DESTRUCTIVE AND USEFUL INSECTS
Their Habits and Control

McGRAW-HILL PUBLICATIONS IN THE AGRICULTURAL SCIENCES

ADRIANCE AND BRISON · Propagation of Horticultural Plants
AHLGREN · Forage Crops
ANDERSON · Diseases of Fruit Crops
BROWN AND WARE · Cotton
CARROLL, KRIDER AND ANDREWS · Swine Production
CHRISTOPHER · Introductory Horticulture
CRAFTS AND ROBBINS · Weed Control
CRUESS · Commercial Fruit and Vegetable Products
DICKSON · Diseases of Field Crops
ECKLES, COMBS, AND MACY · Milk and Milk Products
ELLIOTT · Plant Breeding and Cytogenetics
FERNALD AND SHEPARD · Applied Entomology
GARDNER, BRADFORD, AND HOOKER · The Fundamentals of Fruit Production
GUSTAFSON · Conservation of the Soil
GUSTAFSON · Soils and Soil Management
HAYES, IMMER, AND SMITH · Methods of Plant Breeding
HERRINGTON · Milk and Milk Processing
JENNY · Factors of Soil Formation
JULL · Poultry Husbandry
KOHNKE AND BERTRAND · Soil Conservation
LAURIE AND RIES · Floriculture
LEACH · Insect Transmission of Plant Diseases
MAYNARD AND LOOSLI · Animal Nutrition
METCALF, FLINT, AND METCALF · Destructive and Useful Insects
NEVENS · Principles of Milk Production
PATERSON · Statistical Technique in Agricultural Research
PETERS AND GRUMMER · Livestock Production
RATHER AND HARRISON · Field Crops
RICE, ANDREWS, WARWICK, AND LEGATES · Breeding and Improvement of Farm Animals
ROADHOUSE AND HENDERSON · The Market-milk Industry
STEINHAUS · Principles of Insect Pathology
THOMPSON · Soils and Soil Fertility
THOMPSON AND KELLY · Vegetable Crops
THORNE · Principles of Nematology
TRACY, ARMERDING, AND HANNAH · Dairy Plant Management
WALKER · Diseases of Vegetable Crops
WALKER · Plant Pathology
WILSON · Grain Crops
WOLFE AND KIPPS · Production of Field Crops

Professor R. A. Brink was Consulting Editor of this series from 1948 until January 1, 1961.
The late Leon J. Cole was Consulting Editor of this series from 1937 to 1948.
There are also the related series of McGraw-Hill Publications in the Botanical Sciences, of which Edmund W. Sinnott is Consulting Editor, and in the Zoological Sciences, of which Edgar J. Boell is Consulting Editor. Titles in the Agricultural Sciences were published in these series in the period 1917 to 1937.

DISCARDED

SB
931
M4.5
1962

DESTRUCTIVE AND USEFUL INSECTS

THEIR HABITS AND CONTROL

C. L. METCALF

W. P. FLINT

Revised by
R. L. METCALF

*Professor of Entomology and
Entomologist in the
Agricultural Experiment Station
University of California, Riverside*

FOURTH EDITION

NORMANDALE STATE JUNIOR COLLEGE
9700 FRANCE AVENUE SOUTH
BLOOMINGTON, MINNESOTA 55431

McGRAW-HILL BOOK COMPANY   1962
New York   San Francisco   Toronto   London

DISCARDED

DESTRUCTIVE AND USEFUL INSECTS

Copyright © 1962 by the McGraw-Hill Book Company, Inc.
Copyright renewed 1967 by Mrs. Cleo F. Metcalf

Copyright, 1928, 1939, 1951 by the McGraw-Hill Book Company, Inc. Printed in the United States of America. All rights reserved. No part of this publication may be reproduced, stored in a retrieval system, or transmitted, in any form or by any means, electronic, mechanical, photocopying, recording, or otherwise, without the prior written permission of the publisher. *Library of Congress Catalog Card Number* 61-14049

41658

7 8 9 10 11 12 – MAMB – 1 0 9

To
STEPHEN ALFRED FORBES
DEAN OF AMERICAN ECONOMIC ENTOMOLOGISTS

and
HERBERT OSBORN
MASTER TEACHER OF ENTOMOLOGISTS

this book is affectionately dedicated

PREFACE

Thirty-three years ago, C. L. Metcalf and W. P. Flint wrote in the Preface to the First Edition[1] of "Destructive and Useful Insects": "This book is intended as a text for the beginning student in entomology and also as a guide or reference book for practical farmers, gardeners, fruit growers, farm advisers, physicians, and general readers who desire up-to-date and reliable information about the many kinds of insect pests. . . . It aims to present what the authors believe to be the essentials of economic entomology, in language which any reader can understand.

"As a textbook, it is adapted to the two types of introductory courses commonly given in American universities and colleges. In the first ten chapters, enough of the fundamentals of technical entomology is given to serve as a basis for further special study in the subject or for students of biology who desire an introduction to entomology. The later chapters of the book are devoted to an analysis of the more important insect pests of the major crops in the continental United States and southern Canada. It is hoped that these discussions will be a ready reference for the practical worker who wishes to determine particular pests and learn their control; and serve classes seeking a broad course in applied entomology, as a part of an agricultural education. Material enough is available so that, by selection on the part of the teacher, practical courses adapted to the needs of a great variety of special classes, or different sections of the country, may be given."

In preparing the Fourth Edition, the writer has endeavored to preserve the same point of view as that expressed above. The wide acceptance of the previous editions both as a textbook of economic entomology and as a reference book in its principles and practices has imposed a strict obligation to keep pace with the exponential growth of pertinent scientific information about insects and insect control. This growth is reflected in the present work both qualitatively, in the level of technical competence expected of the reader, and quantitatively, in the substantial increase in the amount of information included. To those critics who object to the former, one can suggest only that they compare a recent text in any scientific field with a counterpart of a generation ago. To those who will inevitably take exception to the latter in pointing out the many other topics which might have been included, one can only plead the stern realities of space conservation in a book which is already bursting its covers.

The text of the Fourth Edition has been completely modernized, and large sections have been rewritten. Chapters 3 and 4 of the previous editions have been combined into a single chapter, Morphology, Physiology, and Biochemistry of Insects, designed to present an integrated

---

[1] The book first appeared in two mimeographed editions, copyrighted in 1925 and 1926.

view of the relations of structure, function, and biochemical processes. Chapters 7 and 8 have also been combined into a single chapter, The Classification of Insects, and this has been expanded to include more descriptive material about the important families of insects, taken largely from the senior authors' "Fundamentals of Insect Life," now out of print. The chapter on Development and Metamorphosis has been enlarged to include discussions of embryology and of growth. The chapters on Insect Control and Application of Insecticides have been almost completely rewritten. All tabular material has been revised, and a number of new tables and illustrations have been incorporated.

The last half of the book has been thoroughly revised to include recent information and references on life histories and current control practices for the insect pests discussed. Wherever compatible with limitations of space, complete descriptions of additional and important insects have been added. These include the beet armyworm, sugarcane borer, sorghum midge, meadow spittlebug, rice stink bug, European chafer, spotted alfalfa aphid, Khapra beetle, face fly, imported fire ant, and various spider mites. It was of course inevitable that, in order to include this large amount of new material, it would be necessary to condense or eliminate some of the least significant of the older tables, charts, and illustrations. From this procedure and by the judicious use of smaller type, it is estimated that the Fourth Edition contains at least 10 per cent more material than the Third Edition. In all this combination of old and new, it is again to be hoped that the result would have been acceptable to the original authors, C. L. Metcalf and W. P. Flint, for this book represents a memorial to them.

In the preparation of the manuscript of this edition, the writer is particularly indebted for stimulating discussions, valuable suggestions, and criticisms of the various chapters to L. D. Anderson, M. M. Barnes, W. H. Ewart, T. R. Fukuto, F. A. Gunther, L. R. Jeppson, D. L. Lindgren, R. B. March, H. T. Reynolds, Harry Shorey, and Vernon Stern. C. P. Clausen not only provided advice and encouragement but also kindly loaned illustrations from his classic, "Entomophagous Insects." My wife, Esther Rutherford Metcalf, has again contributed immeasurably as literary critic, secretary, and cheering section.

Finally, it must be stated that the control of insects is often an exceedingly complex problem. The results obtained are affected by climatic factors, crop varieties, physiological races of insects, and local conditions of many kinds which cannot be anticipated or detailed in this book. Further, the use of chemicals for insect control is regulated by the complex restrictions of the Federal Food and Drug Laws and by additional state legislation. When improperly carried out, it can be hazardous to man and animals. The suggestions made here are summarized from information available in publications of the U.S. Department of Agriculture and state agricultural experiment stations and are presented solely as an indication of current insect-control practices. *The reader is warned, therefore, that any use made of the remedies suggested will be made at his own risk, and is strongly advised to consult his local state experiment station for detailed information as to the proper way to control insect pests in his locality as well as to follow carefully the directions given by the manufacturer on the label of any material to be applied for insect control.*

R. L. Metcalf

CONTENTS

TABLES, SYNOPSES, AND OUTLINES

CHAPTER 1

INSECTS AS ENEMIES OF MAN

"The struggle between man and insects began long before the dawn of civilization, has continued without cessation to the present time, and will continue, no doubt, as long as the human race endures. It is due to the fact that both men and certain insect species constantly want the same things at the same time. Its intensity is owing to the vital importance to both, of the things they struggle for, and its long continuance is due to the fact that the contestants are so equally matched. We commonly think of ourselves as the lords and conquerors of nature, but insects had thoroughly mastered the world and taken full possession of it long before man began the attempt. They had, consequently, all the advantage of a possession of the field when the contest began, and they have disputed every step of our invasion of their original domain so persistently and so successfully that we can even yet scarcely flatter ourselves that we have gained any very important advantage over them. Here and there a truce has been declared, a treaty made, and even a partnership established, advantageous to both parties of the contract—as with the bees and silkworms, for example; but wherever their interests and ours are diametrically opposed, the war still goes on and neither side can claim a final victory. If they want our crops, they still help themselves to them. If they wish the blood of our domestic animals, they pump it out of the veins of our cattle and our horses at their leisure and under our very eyes. If they choose to take up their abode with us, we cannot wholly keep them out of the houses we live in. We cannot even protect our very persons from their annoying and pestiferous attacks, and since the world began, we have never yet exterminated—we probably never shall exterminate—so much as a single insect species. They have in fact, inflicted upon us for ages the most serious evils without our even knowing it."[1]

"It is difficult to understand the long-time comparative indifference of the human species to the insect danger. . . . Men and nations have always struggled among themselves. But . . . there is a war, not among human beings, but between all humanity and certain forces that are arrayed against it. Man . . . has subdued or turned to his own use nearly all kinds of living creatures. There are still remaining, however, the bacteria and protozoa that cause disease and the enormous forces of injurious insects which attack him from every point and which constitute today his greatest rivals in the control of nature. . . . If human beings are to continue to exist, they must first gain mastery over insects. . . . Insects in this country continually nullify the labor of one million men. Insects are better equipped to occupy the earth than are humans, having been on the earth for fifty million years, while the human race is but five hundred thousand years old."[2]

If the reader has never experienced or witnessed any great injury by insects, these statements may sound extreme. Almost everyone, however, has learned to appreciate the destructive capacity of at least a few kinds of insects. Perhaps he has seen a field of corn devoured by army worms or grasshoppers, an orchard killed by scale insects, a building undermined by termites, a bin of grain consumed or contaminated by

[1] FORBES, S. A., "The Insect, the Farmer, the Teacher, the Citizen, and the State," *Bul. Ill. State Lab. Natural History*, 1915.

[2] HOWARD, L. O., "The War against Insects" and other writings.

weevils, or some valuable garment ruined by clothes moths. Few who have not studied the matter carefully will have any idea how many and how varied are the ways in which these minute creatures injuriously affect us.

The awesome capacity of insects for self-preservation and for destruction of human property is perhaps most clearly revealed by the locusts such as *Schistocerca gregaria* and *Locusta migratoria* of Europe, Asia, and Africa. These frequently form vast swarms which may measure several hundred square miles and number 300,000,000 locusts (500 tons) per square mile. Such swarms often travel hundreds of miles a day and, where they alight, devour every blade of edible vegetation (Fig. 1.1).

Fig. 1.1. Aerial view of swarm of locusts, *Schistocerca gregaria*, in Kenya. (*From H. J. Sayer and Shell Chem. Co.*)

The equally incredible industry and persistence of insects may be gaged by the accomplishments of termites such as *Nasutitermes triodiae* of northern Australia, which constructs towering, skyscraper-like mounds, sometimes measuring from 15 to 20 feet high and as much as 12 feet in diameter at the base (Fig. 1.2). These mounds, which contain millions of individuals, are constructed of earth cemented together with saliva and so hard as to defy destruction with pick and shovel.

In assessing the destructive capacity of insects it must be realized that man's advancing civilization has in many ways made him more vulnerable to attack. Modern agriculture with its vast contiguous plantings of standardized crops offers optimum conditions for the development and spread of enormous populations of destructive insects. Thus the Colorado potato beetle, after living precariously for centuries on solanaceous weeds, found the newly introduced potato to be of such superior quality and quantity that it soon became a notorious insect pest. The accumulation and storage of millions of bushels of farm surplus grains has enabled the stored-product insects to multiply on an unprecedented scale. Mountains, deserts, and oceans—geographic barriers which have restricted the dispersal of insects since time began—have lost much of their effectiveness because of modern air transportation. In such a way, the spotted alfalfa aphid was introduced, probably as a single female, into New Mexico in 1954 and, in the absence of the parasites, predators, and diseases of its native home, multiplied so extensively that it blanketed the alfalfa-growing areas of the country in about 3 years. Modern standards of

Fig. 1.2. Mounds of termites, *Nasutitermes triodiae*, in northern Australia. The pickax in the wall of the right-hand termitarium is 3 feet long. (*From Frank Gay, C.S.I.R.O., Canberra.*)

sanitation and food quality have made it impossible to sell agricultural produce blemished by insect attack or contaminated with insect body fragments. Many other examples will occur to the reader of ways in which insects have increased their injurious capacity at the expense of civilized man.

METHODS OF INJURY BY INSECTS

A. *Insects destroy or damage all kinds of growing crops and other valuable plants:*
    1. By chewing leaves, buds, stem, bark, or fruits of the plant.
    2. By sucking the sap from leaves, buds, stem, or fruits.
    3. By boring or tunneling in the bark, stem, or twigs ("borers"); in fruits, nuts, or seeds ("worms" or "weevils"); or between the surfaces of the leaves ("leaf miners").
    4. By causing cancerous growths on plants, within which they live and feed ("gall insects").

5. By attacking roots and underground stems in any of the above ways ("subterranean" or "soil insects").
6. By laying their eggs in some part of the plant.
7. By taking parts of the plant for the construction of nests or shelters.
8. By carrying other insects to the plant and establishing them there.
9. By disseminating organisms of plant diseases (fungi, bacteria, protozoa, and viruses), injecting them into the tissues of the plant as they feed, carrying them into their tunnels, or making wounds through which such disease organisms may gain entrance.
10. By bringing about cross-fertilization of certain rusts, which cause diseases of plants, without which help their aecia will not develop.

B. *Insects annoy and injure man and all other living animals, both domesticated and wild:*
  1. *Causing annoyance:*
    a. By their presence in places where we object to them.
    b. By the sound of their flying about or "buzzing."
    c. By the foul odor of their secretions or decomposing bodies.
    d. By the offensive taste of their secretions and excretions left upon fruits, foods, dishes, tableware.
    e. By irritations as they crawl over the skin.
    f. By chewing, pinching, or nibbling the skin.
    g. By accidentally entering eyes, ears, nostrils, or alimentary canal, causing myiasis.
    h. By laying their eggs on the skin, hairs, or feathers.
  2. *Applying venoms:*
    a. By means of a stinger.
    b. By means of piercing mouth parts.
    c. By the penetration of nettling hairs.
    d. By leaving caustic or corrosive body fluids on the skin, when they are crushed or handled.
    e. By poisoning animals when they are swallowed.
  3. *Making their homes on or in the body, as external or internal parasites, injuring the host animal:*
    a. By causing nervous irritation in crawling about.
    b. By causing inflammation in chewing or piercing the skin.
    c. By contaminating fur or feathers with their eggs and excreta.
    d. By sucking the blood.
    e. By tunneling into muscles, nasal, ocular, auricular, or urogenital passages, causing mechanical injury and promoting infections.
    f. By anchoring to stomach or intestinal lining, mechanically blocking food passages, disturbing nutrition, causing an ulcerous condition, or secreting toxins.
  4. *Disseminating diseases (bacteria, protozoa, parasitic worms, fungi, or viruses) from sick to healthy animals; from some wild animal (the "reservoir") to man or domestic animals; from a diseased parent or antecedent life stage in one of the following ways (the pathogen may merely cling to the body of the insect, it may increase its numbers in the internal organs of the insect, or it may undergo in the insect an essential part of its life cycle that cannot take place anywhere else):*
    a. By accidentally conveying pathogens from filth to food.
    b. By transporting pathogens from filth or diseased animals to the lips, eyes, or wounds of healthy animals.
    c. By the insect host of a pathogen being swallowed by the larger animal, in which the pathogen causes disease.
    d. By inoculating the pathogen hypodermically as the insects bite animals.
    e. By depositing the pathogen upon the skin, in feces, or through its proboscis, or in its crushed body; and the pathogen enters through the bite of the insect, or a scratch, or the unbroken skin.

C. *They destroy or depreciate the value of stored products and possessions including food, clothing, drugs, animal and plant collections, paper, books, furniture, bridges, buildings, mine timbers, telephone poles, telegraph lines, railroad ties, trestles, and the like:*

1. By devouring these things as their food.
2. By contaminating them with their secretions, their excretions, their eggs or their own bodies, even though the product may not be eaten.
3. By seeking protection or building tunnels or nests within or on these substances.
4. By increasing the labor and expense of sorting, packing, and preserving foods.

A. INSECT INJURY TO GROWING PLANTS

Nearly all the injury done by insects results directly or indirectly from their attempts to secure food. They are undoubtedly man's chief rivals for the available food supply of the world. When an insect desires as its food something that man also desires, it becomes his enemy, and we say it is an injurious insect. Because of the great numbers of insects and their unequaled variety we find that there are one or more species adapted to take as food, apparently every kind of organic material in the world—plant or animal, living or dead, dry or decomposing, raw or manufactured, sweet or sour, hard or soft.

Injury by Chewing Insects

Insects take their food in a variety of ways. A primitive and very important method is by chewing off the external parts of a plant, grinding them up, and swallowing them, solids and liquid parts together, very much as a cow or a horse grazes, though of course, taking infinitely smaller bites. Such insects we call *chewing insects* (Fig. 4.3). No one can fail to see examples of this injury (Fig. 1.3). Perhaps the best way to gain an idea of its prevalence is to seek to find leaves of plants absolutely perfect in their freedom from such attack. Cabbageworms, armyworms (Fig. 4.3,*K*), grasshoppers (Fig. 4.3,*C*), the Colorado potato beetle (Fig. 4.3,*J*), the pear-slug (Fig. 4.3,*H*), and the cankerworm are common examples causing injury by chewing. The common, familiar Colorado potato beetles find nearly every potato patch east of the Rocky Mountains every year and, unless checked by insecticide, may soon strip the leaves from the plants and make the carefully planted and cultivated crop a total failure.

Grasshoppers have periodically overwhelmed American farmers from the earliest pioneer days down to the present. In 1923, these insects completely destroyed the crops in one area in Montana larger than an average state in the East.

"Several counties near the Canadian boundary were completely denuded. Many train loads of livestock were shipped out of this territory because of actual lack of forage to keep them alive. Numerous farmers lost everything and moved out. In 1922, the farmers of this state used more than 5,500 tons of poisoned bran mash to destroy the grasshoppers."

Almost the same situation has prevailed in many states and in other years. In 1920 the Canadian entomologists directed the treatment of more than 1,400,000 acres of wheat in Saskatchewan with the saving of $20,000,000 worth of grain, otherwise certain to have been destroyed. Paying a bounty for dead grasshoppers at the rate of 60 cents a bushel, one county in Utah paid out more than $5,000 in a single year, accounting in this way for 274 tons of grasshoppers averaging about 8,000,000 hoppers to the ton. More than 150,000,000 pounds of poisoned bran bait was used against grasshoppers in the United States in 1937, at an estimated cost of $2,000,000. It is believed that over $100,000,000 worth of crops were saved in this way, giving an average return of about $50 for each dollar expended.

Armyworms, like grasshoppers, appear in countless numbers in certain years and practically devastate large areas of the country. Notable outbreaks of this kind have occurred in 1743, 1861, 1896, 1914, 1924, and 1936. The numbers of caterpillars that occur in such outbreaks can hardly be overestimated. Whole fields in which a man could scarcely put his foot to the ground without covering 10 or 12 worms are commonly observed. Since these insects feed chiefly at night, they may, unless noticed in the early stages of the outbreak, destroy a farmer's entire crop before he has time to apply control measures.

In 1868, the gypsy moth, a very destructive leaf-eating caterpillar of shade and forest trees, accidentally escaped into the woodlands of

Fig. 1.3. Two heads of cabbage from adjoining plants. *A*, sprayed to protect it from insects; *B*, not sprayed and badly injured by chewing insects. (*From Wilson and Gentner, Jour. Econ. Ento.*)

Massachusetts and was soon stripping the leaves and killing fruit and forest trees over thousands of square miles. A thousand men at a time have been employed to fight the pest with spray guns, fire, axes, and parasites. By 1927, $25,000,000 had been spent in keeping this pest in control, which at one time could almost certainly have been exterminated at a cost of a few hundred thousands. In spite of efforts to prevent its spread, the gypsy moth has become established in over 40,000,000 acres of northeastern United States. In 1945, this insect caused from 25 to 100 per cent defoliation of more than 800,000 acres of woodland.

INJURY BY PIERCING-SUCKING INSECTS

A second very important way in which insects feed on growing plants is by piercing the epidermis ("skin") and sucking out the sap from the cells within. In this case, only internal and liquid portions of the plant are swallowed, although the insect itself remains externally on the plant.

Such insects we call *piercing-sucking insects*. Their work is accomplished by means of an extremely slender and sharp-pointed portion of the beak (Fig. 4.6) which is thrust into the plant and through which the sap is sucked. This results in a very different-looking, but none the less severe, injury. The hole made by the beak is so small that it is never seen, but the withdrawal of the sap results in either minute spotting of white, brown, or red on leaves, fruit, or twigs; curling of the leaves; deforming of the fruit; or a general wilting, browning, and dying of the whole plant (Fig. 1.4). Aphids, scale insects, the chinch bug, the harlequin cabbage bug, leafhoppers, and plant bugs are well-known examples of piercing-sucking insects (Fig. 4.5).

Aphids (plant lice) (Fig. 4.5,*A,B*) are probably the most universal group of plant-feeding insects. There is scarcely a kind of plant, cultivated or wild, but what supports from one to several species of aphids, and a large percentage of the individual plants will be found infested each summer. The innumerable beaks of these little pests continuously pumping sap from the plants constitute a very severe drain on their vitality. It curtails growth, and interferes with the size and flavor of the fruit developed, if indeed the plant is not killed outright. The pea aphid, for example (Fig. 14.16), caused in 1937, in Wisconsin, a 50 per cent loss of the pea crop over 50,000 acres. Even when the quantity of the yield is not appreciably reduced, the quality and flavor of the peas are depreciated, making it necessary for the commercial canners to add more sugar in an attempt to make up the deficiency. In one experiment nearly 2 million aphids were found to the acre of peas. The spotted alfalfa aphid caused an estimated damage of $42,000,000 in 1956. The corn root aphid, the rosy apple aphid, the woolly apple aphid, the green-bug, and the melon aphid also produce tremendous annual losses.

Another sap-sucking insect, the San Jose scale (Fig. 4.5,*E,F*), has killed tens of thousands of acres of fruit trees since its introduction to the eastern states in 1886 and 1887, and more than one thousand acres of commercial apple orchard were killed by this pest in Illinois alone during the years 1921 and 1922.

The chinch bug (Fig. 4.5,*H*), since its first recorded outbreak in the United States in 1783, has destroyed more than one billion dollars worth of grain crops. In 1914, this insect caused the loss of more than six million dollars worth of corn, wheat, and oats in 13 Illinois counties. There is scarcely a year when the farmers of the Mississippi Valley do not have to reckon with this insect in the production of their crops. In 1934 this pest was so prevalent that a federal appropriation of one million dollars was made to aid farmers in fighting it.

These two groups of insects, the chewing and the piercing-sucking, are the ones for which most spraying is done. It would be difficult to say which group is the more injurious on the whole; but it may be said that the piercing-sucking kinds are generally more difficult to control.

INJURY BY INTERNAL FEEDERS

So long as an insect feeds externally upon crops, it can usually be destroyed by the application of the proper insecticide. But many of our worst pests feed *within* the plant tissues during a part or all of their destructive stages. They gain entrance to the plant either by having the egg thrust into the tissues by the sharp ovipositor of the parent insect or by eating their way in after they hatch from the eggs. In

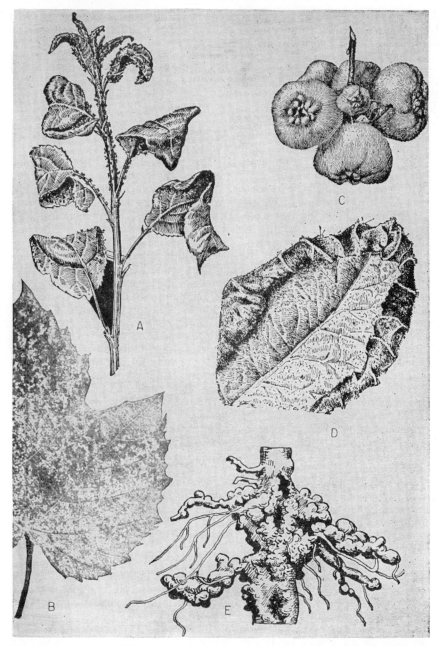

FIG. 1.4. Examples of injury by piercing-sucking insects.   A, curling of leaves and stunting of terminal growth by the green apple aphid (*from Quaintance and Baker, U.S.D.A.*); B, minute white spotting, caused by the feeding of grape leafhoppers (*from Slingerland*); C, aphid apples, the result of the feeding of rosy apple aphids (*from Fulton*); D, hopperburn or tipburn, caused by the feeding of apple leafhoppers on potato (*from Dudley, U.S.D.A.*); E, galls on roots of apple, caused by the feeding of woolly apple aphids (*original*).

either case, the hole by which they enter is almost always very minute, often invisible. A large hole in a fruit, seed, nut, twig, or trunk generally indicates where the insect has come out and not the point where it entered.

The chief groups of internal feeders are indicated by their common group names: (a) "borers," in wood or pith, (b) "worms" or "weevils," in fruits, nuts, or seeds, (c) "leaf miners," and (d) "gall insects." Each group except the third contains some of the foremost insect pests of the world. In nearly all of them the insect is *internal* in only a part of its life stages, sooner or later emerging for a period of free living, usually as adults. This often affords an opportunity to control internal insects by dusting or spraying before their progeny gains entrance to the plant again.

Borers (Fig. 1.5) may attack any plant or part of a plant large enough to contain their bodies. Fruit and shade trees and many herbaceous plants suffer severely in this way. Various bud moths eat out the succulent tissues of swelling buds of trees. The bark beetles, the flatheaded borers, and the peach tree borer work chiefly in the vital cambium layer ("inner bark") of twigs or trunk. The roundheaded borers tunnel through the heartwood, as well as the cambium, greatly weakening the tree and damaging it for lumber.

The European corn borer, which tunnels throughout the stem of the corn plant, was introduced into the United States in about 1908 and by 1950 was found in 36 states and had spread over practically all the major corn-growing areas of the country. In 1949 alone, it was estimated to have caused a total crop loss valued in excess of $349,000,000. The corn

Fig. 1.5. Seven-inch trunk of black oak tree tunneled by larvae of the long-horned beetles. The broad tunnel in the inner bark, below, represents the work of the first year of the insects' life. (*From Ill. Natural History Surv.*)

earworm feeds on growing corn kernels underneath the husks at the tip of the ear. It has been estimated that, in its worst years, this insect attacks more than 70 per cent of the ears of field corn the country over, actually consuming from 1 to 17 per cent of the grain in the infested ears, an annual loss of $140,000,000.[1]

Borers in fruits, including nuts and seeds, are generally called *worms* or *weevils*. Notorious examples are the codling moth (Fig. 4.3,*E*), bean weevils, the cotton boll weevil, the plum curculio, the melonworm, the apple maggot, and the chestnut weevil. Sometimes only one life stage is

[1] See *U.S.D.A., Farmers' Bul.* 872, 1922, and *Yearbook*, 1952.

spent in the fruit, as with the codling moth and apple maggot; in other cases egg, larva, and pupa are all thus concealed from external attack; while in the bean weevil and granary weevil almost the entire life history is spent inside the seeds.

The cotton boll weevil (Fig. 12.2) inserts its eggs into holes made by its long snout in the tissues of the developing bolls from which the cotton lint should later unfold. The grubs that hatch from these eggs devour the immature lint so that no cotton is secured from the infested bolls. Entering this country into southeastern Texas from Mexico in 1890, by 1900 this weevil had increased and spread to such an extent that whole counties were destitute. Their one crop, cotton, having failed, their credit was gone; families became needy; farms were deserted; merchants went bankrupt; and banks failed. The loss increased rapidly in succeeding years until it reached the stupendous sum of $1,000,000,000 in a single year. The beetle spread northward and eastward at an average rate of about 60 miles a year, until all the important cotton-growing states were invaded (Fig. 12.3). Recent discoveries of improved methods of control have greatly checked these losses, but still in certain years the cotton boll weevil harvests more than half of all the cotton planted in the United States. This one insect has collected a toll of nearly $3 a year from every acre of cotton land in the United States.

In April, 1929, the Mediterranean fruit fly, a serious maggot pest of both citrus and deciduous fruits, whose introduction to the United States entomologists had feared for 20 years, was discovered in Florida. The federal Congress appropriated $4,250,000 for its eradication. A strict quarantine preventing the shipment of citrus fruits out of the state was enforced. By the wholesale destruction of every orange, lemon, and grapefruit, both fallen and on the trees, within the infested area, and the spraying of trees of all kinds, the miracle of complete eradication was accomplished. Unfortunately, the pest was rediscovered in Florida in April, 1956, and was eliminated only after spraying an area of 725,000 acres with malathion-protein baits, at a cost of more than $10,000,000.

A number of internal feeders are small enough to find comfortable quarters and an abundance of food between the upper and lower epidermis of a leaf. These are known as *leaf miners*. Surely these are the things Lowell had in mind when he said "There's never a leaf nor a blade too mean to be some happy creature's palace." Among the injurious forms are the apple leaf miners, beet leaf miner, spinach leaf miner, and many others.

The *gall insects* "sting" the plant and make it grow a home for them, within which they find not only shelter but also suitable and abundant food. This is probably the most marvelous instance in biology of the profound influence exerted over one organism by another. We do not know as yet exactly what it is that makes the plants, when attacked by the insect, grow these curious, often elaborate structures (Fig. 1.6) which are absolutely foreign to them in the absence of the gall insect. However, it is clear that the growth of the gall is initiated by the oviposition of the adult and its continued development results from the secretions of the developing larva. A strange feature of the work of gall insects is that the same species of insect on different species of plants causes galls that are similar; while several species of insects attacking the same plant cause galls that are greatly different in appearance. Although the gall is entirely plant tissue, the insect in some unknown manner controls and directs the form and shape it shall take as it grows. There is a marvelous variety of such homes for insects built by the "unwilling" but helpless plants.[1] Many of these galls seem to be practically harmless to the plant that grows them. The wheat jointworm, however, one of our worst pests

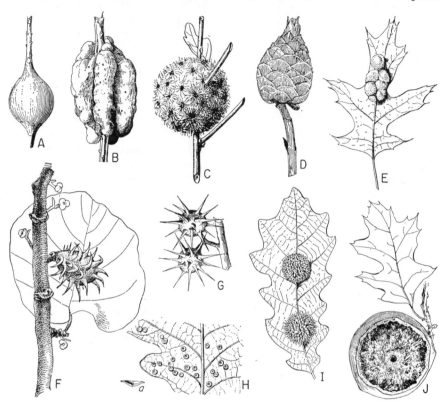

Fig. 1.6. A group of insect galls. *A*, goldenrod ball gall, caused by a fly, *Eurosta solidaginis*; *B*, blackberry knot gall, caused by a gall wasp, *Diastrophus nebulosus*; *C*, wool sower gall on oak twig, caused by a gall wasp, *Andricus seminator*; *D*, pine cone gall, a common growth on willow, caused by a gall fly, *Rhabdophaga strobiloides*, *E*, dryophanta galls on oak leaf, caused by a gall wasp, *Dryophanta lanata*; *F*, spiny witch hazel gall, caused by an aphid, *Hamamelistes spinosus*; *G*, spiny rose gall, caused by a gall wasp, *Rhodites bicolor*; *H*, oak spangles caused by a gall fly, *Cecidomyia poculum*, one gall shown in section at *a*; *I*, spiny oak gall, caused by a gall wasp, *Philonix prinoides*; *J*, large oak apple, caused by a gall wasp, *Amphibolips confluens*. (*From Felt, "Key to American Insect Galls," N.Y. State Museum Bul. 200.)

of wheat, is a gall insect, and the grape phylloxera has destroyed thousands of acres of the most valuable vineyards in Europe and America.

INJURY BY SUBTERRANEAN INSECTS

Almost as secure from man's attack as the internal feeders are those insects that attack plants below the surface of the ground. These include chewers, sap suckers, root borers, and gall insects, the attacks of which

[1] See FELT, E. P., "Key to American Insect Galls," *N.Y. State Museum Bul.* 200, 1917.

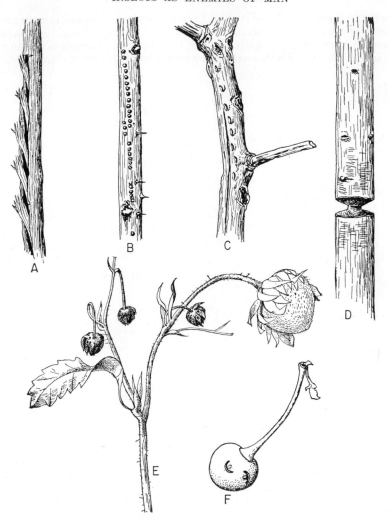

Fig. 1.7. Examples of injury to plants caused by the egg-laying of insects. *A*, twig split by egg-laying of the periodical cicada; *B*, holes in stem of raspberry made by egg-laying of a tree cricket; *C*, slits in bark of apple twig beneath which a treehopper has thrust her eggs; *D*, twig of pecan cut nearly in two by egg-laying of twig girdler; *E*, fruit buds of a strawberry, partially severed by strawberry weevil after laying an egg in the buds; *F*, cherry showing two egg punctures of the plum curculio. (*A–D and F, original; E, from N.J. Agr. Exp. Sta., after U.S.D.A.*)

differ from the aboveground forms just described, only in their position with reference to the soil surface. The subterranean insects may spend their entire life cycle below ground, *e.g.*, the woolly apple aphid. This insect, as both nymph and adult, sucks the sap from the roots of apple, causing the development of ugly tumors (Fig. 1.4,*E*) and the subsequent decay of the roots at the point of attack. More often there is at least one life stage of the insect that has not taken up the subterranean habit, as in the case of the white grubs, wireworms, Japanese beetle, root maggots, and grape and corn rootworms, in all of which the larvae are root feeders while the adults have largely retained the more primitive life above ground. Interesting gradations and adaptations to the subterranean life are seen in the way in which the eggs of these insects are laid and in the place of pupation. In general it may be said that the more of its life stages the insect spends underground, the more difficult it is to control.

Injury by Laying Eggs

Probably 95 per cent or more of the direct injury to plants is caused by insects feeding in the various ways just described. Another instinct, almost as powerful, is the urge to provide for the welfare of the offspring. While, in general, the maternal instinct among insects is not developed to the point where they care for their young after birth, most insects have a marvelously effective instinct to lay their eggs in exactly the right place so that their young will have the best chance to survive; and there are some very striking cases of great effort and care in the preparation of a nest or the deposition of the eggs. Sometimes this provision for the young leads to serious injury to man's possessions. The periodical cicada[1] deposits her eggs in the 1-year-old growth of fruit and forest trees, splitting the wood so severely that the entire twig beyond this point often dies (Fig. 1.7,*A*). The treehoppers and tree crickets split and ruin the bark or twigs of raspberry, currant, and apple in pushing their eggs into the plant tissues (Fig. 1.7,*B,C*). It is interesting to note that these are purely nesting sites. As soon as the young hatch, they desert the twigs and injure the plant no further. In other cases, the young, at least, subsequently feed upon the plant attacked by the egg-laying female; but we wish to emphasize at this point the injury *by the egg-laying act*, quite independent of any subsequent feeding of the young. Thus the plum curculio ruins the fruits of apple, plum, peach, and cherry by her characteristic egg-laying punctures (Fig. 1.7,*F*). The strawberry weevil, after laying an egg in the unopened bud, cuts the blossom stem partly off so that the flower never opens (Fig. 1.7,*E*). One of the most extreme cases of devotion to the welfare of the young is that of the twig girdler. In order that the larvae of this insect may have wood in a suitable condition of moisture and decay, the female laboriously chews off twig after twig of oak, hickory, pecan, elm, persimmon, or other tree in which to lay her eggs (Fig. 1.7,*D*). The severing of a single twig requires several days of work by the female.

The Use of Plants for Making Nests

Besides laying eggs in plants, insects sometimes remove parts of the plant for the construction of nests or for provisioning nests elsewhere, though they do not feed on these materials. This injury is more interesting than it is serious. Leaf-cutter bees thus nip out rather neat circular pieces of rose and other foliage which are carried away and fashioned and cemented together to form thimble-shaped cells one above the other in a tunnel previously made in the stem of a plant. Each cell when completed contains a mass of nectar and pollen and an egg completely surrounded by bits of leaf; in this nest the young bee develops. The tropical leaf-cutting ants strip millions of leaves from trees or herbaceous plants and carry them into their nests where they are cut into fine pieces, sometimes are mixed with bits of their own or other insects' excreta, and form the medium upon which fungi are grown, as the only food of both larvae and adults. Other kinds of ants hollow out the stems or thorns of plants in which they dwell; but this phase of injury is not serious to man.

Insects That Care for Other Insects

Ants and some other kinds of insects, which are not in themselves serious pests, become injurious because they bring to our cultivated crops (corn, asters, citrus fruits) such noxious forms as aphids and mealy bugs, which they care for and protect because they like to eat the honeydew secreted by these pests. In some cases the most intricate and intimate

[1] Also called "seventeen-year locust."

interrelations have grown up between the ants, on the one hand, and the aphids, on the other. In general, ants furnish protection and a feeding place for the aphids, and the aphids furnish food ("honeydew") for the ants. Such cases of mutual dependency of two organisms upon each other are known as *mutualism*. One of the best examples is furnished by the cornfield ant and the corn root aphid. This destructive aphid (Fig. 9.25) has become totally dependent upon the ants, which care for the aphid eggs over winter, and in the spring and throughout the summer carry the young aphids in their mouths through underground tunnels and actually place them on the roots of corn and weeds on which the aphids can feed. The ants are paid for this solicitous care with the sweet honeydew, which the aphids continually excrete and which serves as food for the ants. The cornfield ant is thus a menace to the corn crop although the ants themselves probably never injure the corn plant in any way.

INSECTS AS DISSEMINATORS OF PLANT DISEASES[1]

A serious phase of insect injury and one which may rival in importance the destruction caused by their direct feeding is the connection of insects with the ravages of plant diseases. Everyone has known something about insects that transmit human and animal diseases, but few have realized that other insects are engaged in spreading very disastrous diseases of plants. Since 1892, when it was first proved that a plant disease (fire blight of fruit trees) may be spread by an insect (the honeybee), the knowledge of this subject has grown rapidly, and at present there is good evidence that more than 200 such diseases are disseminated by insects. The majority of them—about 150—belong to the group known as viruses, 25 or more are due to parasitic fungi, 15 or more are bacterial diseases, and a few are caused by protozoa. The essential facts regarding a few of the most important of them are given in Table 1.1.

Insects are responsible for favoring plant diseases in a number of different ways: (a) By feeding or laying eggs or boring into plants, they may make an entrance point for a disease that is not actually transported by them. (b) They have been found actually to disseminate the pathogens on or in their bodies, from one plant to a susceptible surface of another plant, such as a blossom or to a wound made by some other agent. (c) They carry pathogens on the outside or inside of their bodies and inject them hypodermically into the plant as they feed. (d) They harbor the pathogens inside their bodies during adverse periods, as over winter or through a period of drought or host-plant scarcity, protecting them from the adverse climatic condition and from natural enemies. (e) They may be essential hosts for an incubation period, for increase in numbers of the pathogen, or for some part of its life cycle that cannot be completed elsewhere; or they may be essential for dissemination that is not normally accomplished otherwise.

Agents of Invasion and Passive Transmission. The epidermis and bark of plants, like the skin of animals, have a highly protective function. When either is broken, an opportunity is afforded for the ingress of various injurious organisms. The attacks of the corn earworm (Fig. 9.23) are almost always followed by destructive molds and rots which spoil the corn

[1] SMITH, K. M., "A Textbook of Plant Virus Diseases," 2d ed., Churchill, 1957; LEACH, J. G., "Insect Transmission of Plant Diseases," McGraw-Hill, 1940; STEINHAUS, E. A., "Insect Microbiology," Comstock, 1946; SMITH, F., and P. BRIERLEY, *Ann. Rev. Ento.*, **1**:299–322, 1956; SMITH, K., *ibid.*, **3**:469–482, 1958.

and some of which are dangerous to animals that eat them.   Many of these would not normally gain entrance to the ear were it not for the pathways formed by the tunnels of the worms.   The fungus causing early blight of potatoes is similarly favored by the numerous holes made in the leaves by flea beetles (Fig. 4.3,*I*), and chestnut blight by the various bark beetles and borers attacking the bark of that tree.

Active Mechanical Transmission and Inoculation.   In addition to thus passively favoring plant diseases, certain kinds of insects actually carry the pathogens on or in their bodies from plant to plant.   Bacteria are especially likely to be disseminated mechanically, because they adhere readily to the insects.   Some fungus spores are sticky or moist, and some are very fine, dusty, or spiny, or they have an electric charge opposite to that of the insect body, so that they cling well to the hairy bodies of insects.   Bacteria and fungus spores may, therefore, be carried on the outside of the body of almost any insect and, if deposited upon a susceptible plant tissue, may start disease.   Insects that pierce plant tissues with their mouth parts and those which tunnel beneath the surface, when they attack a diseased plant, are very likely to pick up externally, or take internally with their food, the germs causing that disease.   As they pass to fresh plants to feed, the pathogens may be deposited on these plants with their excrement, whence they may enter any wound or the germinating spores may penetrate the plant unaided.   In other cases the disease organisms may actually be injected into the tissues by contaminated mouth parts or ovipositors; or, in the case of borers, by their tunneling into the plant.   The active flying and feeding insects serve admirably to disseminate, widely and rapidly, and to act as agents of ingress for, the inactive disease organisms which by their own efforts could seldom get from one plant to another.   The habit of many insects of visiting only certain kinds of plants and even specific organs of those plants aids in their efficiency as inoculators.   Fire blight of apple and pear is carried by aphids, bees, and other insects, which, as they feed, spread the destructive bacilli.   The Dutch elm disease, which was imported into the United States in 1930 with results alarming to lovers of shade trees, is disseminated from tree to tree by the European bark beetle (page 848).   Since bark beetles prefer to make their breeding tunnels in trees that are in a weak or dying condition, they flock to trees dying of this fungus and later spread the malady far and wide as they seek new, vigorous trees upon which to feed.   The egg punctures and feeding punctures of the plum curculio (Fig. 1.7,*F*) are commonly starting points for the brown rot of the peach, and the amount of brown rot in any season is highly correlated with the abundance of curculios.   The areas around the punctures of aphids, bugs, and leafhoppers frequently thicken, and the leaves become diseased, curl, and drop off, troubles that are often inseparable from and specific for the particular insect.

In feeding upon a plant, the amount of damage that the insect can do is more or less limited by the amount of tissue that it can devour; but if the insect's mouth parts are contaminated with disease organisms, the organisms may be established on the plant and not cease their attack until the entire plant is killed.   In this connection it is noteworthy that much more effective control is needed for these disease-carrying insects than for insects which harm the plant only by their feeding.   Nothing short of absolute control is satisfactory for some of the disease carriers, because a single insect may deal the plant a death blow by inoculating it with a

TABLE 1.1. A SUMMARY OF CERTAIN INSECT-BORNE DISEASES OF PLANTS

A. Fungus Diseases of Plants

Name of disease	Plants affected	Insect carriers	Method of transmission	Causal organism or pathogen
Dutch elm disease.......	Various kinds of elms	The bark beetles, *Scolytus scolytus*, *S. multistriatus*, and *Hylurgopinus rufipes*	Spores introduced into cambium as beetles feed or make tunnels	*Ceratostomella ulmi*
Chestnut blight..........	Chestnut	Various beetles	Spores carried by insects or by birds. Rain washes spores into insect holes	*Endothia parasitica*
Blue stain..............	Norway pine	The bark beetles, *Ips pini* and *I. grandicollis*	Spores introduced into cambium as beetles make egg tunnels	*Ceratostomella ips*
Brown rot.............	Peach, plum, cherry	The plum curculio, *Conotrachelus nenuphar*	Punctures made while feeding or egg-laying inoculate or permit entrance of pathogen	*Sclerotinia fructicola*
Perennial canker........	Apple	Woolly apple aphid, *Eriosoma lanigerum*	Feeding of insect causes cracking of callus and permits entrance of pathogen	*Gloeosporium perennans*
Downy mildew.........	Lima beans	Bees	Transferred to blossoms as bees touch the floral organs	*Phytophthora phaseoli*
Blackleg..............	Cabbage	The cabbage maggot, *Hylemya brassicae*	Maggots carry spores on bodies and make wounds for ingress	*Phoma lingam*
Bud rot.............	Carnations	The mite, *Pediculopsis graminum*	Mites bearing spores crawl down into buds to feed	*Sporotrichum poae*
Ergot...............	Rye, barley, wheat, many grasses	Flies, bees	Carried in alimentary canal and externally by flower-visiting insects	*Claviceps purpurea*
Fusarium wilt........	Cotton	Several species of grasshoppers	Insects spread spores in fecal pellets	*Fusarium vasinfectum*
Plum wilt...........	Plum	The peach tree borer, *Sanninoidea exitiosa*	Fungus enters tree through wounds of borers	*Lasiodiplodia triflorae*

	Potato	Potato flea beetle, *Epitrix cucumeris*	Larva introduces fungus into tuber	*Actinomyces scabies*
Potato scab.............	Potato	Potato flea beetle, *Epitrix cucumeris*	Larva introduces fungus into tuber	*Actinomyces scabies*

B. Bacterial Diseases of Plants

Cucurbit wilt............	Cucumbers, melons, and related plants	The cucumber beetles, *Acalymma vittata* and *Diabrotica undecimpunctata*	Winters in alimentary canal of insects; deposited as they feed or in feces over wounds	*Erwinia tracheiphila*
Bacterial wilt or Stewart's disease	Corn, Job's-tears, teosinte	Corn flea beetle, *Chaetocnema pulicaria*, corn rootworms, *Diabrotica longicornis* and *D. undecimpunctata*, and seed-corn maggot, *Hylemya cilicrura*	Winters in alimentary canal of flea beetles; deposited in wounds made by feeding	*Bacterium stewarti*
Fire blight............	Apple, pear, quince	Bees, wasps, flies, aphids, leaf-hoppers, tarnished plant bug, bark beetles	To blossoms from mouth parts of nectar feeders. From cankers to new growth by direct inoculation	*Erwinia amylovora*
Bacterial soft rot............	Cabbage, potato, and other vegetables	Maggots, *Hylemya brassicae*, and *H. cilicrura*	Maggots inoculate seeds and roots. Persists through metamorphosis and adult flies spread while ovipositing	*Erwinia carotovora*
Olive knot............	Olive	Olive fruit fly, *Dacus oleae*	Persists through metamorphosis and adult flies spread while ovipositing. Infection also passes through egg stage	*Pseudomonas savastanoi*
Bacterial rot............	Apple	Apple maggot, *Rhagoletis pomonella*	Persists through metamorphosis and adult flies spread while ovipositing	*Pseudomonas melophthora*

C. Virus Diseases of Plants

Curly top............	Sugar beet and many other vegetable, ornamental, and wild plants	Beet leafhopper, *Circulifer tenellus*	Virus is inoculated by feeding of insect after a short incubation period	*Ruga verrucosans*

17

TABLE 1.1 (*Continued*)

Name of disease	Plants affected	Insect carriers	Method of transmission	Causal organism or pathogen
Peach yellows	Peach	Plum leafhopper, *Macropsis trimaculata*	Virus is inoculated by feeding of insect after an incubation period	*Chlorogenus persicae*
False blossom	Cranberry	The leafhopper, *Scleroracus vaccinii*	Virus is inoculated by feeding of insect	*Chlorogenus vaccinii*
Aster yellows	Asters and many other plants	The leafhopper, *Macrosteles divisus*	Virus is inoculated by feeding of insect following an incubation period	*Chlorogenus callistephi*
Streak disease of corn	Corn	The leafhoppers, *Cicadulina mbila, C. zeae,* and *C. storeyi*	The same	*Fractilinea maidis*
Dwarf disease of rice	Rice and other grasses	The leafhoppers, *Nephotettix apicalis* and *Deltocephalus dorsalis*	Virus inoculated by feeding of insect. Passes to young through the egg	*Fractilinea oryzae*
Spindle tuber	Potato	Flea beetles, aphids, grasshoppers, tarnished plant bug, Colorado potato beetle	Virus is inoculated by feeding of insect	*Acrogenus solani*
Leaf roll	Potato	*Myzus persicae, M. convolvuli, M. circumflexis, Aphis abbreviata*	The same	*Corium solani*
Sugarcane mosaic	Sugarcane, corn, sorghum	The aphids, *Aphus maidis, Hysteroneura setariae,* and *Toxoptera graminum*	The same	*Marmor sacchari*
Bean mosaic	Beans	*Myzus persicae, Macrosiphum pisi, M. solanifolii, Aphis gossypii, A. medicaginis, A. spiraecola, Brevicoryne brassicae*	The same	*Marmor phaseoli*

Mosaic of crucifers........	Cauliflower, cabbage, turnip, mustard	The aphids, *Myzus persicae* and *Brevicoryne brassicae*	The same	*Marmor cruciferarum*
Cucumber mosaic........	Cucumber and other cucurbits	The aphids, *Aphis gossypii, Myzus persicae, M. circumflexus, M. solani,* and cucumber beetles	The same	*Marmor cucumeris*
Spotted wilt............	Tomato	The thrips, *Thrips tabaci, Frankliniella lycopersici, F. occidentalis,* and *F. moultoni*	Must feed on diseased plants in nymphal stage	*Lethum australiense*
Leaf curl..............	Cotton	The whitefly, *Bemisia gossypiperda*	Virus inoculated by feeding of insect	*Ruga gossypii*
Yellow dwarf..........	Onion	More than 50 species of aphids, especially *Aphis gossypii, A. maidis, Myzus persicae, Brevicoryne brassicae*	The same	*Marmor cepae*
Phony peach disease	Peach	The leafhoppers, *Homalodisca triquetra, Oncometopia undata, Graphocephala versuta, Cuerna costalis*	The same	*Nanus mirabilis*
Peach X-disease........	Peach	The leafhopper, *Colladonus geminatus*	The same	*Carpophthora lacerans*
Peach mosaic..........	Peach and nectarine	The mite, *Eriophyes insidiosus*	The same	*Marmor persicae*
Tristezia.............	Citrus	*Aphis citricidus, A. gossypii, A. spiraeola*	The same	*Corium viatorum*
Elm phloem necrosis	Elm	The leafhopper, *Scaphoideus luteolus*	The same	*Morsus ulmi*
Pierce's disease........	Grape and alfalfa	The leafhoppers, *Heleochara delta, Carneocephala fulgida, Draeculacephala minerva;* the spittlebugs, *Aphrophora annulata, A. permutata, Clastoptera brunnea, Philaenus leucophthalmus*	The same	*Morsus suffodiens*

disease organism, whereas the *feeding* of one insect would be ordinarily insignificant.   It is also extremely important to prevent disease-carrying insects from starting to feed upon a crop.

Biological Transmission.   In most of the above cases, the disease may and probably does survive and spread to some extent without the help of the insects.   In the following cases, however, it seems that the normal means of spread of the disease from one plant to another is always by the intervention of some particular insect.   In at least a part of these cases it appears that the insect is necessary to the continued development and life of the pathogen, some essential part of its life cycle taking place in the insect's body, usually with a concomitant increase in numbers.   The best known case in this category is the cucurbit wilt disease carried by the striped cucumber beetle (Fig. 14.18) and the spotted cucumber beetle (Fig. 9.34).   The causal organism of the disease, a bacterium, spends the winter in the digestive tract of the hibernating beetle.   The infected beetle, when it begins feeding upon a young cucumber plant in the spring, deposits in its feces some of the wilt bacteria.   These are later washed over the surface of the leaf by dew or rain, and wherever there is a fresh wound opening into the vascular system of the leaf, the disease may become established.   Probably the wounds, which are necessary to infection, are chiefly made by the insects in feeding.   After the disease is started in this way, any cucumber beetle feeding on the plant may contaminate its mouth parts and then infect the next plant on which it feeds.   No other means of spread for this disease are known.   A similar relation exists between Stewart's disease or bacterial wilt of corn and the corn flea beetle (Fig. 9.40).   The bacterium of this disease (*Bacterium stewarti*) may be found in winter in the viscera of this beetle, becoming an almost pure culture by spring.   It has been shown that this disease usually becomes more destructive after mild winters but is not important after cold winters when, nor in regions north of an isotherm where, the sum of the mean monthly temperatures for December, January, and February is below 80°F.   It is believed that this results from the effects of cold, not upon the bacterium that causes the disease, but upon the insect carrier.   In the spring the diseased beetles fly to fields of corn and transfer the wilt organism directly to the leaves by their feeding.

The viruses which cause the mosaics and related diseases of plants, are nearly or quite ultramicroscopic and capable of passing through filters fine enough to sieve out the smallest known bacteria.   They are either a precellular form of life, or nonliving gigantic protein molecules close to the border line between living and nonliving.   They are not able to multiply in the absence of living cells and so cannot be grown in artificial media, but they have the ability to cause serious diseases when in contact with living cells and are regenerated and reproduced in the process.   They are not destroyed by being crystallized by chemical treatment.   They have great tenacity and longevity.   It has been claimed that tobacco mosaic has been kept viable in dried tobacco leaves for 24 years and bean mosaic in stored seeds for 30 years.

Insects are known to be the carriers of more of the virus diseases than of bacteria, fungi, and protozoa combined.   Some of these virus diseases appear capable of being transmitted mechanically by almost any insect that will feed upon the affected plant.   It is difficult to distinguish some of them from disease-like injuries by insects in which no virus or other pathogen has been demonstrated.   Thus the psyllid yellows disease of potato and tomato appears upon the plant only when it has been attacked by the psyllid, *Paratrioza cockerelli*. The tipburn of potatoes (one of the most serious of potato diseases, Fig. 1.4,*D*) and similar injury to alfalfa, clovers, soybeans, and peanuts appear only upon leaves that have been pierced by the potato leafhopper (page 643). The disease has been attributed to interference with the translocation of food materials within the plant by the mechanical injuries caused as the leafhoppers feed or to a toxin or enzyme injected into the leaf by the mouth parts of the insect. In some cases the virus disease apparently can be transmitted by only one kind of insect, probably because some stage in its development requires incubation or nourishment in the body of that particular host. Thus the curly top disease of sugar beets is contracted by the plant only when it is punctured by the mouth parts of the beet leafhopper (page 676), which also carries the disease over the winter. False blossom of cranberries, which practically exterminated certain, very susceptible varieties of cranberries in the United States about 1895 to 1905, is transmitted by the leafhopper, *Scleroracus vaccinii*. Control measures for the leafhopper, such as flooding, dusting, and the use of resistant varieties, have greatly reduced the rate of spread of the disease. In some cases, such as the Japanese dwarf disease of rice, a virus disseminated by the leafhopper, *Nephotettix apicalis*, if the insect feeds upon a diseased plant, the eggs become infected, and the young subsequently born may be infective, sometimes for three or four generations.

With reference to the kinds of insects which disseminate plant diseases, it is noteworthy that the piercing-sucking insects are much more important than chewing insects, probably because the superficial wounds made by chewing insects dry out so quickly that the organisms fail to gain entrance to the plant, whereas the slender deep-seated punctures made by the piercing insects protect the pathogen from desiccation until it has established itself in the new host. Only a few Orthoptera and Coleoptera have been incriminated in disease transmission. There are several species of thrips, whiteflies, plant bugs, and mites connected with definite diseases; but the Hemiptera are, so far as we know, much less extensive carriers of plant infections than the closely related order Homoptera (page 213). The explanation suggested for this is that the Hemiptera often kill the cells adjacent to their punctures by toxic saliva, so that the pathogen cannot establish itself. The leafhoppers (page 215) and especially the aphids (page 217) are the known disseminators of more plant diseases then all other kinds of insects combined. The green peach aphid, *Myzus persicae*, has been shown to transmit more than 50 plant viruses. Some of the diseases, such as cranberry false blossom and peach yellows, seem to have a very restricted host range, while others, *e.g.*, aster yellows and curly top of sugar beets, affect a great number of host plants.

Aphid-transmitted viruses have been classified as *nonpersistent* in which the vector becomes optimally infective after feeding on the infected plant for only about 30 seconds, begins to lose its capacity to transmit the virus after starvation for 2 minutes, and loses its infectivity very rapidly when fed on healthy plants. With the aphid vectors of mosaic diseases (Table 1.1), there is no clear-cut distinction between mechanical and biological transmission, and some authorities have described this as modified or delayed mechanical transmission. In the *persistent* viruses the aphid may require a longer period of feeding to become infective and a latent period of a few hours to 10 days or more before it can transmit the infection; and the insect usually remains infective for long periods, often for the remainder of its life. With most or perhaps all of the leaf-hopper-transmitted viruses, there is a pronounced latent period before transmission can occur. This apparently results from the time required for the virus to multiply in the intestine of the insect before it can reach a high enough concentration to penetrate into the salivary glands in sufficient concentration to become infective.

An impression of the highly infective nature of the plant viruses can be gained from the very small amounts of plant juices ingested in feeding: 0.4 microgram per minute by the leafhopper, *Orosius argentatus*, and 1 microgram per minute by the aphid, *Myzus persicae*. The marked specificity in the abilities of aphid or leafhopper species to transmit particular viruses results from such factors as preferred plant tissues for feeding, *i.e.*, xylem or phloem; percentage of feeding on diseased and susceptible host plants; presence of chemical inhibitors in the salivary glands, blood, or intestine of the vector; and permeability of the intestine or salivary glands by the virus.

A new light on the intricate ramifications of insect injuries is the revelation that for black stem rust of wheat (*Puccinia graminis*), growing on barberry, and the related *P. helianthi*, on sunflower, if the sori are protected from the visits of insects, the aecia never produce spores, while the visits of insects usually bring about spore production within a few days. The rusts have two types of sexual cells or pycniospores, + and −, and are incapable of self-fertilization. Insects are almost indispensable in fertilizing these rusts as they forage for nectar exuding from the pycnia and thus favor the ravages of these destructive fungi in much the same way that they pollinate valuable fruits, vegetables, and flowers (page 56). Thus is added new confirmation of Dr. Asa Fitch's surmise: "There is no kind of mischief going on in the world of nature around us but what some insect is at the bottom of it."

B. INSECT INJURY TO MAN AND OTHER LIVING ANIMALS

The second great group of organisms that fall prey to insects is all manner of animal life, from protozoa to man. We find no records of insects that feed upon the marine animals known as echinoderms (starfish, sea cucumbers, etc.). With this exception all the principal branches of the animal kingdom are attacked. Insects in their relation to us make no distinction between man and other animals, and we shall make none in this discussion.

ANNOYING INSECTS

There are, first of all, a number of minor ways in which insects conflict with man's comfort and pleasure. They are annoying by their presence, by their sounds, by the bad odors and tastes of their secretions, by crawling over one's body, by getting into the eyes or ears, or by laying their eggs upon animals. All of us have experienced great annoyance from flying, buzzing, or crawling creatures at times, particularly when we desired to rest or apply ourselves to some exacting task. The unpleasant taste left by certain stink bugs on berries and other fruits, the disgusting odor of cockroaches about the table service of some restaurants, and the sharp pain caused by getting certain minute insects[1] into the eye when driving at night are familiar examples of annoyance by insects. The accidental invasion by living insects of the ears, nostrils, or stomach is usually serious but fortunately a rare experience. Animals suffer from

---

[1] For example, staphylinid beetles and certain small Hemiptera.

the attempts of certain flies (the bot flies) to lay eggs on their bodies. This is sufficient to cause the wildest stampeding of cattle, horses, and deer. While these annoyances constitute the least important of all the phases of insect injury to animals, still they are sufficient to account for a great deal of monetary loss, discomfort, and inefficiency.

VENOMOUS INSECTS

Insects are not popular. The innate abhorrence of "crawling things" possessed by some persons is unfortunate, however, because it prevents them from learning anything about insects and interferes unnecessarily with their enjoyment of out-of-door life. An extreme fear of bugs, caterpillars, spiders, and bees is not warranted by the facts. In temperate climates there are very few kinds of insects that can harm the body seriously. Nevertheless, there are a number of kinds that can bite and sting painfully, and these wounds may become infected with serious results. It is so often true as to be almost a rule that the worst-looking forms are generally harmless; while some of the most painful experiences result from contact with very innocent-looking specimens.

Bodily pain and illness may be caused by the venoms of insects applied to the body in the following ways: (a) by stinging, i.e., penetrating the skin with a defensive and offensive organ located near the tip of the abdomen; (b) by biting, i.e., with the mouth parts—generally inserted to secure food, but sometimes used in a defensive way when certain insects are handled; (c) by nettling with hollow poison hairs located on the bodies of certain caterpillars, which inject venom after the manner of the common nettle plant; (d) by the application of caustic or corrosive fluids to the unbroken skin; (e) by poisoning animals when they are swallowed, accidentally, or with food. Chewing insects are rarely able to cause much pain. The "pinching bugs," which are about the worst of this kind, can scarcely break the skin. The really painful bites are made by insects with piercing mouth parts, are accompanied by the in-

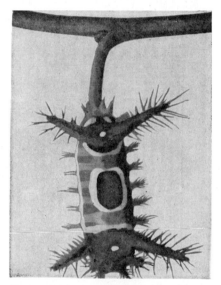

FIG. 1.8. A caterpillar with poisonous hairs, the saddle-back, *Sibine stimulea*, about natural size. (*From U.S.D.A., Farmers' Bul.* 1495.)

troduction of a venom, and are, therefore, a chemical injury. The nature of the venom appears to vary (page 116) but has the common characteristic that it is in some way toxic to animal tissues and so causes pain.

Besides the spiders, ticks, and centipedes (pages 181–184), the following kinds of true insects are notorious for the injury inflicted by their bites: the Diptera or two-winged flies, including mosquitoes, black flies, horse flies, the stable fly, tsetse flies, the sheep tick, etc.; the Hemiptera or true bugs, including the bed bug, assassin bugs, back swimmers, water scorpions, etc.; the Anoplura or bloodsucking lice; the Siphonaptera or fleas.

Many insects regularly feed on animal blood, including that of man, as their only food; and, while so feeding, usually introduce a poison that causes a painful irritation. Many sections of the mountains and woods are rendered temporarily uninhabitable by swarms of black flies in early summer. Forbes[1] quotes from Agassiz's "Lake Superior" as follows: "Nothing could tempt us into the woods so terrible were the

[1] FORBES, S. A., "The Insect, the Farmer, the Teacher, the Citizen, and the State," *Bul. Ill. State Lab. Natural History,* 1915.

black flies. One, whom scientific ardor tempted a little way up the river in a canoe, after water plants, came back a frightful spectacle, with blood-red rings around his eyes, his face bloody, and covered with punctures. The next morning his head and neck were swollen as if from an attack of erysipelas.''

The stable fly (Fig. 20.4), that dreaded but constant companion of horses, mules, cattle, and hogs, all summer long, inflicts such painful bites and withdraws so much blood that animals are sometimes killed outright. Unable to don protective clothing, to retreat into screened houses, or even to "swat" efficiently, domestic animals must suffer beyond our comprehension from these many, bloodthirsty pests. This suffering is translated into losses to the livestock farmer in decreased milk yield, loss of flesh, unsatisfactory growth, inefficiency and unmanageableness of work animals, and in greater susceptibility of the weakened animals to diseases.

The other methods of applying a venom are less important than that by biting. The stinging insects are, so far as man and the larger animals are concerned, largely

Fig. 1.9. An entirely harmless, though evil-looking, caterpillar, the hickory horned devil (*Citheronia regalis*), about ⅕ smaller than natural size. (*From Houser, after Packard.*)

a peaceable and defensive lot, inflicting their punishment almost exclusively on creatures that have injured them or disturbed their nests. However, certain individuals are extremely sensitive to the venom of bees and wasps, and it has been stated that from 1950 to 1954, more people were killed in the United States by the stings of hymenopterous insects than from the bites of venomous snakes.[1] The fire ants (page 896) sting viciously when disturbed and often kill small birds and mammals. Some of the ants possess a venom but have lost the stinger with which to inject it. According to Wheeler, these spray the poison from the tip of the abdomen into a wound made by the mouth parts. Certain beetles have a similar method of defense. The bombardier beetles, *Brachinus*, eject an acrid fluid which is discharged with a distinct popping sound and a small cloud of vapor that looks like the smoke from a miniature cannon (page 115).

Among the most interesting protective structures that insects possess are the nettling hairs of many caterpillars. These structures are similar to the poison hairs

[1] PARRISH, H. M., *Arch. Int. Med.*, **104**:198, 1959.

of the nettle plant.   Not all the hairs of the body are of this type but only certain ones are hollow and connect at their base beneath the cuticle with poison gland cells. When these hairs penetrate the human skin the poison is released at a broken point and may create a serious skin eruption accompanied by intense itching and intestinal disturbance.   The best known of the nettling caterpillars are: the brown-tail moth (Fig. 17.10), the io moth, the saddle-back caterpillar (Fig. 1.8), the puss caterpillar (Fig. 6.59), the hag moth, and the buck moth.   There is nothing distinctive about the appearance of these stinging caterpillars as a group.   One has simply to learn to recognize each of them.   As pointed out already, there are many more formidable-looking kinds that are totally harmless.   For example, the hickory horned devil (Fig. 1.9) with its many thorny spines, some of them ¾ inch long; the common tomato hornworm (Fig. 14.33) and other sphingid larvae with a pointed horn near the tail end of the body; the celery caterpillar (Fig. 6.58,b) with a pair of soft yellow horns near the head that are erected and thrust out when it is disturbed and give off a peculiar odor; these and many other dangerous-looking forms are absolutely incapable of harming a person.

There are certain insects that carry a venomous substance diffusely throughout the body, especially in the blood, rather than confined to particular glands.   In some cases notably the blister beetles[1] (Fig. 14.25), this poison, cantharidin (page 55), possesses caustic or blistering properties when the insects are accidentally crushed on the body.   They are also poisonous if taken internally, as when cattle eat them while grazing.   Chickens are often killed by feeding upon the rose chafer (Fig. 15.72) in localities where it is abundant, and these beetles have also been shown to contain the poison, cantharidin.

External and Internal Parasites

The most loathsome attack we suffer at the hands of insects is their very common habit of taking up their residence on or within our bodies or those of domesticated animals.   Such insects are called *zoophagous parasites*.   Many insects, such as lice, lay their eggs on animals and live continuously on their hosts, generation after generation, never leaving them except as they instinctively transfer from older to younger animals of the same species or as they are forced to migrate when their host dies. Others spend certain life stages or certain parts of the day on the host and are free-living the rest of the time.   Some of these lay their eggs on the host and live there for a time but desert it in their later stages (*e.g.*, bot flies); others lay eggs and develop away from the host and then become parasites only as adults (*e.g.*, fleas).

Three entire orders of true insects, *viz.*, the chewing lice (Mallophaga), the bloodsucking lice (Anoplura), and the fleas (Siphonaptera), a total of more than 4,000 described species, are entirely parasitic, besides hundreds of species from among the flies (Diptera) and true bugs (Hemiptera) and hundreds more of the ticks and mites of the class Arachnida.

The greater number of these species live externally on the surface of the skin.   Their constant crawling about on the skin causes nervousness, restlessness, loss of sleep, failure to feed, and thus a general run-down condition and increased susceptibility to diseases.   Their excreta and eggs mat the coat and create a foul condition that interferes with the excretory function of the skin.   All these external parasites, except the order Mallophaga, feed by inserting their mouth parts and pumping out the blood.   When they are abundant, this irritation, intensified by the animal's rubbing or scratching, may lead to great sores as in the case of the scab mite of sheep (Fig. 20.26) or the body louse ("cootie") of man. In other cases serious constitutional disturbances result.   The mites

[1] Order Coleoptera, Family Meloidae.

known as chiggers or harvest mites often give rise to chills, nausea, and vomiting. The insertion of the mouth parts of the Rocky Mountain wood tick or American dog tick into man or animals may cause tick paralysis. This affliction, which apparently results from the injection of a salivary toxin, is characterized by an ascending motor paralysis involving complete loss of the use of the limbs and finally death, unless the tick is removed.

External parasites that suck the blood of the higher animals include such common pests as the hog louse, the sheep "tick," fleas, bed bugs, and the cattle tick. There is one group of lice that does not suck blood. These are the so-called bird lice or chewing lice,[1] including all common poultry lice (Fig. 20.29) and certain species of lice found on horses, cattle, and other mammals. Their mouth parts are formed to cut off and ingest solid particles rather than to draw blood. They feed upon the

Fig. 1.10. A piece of grubby hide, after tanning, showing holes made by cattle grubs in the most valuable part of the hide, which render the leather practically useless. (*From Hadwen, Can. Dept. Agr., Health of Animals Branch, Sci. Ser. Bul. 27.*)

dry skin, parts of feathers or hairs, clots of blood, and the like, and their injury is probably chiefly due to nervous irritation from nibbling at the skin and running about over it.

Internal insect parasites of the higher animals are of few kinds. But they are so troublesome to our livestock that they constitute a group probably more destructive than the external parasites. These internal parasites are either mites (Acarina) or true flies (Diptera). With a few possible exceptions all internal insect parasites are *transitory; i.e.,* they pass only a part of their life cycle inside the body. For example, in flies it is always the maggot stage that lives within the animal body; the adult, at least, living away from the host. In the mites the young may be internal, but they live as external parasites on the surface of the skin for at least a part of the adult life.

Just as there are borers in plants, so there are insects that live as borers in the animal body. The itch mite, scaly-leg mite, and their

[1] Order Mallophaga, p. 210.

kind dig tunnels into the flesh, in which their eggs are laid (Fig. 20.7). This explains the intolerable itching that is the most prominent symptom of their attack. Before the cause of the disease was known, the "seven-year itch," as it was often called, was a most loathsome and persistent affliction.

We have spoken of the injury insects cause when they try to lay their eggs on the bodies of animals. The larvae from such eggs are very serious parasites. For example, cattle grubs cause serious damage to the hides of cattle which are rendered more or less useless for making leather (Fig. 1.10). The loss in value of the hides (augmented by the pain suffered by the animals, the depreciation in value of the carcass for beef, the loss of milk flow, and decreased growth) amounts to about

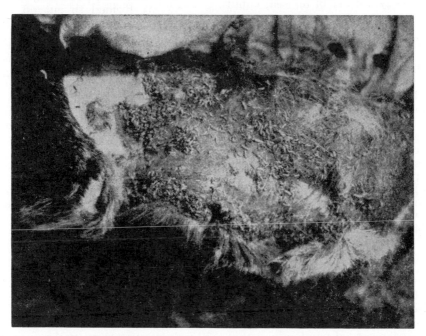

Fig. 1.11. Larvae of a fly (*Phormia terrae-novae*) found under the skin of a reindeer. When heavily infested, the animal usually dies. (*From Hadwen, U.S.D.A., Dept. Bul.* 1089.)

$160,000,000 a year in the United States. In South America, the human bot fly, *Dermatobia hominis*, attacks man in a manner similar to that of the ox warble in cattle. Another species of bot fly causes lumps in the necks of rabbits, and there is a species that emasculates squirrels by living in the scrotum.

The screw-worm and other flies live in the flesh of cattle, horses, hogs, man, and other animals during the maggot stage (Fig. 1.11). These flies are attracted by any wound such as a barbed-wire cut, a dog's bite, dehorning or branding wounds, even the spot where a tick has bitten an animal, or by foul secretions, nasal catarrh, bad breath, and the like. In such situations the eggs are laid, and the larvae tunnel about and feed, greatly aggravating the inflammation and suppuration of the wound and preventing its healing. It is necessary in the southern half of the United States to treat all wounds of animals with a repellent, antiseptic dressing to prevent contamination by this pest.   Fortunately, such attacks upon human beings are relatively rare.

A few kinds of insects, habitually, and some others, accidentally, live in the alimentary canal of animals.   These should not be confused with the intestinal worms,[1] which are not insects.   The best known of intestinal insects are the several species of horse bots (Fig. 20.9).   A horse that is heavily infested with bots generally presents a badly run-down condition. There are no insects that habitually live in the alimentary canal of man, but the eggs or small larvae of bluebottle flies, flesh flies, the house fly, rattailed maggots, and others may be swallowed with impure drinking water, milk, or infested food.   Their presence in the stomach generally causes symptoms of nausea, vomiting, and fever (page 1031).

INSECTS AS CARRIERS OF ANIMAL DISEASES[2]

The most complicated way in which insects injure man and his animals, with the most sinister effects, is as carriers of disease organisms. Because these depredations are commonly associated with primitive living conditions and are not frequently encountered in areas of high standards of living, we tend to forget the importance of insects as vectors of diseases. Nevertheless, the scourge of such insect-borne epidemics as typhus, bubonic plague, yellow fever, African sleeping sickness, and malaria has prevented the colonization and settlement of vast areas of the tropics, decided the outcome of wars, toppled empires, and spread death and disability among billions of the world's inhabitants.   The cost of such depredations is incalculable.

Epidemic typhus is a rickettsial disease transmitted by the bite of the human body louse, *Pediculus humanus humanus*.   Typhus is stated to have killed 2,500,000 or more Russians during the First World War, and millions more died in the Balkans, Poland, and Germany.   Serbia had as many as 9,000 deaths a day and a total of 150,000 victims.   In the Second World War, the disease was again a factor, and an epidemic in Naples in 1944 was stopped only by the mass delousing of 3,000,000 people with DDT powder.

Bubonic plague, the "black death" of the Middle Ages, is a bacterial disease transmitted by bites of the oriental rat flea, *Xenopsylla cheopis*, and other fleas.   This disease caused an estimated 100,000,000 deaths in its first recorded pandemic during a 50-year period of the sixth century and some 25,000,000 victims during a second plague of the middle fourteenth century.   From one-half to two-thirds of the inhabitants of Great Britain perished, and even as late as 1664 to 1666 in London, 70,000 persons of a total population of 450,000 were killed.   In India, almost 10,000,000 deaths from bubonic plague were reported from 1896 to 1917.

Yellow fever is a virus infection transmitted by the bite of the *Aedes aegypti* mosquito.   This disease, the deadly "yellow jack," retarded the development of the American tropics for centuries and nearly prevented the building of the Panama Canal.

African sleeping sickness is a protozoan disease transmitted by the bites

---

[1] Phylum Nemathelminthes.

[2] MATHIESON, R., *Ann. N.Y. Acad. Sci.*, **44**:225, 1943; LINDSAY, D., and H. SCUDDER, *Ann. Rev. Ento.*, **1**: 323, 1956; FULLER, H., *ibid.*, **1**:347, 1956; ADLER, S., and O. THEODORE, *ibid.*, **2**:203, 1957; JELLISON, W., *ibid.*, **4**:389, 1959; RUSSELL, P., *ibid.*, **4**:415, 1959; WEYER, F., *ibid.*, **5**:405, 1960.

of the tsetse flies, *Glossina* spp. (Fig. 1.12). Nearly 500,000 natives perished from this disease between 1896 and 1906, and villages with 30 to 50 per cent infections were found in some areas of Central Africa. This disease has effectively prevented the colonization of large areas of West Africa.

Malaria, a protozoan disease transmitted by the bites of 85 or more species of anopheline mosquitoes, is undoubtedly the most important disease of man. Approximately 200,000,000 clinical cases and 2,500,000 deaths result from it each year. In Ceylon, in 1934–1935, over 66,000 persons were killed in a virulent epidemic, and in Egypt in 1942, 135,000 deaths occurred. An epidemic in Ethiopia, in 1958, resulted in 3,000,000 victims, of which 100,000 died. In areas of the Amazon River basin it is

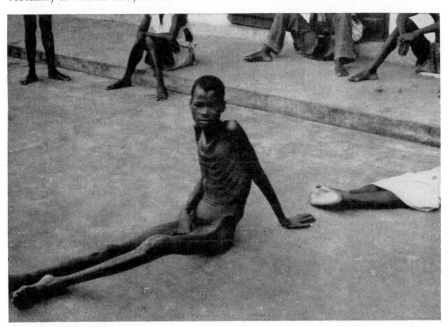

Fig. 1.12. A victim of sleeping sickness, a protozoan disease transmitted by the bite of the tsetse fly, in the French Cameroons. These insects infest about 4,500,000 square miles of land in tropical Africa. (*From World Health Organization, U.N.*)

difficult to find a single individual not chronically infected, and the disease is largely responsible for the lack of colonization of that area. Even in the United States as late as 1935, it is estimated that more than 900,000 cases of malaria occurred, 3,500 deaths being recorded. Malaria has been found in the far north near Archangel, U.S.S.R. (64°N. latitude) and as far south as Cordoba, Argentina (32°S. latitude) and ranges from 1,300 feet below sea level in the Dead Sea basin to as high as 9,000 feet in Bolivia.

Any disease, in order to persist, must continually find new hosts to supplant those lost by death, for the death of a host is calamitous to the parasites that caused it and must often be followed by the death of all the parasites on or in the body. Of the various methods of disease transmission, *i.e.*, bodily contact with infected individuals, contact with contaminated food, water, soil, or clothing or through the air, none is as efficient as the transmission by insects and other arthropods. Most of the pathogens are helpless and inactive, often so delicate that they cannot withstand even exposure to dry air. The insects that carry them are active, hardy, and ubiquitous, and they instinctively seek out for

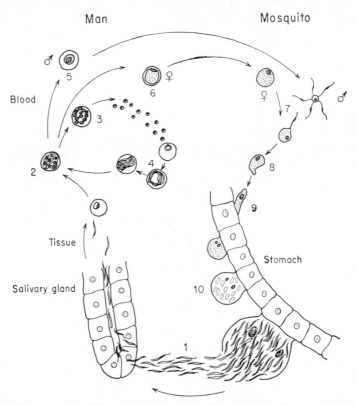

Fig. 1.13. Diagram of the life cycle of the malaria *Plasmodium* in the anopheline mosquito and human blood. (*1*) Sporozoites leave ruptured oöcyst in mosquito stomach wall and migrate to the salivary glands through the mosquito hemolymph. They are injected into the human as the female mosquito feeds, and after an intermediate stage in the liver they infect red blood cells (*2*) and pass through an asexual reproductive cycle, forming (*3*) schizonts, which rupture, liberating merozoites and causing the sporadic fever. The merozoites enter and infect other red cells to form (*4*) trophozoites and repeat the cycle. Eventually, some of the cells develop the sexual forms, the male microgametocytes (*5*) and the female macrogametocytes (*6*), which are taken into the mosquito stomach as the female feeds on the infected human. There they change to the mature sexual forms, the micro- and macrogametes (*7*), which unite to form the fertilized female zygote (*8*). This becomes the motile oökinete (*9*) and bores into the stomach wall of the mosquito, where they produce the oöcyst (*10*).

their own food the very animals upon which alone the disease organisms can survive. The insects thus provide rapid, selective transportation of the pathogens on legs, wings, or mouth parts, and they may shelter them in their bodies—in the alimentary canal, body cavity, salivary glands, muscles, Malpighian tubes, or fat-body. The pathogens may, during the journey to a new host, remain unchanged in the insect body and may overwinter there.   They may increase their numbers without a change in form, or they may increase by a metamorphosis, passing through definite stages of their life cycle in the body of the insect which they cannot undergo in the body of any other animal or in any other situation.   In the latter case the insect is an *essential host* for the pathogen, and the life cycle of the malaria parasite, as shown in Fig. 1.13, is an example of the complexities of such a relationship.

When the infected insect bites man or other animals, the pathogens may pass down the mouth parts into the blood or lymph; they may be deposited on the skin in a receptive place such as the lips, the surface of the eye, or a small wound or a sore; or they may pass out with the insect feces, be regurgitated in a vomit spot, or be swallowed by the new host along with the insect carrier.   A study of Table 1.2 will show the essential facts regarding some of the most important diseases of man and domestic animals which are known to be carried by insects.

The problems of the insect transmission of animal diseases are complicated by the fact that, besides the insect carrier, the disease organism may have at least two other hosts, one in which it causes a definite disease and another in which it appears to be more or less harmless.   The latter animals are known as *reservoirs* of the disease.   Thus in the case of Rocky Mountain spotted fever, the tick, *Dermacentor andersoni*, and other ticks are carriers, and man is the victim of the disease, but a number of other animals, such as snowshoe rabbits, ground squirrels, gophers, woodchucks, and mice, are reservoirs of the infection.   Certain monkeys, in the vast hinterlands of South America and Africa, into which man rarely penetrates, harbor a type of yellow fever known as *jungle fever*, which may be carried to villages and cities by an occasional human traveler.   This inaccessible, permanent source of yellow fever renders the hope of eradicating the disease from the earth (an accomplishment felt, until 1929, to be within the grasp of science) a dream of the remote future.

An additional complication to the problems of arthropod transmission of animal diseases results from the fact that the disease-causing organism may pass from an infected mite or tick, and perhaps an insect, through its eggs to the succeeding generation or generations, to reappear in a virulent form again after a long period of latency.   Thus the infection of a vector may mean that its offspring, in some cases to the fourth generation, will be infective without further exposure to the disease.

It is obvious that in the three-party relationships involving man and animals and the two great unconquered biotic groups, insects and microorganisms, we have a subject that has been of the utmost importance to human welfare for thousands of years.   However, scarcely 60 years have passed since the discovery of the complex life cycles of the malaria parasites and their transmission by anopheline mosquitoes between human hosts was made by the pioneer work of Laveran, Manson, and Ross from 1880 to 1898, and since Smith and Kilbourne in 1893 published the proof that the Texas fever disease of cattle is caused by a protozoan parasite that lives within and destroys the blood cells and is transmitted exclusively by the bites of the cattle tick, *Boophilus annulatus*.   In this period most of the associations between insects and human and animal diseases have been elucidated and satisfactory control measures, mostly aimed at the insect vectors, have been developed.   Yet there remain innumerable unsolved problems of insect biology and host relationships

TABLE 1.2. SOME OF THE MORE IMPORTANT INSECT-

Name of the disease	Animal affected	Vector of the disease	Classification of the vector	Pathogenic organism
Yellow fever........	Man, monkeys, rodents, opossums, and anteaters	The yellow fever mosquito, *Aedes aegypti, A. leucocelaenus, Hemagogus capricornii, H. equinus, H. spegazzinii,* and *H. albomaculatus*	Class Hexapoda, Order Diptera, Family Culicidae	*Charon evagatus*
Malaria............	Man	85 species of *Anopheles* mosquitoes	The same	*Plasmodium vivax, P. falciparum, P. malariae,* and *P. ovale*
Dengue............	Man	The mosquitoes, *Aedes aegypti* and *A. albopictus*	The same	
Encephalitides......	Man, horse, birds	The mosquitoes, *Culex tarsalis, C. pipiens, C. quinquefasciatus, Aedes taeniorhynchus, A. sollicitans,* and others; the ticks, *Ixodes ricinus* and *I. persulcatus*	The same	*Erro scoticus, E. silvestris, E. scelestus, E. equinus*
Filariasis...........	Man	20 or more species of *Aedes, Anopheles, Culex,* and *Mansonia* mosquitoes	The same	*Wuchereria bancrofti* and *W. malayi*
Onchocerciasis......	Man	At least five species of black flies, genus *Simulium*	Class Hexapoda, Order Diptera, Family Simuliidae	*Onchocerca volvulus*
African sleeping sickness	Man	The tsetse flies, *Glossina palpalis* and *G. morsitans,* and at least five other species	Class Hexapoda, Order Diptera, Family Muscidae	*Trypanosoma gambiense* and *T. rhodesiense*
Nagana............	Domestic animals and wild game	The tsetse flies, *Glossina morsitans* and *G. longipalpis,* and other species	The same	*Trypanosoma brucei*
Verruga peruana or Oroya fever	Man	The psychodid fly, *Phlebotomus verrucarum*	Class Hexapoda, Order Diptera, Family Psychodidae	*Bartonella bacilliformis*
Papataci fever......	Man	The psychodid fly, *Phlebotomus papatasii*	The same	Unknown; a virus
Tularaemia.........	Man, rodents, ground birds	Deer flies, fleas, rabbit louse, ticks	Tabanidae, Pulicidae, Haematopinidae, Ixodidae	*Pasteurella tularensis*
Spotted fevers......	Man, rodents	The ticks, *Dermacentor andersoni, D. variabilis, D. occidentalis,* and others	Class Arachnida, Order Acarina, Family Ixodidae	*Rickettsia rickettsii*

BORNE DISEASES OF MAN AND DOMESTIC ANIMALS

Classification of the pathogen	Distribution of the disease	Method of transmission	Other ordinary ways of getting the disease
Virus	The tropics and subtropics of Africa and America	Directly inoculated into blood by mouth parts of mosquito	None. Exclusively insect-borne
Phylum Protozoa, Class Sporozoa, Order Haemosporidia, Family Plasmodidae	In a broad belt around the globe in the tropics and subtropics	The same	None. Exclusively insect-borne
Unknown	Around the globe in the tropics and subtropics	The same	None. Exclusively insect-borne
Virus	United States, Canada, South America, Europe, Asia	The same	None. Exclusively arthropod-borne
Phylum Nemathelminthes, Class Nematoda, Order Spirurida	Around the globe in the tropics and subtropics	The same	None. Exclusively insect-borne
The same	Equatorial Africa, Mexico, Central America	Directly inoculated into blood by mouth parts of the fly	None. Exclusively insect-borne
Phylum Protozoa, Class Mastigophora, Order Protomonadina, Family Trypanosomidae	Equatorial Africa	The same	None. Exclusively insect-borne
The same	Equatorial Africa	The same	None. Exclusively insect-borne
Rickettsiae, uncertain	Peru, Ecuador, Bolivia, and Chile on slopes of the Andes	Directly inoculated into the blood by mouth-parts of the insect	None. Exclusively insect-borne
Unknown	Mediterranean region, India, Ceylon, South China	The same	None. Exclusively insect-borne
Bacteria, Family, Bacteriaceae	General in United States, Canada, Europe, Japan	The same	Butchering rabbits, eating undercooked flesh of rabbits, or from bites of rodents
Rickettsiae, uncertain	Entire United States and Alaska	Directly inoculated into the blood by bite of tick	None. Exclusively tick-borne

TABLE 1.2.

Name of the disease	Animal affected	Vector of the disease	Classification of the vector	Pathogenic organism
Relapsing fevers....	Man, rodents	Many species of ticks of the genus *Ornithodoros* and the louse, *Pediculus humanus*	Class Arachnida, Order Acarina, Family Ixodidae	At least 15 species of *Borrelia*
Leishmaniasis.......	Man, dog, rodents	A number of species of *Phlebotomus* sand flies	Class Hexapoda, Order Diptera, Family Psychodidae	*Leishmania tropica, L. donovani,* and other species
Texas fever or piroplasmosis	Cattle	The cattle ticks, *Boophilus annulatus* and *B. microplus*	The same	*Babesia bigemina*
Fowl spirochetosis...	Chicken, turkey, goose	The fowl tick, *Argas persicus*	Class Arachnida, Order Acarina, Family Argasidae	*Borrelia anserina*
Scrub typhus or tsutsugamushi fever	Man, rodents	The chigger mites, *Trombicula akamushi, T. deliensis, T. minor*	Class Arachnida, Order Acarina, Family Trombidiidae	*Rickettsia tsutsugamushi*
Epidemic typhus fever	Man	The human louse, *Pediculus humanus*	Class Hexapoda, Order Anoplura, Family Pediculidae	*Rickettsia prowazekii*
Endemic or murine typhus fever	Man, rodents	The oriental rat flea, *Xenopsylla cheopis,* and other fleas, lice, mites, and ticks which spread the disease among rodents	Class Hexapoda, Order Siphonaptera, Family Pulicidae	*Rickettsia typhi*
Plague.............	Man, rodents	The oriental rat flea, *Xenopsylla cheopis,* and other fleas which spread the disease among rodents	The same	*Pasteurella pestis*
Chagas' disease.....	Man, rodents	At least 20 species of assassin bugs of the genera *Triatoma, Rhodnius,* and *Eratyrus*	Class Hexapoda, Order Hemiptera, Family Reduviidae	*Trypanosoma cruzi*

between the insects, the pathogens, and the animal reservoirs which require prolonged and intensive study. However, the introduction of DDT and lindane dusting powders for human louse control has virtually eliminated epidemic typhus, and the residual spraying of dwellings with DDT, lindane, and dieldrin has eradicated malaria in Sardinia and almost eliminated it in the Mediterranean area, by killing infected female anophelines before parasite development is complete, when they become infective (Fig. 1.13). Today the World Health Organization is actively engaged in the global eradication of malaria, and similar conquests are being made of most of the other insect-transmitted diseases. The major difficulty apparent is the development of insecticide resistance among the vector species (page 397).

(*Continued*)

Classification of the pathogen	Distribution of the disease	Method of transmission	Other ordinary ways of getting the disease
Bacteria, Family Spirochaetaceae	Africa, Mediterranean region, Asia Minor, North and South America	By bites or from feces or coxal fluid	None. Exclusively tick- or louse-borne
Phylum Protozoa, Class Mastigophora, Order Protomastigida, Family Trypanosomidae	Mediterranean Asia Minor, India, China, South America	By bite	None. Exclusively Insect-borne
Phylum Protozoa, Class Sporozoa, Order Haemosporidia, Family Babesiidae	Southern United States, Central and South America, South Africa, Philippines, and Europe	Directly inoculated into the blood by mouthparts of the tick	None. Exclusively tick-borne
Bacteria, Family Spirochaetaceae	India, Australia, Brazil, North America, Persia, Egypt, etc.	By bites or from feces	None. Exclusively tick-borne
Rickettsiae, uncertain	Australia, Japan, India, Formosa, China, Malay Archipelago, East Indies	Directly inoculated into the blood by bite of mite larvae	None. Exclusively mite-borne
Rickettsiae, uncertain	Europe, Mexico	Deposited in feces by louse and scratched into skin or wounds from bites	None. Exclusively insect-borne
Rickettsiae, uncertain	Southern United States Mexico	Deposited in feces by flea and scratched into skin or wounds from bites	None. Exclusively insect-borne
Bacteria, Family Bacteriacae	Nearly cosmopolitan	From feces or vomit spots, by bites or through abrasions	By breath of plague victim or from fomites
Phylum Protozoa, Class Mastigophora, Order Protomonadina, Family Trypanosomidae	South and Central America	From feces scratched into skin or by passing through mucous membranes of mouth, nose, or around eyes	None. Exclusively insect-borne

C. INSECTS AS DESPOILERS OF STORED PRODUCTS AND OTHER MATERIALS

We have reviewed the ways in which insects are injurious to living plants and to living animals, including man. The third great phase of insect injury arises from the fact that they compete with us for the possession and use of practically all our stored products—both stored foods and the many other articles with which we habitually surround ourselves. We find here, exactly as in the case with living plants and living animals, that the attacks of insects are motivated mostly by hunger. To a lesser degree, damage results from their efforts to make provision for their eggs or young or in seeking shelter or building nests for themselves.

Since hunger is the principal motive involved and since insects eat organic matter of every kind, and practically only organic matter, it follows that the chief stored products to be protected are things of plant or animal origin, including grains, seeds, flour, meal, candies, nuts, fruits, vegetables, meats, fats, milk, cheese, honey, wax, tobacco, spices, drugs, feathers, furs, leather goods, woolens, paper, books, labels, photographs, boxes, furniture, wooden buildings, bridges, piling, mine props, telephone poles, railroad ties and trestles, and collections of insects, plants, and animals.   These are some of the things that must be guarded from insect depredations.

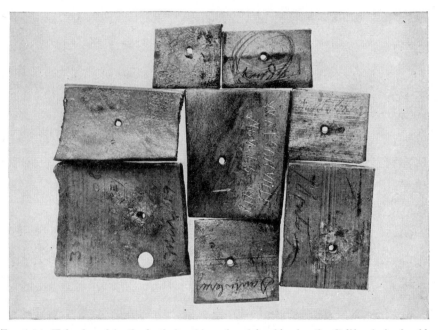

Fig. 1.14. Holes bored in the lead sheathing of aerial cables by the California lead cable borer, *Scobicia declivis*, from various localities in California.   (*From U.S.D.A., Dept. Bul.* 1107.)

Many other things not of an organic origin are generally immune from attack, *e.g.*, jewelry, metals of all kinds, pottery, statuary, brick, stone, and cement work.   Even these inorganic objects are not entirely inviolate by insects.   A species of powder post beetle has been given the name of lead cable borer[1] because of its troublesome habit of eating holes through the lead sheathing of aerial telephone cables (Fig. 1.14).   These holes admit moisture and cause short circuits; often the insulation becomes water-soaked and ruined for an appreciable length, necessitating splicing and resheathing.   In southern California this type of insect injury is reported as causing about one-fifth of all aerial cable troubles. As many as 125 holes to a span of 100 feet have been found.   A single hole may put from 50 to 600 telephones out of use for from 1 to 10 days. Termites similarly have bored through the lead pipe and the cotton

[1] *U.S.D.A., Dept. Bul.* 1107, 1922.

insulation enclosing underground cables, thus ruining them within a year after they were laid down. Beetles of at least a dozen different families, besides the caterpillars of several moths and adult wasps of several kinds, have been recorded as boring through metal. Some years ago, one of the large railroads discovered that a mud-dauber wasp was causing great trouble and expense by building its mud nests in the exhaust port of the pressure-retaining valves of the Westinghouse air brake. It became necessary to change the shape of the valve and the form of its opening to overcome the trouble.

Food-infesting Insects

While the above cases are spectacular, it must be borne in mind that the total injury by all the insects that attack inorganic articles in all time past is probably exceeded in a single year by the injury of any one of a dozen or more pests of stored foods. The pests of stored products are the most expensive of all insects to feed, because they feed upon products that have been grown, harvested, stored, and, in many cases, have incurred further expense through manufacturing, advertising, selling, and distributing processes.

In times of stress, attacks by insects upon man's food supply may mean death to thousands. In earlier times one of the critical duties of sailors upon long ocean voyages was to guard the ship's biscuits from the ever-present and ravenous cockroaches. During the First World War, large quantities of wheat, badly needed by the European nations, were destroyed by weevils in Australia, and entomologists were hurriedly dispatched to check the destruction. Seeds are among the most concentrated foods known, and a large part of the injury to stored products is to seeds of our cereal and leguminous crops. Of the many animals that compete for this valuable food material in any community, such as birds, rats, mice, insects and man, insects probably get the largest share, next to man (Fig. 1.15).

It has been estimated that it costs the American people $500,000,000 a year to feed the insect pests of stored foods.[1] Every individual can recall some case where an article of food had to be discarded because insects "beat man to it." These small losses make in the aggregate a heavy total. To the total of small losses must be added all-too-frequent cases where an entire crop in storage, the contents of a large elevator, or a shipment of foodstuffs has been rendered unfit for human consumption.

The method of attack is varied. Some species hide their eggs in the developing seeds as they grow in the field, and the injury becomes apparent only after the immature insects have been carried into the storehouse. Such is the case with the pea and bean weevils (Figs. 19.25 and 19.26) and the Angoumois grain moth (Fig. 19.22). Other kinds enter by stealth into kitchens, granaries, or factories and deposit their eggs on cured meats, harvested seeds, or any of the products manufactured from the raw-food materials. Some of the grain insects make their homes inside of single whole grains during all their growth; others attack only the broken or ground seeds, roaming about in flour, meal, and other foods and contaminating much more than they eat, with their excreta or the silk that they spin. Many other kinds do not breed in the stored foods, either having nests outside and entering our foods only on foraging expeditions, e.g., the ants, or leading a gypsy life, e.g., the cockroaches,

[1] Haeussler, G. J., U.S.D.A., Yearbook, 1952:141.

which are objectionable more on account of the filth and disease germs that they probably carry than because of the amount of food consumed.

Two orders of insects are of prime importance as pests of stored foods: (a) the Coleoptera, including such notorious pests as the granary weevil (Fig. 19.16), the confused flour beetle, the saw-toothed grain beetle, the pea weevil, the bean weevils, the larder beetle, and many others; and (b) the Lepidoptera, including the Angoumois grain moth, the Mediterranean flour moth (Fig. 19.23), the Indian meal moth, and others.   In the former

Fig. 1.15. Larvae of Khapra beetle, *Trogoderma granarium*, infesting pinto beans.   (*Univ. Calif. Coll. Agr.*)

of these groups both the grubs or larvae and the adult beetles feed on the stored materials, while among the moths only the caterpillars are directly injurious.   In addition to these two groups of most importance, such pests as the book-louse, the cheese skipper, the cheese and ham mites, sporadically destroy large quantities of food (pages 917, 919).

No other economic group is more widely and equitably distributed than these insects of stored products.   Many of the worst kinds are quite cosmopolitan.   Some of them have been said to have spread over an entire continent in two or three years.   They crawl and fly about seeking the concentrations of attractive foods that we bring together; they enter our storehouses on the crops we harvest; they are distributed in the seeds we purchase for planting; they go to market with the grains we sell; and they come back to us in the flour, breakfast cereals, cakes, and crackers from our grocers.

PESTS OF WOOD AND WOODEN ARTICLES, CLOTHING, AND DRUGS

A particularly insidious pest is the termite, or white ant (Fig. 19.4), whose fondness for a diet of woody tissues leads to most surprising invasions of dwellings (Fig. 1.16), libraries, trestles, fence posts, and indeed any article of wooden origin, such as stores of paper stock, cardboard

FIG. 1.16. A public building, the foundation timbers of which have been badly damaged by termites, necessitating extensive repair work (*From photo by Snyder, U.S.D.A.*)

boxes, library books (Fig. 1.17), and the like. Since they avoid exposure to the air, they are seldom seen until great damage has been done. Living in the ground, they make an opening into any timber that touches the soil and, from that tiny entrance, excavate a honeycomb of connecting passageways, working always in the interior of the structure invaded, without breaking the surface, until the timbers are so weakened that they break through before the occupants are aware that an enemy has been at work beneath their feet. Many cases are on record of extensive damage to private and public buildings by this insect, and its numbers seem to be on the increase. The powder post beetles (Fig. 19.5) do similar injury, but their dust-filled tunnels are easily distinguished from the frass-free runways of the termites.

In a somewhat different way the silverfish (Fig. 6.11) usually establishes itself in new quarters before it is suspected, because it is nocturnal and hides in cracks during the daytime.   It has a fondness for starchy material and glue, which leads it to eat at book bindings, photographs, wall paper, and all kinds of labels.   In a large engineering laboratory it was discovered that the silverfish, in order to get the sizing in the paper, had effaced the numbers on inventory cards that gave the only clue to valuable apparatus out on loan.

Everyone has had the discomfiting experience of finding that furs, rugs, upholstered furniture, and winter clothing, stored during the summer, have been so eaten by clothes moths as to render them useless. A few insects have the surprising habit of feeding upon such things as drugs and tobaccos.   The tobacco beetle riddles cigars and cigarettes with fine holes.   It also does great damage to upholstered furniture and

Fig. 1.17. A book from an Arkansas library, ruined by the feeding of termites.   (*From U.S.D.A., Farmers' Bul.* 759.)

in wholesale houses sometimes causes losses by tunneling through the leather soles of boots and shoes.   The drug store beetle is most catholic in its tastes, having been found feeding in at least 45 different kinds of drugs, some of which are poisonous to man, and also on such widely different articles as books, sheet cork, chocolate, red and black pepper, ginger, and yeast cakes.

CONCLUSION

It has been seen how widespread and diverse is the conflict between insects and man.   Our growing crops must struggle against insect attacks from the time the seed is planted until the crop is safely harvested— attacks upon leaves, branches, stems, roots, buds, blossoms, and fruits. Our domesticated animals, and man himself, are harassed and bitten and worried and their bodies infested with maggots and inoculated with

TABLE 1.3. THE INSECTS OF THE UNITED STATES IN ACCOUNT WITH THE
AMERICAN PEOPLE—DEBIT
Most Important Items, 1957

Damage to:		Per cent	Estimated loss due to insects
Quantity	Kind of crop		
160,815,000 lb.	Alfalfa seed	10	$ 3,936,000
505,353,000 bu.	Barley, rye, rice	8	50,564,000
1,577,100,000 lb.	Beans (dry edible)	12	13,336,000
43,000 tons	Broomcorn	10	1,029,000
1,871,000 bu.	Buckwheat	5	104,000
5,932,000 gal.	Cane and sorgo sirup	7	676,000
123,409,000 lb.	Clover seed (red, alsike, sweet, white, Ladino)	28	7,342,000
3,402,832,000 bu.	Corn	9	347,123,000
10,964,000 bales	Cotton lint ⎫	15	278,056,000
4,609,000 tons	Cotton seed ⎭		
481,266,000 bu.	Cowpeas and soybeans	5	50,199,000
25,754,000 bu.	Flaxseed	10	7,557,000
561,977,000 bu.	Grain sorghums	7	37,663,000
121,402,000 tons	Hay	11	246,327,000
40,135,000 lb.	Hops	12	2,575,000
1,308,360,000 bu.	Oats	5	40,024,000
1,445,110,000 lb.	Peanuts	3	4,476,000
3,270,000 bags	Peas (dry field)	8	956,000
239,539,000 cwt	Potatoes	15	69,521,000
15,497,000 tons	Sugar beets	15	26,035,000
6,750,000 tons	Sugarcane for sugar and seed	20	9,286,000
18,053,000 cwt	Sweetpotatoes	5	3,742,000
37,605,000 lb.	Timothy seed	10	306,000
1,660,553,000 lb.	Tobacco	10	93,453,000
107,000 tons	Velvet beans	12	413,000
947,102,000 bu.	Wheat	9	165,273,000

Total estimated damage to staple crops by insects $1,459,972,000

16,400 tons	Artichokes	10	$ 289,000
1,338,000 cwt ⎫ +114,460,000 tons ⎭	Asparagus	5	1,937,000
324,000 cwt ⎫ +92,650 tons ⎭	Lima beans	10	1,602,000
4,921,000 cwt ⎫ +358,950 tons ⎭	Snap beans	8	6,926,000
506,000 cwt ⎫ +159,900 tons ⎭	Beets	5	239,000
19,086,000 cwt ⎫ +169,500 tons ⎭	Cabbage ⎫ Kraut ⎭	20	8,681,000
558,000 tons	Cantaloupes	20	12,215,000
700,000 tons	Carrots	10	4,489,000

TABLE 1.3. (*Continued*)

Damage to:		Per cent	Estimated loss due to insects
Quantity	Kind of crop		
225,200 tons	Cauliflower	10	$ 1,465,000
747,200 tons	Celery	10	6,043,000
11,807,000 cwt ⎱ +1,491,500 tons ⎰	Sweet corn	10	8,004,000
4,284,000 cwt +15,419,000 bu.	Cucumbers ⎱ Pickles ⎰	20	8,261,000
25,000 tons	Eggplant	15	352,000
8,800 tons	Kale	15	106,000
1,693,600 tons	Lettuce	5	7,200,000
1,205,400 tons	Onions	20	12,015,000
288,000 cwt ⎱ +559,890 tons ⎰	Peas	12	6,362,000
137,200 tons	Peppers	10	2,651,000
1,664,000 cwt ⎱ +142,900 tons ⎰	Spinach	10	1,537,000
19,651,000 cwt ⎱ +3,287,800 tons ⎰	Tomatoes	7	16,341,000
1,500,800 tons	Watermelons	15	7,538,000

Total estimated damage to truck crops by insects................ $114,254,000

118,548,000 bu.	Apples	20	$ 43,434,000
190,400 tons	Apricots	10	3,117,000
56,800 tons	Avocados	4	376,000
240,000 tons	Cherries	10	4,884,000
1,050,000 bbl.	Cranberries	6	690,000
2,599,000 tons	Grapes	15	23,979,000
44,700,000 boxes	Grapefruit	10	5,154,000
14,700,000 boxes	Lemons	10	3,528,000
136,190,000 boxes	Oranges	10	25,221,000
62,335,000 bu.	Peaches	20	25,600,000
31,676,000 bu.	Pears	8	5,035,000
141,350,000 lb.	Pecans	10	3,365,000
88,000 tons	Plums	8	1,318,000
231,000 tons	Prunes (dried and fresh)	8	2,956,000
563,832,000 lb.	Strawberries	10	7,948,000
66,600 tons	Walnuts	20	6,235,000

Total estimated damage to fruit crops by insects................ $162,840,000

$136,339,526	Nursery products	12	$ 16,360,000
247,857,089	Flowers	20	49,571,000
69,457,057	Vegetables under glass	15	10,418,000

Total estimated damage to nursery and greenhouse products...... $ 76,349,000

TABLE 1.3.   (*Continued*)

Damage to:		Per cent	Estimated loss due to insects
Quantity	Kind of crop		
77,883,000 tons	Rangeland forage	6	$ 44,376,000
48,840,000,000 bd. ft.	Saw timber	10	73,260,000
10,760,000,000 cu. ft.	Forest growing stock	9	174,312,000
Total estimated damage to forest trees and forest products........			$247,572,000
94,502,000 head	Cattle and calves, including milk cows	5	$432,639,000
3,574,000 head	Horses, mules, and colts	½	1,280,000
30,840,000 head	Sheep and lambs	3	13,840,000
51,703,000 head	Swine	¼	3,187,000
390,137,000 head	Chickens	5	22,851,000
Total estimated loss in livestock production....................			$473,797,000
Total estimated damage to stored grain........................			$500,000,000
Total estimated damage to packaged food products..............			$150,000,000
Total estimated damage to garments, furniture, household goods...			$200,000,000
Injury by noxious insects, including transmission of typhoid, tuberculosis, dysentery, conjunctivitis, cost of mosquito, fly, flea, and bed bug control.......................................			$100,000,000
Grand total...			$3,529,160,000

disease. Our foods are contaminated, our clothing ruined, our books and papers consumed, the wires that carry our messages rendered ineffective, and the timbers of our houses eaten piecemeal.

No one knows how much better off man might be were all his insect enemies destroyed. We can only conjecture and, by piecing together scattered records from many individuals of authentic losses, arrive at estimates of the total. Such an attempt to indicate what our insect enemies cost us is given in Table 1.3. Estimates of this kind have usually been based on the commonly accepted belief that insects destroy, on the average and the country over, at least 10 per cent of every crop every year. Realizing that damage varies greatly from crop to crop and among the different species of animals, the authors have attempted, as a substitute for the flat rate of 10 per cent, to estimate in a more specific manner the damage to the various crops and animals. The percentage estimates given in Table 1.3 are based upon field observations and records kept in the *Illinois Natural History Survey*, the *University of California Division of Agricultural Sciences*, and on U.S. Department of Agriculture statistics.[1] The figures for total crop values were taken from *Agricultural Statistics*, 1957, *Agricultural Census*, 1954, and *Timber Resources Review*, 1955.

[1] See HYSLOP, J., *U.S.D.A., Bur. Ento. Cir.* E-444, 1938; HAEUSSLER, *loc. cit.*

CHAPTER 2

THE VALUE OF INSECTS TO MAN

So much of the writings about insects must necessarily deal with their destructiveness that we may be in danger of forgetting that many insects have beneficial attributes and habits the value of which we can hardly overestimate. It is a little startling to discover that this humble class of animals contributes to the world's commerce products that sell for more than $125,000,000 each year, in the United States alone;[1] or to read from the pen of an American entomologist[2] that:

"Except for the check put upon insect multiplication through warfare within the insect household, by which one species of insect destroys its relatives, no informed naturalist would expect the survival of the human race for a longer period than 5 to 6 years. Not only would man's food supply be appropriated by his insect enemies, but it would be impossible for him to withstand the withering march of malaria, yellow fever, typhoid, bubonic plague, sleeping sickness, and other maladies transmitted by insect carriers."

We need to study these creatures very carefully, in order that we may be able to distinguish insect friends from insect enemies. Almost any entomologist can tell from his own experience of incidents where people have gone to great trouble and expense to destroy quantities of insects, only to learn later that the insect destroyed was not only harmless but was actually engaged in saving their crops by eating the destructive form. Certainly most entomologists have had correspondents send in the larvae of Syrphidae or lady beetles with the complaint that they were injuring plants; at the same time overlooking the smaller aphids which were causing the injury and which these larvae were continually devouring.

Each citizen owes it to himself to know, as well as possible, the sundry ways in which beneficial insects affect the complex currents of plant and animal life to his advantage. The curious facts of honey production, silk production, and shellac production; the wonderfully intricate mechanisms of pollinization; the nature of food and the means of getting it, of insectivorous game, fur-bearing animals, fish, fowl, and songbirds, which subsist so largely on insect food; the possibilities of greatly increasing the quantity of certain fish and game, at present only a delicacy on the tables of the wealthy (or the lucky!); the overwhelming possibilities for human ill or welfare leashed in the prodigious hordes of bugs that eat each other —all these things are entomological topics that deserve our earnest consideration.

[1] See *U.S.D.A.*, *Yearbooks*, statistics on silk, honey, beeswax, shellac, etc.
[2] Gossard, H. A., "Relation of Insects to Human Welfare," *Jour. Econ. Ento.*, **2**: 313–332, 1909.

THE WAYS IN WHICH INSECTS ARE BENEFICIAL OR USEFUL TO MAN

A. *Insects produce and collect useful products or articles of commerce:*
  1. The secretions of insects are valuable:
     a. The saliva of the silkworm is the true silk of commerce.
     b. Beeswax is a secretion from hypodermal glands on the underside of the honeybee's abdomen.
     c. Shellac is the secretion from hypodermal glands on the back of a scale insect of India.
     d. The light-producing secretion of the giant firefly of the tropics is used in minor ways for illumination and may point the way to the synthesis of a substance giving brilliant light with almost no accompanying heat.
  2. The bodies of insects are useful or contain certain useful substances:
     a. Cochineal and crimson lake are pigments made by drying the bodies of a cactus scale insect of the tropics.
     b. Cantharidin is secured from the dried bodies of a European blister beetle known as the "Spanish fly."
     c. Insects, such as the hellgrammite or dobson, are widely used as fish bait, and the best artificial flies are modeled after insects.
  3. Insects collect, elaborate, and store plant products of value:
     a. Honey is nectar assembled from blossoms—concentrated, modified chemically, and sealed in waxen "bottles" by the honeybee.
  4. Insects cause plants to produce galls, some of which are valuable:
     a. Tannic acid from insect galls has been used for centuries to tan the skins of animals for leather or furs.
     b. Many insect galls contain materials that make the finest and most permanent inks and dyes.
B. *Insects aid in the production of fruits, seeds, vegetables, and flowers, by pollinizing the blossoms:*
  1. Most of our common fruits are pollinized by insects. The growing of Smyrna figs is dependent upon a small wasp that crawls into the flower cluster.
  2. Clover seed does not form without the visit of an insect, usually some kind of a bee, to each blossom.
  3. Peas, beans, tomatoes, melons, squash, and many other vegetables require insect visits before the fruits "set."
  4. Many ornamental plants, both in the greenhouse and out-of-doors, are pollinized by insects, *e.g.*, chrysanthemums, iris, orchids, and yucca.
C. *The bodies of insects serve as food for many animals that are valuable to us:*
  1. Many of our food fish subsist largely upon aquatic insects.
  2. Many highly prized song and game birds depend upon insects for a large percentage of their food.
  3. Chickens and turkeys naturally feed upon insects and, under proper conditions, can be raised almost exclusively on such a diet.
  4. Hogs may feed and fatten upon white grubs rooted from the soil.
  5. A few of the wild, fur and game animals eat insects, *e.g.*, skunk and raccoon.
  6. In many parts of the world, from ancient times to the present day, insects have been eaten extensively by human beings. Grasshoppers, crickets, walking-sticks, beetles, caterpillars and pupae of moths and butterflies, termites, large ants, aquatic bugs, cicadas, and bee larvae and pupae are prized as food by most of the more primitive races of men.
D. *Many insects destroy other injurious insects:*
  1. As parasites, living on or in their bodies and their eggs.
  2. As predators, capturing and devouring other insects.
E. *Insects destroy various weeds in the same ways that they injure crop plants.*
F. *Insects improve the physical condition of the soil and promote its fertility:*
  1. By burrowing throughout the surface layer.
  2. Their dead bodies and droppings serve as fertilizer.
G. *Insects perform a valuable service as scavengers:*
  1. By devouring the bodies of dead animals and plants.

    2. By burying carcasses and dung.

H. Certain insects are indispensable in scientific investigations:

    1. The ease of handling, rapidity of multiplication, great variability, and low cost of keeping and rearing make insects ideal experimental animals for the study of physiology, biochemistry, and ecology.

    2. The foundations of modern genetics have been derived from studies of the Drosophila or fruit fly.

    3. Studies of variation, geographical distribution, and the relation of color and pattern to surroundings have been greatly advanced through the study of insects.

    4. Principles of polyembryony and parthenogensis have been discovered by the study of insects.

    5. The behavior and psychology of higher animals have been illuminated by a study of the reactions of insects such as the honeybee, whose behavior can be analyzed into simple tropisms. Valuable lessons in sociology have been deduced from considerations of the economy of social insects.

I. Insects have aesthetic and entertaining value:

    1. Their shapes, colors, and patterns serve as models for artists, florists, milliners, and decorators.

    2. The more highly colored and striking forms are much used as ornaments in trays, pins, rings, necklaces, and other jewelry.

    3. Moths and butterflies are universally admired; those who use the microscope find much to admire in the colors and patterns of many of the smaller insects.

    4. The songs of insects have been found highly interesting.

    5. Insects have served as subject matter for hundreds of poems.

    6. The inimitable variety found among insects and their curious habits afford entertainment and diversion for thousands who collect and study them.

    7. Orientals gamble on crickets trained for fighting, and fleas have often been used for circus stunts.

J. Insects and insect products have limited use in medicine:

    1. The maggots of certain flies, reared aseptically, have been used in the treatment of wounds.

    2. The stings of honeybees have remedial value for rheumatism and arthritis.

    3. Extracts from the bodies of insects and insect products such as royal jelly are used to some extent as medicines.

USEFUL INSECT PRODUCTS

The most obvious and tangible of the benefits that arise from insect activities is the utilization of the things that insects make, collect, or produce, such as silk, honey, beeswax, shellac, paints, dyes, and medicines.

Silk. Preeminent among insects valuable in this way is the silkworm. Very few persons know the silkworm by sight, but everyone knows the product it manufactures. Those who know this product most seldom if ever think of its lowly origin; many doubtless do not know that silk is caterpillar's spittle. The silkworm has been a creature of domestication from time beyond memory and history, a captive and slave to man. For more than 35 centuries it has toiled ceaselessly for man: countless generations laying their eggs, eating the mulberry leaves provided for them in the larval stage, spinning their cocoons, and then dying, a perpetual sacrifice to the demand of men and women for adornment. For more than 2,000 years only the Chinese knew of the origin of silk, the discovery of its usefulness being credited to Lotzu, Empress of Kwang-Ti, in about 2697 B.C. They punished with death anyone who attempted to take the eggs or silkworms out of the country, and it is said that silk was for a time valued at its weight in gold. But in the year A.D., 555 two monks sent as spies to China discovered the nature and source of silk and brought back to Constantinople some eggs of the silkworm moth concealed in a hollow staff. In this humble way silk culture was introduced into Europe.

The parent of the silkworm, *Bombyx mori*, is a creamy white moth (Fig. 2.1) about 2 inches across the open wings. It is fat-bodied and feeble-winged; it scarcely ever flies; it takes no food and lives only 2 to 3 days, but long enough to mature and lay 300 to 500 eggs. These are yellowish white, semispherical, and weigh about 1 milligram. The silkworms or larval stage are best reared at 77°F., and at this temperature they molt at 6, 12, 18, and 26 days after hatching, the entire larval stage requiring about 42 days. During this period, the average larva (Fig. 2.2) increases in weight from about 0.65 milligram to 6 grams or more and, when full grown, attains a length of about 3 inches. During this process of growth it consumes about 90 grams of mulberry leaves. After cessation of feeding, the larva spins a cocoon (Fig. 2.3) formed from a continuous thread of silk from 800 to 1,200 yards long and weighing 0.4 to 2 grams, which requires about 3 days to complete. In this cocoon, the larva pupates and emerges after 10 to 12 days as the adult moth by

FIG. 2.1. The parent or moth stage of the silkworm, about natural size. Male above, female below. (*From Slingerland.*)

softening one end of the cocoon by an alkaline secretion which enables it to break through the strands of silk. There are various races of domesticated silkworms which have from one generation a year in Europe and the Near East, to two in Japan, and as many as six in India and China.

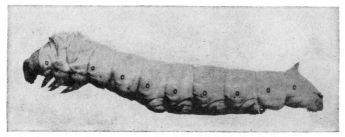

FIG. 2.2. The full-grown silkworm (larva) about natural size. (*From "Silk, Its Origin, Culture, and Manufacture," Corticelli Silk Co.*)

Sericulture, or the commercial production of silk, is an important industry in Japan, China, India, Italy, France, and Spain, where there is an abundance of cheap labor, and from 50,000,000 to 70,000,000 pounds of raw silk are produced annually with a value of $200,000,000 to

$500,000,000. Since as many as 3,000 cocoons are needed to produce a pound of raw silk, this vast production of silk floss demands the sacrifice of at least 100 billion caterpillars each year. Sericulture is a highly organized and specialized profession in the silk-producing countries, and the silkworms are fed and tended from egg to cocoon as carefully as any domestic animal. Because of the danger of incurring devastating epidemics of diseases such as the pebrine disease, an infection by a myxosporidian parasite, *Nosema bombycis*, which is transmitted through the egg, only pedigreed strains of silkworms are propagated from cultures determined to be disease-free. Selected cocoons are allowed to mature, and the moths are bred and the females allowed to lay their eggs in small sterile cloth bags. In the actual production of silk, the cocoons are treated, about 8 days after spinning, by steam or dry heat to kill the insect and prevent the destruction of the continuous fiber by the emergence of the moth. The cocoons are then carefully dried to prevent

FIG. 2.3. Cocoons of the silkworm, from which the true silk of commerce is unwound, about ½ natural size. (*From Garman, Ky. Agr. Exp. Sta.*)

putrefaction of the pupa, assorted according to color and texture, and the loose outer threads removed, and the cocoons are then soaked in warm water to soften the gum that binds the silk threads together and skillfully unwound by expert operators. The threads from several cocoons are wound together on wheels to form the reels of raw silk. Only about one-half of the silk in each cocoon is reelable; the remainder is used as silk waste and formed into spun silk. Subsequently, the raw silk is boiled, scoured, steamed, stretched, purified by acids or by fermentation, washed and rewashed to remove the gum and bring out the much prized luster; finally, combed and untangled, it is ready for spinning into the beautiful fabrics that eventually appear upon the market to beautify our homes and adorn our bodies.

The silk fiber spun by the silkworm is formed from the secretions of the paired salivary (labial) glands (Fig. 3.9), which have a common opening, the spinneret, on the labium. During the growth of the larva, these glands fill with relatively enormous quantities of clear viscous liquid. When extruded from the glands, this secretion forms the two cores of fibroin, a tough, elastic, insoluble protein which comprises 75 per cent of the weight of the fiber, and these are cemented together by sericin from the middle of the glands, a gelatinous-like protein which is readily soluble in warm water. Small amounts of wax and carotenoid pigment are also present. The silk fibers range from 0.0018 to 0.0033 inch in diameter and have a tensile strength of about 64,000 pounds per square inch, or nearly as great as steel, and remarkable elasticity up to 20 per cent.

Several other mulberry-feeding *Bombyx* moths are domesticated, including *textor*, *sinensis*, *croesi*, *fortunatus*, and *arracanensis*, all of which produce cocoons of reelable silk. Wild silk is produced by at least 30 Asiatic species of moths of the family Saturniidae (page 252), especially *Antheraea pernyi* of China, *A. paphia* of India, *A. yamamai* of Japan, and *A. assamensis* of Assam. The Atlas moth, *Attacus atlas*, the Cynthia moth, *Samia cynthia*, and *S. ricini* have also been domesticated. However, none of these species has been able to rival *Bombyx mori* as a commercial-silk producer.

Honey. Man has domesticated many kinds of animals and directed their activities so that the products resulting from their life processes might be available for his use. Of the many species of insects only two, the silkworm and the honeybee (*Apis mellifera*), have been domesticated. These are so remarkably successful that one wonders if there are not many other kinds that could be domesticated with profit. Many kinds of bees and wasps store honey as food for themselves and their young. But here again, as with the silkworm, only one species of outstanding merit has found a place in the husbandry of man.

Plants secrete nectar in profusion, but in such numerous and infinitesimally small portions that man unaided could never afford to collect it. It requires from 40,000 to 80,000 trips of a honeybee, and visits to many times this number of flowers, to find and assemble nectar enough to make a pound of honey. The average trip is thought to be about 1 or 1½ miles. Hence, for a single bee to collect nectar enough for 1 pound of honey would mean traveling at least twice the distance around the world.

Driven by an instinct to provide for their young and to fortify themselves against times of want, especially to lay up a winter supply of food, and operating in almost countless numbers, day after day, trip after trip, bit by bit, the bees gather these sugar-bearing secretions of plants and store their treasures in minute waxen "bottles" of their own making. In this way 200 to 250 million pounds of this product are collected and stored annually in the United States. Billions of pounds of nectar go to waste each year for want of enough bees to gather it.

The honeybee does not make honey in the same sense that the silkworm makes silk. Nevertheless, the bees are more than harvesters, being an essential intermediary between nectar-bearing plants and man. Honey is truly an insect product. The nectar obtained from flowers, after being mixed with saliva and swallowed, is carried in the honey sac (crop) until the bee reaches the hive. The honey sac is a kind of first stomach surrounded by muscles and provided with valves so arranged that its contents may be passed on into the stomach or, by compression, emptied back through the mouth parts into the cells of the hive (Fig. 2.4). At the hive, the collected nectar is masticated thoroughly with saliva containing the enzymes invertase and amylase, and the sucrose is hydrolyzed to glucose and fructose. A large portion of the water content is removed by the air currents produced over the cells by the rapid beating of the wings of the worker bees. When a cell is filled with a properly "ripened" honey, it is capped over with wax and appears in the form familiar to everyone as comb honey.

Chemically, honey is a viscous water solution of sugars, containing 13 to 20 per cent water, 40 to 50 per cent fructose, 32 to 37 per cent glucose, 2 per cent sucrose, traces of maltose, 1 to 12 per cent dextrins and gums, and traces of minerals, free organic acids, vitamins, enzymes, plant pigments, and suspended solids such as pollen grains and beeswax. It has a specific gravity of about 1.45 to 1.48. Honey is extensively used as a natural sweet, as a spread, and in making candies, cakes, and bread, and it is a rich, nutritious food. For example, deep-sea divers find large amounts of honey one of the few foods suitable to their strenuous physical activity.

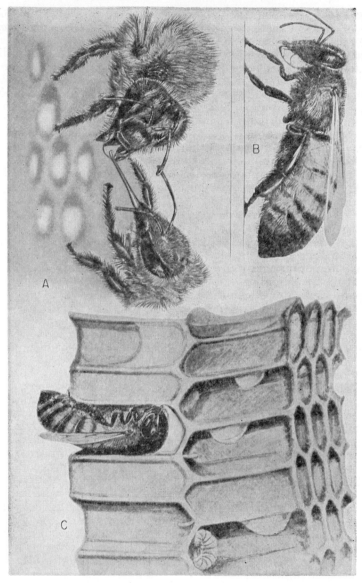

Fig. 2.4. A bit of honeybees' comb in section, showing the waxen "bottles" in which honey is being stored and which serve as cells for the protection of the egg, larval, and pupal stages of the bees. Note the two eggs and the partly grown larva. Above at left a field bee is transferring her load of nectar to a hive bee. At the right a hive bee is ripening the nectar by forcing it in and out of her mouth *(After Park and Joutel, Jour. Econ. Ento., Vol. 18.)*

Beeswax. The life processes of the honeybee also give us the useful material known as *beeswax*. Besides forming the cells in which the honey is stored and serving as "cradles" for the bees during their development (Fig. 2.4), beeswax is extensively used in many arts and trades. From 5 to 15 million pounds of it, worth several million dollars, are used in the United States each year. It is a common mistake to suppose that honeybees convert the pollen they collect from flowers into beeswax or honey. This material is a natural secretion of the worker bees that is poured out in thin delicate scales or flakes from glands that open on the underside of the abdomen[1] (Fig. 2.5). Its production directly follows the digestion of a quantity of honey, a pound of wax resulting from the consumption by the worker bees of from 3 to 20 pounds of honey in about 24 hours' time. About 1 million pounds of the annual crop is pressed into "foundation" and returned to the bee hives as a basis for the comb honey. Several million pounds are used in manufacturing candles, which are nearly smokeless and do not bend over with the heat. Much beeswax is used in shaving creams (to keep the lather from drying quickly), in cold creams, cosmetics, polishes, floor waxes, patterns for castings, models, carbon paper, crayons, and electrical and lithographing products.[2]

Shellac. A tiny species of scale insect (related to our fruit pest, the San Jose scale) yields the substance from which shellac is made. This substance is extensively used in making varnishes and polishes; for finishing woods and metals; for stiffening hat materials; as an ingredient of lithographic ink; as sealing wax; as an insulating material in electrical work;

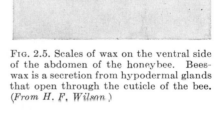

FIG. 2.5. Scales of wax on the ventral side of the abdomen of the honeybee. Beeswax is a secretion from hypodermal glands that open through the cuticle of the bee. (*From H. F. Wilson.*)

and in making phonograph records, airplanes, linoleum, buttons, shoe polishes, pottery, toys, and imitation fruits and flowers. That this substance is derived from an insect is realized by very few persons.

The lac insect, *Laccifer* (= *Tachardia*) *lacca*[3] (Fig. 2.6), lives on native forest trees in India and Burma. The natural function of the lac is to protect the motionless insect from adverse weather and natural enemies. On contact with the air the resinous secretion hardens, and, where the

[1] Chibnall has shown (*Biochem. Jour.*, **28**:2189, 1934) that beeswax and all other insect waxes are complex mixtures of varying proportions of (*a*) even-numbered alcohols ranging from C_{24} to C_{36}, (*b*) even-numbered normal fatty acids from C_{24} to C_{34}, and (*c*) odd-numbered normal paraffins ranging from C_{23} to C_{37}. The various insect waxes differ only in the proportions of these constituents and in the extent to which the alcohols are esterified with the fatty acids.

[2] GROUT, R., "Current Uses of Beeswax," *Amer. Bee Jour.*, September, 1935.

[3] Order Homoptera, Family Coccidae.

insects are closely crowded together, it forms a continuous layer over the branches (Fig. 2.6,*A*). In this condition the substance is known as *stick lac*.

About 40 to 90 million pounds of stick lac are collected annually. It is ground in crude, hand-operated mortars and the resulting material separated into (*a*) granules of lac known as *seed lac*, (*b*) dust, which is used to make toys, bracelets, and bangles, and (*c*) wood, which is used

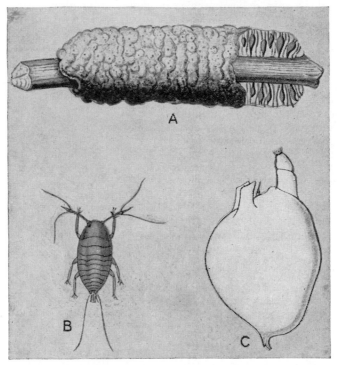

Fig. 2.6. The lac insect. *A*, a piece of twig encrusted with lac. The part on the right is broken open to show the worm-like lac insects in their cells. Each living insect communicates with the exterior through three small apertures, shown on the surface of the incrustation (about natural size). *B*, the young lac insect (first-instar nymph) greatly enlarged. *C*, the body of an adult female lac insect, freed from its resinous secretions, as seen from the side. At the upper end is the segmented abdominal extremity of the insect, terminating in the fringed anus. At the left is the dorsal spine, and overhanging it, the two respiratory processes which guard the spiracles. The mouth parts are at the lower extremity and, during life, are inserted in the bark. About 12 times natural size. (*Modified from Green, "Coccidae of Ceylon."*)

as fuel. The seed lac is next soaked in water and trodden by foot, which crushes and washes out the wine-colored pigments that were formerly sold as dyes but no longer have any commercial value. The granular lac, after drying and bleaching in the sun, is placed in slender cloth bags, 10 to 12 feet long, is heated by open charcoal fires, and, by the twisting of the ends of the bag, is forced out as it melts, dropping upon the floor (Fig. 2.7). Before these pads of melted lac have had time to congeal, they are grasped by natives who stretch them with their hands, teeth, and feet to extremely thin sheets.[1]   After drying, these sheets are broken into thin, small flakes; in this condition they are shipped.   Dissolving the flake lac produces the familiar white or orange liquid shellac.   It requires about 150,000 lac insects to make a pound of lac.

The lac industry is an ancient one, certainly several thousand years old.   The material has been used as a varnish at least since 1590.   In 1709 Father Tachard described the insect that produces lac.   The Mohave Indians used the secretion of a North American lac insect to make baskets watertight for cooking.   The insect is not domesticated or even cultivated but lives in such abundance on the forest trees that millions of the poorer classes of India find in this industry their sole means of livelihood.   No superior modern substitute has been found.   From

Fig. 2.7. Indian natives melting lac.   One step in the preparation of shellac for the market. (*From "Shellac: A Story of Yesterday, Today, and Tomorrow," James B. Day & Co.*)

10 to 20 million dollars worth of this product is used in the United States each year.   Of late, extensive fluctuations in the abundance of the insect have led to the suggestion of simple routine methods of manipulation intended to reduce the extent of parasitism and increase the "set" of young insects on the host trees; and machine methods are now being used to some extent in producing the lac.   Doubtless many other improvements will be initiated from time to time as the demand for the product makes necessary a more economical method of production.   Recently synthetic lacquers have been produced which will doubtless supplant the true shellac for many purposes.

Living Light or Light without Heat.   One of the most remarkable phenomena in nature is the property that many animals possess of emitting light, "phosphorescence"

[1] See "Shellac: A Story of Yesterday, Today, and Tomorrow," James B. Day & Co.; GLOVER, P. M., "Lac Cultivation in India," *Indian Lac Res. Inst.*, 1937; PARRY, E. J., "Shellac," Pitman, 1935.

or *bioluminescence*. This phenomenon is known among the Protozoa, Coelenterata, Mollusca, Annelida, fishes, birds, Crustacea, bacteria, fungi, and others, but doubtless the best-known examples are among the beetles, commonly known as fireflies or lightning-bugs and glowworms[1] (Fig. 6.43).

The substance that gives rise to the light rays has been called *luciferin*. It is formed in certain specialized cells of the insect body. These cells are abundantly supplied with the tracheae or breathing tubes. When air is admitted under the control of the insect, the oxidation of the luciferin takes place under the influence of an enzyme, called *luciferase*, to form *oxyluciferin*, in the cells that produce it, and the production of light is instantaneous. The insect body is generally provided with a reflector for the light, formed by a white material (probably ammonium urate), secreted by the cells directly behind the photogenic tissues. In some organisms this reaction is reversible and the oxyluciferin can be reduced again, with loss of oxygen, and the luciferin used over and over. The following reactions are involved, and firefly tissue is used to detect as little as 0.1 microgram of ATP:

(a) LH_2 (luciferin) + ATP (adenosine triphosphate) → $LH_2 \cdot AMP$ (adenyl luciferin)

(b) $LH_2 \cdot AMP + O_2 \xrightarrow{\text{luciferase}} L \cdot AMP$ (adenyl oxyluciferin)

(c) $L \cdot AMP → L + AMP + hv$ (a quantum of light)

The light emitted by our fireflies is most remarkable in having the maximum of visibility (from 92 to nearly 100 per cent light rays) and practically no heat rays or ultraviolet rays. In the ordinary gas flame, by contrast, only 2 per cent of the energy is converted into light rays, the rest being lost as low heat rays; in the electric arc only 10 per cent of the energy produces light; while sunshine is only 35 per cent light.

Some slight use has been made of these light-giving insects as ornaments, as an artificial illuminant, and in photography. But the significant benefit we may hope to derive from them is guidance to the synthesis of a luciferin in the chemical laboratory which may give to the world an artificial illuminant many times as efficient as the best artificial lights of the present day.

Cochineal. Cochineal, a beautiful carmine-red pigment or dye, is the dried, pulverized bodies of a mealybug, *Dactylopius coccus*, that lives on the prickly pear, *Opuntia coccinellifera*. Cochineal is now used principally as a cosmetic or rouge; for decorating fancy cakes; for coloring beverages and medicines; for dyeing where unusual permanence is desired; and, because of its property of allaying pain, for treating whooping cough and neuralgia.

It was used by the Aztecs in Mexico before the Europeans discovered America and, until the aniline dyes came on the market, was a very important product. The cochineal insect is now cultivated principally in Honduras and the Canary Islands, though still a product from Mexico, Peru, Algiers, and Spain. The insects are carefully cultivated. The Mexicans keep them indoors over winter on branches cut in the fall from the prickly pear. In spring they put the females out in little straw nests fastened to the cacti. The young bugs settle on the cactus, and in 3 months' time the cochineal insects are fully developed and ready to harvest. The branches are broken off, the insects brushed from them into bags and killed by hot water, steam, dry heat, or drying in the sun. The impurities are then removed and the product is ready for the market. It requires about 70,000 insects to make a pound of cochineal, which contains about 10 per cent pure carminic acid.

Insects as Medicine. Many kinds of insects have been reputed to possess medicinal properties. In fact, during the seventeenth century some curative power was attributed to almost every known insect. At that time the belief prevailed that every creature possessed some special usefulness to man. No other virtue being apparent for many of the insects, special therapeutic qualities were ascribed to them. Of course, the vast majority of such recommendations are pure quackery and are founded on the rankest of superstition. Examples are the reputed belief that the bite of a katydid or cricket will remove warts; that cockroaches, crickets, or earwigs,

---

[1] True luminescent organs are found only in the families Lampyridae, Phengodidae, Elateridae, in several Carabidae of the genus *Physodera*, in a species of *Buprestis*, and in the larvae of the genus *Bolitophila*, Order Diptera, Family Mycetophilidae. Several genera of Collembola possess a generalized luminescence.

variously bruised, burned, or boiled and properly compounded and applied, will cure earache, weak sight, ulcers, and dropsy.

Certain insects, however, do have a real medicinal value, notably the maggots of certain flies, the honeybee (*Apis*) and blister beetles.

During the First World War, Dr. W. S. Baer noticed that wounds of soldiers who had been lying on the battlefield for hours did not develop infections, such as osteomyelitis, as did those which had been treated and dressed promptly after they were inflicted. The difference was found to be due to the fact that the older wounds were always infested with maggots developed from eggs laid about the wounds by certain flies. The remarkable discovery that these maggots could clean up the infection in deep-seated wounds much better than any known surgical or medicinal treatment has led to the practice of rearing maggots of the house fly and certain bluebottle flies under sterile conditions and introducing these surgically clean maggots into wounds to eat out every microscopical particle of putrid flesh and bone. In one survey 92 per cent of 600 physicians who had used this treatment reported favorably upon it. Dr. William Robinson[1] has isolated a substance from the secretions of the maggots which has the same property of healing in infected wounds as the maggots themselves. This material, known as *allantoin*, is commercially available and is described as a harmless, odorless, tasteless, stainless, painless, and inexpensive lotion which, when applied to chronic ulcers, burns, and similar pus-forming wounds, stimulates local, rather than general, granulation and so is of especial value in treating deep wounds such as bone-marrow infections, where the internal parts of the wound must be healed first. Robinson does not believe, however, that the allantoin solutions will replace the living maggots in the treatment of bone infections, because the maggots actually eat out the necrotic tissue, kill the pus-forming bacteria by digesting them, and continually apply minute quantities of the allantoin in their excreta to the very depths of the wound, more effectively than can possibly be done by instrumentation.

The best known of the blister beetles is the so-called Spanish fly, *Lytta vesicatoria* that occurs in great abundance in France and Spain. This is a relative of our American blister beetles or "old-fashioned potato beetles." These insects have in their blood and internal organs a substance known as *cantharidin*. It was formerly greatly used as an external local irritant or blister. Lloyd[2] says: "The barbarisms practiced upon the American people during the nineteenth century by the application of cantharis blisters for all sorts of ailments, overtopped the misery endured by those who suffered in the war of the Revolution." This material was also much used as an aphrodisiac before its dangerous nature was appreciated. At present it has a place as an internal treatment in certain diseases of the urinogenital system and in animal breeding.

Carminic acid from cochineal

Cantharidin

Allantoin

While the use of cantharidin appears to be dying out, the use of the honeybee in medicine has increased in the past century. The preparation known as "specific

[1] *Jour. Parasitol.*, **21**:354, 1935.

[2] Lloyd Brothers, *Drug Treatise* XXI, Cincinnati, Ohio.

medicine Apis" is extracted from the bodies of honeybees by killing them in alcohol while they are intensely excited and by digesting their bodies in this medium for a month at a warm temperature. It is finally brought to a strength representing 2 ounces of bees to 1 pint of the medicine. In 1858 this preparation was said to be the most universally useful remedy next to aconite. It has been used by many physicians for the treatment of "hives," diphtheria, scarlet fever, erysipelas, dropsy, urinary irritation, and all kinds of edema accompanied by swelling and burning.[1] In an article entitled "The Remedial Value of Stings," Root[2] summarizes a vast amount of testimony of medical men and others as to the curative properties of bee venom. It appears that this remedy is extensively used especially in the Old World. The bee venom has also been placed upon the market in ampoules to be injected hypodermically, thus providing the same effects as the natural stings without the pain. Bee royal jelly (page 292) has a certain vogue as a "hormone-type" ingredient in cosmetic creams.

Use of Insect Galls. The injury to plants by insect galls was discussed in Chapter 1. These galls contain certain valuable products that have been used in a variety of ways. Many superstitions have attached to these remarkable growths on plants.[3] However, certain galls are reputed to have genuine medicinal or curative properties. The Aleppo gall, or gallnut, of western Asia and eastern Europe has been used in medicine since the fifth century B.C. It is a powerful vegetable astringent, tonic, and antidote for certain poisons.

Other galls have been used as dyes. The African Somali women use them as a tattooing dye. The Turks secure a fine scarlet color from a reddish gall on oak. Turkey red is dyed from the "mad apple" in Asia Minor. The ancient Greeks used the Aleppo gall for dyeing wool, hair, and skins; more recently great quantities of it have been used in dyeing leather and seal skins. Tannic acid occurs in high percentages (30 to 70 per cent) in many of these galls, which are the richest of all sources of this material.

Galls are used in the preparation of very durable and permanent inks. In some countries the laws require that certain records be made with ink compounded of gallnuts.

INSECTS AS POLLINIZERS[4]

The maintenance of plant life generation after generation may be accomplished by *asexual reproduction* (the formation of buds, bulbs, and tubers) or by *sexual reproduction*. In the latter case one specialized reproductive cell, the male gamete (sex cell or sperm), unites with another, the female gamete (sex cell or egg), and from this union a new individual arises. In the higher plants sexual reproduction is made possible by the process known as *pollination*. The essential carrier of the pollen (male sex cells) from the anthers of one flower to the stigma of another is in most cases either the wind or an insect. Well-known examples of wind-pollinated flowers are corn, wheat, and other cereals, nut trees, willows, oaks, pines, etc. Wind-pollinated plants generally have flowers that are small and inconspicuous, with poorly developed petals, unisexual flowers, no nectar, dry and light pollen, and brush-like stigmas.

Most of our fruits and ornamental flowers; many of our vegetables such as beans, peas, tomatoes, melons, squash; and such field crops as clover, buckwheat, cotton, and tobacco depend mainly upon the visits

[1] Lloyd Brothers, *Drug Treatise* XXI, Cincinnati, Ohio.

[2] ROOT, E. R., *Gleanings in Bee Culture*, **75**:16–20, 84–87, 1925; BECK, B. F., "Bee Venom, Its Nature and Effect on Arthritic and Rheumatoid Conditions," Appleton-Century-Crofts, 1935.

[3] See FAGAN, MARGARET M., "The Uses of Insect Galls," *Amer. Naturalist*, **52**: 155–176, 1918.

[4] BOHART, G., *Ann. Rev. Ento.*, **2**:355, 1957, and *U.S.D.A., Yearbook, Agr.*, **1952**: 107; VANSELL, G. H., and W. H. GRIGGS, *ibid.*, p. 88.

of insects to carry the pollen to the stigma and so make possible a fertilization without which no seed or fruit would form.   Flowers that depend upon insects for pollination can be recognized generally by their well-developed corollas of conspicuous size, by showy colors, or by a marked odor.   They have sticky pollen grains, sticky stigmas, and nectaries that secrete a sweet liquid attractive as food for the insects.

Plants do not develop beautiful blossoms and sweet odors to delight the senses of man.   They serve to attract insects.   Plants have many remarkable modifications of their structures which compel the insects that come for nectar to carry away with them to the next flowers visited a load of pollen.   As they crowd their way in and out of flowers, their bodies become covered with the fine pollen dust.   In the honeybee this is removed from the general body hairs by a highly specialized brush on one of the segments of the hind leg (Fig. 3.4,C, *Tarsus*).   When the brush becomes filled, the hind legs are crossed and the pollen grains from one leg are scraped into the pollen basket on another segment (Fig. 3.4,C, *Tibia*) of the opposite leg.   In the pollen baskets it is carried to the hive, where a spine on the end of the middle leg is used to pry it off and it is stored in the cells.   While much of the pollen is thus collected and used as food by the bees, some of it brushes off when the bee or other insect crowds into the next flower visited.   Many flowers are so constructed that an insect can scarcely obtain nectar from them without dusting some pollen from previously visited flowers upon their stigmas.   Therefore, without the valuable work performed by the insect pollinizers, we should have very poor yields of many important crops such as fruits, tomatoes, melons, alfalfa and clover seed, coffee, tea, chocolate, and cotton.   The annual value of insect-pollinated crops to the American farmer is immense (Table 2.1).

Pollination of both alfalfa and red clover depends upon the tripping mechanism whereby pollen is released upon disturbance by insect pollinizers.   In alfalfa, the tripping mechanism is actuated when the bee forces its head into the throat of the flower.   As the flower is tripped, the bee's head is momentarily caught between the petals and the tip of the sexual column, so that the bee is showered with pollen in exactly the proper position to contact the stigma of the next flower visited.   More than 100 species of wild bees are pollinizers of alfalfa in the United States, of which the alkali bee, *Nomia melanderi*, and *Megachile* spp. are especially important.   The honeybee is a relatively inefficient pollinizer because, when seeking nectar, it readily learns to avoid the tripping mechanism by feeding from the side, so that only about 1 per cent of the blossoms are tripped.   However, when honeybees are collecting pollen, about 80 per cent of the blossoms are tripped.   Therefore, the honeybee can become a very valuable alfalfa pollinizer under conditions which favor a high percentage of pollen collectors, as in the arid Southwest or where the density of nectar feeders is very high.

The use of honeybee colonies as pollinizers of legumes has become very important in California, and in 1955, 450,000 hives were employed for this purpose.   The optimum seems to be from 3 to 6 colonies per acre, and such forced pollination has been a major factor in increasing the average alfalfa seed yield from 215 pounds per acre in 1945 to 480 pounds per acre in 1955.

In red clover, tripping takes place through pressure against the standard and wing petals of the flower, which forces the stigma and anthers upward out of the enclosed keel petals.   Nectar cannot be stolen without tripping, and any insect that applies sufficient pressure may become a pollinizer.   The honeybee is not well equipped for red clover pollination because the length of its tongue (Fig. 4.17,*D*) is such that it cannot generally reach the deep-seated nectary of this flower and it is therefore

TABLE 2.1. THE INSECTS OF THE UNITED STATES IN ACCOUNT WITH THE
AMERICAN PEOPLE—CREDIT
Most Important Items, 1957[1]

To products made or collected by or from insects:

13,739,426 lb. silk imported, @$6.35	$ 86,618,921
242,293,000 lb. honey produced, @$0.19	45,551,084
4,476,000 lb. beeswax, @$0.57	2,551,320
39,519,130 lb. shellac imported, @$0.23	9,251,144
Total value insect products utilized in United States	$143,972,469

To pollination of fruits, vegetables, and flowers:
  The following crops depend almost exclusively upon insects for the production of
    fruits and seeds:

Apples	Oranges	Eggplant
Pears	Lemons	Tomatoes
Peaches	Grapefruit	Peppers
Plums	Figs	Clovers
Prunes	Melons	Alfalfa
Cherries	Cucumbers	Soybeans
Strawberries	Pumpkins	Cowpeas
Raspberries	Squash	Sweet clover
Blackberries	Beans	Cotton
Cranberries	Peas	Certain flowers

Total value insect-pollinated crops	$4,534,634,000

To serving as food for other animals:
  Two-fifths of the food of fresh-water fishes is insects.
  One-third of the food of wild song and game birds is insects.
  Hogs, turkeys, certain fur-bearing animals eat insects.
To services in controlling destructive insects:
  Unless checked by parasites and predators, injurious insects would soon make it
    impossible for man to exist.
To services in aerating and fertilizing the soil with their dead bodies and excreta
To services as scavengers:
  Transforming plant and animal refuse into forms available for utilization by grow-
    ing plants
To services in scientific investigations
To services as models, ornaments, and for entertainment and diversion
To services in medicine and surgery:
  It is impossible to estimate in dollars the enormous benefits to mankind from these
    services.

  [1] Figures from *U.S. Dept. Commerce, Quart. Summary Foreign Commerce U.S.*, January–December, 1957, and *U.S.D.A., Agr. Statistics*, 1958.

reluctant to visit this crop.   However, where honeybee populations are concentrated in red clover fields comparatively isolated from competing sources of nectar and pollen, they are satisfactory pollinizers of red clover. The bumblebees, *Bombus* spp., are the most important pollinizers of second-crop clover because of their longer tongues.   The bumblebees winter only as young queens, and it takes several months in spring and summer to build up colonies to a point where they are sufficiently abundant to visit enough flower heads to produce satisfactory set of clover seed. Although bumblebee populations often fluctuate from year to year, making seed production uncertain, unless honeybees are also present, their value is shown by the fact that in New Zealand it was found to be impossible to obtain seed from red clover until bumblebees were imported into that country. Certain wild bees, *Tetralonia* and *Mellissodes* spp., are very efficient pollinizers of red clover, and when they are abundant, the first crop of clover will usually set a large amount of seed, often producing a greater yield than that of the second crop.

It has been suggested that races of honeybees with longer tongues, or strains of red clover with shorter corollas, should be bred, so that pollination of this crop could be accomplished by the honeybee. Since entomologists have learned how to control matings in the honeybee, this and other improvements, such as the capacity to carry greater loads of pollen and nectar and the inability or reluctance to sting, are distinct possibilities.

Too often the impression prevails that bees are the only insects of importance in cross-pollination. Many kinds of flies, butterflies, moths, and beetles also share in this work. Many of these insects, however, produce injurious caterpillars, grubs, or maggots which may largely offset the benefit derived from the parents. Bees and wasps develop no objectionable progeny. The honeybees can be introduced into any neighborhood, orchard, or garden, and their numbers increased as we wish; or should they no longer be desired, the whole population could be exterminated at will. They are under man's control as no other insect is, except the silkworm.

Most people think of honeybees only as a source of honey. It seems that their most important contribution to man is in the production of fruits and seeds. It has been said that every time a colony of bees makes $5 worth of honey, they have made $100 worth of seeds or fruits. In orchards where colonies of honeybees have been placed at blossoming time, the yields of fruits have been increased almost beyond belief. In a Michigan apple orchard where the largest crop in 8 years had been 1,500 bushels, 40 colonies of bees were introduced in 1927, and 5,200 bushels of apples were harvested. In a cherry orchard, the beekeeper made $100 from the honey and rent for his bees, and it was estimated that the owner of the orchard made $10,000 more from his cherry crop than he would have without the bees. In another case, an alfalfa grower harvested alfalfa seed at the rate of 1,200 pounds per acre where bees were abundant, and only 300 to 400 pounds without the bees.

The Fig Insect. One of the most striking illustrations of the dependence of plants upon insects is found in the history of the introduction of Smyrna fig culture into the United States. Previous to 1900 figs grown in this country were of quality and flavor very inferior to those of figs imported from Asia Minor. A careful study of the situation revealed the remarkable fact that the palatability of the Smyrna fig, in Asia Minor, was dependent upon the pollination of the flowers by a small chalcid wasp, *Blastophaga psenes*.[1]

A fig is a hollow pear-shaped receptacle that bears a very large number of minute flowers lining its inner surface. The only entrance to the flowers is a tiny opening at the free end of the fig. If the flowers are not fertilized, the seeds do not form, and

[1] *Calif. Agr. Exp. Sta. Bul.* 319, 1922; *U.S.D.A., Dept. Bul.* 732, 1918.

the fleshy, nearly closed receptacle that bears them does not develop the sweet, nutty flavor that characterizes the perfect fruit. The Smyrna fig is exclusively female and produces no pollen. A pollen-producing variety, known as the *caprifig*, produces inedible fruits but an abundance of pollen. Pollination is performed only by the female of this tiny fig wasp (Fig. 2.8,*a*). These insects lay their eggs in the flowers of wild figs, or *caprifigs*, and their larvae develop in small galls at the base of the caprifig flowers. The males that are formed are wingless (Fig. 2.8,*e,f*) and never leave the wild fig in which they develop. They crawl about, gnaw into the galls in which females are developing, and fertilize the females through the puncture. After mating, the female escapes from the gall, becomes covered with pollen from the stamens of the caprifig, then squeezes her way out of the small opening at the free end of the caprifig, and flies about among the trees. In seeking places to lay their eggs, the females enter

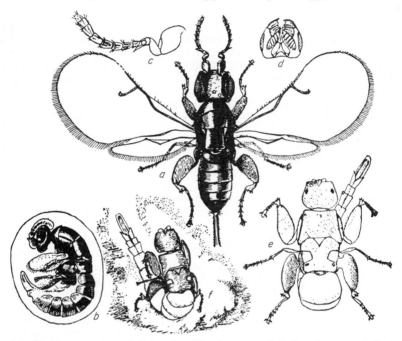

FIG. 2.8. The fig wasp. *a*, adult female with wings spread; *b*, female not entirely issued from pupal skin and still contained in gall of the fig flower; *c*, antenna of female; *d*, head of female from below; *e* and *f*, adult males. All much enlarged. (*From U.S.D.A., Dept. Bul.* 732.)

Smyrna figs as well as the caprifigs, and, although they are said not to lay their eggs in the former, because the ovaries are so deep-seated that their ovipositor cannot reach them, they crawl over the minute flowers and scatter pollen over them from the flowers of the caprifigs in which they developed.

When the part that this minute insect plays in the culture of figs was known, efforts were made to bring some of the wasps from Algeria into California. There were many failures, but a decade of effort finally resulted in the establishment of the insect and the subsequent production in America of figs equal to those grown in Asia Minor. It is necessary to grow caprifigs as well as Smyrna figs to keep up the supply of wasps, since the insect does not reproduce in the edible fig. Figs containing mature fig wasps are removed from the caprifig tree, strung on fibers, and suspended among the branches on the Smyrna fig trees when the latter are ready for fertilization.

Recently the fig wasp threatened ruin to the fig growers in California by spreading an endosepsis or internal rot disease to the Smyrna figs from infected caprifigs in which they develop. The little wasps are being reared by millions in sterile incubators and released in the orchards free from brown rot germs in an attempt to prevent this contamination.

INSECTS AS FOOD

Although of small size, insects, because of their prodigious numbers, probably exceed in weight all other animal matter on the land areas of the earth. This great mass of material possesses genuine food value. Chemical analyses of white grubs and May beetles, for example, have shown[1] that these insects compare favorably with tankage in food value. Turkeys, hogs, and other domestic animals will often fatten on insects.

It has been said that insects make up, on the average, about two-thirds of the food of our common land birds.[2] The extensive investigations carried on by Forbes and his associates in Illinois led him to the conclusion, from the examination of over 1,200 fishes from all kinds of Illinois waters, that fully two-fifths of the food of adult fresh-water fishes is insects.[3] The insects of most importance as fish food are (a) small, slender midge larvae known as bloodworms,[4] (b) May-fly nymphs,[5] and (c) caddice-worms.[6]

In many parts of the world considerable quantities of insects are regularly eaten by human beings. These are generally looked upon as great luxuries by the less civilized races. In Mexico the eggs of certain large aquatic bugs are regularly sold in the city markets. The eggs are about the size of bird shot. The Mexicans sink sheets of matting under water upon which the eggs are laid by millions. These are then dried and placed in sacks, sold by the pound and used for making cakes. The people of Jamaica consider a plate of crickets a compliment to the most distinguished guest. Ox warbles are eaten raw by the Dog Rib Indians. Natives of Australia collect quantities of the bugong moth, *Agrotis infusa*, in bags, roast them in hot coals, and claim that they taste like nuts and abound in oil. The Indians and semicivilized natives of many countries catch quantities of ants, grasshoppers, and the larvae and pupae of bees, moths, crane flies, and wood-boring beetles and eat them raw, dried, or roasted. The manna or sugary honeydew excreted by aphids and scale insects (page 109) is used as a sweet by peasants of Turkey, Iraq, and Iran. According to Bodenheimer and Swirski,[7] about 70,000 pounds are collected and sold in Iraq annually.

From the actions of wild animals and the testimony of those persons who have tried insects as food, it seems that much of this material is palatable. It would, in fact, be difficult to give any sound reasons why we should consume quantities of oysters, crabs, and lobsters and disdain to eat equally clean, palatable, and nutritious insects. Perhaps the economists of the future, if hard pressed to maintain an ever-increasing population, may well turn their attention to the utilization of certain kinds of insects as human food. In any event such gourmet's delicacies as fried grasshoppers, caterpillars, and ants are already available on our supermarket shelves.

PREDACEOUS AND PARASITIC INSECTS[8]

Many of the benefits from insects enumerated above, although genuine, are insignificant compared with the good that insects do by fighting

---

[1] *U.S.D.A., Farmers' Bul.* 940, 1918.

[2] *U.S.D.A., Farmers' Bul.* 630, 1915.

[3] *Bul. Ill. State Lab. Natural History*, **1**:75, and **2**:475–538, 1888.

[4] Order Diptera, Family Chironomidae.

[5] Order Ephemeroptera, Family Ephemeridae.

[6] Order Trichoptera.

[7] "Aphidoidea of the Middle East," Weisman Science Press, Jerusalem, 1957.

[8] CLAUSEN, C. P., "Entomophagous Insects," McGraw-Hill, 1940; BALDUF, W. V., "Bionomics of Entomophagous Coleoptera," 1935, and "Bionomics of Entomophagous Insects," 1939, published by the author, Urbana, Ill.

FIG. 2.9. An insect predator: a robber fly with its prey. (*From Howe, "Insect Behavior," Badger.*)

among themselves. There is no doubt that the greatest single factor in keeping plant-feeding insects from overwhelming the rest of the world is that they are fed upon by other insects. It is easy to see how the industry of insects and their devotion to purpose, when coupled with almost unlimited numbers of individuals, can work a miracle for us when their instincts lead them to seek and devour myriads of pests scattered over a farm or a forest. Man will probably never be able to do so much in controlling his insect enemies as his insect friends do for him.

These insect eaters, or *entomophagous insects*, as they are called, are advantageously considered in two groups known as (*a*) *predators* and (*b*) *parasites*. Predators (Figs. 2.9 and 2.13) are insects (or other animals) that catch and devour smaller or more helpless creatures (called the *prey*), usually killing them in getting a single meal. The prey is generally either smaller, weaker, or less intelligent than the predator. *Parasites* are forms of living organisms that make their homes on or in the bodies of living organisms (called the *hosts*) from which they get their food, during at least one stage of their existence (Fig. 2.10). The hosts are usually larger, stronger, or more intelligent than the parasites and are not killed promptly but continue to live during a longer or shorter period of close association with the parasite. An important difference between these two groups is that, in parasitism, the host makes or determines the *habitat* for the parasite, whereas the prey does not necessarily fix the habitat for the predator, which lives quite independently of its victims during the intervals between meals. Another useful distinction is that generally a parasitic larva requires only a single individual of the host species to nourish it to maturity, while a predaceous insect takes many individual victims in completing its development. Predators are typically very active and have long life cycles; parasites are typically sluggish, often sessile, and tend to have very short life cycles.

FIG. 2.10. Insect parasites: larvae of a parasitic wasp, *Apanteles fulvipes*, leaving the body of the still living, but doomed, host (a caterpillar of the gypsy moth) after having fed within its body for several weeks. (*From U.S.D.A., Bur. Ento. Bul. 91.*)

Phytophagous parasites are those which live upon plant hosts. Insects that parasitize the larger animals (especially domestic animals) may be called *zoöphagous parasites*. Insects that parasitize other insects are called *entomophagous parasites*. Zoöphagous parasites as a group are highly injurious to man (page 939). They are often permanent parasites and rarely kill the host. The entomophagous parasites are largely beneficial to us. They are almost always transitory and generally kill their hosts.

There are numerous kinds and gradations of parasitism. The term *permanent parasite* is used for those parasites (such as the bloodsucking lice) which spend all their time and all life stages on or in the body of the host. *Transitory parasites* are those that pass certain life stages with one host and during other life stages are either free-living, like the horse bot, or parasitic in the body of an alternate host of a different species, such as the protozoan that causes human malaria. The term *intermittent parasitism* is used by some authors for such attacks as those of mosquitoes or bed bugs, which approach the host only at the time of feeding and, after the meal, leave the host for a period of free living. It is probably better to call such an attack *predatism*. *Obligatory parasites* are those which can live only as parasites and usually on only one species of host, *e.g.*, many of the chewing lice. *Facultative parasites* are those which, like the common flea of cats and dogs, can live free from the hosts part of the time and shift successfully from one individual or species of host to others. Parasites that live on the outside of the body are known as *ectoparasites*, while those that enter the body or eggs of their hosts are known as *endoparasites*. *Monophagous parasites* are those which are restricted to one species of host; *oligophagous parasites* those which are capable of developing upon a few closely related host species; and *polyphagous parasites* those which are capable of parasitizing a considerable number of host species. *Simple parasitism* refers to the condition resulting from a single attack of the parasite, whether one or many eggs are laid; *superparasitism* refers to the condition resulting from several different attacks by individuals of the same parasite species; while *multiparasitism* describes the condition of a host which is suffering simultaneously from the attacks of two or more species of primary parasites. A *monoxenous parasite* requires only one host for its complete development; a *heteroxenous parasite* requires several or different hosts for its complete development.

A classification of parasitism that should be kept in mind is that, with respect to a given host, a parasite or predator may be primary, secondary, tertiary, or quaternary. Not all parasites that kill insects are beneficial to man, and indeed it is sometimes almost impossible to tell whether the presence of a parasitic species in a given territory would be beneficial to man or injurious. If we have an injurious insect such as the cotton boll weevil, any parasite attacking it is *primary* for the cotton boll weevil and helpful to man. That parasite may in turn be attacked by a parasite, which is then *secondary* to the cotton boll weevil and inimical to man. A parasite attacking such a secondary parasite would be known as a *tertiary* parasite of the cotton boll weevil and would in this capacity be beneficial to man. All parasites whose hosts are also parasites are collectively known as *hyperparasites*. When abundant, the hyperparasites may completely offset the beneficial work of primary parasites. How complicated the interrelations of good and bad parasites may be is illustrated by the accompanying diagram (Fig. 2.11) of the various parasites and predators associated with the cotton boll weevil.

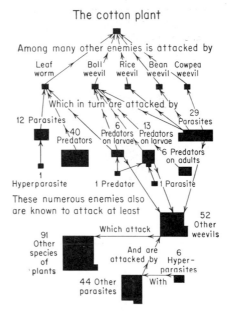

The cotton plant

Among many other enemies is attacked by

Leaf worm · Boll weevil · Rice weevil · Bean weevil · Cowpea weevil

Which in turn are attacked by

12 Parasites · 40 Predators · 6 Predators on larvae · 13 Predators on larvae · 29 Parasites · 6 Predators on adults

1 Hyperparasite · 1 Predator · 1 Parasite

These numerous enemies also are known to attack at least

91 Other species of plants · Which attack · 52 Other weevils · And are attacked by · 6 Hyperparasites

44 Other parasites · With

FIG. 2.11. Diagram illustrating the interrelations of some helpful and harmful parasites and predators associated with the cotton boll weevil. (*From U.S.D.A., Bur. Ento. Bul.* 100.)

The practical man should make an effort to become acquainted with the typical appearance and manner of life of the more important groups of predators and parasites; and those scientifically inclined can hardly find more fascinating groups for study.

Insect Predators.   Among the best-known entomophagous predators are the following:

Dragonflies, Order Odonata (page 205).

Aphid-lions, Order Neuroptera, Family Chrysopidae (page 243).

Ground beetles, Order Coleoptera, Family Carabidae (page 232).

Lady beetles,[1] Order Coleoptera, Family Coccinellidae (page 235).

Flower flies, Order Diptera, Family Syrphidae (page 305).

Tillyard said of the dragonflies: "They are the most powerful determining factor in preserving the balance of insect life in ponds, rivers, lakes, and their surroundings." Although popularly called "snake-feeders," "snake-doctors," "devil's darning

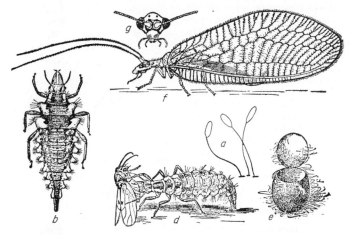

Fig. 2.12. An aphid-lion or lace-winged fly. *a*, Several eggs, showing the long pedicels that elevate them; *b*, larva, dorsal view; *d*, larva feeding on a plant louse; *e*, empty cocoon, showing lid through which the adult has escaped; *f*, adult, side view; *g*, head of an adult, front view. (*From Marlatt, U.S.D.A.*)

needles," and other unsavory names, neither the adults (Fig. 6.19,*5*) nor the nymphs (Fig. 6.19,*1,3*)—which develop for a year or two in the water before the winged stage appears—bite, sting, feed snakes, perform sorcery, or harm man in any way whatever.   They scull, soar, and dart about, near and over ponds and streams in a manner to arouse the envy of the most daredevil aviator.   The swiftest of them attain a speed of nearly 60 miles an hour.[2]   They both catch and eat their prey while on the wing.   Tillyard found over a hundred mosquitoes in the stomach of a dragonfly at one time.   The same author fed mosquito larvae to a hungry dragonfly nymph which swallowed 60 of them in 10 minutes.   The adults also are known to catch flies, beetles, moths, and wasps, many of which are doubtless injurious to man.

Aphid-lions.   These voracious creatures are the young of delicate, gauzy-winged, weak-bodied insects called lace-winged flies or golden-eyed flies (Fig. 2.12,*f*; Chrysopidae, page 243).   In the adult stage (*f*) these insects probably have little importance to man, but their larvae (*b*, *d*) are very beneficial.   Their white eggs (*a*) are curious objects placed on long slender stems attached to the leaves or stems of trees, vegetables,

[1] Also called "ladybirds" and "ladybugs."

[2] TILLYARD, R. J., "The Biology of Dragon Flies," Cambridge, 1917.

or field crops (Fig. 5.1, *R*). The spindle-shaped larvae that hatch from them may be found scurrying about on the plants in search of aphids. They have very long, sharp-pointed jaws or mandibles, with which they grasp and puncture the bodies of aphids or other small, soft insects, or their eggs. These mandibles have grooves along their ventral surface against which the maxillae fit to make two closed tubes through which the juices of the victim are sucked into the mouth (Fig. 4.11). The aphid-lions spend the winter in small, white, silken, spherical or oval cocoons (Fig. 2.12,*e*) about the size

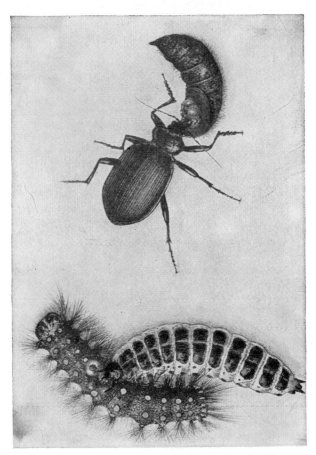

Fig. 2.13. European ground beetle, *Calosoma sycophanta*. Adult, at top, feeding on pupa of gypsy moth, and larva, at bottom, feeding on caterpillar of gypsy moth. (*From Mass. State Forester.*)

of a common elderberry. In the spring the adult cuts off a circular lid through which it makes its escape.

Ground Beetles. The exact food habits of the ground beetles of the Family Carabidae (page 232) are less well known. One finds the general statement "they are beneficial because of their predaceous habits"; or, since "both larvae and adults feed on many of our most noxious insects, ground beetles must rank among the farmer's best friends." At least a few of the species are injurious by feeding on seeds and berries, and the food habits of most of the 1,200 or more American species have never been recorded. But certain species are known to be very valuable; and, in general, these flattened, black or brown, long-legged, swift-running, strongly built "caterpillar hunters" are probably helpful to man. They hide during the day under stones, boards, logs, in the grass, or below the surface of the soil and hunt chiefly at night. The larvae are slender, a little flattened, slightly tapering to the tail, which terminates in two bristly hair-like or spine-like processes. *Calosoma sycophanta* was introduced into New England to control the gypsy moth and is an important natural enemy of this insect and other lepidoptera. A single larva of the predator will destroy at least 50 full-grown gypsy moth larvae during its 2 weeks of development, and the adult beetle which lives from 2 to 4 years will destroy several hundred (Fig. 2.13).

Lady Beetles. This group of the family Coccinellidae (page 235 and Fig. 2.14) needs no introduction to most persons. Their bright bodies, active habits, great abundance and equitable distribution, and their fame in popular songs and stories, all combine to ensure that almost all of us make their acquaintance at an early age. The lady beetles are nearly hemispherical in shape, and the commonest species are red, brown, or tan, usually with black spots; a few are black, sometimes spotted with red. In size they are commonly from $\frac{1}{16}$ to $\frac{1}{4}$ inch long and about two-thirds as broad. Some of the destructive leaf beetles (page 239) look much like lady beetles. The student need never confuse these two groups if he will remember that the lady beetles have three-segmented tarsi, the leaf beetles four segments in each tarsus.

Both the adults and the larvae of the lady beetles feed on scale insects, aphids, or other small, soft-bodied creatures or their eggs. The larvae of lady beetles, are carrot-shaped and resemble somewhat the aphid-lions in their flattened, gradually tapering bodies, distinct body regions, long legs, and warty or spiny backs. They do not have the inordinately long mandibles or extra-wide thoracic segments of the aphid-lions, and are generally more conspicuously colored with patches of blue, black, and orange. The pupae are not enclosed in cocoons but are exposed on the leaf to which the tips of their abdomens are cemented. When disturbed, they have the curious habit (possibly protective) of lifting the body into a vertical position and soon dropping

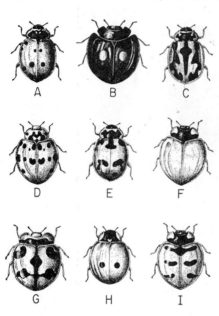

FIG. 2.14. Some common lady beetles. *A*, convergent lady beetle, *Hippodamia convergens; B*, twice-stabbed lady beetle, *Chilocorus stigma; C*, American lady beetle, *Hippodamia americana; D*, ash-gray lady beetle, *Olla abdominalis; E*, Washington's lady beetle, *Hippodamia washingtoni; F*, California lady beetle, *Coccinella californica; G*, vedalia or Australian lady beetle, *Rodolia cardinalis; H*, two-spotted lady beetle, *Adalia bipunctata; I*, eastern lady beetle, *Coccinella transversoguttata.* (*From C. Papp, Bul. S. Calif. Acad. Sci., vol. 56.*)

back again. The orange eggs of many lady beetles are placed in small masses of a dozen or two, the individual eggs standing on end in contact with each other. They should not be destroyed.

Syrphid Flies. Another important group of predators that rival in importance the lady beetles and the aphid-lions are the creatures known as syrphid flies, flower flies, or sweat flies (Fig. 2.15; Syrphidae, page 305). They are predaceous only in the larval stage; the adult flies never attack other insects. It is a rare aphid colony that does not have from one to many of the elongate, footless, slug-like, tan, or greenish maggots of these flies (Fig. 2.15,*C*) preying upon it. Hidden among the aphids or quietly looping about over the surface of the plant, these larvae grasp aphid after aphid by their pointed jaws, raise it in the air, and slowly pick out and suck out all the body contents, finally discarding the empty skin. A syrphid fly larva often destroys aphids at the rate of one a minute over considerable periods of time. It is difficult to measure the great service thus performed. The adults lay their glistening-white elongate eggs (Fig. 2.15,*B*) usually one in a place, among groups of aphids. They themselves feed on nectar and pollen and have considerable value as pollinizers. Adult syrphid flies are often confused with wasps and bees. They are generally banded or spotted with bright yellow or covered with long black and yellow hairs that give them a very striking general resemblance to stinging insects. They can be distinguished by having only one pair of wings, by their hovering or poising flight, and by the presence of the false vein (*x*, Fig. 2.15,*A*) in their wings.

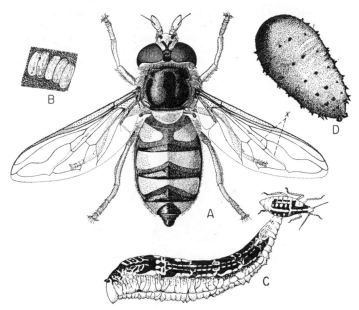

Fig. 2.15. A syrphid fly. *A*, adult of *Didea fasciata*; *x*, the false vein in the wing that characterizes flies of this family (*after Metcalf, from Ohio Naturalist*); *B*, eggs of *Melanostoma mellinum* (*after Metcalf, Maine Agr. Exp. Sta. Bul.* 253); *C*, larva feeding on an aphid (*redrawn after Jones, Colo. Agr. Exp. Sta.*); *D*, puparium of *Didea fasciata* (*after Metcalf, from Ohio Naturalist*).

The number of predators working in a field is often astonishingly large. W. G. Johnson records that in packing peas in southern Maryland in 1899, the separators sieved out in a few days about 25 bushels of larvae of Syrphidae, chiefly one species. They were so abundant that they almost completely destroyed the pea aphids in the fields. J. E. Dudley reports that in experimenting with an aphidozer or trap to collect pea aphids from pea fields in Wisconsin, he collected with the aphids from 2½ acres, 1,523 syrphid fly larvae, 173 adults and larvae of lady beetles, and 42 other predators.

Insect Parasites. The most valuable *entomophagous parasites* are probably contained in the following families:

Tachinid flies, Order Diptera, Family Tachinidae (page 310).

Ichneumon wasps, Order Hymenoptera, Family Ichneumonidae (page 274).

Braconid wasps, Order Hymenoptera, Family Braconidae (page 275).

Chalcid wasps, Order Hymenoptera, Family Chalcididae or Superfamily Chalcidoidea (page 276).

Egg parasites, Order Hymenoptera, Family Scelionidae (page 276).

These families of entomophagous insects differ from the predators just discussed in that they enter the body of the victim and live inside of it, feeding for a period of time on blood or tissues, instinctively avoiding vital parts, usually until the parasite is full grown. By that time or shortly afterward the victim dies and the parasite completes its transformation to an adult either within or without the dead body of the host. Representatives of all insect orders and all life stages from egg to adult may be thus attacked by the various parasites.

Tachinid Flies.   One of the most important families of entomophagous parasites is the group of flies known as Tachinidae (page 310).   Many of the species resemble, superficially, a much overgrown house fly, being very bristly, usually grayish, brownish, or black-mottled flies, without bright colors (Fig. 2.16).   They may be distinguished at once from the house fly family, however, by the entirely bare bristle on the

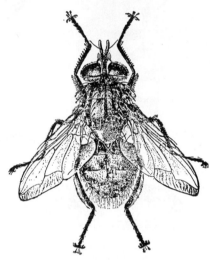

FIG. 2.16. A tachinid fly, *Winthemia quadripustulata*, a fly that lays eggs on armyworms and whose larvae destroy the worms.   (*From U.S.D.A., Farmers' Bul.* 752.)

antenna.   The adults are found chiefly resting on foliage or about flowers upon which they feed, but may often be seen attacking the caterpillars of moths and butterflies which are the commonest hosts of their larvae.

The eggs of the tachinid flies are glued to the skin of the host (Fig. 9.6) or laid on foliage where the host insect may ingest them; or already hatched larvae are deposited on the body of the victim or beneath its skin.   They feed at the expense of muscles and fatty tissues, not absolutely necessary to life, and the host caterpillar may complete its growth and form a chrysalid or cocoon before it dies.   From the cocoon, however, emerges the parasitic fly instead of the moth or butterfly.   Whenever armyworms become abundant, one of the noticeable features of the outbreak is the great number of flies that are buzzing about.   Farmers sometimes think that flies produce the armyworms.   However, the parents of the armyworms are tan-colored moths (Fig. 9.7), and these flies, *Winthemia quadripustulata* (Fig. 2.16), are one of their principal natural enemies.   If you examine the back of an armyworm, you will often find the small white eggs of this fly glued to the skin, in numbers ranging from 1 to 50 (Fig. 9.6, *left*). Few of the eggs are found farther back on the body than the thorax—seemingly a nice protective instinct on the part of the parent fly, since the caterpillar could bite off the eggs deposited near the rear of its body.   Upon hatching, the maggots tunnel directly through the skin and kill the host within several days. In many seasons parasitism of the armyworm may reach 50 per cent.

Another tachinid fly, *Lydella stabulans grisescens*, is an important introduced parasite of the European corn borer (page 415). This parasite, which often attacks 45 to 70 per cent of the corn borer larvae, has two generations each year and passes the winter as a second-instar maggot in the body of the fourth-instar host. The adult female fly oviposits 1,000 or more living larvae (rather than eggs) directly into the burrows of the corn borer, and the young maggots immediately seek the host larva and enter its body through a natural orifice or by burrowing through the skin. The parasite larva attaches to a tracheal trunk of the host and forms a respiratory funnel, where it remains until the host dies. The life cycle requires about 18 days.

The order Hymenoptera contains the most important entomophagous parasites both from the standpoint of the number and diversity of species and the frequency and effectiveness of their attacks upon important insect pests. The families of the parasitic Hymenoptera are extremely large in numbers of species and very difficult to classify. For the practical man it is probably sufficient to know that they are all small to moderate-sized wasps with a sharp ovipositor (homologue of the "stinger" of the honeybee), with which the eggs are generally thrust into the flesh of the host. Most species have a divided or two-segmented trochanter (Fig. 3.4,*D*) and a slender petiole or fore part of the abdomen next to the thorax.

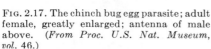

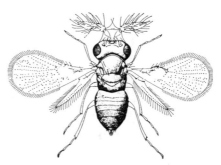

FIG. 2.17. The chinch bug egg parasite; adult female, greatly enlarged; antenna of male above. (*From Proc. U.S. Nat. Museum, vol. 46.*)

FIG. 2.18. The adult male of the egg-parasite *Trichogramma minutum*. (*From Clausen, "Entomaphagous Insects," copyright 1940, McGraw-Hill.*)

The *Scelionidae* or egg parasites of the superfamily Proctotrupoidea (page 276) include some of the smallest of all insects. Many of them are of such a size that they derive all their nourishment and grow to maturity inside of a single egg of another species, especially the Lepidoptera, Hemiptera, Orthoptera, or Diptera. *Eumicrosoma benefica*[1] (Fig. 2.17) is a parasite of chinch bug eggs (page 480) and prefers to deposit each of its average of 22 eggs in host eggs which are 1 to 3 days old. Hatching occurs in a few hours, and the larval stage lasts about 6 to 8 days. The entire life cycle requires 11 to 23 days, and there are 8 or 9 generations a year, 4 or 5 successive generations developing in the eggs of a single host brood.

The *Chalcidoidea* (page 276) includes many thousands of species, mostly parasitic, although some, such as the jointworm and fig wasp, feed on plants. The family Trichogrammatidae is composed entirely of egg parasites, which are of considerable value in natural control and have a wide host range. *Trichogramma minutum* (Fig. 2.18) has been found to attack more than 150 species of Lepidoptera, Coleoptera, Hymenoptera, Neuroptera, Diptera, and Hemiptera and is reared in enormous numbers (page 415) and sold for biological control of such pests as the codling moth, oriental fruit moth, and sugarcane borer. Its life cycle requires about 8 days, and generations are continuous as long as host materials are available.

[1] McColloch, J., and H. Yuasa, *Jour. Econ. Ento.*, **8**:248, 1915.

The *Encyrtidae* are chiefly internal parasites of scale insects of the family Coccidae and have been introduced extensively for biological control. *Aphycus helvolus*, the most successful parasite of the black scale (page 805), lays its egg at either end of the host, the respiratory stalk remaining fixed at the point of insertion. The life cycle is completed in about 13 days, and there may be as many as 8 generations a year, egg

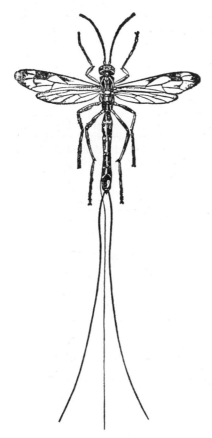

Fig. 2.19. The long-sting, *Megarhyssa lunator*, an ichneumonid parasite of the pigeon tremex. The long appendage is not a stinger but an egg-laying organ for inserting the eggs into the burrows of the pigeon tremex, in a tree trunk. (*From Kellogg, "American Insects."*)

Fig. 2.20. A sphinx caterpillar covered with the cocoons of a braconid parasite. The larvae of the parasite develop inside the body of the caterpillar, which eventually dies as a result of the attack. (*From Clausen, "Entomophagous Insects," copyright, 1940, McGraw-Hill.*)

production ceasing if the host scale is not in a suitable stage for oviposition and for the adult to feed on its body fluids.

The *Aphelinidae* are mostly external parasites of the Homoptera. *Aphelinis mali*[1] is an important parasite of the woolly apple aphid (page 733). The female, which lays an average of about 100 eggs, oviposits on the ventral surface of the aphid and often feeds at the puncture point. The egg hatches in about 3 days, and the larva feeds internally during the next 10 to 12 days. The life cycle requires from 20 to 25 days, and there are 6 to 7 generations per year.

[1] Lundie, A., N.Y. (*Cornell*) *Agr. Exp. Sta. Memoir* 79, 1924.

The *Eulophidae* have been widely imported for biological control and are largely external parasites of leaf-mining or stem-boring Diptera, Lepidoptera, and Hymenoptera and of the Homoptera. *Coccophagus gurneyi*[1] is an internal primary parasite of the citrophilus mealybug, and the female egg hatches in 3 to 4 days, and the life cycle is complete in 3.5 to 6 weeks. However, the male is a hyperparasite, and the egg which is deposited in the body cavity does not hatch until the body fluid of the host has been depleted by the feeding of a primary parasite. The male egg may thus remain alive for up to 86 days if conditions are not favorable for its development. After hatching, the male larva enters the body of the full-grown female primary parasite larva and feeds internally for a time, then completes its development as an external parasite.

The *Pteromalidae* includes many important parasites of major insect pests. *Pteromalus puparum*, which attacks the imported cabbageworm (page 662), may deposit

F꜀ɢ. 2.21. Aphids parasitized by a braconid wasp, *Lysiphlebus testaceipes*. Adult wasps have emerged through the circular holes after having killed the corn leaf aphids by developing in their bodies. Nearly 100 per cent of the aphids were parasitized. (*From Essig, "Insects of Western North America," copyright 1926, Macmillan; reprinted by permission.*)

as many as 700 eggs, and the colony of parasitic larvae in a single cabbageworm may deplete the available food supply, as many as 300 individuals having been reared from a single cabbageworm. The life cycle requires about 21 days, and the generations are continuous as long as suitable hosts are available.

Not all the parasitic chalcids are beneficial to man, because there are many hyperparasites or secondary parasites of beneficial insects.

The *Ichneumonidae* includes the largest of our parasitic species, as well as some very minute ones. They are often brilliantly marked and when active can generally be recognized by their short, jerky flight and constantly vibrating antennae. The very large long-sting, *Megarhyssa lunator* (Fig. 2.19), 1.5 inches long, with a tail-like ovipositor nearly 3 inches long, may often be found fastened by its ovipositor in the trunk of a tree. It thrusts this marvelously thin and strong ovipositor into the wood until it strikes the burrow of a wood-boring larva (the pigeon tremex, Fig. 6.75), when an egg is inserted. The parasitic larva, upon hatching from the egg, crawls along

[1] Fʟᴀɴᴅᴇʀs, S., *Univ. Calif. Pub. Ento.*, **6**:401, 1937.

the tunnel until it encounters the wood-boring larva, which it then attacks.   Possibly the adult long-sting locates the position of the wood-boring larva by the sounds it makes as it tunnels through the wood.   Most ichneumons attack caterpillars, with great benefit to man.

The *Brachonidae* are the parasites responsible for the parasitism of caterpillars such as the tomato hornworm (page 655) or catalpa sphinx (page 825), which are often seen with their backs covered with small elongate silken cocoons (Fig. 2.20).   These enclose the pupal stage of a braconid wasp of the genus *Microgaster*.   The parent wasp had previously thrust many eggs through the skin of the larva; her larvae had fed within the tissues of the caterpillar and, when full grown, had eaten out through the body wall again, to spin their cocoons fast to the host.   Some other species leave the host and form their cocoons in masses on a leaf near their dead victim.

If one examines carefully almost any leaf that is infested with aphids, one will see some specimens of the aphids whose bodies are distended, shiny, and brown and many of them with neat circular holes cut in their backs (Fig. 2.21).   This is the work of the smallest of the braconid wasps of the genera *Aphidius* and *Lysiphlebus* (Fig. 6.76). *Macrocentrus ancylivorus*[1] is one of the best known of the braconids and is an efficient parasite of the oriental fruit moth.   The female locates the host larva in its burrow in the plant by tapping the surface with her ovipositor.   Then she deposits a minute egg into the body cavity of the larva.   This egg begins to grow immediately and forms several embryos by polyembryony.   However, the hatching of the first embryo inhibits the development of the others, and only a single parasite matures.   The third-stage larva emerges from the host and molts simultaneously, so that the fourth-instar larva feeds externally on the host and then spins its cocoon inside the host cocoon.   The entire life cycle requires about 28 days, and the number of generations corresponds to that of the host.   A female wasp lays a maximum of about 500 eggs.

INSECTS AND WEEDS

Whether an insect is injurious or beneficial depends more often than anything else upon the economic status of the thing or material it eats. That great horde of insects that feeds on weeds or attacks noxious animals must perform a very valuable service.   Anyone who stoops to notice the abundance and variety of insects on almost any bit of wild vegetation will need no arguments to convince him that these plants suffer grievously from insect attack.   To a corresponding degree they help the farmer, who has to contend with these undesirable plants.   The biological control of weeds by insect enemies is further discussed on page 417.

This beneficent work is unfortunately not without its drawbacks.   Too often insects that have increased at the expense of weeds subsequently shift their attack to cultivated crops.   This is especially likely to occur when the weeds are botanically close relatives of the cultivated crop.   For example, the writers have often seen numbers of flea beetles and tomato hornworms developing in hog lots, barnyards, fence rows, and fields, where Jimson weed, horse nettle, and morning-glories were allowed to grow in profusion.   A little later these pests shifted to adjoining potato, tomato, or tobacco plants.   Forbes has shown that the corn root aphid is dependent in the early spring upon such weeds as smartweed, foxtail, ragweed, purslane, and crab grass, before corn is planted.   The common stalk borer, *Papaipema nebris* is seldom destructive to corn or other cultivated crops except along the margins of fields where the larvae have begun development on grasses or weeds.   The bean aphid, *Aphis fabae*, is often very abundant on pigweed and dock, from which it easily migrates to beans, dahlias, euonymus, etc.   The spinach flea beetle increases in the early part of the season at the expense of such weeds as chickweed and Chenopodium. The tarnished plant bug occurs in such numbers on ragweed, pigweed, common mallow, goldenrod, evening primrose, and wild asters as to constitute a considerable

[1] DANIEL, D., *N.Y. (Geneva) Agr. Exp. Sta. Tech. Bul.* 187, 1932.

check to their development.   But while they are checking these weeds, they are also increasing their own numbers for a probable later invasion of our nurseries, orchards, and flower gardens.   The thistle butterfly or painted lady, in periods of abundance, is sometimes hailed as a destroyer of Canada thistle.   But it is also likely to feed injuriously upon some cultivated crops.   Even insects that are not at present known to attack any cultivated crop may change their food habits, desert their weed hosts, and adapt themselves almost exclusively to a new and valuable food plant.   Such a shift is known to have occurred in the case of the Colorado potato beetle (page 640).

We may fairly sum up this phase of the subject with the statement that, while we are not unmindful of the vast service performed by insects in checking weed growth, we cannot encourage their development or even look with favor upon the great natural increase of any species, especially in proximity to cultivated crops, because of their potential injury to valuable plantings (pages 417, 418).

Insects as Soil Builders[1]

In the production or maintenance of productive soils insects play an important part.   They help to break up the rock particles and, by bringing them to the surface, expose them to the action of water and other weathering influences.   The numerous tunnels made by the insects facilitate the circulation of air into the soil, so essential to the health of plants.   Insects burrow to depths ranging up to at least 5 feet for white grubs and 10 feet for the nymphs of cicadas.   These burrows doubtless have considerable importance in the movements of capillary water.   Finally, insects are of inestimable value in adding humus or organic matter to the soil.   This is accomplished in several ways.   The dead bodies of the insects themselves accumulating on the surface are a fertilizing element.   The excreta of insects is a rich manure, comparing favorably in chemical content with that of the larger animals and in total amount undoubtedly exceeding the latter.   In burrowing through the ground, insects bring up the subsoil particles, and cover plant and animal materials lying on the surface.   Others carry plant and animal particles into the soil in connection with their feeding or nesting activities: termites, ants, cutworms, burying beetles, dung beetles, and predaceous wasps.   This is somewhat similar to plowing under a cover crop.

Here, as in most of the other relations of insects, their small size is abundantly offset by their unparalleled numbers.   Wheeler says that ants outnumber in individuals all other terrestrial animals.   The earth worm as a soil builder has been brought prominently to our attention by the writings of Darwin; it seems to the writers that insects as a whole must equal or excel earth worms in the formation, fertilization, and renovation of soils.   A great variety of insects is found in the soil, representatives of practically all the natural orders.   The most abundant are ants, bees, wasps, beetles, the larvae of flies, cutworms, the pupae of moths, cicadas, crickets, and springtails.   These creatures have been found in the soil in abundance beyond our comprehension.

Insects as Scavengers

One of the most interesting ecological groups of insects is that group that feeds on the decaying substances of plants and animals.   Their service is twofold: in the first place, they help to remove from the earth's surface the dead and decomposing bodies of plants and animals, converting them into simpler and less obnoxious compounds, and removing what would otherwise be a menace to health; and in the second place, they play a very important part in converting dead plants and animals into simpler substances which can be used as food for growing plants.   Repulsive as they may be, we cannot scorn the beneficent work they do.   In this work termites, the larvae of many flies, and the larvae and adults of beetles are especially important.   A fascinating story of the lives of some of the beetles is told by Fabre in "The Life and Love of the Insect."

[1] McColloch, J., and W. Hayes, "The Reciprocal Relation of Soil and Insects," *Ecology*, **3**:288–301, 1922.

AESTHETIC VALUE OF INSECTS

The aesthetic value of insects is the least tangible of all, and perhaps most readers will think it the least important. Insects rival birds and flowers in beauty; and there are many more of them. Moths and butterflies are universally admired. But there are thousands of kinds of beetles, bees, flies, leafhoppers, planthoppers, dragonflies, mantids, and others that will be found every bit as handsome as moths and butterflies by anyone who will examine them through a mircoscope or magnifying glass.

Much practical use has been made of the beauty of insects. Artists, florists, milliners, and designers have drawn extensively from the Lepidoptera and might profitably draw much more extensively upon the inexhaustible grace of line and beauty of color, so lavishly displayed in the less well-known kinds of insects. Insects are widely used as ornaments. In larger cities one finds stores dealing in trays, pins, necklaces, etc., of which the sole claim to beauty is the actual bodies of preserved insects. The reader should not gain the idea that many insects have a very great commercial value. One of the drawbacks of being an entomologist is the frequent necessity of disillusioning some trusting youngster who has reared his first cecropia moth and has hastened to the nearest "bugman" to sell his valuable (?) find.

In the Bahamas, South Africa, and Australia the natives often wear strings of what are called "ground pearls," which have been found to be the case or shell secreted by the nymphs of a kind of scale insect. They are irregular in shape, of a variety of colors, and often have a beautiful luster or iridescence. In the Caribbean islands living fireflies are enclosed in gauze and worn as hair ornaments.

The songs as well as the form and colors of insects have been of much interest from earliest times. The natives of South America, Africa, Italy, and Portugal cage katydids and crickets for the sake of their songs; and highly ornamented cages containing these little songsters are sold in the streets.[1] One of our entomologists[2] has taken much pains to assemble the poetry based on insects, in the English language alone, from A.D. 1400 to 1900. He records having found over 1,200 separate excerpts, including about 75 complete poems.

The curious habits, structures, sounds, and interactions of insects continually afford profitable diversion or hobbies to a small army of amateur entomologists, ranging from carefree children to staid scientists and millionaires. The Chinese have elaborate cricket fights, contested by champions that are as carefully fed and trained as race horses and that sell for $5, $10, or even $50 each. Many persons have been amused and deceived by the seemingly intelligent performances in trained-flea circuses.

SCIENTIFIC VALUE OF INSECTS

Insects have taught man a great many things. They have helped to solve some of the most puzzling problems in natural phenomena. They have led to some remarkable inventions. They have contributed to the sum of human knowledge in physiology, psychology, and sociology.

The use of insects as an index to stream pollution has been a valuable aid in the science of conservation of natural resources. When toxic factory wastes and large amounts of raw sewage are dumped into streams, aquatic life, such as fish, is destroyed and the pollution also becomes a menace to the health of man and other animals. The degree of pollution

[1] CAUDELL, A. N., "An Economic Consideration of Orthoptera Directly Affecting Man," *Smith. Inst. Ann. Rept.*, 1917, pp. 507–514.

[2] WALTON, W. R., *Proc. Ento. Soc., Washington*, October and November, 1922.

can be measured by the amount and type of insect life that is able to survive in the polluted water. Streams and lakes having few insects are highly polluted. It may eventually be possible to measure the amount of pollution with mathematical accuracy by observing the insect life that is present in any stream.

The great abundance and variety of insects have resulted in their selection for many scientific studies and fundamental biological investigations. Among the many benefits that have accrued from observations of, and experimentation with, these small animals, perhaps the greatest is their contribution to cytology, genetics, and eugenics. The cells of insects are extraordinarily large, so that they are exceptionally good for cytological work. Plant and animal breeders and those interested in the future of the human race, have demanded precise and explicit knowledge of how physical and mental traits are inherited when selected mates of various kinds are crossed. Previous to the discovery of the suitability of the tiny fruit fly, *Drosophila melanogaster*, for this work, genetic experimentation with guinea pigs, rats, pigeons, corn, peas, and the evening primrose involved heavy expenses for feed, cages, and caretakers. Even more serious was the limited number of generations or crosses that one scientist could rear for study during his lifetime. Finally it was realized that the fundamental principles of inheritance, variation, and race improvement could be revealed as clearly by insects as by the more expensive, more cumbersome, slow-breeding plants and animals. Drosophila can be handled with the greatest of ease. Thousands of them can be reared on a bit of fermenting banana in small vials. Little space and few caretakers are required. This little fly is subject to tremendous variations in visible external characters. Its cells have only four pairs of chromosomes (page 152), and its salivary glands have chromosomes large enough that the actual genes (page 152) can apparently be seen under the microscope. Most important of all, this insect will complete a generation in as short a time as 10 days, so that the geneticist can have as many different generations and crosses as he can possibly study during his lifetime. It is safe to say that a large part of our present knowledge of genetics and eugenics would never have been possible without the use of this tiny fly. Cockroaches and stored-product insects have been the subjects of many nutritional studies. The small size and rapidity of reproduction of the insects have made them ideal tools for nutritional work dealing with very small quantities of dietary media and unknown dietary factors. Insects such as the house fly and mosquito larvae frequently have been used for the bioassay of extremely small amounts of insecticide residues on fruits and vegetables.

CHAPTER 3

THE MORPHOLOGY, PHYSIOLOGY, AND BIOCHEMISTRY OF INSECTS[1]

Insects are among the humblest and lowliest of animal creatures; man belongs to a species that has been called the dominant species. Yet it has been shown in the preceding chapters that insects dispute with man for the possession of most of the things he values and often succeed in getting the greater part of the product of his labors. The present chapter is an attempt to analyze the success of insects as a group of animals living in keen competition with thousands of others.

Insects are the most abundant of terrestrial animals in both numbers of species and numbers of individuals. According to the theory of *organic evolution* or *biogenesis,* this has come about because they are better adapted to their surroundings or environment than other groups, many of which have become extinct while the insects have continuously diversified and multiplied. In general, every kind of plant and animal produces more offspring every year than could possibly find food and room to survive (*overproduction*). The competition for every necessity of life is so intense that the great majority of all creatures born into the world die before reaching maturity (*the struggle for existence*). Among the individuals of a species, even those from the same parents, there are generally noticeable variations, sometimes slight (*continuous variations*) and sometimes striking ones (*mutations*). There is good evidence for believing that death due to the struggle for existence is not indiscriminate, but that those individuals possessing variations that are advantageous will in the long run survive to perpetuate their kind, while those less favorably equipped for the struggle of life will be eliminated (*elimination of the unfit* or *survival of the fittest*). It follows that any group that has attained the numerical superiority of the insects, both in species and in individuals, must be unusually well fitted for life. We shall be in a better position to fight those insects that are pests when we understand their structure and the most important characteristics that fit them for life.

CHARACTERISTICS THAT ENABLE INSECTS TO COMPETE WITH MAN

The Size of Insects. One of the greatest factors in the success of insects is undoubtedly their small size. Although the largest insects may reach a length of 10 inches (250 millimeters), as in the giant phasmid, *Palophus titan,* or a weight of 1.5 ounces (42 grams), as in the beetle,

[1] General references: SNODGRASS, R. E., "Principles of Insect Morphology," McGraw-Hill, 1935; WIGGLESWORTH, V. B., "Principles of Insect Physiology," 4th ed., Dutton, 1950; ROEDER, K. D. (ed.), "Insect Physiology," Wiley, 1953; GILMOUR, D., "The Biochemistry of Insects," Academic Press, 1961.

Goliathus goliathus, these are very exceptional and the average insect is probably no larger than 6 to 10 millimeters long and 25 to 50 milligrams in weight. The range in size is, however, astonishing, and W. T. M. Forbes is quoted as saying that the moth having the greatest wing spread is probably *Thysania agrippina*, which measures 11 inches (279 millimeters) from tip to tip of wings, and the smallest probably *Nepticula gossypiella*, whose wing span is slightly over 0.1 inch (2.5 millimeters). The over-all weight range is of the order of 10,000,000 times. The average size and weight of some common insects are given in Table 3.1.

TABLE 3.1. SIZE AND WEIGHT OF SOME INSECTS

Insect	Body length, mm.	Live weight, mg.
Egg parasite, *Trichogramma minutum*	0.37	0.004
Greenhouse thrips, *Heliothrips haemorrhoidalis*	1.2	0.056
Yellow fever mosquito, *Aedes aegypti* ♀	5.0	1.5 (unfed)
		3.6 (fed)
House fly, *Musca domestica*	7.0	20.0
Milkweed bug, *Oncopeltus fasciatus*	15.0	80.0
Honeybee, *Apis mellifera*	15.0	100.0
Squash bug, *Anasa tristis*	17.0	125.0
Japanese beetle, *Popillia japonica*	13.0	150.0
Colorado potato beetle, *Leptinotarsa decemlineata*	10.0	160.0
Grasshopper, *Melanoplus differentialis* ♀	50.0	1,250.0
American cockroach, *Periplaneta americana* ♀	35.0	1,300.0
Saltmarsh caterpillar, full-grown larva, *Estigmene acrea*	40.0	1,580.0
Silkworm, full-grown larva, *Bombyx mori*	75.0	6,000.0
Cockroach, *Blaberus cranifer*	55.0	6,700.0
Goliath beetle, *Goliathus goliathus*	125.0	42,000.0

At first thought, small size might appear to be a disadvantage. However, this characteristic enables insects to live in cracks and crannies of plant and animal communities, where competition is minimal. Many insects can subsist on the portions left from the feeding of larger animals. They can retreat into protected places where larger animals cannot follow. They often escape death because they are completely overlooked; very often the first intimation a gardener has of the presence of scale insects in his orchard or of thrips in the flower garden is the spotting of the fruits at harvest or the withering of the blooms.

A small body makes possible feats of strength and agility that, in comparison to size, seem marvelous. Thus a flea with legs about 0.05 inch long can jump about 13 inches horizontally and 7.75 inches high. If length of legs were the only factor involved, we should expect a human with legs 3 feet long to make a broad jump of 700 feet and a high jump of at least 450 feet. The much larger grasshopper, *Schistocerca*, has been observed to jump 10 times its height and 20 times its body length and to lift a weight of 10 times its body with one leg. The stag beetle, *Lucanus dama*, has been observed to drag a weight of 120 times her body over a short distance and to hold, while suspended by her claws, a weight of

83 times her body, attached to her waist. To equal these feats a man weighing 160 pounds should be able to drag 9.6 tons and hold on by hands and feet with 13.5 tons tied to his waist. Such amazing feats of insect strength are partly the result of the fact that the weight of an organism increases as the cube of its linear dimension, while the strength of muscle increases only as the square of its linear dimension, so that as the body size decreases, the ratio of muscle strength to weight becomes progressively greater.

The Abundance of Insects and Their Rapidity of Reproduction. The great reproductive capacity of insects, in general, is unquestionably responsible in a large measure for their success. Different counts of the number of insects found in and on the soil indicate a frequency of 1,000,000, 3,500,000, and 10,000,000 individuals per acre, under various conditions. Margaret Windsor found insects occurring on and in the forest soil in Illinois, during the winter, to a depth of 18 inches, at an average frequency of 65,000,000 to the acre. H. Elliott McClure, from extensive trapping experiments, concluded that there is frequently an average population of about 3,000 insects in flight over each acre of surface, or about 1,850,000 per square mile in the morning, and about 11,000 per acre, or nearly 7,000,000 per square mile, flying in the evening. By way of comparison, there is an average of one human being to each 16 acres of dry land on the earth's surface; an average of about two head of horses, mules, cattle, sheep, hogs, and chickens, combined, per acre of land in the United States; and an average of three birds per acre in Illinois fields, woods, and orchards during the summer months. The reproductive capacity of insects is further discussed on pages 158 to 159.

The Adaptability of Insects. In the adaptability of insects we see another superior quality. New structures and new habits are continuously developing. New forms are evolving and old forms changing in accommodation to the constantly changing face of the continent. Within the memory of those still living, certain insects have adapted themselves to new host plants, thus changing from insignificant bugs to major pests. Insects have not restricted themselves to one medium, like the fish or the birds, or to one kind of host, like the parasitic worms; they occur on and in the water, in the air and soil, on animals and plants, and within the bodies of both; in houses, ships, and mills, and in almost all sorts of organic and inorganic substances. The problems in insect control are more serious in areas where the biological and physical environment is undergoing rapid changes than they are in areas where conditions have become more stable by a long and gradual development. It is one of the greatest tributes to the remarkable variety and plasticity of the insect class that, in the face of rapid and radical changes, *some* species of insect is certain to adapt itself with surprising promptness to the new opportunity and to accomplish this so successfully that it very soon becomes a serious pest.

The Persistence of Insects. In the instinctive behavior of insects we see an explanation of their apparent fixity of purpose. They lack both reason and judgment. This would seem to put them at a great disadvantage compared to man; and so it does, individual to individual. But considered in the mass, it means on the part of the insect, unfaltering pursuit of the work for which it is adapted. The everyday behavior of the flies about an animal, or in a room when one is trying to sleep, or of chinch bugs trying to cross a barrier from a wheat field to a cornfield, illustrates this point.   They cannot be frightened away or discouraged by repeated assault; they recognize no defeat.   So long as life persists within them they continue unflinchingly to procure a living for themselves and to prepare for the next generation.

The Body Structure of Insects.   A characteristic of insects which has been important in their evolution is the nature of the body wall.   Insects have no bones, but are covered externally with durable, flexible exoskeleton, the *cuticle* or *cuticula*.   This exoskeleton is lighter and stronger than bone and is remarkably resistant to solution or corrosion, not being visibly affected by any of the ordinary chemicals, such as water, organic solvents, strong acids, alkalies, and the digestive fluids of animals.   Even boiling potassium hydroxide, which quickly dissolves flesh and horn, does not destroy or change the appearance of the skin of an insect, unless the treatment is continued for a long time.   Because of the unusually stable

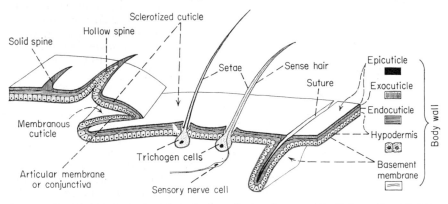

Fig. 3.1. Diagrammatic section of the body wall of an insect.   The living cells, called the *hypodermis*, secrete the three layers of cuticle over their outer (ectal) surface, while internally (entad) the hypodermis is limited by a basement membrane.   The cuticle is usually hardened or sclerotized over most of the body, but at infoldings, called *conjunctivae*, it remains soft and flexible, as membranous cuticle.   A solid and a hollow spine, and ordinary seta, and a sense hair are also shown; at the lower right, subtending a suture, is shown an apodeme, to which muscles may be attached.   (*In part after Comstock and Snodgrass.*)

character of the body wall, insects can be kept, like mummies, for hundreds of years without any preservation, retaining a life-like appearance after death.

The Insect Exoskeleton.   This organ (Fig. 3.1) serves two very important functions: it protects delicate muscles, nerves, and other organs from mechanical injury, and it serves as a framework for the attachment of muscles.   In this latter capacity an exoskeleton seems to offer certain advantages over the bony endoskeleton of the vertebrates.   It gives a vastly greater area for the attachment of muscles and certain opportunities for very effective leverage.   As a protective armor it could scarcely be improved upon, since it prevents the body from being soaked with water, from excessive drying, and from the attacks of many disease organisms and is probably the chief thing that enables insects to live in the greatest variety of conditions.   The typical insect is built with its shell somewhat in the form of a hollow cylinder, which is the strongest type of construction possible with a minimum amount of material.   The ends of the cylinder are closed with more or less convex caps, so that from almost any point of attack the insect body presents an arched construction. This protects insects from injury by ordinary blows or falling, a point of great significance when we consider the very active, apparently reckless sort of lives most of them lead. It should be understood that all insects in all stages, however soft the body wall may be, are completely covered by the sclerotized exoskeleton. The over-all advantages of this type of construction are best typified by the "ironclad" tenebrionid beetles, *Phloeodes* and *Eleodes*, which are virtually indestructible, survive total immersion in liquid hydrogen cyanide, and even after death can be pinned only with the greatest difficulty.

The hardened plates of the insect cuticle are known as *sclerites*. Between these plates, the cuticle is soft and continuous and uninterrupted from one sclerite to another, but remains soft and flexible and is often infolded. Definite impressed lines or internal ridges between sclerites are known as *sutures*. Flexible infoldings of the body wall are called *conjunctivae*, and the definite joints, as between segments of the leg, which permit rotary or hinge-like movements, are called *articulations*. The sclerites may often be moved very freely with reference to each other, by the action of muscles attached to their internal faces.

Formation of the Cuticle. Apart from its vital structural functions, the insect cuticle or body wall is one of the primary organ systems and a site of complex biochemical activity. Morphologically (Fig. 3.1) it is composed of a single layer of epidermal cells, the *epidermis* or *hypodermis*, which is separated from the body cavity by a thin basement membrane. These epidermal cells are the source of secretions which form the outer layers of the cuticle, and also the molting fluid, containing a mixture of the enzymes chitinase and protease, which dissolve the old cuticle (see Development and Metamorphosis, page 165). Above the epidermal cells is the inner colorless and flexible *endocuticle;* above this an amber to black rigid layer, the *exocuticle*, and the thin outer protective layer of the *epicuticle*. The cuticle may contain unicellular hair-like outgrowths known as *setae*, which represent sensory organs of various types and may also take the form of pegs, hooks, or scales. Multicellular outgrowths are known as *spines* if they are immovable and *spurs* if they are articulated and movable (Fig. 3.1). The cuticle is penetrated by the ducts of *dermal glands* and by minute helical pore canals, which in *Periplaneta americana* are from 0.1 to 0.2 micron in diameter and number about 1,000,000 per square millimeter of cuticle.[1]

The outermost layer of the integument, the *epicuticle*, is typically from 1 to 2 microns in thickness, waterproof, very resistant to chemical attack and insoluble in ordinary solvents. The epicuticle is formed by the deposition of a number of components and typically, as in *Rhodnius*, consists of an outer cement layer about 0.1 micron thick, which is a protective coating of shellac-like material that perhaps polymerizes with proteins and waxes. Underneath is the wax layer consisting of oriented lipid molecules of the same general constitution as beeswax, *i.e.*, long chain, even-numbered C_{24} to C_{36} esters of fatty acids and normal alcohols, mixed with odd-numbered C_{23} to C_{37} paraffinic hydrocarbons. The wax layer controls the permeability of the cuticle to water and to a large extent to insecticides. Wigglesworth and Beament[2] have shown that

[1] RICHARDS, A., *Jour. Morphology*, **71**:135, 1942.
[2] BEAMENT, J., *Jour. Exp. Biol.*, **36**:391, 1959.

each species of insect apparently has a characteristic wax with a distinctive melting point and critical temperature above which the impermeability to water is destroyed.   The soft grease of the epicuticle of *Periplaneta* consists of a hard wax dissolved in a slightly volatile hydrocarbon solvent.   The inner cuticulin layer of the epicuticle is apparently a tanned lipoprotein impregnated with lipids, produced as subsequently described.

Underneath the epicuticle lies the chitinous cuticle, which is the principle structural framework of the insect exoskeleton.   *Chitin* is a polysaccharide formed from long chains of *N*-acetyl D-glucosamine units, with the following type of structure, which is analogous to cellulose.   This

material, which is insoluble in all ordinary solvents and dilute mineral acids, constitutes from 30 to 60 per cent of the average weight of the exo- and endocuticles, but is not found in the epicuticle.   Absorbed in the chitinous framework is a typical water-soluble protein, *arthropodin*, which comprises about 35 to 70 per cent of the chitinous cuticle and which is bound to it with weak chemical bonding.   The chitinous cuticle exists in a laminated structure with alternating layers of materials of differing densities.   In *Periplaneta* these are about 0.15 micron thick, and Fraenkel[1] has suggested that they represent alternating monomolecular layers of chitin and arthropodin.

Cuticle formation begins with the secretion of the lipoproteins of the epicuticle as a semifluid continuous layer.   The inner layers of chitin and protein are then formed and impregnated with polyphenols such as 3,4-dihydroxyphenylacetic acid and 3,4-dihydroxybenzoic acid.   This is then waterproofed with a layer of crystalline wax, and within a half hour after molting the wax layer is covered with a secretion of lipoprotein cement.   The newly formed cuticle is soft and colorless, but after 1 to 2 hours, the process of sclerotization occurs in which polyphenol oxidase enzymes form the dark melanin pigment from tyrosine and oxidize the polyphenols in the exo- and epicuticle to orthoquinones, which then polymerize with amino-, imino-, and sulfhydryl groups of the proteins to form the familiar hard and dark or sclerotized cuticle.

Colors of Insects.   Insects are unique among animals for their fascinating array of colors.   Of these, the metallic and irridescent colors are produced by various structural features of the cuticle, which cause interference or diffraction of reflected light.   The classic investigations of Mason[2] have shown that interference colors are produced by the wing scales of Lepidoptera by multiple films separated by material of differing refractive indices.   The lamellae in *Morpho* are oblique and in *Urania* and others are parallel to the scale surface.   The elytra of the beetle *Sericea* contain rows of fine parallel striations from 1 to 2 microns apart which act as diffraction gratings.   Certain other metallic colors appear to be produced by minute particles imbedded in a transparent matrix.

Aside from these purely physical colors, which are limited to a few of the more

[1] Fraenkel, G., *Proc. Roy. Soc.*, **B134**:111, 1947.
[2] Mason, C., *Jour. Phys. Chem.*, **30**:383, 1926; **31**:321, 1856, 1927.

showy Coleoptera, Lepidoptera, and Odonata, the great majority of insect colorations are the result of various pigments located in the cuticle, hypodermis, blood, and fat-body. The most common insect pigment is melanin, which is responsible for the darkening of the cuticle after molting. Melanin is formed from the interaction of a chromogen, such as tyrosine or dihydroxyphenylalanine, with the oxidase, tyrosinase. The elaborate wing patterns of many insects are due to melanin formation, the distribution of the chromogen appearing to control the pattern formation, as tyrosinase is present everywhere in the body.

The pterins are the most numerous group of insect pigments, occurring in Lepidoptera, Neuroptera, Diptera, Hymenoptera, and Homoptera, where they are typically deposited in the hypodermis or in scales or hairs. Chemically, the pterins are related to purines, and it is possible that they may be formed as products of excretion. These pigments may be white as leucopterin, yellow as xanthopterin, or red as erythropterin.

A number of species of coccids contain polyhydroxyanthroquinone pigments such as carminic acid from *Dactylopius coccus*, which has been an important article of commerce (see cochineal, page 55), kermesic acid from *Kermesococcus ilicis*, and laccaic acid from *Laccifer lacca* (see shellac, page 51).

The blood of a number of Aphidae and Phylloxeridae contains a water-soluble magenta pigment, protoaphin, present in concentrations up to 2 per cent in *Tuberolachnus salignus*. In the crushed aphid this is readily converted by an enzyme to the yellow, fluorescent xanthoaphin, which, upon treatment with acid or base, forms orange chrysoaphin and red erythroaphin.[1]

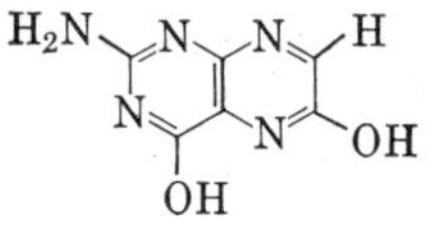

Xanthopterin

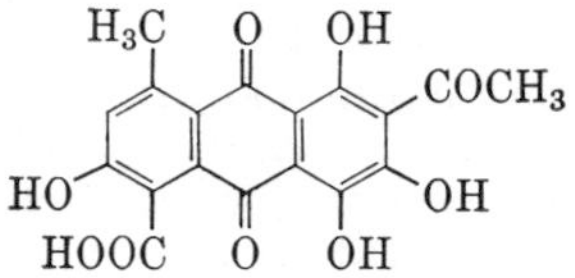

Kermesic acid

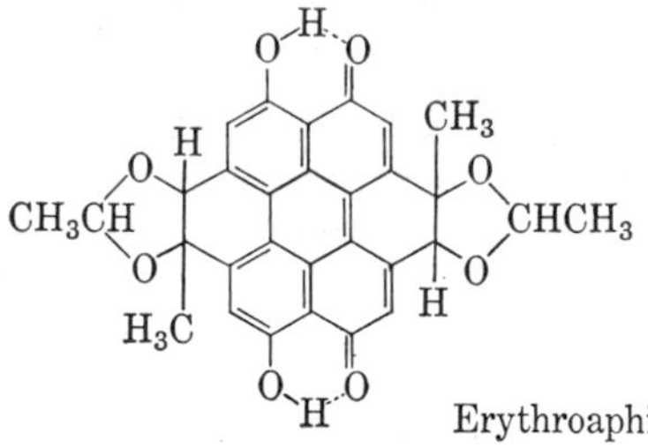

Erythroaphin

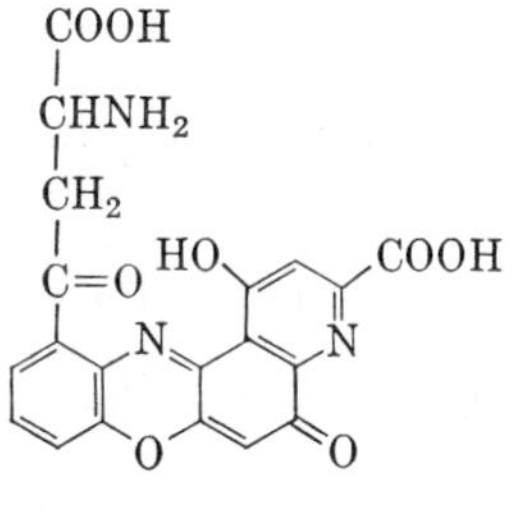

Xanthommatin

Many insect pigments are obtained from their plant foods. The carotenoids are readily absorbed by feeding insects and may occur free or as water-soluble protein complexes. The brilliant wing pigments of the Locustidae are β-carotene and astaxanthin and their protein complexes. β-carotene also occurs in the blood and tissues of Orthoptera, Coleoptera, Lepidoptera, and Hemiptera, along with lutein, which is also the yellow pigment of the silkworm cocoon. Chlorophyll and derivatives are found in the digestive tract and blood of many caterpillars, where they account for the insects' green colors. Chlorophyll derivative products such as phaeophorbides are found in the tissues of plant-feeding Hemiptera and in *Forficula*. Blue and greenish pigments of the biliverdin type are found in locusts and in the blood of Lepidoptera.

[1] CROMARTIE, R., *Ann. Rev. Ento.*, **4**:59, 1959.

A. unique group of pigments, the ommachromes, occur principally as the brown, red, and yellow eye pigments of insects and are also found in the red molting fluid of *Vanessa urticae*. They usually occur as protein complexes and are yellow-brown in the oxidized form and deep red when reduced, *e.g.*, xanthommatin.[1]

Flavones or anthoxanthins from plants are found in the wings of Lepidoptera, and an anthocyanin has been reported from the beetle *Cionus*. Hemoglobin-type pigments are found in *Chironomus* larvae (bloodworms) and in certain aquatic Hemiptera, while *Rhodnius* and other bloodsucking Hemiptera metabolize hemoglobin to form a red hemalbumin in the salivary glands and a green biliverdin-like pigment in the pericardial cells.

The Segmentation of the Body. Many animals that are covered with a hard outer shell are sluggish, inactive, and sedentary in habits, like the clams, snails, and barnacles. This is not true of most insects. The latter have achieved the happy combination of an armor plate and great freedom of movement. This is made possible by the characteristic known as *segmentation*. As one examines the bodies of insects (especially caterpillars), one sees that the external wall does not present a smooth, unbroken surface (Figs. 3.9,*1*, and 3.2). On the contrary, it is divided by constrictions into a series of ring-like pieces, all connected, yet moving rather freely on each other. The word insect is from the Latin *insectum*, "cut into," and refers to the manner in which the parts of the body are separated by constrictions.

Segmentation of the body is an important advantage. If we contrast it with the condition shown in other shelled animals such as snails and clams, we see that it permits great freedom of movement and activity. In the second place, it facilitates specialization, which makes for efficiency. The body so divided into segments may devote one part to securing food, another to locomotion, another to reproduction, another to defense, and so on. It permits *division of labor*, which has always made for success and progress the world over, whether it be in a force of factory workers, in a football team, or in the body of a bumblebee.

The segments of the body are called *somites* or body segments, to distinguish them from the segments of legs, antennae, and other appendages. The term "joint" properly refers to the constriction between two segments, and should not be used for the segment. The flexible portion of the cuticle connecting the hard ring-like portions of any two segments is called a *conjuctiva* or *articular membrane* (Fig. 3.1). Each body segment is made up of (at most) four exposed faces. The dorsal or upper face, above the bases of the legs, is called the *tergum* or *notum;* the ventral face (the part between the legs and next to the ground when the insect is in its normal position) is called the *sternum;* and the lateral face or side piece, containing the base of the leg, is called the *pleurum.* Each of these faces may be made up of several sclerites, which are then collectively called *tergites, sternites,* and *pleurites,* respectively, and given individual names. The pleura, however, are generally nearly or entirely membranous. The typical number of segments in the insect body is about 20 or 21. This number is greatly obscured and reduced in most insects by the fusion of some of the segments and the degeneration of others. So that ordinarily one will recognize only from 8 to 12 obvious body rings in most insects (Figs. 2.15,*A*, and 3.2).

The Body Regions of Insects. Other animals besides insects have the body externally segmented, notably the true worms, thousand-legged worms, crayfish, and their relatives. Members of the phylum Arthropoda,

---

[1] CROMARTIE, R., *Ann Rev. Ento.*, **4**:59, 1959.

to which insects, spiders, crayfish, and similar creatures belong, differ from the true worms (phylum Annelida) by having a pair of jointed appendages on at least a part of the body segments.   In this way we distinguish such things as caterpillars, maggots, and grubs (often called "worms," but really young insects) from the true worms such as the earth worm, and from parasitic worms like the tapeworm.   The latter have no legs.

In the true insects, there is never more than one pair of jointed appendages on any body segment.   Hence, even if the conjunctivae are lost and the segments fused together, as in the head, we can usually tell how many segments are represented by noting the number of paired, jointed appendages present.

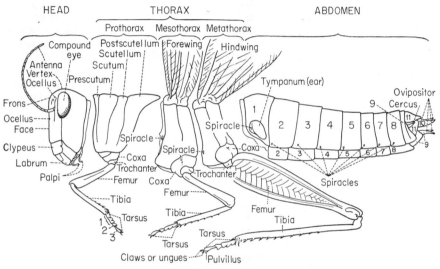

Fig. 3.2. Outline of body of a grasshopper as seen from the side, dissected to show the three body regions and the parts of the body commonly referred to in literature.   (*Slightly modified from Herms, "Medical and Veterinary Entomology."*)

Of the 20 to 21 segments of the insect body, 6 are fused into what is called the head,[1] 3 in front of the primitive mouth and 3 behind it.   The next three segments constitute what is called the *thorax*, which is distinguished as *that part of the insect body which always bears the jointed legs and the wings*, if they are present.   The remaining part of the body (typically 11 or 12 segments) bears no jointed legs and is called the *abdomen*.   The eleventh true abdominal segment bears the appendages called the *cerci* (Figs. 3.2 and 3.8).   Behind this the abdomen is terminated by a twelfth segment that never bears appendages, known as the *telson* or *periproct*. Primitively, the anus is supposed to have opened on the telson, the female reproductive organs between the seventh and eighth segments, on the ventral side, and the male reproductive organs on the posterior border of the ninth abdominal segment, ventrally.   These body openings, however, often become greatly shifted as the body regions become specialized.

[1] There is considerable disagreement among various authorities, who have suggested that from 5 to 8 segments are present (p. 127).

We say, therefore, that the segments of the body of an insect are grouped into three *body regions*, known as head, thorax, and abdomen (Fig. 3.2). This is a further development of the specialization and division of labor spoken of in connection with the segmentation of the body.   The head of the insect takes over the function of locating and taking in food, as well as most of the work of sensing danger and recognizing friends and enemies. The thorax practically always performs locomotion, while the organs of reproduction are borne by the abdomen.   The one thing that marks off the thorax from the head and abdomen is that it bears the legs.   Hence we find the thorax and determine its limits by finding where the six legs are attached, even in such insects as beetles, where the thorax appears to end just back of the front legs.   The three segments of the thorax are given distinctive names: the one bearing the first pair of legs is known as the *prothorax*, the segment bearing the middle pair of legs is the *mesothorax*, and the segment to which the hind legs attach is the *metathorax*.

The Antennae of Insects.   The way in which segmentation of the body facilitates division of labor is well shown in the paired jointed appendages on the heads of insects.   These appendages of the head arise during embryonic development in exactly the same way as the legs and for a time are indistinguishable from them.   But in the active insect these appendages have become greatly differentiated for several distinct functions.   One group of them comprises the mouth parts which are discussed more fully in Chapter 4.   Another pair of appendages, homologous with the legs, has become modified to form the "feelers" or "horns," or, as they are properly called, the *antennae* (Figs. 3.2 and 3.3).

None of the higher animals has anything to compare with the antennae of insects and their relatives.   With them various insects feel their way, detect danger, locate their food, find their mates, and, at least in some cases, use them to communicate with others of their own kind, *e.g.*, the ants; or bear end organs of smell (as in the flies); or use them in hearing, *e.g.*, as the male mosquito; or rarely for grasping a mate or the prey.   A pair of long, flexible, highly sensitive "feelers," with which the insect can sound out the environment ahead, must be of very great advantage to these active animals.   Most insects show great distress and sometimes helplessness when the antennae are removed or injured.

All true insects have one pair of antennae.   They are the appendages of the second head segment (pages 128, 129).   This is a useful distinguishing mark; for the spiders, mites, ticks, and scorpions have no antennae, the crayfish, lobsters, and crabs have two pairs, while the centipedes and millipedes agree with the insects in having one pair.

Types of Antennae.   The antennae of insects vary greatly in size and form, and are much used in classification.   The following special names have been applied to certain of the common types (Fig. 3.3).   A *filiform* or thread-like antenna is one in which all the segments are of about the same thickness and have no prominent constrictions at the joints.   A *moniliform* antenna is one made up of somewhat globular segments, with prominent constrictions between them, the whole suggesting the shape of a string of beads.   A *setaceous* or bristle-like antenna is characterized by a noticeable decrease in the size of segments from the base to the apex, so that the antenna tapers from a rather thick base to a very slender tip.   A *clavate* or club-shaped antenna enlarges gradually toward the tip, the segments near the end being larger than those near mid-length.   An antenna in which the enlargement toward the tip is more abrupt and greater than in the clavate type is called *capitate* or knobbed. If the enlargement at the end is almost entirely toward one side from the axis of the antenna and forms broad, somewhat flattened plates, we call the antenna *lamellate.*

Locomotion of Insects. The success of insects must in large measure be determined by their extraordinary mobility. Anyone who has observed the flight of dragonflies, the swimming of whirligig beetles, or the running of ants is immediately aware of the complete mastery of locomotion which these creatures have attained. Such powers, together

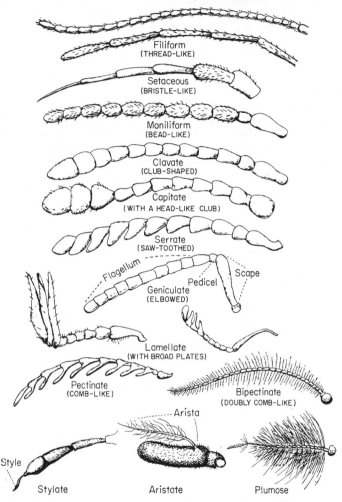

Fig. 3.3. The commoner types of insect antennae, with the names applied to each type. (*Redrawn from various sources.*)

with their small size and durability, enable insects to disperse over almost any geographic barrier and to invade almost every nook and cranny of the living environment.

The Legs of Insects. The most characteristic single thing we can name for insects is the presence of three pairs of jointed legs. These are practically always present in adult or mature insects, and generally present in the other stages. However, a good many maggots, grubs, and other larvae are entirely legless (Figs. 17.20, 19.26,*b*, and 20.19). Some insect larvae, notably the caterpillars, have in addition to the three pairs of jointed legs on the thorax, anywhere from two to eight additional pairs of fleshy, *unjointed* projections of the abdomen, which are used as legs and are known as *prolegs* (Figs. 5.14 and 5.15,*A*,*F*).

Insects are the six-legged Arthropoda. It is this character that has given them their class name Hexapoda (meaning six legs). The spiders, mites, and ticks (Figs. 6.2 and 6.3) have four pairs of legs. The crayfish (Fig. 6.5), lobsters, crabs, and their relatives have five pairs of walking legs. The hundred-legged-worms (Fig. 6.1) have a pair to each body segment, anywhere from 12 to 60 pairs in all, while the thousand-legged-worms (Fig. 14.13) have two pairs to each apparent segment, sometimes as many as 213 pairs.[1]

No one can study active insects long without being impressed by the extensive use they make of their legs. In insects the legs perform many of the functions for which we would use our hands, though sometimes the mouth parts are used for digging, carrying, fighting, and the like. In addition to walking and jumping, insects often use their legs for digging, grasping, feeling, swimming, carrying loads, building nests, and cleaning parts of the body. The cricket and katydid have "ears" on their front legs (Fig. 3.4,*A*).

Perhaps there is something significant in the number of legs that we find in this most abundant group of animals. If we study the movements of the legs, we find that the insect does not move the three on either side together and alternately with those on the other side; nor does it move the two of any pair in unison. Instead they go in tripods, the middle one of either side being raised and advanced about the same time as the front and hind ones of the other side. While these three are being advanced, the other three are supporting the body. Since three supports is the smallest number that will give a stable equilibrium, we see that insects, unlike two-legged and four-legged animals, are in a state of stable equilibrium, whether standing or moving. This requires less muscular effort and practically eliminates that hazardous period that many animals undergo while learning to walk. As to a larger number of legs we need only quote the following ditty:

> A centipede was happy, quite,
> Until a toad in fun
> Said, "Pray, which leg moves after which?"
> Which raised her doubts to such a pitch,
> She fell exhausted in the ditch,
> Not knowing how to run.

The legs of insects are hollow, more or less cylindrical outgrowths or continuations of the body wall at the point where pleura and sterna meet. Nerves, tracheae, blood spaces, and muscles occupy their internal cavities. The coxa of the leg is articulated to pleurites, some of which are considered to have been primitive segments of the leg itself, or to both a pleurite and a sternite; and the segments of the leg are articulated to each other by hinge-like joints or in such a way that some axial revolution is possible. The legs are moved by muscles that extend from one segment of the leg to more distal ones or that enter the base of the leg from adjacent parts of the body wall.

[1] *Parajulus pennsylvanicus.*

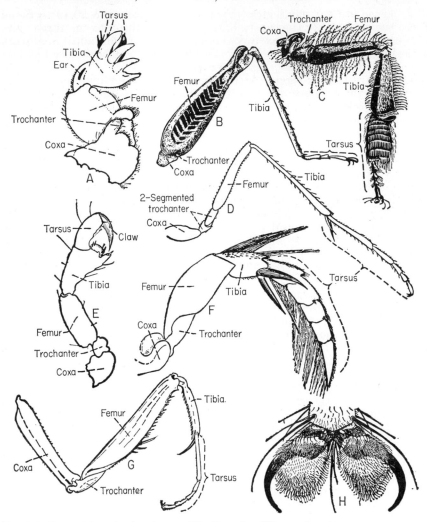

Fɪɢ. 3.4. Legs of insects showing modifications for different functions. *A*, digging leg of a mole cricket. Note the rake-like tibia with the three-segmented tarsus beneath it; also the slit-like "ear" or tympanum toward the base of the tibia. *B*, jumping leg of a grasshopper (*from Univ. of Kansas*). *C*, hind leg of worker honeybee adapted for assembling and carrying food substances; the rows of regular hairs on the basal segment of the tarsus are used for gathering pollen; the large marginal bristles of the tibia form, on the side opposite that shown, the pollen basket for carrying pollen to the hive (*from Cheshire*). *D*, walking leg of an ichneumon wasp. Note the two-segmented trochanter. *E*, clinging leg of the hog louse. Note the one-segmented tarsus with a single claw adapted for clinging around a hair. *F*, swimming leg of a predaceous diving beetle. Note the numerous long hairs used in rowing and the coxa which flattens out on the body wall. *G*, grasping leg of praying mantis, showing the very long coxa to extend the reach and the spiny femur and tibia between which insects are caught to be devoured. *H*, foot of the common house fly, showing claws, pulvilli, and the tenent hairs that make it possible for flies to walk upside down (much magnified). (*H, from Kellogg, "American Insects"; A, D–G, original.*)

All insect legs are made up of five true, independently movable segments, and these parts always occur in the same order. Beginning next to the thorax, the names of the parts are (Figs. 3.2, 3.4, and 3.8):

Coxa (plural coxae).
Trochanter (plural trochanters).
Femur (plural femora).
Tibia (plural tibiae).
Tarsus (plural tarsi).

These names are constantly used in describing and determining insects, and they should be learned once for all. All these parts are single segments except (*a*) the tarsus or "foot," which is commonly divided into from two to five subsegments, besides the claws and pads at the end of the leg which are not counted as segments of the tarsus, but constitute the *pretarsus* (Fig. 3.4,*H*); and (*b*) the trochanter, which in a few cases, notably the parasitic wasps and the Odonata has two segments (Fig. 3.4,*D*). The femur and the tibia, which form, between them, the knee joint of the leg, are usually much longer than any other segments in the leg, and of these the one nearer the body, and usually the thicker one, is the femur; the slenderer, outer one is the tibia. Between the femur and the body there are always two (rarely three) small pieces, the one nearest the femur being the trochanter, and the one next the body the coxa. All that part of the leg beyond the end of the tibia (the tarsus and the pretarsus) is generally placed flat upon the ground when the insect is walking, and the end of the tibia generally has prominent spines or spurs that help to maintain a footing. The pretarsus (Fig. 3.4,*H*) usually bears two sharp curved hooks or claws (the *ungues*), though there may be only one; and some complicated pads (known as *pulvilli, arolia, empodia,* etc.), which are very important in locomotion. For example, in the house fly there are many microscopic hollow hairs on these pads through each of which a sticky substance exudes that enables the fly to walk upside down and up very smooth surfaces.

The Wings of Insects. A characteristic of insects that we may be sure has been of very great advantage in their struggle for existence is the possession of wings. Wings enable insects (*a*) to forage far and wide to find suitable food, (*b*) to flee quickly from enemies and other dangers, (*c*) to disperse widely and intimately to find mates and lay their eggs, (*d*) often to select nesting sites not accessible to many of their animal enemies.

Insects are the only winged invertebrates. That is to say, if one finds an animal that has wings and does not have a backbone one may be sure it is an insect. Wings have been developed also in two groups of vertebrate animals, the birds (Aves) and the bats (Mammalia). But insects were almost certainly the first "flying machines," because we know from fossil records that winged insects were present on the earth in the Carboniferous period and almost certainly in the Devonian and Silurian, long aeons before either birds, flying reptiles, or bats made their appearance upon our globe, in the Jurassic, Cretaceous, and Tertiary periods.

While adult insects regularly have six legs, the number of wings varies among the different kinds. Insects never have functional wings until they are full-grown or adult, and many adult insects do not have wings. Silverfish and springtails (pages 198 and 199) represent wingless-insect groups, whose ancestors apparently never had wings. Others, such as fleas and lice and certain ants and aphids, are considered to be degenerate forms whose distant ancestors possessed wings which have been lost in adaptation to a more quiescent life in the ground or on the bodies of animals. The wings of some beetles have become atrophied and useless, so that they are incapable of flight. Other insects, such as termites and certain ants, break off or tear off their wings after a single nuptial flight and before beginning their life in the soil. No insect has more than four wings, or two pairs. This is the typical number. Some insects have only one pair. A good many are wingless throughout life. Frequently one sex has wings and the other sex is wingless.

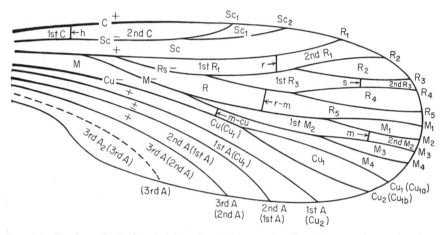

Fig. 3.5. The hypothetical, primitive wing of insects, showing the type of venation from which all modern insects' wings are supposed to have evolved. A slightly different interpretation, applicable especially to the Mecoptera, Trichoptera, Neuroptera, Lepidoptera, and Diptera, is indicated by the abbreviations in parentheses. *The names of veins* are indicated on the veins and are also indicated at the ends of the branches. *The names of cells* are indicated in the spaces: cells take their names from the veins along their anterior margin. C = costa, Sc = subcosta, R = radius, Rs = radial sector, M = media, Cu = cubitus, A = anal. Sc_1 is read first subcostal vein; $1st\ R_1$ is read first radial one cell; $2nd\ M_2$ is read second medial two cell, etc. The cross-veins are named r, radial; s, sectoral; r-m, radio-medial; m, medial; and m-cu, mediocubital. (*Original*.)

The wings, like the legs, are always attached to the thorax, the front pair to the mesothorax and the hind pair to the metathorax (Figs. 3.2 and 3.8). When one pair only is present (Fig. 2.15), it is the first pair, and it is borne by the mesothorax. We see thus that the thorax has all the organs of locomotion. This is ideal because it is near the center of mass of of the body.

The wings of an insect are remarkably different structurally from those of a bird or bat. They contain no bones, muscles, joints, or feathers and nerves and blood vessels are wanting or greatly reduced in activity. They are simply thin sheets of parchment-like cuticle that are moved by the action of muscles attached to the base of the wing, inside the body wall.

The wings develop, as anyone can easily determine by examining a young grasshopper or bug, a moth newly emerged from its cocoon, or a teneral fly, each as a hollow sac folded out from the body wall (Fig. 5.11). Promptly after the insect breaks out of its pupal or last nymphal skin, these wing sacs enlarge, and their walls flatten together and unite, so that the upper and under walls of the sac become fused and indistinguishable from a single membrane. They fuse closely in this way, except along certain lines where the two walls remain separated slightly and become thickened to make a kind of framework of hollow ribs, between which the wing membrane stretches. The hollow linear ribs (Fig. 3.6) which make up the framework of the wing are known as *veins*, though they have nothing to do with the circulation of the blood. The areas, of various shapes, enclosed between the veins are called *cells—closed cells* if the area is entirely surrounded by veins, *open cells* if the area extends to the wing margin without an intervening vein (Fig. 3.6). The exact number, branching, and arrangement of the veins of the wings are extensively used in classifying the various suborders, families, genera, and species of insects. Attempts have been made to homologize the veins of all the orders.[1] What is believed by some authors to be near the condition of the veins in an ancestral form of the winged insects is reproduced in Fig. 3.5. The student should note that each wing cell is given the same name as the wing vein along its anterior border.

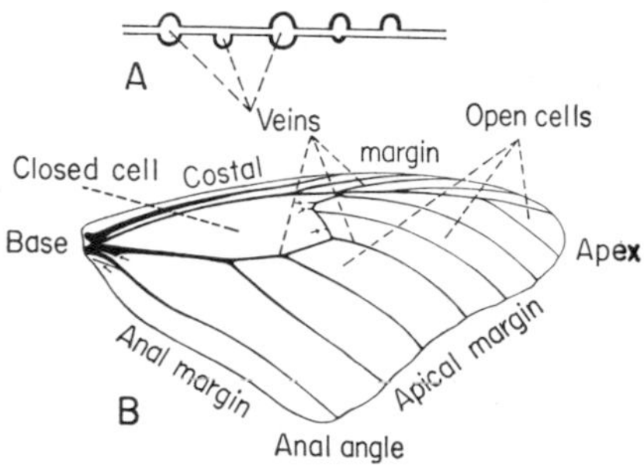

In outline (Fig. 3.6) the wings are often somewhat triangular. The front side of the triangle is known as the *costal margin*, and the outer side

Fig. 3.6. Front wing of the monarch butterfly, showing names applied to various parts of an insect wing (*original*). *A,* diagrammatic cross-section of an insect wing to show how it is formed of two membranes and how the veins develop as hollow rods by the thickening of one or both membranes along certain lines. (*Redrawn after Woodworth.*)

the *apical margin*, while the third side is called the *inner* or *anal margin*. Generally the veins are heavier or more closely placed toward the costal margin, since the greatest stress during flight is on this area of the wing.

Because they vary so widely in form and appearance in different groups, wings are very important structures in the classification of insects. As we shall learn later, most of the order names of insects end in *ptera*, meaning wing. Thus the Diptera (flies) are the "two-winged" insects, the Coleoptera (beetles) are the "sheath-winged" insects, the Lepidoptera (moths and butterflies) are the "scale-winged" insects, the Hemiptera (true bugs) are the "half-winged" insects, the Hymenoptera (wasps, bees) are the "membrane-winged" insects, and the Orthoptera (grasshoppers, etc.) are the "straight-winged" insects (Table 6.3).

Muscular System. The insect muscular system is developed according to the peculiar requirements of the exoskeleton and the segmented construction of the body. Because of these requirements, the number of insect muscles is very large, Lyonnet having described 4,069 in the goat moth caterpillar, *Cossus cossus,* as compared with 529 in man. Two distinct systems are recognizable: the *skeletal* or *voluntary muscles,* which move the body segments and appendages and consist of well-defined bundles of contractile fibers, and the *visceral* or *involuntary muscles,* which form a network about the gut or other organs and in the alary muscles of the heart (Figs. 3.9 and 3.10).

[1] Comstock, J. H., "The Wings of Insects," Comstock, 1918.

Histologically, insect muscles are of the striated type, in which the fibers are composed of a number of fibrillae enclosed in a nucleated sarcoplasm. The fibrillae exhibit a characteristic striated pattern of alternating light (isotropic) and dark (anisotropic) bands (Fig. 3.14,C). The anisotropic areas shorten during contraction of the muscle fiber and are believed to be regions where the molecules of the complex muscle protein, actomyosin, are all oriented in the same direction. Insect muscles are typically yellowish or colorless and, unlike vertebrate muscles, the fibers are not enclosed in a tendinous sheath. The skeletal muscles are attached at the origin to the hypodermis either directly or more generally through a tendinous structure, the *tonofibrilla*, which extends through the hypodermal cells into the endocuticle (procuticle). Special apodemes or invaginations of the integument may serve as points for muscular attachment.

Insect muscles such as those of the grasshopper leg have an extraordinary efficiency and can exert a power of 20,000 times their own weight, and thus outperform human muscle tenfold. The efficiency apparently results from short muscle fibers arranged along the entire length of the exoskeleton of the femur, so that the load is evenly distributed over the entire femur.

Certain intricate nerve pathways are involved in specialized muscular functions. In the grasshopper, Hoyle has shown that the muscles of the jumping legs are provided with a fast nerve fiber which controls the jumping and a slow nerve fiber which controls walking and other leg movements. To jump, the action of the slow nerve fiber must be inhibited and the flexor muscles contracted. Then an impulse in the fast nerve produces the jump, whose intensity varies with the number of impulses received.

Insect Flight. The wings of insects provide lifting force, driving power, and efficient steering apparatus, enabling them to take off with surprising quickness, to change directions with lightning-like rapidity, to dart sidewise, to dip, to bank, and most remarkable of all, to hover in one spot and even to fly backward without changing the position of the body. The continuous down- and upstrokes of the wings produce a propeller-like action which draws air from in front and above and directs it back and down in a relatively concentrated path. The net effect is, as with a fixed-wing aircraft, to create a zone of reduced pressure in front and above and increased pressure in back and below. The partition of the force exerted by the wing beat, between lift and thrust, is controlled by changes in the inclination of the wings, and this determines the air speed, while changes in direction are effected by varying the relative amplitudes of vibration of the right and left wings.

Efficient operation of the insect wing occurs only in smooth air flow, and the turbulence produced by the motion of the forewings adversely affects the performance of the hind wings. Thus primitive insects such as Isoptera and Neuroptera, whose wings move independently, are generally inefficient fliers. The Odonata owe their high degree of flight performance to out-of-phase movements of the forewings and hind wings, so that the latter operate in an undisturbed air stream, and in some Orthoptera, although the forewings and hind wings operate in phase, the latter have a larger amplitude. The Trichoptera, Hymenoptera, and many Lepidoptera and Hemiptera have solved this problem in an evolutionary way by developing various types of hooks or hairs which serve to unite the forewings and hind wings into a single functional aerodynamic unit.   The forewings of Coleoptera form a nonfunctional protective sheath, generally held extended during flight, while in Diptera, the hind wings have become modified into club-shaped balancing organs, the halteres.

The flight performance of certain insects is astonishing, and summary details of wing and flight speeds and range are given in Table 3.2.   These capabilities depend upon the nature of the fuel, the sugars of nectar for Diptera and Hymenoptera, or body-fats for such long-range fliers as the desert locust, monarch butterfly, and the leafhopper, *Circulifer*, and the size of the flight muscles which comprise 7 per cent of the body weight of *Pieris*, 11 per cent of *Musca*, 13 per cent of *Apis*, and 25 per cent of *Aeschna*.

TABLE 3.2. CHARACTERISTICS OF INSECT FLIGHT[1]

Species	Wing beats per second	Flight speed, m.p.h.	Flight range, miles
Apis mellifera............	250	7.5	29
Musca domestica..........	190	4.5	
Schistocerca gregaria.......	20	5.6	217
Danaus plexippus.........	9	6.2	650
Libellula................	20	22.0	

[1] MAGNAN, A., "Le vol des insectes," Hermann, Paris, 1934; HOCKING, B., *Sci. Monthly*, **85:** 237, 1957.

Protection.   Protection from natural enemies and adverse climatic conditions is as necessary to safeguard the individual as reproduction is to perpetuate the species.   The keen struggle for existence among animals has resulted in the perfection and adoption of a great variety of protective structures and devices, which are nowhere better illustrated than among insects.

The methods of protection among insects may be classified into:

1. Protective structures.
2. Protective constructions.
3. Protective size, form, and color.
4. Protective positions.
5. Protective behavior or reactions.

The importance of a sclerotized exoskeleton as a protection to insects has already been emphasized (page 79).   Such a hard body wall is characteristic of the great majority of insects.   The cuticle is often prolonged into bristles, spines, hairs, and scales which further protect the insect from mechanical injury, from excessive heat or evaporation, or from natural enemies, which find the hairy or spiny creatures unpalatable.

In certain caterpillars the protective value of the hairs is further increased by venom which fills them, and which nettles or poisons other animals that touch the hairs (page 24).   The body fluids of blister beetles and some other insects are corrosive or poisonous.   Many insects, especially in the bug order (Hemiptera), have odors that are repulsive (page 115).   Some, like the celery caterpillar, give off odors or bitter secretions from eversible glands (Fig. 6.58,*b*), the sudden erection of which may have value in frightening certain enemies.

Insects build many curious constructions to protect themselves or their young. The best known of these are the cocoons of moths, which are formed of silk secreted from the mouth. A great variety of cases or nests is used to protect especially the motionless pupal stage (Fig. 5.17). Sometimes the larval stage is protected by the cocoon throughout growth, as in the instance of the casemaking clothes moth (Fig. 19.12,*b*), the bagworm (Fig. 17.2), and the caddice-worms. Soil, leaves, small pebbles, shavings of wood, and many other substances are used to cover the body. Or a special secretion or excretion may be poured out through the body wall—as the waxy, woolly covering of mealybugs and many aphids. The social insects build elaborate nests, *e.g.*, the paper globe of the bald-faced hornet, the earthen mounds of ants, the soil pyramids of termites, or the overwintering nests of the brown-tail moth larvae.

A curious method of protection is illustrated by the larvae of the lace-winged flies and the larvae of sweetpotato beetles; these insects pile their own excrement and shed skins on their backs for concealment. In many flies the next to the last exuviae of the larvae are not shed but are retained about the body during the last larval and pupal stages, forming an excellent protective case (Fig. 5.17,*C*). The elevating threads that bear the eggs of the lace-winged fly (Fig. 5.1,*R*) illustrate another device which is probably of protective value.

It is probable that size may have protective value from certain enemies, extremely small insects being overlooked by larger enemies, and unusually large insects appearing too formidable for certain of their smaller enemies. Some insects have a very grotesque appearance, which may well be frightening to some of their enemies. Concealing form and coloration (camouflaging) is pronounced among the insects. It takes two special forms. One condition, known as *protective resemblance*, is well illustrated by walking-sticks (Fig. 6.15), which look so much like the twigs among which they live as to be difficult to detect except when they move. The other condition is known as *mimicry*. Many butterflies, flies, and other insects which are edible to birds and toads resemble in shape or color or both other butterflies which are poisonous or bitter to taste, or wasps which have a sting. It is believed by many naturalists that the palatable kinds gain a valuable protection from their natural enemies by this deceptive appearance. Such resemblance to another animal is known as *protective mimicry*. If the camouflaging or mimicry enables its possessor to stalk its prey or lie in ambush more successfully, it is called *aggressive resemblance* or *aggressive mimicry*. Many insects that have stings or bad tastes, so as to be distasteful to predators, are gaudily marked, *e.g.*, the bright-banded wasps. This is called *warning coloration*.

The surroundings in which an insect normally lives and feeds may make other protection less necessary. Insects which burrow in the soil, or tunnel in trunks of trees, or live in fermenting organic material, or swim in the water, or live on the skins or in the bodies of animals, gain a greater or lesser degree of security from extremes of temperature, excessive evaporation, and storms. They also incidentally gain security from a great many of the parasites and predators that would molest them if they fed in exposed situations.

Some insects which normally feed in exposed places have learned to take shelter or hide at the approach of certain enemies. Others depend upon their legs or wings for escape. Running away, flying away, jumping, swimming, or diving is perhaps the commonest of all protective measures. A few insects carry away their young or eggs when forced to retreat.   An interesting method of escape, rather widely exemplified, is by insects feigning death or "playing 'possum" when danger threatens. Leaf beetles, click beetles, measuring-worms, sphinx larvae, cuckoo wasps, and many curculios are examples of insects that behave in this manner.

In contrast to those just mentioned are the pugnacious insects that stand their ground or take the offensive when danger threatens them or their nests.   The stinging Hymenoptera illustrate this best, although many kinds pinch or pierce with the mouth parts when handled.   Others, which have no weapons, threaten or show fight and doubtless succeed with their bluffing especially if they resemble well-protected kinds.

Just as some kinds enjoy a measure of immunity by living in situations removed from the beaten paths of insect life, so others, by using the less crowded periods of the day, escape attack from some enemies though they expose themselves to others.   Insects which remain quiet during the day and become active at dusk are called *crepuscular* insects; those which confine their activity to the darkness of night are called *nocturnal*.

THE METABOLIC PROCESSES OF INSECTS

Everything about the life and behavior of insects is dependent upon chemical changes within their bodies.   These are organized at the cellular level as biochemical processes and integrated in organ systems as physiological processes.   Insects perform all the functions common to man or other vertebrates and, so far as we can determine, none that are peculiar to insects and, despite their small size, are as intricate and complex in structure as any other animal.

Insects, perhaps more than any other animals, are noteworthy for their great activity, the continual changes in their physical bodies, and their ability to do "work."   This work involves especially moving about to secure food and congenial surroundings, to escape enemies, to find mates, to construct nests, and to produce eggs for another generation.   The ability to do work depends upon the availability of energy.   This is secured from food substances in which the energy from the sun has been stored by green plants in the process known as *photosynthesis*.   *Metabolism* is the term applied to all the chemical changes taking place in the living body and therefore includes all the physiological functions listed above.   Although these functions are accomplished in insects by organs and processes which sometimes differ from similar functions in mammals, it appears that the fundamental biochemical reactions occurring in the cells of the two groups of organisms are very similar.

Digestion and Nutrition.   *Food Habits.*   The food habits[1] of insects are perhaps the dominant characteristic in determining their ecological habitat and their relative importance to man.   For example, the Colorado potato beetle, *Leptinotarsa decemlineata*, originally lived in the eastern slope of the Rocky Mountains on a weed, *Solanum rostratum*, buffalo burr or sand burr, and had it not in the 1850s developed a liking for the newly introduced potato, would even today be only an obscure collector's item, instead of a world-famous enemy of man.   Were the anopheline mosquitoes addicted solely to sucking rodent or avian blood, their role as transmitters of protozoan infections would be of scientific curiosity, rather than the target of a world-wide malaria-eradication campaign.

[1] BRUES, C. T., "The Insect Dietary," Harvard, 1946.

The foods of insects may consist of almost every variety of natural organic substance. Some insects develop on substances such as dry wood (cellulose) and wool (keratin), silk, and feathers which mammals cannot assimilate, while others feed on materials such as black and red pepper, ginger, ergot, tobacco, or even rotenone-bearing roots, which would actually be poisonous to most animals. As with other animals, however, the ultimate energy sources are carbohydrates such as starches and sugar, fats, and proteins. In the cockroaches and termites, the digestion is aided by the presence of intestinal symbiotic bacteria or protozoa. Thus the insect may function to utilize atmospheric nitrogen, an essential element not present in cellulose. These intestinal flora and fauna are of special importance in the production of vitamins and enable many insects to subsist upon foods which are largely vitamin-free. The diversity of insect foodstuffs has resulted in extensive modifications of the alimentary tract and in the digestive processes, in certain species. The enzymes which accomplish the digestion of food vary with the nature of the food. In nectar-feeding insects, only simple sugars are to be digested, while in omnivorous species such as the cockroaches, the variety of foods to be transformed by enzymatic action is virtually as great as in the human.

The *ingestion* or taking in of foods is treated in the chapter on Mouth Parts of Insects (Chapter 4). Most food materials when taken into the alimentary canal cannot pass through its walls into the blood and be carried to the various tissues requiring them until they have undergone certain modifications known as digestion, which render them soluble or absorbable. Much of the material swallowed is indigestible and is *egested* from the alimentary canal, never having been a part of the living body of the insect. This act of voiding material that has not been digested (*egestion* or *evacuation*) should not be confused with *excretion*, discussed later. It is interesting that certain insects such as the honeybee and parasitic wasps, while young, have no connection between the mid-intestine and the hind-intestine and cannot void excreta until they are full-grown larvae, before spinning cocoons, or until, as adults, they leave the body of the host.

Alimentary Canal. The alimentary canal of insects is a tube leading from the mouth to the anus at the tip of the abdomen. In the primitive insect the mouth typically opens between the third and fourth head segments, and the anus opens upon the twelfth abdominal segment (the *telson*). The space between the alimentary canal and the body wall is called the body cavity or *hemocoel* and is largely filled with blood. The length of the alimentary canal and the complexity of its structure are correlated with the food habits of the insect. The simplest type is found in insect larvae which usually feed on solid tissues, such as Lepidoptera, Hymenoptera, Orthoptera, and Diptera. The greatest length and complexity occur in the sap-feeding Homoptera and in bloodsucking Hemiptera and Diptera, where the alimentary canal may be several times as long as the insect's body.

During embryonic development, the alimentary canal forms in three sections (Fig. 3.7). The *fore-intestine* and the *hind-intestine* grow as invaginations from the outside, and consequently have a cuticular lining which is molted each time the external skin is shed. The *mid-intestine* develops internally (Fig. 3.7,*Ment*) and lacks the cuticular lining. In some insects, the fore-intestine consists only of the mouth or *buccal cavity*, the *pharynx* (Fig. 3.7,*Phy*), and a straight thin-walled tube, the *esophagus* (*OE*), leading back to the *stomach* (*Vent*).   It is surrounded by an inner layer of longitudinal muscles and an outer layer of circular muscles.   The pharynx, especially of insects with piercing, siphoning, or sponging mouth parts, is generally developed into a pump or sucking device, closed by the elasticity of its cuticular lining and by muscles in its walls and opened by

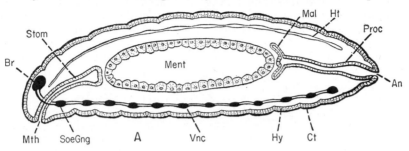

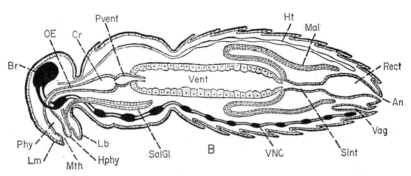

Fɪɢ. 3.7. Diagrammatic sagittal sections of the body of an insect to show the arrangement of the principal organs and the formation of the alimentary canal.   *A*, section of an insect in the embryonic stage, showing the anterior invagination of the ectoderm which forms the mouth (*Mth*) and fore-intestine or stomodeum (*Stom*); the posterior invagination which forms the anus (*An*) and hind-intestine or proctodeum (*Proc*); and the endodermal development of the mid-intestine called the mesenteron (*Ment*).   *B*, section of a mature insect, after fore- and hind-intestines have united with the mid-intestine to form a complete canal.   The fore-intestine has differentiated into small intestine (*SInt*), with the Malpighian tubes (*Mal*) arising from its anterior extremity, and rectum (*Rect*).   The mesenteron has become the stomach or ventriculus (*Vent*) of the adult.   *An*, anus; *Br*, brain; *Cr*, crop; *Ct*, cuticle; *Hphy*, hypopharynx; *Ht*, heart; *Hy*, hypodermis; *Lb*, labium; *Lm*, labrum; *Mal*, Malpighian tubes; *Ment*, mesenteron; *Mth*, mouth; *AE*, esophagus; *Proc*, protodeum; *Pvent*, proventriculus; *Rect*, rectum; *SalGl*, salivary gland; *SInt*, small intestine, *SoeGng*, subesophageal ganglion; *Stom*, stomodeum; *Vag*, vagina; *Vent*, ventriculus; *VNC*, ventral nerve cord.   (*From Snodgrass*, "*Anatomy of the Honeybee*.")

muscles extending from these walls to the inside of the head capsule (Figs. 4.6 and 4.10).   The middle region of the esophagous is often enlarged into a *crop* (Figs. 3.7,*Cr*, and 3.8), where food is held temporarily, and the far or caudal end, especially in chewing insects, is specialized into a *gizzard* (*Pvent*), for grinding and straining the food, and by a sphincter-like action, for controlling its passage into the stomach.   In the Orthoptera, the cuticular lining of the gizzard is developed into strong teeth.

The hind-intestine, which begins near the point where the Malpighian tubes attach, may be a simple tube leading from the mid-intestine to the anus.   In other species it may be variously subdivided into a *pylorus*, containing the bases of the Malpighian tubes, an *ileum*, a *colon*, a *rectal sac*, and a *rectum*.   The rectum is usually provided with four to six inwardly projecting "rectal glands" which are thought to conserve water by reabsorption from the feces.

The mid-intestine or stomach (Figs, 3.7, *Ment, Vent,* and 3.8) has valves near its anterior and posterior ends (called *cardiac* and *pyloric*, respectively) which regulate the passage of food into and from the stomach.

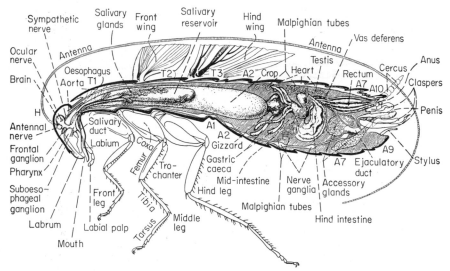

FIG. 3.8. Sagittal section of the body of a cockroach, to show especially the internal anatomy of digestive, reproductive, nervous, circulatory, and excretory systems.   None of the tracheae are shown.   *H*, head; *T₁*, prothorax; *T₂*, mesothorax; *T₃*, metathorax; *A₁* to *A₁₀*, first to tenth abdominal segments.   (*Original; drawing by A. M. Paterno, in part after Miall and Denny.*)

The epithelial portion of the mid-intestine is lined with cells which excrete the digestive enzymes and absorb the products of digestion.   Surrounding the epithelium is an inner layer of circular muscles and an outer layer of longitudinal muscles which keep the food products moving along in the canal.   The surface area of the stomach is often increased by outpocketings of the wall (usually at the esophageal end) into short or long blind tubes, the *gastric caeca* (Fig. 3.8), which vary in number from two to eight or more.

The mid-intestine, unlike the fore- and hind-intestines, does not have a cuticular lining, but in many insects the cells of this organ are protected from abrasion by the gut contents by a delicate membrane less than 0.5 micron thick, the *peritrophic membrane*.   This membrane, like the endocuticle, is composed of chitin impregnated with protein and is readily permeable to the digestive enzymes and the products of their action.   It is secreted as concentric lamellae by cells throughout the length of the mid-gut in Odonata, Orthoptera, larval Lepidoptera, and many Coleoptera, and Hymenoptera.   However, in Diptera and Dermaptera, the peritrophic membrane is secreted as a single layer by a ring of cells at the anterior of the mid-intestine.   The membrane is absent in many insects that ingest only liquid foods, such as Hemiptera, adult Lepidoptera, Thysanoptera, Anoplura, and some Diptera.

Digestive Enzymes.   Enzymes secreted by the salivary glands (page 114) are usually concerned with digestion.   In the cockroaches and honeybee, the saliva contains an amylase which converts starch into maltose and an invertase which reduces sucrose to glucose and fructose. However, most of the digestive enzymes are produced by the epithelial cells of the mid-intestine and the gastric caeca.   The digestive juices are usually either weakly acid or alkaline, the crop in the cockroach being generally acidic due to the presence of lactic acid formed from sugars. In omnivorous insects such as *Periplaneta*, digestive secretions contain maltase, hydrolyzing maltose to glucose; lactase, hydrolyzing lactose to glucose and galactose; amylase; invertase; lipase, which hydrolyzes fats to glycerol and fatty acids; proteinase, which reduces natural proteins to polypeptides and amino acids; and several peptidases, which reduce various polypeptides to free amino acids.   In this insect, the crop is the principal site of digestion, the enzymes of the mid-intestine being regurgitated to this region, where they are mixed with the food.   After digestion, the food is admitted to the mid-intestine, where absorption of the simple sugars, amino acids, and fatty acids into the blood takes place, principally from the gastric caeca.   There is some evidence that final digestion of certain polypeptides takes place in the epithelial cells of the mid-intestine.   In other insects whose foods are of simpler composition, the digestive enzymes vary accordingly.   Protease and lipase predominate in carnivorous forms, whereas some insects eating cellulose, such as Cerambycidae and other wood-boring beetle larvae, secrete cellulase, hemicellulase, and lichenase, which are capable of breaking down cellulose to glucose.   These enzymes, however, are not found in termites or wood-feeding cockroaches.   Insects which feed on wood seem never to digest the lignin fraction.   Carpet beetles, clothes moth larvae, and blow fly larvae, which feed on animal proteins such as keratin and collagen, possess appropriate enzymes, keratinase and collagenase, which reduce these proteins to peptides which can be further digested by proteinase.   A number of insects predigest their food by the injection of enzymes during feeding.   This occurs in plant-feeding Homoptera, predaceous Hemiptera, predaceous and parasitic Hymenoptera and Diptera, and in most predaceous Coleoptera and Neuroptera larvae.   In some cases the enzymes appear to be secreted by salivary glands, but in the beetle larvae these are absent and the enzymes are produced in the mid-intestine and forced out through the perforated mandibles.

Nutrition.   The study of insect nutrition is exceedingly complex because of the vast number of insect species and the diversity of their food habits.   However, certain essential nutritive elements are required by all, either to be supplied in the diet or synthesized by intestinal symbionts.

MINERALS.   The mineral requirements of insects have not been exhaustively investigated, but phosphorus and potassium seem to be limiting growth factors for all species studied.   Calcium is required for the growth and transformation of mosquito larvae.   Cobalt, which is an essential part of the vitamin B_{12} molecule, and magnesium and manganese, which are cofactors in various enzymes, must also be essential.   Very limited data also indicate the importance of sodium, zinc, iron, and copper.

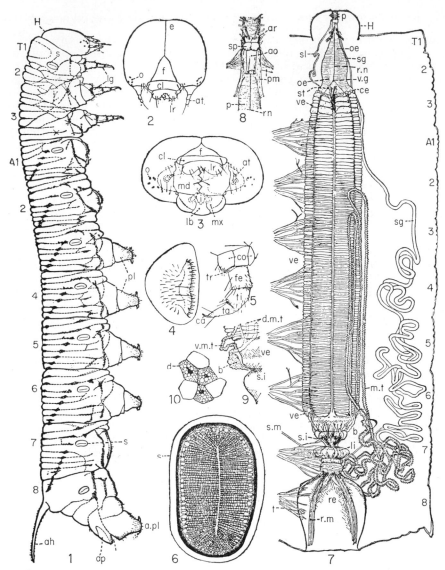

Fig. 3.9. Anatomy of the tomato hornworm. *1*, Side view of entire larva; *2*, front view of head; *3*, ventral view of head to show chewing mouth parts; *4*, a proleg, ventral view, showing crochets; *5*, a true or thoracic leg, front view; *6*, a single spiracle, greatly magnified; *7*, body of the larva opened from the dorsal side to show the alimentary canal. On the left side are shown the salivary gland and the terminal branches of tracheae which enter the canal; on the right, the silk gland and Malpighian tubes of that side of the body are represented; *8*, the pharynx from above in the region of the brain; *9*, enlarged view of the alimentary canal at the point where the bladder of the Malpighian tube is attached; *10*, a bit of the adipose tissue of fat-body. The following abbreviations are used: *A1* to *A8*, abdominal segments 1 to 8; *a*, anus; *ah*, anal horn; *ao*, aorta; *ap*, anal plate; *a. pl*, anal proleg; *ar*, arched nerve; *at*, antenna; *b*, bladder of Malpighian tubule; *ca*, claw; *ce*, caeca; *cl*, clypeus; *co*, coxa; *d*, adipose tissue; *d.m.t.*, dorsal Malpighian tubule; *e*, epicranial suture; *f*, front; *fe*, femur; *H*, head; *i*, spinneret; *lb*, labium; *lg*, leg; *li*, large intestine; *lr*, labrum; *md*, mandible; *mt*, Malpighian tubule; *mx*, maxilla; *o*, lateral ocelli; *oe*, esophagus; *p*, pharynx; *pl*, proleg; *pm*, pharyngeal muscle; *re*, rectum; *rn*, recurrent nerve; *rm*, rectal muscle; *s*, spiracle; *sg*, silk gland; *si*, small intestine; *sl*, salivary gland; *sp*, brain; *st*, part of sympathetic nerve; *T1*, *T2*, and *T3*, thoracic segments, prothorax, mesothorax, and metathorax, respectively; *t*, tracheae; *ta*, tarsus; *ti*, tibia; *tr*, trochanter; *ve*, ventriculus; *v.m.t.*, ventral Malpighian tubule. (*From Peterson, Ann. Ento. Soc. Amer., vol. 5.*)

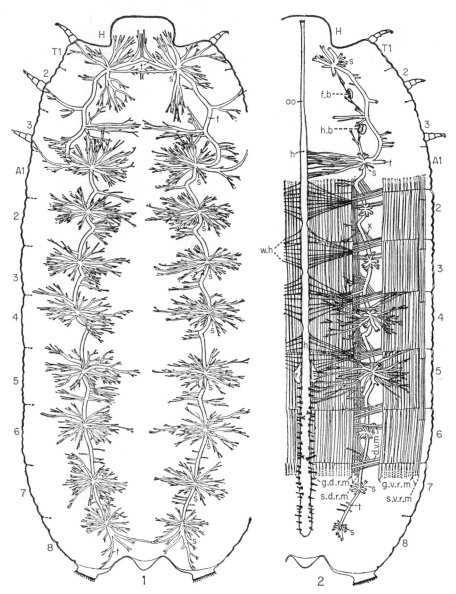

Fig. 3.10. Anatomy of the tomato hornworm. The body wall of the larva has been opened along the ventral side and spread to show, on the left, the principal branches of the tracheae, all finer branches omitted; *s* indicates the points at which spiracles open to the outside through the body wall. On the right are shown the median heart (*h*) and aorta (*ao*) with the wing-shaped muscles (*w.h.*) that hold the heart in place; the longitudinal tracheal trunk of that side (*t.t.*); the buds from which the wings of the adult are subsequently developed—(*f.b.*), the front-wing bud, and (*h.b.*), the hind-wing bud; the reproductive (*r*) and some of the many muscles of the larva (*d.v.m., g.d.r.m., g.v.r.m., s.d.r.m., s.v.r.m., x*). (For explanation of the other letters, see under Fig. 3.9). (*From Peterson in Ann. Ento. Soc. Amer., vol. 5.*)

CARBOHYDRATES.   Carbohydrates form a most important source of cellular energy and are important articles of most insect diets for energy and synthesis of fat and glycogen, although *Aedes aegypti* larvae, *Tribolium*, *Tineola*, *Ptinus*, and *Lasioderma* are able to grow on a carbohydrate-free diet.   However, the larvae of *Tenebrio molitor* did not grow on a diet of less than 40 per cent carbohydrate.   The ability of an insect to utilize complex carbohydrate foods may range from simple sugars such as glucose and fructose to such complex materials as cellulose, hemicellulose, starch, and glycogen.   Of the simple sugars, the honeybee can live on glucose, fructose, maltose, trehalose, melezitose, raffinose, arabinose, xylose, galactose, and the *Calliphora* adult on these sugars as well as melibiose, lactose, mannose, mannitol, and sorbitol.   *Chilo suppressalis*, on an aseptic diet, was able to grow on the monosaccharides fructose, glucose, galactose, mannose, sorbose, rhamnose, xylose, arabinose, and ribose, and on the polysaccharides maltose, sucrose, trehalose, raffinose, melezitose, melibiose, lactose, and cellobiose, in the order named.

LIPIDS.   Insects are known to synthesize fats from carbohydrates, as in *Circulifer tenellus*, the sugar-beet leafhopper, or from protein, as in the german cockroach, *Blattella germanica*.   Dietary fats, however, are utilized by most insects as sources of energy, for metabolic water, and for storage as reserves of fat or glycogen.   The nature of the body-fat deposited (unsaturation) in *Heliothis armigera* and *Lucilia sericata* is dependent upon the composition of dietary fat.   In *Schistocerca gregaria*, fat is the major source of energy during flight.   Fatty acids, especially linoleic acid, are required for normal growth and pupation of *Ephestia*, *Pyrausta nubilalis*, *Pectinophora gossypiella*, *Pseudosarcophaga affinis*, and *Calliphora erythrocephala*.

STEROLS AND OTHER FAT-SOLUBLE VITAMINS.   All insects require a dietary source of sterols, and cholesterol and 7-dehydrocholesterol are about equally effective for *Tribolium*, *Ptinus*, *Stegobium*, *Silvanus*, and *Lasioderma*.   Phytophagous insects such as *Pectinophora gossypiella* grow better on the plant sterols ergosterol, sitosterol, or stigmasterol, but *Dermestes vulpinus* cannot utilize these plant sterols.   No insect has been found which can utilize calciferol or related sterols of the vitamin D group.   Vitamin K and vitamin E also do not play an essential part in insect nutrition.   Vitamin A and β-carotene have not been substantiated as essential for insects, and *Periplaneta americana* grows normally without them, although *Schistocerca gregaria* suffered retarded growth, high mortality, and abnormal development of body colors when reared on a β-carotene-deficient diet.

WATER-SOLUBLE VITAMINS.   All insects apparently require for growth and development those water-soluble vitamins which act as catalysts in cellular metabolism, i.e., thiamin, riboflavin, nicotinic acid (niacin), pantothenic acid, pyridoxin, biotin, folic acid (pteroylglutamic acid), and choline.   These must be either supplied in the diet or synthesized by symbiotic microorganisms.   Thus *Tribolium confusum*, *Tenebrio molitor*, *Blattella germanica*, *Plodia interpunctella*, *Ephestia*, *Aedes aegypti*, *Chilo suppressalis*, *Hylemya antiqua*, *Culex molestus*, and *Pseudosarcophaga* do not grow normally on a purified diet with the omission of any one of these factors (Fig. 3.11), while *Stegobium paniceum* and *Lasioderma serricorne* normally require only thiamin and pyridoxin; when reared under sterile conditions, riboflavin, niacin, pantothenic acid, biotin, folic acid, and choline were also required to replace the factors normally provided by intercellular symbionts.   Nucleic acid increased the growth rate of *Aedes aegypti* and *Drosophila melanogaster*, while carnitine (β-hydroxybutyrobetaine) is essential for the growth of *Tenebrio molitor*, *Tribolium confusum*, *T. castaneum*, and *Palorus ratzeburgi*.   No clear-cut need for inositol, *p*-aminobenzoic acid, vitamin B_{12}, or vitamin C (ascorbic acid) has been demonstrated in insects.

PROTEINS AND AMINO ACIDS.   All insects require complete proteins or their amino acid constituents for growth and development, although many adult species such as Lepidoptera feed only upon nectar, which is almost entirely nitrogen-free, so that their development and that of their genital products are at the expense of stored reserves of protein and fats. Casein has generally proved a complete protein for the growth of species such as *Blattella germanica*, *Anthrenus museorum*, *Tineola bisselliella*, and *Tribolium confusum*.   However, zein, gelatin, silk fibroin, and gliadin are not adequate protein sources for the growth of these insects.   Studies

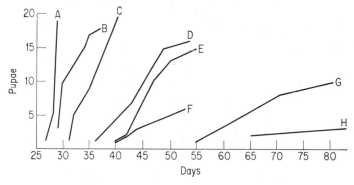

FIG. 3.11. Growth curves showing the number of pupae of *Tribolium confusum* produced from 20 newly hatched larvae on various vitamin-deficient diets.   *A*, whole-wheat flour; *B*, synthetic diet with all vitamins listed; *C*, no choline; *D*, no pyridoxin; *E*, no folic acid; *F*, no biotin; *G*, no thiamin; *H*, no riboflavin.   Without nicotinic acid, one pupa was produced in 114 days, and without pantothenic acid, no pupae were formed.   (*After Fraenkel and Blewett, Biochem. Jour.*, 1943).

with synthetic mixtures of amino acids under sterile conditions have shown that *Tribolium* requires valine, tryptophane, histidine, and leucine for pupation, while lysine, threonine, phenylalanine, methionine, isoleucine, and arginine were essential for normal growth.   *Attagenus*, *Tribolium*, *Drosophila melanogaster*, *Pseudosarcophaga affinus*, and *Apis mellifera* and the phytophagous insects, *Hylemya antiqua*, *Pectinophora gossypiella*, and *Chilo suppressalis* also require these same 10 essential amino acids for normal growth and development, as in the case with the rat and man.   Other amino acids may also be important, since cystine is required for normal pupation of *Lucilia*, *Drosophila*, *Aedes*, and *Blattella germanica*, and glycine is required by *Aedes*, *Blattella*, *Pseudosarcophaga affinus*, and *Calliphora erythrocephala*, as it is for the chick.

Circulation.   The circulatory system in insects, unlike that of higher animals, serves primarily as the medium for all chemical exchanges between body organs and thus functions in the transport of nutritive materials, excretory products, hormones, and respiratory gases and also as a hydraulic fluid for transmitting and maintaining body pressures during respiratory movements, in hatching and molting, and for the function of protrusible organs such as the penis and the ptilinum.

Morphology. Structurally, the circulatory system is very simple, consisting of a single tube lying close under the body wall down the middle of the dorsum. This *dorsal vessel*, the abdominal portion of which is commonly called the "heart" (Figs. 3.7,*Ht*, 3.8, and 3.10,*h*) is closed at the posterior end but has small paired openings (ostia) at regular intervals along the sides. The blood may enter these openings, but an arrangement of *auricular valves* prevents the backward flow of the blood in the heart. The anterior nonpulsating portion of the heart is called the *aorta* (Fig. 3.10,*ao*). The heart is held in place by triangular sheets of *alary muscle* extending to the body walls. A fibromuscular *dorsal diaphragm* usually extends across the abdominal cavity above the alimentary canal and forms a definite *pericardial* or *dorsal sinus* enclosing the heart. A corresponding ventral diaphragm is present in Hymenoptera, Lepidoptera, Orthoptera, Odonata, and Ephemeroptera, which lies above the ventral nerve cord and forms a *perineural ventral sinus*, and in the *visceral sinus* between these two diaphragms are the principal internal organs. The number of chambers and pairs of alary muscles varies from 12 in *Periplaneta* to 1 in *Aeschna* nymphs, while *Musca* has 3 and *Apis*, 4.

In addition to the dorsal vessel, *accessory pulsating organs* aid in the circulation of blood through the appendages. These are sac-like structures often located near the base of the antenna, the legs, and the wings.

Physiology. Circulation in insects is a relatively inefficient process, and as a result insects may live, apparently normally, for some time after the heart has stopped beating. However, under normal conditions the blood is kept in motion by the rhythmic pulsating of the dorsal vessel, which takes place as a wave of contraction, *systole*, proceeding in an anterior direction. This forces the blood to the region of the head, where it leaves the aorta and enters the body cavity, its pressure displacing blood in a posterior direction. As the heart muscle relaxes and the alary muscles contract to expand the heart in *diastole*, the blood enters the heart through the ostia (Fig. 3.12). A low-pressure area is thus formed in the dorsal sinus, and the blood filters back-ward and upward through the vis-

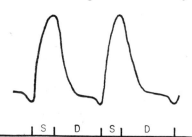

FIG. 3.12. Mechanical recording of heart-beat of *Periplaneta americana: S*, systole, *D*, diastole. (*After Yeager, Jour. Agr. Res.*, 1938).

ceral sinus to equalize this. The circulation is aided by movements of the dorsal diaphragm, which undulates anteriorly, and the ventral diaphragm, which undulates posteriorly, and these aid in circulating the blood through the pericardial and perineural sinuses. Reversal of the direction of the heart pulsation may occur intermittently in some insects, as in *Bombyx mori*, where it is observed before metamorphosis.

The accessory pulsatile organs assist in pumping the blood through perforated vessels which pass through the length of the antennae, into the legs, and between the tracheae and walls of the veins of the wings, where it usually enters by the costa and returns along the posterior margin.

The rate of contraction of the dorsal vessel is directly proportional to environmental temperature and in *Melanoplus* increases over a range of 1.5 to 1.8 times for each 10°C. rise in temperature. Each species also has a characteristic rate of beat which varies with stage of development, metabolism, and activity and at normal temperatures ranges from about 14 beats per minute in *Lucanus cervus* to 150 to 160 in *Campodea*.

The rhythmic muscular contractions of the heart wall are generally considered to be myogenic, *i.e.*, without initiation by the nervous system, although the rate and amplitude are controlled by impulses from the paired cardiac ganglia of the visceral or sympathetic system and the segmental ganglia of the ventral nerve cord, which innervate the heart. The rate of heartbeat is influenced by the neural hormone acetylcholine and in the isolated heart of *Periplaneta*, at 28°C., increases from about 80 beats per minute at 2×10^{-9} M acetylcholine to 140 beats at 1×10^{-8} M.

Blood. The insect blood or hemolymph does not have a major respiratory function because of its low oxygen content (Table 3.3), although it serves in the removal of carbon dioxide, mostly as bicarbonate, and in aiding in the transfer of gases between the ends of the tracheoles and the tissues. Insect blood does not contain hemoglobin respiratory pigment except in the bloodworm or *Chironomus* larvae, where the tracheal system is rudimentary, and the blood, which contains a red hemoglobin pigment, apparently has a respiratory function. Insect blood cells or *hemocytes* are primarily phagocytic and function in engulfing and removing foreign bodies and in wound healing by agglutination. These cells may number from about 1,000 to 100,000 per cubic millimeter and have been classified in 7 to 10 major types, although these may include transitional forms. They include (*a*) proleucocytes, (*b*) oenocytoids, (*c*) plasmocytes, (*d*) podocytes, (*e*) cystocytes, and (*f*) spheriodocytes. The hemocytes may also have a food-storage function for glycogen and fat.

Insect blood is a yellowish to greenish fluid which may comprise 5 to 40 per cent of the total body weight and with the composition typified by the species shown in Table 3.3, where they are compared with human blood. The insect bloods have a water content of about 90 per cent, a pH very close to neutrality, and a total osmotic pressure equivalent to 0.9 to 1.6 per cent sodium chloride. Insect bloods have a wide range of inorganic constituents (sodium, potassium, magnesium, phosphorus, and chlorine and traces of copper, zinc, aluminium, and manganese), and the sodium/potassium ratios range from 10 to 25 for carnivorous species such as *Triatoma*, *Dytiscus*, *Hydrophilus*, and *Cicindela* to 0.1 to 0.3 in phytophagous feeders such as *Bombyx*, *Pieris*, and *Samia*. Insect blood has a free amino acid content 25 to 70 times that of human plasma. Also present are reducing sugars, fats, and proteins and traces of plant and animal pigments. In contrast to mammals, where the blood sugar is glucose, the predominant blood sugar in insects is the nonreducing disaccharide trehalose, and in *Schistocerca* this is rapidly formed from and can be reconverted to glucose.[1]

The insect blood is also an avenue for the removal of nitrogenous organic wastes from various organ systems, and its content of urea and uric acid is reduced by passage about the Malpighian tubules as discussed on page 107.

Fat-body. This organ is composed of masses of cells confined by delicate connective membranes and fibers and occupies much of the

[1] TREHERNE, J., *Jour. Exp. Biol.*, **35**:611, 1958.

space between the larger organs of the body cavity, thus presenting a maximum surface to the circulating blood (Fig. 3.14,*D*).   The cells of this organ are capable of great variations in structure during the life of the insect and have the important function of storing fat, glycogen, and protein not immediately required for sustenance.   In *Aedes aegypti*, and presumably in other insects, the type of nutrition determines the substance stored, and fat is accumulated after feeding olive oil, glycogen after starch or sugars, and protein together with glycogen and fat after

TABLE 3.3. COMPOSITION AND PROPERTIES OF INSECT BLOODS COMPARED WITH HUMAN BLOOD[1]

Constituent	*Hydrophilus piceus*	*Gasterophilus intestinalis* larva	*Apis mellifera* larva	*Prodenia eridania* larva	Man
	Millimoles per liter				
Sodium..............	124	175	6	22	146
Potassium...........	13	12	24	40	5
Calcium.............	11	3	4	9	2
Magnesium..........	20	16	8	7	1
Phosphate..........	4	4	10	34	1
Chloride.............	40	15	33	6	104
Protein, %...........	3.4	10.8	6.6	1.0	7
Amino N, mg., %....	40–80	94	306–385	235	3.9–5.5
Urea, mg., %........	7.4	20.4		6	24–47
Uric acid, mg., %....	12	2.2	5.3	15	2.9–6.9
Reducing sugars, mg., %................	20–104	356	120–258	66	87–119
pH..................	6.7–7.0	6.8	6.8	6.4–6.7	7.4
Oxygen, ml. per 100 ml.................	0.11	0.15		2	19
Carbon dioxide, ml. per 100 ml........	73–89	72.4	25–30	10	45–50
Water, %............	88–92	84			
Specific gravity.......	1.012	1.062	1.045	1.032	1.027

[1] Values from various sources as summarized by BUCK, J., p. 147 in K. D. ROEDER (ed.), "Insect Physiology," Wiley, 1953.

casein.   During larval growth, these reserves may accumulate to enormous proportions, and in the honeybee larvae, the fat-body forms about 65 per cent of the total body weight, most of which is glycogen. This store is drawn upon for the histogenesis necessary when the insect changes from larva to pupa and adult, and as a source of energy to maintain life during hibernation or under prolonged stress, as in extended flight.   During metamorphosis and molting in some insects such as Hymenoptera and Diptera, the fat-body of the larva almost completely disintegrates and a new organ is reformed in the adult.

The fat-body has an accessory role in storage excretion in insects such as Collembola, which have no Malpighian tubes, and in cockroaches and certain Hymenoptera, Diptera, and Lepidoptera, where deposits of uric acid or its salts are formed in special *urate cells* of the fat-body.

Excretion.[1]   The process of excretion serves to eliminate the nitrogenous waste products of protein katabolism from the internal body environment and to regulate the balance of water and ions.   In insects, several organs contribute to this function: (*a*) the *Malpighian tubes,* (*b*) the *fat-body,* and (*c*) the *nephrocytes.*

Malpighian Tubes.   This organ functions in the formation and excretion of urine and consists of a variable number of tubes, closed at the distal end and opening into the hind intestine near its junction with the mid-intestine (Figs. 3.7 and 3.8).   The distal portions of the tubes are generally free in the body cavity, but in many Coleoptera, the larvae of most Lepidoptera, and the larvae of some primitive Hymenoptera and Myrmeleontidae, the distal ends are bound to the hind-intestine by a layer of cells or muscles.   This condition is called cryptonephry and

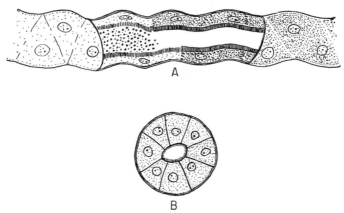

Fig. 3.13. *A,* section of the Malpighian tube of *Rhodnius prolixus* at junction of proximal (left) and distal (right) sections, showing region of uric acid precipitation (*after Wigglesworth, Jour. Exp. Zool.,* 1931).   *B,* cross-section of Malpighian tube of *Periplaneta americana.*

presumably aids the insect in the conservation of water by its removal from the feces.   The number of Malpighian tubes is exceedingly variable, although the primitive number is 6 and they generally occur in multiples of 2.   There are 2 in the Coccidae; 4 in Hemiptera, Anoplura, Thysanoptera, and Siphonoptera; and 4 in most Diptera; but 5 in Culicidae; 6 in Mecoptera, Lepidoptera, and Trichoptera; 4 to 6 in the Coleoptera and Psocidae; 2 to 8 in Isoptera; 4 to 16 in Thysanura; 6 to 8 in Neuroptera; 8 to 20 in Dermaptera; 8 to 100 in Ephemeroptera; 50 to 60 in Plecoptera; 30 to 200 in Orthoptera; 50 to 200 in Odonata; and 6 to more than 100 in Hymenoptera.   The Aphidae and certain Collembola have no Malpighian tubes.   Where the tubes are very numerous, they may unite into groups which discharge through a common duct.   Morphological differences are observed between the various tubes of a species, and in many insects such as *Rhodnius* the individual tubes may show different structures and functions between distal and proximal portions (Fig. 3.13).

In cross-section the tubes are seen to be composed of a single layer of three to eight epithelial cells surrounding a lumen (Fig. 3.13).   The tubes are covered exteriorly by a peritoneal coat which contains tracheoles

[1] CRAIG, R., *Ann. Rev. Ento.,* **5**:53, 1960.

and, except in Thysanura, Dermaptera, and Thysanoptera, by muscle fibers which carry out peristaltic movements of the tubes. The inner surface of the cells is composed of a striated border made up of honeycomb-like rods or brush-like filaments.

Physiology and Biochemistry. Uric acid is the most important nitrogenous excretory product of insects and comprises about 86 per cent of the nitrogenous excreta of *Bombyx mori* and 64 to 84 per cent of the urine of *Rhodnius.* In this, the insects resemble birds and reptiles rather than mammals; in the latter urea is the important excretory product. Uric acid and its sodium, potassium, calcium, and ammonium salts are often present in the lumen of the Malpighian tubes as fine crystals, and where the water intake of the insect is low, the urine becomes semisolid or even a dry powder. In addition to uric acid, the urine may contain ammonia as various salts and smaller amounts of urea and amino acids (Table 3.4). The blow fly maggot *Lucilia* excretes allantoin (page 55) in the urine, and creatine is found in the urine of *Rhodnius* in small amounts. Calcium carbonate and calcium oxalate are found in the urine of plant-feeding insects, often in crystalline form. Pteridines comprise about 0.2 per cent of the dried feces of *Oncopeltus.*

TABLE 3.4. COMPOSITION OF NITROGENOUS EXCRETA OF INSECTS COMPARED WITH THAT OF MAN

Constituent	*Melanoplus bivittatus*	*Aedes aegypti*	*Brevicoryne brassicae*	*Tineola bisselliella*	Man
	Per cent of total nitrogen				
Uric acid............	36.2	47.3	0	47.2	1–3
Urea................	3.7	11.9	0	0.5	80–90
Ammonia............	1.3	6.4	5–8	13.2	2.5–4.5
Amino N............	7.4	4.4	90	....	2–6
Protein..............	0	10.8	...	....	Trace

Wigglesworth has found that in *Rhodnius* granules of uric acid occur only in the proximal third of each Malpighian tube in the area of the brush border (pH of 6.6), while in the distal two-thirds with honeycomb border (pH of 7.2), the urine is a clear fluid (Fig. 3.13). This suggests that the lower area is concerned with reabsorption and the upper with secretion. He has suggested that soluble urate salts are absorbed in the distal area and react with carbon dioxide in the proximal portion to produce relatively insoluble uric acid and bicarbonates which are resorbed along with water. Limited biochemical studies suggest that uric acid synthesis, *e.g.,* in the fat-bodies of *Antheraea* and *Periplaneta*, occurs through the reaction of formate, ammonia, carbon dioxide, and glycine to produce the purine hypoxanthine, which is oxidized to xanthine and then to uric acid. Many Diptera contain the enzyme uricase, which further oxidizes uric acid to allantoin (page 55).

In addition to their excretory function the Malpighian tubules serve as storage depots for water-soluble vitamins. In *Periplaneta*, riboflavin at 840 to 1,000 p.p.m., thiamin at 35 to 50 p.p.m., pantothenic acid at 80 p.p.m., niacin at 200 to 460 p.p.m., and ascorbic acid at 600 to 1,012 p.p.m. were found in the Malpighian tubes. Many other constituents are doubtless present, including plant pigments. The Malpighian tubes of the Cercopidae produce a sticky viscous substance which forms the foam or spittle in which these insects hide. The larvae of Neuroptera, such as *Chrysopa*, and of certain Coleoptera produce a true silk from the modified Malpighian tubes, which is used to construct the pupal cocoon.

The excreta of the plant-feeding Homoptera, particularly aphids and coccids, are especially interesting both because of the relatively enormous volumes produced, as much as several times the weight of the insect in 24 hours, and the extremely high carbohydrate content, which may exceed 80 per cent of the total weight of the fresh excreta. These excreta are very attractive to other insects and are termed honeydews or manna. The composition of these honeydews varies with the seasonal composition of the plant sap, but in addition to the constituents which pass directly through the alimentary tract, there is present a variety of sugars and nitrogenous compounds which are synthesized in the body of the insect.

The amino acid content of the honeydew of the pea aphid, *Macrosiphum pisi*, reached a maximum of 1.83 per cent, and 15 amino acids were present, while that of the cabbage aphid, *Brevicoryne brassicae*, was 2.35 per cent, with 18 amino acids, of which glutamine and asparagine constituted 36 per cent. In the coccids, *Coccus pseudomagnoliarum*, *C. hesperidum*, *Icerya purchasi*, *Saissetia oleae*, and *Pseudococcus citri*, the same 10 amino acids, alanine, arginine, aspartic acid, cysteine, glutamic acid, proline, glutamine, serine, threonine, and valine, were found in the honeydew. In general, the amino acid content of the honeydew reflects that of the host plant sap.

The carbohydrate content of the honeydews represents a more complex situation, and most of the higher sugars present are synthesized enzymatically in the insect from the simple sugars, glucose, fructose, and sucrose, found in the plant sap. Thus *Icerya purchasi* honeydew contained, in addition to small amounts of these, a relatively large amount of the trisaccharide, melezitose, which does not occur in the sap. This sugar was also found in the honeydews of the other coccids mentioned above. In *Coccus hesperidum* honeydew, however, the principal sugar was the trisaccharide, glucosucrose, and small amounts of the more complex sugars, maltosucrose and maltotrisucrose, were also synthesized. The sugar alcohols, ribotol, which is present in the honeydew of *Ceroplastes*, and dulcitol, in *Aphis euonymi*, are believed to be obtained unchanged from the host plant sap.

Respiration. The term respiration encompasses all the processes whereby the oxygen of the atmosphere enters the air passages of the insect, is transported to the various body cells, and participates in the metabolic processes of oxidation and the means by which carbon dioxide is eliminated from the body. Therefore it is important to distinguish between the *ventilation* of the air passages and *cellular respiration*, the fundamental aspects of which are very similar in all animals. Most insects have a special ventilating system, the tracheal system, in which air enters a system of internal air ducts, the *tracheae*, through paired lateral openings, the *spiracles* (Figs. 3.2 and 3.9,*l,s,6*). In many aquatic insects, air enters the tracheal system through *gills* (Figs. 6.18 and 6.47).

Cutaneous Respiration. Many simple animals ventilate the body directly through the body wall, but the evolution of the highly impermeable cuticle has made this an inefficient process in terrestrial insects, and cutaneous respiration, although it may account for as much as 10 to 50 per cent of the respiration, is the major respiratory mechanism only in the Collembola. However, with aquatic insects and certain parasitic larval Diptera and Hymenoptera living in aqueous surroundings, a major portion of the respiratory exchange takes place directly through the integument. Many species of aquatic insects such as nymphal Odonata, Trichoptera, and Ephemeroptera can survive even when the gills are surgically removed, and cutaneous respiration is a major means of gaseous exchange in the Chironomidae.

Insect eggs respire entirely by diffusion, and Krogh has calculated that, assuming a spherical shape, the maximum-sized diameter for which this type of respiration is adequate is 2 millimeters. As in the adult insect cuticle, the waterproofing wax layers of the egg are highly impermeable, and gaseous exchange takes place largely through the micropyle (Fig. 5.2).

Spiracles. These are usually located on the soft membranous areas between segments or sclerites, but the number and arrangement are highly variable and have been used to classify the insect respiratory system as follows:

HOLOPNEUSTIC is the most primitive arrangement in which two pairs of spiracles are present on the thorax and eight pairs on the abdomen.

HEMIPNEUSTIC is the condition where one or more pairs of the primitive system have become nonfunctional. Various arrangements occur, such as the *peripneustic* system, in which the spiracles occur in rows along the body as found in many insect larvae; the *amphipneustic*, in which only prothoracic and posterior abdominal spiracles are functional, as in many immature Diptera; the *propneustic*, with only prothoracic spiracles; and *metapneustic*, where only the last abdominal pair are functional, as in aquatic Diptera.

APNEUSTIC is the condition without functional spiracles. This condition is found in many aquatic and endoparasitic insects, air entering the tracheal system through gills or through the body surface.

HYPOPNEUSTIC is the condition where one or more pairs of the primitive 10 spiracles have completely disappeared, e.g., scale insects, lice, and parasitic Hymenoptera.

The spiracles are extremely variable in form and range from simple invaginations opening directly into the trachea to elaborate chambers equipped with intricate closing mechanisms, so that water loss from the trachea is under muscular control, and provided with hairs, lips, or ridges, which exclude foreign matter (Fig. 3.9,6), or with glands, which secrete a hydrophobic material to prevent wetting of the opening.

Tracheal System. The tracheae form an elaborate series of air ducts which divide and subdivide and continue into minute branches, <0.1 to 0.3 micron in diameter, the *tracheoles* (Fig. 3.14), which reach ultimately to every organ, tissue, and cell of the body. Through the thin walls of the tracheoles, the living cells withdraw the oxygen necessary for respiration, by diffusion, and in the same manner return the carbon dioxide produced in metabolism to the tracheae and so out through the spiracles. The tracheae and tracheoles are lined with a cuticular lining which has regular spiral thickenings, the *taenidia*, which keep the air passages expanded (Fig. 3.14).

The tracheae in primitive insects such as some Thysanura and Collembola arise from each spiracle without interconnections, but in most insects the spiracles are interconnected by longitudinal and transverse tracheal trunks (Fig. 3.10).   In many flying insects certain areas of the tracheae are greatly expanded into air sacs, which increase the volume of respiratory exchange and lower the specific gravity of the insect.

Gills.   Special problems in respiration occur in aquatic insects.   Certain aquatic Coleoptera and Hemiptera carry bubbles of air beneath the water, trapped under the elytra or in hydrofuge hairs, and breathe this air through the spiracles.   A few unique species such as the larvae of the *Mansonia* mosquito and the beetle *Donacia* penetrate the air spaces of aquatic plants by piercing the plant tissue with a modified respiratory syphon.   However, most aquatic insects respire through a combination of cutaneous respiration and tracheal gills.   The gills are pads of very thin membrane covering a network of tracheoles through which respiratory exchange takes place by diffusion.   The gills are of various types;

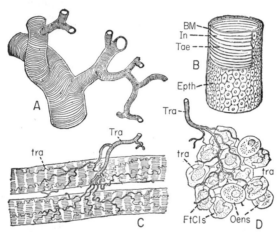

FIG. 3.14. Structure and terminal branches of tracheae.   *A*, a piece of trachea showing characteristic cross-striated appearance, due to spiral taenidia, and method of branching. *B*, basement membrane (*BM*) and inner cuticular lining or intima (*In*) with spiral thickenings of taenidia (*Tae*).   *C*, tracheal branches (*Tra*) ending in tracheoles (*tra*) on muscle fibers.   *D*, tracheation of piece of fat-body, showing tracheoles on fat cells (*Ft Cls*), but not on oenocytes (*Oens*).   (*From Snodgrass, "Anatomy of the Honeybee."*)

in the Ephemeroptera, Plecoptera, Neuroptera, Trichoptera, and Coleoptera, they generally consist of abdominal filaments or sometimes lamellae (Figs. 6.18 and 6.47).   The dragonflies (Anisoptera) have gills located in foldings of the rectal wall, while the damselflies (Zygotera) have three external gills.   Of the Diptera, mosquito larvae have four lamellate anal gills and black fly larvae (*Simulium*) have rectal gills.

Physiology and Biochemistry.   Gaseous diffusion through the tracheae has been shown to be adequate for the normal oxygen consumption of small insects and even of the larvae and pupae of such relatively large forms as *Cossus*, *Dytiscus*, and *Tenebrio*, provided the spiracles remain open.   However, the conservation of water is extremely important to insects, with their high ratio of body surface to volume, and in most species the spiracles are partially or completely closed most of the time. Many adult insects are extremely active, and oxygen consumption during the flight of *Apis*, *Lucilia*, and *Schistocerca* rises about fiftyfold from normal resting values in the range of 33 to 50 microliters per gram per minute. Both the metabolic levels and rates of increase during violent activity are much higher than those recorded for higher animals. Under such conditions mechanical ventilation is variable—in the relatively sluggish *Cossus* larva, 60 per cent of a volume of 107 microliters, and in *Melalontha* adult, 44 per cent of a volume of 630 microliters.

The direction of movement of air is controlled by selective opening and closing of the various spiracles, so that as in the grasshopper *Schistocerca*, inspiration takes place through the four anterior spiracles and expiration through the six abdominal spiracles. Carbon dioxide, which diffuses through animal tissues much faster than oxygen, is eliminated through the body surface as well as from the tracheae.

As we have seen, the regulation of respiration is effected by diffusion control through spiracular closure, and ventilation control through the pumping movements. Both these processes are under the control of the central nervous system, but the directing influence seems to be the concentrations of carbon dioxide and oxygen in the tracheae. Thus either a deficiency of oxygen or an excess of carbon dioxide results in opening of the spiracles and increase in the rate of ventilating movements, but the interaction is complex. It is of interest that high concentrations of

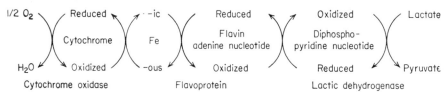

FIG. 3.15. Generalized scheme showing respiratory enzymes involved in the oxidation of lactic acid in cellular metabolism.

carbon dioxide cause complete and prolonged anesthesia in insects, and this is extremely useful in physiological and toxicological manipulations.

Cellular respiration in insects shows no essential differences from that in other lower and higher animals. This is a very complex process whereby the oxygen of air is utilized through a system of enzymes and carriers to oxidize organic foodstuffs such as sugars, fats, and proteins to provide the chemical energy for the processes of metabolism. The system involved in the utilization of oxygen to convert lactic acid, the product of glucose utilization in muscle metabolism, to pyruvic acid is depicted in Fig. 3.15.

Intermediary Metabolism of Muscle. Muscular energy results from the oxidation of carbohydrates from food to form carbon dioxide and water. The cellular reactions involved in this process are many and complex and can be only briefly summarized here (Fig. 3.16). The energy directly utilized for the contraction of the complex muscle protein actomyosin is the approximately 11,500 calories per mole liberated by the dephosphorylation of adenine triphosphate (ATP) to adenine diphosphate (ADP) in response to a nerve impulse. The ATP is then regenerated by a transfer of energy, by phosphorylation, from a phosphagen, probably arginine phosphate (in distinction to the creatine phosphate of vertebrates). Arginine phosphate is resynthesized by energy contributed during the aerobic process of glycolysis, in which glycogen of muscle, which may be formed from dietary fat or protein as well as carbohydrate, is degraded successively to glucose and fructose phosphates.   These are converted by successive phosphorylations and dephosphorylations to glyceraldehyde phosphate and to α-glycerophosphate, and during these operations a net gain of three new high-energy phosphate bonds are formed from each

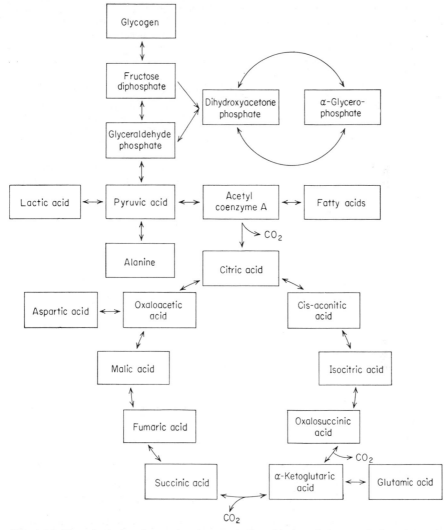

FIG. 3.16. Hypothetical pathways for the metabolic oxidation of glycogen in insect muscle tissue.   The larger circle represents the citric acid cycle, the smaller circle, the glycerophosphate cycle.   Each reaction is catalyzed by a specific enzyme.

molecule of glucose.   These as ATP react with arginine to form a new supply of arginine phosphate.   The final breakdown product of glycolysis under conditions of adequate oxygen supply is pyruvic acid.   However, when very intensive muscle activity occurs, the glycolysis may proceed to lactic acid and an oxygen debt may occur.

Glycolysis in insect muscles is not completely understood. The tracheal system provides an extremely efficient means for the transport of oxygen to the tissues, and in contrast to vertebrate muscles, anaerobic processes are of little importance, especially in flight muscles where the consumption of oxygen is enormous. In such circumstances the oxidation of α-glycerophosphate is the dominant reaction. However, some anaerobic activity occurs in the leg muscles of *Belostoma* and *Melanoplus*, and here the oxidation of lactic acid is an important pathway. Some of the reactions which are believed to occur are diagramed in Fig. 3.16.[1] The citric acid cycle is known to occur in insects and provides for the oxidative decarboxylation of pyruvic acid and the consequent yield of energy. This system also provides for the interconversion of carbohydrate, fat, and amino acids which is known to occur. Each reaction is catalyzed by an entirely specific enzyme system.

The flight muscles of Diptera, Hymenoptera, etc., show astounding frequencies of contraction, sometimes more than 1,000 times a second. These muscles also show amazing endurance, since in migratory locusts, butterflies, etc., these relatively high frequency contractions are maintained for many hours. As a result of this intense muscular activity, the oxygen consumption during flight may reach fiftyfold over the resting level, as compared with only about 5.5 times in the hummingbird. The energy source for flight activity is generally thought to be carbohydrate, especially glycogen or perhaps blood sugar as in *Apis*, and the respiratory quotient is 1.0. However, with the migratory locust *Schistocerca*, some 85 per cent of the flight energy comes from the oxidation of fats, and the respiratory quotient is about 0.7.

Secretion. The metabolism of insects results in the formation not only of the living protoplasm, but also of various other chemical substances that are essential to the body. Such substances are called *secretions*, and they are formed in specialized cells or groups of cells, the *glands*. These are of two general types, the *exocrine glands*, which discharge their secretions outside the insect body or into the viscera through a well-defined duct, and the *endocrine glands*, whose secretions, called hormones, diffuse into the blood stream.

Exocrine Glands. The epidermal or hypodermal cells which form the cuticle and molting fluid (page 80) and the Malpighian tubes (page 107) are examples of this type of gland. Exocrine glands may be single cells or simple aggregations of cells.

LABIAL OR SALIVARY GLANDS. These are found in nearly all insects, with the exception of some Coleoptera, and are highly diversified in various species (Figs. 3.8, 3.9,7, and 4.6) and consist of simple or convoluted tubes which are often branched or lobed and sometimes provided with enlarged reservoirs. The glands usually lie in the thorax but often extend into the abdomen. Their secretions are highly variable but are usually concerned with digestion. In Lepidoptera and Hymenoptera the larval glands are modified into silk-producing organs. Certain blood-feeding insects, such as *Anopheles* and *Glossina*, secrete anticoagulins and agglutinins for mammalian blood, while ants and bed bugs, respectively, secrete formic acid and histamine in the saliva.

WAX GLANDS. The secretion of beeswax and its chemical constitution have already been described (page 51). The secretion of waxes by the epidermal glands is a normal process of cuticle formation in all insects. In Homoptera profuse discharges of wax occur, as in the scale coverings of *Coccidae*, especially in the lac insect (page 51), and in woolly aphids and mealybugs.

SILK GLANDS. The formation of silk for the construction of cocoons and shelters takes place in the labial glands of Lepidoptera and Tri-

---

[1] See BOETTIGER, E., *Ann. Rev. Ento.*, **5**:1, 1960, and SACKTOR, B., *ibid.*, **6**:103, 1961.

choptera, and aspects of silk production in *Bombyx mori* are discussed on page 48.   In Embioptera and certain Carabidae, silk is formed by secretions of the Malpighian tubes and in the Empididae silk is formed by glands of the protarsi.

REPELLENT GLANDS.   Dermal glands which secrete unpleasant odors that act as a protective function are common in the Hemiptera.   In *Anasa tristis* the paired tubular glands (Fig. 3.17) lie in the mesothorax below the digestive tract and empty into a reservoir which opens to the exterior through two ostioles between the middle and hind coxa.   In the nymph the glands open on the dorsal surface of the abdomen.   Similar glands in the rice stink bug, *Oebalus pugnax*, secrete 40 per cent *trans*-2-heptenal, $CH_3CH_2CH_2CH_2CH=CHCHO$, dissolved in the hydrocarbon *n*-tridecane.   The cockroach, *Eurycotis floridana*, secretes a repellent, 2-hexenal, $CH_3CH_2CH_2CH=CHCHO$.   In the Coleoptera, especially the Carabidae and Tenebrionidae, paired pygidial glands opening near the anus produce butyric acid and other evil-smelling substances. Those of *Brachinus*, the bombardier beetle, discharge, with an audible sound, a volatile fluid of corrosive properties which contains benzoquinone and 2-methylbenzoquinone. A number of insects also secrete various quinones that have a repellent effect on other predatory species.   These include *Tribolium*, whose glands are located on the anterior margin of the thorax and on the last abdominal segment, the cockroach, *Diploptera punctata*, whose second abdominal spiracles and attached tracheae have been modified to glands which discharge 2-methyl- and 2-ethyl-1,4-benzoquinone and *p*-benzoquinone.   The

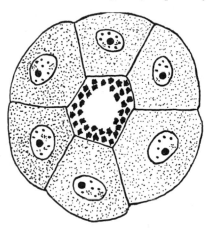

FIG. 3.17. Cross-section of repugnatorial gland of the squash bug, *Anasa tristis*, showing the secretory cells surrounding the lumen, which is partially filled with granules. (*After Moody, Ann. Ento. Soc. Amer.*, 1930.)

two former compounds are also secreted from thoracic glands of the tenebrionid beetle, *Diaperis maculata*. The leaf-cutting ant, *Atta sexdens*, secretes citral as a warning substance, from mandibular glands.

ATTRACTANT GLANDS.   *Pheromones* or ectohormones are substances liberated by external glands which produce specific reactions in other individuals of the same species.   Olfactory pheromones include the sex-attractant substances produced by the lateral glands of the last abdominal segment of virgin female moths.   This has been identified in *Bombyx mori* as 10,12-hexadecadiene-1-ol, $(CH_3CH_2CH_2CH=CH—CH=CH(CH_2)_8-CH_2OH)$, and in the gypsy moth, *Porthetria dispar*, as *d*-10-acetoxy-1-hydroxy-*cis*-7-hexadecene,

$$CH_3(CH_2)_5CHCH_2CH=CH(CH_2)_5CH_2OH.$$
$$O$$
$$O=CCH_3$$

The wings of many male butterflies contain specialized scales or *androconia*, which serve as outlets for sex attractants produced by glands at their bases, and similar glands are found on the legs of other male Lepidoptera. Sex pheromones are produced by glands on the last abdominal segment of both sexes of *Tenebrio molitor* and of female *Periplaneta americana*. The male tropical waterbug, *Lethocerus indica*, secretes 2-hexenyl acetate, $CH_3CH_2CH_2CH=CHCH_2OCOCH_3$, which is apparently a sex pheromone, from dorsal abdominal glands.

Apart from sex attractants, specific marking scents are secreted by the honeybee and bumblebee to identify individual colony members and for orientation of the nest, and the ant, *Formica rufa*, is believed to use formic acid to mark its trails.

Oral pheromones are secreted by queen bees and ants, which, when licked by the workers, inform the latter of the presence of the queen, retard the development of their ovaries, and influence their behavior. Pheromones are also elaborated by king and queen termites that, acting together, repress the production of individuals of the reproductive caste.

Numerous species of Coleoptera that live as symbionts in the nests of ants and termites and Lycaenidae larvae in ant nests owe their congenial reception to the secretions of dermal glands, which are very attractive to the worker insects.

POISON GLANDS. The venoms of many ants, bees, and wasps are secreted by modified accessory glands of the female and injected through the modified ovipositor or stinger. In the honeybee, *Apis*, the toxin appears to be a complex mixture containing histamine, histidine, lecithinase A, hyaluronidase, and phospholipase A, while that of *Formica* is formic acid, which may reach a concentration of 71 per cent and constitute 20 per cent of the body weight. Cantharidin (page 55) is secreted by accessory glands of the genital tract of blister beetles of the family Meloidae. Lepidopterous larvae of several species have poison glands which discharge through hollow poisonous hairs (page 24). The dolichoderine ant, *Iridomyrmex humilis*, contains up to 1 per cent by weight of iridomyrmecin, the lactone of (2-hydroxymethyl-3-methyl cyclopentyl)-propionic acid, which is highly toxic to other insects.

Iridomyrmecin

Endocrine Glands. The endocrine system of insects controls such functions as molting and metamorphosis by the elaboration of *hormones*. These are produced in a variety of neuroendocrine cells, which are neuron in form and secretory in function.[1]

NEUROENDOCRINE CELLS OF THE BRAIN. Molting in *Rhodnius* and in many other insects is controlled by a hormone elaborated by the medial neuroendocrine cells of the pars intercerebralis. The hormone elaborated stimulates the prothoracic glands to produce yet another hormone *ecdyson*, which acts directly on the body tissues to produce the molting

[1] VAN DER KLOOT, W., *Ann. Rev. Ento.*, **5**:35, 1960.

process.   This process has been demonstrated in Hemiptera, Orthoptera, Odonata, Mecoptera, Diptera, and Lepidoptera, but certain exceptions are known to occur.

PROTHORACIC GLANDS.   These paired structures are present in the immature stages of Lepidoptera, Coleoptera, Mecoptera, and Hemiptera. Williams has shown with *Hyalophora cecropia* that the prothoracic gland secretes the growth and molting hormone only in response to the action of a hormone from the brain.   This two-phase molting process has also been demonstrated in Hemiptera, Orthoptera, Odonata, Mecoptera, and Diptera.

CORPORA CARDIACA.   These structures, which occur in the Pterygota, lie on either side of the aorta behind the brain and are connected to the protocerebrum by nerve fibers.   The neurohormone from the brain accumulates in these organs and is liberated into the blood.

CORPORA ALLATA.   These are closely associated with the corpora cardiaca and are generally paired organs, although they are fused into a single body in the Hemiptera.   In *Rhodnius*, Wigglesworth has shown that they produce the juvenile hormone which inhibits the appearance of adult characteristics during development, and thus opposes the action of the molting hormone of the prothoracic gland.   The corpora allata become inactive in the last immature instar, and the change in hormonal balance induces metamorphosis.   This action of the hormone from the corpus allatum in inhibition of metamorphosis has been demonstrated in Lepidoptera and Coleoptera, and it is of general occurrence in the Insecta.

WEISMANN'S RING GLAND.   This organ, which lies behind the brain and encircling the aorta, is found in the larvae of cyclorrhaphous Diptera, where it is the source of the puparium-forming hormone.   This gland contains three types of glandular tissues homologous with the corpus cardiacum, corpus allatum, and prothoracic gland.

Hormones.   Considerable knowledge has been accumulated regarding the nature of the growth and molting hormones.   Ecdyson, the growth and molting hormone of the prothoracic glands, has been isolated in crystalline form by Butenandt and Karlson.   It has the empirical formula $C_{18}H_{30}O_4$ and a molecular weight of about 300.   This hormone, although originally isolated from the silkworm, causes molting and puparium formation in Diptera, Hemiptera, and other Lepidoptera.

The juvenile hormone, *neotenin*, which induces the formation of larval characters by the epidermis, growing under the influence of ecdyson, has been isolated by Williams from the male *Hyalophora cecropia* as a yellow oil which is thought to be a steroid.

CONDUCTION, COORDINATION, SENSATION

It is one of the inherent characteristics of living substance to be sensitive or irritable to the various stimuli that act upon it and to respond by altering its behavior in some way.   In insects, as in all higher animals, this function is accomplished by a nervous system.   Nerves are composed of cells highly specialized for the functions of sensation, conduction, and coordination.   Each nerve cell is called a *neuron* and may be a *sensory* (afferent) neuron, which has its cell body lying in the hypodermis and which conducts impulses inward from a sense organ; a *motor* (efferent) neuron, which has its cell body lying in the central nervous system and which conducts impulses from the ganglia outward to muscles or glands;

or an *association* neuron, which stands between a sensory and a motor neuron and may modify, direct, or harmonize impulses received from one or several sensory cells so as to coordinate the response of the organism as a whole. The association neurons are therefore believed to be the seat of consciousness and intelligent behavior. In insects the nervous system is conveniently characterized into three groups (Fig. 3.8): (*a*) the *central nervous system*, which consists of a series of ganglia and their connectives; (*b*) the *visceral* or *sympathetic nervous system*, which innervates the heart, digestive tract, spiracles, reproductive system, etc.; and (*c*) the *peripheral nervous system*, which is composed of the nerves radiating from the central nervous system, especially the sensory neurons and end organs, and is exceedingly delicate and far-reaching. Thus sensations from the eyes, antennae, palps, etc., are conducted by the sensory peripheral nerves to the ganglia of the central nervous system, where associations take place and the resulting nerve impulses are conducted along the motor nerves to the muscles and glands, which respond by a directed behavior. Meanwhile, the visceral nervous system is carrying out the numerous bodily functions of digestion, circulation, reproduction, molting, etc.

Central Nervous System. The central nervous system of the Insecta consists of a series of ganglia connected by a double ventral nerve cord which passes from end to end of the body (Figs. 3.7,*VNC*, and 3.8). In the primitive arthropod ancestors, there were two paired ganglia to each body segment united by transverse *commissures* and by longitudinal *connectives*. In modern insects, however, a considerable degree of fusion has occurred so that the paired ganglia are coalesced into a single ganglion and the connectives are generally so closely approximated as to appear as a single cord. In the evolutionary scheme of the Insecta, various other fusions have taken place, so that it is only in the primitive *Thysanura* and in many insect larvae that the subesophageal, three thoracic, and eight abdominal ganglia are recognizable. In the Orthoptera, Mecoptera, Trichoptera, and Hymenoptera, the metathoracic ganglion is fused with the first three abdominal ganglia and the seventh and succeeding abdominal ganglia are fused. In such specialized insects as the higher Diptera, only the subesophageal ganglion and a single fused thoracic-abdominal ganglion are found, while in the Coccidae and Aphidae all the ganglia are united into a single mass.

The brain or *supraesophageal ganglion* (Figs. 3.7,*Br*, and 3.8) is the largest ganglionic mass in the insect body and is the chief body center for the regulation of behavior by the association of stimuli received by the sensory organs with reflex responses by the rest of the body. The ganglionic mass is made up of the ganglia from the first three segments: the protocerebrum from the optic segment, the deutocerebrum from the antennal segment, and the tritocerebrum from the intercalary segment innervating the labrum and anterior portion of the gut. The development of the brain is often clearly associated with the complexity of the insect's behavior or sensory processes and ranges from about 0.00024 of the body volume in *Dytiscus* to 0.0058 in *Apis*. The insect brain is principally an association center and is composed of association neuron masses in the cortex and a number of association centers or tracts in which fibers from all parts of the brain converge. The most conspicuous structures in the protocerebrum are the *corpora pedunculata* or mushroom bodies, which surround the *central body*. The *optic lobes* are the association centers from the compound eyes, and their degree of development is correlated with that of the eyes.   The deutocerebrum is composed chiefly of the *antennal* association centers or *olfactory lobes*, and from it arise the antennary sensory and motor nerves.

Immediately behind the tritocerebral lobes are the circumpharyngeal connectives which join the brain and the *subesophageal ganglion* (Figs. 3.7 and 3.8).   This ganglion is formed by the fusion of the ganglia of the mandibular, maxillary, and labial segments and influences motor activity of the entire body.   The other ganglia generally control the movements of the appendages attached to the particular segment, although more complex reflex activity may be controlled by several ganglia.

Nerve Structure.   As in other animals, the basic units of the insect nervous system are the neurons, which consist of a nucleated cell body and a long axonal filament.   The junctions between the neurons are known as *synapses*, and these constitute the integrating mechanism between the transmission system of the nerve fibers.   The sensory nerve fibers are in general small and numerous, while the motor nerve fibers are few and large.   In the cercal nerve of *Periplaneta* there are 140 sensory fibers 7 to 10 microns in diameter, while the leg muscles are supplied by only two motor fibers.   Giant fibers up to 50 microns in diameter are found in the ventral nerve cord.

The nerve axon consists of a cylindrical axoplasmic core surrounded by a thin sheath of highly organized lipoprotein whose ultrastructure resembles that of the heavier myelinated sheaths of vertebrate nerves. In the insect this sheath may be from 1 to 10 per cent of the fiber diameter as contrasted with as much as 42 per cent in the vertebrate nerve.

Physiology and Biochemistry.   From the fundamental level there is little difference between the nervous system of the insect and of higher animals.   The nerve impulse through the axon is electrical in nature, while synaptic transmission appears to be dependent upon the release of acetylcholine and perhaps other substances.   The rate of conduction in *Periplaneta* is of the order of 6 meters per second as compared with 100 meters or more for vertebrate nerves.   From the standpoint of chemical constitution, the insect central nervous system (CNS) of *Apis* is composed of about 39 per cent lipids and 56 per cent protein.   The lipid fraction is nearly half phospholipid and a third cephalin, with traces of neutral fats, sterols, lecithin, and sphingomyelin.   No cerebrosides were found, and the insect tissue was closely comparable in composition with that of the brain of a young vertebrate prior to the appearance of visible medullation.

Insect CNS tissue contains very high amounts of acetylcholine, 10 to 20 times that of the vertebrate CNS, and correspondingly high activities (10 to 50 times vertebrates) of the enzyme cholinesterase.   This system is clearly involved in the mechanism of nerve-impulse transmission, and its interruption by the anticholinesterase organophosphorus insecticides leads to progressive nervous disturbances and death.   The lipid sheath in the insect nerve provides a distinct barrier to the penetration of ionic materials such as potassium ions, acetylcholine, etc., into the neurons, and this is the explanation for the inactivity of externally applied acetylcholine in affecting conduction.

Sense Organs.   The function of a nervous system is to acquaint the insect with changes in its environment.   For this it is necessary to have a variety of *receptors* (sensory end organs) which detect and interpret these changes (stimuli) and translate them into nerve impulses that ultimately produce responses in muscles or glands. The insect sense organs may be most conveniently classified by responses to (*a*) pressure or contact, *mechanoreceptors*, (*b*) sound, *audioreceptors* or auditory organs, (*c*) chemicals in solution or as gases, *chemoreceptors*, (*d*) light, *visual organs*, and (*e*) temperature and humidity. Each type of sensory organ is adapted to receive only a single kind of stimulus, ignoring or excluding all others.

Mechanoreceptors. *Sensory hairs* on the surface of the body differ from ordinary *setae* by having a process from a bipolar neuron in contact with the base. These are widely distributed over the body, especially in the antennae, tarsi, and cerci and, as they arise from very thin rings of cuticle, produce nerve impulses in response to the slightest stimuli, such as air currents and earth vibrations.

CAMPANIFORM SENSILLA are similar organs in which the hair is modified into a thin dome-like covering in contact with the distal portion of the sensory neuron. These are often located in groups on the palps, wings, halteres, basal joints of appendages, gills, cerci, etc., where they respond to stretching or compression of the integument.

CHORDOTONAL SENSILLA OR SCOLOPIDIA are internal sensory rods attached to a flexible region of the cuticle and containing the termination of a bipolar neuron. The sensilla may be attached to the integument at each end, or one end may be free in the body cavity. Such organs occur on the legs, wings, antennae, palps, and epidermis and may be associated with the tympanic membranes of auditory organs or highly specialized as in Johnstone's organ, which consists of a group of these organs located in the second antennal segment of the Pterygota. Johnstone's organ is most highly developed in Chironomidae and Culicidae, where it may fill the large globular second antennal segment (Figs. 3.3 and 4.7). The chordotonal sensilla respond to tension resulting from external and internal pressures, muscular action, and sound.

Auditory Organs. Two types of mechanoreceptors in insects have become specialized for auditory responses to movements of the surrounding air.

TACTILE SENSILLA, such as those of the cerci of *Periplaneta* or *Gryllus*, are long sensory hairs which respond to low-frequency sounds up to 4,000 cycles per second. In male mosquitoes, Johnstone's organ has an auditory function, which apparently results from stimulation of antennal hairs that vibrate the flagellum, which produces a response of the chordotonal sensilla. These hairs are of varying lengths and vibrate to sounds of differing tones, especially those of the same pitch, as the hum of the female mosquito. Thus it has been suggested that the male turns his body until the two antennae are equally stimulated by such a sound, when, by flying straight ahead, he is able to find his mate. It has been found that the "songs" of each species of mosquito are distinctive, and attempts have been made to use amplified recordings of mosquito mating calls to lure the insects to destruction.

TYMPANAL ORGANS are drum-like cuticular membranes and a group of scolopidia or chordotonal sensilla attached to the drum or to associated tracheal structures. Such organs respond to vibrations from about 250 to 45,000 cycles per second, and in Orthoptera the sensitivity at the region of maximum response, *i.e.*, about 10,000 cycles per second, is close to that of man. Tympanal organs may be large conspicuous oval plates on the sides of the first abdominal segment of the Locustidae

(Fig. 3.2) or smaller organs near the base of the front tibia in Gryllidae and Tettigoniidae (Fig. 3.4). They are also found on the second abdominal segment of the Cicadidae, on the metathorax of many Lepidoptera, and the meso- and metathorax of Hemiptera.

Chemoreceptors. Chemoreception or the perception of odor and taste in insects is accomplished by modified sensory hairs or sensilla. These consist of a bipolar neuron associated with a trichogen or glandular cell and covered with a very thin cuticular wall, through which the chemical enters by osmosis, the *sensilla trichordea;* peg-like or cone-like *sensilla basiconica* and *styloconica,* as in the antennae and palps of *Periplaneta* and *Hydrophilus,* the antennae of Diptera, and palps of butterflies; and plate-like *sensilla placodea,* the pore plates of the antennae of *Apis* and other Hymenoptera, Coleoptera, lepidopterous larvae, and Aphidae. The drone bee has about 30,000 of these pore plates per antenna. A generalized chemical sense for the perception of irritants exists independent of the characteristic response to odor and taste.

OLFACTORY RESPONSE OR SMELL of insects is of great importance in locating foods, suitable places for oviposition, members of the same colony or caste of social insects, and in mating, and the nature of this process has been studied in connection with attractants and repellents for insect control (Chapter 7). The degree of perception for certain sex attractants is extremely acute and entirely species-specific. With male *Bombyx mori* moths, it has been estimated that a definite reponse to the female secretion (page 115) can be obtained with as little as 10,000 molecules. A trap baited with live female gypsy moths, *Porthetria dispar,* attracted males from distances greater than 2 miles, and a male *Samia cynthia* is reported to have flown 1.5 miles to reach a female. The male oriental fruit fly, *Dacus dorsalis,* has been attracted to methyl eugenol for a distance of more than $\frac{1}{2}$ mile downwind.

GUSTATORY RESPONSE OR TASTE. The sensilla concerned with the response to liquids or solids generally occur on the epipharynx, hypopharynx, and palps, although they are also found on the antennae of bees and ants, on the tarsi of *Apis,* butterflies, and certain Diptera, and on the ovipositor of parasitic Hymenoptera and *Gryllus. Dytiscus* and *Hydrophilus* have been shown to distinguish the four primary taste qualities of sweet, salt, sour, and bitter. The threshold of sensitivity is extremely variable and depends upon the state of nutrition. The tarsal receptors of the butterfly *Pyrameis* are more than 200 times as sensitive as the human tongue to sucrose, but man is more than 20 times as sensitive to salt and several thousand times as sensitive to quinine as the bee. With regard to sugars, *Apis, Calliphora,* and *Phormia* prefer disaccharides such as sucrose to monosaccharides such as fructose and glucose.

Vision. Insects have eyes of complex structure and of two distinct kinds, ocelli or simple eyes and compound eyes. Yet we believe that vision in most insects is poor and probably subordinate to smell and touch as a guide to them in their reactions to the environment.

OCELLI. The simple eyes of insects differ from compound eyes in having a single lens-like *cornea,* which is usually an arched and thickened area of transparent cuticle. This serves as a fixed focus lens and is supported from beneath by a band of transparent *cornagen cells* and under this a *retina,* which consists of groups of sensory neurons surrounding a *rhabdom* or longitudinal optic rod (Fig. 3.18). *Dorsal* ocelli, innervated from the ocellar lobes of the protocerebrum, are commonly found along with compound eyes in nymphs and adult insects, where they typically occur in groups of three in a triangle on the vertex (Figs. 4.2 and 4.17). However, some insects, including butterflies, have no ocelli, although moths have two near the bases of the antennae. The dorsal ocelli probably serve only for the perception of light, since the image of the lens is focused far behind the retina, but they are very sensitive and have apertures estimated at f 1.0 to 1.8 or equal to the fastest camera lenses. *Lateral* ocelli or stemmata are typically the only eyes of insect larvae and are located in variable numbers, from one to six or more on the gena (Fig. 6.55). They are innervated from the optic lobes of the brain and, where several occur together, are arranged to cover a wide visual field and thus to provide a coarse mosaic vision. Despite their fixed focal length, it has been shown that the image from the lens will fall somewhere along the length of the rhabdom and these ocelli, which in *Isia isabella* have f values of 0.5 to 1.0, probably function in the perception of form and color.

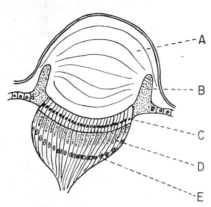

FIG. 3.18. Lateral ocellus of *Cicindella campestris*. *A*, cornea; *B*, pigment; *C*, cornagen cells; *D*, rhabdom; *E*, retinal cells. (*After Friederichs.*)

COMPOUND EYES.   These organs are usually the most conspicuous objects on the insect head and are convex, round, oval, or kidney-shaped areas, one on each side, with a shiny appearance, which suggests their function at a glance (Figs. 3.19 and 4.15). Upon microscopic examination they are found to consist of a number of individual facets, generally hexagonal, which are the exposed faces of individual lenses (Fig. 3.20). The number of facets in a compound eye is quite variable and is related to the habitat and behavior of the insect. Worker *Solenopsis* ants living underground have 6 to 9 facets, while the males which must find the females in nuptial flight have 400. The wingless female of *Lampyris* has 300, and the winged male 2,500. The house fly, *Musca*, and the worker honeybee, *Apis*, have about 4,000, while *Dytiscus* has 9,000, the swallowtail butterfly 17,000, and dragonflies whose visual acuity is extreme, 28,000. The individual facets range in diameter from 16 microns in *Culex* to about 40 microns in dragonflies. In some male Diptera such as *Tabanus*, the facets are larger in the upper portion of the eye, and in *Bibio* and *Simulium* there is distinct demarcation between the two areas of different-sized facets.

FIG. 3.19. Head of a fly, *Didea fasciata*, dorsal view, showing the large compound eyes, which occupy the entire sides of the head; between them the three minute simple eyes or ocelli arranged in a triangle; and the aristate antennae. This is a female fly; many male flies have the compound eyes touching each other on top of the head. (*From Metcalf, Ohio Naturalist.*)

Each facet or *cornea* of the compound eye is the lens of an individual optic unit, the *ommatidium*, which passes inward like a tube or rod toward the center of the head (Fig. 3.21). The light admitted through each facet is focused or concentrated by a cone-shaped body of high refractive index, the *crystalline cone*, although in some species this may be wanting or composed of transparent jelly, and these cones are surrounded by densely pigmented *iris cells*.   The light traversing the cone falls upon the *retinula* or visual end organ, which consists of a group of about seven cells collectively secreting an *axial rhabdom* or optic rod, which is composed of fine nerve fibrillae coalescing at the base of each rhabdom into a single postretinal nerve fiber which passes to the optic lobe of the brain.   The typical compound eye has an aperture of f 2.5 to 4.5 and is thus less sensitive to light than the lateral ocellus.   It is not really known how well insects can see with such eyes.   They are sometimes so big and bulging that the insect must be able to see in front of it, to each side, above and below, and even to some extent behind.   On the other hand, it cannot move its eyes, it cannot focus them upon objects at various distances, and of course it cannot close them.   In all probability, the closer an object, the better it is seen, because more ommatidia will cover an object close at hand than one far away.   It is believed that the compound eye is especially good at detecting objects in motion, because different independent ommatidia are stimulated in succession.   It has been claimed that insects such as the dragonflies may respond to objects in motion as far away as 60 feet.   However, the perception of the shape of objects is probably limited to distances of a few feet.

PHYSIOLOGY OF VISION.   Vision in insects has been demonstrated to be of the mosaic type, in which each ommatidium does not form an image of the entire field of view but only preserves the intensity, pattern, and color of the light from its projection upon the visual field, *i.e.*, at an ommatidial angle of 1 to 3 degrees for *Apis* and *Musca*.   The erect image formed by the entire compound eye is therefore the combined effect of thousands of minute areas of light, shade, and color and thus resembles, in principle, the newspaper photograph or jigsaw puzzle.

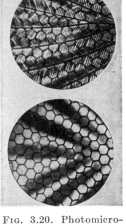

FIG. 3.20.   Photomicrographs taken through the cornea of a fly.   Note the honeycomb-like margins which divide the cornea into a number of hexagonal facets.   In one case the object photographed is extremely close to the eye, in the other case a slight distance farther from the eye. (*From Howe*, "*Insect Behavior*," *Badger*.)

The processes of image formation by the cornea and crystalline cone are complex, since these have a laminated structure with a variable refractive index which decreases toward the periphery.   As a consequence, two distinct types of vision are possible with the compound eye: (*a*) *apposition* eyes, common among most diurnal insects, in which the cones and rhabdoms are surrounded by pigment so that only light rays parallel to the axis of each ommatidium can be focused on its rhabdom, and (*b*) *superposition* eyes of nocturnal insects, such as Noctuidae and Lampyridae, in which the ommatidia are elongated so that the retinulae are separated from the cones by long transparent iris pigment cells.   In this type of eye the pigment is capable of migrating upward around the cones in dim light, so that refracted light from as many as 30 contiguous cones may fall upon each rhabdom, while during daylight the pigment migrates to surround each retinula so that the eyes function as apposition eyes.

The time required for this adaptation is from 30 to 60 minutes in the codling moth.

The visual acuity or resolving power of insect eyes is low and for *Apis* is about one one-hundredth that of man. As a result, the bee is unable to distinguish simple shapes such as solid squares, circles, or triangles, although it can separate these from patterned figures such as rows of stripes, hollow squares, and crosses. However, it is certain that bees, ants, and wasps utilize visual landmarks in finding their way to the nest. Insects have marked perception of motion, and *Aeschna* nymphs can detect the flicker of light intensity at about 60 per second as compared with about 50 for man. Because of the fixed nature of insect eyes, no stereoscopic vision as such can occur, although insects have depth perception resulting from the overlap of the fields of vision of each eye and measured by simultaneous and equal illumination of corresponding ommatidia of each eye.

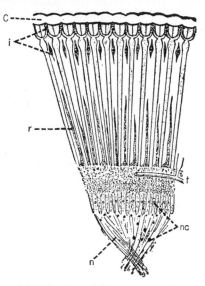

FIG. 3.21. Portion of the compound eye of a blow fly, *Calliphora vomitoria*, in radial section. *c*, Cornea, which is modified cuticle; *i*, iris pigment; *n*, nerve fibers leading to the brain; *nc*, nerve or retinal cells; *r*, retinal pigment; *t*, trachea. (*From Folsom, "Entomology," after Hickson, McGraw-Hill–Blakiston.*)

COLOR PERCEPTION. Insects have a somewhat wider range of color perception than man, *i.e.*, from about 2,500 to 7,000 Angstrom units, and can thus detect ultraviolet radiation. All wavelengths are not equally stimulating, and maximum response is usually found in the ultraviolet at about 3,650 Angstrom units, with other peaks at about 4,920, 5,150, and 5,550 Angstrom units. Color vision has been demonstrated in *Apis*, which distinguishes four regions: 3,100 to 4,000, 4,000 to 4,800, 4,800 to 5,000, and 5,000 to 6,500 Angstrom units. This insect can also perceive the plane of vibration of polarized light, which it uses as a direction-finding mechanism.

THE REPRODUCTIVE SYSTEM OF INSECTS

In some animals each individual may be both male and female or hermaphroditic (as in the common earth worm), but among the insects, male and female reproductive organs are nearly always borne by different individuals. The cottony-cushion scale, *Icerya purchasi*, is the most notable exception and is both hermaphroditic and self-fertilizing. The hermaphrodite has the body form and behavior of a female and may mate with the rare male, but commonly undergoes self-fertilization. Male and female insects are often equally injurious. In some species the females are more injurious, because the adult males are short-lived and do not feed (as in scale insects), because the female inflicts injury in the egg-laying act (*e.g.*, the tree crickets), or because females must consume more food to mature large numbers of eggs. Male insects are generally smaller than the females.   The sexes are sometimes indistinguishable on external appearance but generally show minor differences by which they can be identified.   In many species of flies the sexes can be told apart by the difference in the eyes, as shown in Fig. 3.19.   Frequently an examination of the tip of the abdomen will show differences sufficient to distinguish the sexes.   In some cases the two sexes are so different-looking as to appear like widely separated kinds (*cf.* Figs. 6.45 and 6.46).

The reproductive organs are usually found toward the tip of the abdomen (Fig. 3.8), although, when the eggs are developing, they often

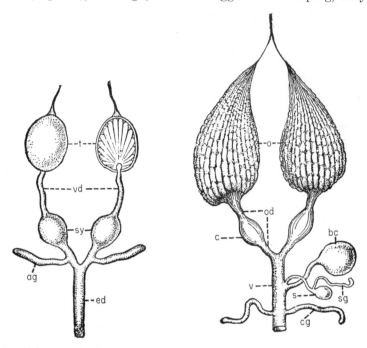

Fig. 3.22. Male reproductive organs, on the left, diagrammatic; right testis is shown in section to expose the testicular follicles. *ag,* Accessory glands; *ed,* ejaculatory duct; *sv,* seminal vesicles; *t,* testes; *vd,* vasa deferentia.   Female reproductive organs, on the right, diagrammatic. *o,* Ovary; *od,* oviduct; *c,* egg calyx; *v,* vagina; *s,* spermathecal gland; *cg,* colleterial gland.   (*From Comstock, "Introduction to Entomology."*)

pack the entire body cavity of the female.   The opening from the reproductive organs is near the posterior end of the body (on the eighth or ninth sternite) in all insects and is commonly surrounded by external genitalia.   In the female, there may be an *ovipositor* (Figs. 3.2 and 6.13), an organ for thrusting eggs into the ground, the tissues of plants, or the bodies of other animals.   In the male, there are *claspers* (Fig. 3.8), used to hold the female during mating, and the *penis.*   The external genitalia are increasingly used very extensively in the determination of species; the small, and often complex, parts showing perhaps more distinct differences between species than any other group of structures.

In either sex the *gametes* or germ cells (eggs and sperms, respectively) are developed from cells of ordinary appearance, and prepared for union in the sex glands.   These glands (Fig. 3.22) are commonly more or less compact bodies lying among the other organs of the abdomen or suspended by filaments or ligaments from the inside of the dorsal wall of the body. The *testes* and *ovaries* differ so much in form that no brief general description can be given. Each consists internally of a number of minute tubes or bead-like strings of cells, the *ovarian tubes* in the female (ranging in number from 1 to as many as 2,400 in each ovary, but commonly 4 to 8), and the *testicular follicles* in the male. In these tubes the sex cells (*gametes*) become more and more highly specialized, changing from primary germ cells to *oögonia* or *spermatogonia* and then to *oöcytes* or *spermatocytes* and, finally, after the reduction division of the chromosomes (page 153), emerging as matured eggs or sperms. The growth of the eggs is made possible by the secretions (yolk) of adjacent cells known as *nurse cells*. The entire ovary or testis may or may not be enclosed in a sheath of connective tissue. As the eggs descend the ovarian tubes, the shell is secreted about them by the small *follicular cells* that form the walls of the tube. From the sex glands the gametes are conveyed toward the exterior of the body through tubes known as *vasa deferentia* or *seminal ducts* in the male and *oviducts* in the female. Pending mating, the accumulated sperms are held in the body of the male in dilatations of the vasa deferentia known as *seminal vesicles*. At the time of mating, the sperms are usually received into a special pouch of the female system known as the *seminal receptacle* or *spermatheca*, which is attached to the oviducts or vagina. Later, as the eggs are being laid, some of the stored sperms are forced out upon them when they pass the opening of the seminal receptacle, thus bringing about fertilization. In the case of the honeybee queen, the sperms retain their viability in the seminal receptacle throughout her lifetime of several years. Eggs may be fertilized by these sperms years after the mating took place. The two oviducts usually unite to form a common duct, the *vagina*, opening ventrally on the seventh, eighth, or ninth abdominal segment. The two seminal ducts usually fuse into a single *ejaculatory duct* opening ventrally on the ninth abdominal segment. *Accessory glands* of the male (Fig. 3.8) secrete the fluid with which the sperms are mixed and sometimes sac-like coverings for packets of sperms known as spermatophores. In the female the accessory glands secrete a substance for cementing eggs together, or fastening them to leaves and other objects; or capsules to enclose a number of eggs; and, in the honeybee, one of the poisons used in the stinger. The correspondence of the organs of male and female is shown in the following table (after Folsom; compare with Fig. 3.22):

Male	*Female*
Paired testes[1]	Paired ovaries
Sperm tubes (testicular follicles)	Egg tubes (ovarian follicles)
Paired vasa deferentia (seminal ducts)	Paired oviducts
Median ejaculatory duct	Vagina (median oviduct)
Penis (phallus or aedeagus) and claspers	Genital chamber (bursa copulatrix) and ovipositor or stinger
Seminal vesicle	Seminal receptacle (spermatheca)
Accessory glands	Accessory glands

[1] Sometimes united into a single testis, as in moths and butterflies.

The growth and development of the insect following the formation and fusion of the egg and sperm are discussed in Chapter 5.

CHAPTER 4

THE MOUTH PARTS OF INSECTS

The most important way in which insects inflict losses and injury upon man and his possessions is by eating or feeding. Since insects feed in various ways, it is evident that a knowledge of insect mouth parts is of prime importance in the study of entomology. By those interested in controlling insects, this part of insect anatomy receives very careful consideration; and the homologies of the regions and appendages of the heads of various insects have presented some of the most interesting and puzzling problems in insect anatomy.

The Segmentation of the Head

The head of an insect is believed to be composed of six segments, or twice the number in the thorax, although it looks like a single segment.[1] The reasons for concluding this are that in the embryos of certain insects, six nerve ganglia, six pairs of rudimentary appendages, and six pairs of coelomic sacs or primary divisions of the body cavity can be recognized in the part that subsequently forms the head. Since the typical arthropod segment bears one ganglion, one pair of jointed appendages, and a pair of coelomic sacs, this can only mean that six such segments have fused to form the head of the insect. At this stage of development, the mouth opening is found about mid-length of the head on the ventral side between the third and fourth segments. Probably in the ancestors of present-day insects these six segments were as distinct as those of the abdomen in the insects living now; and each segment bore a pair of appendages, all very much alike and similar to the legs.

For the function of walking, we concluded (page 87) that it is an advantage to have the segments distinct and the appendages thereof widespread to form two tripods. But for chewing and ingestion of food the appendages must be close together where they can work against each other, to cut off, to hold, and to masticate the food. So the head of the insect (Fig. 4.2) has come to be more and more compact, its six segments have fused together, and almost all trace of their conjunctivae and segmentation has been lost. Its walls have become thick and firm to serve as an adequate base for the attachment of the powerful muscles needed to operate the mouth parts. In this way the skull case (called the *cranium* or *head capsule*, in insects) has been developed. This strong cranium supports the eyes, the antennae, and the mouth parts. It encloses the brain, the mouth or buccal cavity, the pharynx, and the muscles that operate the mouth parts. In insects the size of the head is not an indication of the development of the brain but is correlated largely with the size and strength of the jaws.

Differences in the various appendages, sutures, and sclerites of the head are much used in determining the names of insects. To one who understands these parts, they will tell almost the whole story of the insect's food habits. So it is important to learn the names of the appendages and sclerites that follow.

The Sclerites of the Head

The most evident landmarks on the wall of the head of adult insects are the large compound eyes (Figs. 4.7 and 4.15), described in Chapter 3. These represent the

---

[1] Various authorities have concluded that the head is composed of from five to eight segments. See Butt, F. H., *Biol. Rev.*, **35**:43, 1960, for a comprehensive discussion.

first of the six head segments. In some insects there is an impressed line or suture running over the top of the head, between the compound eyes on the median line, and forking into two branches toward the front, the whole thing being like the letter Y, with its stem toward the thorax and its arms toward the front. This suture is called the *epicranial* or *coronal suture* (Fig. 4.1). The area of the wall of the head between the branches of this suture, or the equivalent area in insects where the suture cannot be found, is known as the *front* or *frons*. It usually bears the median ocellus. Below the front are two other areas or sclerites typically separated from each other by transverse sutures. The upper one of these is called the *clypeus*, sometimes divided

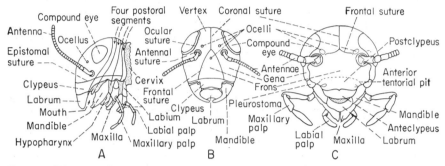

Fig. 4.1. Diagrams illustrating the fundamental structure of the head of insects and the principal named parts. *A* suggests the origin of the head from six primitive segments. (*Redrawn from Snodgrass, "Principles of Insect Morphology," McGraw-Hill.*)

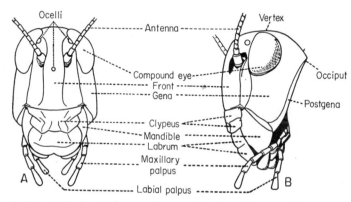

Fig. 4.2. Outline of the head of a grasshopper showing sclerites and appendages. *A*, as seen from in front; *B*, from the side. (*Rearranged from Folsom, "Entomology," McGraw-Hill–Blakiston.*)

into two parts (Fig. 4.1), and the lower one the upper lip or *labrum* (Fig. 4.4). One *condyle* or root of the mandible articulates to the clypeus, while the labrum is hinged along its entire base to the clypeus but is otherwise free and movable as an upper lip. The area between the compound eyes on the upper part of the head is called the *vertex*. Here the other two ocelli (Figs. 4.1 and 4.2) are usually located, and from this area and slightly below the ocelli the antennae (Figs. 4.1 and 4.7) arise. The cheeks or sides of the head below and behind the compound eyes and between them and the mandibles are called the *genae*. Sometimes a narrow strip along their lower margins is marked off by a suture and is known as the *subgena*. The part of the head that abuts against the thorax is called the *occiput* (Fig. 4.2). It generally consists of a somewhat horse-shoe-shaped arch the lower or lateral parts of which (the heels of the horseshoe) are called the *postgenae*. The occiput surrounds the small posterior opening or passageway in the skull case through which all the food, as it is swallowed, and also the nerve cords, dorsal blood vessel, and tracheae, pass from head to thorax. This passageway is called the *occipital foramen* (Fig. 4.9,*B*). The *tentorium* is an internal structure in the head of insects formed by two pairs of ingrowths of the body wall (*anterior and posterior arms of the tentorium*) that unite to form a cuticular framework or plate supporting the pharynx and esophagus and arching over the subesophageal ganglion. The tentorium may be further braced by a pair of *dorsal arms,* extending from the cranium to the anterior arms, above the bases of the antennae. These arms and their intersections brace the cranial walls and serve as a place of origin for muscles that operate the mouth parts and the antennae. The pits or slits where the *anterior arms* of the tentorium invaginated (Fig. 4.1,*C*) lie near or in the lateral ends of the frontoclypeal (epistomal) suture, just above the anterior articulation of the mandibles. The pits or slits where the *posterior arms* of the tentorium invaginated lie at the caudoventral corners of the postgenae, in the lower ends of the suture separating the fifth and sixth head segments. When the dorsal arms of the tentorium are present, their points of origin are marked by spots or depressions somewhat above the base of the antennae and near the arms of the epicranial suture.

The compacting and fusion of the originally separated segments of the head have not always been strictly in the direction of the long axis of the body, but a distortion has sometimes carried the foremost segments, with their appendages, toward the top of the head, leaving the mouth at the very front of the body (*prognathous*); or sometimes, by reason of an opposite tendency, the mouth has been shifted backward and comes to occupy a place at the lowermost part of the head (*hypognathous*).

THE APPENDAGES OF THE HEAD

The appendages of the first one of the head segments have become replaced by, or specialized into, the compound eyes (Figs. 4.7 and 4.15). Those of the second segment are the antennae or feelers, (Fig. 3.3, page 86). The third pair of the original appendages (second pair of antennae) has been lost in the insects, but we have in this region an unpaired sclerite of the head, developed somewhat like an appendage, the *labrum* or upper lip (Figs. 4.2 and 4.4). The appendages of the fourth head segment are the *mandibles* or first pair of jaws (Figs. 4.2 and 4.4); those of the fifth segment, the *maxillae* or second pair of jaws (Figs. 4.2 and 4.4); and those of the sixth segment, the *labium* or lower lip (Figs. 4.2 and 4.4). In addition to the above parts, portions of the wall of the mouth or preoral cavity are sometimes projected outward, or otherwise so specialized that they form an essential part of the mouth parts; the one from the dorsal wall (roof of the mouth) is called the *epipharynx* (Figs. 4.4 and 4.7), the one from the ventral wall (floor of the mouth) is called the *hypopharynx* (Figs. 4.4 and 4.7). The typical arrangement of these parts with reference to the mouth opening may be indicated by the following diagram. It is plotted to represent the arrangement of parts as the

Right compound eye	Vertex	Left compound eye
Right ocellus		Left ocellus
Right antenna	Median ocellus	*Left antenna*
	Frons	
Right gena and	Clypeus	Left gena and
subgena		subgena
	(Labrum)	
	(Epipharynx)	
(*Right mandible*)		(*Left mandible*)
	MOUTH	
	(PREORAL CAVITY)	
(*Right maxilla*)		(*Left maxilla*)
	(Hypopharynx)	
	(*Labium*)	
Right postgena	Occiput	Left postgena

insect would face the reader. The parts that are printed in italics are believed to be homologous to the legs of the thoracic segments. The parts written in parentheses constitute the mouth parts.

The eight parts which are collectively called the mouth parts vary extremely in different insects for different kinds of feeding. They are first described as they are found in an insect that chews solid foods, such as a grasshopper, cricket, beetle, or caterpillar. Insects of this type of mouth parts (Fig. 4.3) inflict losses upon American agriculture in excess of a billion dollars a year.

The Chewing Type of Mouth Parts

The Labrum (Figs. 4.2 and 4.4). This so-called upper lip covers the mandibles and closes the mouth cavity from in front, much as our upper lip covers our teeth. It helps to pull food into the mouth. There is often a notch at the middle of the labrum which is of service in holding the edge of a leaf in position so that the mandibles can bite across it effectively. Very often the insect places itself in such a position at the edge of the leaf that the sagittal plane of its body (the plane that divides the body into two equal halves) is parallel to the plane of the leaf. Some insects, however, can eat directly into the face of a leaf away from its margin.

The Epipharynx (Fig. 4.4). If we had only the chewing insects to consider, it would hardly be worth while to give a separate name to this part. But in some other types of mouth parts it becomes differentiated into an important structure. In the grasshoppers and crickets, the epipharynx is inseparably attached to the labrum forming the inner, under, or posterior face of the labrum and is continuous with the roof of the mouth and thence into the esophagus. It is a sensory area believed to contain end organs of taste.

The Mandibles (Figs. 4.2 and 4.4). The mandibles, teeth, or first pair of jaws are in chewing insects the most important part of the mouth structures. Besides masticating the food, they are the structures that cut it off or tear if off the leaf or other object on which the insect is feeding. In different insects they function to carry things, to fight with, or to mold wax. In the chewing insects there are generally several projections or small teeth on each mandible, which work against those of the opposite side and so make very efficient grinders. It should be noted that the action of the mandibles and maxillae in insects is transverse, or from side to side, instead of longitudinal, or up and down, as in man.

Each mandible is typically a single, nearly solid piece of sclerotized cuticle, roughly shaped like a pyramid with three faces, one of which is continuous with the gena or cheek (Fig. 4.2,*B*). These heavy teeth articulate, in front, by a socket to convex processes from the lateral corners of the clypeus where it joins the genae and, behind, by a rounded process that fits into a socket of the genae, postgenae, or subgenae. Two sets of muscles operate them, one closing them against each other, the other pulling them apart.

The Maxillae (Figs. 4.2 and 4.4). These are the second pair of jaws, much more complicated in structure than the first pair, but working from side to side in a manner much like the mandibles. The exact shape differs with the kind of food and the manner of feeding, but the following parts are typically represented. There is a central body of three or four sclerites (*cardo, stipes, palpifer*), from which three appendages arise. One

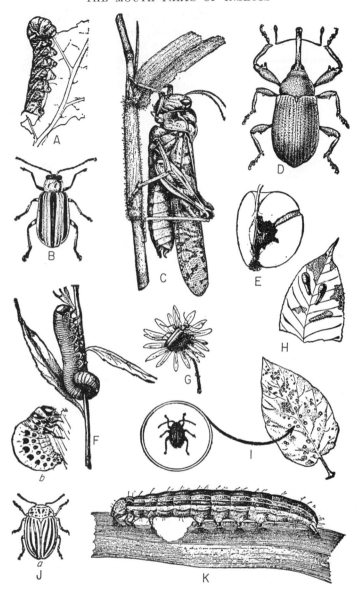

Fig. 4.3. A group of common insects that have chewing mouth parts. *A*, tomato horn-worm (*from Conn. Agr. Exp. Sta.*); *B*, striped cucumber beetle (*from U.S.D.A.*); *C*, grass-hopper (*redrawn after Walton, U.S.D.A.*); *D*, cotton boll weevil (*from U.S.D.A.*); *E*, codling moth larva in injured apple (*redrawn after Conn. Agr. Exp. Sta.*); *F*, elm sawfly (*from Riley*); *G*, black blister beetle (*from Conn. Agr. Exp. Sta.*); *H*, pear-slugs, skeletonizing a leaf (*from Conn. Agr. Exp. Sta.*); *I*, potato flea beetle and characteristic work on potato leaf (*original*); *J*, Colorado potato beetle, *a*, adult, *b*, larva (*from U.S.D.A.*); *K*, armyworm feeding on leaf (*from U.S.D.A.*).

of the appendages is antenna-like in shape, of from one to five or six segments, and is a kind of sense organ bearing tactile hairs and probably also organs of smell or taste.   It is known as the *maxillary palp* or *palpus* (Fig. 4.4) and is regularly the longer of the two pairs of palps found in chewing insects.   The second appendage, called the *galea*, is very variable in form (helmet-like in the grasshopper); and the third, known as the *lacinia*, is the tooth part of the maxilla, modified often for cutting,

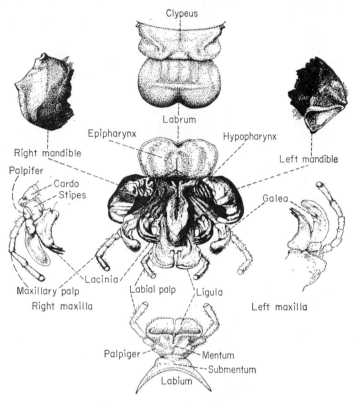

Fig. 4.4. The chewing type of mouth parts as found in a grasshopper.   The central figure is looking into the mouth with all the appendages widespread.   The upper left shows ectal view of right mandible; upper right ental view of left mandible; upper center the labrum with the clypeus to which it attaches; lower left is ectal view of the right maxilla; lower right, ental view of left maxilla; lower center, ectal view of labium.   The hypopharynx is shown near the center of the central figure.   (*Original drawings by Antonio M. Paterno.*)

grasping, or grinding the food.   The central body of the maxilla articulates to the lower part of the posterior wall of the head (the *subgena*) (Fig. 4.2,*B*), and its several parts are freely movable by muscles that run out from inside the wall of the head and the tentorium.

   The Hypopharynx (Fig. 4.4).   This is a tongue-like prolongation of the floor of the mouth or preoral cavity, usually attaching to the inside (anterior) wall of the labium.   It is of interest as the part through which the salivary glands of insects open, these glands being especially significant in the silkworm (see page 48) and also in some disease-carrying insects.

The Labium (Figs. 4.2 and 4.4). A good idea of the function of this part is conveyed by its common name, lower lip. It stands apposed to the upper lip, closing the mouth below or behind. It is the most complicated of all the parts, but we easily analyze its structure when we understand its origin. It has developed from two maxilla-like pieces by the two growing together along the middle line. It therefore consists of a large central body, more or less produced into lobes or unsegmented appendages (typically four) at its free end, and gives off at each side a short antenna-like appendage known as the *labial palp* (Fig. 4.4). The four unsegmented appendages are a median pair of *glossae* and a lateral pair of *paraglossae*. These four parts, when more or less fused, are called the *ligula*. The labial palps are regularly shorter than the maxillary palps, the number of segments varying from one to three or four. Their function is similar to that of the maxillary palps. In some insects care will be needed not to confuse the palps with antennae, especially if the antennae are concealed. The labium attaches to the neck of the insect, articulating with the lower ends of the *occiput*, between the maxillae at the back part of the head.

How Insects Feed

Most of our serious insect pests feed in one of the following ways: (*a*) They tear or pinch off, chew up, and swallow bits of plant and animal tissue, much as a cow eats the leaves off a stalk of corn. In other words they are "chewers," like grasshoppers and caterpillars, and are said to have chewing mouth parts (Fig. 4.3). (*b*) Or they extract from beneath the surface of a plant or animal body the body liquids (without swallowing the tissues), just as a person might insert a straw into a piece of wet sponge and suck out the liquid or into a cocoanut and suck out the "milk." In other words, they are "drinkers," like mosquitoes or aphids, and are said to have piercing-sucking mouth parts (Figs. 1.4 and 4.5).[1]

The condition of the mouth parts in a typical chewing insect has been described in some detail because it is the most primitive type and the kind from which all other mouth parts have been derived. Insects with chewing mouth parts may generally be controlled by spraying or dusting with a stomach poison.

The Piercing-sucking Type of Mouth Parts

The piercing-sucking mouth parts take the most valuable liquids in the world—the sap of growing plants and the blood of living animals. In order to get this food, they puncture or pierce the skin of the animal or the epidermis of the plant, making a very tiny, invisible hole with their

---

[1] In most textbooks these two general feeding habits have been called "biting" and "sucking." A little reflection will show that "biting" is a particularly ambiguous term, which in common usage conveys exactly the opposite meaning to that intended in the classification of insect mouth parts. That is, when one speaks of a *biting insect* the average person thinks—not of a beetle or a caterpillar chewing up leaves—but of a "biting" mosquito or flea or horse fly, whose mouth parts, according to the usual classification, are not *biting* at all but *sucking*. Thus the "biting house fly" would have, according to such a classification "sucking" and not "biting" mouth parts. The term "sucking" is also objectionable because it is too general; as shown in the table at the end of this chapter, there are several, radically different, kinds of mouth parts which are all "sucking" or suctorial in function. We therefore urge the adoption of the term *chewing* instead of "biting" and *piercing-sucking* instead of "sucking," for the two commonest types of mouth parts.

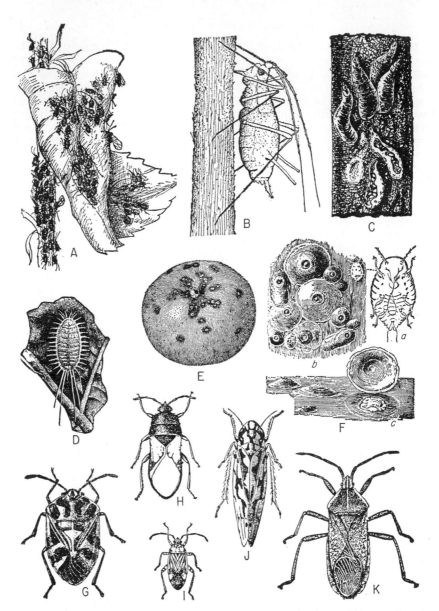

FIG. 4.5. A group of common plant-feeding insects that have piercing-sucking mouth parts. *A*, aphids or plant lice clustered on stem and leaf, about natural size (*original*); *B*, a single potato aphid in feeding position, greatly enlarged; note beak appressed to stem (*from Ohio Agr. Exp. Sta.*); *C*, oyster-shell scales on bark, the lower two turned over to expose eggs from beneath the scale, enlarged (*original*); *D*, long-tailed mealybug on leaf, much enlarged (*from Comstock*); *E*, San Jose scale on fruit of apple showing spotting due to its feeding (*from Fulton*); *F*, San Jose scale, *a*, first-instar nymph or "crawler," greatly enlarged, *b*, a group of scales as seen on the bark, much enlarged; the large round ones cover mature females, the elongate ones cover males, and the small ones cover the second-instar nymphs; *c*, a group of scales from the side, one of them lifted to show the body of the female insect (*from Quaintance, U.S.D.A.*); *G*, harlequin cabbage bug, enlarged (*from U.S.D.A.*); *H*, chinch bug, enlarged (*from S.D. Agr. Exp. Sta.*); *I*, tarnished plant bug, enlarged (*from U.S.D.A.*); *J*, grape leafhopper, enlarged (*from U.S.D.A.*); *K*, squash bug, enlarged (*from Iowa Agr. Exp. Sta.*).

mouth stylets through which they suck the sap or blood from beneath the surface.   There are two very distinct operations involved in this act—(a) piercing and (b) sucking—hence we call these mouth parts piercing-sucking.   The insects that have piercing-sucking mouth parts include both serious animal parasites, such as many flies and mosquitoes, the fleas, the bloodsucking lice, and destructive crop pests, especially the true bugs of the orders Homoptera and Hemiptera (Fig. 4.5).   Insects of this type of mouth parts are therefore pests of importance to both the livestock raiser and the grain farmer or gardener.

The appearance of the mouth parts of this type is totally different from those described above.   We do not find a complex group of appendages surrounding an evident mouth.   In fact it is sometimes hard to tell just where the mouth opening is, it is so small or well hidden.   We do not find hard, tooth-like mandibles for grinding.   What we do find (Figs. 4.6,C, and 4.7) is a long, needle-like beak, slender, cylindrical, usually jointed, which may point forward, downward, or backward, but, when not in use, is generally found laid back on the breast between the front legs. In many cases there are no palps at all, in other subtypes one pair, and rarely both pairs are present.

In the true bugs, such as chinch bugs, cicadas, bedbugs, aphids, and squash bugs, we find a jointed slender beak of three or four segments (Fig. 4.6,C), inside of which lie four extremely slender, pointed stylets (Figs. 4.6,B,C), that normally cling together to form what appears like a single, slender, brown bristle.   The jointed beak is not a closed cylinder but is open down the entire front side and at the end, like a trough, or the handle of a pocketknife.   This largest outside piece is the *labium* (Fig. 4.6,C).   *It has nothing to do with puncturing the plant or sucking up the sap.* Lying inside the groove of the labium are four very sharp, chitinous "stabbers" or "needles," the *stylets*, which do the work of piercing the plant and drawing out the sap.   These four pieces are the two *mandibles* and the two *maxillae* (Fig. 4.6), all extremely modified from their condition in chewing insects.   The *labrum* in this type (Fig. 4.6,B,C) is a short flap that covers the groove in the labium toward the base of the latter.

Sometimes the mandibles and maxillae have little sharp barbs near the apex.   As the stylets are alternately thrust out from the head at a rapid rate, the barbs catch in the leaf tissue or flesh, and, anchored in this way against a backward pull, they help to sink the stylets deeper into the wound at each thrust, until the level of sap or blood is reached.   In other species, such as the scale insects and whiteflies, the barbs are wanting, and the stylets are kept from being pulled out of the wound at each counterthrust by a muscular clamp in the groove of the labium, which alternately grips the stylets and allows them to slip deeper into the wound.   Each maxilla is doubly grooved from end to end along its inner face (Fig. 4.6,A), and these concave faces of the two maxillae fit tightly together to make two closed tubes known as the *food channel* and the *salivary duct*.   A mandible fits closely against each side of the apposed maxillae, and in some species, such as the squash bug, a tongue (*Tn*) on the outer face of each maxilla locks into a groove (*Gr*) on the inner face of each mandible so that they can slide in and out on each other.   In this way the delicate hollow apparatus is greatly strengthened for the thrust into the tissues.   The stylets are sometimes many times as long as the labium and even longer than the entire body.   The extra length needed to reach the deeper tissues of plants or animals is accommodated when the

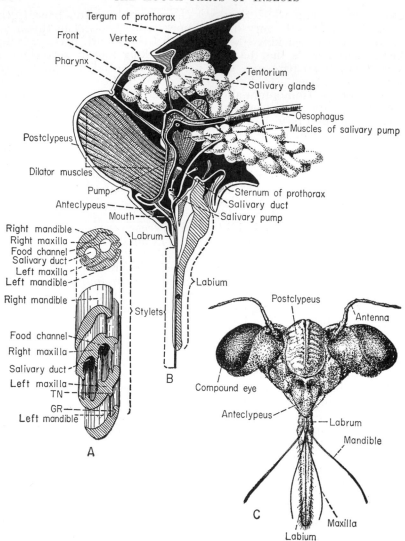

Fig. 4.6. The piercing-sucking type of mouth parts as found in the squash bug and cicada. *A*, cross-section and isometric projection of the stylets as described by Tower, greatly magnified (*original*); *B*, sagittal section of the head of the periodical cicada showing the relation of the stylets to the mouth opening, the pump, the pharynx, and the salivary glands, duct, and pump, much enlarged (*redrawn after Snodgrass, Proc. Ento. Soc. Wash.*); *C*, front or dorsal view of head and mouth parts of a dog-day cicada, much enlarged (*original*).

insect is not feeding by being thrown into loops or coils inside the head, the thorax, or the base of the labium. The slender hair-like stylets, just described, have their bases sunken in the head and enclosed in pockets or pouches, whose lining is continuous with the integument of the cranium. In addition to the highly functional stylet parts, the mandible and the maxilla have each a plate-like basal part on the side of the head capsule,

below the compound eye.   A cuticular lever reaches from each mandibular and maxillary plate to the expanded base of the stylet inside the
head, and muscles originating on the inner faces of the mandibular and
maxillary plates are inserted upon these levers.   As these muscles contract, they force the stylets outward or downward into the food tissue.
Other muscles, originating near the top of the head and inserted upon the
bases of the stylets themselves, retract or pull the stylets upward and out
of the wound.   By rapid alternate protraction of right and left mandibles,
followed by right and left maxillae, the stylets are forced into the food
tissue rapidly, although by very short thrusts.

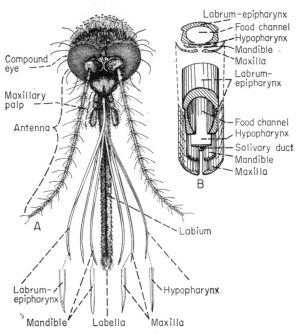

Fig. 4.7. Piercing-sucking mouth parts as found in a female mosquito; common biting
fly subtype.   A, front or dorsal view of head and mouth parts with the stylets spread out
of the labium and their tips more enlarged below. B, cross-section and isometric projection
of the stylets as described by Howard, Dyar, and Knab.   Much enlarged.   (*Original
drawing by Antonio M. Paterno.*)

When the stylets have reached the sap beneath the epidermis of leaf
or bark, or the blood in bloodsucking forms, the liquid food is sucked up
through the microscopic tube called the food channel (Fig. 4.6,*A*) by
expansion of a pump, formed by the walls of the buccal or mouth cavity
and operated by powerful dilator muscles (Fig. 4.6,*B*).   The contraction
of these muscles pulls the dorsal wall of the pump away from the ventral
wall, thus producing the necessary suction clear to the tip of the food
channel in the stylets.   As the muscles relax, the elastic dorsal wall
springs back into place, its lower or anterior end closing first, and so
forcing the sucked-up liquid backward and upward into the *esophagus*
(Fig. 4.6,*B*).   The labium shortens up or bends back out of the way and
does not enter the plant tissues or flesh.   There may be sensory hairs at

the end of the labium that serve to sample the food and select the spot for feeding, and this part may also be used as a kind of fulcrum to steady the head and the stylets while they are piercing.   However, its chief function appears to be to act as a kind of scabbard for the four-parted dagger that we call the stylets.   The saliva that is pumped through the salivary duct into the plant may soften the cell walls and even predigest the liquid food. *Both pairs of palps are wanting in the Homoptera and Hemiptera.*

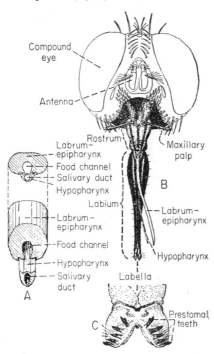

Any insect having mouth parts of this nature is open to suspicion as a serious enemy of crops or animals. If leaves show minute pale or brown spots, or are curled or wilted (Fig. 1.4), this kind of insect should be sought as the probable cause.   Many piercing insects feed on other insects in a manner helpful to us, but as a group they are predominantly destructive.

One important principle should be fixed in mind at this point.   *No insect with piercing-sucking mouth parts can be killed by applying a stomach poison to the plant on which it is feeding,* because the minute stylets may penetrate through the deposit of insecticide on leaf or fruit.   Therefore contact insecticides are commonly employed against such insects.   An exception to this lies in the use of systemic insecticides, which poison plant or animal tissues (page 357), so that the insect with piercing-sucking mouth parts ingests the insecticide as it feeds.

Fig. 4.8. Piercing-sucking mouth parts as found in the stable fly; special biting-fly subtype. *A,* cross-section and isometric projection of the stylets to show the food channel and salivary duct; *B,* front or dorsal view of the head and mouth parts with the stylets spread out from the labium; *C,* the labella more magnified to show the prestomal teeth, which are cutting organs, according to Patton and Cragg. Much enlarged. (*Original drawings by Antonio M. Paterno.*)

VARIETIES OF PIERCING-SUCKING MOUTH PARTS

Besides the piercing-sucking mouth parts of the true bugs, there are several variations of the same functional type, especially among the bloodsucking insects. These are spoken of as subtypes and their chief structural features are indicated in the table of mouth parts below.   For example, in the piercing apparatus of the mosquito (Fig. 4.7), we note that there are six stylets instead of four, the *labrum-epipharynx* and the *hypopharynx* being long and slender like the *mandibles* and the *maxillae.*   These stylets are not protractile and retractile, like those of the Homoptera and Hemiptera, but are imbedded, when the insect bites, by a strong downward or forward thrust of the body.   We find also a pair of palps (entirely wanting in the bugs), which are the *maxillary palps.*   The food channel (Fig. 4.7,*B*) is formed chiefly by the labrum-epipharynx, which is the heaviest of the six stylets.   The groove along its lower or posterior face is closed by the apposition of the hypopharynx or the flattened mandibles.   The hypopharynx carries throughout its entire length, in the *salivary duct,* the saliva that causes the irritation when a mosquito bites and the malaria parasite from an infected mosquito to the blood stream of man.   The labium is not jointed except for the differentiation of a pair of oral lobes or labella (Fig. 4.7) at its tip.

Among the flies of the house fly family are some species, such as the stable fly, that bite, although the house fly never does.   These have mouth parts of the type illustrated (special biting-fly subtype) in Fig. 4.8.   The *labium*, the *labella*, the *labrum-epipharynx*, the *hypopharynx*, and the *maxillary palps* are much like those of the mosquito.   But the mandibles and maxillae are entirely wanting, there being therefore only two stylets.   Patton and Cragg[1] state that piercing by these insects is accomplished by the rapid protraction and retraction of the labella, which bear sharp teeth on their inner faces (Fig. 4.8,*C*), and that the labium as well as the stylets follows these teeth into the flesh.   The food channel is formed by the labrum-epipharynx and hypopharynx and the salivary duct traverses the hypopharynx (Fig. 4.8,*A*).

The mouth parts of the fleas show another variation in the combination of stylets that are used for piercing and sucking (Fig. 4.9).   There are three stylets: the *labrum-*

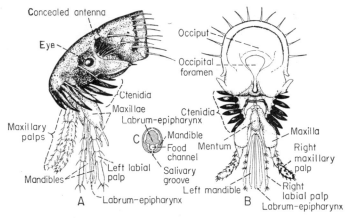

Fig. 4.9. Piercing-sucking mouth parts as found in a flea.   *A*, head and mouth parts in side view.   Note the antenna nearly concealed in a groove in the side of the head, the small eye, the ctenidia or comb of heavy spines, and the way the labial palps shield the stylets.   *B*, caudal view of head and mouth parts of a flea, the head having been removed from the thorax.   *C*, cross-section of mandibles and labrum-epipharynx, as described by Patton and Cragg, to show how the food channel and salivary groove are formed.   All much enlarged.   (*Original drawings by Antonio M. Paterno.*)

epipharynx and the two *mandibles*.   The mandibles have toothed edges and are the chief cutting organs.   The labrum-epipharynx is grooved on its ventral or posterior face, much as in the stable fly (Fig. 4.9,*C*; *cf.* Fig. 4.8,*A*).   In order to make this groove or food channel tight enough to suck up blood, the slit is probably closed by the apposition of the edges of the mandibles.   Each mandible has in its posterior (mesal) edge a tiny groove (Fig. 4.9,*C*); these two grooves when pressed together, form a second tube, the *salivary groove*, to carry the saliva to the tips of the stylets.   The maxillae do not enter the wound but are broad flaps, said to serve as levers or fulcrums to steady the head during the piercing and sucking operations.   They have a pair of prominent *maxillary palps*.   The labium suggests somewhat the condition in chewing insects, consisting of a basal piece called the *mentum* and two slender, jointed pieces called the *labial palps*.   These palps are concave on their sides which are next to stylets, and the two together form a protective sheath for the mandibles and labrum-epipharynx.

The piercing-sucking organs of the bloodsucking lice are anomalous, and the homologies of the parts are not clear.   There is no external evidence of the mouth

[1] PATTON, W., and F. CRAGG, "A Textbook of Medical Entomology," Christian Literature Society for India, London, 1913.

parts except a few teeth (mouth hooks) around the mouth opening (Fig. 4.10), which serve to anchor the head of the louse against the skin as it prepares to bite. Between these teeth an opening leads into the mouth cavity, called the *buccal funnel* or *preoral cavity.* When not in use, the piercing structures are entirely withdrawn into a sac, the *stylet sac,* that branches off ventrally from the floor of the buccal funnel near the front of the head and ends blindly beneath the esophagus near the back of the head. At the caudal end of this sac are attached three slender stylets. They are forked at the caudal (proximal) end and imbedded in the walls of the stylet sac. Their cephalic or distal ends extend nearly to the mouth opening and in the region of the buccal funnel are ensheathed in a trough called the *sac tube.* Although morphologists are not agreed as to their homologies, it is possible that the largest and most *ventral stylet* is the labium and that a delicate hypopharynx surrounds the salivary ducts as a *median stylet,* while the so-called *dorsal stylet* may represent the interlocked maxillae. The mandibles and both pairs of palps are entirely wanting. When the louse is ready to feed, protractor muscles (some having their origin on the inner wall of the cranium

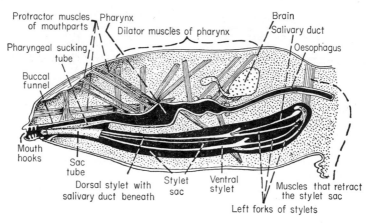

FIG. 4.10. Piercing-sucking mouth parts as found in the human body louse. Note the stylet sac lying beneath the pharynx, in which are two prominent stylets. In order to feed, the stylets are thrust through the mouth opening and forced into the flesh, and the blood as drawn is sucked up through the pharyngeal sucking tube. (*Redrawn after Imms and Peacock.*)

and insertion on the walls of the stylet sac, and others arising on the walls of the stylet sac and being inserted on the basal forks of the stylets) force the stylets out of the mouth and into the flesh. As the blood exudes, it is sucked up through the short, trough-like or funnel-shaped passageway and forced into the esophagus by the action of a powerful pharyngeal pump, operated by the action of dilator muscles. The saliva probably serves to prevent coagulation of the blood.[1]

A CLASSIFICATION OF INSECT MOUTH PARTS

While the chewing and piercing types of mouth parts are the commonest and most important, there are a number of other types found in insects, as indicated in the following table, which is a summary of this chapter.

A. The Chewing Type (Fig. 4.4). Generalized mouth parts (consisting of eight named parts surrounding an evident mouth opening), the essential features of which are two pairs of tooth-like jaws; the mandibles and maxillae, fitted to work transversely (and used for tearing off and masticating food, or for carrying things, for fighting, etc.); and an upper and a

[1] PEACOCK, *Parasitology,* **2**:98–117, 1918.

lower lip. A further characteristic is the presence of *two pairs of jointed palps*.

This is the commonest type of mouth parts and is found in the silverfish, grasshoppers, crickets, earwigs, termites, book-lice, chewing lice, beetles, weevils, some Hymenoptera, and many insect larvae, especially grubs and caterpillars, besides many others of little importance to man.

I. *The Grinding or Masticatory Subtype* (Fig. 4.4). Mandibles provided with a basal, molar, or grinding area suited for masticating solid plant or animal tissues, in addition to the apical cutting edges. EXAMPLES: the majority of chewing insects including nearly all caterpillars, most beetles, and most of the Orthoptera.

II. *The Grasping or Predaceous Subtype* (Fig. 6.44). Mandibles elongate, curved, with one or more sharp distal points for catching and holding the prey, and without a well-developed molar area. EXAMPLES: most predaceous beetles, soldier ants, and such beetles as the stag beetles, in which the mandibles serve for holding the female during mating.

III. *The Grasping-sucking or Mandibulo-suctorial Subtype* (Fig. 4.11). Mandibles long, slender, and sickle-shaped for grasping and piercing the prey and with a groove or closed canal extending from near their tips to near their bases through which the blood of the prey is sucked. The groove may be closed by overgrowth of the mandible itself, by the application of accessory lobes of the mandible, or by the close application of the elongate, flattened, maxillary blades. The mouth opening (preoral cavity) is usually partly or completely closed, the blood of the victim being sucked into the mouth by a pharyngeal pump through its lateral corners, or by secondary "mouths" applied to the opening of the groove near the base of each mandible. Frequently a paralyzing and digestive fluid is pumped into the body of the prey before sucking begins.

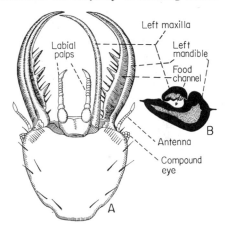

FIG. 4.11. Mandibulo-suctorial subtype of mouth parts, as found in certain Neuroptera, ventral view, showing the greatly elongated, sickle-shaped mandibles and maxillae, which fit together to form the food channels. These extend from the tip of each jaw to its base, where an adventitious mouth leads into the pharynx. *A*, left maxilla has been displaced to show the groove in the mandible. *B*, diagrammatic cross-section of the mandible and maxilla to show how food channel is formed by their close apposition. (*Drawn by Kathryn Sommerman.*)

EXAMPLES: larvae of ant-lions (Fig. 6.48), aphid-lions (page 64), and some adult beetles such as the predaceous diving beetle and fireflies.

IV. *The Brushing, Spatulate, or Scraping Subtype.* Mandibles without incisor or molar teeth, densely covered with stiff hairs; or flat, thin, and spatula-like for molding wax, mud, or dung. EXAMPLES: pollen-feeding beetles and dung beetles.

B. The Rasping-sucking Type (Fig. 4.12). Mouth parts which are somewhat intermediate in structure between the piercing-sucking type and the chewing type, but are rasping and sucking in their action, serving to lacerate the epidermis of plants and to suck up the exuding sap. The right mandible is reduced, making the head and the mouth parts somewhat asymmetrical. The left mandible, the maxillae, and, according to Borden,[1] the hypopharynx are elongate, suggesting the stylets of the piercing type and adapted to move in and out through a circular opening

[1] *Jour. Econ. Ento.*, **8**:354, 1915.

at the apex of the cone-shaped head.   The stylets are each contained in a separate pouch that has invaginated from the surface to give them their internal position.   The stylets apparently do not form a food channel, external to the wall of the head, nor do they enter deeply into the wound. The sap as it exudes on the surface is sucked up by the cone-shaped mouth rather than by the stylets.   According to Snodgrass, the food channel, within the head capsule, lies between the labrum and the hypopharynx, and the saliva passes to the tip of the stylets between the hypopharynx

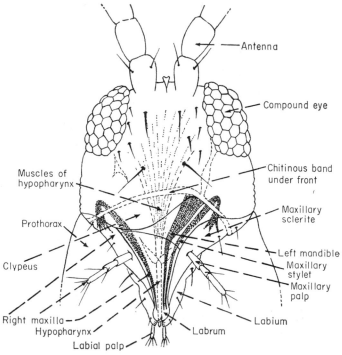

Fig. 4.12. Rasping-sucking type of mouth parts as found in the flower thrips.   View of the head from in front with only the bases of the antennae shown.   The chitinous band serves to connect the mouth parts to the front.   It also sends a branch to the left eye, while the one to the right eye is a triangular rudiment.   Note the two pairs of palps, the three or four stylets, and the cone formed by labrum, maxillae, and labium for sucking up the sap.   (*Redrawn after Borden, Jour. Econ. Ento.*)

and the labium.   Both pairs of palps are present.   The rasping-sucking mouth parts are characteristic of the thrips (page 211).

   C. The Piercing-sucking Type.   Specialized mouth parts characterized by a tubular, usually jointed beak, enclosing several needle-like stylets.   The outer tube is formed by the *labium*, which is simply a protective structure for the other parts and has nothing to do with piercing the tissues or drawing up the liquid food.   The mandibles and maxillae, sometimes supplemented by or replaced by the labrum-epipharynx and hypopharynx, are greatly elongated and slender structures, which serve for piercing the skin of an animal or the epidermis of a plant, and also as the food channel, *i.e.*, an inner tube up which the liquid blood or sap is drawn.

There are several structural variations of this type, of which the following must be noted:

I. *The Bug or Hemipterous Subtype* (Fig. 4.6). No palps. Four stylets: two *mandibles* and two *maxillae*, the latter partly fused. Food channel and salivary duct formed by the maxillae. EXAMPLES: chinch bug, aphids, scale insects, and bedbug.

II. *The Louse or Anoplurous Subtype* (Fig. 4.10). No palps. Two prominent "stabbers" or stylets, and some associated structures. The proper names and homologies are not well understood, but the stylets are believed to represent the maxillae

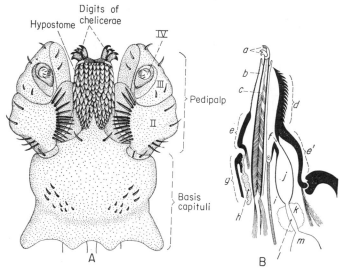

FIG. 4.13. Anchoring subtype of mouth parts, as found in ticks. *A*, ventral view, showing the median *hypostome*, with its many retrorse teeth; projecting slightly beyond the hypostome the two *chelicerae*, each with two articles bearing several *digits* or teeth; and at either side the *pedipalp*, with its four segments marked I, II, III, and IV. All the above structures arise from the sclerotized ring-like false head or *basis capituli*. *B*, diagrammatic sagittal section of the same showing at *d*, the *hypostome;* and *a*, the *digits of the chelicerae; b*, the *sheath of the chelicera*, which encloses the *rod of the chelicera, c*. Within the rod of the chelicera are muscles that operate the tendons extending to the digits and to the base of which other muscles attach that manipulate these cutting teeth; *e* and *e'*, the *basis capituli*, sectioned above, and below, the *mouth opening, f*, which leads to the *pharynx, j; l, esophagus; m, stomach*. The *brain, k*, surrounds the esophagus. The *salivary duct, i*, opens just above the mouth. Between the *scutum, g*, and the basis capituli on the upper side of the body is the opening from the gland called *Gene's organ, h*, the secretion from which is used in laying the eggs. (*Redrawn in part from Nuttall, "Monograph of the Ixodoidea."*)

(interlocked), the hypopharynx, and the labium. This type differs from all other insect mouth parts in that, when not in use, all the parts are entirely withdrawn into a long, slender pocket in the head beneath the pharynx. EXAMPLES: human lice or "cooties," hog lice, and other bloodsucking lice.

III. *The Common Biting-fly or Dipterous Subtype* (Fig. 4.7). Maxillary palps present. Six stylets: two *mandibles*, two *maxillae*, the *labrum-epipharynx*, and the *hypopharynx*, the last two parts forming the food channel and the hypopharynx surrounding the salivary duct. The stylets are not protractile and retractile as in the Hemiptera and Homoptera but are imbedded, when the insect bites, by a strong downward or forward thrust of the body. EXAMPLES: mosquitoes, horse flies, black flies, and "no-see-ums."

IV. *The Special Biting-fly or Muscid Subtype* (Fig. 4.8). So far as structure is concerned, derived, according to Patton and Cragg,[1] from the sponging type (see below) by the reduction in size of the labella and the attenuation and sclerotization of the labium, which in this subtype is rigid and not retractile. The labella are provided with cutting teeth, and this type differs functionally from all other piercing insects in that the *labium* itself enters the puncture.[1] The *labrum-epipharynx* and *hypopharynx* are similar to those of the sponging type, together forming the food channel; the mandibles are wanting and the maxillae represented only by a pair of palps. This type is exemplified by the bloodsucking species of the order Diptera, family Muscidae, such as the stable fly, horn fly, and tsetse flies.

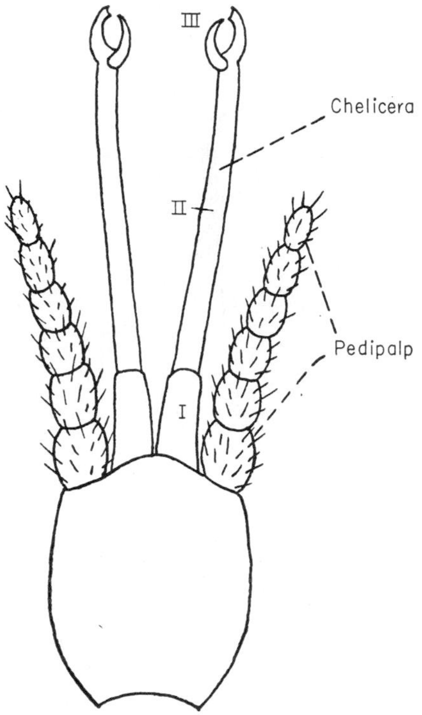

FIG. 4.14. Chelate subtype of mouth parts as found in many mites. I, II, III, the three segments of the chelicerae. (*Drawn by Kathryn Sommerman.*)

V. *The Flea or Siphonapterous Subtype* (Fig. 4.9). Maxillary palps present. Only three stylets (two *mandibles* and the *labrum-epipharynx*) enter the wound. The maxillae are triangular plates that serve as levers while biting. The labium bears two segmented parts, which are probably the *labial palps*. The food channel is formed by the labrum-epipharynx and the apposition of the caudal edges of the mandibles. The latter also form a salivary duct (Fig. 4.9,*C*). This type of mouth parts is found in the fleas.

VI. *The Tick or Anchoring Subtype* (Fig. 4.13). Attached to a false head or *basis capituli* is a pair of four-segmented pedipalps which may function both as sensory organs and for the protection of the more vital parts between them. The latter consist of a labium-like *hypostome* which arises from the basis capituli below the mouth opening. Its ventral surface is usually provided with a number of hooks or teeth projecting backward, and it is the structure with which the tick anchors to the skin of its host. The cutting organs consist of a pair of mandibles or *chelicerae* which arise from the basis capituli above the mouth opening. Each consists of a firm sclerotized rod or shaft enclosed by a sheath and extending back inside the false head where muscles attach that control its movements. The distal end of the shaft protrudes from the sheath and bears two movable digits, each bearing a few very sharp teeth. By movement of these digits the hole is cut in the skin into which the hypostome is thrust and through which blood is sucked. This method of making the wound suggests somewhat that employed by the teeth on the labella of the Special Biting-fly Subtype. This type is found in the cattle tick, the spotted-fever tick, the American dog tick, and their relatives.

VII. *The Chelate or Mite Subtype* (Fig. 4.14). Similar to Subtype VI, but the hypostome is not developed. The chelicerae often much more elongated to form needle-like piercing structures. The third or terminal segment of the chelicera sometimes articulated against the side of the second segment to form a pair of minute pincers; at other times wanting, leaving the chelicerae or mandibles like those of a mosquito, simple slender needles. This type is found in many mites such as the poultry mite, the tropical rat mite, mange mites, scab mites, chiggers, and others.

[1] PATTON and CRAGG, *op. cit.*

D. The Sponging Type (Fig. 4.15). This type of mouth parts is well illustrated by the condition in the common house fly. We find on the lower side of the head a fleshy, elbowed, and retractile proboscis, which is the *labium*. The basal segment of this elbowed proboscis is the *rostrum* (which contains a part of the clypeus and the basal maxillary plates), and the distal segment is called the *haustellum*. The end of the labium is specialized into a large sponge-like organ, the *labella* (Fig. 4.15,*A*). The labella are traversed by a series of furrows or channels, narrowly open all along the exposed edge, the *pseudotracheae* (Fig. 4.15*C*). These insects

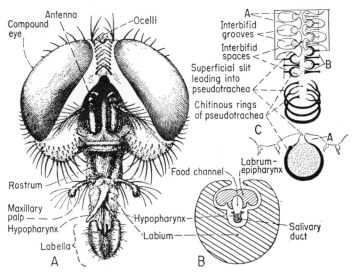

Fig. 4.15. Sponging mouth parts as found in the house fly. *A*, dorsal or front view of head and mouth parts with the proboscis extended and the stylets spread out from the labial gutter. Note the pseudotracheae and sensory hairs on the labella. Much enlarged. (*Original drawing by Antonio M. Paterno.*) *B*, diagrammatic cross-section of the proboscis to show composition of food channel, salivary duct, and labial gutter. *C*, details of a single pseudotrachea, greatly magnified. In the upper part of the figure a surface view of the pseudotrachea shows the superficial slit and interbifid spaces through which liquid foods enter the pseudotrachea, and on the left two interbifid grooves leading to the interbifid spaces. On the right the chitinous rings are shown as seen by transmitted light. At *B* the integument of the oral surface of the labellum has been removed to expose the membrane that lines the interior of a pseudotrachea and stretches between the chitinous rings with their alternate bifid and flattened extremities. *A* is the integument of the oral surface of the labellum; *B* is the membrane lining the interior of the tube. At the center are represented three consecutive chitinous rings to show how their bifid and flattened extremities alternate on each side of the superficial slit. In the lowest part of figure *C* is represented a transverse section of a part of the oral surface of the labellum, cutting across a single pseudotrachea. Greatly magnified. (*B and C, redrawn after Graham-Smith and Hewitt.*)

feed on exposed liquids, such as nectar or sap, or by dissolving solid substances, such as sugar, in their saliva. When the labella are appressed to such liquids, the pseudotracheae fill with the liquid by capillary attraction. These little channels all converge at one point on the labella, and from this point the liquid food is drawn up through the food channel into the esophagus. The food channel is formed by the labrum-epipharynx and the hypopharynx as in other flies (Fig. 4.15,*B*). These are the only two stylets present, and in this type they are entirely incapable of piercing the skin. The mandibles are wanting and the maxillae are represented by only a pair of maxillary palps.

This type is found in the house fly and other nonbloodsucking Muscidae, in the Syrphidae, and in many other of the Diptera.

E. The Siphoning Type (Fig. 4.16). This is a very much specialized type, in which the labrum is greatly reduced, the maxillary palps rudimentary, and the mandibles usually entirely wanting. The labium is represented only by the large, hairy or scaly, three-segmented labial palps and a very small basal plate. The essential working parts are formed by the *maxillae*, parts of which, the *galeae*, are greatly elongated and joined to form a slender hollow tube which is coiled up under the head like a watch spring when not in use. Its structure suggests that of a piece of flexible metal tubing. Innumerable tiny short muscles extend from one ring of the tube to another, inside each half of the proboscis, which Snodgrass believes serve to coil the tube, while it is extended or straightened out to feed, possibly by the pressure of blood forced into it from the body cavity. This proboscis is not capable of piercing the skin of an animal or the epidermis of a leaf or fruit, except in rare instances. Feeding is accomplished by uncoiling this tube and projecting the tip of it into some exposed liquid (commonly the nectar in the nectary of a flower) and then sucking the liquid up through the *food channel* (Fig. 4.16,*B*), which runs full length through the proboscis.

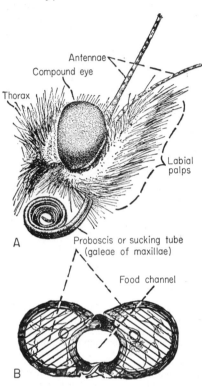

FIG. 4.16. Siphoning type of mouth parts as found in a moth or butterfly. *A*, side view of the head, with the proboscis partly coiled. Note the labial palps, which are so covered with hairs that segmentation cannot be distinguished. *B*, cross-section of the proboscis to show how the right and left galeae lock together to form the food channel. Highly magnified. (*Redrawn after Comstock.*)

This type of mouth parts is found in practically all adult moths and butterflies, the order Lepidoptera.

F. The Chewing-lapping Type (Fig. 4.17). This type, which is so well illustrated by the honeybee or bumblebees, is a kind of combination type in which the labrum and mandibles are of the same structure as in the chewing type, but the maxillae and the labium are elongated, and closely united, to form a sort of lapping tongue (Fig. 4.17*A,D*). Both the maxillae and the labium are suspended from the cranium, and articulated to it, chiefly through the base of the maxillae. Both pairs of palps are present, the *labial palps* are long and conspicuous but the *maxillary palps* (Fig. 4.17,*D*) are very small. The glossae of the labium are greatly elongated to form a hairy, flexible tongue that can be rapidly protracted and retracted to reach deep into the nectaries of tubular flowers. According to Snodgrass, a temporary *food channel* is formed by the concave inner surfaces of the galeae, roofing over the *glossa* and fitting snugly lengthwise against the labial palps, which in turn lie tightly against the sides of the glossa. Through such a complexly formed tube ("held, like a straw in one's mouth, by the mandibles grasping the bases of the galeae while the epipharynx plugs the gap

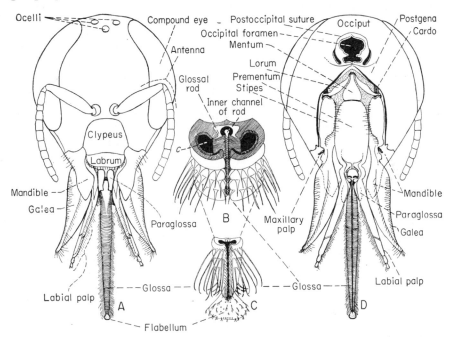

Fig. 4.17. Mouth parts of the chewing-lapping type as found in the honeybee. *A*, cephalic or front view of head and mouth parts. Note that labrum and mandibles are of the chewing type, the latter specially shaped for molding the waxen combs, while the galeae, labial palps, and glossa make a five-part tongue for lapping up nectar. *B*, cross-section and isometric view of a portion of the glossa, greatly magnified, showing the ventral groove which communicates with the canal, *c*, the rings of long hairs, the glossal rod imbedded in its anterior wall, and the hair-guarded inner channel of the latter. The ventral canal may be used to conduct saliva to the tip of the tongue but is not believed to be important in sucking up nectar. *C*, tip of the tongue or glossa, greatly magnified, ventral or posterior view, showing the close-set hairs that guard the ventral canal and the spoon-shaped flabellum at its tip. *D*, caudal view of head and mouth parts, showing the cardo, stipes, maxillary palp, and galea of the maxilla; and the lorum, mentum, prementum, paraglossae, and glossa of the labium. (*Redrawn, in part, from Snodgrass, "Anatomy and Physiology of the Honeybee."*)

where the ends of the galeae diverge toward the head") a drop of honey may be sucked up.

According to George E. King, in securing nectar from the open nectaries of flowers, the bee thrusts out the glossa or tongue and licks the nectar with the tip of it. The glossa, thus smeared with nectar, is rapidly retracted between the labial palps and galeae, and the nectar is squeezed off the tongue by the galeae and deposited so as to accumulate in the small cavity formed by the paraglossae at the base of the glossa. Then by the bending of the labium upward near mid-length, the base of the glossa is brought into close apposition to the mouth cavity, and the accumulated nectar is sucked into the esophagus by the action of a pharyngeal pump. Imms states that the liquid food ascends by capillary action through the ventral canal of the glossa (Fig. 4.17,*B*,*c*), and that the glossa is shortened by a muscular pull on the inner rod, squeezing the nectar upward to enter the space between the glossa and base of the paraglossae (Fig. 4.17,*A*); but Snodgrass finds that the liquid food is sucked up more rapidly than could be accomplished in this way. The nectar thus gathered serves as food for the bees, and the surplus is stored as honey.

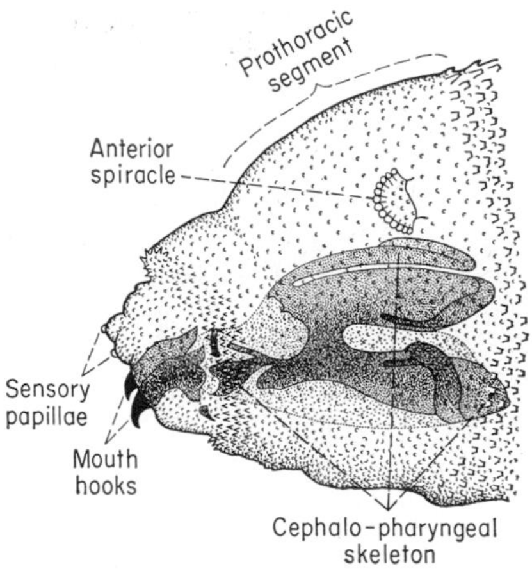

FIG. 4.18. Degenerate head and mouth parts of a dipterous larva showing the mouth hooks, the anterior or prothoracic spiracles, and, inside the spiny body wall, the cephalopharyngeal skeleton or hypopharyngeal sclerite. (*Drawn by Kathryn Sommerman.*)

The inner channel of the rod or the ventral canal of the glossa, or both, may serve as a salivary groove to conduct saliva from the salivary duct, which ends near the base of the glossa, to the tip of the tongue, where it may be used to dissolve solids, such as sugar, preparatory to swallowing. The mandibles are used for carrying things and, in the honeybee, for molding wax into cells.

This type of mouth parts is found in some species, only, of the order Hymenoptera, which includes the bees, wasps, and ants.

G. Degenerate Types. The mouth parts of nymphs are generally very similar to those of the adults. The mouth parts of larvae are fundamentally of the chewing type regardless of the nature of the mouth parts of their parents. These become somewhat reduced in the larvae of certain orders.

I. *Hymenopterous, Trichopterous,* and *Lepidopterous Larvae.* Maxillae, hypopharynx, and labium associated closely or united to form a combined lower lip. Ligula and hypopharynx form a median lobe through which silk for spinning cocoons issues through a *spinneret.* The typical parts are often reduced in size and simplified, the labial palps, especially, often wanting.

II. *Dipterous Larvae.* In the lower (Orthorrhaphous, page 295) Diptera the mouth parts are essentially of the chewing type, but in the higher (Cyclorrhaphous, page 296) Diptera, such as the house fly and flower fly maggots and many others, the true mouth parts are entirely wanting, the true head segments have been invaginated ("sucked down") into the throat, and special mouth parts, which work vertically instead of laterally, special sense organs, mouth lips, and a mouth cavity or atrium have been developed to serve, instead of the true head parts, during larval life. Parts of the true hypopharynx, the clypeus, and associated sclerites of the head form a conspicuous sclerotized organ surrounding the beginning of the alimentary canal which is known as the *cephalopharyngeal skeleton* or *hypopharyngeal sclerite* (Fig. 4.18).

CHAPTER 5

DEVELOPMENT AND METAMORPHOSIS

That "truth is stranger than fiction" is well illustrated by the life cycles of insects. The life cycle or life history of an insect means the record of all that the insect habitually does and all the changes in form and habits that it undergoes from the beginning of its life until its death, including the situations where each life stage and every season is spent and the length of time occupied by each stage. This is an important and fascinating study. No one could possibly predict the life cycle of even the most humdrum bug with half of its manifold interrelations and complications. Each species must be studied *by observation*, to determine what normally happens in every region where it lives, and *by experiment*, to determine the effect upon it of unusual or varying conditions.

Since the death and disintegration of every individual are inevitable, in order to preserve any line of descent or kind of animal or plant from extinction, a certain part of its living stuff must be freed from the individual before death overtakes it; and under such conditions that this bit of living material will not only survive but grow into a whole new individual, capable of repeating the reproductive process when it in turn has reached maturity. Among insects reproduction is accomplished *sexually, i.e.,* by the release of single cells, known as *gametes*. Gametes are of two kinds: *eggs* from females and *sperms* from males. These are the perfect or complete cells from which all other kinds of cells of the entire body may be produced during development.

All insects, like all other animals, begin life from such a single cell known as the egg (Fig. 5.1). They do not appear spontaneously or spring up out of nothing, as people sometimes suppose, but come from eggs previously hidden about us by insects of the same kind. Only occasionally do swarms of insects invade a locality from some distant point.

It is probably safe to say that nine-tenths of the insect troubles of a given farm come from eggs laid on that farm. Each farmer, or certainly each community, raises its own insect pests, with some exceptions. It is part of the function of entomology to teach us to recognize insect outbreaks in their incipiency and the stages of pests which are harmless, as well as those that do damage.

Before development can begin, it is usually necessary that the insect egg be fertilized by union with a sperm from the male insect (page 153). But many cases have been found among insects in which fertilization is not necessary, the female insect producing living, normal young without the necessity of mating. This is known as *parthenogenesis*. In the honeybee the fertilized eggs produce "workers" or "queens" (*i.e.,* females); the unfertilized eggs invariably produce males or "drones." In aphids all summer generations are exclusively females developed from

149

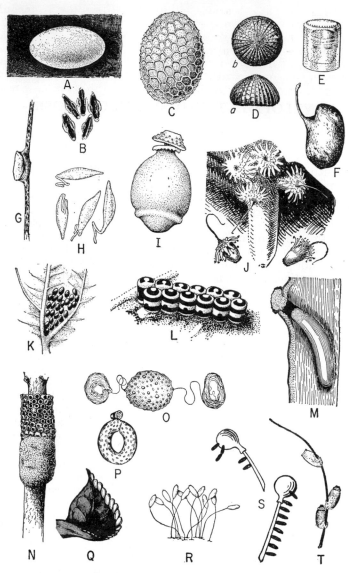

Fig. 5.1. Eggs of various insects to show something of the variety in shape, pattern, sculpturing, and arrangement.   All much enlarged.   *A*, egg of Japanese beetle; *B*, a group of eggs of the malarial mosquito; *C*, egg of honeysuckle miner, *Lithocolletes fragilella* (*from Crosby and Leonard*); *D*, egg of the fall armyworm; *a*, side view, *b*, from above; *E*, egg of southern green plant bug, side view; *F*, egg case of great water scavenger beetle which encloses 50 to 100 eggs (*redrawn after Kellogg*); *G*, egg of a ground beetle, *Chlaenius tricolor*, in its mud cell on the stem of a sedge (*redrawn after King, Ann. Ento. Soc. Amer.*); *H*, eggs of the apple seed chalcid; *I*, egg of a stone-fly, *Perla immarginata* (*from Smith, Ann. Ento. Soc. Amer.*); *J*, eggs of poultry lice (*from Ohio Agr. Exp. Sta.*); *K*, egg mass of the squash bug (*original*); *L*, egg mass of the harlequin bug; *M*, egg of snowy tree cricket (*from Parrott and Fulton*); *N*, eggs of tent caterpillar forming a collar about a twig.   In the upper portion of the mass the eggs have not been covered with the glue-like secretion (*original*); *O*, egg of a May-fly, *Heptagenia interpunctata*, showing skein of thread at each end which anchors

unfertilized eggs, males appearing only in the fall and fertilizing only the overwintering eggs. In a considerable number of species, no males have been found, or in certain generations no males are produced. For example, 98 successive generations of aphids have been produced under observation without a single fertilization. Other things being equal, a parthenogenetic species is likely to be a worse pest than one in which mating is necessary, because the hazard of not finding a mate is removed and successful reproduction is that much more sure. In spite of these exceptions, the normal thing among insects is for fertilization to occur.

Eggs and sperms are living cells highly specialized for the particular function of generating complete new individuals, just as nerve cells are highly specialized for sensation and conduction, or muscle cells for contractility. In order that we may understand the specialization that these particular cells undergo, and their subsequent history, we should examine the essential parts of the insect egg and sperm.

The Egg. The eggs of insects (Fig. 5.1) are not so varied in size, shape, and appearance as the insects that lay them. They are very small; the characters by which they may be distinguished are often most obscure and elusive; and very little time has yet been given by entomologists to the study of insect eggs. Nevertheless, it is often possible to tell from an examination of the egg the exact kind of insect that will develop from it. This may at times be of the greatest importance in forecasting the appearance of the destructive stages of a particular insect pest. For example, grasshopper or fruit tree leaf roller epidemics may be predicted 6 months before any damage will begin, and materials and an organization to combat them may be perfected long before the voracious insects hatch.

In attempting to recognize the kind of insect from its eggs, one should note the size, shape, and color of the egg; the place in which it is found; the way in which it is laid or attached—whether on its end (Fig. 5.1,*S*), its side (*K*), or on an elevating stem (*R*); or, if inserted into the tissues of plants or animals, the kind of scar made by the laying (Fig. 1.7); whether laid singly, in indefinite masses, or in accurately spaced or definitely ranked groups (Fig. 5.1,*LNQ*); and the arrangement of the eggs with respect to each other, whether free, cemented together, or covered over with hairs or a cement-like secretion. Of special usefulness is the sculpturing of the eggshell as seen under the microscope. The exact nature of the surface pattern often serves to distinguish one species from another (Fig. 5.1,*C–E,H–J,O*).

Parts of the Insect Egg. Regardless of shape or size, the following parts of the egg will usually all be represented (Fig. 5.2):

The *chorion* or eggshell is secreted by the cells of the follicular epithelium of the ovary (page 126) and is composed of lipoproteins arranged in a number of layers and is devoid of chitin. In the egg of *Rhodnius*, Beament[1] has recognized the following distinct features, from outside to inside: (*a*) the exochorion, composed of a sculptured surface layer of resistant lipoprotein and an inner layer of soft lipoprotein, and (*b*) the

---

[1] BEAMENT, J., *Bull. Ento. Res.*, **39**:359, 1948; **39**:467, 1949.

---

egg on surface of water by winding about sticks or plants (*from Morgan, Ann. Ento. Soc. Amer.*); *P*, egg mass of a caddice-fly, *Phryganea interrupta* (*from Lloyd*); *Q*, egg mass of angular-winged katydid on edge of leaf (*from Comstock*); *R*, eggs of a lace-winged fly (*original*); *S*, eggs of asparagus beetle; *T*, eggs of little red louse on hair of cow (*A, B, D, E, H, L, S, and T from U.S.D.A.*).

endochorion of five layers of protein impregnated with polyphenols and lipids and tanned as described in cuticle formation (page 81). Inside the endochorion is a waterproofing layer of wax very similar to that of the insect cuticle.

The *micropyle,* one or more small openings through the chorion, usually at one end of the egg, through which the sperm may enter the egg to fertilize it. The micropyles or other very similar openings serve as respiratory channels.

The *vitelline membrane* or cell wall of the egg is a delicate membrane completely lining the shell within and enclosing the following parts:

The *cytoplasm* or general living substance of the egg.

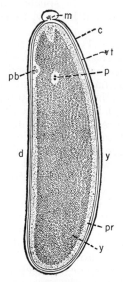

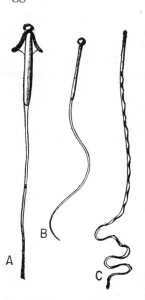

FIG. 5.2. Sagittal section of an egg of the house fly in process of being fertilized. *c,* Chorion; *d,* dorsal side; *m,* micropyle, with exudation; *p,* nuclei from sperm and egg about to unite; *pb,* polar bodies; *pr,* peripheral protoplasm; *v,* ventral side; *vt,* vitelline membrane; *y,* yolk. Greatly magnified. (*From Folsom, "Entomology," after Henking and Blochmann, McGraw-Hill–Blakiston.*)

FIG. 5.3. Sperms of insects. *A,* of grasshopper; *B,* of cockroach; *C,* of a scarabaeid beetle. Greatly magnified. (*From Folsom, "Entomology," after Bütschli and Ballowitz, McGraw-Hill–Blakiston.*)

The *yolk,* deutoplasm or lifeless food material, which consists of carbohydrates, protein, and lipids scattered as globules throughout the reticulum of the cytoplasm.

The *nucleus,* a highly organized dynamic part of the cell, containing the chromatin, which at certain regular times forms the *chromosomes.* These are composed of a large number of giant molecules of deoxyribonucleic acid (DNA), which are the *genes,* the bearers of heredity characteristics. It is believed that the arrangements of the adenine, guanine, thymine, and cytosine fragments of the DNA chain, in which millions of combinations are possible, provide a template for the synthesis of the proteins of the developing organism, thus determining that the insect developing from an egg laid by a bumblebee, for example, shall be a bumblebee and not a grasshopper or a house fly.

The size of insect eggs is largely determined by the yolk content. That of the endoparasitic *Clemelis*, which is swallowed by its host, is practically devoid of yolk and has a volume of only 4,800 cubic microns, while in the related *Hypodermodes*, the egg has a volume of 25,000,000 cubic microns. In many parasitic Hymenoptera whose eggs are deposited internally in the host, the egg often grows rapidly in size after deposition because of absorption of the host fluids, increases of over 1,000 times in volume having been observed for species of *Perilitus* (Brachonidae).

The Sperm. The sperms of insects are in a general way similar to those of other animals. They are elongate, extremely slender cells, with a whip-like, vibratile tail, by which they may swim actively to find an egg. When examined at high magnification, three different parts may be distinguished (Fig. 5.3): (*a*) a slender, rod-like *head*, which contains the nucleus and carries the chromosomes and is the part that bears the hereditary characters of the male parent to the egg; (*b*) a *middle piece*, which is thought to contain an "attraction sphere," of significance in the division and development of the egg after fertilization; and (*c*) the *tail*. Such a curious cell is developed from an ordinary-appearing cell in the testis, during the process known as maturation.

Maturation and Fertilization. The nucleus of the egg and of the sperm is the portion that contains the "germ of life." Every other part of the egg is subservient to the nucleus, serving to protect or nourish this vital part. The maturation of the sperm cells (*spermatogenesis*) occurs through certain complicated changes known as *meiosis*, during which (*a*) the chromatin collects to form the chromosomes, (*b*) these pair, so that only one-half the original number are present, (*c*) the cell divides, so that each daughter cell contains one-half (haploid number) of the original chromosomes. During this process, the sperm cell becomes specialized for locomotion and most of the cytoplasm is discarded.

The entrance of the sperm into the egg (through the micropyle) provides the biochemical stimulus for maturation of the egg (*oögenesis*), which is similar to that of the sperm, except that after the reduction division, one-half the chromosomes are given off in a small polar body while the other half are retained with all the yolk, in the egg. The sperm loses its tail, forming the male pronucleus, and this unites with the nucleus of the egg, in the process of *fertilization*, to form the *fusion nucleus*, in which the original (diploid number) of chromosomes is restored. All cells in the new body of the insect are the direct descendants of the "perfect cell" so formed.

Development. The life cycle begins with fertilization, the fusion of the sperm and egg into a single cell. The life cycle ends with a body composed of millions of cells, highly organized into a complex, living machine. All that takes place between the fertilization of the egg and the perfection of the full-grown insect we call *development* and *growth*. This is sharply divided into two phases by the act of hatching or escape from the eggshell. That part of the development that occurs before hatching or birth is called the *embryology* (embryonic development), and all that takes place after hatching or birth is *postembryonic development*.

Embryology.[1] The embryology of insects is exceedingly intricate and can be presented here only in brief outline. The single cell of the fertilized egg divides mitoti-

[1] JOHANNSEN, O., and F. BUTT, "The Embryology of Insects and Myriapods," McGraw-Hill, 1941; HAGAN, H., "Embryology of the Viviparous Insects," Ronald Press, 1951.

cally, and the succeeding cells divide and further multiply. These *cleavage nuclei* migrate to the periphery of the egg just beneath the vitelline membrane, where they form a continuous layer of cells called the *blastoderm* (Fig. 5.4,*A*). Along the ventral side of the egg some of the blastoderm cells become thicker, forming the *germ band* or ventral plate (Fig. 5.4). Lengthwise of the germ band an invagination or groove forms that carries some of the blastoderm cells to an internal position. These form the mesoderm, while those remaining in a superficial position are known as the *ectoderm*. Folds of the blastoderm now grow over the germ band from each side to form a double protective covering, the embryonic membranes. When these folds meet and fuse, the outer layer of the folds forms the *serosa* and the inner, the *amnion* (Fig. 5.4). Thus

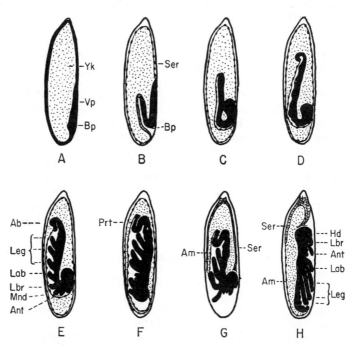

FIG. 5.4. Stages in the embryonic development of the dragonfly, *Agrion*, showing formation of embryonic envelopes and blastokinesis. *A*, formation of ventral plate; *B–D*, invagination of embryo; *E–G*, development of appendages; *G–H*, rupture of amnion and inversion of embryo. *ab*, abdomen; *am*, amnion; *ant*, antenna; *bp*, blastopore; *hd*, head; *lab*, labium; *lbr*, labrum; *mnd*, mandible; *prt*, proctodeum; *ser*, serosa. (*After Brandt.*)

the developing insect becomes separated from the vitelline membrane by two additional membranes.

The growth and movements of the embryo during its development are termed *blastokinesis*. Several generalized patterns have been discerned: (*a*) the embryo develops from the germ band without immersion in the yolk and remains on the ventral surface of the egg during its entire development, the embryonic membranes forming as lateral folds of the germ band, which meet and fuse to form the amnion and serosa. This development is characteristic of the Siphonaptera and nemocerous Diptera. (*b*) The caudal end of the germ band invaginates into the yolk, and the membranes develop across the cavity which is formed. As the embryo grows, it revolves about its longitudinal axis as in some Orthoptera, Lepidoptera, and Diptera, or (*c*) about an axis perpendicular to its sagittal plane as in Orthoptera, Dermaptera, Odonata, Ephemerida, Homoptera, Hemiptera, Anoplura, and Mallophaga. This is illustrated for the dragonfly *Agrion*, in Fig. 5.4. In this insect the single layer of cells of the blastoderm becomes thickened, on the ventral side, into a two-layered ventral plate (Fig. 5.4,*B*). At the *blastopore* the ventral plate sinks into the yolk and migrates anteriorly and dorsally, drawing with it a portion of the blastoderm, which becomes the amnion, while the remainder thins out to form the serosa (Fig. 5,4*C*). As the embryo grows, segmentation develops and the appendages appear (Fig. 5.4,*E*). Subsequently, the embryo withdraws entirely into the yolk and the amnion forms a sac about its head, which fuses in part with the serosa (Fig. 5.4,*F*). The serosa now contracts and thickens, drawing the yolk into the anterior of the egg, and the increased pressure ruptures the amnion at the point of fusion. The head of the embryo protrudes from the opening, and the embryo moves along the ventral side (Fig. 5.4,*G*), so that in its final development it lies again along the ventral surface of the egg (Fig. 5.4,*H*).

Within the sheltered environment of the egg yolk, from which the substance and energy for development are obtained, the embryonic development takes place rapidly and synchronously. Transverse furrows of the germ band divide it into a series of segments. Invaginations or ingrowths from the ectoderm form the fore- and hind-intestine, while internally the mesoderm develops the tube known as the mid-intestine to connect the other two (Fig. 3.7). Outgrowths of the ectoderm also form the paired appendages, such as antennae, legs, and the labrum, mandibles, maxillae, and labium of the mouth parts (Fig. 5.5). Invaginations produce the central nervous system,

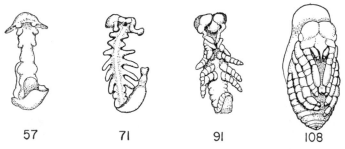

| 57 | 71 | 91 | 108 |

Fig. 5.5. Stages in the embryonic development of the milkweed bug, *Oncopeltus fasciatus;* age in hours. (*After Butt, Cornell Univ. Agr. Exp. Sta. Memoir* 283, 1949.)

tracheal tubes, and salivary glands. The germ band grows up around the sides, gradually eliminating the blastoderm until the edges of the germ band meet and unite, thus completely enclosing the yolk and giving cylindrical form to the body. From the mesoderm develop the muscles, the heart, and the fat-body and the pericardial and blood cells. As the cells increase in number, all take their places in an orderly and definite manner to form the various tissues and organs of the future insect, until, when it hatches, we have within the eggshell (Fig. 5.6,*A,F,K*) an organism capable of discharging all the necessary functions of life.

Something of the chronology of the development of the milkweed bug, *Oncopeltus fasciatus*, at 21°C., is shown in Fig. 5.5 and may be summarized as follows:

First day. Blastoderm forms.

Second day. Germ band and serosa form, amnion appears.

Third day. Stomadeum invaginates, segmentation of thorax is evident; labral, antennal, mandibular lobes form; embryo reaches greatest length, neuroblasts appear.

Fourth day. Proctodeum forms; tracheae, Malpighian tubes, fat-body, muscle tissues differentiate; segmentation of thoracic appendages occurs.

Fifth day. Amnion and serosa fuse and rupture.

Sixth day. Eyes differentiate, and pigment appears.

Eighth day. Hatching occurs.

Physiology of Embryonic Development. The development of the egg is initiated at the *cleavage center*, located where the head of the future embryo will develop. Here the cleavage nuclei form and are induced to begin their peripheral migration. The

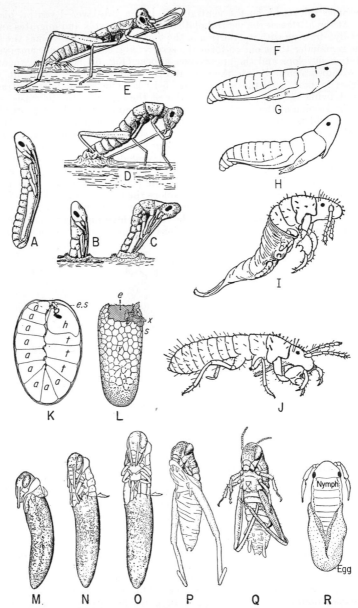

FIG. 5.6. The hatching of insects. *A–E* are of a tree cricket: *A* shows the position of the embryo in the egg; *B–E*, successive stages in the hatching of the nymph from an egg sunken in the wood (*from Parrott and Fulton*). *F–J* are the periodical cicada: *F*, the egg with the eye of the embryo showing through the chorion; *G*, the newly hatched nymph; *H*, the same in motion; *I*, the same, shedding embryonic membrane; *J*, the same, free from embryonic membrane (*from Snodgrass*). *K* shows the embryo peach borer larva in its U-shaped position in the egg before hatching, and *L* the empty eggshell from which the larva has hatched (*from Peterson*). *M–Q* show successive stages in the hatching of a grasshopper (*from S.D. Agr. Exp. Sta.*). *R* is of an apple aphid showing the nymph partly hatched from the egg (*From Peterson*).

cleavage nuclei are isopotent; *i.e.*, they can develop into any portion of the future embryo as determined by their subsequent organization. As these migrating cells reach the posterior pole of the egg, the *activation center* there produces a wave of chemical differentiation which spreads anteriorly toward the thorax of the future embryo and induces the *differentiation center* to begin the accumulation of cells necessary for the formation of the germ band. At about this point determination becomes complete, and the egg is now a mosaic in which the future functions of all the cells are predictable from their geographic location, and destruction of any portion will produce a defective embryo. Determination of the eggs of *Musca* and *Drosophila* is complete at the time of laying, while that of *Bruchus* requires about 6 hours and of *Apis* 24 hours after laying.

The rate of embryonic development is controlled by many factors, of which temperature is most critical. Within the tolerated limits, development increases with increased temperature and will not take place below a lower critical temperature. For example, the eggs of *Anopheles quadrimaculatus* will not hatch below 10°C. or above 35°C., and embryonic development required an average of 492 hours at 10°, 357 hours at 12°, 109 hours at 17°, 54 hours at 23°, 38 hours at 28°, 33 hours at 33°, and 24 hours at 35°. Moisture is also an important factor, and the development of some insect eggs may be delayed under low humidity. Desiccated insect eggs may remain dormant for long periods until moisture is supplied. Those of the floodwater mosquito *Aedes vexans*, which hatch within 1 day when submerged, have remained viable after 2 to 5 years of desiccation, and eggs of the grasshopper, *Locustana pardalina*, which hatch in about 14 days in moist soil, have survived 3.5 years of desiccation.

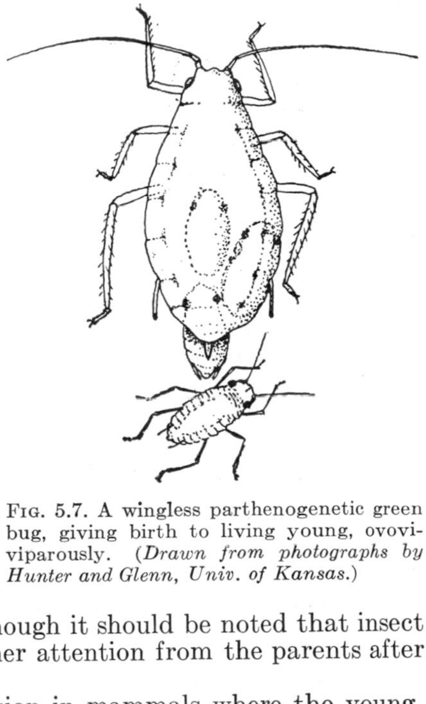

Types of Reproduction. The development within the egg requires food. This may be derived from food material, *yolk*, stored inside the shell by the parent insect before the egg is laid. The parent then nourishes it no further. This method of reproduction is known as *oviparous*, *i.e.*, bringing forth eggs. It is par-

FIG. 5.7. A wingless parthenogenetic green bug, giving birth to living young, ovoviviparously. (*Drawn from photographs by Hunter and Glenn, Univ. of Kansas.*)

alleled by the condition in birds, although it should be noted that insect eggs do not require incubation or other attention from the parents after they are laid.

In contrast with this is the condition in mammals where the young, during embryonic development, establish a definite connection (the *placenta*) with the blood system of the mother and receive their nourishment, moment by moment, from the circulatory system of the parent. Food as well as oxygen for an embryonic mammal passes from mother to young by diffusion or osmosis through the so-called fetal membranes. This condition is known as *viviparous* reproduction, *i.e.*, bringing forth active young.

A somewhat intermediate condition is found among insects, some of which do not lay the eggs but retain them until after they have hatched and then bring forth active young (Fig. 5.7). This is not all equivalent to viviparous reproduction, however, because the young insect receives its nourishment from the yolk of the egg and not from the parent's circulation. No organic connection is established. It is simply premature hatching or delayed oviposition. It is distinguished as *oviviparous reproduction*. The two kinds of reproduction common among insects are oviparous and ovoviviparous. A condition somewhat analogous to that found in man and other mammals (viviparous reproduction) is known only in the case of a few flies (the sheep tick, page 976, Fig. 20.22, and the tsetse flies) in which the young are actually nourished by special nutritive glands in the uterus of the parent fly.

Number of Eggs and Methods of Deposition. Insect eggs (Fig. 5.1) are generally small and consequently are seldom noticed except when laid in masses or groups that are conspicuous. Some insect eggs are so small that several dozen could be placed side by side on the head of a common pin; the largest eggs of our common insects are not over ⅛ inch in diameter.

The number of eggs laid by one insect is as varied as their shape and size. A single female may lay as few as 1 egg (in exceptional cases, like the true females of certain aphids), or at the other extreme 1,000,000 or more. The honeybee queen lays 2,000 to 3,000 eggs a day, actually producing several times her own weight of eggs each day for weeks at a time. The termite queen lays as many as 60 eggs a minute until millions have been produced. The average number for all insects is probably over 100.

The eggs may all be laid at one time, as in the tussock moth; they may be laid a few a day for many days, as in the bloodsucking lice; or there may be a number of successive "batches" of eggs produced at intervals, as in the case of the common house fly, which lays from two to seven lots of eggs at intervals of 2 to 5 days, each lot consisting of about 125 eggs.

Insect eggs are generally laid in such a situation that the young, upon hatching, may find suitable food with the minimum of effort or discriminative action. More often than not, they are simply extruded in a suitable place, which is generally selected with remarkable care. This done, the mother pays no further attention to them, usually dying shortly afterward. There is ordinarily very little of parental care or family life among insects. The eggs hatch (Fig. 5.6) without attention or control by their parents, and from the moment of hatching the young insect must ordinarily lead an independent self-supporting existence.

There are some very interesting adaptations of egg laying to the subsequent life of the young insect. The chestnut weevil has a beak longer than her own body, with which she reaches through the chestnut bur to chew a hole into the developing nut. In this hole is laid the egg from which the chestnut "worm" develops. The plum curculio and the strawberry weevil make remarkable provision to assure the success of the eggs and young (Fig. 1.7,*E*,*F*; see the discussion of egg laying under these insects). The horse bot fly, whose young must reach the stomach of the horse in order to develop, lays its eggs, not in the mouth of the horse, but on the hairs of the legs of the horse. Their success in reaching their feeding grounds is thought to depend upon the activities of another insect, the stable fly. The bites of the stable fly cause the horse to nibble at its legs. The eggs of the horse bot hatch almost instantly when stroked by the moist lips or tongue, to which the small larvae cling, and are subsequently swallowed.

One of the most remarkable cases of provision for the young is exhibited by the twig girdler. The twig girdler lays her eggs one in a place in holes which she chews into the soft bark of the terminal twigs of such trees as hickory, oak, pecan, persimmon and many others. Before the eggs are laid, however, she completely girdles the twig, usually to a depth of ⅛ inch or more, by chewing out the wood, bit by bit, in a band around the twig (Fig. 1.7,*D*).   This girdling requires many hours of work on the part of the female (commonly 40 or 50 hours), and, when it is completed, the insect lays in the partially severed twig perhaps 12 to 20 eggs.   Other twigs are then attacked in a similar manner, and the female busies herself in this way all during the long autumn months.   The twigs subsequently break off in the wind and the larvae develop in the dead, decaying wood at the surface of the soil.   This represents probably the extreme of parental care on the part of an insect, in the mere act of egg laying.   A closely related species in the subtropics often cuts off branches 1 or 1½ inches in diameter, several females working together to sever so large a branch.

Many other insects, especially the Hymenoptera, build elaborate nests and provision them with paralyzed insects or honey to serve as food for the young (Figs. 2.4 and 6.85).   The European earwig actually broods over her eggs and young after the manner of birds (Fig. 6.17).   But the great majority of insects lay their eggs and die without ever seeing them again.   There is in almost every case, a new "crop" of insects each year.   It is the exceptional species, like the white grubs, wireworms, periodical cicada, ants, and honeybee, in which the same individuals live longer than one year; even in these cases the adult is generally short-lived, and only the young or the queens persist into the following seasons.

The Rapidity of Insect Increase.   Insects multiply very rapidly. This great increase may be due to either one or both of the following factors: (*a*) a great number of eggs or young in a family or generation and (*b*) a short life cycle and the rapidity with which generations succeed each other.   Compared with the half dozen or dozen children that characterize the families of man and our domestic animals and fowls, the hundreds of the average insect family are impressive.   Again, the life cycle or generation in the larger animals is from a few months in the case of smaller rodents to as much as 30 years in the case of man, while the shortest known life cycle among insects is about 10 days!

Either one of these factors operating independently may result in a tremendous population of insects from one or a few individuals that may begin the season.   For example, the corn root aphid has a family averaging from 12 to 16 young.   But each of these begins reproducing at the age of 8 days, and a generation may be completed in 16 days.   In this way it would be possible, theoretically, for a single female to produce in 1 year, if all her descendants survived, a chain of these aphids long enough to encircle the earth.   The San Jose scale has fewer generations—from two to four in the northern states—yet, because of the large number of young produced by one female (400 to 500), a single pair might be the progenitors, if all their descendants survived, of more than 1,000,000 in a single season.   In the case of the house fly both of the above factors operate to produce the alarming increase in numbers of flies as summer comes on.   According to Hodge: "A pair of flies beginning operations in April, might be progenitors, if all were to live, of 191,010,000,000,-000,000,000 flies by August.   Allowing ⅛ cubic inch to a fly, this number would cover the earth 47 feet deep."   Needless to say, owing to the factors of natural control, no such rate of increase ever actually occurs.

The first life stage of all insects is the egg (sometimes concealed within the mother's body).   The time spent within the egg may be as short as 8 hours, as in the case of the house fly; is commonly a week or two; and very often insects go through the winter in this stage, all other stages then usually dying off before winter is passed.

Insects without a Metamorphosis.   The egg may lie dormant for a long period of time, but during at least a part of the egg stage there is

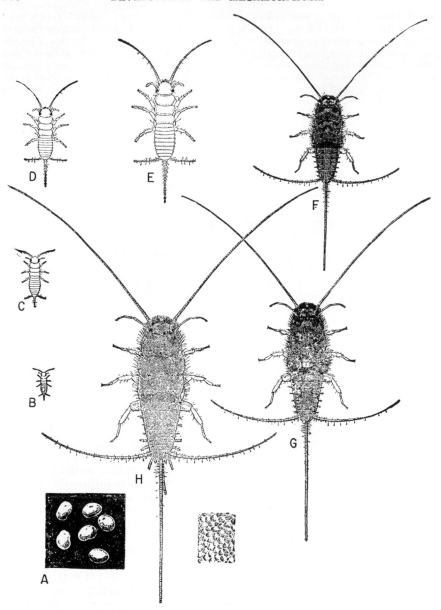

Fig. 5.8. Development of the firebrat, *Thermobia domestica*, illustrating development without metamorphosis.   There are more instars than illustrated, the intervening ones being of intermediate sizes but otherwise very similar to those shown.   The covering of scales first appears at the third or fourth instars.   *A*, a group of eggs; *B–G*, nymphs or young of various sizes; *H*, adult female, as recognized by the styli and ovipositor at tip of abdomen; *I*, sculpturing of the eggshell, as faintly visible under the microscope.   (*Original drawings by R. Slabaugh and K. Sommerman.*)

great activity within the eggshell, as a result of which a perfectly formed young insect is finally disclosed by hatching[1] (Fig. 5.6).   When a chicken or a duckling hatches from its egg, it resembles in most respects (except size) the full-grown chicken or duck.   A few insects, when they hatch, also are so much like their parents that anyone would know they belonged to the same kind of insect (Fig. 5.8).   This is especially true of a small number of wingless insects belonging to the orders Thysanura and Collembola, the fishmoths and springtails.   Such insects are said to undergo *no metamorphosis*, and the two orders just mentioned are collectively called the *Ametabola* (which means *without change*).   Their growth from smallest to largest is hardly accompanied by greater changes in appearance than those that take place from infancy to manhood.

Insects with a Gradual or Simple Metamorphosis.   If the full-grown insect has wings, the young never resemble it completely at hatching, for (unlike birds) *no insect has visible wings when it emerges from the eggshell* (Fig. 5.6).   All winged insects, therefore, undergo a metamorphosis during their development.   *Metamorphosis may be defined as a conspicuous change in the form and appearance of an animal between birth (or hatching) and maturity.*   Frogs or toads, with their curious young tadpole stages, have a metamorphosis, while birds, rabbits, men, and the like, have no metamorphosis during their postembryonic development.

In many insect species the young are very similar to the adult except for the complete absence of wings and genitalia; in other cases there are striking differences in the color or shape, or in the structure of some of the appendages.   In either case, after a period of growth the wings may appear attached to the outside of the body as small *wing pads* which become larger and larger.   The more developed the young insect becomes, the more it resembles its parents (Fig. 5.10).   Such a development is called a *gradual* or *simple metamorphosis*.   The young of such insects are called *nymphs* (Fig. 5.9).   They commonly have the same habits as their parents, and the old ones and young ones may frequently be seen feeding together, not unlike a hen and her chicks.   Grasshopper nymphs and adults both eat grasses and clovers and may be found hopping about together in the pastures.   Squash bug nymphs and adults both suck the sap of the squash plant.   Bedbug nymphs and adults all suck human blood.   This group as a whole is known as the *Heterometabola* (meaning *different change*).   It includes many important insects of the orders Orthoptera, Isoptera, Mallophaga, Thysanoptera, Homoptera, Hemiptera, Anoplura, and others (Table 6.3).

Insects with a Complete or Complex Metamorphosis.   Finally, we have a large group of insects, most of which have different habits when they are young than when they are full grown.   For example, the young may swim in the water and the adults live in the air, like the mosquitoes.   Or the young may tunnel through the soil and eat grass roots like the white grubs, while their parents the May beetles, fly about and feed on the leaves of trees.   Or the young may live in the stomach of a horse like the bots and the parents fly freely in the air.   Obviously these insects could not exist in such different environments unless they were very different in structure when young and when full grown.   Indeed, they are generally so very different in appearance in their several life stages

---

[1] The term *hatching* or *eclosion* properly refers only to breaking out of the eggshell or embryonic membranes.   The escape of a winged insect from its cocoon or pupal case should be called *emergence*, not hatching.

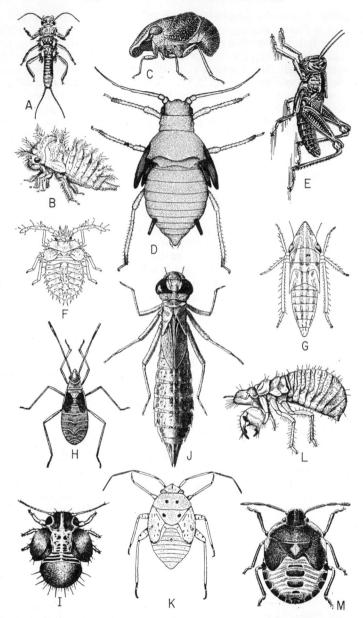

FIG. 5.9. *Nymphs:* examples of insects that develop the wings externally during growing period and transform to the adult usually without a pupal stage. *A*, nymph of a stone-fly, order of Plecoptera (*from Kellogg's "American Insects"*); *B*, nymph of a treehopper, *Ceresa basalis*, order Homoptera (*from N.Y. State Coll. Forestry*); *C*, nymph of a fulgorid, *Bruchomorpha oculata*, order Homoptera (*from N.Y. State Coll. Forestry*); *D*, nymph of an aphid, *Aphis cucumeris*, order Homoptera (*from U.S.D.A.*); *E*, nymph of a grasshopper, order Orthoptera (*from U.S.D.A.*); *F*, nymph of a lace bug, *Corythuca pergandei*, order Hemiptera (*from Ohio Biol. Surv. Bull.* 8); *G*, nymph of a leafhopper, *Draeculacephala mollipes*, order Homoptera (*from U.S.D.A.*); *H*, nymph of cotton stainer, *Dysdercus suturellus*, order Hemiptera (*from Insect Life*); *I*, nymph of the pear psylla, *Psylla pyricola*,

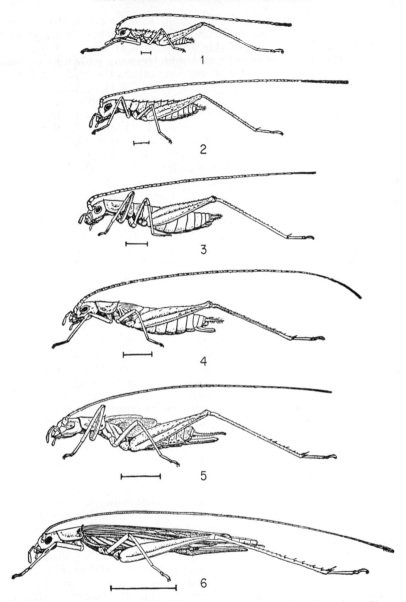

FIG. 5.10. Instars or stages of growth of the snowy tree cricket, *Oecanthus niveus*, illustrating a gradual metamorphosis. *1–5*, The first to fifth nymphal instars, respectively; *6,* the adult. Note the appearance of the wing pads at *4* and their expansion to full size at *6;* also the very gradual assumption of the adult condition. (*From N.Y. (Geneva) Agr. Exp. Sta. Bul.* 388.)

order Homoptera (*from Slingerland*); *J*, nymph of a dragonfly, *Anax junius*, order Odonata (*from Ill. Natural History Surv.*); *K*, nymph of tarnished plant bug, *Lygus oblineatus*, order Hemiptera (*from Ill. Natural History Surv.*); *L*, nymph of periodical cicada, order Homoptera (*from U.S.D.A.*); *M*, nymph of the green plant bug, *Nezara hilaris*, order Hemiptera (*from U.S.D.A.*).

that no one, without previous information, would ever suspect them of being even closely related animals, much less the successive stages of *the same individual*. The young of this group show no trace of wings externally during any period of growth, although the wing buds may be found, by dissection, inside the body wall (Figs. 3.10,*f.b,h.b*, and 5.11,*II,III*). Furthermore, the oldest larva shows no greater resemblance to the adult than the smallest one, except in size (Fig. 5.14). Such young insects are known as *larvae* (singular *larva*), in contrast with nymphs (Fig. 5.15).[1]

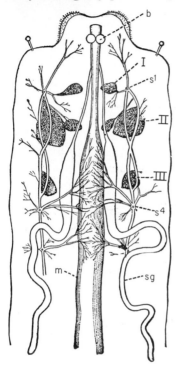

When a larva is full grown, a striking change takes place, and the insect, after shedding its skin, appears with the wings exserted as large wing pads and with the usually long legs and feelers of the adult now recognizable. But legs, wings, and antennae always remain functionless for a definite period of time while the internal organs are being transformed to the adult condition. This transformation stage is known as the *pupa* (plural, *pupae*) (Fig. 5.16). At its completion, the pupal skin is shed and the adult formed by rapid expansion of the wings to full size, the general hardening of the body wall, the development of the color pattern, and numerous other changes. Such development is called a *complete* or *complex metamorphosis*. The largest orders of insects have such a complete metamorphosis, *e.g.*, the Coleoptera, Lepidoptera, Hymenoptera, and Diptera, besides some smaller orders like the Siphonaptera, Neuroptera, and others (Table 6.3). All these insects that have a complete metamorphosis are referred to collectively as *Holometabola* (meaning *complete change*).

Fig. 5.11. Dissection of a full-grown caterpillar, *Pieris* sp., from above, to show the wing buds which are developing inside the body wall. *b*, Brain; *m*, alimentary canal; *sl*, prothoracic spiracle; *sg*, silk gland; *I*, bud of the prothoracic segment; *II*, bud of the front wing; *III*, bud of the hind wing. (*From Folsom, "Entomology,*' McGraw-Hill–Blakiston, after Gonin.*)

Growth. For a further understanding of the growth and metamorphosis of insects, it is essential to have clearly in mind what is meant by life stages and by instars. The *life stages* are those several periods of an insect's life which are radically different from each other in appearance and usually also in behavior or activity. Thus the insects with a complete metamorphosis have four life stages, the *egg*, the *larva*, the *pupa*, and the *adult*. Among insects all increase in size takes place in the life

[1] The authors realize that the term *larva* is often used for the first stage of any insect or other animal having a metamorphosis and wish that a special term could have been employed for the young of insects with a complete metamorphosis. However, the term *larva* is so firmly established as a general term for caterpillars, grubs, maggots, and the like that there would be no hope of any special term for this purpose becoming adopted. *Larva* as used in this book, therefore, means the young of insects that have a complete or complex metamorphosis.

stage that immediately follows hatching, *i.e.*, either as a nymph (if the insect has a gradual metamorphosis) or as a larva (if the insect has a complete metamorphosis). No growth occurs in the adult stage after the insect once acquires functional wings, and none in the pupal stage. Little flies do not ever grow into large flies, or little moths into larger ones. With the appearance of the full-spread wings the size of the insect is fixed for the rest of its life, except as the body expands to accommodate a large meal or developing eggs.

The enormous increase in size of some insects from birth to maturity is astonishing, especially as compared with man, where the increase is approximately 20 to 25 times. The following ratios for increases in the weight of the newly hatched larva to the fully grown stage will indicate the range which exists: *Toxoptera graminum*, 16; *Dytiscus marginalis*, 52; *Tribolium confusum*, 85; *Schistocerca gregaria*, 126; cabbage looper, *Trichoplusia ni*, 1,640; *Ephestia kühniella*, 1,320; *Apis mellifera*, 1,576; *Bombyx mori*, 8,417; salt marsh caterpillar, *Estigmene acrea*, 11,000; and *Cossus cossus*, 72,000.

Since growth occurs exclusively during the nymphal or larval period, it follows that there must be various sizes of nymphs or larvae (Figs. 5.10 and 5.14) in the case of every species and in the development of every individual. An insect does not grow by regular, gradual, imperceptible degrees like a child, just as its sclerotized exoskeleton will not expand like the mammalian skin to permit this. Therefore growth inside this inexpansible shell cannot be regular and continuous. In order to make any considerable increase in size, the shell must be split off. This process is known as *molting*, and the old cuticle so cast off is known as the exuviae (meaning clothes). Before the old skin or cuticle is split off, new epicuticle and exocuticle are secreted inside it by hypodermal cells. Then the molting fluid is poured out by cells in the hypodermis or by special exuvial glands (page 80). This loosens the old cuticle from the new by dissolving the inner part of the old endocuticle. The molting fluid also softens the remainder of the old skin and separates it from the new one by a thin film of fluid. The old skin is then split open by pressure from within, and the insect squirms and crawls out of it, pulling its appendages free (Fig. 15.45). At this point there is a considerable expansion in the size of the insect, before its new exocuticle becomes sclerotized or "set" to the definite size of the next instar. The molting process is completed by the formation of a new endocuticle beneath the hardened exocuticle. Subsequently there is a relatively long period during which the insect is feeding and accumulating reserve materials within its body, but without any noticeable increase in size. This is followed by another molt and period of constancy in size, and so on.

The molts occurring during the growing period divide this life stage (nymph or larva) into a number of sharply separated sizes or steps that are called *instars* (Fig. 5.14). Upon hatching from the egg, the insect is said to be in the first instar. This instar is terminated by the first molt, which ushers in the second instar, distinguished from the first at least by its larger size and often also by differences in structures or coloration. The second molt introduces the third instar, and so on, until commonly 3, 4, 5, 6, and sometimes as many as 20 molts have occurred. However, the number of molts is not always constant and may be hereditary, as in the different races of silkworms (page 47), or influenced by environmental temperature, as in the Khapra beetle, *Trogoderma granarium*, which requires an average of 8 molts at 70°F. and 4 at 90°F. Inadequate nutrition often affects molting, and the clothes moth, *Tineola bisselliella*, may require from 4 to 40 molts under various conditions, while species of *Trogoderma*, under starvation conditions, molt repeatedly for 4 to 5 years, growing smaller with each instar molt until the larva is smaller than after hatching from the egg. The ametabolous Collembola and Thysanura continue to molt throughout life, as many as 60 instars having been recorded for *Thermobia* (Fig. 5.8), but in nearly all other insects molting ceases with adulthood. When growth (increase in size) is completed, if the insect is a nymph, a final molt discloses the adult; if it is a larva, the corresponding molt gives rise to the pupal stage; and when the transformations of this stage are perfected, a final molting of the pupal epidermis discloses the adult.

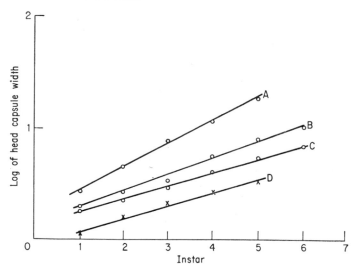

Fig. 5.12. Graph illustrating Dyar's law relating width of head capsule to instar. *A, Trichoplusia ni*, ratio 1.60; *B, Ephestia kühniella*, ratio 1.43; *C, Tribolium confusum*, ratio 1.31; and *D, Oncopeltus fasciatus*, ratio 1.30. (*From Brindley, Beck, and McEwen, Ann. Ento. Soc. Amer.*, 1930, 1958, 1960.)

The increases in size observed after each molt, *i.e.*, with each instar, are surprisingly regular. Dyar[1] showed that for many lepidopterous larvae, the increases in the widths of the head capsule occurred in a regular geometric progression with a ratio varying from 1.1 to 1.9 and averaging about 1.4 (Fig. 5.12). This phenomenon, known as Dyar's law, has been shown to apply to a large variety of cuticular structures and is of value in determining the total number of instars in a life cycle. Later observations suggest that for some insects, at least, the relationship is parabolic rather than linear, but the deviation from a straight line is almost imperceptible and in no way alters the usefulness of the relationship.

Although the growth of the cuticle is discontinuous, growth in weight is essentially a continuous process, as shown by the growth curves in Fig. 5.13, which demonstrate wide variations even in insects with

[1] DYAR, H., *Psyche*, **5**:420, 1890.

short growth periods. Przibram[1] has proposed that the weight of a typical insect is doubled at each instar and the linear dimensions increased by $\sqrt[3]{2}$ (or 1.26) at each molt.   This rule is more applicable to hemimetabola than to holometabola, but is of doubtful value for many insects

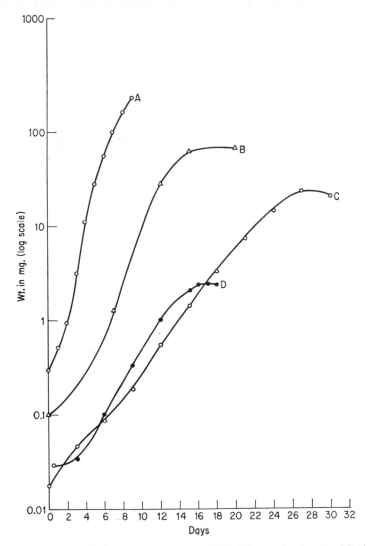

FIG. 5.13. Growth curves showing increase in weight of deve'oping insect with time. *A, Trichoplusia ni; B, Oncopeltus fasciatus; C, Ephestia kühniella; and D, Tribolium confusum. (From Brindley, Beck, and McEwen, Ann. En'o. Soc. Amer., 1930, 19*≤8, 1960.)*

where the weight increases severalfold with each molt (Fig. 5.13).   The growth curves of some insects appear discontinuous, as is the case with *Rhodnius,* where the insect ingests from 6 to 10 times its body weight of

[1] See BODENHEIMER, F., *Quart. Rev. Biol.,* **8**:92, 1933.

blood in feeding after each molt, and in *Notonecta*, where water is absorbed by the newly molted insect, so that its weight is rapidly doubled.

Growth in most insects is a complex phenomenon in which the various appendages have individual growth rates, which differ among themselves and from the body as a whole. Such growth is termed heterogonic or allometric and implies that at any period of development, $x = ky^a$, or that the log of the dimension of the part is proportional to the log of the dimension of the whole.

Diapause.[1] This term refers to a period of arrested development in the ontogeny of the insect, *i.e.*, as egg, larva, pupa, or adult, during which growth, differentiation, and metamorphosis cease. This dormancy, which may normally last several months to a year or more, enables the insect to survive most easily a period of unfavorable conditions. It is important to distinguish diapause, which when initiated proceeds to

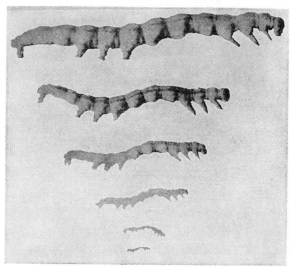

Fig. 5.14. Instars or stages of growth of the green clover worm, *Plathypena scabra*, illustrating a complete metamorphosis. Note the entire absence of wing pads even in the largest instar, that the largest larva is no more like the adult moth in form than the smallest, and that a pupal stage intervenes between the last larva and the adult. (*From U.S.D.A., Farmers' Bul.* 982.)

its predetermined course, from the quiescence resulting from exposure to low temperatures, which is abruptly terminated by a rise in temperature. It is of interest that the remarkable cold-hardiness of some insects such as *Camponotus pennsylvanicus* apparently results from its storage of glycerol during cold periods. The body content of glycerol may reach 10 per cent of body weight, but this disappears with exposure to warm temperatures. Diapause can be induced or terminated by a number of external stimuli such as changes in temperature, photoperiod, or nutrition, by desiccation, or by wounding the cuticle. In *Bombyx mori* the eggs laid in the fall will not develop until exposed to temperatures from about 5 to 7.5°C. for 60 days. The ability of the adult silkworm to produce such diapausing eggs is determined by the genetic constitution of the female, and strains or races range from univoltines, which lay diapausing eggs at all temperatures, through typical bivoltines, which produce nondiapausing eggs at low temperatures and diapausing eggs at high temperatures, to multivoltines, which never produce diapausing eggs.

[1] Lees, A., "The Physiology of Diapause in Arthropods," *Cambridge Monograph in Exp. Biol.*, 4, 1955.

In the oriental fruit moth, *Grapholitha molesta*, diapause of the larva is controlled by temperature and daily exposure to light (photoperiod). Larvae reared in darkness do not enter diapause, and as the photoperiod increases above 3 hours per day, the percentage of diapausing larvae increases, reaching 100 per cent at about 12 hours and abruptly decreasing to zero at 13 hours. Diapause is, however, produced by this photoperiodic mechanism only at moderate temperatures and does not occur below about 12°C. or above 30°C. It is of interest that in *Lucilia sericata*, diapause is controlled only by low temperatures and is not affected by photoperiod. Knowledge of the photoperiodic response, which appears to be rather general in diapausing insects, has proved to be very useful in developing procedures for the continuous rearing of insects such as the codling moth, *Carpocapsa pomonella*.

From a biochemical viewpoint, diapause has been shown to be produced in the pupa of *Hyalophora cecropia* by the failure of the brain to initiate the production of the prothoracic gland growth hormone, ecdyson (page 117). Williams, after intensive study, has concluded that this hormone normally controls the rate of synthesis of the respiratory pigment cytochrome *c*, and that when diapause is broken by chilling the insect, this stimulates the brain to produce an unidentified hormone, which activates the prothoracic glands.

Nymphs vs. Larvae. The best criterion to divide the winged insects into two groups, in respect to metamorphosis, is (*a*) whether the wing pads are borne externally during the growing stage or concealed beneath the body wall. If the wing pads are developed on the outside of the body wall (Figs. 5.9 and 5.10), we call the growing stage a *nymph*, and we say that insect has a *simple* or *gradual metamorphosis*. If the wing pads are developed internally during the growing stage (Figs. 3.10,*f.b*,*h.b*, and 5.11,*II*,*III*), we call that stage a *larva* (Figs. 5.14 and 5.15) and say that insect has a *complete* or *complex metamorphosis*. A nymph, then, is the growing stage of such insects as have a gradual or simple metamorphosis, and it develops its wings (if it has any) on the outside of the body wall as visible pads. Other general differences between nymphs and larvae are that (*b*) the nymph generally has a shape and body construction similar to that of the adult; (*c*) each successive instar usually looks more like the adult than the one that preceded it; (*d*) nymphs have very few organs that are not also possessed by the adult; (*e*) a nymph has compound eyes unless its parents are without compound eyes; (*f*) it always has the same type of mouth parts as the adult; (*g*) it generally occupies the same kind of habitat, takes the same kind of food, and leads the same manner of life as the adult; and, finally, (*h*) the nymphal period generally passes over into the adult period without any prolonged inactive or pupal stage intervening.

In contrast with the nymph, the larva, or growing stage of insects with complete or complex metamorphosis, (*a*) develops its rudimentary wings during this stage inside the body wall of the thorax; (*b*) it generally has a more or less worm-like form of body, often strikingly different from that of the adult; (*c*) the later instars are no more like the adult, as a rule, than the earlier ones; (*d*) the larva often has provisional structures or organs, of use only in this stage, which are lost or supplanted before the adult stage is reached; (*e*) the larva never has functional compound eyes, though it may have simple eyes or ocelli; (*f*) it may occupy the same habitat as the adult but very often lives in a totally different sort of situation; (*g*) the larva commonly has a different type of mouth parts from the adult and often takes a wholly different kind of food; and, finally, (*h*) the larva is always separated from the adult by a pupal stage during which the insect takes no food and is usually quiescent.

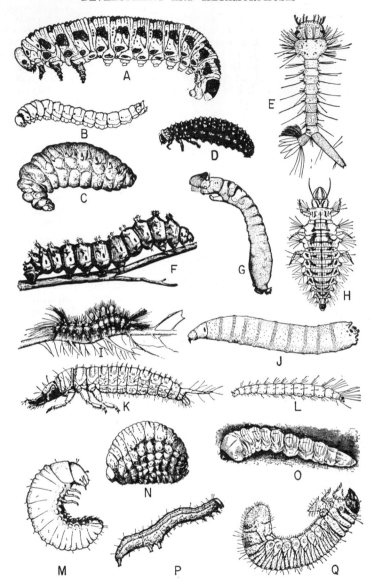

Fig. 5.15. *Larvae:* insects that develop wings internally during the growing period, that are very different in appearance from the adults, and that have a pupal stage intervening between larva and adult. *A,* larva of sawfly, *Neodiprion lecontei,* order Hymenoptera (*from Middleton, Jour. Agr. Res.*); *B,* larva of European wheat stem sawfly, *Cephus pygmaeus,* order Hymenoptera (*from Ries, Jour. Agr. Res.*); *C,* larva of black digger wasp, *Tiphia* sp., order Hymenoptera (*from Davis, Bul. Ill. Natural History Surv.*); *D,* larva of beet leaf beetle, *Erynephala puncticollis,* order Coleoptera (*from U.S.D.A.*); *E,* larva of mosquito, *Culex territans,* order Diptera (*from Bul. Ill. Lab. Natural History*); *F,* larva of cecropia moth, *Hyalophora cecropia,* order Lepidoptera (*from Saunders*); *G,* larva of black fly, *Simulium venustum,* order Diptera (*from H. Garman*); *H,* larva of lace-winged fly, *Chrysopa quadripunctata,* order Neuroptera (*from R. C. Smith*); *I,* larva of western tussock moth, *Hemerocampa vetusta,* order Lepidoptera (*from Volck, Calif. Agr. Exp. Sta.*); *J,* larva of

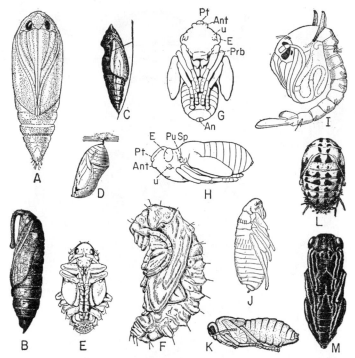

FIG. 5.16. Pupae of various insects: *A–D* of Lepidoptera; *E, F,* and *L* of Coleoptera; *G–I* of Diptera; *J* of Siphonaptera; *K* and *M* of Hymenoptera. *A,* pupa of pink bollworm (*from Heinrich, Jour. Agr. Res.*); *B,* pupa of tobacco hornworm (*from U.S.D.A.*); *C,* naked pupa or chrysalis of alfalfa caterpillar: note how it is suspended by a thread or girdle of silk (*from U.S.D.A.*); *D,* pupa of monarch butterfly as it hangs suspended by posterior end (*from French*); *E,* pupa of beet leaf beetle (*from U.S.D.A.*); *F,* pupa of cherry leaf beetle (*from U.S.D.A.*); *G,* pupa of apple maggot; *H,* the same in side view (*from Snodgrass, Jour. Agr. Res.*); *I,* pupa of house mosquito, a pupa that swims in water (*from U.S.D.A.*); *J,* pupa of dog flea (*from U.S.D.A.*); *K,* pupa of pear-slug (*from Iowa Agr. Exp. Sta.*); *L,* pupa of convergent lady beetle; it is fastened by silk to the leaf and can rise up on the rear end when disturbed (*from U.S.D.A.*); *M,* pupa of hymenopterous parasite, *Pardianlomella ihseni* (*from U.S.D.A.*).

Sometimes it is difficult to decide whether the metamorphosis of a given insect should be called complete or gradual. We must bear in mind that nature does not make sharp division lines; that there is likely to be every conceivable gradation from one condition to another which is remarkably different. It is so with the metamorphosis of insects. The groups described above, however, are very important for convenience of

---

apple maggot, *Rhagoletis pomonella*, order Diptera (*from Pa. State Dept. Agr.*); *K,* larva of ground beetle, *Harpalus pennsylvanicus*, order Coleoptera (*from Davis, Ill. Natural History Surv.*); *L,* larva of flea, order Siphonaptera (*from Bishopp, U.S.D.A.*); *M,* larva of *Colaspis flavida*, order Coleoptera (*from Ill. Natural History Surv.*); *N,* larva of granary weevil, *Sitophilus granarius*, order Coleoptera (*from U.S.D.A.*); *O,* larva of broad-necked root borer, *Prionus laticollis*, order Coleoptera (*from N.J. Agr. Exp. Sta.*); *P,* larva of alfalfa looper, *Autographa gamma californica*, order Lepidoptera; note reduced number of prolegs (*from Hyslop, U.S.D.A.*); *Q,* larva of scarab beetle, *Adoretus caliginosus*, order Coleoptera (*from T. B. Fletcher*).

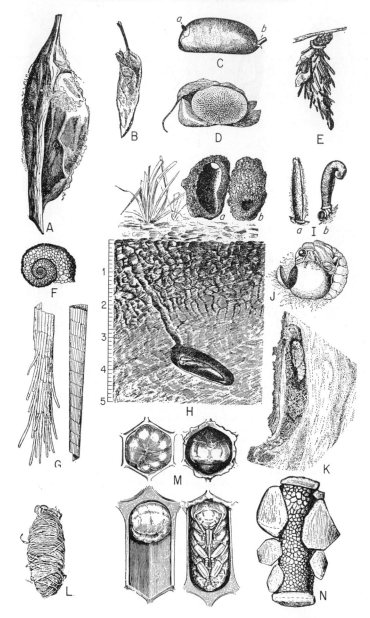

FIG. 5.17. Some methods of protection for insect pupae. *A*, cocoon of the cecropia moth (*from Saunders, "Insects Injurious to Fruits,"*); *B*, folded leaf in which the apple leaf sewer, *Ancylis nebeculana*, feeds as a larva and which also protects the pupal stage; empty pupal shell from which adult emerged projects at upper left (*from U.S.D.A.*); *C*, puparium of *Tropidia quadrata;* a protective case formed from the larval skin (*from Metcalf, Maine Agr. Exp. Sta. Bul. 253*); *D*, cocoon of clover leaf weevil, partly surrounded by clover leaves (*from Tower and Fenton, U.S.D.A.*); *E*, the bagworm, a case of silk covered with spruce needles carried about by larva during its life and later closed to protect pupa (*redrawn after Riley*); *F*, larval case of caddice-fly, *Helicopsyche borealis*, formed in the

study and reference, and most insects can easily be fitted into one or another of them.

The Meaning of a Complete Metamorphosis. The explanation of a complete metamorphosis is probably that growth is all confined to one life stage (the larva) and reproduction to another (the adult). For the functions of eating, growing, and storing up energy a simple cylindrical body with few appendages is well adapted. Since the parent insect generally places the young in the midst of an abundance of food, highly specialized sense organs are not usually required. But to secure a mate, to locate a suitable place to deposit the eggs, and to care for the necessary dispersal of the species, a highly sensitive, active, complex body is required. Hence we have on the one hand the sluggish, stupid, gluttonous caterpillar of simple structure (Fig. 5.15) and, on the other, the alert, highly specialized fly, bee, moth, or beetle. So very different have these stages become in many cases that the change from one form to the other is profound. In some insects nearly all the larval tissues disintegrate, and the corresponding adult tissues and organs are built up anew from small groups of cells (histoblasts) that have remained dormant and rudimentary during larval life but are now able to multiply rapidly by utilizing the nutritive products resulting from the histolysis of the larval cells and from the fat-body. So profound is the change from larva to adult in many cases that the organism can accomplish no other functions while it is going on. Locomotion ceases, feeding is suspended, respiration is reduced, and the insect undergoes a transformation period, externally quiescent but internally probably as active as any period subsequent to embryonic development. All available energy is devoted to the development of the wings, legs, eyes, antennae, mouth parts, and other appendages of the adult and to the maturing of the reproductive system and changes in other internal organs. The insect during this period is known as a *pupa* (Fig. 5.16).

Because it neither feeds nor moves about, the pupal stage is neither injurious nor beneficial to man.

It appears that larvae represent an earlier ontogenetic period of development than do nymphs. The period in gradual metamorphosis corresponding to the larvae of complete metamorphosis is passed in the embryo before the nymph hatches. The pupal period corresponds more nearly with the nymphal stages, and pupae may, in a way, be thought of as nymphs which have lost their activity. Nymphs accom-

shape of a snail shell and covered with grains of sand (*from Lloyd, "North American Caddice Fly Larvae"*); G, larval and pupal cases of a caddice-fly, *Phryganea vestila*, made of slender sticks and bits of leaves arranged in spiral form (*from Lloyd*); H, pupa of tobacco hornworm in its earthen cell formed by the caterpillar. The depth is indicated by the inch marks at the left. Above are shown two earthen cells removed from the ground; note hole by which the larva entered (*from Morgan, U.S.D.A.*); I, casebearers; at left the cigar casebearer, at right the pistol casebearer with head and thorax of larva projecting below (*from N.Y. (Geneva) Agr. Exp. Sta.*); J, cocoon of the California green lace-wing fly; the pupa has just emerged through the circular hole in the cocoon (*from Wildermuth, Jour. Agr. Res.*); K, pupa of the roundheaded apple borer; note the sawdust and shavings with which the larva had closed its tunnel behind the pupa, and also plugged the exit hole for the adult, near the head of the pupa (*drawn from photograph by Slingerland*); L, cocoon of Hawaiian sugarcane borer (*from Van Dine, U.S.D.A.*); M, larvae and pupae of honeybee in the hexagonal cells made of wax by worker bees; at left, end and side views of larvae; at right, end and ventral view of pupae (*from White, U.S.D.A.*); N, pupal case of a caddice-fly, *Neophylax concinnus*, made of minute grains of sand cemented together with a few heavier pebbles for ballast (*from Lloyd*).

plish both growth and transformation to the adult. In complete metamorphosis growth has all been relegated to the larval stage and transformation has been greatly telescoped or abbreviated into a single period, the pupa.

Methods of Protection for the Pupa. The life of the insect during the helplessness of the pupal stage is generally safeguarded by the larva. Sometimes the pupa is found naked and exposed, as with many butterflies, lady beetles, and the like. These (Fig. 5.16,*A,C,D,L*) usually have the tips of their bodies fastened to a leaf but are not covered in any way. Commonly the larva retreats into a protected situation under overhanging bark; into logs, stones, grass or leaf mold; or into the soil before entering upon this defenseless period of its life (Fig. 5.17,*H,K*). Often a case is formed about the larval body before pupation. The case may consist of a folded leaf (Fig. 5.17,*B*), fine pebbles (*F,N*), fine shavings of wood, bits of soil, hairs from the body of the larva, or other materials that surround the larva as pupation approaches (*E,G,K,L*). These are generally cemented or tied together by a silken secretion from the mouth of the larva; in many moths and Hymenoptera the silk is abundant and forms a complete, sometimes dense, case about the pupa, known as a *cocoon* (Fig. 5.17,*A,D,J*).

In many of the flies, instead of spinning a silken cocoon or constructing a case of extraneous material, the larva practices an interesting economy by retaining about itself one of its own cast, dry skins, which is slightly modified by inflation and hardening, to form a waterproof and airtight case known as a *puparium* (Fig. 5.17,*C*). This next-to-the-last larval skin is not discarded at the time of pupation as in most insects but is retained until the adult breaks out of the pupal skin.

It is important to note that the actual change to the pupa does not coincide with the completion of the active larval life or the formation of the pupal case. The plum curculio buries itself in the soil for about 4 weeks; but 2 weeks of the time it remains as a larva before it pupates. The Hessian fly larva may remain many months after the puparium is formed before the pupa is formed. Many moths and sawflies that winter in a cocoon do so in the larval stage, not pupating until spring. The pupal stage may be said to begin when the larval skin has been molted off, but a great portion of the transformation toward the adult has already taken place before this molt occurs, notably the eversion of the wing pads from their position inside the body wall to a position outside the pupal integument but still enclosed by the unshed larval skin. In the Diptera the formation of the motionless puparium may be called *pupariating*, to distinguish it from the shedding of the final larval skin, or true pupating. This period of variable duration between the retreat to the pupal position or formation of the pupal case and the actual change to the pupa is known as the *prepupal period.*

The order to which a given pupa belongs can usually be told by characteristics which are given in connection with the discussion of the orders in Chapter 6, and the actual species of insect may generally be recognized, by an expert, from the pupa.

CHAPTER 6

THE CLASSIFICATION OF INSECTS[1]

No class of objects on earth presents so great and varied an assortment as the living things that inhabit its surface. The biggest job of classification ever undertaken is the systematic study of plants and animals, including insects. From the time of Aristotle (384–322 B.C.) and Pliny (A.D. 23–79) through all the generations since, thousands of men the world over have given their lifetime to the great problems of systematic biology. However, in spite of these immense labors, the task is still very far from completion.

The consensus of opinion of students today is that life probably originated only once upon the earth and therefore that all living things had a common origin and have grown to their present complexity of forms by an orderly and extremely slow process of ascent and differentiation or *evolution* in which the principal potent factors have probably been variation, natural selection, and heredity. If this is the case, it follows that all living organisms must present but one true arrangement with respect to their blood relationships to each other. To discover and record this arrangement and reveal the pathways along which each organism has developed is to make a *natural classification*—the ideal of systematic botanists, zoologists, and entomologists.

The living or organic world is clearly made up of two great categories of organisms, the *animal kingdom* and the *plant kingdom*. If plants and animals all arose from the same original spark of life, the divergence of these two main branches must have occurred very early because they have so many points of difference. Yet in such fundamental matters as respiration, cell structure, mitosis, and reproduction plants and animals are so peculiarly alike that no one who understands these processes can doubt that they are the result of a common ancestry. It is generally easy to distinguish the common plants, which are sessile, require inorganic food, possess chlorophyll, and have cellulose cell walls, from the common animals, which are free-moving, require organic food, do not possess chlorophyll, and have protein cell walls, though it is less easy to point to distinct differences by which we can separate them all. Indeed, there are some minute, intermediate forms of life that partake to some extent of the characteristics of both groups, and this fact adds support to the great truth of evolution and the theory of the common origin of all life.

[1] For a more detailed discussion of the classification of insects see BRUES, C. T., A. L. MELANDER, and F. M. CARPENTER, "Classification of Insects," *Bul. Harvard Museum Comp. Zool.*, vol. 108, 917 pp., 1954; COMSTOCK, J. H., "Introduction to Entomology," Comstock, 1924; ESSIG, E. O., "College Entomology," Macmillan, 1942; BORRER, D. J., and D. M. DELONG, "An Introduction to the Study of Insects," Rinehart, 1954; IMMS, A. D., O. W. RICHARDS, and R. G. DAVIES, "A Textbook of Entomology," 9th ed., Methuen, 1957.

PHYLA, CLASSES, ORDERS, FAMILIES, AND GENERA

Taking the animal kingdom as a whole, nearly a million species or kinds have been discovered and named. Nearly all the known animals fall naturally into about a dozen important branches, which are called by the Latin name that means branch, *i.e.*, *phylum*. The important phyla of animals and the numbers of described species are listed in Table 6.1.

TABLE 6.1. THE ANIMAL KINGDOM

Phylum	Class	Examples	Estimated Number of Living Species Described
		VERTEBRATES	
Chordata			
	Mammalia	Man, cat, horse, bat, whale.............	3,200
	Aves	Birds, fowls..........................	8,600
	Reptilia	Turtles, snakes, lizards, alligators........	5,500
	Amphibia	Frogs, toads, salamanders..............	2,000
	Pisces	Fishes...............................	25,000
		INVERTEBRATES	
	Minor classes	Tunicates, Balanoglossus, etc............	2,500
		Total Chordata........................	44,800
Arthropoda		**(See Table 6.2 for Classes and Examples)**.......	**750,950**
Mollusca		Snails, slugs, clams, oysters...................	80,000
Echinodermata		Starfish, sand dollar, sea urchin...............	5,500
Annelida (Annulata)		Earthworm, leeches.......................	8,000
Bryozoa (Polyzoa)		Moss animals, sea mats......................	3,100
Brachiopoda		Lamp shells................................	500
Nemertinea		Nemertines.................................	600
Nemathelminthes		Roundworms, *Trichina*, filaria................	10,000
Platyhelminthes		Flatworms, flukes, tapeworms................	7,000
Trochelminthes		Rotifers, wheel animalcules...................	1,750
Ctenophora		Sea walnuts, comb jellies.....................	100
Coelenterata		Jellyfishes, coral animals, *Hydra*..............	10,000
Porifera		Sponges...................................	4,500
Protozoa		*Amoeba, Paramecium, Euglena,* malarial organisms, trypanosomes.............	30,000
Minor phyla............			200
		Grand total...........................	957,000[1]

[1] For comparative purposes, the total number of described species of plants, including bacteria, is about 335,000. See *Science*, **94**: 234, 1941

From this it is apparent that the phylum Arthropoda, which contains the insects and allied groups, comprises more than 75 per cent of all the species of animals. Each of the animal phyla includes large numbers of species having common characteristics, *e.g.*, the backbone of the vertebrates. However, it is obvious that certain organisms in any phylum resemble each other more closely than they do the others. To express these differences between animals in the same phylum, secondary groupings known as *classes* are used. The important classes of the phylum Arthropoda are given in Table 6.2 and discussed on page 180.

The members of the class Hexapoda or Insecta are all clearly alike in possessing six legs, but there are still wide differences among them in the type of metamorphosis, mouth parts, number of wings, etc.   These and many other characteristics are used as the basis for third-rate categories known as *orders*.   A synopsis of the 26 orders of insects is given in Table 6.3 and discussed on page 184.   The orders with their vast numbers of species are divided into many smaller groups, the *families*.   Two other categories of successively smaller scope than the family must also be understood, *viz.*, the *genus* and *species*.   Most families of animals have a number of genera.   The *family name* is always formed by adding the suffix *-idae* to the stem of the name of the typical genus, *e.g.*, Musca, Muscidae.

THE SPECIES

The innumerable insects and other living things about us naturally group themselves into kinds or species, as everyone knows who has said "Here is a new kind of flower," or "What kind of weed," or "What kind of bird is this?"   Not only do the progeny of a single pair associate together closely, but many other individuals of the same kind that look exactly like them mingle with them, behaving in the same manner, eating the same kind of food, building nests that are very much alike, mating with each other, and continually bringing forth new individuals like themselves.

These natural kinds which normally associate in nature are known as *species*.   We should recognize that a species is a *real group* just as truly as the individual animal or plant is real.   To define species, however, is difficult.   One important criterion is that the members of a species generally interbreed, the matings producing fertile offspring, while members of different species seldom interbreed and, if they do, the offspring are usually not fertile.   Differences in characteristics between one species and another are greater than those between children of the same parents.   These specific characteristics are constant from one generation to another and throughout the natural range occupied by that species.

The species is the natural reproductive unit among animals.   Species do not originate in the cabinets of the museum worker or under the microscope, though the *names* of species may be originated there.   Species are formed by the hand of nature, and many of them were present on the earth in remote times before systematic zoologists or botanists or entomologists came into being, just as truly as they are today, when systematists have discovered, named, and labeled over three-fourths of a million of them.

Because we often apply inadequate standards in recognizing species, and because resemblances which seem important to one naturalist may seem trivial to another, there are great differences of opinion about how many species exist.   Too often insects have been classified according to their structure alone, and indeed very often according to the structure of only one part or one set of organs.   If we are really to know the groups as nature has developed them, we should study not only the structure of every part, but also their habits, mating, development, distribution, fossil history, physiology, and perhaps even their psychology.

Because of the great number of insects, the species is a fundamental conception without which little progress in economic entomology would be possible.   If we had to cope with each *individual* insect, as a separate problem, first determining its status as friend or enemy, and then working out a scheme of control for it, as determined by its structure and habits, our task would be utterly hopeless. But all the individuals of a species *look alike, act alike, eat the same kind of food in the same manner, and are controllable in exactly the same way.* This is because, for the most part, insects lack initiative or intelligence. That is, they do not have the ability to profit by experience, but depend upon instinct,[1] and do everything in the same way their ancestors have done for thousands of generations. All chinch bugs feed on grass plants, fly to the same type of winter quarters, and crawl on foot from small grains to corn in early summer, and, if a barrier or trap is provided, they fall into it just as their ancestors have done ever since man has been trapping them. In the same way all boll weevils, all codling moths, all Colorado potato beetles present one problem, and not a million different ones, although each species may be represented by millions of *individuals.*

In human relationships we deal with individuals, each of whom has individuality of behavior and therefore must have a distinctive name; but among insects these separate instinctive individuals do not require separate names. In dealing with insects, *the unit is the species* instead of the individual, and consequently every *species* of insect about which we wish to write or speak must have a name. This may be a common name or a scientific name. In the past, common names have been applied by anyone who chose to apply them and without rules or regulations. Accordingly, the same kind of insect may have many different common names (*synonyms*) in different communities, and especially in different countries. Thus the corn earworm is known in different sections of this country as the tobacco budworm, cotton bollworm, tomato fruitworm, and vetchworm. The squash bug is also commonly called pumpkin bug and stink bug. Furthermore, the same common name is often applied to several very different kinds of insects (*homonym*). For example, the *locust* in Biblical times, and at present in Europe, means a grasshopper, while to many of us it means a kind of cicada. Recently certain rules have been formulated for the application of common names of insects, and a list of about 1,400 names, approved by the Entomological Society of America, has been published.[2]

Dr. F. E. Lutz, in his delightful "Fieldbook of Insects," has termed the common names of insects "nicknames." He says:

"There are thousands of kinds of native-born, United States insects which have been really named but not nicknamed. . . . Often real names are no longer or harder than the "common" names. An insect is considered to be christened when some student, who has found a kind which he thinks has never been named, publishes a

---

[1] Instincts, which are so characteristic of insect activities, are described as coordinated reflex behavior. They may be distinguished from intelligence and habits by the following characteristics: (*a*) They are performed in the same essential way by all the individuals of a species; (*b*) they are done in a nearly perfect manner without learning, *i.e.*, essentially as well the first time they are tried as the second or the hundredth; (*c*) they are done without consciousness or awareness of what is being done or knowledge of its purpose; (*d*) the separate acts of a complicated performance must take place in a definite, fixed sequence, each step, except the first, requiring the stimulus of the preceding step to set it going and, once started, tending to go ahead to completion, even though something may have happened in the meantime to make the rest of the performance useless and even ridiculous; (*e*) they are always acts necessary to the life or welfare of the individuals and are done with remarkable perfection so long as everything goes successfully, but the procedure cannot be changed, even slightly, to meet emergencies and save the act from failure or disaster.

[2] *Bul. Ento. Soc. Amer.,* **6,** no. 4, 1960.

description of it, and gives it a properly formed name.   If somebody had previously named the same kind, *the prior name usually holds*."

It is evident that no sound progress could be made by using names as various and ambiguous as the nicknames of insects, because we could never be sure just what kind of insect was being considered by a speaker or writer.   There is at least as much reason for insisting that each important species of insect pass under a single name as for objecting to the use of aliases and pseudonyms by our fellow men.   This points to the necessity for *scientific names*.   Scientific names are simply names applied under a set of rules and regulations to which the majority of systematic workers have agreed.   Scientific names are written in Latin form since Latin is the most nearly universal language.   Also, being a dead language, its form is fixed, and the scientific names are therefore the same in all countries and all languages.   If these names seem strange, unnecessarily long, and difficult, we should remember that one cannot have 686,000 different names (page 186) using only four letters and one syllable! We should be willing to concede something to the necessity for exactness and uniformity and to the importance of making these names available to all nationalities.   In the latter part of this book the scientific name of each economic species discussed is given in a footnote.

By general agreement the system called *binomial nomenclature* has become universal.   By this system every species is given a scientific name consisting of two parts: the first part is the *genus* name, which is common to from one to many similar species in much the same way that our family names or surnames, such as Johnson or Andrews, are common to all members of the same social family; and the second part is the *species* name, which must never be given to more than one kind of insect in the same genus,[1] in the same way that one family would not have more than one child with the same Christian name or given name, such as William or Richard.   By using a Christian name with a surname, we designate each of our friends; by using the correct species name with the correct genus name, each insect kind may be clearly designated.   A third part of the scientific name is usually added for convenience in finding the original description of the insect, *i.e.*, the name of the person, called the *author*, who first applied that particular name to the insect along with a published description of it.[2]

There are a few simple rules about scientific names that all should observe.   The genus and species parts should always be printed in *italics* or, when written, should be underscored once.   The author's name is not printed in italic letters or underlined.   The genus part of the name comes first and is always written with an initial capital letter; the species part of the name is given second and should always begin with a small letter.   The three parts of the name are written without any punctuation.   Thus *Anasa tristis* (DeGeer) is the scientific name of the squash bug.   This name tells us at once that the squash bug belongs in the genus *Anasa* and the species *tristis* and, further, that it was first named and described by DeGeer, who was a Swedish naturalist of the eighteenth century.

[1] Two names applied to the same species of insect are called *synonyms*, and only the first used is valid.   The same name applied to two different species is called a *homonym*, and its second application becomes invalid.

[2] The use of parentheses around the name of the author indicates that the species has been placed in another genus than the one in which it was originally described.

The Phylum Arthropoda

In Chapter 3 the important structural and functional characteristics of insects were described. No single one of these characteristics will define an insect and distinguish all insects from all other kinds of animals. For example, *the segmented body, bilateral symmetry, paired jointed appendages* usually terminating in claws, *chitinous exoskeleton, ventral nervous system, and dorsal heart* are the characteristics of the entire phylum Arthropoda, which includes besides the true insects many other creatures such as crayfish, crabs, lobsters, sowbugs, centipedes, millipedes, spiders, mites, ticks, scorpions, harvestmen, and many others. The *phylum Arthropoda* is the largest phylum in the animal kingdom and, aside from the vertebrates, the phylum of most importance to man. This phylum embraces five important and well-known classes, of which insects (the class Hexapoda) is one (Table 6.2). More than 75 per cent of all the animal kinds hitherto found and named belong in the phylum Arthropoda, and about 90 per cent of these are true insects. The class Hexapoda is further analyzed in this chapter, and, except for a number of references to mites and ticks, the remainder of this book is devoted to a discussion of important insects.

TABLE 6.2. THE PHYLUM ARTHROPODA

Classes	Examples	Estimated Number of Living Species Described
Hexapoda (Insecta)	**All true insects**.....................	**686,000**
Chilopoda	Centipedes or hundred-legged-worms...	1,200
Diplopoda	Millipedes or thousand-legged-worms...	1,300
Arachnida		
Orders of the Class Arachnida		
Scorpionida	Scorpions.........................	600
Phalangida	Harvestmen or daddy-long-legs........	1,900
Araneida	Spiders...........................	22,000
Acarina	Mites and ticks....................	9,200
Minor orders	Pedipalpa, Pseudoscorpionida, Solpugida, Palpigrada...................	2,500
	Total Arachnida....................	36,200
Crustacea	Crayfish, lobster, crab, sowbug, barnacles, water fleas, cyclops............	25,000
Minor classes	Pauropoda, Symphyla, Pycnogonida, Xiphosura, Linguatilida............	1,250
	Total..............................	750,950

The possession of three pairs of legs, three body regions, and wings are characteristic things that mark off the insects from the other arthropods. Each of the four other classes is distinguished by certain structural features not possessed (at least in their entirety) by the rest of the arthropods. In common usage, many of the representatives of these other classes are considered to be "bugs," and the species of economic importance are dealt with chiefly by entomologists. Hence a brief discussion is given here of the classes Chilopoda, Diplopoda, Arachnida, and Crustacea (Table 6.2).

Class Chilopoda: The Centipedes or Hundred-legged-worms (Fig. 6.1). The closest relatives of the true insects are the centipedes. Like the insects, they have a single pair of antennae, they breathe by tracheae, and the reproductive organs open at the posterior end of the body. They differ from insects in having neither thorax nor wings and in the large number of legs, typically one pair to each body segment. Compound eyes are rarely developed. They are worm-like in form but differ from the true worms in having a distinct head and definite jointed legs. They are usually somewhat flattened. There are a pair of poison claws or legs, on the first segment behind the head, that are used to paralyze insects and other prey that they devour. The centipedes may as a group probably be considered beneficial, although some of the species, especially the larger ones, which in the tropics may reach a length of 18 inches, sometimes inflict very painful bites upon man.

Fig. 6.1. A giant centipede from the southeastern United States, feeding on a white grub. Note the size as compared with chair. (*Photograph from life by A. R. Cahn.*)

Class Diplopoda: The Millipedes or Thousand-legged-worms. Millipedes are superficially much like centipedes but differ in the following important respects. The legs are still more numerous than in the centipedes, each apparent body segment having two pairs of legs. The body is typically round in cross-section, not flattened; there are no poison legs; the antennae are short; and the reproductive organs open far forward close to the head. The millipedes generally feed on decaying vegetable matter but some species attack growing crops in damp soil, eating either the roots or the leaves that lie close to the ground. They are sometimes mistaken for wireworms and may be serious pests in fields and greenhouses (page 884). Many of the species have an offensive odor.

Class Arachnida: The Spiders, Ticks, and Their Relatives. Next to the insects, the largest class of Arthropoda is the class Arachnida,[1] to which the spiders, scorpions, mites, ticks, and harvestmen belong. The Arachnida resemble insects in their small size, in their predominantly terrestrial habits, and in the possession of tracheae and Malpighian tubes. They differ radically, however, in having four pairs of legs; in having no antennae, true jaws, or compound eyes; in having only two body regions, the

[1] Care should be taken not to confuse the name of this class, *Arachnida*, with the name of the phylum, *Arthropoda*, since the names somewhat suggest each other to the beginner.

head and thorax being grown together into one region; in the curious "book-lungs" used for respiration; in not having a conspicuous metamorphosis; and in the position of the openings from the reproductive organs which are near the front of the abdomen. There are a number of orders in this class of which the following need mention here.

Class Arachnida: Order Araneida.   This large order includes all the spiders—a group of animals rivaling the snakes in their ability to frighten people.   Comstock, the author of "The Spider Book," has well said: "Few groups of animals are more feared, and few deserve it less."   All spiders have a pair of venomous jaws and live on insects which they poison with their bites.   They can bite; and occasionally such bites may become infected and cause serious results.   But probably in all the world there are not more than a few species, if any, that are capable of killing man by their bites.   The large "tarantula" (Fig. 6.2), which comes into our midst in bunches of bananas, is capable of killing birds and small mammals by its bite.   It apparently cannot kill a man, and besides it seems hard to persuade to bite a person, as has been shown by the experiments of Baerg.[1]   Among our native spiders, the one having the worst reputation is the "hourglass spider" or "black widow," *Latrodectus mactans* (page 1008).

FIG.   6.2.   Tarantula.   About ½ natural size.   (*From Herrick, "Insects Injurious to the Household," copyright,* 1914, *Macmillan. Reprinted by permission.*)

One of the most characteristic things about spiders is their habit of spinning silk.   This is used in a variety of ways. (*a*) Chiefly it serves as a snare to capture food.   It is truly a wonderful thing for a dumb animal to manufacture and set a trap.   We know of none of the higher animals except man that do this, although it is done by some of the insects.   (*b*) It forms tubes or tents for protection.   (*c*) It forms sacs for protection of the eggs and newly hatched young.   (*d*) It is used for locomotion.   Spiders descend from higher to lower levels by spinning out a thread as they let themselves slowly down. Some spiders climb to a high point, and resting on their front legs begin to spin silk, supporting it by the hind legs until the loose end is caught by the breeze. More and more is thrown out until finally this simple kite exerts pull enough to carry the spider away.   This can usually be observed in the open country on any bright autumn day, and the threads of "ballooning" silk, revealed by the descending sun, often seem to carpet the grass.

Class Arachnida: Order Acarina (Fig. 6.3).   Mites and ticks can usually be told at a glance from spiders or insects, because the body is all one region, there being little indication of either body regions or segments.   They are like the spiders in the matter of appendages, and some of them spin silk.   A curious feature is that the newly hatched young have only three pairs of legs.   They breathe either by tracheae or directly through the skin.   The chief difference between mites and ticks is a difference in size, *i.e.*, the larger members of this order (Fig. 6.3,*A*) are called *ticks*, while the smaller ones (*B*) are called *mites*.

The economic importance of the mites and ticks is at least fourfold: (*a*) Some of them injure plants, *e.g.*, the red spider mites (page 616) and the gall mites.   (*b*) A number of species are found on or in the bodies of insects.   Some of them are said simply to be riding upon the insects to a new feeding ground, but at least some species are parasitic upon the insects.   Thus the Isle of Wight disease among honeybees is caused by a kind of mite that lives in the tracheae of the bee.   (*c*) Many species are

[1] *Ann. Ento. Soc. Amer.,* **18:**471, 1925.

parasitic upon other animals, including man.   Here the most notorious examples are the cattle tick (page 960), the Rocky Mountain wood tick (page 1017), the poultry mite (page 991), itch mite (page 1025), scab mite (page 979), and scaly-leg mite (page 993).   (*d*) Many of the parasitic species are to be feared because they are the known and only carriers of some animal diseases.   Thus Texas fever is transmitted chiefly by the bite of the cattle tick, Rocky Mountain fever of man by the wood tick, and fowl spirochetosis by the fowl tick (Table 1.2).   The important species of Acarina are further discussed in the following chapters, in connection with the crops and animals they injure.

Class Arachnida: Order Scorpionida.   The scorpions are common in the south-western part of the United States and other subtropical and tropical regions.   They are well known at least by name to nearly everyone, because of their reputation as stingers.   The sting is borne at the tip of the abdomen.   The latter is unusually long and the terminal half of it much more slender than the basal half.   In addition to the four pairs of walking legs, scorpions have the pedipalps developed to very large size and provided with a pair of pincers so that they appear to have five pairs of legs. The pedipalps are used to grasp prey, and the abdomen is then curled forward over the back and the stinger plunged into the victim to paralyze it.

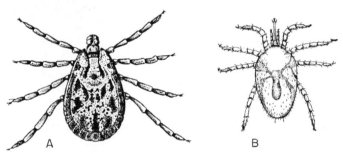

Fig. 6.3.   *A*, the spotted-fever tick (*from Cooley*) and *B*, one of the smaller members of the same order, usually called a mite, the tropical fowl mite.   Much enlarged.   (*From Cleveland.*)

The young are born after hatching from the eggs, and are carried about by the mother for a time after birth, clinging with their pincers to her body.   They are nocturnal creatures that forage about at night, catching and stinging spiders and insects.   Although the stings of most scorpions are scarcely more painful than those of a bee or wasp, the venom of *Centruroides sculpturatus* and *C. gertschi* of the South-west is neurotoxic and may cause severe illness or even death.

Class Arachnida: Order Phalangida (Fig. 6.4).   The harvestmen or daddy-long-legs are familiar to all out-of-door persons.   They look much like very long-legged spiders, but close examination will show that the body is not divided by a slender waist.   The legs are carried with the "knees" high, and the body swung low between them.   The creatures have a noticeable odor that probably discourages many enemies, and Comstock suggests that the ease with which the legs separate from the body is a protective adaptation, enabling them to get away from predators that grasp them by a leg—minus that leg!

The food of the harvestmen is not well known.   Some authors state that they feed largely upon insects, others that they take only dead insects, soft fruits, and other plant tissues.   At any rate they are not known to have any injurious or objectionable habits.

Class Crustacea: The Crayfish, Crabs, Sowbugs, Barnacles, and Their Relatives (Figs. 6.5 and 6.6).   The crayfish, lobsters, and crabs are the largest and best known representatives of this class.   They are primarily aquatic in habit and furthest removed from the insects of any of the classes of arthropods discussed in this book. They have five pairs of walking legs, paired jointed appendages on the abdomen, two pairs of antennae, a pair of compound eyes, and only two body regions. The legs are forked or branched into an outer branch, the *exopodite*, and an inner branch, the *endopodite*. The coxopodite, from which these branches arise, usually bears a gill. Unlike insects they have no tracheae and breathe by blood gills or through the skin. The excretory organs lie in the head, opening at the base of the antennae. The reproductive organs open at the base of the walking legs. The forms most likely to be confused with insects are the small terrestrial sowbugs and pillbugs which abound

FIG. 6.4. A harvestman or daddy-long-legs. Natural size. (*From Slingerland.*)

FIG. 6.5. A crayfish about ½ natural size. (*From Fernald, "Applied Entomology.*")

FIG. 6.6. The greenhouse pillbug. Left, extended; right, rolled into a ball. Enlarged. (*From U.S.D.A., Farmers' Bul.* 1362.)

under boards, logs, in greenhouses, and other damp places. The pillbugs have the habit of rolling themselves into a nearly perfect sphere when disturbed. They are sometimes injurious in greenhouses.

THE ORDERS OF INSECTS

As we have seen in Tables 6.1 and 6.2, about 72 per cent of all known kinds of living animals are insects. About 686,000 different kinds have been discovered, properly named, and described. Thousands of new species are being found every year, indicating that we are far from having reached a full knowledge of the class Hexapoda.   The number of species that are probably living in all parts of the earth, according to H. A. Gossard,[1]

" . . . is variously estimated at from 2,500,000 to 10,000,000, with the probabilities favoring the latter figure as the more nearly correct.   Assuming the maximum figure to be correct, in what a field does the entomologist find himself!   Suppose that he attempts to familiarize himself with each species so that he will recognize it the next time he sees it.   Since his task is obviously great, we will start him at it at the age of 5 years and allot him 5 minutes in which to study each species giving him one-half of the time to a male specimen and one-half to a female.   Lest he should become lazy, we will provide him with electric lights and keep him working day and night and lest he should become fat, we will forbid him to eat except as he is able to snatch mouthfuls from the 5-minute intervals during which he is expected to fix in his memory the anatomical characters, color patterns, etc., which differentiate each species from every kindred one.   Working in this manner and at this rate, the rains of nearly 100 summers will have fallen on his roof before the last representative of the long procession of insects has passed before him."

Obviously none of us shall ever learn to know all of the kinds of insects! Obviously, too, no one need be surprised if an entomologist cannot tell him, offhand, the name of every insect encountered.   The field is so vast that it is, in its finer aspects, beyond the comprehension of any one man.

A good working knowledge of the groups of insects, however, is within the grasp of any earnest student.   The largest groups of insects are known as *orders*.   About 26 orders are commonly recognized by entomologists, and these are listed in Table 6.3.   Of these the Protura, Embioptera, and Zoraptera are comparatively rare and contain so few species that they are unlikely to be encountered by any save the specialist.   They are of no apparent economic importance and will not be discussed further.   A number of other orders contain many common and interesting species, but these are generally of no great importance to man.   From the orders in Table 6.3 there have been selected for special study 13 orders which contain species of importance to man; these are printed in **boldface type.**

KEYS TO THE PRINCIPAL ORDERS OF INSECTS

A. Insects usually provided with 1 or 2 pairs of wings and capable of flying.   Head, thorax, abdomen, and jointed legs nearly always distinct.   Tarsi generally of more than 1 segment: if 1 tarsal segment and 1 claw, see page 226.   Body wall generally firm, hard, often highly colored.   Individuals nearly always active, with little variation in size; often engaged in mating or egg laying.   Mouth parts often modified for siphoning (Fig. 4.16), lapping (Fig. 4.17), or sponging (Fig. 4.15), sometimes entirely wanting or functionless. . . . . **Adult Insects,** see A, couplet 1*a*, 1*b*.

B. Insects without functional wings, either entirely wingless or with short wing pads incapable of transporting the body.   Legs sometimes normally developed, but often short or entirely wanting.   Rarely entirely sessile or incapable of locomotion (page 213).   Tarsi frequently consisting of a single segment and bearing a single claw.   If legs of type shown in Fig. 3.4,*E*, and no exposed mouth parts, see couplet 7. Sometimes without a distinct head, often the thorax not distinct from abdomen. Often enclosed in a case of some kind. . . . .**Immature Insects,** see B, couplet 41*a*, 41*b*, 41*c*.

A. KEY TO ORDERS OF INSECTS IN THE ADULT STAGE

1*a*, Insects with wings.   (If wings very short, try also couplet 41*a* and 41*c*). . . . . . .20.
1*b*—Wingless insects. . . . . .2.

[1] *Jour. Econ. Ento.*, **2**:314, 1909.

TABLE 6.3. THE ORDERS OF THE CLASS HEXAPODA

Types of Mouth Parts	Orders	Examples	Estimated Number of Species Described[1]		Number of Wings in Adult
			World	North America North of Mexico	
		Without Metamorphosis *Primitively Wingless*			
	Protura	Telsontails	90	29	
	Thysanura	Bristletails, silverfish	700	50	No wings
	Collembola	Springtails	2,000	314	
		With Gradual or Simple Metamorphosis *Wings Develop Externally on Nymphs Which Have Compound Eyes*			
Chewing in nymphs and in adults	**Orthoptera**	Roaches, crickets, grasshoppers, katydids, mantids, walking-sticks	22,500	1015	
	Dermaptera	Earwigs	1,100	18	
	Embioptera	Embiids	149	8	Four wings
	Ephemeroptera	Mayflies	1,500	550	(rarely none)
	Odonata	Dragonflies, damselflies	4,870	412	
	Plecoptera	Stoneflies	1,490	340	
	Isoptera	Termites, "white ants"	1,717	41	
	Corrodentia	Book-lice, bark-lice	1,100	120	Four wings or none
	Zoraptera	Zorapterans	19	2	
	Mallophaga	Chewing lice, bird lice	2,675	318	No wings
Rasping-sucking in nymphs and in adults	**Thysanoptera**	Thrips	3,170	606	Four wings
Piercing-sucking in nymphs and adults	**Homoptera**	Aphids, scales, cicadas, leafhoppers	26,500	8,742	Four wings or none
	Hemiptera	True bugs	28,500		
	Anoplura	Bloodsucking lice	250	62	No wings

With Complete or Complex Metamorphosis

Wings Develop Internally in Larvae Which Do Not Have Compound Eyes

Chewing in larvae and in adults	**Coleoptera**	Beetles, weevils	277,000	26,576
	Strepsiptera	Twisted-wing parasites	300	100
	Neuroptera	Lacewings, ant-lions, dobson flies	4,670	338 } Four wings (rarely none)
	Mecoptera	Scorpion-flies	350	66
	Trichoptera	Caddis-flies	4,450	921
Chewing in larvae, siphoning in adults	**Lepidoptera**	Butterflies and moths	112,000	10,300
Chewing or reduced in larvae; chewing or chewing-lapping in adults	**Hymenoptera**	Bees, wasps, ants, sawflies	103,000	14,528 } Four wings or none
Chewing or reduced in larvae; piercing-sucking or sponging in adults	**Diptera**	Flies, mosquitoes, gnats	85,000	16,700 } Two wings or none
Chewing in larvae and piercing-sucking in adults	**Siphonaptera**	Fleas	1,100	238 No wings
	Total insects		686,200	82,394

¹ Modified from Sabrosky, C., *U.S.D.A., Yearbook*, 1952.

187

WINGLESS ADULTS

2a, Tip of abdomen with 2 or 3 prominent appendages which are at least ⅕ as long as the body..3.

2b—Tip of abdomen without any appendages, or with a single prominent appendage, or with a short inconspicuous pair of processes............................4.

3a, With 2 stiff, terminal, abdominal appendages, forceps- or pincer-like in shape (Fig. 6.7). Tough-skinned insects, not covered with scales. Tarsi 3-segmented.. ....*Order Dermaptera*, page 203.

3b—Nearly always with 2 or 3 slender, flexible, antenna-like appendages at tip of abdomen. (If abdominal appendages forceps-like, the tarsi are 1-segmented, Order Thysanura, Family Japygidae.) Thin-skinned, delicate, carrot-shaped insects, covered with scales..............................*Order Thysanura*, page 198.

4a, Very small, delicate insects, rarely over ⅕ inch long, not much flattened, with not more than 6 abdominal segments, the first usually bearing a short, forked "sucker," and the fourth a long forked "spring" used to flip the body in jumping. Mouth parts sunken in the head. Antennae never more than 6-segmented............. ....*Order Collembola*, page 199.

4b—Abdomen of more than 6 segments, though some of them may be obscure; not provided with a terminal, ventral appendage for use in leaping................5.

5a, Mouth parts of the chewing type, with a pair of teeth or mandibles adapted for chewing or pinching transversely, and usually 2 pairs of mouth palps (see also Thysanoptera, 14a)...6.

5b—Mouth parts not adapted for chewing, but generally with an elongate tongue, beak, or proboscis for piercing, lapping, or sucking. At most 1 pair of palps present (except in Thysanoptera, see 14a). (If there are no evident mouth parts see Anoplura, 18a.)...12.

6a, Body louse-like, *i.e.*, rarely over ¼ inch long, flattened, usually somewhat oval in outline, and generally thin-skinned. Legs short. Antennae short, never more than 5-segmented. Live as parasites among clothing, hairs, or feathers of animals...7.

6b—Body not louse-like; if flattened, larger. Antennae always of more than 5 segments. Not parasites on the skin of animals............................8.

7a, Head as wide, or nearly as wide, as the body and broadly rounded or blunt in front. Tarsi with 1 or 2 segments and 1 or 2 claws. Spiracles of thorax on ventral side. From birds or mammals...................*Order Mallophaga*, page 210.

7b—Head narrow, especially toward the front, where it is somewhat pointed. Legs thick, tarsus with a single segment and only 1 claw. Mouth parts piercing-sucking but not visible externally. Thoracic spiracles on dorsal side. From mammals only..*Order Anoplura*, page 226.

8a, Body somewhat ant-like in form, *i.e.*, small, but not flattened; nearly as deep as broad, with moderately long legs and antennae (Fig. 6.8)....................9.

8b—Body not ant-like in shape, generally much larger insects, with firm, dark-colored body wall...11.

9a, Waist slender, *i.e.*, a very slender portion of the body between thorax and abdomen. Mouth parts chewing or chewing-lapping. Tarsi 5-segmented. Generally hard-skinned insects................................*Order Hymenoptera*, page 226.

9b—Waist not conspicuously slender, the body at mid-length nearly or quite as thick as elsewhere. Tarsi 2-, 3-, or 4-segmented. Pale, thin-skinned, delicate insects with large broad heads (Fig. 6.9)...10.

10a, Tarsi 2- or 3-segmented. Very small insects rarely over ⅒ inch long, with compound eyes, very small prothorax, and no cerci. Usually solitary in habit....... ....*Order Corrodentia*, page 209.

10b—Tarsi 4-segmented. Small blind insects up to ¼ inch in length, with a pair of very small cerci. Live in soil or in wood or paper usually in great colonies....... ....*Order Isoptera*, page 207.

11a, Cerci and a conspicuous ovipositor usually present. Prothorax saddle-like. Ocelli present. Antennae usually of many segments, at least 15............... ....*Order Orthoptera*, page 200.

11b—Never with cerci or a firm ovipositor. Ocelli wanting. Antennae of not more than 11 segments.................................*Order Coleoptera*, page 228.

12a, Motionless or sessile insects without distinct head, thorax, and abdomen, and often without legs, eyes, or antennae (see also *B*, Key to Insects in Their Immature Stages, couplet 41a, 41b, 41c)...............................13a, 13b, 13c.

12b—Insects capable of locomotion: jointed legs, antennae, and body regions more or less evident..14.

13a, Body of the insect covered with a mealy powder, or cottony tufts, or a firm, thin, separable shell or scale. Mouth parts piercing-sucking, with long stylets but no encasing, jointed labium or palps. Living on plants. Female scale insects......
....*Order Homoptera*, page 213.

13b—Living on bodies of bees and wasps, partly enclosed by body wall of the host from which they project like a wart or tumor........*Order Strepsiptera*, page 241.

13c—Worm-like or caterpillar-like, sometimes without legs, antennae, or mouth parts. Enclosed in tough silken cases, covered with bits of leaves or evergreen "needles," and suspended from twigs of trees by a silken loop.............................
............................Female bagworms, *Order Lepidoptera*, page 247.

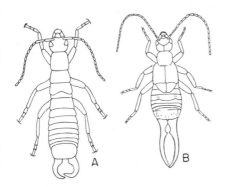
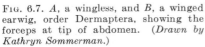

FIG. 6.7. *A*, a wingless, and *B*, a winged earwig, order Dermaptera, showing the forceps at tip of abdomen. (*Drawn by Kathryn Sommerman.*)

FIG. 6.8. A velvet-ant, order Hymenoptera. (*Drawn by Kathryn Sommerman.*)

FIG. 6.9. A termite worker, order Isoptera. (*Drawn by Kathryn Sommerman.*)

14a, Tarsi usually without claws, ending in an inflatable hoof or bladder-like segment (Fig. 6.10,*F*). Head cone-shaped, with 2 pairs of palps, but no segmented beak; the mandibles and maxillae stylet-like and retractile in the head capsule. Very small slender insects, rarely over ⅛ inch long, especially prevalent in flowers.........
....*Order Thysanoptera*, page 211.

14b—Tarsi always with terminal claws, though they may be obscured by surrounding hairs or scales; no protrusible terminal membrane (Fig. 6.10,*A–D*)............15.

15a, Body thickly covered with hairs and flattened scales. Mouth parts, if present, consisting of a proboscis, coiled up under the head like a watch spring, with a pair of palps margining it. Prothorax fused with mesothorax....................
....*Order Lepidoptera*, page 247.

15b—Body not thickly covered with scales and not more than moderately hairy. Mouth parts never a coiled proboscis.................................16.

16a, Tarsi 5-segmented (Fig. 6.10,*A*). Antennae very short, usually completely concealed. Palps present.......................................17.

16b—Tarsal segments fewer than 5. No palps...............................18.

17a, Body strongly compressed or flattened from side to side (thin horizontally). Small, spiny, jumping parasites of animals. Prothorax distinct. Coxae enormously large, those of same pair contiguous. Two pairs of palps..............
....*Order Siphonaptera*, page 312.

17b—Body depressed or flattened from above downward (thin vertically). Body leathery and hairy, but not spiny. Coxae of same pair well separated. Not jumping insects. Abdomen not distinctly segmented. Palps present..........
....*Order Diptera* (in part) *Pupipara*, page 292.
18a, All 3 thoracic segments fused. Legs short, stout, with 1-segmented tarsi bearing a single claw (Fig. 6.10,*E*). Head pointed, narrow, no trace of mouth parts externally; stylets within the head apparently 3 in number. Abdomen broad. Parasitic on mammals (if head broad and mandibles minute, see Mallophaga)........
....*Order Anoplura*, page 226.
18b—Beak evident, segmented. No palps. Four stylets, external to head, usually concealed in a segmented beak..19.
19a, Beak arising from posterior (caudal) part of head. Frequently very small and delicate insects. Prothorax not unusually large except in treehoppers...........
....*Order Homoptera*, page 213.
19b—Beak arising from front (cephalic) part of head. Prothorax large and distinct.
....*Order Hemiptera*, page 218.

WINGED INSECTS

20a, With only 1 pair of wings, the hind wings entirely wanting or represented only by a pair of minute knobbed hairs or balancers or halteres...................21.
20b—With 2 pairs of wings, *i.e.*, 2 pairs of readily visible projections from the thorax in addition to the legs. (The sheath- or shield-like wing covers of beetles and earwigs and the club-like, mesothoracic projections of Strepsiptera are wings)...23.
21a, Tip of abdomen with 1 or more long or prominent processes. Mouth parts wanting...22.
21b—Tip of abdomen without prominent projections. Mouth parts generally with a conspicuous proboscis. Hind wings represented by a pair of halteres..........
....*Order Diptera*, page 292.
22a, Wings netted-veined, with numerous cross-veins. Antennae very short. No halteres......................................*Order Ephemeroptera*, page 204.
22b—Wings with few veins and no cross-veins. Antennae prominent, long. A pair of hook-like halteres present. Males of scale insects...*Order Homoptera*, page 213.
23a, Front wings distinctly thicker, stiffer, and/or less transparent the hind pair; often horny or leathery. If the front pair is only slightly thicker, the hind wings are broad and folded fan-like. Prothorax distinct from rest of thorax, usually large...24.
23b—Front wings of the same texture, stiffness, and color as the hind pair, generally thin and transparent like cellophane.....................................30.
24a, First pair of wings very small, mere short, blunt clubs; the hind pair large, triangular, folding fan-like, and without cross-veins. Minute, rare insects with protruding eyes and short, flabellate antennae. No cerci.....................
....*Order Strepsiptera* (males), page 241.
24b—First pair of wings wide enough to cover the abdomen....................25.
25a, Front wings very stiff and horny, the veins completely obscured (though there may be parallel ridges on them). The hind wings longer than the front pair and *folded crosswise*, to be completely covered by the front ones when at rest......26.
25b—Front wings leathery or parchment-like (at least at the tip), rather than horny. Hind wings not longer and never folded crosswise at rest....................27.
26a, Tip of abdomen without conspicuous appendages. Hind wings folding transversely. Front wings generally covering all or most of the abdomen like a sheath.
....*Order Coleoptera*, page 228.
26b—Tip of abdomen with a pair of prominent pincer-like or forceps-like appendages (Fig. 6.7,*B*). Front wings always much shorter than the abdomen. Hind wings folding radially................................*Order Dermaptera*, page 203.
27a, Front wings stiff and horny or leathery at the base, the tips (distal third or half) abruptly thinner and membranous, and lying flat, one over the other, above the tip of the abdomen, when at rest. Head usually horizontal in position, with a piercing-sucking beak arising near its front end........*Order Hemiptera*, page 218.

27*b*—Front wings almost never stiff and horny, though they may be somewhat leathery and colored; always of about the same texture from base to tip. Head usually vertical in position...28.

28*a*, Mouth parts of the chewing type. Hind wings larger than front wings and folding fan-like. Often with modifications of legs and wings in the males for singing or chirping. Generally large insects with long antennae, distinct large prothorax, and a pair of cerci.......................*Order Orthoptera*, page 200.

28*b*—Mouth parts of the piercing-sucking type, the labium attached near posterior end of head, close to front legs. No cerci.................................29.

29*a*, Head horizontal. Pronotum large and distinct. Beak attached farther forward on the head.....................................*Order Hemiptera*, page 218.

29*b*—Head vertical. Pronotum often short, but enormous in treehoppers (Membracidae)..*Order Homoptera*, page 213.

30*a*, Both pairs of wings largely or entirely covered with minute scales (short flat hairs) on both upper- and undersurface; wings very large and usually varicolored. Mouth parts consisting of an elongate, slender, sucking tube, coiled like a watch spring when not feeding, and a pair of palps margining it; or mouth parts wanting. Prothorax small. Wing vein M_4 distally fused with vein Cu_1.....................
....*Order Lepidoptera*, page 247.

30*b*—Wings not shingled with scales, though they may be hairy. Mouth parts not of the siphoning type...31.

31*a*, Membranous part of both wings very narrow, stick-like, but margined with very long stiff hairs or setae. The tarsi 1- or 2-segmented, without claws, ending bladder-like or hoof-like (Fig. 6.10,*F*). Mouth parts of rasping type. No cerci..........
....*Order Thysanoptera*, page 211.

31*b*—Wings always with broad membranous expansions and bearing veins. Tarsi always with claws (which may be concealed by hairs) (Fig. 6.10,*A–D*).......32.

32*a*, Mouth parts of the piercing-sucking type. No palps. The beak or stylets arising far back on the underside of the head. No cerci......................
....*Order Homoptera*, page 213.

32*b*—Mouth parts not of the piercing-sucking type. At least the mandibles developed to work transversely for chewing, pinching, or grinding; or mandibles absent. Palps always present...33.

33*a*, Antennae very inconspicuous, scarcely as long as the head. Wings netted-veined. Insects found chiefly near water in which their young develop.......34.

33*b*—Antennae usually very well developed. Always longer than the head, or if rarely very small, the wings generally with few cross-veins (the finely reticulated wings of Isoptera are an exception, see 36*a*)................................35.

34*a*, Front wings much larger than hind pair; at rest, folded vertically over the back as in butterflies. Two or three long antennae-like processes at tip of abdomen. Delicate, short-lived insects, found near water in which their young develop......
....*Order Ephemeroptera*, page 204.

34*b*—Front and hind wings of about equal size, with a nodus and pterostigma; held outspread at sides of thorax when insects come to rest. Abdomen generally very long and slender, with only short or inconspicuous appendages at its tip.........
....*Order Odonata*, page 205.

35*a*, Head prolonged downward into a short, thick trunk or beak, 2 or 3 times as long, and about ½ as thick, as the head; bearing chewing mouth parts at the end. Two pairs of wings of about the same size, netted-veined, often spotted. Males with a pair of swollen, pincer-like appendages at tip of abdomen (besides the short cerci), suggesting the sting of a scorpion....................*Order Mecoptera*, page 244.

35*b*—Head not prolonged into a beak. Males without swollen terminal appendages suggesting the sting of a scorpion.......................................36.

36*a*, The two pairs of wings of the same size and shape, with a suture near the base, along which they easily break off; finely but faintly netted-veined. Tarsi 4-segmented. Often appear in swarms from buildings or from wood..................
....*Order Isoptera*, page 207.

36*b*—Front and hind wings rarely of the same shape and size; never with a suture for dehiscence. Tarsi not 4-segmented...37.

37*a*, Tarsi 5-segmented.   No cerci or caudal appendages............38*a*, 38*b*, 38*c*.
37*b*—Tarsi of fewer than 5 segments, 2- or 3-segmented (Fig.6.10,*C*,*D*).........40.
38*a*, Wings generally covered with rather fine, long, silky hairs (*never broad scales*), giving the insects a moth-like appearance.   Wing spread not over 2 inches.   A bare, semitransparent, whitish spot near the center of each wing.   Not many cross-veins along the costal margin.   Wings held roof-like over the back when at rest; vein M_4 of the forewing never fused with vein Cu_1.   Prothorax small, weak. Mandibles reduced or wanting; no long proboscis.   Rather delicate shy insects found about water in which the young develop.......*Order Trichoptera*, page 245.
38*b*—Wings appearing transparent, like cellophane.   Mandibles generally well developed...39.

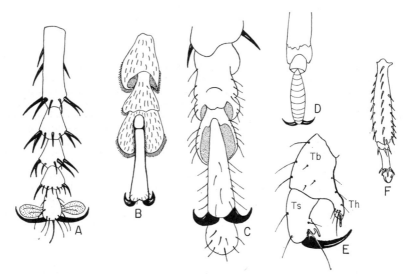

Fig. 6.10. Tarsi of various insects.   *A*, five-segmented tarsus of a robber fly, showing terminal claws, pulvilli, and the median bristle-like empodium; *B*, tarsus of a leaf beetle, showing the condition described as apparently four-segmented, the fourth segment being very small and hidden in the partially divided third segment; *C*, three-segmented tarsus of a grasshopper, with arolium and claws at end; apex of tibia is also shown; *D*, two-segmented tarsus of an aphid, with apex of tibia shown; *E*, one-segmented tarsus (*Ts*) of the human louse, bearing a single claw; the tibia (*Tb*) with its thumb (*Th*) is also shown; *F*, tibia and two-segmented tarsus of a thrips, terminating in a "hoof" or "bladder," without claws. (*Original*.)

38*c*—Wings neither hairy nor transparent, generally colored like the body; hind pair folding fan-like.   Prothorax distinct.   Cerci well developed.   The female with long stiff ovipositor.   Crickets......................*Order Orthoptera*, page 200.
39*a*, Hind wings of much smaller area than the forewings and never folded fan-like in the anal area; frequently fitted against, and hooked to, the front pair.   Wings with few veins and few cross-veins, which are never numerous in the costal area. Prothorax not distinct from mesothorax.   Often slender-waisted insects.   Females often with a stinger or a prominent ovipositor.   Mouth parts often of chewing-lapping type....................................*Order Hymenoptera*, page 266.
39*b*—Hind wings often nearly or quite as large as the front pair; generally with many veins and cross-veins especially in the costal area.   Radial sector of wings generally pectinately branched.   Prothorax well developed.   Many large species.   Mouth parts of chewing type.............................*Order Neuroptera*, page 242.
40*a*, Hind wings always distinctly smaller than front pair; no folded anal area; veins few; wings usually held roof-like over abdomen, at rest.   Prothorax small.   Cerci wanting or very short.   Very small insects...........*Order Corrodentia*, page 209.

40*b*—Hind wings generally as large as or larger than the front pair, the anal area folding fan-like. Wings flat over abdomen at rest. Prothorax well developed. Abdomen generally with two short appendages at tip. Body somewhat flattened. Found chiefly near water in which the young develop..*Order Plecoptera*, page 205.

B. KEY TO THE ORDERS OF INSECTS IN THEIR IMMATURE STAGES

41*a*, **Nymphs:** Wing pads often present in the later instars or larger individuals, but never capable of flight. Compound eyes present, unless the adults are eyeless. Entire body usually fairly well sclerotized, tough-skinned, sometimes brightly colored. Abdomen seldom twice as long as head and thorax together. Shape and appendages much like the adults, with which they are often associated. Legs usually long, rarely wanting, nearly always with two claws on the tarsus. A feeding stage: food and mouth parts of same type as the parents, either chewing or piercing-sucking. Individuals show great variation in size, the largest molting directly to the adult. . . . .**Nymphs,** couplet 42.
41*b*—**Larvae:** Never any trace of wings or wing pads. Compound eyes never present. Abdomen and upper part of thorax, or entire body, often soft, thin-skinned, or weakly sclerotized; often whitish or yellowish, but sometimes brightly colored. Abdomen usually 3 or 4 times as long as head and thorax together. Shape often very different from adults, often cylindrical or spindle-shaped. Legs often very short or wanting. A feeding stage—mouth parts always chewing or reduced, sometimes secondarily modified for sucking. Food and habitat often extremely different from that of adult, very often in the ground, or inside living or dead plant or animal tissues. Individuals show great variation in size, the largest followed by a pupal stage (see 41*c*) before the adult. . . . .**Larvae,** couplet 58.
41*c*—**Pupae:** Functionless legs and wing pads encased in an extra membrane; never used for locomotion; usually incapable of being moved. Sometimes wriggle abdomen and rarely swim actively. Compound eyes visible unless adults are eyeless. Often entirely surrounded or encased by a silken cocoon, a dried larval skin (puparium), or cell of wax, paper, soil, or other material from the environment: then the entire body usually very soft and pale colored. If without a case, the body wall often very hard and dark-, usually somber-colored. Abdomen not unusually elongate; shape and appendages something like adult. A nonfeeding "resting" stage. Mouth parts usually foreshadow the type of the adult. Habitat often very different from both larva and adult. Individuals show little variation in size. Followed directly, upon molting, by the adult stage. . . . .**Pupae,** couplet 70.

NYMPHS

42*a*, Mouth parts chewing in type, frequently sunken into the head, so that the mandibles and maxillae are concealed by the walls of the head as seen from the side. Legs always present, the tarsi usually consisting of a single segment, rarely 3- or 4-segmented, frequently with two claws (Fig. 6.10). Wing pads never present. Thoracic segments larva-like, *i.e.*, pleura and sternum not divided into distinct smaller sclerites (Subclass Apterygota). . . . . .43.
42*b*—Mouth parts chewing in type (Fig. 4.4), greatly elongated (Fig. 4.6), or sometimes wanting. If of chewing type, always exposed, never concealed by the extensions of lateral aspects of the head. Legs usually present, the tarsi variable, consisting of 1 to 5 segments; if only 1 segment, usually with a single claw (Fig. 6.10,*E*). Wing pads frequently present (Fig. 5.9). Thoracic segments usually with pleura and sternum divided into smaller sclerites by distinct sutures (Fig. 3.2) (Subclass Pterygota). . . . . .44.
43*a*, Antennae long, consisting of 10 segments or more (Fig. 5.8). Tip of abdomen with 2 or 3 long, jointed, antennae-like appendages (cerci) (Fig. 5.8), or modified into a pair of forceps-like appendages (Fig. 6.7). (Do not confuse with Dermaptera, page 203.) Abdomen never with a furcula (Fig. 6.12) or a collophore (Fig. 6.12). Prothorax never concealed by the overlapping mesothorax. Slender rapid-running insects. Body covered with scales. . . . .*Order Thysanura*, page 198.
43*b*—Antennae short, never with more than six segments (Fig. 6.12). Cerci always wanting, the tip of the abdomen never with a pair of forceps-like appendages

Abdomen usually with a jointed furcula, used for jumping, on underside of the fourth segment (Fig. 6.12) and a collophore or adhesive organ on the underside of second segment (Fig. 6.12). Prothorax usually small and usually concealed by the mesothorax. Minute, chunky, jumping insects found in damp places . . . .
. . . . Order Collembola, page 199.

44a, Tarsi usually consisting of 2, 3, or 4 segments, rarely of 5 and very rarely of a, single segment. Legs very rarely wanting. Thorax with all three segments exposed and generally different in form (Fig. 5.9); pleural and sternal sclerites usually distinct and never concealed (Fig. 6.29); wing pads usually present (Fig. 5.9) on dorsal and lateral aspects of body. Epicranial suture does not extend to the clypeus. External genitalia may be evident in later instars. (Nymphs) . . . . 45.

44b—Tarsi usually consisting of a single segment (Fig. 3.9,5), or legs wanting, or segmentation of tarsi difficult to determine; more rarely tarsi of 2, 3, or 4 segments. If legs of type shown in Fig. 3.4,E and no exposed mouth parts, see couplet 55. Thorax with all three segments similar in form (Fig. 5.15) and wing pads wanting; or, wing pads present, laterally and ventrally, the thoracic segments not exposed; the pleural and sternal sclerites never distinct, either not differentiated from notum or concealed by legs and wing pads (Fig. 5.16). Epicranial suture usually extends to clypeus. External genitalia not evident. (Larvae and pupae) . . . . 57.

45a, Tarsi almost always without claws, the last segment bladder-like or hoof-like in form (Fig. 6.10,F). Body cylindrical, with the tip of the abdomen pointed (Fig. 6.24,4). Mouth parts with 2 pairs of palps; no segmented beak . . . .
. . . . Order Thysanoptera, page 211.

45b—Tarsi always with 1 or 2 claws, the last segment not bladder-like or hoof-like (Fig. 6.10) . . . . 46.

46a, Mouth parts fitted for chewing (Fig. 4.4), labium never modified into a tube-like beak, always exposed; maxillary and labial palps rarely wanting . . . . 47.

46b—Mouth parts fitted for piercing and sucking (Fig. 4.5); maxillary and labial palps always wanting . . . . 55.

47a, Labium or lower lip 5 or 6 times as long as broad, elbowed at middle, folded beneath the head and over the mouth like a mask when at rest, but capable of being extended a considerable distance beyond the head. Antennae very small. First pair of legs shortest or about equal to the others. Abdomen broad, entirely without appendages or with 3 leaf-like or finger-like gills at the end, never with long antenna-like or bristle-like tails or cerci. Wing pads subequal. Always aquatic . . . .
. . . . Order Odonata, page 205.

47b—Labium normal in form, not greatly elongated, extensible, or folded elbow-like beneath the head . . . . 48.

48a, Abdomen with a series of plate-like or finger-like tracheal gills along each side of the body (Fig. 6.18). Two or three long cerci or antenna-like filaments at tip of abdomen, often "feathered" or fringed with long setae. Tarsi with a single claw. Antennae small, 1 to 3 times length of head. Mesothoracic wing pads larger than the metathoracic pair. Always aquatic . . . . Order Ephemeroptera, page 204.

48b—Abdomen never with a series of plate-like, tracheal gills along each lateral margin. Cerci wanting, or short, or if long not fringed with prominent hairs. The tarsi almost always with 2 claws, except in Mallophaga from mammals . . . 49.

49a, Antennae never consisting of more than 5 segments, short. Body always strongly depressed or flattened dorsoventrally. Head flat, prognathous, broad and rounded in front, eyes simple. Prothorax small, but distinct; meso- and metathorax somewhat fused. Ectoparasites on birds or mammals . . . . . Order Mallophaga, page 210.

49b—Antennae always with more than 5 segments . . . . 50.

50a, Prothorax always much shorter and smaller than the other thoracic segments. Head hypognathous, eyes compound. Body not flattened. Not over ⅛ inch long. Antennae long. Clypeus swollen. (Fig. 6.22) . . . . Order Corrodentia, page 209.

50b—Thorax with the 3 segments about equal, or the prothorax or mesothorax the largest, the pronotum quadrangular or subquadrangular . . . . 51.

51a, Head vertical with the mouth ventral in position (hypognathous). Ligula 4-lobed. Antennae attached on cephalic or front aspect of head. Pronotum quadrangular, pleura on ventral aspect, with the sterna. Segmented cerci usually present. Hind legs often developed for jumping. Wing pads, when present, with the hind pair overlapping the front pair..............*Order Orthoptera*, page 200.

51*b*—Head horizontal, with the mouth cephalic in position (attached to front end of head—prognathous). Antennae usually attached to dorsal aspect of head. Wing pads not as in 51*a*...52.

52*a*, Tarsi with four segments. Head prognathous, usually distinctly longer than broad. Labium very long. Thin-skinned, pale insects. Wing pads, if present, subequal. Cerci of 3 short segments, very small........*Order Isoptera*, page 207.

52*b*—Tarsi with two or three segments. Head distinctly broader than long......53.

53*a*, Basitarsus of first pair of legs about as long as tibia or longer, strongly dilated and provided with openings of silk glands on the ventral surface. Basitarsi of other legs normal in form. Head prognathous.........................*Order Embioptera*.

53*b*—Basitarsus of first pair of legs never so long as the tibia, not different in form from the basitarsi of the other legs...54.

54*a*, Thorax and abdomen never with tracheal gills. Terrestrial insects, usually with cerci consisting of a pair of unsegmented forceps. Head hypognathous. Ligula 2-lobed.....................................*Order Dermaptera*, page 203.

54*b*—Thorax and cephalic (front) segments of abdomen with tufts of slender finger-like tracheal gills near bases of legs. Usually no other appendages on sides of abdomen, but generally a pair of long antennae-like tails at its tip. Legs fringed with strong hairs. Aquatic insects, never with a pair of abdominal pincers or forceps........ ....*Order Plecoptera*, page 205.

55*a*, Wing pads always wanting. Labium never an exposed beak, all the mouth parts withdrawn into head at rest. Head pointed in front. Tarsi adapted for clinging to hairs, 1-segmented, and with a single claw (Fig. 6.10,*E*). Pleura and thoracic spiracles on the dorsal aspect.........................*Order Anoplura*, page 226.

55*b*—Wing pads usually present, labium almost always exposed and segmented. Tarsi never adapted for clinging to hairs. Pleura on ventral or lateral aspect. Thoracic spiracles concealed...56.

56*a*, Labium attached to front end of the head. Pronotum large, distinct.......... ....*Order Hemiptera*, page 218.

56*b*—Labium attached to caudal end of the head, often very close to front coxae. Pronotum often short, but enormous in treehoppers. Body sometimes without trace of legs or wings, flattened, scale-like, covered with a powdery or plate-like secretion, beneath which the insect lives, motionless..*Order Homoptera*, page 213.

57*a*, Thorax never with exposed wing pads present. Tarsi consisting of a single segment. Legs usually short and frequently wanting entirely. *Larvae*...........58.

57*b*—Thorax with 3 segments distinct, usually with wing pads. Legs always present, long, and well developed, folded against sternal and pleural surfaces, sometimes fused with the sternal and pleural surfaces of the body and with each other. Tarsi always consisting of more than a single segment, varying from 2 to 5, but the number of tarsal segments sometimes difficult to determine. Generally quiescent stages, often immobile, and frequently enclosed in silken cocoons or in cases made of soil, plant particles, and the like. Usually no epicranial suture. External, genitalia often indicated. *Pupae*...70.

LARVAE

58*a*, Three pairs of jointed legs on the thoracic segments, the legs often small and inconspicuous, rarely indistinctly segmented...............................59.

58*b*—Thoracic segments never with legs.....................................66.

59*a*, Abdominal segments 1 to 8 without true prolegs or larvapods, except rarely 1 pair at end of abdomen, the other segments at most with ventral folds or wrinkles. First pair of spiracles usually located on the mesothorax. Head often flat and depressed, the mouth directed cephalad (prognathous) (Fig. 6.47).....................60.

59*b*—Abdomen always with true prolegs or larvapods on several or all of the segments 1 to 8. First pair of spiracles located on the prothorax or apparently wanting. Tarsal segments with a single claw. Head globular.........................64.

60*a*, Mesothoracic and metathoracic legs (second and third pairs) noticeably larger than prothoracic (first) pair...61.

60*b*—Legs all about equal in length, rarely, if ever, all 3 pairs directed cephalad. .62.

61*a*, Prothoracic legs directed ventrally, meso- and metathoracic legs much larger, directed laterally; never consisting of more than 4 segments and a single claw. No gills. A cluster of several ocelli on side of head. Pronotal plate wanting. Body strongly curved............*Order Mecoptera* (in part, *Family Boreidae*), page 244.

61*b*—Legs all directed cephalad; consisting of 5 or 6 segments and a single claw. Aquatic insects; abdominal segments usually with hair-like tracheal gills, but without firm, hairy, or feathered filaments. Usually a single pair of abdominal prolegs at tip of abdomen, bearing strong hooked claws. First abdominal segment often with a dorsal, and 2 lateral, fleshy tubercles. Pronotal plate present. Antennae very short. Maxillary palps 4- or 5-segmented. Usually found on bottom of streams or ponds, surrounded by a silken case of varied shapes, covered with pebbles or sticks and open at the head end......*Order Trichoptera*, page 245.

62*a*, Tarsi of thoracic legs with 2 claws (except Sisyridae). Antennae and mandibles usually as long as, or longer than, the head, the latter usually sharp, often sickle-shaped. Abdomen terminating in hooked claws or anal prolegs or an unpaired, median filament, never a pair of cerci. Head usually with a gula. No spinneret. Body usually tapering toward both ends. Often aquatic; then with gills or firm bristle-like or feathered filaments (often segmented) along sides of abdomen, but no prominent terminal appendages. Terrestrial forms without maxillary palps, the maxillae closely fitted to mandibles to form a pair of sucking tubes (Fig. 4.11)....
....*Order Neuroptera*, page 242.

62*b*—Tarsi of thoracic legs usually with a single claw; if with 2 claws, the abdomen has a pair of anal cerci, sometimes retractile. Antennae and mandibles rarely, if ever, longer than the head. Abdomen never with long anal prolegs with hooked claws.63.

63*a*, Head flat, depressed, the mouth directed cephalad (prognathous). Head never with adfrontal sclerites............................*Order Coleoptera*, page 228.

63*b*—Head usually globular, the mouth directed ventrad (hypognathous). Whether flattened or globular, always with adfrontal sclerites (Fig. 6.55), at least in the older larvae. Spinneret present. Thoracic legs with not more than 5 segments, 1 claw. Antennae contiguous to base of mandibles...*Order Lepidoptera*, page 247.

64*a*, Head on each side with a group of 12 to 20 or more ocelli, resembling a compound eye. The thoracic legs 3-segmented with 1 tarsal claw. Six to eight pairs of prolegs, without crochets...
.........*Order Mecoptera* (in part, *Families Panorpidae* and *Bittacidae*), page 244.

34*b*—Ocelli on each side of head never more than 10, usually fewer, and sometimes wanting...65.

65*a*, Prolegs usually 5 pairs, on segments 3 to 6 and 10; sometimes 2 or 3 pairs only. The prolegs always provided with crochets. Ocelli on each side of head either more than one or wanting. Antennae attached to articulating membrane at base of mandibles. Labium usually with a protruding, median spinneret..............
....*Order Lepidoptera*, page 247.

65*b*—Prolegs usually 6 to 8 pairs on segments 2 to 8 and 10, or 2 to 7 and 10, or 2 to 6 and 10. The prolegs never provided with crochets. Only 1 ocellus on each side of head, or none. Antennae attached to frons.................................
..................*Order Hymenoptera* (in part, *Family Tenthredinidae*), page 266.

66*a*, Head always with distinct adfrontal sclerites, at least in the last instar. Labium with a projecting median spinneret.........*Order Lepidoptera* (in part), page 247.

66*b*—Head never with adfrontal sclerites. No projecting spinneret.............67.

67*a*, Head always present, usually much darker in color and easily recognized as a distinct region from the rest of the body; never retracted within the prothorax. Head always with recognizable antennae. Maxillary palps of 2 or more segments.
....68.

67*b*—Head frequently somewhat or completely retracted within the prothorax, often apparently wanting, the body then distinctly tapering and pointed at one end. If the head is exposed, it is usually globular and of the same color as the thorax and abdomen, its transverse width usually much less than that of the prothorax. Antennae and ocelli usually wanting. If antennae distinct, the spiracles of the eighth abdominal segment usually larger than those of other segments and sometimes located at ends of breathing tubes of varying length. Spiracles of all other abdominal segments may be wanting..69.

68a, Body generally short and frequently U-shaped. Ocelli usually present. Antennae and mandibles generally shorter than the head. Each of the principal abdominal segments with a pair of easily recognized spiracles. Abdomen often with fewer than 10 distinct segments and without prominent subanal processes... ....*Order Coleoptera* (in part), page 228.

68b—Body long and unusually slender, less than ⅓ inch long, never U-shaped. Eyes always wanting. Head usually light in color. Antennae distinct, though short, usually of 3 segments. Maxillae brush-like. Spiracles of abdomen minute and inconspicuous or wanting. Abdomen with 10 distinct segments, terminating in a pair of subanal processes; each segment with about 12 stiff, erect setae, longer on the posterior segments...........................*Order Siphonaptera*, page 312.

69a, Abdominal segments usually with several pairs of spiracles, at least in the later instars; the last pair usually the same size as those on other segments; if larger, never situated close together on the dorsomeson. If any abdominal spiracles are wanting, all are usually wanting. Antennae always wanting. Maxillary palps never of more than 1 segment. Mandibles apposable, of chewing type. Generally sluggish, soft, white or yellow worms or grubs, usually tapering somewhat toward both ends, very commonly found in individual cells of wax, in paper, or in bodies of other insects or in galls... *Order Hymenoptera* (in part, bees, wasps, ants, parasites, and gall wasps), page 266.

69b—Frequently only 1 pair of spiracles on the abdomen; usually large, complex, and located adjacent to each other on the dorsomeson of the eighth segment or sometimes at the end of short subconical to long cylindrical tubes. If a number of abdominal spiracles of about the same size are present, the antennae distinct, or a spatula-shaped "breastbone" on the thorax. Mouth parts often consisting of a pair or group of hooks, not apposable for chewing and articulated to a pharyngeal skeleton (Fig. 4.18). Head often greatly reduced or apparently wanting, that end of body pointed. Very often found in dead plant or animal refuse or in bodies of living insects or other animals.........................*Order Diptera*, page 292.

PUPAE

70a, Mandibles, maxillae, and labium recognizable on the head and of the form usually found in the chewing type of mouth parts. (If at end of long prolongation of head, bearing the antennae on its sides, *Order Coleoptera*, in part, *Rhynchophora*.).....71.

70b—Mandibles, maxillae, and labium wanting, or, if present, with some or all of them modified into tubular, piercing or sucking organs. (If a tubular immovable beak-like prolongation of the head with antennae attached on its sides, *Order Coleoptera*, in part, *Rhynchophora*.)...74.

71a, Antennae elongate, always with more than 12 segments; wing pads never clytra-like, *i.e.*, the front pair not unusually thick, generally with a number of veins distinct. ....72.

71b—Antennae either much shorter than the body, with fewer than 12 segments, or much longer than the body, with numerous stout segments. Wing and leg cases rarely fused to the body. The antennae usually lie against the sides of the body curved around above the knees. Wing pads always elytra-like, with few or no veins. Prothorax large, and distinct from mesothorax..*Order Coleoptera*, page 228.

72a, Head normal in form provided with gula. Clypeus or labium and mouth parts not elongated to form a beak or trunk. Prothorax distinct from mesothorax..73.

72b—Head abnormal in form, clypeus or labium and mouth parts greatly elongated to form a trunk-like immovable proboscis............*Order Mecoptera*, page 244.

73a, Mandibles stout, curved, subcylindrical, and overlapping or crossing each other. Thorax and abdomen frequently with finger-like or filamentous tracheal gills. Pronotum small and inconspicuous. Nearly always aquatic; often in silken cases covered with sticks, pebbles, and the like............*Order Trichoptera*, page 245.

73b—Mandibles large and stout, but never overlapping or crossing each other. Pronotum large and quadrangular. Wing and leg cases not tightly folded against body. The antennae not lying against body above the knees.................. ....*Order Neuroptera*, page 242.

74*a*, Mouth parts wanting. Front legs directed forward beneath the head. Body in a cocoon or enclosed in old, molted skins or covered with a waxy separable shell .... *Order Homoptera* (Male Coccidae), page 213.

74*b*—Mouth parts usually present. Front legs not extended forward under head..75.

75*a*, Antennae, mouth parts, legs, and wings usually immovable, firmly grown fast or fused to the pleural and sternal surfaces and to each other. The maxillae extend as a pair of long, slender, adjacent plates along the ventromeson, forming a long proboscis. Pronotum small. Wing pads very large. Antennae usually lie parallel with ventral margin of wing pads. Either in a dense cocoon, in a cave in soil, or attached to some object by tail and girdle of silk around center of body.. ....*Order Lepidoptera*, page 247.

75*b*—Appendages of the head and thorax freely movable without tearing, never grown fast to each other or to pleural and sternal surfaces. The maxillae never forming 2 long, slender plates extending along the ventromeson. Proboscis, if present, shorter than in 75*a*...76.

76*a*, Pupae rarely in silken cocoons, but often enclosed in a firm seed-like case or puparium, completely concealing all appendages. The puparium composed of a dried larval skin and bearing the large adjacent pair of larval spiracles or stigmal plates near one end, and often pupal respiratory horns on thorax. If not in a puparium, or when removed therefrom, wing pads very rarely wanting, consisting of a single pair which, with legs and antennae, are not usually grown fast to the body. Pronotum small, not distinct from mesothorax...*Order Diptera*, page 292.

76*b*—Pupae never enclosed in a puparium formed of the larval skin and bearing the larval spiracles at the end. If enclosed in a cocoon or case of any kind, the latter of silk without a pair of large adjacent spiracles or stigmal plates at one end. Appendages of head and thorax always exposed and movable except as they may be concealed by case or cocoon. Wing pads, if present, four in number......77.

77*a*, Body subcylindrical, often a slender waist between thorax and abdomen. Wing pads usually present, the mesothoracic (first) pair veined and larger than meta-thoracic pair. Antennae always longer than the head. Compound eyes distinct. Mandibles of chewing type, the maxillae and labium often elongate. Prothorax small, fused with mesothorax. Sometimes in cocoons which are usually parchment-like. Often in nests of many individuals.........*Order Hymenoptera*, page 266.

77*b*—Body strongly compressed or flattened from side to side, not over ¼ inch long. Wing pads always wanting. Antennae always minute and shorter than the head. Compound eyes never present; sometimes with simple eyes. Mandibles long and slender, fitted for piercing. Pronotum large and conspicuous. Nearly always in dust- and trash-covered cocoons..................*Order Siphonaptera*, page 312.

ORDER THYSANURA

THE BRISTLETAILS OR SILVERFISH

This is one of the smallest orders in number of known species, but is included here for two reasons. It contains a few species known as silver-fish, fishmoths, slickers, or firebrats that are great household pests.[1] These are further discussed in Chapter 19, on Household Insects, and it need be said here only that these carrot-shaped, swift-running, somber-colored, nocturnal pests (Fig. 6.11) are often injurious to stores of paper stock, book bindings or lettering, card labels of indexes, rayons, wall-paper, and similar starched or sized articles, which they eat. Another reason for placing the Thysanura in the list for special study is that they may represent a very lowly offshoot of the insect family tree. Together with the Collembola, they make a group called the *primitively wingless insects*, which are very different in structure and metamorphosis from

[1] *Ento. News,* **51**:95, 1940.

the higher insects. Many insects such as fleas, lice, some ants, and aphids are wingless throughout all the stages of their life. Study has shown that the kinds just named are wingless by specialization or degeneration. The members of the order Thysanura, however, are believed to be insects that never had wings in their ancestry, having branched off from the insect stock before the latter evolved wings.

Because of the absence of wings in the Thysanura and their direct development, they have no metamorphosis. The young, also called nymphs, grow gradually toward the adult condition without any appreciable change in form or appearance except the change in size (Fig. 5.8). The mouth parts are of the chewing type, sometimes curiously set into the head cavity so that only the tips of the parts project from the surface. Compound eyes are present in some species, degenerate in others, and wanting in some; ocelli are usually wanting. The antennae are long and many-jointed. Most species have at the tail end of the body two or three bristle-like many-jointed appendages something like antennae, from which the common name bristletails is given. Sometimes these appendages are unsegmented and forceps-like. Some of the species have leg-like structures on the segments of the abdomen, a condition unique among insects. The body is very soft but is covered with scales or hairs that give it a shiny appearance and also account for the name fishmoths. Thysanura live a hidden life, being found in cracks and crevices about buildings, under stones, and in the soil among leaf mold. They are active chiefly at night or in darkness. When disturbed they scuttle about with great rapidity. Most of the species are thought to be scavengers.

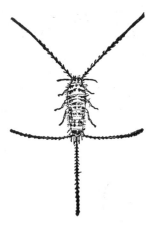

Fig. 6.11. A common household thysanuran (*Thermobia domestica*), a little larger than natural size. (*From Kellogg, "American Insects," after Howard and Marlatt.*)

Primitively wingless insects. Mouth parts chewing. Abdomen of 10 or 11 segments. Antennae very long, many-segmented. Usually a pair of long cerci at the posterior end, and sometimes three such antenna-like tails. Sometimes rudimentary legs on the abdominal segments. Tarsi one-to four-segmented. Malpighian tubes sometimes wanting. No metamorphosis.

Fig. 6.12. Spotted springtail, *Papirius maculosus*, about 6 times natural size. (*From Kellogg, "American Insects."*)

ORDER COLLEMBOLA[1]

The Springtails and Snow-fleas

The springtails are minute insects, rarely ⅕ inch long, often occurring in enormous numbers on the surface and in the soil of woodlands, in decaying vegetable matter, on the surface of stagnant water, on snow, in mushroom houses, and other damp places. They are seldom noticed except by those who seek them. The points of principal interest about them are that they are entirely and primitively wingless, that they generally have a forked muscular appendage at the tip of the abdomen, which is used in springing into the air (Fig. 6.12), and that they occasionally become pests about

[1] MILLS, H. B., "Collembola of Iowa," Iowa State Coll. Press, 1934; MAYNARD, E. A., "Collembola of New York," Comstock, 1951.

maple-sap buckets, in mushroom beds, or on seedlings in greenhouses.   They are often deeply colored.

Primitively wingless insects.   Mouth parts chewing; sunken into the head.   Compound eyes degenerate.   Malpighian tubes wanting and the tracheal system very slightly developed.   Never more than six abdominal segments, the first with forked adhesive organ or ventral tube, shown between the first and second pairs of legs in the figure, and the fourth with a forked spring.   Antennae of few segments.   Tarsi one-segmented.   Development without a metamorphosis.

ORDER ORTHOPTERA[1]

Grasshoppers, Crickets, Katydids, and Others

This large order includes some very primitive insects, such as the roaches; some that are well known to everyone, such as the katydids and crickets; some of the most curious of all insects, such as the walkingsticks and mole crickets; and some that are very destructive to crops, such as the grasshoppers.   They are mostly large insects, and many of the species make sounds, so that they attract a great deal of attention.

Fig. 6.13. A common field cricket, *Acheta assimilis*, female.   (*From Kellogg, "American Insects."*)

The groups that make noises have the hind legs unusually long and powerful, and progress by jumping. The sounds are not voices.   They do not come from the mouth, but are produced by rubbing rough surfaces of the body together.   In the crickets and katydids, specialized parts of the front wings are rubbed together to make the sounds.   In some grasshoppers the front and hind wings are rubbed together; others rub the inner surface of the hind leg (the femur) over the outer edge of the front wing.   It is the males that produce these sounds, the females almost never having special organs for this purpose.   Both sexes, however, commonly have sound-perceiving organs or "ears" (page 120).

All the Orthoptera have well-developed chewing mouth parts.   Many kinds feed on plants, others on small animals, and still others are scavengers.   The metamorphosis is a gradual one, the nymphs generally passing through five instars.   Most species live exposed on plants or hidden on the surface of the ground, but a few burrow into the soil, a few live in houses, and a very few take to the water.   The wings when present are four in number, the front pair narrow and thickened, but with the veins showing, and capable of bending without breaking, somewhat like a piece of leather.   The hind wings are thin, often brightly colored and with many veins, broadly triangular, and, when brought to rest, they fold along radiating straight lines from the base, like a fan, and are laid back over the abdomen, so that they are covered by the front wings (Fig. 9.4).   The wings are often incompletely developed.   The antennae are generally long and prominent, the legs are long, the prothorax is large, and the tip of the abdomen is provided with a pair of

[1] Scudder, S. H., "Catalogue of Orthoptera of United States," *Proc. Davenport Acad. Nat. Sci.*, **8**:1–101, 1900; Blatchley, W. S., "Orthoptera of Northeastern America," Nature Pub. Co., 1920; Hebard, M., "Orthoptera of Illinois," *Bul. Ill. Nat. Hist. Surv.*, 20, pp. 125–279, 1934.

cerci and often with a prominent ovipositor in the female (Fig. 6.13). Here belong some of the largest of all insects: a Venezuelan grasshopper that measures 6½ inches in length, and African walking-sticks 10 inches long. In temperate latitudes most species spend the winter in the egg stage.

Mouth parts chewing. Ligula four-lobed. Prothorax large and distinct. Wings four, sometimes greatly reduced or wanting. The front pair narrow, somewhat thickened, and usually colored like the rest of the body but distinctly veined. The hind pair membranous, broad, and folded fan-like when at rest. Cerci and an ovipositor generally present. Metamorphosis gradual.

Important economic species of this order are grasshoppers (pages 465 and 602), tree crickets (page 788), and roaches (page 906).

THE FAMILIES OF ORTHOPTERA

The families of Orthoptera are so distinct that certain writers have proposed separate orders for some of them. The first three families discussed have been called the singing, jumping Orthoptera (Saltatoria), since most of the males have stridulating organs, and all have the hind legs noticeably longer and stouter than the others. In the females the ovipositor is usually well developed. The tarsi have fewer than five segments in these three families.

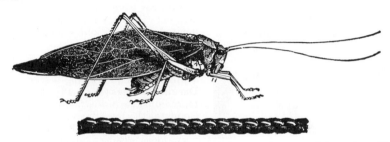

Fig. 6.14. The angular-winged katydid, *Microcentrum retinerve*, and its eggs. Natural size. (*From Sanderson and Jackson, "Elementary Entomology," after Riley.*)

Family Locustidae (Formerly Called Acrididae). The Grasshoppers or Locusts (Figs. 9.1 to 9.4). These are moderately long insects, a little *deeper than wide, of dark colors, variously mottled*, with prominent heads and large eyes, all characteristically active in the daytime. The antennae are always much shorter than the body, the tarsi have three segments, and the "ears" (Fig. 3.2) are found on the sides of the first abdominal segment. The ovipositor consists of four short, finger-like pieces, well separated (Fig. 3.2). These are used to thrust the eggs into the soil or into soft wood, ½ inch or 1 inch below the surface (Fig. 9.1). The eggs are formed into definite masses of 20 to 100 or more each, and are surrounded by a frothy, gummy substance which hardens to form a protective case for them. There is usually only one generation a year.

Family Tettigoniidae (Formerly Called Locustidae). The Long-horned Grasshoppers, Green Meadow Grasshoppers, Katydids (Fig. 6.14), **Cave and Camel Crickets.** These are more delicate and less hardy insects than the true grasshoppers, most commonly green in color, but of somewhat similar form. They are often nocturnal and can be distinguished from the grasshoppers by their very long antennae, often longer than the body, and by having four segments in each tarsus. The "ears" are on the base of the front tibiae (Fig. 3.4,*A*), and the ovipositor is *sword-shaped*, the four pieces being flattened and closely appressed. The eggs are laid singly or in rows often on or in leaves, stems, twigs, or sometimes in the soil. There is only one generation a year. As a rule they eat either plants or small animals. Only a few of them are serious pests.

Family Gryllidae. Crickets (Fig. 6.13), **Tree Crickets** (Fig. 15.87), **Mole Crickets.** These are usually somewhat short dark-colored Orthoptera with the tarsi three-segmented like the grasshoppers, but the antennae very long. In those kinds that have wings and produce sounds, the "ears" are found on the front tibiae (Fig. 3.4,*A*). The ovipositor is a long *spear-shaped* tube. The front margin of the wings is bent sharply down over the sides of the abdomen, like the edge of a box lid. The eggs are laid in groups in the soil or inserted into the stems of plants.

Crickets are nocturnal and negative to light. They feed upon a great variety of substances. The tree crickets are slender greenish insects that live among tall weeds, trees, or bushes and sometimes cause damage by slitting twigs and depositing their eggs in them (Fig. 1.7,*B*). A curious fact is that the male has glands opening on the upper side of the thorax from which the female feeds at the time of mating. The mole crickets often burrow and make nests in the soil in the vicinity of water. They eat plant roots, other insects, and earthworms. The body is covered with fine, brown, velvety hairs. The front legs (Fig. 3.4,*A*) are remarkably developed, both for digging and to act like a pair of scissors for cutting off small roots that are in their way. The ordinary black field crickets and the "cricket of the hearth" are other well-known representatives of this family. Crickets often do damage in grain fields by cutting the twine used to bind sheaves.

The last three families of Orthoptera have been called the mute nonjumping Orthoptera, since they make no particular sounds and do not have the hind legs enlarged for leaping. The tarsi are always five-segmented in these families, and the ovipositor is concealed or wanting.

Family Phasmidae (Sometimes Classified as a Distinct Order, Phasmoidea). Walking-sticks, Walking-leaves, Devil's Darning Needles (Fig. 6.15). Our common representatives of this family are extremely elongate (the mesothorax especially long), cylindrical, wingless, with long stiff legs and very long, slender antennae. They are found feeding upon the foliage of trees, but also often resting about buildings. They move but little and in a very stealthy manner; they are doubtless often overlooked by their enemies because they look so much like slender sticks, being good examples

FIG. 6.15. A walking-stick in its usual environment among leaves and twigs. The head is just above the center of the picture, with the long front legs projecting straight forward toward the upper right. (*From Slingerland.*)

of the phenomenon known as *protective resemblance*. The eggs are simply dropped by the females, one at a time, as they rest among trees, and have often been said to make a noise like the patter of rain drops, when the insects are abundant. They are harmless to man except that they may injure trees by eating the leaves. Some tropical species have well-developed wings, and some of them are broad, flat, green insects that look astonishingly like the leaves among which they live.

Family Blattidae (Sometimes Classified as a Distinct Order, Blattoidea). The Roaches or Cockroaches (Fig. 19.6). This family has been called the running Orthoptera. The body is flattened, the head is bent downward and backward and is not prominent. The prothorax is very large, the legs are long and bristly, the hind pair only moderately larger than the others, but the coxae are all very large. The wings may be well developed, short, or wanting; they lie flat over the back, crossing over somewhat toward the tip.

Cockroaches usually have a bad odor, and they frequent all sorts of filthy places and dusty crevices, so that they are inexpressibly dirty. They are common in kitchens, bakeries, and restaurants and may be carriers of disease germs. The eggs are formed into packets of 16 to 40, enclosed in seed-like cases, and these so-called oöthecae are sometimes carried about by the female partly extruded from the abdomen until the nymphs hatch from them (Fig. 19.6).

Cockroaches are very sensitive to cold. Many species live out-of-doors and in moist tropical countries are very abundant. A half dozen species have the habit of living in dwellings and other buildings. These will be further considered under the discussion of household insects.

Family Mantidae (Sometimes Classified as a Distinct Order, Mantoidea). Praying Mantes, "Mule Killers" (Fig. 6.16). These remarkable creatures have curious habits and odd structures. The common name comes from the manner in which they hold up the fore part of the body, with its enormous front legs, as though in an attitude of prayer. They might also be called *preying* mantes, for they are the only family of Orthoptera that seems to be exclusively carnivorous, eating other insects.

Fig. 6.16. A praying mantis, *Stagmomantis carolina*. About natural size. (*From Comstock, "Introduction to Entomology."*)

The body is elongate, the prothorax and fore coxae especially long, and the front legs so modified that they can grasp small insects between the spiny tibiae and femora, the former closing against the latter like a knife blade against its handle (Fig. 3.4,*G*). The wings are usually well developed, but the mantes commonly remain quiet in one place until some insect comes within reach. They sometimes cautiously stalk their prey. The eggs are laid in large masses an inch or so long, in a frothy, gummy substance, on the twigs of trees. These ferocious but beneficial insects are not commonly found north of the fortieth parallel of latitude. They never injure man or the large animals. Some tropical species of this family are very broad and have the forewings so modified as to resemble leaves or flowers in shape, color, and venation.

ORDER DERMAPTERA (EUPLEXOPTERA)

The Earwigs

The earwigs are beetle-like insects easily distinguished from the Coleoptera by the prominent forceps at the rear end of the body and by their gradual metamorphosis. The mother broods over her nest of eggs in the soil and guards the young nymphs (Fig. 6.17). The food is variable, some species attacking plants in an injurious manner, others catching insects, and others feeding on decaying matter. Sometimes they become serious pests in and about houses, although the superstition that they attack people's ears is absurd.

Front wings horny, veinless, beetle-like, meeting in the middorsal line, but very much shorter than the abdomen. Hind wings membranous, ear-shaped, the veins radiating from the middle of the costal margin, folding both radially and transversely. Often wingless. A conspicuous pair of hooks or forceps at the end of the abdomen. Mouth parts of the chewing type; ligula two-lobed. Tarsi three-segmented. Compound eyes present; ocelli absent. Metamorphosis gradual.

Fig. 6.17. A female earwig brooding over her nest of eggs. (*From Fulton, Ore. Agr. Exp. Sta. Bul.* 207.)

ORDER EPHEMEROPTERA (EPHEMERIDA, PLECTOPTERA)

The May-flies, Lake-flies, or Shad-flies

These delicate, defenseless creatures (Fig. 6.18) often appear in surprising myriads in cities near lakes or streams, being strongly attracted by lights. They live but a few hours or a few days as adults, but this is often preceded by 1, 2, or 3 years of life beneath the water as nymphs. This ephemeral adult life is responsible for the order name.

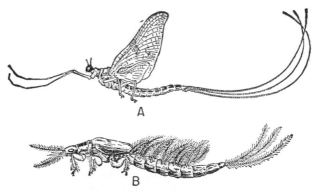

Fig. 6.18. A May-fly, *Ephemera varia. A*, adult; *B*, nymph, about natural size. (*From Comstock, after Needham, Comstock Pub. Co.*)

The nymphs (Fig. 6.18,*B*) feed mostly on living or dead aquatic vegetation and breathe through tracheal gills. They are important as food for fishes. The adults have rudimentary mouth parts and take no food but are frequently a nuisance because their dead bodies accumulate in windrows on streets and about watering places and have an offensive smell.

Slightly chitinized adults, living but a short time, and molting once after reaching the winged, adult stage. Four, triangular, net-veined, gauzy wings, folding vertically over the back as in butterflies, when at rest; the hind pair much smaller, rarely wanting. Mouth parts of chewing type, but degenerate or wanting in adults. Antennae very short. Compound eyes and ocelli present. Mesothorax large. Two or three, very long, slender, many-jointed "tails." Genital openings double. Metamorphosis gradual, sometimes as many as 20 molts, the nymphs aquatic, elongate; the abdomen with two or three slender tails and seven (or fewer) pairs of tracheal gills along the sides and on the back of the abdomen. Tarsi one- to five-segmented. A single claw on each nymphal tarsus.

ORDER ODONATA

The Dragonflies (Fig. 6.19) and Damselflies

The adults are aerial; expert flyers; abounding about ponds and streams; catching and eating other insects on the wing (page 64). The nymphs (Fig. 6.19,*1,2,3*) are aquatic, walking or hiding on the bottom of ponds and streams and catching other small animals for food. The development of a generation generally requires about a year, but some require several years. Odonata are of value to man by their feeding on horse flies and mosquitoes and as food for fish.

Large, often beautifully colored insects. Head vertical. Chewing mouth parts. Four membranous, slender, finely net-veined wings of about equal size, often not laid over the back when at rest. Near the middle of the front margin is a short, heavy cross-vein and a slight notch, like a joint; and near the tip of the front wings is a dark stigma. Antennae very small. Compound eyes large, ocelli present. Tarsi three-segmented. Abdomen very long and slender. Copulatory organs of male on the second abdominal segment, separate from the openings of the vasa deferentia. Metamorphosis gradual, the nymphs developing in the water and being provided with a very long extensible labium, used in capturing prey and folding like a mask over the face when not in use (Fig. 6.19,1, cf. 4). Molts numerous, 10 to 15.

There are two suborders differing as follows:

The Anisoptera or Dragonflies	*The Zygoptera or Damselflies*
Hind wings broader at base, not folded but held in a horizontal position at sides of the body when at rest. Strong flyers	Two pairs of wings of same size and shape, narrow at base; folded back over the abdomen or up over the back, like those of a butterfly, when at rest. Feeble flyers
Eyes do not project from the side of the head	Eyes projecting, constricted at the base
Eggs laid on the water or on aquatic plants or rarely in their stems	Eggs thrust into the stems of aquatic plants, often beneath water
Nymphs respire through tracheal gills inside of the rectum, and the forcible ejection of water from the anus propels the nymphs forward	Nymphs respire by three leaf-like tracheal gills, projecting from the end of the abdomen

ORDER PLECOPTERA

Stone-flies

These retiring insects are seldom seen except by those who seek them. They live near streams, resting on stones, trees, and bushes, or flying over the water. Some species are not attracted to lights. The adults (Fig. 6.20) have rather long antennae, but the tail-like cerci are much shorter than in the May-flies; and the hind wings are much broader than the front pair. The wings fold flat over the back, giving the insects a straight-sided, square-shouldered appearance when resting. The nymphs develop in the water, being common on the surface of stones in swift streams. They feed mostly on diatoms, algae, and other small plant and animal forms. In contrast with May-fly nymphs, they usually bear their gills on the thorax and have two claws on each leg (Fig. 5.9,*A*). Their only importance to man is as food for fishes.

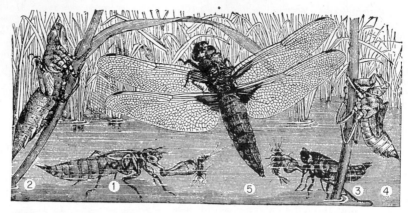

FIG. 6.19. A dragonfly, adult and nymphs. At *1* and *3*, two nymphs are shown catching prey with the extended labium. *2*, a mature nymph ready to change to the adult. *4*, the shed skin (exuviae) from which the adult, *5*, has emerged. (*From Sanderson and Jackson, after Brehm.*)

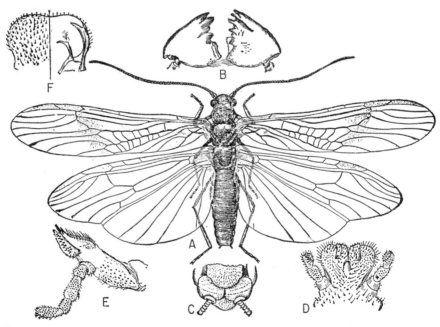

FIG. 6.20. A stone-fly, *Taeniopteryx pacifica*, much enlarged. *A*, dorsal view of adult about 4 times natural size; *B*, mandibles; *C*, last segment of female from below; *D*, labium; *E*, maxilla; *F*, labrum, dorsal half at left, ventral half (epipharynx) at right. (*From Newcomer, Jour. Agr. Res., vol. 13.*)

Wings four, netted-veined; front pair narrow, hind pair very broad, folding like a fan; abdomen thinly chitinized. Compound eyes and ocelli present. Antennae long, filiform. Tarsi three-segmented. Mouth parts of the chewing type but often reduced and weak. Metamorphosis gradual. Nymphs with long antennae, and usually long cerci. Tracheal gills on the thorax. Tarsi with two claws.

ORDER ISOPTERA[1]

Termites or "White-ants" (Fig. 6.21)

These are the yellowish-white, soft-bodied "wood-ants" that are seen so often in countless numbers in logs, stumps, timbers of buildings, or wood lying in contact with the soil. They are not ants at all, being very different from ants in structure and in metamorphosis. One easy way to distinguish them from ants (see order Hymenoptera) is to note that the base of the abdomen is broadly joined to the thorax and not by a slender petiole.

Their chief resemblance to ants is in their colonial or social life. This is a curious condition, found in this order and in some of the Hymenoptera, in which there are individuals of several *castes*, that differ in structure and in duties in the same species and only a few of which become parents, all the others devoting themselves to the care of the thousands of offspring from the few "kings" and "queens."

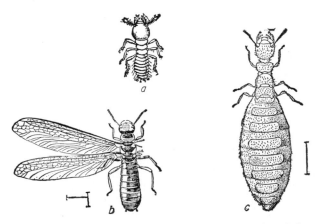

Fig. 6.21. Common North American termite. Above left, a worker; below, a male with right wings removed; right, a queen after she has lost her wings and her eggs have developed. Note the thick-waisted condition of all. (*From Kellogg, "American Insects."*)

In the termites the castes consist of: (*a*) dark-bodied males and females with four long wings, known as kings and queens; (*b*) short-winged males and females; (*c*) wingless males and females; (*d*) wingless workers; and (*e*) wingless soldiers. The kings and queens can reproduce all the castes, including others like themselves, which at certain seasons swarm out of the nest in enormous numbers, often appearing in or about infested buildings and giving the first warning or indication that termites are present. Their wings are long, narrow, whitish, or semitransparent, with many indistinct veins; and the two pairs are almost exactly alike in size and appearance. The wings are used only for a single, wedding flight, after which they are broken off along a suture of weakness near the base. Then a nest may be started, copulation takes place, and the queen subsequently becomes enormously enlarged with developing eggs.

[1] Snyder, T. E., "Our Enemy the Termite," 2d ed., Comstock, 1948; "Catalogue of Termites of the World," *Smithsonian Inst. Misc. Collections*, no. 112, 1949.

These swarming, winged kings and queens are the only termites that normally appear in the open air. The young and all the other castes stay in the same nest throughout their lives and are very delicate, thin-skinned, pale-colored, and blind. The kings and queens are not rulers in any sense, but simply reproductive individuals that do not even feed themselves. Among the termites, in contrast with the ants, bees, and wasps, the males live a long time and mate repeatedly with the queens, and all termite eggs are fertilized. The nymphs are all similar in appearance to workers but they early begin to show minor differences and develop into the different castes at the final molt.

Unlike the ants, bees, and wasps in which workers and soldiers are exclusively females, the termite soldiers and workers are of both the male and female sex but have the reproductive organs underdeveloped, and they never mate or lay eggs. One might almost say that they never pass beyond the nymphal stage. The short-winged and wingless males and females, which are sometimes present, may reproduce their own kind and workers and soldiers, but never the long-winged kings and queens, and so they cannot start new colonies. All the wingless castes have a strong aversion to being exposed to the open air, and rarely show themselves outside of the nest, the soil, or the wood in which they are working. This has usually been attributed to a negative reaction to light, but they appear to have little aversion to light, if they are completely protected from the open air. If they must cross an exposed place, a covered runway of soil and body secretions is built. The workers build nests, supply food, care for the eggs, feed the extremely young nymphs and the queens and kings, and perform all other duties for the colony, except reproduction and defense. The mouth parts are of the chewing type; the mandibles of the soldiers become enormously enlarged, and are borne on very large heads. Compound eyes and ocelli are wanting or greatly reduced, except in the kings and queens, and in the workers of certain tropical species, which forage in the open air for food.

The food of our common termites is primarily wood, and often dead, hardwood, in which there is apparently little except cellulose, a material not usually digested by the larger animals. The digestive tracts are packed with protozoa, and it is believed that the termites live on the products of the digestion by the protozoa, and not on the cellulose directly. Termites are great pests of all kinds of wood and products of wooden origin (page 39 and Figs. 1.16 and 1.17). They also attack living plants, hollowing out the stems.

Termites are really a tropical group. In Africa, Australia, and other tropical lands they build enormous nests, 12 to 20 feet tall, containing incalculable numbers of workers (Fig. 1.2). The queens of tropical species may reach a length of 4 or more inches, and they are said to produce eggs at the rate of 60 or more a minute. Many tropical species cultivate mushroom beds or fungus gardens in their nests, which furnish their food. The important species of Kalotermitidae or dry-wood termites and Rhinotermitidae or subterranean termites are discussed on page 898.

Moderate-sized, thin-skinned, slender, social insects, consisting of several castes, living together in great nests or colonies, like ants. Tarsi nearly always four-segmented. Abdomen wide where it joins the thorax and has a pair of small cerci at the end. Mouth parts chewing. Metamorphosis simple. Only the kings and queens normally reproduce; they are four-

winged, often dark-colored. The wings are equal in size, long, narrow, with membrane somewhat opaque and the veins indistinct (except along the costal margin and in the anal region), laid flat over their backs when not in flight, and usually broken off after the pairing flight, along a joint of weakness near the base. The workers and soldiers are of both sexes, but do not regularly reproduce. They are wingless, usually pale-colored, soft-bodied, and eyeless.

ORDER CORRODENTIA (PSOCOPTERA, COPEOGNATHA)[1]

Book-lice, Dust-lice, Bark-lice, Deathwatches

Often, as one opens an unused book or disturbs some old papers, a very small yellow insect runs across the page. If one examines such an animated speck under a lens, one finds a wingless, soft-bodied insect with well-developed head, chewing mouth parts, small compound eyes, antennae nearly as long as the body, and six large legs (Fig. 6.22). Relatives of this book-louse live on the bark of trees, vegetation such as corn stalks and greenhouse plants, and some of them have four membranous wings (rarely covered with scales like moths). They look like aphids but have the veins of the wing peculiarly kinked as shown in Fig. 6.23. At rest the wings are held roof-

Fig. 6.22. A book-louse, greatly enlarged. (*From Comstock, "Introduction to Entomology."*)

Fig. 6.23. A winged bark-louse, 13 times natural size. (*From Kellogg, "American Insects."*)

like over the back. A structural peculiarity of this order is the "pick" or rod attached to each maxilla and working in and out of the softer part of the maxilla like "a piston sliding to and fro in its cylinder."[2]

The known importance of this order is not very great. Those that live indoors may occasionally become pests by feeding on paper, starch, grain, and other substances in damp places, and one species has been accused of spreading plant diseases. Some species are active at very low temperatures.

Minute insects; wingless or with four membranous wings with few veins; the first pair larger and at rest held roof-like over the abdomen. Mouth parts chewing, with a curious rod in the maxilla. Compound eyes present, but ocelli wanting in the wingless forms. Prothorax small, tarsi two- or three-segmented, and cerci wanting. The metamorphosis is very simple.

The members of the family Psocidae (Fig. 6.23) have well-developed wings. With their young they inhabit the trunks of trees, stone walls, and other dry locations, where lichens are available as food. The family Atropidae includes the book-lice of dark, little used, damp buildings (Fig. 6.22). *Atropos pulsatoria* is a cosmopolitan species found in libraries, museums, and in deserted beehives and wasp nests. The Liposcelidae include both winged and wingless species. The wingless book-louse or cereal psocid, *Liposcelis divinatoria*, is ubiquitous in old wooden houses, warehouses, granaries, museums, and libraries, where it is a pest chiefly because of its annoying presence, although it may injure books and specimens.

[1] CHAPMAN, P. J., "Corrodentia of the United States," *Jour. N.Y. Ento. Soc.*, **38**:219–280, 319–402, 1930.

[2] LEFROY, M., "Manual of Entomology," Longmans, 1923.

ORDER MALLOPHAGA
Chewing Lice or Bird Lice[1]

There are two groups of lice that make their homes continuously on the bodies of warm-blooded animals. These two orders (Mallophaga and Anoplura) agree in never showing any trace of wings; in being flattened, oval, tough-skinned, external parasites; in gluing their eggs to the hairs or the feathers of the host; and in spending all their lives generation after generation on the same host animal. The members of the other order, the Anoplura, suck blood; but the Mallophaga have chewing mouth parts and subsist on bits of hair or feathers, skin scales, or the dried blood from scabs. Another noteworthy difference is that, whereas the Anoplura are confined to the mammals or hair-bearing animals, the Mallophaga occur on both birds and mammals. By far the greatest number, however, are found on birds, nearly every kind of wild or domesticated bird being attacked by one or more kinds of chewing lice. As a general rule, one species of louse will seldom live on more than one species of bird; the chicken, however, has seven common species. Cattle, horses, sheep, dogs, and cats also are attacked, but none of the chewing lice live on hogs or man. The species that live on birds normally run very rapidly when they are exposed. But the ones that live on mammals have the tarsus highly modified into a clamp for clinging to the hairs, and move about but awkwardly. The nature of this clamp is explained in the discussion of the Anoplura, which have a similar structure (Fig. 3.4,*E*). The metamorphosis is a very simple one, almost the only change from hatching to maturity being an increase in size, in thickness of the body wall, and in darkness of coloring.

The chewing lice are small, wingless, oval or elongate, flattened insects (Figs. 20.28 and 20.29), *mostly* ⅕ *to* ¹⁄₂₅ *inch long, with large, broad heads, rounded in front, and bearing short, three- to five-segmented antennae often hidden in grooves of the head. The eyes are degenerate. The legs are not very large, tarsi one- or two-segmented, with one or two claws. The prothorax is distinct, the meso- and metathorax often more or less united. Thoracic and abdominal spiracles located on the ventral side. The skin is tough, often with heavily sclerotized plates of dark color and with scattered hairs. Mostly parasitic on the bodies of birds, but some species on mammals. Mouth parts chewing; not bloodsuckers; the labial palps, sometimes all four palps, wanting. No cerci. Metamorphosis gradual or wanting.*

The suborder Amblycera includes species in which the antennae are concealed in grooves of the head and are usually four-segmented and capitate. The maxillary palps are also four-segmented and are easily confused with the antennae. The mandibles are turned forward to work in the same plane as the long axis of the body. The family Gyropidae includes chewing lice of the guinea pig and rodents, with a single tarsal claw for clinging to the hairs. The family Menoponidae, with two-clawed tarsi, includes a number of important pests of birds (page 988).

The suborder Ischnocera includes species in which the antennae are normally exposed, filiform, three- or five-segmented; the maxillary palps are wanting, and the mandibles are turned downward to work in a plane at right angles to the long axis of

[1] These lice have often been called *biting lice* in an attempt to distinguish them from the Anoplura. This seems absurd, for, if the victims of the attacks were asked which kind of lice "bite," they would undoubtedly say the Anoplura, which insert sharp stylets into the skin to suck the blood, and not these forms, which, at most, only nibble at the skin.

the body. The family Trichodectidae includes the common chewing lice found on horses, cattle, sheep, and dogs (page 957). These have tarsi with single claws and the antennae are three-segmented. The family Philopteridae have five-segmented antennae and normal two-clawed tarsi. They live only on birds and include several important species (page 988).

ORDER THYSANOPTERA (PHYSOPODA)[1]

THE THRIPS (FIG. 6.24)

The thrips are minute, slender, and agile, rarely as long as ⅛ inch. They live in flowers or on other parts of plants, feeding on the sap. Many species are serious pests of fruits, vegetables, flowers, and field crops.

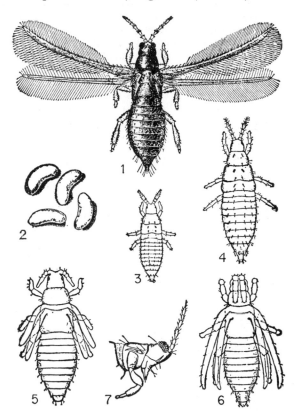

FIG. 6.24. Pear thrips, *Taeniothrips inconsequens*. *1*, Adult; *2*, eggs; *3–6*, first-, second-, third-, and last-instar nymphs; *7*, head of adult from the side. All greatly enlarged. (*From Moulton, U.S.D.A., Dept. Bul. 173.*)

With the order Thysanoptera we take up the first insects with other than the chewing type of mouth parts. The mouth parts of thrips (Fig. 4.12) are unique, in some respects being intermediate between the chewing and the piercing-sucking types. They have two pairs of palps,

[1] HINDS, W., "Monograph Thysanoptera of N. America," *Proc. U.S. National Museum,* **26**:79–242, 1902; WATSON, J., "Synopsis and Catalogue of Thysanoptera, N. America," *Fla. Agr. Exp. Sta. Bul.* 168, pp. 1–100, 1923.

like the former, but the mandibles and maxillae are suggestive of the form in the Hemiptera and Homoptera. The head bears well-developed compound eyes and ocelli, and well-developed but not extremely long antennae of six to nine segments. The head capsule tapers downward in the shape of a cone to a small mouth opening at its lowermost part (Fig. 6.24,7). Around this opening are the two maxillary and the smaller two labial palps. The labium is not elongated into a beak, but fused into the head cone as are the bases of the maxillae and the labrum. In and out of this funnel-like opening three slender jabbers or stylets operate by end thrusts to lacerate the epidermis of the plant. The sap which exudes is then sucked up through the mouth of the cone, there being no long food channel. The three stylets consist of the left mandible (the right being degenerated) and the two maxillae.

Two other structural peculiarities of thrips, the wings and the tarsi, need explanation. In some species one of the sexes and in other species both sexes, are wingless. Many of the species have four wings which are extremely narrow, almost without veins and laid back over the abdomen at rest. The wing membrane would scarcely be sufficient to sustain the insects in flight. The wing, however, is fringed with close-set long hairs to furnish resistance to the air in flight, somewhat like the long feathers on the wing of a bird. This gave the order the name Thysanoptera, which means *bristle wings*. The other peculiarity is in the foot. The tarsus has one or two segments but usually *no claws*. It ends in a hoof-like or cup-like depression surrounding a small bladder that can be protruded or withdrawn by the insect. This characteristic gave the insects the name Physopoda (*bladder-footed*), which is sometimes used instead of Thysanoptera.

Thrips are very active insects, at least when disturbed. They spring or fly readily, they turn up their tails at one, as if to sting, and, according to one writer, they spend most of their time combing the hairs of the body. The compound eyes are small; ocelli are usually present.

The eggs are laid on the tissues of plants or, in some species, inserted into slits made by a sharp ovipositor. Parthenogenesis is common. There are four or more nymphal instars. The last two do not feed and may be quite inactive—a foreshadowing of the complete metamorphosis of the higher orders.

Small to minute, mostly phytophagous, slender-bodied insects, wingless or with two pairs of very slender, nearly veinless, equal wings fringed with long hairs and laid longitudinally over the back when not in use. Mouth parts rasping-sucking, asymmetrical, with two pairs of palps but only one mandible. The tarsus consists of one or two segments and terminates in a protrusile bladder. Cerci wanting. Metamorphosis gradual, but the larger nymphal stages quiescent.

The suborder Terebrantia includes very agile, rapidly running and jumping species. Wings are usually present, with microscopic hairs on the membrane and more than one vein on the front wing. In the female, the last segment is cone-shaped and bears a saw-like ovipositor for inserting eggs into the tissues of plants. The last segment in the male is bluntly rounded. The family Thripidae contains a number of injurious species such as the onion thrips (page 659), the greenhouse thrips (page 863), the pear thrips (page 742), the citrus thrips (page 800), and the gladiolus thrips (page 864). Several species are vectors of the spotted wilt virus of tomatoes (page 19). The onion thrips, the corn or oat thrips, *Limothrips cerealium*, the blossom thrips, *Thrips marginalis*, and other species often swarm in enormous numbers on hot dry days and plague human beings by getting into their eyes, ears, nose, mouth, and clothes and sometimes by biting as they probe for moisture, so as to produce a skin rash.

The suborder Tubulifera includes less active species whose wings, if present, are without pubescence and without veins or with a single longitudinal vein. The last abdominal segment is tubular in both sexes, and the female has no ovipositor, eggs being laid on the surface of plants.

ORDER HOMOPTERA (HEMIPTERA OR RHYNCHOTA, IN PART)[1]

The Cicadas, Aphids, Scale Insects, Leafhoppers, Treehoppers, and Others

For many years authors included the above kinds, together with the leaf bugs, stink bugs, bed bugs, squash bugs, aquatic bugs, and their relatives, all in the order Hemiptera. They have important points in common. *They have a gradual metamorphosis. The mouth parts of all of them are of the piercing-sucking type, with four stylets, and without palps. There are four wings or none. The antennae are of few segments. Compound eyes and ocelli are usually present and cerci wanting.*

They have such important differences, however, that most recent authors separate the above groups into two orders, Homoptera and Hemiptera. Except for the cicadas, the Homoptera are mostly small, inconspicuous insects. Some, however, are brilliantly colored and many are of grotesque shapes. Many of them are wingless, at least in the female sex or under certain conditions. *When wings are present, they are four in number, of a nearly uniform, membranous, or sometimes somewhat leathery texture; the front pair is longer, the hind pair often wider; they usually stand sloping roof-shaped over the abdomen when at rest; their bases are never abruptly thicker than their tips; and they do not overlap much at the tip. An ovipositor sometimes well developed. All terrestrial in habit.* Male scale insects are an exception, having only one pair of wings (Fig. 15.20) and a complete metamorphosis. Some are very degenerate in form, the female scale insects having neither body regions, eyes, wings, nor legs (Fig. 4.5,*Fc*).

The labium or beak of the Homoptera attaches to the head near its hinder part, often seeming to arise from between the front legs, which touch the head; sometimes, in sessile forms, the labium is very short, apparently wanting. The food is exclusively the sap of plants. There are many extremely destructive species, some carrying diseases from plant to plant.

The important families are listed in the following synopsis:

SYNOPSIS OF THE ORDER HOMOPTERA

A. **Suborder** *Auchenorhynchi.* The Free Beaks. Labium plainly attached to head. Tarsi three-segmented. Active, free-moving insects. Antennae very small, ending in a bristle. Medium-sized to very large insects. Females generally with a stiff ovipositor.

Family 1. The cicadas ("locusts"), Family Cicadidae.

Family 2. The spittle bugs or froghoppers, Family Cercopidae (Figs. 6.25 and 6.26).

Family 3. The treehoppers, Family Membracidae (Fig. 6.27).

[1] Van Duzee, E. P., "Catalogue of Hemiptera," *Univ. Calif. Publ. Ento.*, vol. 2, 902 pp., 1917; Britton, W. E. (ed.), "Hemiptera of Connecticut," *Conn. Geol. Natural History Surv. Bul.* 34, pp. 24–382, 1923.

Family 4. The leafhoppers, Family Cicadellidae (Jassidae).

Family 5. The planthoppers, Family Fulgoridae (Fig. 6.28).

B. Suborder *Sternorhynchi.* The Fused Beaks. Labium appears to attach to the thorax between the front legs. Tarsi one- or two-segmented. Females sluggish or sedentary, without a stiff ovipositor. Antennae larger, not ending in a bristle, sometimes wanting. Nearly all very small insects.

Family 6. The jumping plant lice, Family Chermidae (Psyllidae).

Family 7. The plant lice or aphids, Family Aphidae.

Family 8. The whiteflies, Family Aleyrodidae.

Family 9. The scale insects, Family Coccidae.

THE FAMILIES OF HOMOPTERA[1]

Family Cicadidae.[2]   The Cicadas (Also Wrongly Called Locusts). These are the largest of all the Homoptera, certain tropical species reaching a length from head to tip of wings of nearly 4 inches; some species, however, are only ½ inch long. The bodies are wedge-shaped, the heads very broad, eyes slightly bulging at the sides, the abdomen tapering rapidly behind. The forewings are longer than the body, lens-shaped, glistening, prominently veined; the hind pair half as long. The face is prolonged downward and backward V-shaped to the base of the slender, stiff labium (Fig. 4.6). The body is generally patterned with olive-green, reddish-yellow, or whitish pruinose, irregularly shaped spots.

The cicadas are best known from the loud, shrill calls made by the males; the females have no sound-making apparatus.

Happy are cicada's lives
For they all have voiceless wives.

The sounds are produced by shiny, tense membranes on the sides of the body just above the hind legs. Powerful muscles attach to the inside of these membranes and vibrate them so rapidly as to give the high-pitched sound. No obvious ears are possessed by these insects, and the use of the sound, if any, is not known. Another noteworthy thing about the cicadas is the very long life of the young in one North American species, the periodical cicada, *Magicicada septendecim* (Fig. 15.46), which spends either 13 or 17 years as slowly growing nymphs underground, sucking the sap from the roots of trees. After this long subterranean existence they emerge in the form best known, the winged adults. The adults sometimes damage trees by splintering the twigs as they lay their eggs (Fig. 1.7,*A*).

FIG. 6.25. Adult spittle bug, parent of the nymphs which make the masses of spittle. (*From Garman, Conn. Agr. Exp. Sta. Bull.* 230.)

Family Cercopidae.[3]   The Spittle Bugs. The spittle bugs or froghoppers are ⅛ to ½ inch (typically ¼ inch) long and of inconspicuous brown color and broad, suboval shape (Fig. 6.25). A few species are brightly marked with yellow or red bands or spots. The wings are not clear but conform in color to the body, about which they fit closely. The head is somewhat like that of the cicadas but flat on top, with a sharp angle between vertex and front, the eyes not projecting, antennae inserted beneath the margin of the vertex and between the eyes. The thorax is simple, and the hind tibiae are armed with one or two stout spurs along their length and an incomplete circle of similar spines at the end, which is characteristic for the family. They are most remarkable for the frothy, spittle-like masses of sap that the nymphs whip up and beneath which the young rest, feeding on the plant (Fig. 6.26).

---

[1] These families have been raised to superfamily status by some authors.

[2] LAWSON, P. B., "Cicadidae of Kansas," *Univ. Kans. Sci. Bul.* 12, pp. 307–352, 1920.

[3] DOERING, K. C., "Synopsis of the Family Cercopidae in North America," *Jour. Kans. Ento. Soc.,* **3**:53–108, 1930.

Family Cicadellidae[1] (Also Called Jassidae). The Leafhoppers. The leafhoppers are similar to spittle bugs in general shape (Fig. 4.5,*J*) and size, the largest North American species being scarcely ½ inch long. They are, however, generally more slender and more pointed behind. The average size is close to ¼ inch. They are often brilliantly colored. Greens, yellows, blues, and reds of various shades are arranged in the most startling patterns of stripes, spots, and bands, which rival in beauty that of the birds and butterflies. Many others which are of more somber brown, yellow, black, white, or greenish hues have, nevertheless, when magnified so that they can be appreciated, very beautiful patterns. The wings conform in color to the body and are often most brilliantly colored of all. There is a distinct tendency for the head to be prolongated forward as a smooth, flat, triangular, or shovel-shaped process. The thorax is simple in form. The hind legs have two parallel rows of spines along the tibia instead of the circlet of spines at the end of the tibia which are found in the spittle bugs. The antennae are placed as in the cercopids. Leafhoppers are often seen in abundance in grassy fields and at night swarming about lights. They are specially interesting to the economic entomologist because the feeding of some species causes certain destructive plant disorders. Among the very important species are the apple and potato leafhopper (page 643 and Fig. 14.27), the cause of potato tipburn; the beet leafhopper (page 676), which causes the destructive curly top of sugar beets; and the grape leafhopper (page 777).

Family Membracidae.[2] The Treehoppers. The treehoppers (Fig. 6.27) are grotesque, freakishly shaped, and among the most interesting of insects, structurally. The prothorax is enlarged, inflated, and prolongated into the most outlandish

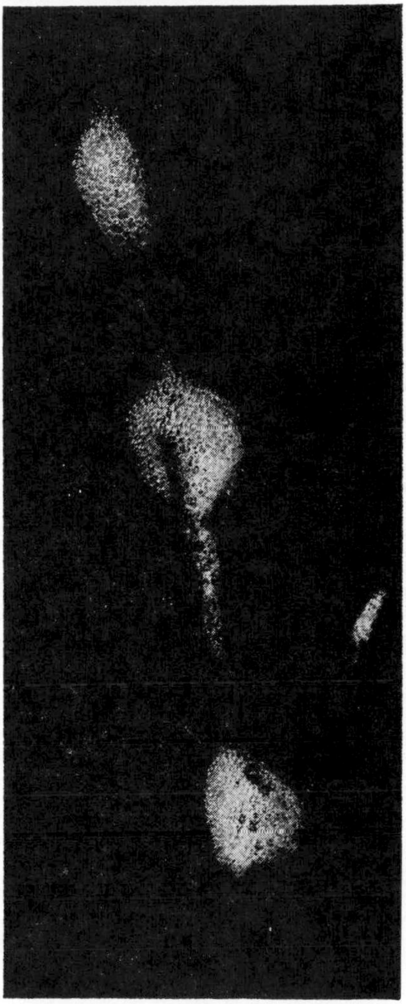

FIG. 6.26. Masses of spittle made by the grass-feeding spittle bug, *Philoenus lineatus*, as a protection for the nymphs which live within it. About 3 times natural size. (*From Osborn, Maine Agr. Exp. Sta. Bul*. 254.)

[1] OSBORN, H., "Leafhoppers Affecting Cereals, Grasses and Forage Crops," *U.S.D.A., Bur. Ento. Bul.* 108, p. 123, 1912, and "Cicadellidae of Ohio," *Ohio State Univ. Bul.* 32, pp. 199–374, 1928; DELONG, D. M., "Cicadellidae of Illinois," *Bul. Ill. Natural History Surv.* 24, pp. 97–376, 1948; LAWSON, P. B., "Cicadellidae of Kansas," *Univ. Kans. Sci. Bul.* 12, 306 pp., 1920; MEDLER, J. T., "Cicadellidae," *Tech. Bul. Minn. Agr. Exp. Sta.* 155, 196 pp., 1942.

[2] FUNKHOUSER, W. D., "Biology of the Membracidae of the Cayuga Lake Basin," *Cornell Univ. Memoir* 11, pp. 177–445, 1917; GODING, F. W., "Catalogue of Membracidae of North America," *Bul. Ill. State Lab. Natural History* 3, pp. 391–482, 1892; VAN DUZEE, E. P., "Review of North American Membracidae, *Bull. Buffalo Soc. Natural Sci.* 9, pp. 29–129, 1908.

shapes, many of which have no known use or adaptive explanations. It is impossible to describe the shapes of these insects, except by illustrations. Some are paper-thin from side to side, some are plump and smoothly rounded above, others are provided with horns, warts, spines, and humps, the variety of which defies description. It is usually the pronotum, or upper part of the thorax, that is responsible for these grotesque shapes. It usually extends backward to cover most, if not all, of the abdomen

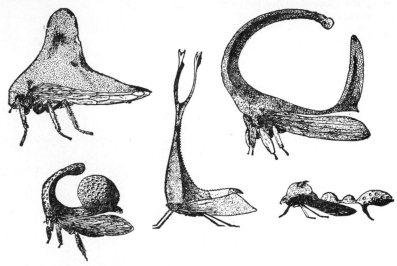

FIG. 6.27. A group of treehoppers or membracids, showing the remarkable development of the pronotum. (*From Funkhouser, Cornell Agr. Exp. Sta.*)

and wings. The colors are usually browns, yellows, and greens, sometimes splotched with white, yellow, or darker shades of the prevailing color. The beauty of the treehoppers is due, however, chiefly to shape rather than color. The average size is between ¼ and ⅓ inch. The nymphs are often very spiny, and both they and the adults hop vigorously when disturbed. As in the cicadas, injury is chiefly by egg-laying in the twigs of woody plants. The buffalo treehopper (page 739) is a species of economic importance.

Family Fulgoridae.[1] The Planthoppers or Fulgorids. Many of the planthoppers have "run to head" much as the treehoppers have "run to thorax." This is a very large family of very diverse forms, agreeing in having the antennae directly beneath the eyes on the side of the head (Fig. 6.28). The first two segments of the antennae are large, the remaining ones very slender. Many of the species are shaped like spittle bugs and leafhoppers; others are very broad, short, seed-like. A common shape is with an upwardly curved, slender, horn-like projection of the head. The wings are variable in shape, venation, and color, frequently short, leaving much of the abdomen exposed. They are often partly or wholly transparent, but more generally conform to the color and pattern of the body and are sometimes greatly expanded and

FIG. 6.28. A fulgorid, *Stoboera tricarinata*. (*From Z. P. Metcalf, Jour. Elisha Mitchell Sci. Soc., vol. 38.*)

[1] METCALF, Z. P., "A Key to the Fulgoridae of Eastern North America," *Jour. Elisha Mitchell Sci. Soc.*, **38**:139–230, 1923; OSBORN, H., "Fulgoridae of Ohio," *Bul. Ohio Biol. Surv.* 35, pp. 283–349, 1938.

look like the wings of moths. Although somber hues predominate, many species are most brilliantly and strikingly colored. There are no very destructive species known in the United States, although they occur on many kinds of plants. The nymphs are frequently covered with filaments of white wax.

Family Chermidae[1] (Also Called Psyllidae). The Jumping Plant Lice. The jumping plant lice differ from true plant lice especially in the development of the hind legs for leaping, in the venation of their wings, the much firmer bodies, and the usually very flat nymphs (Fig. 15.49). A great many species, unlike the aphids, have the wings dark-colored and patterned like the body and superficially resemble leafhoppers. The prevailing colors are browns and yellows and black, the body frequently spotted with yellow. Structurally they are close to the Cicadidae, except for the long filiform antennae which project horn-like in front of the eyes, and terminate in two rather conspicuous, short hairs. Most of the species are about ⅛ inch long. The pear psylla (page 743) and the apple sucker, *Psylla mali*, are two species of importance in America, although of somewhat limited distribution.

Family Aphidae.[2] The Aphids or Plant Lice. The aphids or plant lice (Fig. 18.12) are among the smallest, most defenseless, and most preyed upon of all insects; yet because of their immense vitality and extraordinary fecundity, due to the shortness of the life cycle and their ability to reproduce parthenogenetically, they cover the earth with an enormous assemblage of species and tons of individuals affecting nearly every kind of green plant. They can usually be distinguished by the pair of oil- or wax-secreting tubes (cornicles) on the upper side of the fifth or sixth abdominal segment (Fig. 4.5,*B*) and by their two-segmented tarsi. The majority of adults are wingless, but the males and certain agamic females (migrants) have four clear wings (Fig. 9.25). The head and thorax are short, the abdomen swollen and very soft-walled. The antennae are long, slender, and provided with many complicated sensory pits. While the prevailing color is green, many species are black, rosy, yellow, or bluish. The average length is close to 1/10 inch.

The species are for the most part societies of females, which are of distinct types and are interwoven into remarkable life cycles, parthenogenetic throughout most of the year, and often with a remarkable alternation of generations migrating from one kind of plant to another. Typically, males appear only in the fall to fertilize the overwintering generation of eggs. In a number of species males have never been found. Practically all aphids feed on the sap of plants and may attack either leaves, twigs, fruits, or roots, often causing abnormal growths or deformities. Much of the sap ingested is excreted as honeydew (page 109). Among the hundreds of destructive species are the cabbage aphid (page 667), the corn root aphid (page 501), the greenbug (page 527), the green peach aphid (page 754), the melon aphid (page 631), the pea aphid (page 626), the rose aphid (page 873), the rosy apple aphid (page 710), the spotted alfalfa aphid (page 561), and the woolly apple aphid (page 733).

Family Aleyrodidae.[3] The Whiteflies. Whiteflies differ from aphids in having body and wings covered with a white dust or powder; in that all adult individuals are winged; and that the nymphs, after the first instar, are flattened, sessile, and scale-like (Fig. 18.13). They differ from scale insects in that both sexes are active and injurious; that they have four wings; and that they feed exclusively on leaves. The nymphs are never covered with a separable scale or shell and are often surrounded by a fringe of fine, radiating, waxy, white threads or rods. The species rarely exceed ⅛ inch across the wings, and the adults look like tiny white moths. The eggs are stalked, and,

---

[1] CRAWFORD, D. L., "A Monograph of Psyllidae of the World," *U.S. National Museum Bul.* 85, 1914.

[2] BAKER, A. C., "Generic Classification of the Hemipterous Family Aphididae," *U.S.D.A., Dept. Bul.* 826, 1920; ESSIG, E. O., "Aphididae of California," *Univ. Calif. Publ. Ento.*, **1**:301–346, 1917; GILETTE, C. P., and M. A. PALMER, "Aphididae of Colorado," *Ann. Ento. Soc. Amer.*, **24**:827–943, 1931; HOTTES, F. C., and T. H. FRISON, "Aphididae of Illinois," *Bul. Ill. Natural History Surv.* 19, pp. 121–447, 1931.

[3] QUAINTANCE, A. L., and A. C. BAKER, "Classification of the Aleyrodidae," *U.S.D.A., Bur. Ento. Tech. Bul.* 27, 1913; SAMPSON, W. W., "Generic Synopsis of Aleyrodoidea," *Ento. Amer.*, **23**:173–223, 1943.

to fertilize the eggs, the sperms are said to migrate through the stalk which subsequently shrivels. A very characteristic structural feature is a vase-shaped orifice within which the anus opens, beneath a strap-shaped organ and broad operculum, the whole structure being known as a *vasiform orifice*. An abundance of honeydew is secreted through this opening. Among the most destructive species are those attacking citrus (page 808) and greenhouse plants (page 874).

Family Coccidae.[1] **The Scale Insects and Mealybugs.** The best known of the scale insects are the females of the armored scales (Fig. 4.5,*E,F*), which are legless, wingless, eyeless, degenerate creatures, covered with a convex shell of secretory material beneath which they live (Fig. 18.4) and suck the sap from various parts of plants. The unarmored scales are similar-looking and likewise sessile in all but the first instar. They usually retain legs throughout life and, although generally provided with a hard cuticle, do not form a separable shell over the body. The mealybugs (Fig. 4.5,*D*) are dusted all over with white powder and often have long radiating threads of the same secretion around the margin of the body. All females are wingless, and consequently the dispersal is accomplished by the crawling first-instar nymphs and by the transportation of infested plants. The males of all species are insect-like, minute "gnats," sometimes wingless but usually provided with a single pair of wings and without mouth parts (Fig. 15.20). The males and females thus exhibit a remarkable sexual dimorphism, being so different as to appear like creatures of different orders or classes (*cf*. Figs. 4.5,*Fc*, and 15.20). A structural characteristic of this family is the one-segmented tarsi with a single claw.

The females of the armored scales, although legless, rotate beneath the shell, placing the waxy secretion from their tail ends about them in concentric rings to enlarge the shell or in elongate forms, such as the oyster-shell scale (Fig. 4.5,*C*), in crescent-shaped additions toward the posterior end by swinging the body from side to side in a short arc.

Among the most important of the several hundred destructive species in North America are the black scale (page 805), the brown scale (page 869), California and Florida red scales (page 803), citricola scale (page 806), cottony maple scale (page 837), European elm scale (page 836), oyster-shell scale (page 835), pine needle scale (page 835), purple scale (page 802), San Jose scale (page 705), and many other scales (page 866) and mealybugs (page 869) attacking ornamental plants.

ORDER HEMIPTERA (HETEROPTERA, OR RHYNCHOTA, IN PART)[2]

THE TRUE BUGS

These bugs (Fig. 6.29) are like the Homoptera in having a gradual metamorphosis; two pairs of wings; antennae generally of five or fewer segments, though they may be long, often concealed in grooves; piercing-sucking mouth parts without palps; and the hind wings usually shorter and wider than the front pair. They are distinguished from the Homoptera by having the front pair of wings thickened and quite stiff about the basal half, the distal half abruptly thinner, usually membranous. When folded back they lie horizontally, or flat, over the back, and the membranous tips of the front pair overlap. There are many wingless forms and some adults with short wings often in the same species with long-winged ones.

[1] DIETZ, H. F., and H. MORRISON, "Coccidae of Indiana," Office of the State Entomologist, Indianapolis, pp. 195–321, 1916; MacGILLIVRAY, A. D., "The Coccidae," Scarab Co., 1921; FERRIS, G. F., "Atlas of Scale Insects of North America," ser. 1, 1937; ser. 2, 1938; ser. 3, 1941; ser. 4, 1942; ser. 5, 1950; ser. 6, 1953; ser. 7, 1955, Stanford Univ. Press.

[2] VAN DUZEE, E. P., "Catalogue of Hemiptera," *Univ. Calif. Publ. Tech. Bul.* (*Entomology*) 2, 1917; BRITTON, W. E. (ed.), "Hemiptera of Connecticut," *Conn. Geol. Natural History Surv. Bul.* 34, pp. 383–807, 1923; BLATCHLEY, W. S., "The Heteroptera or True Bugs of Eastern North America," Nature Publ. Co., 1926; TORRE-BUENO, J. R., "Synopsis of the Hemiptera Heteroptera of North America," *Ento. Amer.*, **19:**141; **21:**41; **26:**1; 1939–1946.

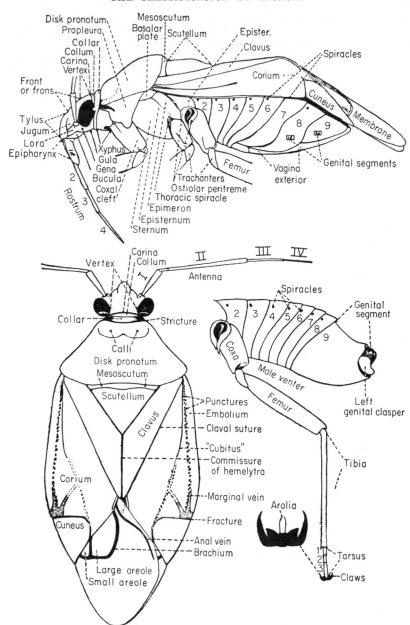

Fig. 6.29. A bug of the order Hemiptera, *Deraeocoris fasciolus*, with the principal parts of the body named. The part labeled epipharynx is called the *labrum*, and the rostrum, the *beak*, by some authorities. The parts of the wing labeled *clavus*, *corium*, and *cuneus* are thick and horny; the part labeled *membrane* is thin and transparent. (*From Knight, Minn. Agr. Exp. Sta. Tech. Bul. 1.*)

The labium or beak of the Hemiptera attaches well forward near the ront end of the head, and the head is free from the coxae and longer than it is in the Homoptera.   The prothorax is large and distinct, and the triangular scutellum (Fig. 6.29) of the mesothorax separates the bases of the wings and is sometimes very large.   The tarsi are three-segmented.   All this order are insect-like in form, typically flattened, there being no extremely degenerate species.   *They commonly possess scent glands that give them a distinct odor*, usually offensive to man and probably defensive against their natural enemies.   The various species attack a wide variety of both plants and animals, always feeding on liquid parts only.   Those that attack plants generally have the labium long, cylindrical, and straight; those that prey on other insects or on larger animals have a short, curved, pointed labium. Many are adapted to live in the water.

The important families are listed in the following synopsis.

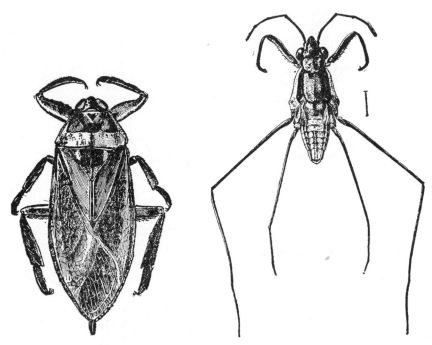

FIG. 6.30. Giant water bug or electric-light bug, *Lethocerus americanus.* Natural size.   (*From Kellogg, "American Insects."*)

FIG. 6.31. A water strider.   Line shows actual length of body.   (*From Osborn, "Agricultural Entomology."*)

SYNOPSIS OF THE ORDER HEMIPTERA

A. **Suborder** *Cryptocerata.*   The Short-horned Bugs.   Antennae shorter than the head, nearly concealed on the underside of the head.   Stink glands wanting.
   1. *Aquatic Predaceous Bugs.*
      Family   1. Water boatmen, Family Corixidae (Fig. 6.32).
      Family   2. Back swimmers, Family Notonectidae (Fig. 6.32).
      Family   3. Water-scorpions, Family Nepidae (Fig. 6.33).
      Family   4. Giant water bugs, Family Belostomatidae (Fig. 6.30).
B. **Suborder** *Gymnocerata.*   The Long-horned Bugs.   Antennae at least as long as head and plainly visible at its sides.

2. *Semiaquatic Predaceous Bugs.*
Family 5. Water striders, Family Gerridae (Fig. 6.31).
3. *Terrestrial Predaceous Bugs.*
Family 6. Assassin bugs, Family Reduviidae.
Family 7. Bed bugs, Family Cimicidae.
Family 8. Damsel bugs, Family Nabidae.
Family 9. Ambush bugs, Family Phymatidae (Fig. 6.34)
Family 10*a*. Stink bugs, Family Pentatomidae (in part).
Family 11*a*. Leaf bugs, Family Miridae (Capsidae) (in part).
4. *Terrestrial Plant-eating Bugs.*
Family 10*b*. Stink bugs, Family Pentatomidae (in part).
Family 11*b*. Leaf bugs, Family Miridae (Capsidae) (in part).
Family 12. Lace bugs, Family Tingidae (Fig. 6.35).
Family 13. Chinch bugs and others, Family Lygaeidae.
Family 14. Squash bug, leaf-footed bugs, and others, Family Coreidae (**Fig. 14.20**).

The Principal Families of Hemiptera

Family Corixidae.[1] **The Water Boatmen.** These chunky, obscurely mottled bugs, which are $\frac{1}{8}$ to $\frac{3}{8}$ inch long, are equally proficient at swimming and flying. They drive their bodies through the water in an upright position by strokes of the oar-like hind legs (Fig. 6.32,*B*). The tarsi of the front legs are rake-like and are used for securing food, and the middle pair serve for clinging to some object under water while they feed. Some species are said to eat larvae of mosquitoes, but probably their chief food is algae or the ooze from the bottom of ponds. Air for breathing is carried under water as a bubble beneath the wings, and a silvery film covers much of the body. The adults fly from pond to pond and are frequently attracted to artificial lights. The head is very broad, beautifully rounded in front, and overlaps the prothorax above. The labium is very short and flat, hardly separated from the head. The antennae are concealed beneath the edge of the head. The front wings are leathery throughout, usually cross-barred with irregular bands of brown and yellow, and rounded at apex to correspond with the rounded head, between which extremities the sides of the body are nearly parallel.

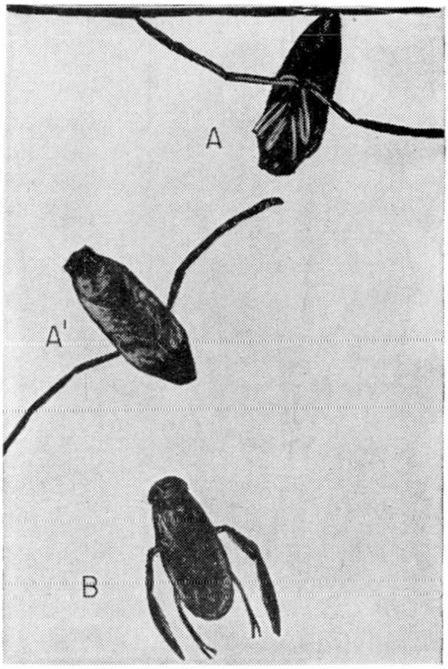

FIG. 6.32. Notonectids and corixid. *A*, notonectid at the surface of water, showing undersurface; *A'*, swimming, showing upper surface; *B*, corixid swimming. Somewhat enlarged. (*From Linville and Kelly, Textbook in General Zoology.*)

Family Notonectidae.[2] **The Back Swimmers.** These bugs (Fig. 6.32,*A,A'*) superficially resemble the water boatmen in shape and long oar-like hind legs, but have the remarkable habit of swimming upside down. The back is convex, like the bottom of a boat. On the underside of the abdomen is a median keel from which a fringe of hairs extends outward toward each side, meeting an inwardly directed fringe from the side of the body and thus enclosing two longitudinal channels or troughs beneath the abdomen, in which air is carried and into which the spiracles open. They suck the

---

[1] HUNGERFORD, H. B., "Corixidae of Western Hemisphere," *Univ. Kans. Sci. Bul.* 32, 827 pp., 1948; and "Biology and Ecology of Aquatic and Semiaquatic Hemiptera," *Univ. Kans. Sci. Bul.* 11, pp. 3–265, 1919.

[2] BLATCHLEY, W. S., *op. cit.,* pp. 1048–1061.

blood of small animals, especially insects, and do not hesitate to bite man severely when they are handled. The front and middle legs are stout, the hind pair elongated and working in a horizontal plane past the sides of the body as oars. The antennae are concealed beneath the eyes. The beak is distinct but not very long. In shape the bodies are oblong oval, a little more pointed behind. In size the back swimmers range from the tiny, convex, beetle-like *Plea* up to *Notonecta* more than ½ inch long. The colors range from pure buff to nearly black, several species being strikingly splotched with white.

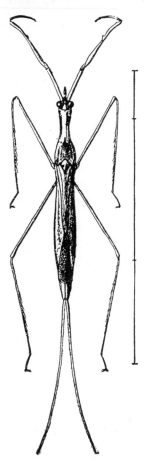

Family Nepidae.[1] **The Water Scorpions.** The members of this family are elongate, stick-like bugs, an inch or two in length, some species (*Nepa*) very slender and cylindrical (Fig. 6.33), others (*Ranatra*) flattened, and usually provided with a tail tube from one-half as long to nearly as long as the body, which is thrust up to the surface film and conveys air to the spiracles at the tip of the abdomen. From the tip of the extended front legs to the apex of the tail tube may exceed 3 inches. Water scorpions are not good swimmers but crawl about in shallow stagnant water among sticks and bottom rubbish and grab with their raptorial front legs small animals upon whose blood they feed. The middle and hind legs are slender, the front pair thickened, their coxae very elongated, the femur grooved to receive the tibia as the handle of a pocketknife receives the blade. The very short beak is directed forward, and the antennae are concealed beneath the eyes. The colors are plain browns, which help to conceal them from their prey.

Family Belostomatidae.[2] **The Giant Water Bugs.** Measuring from 1 to 4 inches long, by one-third to one-half as broad, these hard, broad, flat bugs (Fig. 6.30) never fail to excite curiosity when, as frequently happens, they are found about artificial lights. It is only the adults that are thus lured to lights as they fly from pond to pond, for their natural habitat is on the bottom of ponds, where the eggs of some species are laid on the stems of water plants and where the young grow to maturity. When in the water they carry air under the wings and also get a supply through a tube on the tip of the abdomen, which is, however, much shorter than in the water scorpions. The middle and hind legs are flattened and oar-like, the front pair specialized for grasping. They catch tadpoles, small fishes, and insects, from which they suck the blood.

FIG. 6.33. Water scorpion, *Ranatra americana*. Line shows natural size. (*From Osborn, "Agricultural Entomology."*)

They are capable of biting severely when handled. An amusing habit of certain females is to glue their eggs all over the backs of the males, who are obliged to carry them about until the nymphs hatch. The body is broadest near the middle, the wings tapered to a point behind, the thorax and head correspondingly narrowed in front. The head is prolonged forward and downward, between the eyes. The beak is distinct, and the antennae completely hidden under the eyes. The colors are obscure dull brown.

Family Gerridae[3] (Also Called Hydrobatidae). **The Water Striders.** This is the largest of a group of several families (Veliidae, Mesoveliidae) of similar habits and

[1] HUNGERFORD, H. B., "Nepidae of North America North of Mexico," *Univ. Kans. Sci. Bul.* 14, pp. 425–469, 1922.

[2] CUMMINGS, C., "Belostomatidae," *Univ. Kans. Sci. Bul.* 21, pp. 197–219, 1934.

[3] DRAKE, C. J., and H. M. HARRIS, "Gerrinae of Western Hemisphere," *Ann. Carnegie Museum*, **23**:179–240, 1934.

appearance, whose members walk, skip, or glide over the surface of the water, supported by the surface film. They may even jump from the surface of the water into the air to catch small insects. The body is light in weight and usually slender, four or five times as long as broad (Fig. 6.31), and covered with a waterproof coat of very delicate hairs. The front legs are modified to hold prey; the second and third pairs are very long and slender, often twice as long as the body. The tarsi are covered with fine hairs, which prevent wetting, and these indent the surface film without breaking through it. According to Blatchley, it is the middle legs alone that propel the body, the others dragging or skating on the surface film. They are common in small streams and the margins of lakes and ponds. The eggs are fastened to aquatic vegetation. The food of nymphs and adults consists of living or dead floating insects and other animals. In contrast with all the aquatic insects described above, the antennae are long, slender, and not concealed. Wings may be wanting, fully developed, or half size, even in members of the same species. The head is directed forward, the slender pointed beak descending from its front end. The claws of the hind tarsi are attached a short distance before the apex of the segment. The coxae are very far apart, attached toward the sides of the body. While typical species average about $\frac{1}{2}$ inch long by one-fourth or one-fifth as broad, some are short and broad and not over $\frac{1}{8}$ inch long. The colors are obscure brown, except for white patches formed by the coat of hair and occasional yellow spots.

Family Reduviidae.[1]   **The Assassin or Kissing Bugs.** This is a very large family of mostly flattened, oval bugs (Fig. 21.8) with narrow heads, the beak rigid and bowed away from the front of the head, its tip resting in a groove on the underside of the prothorax. There are two or three large basal cells in the membrane of the wings. The assassin bugs catch small insects and suck their blood as food. Species of *Triatoma* and *Reduvius* readily enter human habitations and inflict painful bites (page 1012). In Central and South America they are the vectors of a serious trypanosomiasis, Chagas' disease (Table 1.2). Other species bite severely when handled. Common species are $\frac{1}{2}$ to $\frac{2}{3}$ inch long, others from $\frac{1}{3}$ to $1\frac{1}{2}$ inch, black or brown, and of such an appearance as easily to be confused with spiders by those not too observant. Many of the reputed spider bites are believed to be due to these insects. While the colors are usually uniform brown or black, some species are marked with red, white, or black spots or bands in a manner supposed to warn predaceous animals that they are vicious biters and thus gain for them a measure of freedom from attack. Two of the most aberrant species are the wheel bug, *Arilus cristatus*, with a crest, like an erect, half cogwheel on the back of the thorax, and the thread-legged bug, *Emesaya brevipennis*, which is extremely slender, elongate, and looks like a walking-stick but has grasping front legs and, of course, piercing-sucking mouth parts. Typically, the abdomen is somewhat expanded at the middle, the margins thin, sharp, and exposed beyond the sides of the wings. The thorax is often smaller and armed with spines, divided by a transverse groove, in front of which it narrows and behind which it widens gradually to the base of the wings. The head is usually very narrow and elongate, forming a slender "neck" behind the eyes. There is an impressed groove across the top of the head between the eyes. The antennae are slender, often nearly as long as the body.

Family Cimicidae.   The Bed Bugs. Besides the common bed bug (page 1011), there are several other species in various parts of the world that attack man, and others that feed on bats, swallows, and poultry. All are wingless or nearly so, without ocelli, and are flattened, bad-smelling, tough-skinned bloodsuckers. The thorax is deeply notched in front to receive the short head, up to the bulging eyes. The third and fourth segments of the antennae are conspicuously more slender than the first and second. The abdomen is subcircular in outline, very flat, prominently segmented, and entirely unprotected by wings. The wings are represented only by the greatly atrophied, short, oval pads of the first pair. The body is covered with thick-set short hairs. The common bed bug reaches a length of about $\frac{1}{5}$ inch.

[1] FRACKER, S. B., "Reduviidae of North America," *Proc. Iowa Acad. Sci.*, **19**:217–47, 1912; READIO, P. A., "Biology of Reduviidae of America North of Mexico," *Univ. Kans. Sci. Bul.* 17, pp. 1–248, 1927.

Family Nabidae.[1]   **The Damsel Bugs.** This is a small family of predaceous bugs, mostly beneficial because they feed on other insects.   They resemble small assassin bugs but have four long veins in the wing membrane, which are frequently subdivided by short branches, especially around the margin, and there are four segments in the beak instead of three.   Damsel bugs are smooth-looking, rather straight-sided and slender bugs, with long, slender legs and very slender antennae.   The front legs are somewhat modified for grasping prey, the tibiae being spiny and the femora swollen, but not greatly so, since their prey is mostly soft-bodied caterpillars and insects such as aphids.   The prothorax narrows strongly and evenly from its base to the head, which is slender, rather long, with bulging eyes.   The colors are never conspicuous, mostly brown or black, sometimes finely lined and spotted with black or light gray, which spotting often extends upon the legs.   Most of our species range from $\frac{1}{4}$ to $\frac{3}{8}$ inch long.

Family Phymatidae.[2]   **The Ambush Bugs.** These bugs, of which *Phymata erosa*, a species about $\frac{1}{2}$ inch long, is common throughout the country, will well repay careful study, for their greenish-yellow bodies, irregularly marked with brown, are excellent examples of aggressive resemblance (Fig. 6.34).   Their extraordinarily thickened front legs, with the tarsi neatly packed out of the way in grooves of the tibiae, are marvels of specialization for prehension.   They hide in the heads of flowers and grasp spiders and insects, such as honeybees, which come close enough, and leisurely suck their blood.

The body is somewhat 8-shaped in outline, the abdomen prolonged sideways at mid-length, nearly diamond-shaped, and the basal part of the pronotum also expanded and upturned sideways.   The abdomen is triangular in cross-section, with a sharp, keel-like, median angle on the underside.   The body wall is usually granulated and sometimes bears short spines.   The antennae are short, the terminal segment is enlarged, sometimes more or less concealed in grooves along the sides of the head and prothorax. The base of the beak is guarded by downward-extending expansions of the head.   The wings are narrow, leaving much of the margins of the abdomen uncovered, their membranes large, provided with numerous veins.   In one genus (*Macrocephalus*) the scutellum is greatly expanded backward, leaf-like, to cover nearly all of the wings. The species are from $\frac{1}{4}$ to nearly $\frac{1}{2}$ inch long, by about half as broad.

Family Pentatomidae.[3]   **The Stink Bugs.** Since all Hemiptera have stink glands, "stink bugs" may hardly seem a distinctive enough name for any certain family, yet the members of this family pretty well make it so.   They are among the handsomest of the Hemiptera, having graceful, shield-shaped bodies and often brilliant color markings.   The very large scutellum which is narrowed behind and reaches back to the membrane of the wings is distinctive.   Although primarily sapsuckers, a number of species suck the blood of caterpillars, beetles, and other pests.   The body is broad, flat, generally smooth, but frequently spiny, rugose, or punctured.   The prothorax is broad, often provided with spines, usually broader than the abdomen, and strongly narrowed in front.   The head is narrow, frequently somewhat prolonged forward in front of the eyes above the base of the beak and antennae, as a flattened hood.   The antennae are five-segmented, uniformly slender and filiform, or somewhat clavate. The legs are rather short and generally not spiny.   The membrane of the wings usually has numerous veins, but few or no closed cells.   The size range is from $\frac{1}{4}$ to 1 inch long by about half as broad.   The colors are usually obscure or concealing greens, yellows, or browns, but some of the bad-tasting species are glaringly spotted with red or yellow.   The eggs, which are laid in groups, often have a fringe of beautiful spines surrounding the lid through which the nymph emerges, sometimes resembling grotesquely painted kegs (Fig. 5.1,*L*).   The harlequin bug (page 668) is a very destructive pest of cruciferous plants in the southern states.

[1] Harris, H. M., "A Monographic Study of the Hemipterous Family Nabidae in North America," *Ento. Amer.*, **9**:1–90, 1928.

[2] Evans, J. H., "Revision of North American Phymatidae," *Ann. Ento. Soc. Amer.*, **24**:711–36, 1931.

[3] Van Duzee, E. P., "Pentatomidae of America North of Mexico," *Trans. Amer. Ento. Soc.*, **30**:1–80, 1904; Hart, C. A., "Pentatomoidea of Illinois," *Ill. Natural History Surv. Bul.* 13, pp. 157–223, 1919.

Family Miridae[1] (Also Called Capsidae). The Leaf Bugs. Anyone who sits down in a grassy, weedy spot in early summer, and has eyes to see can scarcely fail to make the acquaintance of some of the hundreds of kinds of mirids that crawl about over the vegetation and feed on its sap. They are $\frac{1}{10}$ to $\frac{1}{4}$ inch long (exceptionally longer), rather soft-skinned, and usually pleasingly colored. They are technically characterized by having a cuneus (Fig. 6.29, an illustration of a common mirid), one or two large cells (areoles) in the membrane of the wing, both antennae and labium four-segmented, and no ocelli. The body is generally elongate oval in outline, rather straight-sided, usually three or four times as long as broad, often clothed with fine hairs. The pronotum is smooth, evenly narrowed to the head, sometimes constricted collar-like in front. The head is short and downbent. The antennae are long, the first segment often thickened. The hind legs are long. The elytra are longer than the abdomen, more or less downbent from the base of the membrane to the tip. The prevailing colors are green, black, or red, often flecked, spotted, or striped with black, yellow, red, white, or hyaline. The lygus bugs (page 565), the tarnished plant bug (Fig. 14.9), and apple redbugs (page 714) are very destructive species. A few kinds feed on insects.

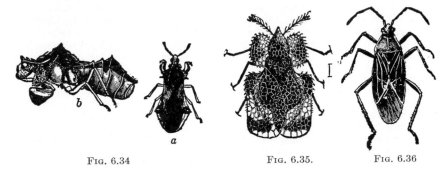

FIG. 6.34   FIG. 6.35.   FIG. 6.36

FIG. 6.34. Ambush bug, *Phymata erosa.* *a*, Adult from above, about twice natural size; *b*, the same from the side, more enlarged. (*From Fernald, "Applied Entomology," after Riley.*)
FIG. 6.35. Lace bug, *Corythuca crataegi.* Line shows actual length. (*From Comstock, "Introduction to Entomology," Comstock Pub. Co.*)
FIG. 6.36. Boxelder bug, *Leptocoris trivittatus.* Twice natural size. (*From Kellogg, "American Insects."*)

Family Tingidae.[2] The Lace Bugs. Lace bugs (Fig. 6.35) look as though they were cut out of fine gauze and will abundantly repay study under a microscope as objects of beauty. Not only the wings but the head and thorax, and sometimes broad, wing-like expansions of the latter, are reticulated with veins and small cells. The upper part of the prothorax often bears sharply elevated longitudinal ridges and swollen knobs. The head is short, and the wings much wider than the abdomen. They are mostly under $\frac{1}{8}$ inch long and very broad and flat. They feed, often in enormous numbers, on the underside of leaves on which their eggs are laid, each egg being covered with a little cone of dark, sticky excrement. The nymphs are usually extremely spiny.

Family Lygaeidae.[3] The Chinch Bug Family. Best known from the very destructive chinch bug, (Fig. 4.5,*H*), this family nevertheless includes several hundred other

[1] Van Duzee, E. P., "Keys to Genera of North American Miridae," *Univ. Calif. Publ. Ento.,* **1**:199–216, 1916; Knight, H. H., "Miridae of Illinois," *Bul. Ill. Natural History Surv.* 22, pp. 1–234, 1941.

[2] Poor-Hurd, M., "Generic Classification of North American Tingidae," *Iowa State Coll. Jour. Sci.,* **20**:429–489, 1946.

[3] Torre-Bueno, J. R., de la, "North American Lygaeidae," *Ento. Amer.,* **26**:1–141, 1946.

American species. All have ocelli, four or five long veins in the wing membrane, which are usually not branched, and the antennae attached low on the head. All are sap-suckers. The body is rather narrowly oval or straight-sided, about three times as long as wide, nearly an equilateral triangle in cross-section. The head is moderately broad and short, the pronotum smooth, nonspecialized, only slightly narrowed toward the head, sometimes strongly constricted at the middle. The antennae and beak are four-segmented, the antennae slightly clavate, the beak often very slender. The front femora are generally swollen, often spiny. The wings cover nearly or quite the full width of the abdomen and often exceed its length, but are sometimes short. The membrane is rather large, usually distinct in color and appearance from the rest of the wings. The scutellum is moderately large. The scutellum, membrane, and corium of the wing are often margined with a distinct color, forming an x on the back of the insect. The Lygaeids are of moderate size, ranging from about $\frac{1}{6}$ to $\frac{5}{8}$ inch long. The colors are often dull grayish brown or black, but many are beautifully marked with red, black, or white lines, spots, or flecks.

Family Coreidae.[1] **The Squash Bug and Its Relatives.** These insects resemble the Lygaeidae but are mostly much larger, with the antennae attached higher on the head in front of the eyes, the pronotum more strongly narrowed to the head, and the wing membrane with from six to many veins, often branching (Fig. 4.5,K). Some species are broad, others very slender. Many forms have parts of the legs greatly swollen or flattened like narrow leaves. They are mostly dull-colored and foul-smelling. The head is small, directed forward, the eyes bulging, behind which the head is constricted neck-like. The antennae are long, four-segmented, frequently ornamented by swellings or small leaf-like expansions. The beak is short, four-segmented. The pronotum is often strongly narrowed and sharply downbent from the base of the wings to the head, the margins often upturned and spiny. The scutellum is not large. The wing membrane is larger, distinct in color and texture. The margins of the abdomen are narrowly exposed beyond the sides of the wings. In some species the base of the abdomen is so constricted as to give a striking resemblance to ants. The colors are mostly browns, sometimes conspicuously banded or striped with yellow or red or with the margins of the abdomen so spotted. The length is commonly $\frac{3}{8}$ to 1.5 inch. The squash bug (page 532) is the most destructive American species of the family. The boxelder bug (page 839), a common black and red species, is sometimes placed in a separate family, Corizidae.

ORDER ANOPLURA (SIPHUNCULATA OR PARASITA)[2]

The Bloodsucking Lice

These lice should be distinguished from the chewing lice. These are much the more serious kind. They live exclusively by sucking the blood of the host, and in so doing are very likely to be carriers of diseases. The members of this order are small, wingless, tough-skinned, flattened, usually dark-colored, external parasites. They attack all kinds of wild and domesticated mammals, but none are known to live on any other class of host. Two species attack man—the cootie, body louse or head louse (page 1026), and the crab louse (page 1029). The first of these is the carrier of epidemic typhus, trench fever, and European relapsing fever. The largest species are about $\frac{1}{4}$ inch long. They can be distinguished from Mallophaga by the difference in mouth parts and by the relatively narrower, more pointed head. The antennae are short, but usually conspicuous, three-, four-, or five-segmented. The eyes are very degenerate or wanting and ocelli absent. The three thoracic segments are fused together and not well separated from the abdomen. The pleura are usually more strongly sclerotized than the rest of the segment. The legs

[1] Deay, H. O., "Coreidae of Kansas," *Kans. Univ. Sci. Bul.* 18, pp. 371–415, 1928.
[2] Ferris, G. F., "The Sucking Lice," *Memoir Pacific Coast Ento. Soc.*, 1, 320 pp., 1951.

are relatively very heavy and are highly specialized to enable the insects to hold on with a death-like grip to the hairs. The tarsus is a single segment with one enormous claw (Fig. 3.4,*E*). This is drawn around by muscles to clamp against a thumb-like projection from the end of the tibia, gripping the hair between the claw, the tarsal segment, the end of the tibia, and its thumb. In some species a spiny pad is forced out from the end of the tibia, further to strengthen the grip.

The mouth parts (Fig. 4.10) are unique. They are of the piercing type, but when not in use are completely withdrawn into the head of the louse. All one can usually see externally is a fringe of minute teeth at the foremost part of the head. At a short distance inside this fringe the mouth cavity or *buccal funnel* gives rise to two tubes. The upper tube is the pharyngeal duct, which leads to the stomach; the lower tube is only about as long as the head and ends blindly like an introverted glove finger. In this blind tube or *stylet sac* are three slender stylets believed to represent a pair of maxillae fused together, the hypopharynx, and the labium. They are all attached near the blind posterior end of the stylet sac, and their free ends reach to the mouth opening. Muscles are arranged to pull the stylet sac forward and drive the stylets into the flesh. Saliva is carried into the wound through the stylets, and blood is sucked up through the stylets, aided by a short trough-like passageway inside the mouth, and so to the pharynx and esophagus.

The eggs of Anoplura, often called "nits," are glued fast to the hairs or, in one species, laid in seams of clothing while it is being worn. The metamorphosis is very simple. The entire life is spent on the body of the host. For the most part, each species attacks one species of host and can live on no other.

Small, wingless, flattened, external parasites of mammals. Mouth parts retractile, piercing-sucking. Head narrow, pointed in front; eyes wanting or degenerate. All three thoracic segments fused; tarsi one-segmented, with a single, grasping claw. Thoracic and abdominal spiracles located on the dorsal side. Cerci wanting; ovipositor not elongate. Metamorphosis simple.

Only about 250 species are known, but their habits give to this order an importance out of all proportion to its size. The species attacking man belong in the family Pediculidae (page 1026). The family Haematopinidae includes the species attacking domestic animals. Important species are the hog louse (page 971), the bloodsucking horse louse (page 945), the short-nosed and long-nosed ox lice (page 958), and the bloodsucking sheep lice (page 975).

INSECTS WITH A COMPLETE METAMORPHOSIS

With the Anoplura we completed our discussion of the orders that have a gradual or simple metamorphosis. All the orders that follow have a complete or complex metamorphosis. A glance at Table 6.3 will show that the four largest orders are in this group, and a little checking of the species of economic importance, discussed in the following chapters, will show that most of our destructive insects are representatives of one of the orders with a complete metamorphosis. Apparently the complete metamorphosis is a very successful thing, in spite of the fact that insects undergoing such transformations must usually become adapted to three distinctly different environments during the course of the life cycle and have a helpless pupal stage. In the case of gradual metamorphosis, the same environment usually serves for both nymphs and adults; but in complete metamorphosis the larva must find suitable food, the pupa must be protected from enemies and adverse physical conditions, and the adult must again find food (generally different from that of the larva) or, if it does not feed, at least a suitable place to deposit eggs.[1]

ORDER COLEOPTERA[2]

The Beetles and Weevils

This is the largest of all orders of insects. Two out of every five kinds of insects that have been discovered and named are beetles. This may be due in part to the fact that most of the orders have not been studied so extensively as the beetles have. Nevertheless ordinary observation will show that these insects are so numerous and ubiquitous, and of such diverse form, habit, and appearance, that not to know something about the beetles is to remain in ignorance of a large and very interesting part of our environment. It is well to study beetles also in self-defense, for they attack us at many points, feeding on growing crops of all kinds, from forest trees to greenhouse plants, as well as on stored foods and other possessions. It is noteworthy that there are practically no beetles that attack the larger animals, and very few *parasites* on any group, although many of them are predaceous on insects and other small animals.

The most characteristic thing about beetles is their wings. The mature insects have the front wings specialized into what are called *elytra* (pronounced ell'-it-ra). These are generally thickened by chitin so that they usually show no veins, and they are not flapped in flight, but serve as a pair of convex shields to cover the hind wings and the rather delicate-walled abdomen from above. When the insect is not flying they lie close over the abdomen, the inner edges of the two coming together to form a straight line from the prothorax back to the tip of the wings, neatly covering the mesothorax, the metathorax, and at least part of the abdomen (Figs. 6.39 to 6.43). A small part of the mesothorax, known as the *scutellum*, remains exposed as a little triangle between the bases of the elytra but is never so large as it is in many of the Hemiptera. On account of the nature of the front wings, the prothorax is unusually distinct from the rest of the thorax and is often wrongly called "the thorax."

The hind wings, or second pair, are the real organs of flight. When a beetle flies, the front wings are held stiffly out at the sides of the body, while the hind wings, only, beat so rapidly that one can scarcely see them. These wings are about as wide as the elytra but commonly one-fourth or one-third longer. They do not project beyond the elytra when the insect comes to rest, but by a remarkable automatic "joint" the distal third or fourth is bent under, folding transversely but not longitudinally, as the wings are laid back to the resting position. The second pair of wings are thin and membranous, with a few veins. Sometimes these hind wings, and rarely both pairs, are wanting, or the wings may be grown together so they cannot be moved.

[1] See "The Life Cycle in Insects," *Ann. Ento. Soc. Amer.*, **13**:133, 1920.

[2] Blatchley, W. S., "Coleoptera or Beetles of Indiana," Nature Publ. Co., Indianapolis, 1910; Blatchley, W. S., and C. W. Leng, "Rhynchophora or Weevils of Northeastern America," Nature Publ. Co., 1916; Leng, C. W., "Catalogue of Coleoptera of America North of Mexico," Sherman, 1920; Bradley, J. C., "A Manual of the Genera of Beetles of America North of Mexico," Daw, Illston & Co., 1931.

The mouth parts of beetles are of the chewing type in both larvae and adults, the same parts being recognizable as in grasshoppers or crickets (page 130). One can usually tell something of the habits of a beetle by examining its teeth. If the mandibles are short, chunky, and with a small number of blunt denticles on the mesal face, the grinding subtype (page 141), it indicates a species that takes plant food. If the mandibles are elongate, come out to one or two sharp points toward the end, or have the inner edges sharp for cutting, the grasping subtype (page 141), the insect is carnivorous and probably benefits us by eating other insects. If the mandibles lack distinct teeth and are covered with stiff hairs, the brushing subtype (page 141), the insect is a harmless pollen feeder.

In one group of this order (the snout beetles) the head is prolonged forward and downward into a cylindrical snout (Fig. 15.36) that varies in length from shorter than the rest of the head to several times the length of the whole body. The student must be careful not to confuse such a snout with the beak of the Homoptera or Hemiptera. This snout is not the mouth parts, but a part of the cranium, as indicated by the attachment of the antennae to it. It is not jointed, it is not furrowed down the front, and it, of course, contains no stylets. The mandibles, maxillae, and other mouth parts are found on the end of the snout, being very small but functioning as chewing mouth parts. This long snout is used for making a hole deep into the tissues of plants, in which the eggs are laid. It also enables its possessor to eat the tissues beneath the surface.

Beetles generally have no ocelli in the adult stage, though the compound eyes are well developed. The reverse is true of the larvae, which have a small group of ocelli at each side of the head but never compound eyes. The coxae of the hind legs (Fig. 3.4,F) are flattened out like a part of the body wall, instead of articulating into a socket of the latter, so that these legs at first sight appear to lack the coxae. There are no cerci. The Coleoptera embrace many very small insects and, from these, range upward in size to some tropical kinds that are several inches in length. There are very few insects likely to be confused with the Coleoptera. A few kinds of Hemiptera (Fig. 14.44) have a superficial resemblance to beetles, but the shield over the back is not formed of two wings but is a single piece, an overgrown thorax (*scutellum*). The earwigs look much like the rove beetles (*cf.* Figs. 6.17 and 6.39), but always show at the tip of the abdomen a pair of heavy forceps or pincers.

Beetles all have a complete, sometimes a very complex, metamorphosis, known as a hypermetamorphosis (page 643). The larvae of beetles (Figs. 9.26, 9.31, 14.24,*b*, and 19.26) are generally called *grubs*, sometimes *borers*. They are very diverse in shape but can be recognized by the following characteristics: They commonly have six thoracic legs each ending in one or two claws, but some are entirely legless. There is never a series of prolegs, at most one pair at the tip of the body, and these have no crochets on the end of them. The head never has the *adfrontal area* that characterizes lepidopterous caterpillars (Fig. 6.55). The spinneret at the middle of the labium also is wanting. The head is always distinct, usually dark-colored, bearing definite though often minute antennae. There are always a number of pairs of spiracles on the abdominal segments.

The pupae of the beetles (Fig. 9.26 and 12.2,*h*) can be distinguished from other pupae by noting that all the mouth parts are of the chewing type and that the antennae have fewer than 12 segments. The membranous sacs that enclose the antennae, legs, and wings are not grown fast ("cemented down") to the sides of the body but are free and can be moved about. Beetle pupae are not protected by the dense silken cocoons characteristic of moth pupae. They may have thin cocoons but are often openly exposed on leaves; at other times hidden in the soil, in burrows in wood, or covered over with foreign material accessible to the larva as it prepares for pupation.

In this order both larvae and adults are commonly injurious and in the same manner. Sometimes, however, the habits of larvae and adults are different, so that they may be injurious in two totally different ways; or one or the other, only, may be destructive. In a few cases one life stage is harmful to us and the other beneficial; thus the blister beetles as adults feed on the foliage of potatoes, chard, asters, and other plants, while their larvae are helpful to man by eating the eggs of grasshoppers in the soil. The lady beetles, except for a few species, and many species of ground beetles are highly beneficial, both as larvae and as adults, by devouring injurious insects (page 66). There are numerous aquatic species.

The Coleoptera are minute to very large, usually heavily sclerotized, robust insects. The front wings are much thickened, veinless, and meeting in a middorsal straight line; the hind wings membranous, with few veins, and the apex folded under transversely when at rest; sometimes wanting. Mouth parts of typical chewing type, although in the snout beetles they are reduced and placed at the end of a slender trunk-like snout easily mistaken by the tyro for piercing mouth parts. Ocelli generally wanting, antennae mostly of 10 or 11 segments. Prothorax very distinct from meso- and metathorax, and freely movable against the mesothorax; meso- and metathorax somewhat fused and united with the abdomen. Tarsi mostly of four or five segments, rarely three. Hind coxae plate-like, immovable. Metamorphosis complete. Larvae worm-like or shaped like Thysanura, sometimes with prominent cerci; usually with six thoracic legs and not more than one pair of prolegs; rarely apodous; the legs in most species with a single tarsal claw. Spiracles on principal segments. No adfrontal area. Pupae with appendages nearly always free (obtect in Staphylinidae and Coccinellidae); the body wall generally thin, soft, and pale-colored; rarely in cocoons. Cerci wanting in adults and often in larvae also. No firm ovipositor.

Among the many very destructive species are the following, to which the reader should refer for figures and descriptions of typical species: white grubs (page 503), wireworms (page 506), billbugs (page 487), corn rootworms (page 509), lesser clover leaf weevil (page 553), alfalfa weevil (page 550), cucumber beetles (page 629), Colorado potato beetle (page 640), flea beetles (page 514), Mexican bean beetle (page 622), bean and pea weevils (page 935), sweetpotato beetles (page 651), asparagus beetle (page 680), bark beetles (page 846), flatheaded and roundheaded apple borers (page 718), Japanese beetle (page 749), strawberry weevil (page 797), plum curculio (page 724), tobacco beetle (page 910), buffalo beetles or "moths" (page 912), grain beetles (page 926), granary and rice weevils (page 920), mealworms (page 922), and the larder beetle (page 917).

Only the most important families are listed in the following synopsis.

SYNOPSIS OF THE ORDER COLEOPTERA

A. Suborder *Adephaga.* The Predaceous Beetles. Mostly feed on other insects. Largely beneficial to us. Tarsi of five segments. Antennae generally filiform. Hind wings with most of the typical veins preserved, and with some cross-veins.

Ventral part of first segment of abdomen divided into three areas by the hind coxae—a very small median piece between the coxae and two large side pieces. Larvae shaped like Thysanura, active, carnivorous, with six segments in each leg and two claws at the end of the leg (Fig. 5.15,*K*).

1. *Terrestrial Predaceous Beetles.*
    Family   1. Tiger beetles, Family Cicindelidae (Fig. 6.37).
    Family   2. Ground beetles, Family Carabidae.
2. *Aquatic Predaceous Beetles.*
    Family   3. Predaceous diving beetles, Family Dytiscidae (Fig. 6.38).
    Family   4. Whirligig beetles, Family Gyrinidae.

B. **Suborder** *Polyphaga.* Beetles of a great variety of form, habit, and economic importance. All have the first ventral segment of the abdomen in a single piece, not divided by the hind coxae. Hind wings with the venation much reduced; cross-veins often wanting. The legs of the larvae always end in a single claw and have five segments or fewer. Larvae of extremely varied habits, often worm-like in shape.

3. *The Short-winged Beetles or Brachelytra.* Elytra short, exposing much of the abdomen. Tarsi five-segmented.
    Family   5. Rove beetles, Family Staphylinidae (Fig. 6.39).
    Family   6. Carrion or burying beetles, Family Silphidae (Fig. 6.40).
4. *The Club-horned Beetles or Clavicornia.* Antennae clavate. Tarsi five- or three-segmented.
    Family   7. Water scavenger beetles, Family Hydrophilidae (Fig. 6.41).
    Family   8. Flat bark beetles, Family Cucujidae.
    Family   9. Lady beetles, Family Coccinellidae (Tarsi three-segmented).
    Family 10. The skin beetles, Family Dermestidae.
5. *The Saw-horned Beetles or Serricornia.*—Antennae serrate. Tarsi five-segmented.
    Family 11. Checkered beetles, Family Cleridae (Fig. 6.42).
    Family 12. Firefly or lightning beetles, Family Lampyridae (Fig. 6.43).
    Family 13. Soldier beetles, Family Cantharidae (Fig. 6.43).
    Family 14. Metallic wood borers or flatheaded borers, Family Buprestidae.
    Family 15. Click beetles, Family Elateridae.
6. *The Beetles with Different-jointed Tarsi or Heteromera.*—Five segments in tarsi of front and middle legs; four segments in hind tarsi.
    Family 16. Darkling beetles, Family Tenebrionidae.
    Family 17. Blister beetles, Family Meloidae.
7. *The Leaf-horned Beetles or Lamellicornia.*—Tarsi five-segmented throughout. Antennae lamellate, a cylindrical basal part and a number of flattened leaf-like segments at the tip.
    Family 18. The stag beetles, Family Lucanidae (Fig. 6.44).
    Family 19. The lamellicorn beetles, Family Scarabaeidae.
8. *The Plant-eating Beetles or Phytophaga.*—Tarsi apparently four-segmented.
    Family 20. The long-horned beetles, Family Cerambycidae.
    Family 21. The leaf beetles, Family Chrysomelidae.
    Family 22. The pea and bean weevils, Family Mylabridae (Bruchidae).
9. *The Snout Beetles or Rhynchophora.*—Head often produced into a long snout. Tarsi apparently four-segmented. Antennae clubbed and elbowed.
    Family 23. Typical weevils, Family Curculionidae.
    Family 24. Engraver or bark beetles, Family Scolytidae (Ipidae).

THE PRINCIPAL FAMILIES OF COLEOPTERA

Family Cicindelidae.[1] **The Tiger Beetles.** On sandy beaches and barren pathways are often seen these bright blue or green iridescent beetles which have the

[1] HORN, W., "List of Cicindelidae of North America," *Trans. Amer. Ento. Soc.*, **56**:76–86, 1930; LENG, C. W., "Cicindelidae of Boreal North America," *Trans. Amer. Ento. Soc.*, **28**:93–186, 1902; HAMILTON, C. C., "Morphology, Taxonomy, and Ecology of Larvae of Holarctic Tiger-beetles," *Proc. U.S. Natural Museum*, vol. 65, 87 pp., 1925.

tantalizing habit of flying up just before they are within reach of the collector's net and alighting only a little farther away. Their quick movements enable them to catch other insects, which they devour. The larvae make deep slender holes in the soil and lie in the mouth of these burrows ready to grab any insect that wanders within reach. (Fig. 6.37). A hump, bearing hooks, on the fifth abdominal segment of the larva prevents it being pulled out of its burrow by the struggles of its victim.

The adults are very hard-shelled, long-legged, exceedingly alert and active beetles, black, brown, white, or metallic blue, green, or bronze, mostly with thin, whitish or blackish curved or hooked lines or bars across the wing covers. The head and thorax are distinctly narrower than the wings. The antennae are filiform. The mandibles are very sharp-pointed, sickle-shaped, with several additional teeth on the inner face. The average length is about ½ inch.

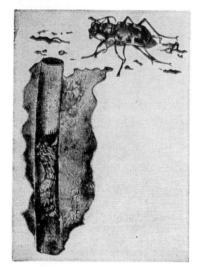

FIG. 6.37. A tiger beetle and, in the tunnel, a larva of the same. About natural size. (*From Sanderson and Jackson.*)

Family Carabidae.[1] **The Ground Beetles.** This enormous family of mostly blackish, somber-looking, long-legged beetles is believed to be largely predaceous as both larvae and adults and very beneficial. They hide under surface trash during the day and hunt at night. Some of the species, known as bombardier beetles, shoot out from the anus a very acrid fluid which volatilizes to form a cloud and makes a faint popping sound as it is discharged so as to suggest a miniature cannon in action. A few kinds climb trees and other vegetation, but most of them live on the surface of the ground and rarely take flight. Most of the species have very broad, hard wing covers, with numerous, fine, parallel, longitudinal ridges, a narrower prothorax and a still narrower head (Fig. 2.13). The size ranges from 1/16 inch, or less, to over 1 inch long, with many species measuring close to ⅔ inch. Sometimes the colors are brilliant metallic greens, blues, or purples, occasionally spotted with iridescent dots or pits of gold, and the head and thorax may be colored contrastingly with the wings, but there are almost never any spots or bands and usually the entire upper surface is uniform in color. The head is long, directed forward. The eyes are small, and the antennae rather thick toward the base and attached between eyes and mandibles on the side of the head. The species can be separated only by very careful study.

Family Dytiscidae. The Predaceous Diving Beetles. The habits of these smooth, shiny, black or brown beetles are well told by the common name of the family. They parallel the adaptations of the aquatic Hemiptera in having flattened and fringed legs that are used for oars and a space beneath the wings that is used as a storehouse for air while under water. Both adults and larvae catch other insects, small fish, and the like for food (Fig. 6.38). In the case of the large *Dytiscus marginalis* the larva digests its food before swallowing it, by forcing digestive secretions out of its mouth parts and into the body of the worm, insect, or snail which it has attacked and then sucking in the liquefied food. The males of many species have suckers on the front tarsi that work on the principle of vacuum cups to hold the body of the female. Head, thorax, and abdomen are compactly joined in one smooth, even, oval outline. The body is flattened, the antennae very slender and filiform, the hind legs placed far back near or behind the middle of the body and remote from the middle pair. In swimming, the Dytiscids strike with both hind legs at the same time, which gives them a smooth, even progress through the water. The size varies greatly from

[1] LENG, C. W., and W. BEUTENMULLER, "Handbook of Cicindelidae and Carabidae of Northeastern America," *Jour. N.Y. Ento. Soc.*, **2**, 4:1894–1896.

minute species only $\frac{1}{12}$ inch long to the relatively enormous *Dytiscus* species, which exceed $1\frac{1}{2}$ inch in length by $\frac{3}{4}$ inch broad.

Family Gyrinidae. The Whirligig Beetles or Lucky Bugs. These beetles, like the water striders (page 222), occupy the surface of the water and feed largely on insects that fall into it. Although they move very close to the water, they are really skating or walking upon it, not swimming in it. They also readily dive beneath the surface. The body is compact, oval as in the Dytiscidae, a little more convex above, bright shining blue-black in color. The hind legs are attached behind the middle of the body, and the middle pair are nearer to the hind pair than to the front pair. The front ones are slender for grasping the prey; the middle and hind pairs form broad, thin oars. Common species range from $\frac{1}{6}$ to $\frac{2}{3}$ inch long. They are best known from

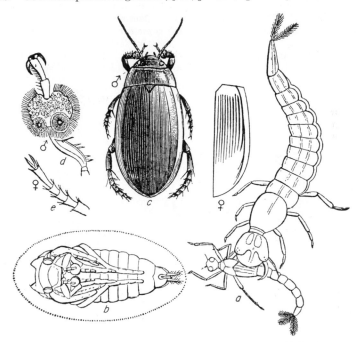

FIG. 6.38. A predaceous diving beetle. *a*, Larva of *Dytiscus marginalis*, known as a water tiger, devouring a May-fly nymph; *b*, the pupa; *c*, adult of *D. fasciventris*, and a single wing cover of the female; *d*, front tarsus of male, showing suckers used in holding female; *e*, corresponding tarsus of female. *a*, *b*, and *c*, about natural size. (*From Riley.*)

their remarkable gyrations, as they circle crazily round and round each other, usually in groups, on the surface of ponds. The antennae are very short, irregular in shape, and concealed in grooves in front of the eyes. They appear to have two pairs of compound eyes. Each eye is divided at the water line, presumably so that one half may look into the water and the other half into the air. The eggs are fastened in rows to aquatic plants, and the larvae, which are predaceous, are provided with long, feathered tracheal gills on their tails.

Family Staphylinidae.[1] The Rove Beetles. This is an enormous assemblage of mostly small or minute forms, many species measuring less than $\frac{1}{8}$ inch long, and being very difficult to classify; exceptional species are nearly 1 inch long. The best

[1] BLACKWELDER, R. E., "Morphology and Subfamilies of Staphylinidae," *Smithsonian Inst. Misc. Collections*, no. 94, 102 pp., 1936; *Proc. U.S. National Museum* **87**:93–135, 1939; CASEY, T. L., "Scaphidiidae, Scydmaenidae, and Pselaphidae," *Ann. N.Y. Acad. Sci.*, **7**:281–606, 1893, and **9**:285–548, 550–630, 1897.

characteristic is the very short front wings, which are usually not longer than their combined width and leave much of the abdomen uncovered (Fig. 6.39). Large, full-length hind wings are completely folded and concealed beneath these stubby elytra. These insects and their larvae abound where there is decaying animal and vegetable matter on the soil. They are scavengers or predators upon other insects found in filth. The more typical species are slender, parallel-sided, and covered with fine hairs, with most of the abdomen exposed, prominently segmented, tapering at the end, and very flexible. When disturbed, they usually turn up the tip of the abdomen as though to sting. The antennae are variable, filiform, geniculate, moniliform, or clavate. The head is directed forward, the mandibles sharp-pointed. The colors are usually brownish or black, rarely metallic blue, almost never spotted or striped, although rarely the wing covers are of a contrasting color or spotted.

Family Silphidae.[1] **The Carrion or Burying Beetles.** This family includes many species from $\frac{1}{2}$ to $1\frac{1}{2}$ inches long, although some are minute. The bodies of some species are brilliantly colored. They are mostly flattened, sometimes greatly so, which enables them to squeeze beneath the bodies of dead animals. Perhaps the most interesting are the burying or sexton beetles, which dig the soil from beneath

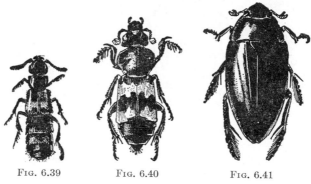

| Fig. 6.39 | Fig. 6.40 | Fig. 6.41 |

Fig. 6.39. Rove beetle *Creophilus maxillosus.* Enlarged $\frac{1}{2}$. (*From Kellogg, "American Insects."*)

Fig. 6.40. Burying beetle, *Necrophorus marginatus.* Enlarged $\frac{1}{2}$. (*From Kellogg, "American Insects."*)

Fig. 6.41. Giant water scavenger beetle, *Hydrous triangularis.* Natural size. (*From Kellogg, "American Insects."*)

birds, mice, snakes, and other small animals, until the latter are completely buried, and then lay eggs upon them so the larvae may feed in seclusion. The two commonest types are the elongate *Necrophorus*, with two or three gaudy, reddish, zigzag bands across the black body (Fig. 6.40), and the broader, much flattened *Silpha*, with prominently grooved or warty elytra, and the prothorax often margined with reddish yellow. The prothorax is very large, the antennae clavate, the wing covers usually too short to cover the entire abdomen.

Family Hydrophilidae. **The Water Scavenger Beetles.** Many of these species resemble the predaceous diving beetles very closely (*cf.* Figs. 6.38 and 6.41) but live on decaying organic matter in the water or in very moist places. The careful observer will find the concealed, club-shaped antennae lying beneath the large compound eyes (Fig. 6.41), which is the best way to distinguish them from Dytiscidae, and will not be deceived by the unusually long palps which look like antennae. The antennae have taken up a peculiar function in helping to exchange the stale air in the space beneath the wings and in the hair-covered troughs which extend to the spiracles

[1] HORN, G. H., "Silphidae and Dascyllidae of United States," *Trans. Amer. Ento. Soc.,* **8**:219–322, 73–114, 1880; ARNETT, R. H., "Nearctic Silphini and Necrophorini," *Jour. N.Y. Ento. Soc.,* **52**:1–24, 1944.

along the sides of the body for fresh air each time the beetles come up to the surface. The maximum size is $1\frac{1}{2}$ inches long, but there are a number of species only about $\frac{1}{8}$ inch in length and some only $\frac{1}{16}$ inch. The beetles are less flattened than Dytiscids or Gyrinids, the legs less perfectly adapted for swimming and more evenly spaced on the front half of the body. In swimming, the oar-like hind legs strike the water alternately. Some of the smaller species do not live in the water. The color is usually black, very brilliantly shining, but a number of species have yellowish stripes along the margins of the body. There is a prominent keel on the underside of the body extending between the bases of the legs and terminating in a sharp spine behind the hind coxae.

Family Cucujidae.[1] **The Flat Bark Beetles.** These insects mostly live under the bark of trees or in burrows in wood, where they feed, not on the decaying wood, but upon wood-boring insects which they encounter. Adults and larvae are often much flattened. The beetles are rather long, trim, and neat, the wing covers straight-sided, rounded at the end, and very flat. The head and prothorax are long, the head directed forward, the antennae large. The colors range from pinkish red through yellows and browns, and they are rarely spotted. The size ranges from $\frac{1}{16}$ to $\frac{2}{3}$ inch in length. A number of important species feed on stored grains and other products; for example, the saw-toothed grain beetle (page 926), the flat grain beetle *Laemophloeus pusillus*, and the foreign grain beetle, *Cathartus advena*. A very common species under the bark of trees is the pinkish red *Cucujus clavipes*.

Family Coccinellidae.[2] **The Lady Beetles.** These short-legged, convex, usually brightly marked beetles are discussed on pages 66 and 414, and the commonest species are shown in Fig. 2.14. They are well characterized by the three-segmented tarsi. Most of the species are predaceous on aphids. The Mexican bean beetle (page 622) is a very destructive species that feeds on foliage of all kinds of beans. Lady beetles are very convex above, flat beneath, almost hemispherical, though some are more elongate and oval in outline. Their colors are tan or brown, usually spotted with black, or black, usually spotted with red, white, or yellow. Most of the species are between $\frac{1}{8}$ and $\frac{1}{4}$ inch long. The head is small, turned downward, and received into a prominent notch of the prothorax. The elytra closely enclose the abdomen, but their margins are prolonged beyond its width. The legs are short, and the antennae short and clavate.

Family Dermestidae.[3] **The Skin Beetles.** An insatiable taste for dead animal matter governs the habits of most of this family and leads to serious injury to furs, woolen goods, hides, museum specimens, stored meats, grain, and many other products. They are small, convex, dark-colored, hairy or scaly, and the larvae, which have food habits similar to those of the adults, are active and covered with long, dense, barbed hairs (Fig. 19.10). The pupal stage is passed inside the ruptured larval skin. The body is short, oval in outline, plump, and closely compacted. The head is very short and directed downward. All the appendages are small, the short, clavate antennae concealed in grooves below and behind the eyes. The elytra are convex, covering the abdomen, and not striate. The short flattened hairs or scales which cover the body sometimes form patterns of white or brick red. Our largest species are barely $\frac{1}{2}$ inch long. Members of this family that are important pests include the carpet beetles (page 912), the larder beetle (page 917), and the Khapra beetle (page 925).

Family Cleridae.[4] **The Checkered Beetles.** This is a fairly large family of no pronounced importance to man but which contains strikingly marked species likely to command notice when they are encountered. They are found chiefly in and about woodlands, feeding, as both adults and larvae, mostly upon other beetles, such as the

[1] Casey, T. L., "Cucujidae of North America," *Trans. Amer. Ento. Soc.*, **11**:69–112, 1884.

[2] Casey, T. L., "American Coccinellidae," *Jour. N.Y. Ento. Soc.*, **7**:71–163, 1899.

[3] Jayne, H. F., "Dermestidae of United States," *Proc. Amer. Phil. Soc.*, **20**:343–77, 1882; Mutchler, A. J., and H. B. Weiss, "New Jersey Dermestidae," *N.J. Dept. Agr.*, *Cir.* 108, 31 pp., 1927.

[4] Wolcott, A. B., "Catalogue of North American Cleridae," *Fieldiana, Zool.*, **32**:61–105, 1947; Knull, J. N., "Cleridae of Ohio," Columbus, Ohio, 1951, 83 pp.

Scolytidae, which make burrows in the wood of trees.   The name "checkered beetles" refers to the conspicuous bands of white, black, and brown, which cross the body (Fig. 6.42).   In many species, there are three subequal, transverse bands or spots of light color on the wings, separated by dark areas of about equal width.   Most of the species are subcylindrical, often hairy.   The prothorax is usually widest in front and constricted behind to form a prominent joint, before the base of the wing covers. Some of the species are distinctly ant-like in form.   The head is broad; the eyes small. The legs are rather large, the front pair often unusually strong.   The common species are between ¼ and ½ inch in length.

Family Lampyridae.[1]   The Fireflies or Lightning Beetles.   Some of these elongate, narrow, and straight-sided beetles with soft, flexible wing covers (Fig. 6.43) have the astonishing property of producing light within their bodies, as discussed on page 54. The flashing is supposed to be a mating signal.   Great numbers of individuals often flash in unison.   The larvae are predaceous, and some of them produce light.   Most of the species range around ½ inch in length.   The prevailing colors are yellow and black, often arranged in spots on the prothorax or slender lines on the wing covers.

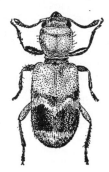

Fig. 6.42. Checkered or clerid beetle, *Enoclerus quadriguttatus*. About 6 times natural size. (*From Ill. Natural History Surv.*)

Fig. 6.43. Firefly or lightning beetle, *Photinus scintillans*. 3 times natural size. (*From Kellogg, "American Insects."*)

The prothorax is nearly semicircular and has its margins extended in a thin, shelf-like manner at the sides and in front, beneath which the head is turned downward and partially hidden.   The wing covers are not very convex but stand out roof-like above the sides of the abdomen.

Family Cantharidae.   The Soldier Beetles.   These are mostly brightly colored beetles very similar in shape to the Lampyridae but differing in not having the head hooded over by the prothorax, and the antennae are attached farther apart; none of the species is light-producing.   The species average about ½ inch in length and are slender, parallel-sided, with narrow, nearly flat, and not very hard wing covers.   The colors are yellow, brown, and black, often arranged in contrasting spots or lines on prothorax and wing covers.   A very common species about flowers in autumn is *Chauliognathus pennsylvanicus*, which is tan-colored with an elongate oval black spot on each wing near the tip.

Family Ptinidae[2] (Including Anobiidae, Bostrichidae, and Lyctidae).   Powder-post Beetles, Deathwatch Beetles, Drug Store Beetles.   Small, brown, or black beetles (mostly ⅛ to ⅜ inch long), with small downbent heads, hooded over by the prothorax, which is frequently swollen or warty.   Some of the species are very hairy, some are

[1] Leconte, J. L., "Lampyridae of United States," *Trans. Amer. Ento. Soc.*, **9**:15–72, 1881.

[2] Fisher, W. S., "North American Bostrychidae," *U.S.D.A. Misc. Publ.* 698, 157 pp., 1950; Kraus, E. J., and A. D. Hopkins, "Holarctic Lyctidae," *U.S.D.A., Bur. Ento. Tech. Bul.*, ser. 20, pp. 111–138, 1911.

glabrous. The elytra usually cover the abdomen completely. Some of the species look like small spiders, with globular, translucent elytra, very small eyes, small head and thorax, and relatively large antennae and legs. Many are nearly cylindrical, with thorax and elytra of about the same width. These insects live for the most part upon dry or slightly decaying vegetable or animal matter and are frequently pests in stores and dwellings. One species, *Xestobium rufovillosum*, has received the name of deathwatch from its habit of banging its head against the sides of its burrows in the wood of houses, which is often a sound impressive and ominous in the stillness of sickrooms. The powder-post beetle (page 903) is one of the worst pests of seasoned stored wood, while the drug store beetle (page 911) and the cigarette beetle (page 910) are serious pests of stored plant materials and upholstered furniture.

Family Buprestidae.[1] Metallic Wood Borers. The adults are hard-shelled, elongate, and somewhat flattened, often large beetles (Fig. 15.29), with the wing covers usually irregularly roughened and the body beautifully shiny and iridescent, as though plated with metal. The common colors are metallic black, brown, coppery, green, or blue. Although most species are of uniform color all over, a number are beautifully splotched or spotted with yellow or white. They are very active and are lovers of the sunshine. All the body regions are very closely fitted together, in contrast with the click beetles, which they somewhat resemble. The prothorax is swollen, the head short, sunken in the thorax. The wing covers often taper curiously and are slightly prolongated behind. The antennae and legs are small. The length ranges from ⅜ to 1¼ inches. The larvae are nearly or quite legless, with a broad, flat enlargement of the thorax, which has given them the name of flat-headed borers. They work beneath the bark of various trees and shrubs. The cane borers (page 787) and the bronze birch borer (page 840) are very injurious, as is the flat-headed apple tree borer (page 718).

Family Elateridae.[2] The Click Beetles. The adults, which are hard, smooth, streamlined, narrow, and mostly brownish or blackish in color (Fig. 9.29), are popularly interesting because of their ability to snap into the air when placed on their backs. This is accomplished by striking hard with the base of the wing covers against the ground. Economically, the family has great prominence because of their destructiveness and the difficulty of controlling the larvae, which are the well-known brownish, tough-skinned wireworms that feed on planted seeds and the roots of field and garden crops (Fig. 9.28). The terminal abdominal segment of the larva is often ornamented and characteristic of the species. The largest species in our central states is the black, white-specked, "pepper and salt," eyed elater, *Alaus oculatus*, which has two prominent eye-like spots on the pronotum. It is found about rotting wood, in which its larvae live as predators.

The prothorax in this family is large, shield-shaped, rounded toward the small head, which fits into its contour without a break. The wing covers are fully as long as the abdomen, gracefully narrowed toward the tip, but not very convex, usually ridged lengthwise. Antennae and legs are rather short, the antennae serrate or sometimes beautifully pectinate. There is a conspicuous break, or hinge joint, between prothorax and wings, and on the underside of the body a stout spine projects back into a socket on the mesothorax and is the organ used in snapping. In size these insects range from ⅛ inch to over 1½ inches long by one-fourth as wide. Color markings are uncommon, but sometimes the wing covers are margined, striped, or contrastingly colored, and one group has the conspicuous eye-like spots on the prothorax. A number of species of *Pyrophorus* have brilliant luminescent organs on either side of the thorax and at the base of the abdomen.

[1] CHAMBERLAIN, W. J., "Catalogue of Buprestidae of North America North of Mexico," Corvallis, Ore., 1926, 289 pp.; FISHER, W. S., "North American Buprestidae of Tribe Chrysobothrini," *U.S.D.A.*, *Misc. Publ.* 470, 274 pp., 1942; KNULL, J. N., "Buprestidae of Pennsylvania," *Ohio Univ. Studies*, no. 2, 71 pp., 1927.

[2] THOMAS, C. A., "Elateridae of Pennsylvania," *Jour. N.Y. Ento. Soc.*, **49**:223–263, 1941; DIETRICH, H., "Elateridae of New York," *Cornell Univ. Agr. Exp. Sta. Memoir* 269, 79 pp., 1945.

Family Tenebrionidae.[1] **The Darkling Beetles.** Superficially resembling the ground beetles in their general appearance and ground-loving habits, these beetles (Fig. 19.17) can at once be distinguished by the four-segmented hind tarsi. The body is frequently curiously shaped, sometimes very humpbacked, or the prothorax flaring upward at the sides or serrate along the margins. The body wall is usually bare and polished, the wing covers often wrinkled or pimply, sometimes strongly angulated at the sides like a box lid. The antennae are moniliform. Color markings are rare, most of the species being dull black or brown. The larvae resemble wireworms, from which they can be distinguished by their distinct labrum and the less specialized apical segment of the abdomen. Both larvae and adults are vegetarians. A very curious species, *Boletotherus bifurcus*, is often taken in the bracket fungi that grow on tree trunks. It is very warty all over, and the prothorax bears two erect horns a third as long as the body. Some of the best-known species are the confused flour beetle (page 925) and the light and dark meal worms (page 922), all three serious pests of stored grains.

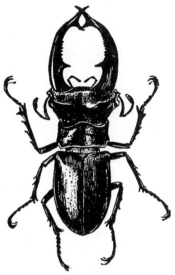

Fig. 6.44. The giant stag beetle, *Lucanus elaphus*, male. Natural size. (*From Kellogg, "American Insects."*)

Family Meloidae.[2] **The Blister or Oil Beetles.** The beetles of this family are mostly clumsy, good-sized (½ to 1 inch long), and straight-sided or cylindrical, with a narrow prothorax forming a distinct neck between the wider head and wing covers (Fig. 14.26). The wing covers are unusually independent of the abdomen. The hind wings are often wanting, and the forewings are frequently short and often soft or leathery. The tarsi have the same number of segments as in Tenebrionidae, five on the front and middle legs and four behind. Each tarsal claw appears as though split lengthwise. The adults feed on foliage and flowers, and are pests of vegetable and ornamental crops (page 642). The colors are often very beautiful, metallic green, blue, coppery, or rose, or nicely spotted or striped yellow and black, while some are unmarked gray, brown, or black. The head is directed downward, the eyes small, the 11-segmented antennae inserted far apart.

These insects are of especial interest on account of the caustic or blistering nature of their blood, which, in the case of a European species, the commercial "Spanish fly," is due to a substance called cantharidin (page 55). They also have remarkable life histories, which, in the species known, involve a hypermetamorphosis; that is, there are additional life stages beyond those recognized in other insects—several distinct kinds of larvae, and also a propupa or pseudopupa stage (page 643). The known species live as larvae in the nests of wild bees or in the soil where they eat the eggs of grasshoppers.

Family Lucanidae. The Stag Beetles. Some of the largest known beetles belong to this family, certain tropical species, such as *Odontolabis alces* of the East Indies, measuring over 4 inches in length. Comparatively few species live in temperate climates. They are usually black or brown, bare, and either dull or shiny, rather broad and long-legged (Fig. 6.44). The legs are modified for digging in the soil, the front tibiae being broadened and provided with spines. The head is large, sometimes very large, and directed forward. The antennae are geniculate and lamellate, the plates or leaves of the club standing distinct from each other, not fitted tightly together as in the Scarabaeidae. The common name refers to the very large, formidable-

[1] HORN, G. H., "Tenebrionidae of North America," *Trans. Amer. Phil. Soc.*, **14**: 253–404, 1870; CASEY, T. L., "Tentyrinae and Coniotinae of American Tenebrionidae," *Proc. Wash. Acad. Sci.*, **9**:275–522, 1907, and **10**:51–166, 1908.

[2] VAN DYKE, E. C., "Genera of North American Meloidae," *Univ. Calif. Publ. Ento.*, **4**:395–474, 1928.

looking mandibles of the males; they are on the same account called "pinching bugs." In spite of their formidable appearance, the insects are not predaceous and are said to feed on honeydew, plant exudates, and wood. The larvae live in rotting trees, frequently below ground, and on the roots of plants.

Family Scarabaeidae.[1] **The May Beetles or June Bugs.** This is an enormous family of small to very large species, usually broad, chunky, and convex, often half as broad as long (Fig. 9.26). As a rule they are not brilliantly colored, although a number of species are metallic green, blue, or rose-coppery, and some are spotted or striped. The legs are long and spiny, and the front tibiae broad, flat, and provided with rake-like teeth on the outer edge. The antennae are lamellate, the three to seven plates of the antennal club being closely compacted. The prothorax is large, as broad as the elytra, frequently with powerful horn-like projections. The tip of the abdomen is usually left exposed by the wing covers, which otherwise fit closely over it. *Dynastes tityus* of southeastern United States may measure up to 2.5 inches long. *Goliathus goliathus* of Africa, 5 inches long, is the world's bulkiest insect. The larvae are typically thick, cylindrical white grubs with well-developed legs, which live in the soil, usually lying in a U-shaped position.

Food habits divide the family into two groups: in one group, including the dung beetles and tumble bugs, both larvae and adults feed on dung, fungi, or decaying vegetable matter, often exhibiting remarkable care in preparing nests for the larvae. This is the only known case among insects where the male aids in providing for the young. In the other group the adults feed on foliage or flowers, while the larvae feed on the roots of grasses and cultivated crops or in rotting wood. These include the destructive white grubs (page 503), the Japanese beetle (page 749), the rose chafer (page 772), and bumble flower beetles.

Family Cerambycidae.[2] **The Long-horned Beetles.** These are perhaps the handsomest of all beetles (Fig. 17.19). Their long cylindrical bodies (from ½ to 2 or 3 inches long) are often beautifully colored, striped, and spotted; and the long antennae, which are frequently several times the length of the body, especially in the males, add to their attractiveness. The antennae are partly surrounded at the base by the compound eyes. As in all the remaining families of beetles, all the tarsi apparently have only four segments. The third segment is bilobed and brush-like beneath. The prothorax and head are generally narrower than the wings, the head downbent, the prothorax often with one or more marginal spines on each side. The wing covers are generally parallel-sided and cover the abdomen completely from above. Sometimes they are concave at the sides or narrowed behind. Although many species are somber brown or black, there are more often some striking spots or bands of white, yellow, or red. The color markings are frequently formed by fine, decumbent hairs.

The adults feed on foliage or tender bark. The larvae, which are called round-headed borers, in contrast to the flat-headed Buprestidae, bore in the solid wood of trees. They are wrinkled, cylindrical, whitish grubs, with a small head and swollen thorax, and are often entirely legless. Larval life may extend over several years, and sometimes in seasoned wood it may be prolonged to 10 or 20 years. The family is a very large one of rather uniform habits and appearance. Among the many very destructive species are the round-headed apple tree borer (page 720), the locust borer (page 843), and the poplar borer (page 842).

Family Chrysomelidae.[3] **The Leaf Beetles.** The leaf beetles rival the Cerambycidae in their beauty of coloring and striking patterns (Fig. 4.3,*B*,*I*,*J*). The colors are often contrasting and frequently metallic. Yellow spots, bands, and stripes

[1] RITCHER, P. O., *Bull. Ky. Agr. Exp. Sta.*, "Anomalini," no. 442, 1943; "Dynastidae," no. 467, 1944; "Copridae," no. 477, 1945; "Cetoniidae," no. 476, 1945; and "Rutelidae," no. 471, 1945.

[2] LINSLEY, E. G., "Pogonocherini and Necydalini of North America," *Ann. Ento. Soc. Amer.*, **28**:73–103, 1935, and **33**:269–81, 1940; KNULL, J. N., "Cerambycidae of Ohio," *Ohio Biol. Surv. Bul.* 39, pp. 133–354, 1946; CRAIGHEAD, F. C., "Larvae of North American Cerambycidae," *Bul. Can. Dept. Agr. Ento.* 23, 238 pp., 1923.

[3] FATTIG, P. W., "Chrysomelidae of Georgia," *Emory Univ. Museum Bul.* 6, 47 pp., 1948.

predominate and are almost always formed of the ground color, not of fine hairs (Fig 9.34). The insects are mostly bare. Their antennae are short or moderate in length (commonly half the length of the body) and do not emarginate the eyes. They are moderate to very small (the largest perhaps $\frac{1}{2}$ inch, the smallest not over $\frac{1}{20}$ inch). The bodies are short, oval, and convex. The legs are short, not spiny, the tarsi apparently four-segmented, the third segment bilobed, the fourth minute, the fifth slender but long. The hind femora are often enlarged for jumping. Most of the adults feed exposed on plant leaves, stems, or flowers. The larvae feed either on the leaves with the adults, or underground on roots, or bore through roots or stems. Some of the tortoise beetles look, while alive, as though they were plated with gold. The larvae of the tortoise beetles cover their bodies with their own dung, shed skins, and other trash, attached to forked spines carried over the back. The larvae of *Donacia* feed on submerged roots or stems of water plants but do not have to come to the surface to breathe because they have learned to tap the plant stems and roots and take their air from the air spaces of the plant. Of the very many destructive beetles of this family we may mention the small jumping flea beetles (page 603), the Colorado potato beetle (page 640), the asparagus beetle (page 680), the elm-leaf beetle (page 821), the strawberry rootworm (page 794), the corn rootworms (page 509), and cucumber beetles (page 629).

Family Mylabridae[1] (Also Called Bruchidae). **The Bean and Pea Weevils.** This is a small family, structurally close to the Chrysomelidae, but having the head produced downward into a very short, broad snout. The wing covers stop two-thirds of the way to the end of the abdomen and do not extend much down the sides of the body. The adults are under $\frac{1}{4}$ inch long, dull-colored, oval, and chunky. The body is covered with fine hairs which often form small spots or flecks. The antennae are short, clavate; the legs stout; the tarsi four-segmented as in the Chrysomelidae and Cerambycidae.

The larvae all live in the seeds of legumes and include species very destructive to stored peas and beans. They have legs in the first instar but lose them after getting into the seeds, where they live until adult and grow fat and humpbacked to conform to the shape of the enclosing seed. The most destructive species include the common bean weevil (page 937), the pea weevil (page 935), and the cowpea weevil (page 936).

Family Curculionidae.[2] **The Curculios, Weevils, or Snout Beetles.** Imms calls this the largest natural family in the animal kingdom. Although not all the snout beetles belong in this huge family, the best-known ones do. These can usually be recognized by the "elephant-trunk-like" snout or prolongation of the head, which has a nearly full set of chewing mouth parts on the end of it, and the elbowed and clubbed antennae attached near its mid-length (Figs. 4.3*D* and 12.2). The snout is usually slender, sometimes very long, usually curved. The snout enables the adults to feed beneath the epidermis of plants and is also, in the females, used in making a cavity for the eggs. The majority of the species are $\frac{1}{2}$ to $\frac{1}{8}$ inch long, but some species may be nearly 2 inches long and some are minute. The colors are generally dull, mostly brown or black, frequently relieved by flecks of white, yellow, or red. The body wall is exceedingly hard and often warty, and they often feign death at the least disturbance, dropping among surface trash, where their shape and color make them very hard to detect. The body narrows in front and is usually covered with short, decumbent hairs or flat scales, which sometimes form color markings. The elytra or wing covers are frequently rugose or warty and embrace the dorsal part of the abdomen very completely, the lateral margins of the abdomen fitting into deep grooves near the edge of the wing covers. The femora are generally swollen. The larvae have no legs, are typically humpbacked, white, and soft-bodied (Fig. 12.2,*e*), and generally feed within nuts, seeds, fruits, buds, stems, or roots.

There are a very large number of serious pests in this family, including the cotton boll weevil (page 582), strawberry weevil (page 797), clover-bud weevil (page 553),

---

[1] HORN, G. H., "Bruchidae of United States," *Trans. Amer. Ento. Soc.*, **4**:311–42, 1873; BRIDWELL, J. C., "Genera of Bruchidae in North America," *Jour. Wash. Acad. Sci.*, **36**:52–57, 1946.

[2] BLATCHLEY, W. S., and C. W. LENG, "Rhynchophora or Weevils of Northeastern America," Nature Publ. Co., 1916.

alfalfa weevil (page 550), plum curculio (page 724), cabbage curculio, *Ceutorhynchus rapae*, chestnut weevils (*Balaninus* spp.), billbugs (page 487), sweetpotato weevil (page 653), and the granary weevil and rice weevil (page 920).

Family Scolytidae[1] (also called Ipidae). The Bark Beetles and Ambrosia Beetles. These are "snout beetles" with a very short snout or none at all. It should be noted that the real characteristic of the Rhynchophora is the absence of the gula, resulting in a single suture on the underside of the head, and not necessarily the presence of a snout. The Scolytidae are very small, short, cylindrical, with short, elbowed antennae terminating in large clubs. The front tibiae are serrate or provided with teeth along the outer edge. The vast majority of the species live between the outer bark and the solid wood of fruit, shade, and forest trees, both deciduous and coniferous. All life stages are passed under the bark, where their tunnels make the curious hieroglyphics often noticed when the bark is removed from dead wood (Fig. 15.32,*C*). Male or female may start the tunnel; the egg gallery is made by the female as she lays her eggs, each in a little niche at intervals along its sides, while the tunnels that radiate from it are larval burrows, which increase in diameter outward and are made by the larvae as they feed and grow. Pupation takes place at the end of the larval burrows, and the new adult finally eats a small "shot hole" out to the surface. The patterns formed by these "family tunnels" are variable and are characteristic for each species. In some species a single male has a harem of from 2 to 50 or more females, each of which makes an egg tunnel. The ambrosia beetles differ from the bark beetles in tunneling into the hardwood at various angles instead of paralleling the bark. They also grow mushrooms on the walls of the tunnels as a source of adult and larval food. The larvae are blind and legless and in some of the ambrosia beetles are fed by the adults. The important species include the clover root borer (page 573), the shot-hole borer (page 755), the peach-tree bark beetle (page 755), the black hills beetle, various pine beetles, turpentine beetles, engravers, spruce beetles, and fir beetles.

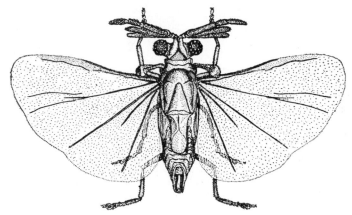

FIG. 6.45. Adult male strepsipteron, *Muirixenos dicranotropidis*. Note the flabellate antennae, club-like front wings, and veinless hind wings. (*From Pierce, Proc. U.S. Nat. Museum, Vol.* 54.)

ORDER STREPSIPTERA[2]

STYLOPS OR TWISTED-WING PARASITES

This is a small order of minute, internal parasites of insects, of little or no importance to man because they destroy mostly wild bees, wasps, and a few leafhoppers

[1] *Ibid.*, pp. 576–669; CHAMBERLAIN, W. J., "Bark and Timber Beetles of North America," 513 pp., Oregon State Coll., 1939.

[2] PIERCE, W. D., "Monographic Revision of the Twisted-winged Insects," *Bul. U.S. National Museum*, 66, 1909; BOHART, R. M., "Revision of Strepsiptera, with Special Reference to North America," *Univ. Calif. Publ. Ento.*, **7**:91–160, 1941.

and planthoppers. They are so abnormal in structure as to constitute an order of very distinct characteristics.

Males (Fig. 6.45) with stalked eyes, flabellate (branched) antennae, and degenerate chewing mouth parts. No ocelli. Wings four, but the first pair reduced to mere short clubs, the hind pair large, triangular, folding fan-like, and without cross-veins. Tarsi often without claws. Metathorax extraordinarily large. No cerci. Female (Fig. 6.46) without legs, wings, eyes, or antennae, and the mouth parts mere vestiges; worm-like and living in the interior of insects throughout life except for the fused head and thorax, which project like a lump, or tumor, between two segments of the host. Her body is enclosed in the last larval skin, which is open just behind the head into the brood chamber. This communicates with the unpaired genital pores on several abdominal segments, and serves for fertilization and also for the escape of the larvae. Metamorphosis is complete, with a hypermetamorphosis.

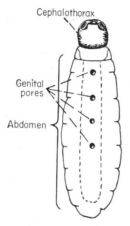

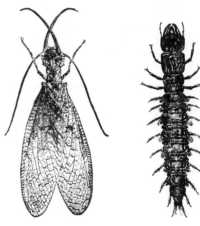

FIG. 6.46. A full-grown or adult female strepsipteron, *Xenos vesparum.* (*Redrawn from Imms, "General Textbook of Entomology."*)

FIG. 6.47. Dobson fly, *Corydalus cornutus;* adult male on left and its larva, the hellgramite, on the right. ½ natural size. (*From Sanderson and Jackson, "Elementary Entomology."*)

ORDER NEUROPTERA (MEGALOPTERA AND PLANIPENNIA)

APHID-LIONS, DOBSON FLIES (FIG. 6.47), ANT-LIONS (FIG. 6.48), AND OTHERS

This order includes species which are mostly predaceous and beneficial to man. The aquatic forms serve as food for fish, and the terrestrial ones are very beneficial in their larval stages by preying upon aphids, ants, and other small insects (page 64).

Insects of variable size, large or small, soft-bodied, with four large leaf-like wings of nearly equal size and texture; often colored; generally finely net-veined and held roof-like over the abdomen at rest. Often with many extra forked veins around the margin of the wings, especially in the costal region, and with the radial sector pectinately branched. Tarsi five-segmented. Mouth parts of the chewing type, though secondarily adapted in many of the larvae (Fig. 2.12,b,d) for sucking the blood of other small animals (graspingsucking subtype) (Fig. 4.11), each mandible having a ventral groove closed by the maxilla to form a sucking tube. Antennae generally long, cerci wanting, tarsi of five segments. Larvae spindle-shaped, much like Thysanura, carnivorous, many of them aquatic, with tracheal gills and sometimes with paired, lateral, often jointed, filaments on many of the abdominal segments (Fig. 6.47). No cerci. Aquatic or terrestrial in habit. The pupae with appendages free; sometimes in cocoons of silk spun from the anus of the larvae.

The Principal Families of Neuroptera[1]

Family Sialidae.[2] **Dobson Flies, Alder Flies, Fish Flies.** This family includes the largest of the Neuroptera as well as some of moderate size. The wingspread ranges from ¾ inch to 5 inches. The head and prothorax are large, squarish, the antennae long, the mandibles enormously elongated in the male dobson fly, which species is over 3 inches long (Fig. 6.47). The wings are ample, often dark-colored, frequently spotted, without a stigma, veins not extensively forked at the wing margins. The hind wings are very broad at the base, this area folding fan-like when at rest. The adults are mostly rather sluggish, clinging to vegetation near the water, but sometimes attracted to artificial lights a considerable distance from water. The larvae are aquatic, with seven or eight pairs of filamentous or tufted, segmentally arranged gills on the abdomen. The abdomen is soft-skinned, the head and prothorax heavily chitinized, legs well developed. The mouth parts are of the chewing type, the mandibles usually long, sharp-toothed. The larvae live underwater in ponds or streams, often under stones in the swiftest current, and devour small insects, worms, and other animal life. Pupation usually takes place in the soft earth of the adjoining shore. The family is of little economic importance, except as fish food.

Family Chrysopidae.[3] **Lace Wings, Aphid-lions, Golden-eyed Flies.** These graceful, weak, slender-bodied insects (Fig. 2.12) are generally saturate leaf-green in color, with beautifully burnished golden eyes. The antennae are long and very slender, the wings of about equal size, the hind pair narrower at the base, lacy, with the veins green, like the body, and forked near the margin. The adults probably take no food. Their fluttering flight suggests that of a butterfly. The eggs are remarkable in being supported or elevated, each at the end of a long slender pedicel. They are often placed in small groups. The larvae are known as aphid-lions from their habit of devouring aphids. They are spindle-shaped, tapering almost as much toward the head as toward the tail, provided with sickle-shaped, bloodsucking mouth parts, slender filiform antennae, slender hairy legs, and a row of small, spine-bearing tubercles down each side of the body. They are usually mottled with green, red, black, gray, and yellow. The pupa is curled up in a perfectly spherical, white, silken cocoon, attached to a tree or other object. The adult emerges from the cocoon through a circular lid. The insects are no doubt highly beneficial in checking aphids and some related insects (page 64).

Family Myrmeleonidae.[4] **Ant-lions or Doodle Bugs.** Most interesting of all the Neuroptera are the ant-lions, which excavate, in sheltered, sandy places, conical or funnel-shaped craters ranging up to 2 inches in diameter and nearly an equal depth. These curious traps (Fig. 6.48) are the work of the larvae, which use them as a means of securing food. Insects such as ants which stumble over the edge of the crater generally slide to the bottom, where the larva lies concealed, except for its sickle-shaped jaws. With the latter it seizes the luckless prey and sucks its body fluids. If the prey regains its footing, it may be confused or brought down again by a shower of sand thrown upon it by the doodle bug. These relatively enormous pits are dug by the larvae chiefly with the head, which is used as a shovel to throw the sand to one side, plowing round and round in circles, terminating at the center. The larva is an uncouth-looking thing, with narrow head and prothorax behind the powerful spiny jaws, followed by a broad plump abdomen. The bodies are covered with hairs to

[1] Hagen, H. A., "Neuroptera of North America," *Smithsonian Inst. Misc. Collections*, no. 4, 1862; Banks, N., "Catalogue of the Neuropteroid Insects of the United States," *Amer. Ento. Soc.*, 1907.

[2] Davis, K. C., "Sialidae of North and South America," *N.Y. State Museum Bul.* 68, pp. 441–487, 1903.

[3] Banks, N., "Nearctic Chrysopidae," *Trans. Amer. Ento. Soc.*, **29**:137–162, 1903; and "Catalogue of the Neuropteroid Insects (except Odonata) of the United States," American Entomological Society, pp. 1–53, 1907; Smith, R. C., "Biology of the Chrysopidae," *Cornell Univ. Memoir* 58, 1922.

[4] Banks, N., "North American Myrmeleonidae," *Can. Ento.*, **31**:67–71, 1899; and *Trans. Amer. Ento. Soc.*, **30**:104–106, 1904.

which the sand particles readily adhere.   The adults are frail, weak-flying.   They have long slender abdomens, like dragonflies, frequently exceeding the wings in length. The antennae are short and clubbed, and the legs short and weak.   There are four nearly equal, delicate, net-veined, black-spotted wings, which narrow evenly from beyond the middle toward the base.   The wing veins are often ornamented with fine, short hairs, the veins branching extensively near the margins.   Pupation takes place in a spherical, silken, sand-covered cocoon.

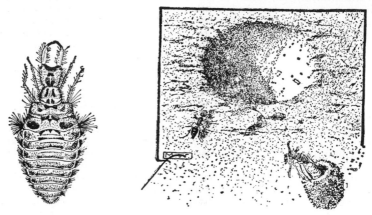

Fig. 6.48. The ant-lion or "doodle bug," *Myrmeleon* sp.; larva on left, 3 times natural size. Note elongate mandibles adapted for sucking.   On right is shown the sand pit or trap made by the ant-lion to catch ants, at the bottom of which the larva lives; about ½ natural size. Below is empty pupal skin and sand-covered cocoon from which the adult has escaped. (*From Kellogg, "American Insects."*)

Family Hemerobiidae.   Brown Lacewings.   These insects are very similar in appearance to the common lacewings or Chrysopidae, but are considerably smaller and brownish in color.   Both larvae and adults are predaceous on aphids, mealybugs, white flies, and other small insects and mites.

Family Rhaphidiidae.   Snake Flies.   This group is a very small one, considered by some to be a separate order, Rhaphidoidea.   The adults are characterized by a long snake-like neck, long antennae, and two pairs of very similar net-veined wings.   The larvae live under the bark of trees and, like the adults, are predaceous.

Family Mantispidae.   Mantid Flies.   These insects have a remarkably similar appearance to the mantids and are predators.   They have grasping front legs with greatly enlarged coxae, femora, and tibiae, and these are attached at the anterior of a very elongated pronotum.   The first-instar larvae actively hunt for the nests of *Lycosa* spiders or *Polybia* wasps, where they assume a parasitic existence and become grub-like in form.   The common American species is *Mantispa brunnea*.

ORDER MECOPTERA (PANORPATAE)[1]

The Scorpion-flies

This order includes some seldom-noticed terrestrial and predaceous insects of little or no economic importance.   The common name is given because the males of some of the species (*Panorpa*) have the tip of the abdomen swollen and carry it curved upward in a manner suggestive of the sting of a scorpion (Fig. 6.49).

Moderate to small insects with elongate bodies and long, many-jointed antennae.   The most distinctive characteristic is the prolongation of the head, maxillae, and labium into

[1] CARPENTER, F. M., "Revision of Nearctic Mecoptera," *Bul. Museum Comp. Zool., Harvard Univ.*, 72, pp. 205–277, 1931.

a broad snout two or three times as long as the width across the eyes. At the end of this snout are the chewing mouth parts. The four, long, rather narrow wings (rarely wanting) have rather numerous cross-veins, are often spotted, and are laid over the abdomen at rest. Large compound eyes and three ocelli usually present. Tarsi five-segmented. Cerci small. The larvae are caterpillar-like but are distinguishable from Lepidoptera because some of them have eight pairs of prolegs without crochets, while others have no prolegs. They live in soil or in moss. The pupae have the wing cases and leg cases free from the sides of the body.

The family Panorpidae has the wings moderately short and usually conspicuously spotted or banded with dark brown. The abdomen of the males is conspicuously swollen and upturned at the end, terminating in a pair of mandible-like claspers, which are suggestive of a scorpion's sting; hence the name of the family. The tarsi terminate in two claws. In the family Bittacidae, the wings, legs, and abdomen are longer than in the Panorpidae, superficially resembling crane flies. The wings are either plain yellow or less conspicuously spotted. The legs are very slender, the tip of the tibiae bearing a pair of very long spines, the last tarsal segment modified to be bent back against the next to last segment and provided with a single claw. The tip

Fig. 6.49. A scorpion-fly, *Panorpa rufescens*, twice natural size. (*From Kellogg, "American Insects."*)

of the abdomen in the male is often enlarged, but not swollen like the Panorpidae. These insects are predaceous upon other insects, commonly hanging down from vegetation with their front legs and grasping their prey with middle or hind tarsi as readily as with the forelegs.

ORDER TRICHOPTERA[1]

CADDICE-FLIES (FIG. 6.51) AND CADDICE-WORMS (FIG. 6.50)

This order includes species of soft-bodied, weak-flying moth-like insects (Fig. 6.51) that are seldom seen except along streams or lakes, in the water of which the larvae

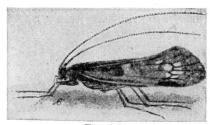

Fig. 6.50   Fig. 6.51

Fig. 6.50. Caddice-fly larvae; above, larva in its case with head and legs projecting in normal position; below, larva removed from its case to expose the tracheal gills on the abdomen. (*From Fernald, "Applied Entomology."*)

Fig. 6.51. An adult caddice-fly. (*From Kellogg, "American Insects."*)

develop. The adults are commonly less than 1 inch in length and have four brownish wings covered with hairs and held sloping roof-like at the sides of the body. The mouth parts are greatly reduced. There are no mandibles and probably most adults take no food.

[1] LLOYD, J. T., "Biology of North American Caddis-fly Larvae," *Bul. Lloyd Library, Ento. Ser.* 1, 1921; BETTEN, C., "Trichoptera of New York," *N.Y. State Museum Bul.* 292, 1934; Ross, H. H., "Trichoptera of Illinois," *Ill. State National History Surv. Bul.* 23, 1944.

The eggs (Fig. 5.1,*P*) are laid in ropes or masses of gelatin-like or cement-like substance in or near the water, often under stones in streams. The larvae are better known than the adults. They make cases that cover their bodies, except the head and legs (Fig. 5.17,*F,G,N*), and often drag these protective cases about with them on the bottom of ponds or streams. They feed on small animal life or bits of vegetation and are among the commonest of insects in the water. The cases are very curious, variable, and characteristic for the species. They are lined with silk and more or less open at each end; in shape cylindrical, ovoid, or spiral, and covered with pebbles, sticks, or pieces of leaves, or tiny snail shells. A few kinds make miniature "fish nets," which are constructed in swift-flowing water and which strain out small particles of food from the current.

Medium-sized insects with four similar membranous wings clothed with rather long hairs, the hind pair usually shorter and broader. The wings stand roof-like over the abdomen in repose; longitudinal veins numerous, but cross-veins few. A semitransparent, whitish spot, devoid of hairs, near the center of each wing. Compound eyes generally small; ocelli three or none. Mouth parts modified from chewing type by reduction, the mandibles often being absent, but the palps well developed. Antennae long, filiform; legs long, coxae large, tarsi five-segmented, tibiae with spurs. Metamorphosis complete; larvae and pupae aquatic, living in cases and breathing by abdominal gills. Larvae worm-like, with three pairs of long thoracic legs and one pair of hook-like prolegs on the last segment of the abdomen. Larval tarsi one-segmented with a single claw. Head and thorax more sclerotized than the abdominal segments. Legs rather long. Antennae and labial palps very small. Most larvae with thread-like tracheal gills on abdominal segments. Pupae with antennae, legs, and wings free, and a pair of large mandibles.

The principal importance of the Trichoptera is as food for fish. Both larvae and adults are of great value as food for trout. One species is said to be a pest of water-cress beds in England.

The Principal Families of Trichoptera

The adults of Trichoptera are separated by structural characters too involved to be of interest to the beginning student. There are, however, some very interesting differences in larval habits, fully discussed by Lloyd, which may be briefly stated as follows.

Family Phryganeidae. These are, according to Lloyd, the best known of all caddice-worms. The larvae are caterpillar-like, averaging about 1 inch long, and live on the bottoms of clear ponds and slow streams, feeding at night upon almost all kinds of aquatic plants and decomposing vegetation. The head and prothorax are conspicuously marked with brown or black on a yellow ground color. The cases are cylindrical, lined with tough silk, and covered with rectangular pieces of leaves fastened in ring-like segments or arranged in spirals (Fig. 5.17,*G*).

Family Limnephilidae. The cases of this family are very diverse, but according to Lloyd, never spiral nor made of segment-like rings, not four-sided, nor tapering, nor entirely of silk. Some are made of sticks arranged crisscross like a log cabin, others of sand with heavy ballast pebbles (Fig. 5.17,*N*) or two long ballast sticks along the sides. Still others use leaf fragments arranged in a two-sided or three-sided case, and a few have curved sand-covered cases. The larvae vary from ⅓ to 2 inches long. Sheltered by their portable cases, they hide in protecting crevices among sticks and stones in the swift current or crawl leisurely about the bottoms of slow brooks or ponds.

Family Sericostomatidae. In this family one species makes a case like a very long, truncated, four-sided pyramid made of minute fragments of wood very regularly placed transversely around the case. From the large open end the legs are protruded spider-like to grasp small animals or plant fragments for food. Another species makes a spiral case of small sand grains, which so closely resembles a snail shell that zoologists have several times described it as a mollusc.

Family Calamoceratidae. One species of this family makes a unique case by boring a cylindrical hole lengthwise through a piece of fallen, water-soaked twig at the bottom of a stream. The hole is lined with silk, and the section of twig is dragged about or allowed to go drifting or rolling downstream with the current. The larvae, which do not exceed ⅔ inch in length, are active in fall and winter as well as during the summer.

Family Leptoceridae. Most of the known larvae of this family live in lakes and ponds where the water is still. Only a few kinds have been reared; these construct variable cases, often tapering, composed in one case of beautiful translucent silk, in others covered with sand or slender bits of leaves, and either straight or shaped like a cornucopia.

Family Hydropsychidae. The best-known species live only in the swift current of streams or on the shores of lakes, where the waves wash over them. They are most remarkable for their habit of building a kind of "fish net," adjacent to which is a nonportable tube in which the larva lives. The net is somewhat funnel-shaped, the open end upstream. As the water flows through the fine mesh of these tiny bags, small insects and other organisms are caught by the net. They are then swallowed whole by the larvae.

Family Polycentropidae. The larvae that are known of this family do not drag their cases about but build them partially submerged in the mucky bottoms of streams or fastened beneath stones by tangled strands of silk. The tubes may be from 1 to 4 inches long, and some of them are branched.

Family Philopotamidae. These larvae make delicate silken nets shaped like glove fingers and usually several of them side by side in the water of swift streams. The nets are attached at the large upstream end, the rest floating in the current. There is a small slit in the downstream end, just big enough for the larva to escape in case of attack. The nets, which are from 1 to 1½ inches long, serve to strain from the water various microscopic organisms, which are used as food.

Family Rhyacophilidae. These caddice-worms live in swift, stony streams. Some of them crawl about under or beneath stones, without a case, until nearly ready to pupate. They then crawl between two large stones and pile up a barricade of tiny pebbles all about them and spin their cocoons. Others form portable cases. All of them, before pupating, spin a tough, rubbery or parchment-like cocoon about themselves, which differs from all other pupal cases of Trichoptera in that it has no openings for the circulation of water.

ORDER LEPIDOPTERA[1]

MOTHS, MILLERS, BUTTERFLIES, AND SKIPPERS

This is the second largest order of insects and one of the most destructive. Its members are well marked, both as adults and as larvae, and scarcely likely to be confused with any other order. Nor are the insects of another order likely to be classed as Lepidoptera by mistake. The name means *scaly-winged*, and the most characteristic thing about these insects is the layer of short, flattened hairs or scales that typically covers both surfaces of the wings and practically all other parts of the body. The scales (Fig. 6.53) are of many shapes, and one can find almost every gradation from broad, flat, plate-like scales to slender, cylindrical hairs. All of them have a projection or pedicel at one end which fits into a cuplike cavity in the cuticle of the wing membrane. In the lower moths they are irregularly scattered over the wings, but in the most specialized ones

[1] BARNES, W., and J. H. McDUNNOUGH, "Check List of Lepidoptera of Boreal America," 392 pp., Decatur, Illinois, 1917; DYAR, H. G., "List of North American Lepidoptera," *U.S. National Museum Bul.* 52, 733 pp., 1902; HOLLAND, W. J., "The Moth Book," Doubleday, 1913, and "The Butterfly Book," Doubleday, 1931; FORBES, W. T. M., "Lepidoptera of New York and Neighboring States," *Cornell Univ. Memoir* 68, 1923; McDUNNOUGH, J. H., "Check List of Lepidoptera of Canada and United States of America," *Memoir Southern Calif. Acad. Sci.* 1, pp. 1–271, 1938, and 2, pp. 1–171, 1939; PETERSON, A., "Larvae of Insects," pt. 1, "Lepidoptera and Plant Infesting Hymenoptera," 315 pp., Columbus, Ohio, 1948.

they present a very perfect arrangement, overlapping on both the sides and the ends, like slates or shingles on a roof (Fig. 6.52).

The function of these scales is primarily to strengthen the wing membrane and make it stiff enough for rapid flight. In some other orders, this has been accomplished by a great increase in the number of veins. In the Lepidoptera there are not many veins, but the scales on both surfaces of the wing membranes give them sufficient rigidity for flight. The swiftest flying moths show the most perfect arrangement of the scales, and the front part of the wing where greatest stress comes has them arranged most regularly.

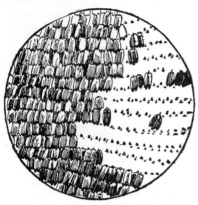

FIG. 6.52. A piece of the wing of a butterfly, *Danaus plexippus*, showing how the scales overlap at sides and ends. On the right some of the scales have been removed to show the cup-like pits out of which the pedicels of the scales grow. Greatly magnified. (*From Kellogg, "American Insects."*)

In addition to strengthening the wing membrane, the scales give protection to most parts of the body, and in them are resident the colors for which moths and butterflies are justly celebrated. If the "dust" (scales) is rubbed from the wings of a moth or butterfly, its characteristic color is lost. The scales are often ornamented with longitudinal ridges or striae, which may occur as closely as 35,000 ridges to the inch. They are very regular, and many of the brilliant colors of these insects are produced by the diffraction of light rays by the striae or by lamellae in the scales, rather than by pigments.

Aside from the possession of these scales, the wings of Lepidoptera are not highly specialized (Fig. 3.6). They are usually very broad, and subtriangular in outline, the front pair somewhat larger. They are often too large to be very effective in flight, since they cannot be flapped up and down rapidly enough. In general we note that the swiftest flying insects

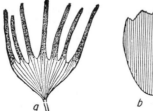

FIG. 6.53. Scales from three species of moths and butterflies. Note the pedicels that attach to cup-like sockets of the wing cuticle and the more or less parallel striae or ridges, which often produce brilliant colors. (*From Kellogg, "American Insects."*)

are those with small wings, and, in this order, the swift-flying ones are those having rather narrow wings, *e.g.*, the hawk moths (Fig. 6.54) and clear-winged moths (Fig. 15.61).

The most highly specialized thing about the Lepidoptera is the mouth parts of the adults. These are probably the most highly specialized of all insect mouth parts.   One finds as he examines the head (Fig. 4.16) a pair of very hairy or scaly palps (the labial palps) projecting forward at the sides of the mouth.   Between them arises a long slender tube for sucking up liquid foods.   Its real make-up is evident from a cross-section (Fig. 4.16,*B*).   From the figure it can be seen that it is double in nature, each half having a groove along its inner (mesal) face, and the two halves are so closely locked together that they form an airtight tube up which exposed liquids can be drawn by suction.   This is called the *siphoning type* of mouth parts (page 146).   When not in use, this tube is carried coiled up like a watch spring, so closely beneath the head that it is inconspicuous.   Some moths take no food during the adult stage, and in these cases the tube is wanting.

It must be emphasized that this proboscis is not a *piercing* structure. It is too flexible to be thrust into plant or animal tissues.   This means that these insects must satisfy themselves with liquids freely exposed. They have become adapted to taking a very special type of food—the nectar concealed in open cups in the corollas of flowers, particularly those

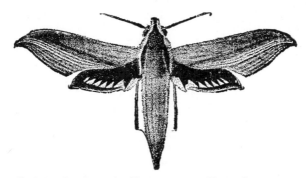

Fig. 6.54. A swift-flying hawk moth, *Theretra tersa*.   Natural size.   (*From Ill. Natural History Surv.*)

too deep to be reached by other kinds of insects.   The exact length of it is correlated with the kind of flowers the particular moth or butterfly visits. It is often longer than the body, and in a number of cases measures 5 or 6 inches.   *Angraecum*, a Madagascar orchid, has a corolla 10 inches deep, and it was predicted that a moth must exist with a proboscis long enough to reach the nectary.   The moth *Macrosilia morgani predicta* was subsequently found.   The removal of this nectar does not injure the plant; in fact, the plant usually benefits by the visits of these insects to its flowers, since in this way cross-pollination is generally brought about.

In rare cases, the tip of the proboscis is provided with stiff spines, sharp enough to lacerate the skin of a ripe fruit.   Thus cotton leafworms (Fig. 12.1) develop only on cotton, but in late summer the adults sometimes fly northward in great swarms, at least as far as Canada.   If they alight in orchards or vineyards, they may do much damage by puncturing the ripe fruits with their "tongues."   This is very exceptional, however, and usually moths and butterflies can do no injury in the adult stage.   When we speak of an injurious moth, such as the clothes moth or the gypsy moth, we refer to the injury caused by the larva or caterpillar of that species.   The economic importance of the Lepidoptera arises almost entirely from the activities of the larva.   This simplifies the problem of control somewhat, as compared with beetles, for example, since only one injurious stage need be considered.

The larvae of Lepidoptera are called caterpillars, and very often "worms." They have chewing mouth parts and are among the world's greatest pests. All the thousands of kinds known are remarkably similar in structure. The shape is generally nearly cylindrical (Figs. 3.9 and 11.8). The body is composed of 13 segments besides the head. Of these the first three, or thoracic, segments have each a pair of jointed legs, terminating in a single claw. The abdominal segments bear unjointed, soft, fleshy projections of the body called *prolegs*, typically one pair each on the

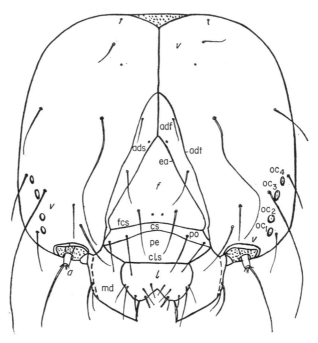

FIG. 6.55. The head of the armyworm, larva. *a*, Antenna; *adf*, adfrontal sclerite; *adt*, adfrontal suture; *cls*, clypeolabral suture; *cs*, clypeal suture; *ea*, epicranial arm or frontal suture; *es*, epicranial stem or coronal suture; *f*, front; *fcs*, frontoclypeal suture; *l*, labrum; *md*, mandible; oc_1 to oc_4, ocelli; *pe* and *po*, the clypeus; *v*, vertex. (*Adapted from Ripley, Ill. Biol. Mono., Vol. 8.*)

third, fourth, fifth, sixth, and tenth segments of the abdomen. Frequently some of these pairs and, in some cases, all the prolegs are wanting (Fig. 15.1). Insects of several other orders have larvae with prolegs, but these are the only insects that have the prolegs armed with a number of fine hooks known as *crochets* (Fig. 3.9,*4*). These crochets, which enable the insect to hold on so tenaciously to a leaf or twig, are arranged in circles or rows across the apex of the proleg.

The head of a caterpillar is usually well developed (Figs. 3.9 and 6.55). There is a group of simple eyes at each side of the head (Fig. 6.55,oc_1 to oc_4), the number varying from two to six pairs. The antennae are very small. A very characteristic thing about the larvae of this order is the presence, in at least the last instar, of the *adfrontal areas* (*adf*)—slender sclerites bordering the epicranial suture (*es*, *ea*) that do not occur in insects of other orders. Another characteristic of Lepidopterous larvae is the presence, near the end of the labium, of the spinneret from which the silk exudes.

The pupae of moths are typically enwrapped in a silken case, called a *cocoon*, which is made from saliva secreted by the full-grown caterpillar (Fig. 5.17,*A*,*E*,*I*). Some of them lie buried in the soil, and a few are formed in tunnels in wood or in other larval habitats. The pupae of butterflies are generally naked "chrysalids" fastened to plants or other supports by a small pad of silk or a girdle (Figs. 5.16,*C*, and 11.5,*C*). Regardless of the nature of protection, the pupae of Lepidoptera (Figs. 5.16,*A*,*B*,*C*,*D*) can be recognized from other orders by these two features: (*a*) the leg cases, antennal cases, and wing pads are fastened down to the sides of the body and immovable; (*b*) the long maxillae in the pupal stage appear as two long slender sclerites along the mid-ventral line.

Adult Lepidoptera are minute to very large, soft-skinned, fragile insects, well characterized by the coiled, siphoning mouth parts (sometimes absent) and the scaly wings (sometimes wanting in females). Wings enormous in proportion to size of body; membranous but not transparent, because characteristically covered on both upper and under sides by minute overlapping scales which often form beautiful color patterns and which easily rub off. Similar scales on most other parts of the body. Wings broad, triangular, the front pair larger, few cross-veins; not folding much at rest. Compound eyes large. Ocelli two or none. Legs relatively small, tarsi five-segmented. Prothorax very small. No cerci. A firm ovipositor very rarely present. Metamorphosis complete. The larvae typically worm-like, with chewing mouth parts; paired spiracles on the prothorax and first eight abdominal segments; two to five pairs of prolegs, with crochets, on the abdomen; and adfrontal areas on the head (Fig. 6.55 adf). Pupae with appendages usually fast to the body wall, often in cocoons. Abdominal segments often capable of much wriggling.

For our purposes, probably the most useful division of the order is the popular one into moths and butterflies. The principal families of each suborder are outlined in the following synopsis.

SYNOPSIS OF THE LEPIDOPTERA

A. **Suborder** *Jugatae or Homoneura*. *Moths or Millers* in which the two wings of each side are held together and made to work as a single wing, in flight, by an angular or finger-like process which projects from the base of the *front* wing over the front margin of the *hind* wing. The venation of the hind wing is very similar to that of the front wing. Mouth parts never of the siphoning type with a coiled proboscis.

  Family 1. Mandibulate moths, Family Micropterygidae.

  Family 2. Swifts, Family Hepialidae.

Suborders B and C are alike in the following characters, by which they differ from suborder **A**. They differ from each other in the characteristics given under **B** and **C**. Lepidoptera in which the action of the two wings on each side is synchronized by one or several strong bristles that project from the base of the *hind* wing over the anal margin of the *front* wing. Such bristles are called a *frenulum*. Sometimes the frenulum is supplanted by an expansion of the base of the hind wing membrane which has taken the place of the bristles. The venation of the hind wing is different from that of the fore wing. Mouth parts siphoning except where they have been reduced by degeneration.

B. **Suborder** *Heterocera* (*Frenatae or Heteroneura*, in Part). *Moths and Millers*. Mostly night fliers. Antennae of varied form, filiform, pectinate, or otherwise, but never enlarged at the tip to form a club. Abdomen heavy. Wings usually

FIG. 6.56. The cecropia moth, *Hyalophora cecropia*, a common and beautiful giant silkworm moth. (*From Fernald, "Applied Entomology."*)

lie horizontally or roof-like at sides of the abdomen or wrapped around the abdomen when at rest. Often have two ocelli. Pupae very often protected by cocoons.

Family 3. Yucca moths and others, Family Incurvariidae.
Family 4. Nepticulid moths, Family Nepticulidae.
Family 5. Carpenter moths, Family Cossidae.
Family 6. Flannel moths, Family Megalopygidae.
Family 7. Slug-caterpillar moths, Family Eucleidae.
Family 8. Clothes moths and others, Family Tineidae.
Family 9. Bagworm moths, Family Psychidae.
Family 10. Ribbed cocoon makers and others, Family Lyonetiidae.
Family 11. Leaf blotch miners and others, Family Gracilariidae.
Family 12. Casebearers, Family Coleophoridae.
Family 13. Parsnip webworm and others, Family Oecophoridae.
Family 14. Gelechiid moths, Family Gelechiidae.
Family 15. Yponomeutid moths, Family Yponomeutidae.
Family 16. Clear-winged moths, Family Aegeriidae (Sesiidae).
Family 17. Leaf roller moths, Family Tortricidae.
Family 18. Olethreutid moths, Family Olethreutidae.
Family 19. Pyralid or snout moths, Family Pyralididae.
Family 20. Hawk moths or sphinx moths, Family Sphingidae.
Family 21. Measuring-worm moths, Family Geometridae.
Family 22. The prominents, Family Notodontidae.
Family 23. Tussock moths, Family Lymantriidae.
Family 24. Owlet or cutworm moths, Family Noctuidae.
Family 25. Tiger moths, Family Arctiidae.
Family 26. Lappet moths, tent caterpillars, and others, Family Lasiocampidae.
Family 27. Royal moths, Family Citheroniidae.
Family 28. Giant silkworm moths, Family Saturniidae.

C. **Suborder** *Rhopalocera (Frenatae or Heteroneura, in Part)*. *Butterflies and Skippers.* Day fliers. Antennae clubbed or enlarged near the tip. Wings usually held vertically above the body when at rest with the upper surfaces of the two pairs in contact. When in flight, the wings of the same side are held together usually by an expansion of the membrane of the hind wing near its base. No ocelli.
The Skippers. Abdomen heavy. Antennae with a recurved hook at the tip, their bases wide apart. Larvae with a distinct neck-like constriction just behind the head.
Family 29. The common skippers, Family Hesperiidae.

The Butterflies. Abdomen slender. Antennae without a recurved hook on the terminal club, their bases inserted close together. Pupae never protected by cocoons.

Family 30. Swallowtails, Family Papilionidae.
Family 31. White and sulfur butterflies, Family Pieridae.
Family 32. Four-footed butterflies, Family Nymphalidae.
Family 33. Hair-streaks or gossamer wings, Family Lycaenidae.

FIG. 6.57. A skipper butterfly, *Epargyreus tityrus*, natural size. Note hooked antennae that characterize skipper butterflies. (*From Fernald, "Applied Entomology."*)

THE PRINCIPAL FAMILIES OF LEPIDOPTERA

Family Micropterygidae. Mandibulate Moths. Although this is a small family of very small, day-flying, metallic-colored moths of no known economic importance, it is of very great scientific interest because it includes the only moths with the chewing type of mouth parts. The proboscis is wanting or rudimentary. The adults feed on pollen of flowers, the larvae on mosses and liverworts.

Family Hepialidae. Swifts. Medium to large swift-flying moths, one Australian form measuring nearly 8 inches from tip to tip of wing. The wings are ample, the head is small, the antennae short, and the mouth parts are poorly developed. The body, legs, and wings are often very hairy. The known larvae are mostly root borers, but few have any recognized economic importance.

FIG. 6.58. Celery caterpillar or black swallow tail butterfly, *Papilio polyxenes*, and its life stages. *a*, Full-grown caterpillar from side; *b*, the same from in front showing osmeteria protruding; *c*, the male butterfly; *d*, outline of egg, greatly enlarged; *e*, a young larva; *f*, pupa or chrysalis. The artist has shown the pupa with the wrong side toward its support. All about natural size except *d*. (*From U.S.D.A., Farmers' Bul. 856.*)

Family Incurvariidae. Yucca Moths and Others. This family includes a moderate number of small moths with very complete, primitive wing venation, sharp, cutting ovipositors, and microscopic fixed hairs scattered over the wing membrane beneath the scales. It includes the yucca moth, *Pronuba yuccasella*, and some tiny leaf miners, whose larvae later crawl about over the foliage carrying a circular bit of leaf over their backs like a turtle's shell.

Family Nepticulidae.[1] Nepticulid Moths. Nearly all the larvae of these small moths are leaf miners, usually in trees or shrubs. The eggs are cemented to the leaves, and the nearly legless larvae make very slender, linear mines between the two leaf surfaces, gradually or abruptly widening as the larvae grow. The full-grown larvae desert their mines and drop to the ground, where they pupate in dense, flat cocoons spun among the surface litter. The species, of which less than 100 are known, can usually be recognized by the shape and appearance of their mines. The adults are among the smallest of all Lepidoptera, some measuring only ⅛ inch from tip to tip of wings. They are seldom seen unless reared from the larvae. The wings, as in the Incurvariidae, have fixed hairs distributed among the scales. They differ from that family in having many fewer veins in the wings, the females lack an ovipositor, and an "eye cap" is formed by the enlarged, concave, first segment of the antenna. The tongue is rudimentary, but the maxillary palps are long.

Family Cossidae.[2] The Carpenter Moths. This is a very small family, the larvae of which are borers in the solid wood of oaks, locusts, elms, maples, and many other deciduous trees and in herbaceous plants, an unusual habit for Lepidoptera. They have a life cycle extending over several years and reach a maximum size of 2 or 3 inches long as larvae, or 2 to 4 inches from tip to tip of wings in the adult females. The wings and body are shaped like those of sphinx moths. The front wings are finely reticulated or spotted with many rounded spots of dark color. The hind wings are less extensively marked. The antennae of the males are bipectinate for at least a part of their length. The proboscis is rudimentary. Our common species are the carpenterworm (page 850) and the leopard moth (page 851).

FIG. 6.59. Empty cocoon of the puss-caterpillar, *Megalopyge opercularis*, showing the hinged lid through which the moth emerged. (*From Comstock, "Manual for the Study of Insects."*)

Family Megalopygidae. The Flannel Moths. This is a very small family of moderate-sized moths that have a dense coat of fine curly hairs intermixed with the scales of the wings and covering body and legs, a condition which resembles the hairy coat of some mammals and has suggested the common name. The proboscis is rudimentary. The larvae, which have two pairs of sucker-like prolegs, without crochets, in addition to the five pairs of true prolegs, feed on the foliage of woody plants. Some of the species have nettling hairs concealed among their long silky coats and may cause a very severe sting. Their curious cocoons (Fig. 6.59), which are often found on the twigs of trees, are very tough, have bumps on the back, and are provided with a very definite lid or trapdoor at one end for the adult to squeeze out of.

Family Eucleidae (Also Called Limacodidae or Cochlidiidae). Slug-caterpillar Moths. The adults are small, hairy, robust moths, without a proboscis, ⅝ to 1 inch across the wings. The prevailing color is some shade of brown, upon the front wings of which is often superimposed, as though washed in water color, a large irregular spot of apple-green, rich brown, or silvery white. The larvae are short, with heads concealed in the thorax, fleshy, bare, or spiny, and often of very curious shapes and striking colors (Fig. 1.8). Some species are nettling. The underpart of the body forms a sole on which the larvae seem to slide along, the thoracic legs being very small and the

[1] BRAUN, A. F., "Nepticulidae of North America," *Trans. Amer. Ento. Soc.*, **43:** 155–209, 1917.

[2] BARNES, W., and J. H. McDUNNOUGH, "Cossidae of North America," *Contribution to Natural History Lepidoptera N.A.*, vol. 1, no. 1, 35 pp., 1911.

prolegs wanting. They feed on foliage and spin oval cocoons that are provided with definite lids, much as in the flannel moths.

Family Tineidae. Clothes Moths and Others. This family name has been used by different writers to cover a varying range of mostly small moths, centering about those whose larvae eat clothing (page 915) and other dried animal matter. Most of the species have long, slender, dull-colored wings, mottled with brown, yellow, or silvery spots and margined with a long fringe on the posterior margin, which is especially pronounced on the hind wings. The antennae are long. Both pairs of palps are usually present in the adult stage. Our common species mostly range from ½ to 1 inch from tip to tip of wings. The most important species are three, which have acquired the habit of living in dwellings and storehouses and eating woolen goods, fur, and feathers. The case-making clothes moth (page 915) makes a portable bag of silk and particles of its food, which covers its body like a sleeping bag and which it drags about wherever it goes. The carpet moth (page 915) makes a long, winding, silk-lined tunnel through the mass of fabric within which it forages, while the webbing clothes moth (page 915) spins some silk but does not make definite cases or tunnels.

Family Psychidae.[1] **The Bagworm Moths.** This small family is of particular interest on account of the larval cases, formed of silk and pieces of foliage, which are dragged about by the larvae as they feed on leaves (Fig. 5.17,E); and the remarkable degeneration of the female adults. Although the males have well-developed, dull-colored wings, the more degenerate females are entirely wingless and spend their adult life in the cases made while they were larvae. Some species have lost also their antennae, eyes, legs, and mouth parts. After a pupal period spent in the case or bag, the female, having been impregnated by a winged male, lays her eggs in the bag and dies. The bag, which is begun by the newly hatched larva and enlarged as the larva grows, thus serves to protect every life stage of the female sex and all but the adult of the males. The bagworm (page 823) is very common on arbor vitae, cedar, poplar, and other trees over much of the eastern United States.

Family Lyonetiidae. Ribbed Cocoon Makers and Others. This is one of several small families of mostly minute, leaf-mining moths, ¼ to ⅓ inch from tip to tip of

FIG. 6.60. Cocoons of the ribbed-cocoon maker, *Bucculatrix pomifoliella*. About 4 times natural size. (*From Slingerland and Crosby, "Manual of Fruit Insects," copyright, 1914, Macmillan. Reprinted by permission.*)

wings. They have been variously classified by different writers. The wings are very narrow, lanceolate, their tips turned up or down, and the margins of the wings fringed with extremely long hairs. The antennae are long, but the palps are small. The most interesting species are those which, after mining the leaves of apple, oak, and other plants, as larvae, construct for the pupal stage elongate, slender cocoons of white silk fastened to the twigs and provided with longitudinal ridges and furrows which make them objects of beauty (Fig. 6.60). The ribbed cocoon maker of the apple is *Bucculatrix pomifoliella*, and the cotton leaf perforator, *B. thurberiella*, often defoliates cotton in the Southwest.

Family Gracilariidae.[2] **Leaf Blotch Miners and Others.** The larvae of this family are miners between the leaf surfaces of woody plants, especially deciduous trees, or feed on the surface of foliage or fruit. The antennae of the adults are exceptionally long, both pairs of palps are present, the membrane of the wings is very narrow and pointed in both wings and, as in the preceding family, often provided with a very long

[1] GAEDE, M., "American Psychidae," *Seitz, Macrolepidoptera of World*, **6**:1177–1186, 1936.

[2] ELY, C. R., "North American Gracilariidae," *Proc. Ento.Soc.Wash.*, **19**:29–77, 1917.

posterior fringe of hairs.   The adults often rest with the front of the body reared up and the tips of the wings touching the surface.   They range from ¼ to ½ inch from tip to tip of wings.   The very flat, young larvae, whose shape is a nice adaptation to life inside the leaf, feed by cutting open the cells on the inside of the leaves and sucking up the sap.   Later they eat the tissues inside the leaf, leaving a very thin epidermis, which is folded up or puckered by the growth of the leaf.

Family Coleophoridae.   (Also Called Haploptiliidae).   Casebearers.   The larvae of this family live in portable cases (Fig. 5.17,*I*).   Some of them are leaf miners when very young, and then emerge and form about their bodies cases made of silk and bits of foliage.   The best-known species are the cigar casebearer (page 698) and the pistol casebearer (page 698), which crawl about over apple and other foliage with their tiny bags held erect above the thorax or fastened to the bark as a protection during the winter.   The adults, which measure from ¼ to ⅝ inch across the wings, have, like those of the preceding families, very slender fringed wings, and both pairs of palps are present.

Family Oecophoridae.[1]   Parsnip Webworm and Others.   The adults of this large family have fairly broad wings with the apices rounded, with much less pronounced fringes than the preceding families, and with rather complete venation.   The labial palps are long and prominent, the antennae of moderate length.   The size ranges, in common species, from about ¾ inch to 1 inch across both wings, and the colors are not very striking.   Most of the larvae web together or roll the leaves and flower heads of Compositae and Umbelliferae.   Others feed beneath the bark of trees or in decayed wood or in stored foods.   A very common species in the East is the parsnip webworm (page 674).

Family Gelechiidae.[2]   Gelechid Moths.   These are moderately small moths, averaging about ¾ inch across both wings.   As a family, they have not received a good common name, although the family is a very large one and embraces some of the most destructive insects of the entire order.   The adults have the front wings of nearly equal width from base to apex, the hind wings moderately broad, with the apical angle often prolonged and pointed.   The proboscis and the labial palps are well developed. The colors are browns with grayish or silvery markings.   Larval habits are varied, but they are nearly all plant eaters and most of them web together the leaves, shoots, or seed heads of herbaceous or woody plants.   A few are leaf miners or leaf rollers, borers, gallmakers, or feed below ground.   Among the most destructive and best-known species are the pink bollworm of cotton (page 587), the Angoumois grain moth (page 929), the peach twig borer (page 761), and the potato tuberworm (page 648).

Family Yponomeutidae (Including Plutellidae and Scythrididae).   This family, the exact scope of which is not a matter of agreement among specialists, includes a hundred or more species of small moths, ranging in size from ⅓ to about 1 inch from tip to tip of wings.   They are diverse in appearance and habits.   The wings are rather broad, sometimes fringed with long hairs, the tongue is usually present, and the eyes are rather conspicuous.   The wing veins are but little reduced and well separated from each other.   In the known species, the eggs are flattened.   The larvae are of various forms and habits, some feeding exposed on foliage and others mining in leaves or boring in buds, fruits, or twigs; and some are social, living in loose webs spun among the foliage of trees.   Some of the pupae form beautifully woven openwork cocoons. Perhaps the best-known species is the diamond-back moth (page 666), the smallest of the common worms on cabbage and related plants.   The apple fruit moth, *Argyresthia conjugella*, bores in the fruits of apple in the Northwest.

Family Aegeriidae.[3]   (Also Called Sesiidae).   Clear-winged Moths.   This family is well characterized by the absence of scales over a considerable part of one or both

---

[1] CLARKE, J. F. G., "North American Oecophoridae," *Proc. U.S. National Museum,* **90**:33–286, 1941.

[2] BUSCK, A., "American Gelechiidae," *Proc. U.S. National Museum,* **25**:767–938, 1903.

[3] ENGLEHARDT, G. P., "North American Aegeriidae," *Bul. U.S. National Museum,* **190**, 222 pp., 1946.

pairs of wings, so that the wings are transparent, and by the habits of the larvae, which are borers in the trunks or roots of trees and shrubs. Usually, the margins of the wings, and frequently a transverse bar across the front wing beyond the middle, retain the scales in lines that look much like thick veins. The front wings are narrow, and the body very slender, polished-looking. The insects fly with great rapidity and during the daytime, in contrast with most moths. They are often colored with blue and yellow or orange, especially as bands across the slender abdomen, so as to possess a striking resemblance to wasps (Fig. 15.61). These forms are believed to be mimics of the stinging Hymenoptera, which they resemble, and to gain protection because they look like stinging species. In the adults the antennae are generally slightly hooked and thickened toward the tip, the proboscis is usually well developed, and there is generally a handsome tuft of long scales at the end of the abdomen. The larvae are nearly bare and white and have the crochets of the prolegs arranged in two transverse rows. The common American species average close to 1 inch across the wings. The peach tree borer (page 757), the lesser peach tree borer (page 760), the squash vine borer (page 635), and the currant borer (page 785), are among the very destructive American species.

Family Tortricidae.[1] **Leaf-roller Moths.** Like many another ancient family name, this one has been variously restricted by different writers, as numerous small groups of genera and species have been taken out and placed in separate families. As so restricted, the Tortricidae includes small dull-colored brown or yellow moths, the most characteristic feature of which is the larval habit of concealing themselves by rolling the leaves of plants and feeding within such a nest. Often a number of larvae work together in this way to make rather large, ugly nests of webbed and chewed leaves containing the larvae and their unsightly excrement. The best-known species are named from this habit, as the cherry tree ugly-nest caterpillar, *Archips cerasivoranus*, and the fruit tree, oak, and rose ugly-nests. The wings are rather broad, the front pair square, truncate, or slightly undulating at the distal end, the marginal fringe never as long as the width of the membrane, usually much shorter. The wings are wrapped about the abdomen or held roof-like when the insect comes to rest. The antennae, the palps, and the proboscis are rather small. The size range in common species is from about ½ to 1¼ inches across both wings. The fruit tree leaf roller and the red-banded leaf roller (page 701) are often very destructive pests of deciduous fruit trees, and the spruce budworm (page 822) is one of the most injurious of forest insects.

Family Olethreutidae[2] **(Also Called Eucosmidae). The Codling Moth Family.** One of the groups separated from the former large family, Tortricidae, on the basis of wing venation, is this great assemblage of moderately small moths, common species of which average about ¾ inch across the wings. Many of them feed as larvae inside the fruiting bodies of plants and have astonishing powers of destruction. They have structural characters similar to the preceding family, from which they are separated by a row of long hairs margining a vein (the cubitus) about the middle of the hind wing on the upper surface, which the other Tortricids lack (Comstock). The apical margin of the front wings is generally more oblique than in the Tortricidae. The colors are mostly various shades of brown with subdued markings. The antennae and palps are small. In this family are placed the codling moth (page 727), the bud moths (page 699), the clover head caterpillar (page 567), the grape berry moth (page 775), the strawberry leaf roller (page 792), the lesser appleworm (page 771), and the oriental fruit moth (page 763). The curious "Mexican jumping beans" are seeds of *Sabestiania* plants, inside of which a caterpillar of *Laspeyresia saltitans* is developing. The jumping of the seed is caused by the impact of the caterpillar against the thin wall, as it threshes about inside the seed.

---

[1] FERNALD, C. H., "Genera of Tortricidae," Amherst, Mass., 69 pp., 1908; HEINRICH, C., "North American Laspeyresiinae and Olethreutinae," *Bul. U.S. National Museum* 132, 216 pp., 1926.

[2] HEINRICH, C., "North American Moths of Subfamily Eucosminae of Family Olethreutidae," *Bul. U.S. National Museum* 123, 298 pp., 1923.

Family Pyralididae.[1] **Pyralid or Snout Moths.** Small to moderate in size, averaging about ¾ to 1 inch across both wings, the moths of this great and destructive family have no common easy recognition mark. The adults have broad, thin, entire hind wings without a long fringe, often distinctly paler-colored than the front pair but in many species colored and patterned like the front ones, and almost an equilateral triangle in shape. The apical margin of the front wings is still more obliquely cut than in the Olethreutidae. A good many species are conspicuously and beautifully spotted or banded. Both pairs of palps are usually present, the labial pair often projecting to form a prominent snout in front of the head. The antennae are moderate in size, the proboscis sometimes greatly reduced. The body is rather slender. The larvae vary greatly in habits. They mostly feed concealed, either in rolled leaves, as borers in the stems, roots, or fruits of plants, or inside stored materials such as grain products and beeswax. They are nearly bare, white or grayish, and remarkably active when disturbed. Destructive species in America include the European corn borer, (page 490), the webworms (page 609), the melonworm (page 638), the pickleworm (page 637), the greenhouse leaf-tier (page 861), the corn stalk borers (page 495), the sugarcane borer (page 520), the Mediterranean flour moth (page 929), the meal snout moth, the Indian-meal moth (page 931), and the wax moth, which feeds on the wax in the hives. A few species are aquatic, with true tracheal gills in the larval stage and at least one species is predaceous in the larval stage upon unarmored scale insects.

Family Sphingidae.[2] **Hawk Moths or Sphinx Moths.** With this family we begin the discussion of those Lepidoptera which, because of their large size, more brilliant colors, or more conspicuous larval forms or habits, the average person is likely to notice. Many common species are from 2½ to 5 inches across the outspread wings. The adults are obviously built for speed. The front wings (Fig. 6.54) are very long and narrow, the outer end very oblique, the margins beautifully curved or undulated. A few species are clear-winged (*Haemorrhagia* spp.) and must not be confused with Aegeriidae. The body presents a distinctly streamlined or spindle-shaped appearance. Broad at the middle, it tapers almost to a point at tail and head. Hawk moths are free of long, wooly hairs and look sleek, polished, and well-groomed. They fly very rapidly, and many species hover like hummingbirds in front of flowers as they probe with their long beaks for the nectar. Some sphinx moths have the longest known proboscides of any insect, one tropical American species having a proboscis nearly a foot long. The antennae are rather thick and nearly always hook slightly at the end. There are no ocelli. For the most part the colors of the adults are quiet, though sometimes exquisitely beautiful. Many species are washed with beautiful tints of rose, green, yellow, or rich brown. Both adults and larvae present good examples of protective resemblance to the bark or foliage against which they rest. A few species are remarkable for their ability to produce sounds both as adults and as larvae. The larvae are not hairy but are marked by a prominent horn projecting obliquely upward at the tail end of the body. It is never provided with venom and is sometimes replaced by a polished spot or button. The name sphinx moths is said to refer to the habit of some of the larvae of resting with the head end reared up, snake-like and rigid. The resemblance to the Sphinx is not very obvious. The pupae generally lack cocoons and lie in cells in the soil. A few species have the tongue case bowed away from the body like a pitcher handle (Fig. 5.16,*B*). Although there are many species, only a few are first-rate pests. The best-known species are the northern or tomato hornworm (page 655), and the southern or tobacco hornworm (page 655, Fig. 14.33).

Family Geometridae.[3] **Measuring Worm Moths.** The adults are moderate-sized moths, averaging close to 1 inch across both wings, though a few species exceed 2 inches. They have, for the most part, slender bodies and thin broad wings, with a suggestion of delicacy, which give them a superficial resemblance in form to butterflies (Fig. 6.61).

---

[1] HEINRICH, C., "American Moths of the Subfamily Phycitinae," *Bul. U.S. National Museum* 207, 1956.

[2] DRAUDT, M., "American Sphingidae," *Seitz, Macrolepidoptera of World*, **6**:841–896, 1931.

[3] PROUT, L. B., "American Geometridae," *Seitz, Macrolepidoptera of World*, **8**: 1–104, 1938.

The wings, however, are generally not folded above the back when the moths are at rest. Many species have wavy color lines diagonally across the wings, but the colors are mostly plain, with whites, yellows, and pale browns predominating. The females are sometimes entirely wingless (Fig. 15.2). In a number of species the apical margins of the wings are cut in angular undulations. The mouth parts are weak to well developed, the palps small, the antennae frequently plumose. The larvae, which are popularly known as inchworms, loopers, or measuring worms, feed exposed on the foliage of plants, but few of them are serious pests. The larvae are naked, slender, greenish, or brownish, and their bodies are wrinkled or otherwise roughened. They get their names from the fact that the prolegs are usually wanting from the middle segments of the body, being present only on the ninth and tenth abdominal segments. Consequently, the larva walks on the two ends of the body, alternately grasping the leaf or twig with the legs at either end of the body while the legless middle region humps or bends upward in a loop. Not infrequently a larva stands erect and rigid on its hinder legs, the body projecting obliquely outward from a twig of about the same diameter and color. In such a case the resemblance to a short stub of a twig is astonishingly accurate. The spring cankerworm (page 689), the fall cankerworm (page 691), the currant spanworm, *Itame ribeuria,* and the snow-white linden moth, *Erannis tiliaria,* are among the most destructive species.

FIG. 6.61. Two species of measuring worm moths. Above, *Ectropis crepuscularia;* below, *Ennomos magnarius.* Natural size. (*Original.*)

FIG. 6.62. Adult of a notodontid moth, *Nadata gibbosa.* (*From Ill. Natural History Surv.*)

Family Notodontidae. The Prominents. This family includes medium-sized, somewhat hairy-bodied moths of dull brown, yellow, or gray color, with inconspicuous spots or bars on the wings (Fig. 6.62). The abdomens are rather long and moderately broad. The front wings are rather narrow, often with a V-shaped tuft of scales extending backward from the anal margin of each front wing. The adults are strictly nocturnal and not often seen. The size range is mostly between 1 and 2 inches across both wings. The legs are very hairy, the antennae sometimes short-plumose, the head small and set close against the broad thorax. The larvae usually feed exposed on the surface of plants and are very frequently noted on account of their curious shapes or striking color markings and also because of peculiar behavior when disturbed and the fact that some species are gregarious and feed close together in flocks. Some of the caterpillars are very hairy, others nearly bare. There are various wart- or thorn- or collar-like processes on the body, and frequently the tail end is prolonged with two flexible, erectile processes that can be thrown out very suddenly, features often supposed to be of value to the larvae as frightening organs. Whether or not they have much effect on any of their enemies except man is not known. These processes or those of the adults' wings probably may justify the common name given to the family. Some species, when disturbed, rear the head and tail into the air (Fig. 15.8) and some of them spray out an acrid, irritating fluid, containing formic acid, from beneath the prothorax. The best-known species are the red-humped caterpillar (page 694), the yellow-necked caterpillar (page 694), and the walnut caterpillar (page 826). The last species has the curious habit of collecting in masses on the trunk of the tree for each larval molt; the skins remain until worn or washed away.

Family Lymantriidae (Also Called Liparidae). The Tussock Moths. The adults of this small family are not very strikingly characterized. They resemble cutworm moths in size and structure but are perhaps more like Geometrids in superficial appearance. They may be separated from the Noctuidae by the absence of ocelli and by the bipectinate antennae of the males. The proboscis is wanting, and in some species the females are wingless or have the wings reduced to mere useless stubs. The hairy front legs are held out in front of the head when the moths are at rest. The size range is between 1 and 2 inches in wingspread. The colors are unattractive browns and whites. The larvae make up for the plainness of their parents. They are frequently of striking or beautiful appearance, bearing tufts or pencils of beautifully cropped, close-set hairs of gaudy colors or brilliantly colored tubercles partially concealed by the hairs of their bodies. Hairs play an important part in the economy of life of these insects. Some of the hairs of the larvae are nettling and no doubt serve to protect their lives from certain enemies; the eggs, which are usually laid in compact masses, are often covered with a mat of hairs derived from the tip of the abdomen of the female; the cocoon, which protects the pupal stage, is woven partly of larval hairs; and the pupae are clothed with hairs, which is very unusual. The white-marked tussock moth (page 828), and various other tussock moths, the satin moth, *Stilpnotia salicis* (page 818), the gypsy moth (page 830), and the brown-tail moth (page 832), are very destructive to shade or fruit trees. The nettling hairs of the brown-tail larvae (Fig. 17.10) are very severely irritating when they come in contact with the skin (page 832).

Family Noctuidae[1] (Also Called Phalaenidae). Owlet or Cutworm Moths. This is the largest and probably the most destructive family of the Lepidoptera. Nearly all are of moderate size, 1 to 2 inches across both wings, but the Catocalas average 3 inches, and a few like the black witch, *Erebus odora*, are enormous moths 4 to 6 inches in wingspread. They are of generally somber brown or gray colors. The front wings are moderately narrow and variously splotched and mottled with silvery white, black, or subdued shades of brown and gray, so that when at rest the insects are protectively colored. The hind wings are most often lighter in color and unmarked and form almost an equilateral triangle. Sometimes they are colored like the front pair and, in the *Catocala*, they are gorgeously colored with concentric bands of black and yellow or red. The body is stout, and the scales of the thorax often appear rumpled or ruffled. The antennae are usually simply filiform, the proboscis generally present. The adults are nocturnal and very common about artificial lights at night. The larvae are fat, unattractive-looking caterpillars of moderate size, either smooth or lightly clothed with hairs, and dull green, brown or gray, striped or spotted with black or slightly contrasting colors. Many, like the cotton leaf worm (page 579) and the cabbage looper (page 664), feed exposed on foliage. Others hide in the soil and eat off roots or stems at or below the ground. These are known as cutworms (page 476) and are numerous and very destructive. A few, such as the corn earworm (page 498), are borers, and some eat into the fruiting bodies of plants. Pupation usually takes place without a cocoon, in the soil. A number of kinds have the armyworm habit. Occurring in enormous abundance when environmental resistance is low and being somewhat gregarious, they march over the ground in devastating armies. Some of the larvae have the first two pairs of prolegs wanting and consequently crawl like loopers or measuring worms. In some species the earliest instars are loopers, later instars having

[1] SMITH, J. B., "Contribution towards a Monograph of the Noctuidae," *Bul. U.S. National Museum* 38, 231 pp., 1890; FORBES, W. T. M., "A Table of the Genera of Noctuidae of Northeastern North America," *Jour. N.Y. Ento. Soc.*, **22**:1–33, 1914; CRUMB, S. E., "Larvae of the Phalaenidae," *U.S.D.A.*, *Tech. Bul.* 1135, 1956.

the full set of legs.   Besides those named above, the following are first-rate pests: the fall armyworm (page 473), the variegated cutworm (page 478), the bronzed cutworm (page 478), the black cutworm (page 477), and the spotted cutworm (page 478).

Family Arctiidae.   Tiger Moths.   The name of this family does not signify carnivorous habits, for the insects are all leafeaters, but refers to the colors of the adults. In contrast with their somber-colored relatives, the owlet moths, the tiger moths are generally brightly marked with contrasty colors, splashed on as by irresponsible fancy in quaint futuristic patterns of geometric lines and spots.   The prevailing colors are white, black, and various shades of pink, red, orange, and yellow (Fig. 6.63).   They have a somewhat delicate appearance, the wings being thin, fine scaled, graceful in shape, and the body free from long coarse hairs.   In size (¾ to 3 inches across the wings), shape of wings, and body build they are similar to the Noctuids, from which

Fig. 6.63. The hickory tiger moth, *Halisodota caryae*, adult.   About natural size.   (*From U.S.D.A., Farmers' Bul.* 1270.)

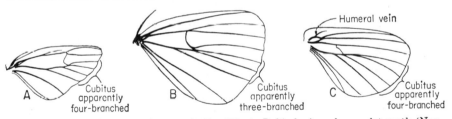

Fig. 6.64. *A*, hind wing of a tiger moth (Arctiidae), *B*, hind wing of an owlet moth (Noctuidae), and *C*, hind wing of a lasiocampid moth, to show characteristic differences in venation.   (*Original.*)

they are distinguishable by the course of the veins in the hind wings: in the tiger moths there is only one vein in front of the large, closed, discal cell for a considerable part of its length; in the owlet moths two veins stand between the discal cell of the hind wing and the costal margin, throughout most of its length (Fig. 6.64).   The larvae are woolly caterpillars, most of them completely covered with close-standing, erect, stiff hairs, which sometimes take the form of long slender pencils or brushes and which, as in the tussock moths, are utilized in weaving their cocoons.   Their hairiness is believed to be good protection from most birds, but the cuckoos are said to eat them readily. Among the few which become abundant enough to cause economic concern is the fall webworm (page 692), whose gregarious larvae make large indefinite nests of dirty silk, covering the leaves of entire twigs of walnut, apple, ash, wild cherry, and many other deciduous trees (Fig. 15.3).   The banded woollybear, yellow woollybear, and saltmarsh caterpillar (page 611) are very hairy, woolly caterpillars, which often become abundant in summer and autumn and attack many garden and field crops.

Family Lasiocampidae. Lappet Moths, Tent Caterpillars, and Others. These moths range in size from moderate to large (1 to 3 inches). They are velvety or wooly-looking, with broad abdomens and body, legs, and wings thickly covered with rather long fine hairs (Fig. 6.65). The adults take no food, their proboscides being atrophied. The palps are very short, the eyes are small, and they have no ocelli. The colors are grays, yellows, and browns, without striking markings. The base of the hind wing is expanded into a large shoulder lobe with one or two small veins extending into it, which locks beneath the front wing, in lieu of a frenulum. The vein just behind the discal cell (cubitus) is apparently four-branched, having appropriated two of these branches from the degenerated medius vein in front of it (Fig. 6.64,C). Both adults and larvae favor browns in their color scheme. The larvae are all more or less hairy, and some species, known as lappet caterpillars, have brush-like clumps of hairs extending outward and downward from the sides of each segment against the leaf or bark on which they rest. Their colors are such as to help make them difficult to distinguish from the background. The family is a small one, the best-known species being the gregarious forest tent caterpillar (page 828) and the eastern tent caterpillar (page 693), which make cooperative nests in fruit and shade trees, like the webworms of the preceding family. The tent caterpillars, however, do not include the foliage in their nest but construct a home in a crotch of the limbs (Fig. 15.5) and, from that, range out in the

Fig. 6.65. Apple tree tent caterpillar moths, male at left, female at right. About natural size. (*From U.S.D.A., Farmers' Bul.* 1270.)

daytime to feed at the ends of the branches. It is said that each caterpillar maintains a silken life line back to the nest, to which all retreat at night. The egg masses of the tent caterpillars form broad rings about the small twigs of the food trees in winter, and are covered with a hard varnish-like material from the female (Fig. 5.1,N).

Family Citheroniidae (Also Called Ceratocampidae). Royal Moths. The large beautifully colored moths, which so many novices wrongly call butterflies, have been variously grouped into families in the past, but are now generally considered to represent two families, the Citheroniidae and the Saturniidae. As in the Lasiocampidae, a humeral expanse of the wing membrane has supplanted the frenulum, but in these families there are no veins extending into it. In both these families, the vein just behind the discal cell (cubitus) is apparently three-branched. The heads of the adults are not prominent, being short, almost hidden beneath the broad thorax. The abdomen is broad, short-haired. These two families have little in common except their large size and showy colors. The Citheroniidae have the mouth parts present, though small, and two large spurs at the end of middle and hind tibiae. In the males the antennae are feathered (bipectinate) only about halfway to the tip, the apex being contrastingly filiform. The larvae usually have prominent horns on the second and third segments behind the head. They make no cocoons for the pupal stage, which is passed in the ground. The most striking species include the regal walnut moth, *Citheronia regalis* (Fig. 6.66), which measures 4 to 6 inches from tip to tip of wings and is colored olive brown with large oval yellow spots and with the veins bordered with red. Its larva is the terrifying-looking hickory horned devil (Fig. 1.9). The imperial moth, *Basilona imperialis* (Fig. 6.67), is a robust, velvety species about as large as the regal moth, but its ground color is yellow, speckled with purplish brown. There are a round spot near the middle of each wing and a narrow band of the same brown color extending diagonally across each wing. The larva is sparsely covered with long stiff white hairs, among which are reared, just behind the head, four stout spiny horns. The only other species commonly noted in this small family are certain oak and maple

FIG. 6.66. The regal walnut moth, *Citheronia regalis*. About ¾ natural size. (*From Ill Natural History Surv.*)

FIG. 6.67. The imperial moth, *Basilona imperialis*. About ¾ natural size. (*From Ill Natural History Surv.*)

worms, which are distinctly smaller in size and the adults of which are brown or yellow suffused with pink or purple.

Family Saturniidae.[1] **Giant Silkworm Moths.** In contrast with the other very large moths, the royal moths, the giant silkworm moths lack mouth parts and have no

[1] MICHENER, C. D., "Saturniidae of Western Hemisphere," *Bul. Amer. Museum Natural History* 98, pp. 335–502, 1952.

spurs at the end of the tibiae.   There is usually a spot at the center of each wing or at least of the hind wings which is contrastingly colored or lacks scales and is transparent and ringed, called an "eye-spot."   The wings are very broad, forming nearly equilateral triangles.   The colors are often beautiful browns and yellows, but for the most part subdued rather than gaudy.   As in many other moths the sexes can be distinguished by the greater length of the feather-like barbs on the pectinate antennae in the male.   In contrast with the royal moths, the antennae of the males are feathered clear to the tip.   The cocoons are very dense, and many of the Oriental species yield silk in commercial quantities.   Shantung silk, tussah silk, and muga silk are derived from different caterpillars of this family (page 48).   The best-known American saturniid is the Cecropia moth, *Hyalophora cecropia* (Fig. 6.56).   It would seem that nearly everyone should know this beautiful moth, but every year in every community, entomologists are regularly sought out by those who have for the first time reared it

Fig. 6.68. The luna moth, *Tropaea luna.*   About ⅗ natural size.   (*From Felt, N.Y. State Museum Memoir 8.*)

from the large overwintering cocoons and, charmed by its beauty, believe they have something unique and commercially valuable.   The beautiful pale-green luna moth, *Tropaea luna* (Fig. 6.68), has the hind wings produced into handsome tails like the swallowtail butterflies.

Family Hesperiidae.[1]   The Common Skippers.   These insects appear to be somewhat intermediate between the moths and the butterflies.   They have heavy, hairy, moth-like bodies (Fig. 6.57), the wings are partially hairy, and the pupae look more like those of moths than butterflies, but most of the species fold their wings like butterflies, though some horizontally like moths.   They have knobbed antennae, lack ocelli, and are diurnal in habits.   The name refers to their short, rapid, darting flight, which is rarely sustained like the moths or butterflies.   The widely separated antennae with their recurved tips and the fact that all wing veins arise separately from the discal cell,

[1] COMSTOCK, J. H., and ANNA B. COMSTOCK, "How to Know the Butterflies," pp. 260–301, Comstock, 1920; DYAR, H. G., "A Revision of the Hesperiidae of the United States," *Jour. N.Y. Ento. Soc.*, **13**:111–141, 1905; LINDSEY, A. W., E. L. BELL, and R. C. WILLIAMS, "Hesperioidea of North America," *Jour. Sci. Lab. Denison Univ.*, **26**:1–142, 1931.

so that there is no forking of veins, distinguish them from all other Lepidoptera. They are small to moderate-sized (1 to 2 inches), mostly very dark, sooty-brown in color, with angular or indefinite-shaped white spots on the wings. The head is very broad between the eyes. The eyes have a fringe of hairs like eyelashes. The palps and beak are well-developed. The larvae are curious-looking, with their large heads borne on a much constricted neck, like a very tight collar. The body is bare, stout at the middle. They live in folded leaves or webs and make a thin cocoon for the pupal stage. They have very little importance.

Family Papilionidae.[1] Swallowtail Butterflies. These are the largest and most beautiful of the butterflies, as the Saturniidae are among the moths. Our common species range from 2 to nearly 5 inches from tip to tip of the front wings. They have a trim, graceful, clean-cut appearance; the very broad wings are paved with perfectly arranged, fine scales. They are magnificently colored, and the hind wings have the anal margin very long and concave and usually prolonged into tail-like extensions (Fig. 6.58,c). The prevailing colors are blacks and yellows with angular splotches of yellow, red, and blue. Males and females often differ extremely in color, and there may be in the same species several kinds of females or males that are very different in appearance. The larvae are bare, generally strikingly colored, and frequently possess a pair of soft, retractile horns which can be erected from the upper part of the segment just behind the head, and give off a sickening sweet scent (Fig. 6.58). They feed mostly on the leaves of various trees but are rarely abundant enough to do any damage. The pupae are angular, with two strong projections from the head, and are suspended by the tail and a girdle of silk about the waist (Fig. 6.58,f), never being enclosed in cocoons.

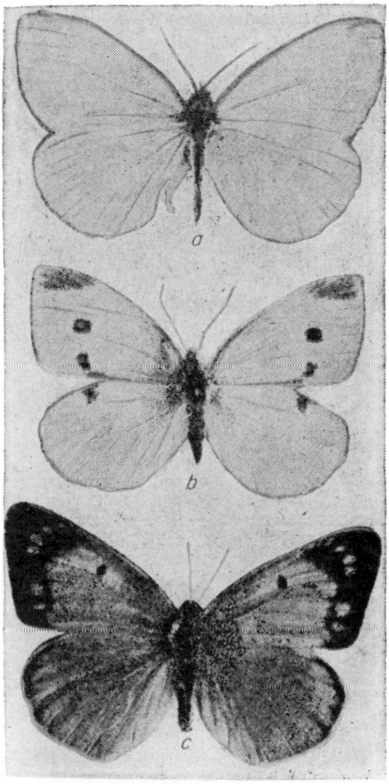

Family Pieridae.[2] White and Sulfur Butterflies. These are common medium-sized butterflies, white, yellow, or orange in color, sometimes margined or spotted with black and silvery white. The wings have a compact, rounded appearance, the apical margins being distinctly convex and the hind wings nearly circular in outline. The antennae are long and strongly clavate, and the abdomen very slender. The size range is mostly from $1\frac{1}{2}$ to $2\frac{1}{2}$ inches. The larvae are very plain, unattractive, nearly cylindrical, and covered with very fine hairs, but without spines or horns of any kind. The pupae are suspended like those of the swallowtails but have only one projection from the head (Fig. 5.16,C). The alfalfa caterpillar (page 555) and the imported cabbageworm (page 662) are serious pests, and the moths closely resemble one another (Fig. 6.69).

Fig. 6.69. Three common pierid butterflies. *a*, Native cabbage butterfly, *Pieris oleracea*, male; *b*, imported cabbage butterfly, *Pieris rapae; c*, clouded sulfur butterfly, *Colias philodice*. (*From Sanderson and Jackson, "Elementary Entomology," after Fiske, Ginn & Co.*)

Family Nymphalidae. Four-footed Butterflies. All adult insects have six legs, except a few degenerated females like the bagworm moths and scale insects, but this

[1] COMSTOCK and COMSTOCK, *op. cit.*, pp. 35–68.

[2] *Ibid.*, pp. 69–101; KLOTZ, A. B., "Generic Revision of the Pieridae," *Entomologica Americana*, **12**:139–242, 1933.

entire family has lost, in part, the feet or tarsi of the front legs. The females have all the tarsal segments but no claws. The males have only one tarsal segment remaining on each front leg. The legs are therefore useless for walking, but may serve to clean the antennae and, rarely, for making sounds. On the average the nymphalids are intermediate in size between the Pieridae and the swallowtails, mostly 1½ to 3½ inches across the wings. This is a very large family, the many species of great variety in size and appearance. The colors on the wings are almost always contrasted, and the spots or bands often sharply angular. Many of the species have a distinctly checkered appearance of many rather small, rounded, or angular spots of contrasting colors. The commonest colors are yellowish brown, black, and silvery white. The shape of the wings is much like that described for the Pieridae, but in some of the species the outer margins of the wings are irregularly notched or cut, as though they had been raggedly torn. The underside of the wings, which alone is visible when the butterflies alight to rest, is often protectively colored. Many of these butterflies are very bad-tasting, and birds readily learn to recognize their appearance and to leave them alone (Fig. 6.70). The larvae are remarkable creatures, often gaudily colored, with fleshy processes or branching spines on head, tail, thorax, or rarely all over the body. The pupae hang head downward, suspended only by the tail (Fig. 5.16,*D*). They have many curious projections and are often metallic-colored and objects of great beauty. The family has little or no economic importance.

FIG. 6.70. An example of protective mimicry. Above, the monarch butterfly, *Danaus plexippus*, a species distasteful to birds. Below, the viceroy, *Basilarchia archippus*, an edible species which, although belonging in a different subfamily, resembles the monarch so closely that it is supposed to be avoided by birds on that account, although it is edible. The hind wings of the monarch show *androconia* as small black spots just behind their centers. ½ natural size. (*Original.*)

Family Lycaenidae. Blues, Hairstreaks, and Gossamer Wings. This family includes the smallest common American butterflies, although many of the micromoths are much smaller. These range from ¾ to 1½ inches across the wings. Though the prevailing color is dark brown, these little butterflies are often exquisitely beautiful. The antennae are ringed with white on nearly every segment, and the eyes are margined with white scales. A good many species are overcast with a beautiful blue or purple sheen, but many are brown-flecked like the fritillaries of the Nymphalidae. The males have the foretarsi without claws. Both palps and proboscis are long and slender. The larvae are somewhat like the slug caterpillars, all the legs being short, and the head is inconspicuous. Some of them have posterior osmeteria, from which ants have learned to lap up the secretion. One of the species, *Feniseca tarquinius*, is remarkable for a butterfly larva in being predaceous. It feeds on woolly aphids. Its pupa looks astonishingly like a skull or a monkey's face.

ORDER HYMENOPTERA[1]

BEES, WASPS, ANTS, SAWFLIES, PARASITES, AND OTHERS

Everyone knows at least three kinds of Hymenoptera—bees, wasps, and ants. But not only are there numerous families of bees, wasps, and

[1] CRESSON, E. T., "Synopsis of the Families and Genera of Hymenoptera of America North of Mexico," with a Catalogue and Bibliography, *Trans. Amer. Ento. Soc.*,

ants; in addition to them the order includes a much greater number of species of other habits, such as the parasitic wasps, the gall wasps, and the sawflies, which are fully as important to us and altogether as interesting as the better-known kinds.

This order is listed as third in point of size (Table 6.3), but it has been so little studied in comparison with the beetles, moths, and butterflies that it would not be surprising to see it surpass these orders when its unnumbered species are once thoroughly studied.

Most authors place the Hymenoptera at the top of the list of orders, in much the same way that man is placed at the pinnacle of the vertebrate animals; and for much the same reason. This order appears to exhibit instinctive behavior in its highest state of perfection. In some of its representatives is found at least a low grade of intelligence, *i.e.*, the ability to learn or profit by experience, to choose and to form concepts.

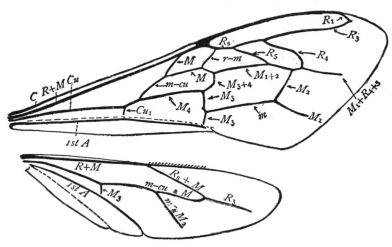

FIG. 6.71. Wings of the honeybee showing the minute hooks or hamuli that serve to lock the two wings together. (*From Comstock, "Introduction to Entomology."*)

As a basis for the social organization which is so elaborately perfected in the better-known Hymenoptera, *care of the young* is widespread and often solicitous, in contrast with most insects which ordinarily pay no attention to their young after the eggs are laid. The larvae of many groups of this order are completely dependent upon their parents (or other adults) for food. Among the parasitic and gall-making species this obligation is discharged by the females when they lay their eggs in the midst of an abundance of food. The solitary wasps and bees generally gather and store a quantity of food of a suitable kind, available to the

Supplementary Volume, Philadelphia, 1887; GODMAN and SALVIN, "Hymenoptera," *Biologia Centrali Americana*, 1883–1900, *Genera Insectorum*, P. Wytsman, Brussels, 1903–1923; BRITTON, W. E., H. L. VIERICK, *et al.*, "The Hymenoptera or Wasp-like Insects of Connecticut," *Conn. State Geol. Natural History Survey Bul.* 22, 1916; MUESEBECK, C. F. W., R. V. KROMBEIN, H. K. TOWNS, *et al.*, "Catalogue of Hymenoptera of America, North of Mexico," *U.S.D.A., Agr. Monograph* 2, 1420 pp., 1951; PETERSON, A., "Larvae of Insects: An Introduction to Nearctic Species, Lepidoptera and Plant-infesting Hymenoptera," 315 pp., Columbus, Ohio, 1948.

larvae in the nest when they hatch from the eggs. But some of the social wasps and bees, and the ants, bring food to the larvae day by day, during their entire lives, and often feed, clean, guard, and care for the young in a manner highly suggestive of the maternal care that is general among the higher vertebrate animals. All the ants and many of the wasps and bees live together in great colonies, leading a complex social or cooperative life, the wonders of which increase, the more intimately man exposes their details to common knowledge.

The social life of the Hymenoptera is rivaled by that of the termites already described. As in that case, the reproductive function is limited to a few specialized individuals (kings and queens), while the vast

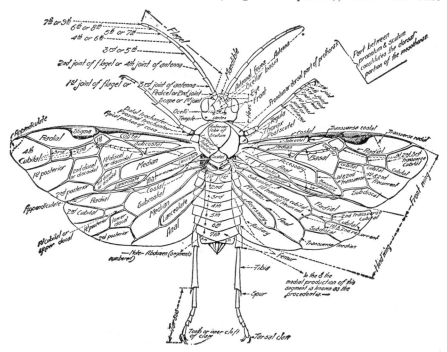

Fig. 6.72. Dorsal view of a sawfly to show names of the principal parts of the body which are used in classification. (*From Conn. Geol. and Natural History Surv. Bul.* 22.)

majority of the adults remain infertile. These barren individuals work, not for themselves, but for the common good, building and cleaning the nests, foraging for food, fighting off invaders, and taking complete care of the prodigious number of young that hatch from eggs laid by the queens. In contrast with the termites, the workers and soldiers of the Hymenoptera, when these castes are developed, are *exclusively females*. The males accordingly have earned the name *drones*, their sole usefulness to the species apparently being to ensure the fertility of the queen's eggs.

The unusual mental development of the higher Hymenoptera is a strong reason for listing this order as the most specialized. But it should be considered as only one criterion and not be allowed to overshadow other considerations. In the matter of wing specialization the Hymenoptera are surpassed by the Diptera, the Coleoptera, and the Hemiptera.

In the specialization of the mouth parts, the Diptera, Siphonaptera, Anoplura, Hemiptera, Homoptera, and especially the Lepidoptera, have gone much farther.   These are some of the reasons we have for giving the Hymenoptera a middle position in the series of insects with a complete metamorphosis.

The Hymenoptera have a complete or complex metamorphosis.   The larvae of this order vary much more in form than those of the beetles or moths, ranging from caterpillar-like sawflies, with distinct head, well-developed legs and prolegs, and independent active habits, to the legless and practically helpless progeny of bees, wasps, and ants.   The sawfly larvae (Fig. 5.15,*A*) can be distinguished from caterpillars (lepidopterous larvae), which they most resemble, by the number of prolegs, which is from six to eight pairs, whereas lepidopterous larvae (Fig. 3.9) never have more than five pairs; also the prolegs are not provided with crochets like those of Lepidoptera; the adfrontal sclerites (Fig. 6.55) are never present in Hymenoptera; and the ocelli are at most one pair in Hymenoptera, always more than one pair, or none, in lepidopterous larvae.   The more

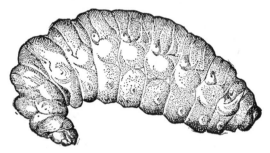

Fig. 6.73. Larva of a digger wasp, *Tiphia* sp.   Note reduced head, spiracles on the sides of the body, and absence of legs.   (*From Ill. Natural History Surv.*)

specialized larvae (Fig. 6.73) differ from dipterous larvae, with which they are most likely to be confused, in having a recognizable head (although it may be reduced in size), with distinct mouth parts.   Also, in contrast with most Diptera, the larvae usually have a pair of small spiracles on each of the principal abdominal segments, rather than a large complex pair close together on the last segment.   The antennae are usually wanting, and the ocelli wanting or a single pair.   In the higher families the head of the larva is opaque white like the rest of the body and often very small, but the true mandibles are retained.

The pupae (Figs. 5.16,*K*,*M*, and 5.17,*M*) resemble beetle pupae in having the appendages not immovably fastened to the general body wall. The antennae of the pupa are always longer than the head, and the mandibles can be recognized as of the chewing type.   The labium and maxillae are often elongate, the compound eyes show distinctly, and the pupa is generally surrounded by a silken cocoon.   In some of the Hymenoptera, the metamorphosis is complicated by hypermetamorphosis (page 643) and by *polyembryony*.   Polyembryony is a remarkable condition in which anywhere from 2 to over 150 individuals develop from a single egg by the splitting up of the egg at an early stage of development to form a number of embryos.   The life cycle may be further complicated by *parthenogenesis* and an *alternation of generations*.   In some species a given generation may be all females; these lay unfertilized eggs on a kind of plant, say an oak, and a characteristic gall grows on the plant to shelter the insect.   When this generation of insects is mature, they may be of both sexes, and the females lay fertilized eggs on another kind of plant, for example a rose.   As a result, the rose develops a gall altogether different in appearance from that of its parents on oak.   When these young are mature, all prove to be parthenogenetic females that seek the oak again and produce galls like those in which their grandparents developed.   Until such species are carefully watched through several generations, they are sure to be described as two distinct species, so different are they, both in habitat and in appearance.

The wings in this order (Fig. 6.71) are four in number.   They are generally small, transparent, and with comparatively few veins.   The front pair is distinctly larger.   The hind wing usually fits exactly against the hind margin of the front wing, to which it is fastened by a row of very minute hooks.   The student should note this carefully, otherwise he may mistake certain ones of this order for Diptera, since the two pairs of wings are so closely fitted together as easily to be mistaken for a single pair.

The mouth parts vary from the chewing type to a combination of chewing and lapping structures (Fig. 4.17).   In all cases the labrum and the mandibles are essentially like those of Orthoptera and Coleoptera. The maxillae and labium are also essentially of the chewing type in the less specialized families, but in the bees, wasps, and ants, these two paired structures become progressively longer and longer, to form a hairy, lapping tongue by means of which liquids are lapped up as the insect feeds (page 146 and Fig. 4.17).   The maxillae are more or less united with the labium in both adults and larvae.

One specialization peculiar to the Hymenoptera is the modification of the ovipositor into a defensive and offensive weapon known as a "stinger." This organ of defense and offense is found only among the insects of the order Hymenoptera—the bees, many of the wasps, and certain kinds of ants, besides the distantly related scorpions.

The stinger of a bee or wasp is a very complex and beautifully adapted organ.   It is known to be the equivalent (homologue) of the egg-laying organ in other insects, and it follows therefore that *only the females of insects can sting*.   In many of the Hymenoptera (Fig. 2.19), a greatly elongated, sting-like organ serves the function of thrusting eggs into plant parts but is never used in defense.   The stinger consists of a similar mechanism for penetrating the skin to a depth of perhaps $\frac{1}{10}$ inch and of a system of glands to secrete the venom that is injected into the wound. In the honeybee the sting proper consists of two extremely sharp, highly polished, brown spears or darts which appear as one.   Their concave inner surfaces make, between them, a fine tube down which the venom is forced, to emerge at their tips.   As the insect stings, these two darts are alternately and very rapidly thrust outward or downward on guide rails of a surrounding sheath.   Each dart has near its tip 9 or 10 recurved hooks which hold it firmly until the next thrust carries it still deeper. Because of these hooks, the honeybee can seldom remove her stinger, and it, with more or less of the viscera, is torn away as she escapes. Other bees, wasps, hornets, and ants may sting repeatedly.

The pain of the sting is due to the venom.   In some of the Hymenoptera this venom is deadly to other insects and small animals receiving it.   In others it has only a paralyzing effect and is used to stupefy flies, spiders, crickets, caterpillars, or beetles upon which eggs are laid and the helpless plunder is then sealed up in their nests as food for the forthcoming young (Fig. 6.74).

Another peculiarity of the adults of this order is that one abdominal segment is fused with the thoracic mass, so that what appears to be the first abdominal segment is in reality the second.

Wings typically four, small, membranous, with few veins; hind wings smaller, often hooked to front pair; females sometimes wingless, the males rarely so (Fig. 2.8). Venation highly specialized; often much reduced. Mouth parts chewing or chewing and lapping, the mandibles always of the chewing type; but the maxillae and labium often elongate (page 146). Compound eyes usually well developed; three ocelli usually present. Pronotum fused to the mesothorax. Tarsi usually five-segmented. Abdomen often with a slender waist and its first segment united with the thorax. Female provided with an ovipositor or stinger. No cerci. Metamorphosis complete. Larva either caterpillar-like or legless; with distinct head, and spiracles on the principal segments. Prolegs, when present, usually more than five pairs and without crochets. No adfrontal areas. Pupae with appendages free, commonly encased in cocoons. Many species live in societies.

Fig. 6.74. A sphecid wasp, *Ammophila* sp., putting a paralyzed measuring-worm into her burrow to serve as food for her young.   When the nest is completed and the eggs laid, the wasp takes a small pebble in her mandibles and packs down the earth, with which she closes her burrow.   (*From Kellogg, "American Insects."*)

The classification of the Hymenoptera appears to be in a very unsettled condition.   The following outline will indicate the principal families and something of their natural grouping.   We have followed Comstock's[1] account in many respects but have varied from the opinions of specialists, where by so doing it seemed possible to give the nontechnical student a more useful working conception of these insects.

A SYNOPSIS OF THE HYMENOPTERA

A. Suborder *Chalastogastra, Symphyta,* or *Sessiliventres.*   Foremost segments of abdomen as broad as the following segments and joined to the thorax by their full width (*i.e.*, without a slender waist).   Trochanters two-segmented.   Ovipositor adapted for sawing or boring, never a sting.   The larvae feed on or in plants and are caterpillar-like in appearance.   They have a single pair of ocelli and usually well-developed legs and prolegs, the latter without crochets.

Family 1. The horn-tails, Family Siricidae.
Family 2. The stem sawflies, Family Cephidae.
Family 3. The sawflies, Family Tenthredinidae.

B. Suborder *Clistogastra, Apocrita,* or *Petiolata.*   Foremost segments of abdomen narrower than those which follow, making the connection to the thorax a slender petiole or "waist."   The thorax appears to bear 3 pairs of spiracles, but the last

[1] COMSTOCK, J. H., "An Introduction to Entomology," Comstock, 1924.

pair is really on the first abdominal segment which is completely fused with the thorax. The larvae are of varied habits, always legless and grub-like, with head and mouth parts reduced, antennae and palps at most of one segment, and ocelli usually wanting.

WASPS OR WASP-LIKE FORMS

Hairs of body not branched. First segment of hind tarsus usually slender, cylindrical. No tubercles on petiole of abdomen.

a. *Solitary species consisting of only males and females;* each female provides for her own young.

    1. Larvae mostly parasitic on other insects. Eggs generally laid on or through the body wall of the active host. Trochanters two-segmented.

        Family 4. Ichneumon wasps, Family Ichneumonidae.
        Family 5. Braconid wasps, Family Braconidae.
        Family 6. Ensign wasps, Family Evaniidae.
        Family 7. Chalcid wasps, Family Chalcididae.
        Family 8. Egg-parasite wasps and others, Family Proctotrupidae (also called Proctotrupoidea and Serphoidea).

    2. Larvae mostly live in galls which they cause to grow on plants (Fig. 1.6). Trochanters apparently two-segmented.

        Family 9. Gall wasps, Family Cynipidae.

    3. Larvae mostly parasitic on other insects or spiders. Eggs laid on paralyzed ("stung") caterpillars, grubs, or spiders, which are buried in the ground or stored in mud cells, burrows, tunnels, mines, or natural cavities; or on active leafhoppers, etc.; or in nests of other Hymenoptera. Or rarely food is brought to larvae in the nest from day to day, by the parent wasps.

        Family 10. Digger wasps, mud-daubers, and thread-waisted wasps, Family Sphecidae.
        Family 11. Dryinid wasps, Family Dryinidae.
        Family 12. Spider wasps, Family Pompilidae.
        Family 13. Cuckoo wasps, Family Chrysididae.
        Family 14. Velvet "ants," Family Mutillidae.
        Family 15. Vespoid digger wasps, Family Scoliidae.
        Family 16. Mud wasps, Family Eumenidae.

b. *Social wasps with a sterile worker caste in addition to both males and females,* the workers taking most care of the females' young. Eggs laid in cells of nests composed of paper, which the adults make of wood. Larvae are fed from day to day on juices of insects or sweets. Wings folded lengthwise when at rest. Trochanters one-segmented.

        Family 17. Hornets, yellow jackets and paper-nest wasps, Family Vespidae

ANTS

Hairs of body not branched. Petiole of abdomen with one or two swellings or tubercles (Fig. 6.87). Often wingless. Social insects with a sterile worker caste, in addition to males and females. Nests in soil, wood, or stems of plants, without well-defined cells for each larva to live in. Larvae fed on regurgitated food from adults, or on bits of insects, seeds, fungi, etc. Trochanters one-segmented.

        Family 18. Ants, Family Formicidae.

BEES

Broader-bodied and more hairy than ants or wasps. The hairs of the thorax branched or plumose (Fig. 6.95). First segment of hind tarsus often broad, flattened, and brush-like for assembling pollen, and hind tibia often specialized as a "pollen basket" (Fig. 3.4,*C*). The pronotum does not extend backward on the sides to the tegulae at base of wings. The mouth parts have the glossae well developed to form chewing-lapping type of mouth parts. Trochanters one-segmented. All individuals winged. Nests provisioned with nectar and pollen from flowers, as food for the young.

a. *Solitary bees consisting of only males and females.*
  1. Tongues (labia) short and broad. Eggs laid' in nests burrowed in ground, in pithy plants, or in crevices of walls or buildings. Cells separated by a silky secretion.
    Family 19. Bifid-tongued bees, Family Prosopidae.
    Family 20. Colletids, Family Colletidae.
  2. Tongues long and slender (Fig. 4.17).
    i. Nests in burrows in plant stems, in cavities about buildings or in the soil. Eggs laid in cells made of pieces of leaves cut from growing plants, or of plant fibers or of clay mixed with saliva. Pollen-collecting brushes on underside of abdomen of females.
      Family 21. Leaf-cutting bees, Family Megachilidae.
    ii. Nests in tunnels cut in the solid wood of buildings or trees. Eggs laid in cells separated by cemented sawdust.
      Family 22. Large carpenter bees, Family Xylocopidae.
    iii. Nests in tunnels in pithy plants. Cells separated by plugs of plant fiber.
      Family 23. Small carpenter bees, Family Ceratinidae.
b. *Parasitic or "cuckoo" bees* that lay their eggs in nests of other bees and so steal the food or parasitize the rightful owners. Tongues long. Legs not adapted for collecting pollen. Males and females only, no worker caste.
    Family 24. Cuckoo bees, Family Nomadidae.
c. *Gregarious bees consisting of only males and females.* Nests placed near together in the soil or in the face of cliffs, often with a common entrance or corridor. Tongues short or long and pointed.
    Family 25. Mining bees, Family Adrenidae (Halictidae).
    Family 26. Anthophorids, Family Anthophoridae.
d. *Social bees, with a worker caste in addition to males and functional females.* Eggs laid in cells made of wax secreted by worker bees (Fig. 2.4).
  1. Nests commonly in deserted mouse nests on the ground (Fig. 6.94). A number of eggs usually laid together in one waxen cell.
    Family 27. Bumblebees, Family Bombidae (Bremidae).
  2. Nests built in trees or in hives provided by man. A single egg laid in each cell (Fig. 2.4).
    Family 28. Honeybees, Family Apidae.

As a group, the Hymenoptera may be considered more beneficial than injurious. There are, to be sure, a number of serious pests. But the work of the very many insect parasites and predators, the activities of the bees in pollinizing plants, and the production of honey and wax, undoubtedly offset manyfold the injuries inflicted by members of this order. As in the Coleoptera and Lepidoptera, there are no parasites of the larger animals.

THE PRINCIPAL FAMILIES OF HYMENOPTERA

Family Siricidae.[1] **The Horn-tails.** The adult horn-tails are moderately large wasps, with a goodly number of veins and cells in the wings. The body is cylindrical, very straight-sided or of nearly equal width, bare, hard, and polished (Fig. 6.75). The colors are black, brown, or blue, usually with yellow bands or spots. The female has a stout ovipositor projecting straight behind, which looks like a sting but is not. It arises about the middle of the abdomen on the underside. Above it is a short, tail-like, spear-shaped projection. The ovipositor is used for drilling into the hardwood of standing trees and laying eggs therein. The eyes are small, and the head curiously swollen behind them and set close against the thorax. The antennae arise low on the

---

[1] BRADLEY, J. C., "Siricidae of North America," *Pomona Coll. Jour. Ento. Zool.*, **5**:1–30, 1913; MACGILLIVRAY, A. D., "Siricidae, Hymenoptera of Connecticut," *Conn. Geol. Natural History Surv. Bul.* 22, pp. 169–172, 1916; Ross, H. H., "Generic Classification of Nearctic Sawflies," *Ill. Biol. Monograph* 34, 1937.

head between the eyes.   There is only one spur at the end of the front tibia.   The white, cylindrical, wrinkled larvae bore through the wood of trees, which they eat. The legs are small and prolegs wanting.   Pupation takes place in the larval burrow, surrounded by a cocoon of silk and sawdust.   The total damage done by this small family is slight.   Our commonest species is the pigeon tremex, *Tremex columba*, frequently found trapped by the ovipositor, which she has been unable to extricate, and hanging dead or dying on the tree trunk.   This species is about 1½ inches long, exclusive of the ovipositor.

Family Cephidae.[1]   **The Stem Sawflies.**   These are medium-sized, soft-bodied, delicate-looking wasps, with slender bodies and very short ovipositors.   The abdomen is compressed, much deeper than wide.   The head, which is much like that of the Siricidae, is very distinct, the prothorax is very long, subconical in front and straight behind and movable against the mesothorax.   There is a single spur at the end of the front tibia.   The larvae (Fig. 5.15, *B*) bore in the stems of herbaceous plants or the tender shoots of woody plants and are almost totally legless.   Important American species include the wheat stem sawfly (page 535), and the currant stem girdler, *Janus integer*.

FIG. 6.75. A horn-tail, the pigeon tremex, *Tremex columba*.   Natural size. (*From Kellogg, "American Insects."*)

Family Tenthredinidae.[2]   **The Sawflies.** The females of sawflies have short, stout, sharply saw-toothed ovipositors, mostly concealed on the underside of the abdomen near the end.   With these ovipositors they cut, like a saw, into plant tissues to deposit the eggs. The species are medium to rather large (¼ to 1 inch long), one of our largest species, the elm sawfly, *Cimbex americana*, having a wing expanse of 2 inches.   The body is broad; head, thorax, and abdomen of nearly equal width, short and compactly joined (Fig. 6.72).   Each anterior tibia has two apical spurs.   The colors are somber, the wings sometimes clouded, broad, well provided with veins, and usually more or less wrinkled.   The antennae vary greatly in form and in number of segments. The colors are black, brown, and yellow—the yellow often in abdominal crossbands. Parthenogenesis is very common in this family; males are generally rare, and in many species no males are known.   Either males or females may be produced from the unfertilized eggs.   The larvae look much like those of Lepidoptera, being cylindrical, provided with both jointed legs and prolegs and often hairy or spiny.   They feed on foliage, especially of woody plants, or bore in stems, fruits, or leaves.   Some species are slug-like and covered with slime.   The larvae can always be distinguished from Lepidoptera by the absence of crochets on the prolegs and by the larger number of prolegs, which are six to eight pairs in the sawflies and from two to five pairs in the Lepidoptera.   Other differences are noted on page 269.   This is a very large family, and a good many species are primary pests, as, for example, the imported currantworm (page 782), the rose-slug (page 857), the pear-slug (page 767), the elm sawfly, and the raspberry, violet, and fern sawflies.

Family Ichneumonidae.[3]   **Ichneumon Wasps.**   This family includes the largest of the wasps whose larvae develop in other insects (especially in caterpillars), as well as

[1] MacGillivray, A. D., "Cephidae, Hymenoptera of Connecticut," *Conn. Geol. Natural History Surv. Bul.* 22, pp. 172–174; Ries, D. T., "Revision of Nearctic Cephidae," *Trans. Amer. Ento. Soc.,* **63**:259–324, 1937.

[2] MacGillivray, A. D., "Tenthredinidae, Hymenoptera of Connecticut," *Conn. Geol. Natural History Surv. Bul.* 22, pp. 41–167, 1916; Norton, E., "Catalogue of the Described Tenthredinidae and Uroceridae of North America," *Trans. Amer. Ento. Soc.,* **1**:31–84, 193–280, 1867, and **2**:211–242, 321–367, 1868–1869.

[3] Ashmead, W. H., "Classification of Superfamily Ichneumonoidea," *Proc. U.S. National Museum,* **23**:1–220, 1901; Vierick, H. L., "Ichneumonidae, Hymenoptera of Connecticut," *Conn. Geol. Natural History Surv. Bul.* 22, 243–360, 1916.

many very small kinds.   The commonest colors are black and bright yellowish-brown with reddish or yellow spots or bands.   The abdomen is often very long and frequently strongly compressed.   The trochanters are two-segmented.   The ovipositor, which always shows behind the abdomen and may reach the phenomenal length of 4 or 5 inches (Fig. 2.19), attaches to the underside of the abdomen some distance in front of the apex.   The antennae are long, filiform, not elbowed, and almost constantly vibrating in life.   The pronotum, or collar behind the head, reaches back on the sides to the tegulae or scales at the base of the wings.   The wings have a well-developed stigma, behind which the closed cells are arranged in a double row, *i.e.*, one behind the other. There is no slender (costal) cell between the stigma and the base of the wing.   There are only two closed cells between the stigma and the anal margin.   The base of the abdomen bows upward from the rear of the thorax.   The middle and hind coxae are

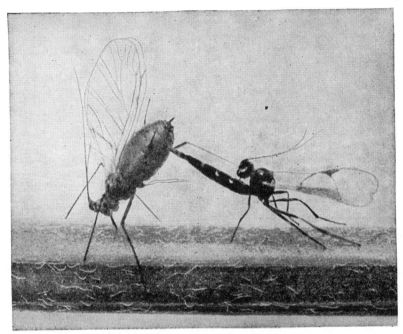

FIG. 6.76. A wasp, *Lysiphlebus testaceipes*, depositing its eggs in the body of the alfalfa aphid.   Much enlarged.   (*From Essig, "Insects of Western North America," copyright, 1926, Macmillan.   Reprinted by permission.*)

enormously large.   The adults feed on flowers.   The larvae have a distinct head and nine pairs of spiracles but no legs.   The larvae occasionally attack the host from the outside, but almost always spend their entire period of growth plowing about inside the body, instinctively not eating any organs that would cause death promptly.   Their growth is concurrent with the host, and they often pupate inside the pupal case of the host insect.   The host insect then dies, and the parasite emerges in the winged form. A cocoon is usually made for the pupal stage.   This is an enormous family the value of which as destroyers of crop pests can hardly be comprehended (page 71).

Family Braconidae.[1]   **The Braconid Wasps.**   Averaging much smaller in size than the ichneumons, but rivaling them in importance by similar parasitic habits, are the closely related braconid wasps.   Although a few species range up to $\frac{1}{2}$ inch long, the great majority are under $\frac{1}{8}$ inch in length.   In most features they are like the ichneumons—the antennae not elbowed, the trochanters two-segmented, the pronotum

[1] VIERICK, H. L., "Braconidae, Hymenoptera of Connecticut," *Conn. Geol. Natural History Surv. Bul.*, pp. 216–239, 1916.

reaching to the wing bases. The colors are generally dark or dull, the wings some-times banded or spotted, but the abdomen almost never so. The abdomen is short, not compressed, the second and third segments firmly fused together. The antennae are rather thick. There are usually three closed cells between the stigma and the anal margin of the wing, and all the closed cells in the wing may be arranged in a single curved row. As in the Ichneumonidae there is no slender (costal) cell between the stigma and the base of the wing. The ovipositor of the female is usually long and exposed. The larval habits are similar to those of the ichneumons, but they more often pupate outside the host's body, attached to leaf or stem, often a number of cocoons more or less webbed together. The braconids most commonly noted are *Apanteles congregatus*, which spin their white cocoons over the backs of tomato horn-worms and other caterpillars (Fig. 2.20 and page 72), and *Lysiphlebus testaceipes*, which live inside an aphid and, in escaping, chew a circular hole in the back of the mummified aphid, whose body contents have nourished them to full development (Figs. 2.21 and 6.76 and page 72).

Family Evaniidae.[1]   **The Ensign Wasps.** The common name of this family refers to the curious appearance of the abdomen: short, compressed, or flattened from side to side and held aloft on the slender basal segments which arise from the upper corner of the thoracic mass, the whole suggesting the shape of a flag. The front wing has a slender (costal) cell extending from the stigma to the base of the wing. The hind wing has no closed cells and only two veins, one along the costal margin and one across the disk. The trochanters are two-segmented, and the ovipositor hidden. The larvae are parasitic in the eggs of cockroaches.

Family Chalcididae[2] **(Also Called the Superfamily Chalcidoidea). The Chalcid Wasps.** This is probably the largest family of the entire order Hymenoptera, con-taining thousands of species, most of them very small (under $\frac{1}{8}$ inch long) and many minute, some measuring only $\frac{1}{60}$ inch in length. The study of the classification and life habits of such minute creatures is extremely difficult. In contrast with the ich-neumons and braconids, the antennae are elbowed and the pronotum does not extend backward on the sides to reach the base of the wings (Fig. 2.18). The body is nearly or quite bare, metallic-colored, the abdomen short, frequently consisting of a very slender stem and a more or less globular gaster. The wings are sometimes banded or spotted, but they have no closed cells, no stigma, and are almost without veins. Small as they are, the chalcids are among the most important to man of all insects. There are some very serious crop pests which develop in the seeds or stems of herbaceous plants, such as the clover seed chalcid (page 568), wheat jointworm (page 536), wheat strawworm (page 538), and cattleya or orchid wasp, *Eurytoma orchidearum*. There are the fig insects discussed on page 59, and thousands of kinds which live within and devour other insects piecemeal. Although overwhelmingly beneficial, nevertheless some parasitic chalcids are detrimental to man by living upon beneficial parasites. Among the outstanding beneficial parasites are *Trichogramma minutum* (page 69), *Aphycus helvolus* (page 70), *Aphelinis mali* (page 70), *Coccophagus gurneyi* (page 71) and *Pteromalus puparum* (page 71).

Family Proctotrupidae[3] **(Also Called Superfamily Proctotrupoidea or Serphoidea). Proctotrupids, Egg-parasite Wasps, and Others.** By some authors this great assembly of small slender-bodied wasps is considered a superfamily of 5 to 10 similar families.

[1] *Ibid.*, "Evaniidae, Hymenoptera of Connecticut," pp. 239–243; BRADLEY, J. C., "The Evaniidae," *Trans. Amer. Ento. Soc.*, **34**:101–194, 1908; TOWNES, H. K., "Nearc-tic Species of Evaniidae," *Proc. U.S. National Museum*, **100**:525–539, 1949.

[2] ASHMEAD, W. H., "Classification of Chalcidoidea," *Memoir Carnegie Museum*, 1(4): 225–393, 1904; VIERICK, H. L., "Chalicidoidea, Hymenoptera of Connecticut," *Conn. Geol. Natural History Surv. Bul.* 22, 443–528, 1916; BURKS, B. D., "Revision of Tribe Chalcidini of North America," *Proc. U.S. National Museum*, **88**:237–354, 1940; GAHAN, A. B., and M. M. FAGAN, "Type Species of Genera of Chalcidoidea," *Bul. U.S. National Museum* 124, 1923.

[3] ASHMEAD, W. H., "Monograph North American Proctotrypidae," *Bul. U.S National Museum*, 45:1–472, 1893; BRUES, C. T., "Serphoidea, Hymenoptera of Connecticut," *Conn. Geol. Natural History Surv. Bul.* 22, pp. 529–576, 1916.

They have few veins in the wings or none at all (Fig. 2.17), and some forms are wingless. The prothorax reaches back on the sides to the base of the wings, and the ovipositor is attached at the tip of the abdomen. They lack beautiful colors or markings, the antennae are generally long but vary in shape, and the trochanters are, as in the above families, two-segmented. Practically all known species are parasites of insects or spiders, and very many are small enough to bring themselves to full growth inside the egg of another insect. Pupation generally takes place inside the host. Although some parasitize beneficial insects and a good number are hyperparasites that develop within another insect, while that insect is living as a parasite, they must exert a tremendous influence in keeping destructive insects in check (page 69).

Family Cynipidae.[1] **Gall Wasps.** These are all small, mostly minute wasps, unattractive in color but presenting the most interesting habits and biology. The body is broad, very compact, nearly bare, the abdomen generally flattened sidewise so that many of the species suggest superficially a winged flea (Fig. 6.77). The antennae are straight, rather long, and the pronotum reaches back to the base of the wings. The trochanters appear to be two-segmented, but Keefer claims the additional segment is a part of the femur. There is no stigma in the wings, but usually a few closed cells near the center, and the ovipositor is attached some distance before the extremity of the abdomen. The second abdominal segment is often very long. These insects are the cause of a very large proportion of the gall-like growths found on plants (page 10 and Fig. 1.6), especially on oak and rose. The eggs are inserted in the growing tissues of some part of the plant, and the gall grows as the larva develops. Its characteristic shape and appearance are in some mysterious way determined by the insect, so that the kind of insect can be identified usually by the nature of the "house" the plant grows for it. The larvae are legless and feed on the tissues of the galls which house them. They make no cocoons, but pupate in the gall. The males are generally rare, and parthenogenesis is common. The alternating generations arc frequently remarkably different from each other in appearance and in galls formed, the grandchildren resembling their grandparents but differing strikingly from their immediate parents. The economic importance of this large family is not great.

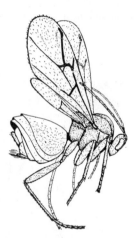

Fig. 6.77. A cynipid or gall wasp, *Cynips plumbea.* (*From Kinsey.*)

Family Sphecidae.[2] **Digger Wasps, Mud-daubers, and Thread-waisted Wasps.** These are generally beautifully colored, extremely active, and graceful wasps, in which the pronotum does not reach back to the wing bases but has at each side a rounded lobe called the tubercle or posterior lobe. An area of the side of the mesothorax below the tegula, known as the *prepectus*, is marked off by definite sutures. In the various groups, sometimes each called a separate family, the shape and appearance vary greatly. In a number of species the last segment of the thoracic mass (really the first abdominal segment, called the *propodeum*) is usually extremely slender, this condition being responsible for the common name, thread-waisted wasps (Fig. 6.78). In other species the body is more compact without the slender waist. In general, the head is rather broad, and antennae and legs are of moderate size. All species and all individuals are winged, the wings not longitudinally folded. The size ranges from

[1] VIERICK, H. L., "Cynipoidea, Hymenoptera of Connecticut," *Conn. Geol. Natural History Surv. Bul.* 22, pp. 361–422, 1916; FELT, E. P., "Key to American Insect Galls," *N.Y. State Museum Bul.* 200, 310 pp., 1918; KINSEY, A. C., "The Gall Wasp Genus Cynips," *Indiana Univ. Studies*, **26**(84–86): 1–577, 1929.

[2] ROHWER, S. A., "Sphecidae, Hymenoptera of Connecticut," *Conn. Geol. Natural History Surv. Bul.* 22, pp. 652–691, 1916; PATE, V. S. L., "Generic Names and Type Species of Sphecoidea," *Memoir Amer. Ento. Soc.* 9, 1937.

less than ¼ to 1¼ inches long. The typical habit is to deposit the eggs, each on a caterpillar or other insect which has been paralyzed by stinging and stored in a nest dug in the ground (Fig. 6.74) or in the hollowed stem of a plant or fashioned of mud and suspended from the underside of a bridge or the ceiling of an attic or shed. The nest is closed, the egg hatches, and the larva feeds to full size on the flesh of the stored insect, which is kept fresh either because it is still alive or because of antiseptic properties of the venom injected by the sting. One of the best-known members of this family is the cicada killer, *Sphecius speciosus* (Fig. 6.79), which provisions its burrow with cicadas.

Family Dryinidae.[1] **Dryinid Wasps.** Collectors and observers of leafhoppers and related Homoptera often find specimens which look as though the abdomen had been ruptured, a hernia-like swelling projecting from between certain segments. This is the habitat of the larva of a dryinid wasp that forms and lives within the cyst, which is composed of its own shed skins. When full-grown the larva leaves the mortally wounded host and spins a cocoon, usually attached to a convenient plant. The

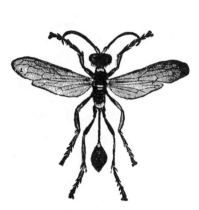

Fig. 6.78. A mud-dauber, *Sceliphron cementarius*. (*From Sanderson and Jackson, "Elementary Entomology," after S. J. Hunter.*)

Fig. 6.79. The female cicada killer, *Sphecius speciosus*, carrying a cicada to her nest to serve as food for her young. About natural size. (*Original.*)

females of the wasps, which finally emerge from the cocoons, are remarkable in the possession of a pincer at the end of each front leg, which is formed by the elaboration of the fifth tarsal segment. With these pincers the wasps catch and hold the small hoppers while they thrust their eggs into the latter's body. The front wings have a well-developed stigma and are often pigmented, but most of the veins are faint or obsolescent. The hind wings are almost veinless. In some species all trace of wings has been lost. The head is broad, the body slender, the prothorax remarkably elongate and drooping. These wasps are very important to man, since the adults, as well as the larvae, may destroy leafhoppers.

Family Pompilidae[2] **(Also Called Psammocharidae). Spider Wasps.** The habit of these wasps of provisioning their individual nests with prey that has been paralyzed by stinging is very similar to that of the Sphecidae, described above, but these insects almost always use spiders as the stored larval food. Here are classified large polished, but not very shiny, mostly black, but sometimes banded wasps, with especially long

[1] Kieffer, J. J., "Dryinidae," *Genera Insectorum*, **54**:1–33, 1906; Fenton, F. A., "Parasites of Leafhoppers," *Ohio Jour. Sci.*, **18**:177–212, 243–278, 285–296, 1918.

[2] Vierick, H. L., "Psammocharidae, Hymenoptera of Connecticut," *Conn. Geol. Natural History Surv. Bul.* 22, pp. 625–634, 1916; Bradley, J. C., "Revision of American Pompilinae," *Trans. Amer. Ento. Soc.*, **70**:23–157, 1944.

hind legs. The hind femora reach beyond the middle of the abdomen. The pronotum reaches the tegulae at the base of the wings, and they differ from the Sphecidae
also in lacking the slender "waist," the petiole being very short. The thorax is
straight-sided and very long, the wings well veined and almost always stained brown

FIG. 6.80. A cuckoo wasp, *Stilbum cyanurum*, showing how the female rolls into a ball to
protect her body from attack. (*From Maxwell, Lefroy, and Howlett, "Indian Insect Life."*)

or black. The mouth parts are elongated. Some species are less than ¼ inch long,
but they range upward to species 1½ inches long, with a wing spread of more than
3 inches. They are good runners, powerful in flight, always on the go, running, flying,
quivering, as they prosecute their relentless search for spiders. They frequent sandy

soil in which the spiders, with the wasp's
eggs attached, are buried. Some large
species of the genus *Pepsis* do not hesitate
to attack the big tarantula spiders, when a
battle royal takes place between the venomous jaws of the tarantula and the venomous tail of the wasp. The economic
importance of this family cannot be properly evaluated until we know more about
the food habits of their prey, the spiders.

Family Chrysididae.[1] **Cuckoo Wasps.**
These insects have received their common
name from their resemblance in habit to
the cuckoo birds, that is, they make no
nests of their own, but stealthily seek out
the nests of other solitary wasps and bees,
in which they lay their eggs. The cuckoo
wasp larva eats either the larva of the
wasp that built the nest or the caterpillars
stored by the latter for her young. In
adaptation for the hazardous life of in-
vading the homes of solitary stinging
wasps and bees, their bodies are heavily
armor-plated and the abdomen is so modified that it can be bent far under the
body to cover legs and mouth parts completely. When so rolled into a ball (Fig.
6.80) only the wings are exposed and the
very heavy dorsum of the body is almost

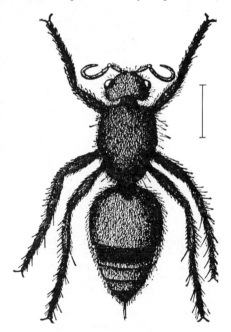

FIG. 6.81. A velvet ant, *Dasymutilla* sp.,
female. (*From Lugger.*)

wound-proof against attacks by the own-
ers of the nest. The integument is typically strongly pitted or sculptured and beautifully metallic green, blue, or red in color. The abdomen has only three or four broad
segments, the remaining ones forming an extensile tube surrounding the ovipositor.

[1] AARON, S. F., "North American Chrysididae," *Trans. Amer. Ento. Soc.*, **12**:209–48,
1885; VIERICK, H. L., "Chrysidoidea, Hymenoptera of Connecticut," *Conn. Geol.
Natural History Surv. Bul.* 22, pp. 602–605, 1916; BISCHOFF, H., "Chrysididae,"
Genera Insectorum, **151**:1–86, 1913.

The pronotum does not reach the wing bases. The hind wings have no closed cells; the antennae are composed of 13 segments. The family is of very little importance to man.

Family Mutillidae.[1] **Velvet Ants.** The velvet ants, like the "white ants," are not true ants. These are wasps, but the ground-dwelling, active females are entirely wingless, ant-like in form, with the three body regions well separated, plump, and rounded (Fig. 6.81). The body is often covered with long, sometimes very long, stiff hairs and usually conspicuously colored black, scarlet, or reddish brown, with reddish, whitish, or yellowish bands, spots, or rings of hairs. Velvet ants can readily be separated from ants by the absence of any tubercles or swellings on the petiole of the abdomen. They are said to be warningly colored, the supposition being that natural enemies more quickly learn not to attack these brightly marked vicious stingers, much as human travelers associate red lights with danger. The males are usually winged and often very different in appearance from the females. The wings are

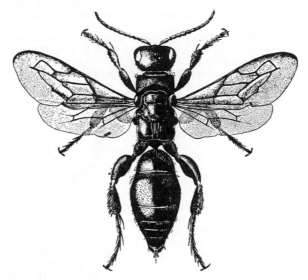

FIG. 6.82. A digger wasp, *Tiphia transversa*, female. About 3 times natural size. (*From Ill. Natural History Surv.*)

stained deep brown or black. They differ from the related Scoliidae in having the bases of the two middle legs touching each other and two spines at the end of the middle tibia. The pronotum extends back to the tegulae at the wing bases. The eyes are small and rounded, the ocelli wanting in the female, usually present in the male, the antennae rather thick, both pairs of palps long, and the coxae small. The females are found especially on hot, dry, sandy soil. So far as known, they lay their eggs in the nests made by other stinging Hymenoptera, and the larvae are parasites on the larvae of the owner of the nest. Some species, because of their extremely painful stings, have been called cow killers or mule killers, but of course they do not kill these animals. Two species are parasites of the African tsetse flies.

Family Scoliidae.[2] **Vespoid Digger Wasps.** These are large, strong-bodied, some-what hairy wasps, with a constriction between the first and second abdominal segments (Fig. 6.82). The color is usually blackish, often with a number of yellow or red crossbands. The legs are short and thick. The abdomen is often (*Elis* spp.) very

[1] BRADLEY, J. C., "Mutillidae and Allies of North America," *Trans. Amer. Ento. Soc.*, **42**:187–214, 309–336, 1916; MICKEL, C. E., "Biological and Taxonomic Investigations of Mutillidae," *Bul. U.S. National Museum* 143, pp. 1–340, 1928.

[2] ROHWER, S. A., "Scoliidae, Hymenoptera of Connecticut," *Conn. Geol. Natural History Surv. Bul.* 22, pp. 616–620. 1916.

long.   The antennae and mouth parts are short.   Both males and females are winged, the bases of the middle legs do not touch, and the middle tibiae have a single spur at the end of each.   The veins of the wings often stop remote from the tip of the wing.   Many of our species range from ¾ to 1 inch long, a few are very large.   The vespoid "diggers" are more direct in their preparations for their larvae than the sphecoid digger wasps.   They make no nest to which larval food is carried, but locate white grubs or other beetle larvae in the soil, dig down to them, sting them, and, depositing an egg upon each, leave the doomed beetle grub to serve as food for their young until full-grown.   The cocoons enclosing the pupae of digger wasps are often found in considerable numbers in soil where grubs abound.   Because of their habit of destroying soil-infesting pests, which are hard for man to control, the scoliids are rated as very beneficial to agriculture.

Family Eumenidae.[1] Mud or Potter Wasps. These are moderate-sized, brightly colored wasps, rather short-bodied and compact, with a single spur at the end of each middle tibia.   The body, including the wings, is black with a few yellowish crossbands or spots on abdomen and thorax.   The legs are small, the eyes large, the mouth parts short or elongated, the antennae rather short and thick.   The wings, which are present in both sexes, fold longitudinally when at rest, a characteristic that is believed to indicate a close relationship to the following family.   The abdomen is either sessile or very slender next to the thorax, often abruptly swollen behind and with a slight to prominent constriction between the first and second segments (Fig. 6.83).   The body is nearly bare, often strongly sculptured or punctate.   The size range is mostly between ⅜ and ¾ inch long.   Nests are often exquisitely molded of clay, attached to a twig or other object, and provisioned with several stung caterpillars and an egg,

Fig. 6.83. A potter wasp, *Eumenes fraterna*, and its clay nest. (*From Sanderson and Jackson,* "*Elementary Entomology,*" *after Weed, Ginn & Co.*)

laid within the nest, before it is sealed.   The eggs, in the known cases, are suspended on slender threads, like a ceiling droplight, just over the stored food.   Other species construct nests in hollowed-out plant stems, mine in the soil, or utilize nests deserted by other wasps or cavities such as keyholes that they find about buildings.   Clay is frequently used to chink or partition such nests.   The adults are common visitors to flowers.

Family Vespidae.[2] Social Wasps, Paper-nest Wasps, Hornets, and Yellow Jackets. Like the Eumenidae, the wings of this family fold longitudinally, but there are in the vespids two spurs at the end of each middle tibia.   Almost all the species are warningly banded with yellow on a black or brown ground color, and they are such severe stingers, especially when a nestful of individuals cooperate, that their

Fig. 6.84. The bald-faced hornet, *Vespa maculata*. About natural size. (*From Sanderson and Jackson,* "*Elementary Entomology.*")

general appearance has come to suggest to us the danger of being stung.   The Vespidae are rather large wasps (mostly ½ to 1 inch long), rather robust, without slender waists, either bare or hairy, and with a pair of obliquely placed yellow stripes on the front of

[1] DALLA TORRE, K. W. VON, *Genera Insectorum,* **19**:9–61, 1904.

[2] VIERICK, H. L., "Vespidae, Hymenoptera of Connecticut," *Conn. Geol. Natural History Surv. Bul.* 22, pp. 640–644, 1916; DALLA TORRE, *op. cit.*, pp. 1–108.

the thorax, which diverge from a point behind the head to the base of the wings (Figs. 6.84, 6.86,*e*). The mouth parts of these wasps are of the chewing-lapping type, although the tongue is not so long as in the honeybee. The pronotum reaches the tegulae at the wing bases. The wings are well provided with veins, and the hind wings are very narrow. Some of these species differ from all the other wasps in having developed a distinct infertile caste, known as workers, which are undernourished females that live in the home of their mother, the queen, and perform for her family all kinds of domestic duties, but have no children of their own. The workers, unlike those of ants and termites, are winged and in general appearance like the kings and queens, but usually smaller. The nest, unlike that of the honeybee, is an annual affair, none of the individuals living more than one season. Only newformed, impregnated females winter over, usually in crevices about trees, the nest

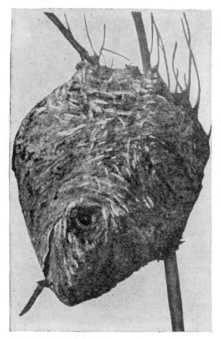

FIG. 6.85. Paper nest of the bald-faced hornet, *Vespa maculata*. Actual size, 10 to 15 inches in diameter. (*From Kellogg, "American Insects."*)

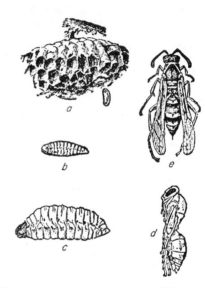

FIG. 6.86. A paper-nest wasp, *Polistes* sp. *a*, Nest and egg; *b*, young larva; *c*, older larva; *d*, pupa; *e*, adult. All enlarged ½, except nest, which is reduced. (*From Kellogg, "American Insects."*)

being completely deserted before the following summer. Adult food is nectar or other sugary solutions such as honeydew and the juices of ripe fruits. They feed the young on bits of caterpillars or flies which they catch and partially chew before presenting to their young. Hornets may be seen almost any summer day engaged in their winged pursuit of flies. For so ephemeral a structure, the nests (Figs. 6.85, 6.86,*a*) are admirably constructed of "paper" which is made by masticating weathered particles of wood from old boards of fences or buildings, mixing with saliva, and drawing out the mixture into thin, strong sheets. Both the envelopes or walls covering the outside or top of the nest and individual cells for the larvae are made of this paper. The cells differ also from those of the honeybee in lying in a vertical position rather than horizontally. Food is not stored in the cells with the eggs, but when the larvae hatch they are fed from day to day by the workers, at first upon sweet solutions, but for the most part upon particles of meat from insects caught by the workers. The cells are not closed over until the larvae are full-grown. Then they are capped and serve as pupal chambers. Nests are attached to twigs or branches of trees or in hollows of trees (hornets) or are constructed beneath the ground (yellow jackets). In *Polistes* (Fig. 6.86) the nest is the work of a single female, has a single layer or tier of cells, and is not enclosed by envelopes. In *Vespa* (Fig. 6.85) the nests usually consist of a number of tiers or "stories," one below the other and completely enclosed by spherical walls. Each cell may be used for two or three successive batches of brood. The young developed early in the season are workers. In late summer queens and males are developed, the latter developing from unfertilized eggs, but in larger cells and as a result of better feeding by the workers. The larvae are entirely legless, and the head is reduced almost to disappearance, so that they resemble maggots, but there are a number of serially arranged spiracles on the sides of the body. These larvae feed the adults a sweetish secretion from the mouths.

Family Formicidae.[1] **Ants.** Ants are said by Wheeler to be the most abundant terrestrial animals. They have perhaps the most highly organized social life of all insects. There is nearly always a wingless worker caste, in addition to the winged males and females, and frequently the workers and more rarely the males or the queens are *polymorphic*, that is, there are different sizes or different shapes in the same species. The males and females, called "kings" and "queens," are concerned chiefly with reproduction and are not rulers in any sense. The workers, which are underdeveloped females, rarely lay eggs, but do all the work of the colony and, in their instinctive way, collectively rule the colony by managing all its essential activities. The food of ants, like their nest-building materials, is extremely diverse. The various species take the nectar of plants, the honeydew of aphids and other Homoptera, seeds, fruits, insects, and other small animals, and secretions of their own larvae. The young are fed on either bits of seeds, pieces of insects or fungi which the workers cultivate, or liquid or solid food regurgitated by the workers. The mouth parts are chewing in type but perform a great variety of different functions, such as fighting, building nests, carrying the young, cleaning the body, weeding fungus beds. The antennae are sharply elbowed, with a very long first segment. Head, thorax, and abdomen are very well separated, the thorax usually distinctly slenderest of the three regions. The pronotum is prolonged backward at the sides. The swollen part of the abdomen (the *gaster*) is attached to the thorax by a short slender petiole which always bears one or two swellings or tubercles (Fig. 6.87). This is the distinguishing mark for the family. In size our species range from the very small *Monomorium minimum*, which is only $\frac{1}{12}$ inch long, to the big carpenter ants, the queen of which is nearly $\frac{3}{4}$ inch long. In some ants, such as the imported fire ant (page 896), the females (queens and workers) can sting, but most of our common species are harmless in this respect. Hard as housewives and others who have been pestered by noxious species would find it to believe, ants as a whole are doubtless beneficial as scavengers and destroyers of many noxious insects. However, their invasion of our houses, the nests built in lawns, their cultivation of troublesome mealybugs and aphids, on corn, in orchards, and in greenhouses make these species pests of first rank. The most intolerable household species, wherever it occurs, is the Argentine ant (page 894).

In ants the wings appear to function only to disperse the new colonies and to prevent close inbreeding. Mating generally takes place on the wing; at all other times ants are crawlers. The workers hold back the newly developed males and females in their nest, until, as though by some prearranged signal, but probably because of a definite meteorological condition, all the individuals from neighboring nests swarm out at the same time and there is considerable mixing of the sexes from different nests. The male dies after mating, the female flies or crawls to some attractive nesting site, breaks off her wings, crawls into a natural cavity or makes a small cave in the soil, and seals herself up for weeks or months until her eggs have developed and hatched and the larvae have been nurtured to the adult stage by a salivary secretion from the female's mouth. This food supply is derived largely from the now useless wing muscles of the

[1] WHEELER, W. M., "Ants: Their Structure, Development and Behavior," Columbia Univ. Press, 663 pp., 1910; EMERY, C., "Formicidae," *Genera Insectorum*, **174A**:1–94, 1921; **174B**: 55–206, 1922; **174C**:207–397, 1922.

queen's body. These first young become workers, establish a connection between the cave and the outside world, and take over all domestic duties for the nest. The female now becomes a queen, is fed by liquids brought in by her daughter workers, and henceforth does nothing but eat and lay eggs. The supply of sperms, received at her initial flight, suffices throughout her life, which may extend over 10 to 15 years, to produce, by fertilizing the eggs, either queens or workers, depending upon the kind of food they receive. Other eggs are unfertilized and produce occasionally a crop of males. Nests are most often built in the soil, especially beneath stones, logs, or piles of rubbish. Many, however, nest in hollowed-out parts of trees or other plants, and a few construct dwellings made of paper and suspended from trees. Unlike the nests of social wasps and bees, no cells or combs are built and a number of eggs, larvae, and

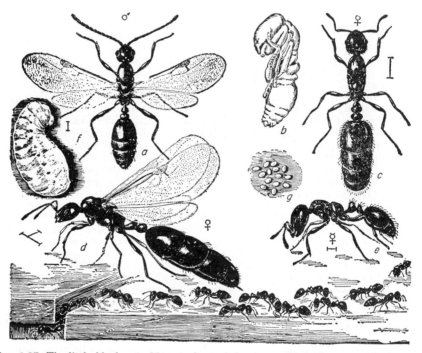

FIG. 6.87. The little black ant, *Monomorium minimum*. *a*, Male; *b*, pupa; *c*, female after losing wings; *d*, winged female; *e*, a sterile female or worker; *f*, larva; *g*, eggs; below, a group of workers in line of march. Lines indicate natural size. (*From Marlatt, U.S.D.A., Farmers' Bul.* 740.)

pupae are generally piled up together in special chambers or rooms of the intercommunicating, underground galleries and are moved about from place to place by the workers. This contact between workers and larvae is given significance and intensified by mutual feeding (trophallaxis), the workers bringing food of varied sort to the larvae and the larvae responding with sweetish secretions sometimes from special glands on the thorax. The larvae are most helpless, legless, and nearly headless creatures, the anterior end typically bent over hook-like. The pupae are often naked but sometimes in whitish silken cocoons.

Family Prosopidae[1] (Also Called Hylaeidae, in Part). Obtuse-tongued Bees or Wasp-like Bees. This small family has certain primitive characteristics in the absence of pollen-collecting structures, in the chewing-type mouth parts, and in the

---

[1] VIERICK, H. L., "Hylaeidae, Hymenoptera of Connecticut," *Conn. Geol. Natural History Surv. Bul.* 22, pp. 737–793, 1916.

very slightly developed pubescence. They are mostly nearly bare, small, rather slender black species with a few yellow markings on head, thorax, and legs. The labium is truncate or obtuse at the end; hence the common name. The antennae and legs are small. Most of our species are only a little more than ¼ inch long. They make nests in the stems of pithy plants or in cavities in the ground or in buildings. The nests consist of cells separated with silky partitions and provisioned each with an egg and with honey and pollen as food for the larva. Lacking pollen-collecting hairs, they swallow pollen along with nectar to transport it to the nest, in their crops.

Family Colletidae[1] (Also Called Hylaeidae, in Part). Bifid-tongued Bees or Plasterer Bees. There is much diversity in the classification of the bees. This family, which is often united with the preceding under one or another joint name, appears to deserve separate rank because of the development of pollen brushes on the

Fig. 6.88. A leaf-cutting bee, showing rose leaves from which she has cut pieces to line her cells, entrance of her nest burrow in the ground, and a single leaf-covered cell removed from the nest. About natural size. (*From Linville, Kelley, and Van Cleave, "General Zoology," Ginn & Co.*)

hind femora of the females, the much greater hairiness of the body and legs, and differences in the wing venation. These bees, whose light crossbands are formed by fine white hairs, rather than by colors on the body wall, make extensive burrowings in soil, the long cylindrical tunnels being lined with silk and divided by silky partitions into separate cells which are provisioned as in the Prosopidae. Sometimes the same clay bank is chosen by very large numbers of females to make a veritable city of nests. They are much larger than the Prosopidae, averaging about ½ inch long and being distinctly broader than members of that family.

Family Megachilidae.[2] Leaf-cutting Bees and Mason Bees. The habits of these bees are most interesting. As the name suggests, the leaf cutters cut, with their scissors-like mandibles, circular or oval pieces from the leaves of rose and other plants and use them to fashion thimble-shaped cells placed end to end in a hollowed bramble

[1] *Ibid.*, "Colletidae, Hymenoptera of Connecticut," 739–741.
[2] *Ibid.*, "Megachilidae, Hymenoptera of Connecticut," pp. 741–753.

or cylindrical burrow in the soil (Fig. 6.88). A bit of leaf ½ to ¾ inches in diameter may be clipped off in 4 to 10 seconds, and the bee flies away bearing it in its mouth. The mason bees seek a cavity in wood, soil, masonry, or pithy stems of plants, a key-hole, or a snail shell and line it with cement made of clay, sand, and a sticky secretion from their mouths, instead of leaves. In either case the egg placed in each cell is accompanied by a mass of pollen and nectar or honey, the pollen having been carried to the nest on the underside of the female's abdomen. These are long-tongued bees of the general appearance of honeybees, but with very large heads, highly polished black or metallic blue or green bodies sometimes banded with thin lines of white hairs across the abdomen. In size they range from that of the honeybee to less than half as large. Pollen is collected by a dense brush of stiff hairs on the underside of the abdomen. Some species in this family (*Coelioxys* spp.) are parasitic in the larval stage upon larvae of other Megachilidae. These parasitic species are slender, with tapering, pointed abdomens banded with white hairs, and the females lack the pollen brushes on the abdomen. The only economic importance of the family is as pollinizers (page 56).

Fig. 6.89. A carpenter bee, *Xylocopa varipuncta*, adult female. About 1⅓ times natural size. (*From Essig, "Insects of Western North America," copyright, 1926, Macmillan. Reprinted by permission.*)

Family Xylocopidae.[1] **Large Carpenter Bees.** These are very large, beautifully colored, blackish or bluish-brown or violaceous bees, resembling bumblebees, but somewhat more flattened and less hairy and often without the yellow bands of hair (Fig. 6.89). The pollen brush of bumblebee or honeybee workers is supplanted by a dense brush of hairs covering the surface of the hind tibia. There is at the tip of the abdomen in the females a short stout spine concealed among the hairs. These bees often become destructive by tunneling into the solid wood of poles, bridges, buildings, fences, or fallen timbers. The tunnels may extend to a depth of a foot or more and are divided into a linear series of cells by partitions made of sawdust or plant fragments, cemented together by their saliva. The food for the young is said to be predominantly pollen. The family is a very small one.

Family Ceratinidae.[2] **Small Carpenter Bees.** These are much smaller than the xylocopids, superficially more like Andrenidae, though structurally agreeing with the xylocopids except for slight differences in wing venation. The body regions are well separated, the color black or metallic blue-green, with almost no markings and very little pubescence. Common species range from ³⁄₁₆ to ⅜ inch long and are only about ⅕ as broad. They choose the soft pith of reeds or brambles in which to tunnel their nests. The females show remarkable maternal care, for a solitary bee, by waiting in the tunnel above the completed cells until the young have become adult. The latter eat out the partitions above their heads and, when all are mature, follow their mother on their first flight. The daughters then assist their mother to clean out the old nest and one of them uses it again.

Family Nomadidae.[3] **Cuckoo Bees.** Most of the species of this small family live, like the cuckoo wasps, by stealthily laying their eggs in the nests already prepared and provisioned by other solitary bees, especially Andrenidae. Their young devour

[1] ACKERMAN, A. J., "United States Species of Xylocopa," *Jour. N.Y. Ento. Soc.*, **24**:196–232, 1916.

[2] SMITH, H. S., "Ceratinidae of North and Middle America," *Trans. Amer. Ento. Soc.*, **33**:115–124, 1907.

[3] VIERICK, H. L., "Nomadidae, Hymenoptera of Connecticut," *Conn. Geol. Natural History Surv. Bul.* 22, pp. 722–730, 1916; LINSLEY, E. G., and C. D. MICHENER, "Generic Revision of North American Nomadidae," *Trans. Amer. Ento. Soc.*, **65**:265–305, 1939.

the food intended for those of the owner of the nest, although it is said that there is sometimes enough for both.   Unlike the cuckoo wasps, these bees are apparently not opposed by the victimized owners of the nests.   These are small (mostly about ½ inch long), but ranging from ¼ to ¾ inch, slender, nearly bare, with a pointed abdomen (Fig. 6.90). They are brightly colored, black or reddish brown, banded or spotted with yellowish or white either of the ground color or of fine, procumbent, short hairs.   The females, in correlation with their lazy habits, have no pollen-carrying apparatus, the legs being nearly bare.

Family Andrenidae[1] (Also Called Halictidae). Mining Bees, Sweat Bees.   The habits of many of these species form an interesting transitional step toward the social life of the Apidae and Bombidae, discussed below.   Although each female produces young and, lacking a worker caste, builds her own nest and stores the necessary honey-and-pollen provender for the larvae, they are distinctly gregarious.   Hundreds, if not thousands, of females dig their cylindrical burrows close together in the same small area of soil, and in some species (*Halictus*), females

Fig. 6.90. A cuckoo bee, *Melecta californica. (From Essig, "Insects of Western North America," copyright, 1926, Macmillan. Reprinted by permission.)*

cooperate to form a common entrance tunnel and vertical corridor from which the private corridor of each female, with its separate nest cells, branches off (Fig. 6.91).

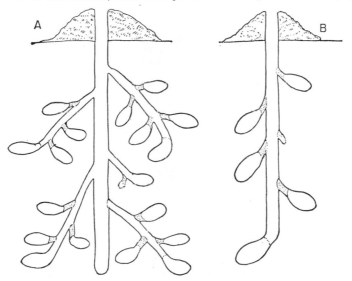

Fig. 6.91. Diagrams of nest burrows of short-tongued mining bees.   *A*, the compound or "apartment-house" nest of *Halictus*, the work of several females that use the common entrance tunnel; *B*, nest of *Andrena*.   (*From Kellogg, "American Insects."*)

Many such cooperative tunnels may occur together, forming what Comstock happily called a city composed of apartment houses, because a number of families use the same

---

[1] VIERICK, H. L., "Andrenidae, Hymenoptera of Connecticut," *Conn. Geol. Natural History Surv. Bul.* 22, pp. 709–720, 1916; VIERICK, H. L., and T. D. A. COCKERELL, "North American Bees of the Genus Andrena," *Proc. U.S. National Museum*, **48**:1–58, 1915.

"front door." A sentinel whose hard head just fits the entrance is constantly on guard. Whether this gregariousness is determined by a social instinct and fondness for others of their kind or whether these bees are very exacting in their requirements for a nesting site so that a just-right spot attracts large numbers of the females to it seems not to have been determined. Fabre claimed that the apartment-house arrangement is the work of the daughters of a single bee; they clean out their old original home, and each builds her nest opening into it. This is a very large family, including many very common species, and aside from the honeybee, these are the bees most commonly seen. Some of them are very small, ranging from ⅓ to ⅔ inch long, but most are from ⅛ to ½ inch (Fig. 6.92). They are metallic red or black or beautiful blue, green, or coppery in color, usually not banded or spotted. A number of species (*Agapostemon*), however, have the head and thorax a beautiful, delicate, metallic green or blue and the abdomen banded with black and yellow. They are not very hairy, and the ground color usually shines through. The females have pollen brushes on the hind femur and tibia, and the hind metatarsus is narrower than the hind tibia. The tongue is slender and pointed, sometimes short, sometimes long. The Andrenidae are considered to be of genuine value as pollinizers of fruit blossoms.

FIG. 6.92. A common short-tongued bee, *Andrena* sp., slightly enlarged. (*From Sanderson and Jackson, "Elementary Entomology," Ginn & Co.*)

Family Anthophoridae.[1] **Anthophorid Bees.** This family, which is sometimes included in the Andrenidae, may be distinguished from the latter by the fact that the labrum of the andrenids is concealed by the clypeus and mandibles, while in the Anthophoridae it is mostly exposed. They are more bee-like and less wasp-like in appearance, usually very densely pubescent, the body regions closely compacted and without bands or spots except for some contrast in the color of the pubescence (Fig. 6.93). The habits are similar to the andrenids, the nests being gregariously built in clay banks, but the entrance to each cluster of nest cells is surmounted by a cylindrical

FIG. 6.93. An anthophorid bee, *Diadasia australis*, adult male. (*From Essig, "Insects of Western North America," copyright, 1926, Macmillan. Reprinted by permission.*)

chimney, extending outward and downward from the vertical bank in which they mine. In size they are between honeybees and bumblebees. *Melissodes* and *Tetralonia* are solitary forms which nest singly in the soil and are sometimes placed in a separate family, the Euceridae. They are of great value as pollinizers of red clover, early in the season (page 59).

Family Bombidae[2] **(Also Called Bremidae). Bumblebees.** Nearly every one knows that these big, buzzing, furry, black and yellow or black and reddish-haired

[1] COCKERELL, T. D. A., "North American Bees of the Family Anthophoridae," *Trans. Amer. Ento. Soc.*, **32**:63–116, 1906.

[2] FRANKLIN, H. J., "Bombidae of the New World," *Trans. Amer. Ento. Soc.*, **38**: 177–486, 1912.

bees can sting severely, that they have their nests in cavities in the soil (Fig. 6.94), especially in a site previously used by field mice as a nest, and that this nest contains a bit of honey stored in irregularly rounded, dark, waxen cells.   Too late for the good of the bees, most boys learn that they are extremely valuable to farmers in pollinizing red clover.   But many interesting features of bumblebee life and habits are known only to the careful student.   For example, the wax used in comb building is a secretion from the underside of the abdomen, which has been worked up in the mandibles and shaped by the mouth parts.   The bees of any one summer do not live over winter. Only the new crop of queens, after mating, crawl away in the shelter of loose bark or hollow trees or other dry, protected places and hibernate.   It is these new queens that come droning about meadows and flower gardens in the early spring.   Having

Fig. 6.94. Nest of a bumblebee, *Bombus auricomus*, in a deserted mouse nest.   Note *A*, the wax-pollen lining material for protection of the nest; *B*, four empty cocoons from which adult bees have emerged and which are used later for the storage of honey and pollen; *C*, a wax-pollen mass enclosing eggs; *D*, cocoons containing larvae, pupae, or adult bees not yet emerged.   The remnants of the old mouse nest are below.   (*From photo by T. H. Frison.*)

found a nesting site or dugout and formed a spherical mass of fine dry grass or moss more or less intertwined, to shelter the comb, the queen next collects a mass of pollen, which is transported to the nest on the hairy-bordered concavities of her hind tibiae. Moistened with nectar, the pollen paste, called "bee bread," is surmounted by a circular wall of wax within which the queen lays her first batch of eggs, commonly 5 to 20 in number, and covers them over with wax.   Between the nest of eggs and the entrance to the nest she makes an open jar or pot which she fills with honey.   Thus she is able to feed herself without losing her watchful guard over the eggs and the nest entrance.   Having thus provided as well as may be for her comfort and safety, she climbs astride the nest of eggs and sits over them for 3 or 4 days or occupies herself by making other masses of pollen paste and laying additional eggs thereon, until the first larvae hatch.   The latter find themselves lying upon a bed of the nutritious bee bread.   From time to time as they grow, the queen gnaws open the waxen roof of their nest sufficiently to insert with her tongue some honey and pollen regurgitated from her honey stomach.   All the larvae may feed at first on the honey dropped in their midst.   Each larva, however, soon eats out a separate cavity in the pollen mass and, as it grows, is fed separately and kept covered with wax by the queen.   By the end of about 10 days the now full-grown larva spins a tough silken cocoon and transforms into the pupal stage. The wax is removed from over the cocoons, which are further incubated by the queen, and in 2 weeks more the young emerge as adults of the female sex and worker caste. The separate cells are thus formed by the spinning larvae and not by the adult bees. After the adults emerge, the empty cocoons may be filled with pollen or nectar, or special wax cells may be made for the storage of food supplies. The workers take over all nest duties, and the queen devotes herself to laying eggs. After a number of broods of workers have been developed, usually late in the summer, functional females and males are developed, the latter from unfertilized eggs. These mate on the wing, the workers and males die off as cold weather sets in, and only the young matrons live into the next year. It is thus seen that the social organization of the bumblebees is about like that of the hornets and yellow jackets—an annual colony—with workers that are winged and differ from the queens only in size and in being sterile. The bumblebees differ from the wasps in kind of larval food and nest-building materials, as well as in the structural features listed on page 273. This is very much less specialized than the social life of the termites and ants or of the honeybee, described below.

Bumblebees have tongues similar to, but longer than, the honeybees' (Fig. 4.17). They have two stout spurs among the hairs at the end of each hind tibia, and they differ in many technical structural features. Along with the true bumblebees, which number about 50 North American species, are some related renegade forms known as *Psithyrus* bees, which have the cuckoo habit, like the Nomadidae and Chrysididae. These are particularly ingrate because they prey upon their very close relatives. Their resemblance in appearance no doubt aids them to sneak into the nests of the queen bumblebees, where, according to some observers, they sting the latter to death; at any rate, utilizing the workers, the food, and the organization of the rightful owner, they lay their eggs in the nest and make use of the gullible bumblebee workers to rear their young. The females of *Psithyrus* are readily distinguished from the true bumblebees by the lack of the pollen basket, the hind tibiae being convex and evenly covered with hairs.

Family Apidae.[1] **The Honeybee Family.** This family, as recently restricted, includes moderate-sized bees, with a very highly developed social life, with hairy eyes, and without spurs at the end of the hind tibiae (Fig. 2.4). The workers have pollen baskets on the hind legs, but the queens, unlike those of bumblebees, are strictly queens, no longer able to do the ordinary work of the colony. They never perform domestic duties and consequently lack the pollen-collecting structures (Fig. 3.4,*C*) and secrete no wax. Except for the activities of man, this family would not be represented in America, our only species, the honeybee, *Apis mellifera*, having been introduced from the old world in early colonial times. The honeybees that occupy bee trees are swarms that have escaped from hives and reverted to their primitive nesting habits.

Three other species of *Apis* live wild in India. The genera *Melipona* and *Trigona* are sometimes included in this family. They are stingless bees, since the sting has become vestigial. Some of the species are very small. They nest in natural cavities of rocks, soil, or trees, using wax, mixed with soil and resin, as a comb-building material. Curiously enough, both males and females may secrete wax, and it is said to flow out upon the dorsal side of the abdomen. The brood cells are kept separate from those used for food storage, and the cells open upward. Food enough for the larva is sealed in the cell with the egg when it is laid.

The most remarkable of all bees is the well-known honeybee, the only American representative of this family. All know the products, honey and beeswax, that are stolen from these industrious bees, but few know the marvels of the association which has been called a "female monarchy." It is indeed a government of females, the males have neither suffrage, authority, nor any usefulness in the organization, except

[1] Root, A. I., and E. R., "The ABC and XYZ of Bee Culture," A. I. Root Co., 1929; Phillips, E. F., "Beekeeping," Macmillan, 1921; Snodgrass, R. E., "Anatomy and Physiology of the Honeybee," McGraw-Hill, 1925; "Anatomy of the Honey Bee," Cornell Univ. Press, 1956.

that one in a thousand may mate with a queen. It must not be supposed, however, that the queen is ruler of the hive. She does nothing except lay innumerable eggs. There is only one queen in a hive, the queens being antagonistic and stinging other queens to death. Ordinarily, the old queen leaves, with a large retinue of workers, called a swarm, about the time a new queen is ready to emerge. In this way reproduction of colonies is provided for. A queen is normally the mother of all other individuals in the nest. It has been estimated that she may lay 1,500,000 eggs and may have as many as 100,000 living children at one time. The queens mate but once and may live from 3 to 5 years. The sperms received at this mating are retained alive in the seminal receptacle in the queen's body as long as she lives, nurtured with a special secretion. We have already noted how the queen determines the sex of her progeny by permitting sperms to unite with some of the eggs, thus producing females, and withholding sperms from other eggs as they are laid, thus producing males. Since the

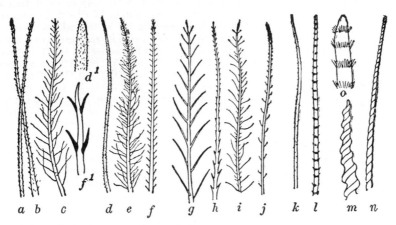

a b c d e f g h i j k l m n

Fig. 6.95. Branched, plumose, and threaded hairs, characteristic of bees: a–f are from bumblebees; g–j of *Melissodes* sp.; k–o of the leaf-cutting bee, *Megachile* sp. (*From Comstock, "Introduction to Entomology," after J. B. Smith.*)

eggs that produce drones are not fertilized, it has been truthfully said that the drone bees have a grandfather but no father. The males do no work, even feeding upon the provisions brought in by the workers. They have broader, blunter bodies than workers; their eyes meet on top of the head; they have no pollen-collecting structures, secrete no wax, and of course do not have a sting. After the swarming season is over, they are often driven out of the hive by the workers and die of exposure and starvation.

The body of 20,000 to 90,000 workers is the mainspring of the hive. Their most important duties are:

1. To collect food, which consists of nectar from a great variety of flowers and provides carbohydrates, and pollen from the same source, which provides proteins. Nectar is swallowed and carried to the hive in the crop, where it is regurgitated and, having been converted into honey, is stored in cells for future use. Pollen is combed off the hairs of the body by a special brush on the hind leg (Fig. 3.4,C), packed into the pollen baskets, and carried to the hive clinging to the hind legs. Water is brought to the nest, and propolis, a resin-like substance, is collected from the buds of trees, such as the poplar, and used for chinking up the hive.

2. To build the nest. The wax passes through the wall of the body from one-celled glands on the underside of the abdomen (Fig. 2.5). The plates or flakes are speared by the hind legs and passed forward to the jaws, where the wax probably receives a secretion from the mouth and is worked to the proper degree of plasticity. It is then built into beautifully hexagonal, thin-walled cells (Fig. 2.4) so arranged that a single bottom and each side wall serve for two cells. The bottom is composed of three parallelograms, the side wall of one cell meeting the center of the one opposite it.

The cells are just large enough to admit the body of the bee, and unlike those of the wasps, they lie horizontally.   They are used both as cribs or cradles for the developing young and as bottles for the storage of honey.   Cells to be used for rearing drones are about one-third larger in diameter than worker cells, while those in which queens develop are very large, bag-shaped, and placed vertically, open end downward.   They have the walls dimpled like a peanut shell.   The queen thrusts her abdomen into the cells made by the workers and deposits a single egg on end, in the bottom of each.

3. Another important duty of the worker bees is to feed the young.   The cells remain open, and when the larvae hatch in about 3 or 4 days, they are fed by the workers day by day.   The food is not put into their mouths, but is regurgitated into the bottom of the cell (Fig. 2.4).   The haploid males differ inherently from the diploid females, since they develop from unfertilized eggs and every cell of their bodies has only half as many chromosomes as the cells of the workers or queens.   The difference between queens and workers is apparently determined by the way in which they are fed.   The eggs are identical, for if one transfers newly hatched larvae from worker to queen cells, the nurse bees will feed the transposed grubs as queens and they will develop into queens.   Similarly, an egg laid in a queen cell, if transferred to a worker cell, is nourished and developed into a worker.   Queen larvae are fed solely on "royal jelly," an oral secretion of the workers, which is rich in vitamins, especially pantothenic acid and biotin.   It also contains ω-hydroxydecenoic acid, $HO(CH_2)_7CH\!\!=\!\!CHCOOH$, which is thought to be a preservative, and biopterin, which may be a pheromone or ectohormone that produces queen differentiation,[1] although some authorities believe that this results from better nutrition.   After about 6 days as larvae, the queens pupate and emerge as adults, completing a total developmental period of $15\frac{1}{2}$ days. Worker and drone larvae are fed royal jelly for the first 3 days and then honey, or mixtures of honey and partially digested pollen called "bee bread."   Their development requires a longer period, 21 days for the worker and 24 for the drone.   When larval growth and the necessity for feeding are completed, the worker bees cover the larvae with a somewhat porous cap of wax over the open end of the cell, and the larva spins an imperfect cocoon and changes to the pupal stage.

Biopterin

4. Other duties performed by the worker bees include the care of the queen—feeding her and brushing her body with their tongues, evaporating water from the nectar, cleaning the hive of dirt, debris, and dead bees, and guarding the nest from attack. The workers are also responsible for maintaining the temperature in the hive, and during the winter when it falls to 57°F., they form a cluster in the center, where they generate heat by muscular activity and fanning of the wings.   Thus the hive may be heated as high as 94°F., at which point the cluster loosens to permit the warm air to escape.

ORDER DIPTERA[2]

FLIES, MOSQUITOES, GNATS, MIDGES

The Diptera are a well-marked group in respect to the condition of the wings, and fairly homogeneous in general appearance; but in habits and in most other characteristics the order presents great diversity.

They are set apart from all other orders of insects by having a single

[1] KARLSON, P., and A. BUTENANDT, *Ann. Rev. Ento.*, **4**:39, 1959.

[2] ALDRICH, J. M., "Catalogue of North American Diptera," *Smithsonian Misc. Collections*, vol. 46, no. 1444, 680 pp., 1905; WILLISTON, S. W., "Manual of North American Diptera," Hathaway, 405 pp., 1908.

pair of wings (the front pair) developed for flight, and each of the hind wings reduced to a short, slender thread, with a knob at the end of it (Fig. 6.96). These rudimentary second wings are called *halteres* or balancers. There is some evidence that they are orienting organs, serving to keep the insects balanced. A few other insects have a single pair of wings, such as certain beetles and May-flies, but these never possess halteres. Many of the Diptera are wingless, having lost the first pair of wings; but even then the halteres usually remain, so that the possession of halteres is perhaps the most distinctive thing about this order. The front pair of wings is similar to those of bees and wasps in texture, *i.e.*, transparent and with comparatively few veins, as a rule. They are small in comparison with the size of the insect, a condition associated with very swift flight. While usually clear, they may have a color pattern (Fig. 6.108), and sometimes the veins are bordered with scales.

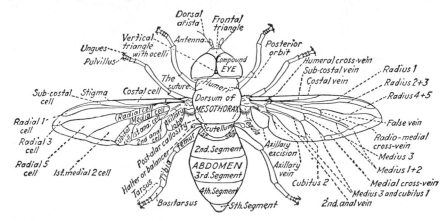

FIG. 6.96. Dorsal view of a male syrphid fly to show names of the principal parts of the body. Some authors use *alula* or *squama* instead of tegula. (*From Metcalf, Ohio Biol. Surv. Bul. 1.*)

The three body regions are very distinct in Diptera. The head is large, often hemispherical, and attached to the thorax by a very slender stem or neck. By reason of the fact that only the front wings are functional, the thoracic mass is largely made up of the mesothorax. A small, distinct, semicircular part of the mesothorax overhanging the base of the abdomen is called the *scutellum*. The abdomen is of varied shape, usually shows from four to nine segments, and the cerci, ovipositor, and male genitalia are normally introverted or retracted, so as to be invisible without special preparation.

The mouth parts of adult Diptera are rather varied in form. Two distinct types are represented; the piercing-sucking type and the sponging type (pages 142, 145). There are several varieties of the former, called subtypes (Figs. 4.7, 4.8, and 4.15). So far as known, no adult fly masticates solid foods; and few, if any, pierce plants to suck the sap. The majority probably feed upon the nectar and pollen of flowers; many others depend upon liquid organic matter such as that from decomposing plant or animal bodies, flowing sap, and honeydew; or they dissolve solid substances in their saliva, *e.g.*, sugar, and sponge up the solution. A number of species are predaceous on other insects, sucking the juices from their bodies.

The females of hundreds of species, representing at least eight families, suck the blood of warm-blooded animals, and, in the Muscidae and Hippoboscidae, the males also have this bad habit. There are many adults that take no food whatever, this life stage generally being short and occupied almost exclusively with the business of getting the eggs developed, fertilized, and laid.

The metamorphosis is complete or complex. The larvae are well separated from the adults both structurally and in habits and specialized to a more extreme degree than the larvae of any other order. There are very few cases where the larvae and adults live together and partake of the same kind of food, as is so common among the beetles. The larvae are always legless and in the larger part of the order have no distinct head (Figs. 2.15,C, and 4.18). In those species where the head is distinct, mosquitoes, for example (Figs. 5.15,E,G, and 21.1), the mouth parts of the larvae are of the chewing type; but in the great majority of species the body tapers gradually to the front end and terminates in a small conical segment which can be protruded or retracted.

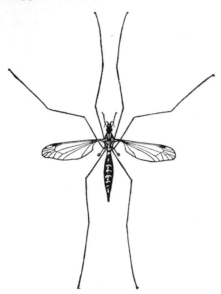

FIG. 6.97. A crane fly or tipulid, male. (*From Sanderson, "Insect Pests," after Weed.*)

This head segment bears no eyes, and no true mouth parts. There is a pair of minute rudimentary sense organs and a pair of prominent mouth hooks which work vertically to tear the tissues upon which the larva feeds or into which it tunnels. The larvae typically have a complex pair of spiracles on the truncate last segment of the abdomen (Fig. 20.10,C), and, sometimes at least, another pair near the front end of the body; but commonly none along the sides of the body on the other abdominal segments. Such larvae are called *maggots*. They live mostly buried or hidden in decaying animal or vegetable matter, in water or mud, or inside the bodies of plants, insects, and other animals. There are strikingly few that feed externally upon plants and comparatively few crop pests of any kind. The most serious are certain gall flies, *e.g.*, the Hessian fly (page 531), the many fruit flies of the family Trypetidae (page 812), some leaf miners, root maggots, and borers in the stems of plants. The attacks of the larvae and adults upon animals are much more serious. This is the most dangerous order for the carrying of human and animal diseases (pages 28, 32). Many species

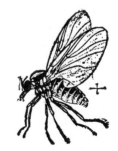

FIG. 6.98. A black fly, *Simulium* sp., female enlarged. (*From Osborn, "Agricultural Entomology," after U.S.D.-A.*)

suck the blood of animals as adults or live as larvae in their bodies. On the other hand, we must note the great benefit that accrues to us from the work of scavenger larvae and from those that are predaceous or parasitic on various insects and from the employment of Drosophila in genetic experiments (page 75).

The pupae of Diptera (Fig. 5.16,*G,H*) have the appendages free from the body wall and can be distinguished from the pupae of all other orders by having a single pair of wing pads. In the higher families the pupa is protected in a unique manner. Instead of shedding the skin of the last active larval instar, when the pupa is formed, this dried larval skin is retained about the pupa and serves as a cocoon. It is often inflated like a large seed, and its walls become very hard and thickened to form an airtight and watertight case. Such a case about a larva or pupa is called a *puparium* (Figs. 2.15,*D*, and 5.17,*C*). All flies of the suborder Cyclorrhapha spend the pupal stage in a puparium; most of the suborder Orthorrhapha do not. In the latter group the pupa commonly has no particular protective covering; rarely a cocoon is formed.

A good many of the flies are ovoviviparous, and in this order we have a few groups in which a viviparous reproduction occurs, *e.g.*, the Hippoboscidae and the tsetse flies of the family Muscidae. Another remarkable method of reproduction is the production of young by larvae and pupae, to which the term *paedogenesis* is applied. This occurs in certain Cecidomyiidae.

It is easy to mistake many of the flies for other kinds of insects. If the observer is not careful to note the number of wings and especially the presence of halteres, he will be likely to place many of his flies as Hymenoptera, because they are commonly banded with yellow and black, like the wasps, or densely covered with hairs like bees. A few species are louse-like or tick-like in form but can be distinguished from ticks by the number of legs and from lice by the nature of the mouth parts, which are exposed piercing organs (Figs. 20.22, 20.23).

Small- to medium-sized, soft-bodied insects. Adults mostly diurnal, with only one pair of wings, the front pair, which are narrow, membranous, with few veins. The hind wings are modified into halteres. Head, thorax and abdomen very distinct. Head vertical, very free-moving, subhemispherical. Compound eyes large; three ocelli usually present. Mouth parts sponging or piercing-sucking; labial palps always wanting. Prothorax and metathorax fused with the mesothorax, but scutellum distinct. Tarsi generally five-segmented. Metamorphosis complete. Larvae legless, usually maggot-like with head greatly reduced; the mouth parts in these cases replaced by a pair of mouth hooks articulating to an internal cephalopharyngeal skeleton. Larval spiracles generally restricted to a small pair on prothorax and a large group on last segment of abdomen. Pupa generally with free appendages; sometimes obtect in the Orthorrhapha; very rarely in a cocoon, but often enclosed in a puparium.

Specialists are fairly well agreed about the classification of the Diptera, at least in its major aspects.

A SYNOPSIS OF THE ORDER DIPTERA

A. **Suborder** *Orthorrhapha*. Straight-seamed Flies. The adult insects escape from the pupal skin or pupal case through a T-shaped or straight split down the back or a transverse split between the seventh and eighth segments of the abdomen. Pupa usually naked, sometimes in a cocoon, rarely in a puparium. Adults do not have a small, lunate-shaped sclerite above the antennae known as the frontal lunule. Larvae often with a distinct head.

1. *Nemocera*. *The Long-horned Flies.* Antennae usually long and slender, of 6 to 39 similar segments. Palps usually four- or five-segmented, generally pendu-

lous. Larvae have a distinct head, eyes, and true mandibles working transversely. First anal cell of wings almost never narrowed toward the wing margin. Discal cell generally absent. Mostly very slender flies.

Family 1. Crane flies, Family Tipulidae (Fig. 6.97).
Family 2. Moth flies and sand flies, Family Psychodidae.
Family 3. Mosquitoes, Family Culicidae.
Family 4. Midges, Family Chironomidae.
Family 5. Gall gnats, Family Cecidomyiidae (Itonididae).
Family 6. Fungus gnats, Family Mycetophilidae.
Family 7. Buffalo gnats or black flies, Family Simuliidae (Fig. 6.98).

2. *Brachycera. The Short-horned Flies.* Antennae usually short, of three segments, last segment sometimes annulate or with a style in addition, like a small whip of withered segments at the end. Palps one- or two-segmented, porrect. Larvae often have the head reduced in size and invaginated, and mouth hooks working vertically, instead of mandibles. Discal cell usually present. First anal cell always closed or narrowed toward the wing margin.

Family 8. Net-winged midges, Family Blepharoceridae.
Family 9. Horse flies, Family Tabanidae.
Family 10. Soldier flies, Family Stratiomyiidae.
Family 11. Snipe flies, Family Rhagionidae (Leptidae).
Family 12. Robber flies, Family Asilidae (Fig. 6.102).
Family 13. Bee flies, Family Bombyliidae (Fig. 6.103).
Family 14. Long-legged flies, Family Dolichopodidae.

B. **Suborder** *Cyclorrhapha.* Circular-seamed Flies. The adults escape from the pupal case through a split that runs round the end of the case and releases a circular lid that is pushed off or aside. Pupa always enclosed by the skin of the last active larval stage, which hardens to form a puparium. Adults with a frontal lunule and antennae generally of three segments, the third bearing an arista or style. Head of larvae always greatly reduced and invaginated into the pharynx. First anal cell always closed.

3. *Aschiza. Flies without a Frontal Suture.* Cap of puparium pushed off by expansion of the face of the adult, when it is ready to emerge; therefore there is no frontal lunule.

Family 15. Humpbacked flies, Family Phoridae.
Family 16. Flower flies or hover flies, Family Syrphidae (Fig. 2.15).

4. *Schizophora. Flies with a Frontal Suture.* A line or seam circles round above the base of the antennae and sometimes extends down nearly to the mouth on either side of the face. This is the vestige of a crack in the head through which a membranous, expansible, bladder-like structure, known as the *ptilinum*, is forced out when the adult is ready to emerge. By inflating the ptilinum with body fluids, the cap of the puparium is forced off. The bladder is then withdrawn into the head and is seen only if one catches the adult very shortly after its emergence.

a. Acalyptratae. Flies with small *tegulae*, *i.e.*, small, flat, membranous expansions connecting the base of the wing, behind, to the thorax. They do not have a complete transverse suture across near the middle of the thorax. They are all small flies, some very small. The eyes of males do not come together on top of the head.

Family 17. Thick-headed flies, Family Conopidae (Fig. 6.106).
Family 18. Ortalid flies, Family Ortalidae.
Family 19. Fruit flies, Family Trypetidae (Fig. 6.108).
Family 20. The frit fly and others, Family Chloropidae.
Family 21. Shore flies, Family Ephydridae.
Family 22. The pomace fly and others, Family Drosophilidae.

b. Calyptratae. Flies with well-developed *tegulae* or *squamae*, *i.e.*, thin, subcircular membranes just behind the base of the wing close against the thorax (Fig. 6.96). The thorax has a complete transverse suture near midlength, above. This division includes our commonest and best-known flies. They are all medium to large in size. The males can often be distinguished from the females, by having the eyes contiguous, at least for a short distance, at the top of the head.

Family 23. Anthomyid flies, Family Anthomyiidae.

Family 24. House fly family, Family Muscidae.

Family 25. Flesh flies, Family Sarcophagidae.

Family 26. Tachina flies, Family Tachinidae.

Family 27. Bot flies, Family Oestridae (Figs. 20.9 and 20.17).

5. *Pupipara.* Louse-like, often wingless flies, with a very tough skin, indistinctly segmented abdomen, and legs inserted far apart on the sternum. External parasites on mammals (including bats), on birds, or on insects. The larvae develop viviparously until full grown and are born shortly before pupation, all growth taking place at the expense of the mother fly, which nourishes the larva from special uterine glands (Fig. 20.22). Antennae one- or two-segmented.

Family 28. The sheep-tick and louse flies, Family Hippoboscidae (Fig. 20.22).

Family 29. Bee lice, Family Braulidae.

THE PRINCIPAL FAMILIES OF DIPTERA

Family Tipulidae.[1] **Crane Flies or Daddy-long-legs.** The tales sometimes told of extremely large mosquitoes, when not pure fabrication, may be founded on the observance of crane flies. Their shape is similar to that of mosquitoes—long, the abdomen cylindrical, the wings narrow, the antennae and legs very long and slender (Fig. 6.97). The wings are very narrow at the base, almost petiolate. The cells on the basal half of the wings are unusually long and slender, and the costal vein continues around the posterior margin of the wing. Many species have the wings spotted or mottled. These flies lack the scales on wing veins that characterize mosquitoes, and of course they cannot bite. Many of the species approach 1 inch in length. Some are nearly 2 inches long, while the smallest measure only ⅛ inch. The legs are two or three times as long as the body and are very easily broken off. The halteres are very long. A good recognition character is an impressed linear mark, in the shape of a V, that extends from the base of one wing to the base of the other across the back of the thorax, the point of the V directed backward. Crane flies are the tan-colored, weak, clumsy flyers, found flying up and dropping into the grass as one walks through low meadows, forests, or damp places where there is an abundance of grass or weeds. The eggs are most commonly thrust into the soil. The cylindrical, brownish or olive-gray larvae are very tough skinned and have very poorly developed heads, and fleshy fingers of varied design surround the tail end. These may be protruded or withdrawn. So far as known, the adults have no economic importance. The larvae usually live in moist soil, muddy water, or decomposing material, rarely on foliage. Occasionally, the larvae cause appreciable injury by eating the roots of grass or corn. They are often called meadow maggots or leatherjackets (page 521).

Family Psychodidae.[2] **Moth Flies and Sand Flies.** Sometimes about sinks and basement drains, and almost always where there is sewage, small, extremely wooly-looking, velvety-winged, short, broad-bodied flies, not over ⅕ or ⅙ inch long (frequently only 1/16 inch) occur in considerable numbers. They look much more like little moths than like flies. Their flight is soft and weak, but they are active on their legs. The legs are short; the wings broad and oval, rather evenly divided by 10 straight parallel long veins but with very few closed cells. The venation is obscured by the hairs. The larvae live in water or moist organic material, feeding on algae, microorganisms, or filth. In the tropics and subtropics, species of the genus *Phlebotomus*, known as sand flies, have piercing-sucking mouth parts and are seriously annoying, besides being carriers of certain human diseases, such as Verruga in Peru, Papataci

[1] ALEXANDER, C. P., "The Crane-flies of New York," pt. I, *Cornell Univ. Memoir* 25, pp. 767–993, 1919; pt. II, 38, pp. 691–1133, 1920; "Crane-flies of Connecticut," *Conn. Geol. Natural History Surv. Bul.* 64, pt. 4, pp. 183–486, 1942; "Crane-flies of California," *Bul. So. Calif. Acad. Sci.*, **44**:33–45, 1945; **45**:1–16, 1946; and **46**:35–50, 1947.

[2] HASEMAN, L., "North American Psychodidae," *Trans. Amer. Ento. Soc.*, **33**: 299–333, 1907.

fever in the Mediterranean area, and Leishmaniasis in Asia and South America (Table 1.2). The only other importance is as a nuisance about habitations of man.

Family Culicidae.[1] **Mosquitoes.** Mosquitoes, because of their aggressiveness toward man, are among the best known of all insects. They are delicate, fragile, soft-skinned, very slender flies (Fig. 21.3). The body is strongly humped, both head and abdomen drooping downward from the thorax, which is much deeper than wide, and lacks the V-shaped suture of the crane flies. The legs are two or three times as long as the body, the head small, the mouth parts long, stiff, and straight and in the female used for sucking blood (Fig. 4.7). The antennae are filiform, about as long as the beak, slender, with enlarged basal segment containing sound-receiving organs (page 120). In the male they are generally densely and long plumose. The body of a true mosquito probably never exceeds ½ inch long, nor the wing spread ¾ inch, and the vast majority are at least one-third smaller than this. The complete fringe of large scales around the wing margin and along each of the veins gives the wings a peculiarly characteristic appearance. The costal vein encircles the wing as in Tipulidae. The mouth palps lie parallel with the beak, like an additional pair of antennae. The eyes are large, but there are no ocelli.

All mosquitoes develop in water. The eggs are usually laid singly in water or depressions where water may accumulate, but in *Culex* they are fastened together in a vertical position to form a raft. They generally float on the surface, and soon the larvae issue from the eggs into the water. The larvae of mosquitoes (Fig. 21.1) are the common "wigglers" of rain barrels and quiet pools. The head is large and has complex mouth brushes that, constantly in motion, waft food into the mouth. The mouth parts of the larvae are of the chewing type, and they feed on algae and other small plant or animal life either living or dead. The thorax is swollen and appears as one segment, but has no trace of legs in this stage. The abdomen is slenderer and bears on the eighth or next-to-last segment a short tube, known as a siphon, which the larva must thrust up into the air at intervals to breathe. This supply of air is supplemented by four finger-like tracheal gills attached to the last segment on the body, by which oxygen is taken from that dissolved in water. The gills alone will not keep the larvae alive, and they must come often to the surface to breathe. Indeed, they usually lie with the siphon projecting up through the surface film and the rest of the body hanging down at an angle in the water. In the larvae of anopheline mosquitoes, the siphon is very short and the larvae generally lie parallel to the surface just below the surface film. When disturbed, mosquito larvae swim down into the water by lashing the abdomen from side to side.

In as short a time as 2 days to 2 weeks, the larvae may be full-grown, about ⅜ inch long in common species. The change to the pupal stage takes place quickly at the fourth molt. This is a very unusual kind of a pupa (Fig. 21.1). It swims about actively in the water, avoids enemies, and does nearly everything the larva does except feed. It breathes through two trumpet-like tubes on the thorax. The eyes, legs, and wings can be seen developing through the body wall on the large combined head and thorax. The pupal stage is often called a "tumbler." After a few hours to a few weeks in this condition, the insect splits its skin down the back and the adult crawls out, balances for a few moments on the empty pupal shell until its wings spread and dry, and then flies away. Most species pass through a number of generations each year.

No family of insects is of greater importance to man. The transmission of malaria, yellow fever, dengue, filariasis, and encephalitis (Table 1.2) constitutes a total injury against the human race that is unequaled by any other family of insects. In addition to the transmission of deadly disease organisms, the irritation, annoyance, and interference with work and pleasure caused by their bites and buzzing render mosquitoes

---

[1] HOWARD, L. O., H. G. DYAR, and F. KNAB, "The Mosquitoes of North and Central America and the West Indies," *Carnegie Institute of Washington, Publ.* 159, vols. I–IV, 1912–1917; MATHESON, R., "Mosquitoes of North America," Comstock, 1944; HORSFALL, W. R., "Mosquitoes: Their Bionomics and Relation to Disease," Ronald, 1955; FOOTE, R. H., and D. R. COOK, "Mosquitoes of Medical Importance," *U.S.D.A. Agr. Handbook* 152, 1959; CARPENTER, S. J., and W. LACASSE, "Mosquitoes of North America," Univ. California Press, 1955.

one of the greatest plagues of man. The life histories and control of mosquitoes are discussed on page 998.

Family Chironomidae.[1] **Midges, Gnats, and Punkies.** In general build (Fig. 21.5) and habits the midges resemble mosquitoes and crane flies, but the members of this family are very small, typical species being about $\frac{1}{10}$ inch long, rarely exceeding $\frac{2}{5}$ inch in length, and often minute. The wings are either bare or, if hairy, the hairs are not confined to the veins and wing margins. The costal vein does not continue around to the posterior margin of the wing, and the mouth parts are short, generally non-piercing. The veins on the anterior margin of the wing are thicker than the ones behind. The antennae of the males are very bushy. The body is generally stouter and the wings shorter and broader than in mosquitoes. In many species the wings are pictured with white and black spots and there are very few closed cells. Swarms of these gnats dancing in the air near swamps are often mistaken for mosquitoes. One group of minute midges are bloodsucking and include the "punkies" or "no-see-ums," which in many sections of American woods and mountains become almost intolerable at dusk and dawn during the summer, because of their hot painful bites. The eggs are usually laid in the water or moist decaying organic matter and are often held together in masses or strings by a gelatinous secretion. The larvae are cylindrical and thin-skinned, with a distinct head. Some of the pupae are active, like those of mosquitoes. Most species develop in standing water of either ponds, ditches, or lakes. Some have been dredged from depths of 1,000 feet in lakes. They often make tiny tubes of the surrounding material, inside which the larvae live. Some of the best known are called bloodworms because of their red color, which is due to hemoglobin in the blood. Others are green or transparent. Most of them are probably scavengers. They are rated as very important as food for fresh-water fishes, but the adults often become dreadful nuisances because of their attraction to light.

Family Cecidomyiidae[2] **(Also Called Itonididae).** **Gall Gnats.** In this family, also, as in the midges and mosquitoes, the body is slender, somewhat mosquito-like, with long slender legs, decumbent head, and filiform antennae (Fig. 18.16), but the wings are broader and with fewer veins, usually only three weak longitudinal veins and no apparent cross-veins. The wings are hairy, and the costal vein encircles the wing; the hairs which cover the wing membrane generally readily rub or wash off. The head is small; the antennae long, usually moniliform, with a whorl of hairs on each segment. The antennae are unique in possessing delicate looped threads among the whorls of hairs on the antennal segments. The abdomen of the female is often tele-scopic terminally and much elongated as an ovipositor. The mouth parts are short, never piercing-sucking, frequently not functional in the adults. The wing spread rarely exceeds $\frac{3}{8}$ inch. These frail gnats or midges have an importance to man and biological interest surprising for creatures so small and delicate. Most of the species are gallmakers. A very common example is the pine-cone willow gall, *Rhabdophaga strobiloides* (Fig. 1.6,*D*). Some very destructive species, such as the Hessian fly (page 531), chrysanthemum gall midge (page 877), wheat midge (page 525), pear midge (page 742), and cloverseed midge (page 570) feed on cultivated crops, sometimes not forming galls. Some are scavengers, and a few are predaceous or parasitic. The larvae have neither legs nor evident head. They usually possess, when full-grown, on the underside of the first thoracic segment, a unique, hard, chitinized, dark-colored plate, more or less forked in front, called the breastbone. Its function is not certainly known, but it is useful in recognizing the larvae of this family. Some of the species increase their numbers in the larval stage, a phenomenon called paedogenesis. The pupae are either naked or protected by a puparium or by a cocoon.

[1] MALLOCH, J. R., "The Chironomidae or Midges of Illinois," *Bul. Ill. State Lab. Natural History*, vol. 10, art. VI, pp. 275–543, 1915; JOHANNSEN, O. A., "Chironom-idae," *Bul. N.Y. State Museum* 86, pp. 76–316, 1905, and 124, pp. 264–85, 1908; *Cornell Univ. Agr. Exp Sta. Memoir* 205, 83 pp., 1937, and 210, 52 pp., 1937.

[2] FELT, E. P., "Key to American Insect Galls," *N.Y. State Museum Bul.* 200, 310 pp., 1917; "A Study of Gall Midges," *N.Y. State Museum Buls.* 165, 175, 180, 198, and 257, 1913–1925, and "Plant-galls and Gall-makers," Comstock, 1940; KIEFFER, J. J., *Genera Insectorum*, vol. 152, 1913.

Family Mycetophilidae.[1] **Fungus Gnats.** This is a large family of delicate, obscure, dull-colored flies of very little importance to man. They are of a general mosquito-like form, but have very long coxae, spurs at the end of the tibiae, two or three ocelli, and a short proboscis. The thorax is usually arched and the antennae long, without whirls of hairs. The wings have a moderate number of veins and fine tangled hairs on the membrane but no scales along the veins. Fungus gnats frequent dark, dank shelters such as basements, stables, and low-lying woods. The eggs are laid in fungi, decaying organic matter, or other damp, dark situations. The economic importance, which is slight, arises from injury to mushroom beds or to roots and planted seeds of crops. One species causes a kind of potato scab. The larvae are soft and whitish, with a small distinct head. Most species develop in wild fungi, especially mushrooms. Others that live in decaying wood, manure, rotting fruits, and vegetables possibly also feed on minute fungus growths. The most interesting thing about fungus gnats is the curious behavior of some gregarious larvae that

Fig. 6.99. Black fly larvae as they are found clinging to stones removed from the swift current of cold mountain streams. (*Original.*)

congregate in large masses several inches wide and 10 or more feet long, crawling over each other, two or three deep, and very slowly advancing.

Family Simuliidae.[2] **Black Flies, Buffalo Gnats.** Black flies are short, chunky, humpbacked flies (Fig. 6.98), not exceeding ¼ inch in length and more often about ⅛ inch, with very broad, short wings, in which the anterior veins are stout, those behind very weak. There are no closed cells. The short, stout, horn-like antennae, composed of 11 closely compacted segments, scarcely exceed the width of the head. There are no ocelli. The legs are stout; the head large but set low against the arched thorax, giving the humpbacked appearance. The abdomen is short, stout, and thin-walled, so that it usually collapses badly in preserved specimens. The wings are free from hairs or scales, and the body has only slight pubescence. The females are among the most insatiable bloodsuckers and vicious of biting insects. Their life histories and control are described on page 1005. The eggs of black flies are fastened to the surface of stones, sticks, or vegetation that breaks the current of streams into ripples or rapids. The larvae (Figs. 5.15,*G*, and 6.99) lead a sporty life, clinging by means of sucker-like groups of minute hooks, one at each end of the body, to the surface of rocks in the swiftest parts of streams, the brinks of waterfalls, and other situations where the water is

[1] JOHANNSEN, O. A., "The Fungus Gnats of North America," *Maine Agr. Exp. Sta. Buls.* 172, 180, 196, 200, 1909–1912.

[2] DYAR, H. G., and R. C. SHANNON, "North American Two-winged Flies of the Family Simuliidae," *Proc. U.S. National Museum*, vol. 69, art. 10, pp. 1–54, 1927; MALLOCH, J. R., "American Black Flies or Buffalo Gnats," *U.S.D.A., Bur. Ento. Tech. Bul.* 26, 1914; METCALF, C. L., "Black Flies," *Bul. N.Y. State Museum* 289, 78 pp., 1932.

highly aerated. Clinging by the tail the body sways freely in the water, and remarkable strainer-like mouth brushes sift minute organic matter from the water and force it into the mouth. Respiration is by means of short retractile gill filaments near the anus. To anchor the pupae in the swift current a slipper- or vase-shaped cocoon is fastened to the surface of the rock, and in this the pupa is fastened by hooks on its abdomen. Long, branching, finger-like tracheal gills attached to the thorax accomplish respiration. The adults emerge under water, float to the surface in a bubble of air, and are on the wing before the swift current can drown them.

Family Blepharoceridae.[1] **Net-winged Midges.** This family of moderate-sized, long-legged, mosquito- or midge-like gnats (¼ to ½ inch long) can be distinguished

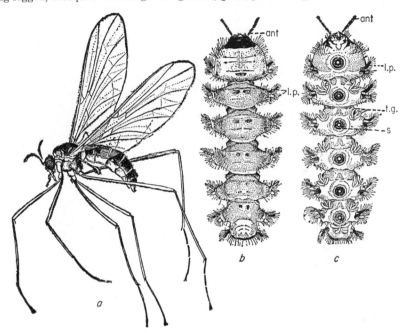

FIG. 6.100. Net-winged midges. *a*, The adult female of *Bibiocephala elegantulus*, 2½ times natural size; *b* and *c*, upper- and undersides of the larva of *B. comstockii*, about 5 times natural size. (*From Sanderson and Jackson, "Elementary Entomology," after Kellogg, Ginn & Co.*)

from all others by the "fine spider-web-like network of lines" which traverses the wings, crisscrossing the true veins. These are creases left by the folding of the wings in the pupal stage (Kellogg). There is no closed discal or first-medial-2 cell near the middle of the wing, and the anal margin of the wing is remarkably excised or notched to fit around the end of the halteres. The eyes of both sexes touch on top of the head. The mouth parts are long and adapted for piercing. According to Comstock, there may be two structurally different kinds of females in the same species, one blood-sucking, the other feeding on nectar. The adults are much less noticed than their curious larvae. The larvae and pupae are common in swift mountain streams in the same situations with black fly larvae and pupae, that is, clinging to the surface of stones and sticks in the swiftest current. They attract attention because of their pronounced segmented appearance (Fig. 6.100). Alternating segments are constricted to less than half the width of the others. If one tries to pick them up, one is amazed at how tightly they cling to the rock. This is accomplished by a median ventral row of

[1] KELLOGG, V. L., "The Net-winged Midges of North America," *Proc. Calif. Acad. Sci., Zool.,* **3**(3):187–226, 1903.

suckers. There are short tracheal gills for respiration. These curious larvae are said to eat algae and diatoms. The pupae are black. Their importance to man is apparently negligible.

Family Tabanidae.[1] **Horseflies, Deerflies.** This family includes the largest blood-sucking flies (Fig. 20.2). They range in size from about ¼ to 1 inch long. The body regions are closely compacted. They are very firm-bodied and hairy but not bristly flies, the hairs short and fine. They have strong legs and wings, the latter well provided with veins and often colored or pictured with bands and spots. The middle tibiae always have two spurs at the end. The costal vein encircles the entire wing margin. The head is large, a little wider than the thorax, short, concave behind. The eyes are often brilliantly colored during life with green, purple, or ruby, frequently in bands. The eyes cover nearly all the upper part of the head, but do not extend close to the mouth parts on the lower side of the head, being peculiarly cut away below. The thorax and abdomen are usually faintly spotted or banded. The antennae are shorter than the head, stiff, swollen at the base of the long third segment, and in the genus *Tabanus*, with a projecting spur on the upper side, like a thumb. The mouth parts are rather short and thick, piercing-sucking in the female, but not capable of penetrating flesh in the males, which feed on the nectar of flowers, honeydew, and other exposed sweets and are not found about animals. The mouth parts have also fleshy labella, as in the sponging mouth parts of the house fly. Horseflies are active in hot clear weather, though often most abundant in low woods along streams and about marshy lakes. They are very strong fliers, easily keeping up with a running horse or an automobile, and make a very loud hum. They bite all sorts of warm-blooded animals, including man; their life histories and economic importance are discussed on page 940. The eggs are typically laid upon leaves or stems of water plants that project above the surface of the water, often in wedge-shaped masses. The larvae tunnel through moist soil or swim in the water, devouring small animal life that they encounter. They are whitish, sometimes ringed with black, taper to both ends, and have elevated rings about the body on the principal segments. The head is small, the skin extremely tough. They respire through terminal posterior spiracles.

Fig. 6.101. A stratiomyid or soldier fly, *Stratiomys potamida*, female. About twice natural size. (*From Verrall, "British Flies."*)

Family Stratiomyiidae.[2] **Soldier Flies.** These are rather large to rather small flies, usually brightly marked, bare of bristles, and not very strong fliers. The antennae have three segments, the third segment with either a terminal style or a dorsal arista or ringed so as to suggest additional segments. The antennae are often carried in the form of a Y, the stem formed by the approximated, very slender, long, first segments of both antennae (Fig. 6.101). The mouth parts are short, of the sponging type, the face generally retreating. The tibiae are without spurs at the end. The scutellum is often ornamented with spines or projections on its margin. The abdomen is often characteristically broad and flattened, but sometimes long. Thorax and abdomen are

[1] HINE, J. S., "Tabanidae of Ohio," *Ohio Acad. Sci. Spcl. Paper* 5, 63 pp., 1903, and "Tabanidae of the Western United States and Canada," *Ohio Naturalist*, **5**: 217–248, 1904; STONE, A., "Bionomics of Tabanidae," *Ann. Ento. Soc. Amer.*, **23**: 261–304, 1930, and "Nearctic Tabaninae," *U.S.D.A. Misc. Publ.* 305, 171 pp., 1938; PHILLIP, C. B., "Catalogue of Nearctic Tabanidae," *Amer. Midland Naturalist*, **37**:257–324, 1947, and **43**:430–437, 1950.

[2] JOHNSON, C. W., "Stratiomyia and Odontomyia of North America," *Trans. Amer. Ento. Soc.*, **22**:227–278, 1895; JAMES, M. T., "Stratiomyiidae," *Jour. Kans. Ento. Soc.*, **9**:33–48, 1935; JOHANNSEN, O. A., "Stratiomyia Larvae and Pupae," *Jour. N.Y. Ento. Soc.*, **30**:141–153, 1928.

typically brightly banded with yellow and pale green especially developed on the margins, but sometimes uniformly yellow, black, or brilliant metallic green or blue. The wings are rather small, laid over each other at rest; the discal cell generally subrotund, not much longer than wide; the veins of the wing close together in front, sparse or faint behind; the squamae very small. The adults are common about flowers but do not hover before them as the syrphid flies do. They favor low wet situations. The eggs are laid in decaying, sappy wood, on or near water, on the soil, or in dung. Some of the larvae are predaceous, others are scavengers. They are very thick-skinned, usually flattened. The head is small, the tail end often tapered, sometimes forming a tubular, breathing organ with a chambered concavity or cleft at the end. Pupation takes place in the last larval skin. No important economic relation to man is known.

Family **Rhagionidae**[1] (Also Called Leptidae). Snipe Flies. These are rather variable, weak-skinned, sluggish flies of moderate to large size, with long tapering abdomens, without bristles, but often densely covered with fine short silky hairs. Many species suggest small robber flies in appearance. The wings are ample, with the wing veins normally distributed, the squamae very small or wanting, halteres large. The legs are long, the tibiae, at least the hind pair, with terminal spurs. The head is large, the face usually retreating. The antennae are three-segmented, with an arista or a style, and usually very short, sometimes long. The tapering abdomen is frequently banded with half a dozen white or yellow crossbands, formed either by close-set silky yellow hairs or by the ground color. Our common species range from $\frac{1}{3}$ to $\frac{2}{3}$ inch long. Adults of certain species are gregarious at egg-laying time and pile up their eggs and their own dead bodies on certain branches of trees or on rocks to form enormous masses as big as a football. Aldrich discusses the habit of the Indians, who collected them by bushels and cooked them for food. The larvae of varied species live in the water or in soil or rotting wood. A few species make sand traps, like the ant-lions, and eat the insects caught. All are supposed to be predaceous. The family has little economic importance. Some adults suck blood, but are not often encountered.

Family **Asilidae**. Robber Flies. Robber flies are moderate-sized to very large, often slender-bodied, but sometimes resembling bumblebees, very bristly or hairy, and often humped-bodied (Fig. 6.102). They got their common name from their ferocious manner of pouncing from the air upon their prey, which con-

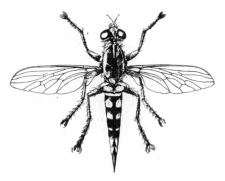

Fig. 6.102. A robber fly, *Erax maculatus*, female. About 2 times natural size. (*From Ill. State. Natural History Surv.*)

sists of all sorts of insects. The head is broad and short and well separated from the thorax by a slender neck, the front characteristically excavated or depressed between the eyes on top. Ocelli are present, the antennae three-segmented, with either a terminal style or an arista. The mouth parts are piercing-sucking, pointed, and stiff, but not very long. The abdomen is most often slender, tapering backward, often longer than the wings. The legs are very stout, long, and bristly, the squamae small, the wings well provided with veins, frequently stained brownish but not pictured. The adults are found in open sunny fields such as stubble fields. When disturbed or attacking prey they fly but a short distance, with a loud buzzing, before alighting. The prevailing colors are gray, brown, or black, but a few are marked with red or golden-yellow pile. The larvae of robber flies are also predaceous. They are found among decaying organic matter, under surface litter, in rotting wood, or in soil where they attack other insect larvae, especially beetle grubs. The pupae, which are not enclosed in a puparium, are very spiny.

[1] LEONARD, M. D., "Rhagionidae (Leptidae) in the United States and Canada," *Amer. Ento. Soc. Memoir* 7, pp. 1–181, 1930.

Family Bombyliidae.[1] **Bee Flies.** The common name, bee flies, is not very distinctive, since many other flies (Syrphidae, Asilidae) have also developed a protective mimicry of their stinging cousins, the bees and wasps. These medium-sized, elusive, handsome flies hover before flowers or stand still in the air or dart swiftly from spot to spot. Their bodies (Fig. 6.103) are usually very light, soft-walled, commonly densely covered with hairs but not usually bristly. The head is tucked close against the thorax, round, nearly covered by the compound eyes. The antennae are usually simply three-segmented, without arista or style, of variable length. Ocelli are present, mouth parts long or short, often very long and slender, adapted for securing nectar and pollen. Head, thorax, and abdomen are usually successively broader, rarely the abdomen slender, thread-waisted, but usually very short. The legs are not very stout, the squamae small. The wings are well traversed by veins, very often beautifully spotted or patterned. The colors are mostly black or dark brown, the wing markings the same, and the body covered with fine long yellowish, brownish, or white hairs, like a halo. While a few species are less than ⅛ inch long, the average is around ½ to ¾ inch long and the wing expanse ¾ to 1½ inches. The larvae which are known are predaceous or parasitic on bee or wasp larvae, cutworms, or the eggs of grasshoppers. They have very small heads. They have a complex hypermetamorphosis. The pupa is not enclosed. Their economic value is not great since some attack other parasites and tend to offset the good done by destroying crop pests.

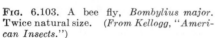
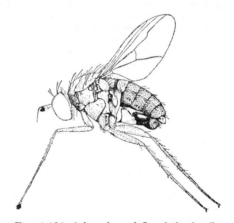

FIG. 6.103. A bee fly, *Bombylius major.* Twice natural size. (*From Kellogg, "American Insects."*)

FIG. 6.104. A long-legged fly of the family Dolichopodidae, male. About 3 times natural size. (*From Woodworth.*)

Family Dolichopodidae.[2] **Long-legged Flies.** Small (⅛ to ½ inch), slender, usually shining green, sometimes blue or yellow, nearly bare but somewhat bristly flies, with the discal (first medial 2) and second basal (medial) cells united, forming a very long cell lengthwise through the middle of the wing (Fig. 6.104). All but one of the large cells are open to the wing margin. The wings are large, sometimes patterned. The mouth parts are short and fleshy, used to envelop small soft-bodied flies and extract their body fluids or to draw nectar. The antennae are placed rather high on the front, not long, with two short and a third long segment, the latter with a dorsal arista or terminal style, which is frequently prominent. The face is not prominent. The abdomen is slender, tapering to the apex, legs sometimes rather long, usually bristly, provided in some males with greatly enlarged ornamental tarsi. The genitalia of the males are often conspicuously large. The adults are often found running over leaves or grass in moist habitats. The larvae, so far as known, are either scavengers or

[1] PAINTER, R. H., "Bombyliidae," *Trans. Kans. Acad. Sci.,* **42**:267–301, 1939.

[2] ALDRICH, J. M., "North American Dolichopodidae," *Trans. Amer. Ento. Soc.,* **30**:269–286, 1904; VAN DUZEE, M. C., F. R. COLE, and J. M. ALDRICH, "Dolichopus in North America," *Bul. U.S. National Museum* 116; 304 pp., 1921.

predaceous, and they live in soil, decaying vegetation, or in the water. The pupae, so far as known, are generally unenclosed and have long breathing horns on the thorax. The economic importance of the family is nearly negligible.

Family Phoridae.[1] **Humpbacked Flies.** In these flies the third segment of the antennae is so much larger than the others that the antennae appear to be one-segmented, with a long stout bristle or arista. The body is rather short ($\frac{1}{20}$ to $\frac{1}{6}$ inch long in common species); the head and thorax are frequently provided with scattered, large bristles. The wings are often wanting, especially in the females. When wings are present, the two veins near the front of the wing are very heavy and terminate before the tip of the wing; the remaining veins are finer, running diagonally, and not forming any closed cells. The costal margin of the wings is often very bristly on the basal half. The eyes are small, the ocelli usually present. Mouth parts are of the sponging type. The thorax is arched, and the abdomen short, narrowed, and drooping behind, to give the flies a strongly humpbacked appearance. The legs are ample and often adapted for jumping. These flies, although small to minute and obscurely colored, are of the most remarkable habits of all the insects. A number have been found to dwell in the nests of ants, bees, and wasps. Others live in decaying vegetable matter, fungi, the carcasses of animals, or decaying insects. The pupa is formed inside the larval skin and breathes through a pair of long horns.

Family Syrphidae.[2] **Flower Flies, Hover Flies, Syrphids.** Although very diverse in form and color, a very large number of the species of flower flies fall into two general types, some resembling wasps, with bare, slender bodies, often slender-waisted at the base of the abdomen, and another lot being broad, very hairy, and mimicking bumble-bees. The prevailing colors are yellow spots or crossbands or fasciae of yellow hairs upon a ground color of polished black or metallic blue, green, or violet. Most species may be found in bright sunshine about flowers, upon the nectar and pollen of which the adults feed. They are expert on the wing, flying with extreme swiftness or hover-ing in the air, with their wings fanning like a haze, but without any visible movement of the body. They are medium to large-sized, hard-bodied, very vigorous flies, mostly $\frac{1}{2}$ to 1 inch long, although a few are only $\frac{1}{8}$ inch. They may be distinguished from all other flies by the presence of a false or spurious vein (Fig. 6.96), which is a vein-like thickening of the wing between radius 4 and 5 and media 1 and 2 and bisecting the radiomedial cross-vein. The head is at least as broad as the thorax, the face usually projecting at the middle or next the mouth; antennae variable, generally short, with an arista on the third segment. There are no large bristles on the head and usually not elsewhere. The adults are important pollinizers, but the economic importance of the family arises largely from the predaceous habits of the larvae of many species, especially *Syrphus*, *Sphaerophoria*, and *Allograpta*. These feed on aphids and other small soft-bodied insects (page 66 and Fig. 2.15,*C*). The larvae of other species are scavengers in all kinds of decaying organic matter, or live in the nests of ants, bees, wasps, or termites, or are pests feeding in bulbs, fungi, cacti, or the trunks of trees. The bulb fly (page 880) and the lesser bulb fly (page 881) are serious pests of narcissus, amaryllis, and onions, in many parts of the United States. The rattailed larvae of *Eristalis*, *Helophilus*, and other genera (Fig. 6.105) have a very remarkable posterior telescopic tube for conducting air down to the larva as it feeds in sewage and other filthy liquids. The larvae of *Microdon* (Fig. 6.105), which are often found in ants' nests, are curious slug-like things that zoologists have several times described as new Mollusca. The pupal stage is always protected in a puparium formed of the larval skin and is to be found on either plants or the soil or in filth. Several species have been recorded as occasional intestinal parasites of man.

[1] BRUES, C. T., "North American Phoridae," *Trans. Amer. Ento. Soc.*, **29**:331–404, 1903; MALLOCH, J. R., "Phoridae of the United States," *Proc. U.S. National Museum*, **43**:411–529, 1913.

[2] WILLISTON, S. W., "Synopsis of North American Syrphidae," *U.S. National Museum Bul.* 31, pp. xxx and 335, 1886; METCALF, C. L., "Syrphidae of Ohio," *Ohio Biol. Survey Bul.* 1, 1913, and "Syrphidae of Maine," *Maine Agr. Exp. Sta. Bul.* 253, 1916, and 263, 1917; CURRAN, C. H., "Syrphidae of North America," *Bul. Amer. Museum Natural History* 78, pp. 243–304, 1941; HEISS, E. M., "Syrphidae, Larvae and Pupae," *Univ. Ill. Bul.* 36, 142 pp., 1938.

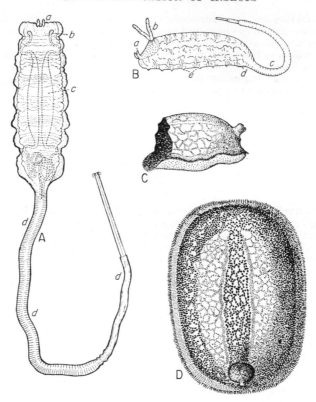

FIG. 6.105. Two curious larvae of Syrphidae or flower flies. *A*, the rattailed maggot of *Eristalis*, showing long extensile breathing tube at *d*; *B*, puparium of the same showing at *b* the cornua through which the pupa breathes; *C*, the empty puparium of *Microdon* from which the adult has escaped; *D*, the larva of *Microdon*, dorsal view. All enlarged 5 to 10 times. (*From Metcalf, Ohio Biol. Surv.*)

Family Conopidae.[1] **Thick-headed Flies.** These are medium-sized ($\frac{1}{4}$ to 1 inch long), slender-bodied, very firm-skinned flies (Fig. 6.106), frequently with the basal part of the abdomen slender and the apex downbent and swollen by the large genitalia in the male or extended into a long ovipositor in the female. Their build and yellowish-color markings often strikingly suggest wasps, but they are not so powerful in flight.

FIG. 6.106. A thick-headed fly, *Physocephala affinis*. Enlarged $\frac{1}{2}$. (*From Kellogg, "American Insects."*)

They are never very hairy, often bare. The eyes are well separated on top of the head in both sexes; the antennae average about as long as the head and are provided with either a terminal style or a dorsal arista. The head is often produced downward far below the eyes. The mouth parts are very long, sometimes as long as the entire body, slender, rigid, often sharply elbowed about mid-length. They are not used for piercing, however, since the adults, so far as known, feed on flowers. The wings are normally veined, sometimes patterned with brown, the cubital cell often unusually long. The adults lay their eggs on the bodies of

[1] VAN DUZEE, M. C., "Conopidae of North America," *Proc. Calif. Acad. Sci.*, **16**:573–604, 1927; PARSONS, C. T., "Conopidae of North America," *Ann. Ento. Soc. Amer.*, **41**:223–246, 1948.

wasps, bees, and grasshoppers, sometimes accomplishing this feat while both insects are in flight. The larvae are internal parasites in such insects. They appose their posterior spiracles to a large trachea inside the host insect, presumably to secure air. The pupal stage is spent within the body of the host. The family is neither large nor of much importance to man.

Family Ortalidae. Ortalid Flies. This is one of two families that have the wings beautifully pictured with blackish spots or bands; the other family is called Trypetidae.

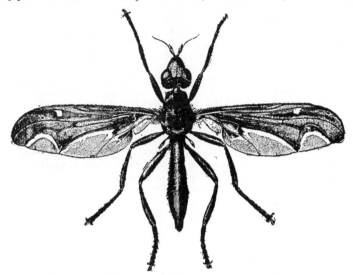

FIG. 6.107. An ortalid fly, *Pyrgota undata*, the larva of which is a parasite on June beetles. About 3 times natural size. (*From Ill. State Natural History Surv.*)

The Trypetidae have the tip of the subcostal vein imperfect, faint, or broken or bent at a sharp angle. In the Ortalidae it is sharp, clear, and normal, meeting the costal margin at an acute angle. There are no prominent bristles ("vibrissae") just above the mouth opening on the lower edge of the face in the ortalids. The head is rather large, the mouth parts sponging in type. The legs are usually stout with spurs at the apex of the middle tibiae only. The venation of the wings is simple; the subcosta and radius are close together but separate and distinct full-length. This is a rather large family of medium-sized flies, ¼ to ½ inch long. Not enough is yet known about the larval habits in this family to estimate their importance to man. The larva of the large wasp-like species, *Pyrgota undata* (Fig. 6.107), is a parasite of adult May beetles. The adult fly lays her eggs beneath the wings of May beetles when they are in flight at night.

FIG. 6.108. A trypetid fly, the white-banded cherry fruit fly, *Rhagoletis cingulata*, female. Enlarged. (*From Lochhead, "Economic Entomology," after Caesar.*)

Family Trypetidae.[1] Fruit Flies. Superficially resembling the Ortalidae, these flies may be distinguished by the differences in wing venation described under that family (Fig. 6.108). The wing markings are in general of finer spots and more complex than in the Ortalidae; the colors are often beautifully intricate,

[1] BATES, M., "American Trypetidae," *Psyche*, **40**:48–56, 1933; BENJAMIN, F. H., "Trypetid Flies," *U.S.D.A. Tech. Bul.* 401, 95 pp., 1934; PHILLIPS, V. T., "Trypetidae of North America," *Jour. N.Y. Ento. Soc.*, **31**:119–154, 1923, and "Trypetid Larvae," *Memoir Amer. Ento. Soc.* 12, 161 pp., 1946.

mostly pale brown, the spotting confined to the wings, the body itself usually plain. The venation is normal except for the characteristic degeneration of the apex of the subcostal vein. The first anal cell is sometimes prolonged into an acute extension at the distal angle. Vibrissae (large hairs of the upper "lip") are wanting. A row of bristly hairs margins the eyes at the upper front corner. The antennae are short, and the eyes rather small, not meeting above in either sex. The eggs of these flies are usually inserted into the tissues of plants. The larvae of some species bore in the stems of plants; some produce galls, some mine in leaves, and most important of all are those which bore into the flesh of fruits and vegetables. The latter include some of the most important of all economic insects: the apple maggot (page 732), the cherry fruit flies (page 769), the walnut husk fly (page 812), the Mexican fruit fly (page 812), the Mediterranean fruit fly (page 812), the oriental fruit fly (page 812), the melon fly (page 812), the olive fruit fly, *Dacus oleae*, and the currant fruit fly, *Epochra canadensis*.

Family Chloropidae[1] (Also Called Oscinidae). **Frit Flies, Grass Stem Maggots, and Eye Gnats.** These are small ($\frac{1}{16}$ to $\frac{3}{16}$ inch long), short-winged, bare flies, abounding in rank vegetation. Like the Ephydridae, the wings are devoid of closed cells except close to the base, the anal cell being absent and the medial not separated from first medial 2. The subcostal vein is also wanting. They may be distinguished from the ephydrids by their small mouths, the short antennae with rounded third segment and bare arista, the occasional presence of vibrissae, the broad front and retreating face, and by the fact that they are not usually entirely black but are generally yellow-spotted or banded. The larvae of many species are phytophagous and include a number of very destructive pests such as the frit fly, *Oscinella frit*, and the gout fly, *Chlorops taeniopus*, which are pests of cereals in Europe. The wheat stem maggot (page 540) mines in the upper part of wheat stems, causing blighted heads. The larvae of *Hippelates* breed in decaying vegetation, and the adults are the vexatious eye gnats (page 1035), which are vectors of acute conjunctivitis.

Family Ephydridae.[2] **Shore Flies.** These flies are similar structurally to the Drosophilidae, and especially to the Chloropidae, from which their uniform dark-brown or black color, without light markings, the frequent pubescence of the arista, the extremely large mouth, the entire absence of vibrissae, and usually swollen, convex face will help to separate them. The entire underside of the head forms a great mouth cavity. As in the Drosophilidae, there are two microscopic "breaks" in the costal vein on its basal half. These flies are commonly $\frac{1}{16}$ to $\frac{3}{8}$ inch long. A number of species of the genus *Ephydra* abound about salt, briny, or alkaline water, in which the larvae live and from which the pupae are frequently washed ashore to form great windrows. On account of their abundance, these have been collected, dried, and used as food by the Indians of the Southwest. The larvae are sometimes rattailed, like certain Syrphidae, but the tail in this case is forked or split toward the end. One remarkable species, called the petroleum fly, *Psilopa petrolei*, lives as larvae in pools of crude oil feeding upon other insects trapped in the oil. These larvae are about $\frac{1}{3}$ inch long, when full-grown.

Family Drosophilidae.[3] **The Pomace or Fruit Fly and Others.** These are small, somewhat bristly flies, commonly $\frac{1}{16}$ to $\frac{1}{4}$ inch long, and of obscure coloration. The frons and face are broad. The face is nearly straight down below the antennae with prominent bristles (vibrissae) usually present at its lower margin. There are long, stout bristles on the upper part of the head. The arista is usually plumose. The abdomen is generally short and rather soft-walled. The wings are broad, with comparatively few veins and closed cells, the venation much as in the Ephydridae. The subcostal vein is degenerate, incomplete, or entirely wanting, and the costal vein has

[1] MALLOCH, J. R., "Genera of Chloropidae of North America," *Can. Ento.*, **46**:113, 1914; SABROSKY, C., "Chloropidae of Kansas," *Trans. Amer. Ento. Soc.*, **61**:207–268, 1935.

[2] JONES, B. J., "Catalogue of Ephydridae," *Univ. Calif. Publ. Ento.*, **1**:153–98, 1906.

[3] STURTEVANT, A. H., "The North American Species of Drosophila," *Carnegie Institution of Washington, Publ.* 301, 150 pp., 1921, and *Univ. Texas Publ.* 4213, 51 pp., 1942; PATTERSON, J. T., "Drosophilidae of Southwest," *Univ. Texas Publ.* 4313, pp. 7–216, 1943.

two distinct interruptions in its basal half as though it had been broken. These flies are found among grass and weeds and about all kinds of overripe, fermenting, or decaying fruits, apparently being attracted chiefly by acetic acid (Barrows). The family is chiefly known from its very common representative, the banana fly, *Drosophila melanogaster* (Fig. 6.109), whose structure, habits, and especially its genetics have been most intensively studied.[1] A very large part of what we know about

inheritance has been determined by laboratory studies of this insignificant-looking little fly.

Family Anthomyiidae.[2] Root Maggot Flies. These are soft-skinned, bristly, grayish to blackish, homely-looking flies (Fig. 14.42) of moderate size ($\frac{1}{4}$ to $\frac{3}{8}$ inch long) and of much the same general build as the house fly. Vein *medius* 1 and 2 is straight or nearly so, and consequently *radial* 5 cell is always widely open; the eyes of the male usually touch on top of the head; and the tegulae are large. The body is usually obscurely spotted or striped with white or grayish pollen, rarely shining. The abdomen is rather short, broad at base, tapering to a pointed tip. The antennae are short, appressed into vertical grooves on the face, the arista varying from plumose to entirely bare. The face is concave, not prominent, provided with vibrissae. This is a large family, the larvae of which have a tendency to feed upon decaying vegetable matter from which a number of species have adopted the habit of attacking the roots of vegetables. The cabbage maggot (page 670), the onion maggot (page 660), the seed-corn maggot (page 518), and the spinach leaf miner (page 677) are all very destructive, because of the practical difficulty of destroying them beneath the soil where the larvae live.

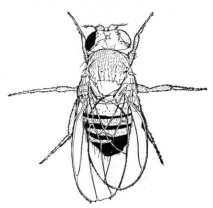

FIG. 6.109. A specimen of the common fruit or pomace fly, *Drosophila melanogaster*, that is half female and half male, a gynandromorph. The dark eye, notched wing, and shape of the abdomen on the left side are those of a female, while the broad wings, color and character of the eye, gnarled or twisted bristles, and other characteristics of the right side show that this half is male. (*From T. H. Morgan, C. B. Bridges, and A. H. Sturtevant, "The Genetics of Drosophila.*")

Family Muscidae[3] (Including Calliphorinae). The House Fly Family. The best-known and most important representative of this family is the common house fly (page 1031), which is fairly typical of the entire family. These flies are rather short, weak-skinned, not very bristly but never bare, grayish or metallic blue or green, with ample wings, in which the *radial* 5 cell is either closed or narrowed at the wing margin by the forward-bent *medius* 1 and 2 vein (Fig. 20.10). Although many species have grayish or brownish pollinose markings, there are a good many in which the abdomen

---

[1] MORGAN, T. H., C. B. BRIDGES, and A. H. STURTEVANT, "The Genetics of Drosophila," *Bibliographia Genetica*, II, 262 pp., 1925; DEMEREC, M. (ed.), "Biology of Drosophila," 632 pp., Wiley, 1950; STURTEVANT, A. H., C. B. BRIDGES, T. H. MORGAN, L. V. MORGAN, and JU CHI LI, "Contributions to the Genetics of *Drosophila simulans* and *Drosophila melanogaster*," *Carnegie Institution of Washington, Publ.* 399, 1929.

[2] JOHANNSEN, O. A., "Anthomyiidae of Eastern United States," *Trans. Amer. Ento. Soc.*, **42**:385–98, 1916; MALLOCH, J. R., "Anthomyiidae of North America," *Trans. Amer. Ento. Soc.*, **46**:133–196, 1920.

[3] HOUGH, G. DE N., "Some Muscinae of North America," *Biol. Bul.* 1, pp. 19–33, 1899; SHANNON, R. C., *Insecutor Inscitiae Menstruus*, **11**:101–118, 1923, and **12**:67–81, 1924; HEWITT, C. G., "The House Fly: A Study of Its Structure Development, Bionomics and Economy," Manchester Univ. Press, 1910; WEST, L. S., "The House-fly," Comstock, 1951.

or thorax and abdomen both are shiny green or blue. The arista is generally plumose full-length, which, with the absence of large bristles on the abdomen except at the tip, will separate them from the Tachinidae. The eyes of the males are close together or touching on top of the head. The mouth parts are of distinctly different types, most of them being sponging in type (Fig. 4.15). A number of species, however, have developed a bloodsucking habit with mouth parts adapted for piercing animal flesh and sucking blood (Fig. 4.8). In these species both males and females feed on blood. These include the stable fly (page 942), the horn fly (page 954), and the tsetse flies of Africa (*Glossina* spp.),[1] carriers of the fatal sleeping sickness (page 29). The size range is a little smaller than the Sarcophagidae. Most of the larvae develop in decomposing animal or vegetable refuse. Some attack living animals, especially living in wounds, as the screw-worm flies (page 968), or upon helpless young, like the nestlings of birds. The bluebottle flies, blowflies, and the cluster fly, *Pollenia rudis*, are very common in and about human habitations.

Family Sarcophagidae.[2] **The Flesh Flies.** Although as a group the flesh flies are rather distinct in appearance from the house fly group, the best technical distinction from the Muscidae appears to be that in this family the arista is plumose only on the basal half, the tip being bare, whereas the Muscidae have it plumose to the tip. Except in the matter of the elbowed *medius* 1 and 2 vein, these flies are very similar also to the Anthomyiidae. The typical color is grayish, somewhat longitudinally striped with white on the thorax, and spotted with the same gray or white pollen on the abdomen. These whitish pollinose markings are generally more prominent than in either the Muscidae or the Anthomyiidae. The abdomen tends to be longer than in the Muscidae (Fig. 6.110). Vein *medius* 1 and 2 bends sharply forward, nearly or entirely closing cell *radius* 5. The size range of common species is from ³⁄₁₆ to ½

FIG. 6.110. *Sarcophaga kellyi*, male, a parasite in the larval stage in grasshoppers. About 6 times natural size. (*From U.S.D.A., Farmers' Bul.* 747.)

inch long. Many of the larvae are parasitic upon such insects as grasshoppers, caterpillars, and beetles. Some of the species attack their victims while both insects are in flight and deposit already hatched larvae on their bodies—an ancient example of aerial warfare. Others develop in the carcasses of animals or in dung or garbage or live in wounds or in the nasal passages or stomachs of larger animals.

Family Tachinidae.[3] **Tachinid Flies.** Superficially resembling the house flies and flesh flies in wing venation and the presence of large tegulae, these may best be distinguished by the entirely bare arista and the presence of large bristles on the base of the abdomen (Fig. 2.16). They are in fact very bristly, usually robust but rarely slender flies, common on leaves or about flowers, from which the adults get their food. They fly swiftly and in general give the impression of being abundantly able to take care of themselves. The veins tend to retreat a little from the posterior margin of the wings. The legs are stout and bristly. They are much more diverse in the structure of the head and antennae and more interesting to the systematist than the Muscidae or Anthomyiidae, some of them being handsomely marked with yellow or with whitish

[1] Austen, E. E., "Monograph of Tsetse-flies," British Museum, 319 pp., 1903; Newstead, R., A. M. Evans, and W. H. Potts, "Tsetse-flies," *Memoir Liverpool School Tropical Medicine* (n.s.), vol. 1, 268 pp., 1924.

[2] Aldrich, J. M., "Sarcophaga and Allies," Thomas Say Foundation, 302 pp., 1916, and *Proc. U.S. National Museum*, vol. 78, no. 2855, 1930; Hall, D. G., "Blowflies of North America," Thomas Say Foundation, vol. 4, 477 pp., 1948.

[3] Coquillett, D. W., "Tachinidae of America, North of Mexico," *U.S.D.A., Div. Ento. Tech. Ser. Bul.* 7, 1897; Curran, C. H., "Tachinidae," *Bul. Amer. Museum Natural History* 89, art. 2, 122 pp., 1947.

or golden pollen.   The family is considered to be one of the most helpful allies of man in his combat with crop pests.   Most of the larvae live in and destroy caterpillars, although various beetles, Hymenoptera, flies, true bugs, and Orthoptera are also attacked (page 68).   The eggs are glued to the skin of the host or laid upon foliage, where they may be eaten by the host caterpillar when it feeds, or larvae already hatched are thrust beneath the skin of the victim.   The larvae are remarkable in forming an organic connection with the host, consisting of a sheath formed of the host tissues, which grow inward from the body wall or from a trachea and completely surround the larvae for at least a part of their life.   For pupation the larvae usually desert the mortally injured host and enter the soil.

Family Oestridae[1] (Including Gastrophilidae and Cuterebridae).   Bot-flies.   The bot-flies are well known to all who work with domestic animals as the large, very hairy, bee-like flies that chase or irritate cattle, horses, and sheep on hot summer days by dashing against their legs, throats, or mouths, in their attempts to glue eggs fast to the hairs in these regions.   The body is very broad.   The head hugs the thorax closely and is nearly hemispherical, with a broad retreating hairy face (Fig. 20.17).   The insects are not bristly, but are usually densely covered with short fine hairs often of contrasting colors.   The antennae are short, almost hidden; the mouth parts degenerate, never large.   The eyes are rather small, well separated in both sexes.   Wings are as in Muscidae, except in the horse bots, where cell *radius* 5 is wide open because vein *medius* 1 and 2 runs straight to the wing margin.   The tegulae are also much smaller in the horse bots than in the other Oestridae.   The larvae of bot flies (Fig. 20.9) are large, very spiny, fat maggots that live in some part of the body of vertebrate animals, especially mammals.   The best-known species are the horse bots (page 948), that live as maggots in the alimentary canal of horses, mules, and donkeys; the ox warbles or cattle grubs (page 963), that live in tumors which they produce under the skin of the back of cattle and bison; and the sheep bot (page 982), which mines in the heads of sheep.   The rabbit bot, *Cuterebra cuniculi*, forms lumps in the necks of rabbits, and the emasculating bots live in the scrotum of squirrels.   The human bot fly (page 1031) and other species occasionally attack man.   Pupation invariably takes place in the soil, and there is usually one generation a year.

Family Hippoboscidae.[2]   Louse-like Flies.   The adults of this family live like lice, upon the skin, among hairs or feathers of mammals and birds, sucking the blood.   The best-known is the sheep tick (page 976 and Fig. 20.22), which is common among the wool of sheep.   Less well-known species are found upon horses, deer, hawks, owls, and other birds.   Some have completely lost their wings; others shed the wings after a time.   A few have them present as vestiges, but most of the species retain ample wings which have the veins concentrated and heavy in front but faint behind, with which they make quick short flights in search of a hairy or feathery host.   The form of the body is little fly-like.   The thorax and abdomen are very broad and flattened, the legs widespread, with the coxae far separated.   The marks of abdominal segments have nearly or quite disappeared, and the skin has become very tough and rubbery, like that of a tick.   The head is closely packed against a concavity of the prothorax; the eyes are small, far separated; the antennae, which are sunken into cavities above the mouth, seem to consist of a single segment with a dorsal or terminal bristle.   The retractile, piercing mouth stylets are shielded, when protruded, by the stiff, elongated palps.   There is a broad band of sternum between the right and left coxae; the legs are strong and bristly; the claws very well developed, each one sometimes double or triple.   Even the halteres have become rudimentary or, in some species, entirely lost.   The family is a small one but possesses great biological interest because of the structural degeneration that has apparently resulted from their parasitic habits and because of their peculiar method of reproduction.   No eggs are laid, but the larvae are developed, one at a time, in the uterus of the female fly, being nourished by a milk-like

[1] HERMS, W. B., "Medical Entomology," 4th ed., Macmillan, 1947.

[2] SWENK, M. H., "North American Hippoboscidae," *Jour. N.Y. Ento. Soc.*, 24: 126–136, 1916; FERRIS, G. F., and F. R. COLE, "Contribution to Knowledge of Hippoboscidae," *Parasitology*, 14:178–205, 1922; MACARTHUR, K., "Hippoboscidae of Wisconsin," *Bul. Publ. Museum Milwaukee* 8, pp. 373–440, 1948.

secretion of the accessory reproductive glands.　When full-grown they are extruded and glued fast to the hairs or feathers of the host (Types of Reproduction, page 157). The larvae are yellowish white, without legs, distinct segments, or mouth parts, with a black "button" surrounding the spiracles at the posterior end of the body.　They take no food after birth but are cemented to the hairs or feathers of the host and very soon form a brown seed-like puparium.　Within the puparium the pupal stage is passed, and from it an adult presently clambers out.　The pigeon louse-fly, *Lynchia brunnea*, is a winged species, imported from Europe, but now widely distributed in North America.　It is the carrier of a protozoan disease, pigeon malaria.

Family Braulidae.[1]　**The Bee Louse.**　These insects are so remarkable that they are included for description, briefly, although there are only two known species in the family and they are only about $\frac{1}{16}$ inch long.　The bee louse, *Braula coeca*, is a tiny, wingless, big-legged, louse-like, degenerate fly that develops in the nests of the honeybee.　They live as adults, upon the body of the queen and drone bees, securing, as food, honey from the bee's mouth.　The eyes are very small; the head very broad between them.　There are no ocelli, and the mouth parts are similar to those of the Hippoboscidae.　There are neither halteres nor any trace of wings.　The thick legs bear on the last tarsal segment a pair of comb-like structures.　The insect lays its eggs among the bees' comb, and the larvae, which are typical maggots, tunnel through the wax caps of comb honey and feed upon the food stores of the bee.

ORDER SIPHONAPTERA (APHANIPTERA OR SUCTORIA)[2]

THE FLEAS

This is a very small order, very well defined, and not closely related to any other group of insects.　All the species are wingless, and all are, in the adult stage, external parasites on warm-blooded vertebrates. They are facultative rather than obligatory parasites (page 63). They are almost unique among insects in being flattened, or thin, from side to side, like a sunfish.　The legs are long and adapted for jumping, and the coxae are abnormally large, often being actually the largest segment of the leg (Fig. 21.9).

The body wall is hard or tough, polished, and provided with many backwardly directed hairs and with short, stout spines, often arranged so regularly as to resemble combs.

There are no compound eyes, and often the simple eyes also are wanting. The three-segmented antennae are concealed in grooves just behind the eyes (Fig. 4.9,*A*).

The mouth parts (Fig. 4.9) are piercing-sucking in type, but differ from any of the other subtypes in having two pairs of palps and the maxillae not stylet-like.　The mandibles and labrum-epipharynx are adapted as stylets for piercing and forming the salivary duct and food channel; the maxillae are broad, triangular plates that do not enter the wound but do bear segmented palps.　Another unusual feature of the fleas is that the three segments of the thorax are very distinct and free from each other.

These are almost the only external insect parasites of our larger animals that have a complete metamorphosis.　The adult stage is the only stage that is recognized generally.　The eggs are not fastened to the host, like the eggs of lice, and so are not noticed.　They drop off the host or are laid on the floor of the nest, kennel, or dwelling of the host.

[1] PHILLIPS, E. F., "The Bee-louse in the United States," *U.S.D.A. Cir.* 334, 1925.

[2] Fox, I., "Fleas of Eastern United States," Iowa State Coll. Press, 1940; Fox, I., and H. E. EWING, "Fleas of North America," *U.S.D.A. Misc. Publ.* 500; 128 pp., 1943; HUBBARD, C. A., "Fleas of Western North America," 533 pp., Iowa State Coll. Press, 1947.

The larvae that hatch from the eggs (Fig. 5.15,L) are very slender, cylindrical, whitish maggots with a distinct head, antennae, and mouth parts, but no ocelli and no legs.   There are 12 well-marked body segments in addition to the head.   There are usually minute paired spiracles on the thoracic and abdominal segments, and each segment bears a transverse band of very long, stiff hairs.   The larvae, which are very active, live on such dead animal and vegetable matter as they find in the cracks of floors or in the dirt about the sleeping quarters of their hosts.   When full-grown they spin a cocoon of silk, covered with particles of dirt, inside of which the pupa is formed.   The pupa may be recognized by its compressed body, absence of wings, inconspicuous eyes, antennae which are shorter than the head, and mandibles elongated for piercing.   The appendages of the pupa are free from the body wall (Fig. 5.16,J).

Small, laterally compressed, jumping insects, entirely wingless, usually spiny, and in the adult stage living as external parasites on warm-blooded animals.   Compound eyes small or absent; antennae short, and concealed in grooves; no neck.   Mouth parts piercing-sucking, with two pairs of palps.   Thoracic segments distinct.   Coxae very large, those of each pair nearly contiguous; hind legs fitted for jumping; tarsal segments five; claws strong.   Cerci minute.   Metamorphosis complete.   Larvae scavengers; slender, cylindrical, without legs or eyes, but with well-developed head, chewing mouth parts, and spiracles on the principal body segments.   Larvae scarcely ⅕ inch long, with transverse rows of very prominent setae.   Pupae without wings, enclosed in a cocoon.

The only families of much importance are:

The common fleas, Family Pulicidae.
The sticktights or chigoes, Family Echidnophagidae.

The fleas of the first family are important, not only as very universal pests of cats, dogs, and hogs, often infesting houses where pets are kept and seriously annoying the persons that live there, by biting them; but also especially as the known carriers of the dreaded human diseases, plague and endemic typhus (page 1013).   The second family includes the sticktight flea, that attaches to the head of chickens, the females remaining fixed in this position sucking the blood; and the true chigoe or jigger of the tropics,[1] the females of which, after mating, burrow into the flesh of man, especially about the feet, and, as the eggs develop, increase enormously in size, causing bad sores.

[1] Not to be confused with the chigger mite (p. 1022).

CHAPTER 7

INSECT CONTROL

To a large extent the value of entomology is based upon insect control. To a still greater extent the support given to this branch of science by the public is in direct proportion to the efficiency of measures for insect control which have been developed by entomologists. While the lessening of insect damage or the control of insect outbreaks is not the only aim of insect study, it is the most important. However, the study of insects is also of great value in increasing our fund of basic knowledge of general biology and in aiding the understanding of the natural laws governing the development and abundance of plants and animals.

Insect control in its broadest sense includes everything that makes life difficult for insects: that kills them or prevents their increase and makes it laborious for them to spread about the world. The control of insects can be accomplished in many ways, as outlined in this chapter, and may broadly be subdivided into (a) *applied-control measures*, which depend upon man for their application or success and can be influenced by him to a considerable degree, and (b) *natural-control measures*, which do not depend upon man for their continuance or success and cannot be influenced greatly by him.

APPLIED CONTROL

Applied control includes those methods, under the control of man, which it is necessary to use when harmful insects have not been held in check by natural factors. Under this heading we have (a) chemical control by the use of insecticides, repellents, attractants, and auxiliary substances; (b) physical and mechanical control by specially designed machines or other devices, and the special manipulation of physical factors of the environment; (c) cultural control by variations in the usual farm operations; (d) biological control by the introduction and establishment of insect enemies; and (e) legal control, by regulating commerce, farming, and other human activities that affect the prevalence and distribution of dangerously destructive insects, the success of insect-control operations, or the health of man.

Applied control of insects is, as a rule, expensive, and the amount which one can reasonably expect to save by the control applied must be weighed against the expense involved. When insects are present in numbers sufficient to cause a heavy loss of property, there is usually a strong desire on the part of the property owner to stop this loss. The cost of killing or checking the insects causing the trouble must be carefully considered, and the most economical means of efficient control employed.

Practical applied methods of control have not been worked out for many of the insects attacking field and forest crops and certain classes of those attacking livestock and man. Applied-control measures have been developed for most of the orchard, truck-crop, greenhouse, stored-grain, shade-tree, and household insects. In general, crops or articles of high value can be protected by applied-control measures, while those of low value per acre will not justify the expense of this type of control.

CHEMICAL CONTROL: INSECTICIDES

Insecticides are those substances which kill insects *by their chemical action*. Insecticides may be grouped into three general classes: (*a*) stomach poisons, (*b*) contact poisons, and (*c*) fumigants. Fumigants or poison gases are generally the most effective insecticides to use when the insects and the products they are damaging are in a tight enclosure such as a house, storeroom, or greenhouse. Sometimes fumigants are used to destroy insects in their burrows in the soil or in wood, and sometimes portable enclosures are placed over plants out-of-doors to fumigate them. Generally when plants, animals, or products in the open are to be treated, sprays or dusts are applied. These are of two fundamentally different kinds known as *stomach poisons* and *contact poisons*.

Another way of classifying insecticides is by their chemical nature and source or supply, as (*a*) inorganic compounds, (*b*) synthetic organic compounds, and (*c*) organic compounds of plant origin. In general the inorganic insecticides are effective only as stomach poisons, and the plant insecticides are largely employed as contact poisons. Most of the synthetic organic insecticides may act as contact and stomach poisons and in certain cases as fumigants. The contact action predominates in most cases, and these materials will be discussed as contact insecticides.

STOMACH POISONS

Insecticides of this class are generally applied against chewing insects but may also be used for insects with sponging, siphoning, lapping, or sucking mouth parts under certain conditions. There are four principal ways of using stomach poisons:

1. The natural food of the insect, *e.g.*, the foliage of plants or the feathers of birds, is covered with the poison so thoroughly that the insect cannot feed without getting some of the poison.

2. The poison is mixed with a substance (an *attractant*, page 394) that is very attractive or tasty to the insect—if possible, more attractive than its normal food—and the poison-bait mixture is placed where the insects can easily find it.

3. Certain poisons may be sprinkled over the runways of insects, so that they get it upon their feet or antennae. In cleaning their appendages with their mouth parts, some of the poison will be swallowed, especially if it is a substance irritating to the feet or antennae.

4. Systemic insecticides, which are readily absorbed and distributed throughout living organisms, may be used to poison the tissues of plants and animals, so that insects feeding thereon are killed. By this means sucking insects may be controlled with stomach poisons.

A satisfactory stomach poison must be sufficiently active to kill quickly, must be inexpensive, and must be available in large quantities. It must not be distasteful to the insects against which it is used, so as to repel them. It must be a sufficiently stable chemical so that it will not undergo chemical changes destroying its toxicity or making it harmful to plants, during shipment and storage or when it is mixed with other chemicals as a spray or dust, or simply from being exposed to the moisture and gases of the atmosphere. If the poison is to be applied to living plants, it must also meet the following requirements: It must not kill or burn the plant to which it is applied. In general this means that stomach poisons applied to foliage *must be insoluble in water,* for soluble substances will generally be taken into leaf or root in solution and so poison the plant. The margin of safety between a dose of poison necessary to kill the insect and the slightly larger dose that will burn or damage the plant is, with many insecticides, very slight. It must spread uniformly and adhere well to the plant surfaces to which it is applied. Since most stomach poisons are not soluble in water, they should be of such a physical nature that they will remain in suspension well in the spray liquid. Great fineness of particles in substances is important, because such substances will cover the plant better, either as a spray or as a dust, and because in general the finer the particles the greater their effectiveness against the insect. Finally, the ideal insecticide should not leave any residue dangerous to the health of man or animals on the food parts of the plants treated; very few stomach poisons meet this requirement.

Some idea of the relative toxicity to insects of the various stomach poisons may be gained from the figures for median lethal dosage determinations (LD_{50}) presented in each section.[1] It should be clearly understood, however, that there are astonishing variations in the toxicity of the same poison to different insects and that only a test of each poison against each pest under consideration can be depended upon to determine its usefulness.

The Arsenicals. The various compounds of the element arsenic, collectively called arsenicals, comprise important and widely used stomach poisons for insects. Certain of these arsenicals meet nearly all the requirements for an ideal stomach poison, listed above, except that they leave residues upon the treated plants that may be dangerous to man and animals.

The element arsenic is apparently not poisonous, but many of its compounds are very toxic. It occurs uncombined in the earth in scattered particles but is more commonly found in combination with sulfur or with metals such as iron and copper. The arsenic used in insecticides is secured largely from the flue dust and fumes from the smelting of various metal ores. The element forms two oxides: As_2O_3, known as "white arsenic," arsenious oxide, or arsenic trioxide; and As_2O_5, known as arsenic oxide, or arsenic pentoxide. These oxides are the anhydrides of arsenious acids and arsenic acids respectively. Three series of arsenites exist, derived from acids which apparently cannot be isolated in the free state: the orthoarsenites, as Na_3AsO_3; the pyroarsenites, as $Na_4As_2O_5$; and the metarsenites, as $NaAsO_2$. The corresponding arsenic acids, orthoarsenic acid, H_3AsO_4; pyroarsenic acid, $H_4As_2O_7$; and metarsenic acid, $HAsO_3$, readily form metallic salts. In most cases the arsenites are less stable and more toxic to both insects and plants, while the arsenates, although less violently poisonous to insects, are more stable and safer to use on plants. Consequently, the arsenates, such as lead arsenate and calcium arsenate, are chiefly used for spraying upon plants but are not toxic enough to be effective in most poison baits. The arsenites, such as paris green and sodium arsenites, however, are used in

---

[1] The median lethal dosage of a poison is the amount required to kill 50 per cent of a test population.

poison baits but cannot be sprayed upon foliage for they "burn" it severely.

The two points of greatest significance about the arsenicals are (a) the percentage of total arsenic in the particular insecticide and (b) the proportion of the arsenic which is soluble in water.

Generally speaking, the killing power of an arsenical is in direct ratio to the percentage of metallic arsenic it contains, although the metal cation with which the arsenic is combined may have toxicity of its own, as in the case of lead or copper arsenate and others. The danger of "burning" or injury to plants is generally in direct ratio to the percentage of its arsenic that is present *in water-soluble form,* since such water-soluble arsenic can enter the living parts of the plant foliage and poison them. The ideal arsenical would be one having a very high arsenic content, none of which should be soluble in water but all of it readily soluble in the digestive juices of the insect.

Lead Arsenate. Lead arsenate was first used as an insecticide in Massachusetts, in 1892, and was developed by the Federal Bureau of Entomology in connection with its work of controlling the gypsy moth, in response to the need of a stomach poison that could be applied at greater strength than paris green without burning. It may be used to destroy or prevent the attacks of most kinds of chewing insects which leave definite holes in the leaves except (a) on very tender plants; (b) for a few insects that are repelled by it or are very hard to kill; and (c) upon parts of plants to be used as human food or feed for animals. It is also used to destroy soil-infesting insects by working it into the topsoil at the rate of 1 or more pounds per 100 square feet.

Of the numerous forms of lead arsenate known to chemists, the acid orthoarsenate, $PbHAsO_4$, and the basic orthoarsenate, $Pb_4(PbOH)$- $(AsO_4)_3$, are the ones chiefly used as insecticides. Most commercial products are mixtures of the two, in which the acid form predominates. The acid lead arsenate contains about 20 per cent metallic arsenic equivalent, much less than $\frac{1}{4}$ per cent of which is usually soluble in water.

The basic lead arsenate is not generally on the market in all areas but may be obtained for special purposes. It has been used for the control of chewing insects upon tobacco and Persian walnut and in regions of very high humidity where the acid lead arsenate may break down and cause burning. It is more stable than the acid form and on this account will not burn even very tender foliage. Because it contains only about two-thirds as much arsenic (about 14 per cent metallic arsenic equivalent) and is not so easily acted upon by the digestive fluids of insects, it requires a larger dosage and a longer time to kill insects.

Compared with the other arsenicals, lead arsenate is a relatively weak poison; but it is as nearly insoluble in water and as stable as any arsenical spray material. This high degree of safety to foliage is its greatest merit. Once the spray has dried upon foliage it adheres, like paint, for a long time. As manufactured it is very finely divided, light, and fluffy, commonly having a volume of 50 to 100 cubic inches per pound (from 350 to 400 grams per 1,000 cubic centimeters). It consequently handles nicely as a dust and also remains in suspension in spray liquids fairly well. Spreaders or deflocculators, such as calcium caseinate, soybean flour, sulfated alcohols, or sulfonated compounds, are frequently added to improve wetting, spreading, and adhesive properties; or the lead arsenate may be made to adsorb gelatin or gums to form colloidal lead arsenate. The alkalies in some soaps and very hard alkaline waters may decompose the acid lead arsenate, forming water-soluble arsenic, but basic lead arsenate is not affected in this way. Both forms of lead arsenate are of doubtful compatibility with the dinitrophenols and tetraethyl pyrophosphate.

Lead arsenate may be applied either as a dust or as a spray. As a dust it is usually diluted with from 2 to 20 parts of some inert carrier such as dusting sulfur, talc, hydrated lime, or gypsum. As a spray, lead arsenate powder is mixed with water or other spray solutions such as bordeaux and is kept from settling out by agitation while it is applied. A standard dosage is 2 to 3 pounds to 100 gallons, but for some insects it is necessary to use it twice that strong. It is very extensively used to protect deciduous fruits, garden and truck crops, ornamental plants, and forest and shade trees from injury by chewing insects. The great advantages of lead arsenate are its high degree of safety to plants, its stability in storage and after it has been applied, and its spreading and adhesive qualities. It may be still further safened for use on delicate foliage by the addition of equal quantities of hydrated lime or one-quarter to one-half as much zinc sulfate, which reduce the amount of water-soluble arsenic formed in solution. Lead arsenate sprays will spot or discolor white painted surfaces.

The LD_{50} values for lead arsenate administered orally to rabbits, rats, and chickens are 125, 825, and 450 milligrams per kilogram, respectively. The LD_{50} values in micrograms per gram of body weight, for oral administration to several species of insects, are *Bombyx mori*, 90; *Melanoplus differentialis*, 2,000 to 4,000; *Leptinotarsa decemlineata*, 140 to 240; and *Protoparce sexta*, 85.

Calcium Arsenate. Although calcium arsenates were used as insecticides at least as early as 1907, the earlier compounds contained or produced so much water-soluble arsenic that they burned foliage badly. The high cost of lead arsenate during the First World War led to renewed efforts to produce a satisfactory calcium arsenate, and by 1915 the federal Bureau of Entomology demonstrated that, when properly manufactured, this insecticide may be satisfactory as a spray for many purposes. The discovery about 1920 that undiluted calcium arsenate applied as a dust to cotton will control the cotton boll weevil, for which other arsenicals had been useless, greatly stimulated interest in the perfection of the product and led to the use of as much as 40,000 tons of it a year. It is also used to some extent on potatoes, tomatoes, and some shade trees. It is considerably cheaper per lethal unit than lead arsenate.

The commercial calcium arsenates are a mixture of tricalcium arsenate, $Ca_3(AsO_4)_2$, and acid calcium arsenate, $CaHAsO_4$ (in which the former predominates), with a considerable excess of lime and calcium carbonate. Basic calcium arsenate, $[Ca_3(AsO_4)_2]_3 \cdot Ca(OH)_2$, is apparently the principal constituent of the "safe" calcium arsenates. It contains the equivalent of 0.4 to 0.5 per cent water-soluble arsenic oxide. The calcium arsenates are very fluffy, white powders, having a volume of 75 to 100 cubic inches per pound (about 400 grams per 1,000 cubic centimeters) and are indistinguishable in appearance from lead arsenate. The commercial products are distinctly colored to prevent them from being mistaken for flour or baking powder. These insecticides contain from 25 to 30 per cent metallic arsenic equivalent and are therefore about one-third more toxic than lead arsenate. When freshly manufactured, they have as small a percentage of water-soluble arsenic as lead arsenate, but the action of carbon dioxide and ammonia in the atmosphere and dilute alkaline solutions decomposes the material, liberating water-soluble arsenic. Buyers should insist upon material manufactured the year in which it is to be used. The suspension is a little better, adhesiveness slightly less, and dusting qualities better than that of lead arsenate.

Calcium arsenates are incompatible with soaps, nicotine sulfate in sprays but not in dusts, flotation sulfur, fluosilicates, cryolite, benzene hexachloride, and tetraethyl pyrophosphate; and are of doubtful compatability with rotenone, pyrethrum, oils, dinitrophenols, chlordane, toxaphene, parathion, and dithiocarbamates.

Calcium arsenate may be applied in the same ways as lead arsenate (page 318). A standard dosage for a spray is $1\frac{1}{2}$ to 3 pounds to 100 gallons of water or other spray liquid. An equal amount of freshly slaked lime or a third more of hydrated lime should be added to the water, unless it contains an excess of lime as in the case of bordeaux.

Calcium arsenate is distinctly more poisonous to insects than lead arsenate but also more toxic to plants, especially when it is not fresh. Only the more resistant plants, such as cotton, potato, and apple, will stand it, and it must not be applied to stone fruits or many vegetables and flowers. It is likely to cause very severe burning when used upon plants attacked by diseases. Its advantages are its moderately high arsenical content, its physical fineness, its cheapness, and the fact that it is compatible with lime-sulfur. Its disadvantages are its instability, the danger of burning foliage, its toxicity to higher animals, and its incompatibility with nicotine sulfate in sprays.

The following oral LD_{50} values in micrograms per gram of body weight have been obtained with calcium arsenate: *Pieris rapae*, 740; *Leptinotarsa decemlineata*, 70 to 140; *Trichoplusia ni*, 500.

White Arsenic, or Arsenic Trioxide, or Arsenious Oxide, As_2O_3. This compound is a very active poison and is one of the cheapest arsenicals. It contains about 75 per cent arsenic. Because it is soluble in water to the extent of about 1.2 grams per 100 cubic centimeters at 20°C., it is so likely to burn foliage that it is not safe to use in spraying plants, but it is used in poison baits for grasshoppers, armyworms, ants, and cockroaches. The oral lethal dosage range in milligrams per kilogram is 60 to 150 for the chicken and 10 to 30 for the rabbit.

Sodium Arsenite, $NaAsO_2$ and Na_2HAsO_3. This is a very highly soluble form of arsenical used mainly in poison baits and stock dips and generally sold in liquid form. It is extremely toxic to plants and is extensively used as a weed killer. The metallic arsenic equivalent is 44 to 57 per cent.

Basic Copper Arsenate, $Cu(CuOH)AsO_4$. This compound was developed as a safe and stable arsenical for foliage applications. It has a metallic arsenic equivalent of about 26 per cent, a copper content equivalent to about 56 per cent copper oxide, and is about 0.1 per cent water-soluble. The material does not hydrolyze in water and is only slightly affected by exposure to carbon dioxide. It appears to be about as toxic as lead arsenate to a variety of chewing insects.

Paris Green. This compound, copper acetoarsenite, $Cu(C_2H_3O_2)_2 \cdot 3Cu(AsO_2)_2$, was the first important stomach poison and was developed in west central United States about 1865 for the control of the Colorado potato beetle. Paris green contains about 33 to 39 per cent metallic arsenic equivalent, 2 to 3 per cent of which is water-soluble, and is manufactured as a brilliant-green powder composed of minute spheres of various sizes which suspend poorly in water and do not dust well. The material is expensive because of its high copper content and burns foliage very easily because of the relatively large amount of water-soluble arsenic, but is more toxic to insects than the other arsenicals. It has limited use as a mosquito larvicide.

Miscellaneous Arsenicals. The *magnesium arsenates*, $Mg_3(AsO_4)_2$ and $MgHAsO_4$, containing 42 to 46 per cent metallic arsenic equivalent, have been used for the control of the Mexican bean beetle. *Zinc arsenate*, $Zn_3(AsO_4)_2$, metallic arsenic equivalent 32 per cent, has been used to control the codling moth with the purpose of eliminating the lead residues from lead arsenate. The *manganese arsenates*, $Mn_3(AsO_4)_2$ and $MnHAsO_4$, containing 34 to 39 per cent metallic arsenic equivalent, and *zinc arsenite*, $Zn(AsO_2)_2$, 54 per cent arsenic equivalent, have had limited insecticidal usage. However, these compounds frequently burn foliage. *Sodium arsenate*, Na_3AsO_4 and Na_2HAsO_4, is a highly soluble form of arsenic, containing about 36 to 40 per cent metallic arsenic equivalent, which is used in poison baits. *Cupric metarsenite*, $Cu(AsO_2)_2$, has been used on a limited scale as a mosquito larvicide.

Fluorine Compounds.[1] Compounds of this highly active element were developed as insecticides largely in an effort to provide substitute stomach poisons for the arsenicals which would not leave highly poisonous residues on edible food crops. Our supply of fluorine is derived from the widely distributed minerals, fluorspar, CaF_2, and cryolite, Na_3AlF_6. The insecticides are derivatives of hydrofluoric acid, HF, especially sodium fluoride; of fluosilicic acid, H_2SiF_6, especially sodium and barium fluosilicates; and of fluoaluminic acid, H_3AlF_6, especially sodium fluoaluminate, commonly known as cryolite.

The principal difficulty with this group of stomach poisons has been to manufacture them in a sufficiently fine, light, and flocculent form to cover foliage as well as the arsenicals do. The fluorine compounds in general kill more quickly than the arsenicals; they are often cheaper; they are less toxic to the higher animals; in certain tests they have proved to be safer to use on plants; they act both as stomach poisons and to a limited extent as contact poisons and are irritating to the appendages of insects, causing them to ingest the poison even if they do not feed upon the treated foliage; and in some cases they have a valuable repellent effect upon the pests.

Sodium Fluoride, NaF. This poison has come into general use since 1915 as a means of combating chewing lice on animals and poultry and for the control of household pests, particularly cockroaches. While it is primarily a stomach poison, it also acts to a slight extent as a contact poison when the powder or solution is taken into the tracheae or directly through minute pores in membranous areas of the body wall, as at the junction of head and thorax or the attachment of the legs to the thorax. It is believed, however, that sodium fluoride kills cockroaches, ants, and possibly chewing lice largely from ingestion during the cleaning of appendages with the mouth parts; consequently, it is necessary that the powder be kept dry and finely pulverized in order to be effective. This insecticide has also been highly recommended as a poison for baits for grasshoppers, cutworms, and earwigs. The commercial form of sodium fluoride is used undiluted as a powder or mixed with other dusts, and in solution. It should not be sprayed on the foliage of plants. The pure material is a

[1] *Univ. Tenn. Agr. Exp. Sta. Bul. 182, 1942.*

white powder, soluble in water to about 4 grams per 100 cubic centimeters at 18°C. and containing 45 per cent fluorine. The insecticidal product is usually colored green to prevent its mistaken use as flour. Sodium fluoride is poisonous to man and other warm-blooded animals, and is somewhat irritating to the skin and respiratory passages. The minimum lethal dosage to the rabbit is about 200 milligrams per kilogram when fed orally, and the following oral LD_{50} values in micrograms per gram of body weight have been recorded for insects: *Bombyx mori*, 110 to 150; *Melanoplus differentialis*, 110; and *Leptinotarsa decemlineata*, 130 to 220.

Sodium Fluosilicate, Na_2SiF_6. This material is a white, granular, odorless powder, containing 60.6 per cent fluorine. It is soluble in water to about 0.65 gram per 100 milliliters at 17.5°C., and a saturated solution has a pH of 3.6. Sodium fluosilicate hydrolyzes in water to form sodium fluoride and hydrogen fluoride and reacts with alkalies to produce sodium fluoride. Therefore it should not be used on plants, as considerable damage sometimes results. Sodium fluosilicate has been widely used as a toxicant for poison baits for grasshoppers, crickets, cutworms, and weevils, and as a saturated water solution for mothproofing woolen goods. It is not as effective against cockroaches as sodium fluoride. The oral minimum lethal dosage to the rabbit is about 125 milligrams per kilogram, while the following oral LD_{50} values in micrograms per gram of body weight have been recorded for insects: *Bombyx mori*, 100 to 130; *Melanoplus femur-rubrum*, 160; *Leptinotarsa decemlineata*, 110; and *Protoparce sexta*, 230 to 420.

Barium Fluosilicate, $BaSiF_6$. This compound, containing 40.8 per cent fluorine, is the least soluble of the inorganic fluorine insecticides, 0.026 gram dissolving in 100 milliliters of water at 17°C. Like sodium fluosilicate, barium fluosilicate hydrolyzes in water to give an acid reaction, the pH of a saturated solution being 3.4. Barium fluosilicate has been used as a spray, at 3 to 8 pounds per 100 gallons of water, or as a dust (diluted 1:3 with tobacco dust, cheap flour, clay, or talc), using 5 to 15 pounds per acre for the control of such insects as blister beetles and flea beetles. Slight foliage injury has sometimes resulted from its application to plants, and it has been reported to be injuriously corrosive to parts of spray machinery. The compound is incompatible with nicotine, calcium arsenate, lime, lime-sulfur, bordeaux mixture, and alkaline soaps. It is best used with common diluents such as talcs, clays, and sulfur. The commercial material contains about 72 per cent barium fluosilicate, 8 per cent cryolite, and 20 per cent inert materials. The oral minimum lethal dosage to the rabbit is about 175 milligrams per kilogram, and the oral LD_{50} to *Bombyx mori* larvae is 90 to 120 micrograms per gram of body weight.

Sodium Fluoaluminate, or Cryolite. This substance, which occurs in Greenland as a mineral, Na_3AlF_6 or $AlF_3 \cdot 3NaF$, and is also manufactured as a synthetic form, was first used as an insecticide by Marcovitch in 1929. It contains 54.2 per cent fluorine. The mineral form consists of 90 to 98 per cent sodium fluoaluminate but when ground, forms a heavy powder which is not entirely suited for insecticidal use. It has a water-solubility of 0.035 gram per 100 milliliters as compared with the solubility of the synthetic form, 0.061 gram per 100 milliliters. The synthetic cryolite is lighter and fluffier than the natural mineral and contains about 86 per cent sodium fluoaluminate; the remainder is principally aluminum hydroxide. Cryolite is soluble in dilute acids and alkalies and reacts with lime to form soluble fluorides which may cause foliage injury.   The pH of a saturated aqueous solution is about 6.2.   Cryolite is incompatible with lime, lime-sulfur, bordeaux mixture, nicotine, paris green, and calcium arsenate and of doubtful compatibility with dormant oils, dinitrophenols, tetraethyl pyrophosphate, parathion, and fixed copper fungicides.   It is best used with inert diluents such as clays and talc and with sulfur.   Cryolite is used as a spray at 3 to 8 pounds per 100 gallons and as a 20 to 80 per cent dust diluted with carriers such as clays and talcs, for the control of many chewing insects such as codling moth, walnut husk fly, Mexican bean beetle, orange tortrix, flea beetles, and many others.   It has the advantage of being of low toxicity to mammals, the oral minimum lethal dosage to the rat being greater than 13.5 grams per kilogram.   The oral LD_{50} values to certain insects in micrograms per gram of body weight are *Bombyx mori*, 60; *Pieris rapae*, 680; *Hyphantria cunea*, 450; *Heliothis armigera*, 70; *Leptinotarsa decemlineata*, 100 to 200; and *Protoparce sexta*, 110 to 360.

Organic Stomach Poisons.   A number of the substances whose most important action is as contact insecticides also are effective as stomach poisons when they are eaten by insects.   These materials include the plant poisons rotenone, nicotine, the veratrine alkaloids of sabadilla and hellebore, the ryania alkaloids, and most synthetic organic insecticides. These materials will be fully discussed under Contact Poisons.

Minor Stomach Poisons.   Of the many substances that have been tested as stomach poisons for insects, the following have proved their usefulness in certain limited fields.   A form of *phosphorus*, known as white or yellow phosphorus, incorporated in a sweet sirup, forms a valuable bait for cockroaches, mice, and rats.   The phosphorus must be handled with great caution, not only because of its toxicity, but also because it ignites spontaneously at temperatures above 95°F.   The compound *thallium sulfate*, Tl_2SO_4, is combined with a sweet or fatty carrier to form a valuable poison for ants.   *Borax*, or sodium tetraborate, $Na_2B_4O_7$, has been used to kill house fly maggots in manure or other refuse; to prevent the breeding of mosquitoes in water to be used only for laundering purposes; and as an emulsion with oil, or as glyceroboric acid, to treat maggot-infested wounds of animals and repel the flies.   *Boric acid*, H_3BO_3, has been used as a stomach poison for cockroaches.   *Form-aldehyde* (formalin, 40 per cent aqueous solution) is valuable as a house fly poison in baits and to destroy the potato scab gnat in seed potatoes. Metallic *mercury*, incorporated in a heavy oil such as vaseline, makes an effective ointment for lice on poultry or on man.   *Mercuric chloride*, or "corrosive sublimate," $HgCl_2$, and *mercurous chloride*, or "calomel," Hg_2Cl_2, are specific remedies for certain root-infesting insects, such as the cabbage maggot (page 670), the onion maggot (page 660), and the larvae of flea beetles and fungus gnats.   Mercuric chloride may also be used in book bindings and ant tapes as a repellent to ants, cockroaches, and termites.   *Potassium antimonyl tartrate*, or tartar emetic, OSbOOC-$(CHOH)_2COOK$, a water-soluble salt, is used as an ant poison and in combination with sugar as an attractant, as a stomach poison for various species of thrips and fruit flies.   *Cuprous cyanide*, CuCN, and *zinc phosphide*, Zn_3P_2, have been used at various times as stomach poisons for mosquito larvae and agricultural pests.

Systemic Poisons.   Systemic insecticides absorbed and generally distributed through plant or animal tissues act as stomach poisons in the control of sucking insects and as contact poisons against leaf miners or cattle grubs. These materials are largely discussed under Contact Poisons because of their chemical similarity. However, *sodium selenate*, Na_2SeO_4, is a water-soluble inorganic salt which has been successfully applied to greenhouse soil for the control of red spider mites and aphids infesting greenhouse plants.[1] This material is a dangerous cumulative poison to warm-blooded animals and should never be used in soil which might be used to grow foods.

Poison Baits. In certain cases where spraying is not practicable, as for certain household insects, insects living inside fruits or vegetables or underground, and chewing insects attacking areas of crops too great to be protected by spraying, stomach poisons may be mixed with materials known to be attractive to the particular pest, and such *poisoned bait* exposed where the insects may get it. Poison baits have been extensively used for grasshoppers and armyworms, especially since about 1914. The success with these pests has stimulated efforts to control many other pests by this method. Success with a poison bait will depend upon many considerations such as attractivity, palatability, toxicity, stability, physical condition, and time, place, and method of exposure. Baits with a dry carrier have been used extensively for grasshoppers, crickets, cockroaches, silverfish, earwigs, cutworms, armyworms, white grubs, and wireworms. Baits in a liquid carrier have been widely employed for ants, house flies, fruit flies, root-maggot flies, and adult oriental fruit moth, greenhouse leaf tier, and corn earworm.

Poison Baits for Grasshoppers, Cutworms, Armyworms, Crickets, and Earwigs. For these insects the most common carrier is wheat bran, moistened with enough water to make it thoroughly wet but not enough to drip without being squeezed. Sometimes hardwood sawdust, fresh horse droppings, ground corncobs, or the hulls of seeds are used as the carrier. "Blackstrap" molasses is usually added, and sometimes other attractants, such as chopped whole oranges or other fruits, oils, amyl acetate, or salt, are recommended. Instead of all bran, up to 50 per cent hardwood sawdust, middlings, cheap flour, whey, or corncobs, ground in a hammer mill using a $\frac{1}{4}$-inch screen, may be used. The bran should be coarse-flaked and free from shorts which cause lumpiness. For poison one may choose white arsenic, paris green, sodium pyroarsenate, dry sodium arsenite, sodium fluosilicate or barium fluosilicate, liquid sodium arsenite, or benzene hexachloride, chlordane, DDT, toxaphene, aldrin, dieldrin, parathion, or Dipterex. If a dry poison is used, it should be mixed thoroughly with the bran, then the water and other ingredients or the oil added and mixed until every flake of bran is moistened. Liquid sodium arsenite may be used by mixing with the water and the poisoned water added to the bran, but liquid poison has not proved to be satisfactory with the oil baits. The poisoned bran is then scattered broadcast over fields to be protected and adjacent wasteland at the rate of 8 to 10 pounds per acre. It may be applied by hand from pails or bags or by the use of an oat seeder, limestone seeder, or special poison-bait spreaders, and it has been distributed over vast areas by airplane. Every effort should be made to scatter the poison thinly so that no lumps or piles of the bait are left on the plants or the ground. If this is done, there is no danger of poisoning livestock or wild birds. For grasshoppers (page 468), the bait should be distributed in the morning so as to be fresh when

[1] NEISWANDER, C., and V. MORRIS, *Jour. Econ. Ento.*, **33**:517, 1940.

the hoppers begin feeding. For cutworms, armyworms, and earwigs the bait is best distributed in the evening, since these insects are nocturnal. See also the baits suggested for cockroaches (page 909) and for silverfish (page 905).

Poison Baits for Ants. These have been extensively used for the Argentine ant and to a lesser extent for other species. For species that prefer fatty foods, the carrier must be a fat, such as ground meat or grease, while for the sweet-eating ants a sirup, sugar, or honey is used. For the latter species, at least, it has been found desirable to use sirup very weakly poisoned with pure sodium arsenite, in order that the worker ants, which alone forage in houses and orchards, may have time after eating the poison bait to carry it back to the nest and feed some of it to the queens and the larvae. In this way the entire colony can be killed. Several formulas found effective for ants are given on page 897. These sirups are put in soda-fountain straws cut into short pieces or in aluminum or tin boxes or cans or paperoid cups, having a piece of sponge or blotting paper in the bottom as a footing for the insects and the sides or covers having holes punched in them or fitting on so as to leave passages through which the ants can enter the boxes. These containers are placed along the runways of the ants out of the reach of children and kept filled so that a part of the sponge is above the sirup. From 75 to 150 such cans are used in each city block, or 5 to 10 around each dwelling, when fighting the Argentine ant, and entire municipalities have been freed of this pest by this method. Fall and early spring are the best times to use this control.

Poison Baits for Flies and Moths. Baits and bait sprays have been successfully used to control the various fruit flies of the family Trypetidae, especially the Mediterranean, Mexican, and oriental fruit flies (page 812). A widely used bait of sugar and molasses employed sodium fluosilicate, lead arsenate, copper carbonate, or tartar emetic as the toxicant. However, this type of bait has been largely replaced by fermenting protein baits, using yeast hydrolysate or corn protein with malathion as the toxicant, and spraying these on vegetation in the infested areas (page 814). The males of the oriental fruit fly have been virtually annihilated in restricted areas by the use of 10-inch fiberboard squares treated with the specific attractant methyl eugenol together with a quick-acting toxicant such as Isolan or DDVP.

The house fly and other domestic flies can be successfully controlled in many situations by bait sprays of sugar and/or molasses with toxicants such as malathion, Diazinon, Dipterex, DDVP, and Dibrom and by baits of dry sugar containing 1 to 2 per cent of these toxicants.

Liquid baits containing fermenting molasses or sugar or with a variety of specific attractants (page 394) have been used with water-soluble toxicants such as sodium arsenite, sodium fluosilicate, Isolan, or Dipterex to control moths such as the oriental fruit moth, tomato hornworm, and corn earworm. These baits are usually exposed in cans or pails suspended among the crops to be protected.

Poison Baits for Subterranean Crop Pests. Sometimes poison baits are buried in the soil to destroy wireworms, white grubs, and similar root-feeding insects. Fresh vegetables, green foliage, germinating seeds, or balls of dough made from scorched flour have been used as carriers for the poison. The soil must be free of growing or rotting vegetation, and the baits must be planted about every 3 feet.

Contact Poisons

In order to kill an insect with stomach poisons, the insect must swallow the poison. Insects with piercing mouth parts take their food from beneath the surface and consequently get none of the poisons applied to the surface of foliage or fruits. Consequently, for the piercing-sucking plant pests we must use a contact poison.

Insecticides of this class kill insects by contacting and entering their bodies either directly through the integument into the blood or by penetrating the respiratory system through the spiracles into the tracheae. These materials may be applied directly to the insect body in a spray or dust or as a residue on plant surfaces, animals, habitations, or other places frequented by insects. There are many types of contact poisons with varying properties, and their manner of use depends upon their stability upon exposure to light, moisture, and air; their toxicity to plants and animals; and their appearance, taste, and odor. Contact insecticides may be classified as (a) plant poisons such as nicotine, anabasine, rotenone, pyrethrum, sabadilla, and ryania; (b) synthetic organic compounds such as DDT, benzene hexachloride, toxaphene, chlordane, organic thiocyanates, dinitrophenols, and organic phosphates; (c) oils and soaps; and (d) inorganic compounds such as sulfur, lime-sulfur, and to a limited extent sodium fluoride and arsenic trioxide.

The action of these materials is very imperfectly understood. It has been shown that the insect cuticle possesses remarkably efficient absorptive properties for most contact insecticides, so that the lethal dosage applied externally is almost equivalent to that when the poison is injected into the body cavity. It is largely this factor of selective absorption rather than any specific toxicity that distinguishes the action of most contact insecticides on insects from their action on higher animals. The penetration of contact insecticides through the insect cuticle is apparently largely controlled by the properties of the lipoids comprising the epicuticle, and this may be a factor in explaining the resistance of certain insects to various insecticides. Residual films of contact insecticides may kill insects solely by action upon the sensory organs present in the insect tarsi. The specific action of contact insecticides will be discussed under the sections devoted to each material.

Many of the contact insecticides are also toxic to plants, and the margin between the concentration required to kill the insect and the amount which is phytotoxic is often small. For this reason, one must use caution in applying these materials. For some of the most resistant insects it is best to apply the spray in winter or before the plants have started active growth, because they will stand stronger applications at that time. Such sprays are called *dormant sprays*, to distinguish them from the *summer sprays* and *dusts* which are applied while the trees and other plants are in foliage. The principal dormant sprays are oil emulsions, lime-sulfur, and the dinitrophenols.

Nicotine Alkaloids.[1] Tobacco was one of the first materials to be used as an insecticide. As early as 1763, a French paper recommended the use of finely powdered tobacco mixed with water, with the addition of some lime, to destroy plant lice without injury to the foliage. It was not

[1] See *U.S.D.A., Bur. Ento. Plant Quar., Cirs.* E-384, E-392, 1936; E-597, 1943; E-646, 1945; E-709, 1946; E-720, 1947; and E-725, 1947; *Chem. Rev.*, **29**:123–199, 1941.

until 1809, however, that the presence of a volatile poisonous substance in tobacco was discovered, and in 1828 this was recognized as an alkaloid and named nicotine. It was synthesized by Pictet and Rotschy in 1904. Tobacco belongs to the same family of plants, Solanaceae, as the potato and tomato. Nicotine occurs in at least 15 species of Nicotiana and in *Duboisia hopwoodii* and *Aesclepias syriaca;* however, the commercially important species are *Nicotiana tabacum* and *N. rustica.* Although 12 alkaloids have been isolated from tobacco, including *l*-nornicotine, anabasine, nicotimine, and nicotelline, nicotine is present in by far the largest portion, comprising approximately 97 per cent of the total alkaloids. From the insecticidal standpoint, only two others, anabasine and nornicotine, are of importance (page 328). Nicotine commonly occurs in *N. tabacum* plants to the extent of 2 to 5 per cent in the leaves, while stalks, pods, and roots have greatly decreasing amounts and the seeds only a trace. *N. rustica* has a nicotine content ranging from 5 to 14 per cent and reaching 20 per cent in selected varieties.

The alkaloid nicotine is *l*-1-methyl-2-(3′-pyridyl)-pyrrolidine, which when freshly distilled is a colorless, nearly odorless liquid, b.p. 247°C., sp. gr. 1.00925 at 20°C., and v.p. 0.0425 mm. Hg at 25°C.

Nicotine

Nicotine can be purified by vacuum distillation and readily steam-distilled. Upon exposure to air it darkens, changing from red to brown to black, becomes more viscous, and develops a disagreeable odor. Nicotine is soluble in alcohol, ether, or petroleum ether and is miscible in all proportions with water below 60°C. and above 210°C. Because of its basic nature ($Kb_1 = 1 \times 10^{-6}$, $Kb_2 = 1 \times 10^{-11}$), nicotine readily forms salts with acids and dibasic salts with many metals and acids, nicotine sulfate, $(C_{10}H_{14}N_2)_2 \cdot H_2SO_4$, being widely used as an insecticide because it is much less toxic to warm-blooded animals than free nicotine and is less volatile and consequently more stable.

Since its early discovery the principal use of nicotine has been for various piercing-sucking insects which cannot be destroyed by stomach poisons. It has been used in a variety of ways, which may be summarized as follows:

1. As a liquid extract: (*a*) standardized 40 per cent *free nicotine* alkaloid, sp. gr. 1.013, which is volatile and is used chiefly for fumigant action, (*b*) standardized 40 per cent *nicotine sulfate*, sp. gr. 1.188, which remains as a toxic residue for a short time after application. To bring about a more rapid liberation of nicotine from nicotine sulfate, soap, hydrated lime, lime-sulfur, or ammonium hydroxide should be added. Diluted nicotine sprays should ordinarily contain 0.05 to 0.06 per cent nicotine.

2. As a dust of concentrated nicotine extract mixed with some light, very finely divided carrier or diluent. Nicotine sulfate should be used in making dust except in extremely dry atmospheres, when free nicotine

should be employed. The type of carrier has an important effect upon the dust, and these have been classified as (*a*) adsorbent carriers such as bentonite, talc, kaolin, and other colloidal materials, which prevent the volatilization of the nicotine; (*b*) inert carriers such as gypsum, clays, sulfur, slate dust, etc., which have little or no effect upon volatilization; and (*c*) active carriers such as hydrated lime and carbonate of lime, which convert nicotine sulfate into free nicotine in the presence of moisture.

3. As a fumigant (page 380): (*a*) by vaporizing the concentrated extracts with heat, and (*b*) by burning papers, punks, or powders soaked in the extracts.

4. As fixed or stabilized nicotine salts or compounds such as nicotine-oil combinations, nicotine tannate, nicotine silicotungstate, nicotine cuprocyanide, nicotine Reineckate, and nicotine bentonite. Nicotine bentonite has been used as a substitute for lead arsenate and other stomach poisons, and residues of this material may retain their toxicity on plant surfaces for 1 to 2 weeks. Nicotine bentonite may be prepared as a tank-mix spray according to the following formula:

Nicotine sulfate (40%)............	1 pt.
Bentonite (Wyoming).............	5 lb.
Soybean oil.....................	1 qt.
Water, to make.................	100 gal.

Nicotine is incompatible with calcium cyanide dust, barium fluosilicate, and cryolite, and of doubtful compatibility with chlordane, benzene hexachloride, dinitrophenols, and tetraethyl pyrophosphate.

Nicotine is a violent poison to warm-blooded animals either orally or through skin absorption. The oral LD_{50} values in milligrams per kilogram are 30 for the rat and 23 for the guinea pig. As a toxicant to insects, nicotine penetrates directly through the integument and through the spiracles. Nicotine acts directly on the ganglia of the insect central nervous system, producing excitation at low concentrations and paralysis at high concentrations, probably due to a direct action at the synapses. The following LD_{50} values have been recorded in micrograms per gram of body weight: by contact, *Aphis fabae*, 48, *Periplaneta americana*, 500; orally, *Bombyx mori*, 10, *Protoparce quinquemaculata*, >4,650, *Pieris rapae*, >3,640, *Leptinotarsa decemlineata* adult, >570.

Anabasine.[1] This alkaloid is closely related to nicotine and occurs in a small woody perennial, *Anabasis aphylla*, Family Chenopodiaceae, which grows wild in Central Asia, Iran, Armenia, Turkey, and North Africa. The alkaloid content varies from a fraction of 1 per cent in the old twigs to more than 2 per cent in the young shoots. Anabasine is also found in the tree tobacco, *Nicotiana glauca*, which grows wild in the southwestern United States. It is commonly present to about 1 per cent, but experimental hybrids have contained up to 8 per cent. Anabasine is *l*-2-(3'-pyridyl)-piperidine and when freshly distilled is a water-white, viscous liquid, b.p. 280.9°C., sp. gr. 1.0481 at 20°C., and v.p. 2.5 mm. Hg at 79°C. The compound is soluble in water in all proportions and darkens upon exposure to air. Like nicotine it is strongly basic and forms salts with a variety of acids and metals. Anabasine is extracted from the

[1] *U.S.D.A., Bur. Ento. Plant Quar., Cirs.* E-537, 1941, and E-636, 1945.

$$H_2C \quad \text{(ring structures)} \quad H_2C$$

Anabasine   Nornicotine

plant by water or dilute acid and is commercially available as crude anabasine sulfate, containing approximately 40 per cent total alkaloids, of which 70 per cent is anabasine, 18 per cent lupinine, and the remainder aphylline, aphyllidine, and extraneous plant material.   Anabasine is the only alkaloid present which is of insecticidal importance.   The use of anabasine sulfate has been largely confined to Russia, although a small amount has been sold in the United States.   It is principally used for the control of aphids and seems to be more effective for this purpose than nicotine.   Anabasine is used as a spray at 1 part in 2,000 parts of water with soft soap, or as a 5 per cent dust with an alkaline carrier such as slaked lime, or with chalk or talc.   It is less volatile than nicotine and is thus less effective as a fumigant but more persistent as a contact poison. It is slightly more toxic than nicotine to mammals but acts in a similar manner, as a nerve poison, to both insects and mammals.   The subcutaneous LD_{50} to the guinea pig is about 22 milligrams per kilogram.

Nornicotine.[1]   This   alkaloid,   2-(3'-pyridyl)-pyrrolidine,   is   closely related to nicotine and anabasine.   It has a boiling point of 270°C., sp. gr. 1.07 at 20°C., and in the pure state is a colorless, viscous liquid which appears to be more stable than nicotine and does not darken readily when exposed to light and air.   *L*-nornicotine comprises 95 per cent of the alkaloid content (1 per cent) of *Nicotiana sylvestris*, and *d*-and *dl*-nornicotine are found in the Australian plant *Duboisia hopwoodii*.   Nornicotine is also present in certain strains of tobacco and may occur in commercial nicotine sulfate preparations to as high as 12 per cent.   It appears to be a somewhat better contact poison but a poorer fumigant than nicotine.

Pyrethroids.[2]   The use of pyrethrum as an insecticide originated in the Transcaucasus region of Asia about 1800.   For many years its nature was kept a secret by these Asiatic countries, and the product was sold very extensively, as a flea and louse powder, at exorbitant prices.   Following the publication of the nature of the product about 1850, its use became world-wide, and it was the most important export of Dalmatia and adjacent countries until the time of the First World War, when Japan became the leading producer.   The industry began to flourish in Kenya about 1932, and by 1940 that country was the principal supplier of pyrethrum to the United States.   Brazil and the Congo have more recently begun the export of pyrethrum.   The researches of Staudinger and Ruzicka in 1924 resulted in the partial identification of the active chemical constituents of pyrethrum, but new discoveries as to the exact compounds concerned were made as late as 1947, due principally to the

[1] *U.S.D.A., Bur. Ento. Plant Quar., Cir.* E-561, 1942.
[2] GNADINGER, C. B., "Pyrethrum Flowers," McLaughlin, Gormley, King & Co., 2d ed., 1936, and Suppl. to 2d ed., 1945.

work of La Forge and Barthel. The active toxicants are four esters, pyrethrins I and II and cinerins I and II, formed from the alcohols, pyrethrolone and cinerolone, and chrysanthemum monocarboxylic acid and chrysanthemum dicarboxylic acid-monomethyl ester. The cinerins are more stable than the pyrethrins, and pyrethrin I and cinerin I seem to be somewhat more toxic than pyrethrin II and cinerin II. The structural formulas for these constituents are given below:

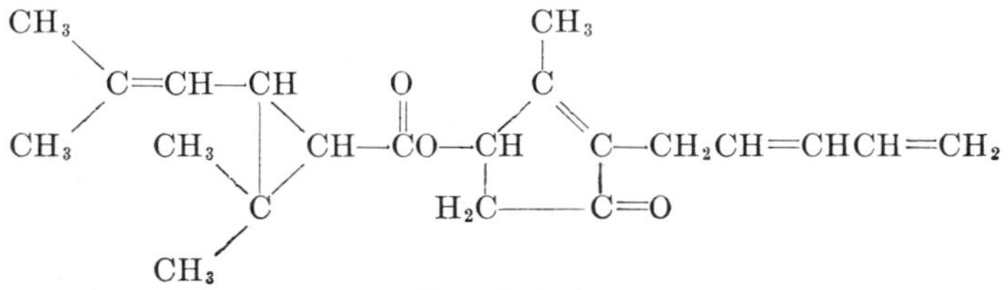

$$CH_3$$
$$\diagdown$$
$$C=CH—CH \quad O \quad C$$
$$\diagup \qquad \diagdown \quad \| \quad \diagup \diagdown$$
$$CH_3 \quad CH_3 \,|\quad CH—CO—CH \qquad C—CH_2CH=CHCH=CH_2$$
$$\diagdown\diagup$$
$$C \qquad\qquad H_2C \quad\text{——}\quad C=O$$
$$\diagup$$
$$CH_3$$

Pyrethrin I

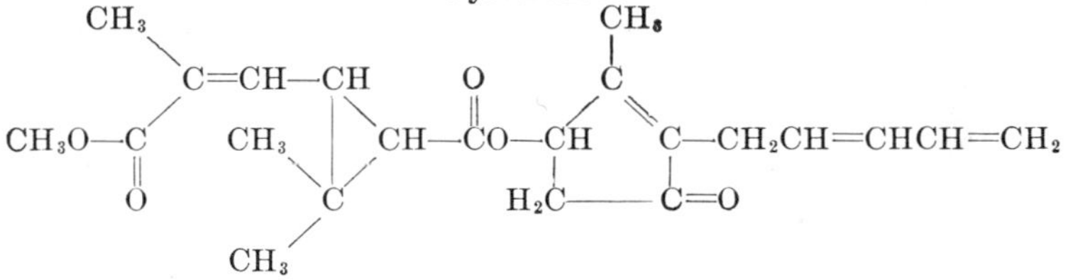

$$CH_3$$
$$\diagdown$$
$$C=CH—CH \quad O \quad C$$
$$CH_3O—C \quad CH_3 \,|\quad CH—CO—CH \qquad C—CH_2CH=CHCH=CH_2$$
$$\|$$
$$O \qquad\qquad C \qquad H_2C \quad\text{——}\quad C=O$$
$$\diagup$$
$$CH_3$$

Pyrethrin II

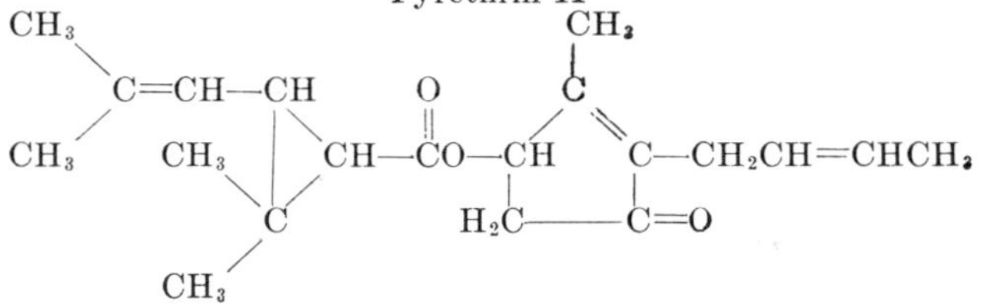

$$CH_3$$
$$\diagdown$$
$$C=CH—CH \quad O \quad C$$
$$\diagup \qquad\qquad \| \quad \diagup\diagdown$$
$$CH_3 \quad CH_3 \,|\quad CH—CO—CH \qquad C—CH_2CH=CHCH_3$$
$$\diagdown\diagup$$
$$C \qquad\qquad H_2C \quad\text{——}\quad C=O$$
$$\diagup$$
$$CH_3$$

Cinerin I

$$CH_3$$
$$\diagdown$$
$$C=CH—CH \quad O \quad C$$
$$CH_3O—C \quad CH_3 \,|\quad CH—CO—CH \qquad C—CH_2CH=CHCH_3$$
$$\|$$
$$O \qquad\qquad C \qquad H_2C \quad\text{——}\quad C=O$$
$$\diagup$$
$$CH_3$$

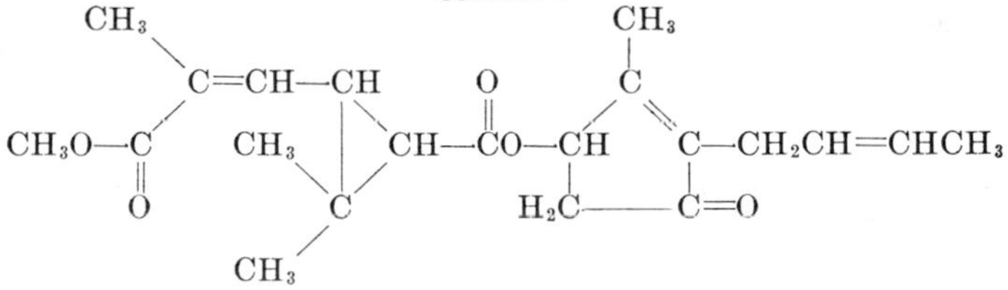

Cinerin II

In addition, esters of pyrethrolone and cinerolone with palmitic and linoleic acids also occur in oleoresin of pyrethrum.

The pyrethrins[1] are highly unstable in the presence of light, moisture, and air. Whole flowers decompose more slowly than ground flowers or dusts, and the potency of the material can best be preserved in sealed, light-proof containers at low temperatures. Various antioxidants such as hydroquinone, pyrogallol, and pyrocatechol have been shown greatly to retard the destruction of the pyrethrins in storage but have not proved of value in preserving insecticidal residues.

[1] To avoid confusion with older terminology, the term pyrethrins will henceforth be used to indicate all the toxic constituents of the pyrethrum flowers.

Pyrethrum occurs only in plants belonging to the genus *Chrysanthemum* (= *Pyrethrum*), Family Compositae. The two species which possess a sufficiently high-toxic content to be suitable for the manufacture of insecticides are *C. cinerariaefolium* and *C. coccineum* (= *roseum*), although the former is at present the only species of commercial importance. Kenya flowers contain an average of 1.3 per cent pyrethrins (maximum of 3 per cent in selected strains), while Japanese flowers average about 1.0 per cent and Dalmatian flowers 0.7 per cent. The mature fully opened flower head contains the majority of the active constituents, the stem having only about one-tenth as much. The achenes of the flower head contain about 90 per cent of the pyrethrins in the flower. Pyrethrum concentrates are prepared by extracting the ground flowers with solvents such as petroleum ether, ethylene dichloride, glacial acetic acid, methyl alcohol, or acetone. These solvents also extract considerable quantities of waxes and coloring matter which may be removed by precipitation, adsorption on charcoal, or by cooling and filtration. By extraction of such a concentrate using nitromethane, followed by adsorption on activated carbon, concentrates containing 90 to 100 per cent pyrethrins may be prepared.

The pyrethrum products are commonly applied as oil- or water-based sprays containing from 0.03 to 0.1 per cent pyrethrins, usually with 5 to 10 times this amount of synergist or activator (page 332). Dusts may be made by mixing pyrethrins extracts with appropriate clay or talc carriers and should be called *pyrethrum dusts* to distinguish them from the powdered flowers or *pyrethrum powders*. The dusts deteriorate rapidly and should be prepared shortly before use.

There are four principal fields of usefulness for the pyrethrum products:

1. As household insecticides. The pyrethrins, because of their rapid paralytic or knockdown activity and their very low toxicity to mammals, are the most widely used ingredients in thousands of commercial insecticides used to control flies, mosquitoes, bed bugs, ants, clothes moths, silverfish, and cockroaches. For this purpose, the extracts are incorporated in deodorized, volatile petroleum distillate that will evaporate without leaving a stain upon furniture, draperies, clothing, or wallpaper. For use in aerosol bombs, a 20 per cent pyrethrins concentrate which is nearly wax-free has been developed. This is used to give a final concentration of 0.04 to 0.25 per cent pyrethrins, usually with 5 to 10 times this amount of synergist, in the aerosol formulation.

2. As livestock or cattle sprays. The pyrethrins as oil- or water-based sprays are particularly effective in protecting animals from the annoyance caused by biting and nonbiting flies and have considerable repellent effect, especially when appropriate synergists or extenders are used.

3. As sprays for mills, warehouses, and for grain in storage. The pyrethrins with appropriate synergists are especially suited because of their low toxicity for the control of stored-product insects. Oil- or water-based sprays are used as residual or space applications and for impregnating sacks and packages containing foodstuffs. A dust containing 0.05 per cent pyrethrins with 0.8 per cent piperonyl butoxide synergist can be mixed directly with stored grain.

4. As dusts and sprays for vegetables and fruits. The very low mammalian toxicity of the pyrethrins makes them useful for applications to truck crops to control aphids and chewing insects shortly before harvest.

Pyrethrum extracts are incompatible with calcium arsenate, lime-sulfur, and hydrated lime and are of doubtful compatibility with dinitro-phenols and tetraethyl pyrophosphate. As already indicated, the principal advantages of these insecticides are their high toxicity to insects and their very slight toxicity to both plants and mammals. They are, however, expensive and deteriorate very rapidly after being applied to plants and otherwise exposed to air, and it is difficult to keep the stored concentrates for more than one season.

The pyrethrins affect insects by a very rapid paralytic action, which makes them especially valuable in household formulations because of their rapid knockdown. The nervous system of the insect is affected and violent convulsions occur before death, which may follow after several days' paralysis. Characteristic vacuolization and degeneration of the central nervous system is found in insects overcome by pyrethrins poisoning. Recovery from sublethal dosages is very common. The pyrethrins are readily absorbed through the insect cuticle or through the spiracles. The following contact LD_{50} values in micrograms per gram of body weight have been recorded for several insects: *Cimex lectularius*, 5; *Pediculus humanus*, 42; *Musca domestica*, 31 to 38; *Aedes aegypti*, 0.5 to 1.0. The pyrethrins are practically nontoxic to mammals by contact or ingestion, but when introduced intravenously they possess a pronounced toxic effect. Persons handling pyrethrum are occasionally subject to dermatitis caused by a volatile oil present in the flowers.

Synthetic analogues of the naturally occurring pyrethroids have become useful insecticides.[1] *Allethrin*, dl-2-allyl-3-methyl-cyclopent-2-en-4-ol-1-onyl dl-*cis-trans*-chrysanthemate was synthesized by La Forge and associates in 1949 and differs from cinerin I in the replacement of the 2-butenyl side chain by an allyl group and in the lack of optical activity. The commercial product is a clear brownish viscous liquid, sp. gr. 1.005 to 1.015, containing 75 to 95 per cent of a mixture of eight optical and geometric allethrin isomers. Allethrin has similar knockdown and toxic properties to the natural pyrethrins and has partially supplanted them, especially in sprays and aerosols for household pests. However, it is generally less effective against agricultural pests and is less effectively activated or synergized by the common activators (page 332). It has an oral LD_{50} to the rat of 920 milligrams per kilogram.

Allethrin

Cyclethrin, dl-2-(2-cyclopentenyl)-3-methyl-2-cyclopenten-4-ol-1-onyl dl-*cis-trans*-chrysanthemate, is a brownish viscous oil, sp. gr. 1.020, with properties very similar to those of allethrin. The technical product consists of 95 per cent of a mixture of eight isomers and is soluble in petroleum solvents and Freon. It has an oral LD_{50} to the rat of about 1,400 milligrams per kilogram. Cyclethrin has about the same insecti-

[1] *Agr. Chem.*, **4**(6):57, 89, 1949; *Jour. Econ. Ento.*, **42**:532–536, 1949; *U.S.D.A.*, *Bur. Ento. Plant Quar.*, *Cir.* E-846, 1952.

cidal activity as allethrin but is synergized more strongly by the available pyrethrins synergists.

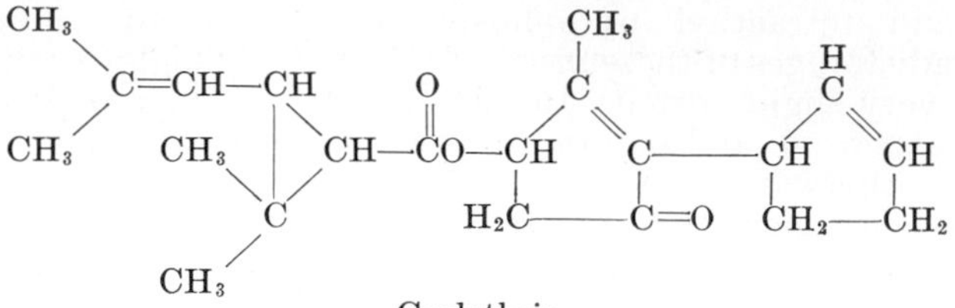

Cyclethrin

A more radical departure from the pyrethrins-type molecule is provided by *barthrin*, 6-chloropiperonyl *dl-cis-trans*-chrysanthemate, which has pronounced insecticidal properties.

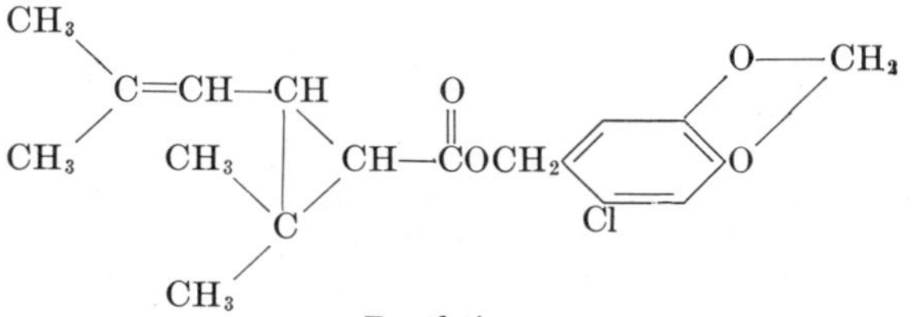

Barthrin

Synergists or Activators. A large number of chemicals are known which, although generally nontoxic or only slightly toxic in themselves, when used in conjunction with the pyrethroids, rotenoids, nicotine, and certain organophosphorus and carbamate insecticides, markedly increase the toxicity of the mixture over the sum of the toxicities of the components. This phenomenon is commonly termed *synergism* or *activation* and has had its most practical application with the pyrethroids, where the use of formulations containing appropriate synergists has resulted in substantial savings in the amounts of the expensive pyrethroids necessary for insect control.

Substances which have had practical application as synergists include N-isobutyl undecyleneamide, $CH_3(CH_2)_7CH{=}CH\overset{\overset{\displaystyle O}{\displaystyle \|}}{C}NHCH_2CH_2(CH_3)_2$; *sesamin*, 2,6-bis-(3,4-methylenedioxyphenyl)-3,7-dioxabicyclo-[3.3.0]-octane, present to about 0.25 per cent in the sesame oil from *Sesamum indicum*, together with *sesamolin*, which differs in structure from sesamin in that one of the 3,4-methylenedioxyphenyl groups is attached to the central cyclic structure through an oxygen atom and is about four times as active as a synergist.

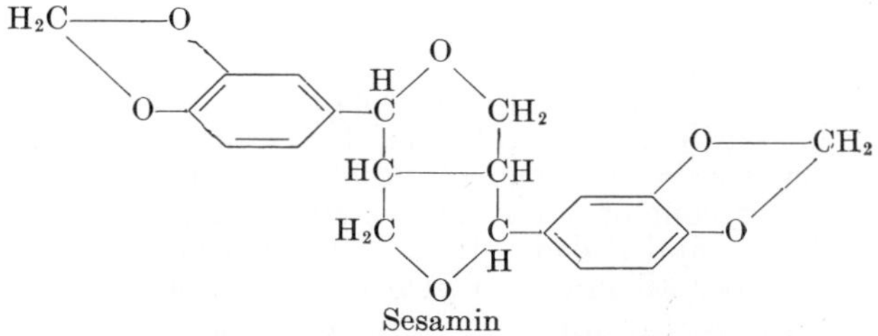

Sesamin

Piperonyl butoxide, 3,4-methylenedioxy-6-propylbenzyl butyldiethylene glycol ether, is a yellowish liquid soluble in petroleum hydrocarbons and dichlorodifluoromethane, sp. gr. 1.06, b.p. 180°C. at 1 millimeter, which has been widely used as a synergist in aerosol bombs.

$$CH_2CH_2CH_3$$

$$O \overbrace{\qquad} CH_2OCH_2CH_2OCH_2CH_2OC_4H_9$$

$$H_2C \overline{\qquad} O$$

Piperonyl butoxide

The related compound, *piperonyl cyclonene*, is a mixture of 3-isoamyl-5-(3,4-methylenedioxyphenyl)-2-cyclohexenone and its 6-carbethoxy derivative and has been used to synergize agricultural sprays.

$$H_2C\!\!-\!\!C\!\!=\!\!O$$

$$O \overbrace{\qquad} CH \qquad CH$$

$$H_2C \overline{\qquad} O \qquad H_2C\!\!-\!\!C\!\!-\!\!CH_2CH_2CH(CH_3)_2$$

Piperonyl cyclonene

Propyl isome is di-*n*-propyl-2-methyl-6,7-methylenedioxy-1,2,3,4-tetrahydronaphthalene-3,4-dicarboxylate, sp. gr. 1.14. It is soluble in aromatic solvents and has an oral LD_{50} to the rat of 15,000 milligrams per kilogram. *Sulfoxide* is 1,2-methylenedioxy-4-[2-(octylsulfinyl)-propyl]-benzene, sp. gr. 1.07.

$$H_2$$
$$H$$
$$CH_3$$
$$H_2C \overbrace{\qquad} COOC_3H_7$$
$$H$$
$$HCOOC_3H_7$$
Propyl isome

$$H_2C \overbrace{\qquad} \overset{CH_3}{\underset{O}{CH_2CHSC_8H_{17}}}$$
Sulfoxide

Sesoxane is 2-(3,4-methylenedioxyphenoxy)-3,6,9-trioxaundecane, b.p. 137 to 141°C. at 0.08 mm. Hg, oral LD_{50} to the rat 2,000 milligrams per kilogram.

$$H_2C \overbrace{\qquad} \overset{OCHOCH_2CH_2OCH_2CH_2OCH_2CH_3}{\underset{CH_3}{}}$$

Sesoxane

A synergist of entirely different chemical type is *N*-(2-ethylhexyl)-bicyclo-[2.2.1]-5-heptene-2,3-dicarboximide or MGK 264, b.p. 158°C. at 2 millimeters, sp. gr. 1.05.

$$H$$
$$C \qquad O$$
$$HC \overset{H}{C\!\!-\!\!C}$$
$$HCH \qquad N\!\!-\!\!CH_2CH(C_2H_5)C_4H_9$$
$$HC \qquad C\!\!-\!\!C$$
$$C \qquad O$$
$$H$$

MGK 264

Activating effects with these materials have been demonstrated for sprays and aerosols, dusts, and residual films. In general, best results have been obtained using 5 to 20 parts of synergist to 1 part of toxicant.

The activation obtained with such mixtures has resulted in equivalent effectiveness being obtained by as little as one-third to one-fifth as much pyrethrins in fly sprays to one one-hundredth as much in the MYL louse powder.   In addition, the synergists apparently stabilize the pyrethrins to the action of light and air so that longer residual action is obtained. Synergism has not been adequately explained from a physiological viewpoint, but may result from such factors as better penetration of the insect cuticle and nervous system and protection of the pyrethroids from enzymatic detoxication.   The combinations of pyrethroids and activators are relatively harmless to mammals and may be applied safely to edible produce within a short time of harvest.

Rotenoids.[1]   The earliest recorded insecticidal use of rotenone was against leaf-eating caterpillars in 1848.   However, plants containing these materials have been used as fish poisons by tropical natives for many centuries.   It was not until 1902 that the active principle, rotenone, was isolated.   The use of rotenone-bearing roots as insecticides in the United States was developed as a result of federal laws against residues of lead, arsenic, and fluorine upon edible produce.   Rotenone is harmless to plants, highly toxic to many insects, and relatively innocuous to mammals. Sixty-eight species of plants, including 21 species of *Tephrosia*, 12 of *Derris*, 12 of *Lonchocarpus*, 10 of *Millettia*, and several of *Mundulea*, have been reported to contain rotenone or rotenoids.   The principal economic species are the various kinds of derris, *Derris elliptica* and *D. malaccensis* from Malaya and the East Indies, and cubé or timbo, *Lonchocarpus utilis* and *L. urucu* from South America.   *Tephrosia* (= *Cracca*) *virginiana* has been studied as a possible native source of rotenone.

A number of toxic constituents have been isolated from the roots and seeds of these plants, the most important of which is *rotenone*, m.p. 163°C. (a dimorphic form exists, m.p. 181°C.).   Rotenone has the following chemical structure:

The other naturally occurring rotenoids are *elliptone*, m.p. 159°C., which

has a furan ring, in place of ring *E* of rotenone; *sumatrol*,

[1] *U.S.D.A.*, *Misc. Publ.* 120, 1932; *U.S.D.A.*, *Bur. Ento. Plant Quar.*, *Circs.* E-402, 1937; E-367, 1936; E-453, 1938; E-514, 1940; E-563, 1942; E-571, 1942; E-579, 1942; E-581, 1942; E-593, 1943; E-594, 1943; E-603, 1943; E-625, 1944; E-652, 1945; E-654, 1945; E-655, 1945; E-656, 1945; E-706, 1946; and E-713, 1947; *Chem. Rev.*, **12**:181–213, 1933; and **30**:33–48, 1942; *Jour. Econ. Ento.*, **34**:684–691, 1941.

m.p. 188°C., which is 15-hydroxyrotenone; *malaccol*, m.p. 244°C., which is 15-hydroxyelliptone; *l-α-toxicarol*, m.p. 101°C., which has a hydroxy

group at carbon 15 and the ring,

$$\begin{array}{c} C-O \\ \diagdown\ \diagup\ \diagdown \\ C \qquad\qquad C(CH_3)_2 \\ \diagdown\ \diagup \\ C=C \\ H \quad H \end{array}$$

in place of ring E

of rotenone; and *deguelin*, m.p. 165 to 171°C., which has a hydrogen atom on carbon 15 in place of the hydroxy group of toxicarol. A related material, *tephrosin*, m.p. 197 to 198°C., which has a hydroxy group on one of the carbon atoms between rings B and C, does not appear to occur naturally in derris resins but is an oxidation product of deguelin. All the naturally occurring rotenoids appear to exist as levo forms. The role of these rotenoids in determining the toxicity of crude rotenone preparations is largely unexplored, but individually rotenone is from five to ten times as effective insecticidally as the other rotenoids. The content of these materials in the various commercial plant species and varieties is variable; roots of *D. elliptica* averaging from 5 to 9 per cent rotenone (maximum 13 per cent) with up to 31 per cent ether extractives, while *D. malaccensis* contains up to 4 per cent rotenone and up to 27 per cent extractives. *L. utilis* averages 8 to 11 per cent rotenone as a maximum and 25 per cent total extractives. Toxicarol and deguelin make up the bulk of the total extractives. When exposed to light and air, rotenone decomposes, changing from colorless through yellow to deep red and resulting in noninsecticidal products. This change occurs in about 10 days in insecticidal residues on plants. Rotenone preparations should therefore be protected from light during storage. Rotenone is readily oxidized in the presence of alkali to dehydrorotenone by elimination of two hydrogen atoms to form a double bond between rings B and C. This material is much less toxic than rotenone, which should therefore be considered incompatible with alkaline dusts such as lime or with soaps and other alkaline wetting and spreading agents. Pure crystalline rotenone is prepared by extracting powdered rotenone-containing roots with a solvent such as ether or carbon tetrachloride and concentrating the solution to produce crystallization. The pure rotenone has the following solubilities in grams per 100 cubic centimeters of solution at 20°C.: water about 0.00002; ethyl alcohol, 0.2; carbon tetrachloride, 0.6; amyl acetate, 1.6; xylene, 3.4; acetone, 6.6; benzene, 8.0; chlorobenzene, 13.5; ethylene dichloride, 33; and chloroform, 47.

Rotenone-containing insecticides are used in the following forms: (*a*) Dusts of ground roots are mixed with 3 to 7 parts of carrier such as talc, clay, gypsum, sulfur, tobacco, or walnut shells ground to pass a 250- to 350-mesh sieve. Impregnated dusts are also used, which are produced by mixing an extract of ground roots in a volatile solvent with an adsorptive carrier. The solvent is then evaporated, leaving each dust particle coated with insecticide. Such preparations are more uniform in particle size than are the ground roots. Dusts containing 0.5 to 1 per cent rotenone and 1.75 to 3.5 per cent total extractives are effective against most insects controlled by rotenone and should be used at 15 to 25 pounds per acre on such crops as cabbage and celery. *Alkaline carriers such as lime should not be used with rotenone.* (*b*) Dispersible powders may be made from finely ground derris roots. Two to five pounds of derris powder with 2 pounds of a neutral soap or the equivalent of a sulfonated oil will make 100 gallons of spray. For small amounts, an ounce of a 4 or 5 per cent rotenone dust and a teaspoonful of spreader should be used in 2 gallons of water. (c) Extracts of derris are widely used and are available as dried resins containing 25 to 35 per cent rotenone and the remainder ether extractives or as solutions in refined petroleum oils. Because of the limited solubility of rotenone in spray oils (about 0.05 per cent), mutual solvents are generally employed to increase the solubility to practical limits. Materials which have been employed for this purpose are dibutyl phthalate, methylated naphthalenes, alkylated phenols, and high-boiling ethers. The concentrated solutions may be either used as fly and cattle sprays or emulsified in water as agricultural sprays. Rotenone concentrates containing 1 per cent rotenone and 3.5 to 4 per cent total extractives may be diluted 1 part to 600 or 800 parts of water, for aphids; the rotenone content in the diluted spray being about 0.00125 per cent.

Crystalline rotenone is also available commercially and is used for mothproofing or wherever staining from insecticides is to be avoided. *Dihydrorotenone*, m.p. 216°C., in which the double bond in the isopropenyl side chain of rotenone is saturated, is about as toxic as rotenone to many insects and is more resistant to decomposition by sunlight.

Many uses have been found for these insecticides. In addition to their effectiveness for both piercing-sucking insects, such as aphids and redbugs, and chewing insects, especially caterpillars upon plants, they make excellent dusts for external parasites of animals, such as fleas and lice. They have become the most widely used remedy for ox warbles. They are, however, ineffective for many kinds of pests such as mealy bugs, squash bugs, scale insects, and many beetles and caterpillars. They act very slowly, often requiring several days to kill. The toxic principles all deteriorate rapidly when exposed to sunlight and air; sprays and dusts usually lose their effectiveness within a week after application. They are more effective when they can be used during periods of cloudy weather, and the dusts should be applied to foliage moist with dew. The outstanding advantages of this group of poisons are that they are harmless to plants, relatively nontoxic to man, and act as both contact and stomach poisons to insects. Rotenone is incompatible with lime and other alkaline materials and with paris green and of doubtful compatibility with calcium arsenate, dinitrophenols, and tetraethyl pyrophosphate.

Rotenone may enter the insect body through the alimentary canal, tracheae, or integument. It appears to kill insects by specific inactivation of the respiratory enzyme, glutamic acid oxidase, resulting in death through oxygen starvation. The following oral LD_{50} values in micrograms per gram of body weight have been recorded for insects: *Bombyx mori* larvae, 3; *Vanessa cardui* larvae, 30; *Heliothis armigera* larvae, >490; *Melanoplus femur-rubrum*, 4,700 to 7,000. The oral LD_{50} values in milligrams per kilogram are white rat, 25 to 75; guinea pig, 50 to 200; and dog, 140 to 200.

Sabadilla and Hellebore.[1] The veratrine alkaloids of the roots of white and green "false hellebores," *Veratrum album* (Europe) and *V. viride* (America), have been used as insecticides for more than 100 years but are

[1] *Jour. Econ. Ento.*, **37**:400–408, 1944; and **38**:293–296, 1945; *Jour. Biol. Chem.*. **159**:517–524, 1945; and **160**:555–565, 1945; *Agr. Chem.*, **1**(4): 19–21, 1946.

at present of little importance because of high cost and lack of standardization of toxic content and because only freshly ground material is fully effective. Recently the seeds of a related plant, sabadilla, *Schoenocaulon officinale*, Family Liliaceae, of South and Central America, have attained commercial importance as an insecticide. The related species *S. Drummondii* and *S. texanum*, indigenous to the United States, also contain the toxic principles. Certain of these plants have been used as native louse powders for centuries. The insecticidal constituents of these two genera of plants are a complex group of alkaloids. *V. viride* (powdered drug) contains about 1.3 per cent total alkaloids, composed of jervine (17 per cent), pseudojervine (3.3 per cent), germine (1 per cent), and traces of rubijervine, protoveratridine, cevadine, veratridine, and veratrimine. Neither jervine nor pseudojervine is toxic to insects, but a crude fraction of the drug suspected of containing *germerine*, $C_{36}H_{57}O_{11}N$, m.p. 193 to 195°C., was highly toxic to *Periplaneta americana*. *V. album* is reported to contain jervine, pseudojervine, rubijervine, germerine, protoveratrine, and protoveratridine. Protoveratrine and germerine are the most pharmacologically active of these alkaloids to mammals.

The crude mixture of alkaloids from sabadilla is termed *veratrine*. It is present in the seeds in yields of 2 to 4 per cent, from which about 13 per cent crystalline cevadine, 10 per cent veratridine, and lesser amounts of cevadilline, sabatine, and sabadine have been isolated. The crude alkaloids are soluble in the following amounts, grams per 100 milliliters of solvent: cold water, 0.055; boiling water, 0.1; alcohol, 35; chloroform, 144; ether, 24; kerosene, 0.1 to 0.2. Sabadilla preparations can be markedly activated by heating to 75 to 80°C. for 4 hours, moistening with 10 per cent sodium carbonate solution, or milling with hydrated lime, which is most effective. The toxic principles are rapidly destroyed by the action of light. *Cevadine*, $C_{32}H_{49}O_9N$, m.p. 205°C., the angelic acid ester of cevine, and *veratridine*, $C_{36}H_{51}O_{11}N$, m.p. 160 to 180°C., the veratric acid ester of cevine, are apparently the alkaloids in sabadilla which are responsible for the insecticidal action, and they act as both contact and stomach poisons. These compounds are highly poisonous to mammals, and they may cause irritation of the eyes and respiratory tract and violent sneezing.

Cevine

Hellebore has been suggested for use as a tea at the rate of 1 to 2 ounces to 1 gallon of hot water or as a 10 to 20 per cent dust with flour, talc, or lime as a diluent. A special use for it is as a larvicide for house flies in manure. One-half pound in 10 gallons of water should be applied to 8 bushels of manure. Unlike borax-treated manure there is no limit to the amount of such hellebore-treated manure that may be applied to the soil.

Sabadilla is generally used as a 5 to 20 per cent dust of ground seeds with lime, pyrophyllite, or sulfur as a carrier or sprayed as a 50 per cent wettable powder. It has shown especial promise for the control of plant-feeding Hemiptera and as a stomach poison for thrips, used with sugar as an attractant. Its rapid deterioration upon exposure to light makes it safe to use on food crops shortly before harvest.

Quassia.[1] The woods of two trees of the family Simarubaceae, *Picrasma* (= *Picroena*) *excelsa* of Jamaica and *Quassia amara* of Surinam, have been used as insecticides since about 1850. The active constituents are intensely bitter, water-soluble compounds of unknown structures. Surinam quassia contains about 0.16 per cent of two isomers of the formula, $C_{22}H_{30}O_6$, *quassin* (m.p. 205 to 206°C.) and *neoquassin* (m.p. 225 to 226°C.), while Jamaican quassia contains about 0.1 per cent of *picrasmin* (m.p. 218°C.) having the same empirical formula. Decoctions of quassia chips in water with soft soap have been used to control aphids and saw-flies. Quassia is marketed in the form of chips and as extracts incorporated into proprietary insecticides.

Ryania.[2] The root and stem of the plant *Ryania speciosa*, family Flacourtiaceae, a native of South America, have been found to contain from 0.16 to 0.2 per cent of insecticidal components, the most important of which is the alkaloid *ryanodine*, $C_{25}H_{35}O_9N$, m.p. 219 to 220°C. This compound, which is of unknown structure, is effective both as a contact and as a stomach poison. Ryanodine is soluble in water, methyl alcohol, and most organic solvents, but not in petroleum oils. It is more stable to the action of air and light than pyrethrum or rotenone and has considerable residual action. Ryania insecticides are applied as 20 to 50 per cent dusts, as suspensions of the ground stems in water, at 3 to 6 pounds to 100 gallons, and as methanolic extracts which form suspensions in water. The material has shown considerable promise in the control of the European corn borer and other crop pests. It appears to be less toxic to mammals than rotenone. Ryania is compatible with other insecticides and fungicides.

Synthetic Organic Insecticides.[3] Recent trends in the development of new insecticides have been almost entirely toward synthetic organic materials. Perhaps the first use of such a chemical was in 1892, when the potassium salt of 4,6-dinitro-*o*-cresol was marketed in Germany as an insecticide. The first large-volume use of a synthetic organic insecticide began in 1932 with the advent of β-butoxy-β'-thiocyanodiethyl ether. However, it was the development of DDT during the Second World War which conclusively proved that a synthetic compound could prove superior to a variety of inorganic and natural products for many insecticidal uses. Since that time the United States chemical industry has expanded until, in 1959, more than $200,000,000 worth of insecticide chemicals were sold.

The development of new organic insecticides has been the result of the

---

[1] *U.S.D.A., Bur. Ento. Plant Quar., Cir.* E-483, 1939.

[2] "Ryania Insecticides," Merck and Co. Entomological Laboratory, May, 1947; *Jour. Amer. Chem. Soc.,* **70**:3086–3088, 1948.

[3] *Ind. Eng. Chem.,* **39**:467–473, 1947; *Chemical Ind.,* June, 1948, 2–9; FREAR, D. E. H., "A Catalogue of Insecticides and Fungicides," vol. I, Chronica Botanica, Waltham, Mass., 1947; *U.S.D.A., Bur. Ento. Plant Quar., Cirs.* E-634, 1945; E-729, 1947; E-730, 1947; E-733, 1947; and E-738, 1947; METCALF, R. L., "Organic Insecticides," Interscience, 1955.

cooperative work of entomologists, chemists, and pharmacologists, from both industry and governmental agencies. Often hundreds and even thousands of compounds were made and tested before the most satisfactory material was found. Then the compound was given exhaustive field tests over a period of several years before it could be approved as an effective insecticide for various pests, harmless to plants and safe to the user and the consumer of treated produce.

The synthetic organic insecticides described in detail below are one of the remarkable triumphs of modern chemical industry. The public health uses of DDT and its relatives are estimated to have saved no less than 5 million lives and to have prevented no less than 100 million illnesses, from 1942 to 1952,[1] and their employment in the world-wide eradication of malaria will save countless millions more. The agricultural applications of the synthetic insecticides have also resulted in dramatic gains. United States Department of Agriculture statistics show that since the advent of DDT and other organics in 1945, the average yield of potatoes has increased about 66 per cent; onions, 46 per cent; sweet corn, 25 per cent; tomatoes, 100 per cent; Lima beans, 29 per cent; cauliflower, 23 per cent; celery, 56 per cent; and milk production, 10 per cent per cow. Many other examples could be cited. These increased yields are attributable almost entirely to the prevention of insect injury, since yields of wheat and oats to which insecticides are not commonly applied were almost unchanged over the same period.

Dinitrophenols.[2] The earliest recorded use of the dinitrophenols as insecticides was in 1892, when a mixture containing potassium dinitro-*o*-cresylate was marketed in Germany. Subsequent work by Tattersfield, Kagy, and Boyce resulted in the modern development of these materials. Two dinitro compounds have received considerable attention as insecticides. *DNOC* is 4,6-*dinitro-o-cresol*, m.p. 85°C., a yellow odorless solid, soluble in water to 0.014 per cent at 15°C. The compound readily forms salts with organic and inorganic bases, and the ammonium, sodium, potassium, calcium, and barium salts are water-soluble. *DNOCHP* is 4,6-*dinitro-o-cyclohexylphenol*, m.p. 106°C., a yellowish-white, odorless, crystalline solid which, like DNOC, forms salts with bases. The water-solubility of DNOCHP varies from about 1.8 milligrams per liter at pH of 1, to 15 milligrams per liter at pH of 6.5. At 30°C. its solubility in grams per 100 grams of solvent is acetone, 40; benzene, 109; carbon tetrachloride, 23; ethyl acetate, 45; ethylene dichloride, 64; xylene, 73; kerosene, 2; light, medium, and heavy spray oils, 2 to 3; dormant oil, 16. *DNOSBP* or 4,6-*dinitro-o-sec-butylphenol*, m.p. 42°C., has been marketed for use as an insecticide and fungicide, used as the triethanolamine salt. These compounds have the following formulas:

The phase distribution of DNOCHP in oil-water emulsions varies from more than 90 per cent of the toxicant in the oil phase at pH of 3.5 to 5.0

[1] KNIPLING, E., *Jour. Econ. Ento.*, **46**:1, 1953.
[2] *Jour. Econ. Ento.*, **32**:432, 1939; **34**:660, 1941.

to less than 5 per cent at pH of 8 to 11.   This phase distribution greatly affects both the ovicidal action and phytotoxicity of the emulsion, optimum conditions resulting from strongly acid mixtures.   Therefore the addition of oxalic acid at 3 to 5 ounces to 100 gallons has been recommended to decrease the amount of phenol in the aqueous phase.   Because of the free phenolic groups of the dinitrophenols, which behave as pseudo acids, they are often highly toxic to plants, and the sodium salt of DNOC is used as a herbicide.   The triethanolamine and dicyclohexylamine salts have proved much less toxic to plants and are marketed for foliage applications.   The dicyclohexylamine salt of DNOCHP is widely used for this purpose, 1.7 parts being equivalent to 1 part of the free phenol.   Because of this ready salt formation, dusts of the dinitrophenols are best prepared with nonionic acid diluents such as walnut-shell and redwood-bark flours. DNOC is formulated as a 20 to 33 per cent water paste of the sodium salt and as a 40 per cent wettable powder for application as a dormant ovicide to control aphids, European red mite, and various scale insects on fruit trees, as a toxicant for grasshopper baits, dusts, and sprays, and as a 2 to 10 per cent dust for chinch bug control, where it is used as a toxic barrier. DNOCHP is formulated as a 20 per cent wettable powder or a 1.5 per cent dust of the dicyclohexylamine salt and as a 1 per cent dust of the free phenol, for the control of red spider mites on apples, citrus, walnuts, beans, and other crops.   The dinitrophenols are incompatible with summer oils and of doubtful compatibility with the arsenicals, cryolite, rotenone, pyrethrum, nicotine, tetraethyl pyrophosphate, parathion, lime, lime-sulfur, bordeaux, sulfur, fixed coppers, and quinone fungicides. They should be used with caution upon all plant foliage because of the injury which may result from overdosage or unusual weather conditions of high temperature and humidity.

DNOC and DNOCHP have been shown to increase greatly the rate of oxygen consumption in both insects and mammals by inhibiting oxidative phosphorylation, and this accelerated metabolic activity is responsible for the toxic action of the compounds.   The following oral LD_{50} values in micrograms per gram of body weight have been determined for insects: DNOC: *Bombyx mori* larvae, 49; *Apis mellifera*, 20; DNOCHP: *Bombyx mori*, 7; *Heliothis armigera*, 87; *Cirphis unipuncta*, 15; *Melanoplus femurrubrum*, 65; *Leptinotarsa decemlineata* larvae, 16.   The oral LD_{50} of DNOCHP in olive oil to mice and guinea pigs is from 50 to 125 milligrams per kilogram of body weight.

A related material, 2,4-*dinitroanisole*, $(NO_2)_2C_6H_3OCH_3$, m.p. 268°C. is a yellow crystalline compound which has been used as an ovicide in the army MYL louse powder and as a toxicant in certain proprietary roach and ant powders.   The compound 4,6-*dinitro-2-caprylphenyl crotonate*,

$$O_2N \underset{CH(CH_3)C_6H_{13}}{\overset{NO_2}{\bigcirc}} O\overset{O}{\overset{\|}{C}}CH=CHCH_3$$

is a fungicide and acaricide.   It is a dark-brown liquid which is formulated as a 25 per cent wettable powder.[1]

Organic Thiocyanates.[2]   A number of synthetic organic thiocyanates have appeared on the market since their first development in 1932.

[1] Karathane.

[2] *Del. Agr. Exp. Sta. Bul. 253, Tech. 33*, 1945; "Chemicals for Industry," Rohm and Haas, 1945.

β-Butoxy-β'-thiocyanodiethyl ether, $C_4H_9OCH_2CH_2OCH_2CH_2SCN$, and *β-thiocyanoethyl laurate*, $C_{11}H_{23}COOCH_2CH_2SCN$, are marketed as 1:1 mixtures with petroleum distillate.[1] *β, β'-Dithiocyanodiethyl ether*, $NCSCH_2CH_2OCH_2CH_2SCN$, is available as a 90 per cent technical material[2] and as a 13.5 per cent dust and wettable powder. Technical *lauryl thiocyanate*, $C_{12}H_{25}SCN$, is marketed with a wetting agent.[3] A mixture of about 80 per cent *isobornyl thiocyano acetate* and the remainder the closely related bornyl and fenchyl thiocyano acetates is available as a technical product and as a 20 per cent solution in deodorized kerosene.[4]

$$
\begin{array}{c}
CH_3 \\
C \\
H_2C \quad\quad H \\
\quad\quad COOCCH_2SCN \\
CH_3CCH_3 \\
H_2C \quad\quad CH_2 \\
C \\
H
\end{array}
$$

Isobornyl thiocyano acetate

β-Butoxy-β'-thiocyanodiethyl ether, β-thiocyanoethyl laurate, and isobornyl thiocyano acetate are used at 2.5 to 5 per cent in deodorized kerosene as fly and cattle sprays and as household insecticides. The latter two materials have milder odors and are better suited for inside use. The materials have a very rapid paralytic action and give good knockdown of flying insects. The addition of 0.5 to 1 per cent of a chlorinated hydrocarbon insecticide to such sprays increases the toxicity and residual action. Properly used these thiocyanates will not stain or corrode household furnishings. Lauryl thiocyanate, β-thiocyanoethyl laurate, and β, β'-dithiocyanodiethyl ether are used for the control of aphids, whiteflies, thrips, mealy bugs, and leafhoppers in greenhouses and on garden and truck crops. They are compatible with and often used with pyrethrum and rotenone.

These materials are all contact poisons acting on the insect nervous system. The following LD_{50} values in per cent of spray solution needed for contact action to *Pediculus humanus* and *Cimex lectularius*, respectively, have been determined: β-butoxy-β'-thiocyanodiethyl ether, 1.5 and 4.0; β-thiocyanoethyl laurate, 8.1 and 32; lauryl thiocyanate, 6.0 and 19.5; and isobornyl thiocyano acetate, 3.2 and 75. These compounds are not highly toxic to mammals, the LD_{50} in milligrams per kilogram to the rat for β-butoxy-β'-thiocyanodiethyl ether being, oral, about 250, contact, about 3,000. However, spraying with these materials should always be carried out with good ventilation.

DDT and Derivatives.[5] *DDT* or *dichlorodiphenyltrichloroethane* was first synthesized by Zeidler in 1874, but its insecticidal properties were not discovered until the work of Müller in 1939. The proper chemical

[1] Lethane 384 and Lethane 60.

[2] Lethane A-70.

[3] Loro.

[4] Thanite.

[5] *Jour. Amer. Chem. Soc.*, **67**:1591–1602, 1945; *Jour. Chem. Soc.*, **1946**:333–339; *U.S.D.A.*, *Bur. Ento. Plant Quar.*, *Cirs.* E-631, 1944; E-660, 1945; E-687, 1946; E-674, 1947; and E-728, 1947; WEST, T., and G. CAMPBELL, "DDT the Synthetic Insecticide," Chapman and Hall, 1946.

name of the compound is 1,1,1-trichloro-2,2-bis-(p-chlorophenyl)-ethane, and it has the following structural formula:

Technical DDT is a white to cream-colored amorphous powder produced by reacting chloral (or its alcoholate or hydrate) with monochlorobenzene in the presence of concentrated sulfuric acid. The technical product has a rather variable composition and is composed of up to 14 chemical compounds. These include, in addition to from 65 to 80 per cent of the active p-p' DDT, 15 to 21 per cent of nearly inactive 1,1,1-trichloro-2-(o-chlorophenyl)-2-(p-chlorophenyl)-ethane or o-p' DDT, m.p. 73 to 74°C.; up to 4 per cent of 1,1-dichloro-2,2-bis-(p-chlorophenyl)-ethane, DDD or TDE (page 343); up to 1.5 per cent of 1-(p-chlorophenyl)-2,2,2-trichloroethanol; and traces of 1,1,1-trichloro-2,2-bis-(o-chlorophenyl)-ethane or o-o' DDT; and bis-(p-chlorophenyl)-sulfone. Technical DDT should have a setting point of 89°C. or above and should contain 9.5 to 11 per cent hydrolyzable chlorine and less than 1 per cent volatile and alcohol-insoluble material. It melts over a range of 80 to 94°C. A partially purified or aerosol grade of DDT is available having a setting point of not less than 103°C., while chemically pure p-p' DDT, recrystallized from ethyl alcohol, consists of white needles which melt at 109°C. and distill at 185°C. at 1 mm. Hg. DDT has a density of 1.6, v.p. about 1.5×10^{-7} mm. Hg at 20°C. In alkaline solution, DDT is readily dehydrochlorinated to form a noninsecticidal product, 1,1-dichloro-2,2-bis-(p-chlorophenyl)-ethylene, m.p. 85°C. This compound may be oxidized to p,p'-dichlorobenzophenone, a reaction which is catalyzed by ultra-violet light. These two reactions apparently account for the decomposition of DDT residues. The dehydrochlorination reaction of DDT is readily catalyzed by traces of iron, aluminum, and chromium salts, but DDT is stable in the presence of aqueous alkali. Highly purified DDT is quite stable to heat, not decomposing below 195°C. but the technical material readily decomposes at about 100°C., probably due to the presence of impurities such as iron. Most petroleum solvents inhibit the decomposition of DDT, but chlorinated and nitrated solvents accelerate it. Pure DDT has the following approximate solubilities in grams per 100 cubic centimeters of solvent at 27°C.: water, 0.00001; xylene, 60; carbon tetrachloride, 47; cyclohexanone, 100; ethyl acetate, 68; ethylene dichloride, 60; dioxane, 100; o-dichlorobenzene, 68; tetralin, 68; crude kerosene, 8; refined kerosene, 4; methylated naphthalenes, 40 to 60; mineral oil, 5; dichlorodifluoromethane, 2; dibutyl phthalate, 33; and acetone, 50.

DDT is the most permanent and durable of all the commonly used contact insecticides, due to its insolubility in water, its very low vapor pressure, and its resistance to destruction by light and oxidation. Residues of DDT applied to indoor locations may remain effective for periods as long as a year, becoming ineffective only when covered by accumulations of grease and dirt. Outdoor applications on foliage are less durable, owing to weathering and the slow decomposition of DDT catalyzed by light and traces of metals. Such deposits may remain effective from 1 month to an entire growing season. This unusual stability has resulted not only in improved control methods for many insects but also in difficulties in residue removal from fruits and in possibilities of toxic contamination of soil and foodstuffs not hitherto experienced with contact insecticides.

DDT insecticides are used in the following forms: (a) 2 to 10 per cent dusts of DDT prepared by grinding with talc, pyrophyllite, clays, sulfur, and other diluents, used for the control of mosquito larvae, human and cattle lice, roaches, ants, and a variety of agricultural pests; (b) solutions of 0.5 to 20 per cent DDT in kerosene, mineral oils, methylated naphthalenes, xylene, cyclohexanone, and other solvents, employed in aerosols, household insecticides, as mosquito larvicides, and for mothproofing; (c) emulsions of 0.1 to 5 per cent DDT in water prepared from concentrates of DDT in kerosene or aromatic solvents plus an emulsifier, used in applying residual treatments of DDT to plants for the control of a variety of plant pests, to animals for the control of biting flies, ticks, and lice, to habitations for the control of flies and mosquitoes, and to water as a larvicide for mosquitoes and other biting flies; (d) water suspensions of 0.1 to 5 per cent DDT prepared from wettable powders containing 50 to 95 per cent DDT formulated with a carrier such as clay and small amounts of wetting and dispersing agents, most commonly used for the residual application of DDT to control a great variety of plant-feeding insects, insects attacking livestock, and insects infesting habitations; (e) many miscellaneous forms such as incorporation into water-based paints and wall paper and into clothing and fabrics for mothproofing, use as a soil insecticide for control of wireworms and white grubs, and as a toxicant for poison baits. DDT is compatible with all the commonly used insecticides and fungicides, with the possible exception of lime-sulfur.

DDT acts as either a contact or a stomach poison to insects, affecting the sensory organs and nervous system and causing violent agitation at first, followed by paralysis and death. It is a relatively slow-acting material, in some cases 2 to 3 or more days elapsing before death occurs. The contact LD_{50} values to several insects in micrograms per gram of body weight are *Periplaneta americana*, 10; *Musca domestica*, 6 to 8; *Aedes aegypti* adult, 5 to 8; *Pediculus humanus*, 27; and *Cimex lectularius*, 63. Pure DDT is relatively nontoxic to mammals, the oral LD_{50} to rats being about 200 and by skin contact about 3,000 milligrams per kilogram. DDT in oil solutions, however, is absorbed by the skin, and long exposures to such formulations should be avoided. The usage of DDT on crops intended for animal forage has been discouraged because of the accumulation of the insecticide in animal fat and its secretion in the butterfat of lactating animals. DDT has proved unusually safe when applied to plant foliage, only a few sensitive species such as cucurbits having shown severe injury.

Several analogues of DDT have attained commercial importance as insecticides. *TDE* or *DDD* is 1,1-dichloro-2,2-bis-(p-chlorophenyl)-ethane and is almost indistinguishable from DDT, in which it occurs as an impurity. The technical product consists largely of the p-p' isomer, m.p. 110°C., and 7 to 8 per cent of the o-p' isomer, m.p. 76°C. In general, TDE has proved slightly less effective against most insects than DDT, but it has the advantage of being from one-fifth to one-tenth as acutely toxic to warm-blooded animals and fish, oral LD_{50} to the rat 3,400 milligrams per kilogram. TDE is superior to DDT for the control of such insects as mosquito larvae, tomato hornworms, and the red-banded leaf roller.

The very closely related compound 1,1-dichloro-2,2-bis-(*p*-ethylphenyl)-ethane or Perthane, m.p. 56°C., oral LD_{50} to the rat 8,170 milligrams per kilogram, has been developed for applications where very low acute and chronic toxicity are desired.

TDE

Perthane

Methoxychlor or dianisyl trichloroethane is 1,1,1-trichloro-2,2-bis-(*p*-methoxyphenyl)-ethane. The technical product is a white powder consisting of about 88 per cent of the *p-p'* isomer, m.p. 89°C. (a dimorphic form melts at 78°C.). This compound gives a more rapid knockdown of many insects than does DDT and is of lower chronic and acute toxicity, oral LD_{50} to the rat 6,000 milligrams per kilogram. It does not accumulate in animal fats as does DDT and is therefore favored for use on animals and animal forage. It has shown special effectiveness in the control of the Mexican bean beetle.

Two nitroparaffin analogues of DDT are 1,1-bis-(*p*-chlorophenyl)-2-nitropropane or *Prolan*, m.p. 80°C., and 1,1-bis-(*p*-chlorophenyl)-2-nitrobutane or *Bulan*, m.p. 66°C. These two compounds are white crystalline powders with fruity odors, insoluble in water and soluble in organic solvents. They have been marketed as a mixture of 52.2 per cent of Bulan and 26.2 per cent of Prolan, together with their *o-p'* isomers, under the name of Dilan. The oral LD_{50} values to the rat are Prolan, 4,000, and Bulan, 330 milligrams per kilogram. They have proved to be more effective than DDT to certain insects.

Methoxychlor

Prolan

Bulan

DFDT or 1,1,1-trichloro-2,2-bis-(*p*-fluorophenyl)-ethane, $(p\text{-}FC_6H_4)_2\text{-}CHCCl_3$, was developed in Germany as Gix and marketed as an emulsifiable concentrate. The technical material is a colorless liquid containing about 10 per cent *o-p'* isomer, and the pure *p-p'* isomer melts at 45°C. This material is about 20 times as soluble as DDT in kerosene and spray oils and has a much less persistent residual action.

All the above-mentioned DDT analogues are formulated and used in about the same manner as DDT, and like DDT they are decomposed in strongly alkaline solutions.

Benzene Hexachloride and Lindane.[1] The active constituent of this insecticide is γ-1,2,3,4,5,6-hexachlorocyclohexane or *lindane*, which has the following structural formula:

Lindane

This compound was first prepared by Faraday in 1825, but its insecticidal properties were not discovered until 1940 to 1942, when British and French workers independently developed the material as an insecticide. Benzene hexachloride is prepared by the chlorination of benzene in the presence of sunlight. The crude product is a greyish or brownish amorphous solid with a very characteristic musty odor and begins to melt at 65°C. It consists of 10 to 18 per cent of the active γ-isomer, m.p. 112°C., together with at least four other nearly inactive stereoisomers: α-isomer, m.p. 157°C., 55 to 70 per cent; β-isomer, m.p. 309°C., 5 to 14 per cent; δ-isomer m.p. 138°C., 6 to 8 per cent; and ϵ-isomer, m.p. 219°C., 3 to 4 per cent. Also present are a heptachlorocyclohexane, up to 4 per cent, and a trace of an octachlorocyclohexane, which are insecticidally inactive. The γ-isomer is by far the most toxic of the isomers, being from 500 to 1,000 times as active as the α-isomer and about 5,000 to 10,000 times as active as the δ-isomer, while the β- and ϵ-isomers are nontoxic. The various isomers differ greatly in their solubilities, and pure γ-isomer can be prepared by treating the crude product with methyl alcohol or acetic acid, in which the α- and β-isomers are nearly insoluble (leaving a marketable product containing 30 to 40 per cent γ-isomer), and then fractionally crystallizing the alcohol-soluble fraction from chloroform; or by chromatographic adsorption. The pure γ-isomer has a slight aromatic odor and a crystal density of about 1.85, v.p. 9.4×10^{-6} mm. Hg at 20°C. It is very stable to the action of heat, light, and oxidation and can be burned without appreciable decomposition, but it is readily decomposed by alkaline materials to form principally 1,2,4-trichlorobenzene and 3 moles of hydrogen chloride. The solubilities of the γ-isomer in grams per 100 grams of solvent at 20°C. are acetone, 43; benzene, 29; carbon tetrachloride, 7; cyclohexanone, 37; ethyl acetate, 36; ethyl alcohol, 6; ethylene dichloride, 29; diesel oil, 4; deodorized kerosene, 2; dioxane, 31; mineral oil, 2; xylene, 25; and water, 0.001.

Benzene hexachloride is generally formulated as a wettable powder containing from 5 to 25 per cent γ-isomer or as a dust containing 0.5 to 2 per cent γ-isomer, for agricultural uses. One-half to five per cent oil solutions in kerosene, xylene, or other solvents have been used for mosquito and fly control out-of-doors, but the persistent odor of the crude material makes it unsuitable for a household insecticide. The crude product has also been extensively used as a soil poison, as a toxicant for grasshopper control, and against cotton insects. Great care should be exercised in applying benzene hexachloride to edible crops or to soil in

[1] *Chem. & Ind.* (London), **1945**:314; *Ind. Eng. Chem.*, **39**:1335, 1947; *U.S.D.A.*, *Bur. Ento. Plant Quar.*, *Cir.* E-731, 1947.

which such crops are to be grown, as such applications have resulted in severe tainting of potatoes, lettuce, apples, and other crops which rendered them completely unpalatable. This difficulty has been partially overcome by the marketing of a 99 per cent pure γ-isomer preparation, called *lindane*, as a 20 per cent emulsive concentrate and as a 25 per cent wettable powder, for use on food crops and as a household insecticide. Benzene hexachloride has also been successfully employed as a mosquito larvicide and for the control of chiggers, cattle lice, and ticks. It is incompatible with calcium arsenate, lime, lime-sulfur, and bordeaux, and of doubtful compatibility with nicotine and fixed coppers. Residues of benzene hexachloride have remained active for several months under indoor conditions, but foliage applications outdoors usually become ineffective within a few days. The material is unusually applicable in that it may kill as a contact or a stomach poison or as a fumigant. γ-Benzene hexachloride is somewhat more toxic to warm-blooded animals than DDT. The approximate LD_{50} to the rat is about 200 milligrams per kilogram orally and about 500 milligrams by contact with the skin. The crude product is very irritating to the eyes and mucous membranes when sprayed and dusted. Oil solutions are readily absorbed through the skin, and undue contact with such preparations should be avoided and the skin thoroughly washed with soap after their use. Like DDT it is a nerve poison to insects, and the following LD_{50} values in micrograms per gram of body weight have been recorded for contact action: *Periplaneta americana*, 5; *Musca domestica*, 1 to 3; *Aedes aegypti* adult, 3; *Pediculus humanus*, 1.5; *Cimex lectularius*, 6; and *Melanoplus differentialis*, 4.

Chlorinated Terpenes. A group of incompletely characterized insecticidal compounds has been produced by the chlorination of the naturally occurring terpenes. *Toxaphene*[1] is prepared by the chlorination of the bicyclic terpene, camphene, to contain 67 to 69 per cent chlorine, and it has the empirical formula $C_{10}H_{10}Cl_8$. The technical product is a yellowish, semicrystalline gum, m.p. 65 to 90°C., d 1.64 to 1.66 at 27°C., and is a mixture of isomers. Toxaphene is unstable in the presence of alkali, upon prolonged exposure to sunlight, and at temperatures above 155°C., liberating hydrogen chloride and losing some of its insecticidal potency. It is very soluble in organic solvents, the solubilities in grams per 100 milliliters of solvent being benzene, carbon tetrachloride, ethylene dichloride, and xylene, >450, kerosene, 280, mineral oil, 55 to 60, but it is insoluble in water.

Toxaphene is formulated as a 25 to 40 per cent wettable powder, as an emulsive concentrate, as a kerosene solution, and as a dust. Although it is effective against many of the insects controlled by DDT, it has proved especially useful in the control of grasshoppers, cotton insects, and pests of livestock. It has somewhat less residual action than DDT and is of doubtful compatibility with calcium arsenate, bordeaux, lime, and lime-sulfur. The acute oral LD_{50} to the rat is 69 milligrams per kilogram.

Toxaphene

[1] *Univ. Del. Bul. 264, Tech. 36, 1947.*

A product of similar composition, Strobane, is prepared by chlorinating a mixture of camphene and pinene to contain 66 per cent chlorine. The technical material is a straw-colored liquid, d 1.60, v.p. 3×10^{-7} mm. Hg, oral LD_{50} to the rat, 200 milligrams per kilogram.

Cyclodiene Insecticides. The compounds of this group are highly chlorinated cyclic hydrocarbons with "endomethylene-bridged" structures, prepared by the Diels-Alder diene reaction.

Chlordane and Heptachlor.[1] Chlordane was discovered by Hyman in 1945. The technical material is a brown, viscous liquid with an odor resembling cedar. Chemically it is composed of up to 60 per cent of 2,3,4,5,6,7,8,8-octachloro-2,3,3a,4,7,7a-hexahydro-4,7-methanoindene, which is present as two isomers, α, m.p. 106.5°C., and β, m.p. 104.5°C., the β form being the more toxic; of variable amounts of heptachlor or 1,4,5,6,7,8,8-heptachloro-3a,4,7,7a-tetrahydro-4,7-methanoindene, m.p. 95°C.; and of 4,5,6,7,8,8-hexachloro-3a,4,7,7a-tetrahydro-4,7-methano-indene, m.p. about 154°C., which is only slightly toxic.

Chlordane

Heptachlor

Chlordane is prepared from hexachlorocyclopentadiene by forming an adduct with cyclopentadiene and chlorinating. The technical material has a specific gravity of 1.61 and is unstable in the presence of alkaline materials, liberating hydrogen chloride and forming nontoxic products. Chlordane is of doubtful compatibility with lime, lime-sulfur, bordeaux, nicotine, calcium arsenate, basic lead arsenate, and dithiocarbamates. It is completely soluble in common solvents such as kerosene, xylene, methylated naphthalenes, esters, ketones, ethers, and in spray oils, but is insoluble in water. Chlordane is slowly volatile and has a slight fumigant action, but also possesses some residual activity.

Chlordane is formulated as a 40 to 50 per cent wettable powder by adsorption on clays and as a 2 to 4 per cent dust with nonalkaline diluents. Oil solutions and emulsive concentrates can be prepared in any desired concentration, using kerosene as the preferred diluent. A stable colloidal dispersion of chlordane in water may be prepared by mixing the technical material with equal parts of a nonionic emulsifier[2] and diluting with water. Chlordane has proved most useful against (a) pests of man and domestic animals such as lice, ticks, fleas, and biting flies; (b) household insects such as ants, roaches, flies, silverfish, clothes moths, and carpet beetles; (c) specific agricultural pests such as grasshoppers, cotton insects, soil insects, leaf miners, squash bugs, plum curculio, and various garden insects.

Heptachlor is in general about three to five times as toxic to insects as chlordane and is used in a similar manner. The pure crystalline

[1] *Jour. Econ. Ento.*, **38**:661, 1946; **44**:910, 1951; **45**:452, 1952.
[2] Trex 80.

material, v.p. 3×10^{-4} mm. Hg at 25°C., has the following solubilities in grams per 100 milliliters at 27°C.: acetone, 75; benzene, 106; carbon tetrachloride, 112; cyclohexanone, 119; *ortho*-dichlorobenzene, 100; ethyl alcohol, 4.5; deodorized kerosene, 18.9; methylated naphthalenes, 82.5; and xylene, 102. The technical material is a waxy solid containing about 67 per cent heptachlor and is formulated as a 25 per cent wettable powder and 25 per cent emulsive concentrate and 1 to 2 per cent dusts. Heptachlor residues in and on plants and in animal tissues are slowly converted to heptachlor epoxide,[1] m.p. 159 to 60°C., which has about the same insecticidal activity and mammalian toxicity.

Heptachlor epoxide

Chlordane and heptachlor appear to be specific nerve poisons, affecting insects with violent convulsions. The respective contact LD_{50} values in micrograms per gram are *Musca domestica* chlordane, 4.0; heptachlor, 1.6, heptachlor epoxide, 1.0; *Periplaneta americana* chlordane, 10, heptachlor, 1.0; and *Oncopeltus fasciatus* chlordane, 145, heptachlor, 31. The oral LD_{50} values to the rat are chlordane 450 and heptachlor 90 milligrams per kilogram, and the dermal LD_{50} values to the rabbit are chlordane about 750 and heptachlor 2,000 milligrams per kilogram.

Aldrin or 1,2,3,4,10,10-hexachloro-1,4,4a,5,8,8a-hexahydro-1,4-*endo,-exo*-5,8-dimethanonaphthalene is a white crystalline compound, m.p. 104°C., which is very similar to heptachlor in insecticidal activity. The pure compound has the following solubilities at 25°C. in grams per 100 milliliters of solvent: acetone, 66; amyl acetate, 30; benzene, 83; carbon tetrachloride, 105; ethyl alcohol, 5; ethylene dichloride, 104; deodorized kerosene, 24; methyl ethyl ketone, 24; and xylene, 92. The technical material contains about 78 per cent of the active aldrin, and impurities present may slowly form hydrochloric acid in storage so that an acid inhibitor such as epichlorohydrin is of value as a stabilizer. Aldrin is stable in strong alkali and is compatible with all the commonly used insecticides. It is formulated as a 25 per cent wettable powder, 25 to 50 per cent emulsive concentrate, dusts, and granules. Aldrin is especially useful for the control of grasshoppers, ants, cotton insects, and as a soil poison. Its contact LD_{50} values are *Musca domestica*, 1.6; *Periplaneta americana*, 1.0; *Oncopeltus fasciatus*, 10.3; and *Melanoplus differentialis*, 1.8 micrograms per gram. The oral LD_{50} to the rat is 67 milligrams per kilogram, and the dermal LD_{50} to the rabbit about 150 milligrams per kilogram.

[1] *Jour. Econ. Ento.*, **51**:1, 1958.

```
   H      Cl              H      Cl
   C      C               C      C
     \  H/  \               \  H/  \
HC    C     CCl        H   C    C     CCl
                         \ /   H/
‖  HCH │ ClCCl ‖      O    │ HCH │ ClCCl ‖
     \            \         \
HC    C     CCl        C    C     CCl
   \ /H\  /           H\  /H\  /
   C      C               C      C
   H      Cl              H      Cl
      Aldrin                 Dieldrin
```

Dieldrin is the epoxide of aldrin or 1,2,3,4,10,10-hexachloro-6,7-epoxy-1,4,4a,5,6,7,8,8a-octahydro-1,4-*endo,exo*-5,8-dimethanonaphthalene, m.p. 176°C. Technical dieldrin contains about 76 per cent of this compound. The pure compound has the following solubilities in grams per 100 milliliters of solvent at 25°C.: acetone, 26; amyl acetate, 32; benzene, 56; carbon tetrachloride, 48; ethyl alcohol, 4; ethylene dichloride, 70; methyl ethyl ketone, 39; methylated naphthalenes, 37; and xylene, 52. Dieldrin is a very stable material with a long residual action and is compatible with all the commonly used insecticides and fungicides. It is formulated as a 50 per cent wettable powder, 18 per cent emulsive concentrate, dusts, and granules. Dieldrin is widely used for the control of cotton insects, pests of fruit, field, and vegetable crops, as a soil insecticide, as a residual house spray for flies, mosquitoes, and bed bugs, and for mothproofing. Its contact LD_{50} values in micrograms per gram are *Musca domestica,* 1.1; *Oncopeltus fasciatus,* 15; *Melanoplus differentialis,* 1.4; *Periplaneta americana,* 1.5. The oral LD_{50} to the rat is 87 milligrams per kilogram, and the dermal LD_{50} to the rabbit is about 150 milligrams per kilogram. It is of interest that aldrin is rapidly converted to dieldrin in the living tissues of plants, mammals, and insects.

Endrin is 1,2,3,4,10,10-hexachloro-6,7-epoxy-1,4,4a,5,6,7,8,8a-octahydro-1,4,*endo,endo*-5,8-dimethanonaphthalene, m.p. 245°C. dec. Endrin is thus the *endo,endo*-isomer of dieldrin, differing only in the spatial arrangement of the two rings, which are bent toward one another rather than away as in dieldrin. The technical endrin is less stable than dieldrin, and formulations are stabilized with small amounts of hexamethylene tetramine. Endrin has the following solubilities in grams per 100 milliliters of solvent at 25°C.: acetone, 17; benzene, 13.8; carbon tetrachloride, 3.5; and xylene, 18.3. It is formulated as a 25 per cent wettable powder, 19 per cent emulsive concentrate, and dusts. Endrin is among the most toxic of the chlorinated hydrocarbon insecticides, with an oral LD_{50} to the rat of 10 milligrams per kilogram. Because of its toxicity it is not useful for direct application to the edible portions of crops. However, it has a surprisingly different spectrum of activity from dieldrin and is especially useful for the control of aphids and lepidopterous larvae.

Thiodan[1] is 6,7,8,9,10,10-hexachloro-1,5,5a,6,9,9a-hexahydro-6,9-methano-2,4,3-benzodioxathiepin-3-oxide. The technical material is a brownish crystalline solid melting from 70 to 100°C., insoluble in water but soluble in xylene to about 50 grams in 100 grams at 20°C. Technical Thiodan is composed of two stereoisomers; α, m.p. 108°, and β, m.p. 206 to 8°, together with unreacted 1,4,5,6,7,7-hexachloro-2,3-bis-(hydroxy-

---

[1] *Jour. Econ. Ento.,* **50**:483, 1957.

methyl)-bicyclo-[2.2.1]-heptene-5.   Thiodan is a general-purpose insecticide which has proved especially effective against aphids, leafhoppers, spittle bugs, and other Homoptera.   The two isomers are of comparable effectiveness, the topical LD_{50} values to *Musca domestica* being α 6.2, and β 8.5 micrograms per gram.   The acute oral LD_{50} to the rat for the technical Thiodan is about 110 milligrams per kilogram, and the dermal LD_{50} to the rabbit about 360 milligrams per kilogram.   Thiodan is formulated as a 25 and 50 per cent wettable powder, a 25 per cent emulsive concentrate, and as a dust.

Thiodan

Kepone is 1,2,3,5,6,7,8,9,10,10-decachlorotetracyclo-[5.2.1.0^{2,6}.0^{3,9}.0^{5,8}]-decane-4-one or decachlorotetracyclodecanone.   The compound is a white to tan solid which sublimes with decomposition at 300°C. and is readily soluble in acetone and lower aliphatic alcohols and soluble in benzene and toluene.   Kepone is deliquescent and has an oral LD_{50} to the rat of 95 milligrams per kilogram.   It appears to be most effectively used as an organic stomach poison for ants, grasshoppers, and house fly larvae.

Kepone

Organophosphorus Insecticides.   The development of the organophosphorus insecticides resulted from the researches of Schrader in Germany just prior to the Second World War[1] and has had a profound influence upon the chemical control of insects.   Extensive research in this field has resulted in the discovery of thousands of compounds, with insecticidal properties of every description.   Thus there are available materials with very short residual action such as TEPP and Phosdrin or with prolonged activity such as diazinon and Guthion.   There are broad-spectrum insecticides such as parathion and materials with highly selective action such as schradan.   The unique properties of compounds such as Systox have resulted in successful plant systemic insecticides, and this has been still further refined into seed and soil treatments with materials such as Thimet, which will protect newly developed seedlings from insect attack. Compounds such as ronnel can be fed to cattle and will kill cattle grubs living in the animals' bodies, while others such as Dipterex have pro-

[1] SCHRADER, G., *Angew. Chemie. Monograph* 62, 1952.

nounced stomach-poison action but virtually no contact activity and are especially useful in poison baits.   The organophosphorus insecticides as a class range from materials of high toxicity to warm-blooded animals to those with lower toxicity than the chlorinated hydrocarbon insecticides. The organophosphorus insecticides have the decided advantage, however, of being detoxified very rapidly in animal tissues and eliminated, rather than of being stored in fatty tissues, as are most of the chlorinated hydrocarbon insecticides.   All the organophosphorus insecticides appear to have a common mode of action as irreversible inhibitors of the cholinesterase enzymes of the neuromuscular system.

Tetraethyl Pyrophosphate.[1]   This compound, generally called TEPP, was first synthesized by Clermont in 1854, but its insecticidal properties were discovered by Schrader in 1939.   TEPP is produced in 20 to 40 per cent yield from the reaction of phosphorus oxychloride or phosphorus pentoxide with triethyl phosphate, together with other inactive ethyl polyphosphates.

$$\underset{\text{Tetraethyl pyrophosphate}}{(C_2H_5O)_2\overset{\displaystyle O}{\overset{\|}{P}}-O-\overset{\displaystyle O}{\overset{\|}{P}}(OC_2H_5)_2}$$

Pure tetraethyl pyrophosphate, TEPP, can be separated from this mixture by vacuum distillation and is a water-white liquid, b.p. 104 to 110°C. at 0.08 mm. Hg, sp. gr. 1.1845 at 25°C., weighing about 10 pounds per gallon.   The pure compound decomposes above 200°C. to produce ethylene and metaphosphoric acid, and various technical products undergo this decomposition above 140°C.   TEPP is miscible in water, ethyl alcohol, ethyl acetate, carbon tetrachloride, benzene, xylene, and methylated naphthalenes but not in kerosene or mineral oils.   Tetraethyl pyrophosphate hydrolyzes in water to produce two equivalents of nontoxic diethyl-*o*-phosphoric acid, the times for 50 and 99 per cent hydrolysis at 25°C. being 6.8 hours and 45 hours, respectively.   Such solutions are strongly acid (pH of hydrolyzed 1 per cent solution is about 1.5) and will corrode black iron, zinc, tin, and some porcelains.   All the insecticidal products containing TEPP are hygroscopic and should be protected from moisture in order to preserve the insecticidal activity.

TEPP may be formulated as a spray by adding the material directly to water with a wetting agent or by dissolving it in an organic solvent such as xylene or amyl acetate and emulsifying the solution in water. Dusts are prepared by adding the material to an inert carrier such as gypsum or volcanic ash, which should be as moisture-free as possible. Because of the rapid hydrolysis of the active ingredient into nontoxic products, both sprays and dusts should be applied as soon as possible after mixing.   This rapid decomposition eliminates any danger of toxic residues remaining on food crops.   Spray equipment should be well washed after applications of TEPP to prevent destructive corrosion. Tetraethyl pyrophosphate is incompatible with arsenicals, lime, lime-sulfur, and bordeaux and of doubtful compatibility with nicotine, pyrethrum, rotenone, fixed coppers, cryolite, dinitrophenols, and dithiocarbamates.   Because of the variable content of TEPP in proprietary products, care should be taken to follow the manufacturers' directions, as plant injury may result from overdosage.   The principal uses of TEPP

---

[1] *Jour. Econ. Ento.*, **40**:97, 1947; *Ind. Eng. Chem.*, **40**:694, 1948; *Jour. Amer. Chem. Soc.*, **70**:3882, 1948.

have been to control aphids and red spider mites on agricultural and ornamental crops and in greenhouses. It is effective at dilutions of 1:5,000 to 1:20,000.

Tetraethyl pyrophosphate is extremely toxic to mammals. The LD_{50} to the rat is about 0.85 milligram per kilogram of body weight intraperitoneally, 2 milligrams orally, and 40 milligrams dermally. The material should be handled with extreme care and instantly removed by washing if allowed to contact the skin. A respirator should be worn at all times when spraying or dusting with this material. In cases of acute poisoning, treatment with atropine has been of benefit. The toxic action of TEPP to insects and mammals occurs as a result of its specific inhibition of the enzyme, cholinesterase, which is a vital component of the mechanism of nerve-impulse transmission (page 119). The following LD_{50} values have been recorded for insects: *Periplaneta americana*, intra-abdominal, 1; *Melanoplus differentialis*, contact, 4.4; and *Apis mellifera*, contact, 1.2 micrograms per gram of body weight.

Tetraethyl dithionopyrophosphate or *sulfotepp* is a yellowish liquid with a garlic odor, b.p. 110 to 13°C. at 0.2 mm. Hg, d 1.196. It is only about 0.002 per cent soluble in water, is soluble in organic solvents, and hydrolyzes only in alkaline solution. The compound has an oral LD_{50} to the rat of 5 milligrams per kilogram and is especially useful where a more stable product than TEPP is wanted, especially in aerosols or smokes for the control of greenhouse pests.

$$\underset{\text{Sulfotepp}}{(C_2H_5O)_2\overset{\overset{\text{S}}{\|}}{P}-O-\overset{\overset{\text{S}}{\|}}{P}(OC_2H_5)_2}$$

Parathion or *O,O*-diethyl *O-p*-nitrophenyl phosphorothionate was also discovered by Schrader and has become the most widely used of all the organophosphorus insecticides. The technical material is a dark-brown liquid, d 1.265, which has an unpleasant garlic odor and is about 98 per cent pure. The pure compound, m.p. 6° and calculated b.p. 375°, has a v.p. 0.00004 mm. Hg at 27°C. Parathion is very resistant to aqueous hydrolysis but is hydrolyzed by alkali to form inactive diethylphosphorothioic acid and *p*-nitrophenol. The times for 50 per cent hydrolysis at 25°C. are 120 days for a saturated aqueous solution and 8 hours for a solution in limewater. At temperatures above 130°C., parathion slowly isomerizes to form *O,S*-diethyl *O-p*-nitrophenyl phosphorothioate, which is much less stable and less effective as an insecticide. Parathion is readily reduced to the nontoxic *O,O*-diethyl *O-p*-aminophenyl phosphorothionate and oxidized with difficulty to the highly toxic diethyl *p*-nitrophenyl phosphate. Parathion is completely miscible in benzene, xylene, phthalates, and glycols, but is almost insoluble in kerosene and mineral oils and is soluble in water to about 0.00002 per cent.

$$\underset{\text{Parathion}}{(C_2H_5O)_2\overset{\overset{\text{S}}{\|}}{P}O-\!\!\left\langle\right\rangle\!\!-NO_2} \qquad \underset{\text{Methyl parathion}}{(CH_3O)_2\overset{\overset{\text{S}}{\|}}{P}O-\!\!\left\langle\right\rangle\!\!-NO_2}$$

Parathion has proved effective against a wider variety of insects than any other insecticide and is generally applied at concentrations of 0.01 to 0.1 per cent. It has a relatively short residual action and may be safely used on edible produce if applied 15 to 30 days before marketing. Parathion is formulated as a 25 per cent wettable powder, 50 per cent emulsive concentrate, and as 0.25 to 2 per cent dusts. It is incompatible with lime and of doubtful compatibility with the arsenicals, cryolite, dinitrophenols, bordeaux, summer oils, and lime-sulfur.

The oral LD_{50} of parathion to the rat is about 6 to 15 milligrams per kilogram, and the dermal LD_{50} to the rabbit about 40 to 50 milligrams per kilogram. The compound is extremely toxic to humans, and several fatalities have resulted from its careless use. It may enter the body in toxic amounts through the mouth, nose, or skin, and great care should be exercised in handling the concentrate materials. The use of rubber gloves, goggles, a respirator, and other protective clothing is advisable. Any material spilled on the skin should be immediately removed with soap and water. When spraying and dusting with parathion, contaminated clothing should be changed frequently. Periodic estimation of blood cholinesterase levels is of value in the early detection of overexposure. In the case of acute poisoning by parathion or other organophosphorus insecticides, the victim should be given prompt medical attention, with the administration of therapeutic doses of atropine, which is a specific antidote. Parathion is of moderate chronic toxicity and is not stored in the mammalian body. Parathion is toxic to insects and mammals alike by its *in vivo* conversion to *O,O*-diethyl *O-p*-nitrophenyl phosphate or paraoxon, which is a powerful irreversible inhibitor of the cholinesterase enzymes of the neuromuscular system. The following contact LD_{50} values have been reported for insects: *Periplaneta americana*, 1; *Melanoplus differentialis*, 0.7; *Apis mellifera*, 3.5; and *Musca domestica*, 1 microgram per gram.

A limiting factor in the use of parathion has been its high toxicity to mammals, and much effort has been devoted to finding safer substitutes. Several closely related compounds have proved especially useful.

Methyl parathion is *O,O*-dimethyl *O-p*-nitrophenyl phosphorothionate, a white solid, m.p. 36°, sp. gr. 1.358. This compound is about as widely used as parathion and is very similar in properties, but hydrolyzes and isomerizes more easily and is therefore less stable both in storage and as an insecticide residue. Methyl parathion is more effective than parathion against aphids and beetles and is somewhat less toxic to mammals, oral LD_{50} to the rat 14 to 42 milligrams per kilogram.

Two chlorinated analogues of methyl parathion have greatly reduced acute mammalian toxicities and are useful as household insecticides and for certain agricultural pests. *Dicapthon* is *O,O*-dimethyl *O*-2-chloro-4-nitrophenyl phosphorothionate, a solid, m.p. 53°, oral LD_{50} to the rat 500 to 1,000 milligrams per kilogram. *Chlorthion* is *O,O*-dimethyl *O*-3-chloro-4-nitrophenyl phosphorothionate, an oil, sp. gr. 1.4330 at 20°C., oral LD_{50} to the rat 1,500 milligrams per kilogram.

$$(CH_3O)_2 \overset{\overset{\textstyle S}{\|}}{P}O \langle\rangle NO_2$$
$$Cl$$
Dicapthon

$$(CH_3O)_2 \overset{\overset{\textstyle S}{\|}}{P}O \langle\rangle NO_2$$
$$Cl$$
Chlorthion

Baytex is *O,O*-dimethyl *O*-3-methyl-4-methylthiophenyl phosphorothionate, a brownish liquid, b.p. 105°C. at 0.01 mm. Hg, sp. gr. 1.245 at 20°C., v.p. 2.15×10^{-6} mm. Hg at 20°C. It is soluble in organic solvents and insoluble in water and has an oral LD_{50} to the rat of 325 milligrams per kilogram. Baytex is a persistent, general-purpose insecticide, especially useful as a residual spray for fly and mosquito control and as a mosquito larvicide.

Ronnel is *O,O*-dimethyl *O*-2,4,5-trichlorophenyl phosphorothionate, a white crystalline solid, m.p. 35 to 7°C., which is soluble in organic solvents and to about 0.008 per cent in water. It has an oral LD_{50} to the rat of 1,740 milligrams per kilogram. It is a persistent, general-purpose insecticide and is also an animal systemic insecticide (page 360).

$$(CH_3O)_2\overset{\overset{S}{\|}}{P}O\!\!-\!\!\langle\rangle\!\!-\!\!SCH_3$$
$$CH_3$$
Baytex

$$(CH_3O)_2\overset{\overset{S}{\|}}{P}O\!\!-\!\!\langle\rangle\begin{matrix}Cl\\[-2pt]\\[-2pt]Cl\end{matrix}Cl$$
Ronnel

Diazinon is *O,O*-diethyl *O*-2-isopropyl-4-methylpyrimidyl-(6) phosphorothionate, a brownish liquid, b.p. 83 to 4°C. at 0.002 mm. Hg, sp. gr. 1.11, which is soluble in most organic solvents but only to about 0.004 per cent in water. It has an oral LD_{50} to the rat of 100 to 150 milligrams per kilogram and a dermal LD_{50} to the rabbit of about 1,000 milligrams per kilogram. Diazinon is a persistent, general-purpose insecticide and is formulated as a 25 per cent wettable powder and 25 per cent emulsive concentrate.

Guthion is *O,O*-dimethyl *S*-4-oxo-1,2,3-benzotriazin-3-(4-*H*)-yl-methyl phosphorodithioate, a white solid, m.p. 73 to 4°C., sp. gr. 1.44 at 20°C. It is soluble in organic solvents but only to about 0.003 per cent in water. It has an oral LD_{50} to the rat of about 18 milligrams per kilogram. Guthion is a persistent, general-purpose insecticide and is formulated as a 15 per cent wettable powder and 15 per cent emulsive concentrate.

$$\begin{matrix}CH_3\\(CH_3)_2CH\!\!-\!\!\langle\rangle\!\!-\!\!O\overset{\overset{}{\|}}{P}(OC_2H_5)_2\end{matrix}$$
Diazinon

$$\overset{O}{\underset{}{\|}}C\!-\!NCH_2S\overset{\overset{S}{\|}}{P}(OCH_3)_2$$
Guthion

Malathion is *O,O*-dimethyl *S*-(1,2-dicarbethoxyethyl) phosphorodithioate, a brownish liquid, b.p. 156 to 7°C. at 0.7 mm. Hg. The technical material is 95 to 98 per cent pure, with an unpleasant odor and a sp. gr. 1.23 at 25°C. It is soluble in most organic solvents, slightly soluble in mineral oils, and soluble in water to 0.0145 per cent. Malathion is easily hydrolyzed above pH 7.0 and below pH 5.0 and is incompatible with alkaline materials. The oral LD_{50} of malathion to rats varies from 1,400 to 5,800 milligrams per kilogram, making it one of the safest of all insecticides. It is a persistent, general-purpose insecticide, especially suited for household, home garden, vegetable, and fruit insect control. It is formulated as a 25 per cent wettable powder, 50 per cent and 95 per cent emulsive, dusts, granulars, and aerosols.

$$(CH_3O)_2\overset{\overset{S}{\|}}{P}SCH\overset{\overset{O}{\|}}{C}OC_2H_5$$
$$CH_2\overset{}{C}OC_2H_5$$
$$\overset{}{\underset{O}{\|}}$$
Malathion

Trithion is *O,O*-diethyl *S-p*-chlorophenylthiomethyl phosphorodithioate, a light-amber liquid, sp. gr. 1.265 to 1.285, insoluble in water and soluble in kerosene, xylene, alcohols, ketones, and esters.   It has an oral LD_{50} to the rat of 28 milligrams per kilogram, and is a general-purpose, persistent insecticide.   It is formulated as a 50 per cent emulsive concentrate and 25 per cent wettable powder.

EPN is *O*-ethyl *O-p*-nitrophenyl phenylphosphonothionate, when pure a yellowish, crystalline solid, m.p. 36°C.   The technical material is a brown liquid, sp. gr. 1.27 at 25°C., v.p. 0.03 mm. Hg at 100°C., insoluble in water and soluble in organic solvents.   It has an oral LD_{50} to the rat of 35 to 45 milligrams per kilogram.   EPN is a general-purpose, persistent insecticide and is formulated as a 25 per cent wettable powder.

Trithion

EPN

Delnav is 2,3-*p*-dioxanedithiol *S,S*-bis-(*O,O*-diethylphosphorodithioate), a brown liquid, sp. gr. 1.257 at 26°C., insoluble in water and soluble in organic solvents.   It is a persistent acaricide and insecticide and is formulated as a 25 per cent wettable powder and emulsive concentrate. It has an acute oral LD_{50} to the rat of 110 milligrams per kilogram and an acute dermal LD_{50} to the rabbit of 107 milligrams per kilogram.

Delnav

Dylox or *Dipterex* is dimethyl 1-hydroxy-2-trichloroethyl phosphonate,

$$(CH_3O)_2PCH(OH)CCl_3$$

a white crystalline solid, m.p. 83 to 4°C., b.p. 120°C. at 0.4 mm. Hg, sp. gr. 1.73 at 20°C.   It is soluble in water to 15 per cent and in aromatic hydrocarbons, alcohol, and acetone.   It has an oral LD_{50} to the rat of 450 milligrams per kilogram.   This compound is a stomach poison useful in dry sugar bait for flies and on foliage for chewing insects.   Above pH 6, it is rapidly converted to *DDVP* or dimethyl dichlorovinyl phosphate, $(CH_3O)_2POCH{=}CCl_2$, a reaction which accounts for its toxic action.   DDVP is a colorless liquid, b.p. 120°C. at 14 mm. Hg, sp. gr. 1.415 at 25°C., with an oral LD_{50} to the rat of 80 milligrams per kilogram.   DDVP is a highly volatile compound, very effective in baits and in aerosol formulations for the rapid knockdown of flies, mosquitoes, moths, etc.

Dibrom is brominated DDVP or dimethyl 1,2-dibromo-2,2-dichloroethyl phosphate, $(CH_3O)_2POCHBrCBrCl_2$, m.p. 26°C., b.p. 110° at

0.5 mm. Hg. It is insoluble in water and soluble in aromatic solvents and has an oral LD_{50} to the rat of 430 milligrams per kilogram. Dibrom is destructively hydrolyzed in water in about 2 days. It provides a short-lived residual insecticide on plants and is a household insecticide.

Ethion or O,O,O',O'-tetraethyl S,S'-methylene-bis-phosphorodithioate is a liquid, m.p. −12 to −15°C., insoluble in water and soluble in most organic solvents, aromatic and aliphatic hydrocarbons, including spray oils. It is formulated as a 25 and 50 per cent wettable, 50 per cent emulsive concentrate, and as a 2 per cent solution in spray oil. The compound has an oral LD_{50} of 96 milligrams per kilogram to the rat and an acute dermal LD_{50} to the rabbit of 915 milligrams per kilogram. It is a promising general-purpose insecticide.

$$C_2H_5O \quad S \qquad\qquad S \quad OC_2H_5$$
$$\diagdown \parallel \qquad\qquad \parallel \diagup$$
$$P{-}S{-}CH_2{-}S{-}P$$
$$\diagup \qquad\qquad\qquad \diagdown$$
$$C_2H_5O \qquad\qquad\qquad OC_2H_5$$

Ethion

Carbamates. The carbamate insecticides are a comparatively recent development, representing a unique class of insecticidal compounds of considerable diversity. These apparently owe their activity to action as competitive inhibitors of the cholinesterase enzymes of the neuromuscular system. They are rapidly detoxified and eliminated from animal tissues and thus are not accumulative in fats or excreted in milk. The carbamates are unstable in alkaline solutions.

Isolan is 1-isopropyl-3-methylpyrazolyl-(5) N,N-dimethyl carbamate, a colorless liquid, b.p. 105 to 7°C. at 0.3 mm. Hg, soluble in water and organic solvents. It has an oral LD_{50} to the rat of 54 milligrams per kilogram and is especially effective as an aphicide with systemic action and as a bait for flies.

Sevin is 1-naphthyl N-methylcarbamate, a white crystalline solid, m.p. 142°C., sp. gr. 1.232 at 20°C., v.p. <0.005 mm. Hg at 26°C. The technical material is about 95 per cent pure and has the following per cent solubilities: water 0.004; carbon tetrachloride, 10; methyl isobutyl ketone, 20; petroleum ether, 20; and xylene, 10. Sevin has an acute oral LD_{50} to the rat of 540 milligrams per kilogram and a dermal LD_{50} to the rabbit of >5,000 milligrams per kilogram. It is a general-purpose insecticide, especially effective for fruit, vegetable, and cotton insect control, and is formulated as a 50 per cent wettable powder, 24 per cent emulsive concentrate, and dusts and granulars.

$$CH_3C{-}{-}{-}{-}CH$$
$$\parallel \qquad\quad \parallel \quad O$$
$$N \qquad COCN(CH_3)_2$$
$$\diagdown N \diagup$$
$$CH(CH_3)_2$$

Isolan

$$\overset{O}{\overset{\parallel}{OCNHCH_3}}$$

Sevin

Other carbamates which appear to be especially useful include *Zectran*, 3,5-dimethyl-4-dimethylaminophenyl N-methylcarbamate(I), m.p. 85°C.

and 2-isopropoxyphenyl N-methylcarbamate(II), m.p. 88 to 9°C., which are general-purpose insecticides.

$$(CH_3)_2N \underset{CH_3}{\overset{CH_3}{\diagdown}} \!\!\!\!\!\!\!\!\! \left\langle \right\rangle \!\!\!\! O\overset{\overset{O}{\|}}{C}NHCH_3 \qquad \qquad \left\langle \right\rangle \!\!\!\! O\overset{\overset{O}{\|}}{C}NHCH_3$$

I   O
CH(CH_3)_2
II

SYSTEMIC INSECTICIDES FOR PLANTS

These toxicants, when applied to seeds, roots, stems, or leaves of plants, are absorbed and translocated to the various plant parts in amounts lethal to insects feeding thereon. This method of plant protection has the decided advantages of (a) minimizing to some extent the inequalities of spray coverage, (b) increasing the length of residual control by protection of the spray residue from attrition by weathering, (c) protecting new plant growth formed subsequent to application, and (d) having less damaging effects on beneficial predatory and pollinating insects. Systemic compounds may be applied as direct sprays to the foliage by drenching the soil as in the irrigation water, by direct application or injection of concentrates to the trunk or stem, by implantation of encapsulated material about the roots or into the stem, or by treatment of seeds before planting. Because of the possible contamination of edible portions of the plant, the use of systemics on food crops must be predicated on a thorough knowledge of the nature and magnitude of toxic residues. The inorganic compound *sodium selenate* has systemic properties and is discussed on page 323. Most of the compounds in practical use, however, are organic stomach poisons which are sufficiently water-soluble to be translocated in toxic amounts.

Schradan or octamethyl pyrophosphoramide,

$$[(CH_3)_2N]_2\overset{\overset{O}{\|}}{P}\!-\!O\!-\!\overset{\overset{O}{\|}}{P}[N(CH_3)_2]_2$$

was among the first practical systemics. The technical material is a brown liquid, b.p. 154°C. at 2 mm. Hg, sp. gr. 1.13 at 25°C., which contains about 40 per cent of the octamethyl pyrophosphoramide and an equivalent amount of decamethyl triphosphoramide,

$$[(CH_3)_2N]_2\overset{\overset{O}{\|}}{P}\!-\!O\!-\!\overset{\overset{O}{\|}}{P}\!-\!O\!-\!\overset{\overset{O}{\|}}{P}[N(CH_3)_2]_2$$
$$\underset{N(CH_3)_2}{\big|}$$

which is of about the same activity. The product is water-miscible and also soluble in most organic solvents, but only slightly soluble in petroleum hydrocarbons. Schradan is remarkably stable to alkaline hydrolysis but hydrolyzes much more rapidly in acids to form dimethyl amine and phosphoric acid. It possesses only a weak contact action to most insects, although it is fairly toxic to Hemiptera and Homoptera and has been shown to be slowly converted in plants and rapidly in animals to the monoamide oxide, which is the actual toxicant. When absorbed into plants, schradan provides protection from aphids and mites lasting up to several months.   The oral LD_{50} to the rat is about 10 milligrams per kilogram, and the material should be handled with the same precautions as described for parathion.

Dimefox or *Hanane* is bis-(dimethylamino) phosphoryl fluoride

$$[(CH_3)_2N]_2\overset{\overset{\displaystyle O}{\parallel}}{P}F,$$

which is a colorless liquid, b.p. 67°C. at 4 mm. Hg, sp. gr. 1.12., v.p. 0.4 mm. Hg at 30°C.   This compound is water-soluble and has been used to control the mealybug vector of the swollen shoot disease of cacao by implanting soluble capsules about the roots.   Because of its volatility dimefox has a very short residual activity and is not suited for foliage application.   The oral LD_{50} to the rat is 3.5 milligrams per kilogram.

Demeton or *Systox* is a mixture of 2 parts of *O,O*-diethyl *O*-2-(ethyl-

$$\text{thio})\text{-ethyl phosphorothionate, } (C_2H_5O)_2\overset{\overset{\displaystyle S}{\parallel}}{P}OCH_2CH_2SC_2H_5, \text{ b.p. } 94°C.$$

at 0.4 mm. Hg, sp. gr. 1.119 at 20°C., water-solubility 0.0066 per cent, and 1 part of *O,O*-diethyl *S*-2-(ethylthio)-ethyl phosphorothiolate,

$$(C_2H_5O)_2\overset{\overset{\displaystyle O}{\parallel}}{P}SCH_2CH_2SC_2H_5,$$

b.p. 110°C. at 0.4 mm. Hg, sp. gr. 1.132 at 20°C., water-solubility 0.2 per cent.   The technical material is a yellowish liquid, sp. gr. 1.183, soluble in organic solvents, and unstable in alkaline solution.   It has an oral LD_{50} to the rat of 9 milligrams per kilogram and a dermal LD_{50} to the rabbit of 27 milligrams per kilogram and is formulated as a 25 per cent emulsive concentrate.   Systox provides a long-lasting systemic insecticide rapidly absorbed by roots, stems, or foliage. Internally in plant tissues, both isomers are rapidly oxidized to the sulfoxide and sulfone derivatives as shown with phorate (page 359).

Meta Systox is *O,O*-dimethyl *S*-2-(ethylthio)-ethyl phosphorothiolate,

$$(CH_3O)_2\overset{\overset{\displaystyle O}{\parallel}}{P}SCH_2CH_2SC_2H_5,$$

which is widely used in Europe as a systemic insecticide.   It has very similar properties and behavior to demeton and an oral LD_{50} to the rat of 80 milligrams per kilogram.   *Meta Systox*

sulfoxide, $(CH_3O)_2\overset{\overset{\displaystyle O}{\parallel}}{P}SCH_2CH_2\overset{\overset{\displaystyle O}{\uparrow}}{S}C_2H_5$, is more water-soluble, and is a very active systemic, especially suited for granular applications to the soil.

Dimethoate is *O,O*-dimethyl *S*-(*N*-methylcarbamoyl)-methyl phos-

$$\text{phorodithioate, } (CH_3O)_2\overset{\overset{\displaystyle S}{\parallel}}{P}SCH_2\overset{\overset{\displaystyle O}{\parallel}}{C}NHCH_3,$$

a white crystalline solid, m.p. 51°C.   It is soluble in organic solvents and soluble in water to about 7 per cent.   Dimethoate has an oral LD_{50} to the rat of 245 milligrams per kilogram.   Because of its water-solubility it has proved especially useful as a persistent systemic for fruit fly larvae and for side-dressing of soil about plants.

Phosdrin is dimethyl 2-carbomethoxy-1-methylvinyl phosphate,

$$(CH_3O)_2\overset{\overset{\displaystyle O}{\parallel}}{P}OC(CH_3)\!=\!CH\overset{\overset{\displaystyle O}{\parallel}}{C}OCH_3.$$

The technical material is a colorless liquid, b.p. 106 to 7.5°C. at 1 mm. Hg, sp. gr. 1.25 at 20°C., v.p. 0.0029 mm. Hg at 21°C.   It is miscible in water, xylene, and acetone, but is less than 5 per cent soluble in kerosene. The technical product is composed of about 2 parts of *trans*- and 1 part of the *cis*-isomer, the former of which is about ten times as active as the latter.   The oral LD_{50} to the rat is about 6.8 milligrams per kilogram. Phosdrin is especially useful for the treatment of edible produce close to harvest since it is rapidly dissipated by volatilization and enzymatic decomposition in the plant.   It is formulated as a 25 per cent emulsive concentrate, a 10 per cent wettable powder, and as a granular.

Phosphamidon is dimethyl 2-chloro-2-diethylcarbamoyl-1-methylvinyl

$$(CH_3O)_2\overset{O}{\overset{\|}{P}}OC(CH_3){=}CClC\overset{O}{\overset{\|}{N}}(C_2H_5)_2$$

phosphate, $(CH_3O)_2POC(CH_3){=}CClCN(C_2H_5)_2$, a colorless liquid, b.p. 160°C. at 1.5 mm. Hg, which is miscible in water and organic solvents. It has an oral LD_{50} to the rat of 17 milligrams per kilogram.   Phosphamidon is rapidly absorbed by plant surfaces and quickly decomposed in the plant to provide a short-lived systemic.

Seed and Soil Treatments.   Several systemic compounds of very low water-solubility are especially suited for the protection of young plants from attack by mites, thrips, aphids, and leafhoppers by application (*a*) as a 50 per cent charcoal powder to the seeds of cotton, alfalfa, or sugar beets before planting, (*b*) by granular application at time of transplanting, or (*c*) as a granular side-dressing applied at planting.

Phorate or *Thimet* is 0,0-diethyl *S*-2-(ethylthio)-methyl phosphoro-dithioate, a yellowish liquid, b.p. 118 to 20°C. at 0.8 mm. Hg, sp. gr. 1.167 at 25°C.   It is soluble in organic solvents and to 0.0085 per cent in water.   The oral LD_{50} to the rat is 3.7 milligrams per kilogram.   In plant tissue, phorate is rapidly oxidized as shown below:

$$(C_2H_5O)_2\overset{S}{\overset{\|}{P}}SCH_2SC_2H_5 \qquad \text{Phorate}$$

$$(C_2H_5O)_2\overset{S}{\overset{\|}{P}}SCH_2\overset{O}{\overset{\uparrow}{S}}C_2H_5$$

$$(C_2H_5O)_2\overset{O}{\overset{\|}{P}}SCH_2\overset{O}{\overset{\uparrow}{S}}C_2H_5 \qquad (C_2H_5O)_2\overset{S}{\overset{\|}{P}}SCH_2\overset{O}{\overset{\uparrow}{S}}C_2H_5$$

$$(C_2H_5O)_2\overset{O}{\overset{\|}{P}}SCH_2\overset{O}{\overset{\uparrow}{S}}C_2H_5$$

These oxidative sulfoxide and sulfone metabolites are responsible for the systemic toxicity of the compound.   They are more water-soluble and less stable than the parent material and are decomposed in the plant to nontoxic hydrolysis products.

Di-Syston is 0,0-diethyl *S*-2-(ethylthio)-ethyl phosphorodithioate and is thus the dithioanalogue of demeton (page 358).   This compound,

$$(C_2H_5O)_2\overset{S}{\overset{\|}{P}}SCH_2CH_2SC_2H_5$$

$(C_2H_5O)_2PSCH_2CH_2SC_2H_5$, is a pale-yellow liquid, b.p. 62°C. at 0.01 mm.

Hg, sp. gr. 1.144 at 20°C.   It is soluble in organic solvents and to 0.0066 per cent in water.   The oral LD_{50} to the rat is from 2 to 12.5 milligrams per kilogram.   Di-Syston is oxidized in plant tissue in the same manner as shown for phorate above, the final active metabolites being the same as the sulfoxide and sulfone metabolites of demeton thiol-isomer.

Ekatin is *O,O*-dimethyl *S*-2-(ethylthio)-ethyl phosphorodithioate,

$$(CH_3O)_2\overset{\overset{\displaystyle S}{\|}}{P}SCH_2CH_2SC_2H_5,$$ and is the dithioanalogue of Meta Systox (page 358).

Systemic Insecticides for Animals

Several insecticides, when fed or topically applied to domestic animals, have been found to move through the body tissues in quantities lethal to such internal parasites as cattle grubs, screw-worm larvae, and helminths and over a shorter period to external parasites such as horn and stable flies, mites, lice, and ticks, without injury to the host animal.   The insecticides are slowly destroyed by enzymatic action in the animal body, so that after a safe period of 60 days, the animal can be used for milk production or slaughtered for meat.

Ronnel or *Trolene*, *O,O*-dimethyl *O*-2,4,5-trichlorophenyl phosphorothionate (page 364), is effective when fed at 100 milligrams per kilogram of animal weight.

Co-Ral is *O,O*-diethyl *O*-(3-chloro-4-methyl-2-oxo-2*H*-1-benzopyran-7-yl) phosphorothionate, a tan crystalline solid, m.p. 90 to 2°C., v.p. 10^{-7} mm. Hg at 20°C.   It is insoluble in water and soluble in organic solvents and has an oral LD_{50} to the rat of 56 to 230 milligrams per kilogram.   Co-Ral is applied as a 0.25 to 0.5 per cent spray.

Ruelene is *O*-methyl *O*-(4-*tert*-butyl-2-chlorophenyl) methylphosphoramidate, m.p. 61°C., soluble in most organic solvents and insoluble in water.   It is effective either as an 0.75 per cent spray or when fed at 20 to 25 milligrams per kilogram of animal weight.   Ruelene has an oral LD_{50} to the rat of 1,000 milligrams per kilogram.

Co-Ral   Ruelene

Acaricides

Chemicals which are especially effective in controlling the mites and ticks of the order Acarina are termed acaricides.   Most of the insecticidal chemicals discussed elsewhere in this chapter, with the exception of the dinitrophenol and organophosphorus insecticides, are not of practical value as acaricides.   Applications of many of the chlorinated hydrocarbon insecticides have no effect on the phytophagous spider mites, but often kill their predators and thus result in abnormally large populations of the spider mites.   A number of acaricidal chemicals have come into widespread use which have almost specific toxicity to the mites but are inactive against insects. In general, these acaricides are highly stable compounds with comparatively prolonged residual action and low mammalian toxicity. Certain of the compounds described below are effective only as ovicides, killing the eggs and sometimes the newly emerged nymphs, while others are active against all the stages of the mites. These acaricides exhibit a considerable degree of specificity for various species of acarina and have been found most useful for the phytophagous Tetranychidae and Eriophyidae. Other acaricides which repel or kill the mites and ticks which attack man and animals are described under repellents (page 389).

DMC or *Dimite* is di-(*p*-chlorophenyl) methyl carbinol or 1,1-bis-(*p*-chlorophenyl)-ethanol, a white solid, m.p. 69.5 to 70°C. It is insoluble in water and has the following solubilities in grams per 100 milliliters of solvent at 25 to 30°C.: petroleum ether, 4.3; ethanol, 125; and toluene, 110. The compound is readily dehydrated upon heating or in the presence of strong acids and forms the inactive 1,1-bis-(*p*-chlorophenyl)-ethylene. Technical DMC contains small amounts of the *o-p'* and *o-o'* isomers and traces of isomeric dichlorobenzophenones. DMC is formulated as a 25 per cent emulsive concentrate and is active against all stages of mites. It has an oral LD_{50} to the rat of about 200 milligrams per kilogram.

Chlorobenzilate is ethyl *p,p'*-dichlorobenzilate, a yellowish viscous oil, b.p. 141 to 2°C. at 0.06 mm. Hg. The technical material, d 1.2816, contains about 90 per cent of the active compound, is insoluble in water, and soluble to more than 40 per cent in deodorized kerosene, benzene, and methyl alcohol. Chlorobenzilate is hydrolyzed in alkali and in strong acids to the inactive *p,p'*-dichlorobenzilic acid and ethanol. The compound is active against all stages of mites and is formulated as a 25 per cent wettable powder and emulsive concentrate. It has an oral LD_{50} to the rat of about 700 milligrams per kilogram.

Kelthane is 1,1-bis-(*p*-chlorophenyl)-2,2,2-trichloroethanol, a white crystalline solid, m.p. 78.5 to 79.5°C. This compound is insoluble in water and soluble in organic solvents and in the presence of alkali forms the inactive *p,p'*-dichlorobenzophenone and chloroform. Kelthane is a long-lasting residual acaricide formulated as a 25 per cent wettable powder and emulsive concentrate and is active against all stages of mites. It has an oral LD_{50} of about 700 milligrams per kilogram to the rat.

DMC   Chlorobenzilate   Kelthane

Tedion is 2,4,5,4'-tetrachlorodiphenyl sulfone, a crystalline solid, m.p. 148°C. It is insoluble in water and has the following solubilities in grams per 100 grams of solvent at 18°C.: petroleum ether, 0.4; ethyl acetate, 7.1; carbon tetrachloride, 1.6; methyl ethyl ketone, 10.5; xylene, 11.5. Tedion is stable to the action of acids and alkalies, light, and temperature and has a very prolonged residual action. It is formulated as a 25 per cent wettable powder and emulsive concentrate and is active against all stages of mites.   Tedion has an oral LD_{50} to the rat of $> 14,700$ milligrams per kilogram.

Sulphenone is *p*-chlorophenyl phenyl sulfone, a white solid, m.p. 98°C. The technical material consists of about 80 per cent of this compound, with small amounts of *o*- and *m*-isomers, bis-(*p*-chlorophenyl) sulfone, and diphenyl sulfone.   Sulphenone is insoluble in water and has the following solubilities in grams per 100 grams of solvent at 20°C.: hexane, 0.4; xylene, 18.2; carbon tetrachloride, 4.9; and acetone, 74.4.   Sulphenone is formulated as a 50 per cent wettable powder and is effective against all stages of mites.   It has an oral LD_{50} to the rat of $> 2,000$ milligrams per kilogram.

Tedion   Sulphenone

Ovex or *Ovotran* is *p*-chlorophenyl *p*-chlorobenzene sulfonate, a white solid, m.p. 86.5°C.   It is insoluble in water and has the following solubilities in grams per 100 grams of solvent at 25°C.: kerosene, 2; carbon tetrachloride, 41; cyclohexanone, 110; ethyl alcohol, 1; ethylene dichloride, 110; and xylene, 78.   Ovex is formulated as a 50 per cent wettable powder and is effective only as an ovicide.   It has an oral LD_{50} to the rat of 2,000 milligrams per kilogram.

Two analogues of ovex have similar solubilities, ovicidal properties, and like ovex are hydrolyzed in alkali to form the phenol and benzene sulfonate salt.   *Genite* is 2,4-dichlorophenyl benzene sulfonate, m.p. 45 to 7°C., v.p. 2.7×10^{-4} mm. Hg at 30°C.   The technical product is about 97 per cent pure and has an oral LD_{50} to the rat of 1,400 milligrams per kilogram.   It is formulated as a 50 per cent emulsive concentrate.   *Fenson* is *p*-chlorophenyl benzene sulfonate, m.p. 61 to 2°C., d 1.33.   It is formulated as a 20 per cent wettable powder and emulsive concentrate.

Ovex   Genite   Fenson

Chlorbenside is *p*-chlorobenzyl *p*-chlorophenyl sulfide, a white crystalline solid, m.p. 74°C., v.p. 2.6×10^{-6} mm. Hg at 20°C.   It is insoluble in water and has the following solubilities in grams per 100 grams of solvent at 20°C.: kerosene, 5 to 7.5; methyl ethyl ketone, 137; xylene, 93. The technical product contains about 90 per cent *p*-*p'* isomer, 5 per cent *o*-*p'* isomer, and 2.5 per cent *m*-*p'* isomer.   Chlorbenside is unaffected by reduction and by acid and alkaline hydrolysis but is readily oxidized to *p*-chlorobenzyl *p*-chlorophenyl sulfoxide m.p. 125°, and more slowly to *p*-chlorobenzyl *p*-chlorophenyl sulfone, m.p. 150°.   These reactions occur on the leaf surface, and the oxidation products are acaricidal, but do not penetrate locally into the leaf tissue as does chlorbenside.   Chlorbenside is formulated as a 20 per cent wettable powder and 25 per cent emulsive concentrate and is active only against eggs and immature mites.   The oral LD_{50} to the rat is $> 10,000$ milligrams per kilogram.

Fluorbenside or *p*-chlorobenzyl *p*-fluorophenyl sulfide, m.p. 36°C., v.p. 8.0 × 10⁻⁵ mm. Hg at 20°C. is used in aerosol formulations for the control of greenhouse mites. It is about 6 times as soluble in petroleum solvents and 20 times as volatile as chlorbenside. The technical material consists of 95 per cent *p-p'* isomer, and the compound undergoes the same oxidative reactions as chlorbenside. Fluorbenside has an oral toxic dose to the rat of 3,000 milligrams per kilogram.

Cl⟨ ⟩CH₂S⟨ ⟩Cl   Cl⟨ ⟩CH₂S⟨ ⟩F

Chlorbenside   Fluorbenside

Aramite is 2-(*p-tert*-butylphenoxy)-isopropyl 2'-chloroethyl sulfite. The technical material is a brownish oil, b.p. 175°C. at 1 mm. Hg, d 1.148, and contains at least 90 per cent of the active ingredient. It is insoluble in water and soluble in aliphatic solvents and miscible in aromatic solvents. Aramite hydrolyzes in alkali to form 1-*p-tert*-butylphenoxypropan-2-ol, ethylene oxide, and inorganic sulfite and under strong sunlight liberates SO₂. Aramite is formulated as a 15 per cent wettable powder and 25 per cent emulsive concentrate and has an oral LD₅₀ to the rat of 3,900 milligrams per kilogram.

$$\underset{\text{Aramite}}{ClCH_2CH_2O\overset{O}{\underset{\;}{S}}OCHCH_2O\langle\;\rangle C(CH_3)_3}$$
CH₃

Several other acaricides have more selective action or specific use. *Neotran* is bis-(*p*-chlorophenoxy) methane, m.p. 70 to 2°C., which is formulated as a 40 per cent wettable powder and has an acute oral LD₅₀ to the rat of 5,800 milligrams per kilogram. It is effective against all stages of the citrus and European red mites.

Cl⟨ ⟩OCH₂O⟨ ⟩Cl

Neotran

Azobenzene, C₆H₅N=NC₆H₅, is an orange dyestuff, m.p. 68°C., b.p. 297°, d 1.203, which is effective as an ovicide for spider mites in greenhouses as a spray or when volatilized by burning 45 per cent pyrotechnic mixture or from a 70 per cent paste applied to steam pipes. Eight grams are used per 1,000 cubic feet. Azobenzene is insoluble in water and soluble in most organic solvents.

The fungicide *zineb* or zinc ethylene-bis-dithiocarbamate has proved to be especially effective for the control of the citrus rust mite.

$$\begin{array}{c}\text{S}\\ \text{H} \quad \| \\ CH_2N-C-S \\ \;\;\;\;\;\;\;\;\;\;\;\;\;\searrow Zn \\ CH_2N-C-S \nearrow \\ \text{H} \quad \| \\ \text{S} \end{array}$$

Zineb

Miscellaneous Organic Insecticides. *Phenothiazine*[1] or thiodiphenyl-
amine is a yellowish to green greasy powder, m.p. 185°C.

Phenothiazine

It has the disadvantage of being readily oxidized in light and air to form
the related compounds phenothiazone and thionol, which are less effective
to insects. For this reason spray deposits of phenothiazine have not
weathered well and the material has given erratic performances. Phenothiazine is insoluble in water but is slightly soluble in organic solvents.
It has been used for the control of codling moth larvae, as a mosquito
larvicide, and as an internal parasiticide for warm-blooded animals. An
interesting use for it has been the demonstration that feeding it to cattle
at the rate of 0.1 gram per kilogram of body weight will prevent the
development of horn fly larvae in the dung of the animals.

Diphenylamine,[2] , m.p. 53°C., is used for the control of screw-worm larvae in wounds.

Soil Poisons and Wood Preservatives. Several chemicals are widely
used as soil poisons for subterranean termites and as wood preservatives
against termites, powder post beetles, and carpenter ants. *Pentachlorophenol*, C_6Cl_5OH, is a brownish flaky material, m.p. 175 to 180°C., which
has the following solubilities in grams per 100 grams of solvent at 20°C.:
water, 0.0014; pine oil, 32; *o*-dichlorobenzene, 8.5; Stoddard's Solvent,
1.5; diesel oil, 3.1; and carbon tetrachloride, 2. The sodium salt is soluble
in water to 26 grams per 100 cubic centimeters and is converted to the free
phenol below pH of 6.8. The material is highly irritating to the skin,
nose, and eyes and should be handled with care. *o-Dichlorobenzene*,
$C_6H_4Cl_2$, is a colorless to yellowish liquid, b.p. about 179°C., sp. gr. 1.305
to 1.313 at 15°C. It crystallizes at 7°C. The technical material is a
mixture of 75 to 85 per cent *o*-isomer and 15 to 25 per cent of dissolved
para-dichlorobenzene (page 386). *Trichlorobenzene*, $C_6H_3Cl_3$, is a clear
liquid, b.p. 205 to 250°C., sp. gr. 1.460 to 1.477, which crystallizes at
10°C. The technical product is a mixture of the 1,2,3 and 1,2,4 isomers.
Creosote and *carbolineum* are crude coal-tar distillates.

Sulfur, Lime-sulfur, and Other Sulfur Compounds. Sulfur and its
compounds are among the most important contact poisons. The following list will indicate some of its important uses:

A. Sulfur Dusts:
1. Especially toxic to mites, such as chiggers and red spider mites, for thrips, for
   newly hatched scale insects, and as a stomach poison for some caterpillars.
2. Important as a fungicide carrier for other dust insecticides. The following
   mixture or some modification of it is extensively used on cotton.

   DDT. . . . . 5 pt.
   Toxaphene . . . . 5 pt.
   Dusting sulfur . . . . 50 pt.
   Dusting talc . . . . 40 pt.

[1] *U.S.D.A., Bur. Ento. Plant Quar., Cir.* E-480, 1939.
[2] *Jour. Econ. Ento.*, **27**:1176–1185, 1934; **28**:727–728, 1935; **37**:796–808, 1944.

B. *Mechanical Spray Mixtures of Sulfur with Other Substances:*
1. Wettable sulfur: sulfur so treated as to be readily miscible in water is widely used in spray mixtures, especially for its fungicidal effect.
2. Flotation sulfur is an extremely finely powdered sulfur which is more active chemically than flowers of sulfur.
3. Self-boiled lime and sulfur:
   Sulfur .... 16 lb.
   Unslaked lime (high-grade quicklime) .... 16 lb.
   Water... 100 gal.
   The sulfur is intimately mixed with the lime by adding it as the lime begins to slake. The mixture is cooled before a chemical combination of the lime and sulfur takes place.
4. Colloidal sulfur: made by passing sulfur fumes into soap and water or glue and water or by passing hydrogen sulfide (H_2S) gas into a solution of sulfur dioxide (SO_2).

C. *Chemical Compounds of Sulfur with Calcium, Ammonia, Potassium, and Barium:*
Used as a dormant spray for many scale insects and fungus diseases and as a foliage spray especially for fungus diseases, thrips, and various mites.
1. Liquid, fire-boiled lime-sulfurs.
2. Dry lime-sulfur.

Sulfur and lime are both practically insoluble in water, but when boiled together, they combine to form a series of salts, some of which are soluble and others insoluble. Lime-sulfur is a mixture of the following compounds:

CaS_5—Calcium pentasulfide
CaS_4—Calcium tetrasulfide
CaS_2O_3—Calcium thiosulfate
$CaSO_3$—Calcium sulfite

The first three are soluble in water. The calcium sulfite is comparatively insoluble, together with any uncombined sulfur or excess lime, magnesium, or iron from impure lime, it settles out and forms the sediment or "sludge" in home-boiled lime-sulfurs. In the commercial lime-sulfurs the sludge is removed before shipping. Of the first three compounds the pentasulfide and tetrasulfide, together called polysulfides, are believed to be the valuable ingredients, and it is to them that the lime-sulfur owes its killing power as well as its characteristic red color.

Solutions of lime-sulfur were first used as stock dips; they were tried as insecticides in California in 1886, and have, since about 1902, been a standard remedy for controlling certain scale insects, particularly the San Jose scale. These solutions are also efficient fungicides and are used in a large number of combinations. They are very disagreeable to use because of their caustic properties, which cause them to burn the face, eyes, and hands of the operator, and because of their odor. Lime-sulfur solutions are compatible with calcium arsenate, nicotine, lime, and sulfur but are incompatible or of doubtful compatibility with all other commonly used insecticides and fungicides.

Lime-sulfur solutions can be made at home by cooking together fresh or stone lime and sulfur. However, the material in use today is almost all of standardized commercial production. A number of formulas are in use, of which the following is typical:

Lump or stone lime................ 100 lb.
Commercial ground sulfur.......... 200 lb.
Water, enough to make............ 100 gal.

Heat about one-third of the required amount of water and to this add the lime. As soon as the lime starts slaking, add the sulfur, which should have been previously mixed thoroughly with enough water to make a thick paste.   Then add the remainder of the water and boil for from 45 minutes to 1 hour, adding more water as necessary to keep up to the original level.   If cooked by steam, as is usually done, a mechanical agitator should be provided.   When the free sulfur has all disappeared, the mixture should be strained and may be stored in barrels, tanks, or cisterns.[1]

In order to use lime-sulfur solutions intelligently, their strength or concentration must be tested.   This is done by means of a hydrometer.   The concentrated home-boiled and commercial lime-sulfurs must always be diluted.   Different batches of the home-boiled preparations will vary somewhat in strength.   The many brands of commercial lime-sulfur on the market usually test about 33° Bé.   The 33° material should be diluted, 1 part to 7 or 8 of water, for use as a dormant or winter spray on fruit trees; or about 1:49 for a summer spray.   Lime-sulfur which has been drawn from different levels in storage tanks or barrels will give varying readings on the hydrometer.   It is better to test the diluted spray mixture when it is ready to be applied rather than to depend on a dilution table.   When the application is to be made to dormant trees, *the diluted lime-sulfur mixture should give a reading of about 5° on the Baumé hydrometer, or 1.035 specific gravity.   For summer spraying, the reading should be about 1° Bé., or about 1.005 specific gravity.*   However, since conditions vary so widely in different localities, the state experiment station should be consulted regarding the best dilution to use for various insects and diseases.

Dry Lime-sulfur.   There are a number of dry powdered materials on the market as substitutes for the liquid lime-sulfur.   Dry lime-sulfur is liquid lime-sulfur with nearly all the water removed.   The chemical content of the material is slightly changed during the drying.   Tests by several experiment stations indicate that these materials are about as effective as liquid lime-sulfur when mixed so as to give the same content of sulfur in the dilute spray that occurs in liquid lime-sulfur.   If made to contain the same amount of sulfur as liquid lime-sulfur, the materials are much more expensive.   They are also troublesome because of the amount of sludge in the dilute material.   On the other hand, they are much more convenient to ship, store, and handle than liquid lime-sulfur.   A typical analysis of a dry lime-sulfur shows active ingredients CaS_5 and CaS_4, about 65 to 70 per cent; CaS_2O_3, about 5 per cent; free sulfur, about 5 to 10 per cent; and inert ingredients, about 20 per cent.

Oils and Emulsions.   Although recommended as insecticides as early as 1763, oils were probably little used until well along in the nineteenth century.   At first petroleum, turpentine, and kerosene were applied, unmodified, and, while very toxic to the insects, they also sickened or killed the plants.   The next attempt was to mix the kerosene mechanically with water, but such mixtures were very unstable and often resulted in disastrous destruction of trees and other foliage.   Consequently oils

[1] *Ill. Agr. Exp. Sta., Cir.* 492, 1939.

did not become satisfactorily useful until entomologists learned how to overcome the severe effects upon plants by emulsifying the oils in water. The first of such emulsions were made from kerosene, about 1870. In 1874 a good formula for a kerosene, soap, and water emulsion was discovered; from that time on, oils have become increasingly important and formulas and uses have multiplied continually. In 1904 the first commercial emulsion or miscible oil was placed on the market. In 1919 to 1923 lubricating-oil emulsions were found to have great efficiency in killing San Jose scale. About 1930 certain highly refined neutral or white oils, free from unsaturated hydrocarbons, acids, and highly volatile elements, were found safe to use upon plants in foliage, and thus the field for oil sprays was greatly enlarged.

The more important fields of usefulness for oil insecticides are:

1. As dormant (winter) sprays for scale insects, mites, insect eggs, and some hibernating caterpillars.

2. As summer or foliage sprays for aphids, mealybugs, mites, thrips, psyllids, whiteflies, and scale insects.

3. As parasiticides for lice, fleas, and mites on animals.

4. As a carrier for contact insecticides to increase their effectiveness.

5. Mixed with lead arsenate and other stomach poisons to increase their effectiveness.

6. As attractants in poison baits (page 323).

The term *oil* is a generic expression for a great variety of chemical substances which are composed principally, if not exclusively, of carbon and hydrogen. At ordinary temperatures they are greasy fluids, insoluble in water but readily soluble in ether, chloroform, carbon disulfide, carbon tetrachloride, and the like. They are readily inflammable, are slightly lighter than water, and have a peculiar property of penetrating readily into the pores of dry substances.

There are three great classes of these oils: (*a*) The *fixed oils*, such as linseed, soybean, castor, neat's-foot, fish oil, and many others, derived from both plants and animals, which are essentially glycerides that saponify or form soaps with alkaline bases, setting free glycerin. Fish oil, which is important in making insecticidal soaps, and soybean oil are those which have been most used for insecticides. (*b*) The *volatile* or *ethereal oils*, which differ from the fixed oils in not being greasy or viscous and not being capable of saponification. They are derived from special glands of plants and usually have a very pungent odor characteristic of the plant. Common examples are menthol, camphor, eugenol, oil of peppermint, wintergreen, and citronella. Their chief use in entomology is as attractants in baits and as repellents. (*c*) The *petroleum oils*, which are derived from sedimentary rocks, are, for the entomologist, by far the most important class. They are complex solutions of hundreds of hydrocarbons from which are derived natural gas, kerosene, gasoline, lubricating oils, asphalt, tar, and many other mixtures.

In order to understand the effects of different oils upon insects and upon plants, the following properties must be considered:

1. *Volatility or distillation range*, which is measured by the temperature at which the particular fraction distills. The lower the volatility or "heavier" the oil, *i.e.*, the higher the boiling point, other things being equal, the more effective oils are in killing insects. For dormant spraying, the oil should distill, at an even rate, about 90 per cent of its volume

between 590 and 700°F., and not over 2 per cent at 230°F., for 4 hours. Unfortunately, the heavier oils are also the more toxic to plants, so that the practical problem is to find the lightest oil that will kill the insect or the heaviest that can be used with safety on the particular plant.

2. The second important property to consider concerning spray oils is their *viscosity*, which is defined as resistance to flowing—the opposite of fluidity. It is measured in arbitrary units, which are the number of seconds required for a given volume (60 cubic centimeters) of the oil to flow through a standard orifice at a definite temperature (100°F.) (Saybolt test). While it is customary to speak of oils of low viscosity as "thin" and those of high viscosity as "heavy," it should be understood that viscosity is a very different property from specific gravity. Other properties being equal, oils of low viscosity are safer to use upon foliage than those of high viscosity. For dormant sprays on deciduous trees, an oil having a viscosity between 100 and 200 seconds (Saybolt), at 100°F., is considered satisfactory, a lower range often being used in more northern areas and a higher range in warmer territory.

3. The third property of great importance in spray oils is their *purity* or *degree of refinement*. An oil of the proper volatility and the right viscosity for spraying may still be quite unsuitable because of the presence of impurities, especially unsaturated and aromatic hydrocarbons. These compounds are chemically active and easily oxidized, causing the oils to become turbid and acid in reaction. It is generally believed necessary to remove 85 to 100 per cent of these unsaturated hydrocarbons, if the oil is to be used as a foliage spray; whereas, for dormant spraying, the removal of 65 to 75 per cent of the unsaturated hydrocarbons is satisfactory. The refining is brought about chiefly by treating with sulfuric acid and washing out the resulting sludge. For any oil of unknown purity, the degree of refinement may be determined by treating a sample with sulfuric acid. If the unsaturated hydrocarbons have already been removed, this treatment will not remove any portion of the oil. Such an oil is said to be "100 per cent unsulfonatable" or to have a "purity of 100 per cent." "Superior" dormant oils, which are about 25 to 30 per cent more efficient than the "regular" dormant oils, have been developed in New York State. The comparable specifications for these two grades are as follows:

Property	Regular	Superior
Viscosity, Saybolt sec. at 100°F.	90–120	90–120
Viscosity index, kinematic (minimum)	65	100
Gravity, A.P.I. deg. (minimum)	28	31
Unsulfonated residue, per cent	78–85	90–92
Pour point	Not greater than 30°F.	
Homogeneity	Relatively narrow boiling petroleum distillate	

The following classification of spray oils for foliage application is used in California:

Grade	Distillation at 636°F., %	Unsulfonated residue (U.R.), %	Viscosity, Saybolt sec. at 100°F.
Light...............	64–79	90	55–65
Light-medium.......	52–61	92	60–75
Medium...........	40–49	92	70–85
Heavy-medium......	28–37	92	80–95
Heavy.............	10–25	94	90–105

Spray oils are generally applied as mechanical mixtures of oil and water in which the oil is very evenly dispersed as small droplets in the water and kept so, at least until it has been applied to the plant, by the use of the *emulsifier* (page 400). Almost any degree of permanency or quick breaking desired can be produced by using oils of different viscosity, by varying the nature or amount of the emulsifier used, or by varying the size of the droplets of oil produced at the time of dispersal. The nature and amount of emulsifier also determine the amount of oil retained or "built up" on the plant surface and hence must be gaged carefully to give just enough deposit to kill the insect but to avoid the excess which will kill the bark or spot the foliage or fruit. Occasionally, oils may be applied directly to plants as fine mists or aerosols.

Oil sprays kill insects and mites and their eggs by enveloping them in a continuous film of oil which interferes with their respiration and ultimately causes death by suffocation. Spray oils prepared from paraffinic-base crudes have, in general, proved more effective than those derived from naphthenic-base crudes.

Three types of spray oil formulations are in general use: (*a*) *oil emulsions* are fluids or pastes consisting of 80 to 90 per cent oil emulsified with a small amount of water to form stock preparations which mix readily when added to water in the spray tank; (*b*) *emulsive oils* contain 97 to 99 per cent oil and a dissolved emulsifier but no water and are emulsified in the spray tank by agitation and pumping; the *miscible oils* are preparations of this type which contain sufficient emulsifier to form stable emulsions with very slight agitation; (*c*) *tank-mix oils* are prepared immediately before use by adding the oil and the emulsifier and spreader[1] separately to the water in the spray tank and emulsifying by agitation and pumping. This type of formulation forms an especially quick-breaking emulsion. The oil content of an emulsion is always expressed in per cent by volume. The following rule may be used to calculate readily how to dilute a stock emulsion with water to give any desired amount at any desired concentration: *multiply the amount of spray desired by the per cent of oil desired and divide the product by the per cent oil in the stock or concentrate.* Thus, if one desires to make 50 gallons of a 3 per cent oil spray from a 90 per cent oil stock, multiply 50 by 3 and divide by 90, which gives a quotient of 1⅔, the number of gallons of concentrated stock needed to make the 50 gallons of spray. If hard water is to be used for diluting the stock emulsion, a stabilizer such as caustic soda, casein, glue, or gelatin is often

[1] Blood albumin 1 part to fuller's earth 3 parts is commonly used at 4 to 8 ounces to 100 gallons of spray.

added to the stock emulsion before dilution to prevent the breaking of the emulsion.

The advantages of oils may be stated briefly as low cost compared to most other insecticides, their "creeping" or covering capacity, ease of mixing, and pleasantness of handling. Their disadvantages are their relatively low toxicity to most insects, their lack of stability in storage, danger of injuring plants, injury to spray hose, and the lack of fungicidal value as compared with lime-sulfur. Soap-oil emulsions are generally not compatible with lime-sulfur, the arsenicals, or fluosilicates; while those made with inert emulsifiers are compatible with nearly all standard fungicidal and insecticidal materials. Among the many kinds of emulsions recommended for various purposes, some of the better known are the lubricating-oil emulsions, the white- or summer-oil emulsions, coal-tar-oil emulsions, distillate-oil emulsions, kerosene emulsions, carbon bisulphide emulsions, and the miscible oils. All of them can be purchased already emulsified, under various trade names.

Soaps. A soap is a salt of a fatty acid, such as oleic, stearic, or palmitic acid derived from animal or vegetable oils, and an alkali-metal base such as sodium or potassium hydroxide. The sodium base forms hard soaps, the potassium base, soft soaps. Most soaps when dissolved in water at sufficient strengths have value as contact insecticides. They have been used against insects since 1787. Potash fish-oil soaps are the most widely used for spray purposes. Potash vegetable-oil soaps of cocoanut, linseed, soybean, or corn oil are of equal merit and do not have such a disagreeable odor. Rosin fish-oil soap at 1 ounce to the gallon or 6 pounds to 100 gallons, or a good commercial insecticide soap, applied on calm humid days or during a light rain or mist, when the evaporating power of the atmosphere is very low, is surprisingly effective against many kinds of pests. Sodium soybean-oil soap has been found of value in the control of insects. Soaps are used chiefly in the preparation of emulsions and as spreaders, wetting agents, and stabilizers for nicotine and pyrethrum sprays. The potash soaps are readily soluble in alcohol or cresol and are used in solution in oils to form the miscible oils. Soaps may be used in sprays containing lead arsenate, especially if hard water has to be used, in which case the soap may free the hard water of soluble calcium and magnesium salts.

FUMIGANTS

Gaseous poisons used to kill insects are called fumigants. Their application is generally limited to plants or products in tight enclosures or those which can be enclosed in relatively gastight tents or wrappings or to soil. Fumigants may be used to combat all kinds of insects, regardless of the type of mouth parts, since the gas enters the insect body through the spiracles during respiration.

The necessity for a successful fumigant to vaporize readily at room temperature greatly limits the number of insecticides which may be usefully employed for fumigation. In general, compounds boiling at about room temperature such as hydrogen cyanide, methyl bromide, and ethylene oxide are the most useful. However, for soil fumigation, the slower release of vapor from substances boiling as high as 180°C. has proved most effective. Substances of relatively high vapor pressure such as naphthalene and *para*-dichlorobenzene exert fumigant action in tight containers, and azobenzene, lindane, DDVP, Phosdrin, and other substances discussed under Contact Insecticides may sometimes kill by vapor action. The use of the more difficultly vaporizable insecticides as fumigants is enhanced by the use of atomization, volatilization from thermostatically controlled heating devices (page 447), or by burning in pyrotechnic mixture (page 446).

The various fumigants often exhibit a marked degree of specificity in their action against various insect pests, as shown in Table 7.1. The choice of the proper fumigant for any specific purpose is determined not only by the relative effectiveness but also by cost; safety to human beings, animals, and plants; flammability; penetrating power; effect on germination of seeds; reactivity with household furnishings; etc.

Fumigants are employed to control a great variety of insects under many different conditions.

A. Not Involving Living Plants:
  1. Mills, Factories, Storerooms, Packing Plants, Warehouses, Groceries, Museums, Insect Collections, and Herbariums:
      a. For pests of food products, including seeds, cereal products, candies, dried fruits, meats, cheese, nuts, and tobaccos.
      b. For pests of carpets, rugs, upholstery, woolens, furs, skins, dried plants, wood, paper, and leather goods.
  2. Human Habitations, Hotels, Prisons, Barracks, Hospitals, Theaters, and Camps:
      a. For any of the pests mentioned in 1.
      b. For bed bugs, lice, fleas, and mosquitoes.
  3. Ships, Railway Passenger, Freight, and Bunk Cars, and Automobiles:
      a. For any of the pests in 1 or 2.
      b. For plant pests which are under quarantine and likely to be transported by rolling stock.
      c. For rats which are hosts of fleas, the carriers of bubonic plague.
  4. Vacuum and Industrial Fumigation in Especially Constructed Chambers or Vaults, Usually in Partial Vacuum:
      a. For bales of cotton, human baggage, bags of flour, and other materials requiring great penetration.
      b. For packages of cereals, tobaccos, candies, and the like, to make sure they leave the factory free from living insects.
  5. Fumigation of Soil: for ants, grubs, wireworms, nematodes, and many other subterranean pests.
  6. Fumigation of the Intestinal Tracts of Animals: for bots and intestinal worms.

B. Involving Living Plants:
  7. Fumigation of Nursery Stock:
      a. For subterranean pests.
      b. For scale insects.
      c. For plants being imported from foreign states or countries.
  8. Fumigation of Cavities in Trees or Buildings: for borers, nests of bees, and the like.
  9. Fumigation of Citrus Trees and Other Plants Involving the Use of Portable Tents or Boxes: for scale insects and other pests not controlled by sprays or dusts.
  10. Greenhouse Fumigation: for all kinds of greenhouse pests.
  11. Open Field Fumigation: rarely used for cinch bugs, grasshoppers, and other pests.

General Directions for Fumigating. The first law of fumigation is to safeguard human lives. No one should undertake the use of these dangerous chemicals unless he has been thoroughly trained in the procedure

TABLE 7.1. COMPARATIVE TOXICITY OF FUMIGANTS TO STORED-PRODUCTS INSECTS[1]
LD$_{50}$, Milligrams per Liter, for 6 Hours

Fumigant	Sitophilus granarius	Stegobium paniceum	Tribolium confusum	Acanthoscelides obtectus	Oryzaephilus surinamensis	Rhyzopertha dominica
Acrylonitrile	2.0	1.7	3.0	1.1	0.8	0.8
Carbon bisulfide	43.0	42.0	75.0	29.0	40.0	31.0
Chloropicrin	3.4	1.9	6.4	<1.5	<1.5	<1.5
Ethylene dibromide	3.0	2.8	3.4	10.2	0.9	3.0
Ethylene dichloride	127.0	77.0	53.0	49.0	39.0	65.0
Ethylene oxide	13.5	9.0	27.5	10.5	4.0	6.2
Hydrogen cyanide	4.6	<0.4	0.8	0.9	<0.4	0.8
Methallyl chloride	25.0	10.0	27.0	18.0	19.0	25.0
Methyl bromide	4.8	4.4	9.2	4.2	4.4	3.4

[1] LINDGREN, D., L. VINCENT, and H. KROHNE, Jour. Econ. Ento., **47**:923, 1954.

and has adequate gas masks to protect the operators themselves. It is not safe to fumigate one room or apartment while other parts of the building are being occupied. Buildings being fumigated should be plainly placarded and securely locked or guarded, so that no one may enter until they have been thoroughly ventilated. In general it will be well to avoid windy or cold weather. Regardless of the kind of fumigant to be used, great care is required to make the enclosure to be fumigated as nearly airtight as possible. In dwellings, greenhouses, mills, and storerooms all cracks, broken glass, chimney holes, ventilators, and other openings must be very tightly closed. Strips of gummed paper or even 4-inch strips of newspaper, soaked in water or flour paste or smeared with heavy grease, will be found useful to paste over cracks; pieces of wallboard, cut to fit, for large openings; and rags or cotton waste for small holes. For large cracks up to 5 or 6 inches wide, a paste made by mixing ground asbestos with calcium chloride 3:1 and enough water to make a thick "putty" is excellent, as the hygroscopic calcium chloride keeps the paste from drying out and cracking away. Masking or scotch tape, plastic clays, and heavy sheets of kraft paper sealed with Plastic Elastic are also excellent aids in sealing a building. More failures in fumigation are due to insufficient preparation of the building than to any other cause.

Where living plants are being treated, especially greenhouse and citrus plants, the dosage used will have to be very carefully determined; the duration of exposure is rarely over 1 hour; and ventilation must be carefully timed. The temperature must be between 40 and 80°F., and for greenhouse plants, between 55 and 68°F., and it should *rise slowly* to prevent condensation of moisture upon the leaves and resultant burning by the gas. Generally treatment must be made at night or in darkness, since sunlight during or within an hour before or after fumigation is likely to cause injury. Where no living plants are involved, the aim will be to use a slight excess of the fumigant; to leave the materials exposed as long as convenient; or, if the building does not need to be entered, to neglect ventilation altogether; to have the temperature high— between 70 and 100°F.; and to neglect the trend of temperature and the amount of light as of little importance. Dry food substances may be used after treatment with the common fumigants, but those containing moisture may be poisoned.

Before fumigating any building, provision should be made for ventilating it after the fumigation is complete. If a gas mask, correctly charged for the fumigant being used, is available, it is exceedingly useful in setting off the charge and for entering the building afterward to open doors and windows. Otherwise, several doors or windows must be arranged so that they can be opened *from the outside.*

Dosage. The dosage for fumigants is commonly expressed in terms of pounds per 1,000 cubic feet. Successful fumigation is dependent upon the attainment of a critical Ct value, which will kill at least 99 per cent of the insect population (the product of the concentration of gas per unit volume times the duration of the exposure). Therefore it is necessary to know the volume of the enclosed space to be treated and the rate of release of the fumigant vapor. It is apparent that the longer the period of exposure, the lower the concentration of gas required. However, in most cases the exposure time is limited because of escape of gas from the enclosure and for convenience in treatment. Dosage and the attainment of the critical Ct value are affected by the sorption of the fumigant by the commodity being fumigated. Methyl bromide, for example, reacts with proteins and is rapidly sorbed by grain, flour, and seeds, etc. The rate of sorption also depends upon the moisture content of the product. Between the temperatures of about 50°F. (10°C.) and 90°F. (32°C.), the Ct value decreases by approximately one-half for each 10°C. rise in temperature. These relationships are shown in Fig. 7.1. In practice, the calculation of dosage is usually based upon the amount of fumigant which experimentation has proved to be necessary to kill the pest over a standard period of exposure. Where living plants are involved, the dosage is limited to the amount they can tolerate without injury.

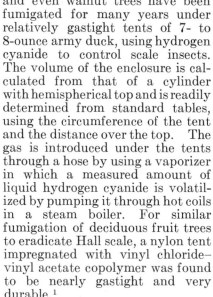

Tent Fumigation. Citrus trees and even walnut trees have been fumigated for many years under relatively gastight tents of 7- to 8-ounce army duck, using hydrogen cyanide to control scale insects. The volume of the enclosure is calculated from that of a cylinder with hemispherical top and is readily determined from standard tables, using the circumference of the tent and the distance over the top. The gas is introduced under the tents through a hose by using a vaporizer in which a measured amount of liquid hydrogen cyanide is volatilized by pumping it through hot coils in a steam boiler. For similar fumigation of deciduous fruit trees to eradicate Hall scale, a nylon tent impregnated with vinyl chloride-vinyl acetate copolymer was found to be nearly gastight and very durable.[1]

FIG. 7.1. Influence of sorption, moisture content, and temperature on Ct of methyl bromide to *Tribolium confusum.* (*Data from Lindgren and Vincent, Jour. Econ. Ento.*, 1960.)

Methyl bromide fumigation is very widely used for the control of termites and powder post beetles in lumber and dwellings, for stored-product insects in mills and warehouses and in bagged commodities, and for soil insects. For this type of fumigation, the structure or commodity is covered with plastic sheeting, rolled and clamped at the edges and sealed at the bottom with soil or lengths of canvas tubing filled with sand. The gas is introduced through tubing from cylinders. Polyethylene or polyvinyl chloride plastic sheeting 0.004 to 0.006 inch thick is very effective in retaining methyl bromide[2] and is light, flexible, and reasonably durable and inexpensive.

Vacuum Fumigation. The lack of penetration of fumigants under atmospheric pressure is overcome and the time required for effective fumigation is greatly reduced by using the gas in a partial vacuum. This method is used particularly with hydrogen cyanide, carbon bisulfide,

[1] *U.S.D.A., Cir.* 978, 1955.
[2] PHILLIPS, C., and H. NELSON, *Jour. Econ. Ento.*, **50**:452, 1957.

ethylene oxide, methyl bromide, and methyl formate in industrial fumigation. The products to be treated are placed in a tight-sealing steel chamber or vault. The air is pumped out until a 27- to 29-inch mercurial vacuum (686 to 737 millimeters), or 1 to 3 inches absolute pressure is reached. The fumigant, thoroughly mixed and heated to about 120°F., is then introduced until atmospheric pressure is attained, penetrating very completely through the bales, bundles, sacks, or packages to effect complete destruction of insect life in 1 to 2 hours. The gas is then pumped out, the treated materials are "air washed" by admitting air, and they are ready immediately for storage or shipment in a clean sanitary condition. The reduction in oxygen content resulting from the removal of the air also makes the insects much more susceptible to the effects of the gas; carbon dioxide is frequently mixed with the toxic gas (except in the case of hydrogen cyanide), and this further stimulates the pests to respire the poison rapidly. This method is used especially for imported products likely to be infested with dangerous foreign insects; for packaged foods, confections, and tobaccos; and for eradicating human lice from clothing, bedding, and baggage of soldiers, hobos, and the inmates of public institutions. All cotton and cotton waste must be vacuum fumigated to meet quarantine regulations for entry into the United States to prevent the introduction of the pink bollworm. Potatoes shipped out of certain states are vacuum fumigated to prevent spread of the potato tuber moth. Nursery stock is often so treated to assure its freedom from scale insects. Thousands of tons of dried fruits, nuts, candies, coffee, cereal, seeds, tobaccos, vegetables, furniture, books, furs, mattresses, and other articles of commerce, packed in cellophane or other cartons, boxes, bags, barrels, or bales, are vacuum fumigated every year. This process does not leave any distasteful or poisonous residue, nor does it bleach, stain, or corrode containers. In this way is avoided the great loss of prestige and patronage that inevitably results when such products are received infested with worms, moths, or weevils. The special chambers are available in sizes ranging from a few cubic feet to fumigators large enough to treat an entire railway car. The cost of treatment is generally less than $\frac{1}{4}$ cent a pound.

Hydrogen cyanide or hydrocyanic acid, HCN. Of the chemicals employed as fumigants for insects, hydrogen cyanide is the most extensively used. It was originally employed to protect insect collections from destruction by museum beetles and was first widely employed against the cottony-cushion scale (page 414) on citrus trees in California in 1886–1887. It was used for nursery stock in 1890, for greenhouse insects in 1894, for household insects in 1898, and mill insects in 1899. Vacuum fumigation was perfected in 1913, liquid hydrogen cyanide was first recommended in 1915, and calcium cyanide became available in 1916.

Hydrogen cyanide is a volatile, colorless liquid, b.p. 26°C., m.p. −14°C., sp. gr. 0.699 at 20°C., which has an odor of bitter almond. The gas has a specific gravity of 0.9483, and 1 pound occupies about 14 cubic feet at atmospheric pressure. Hydrogen cyanide burns freely in air and is dangerously inflammable and explosive in mixtures with air above 5.6 per cent or above 4 pounds per 1,000 cubic feet. Hydrogen cyanide is soluble in water and in alcohol and ether. As a fumigant the gas diffuses rapidly, but because it is lighter than air, the diffusion is mainly upward and outward. Under normal atmospheric pressure, the gas does not readily penetrate closely packed materials such as piles of grain, sacks of flour, or bales of cotton. This lack of penetration may be overcome by using the fumigant in a partial vacuum.   Hydrogen cyanide will tarnish metals such as brass, gold, or nickel, but the tarnish is easily removed by rubbing with a polishing cloth, or it may be prevented by coating the metal with grease.   It has no effect on the colors of most fabrics or papers.

Hydrogen cyanide is formed from the cyanides of alkali metals, especially sodium cyanide, calcium cyanide, and potassium cyanide. Calcium cyanide is sufficiently unstable that it yields its gas simply upon exposure to moist atmosphere, but the other salts require treatment with an acid such as sulfuric acid in order to liberate the gas rapidly. The reaction when sodium cyanide, sulfuric acid, and water are mixed together is

$$NaCN + H_2SO_4 \rightarrow NaHSO_4 + HCN\uparrow$$

The principal present uses of this powerful insecticide cover nearly all the conditions listed on page 371.

There are four principal methods of applying this gas in any enclosure:

1. By *the pot method*, in which the gas is generated from earthenware jars or paraffined barrels placed within the enclosure to be fumigated. This is done by mixing water and sulfuric acid in the container and, while this mixture is still hot, as a result of the chemical action, adding the proper amount of sodium cyanide.   This method, which at one time was used almost exclusively, has been largely supplanted by the other methods, but it is still often the best and cheapest method to use, especially where great accuracy of dosage and small amounts of the gas are needed.   A number of containers, of sufficient capacity that the contents do not come within 4 inches of the top, must be provided as generators. For greenhouse work, ½-gallon glazed earthenware jars are excellent; for household fumigation, 3- or 4-gallon jars are generally used; and for very large jobs, paraffined wooden barrels have been used.   Not more than 5 pounds of cyanide should be used in a 3- or 4-gallon generator, and there must be at least one for each room.   The generators should be placed at intervals over the enclosure, depending on its size, and the proper amount of water should be placed in them.   The correct amount of acid should then be carefully poured into the water.   Considerable heat is generated by the reaction of the acid and water, and, unless the generators are of good grade, they may break, which may result in the burning of floors or injury to the operator.   To protect rugs and floors, the generators may be set in boxes or tubs of soil or ashes.   The cyanide should previously have been accurately weighed into paper bags and placed beside the containers.   Having everything in readiness, the operator should quickly drop each bag into its generator, starting with those farthest from the exit, and at once leave the room.   If more than one floor of a building is to be treated, the fumigation should start on the upper floors.   The gas soon forms an equal concentration in all parts of a building.   Following fumigation and immediately after ventilating, the residue remaining in the containers should be removed and buried or poured into the sewer.   The cyanide used should be of C.P. grade.   Cyanide containing more than a trace of sodium chloride, or sodium nitrate, is not suitable for fumigation purposes.   The dosage of hydrogen cyanide gas required for different insects will vary with the insect and with the conditions of treatment.   For fumigation where the treatment of living plants is not involved, the usual proportions are *for each* 1,000 *to* 1,500 *cubic feet of space:*

Water...................................... 3 pt.
Commercial sulfuric acid (sp. gr. 1.83).......... 1½ pt.
Sodium cyanide[1] (98% pure)................... 1 lb.

In the treatment of living plants the proportions used are about the same, but the dosage is greatly reduced. For greenhouse plants the following formula is generally safe to use, *for each* 1,000 *cubic feet*. This dosage will need to be increased for the more resistant insects or may have to be reduced if very tender plants are to be treated.

Water.. ⅜–¾ fl. oz. (11 to 22 cc.)
Commercial sulfuric acid (sp. gr. 1.83).......... ³⁄₁₆–⅜ fl. oz. (5.5 to 11 cc.)
Sodium cyanide[1] (98% pure)................... ⅛–¼ oz. (3.5 to 7 g.)

2. In *the machine method*[2] a carefully measured quantity of sulfuric acid is introduced to a solution of sodium cyanide in a special chamber or machine, or sodium cyanide in solution may be admitted to a larger quantity of sulfuric acid; thus successive charges of the gas are generated and led from the machine into the enclosure to be fumigated through pipes leading from the machine. This method has been used chiefly in fumigating citrus trees covered with portable tents.

3. In *the liquid method*[3] hydrogen cyanide, stored in the liquid form in steel containers, is pumped out in the desired quantity and liberated through pipes and nozzles to the area to be treated. Twenty cubic centimeters (13 grams) of liquid hydrogen cyanide is equivalent to 1 ounce of sodium cyanide by the pot method, or 1 pound is considered equivalent to 2¼ pounds of sodium cyanide and sufficient for about 3,000 cubic feet in well-built mills or dwellings. This method is limited to cases where large amounts of the gas are needed at one time. It has been used extensively in sterilizing warehouses and citrus trees.

4. In *the dry method* calcium cyanide or liquid hydrogen cyanide, absorbed in inert earths or crude paper disks, is hermetically sealed in cans, and the gas is liberated by opening the tins and scattering the dry material over the floor or blowing it into the atmosphere. This, the newest method, is also the simplest. When these calcium cyanides are exposed to the atmosphere, the following reaction takes place:

$$Ca(CN)_2 + 2H_2O \rightarrow Ca(OH)_2 + 2HCN\uparrow$$

or

$$CaH_2(CN)_4 + 2H_2O \rightarrow Ca(OH)_2 + 4HCN\uparrow$$

It will therefore be evident why the atmosphere must be moist and also that the residue remaining after the reaction has gone to completion is harmless calcium hydroxide. There are two principal types of calcium cyanide on the market. One form in common use[4] contains from 40 to 50 per cent calcium cyanide and gives off its gas very slowly. It will give off about half as much hydrogen cyanide as an equal amount by weight of sodium cyanide. Another form of dust[5] contains 88

---

[1] Potassium cyanide is now little used in fumigation. About one-fourth less hydrogen cyanide is given off where this form of cyanide is used.

[2] See *U.S.D.A., Farmers' Bul.* 1321, 1923.

[3] See *Calif. Agr. Exp. Sta. Bul.* 308, June, 1919; and *U.S.D.A., Farmers' Bul.* 1321, 1923.

[4] Cyanogas.

[5] Calcyanide.

per cent calcium cyanide and will yield about 90 per cent as much gas as an equal weight of sodium cyanide. In this form of dust the gas is generated very rapidly upon exposure to the moist air. *No acid is required to generate the gas from calcium cyanide dusts, the necessary reaction taking place when it is simply exposed to the atmosphere.* Calcium cyanide has been used for killing scale insects on citrus trees by spreading it on the ground or by blowing the calcium cyanide dust into tents placed over the trees. For household insects, mill insects, and greenhouse insects the proper amount of the dust is spread out, not over ⅛ inch thick, on papers, on the floor, or on the walks in the greenhouse. It is used very extensively for the control of rodents by placing or blowing a small amount of material into their burrows. Calcium cyanide in flake form has also been used in combating chinch bugs, to kill the bugs trapped by barriers. The dosage to be used will depend upon the nature of the materials to be treated, the tightness of the building or other enclosure, and the kind of cyanide dust used. With an 88 per cent calcium cyanide dust, ¾ pound to each 1,000 cubic feet of space is recommended for dwellings, mills, etc., and ⅛ ounce per 1,000 cubic feet is about a minimum dosage for ordinary greenhouse fumigation. With a 48 per cent dust the dosage should be about twice the above recommendation. In greenhouse fumigating it is recommended that the above dosage be tried in an experimental way. If a satisfactory kill is not secured, the dosage should be gradually increased, provided no burning results. If the plants are injured, the dosage must be reduced. The correct dosage for particular plants and insects can be determined by consultation with a trained entomologist. In any case, the plants should not be watered for some hours previous to the fumigation. Hydrogen cyanide appears not to conflict with the use of any other insecticides except bordeaux mixture, which should not be used on plants either before or after fumigation with this gas.

The other type of dry cyanides[1] undergoes no chemical change when exposed, but the absorbed liquid hydrogen cyanide, when exposed to the air, undergoes a change of state, simply evaporating to form the deadly gas, leaving the harmless, inert, carrier material. Some commercial cyanide products contain a small per cent of another gas, such as chloropicrin, as a warning gas.

It has been discovered that for some insects the effectiveness of hydrogen cyanide can be increased about 30 per cent by applying the required amount in fractional or tandem doses, rather than all at once. For example, the required dose of liquid hydrogen cyanide may be divided into four equal parts and one such part applied every 2 hours, so as to keep the concentration of the gas at about 3.5 milligrams per liter of air in the enclosure.

Hydrogen cyanide is one of the most deadly gases, quickly killing animals and also killing plants when used in too large doses or for too long exposures. Its toxic action to insects and warm-blooded animals results from its ability to combine with the iron atoms present in iron-containing respiratory enzymes such as cytochrome oxidase (page 112). Extreme care is necessary when fumigating with hydrogen cyanide to prevent injury to the operator or to others who may be exposed. Its use should never be undertaken by inexperienced persons. Operators working with it should be protected by gas masks especially charged to

[1] Zyklon products.

absorb hydrogen cyanide. All persons must be out of the building treated; it is not safe to fumigate certain rooms or apartments while other parts of the building are occupied. Occupants of adjoining buildings should be notified to keep their windows closed. The building being treated should be securely locked and placarded and all fires extinguished. Liquid foodstuffs in unsealed containers must be removed before treating. The building and contents must be thoroughly aired before it is reoccupied. This requires special care in cold weather, including a thorough airing after being heated. The safe limit for prolonged human exposure to hydrogen cyanide is 10 p.p.m. (0.011 milligrams per liter).

Carbon bisulfide, CS_2. This chemical, also called carbon disulfide, was first used in France in 1858 to control grain-infesting insects, and its early development was in connection with the fight against the grape phylloxera (page 779).

Carbon bisulfide is a nearly colorless liquid, b.p. 46°C., m.p. −108.6°C., sp. gr. 1.263 at 20°C., v.p. 360 mm. Hg at 25°C., and flash point about −22°F. (−30°C.). It has a disagreeable odor due to traces of hydrogen sulfide and is slightly soluble in water (0.22 gram per 100 grams at 22°C.) and soluble in organic solvents. It is an excellent solvent for rubber, waxes, varnishes, oils, and plastics. The gas has a specific gravity of 2.63 and is highly inflammable and explosive in mixtures above 1 per cent in air. A flame of any kind, a lighted cigar, or even a spark from metal striking metal or from an electric switch may cause an explosion of the gas. Because of the fire hazard this chemical cannot be shipped by mail or express, and in some cases its use in buildings invalidates insurance. The gas does not discolor fabrics, but the impure commercial liquid leaves a tenacious yellowish residue.

The chief fields of usefulness for carbon bisulfide, at present, are (a) for insect pests in stored grains and seeds, in granaries, mills, and storerooms, where there is no fire hazard; (b) for insect pests of clothing, fabrics, furniture, etc., in houses, warehouses, etc.; (c) as emulsions, for soil insects and other vermin in the soil; (d) for borers in the wood of trees and insects such as honeybees in buildings; (e) for bots and intestinal worms in the stomachs of animals.

To use this gas, the sides and bottoms of the bins, rooms, or containers to be fumigated must be made as nearly airtight as possible. Where there is sure to be some leakage, the amount of carbon bisulfide must be increased above that given below. One should never attempt to fumigate a room or bin with large cracks or openings in the bottom or sides. The best results will be obtained at temperatures from 75 to 90°F. Do not fumigate when the temperature is below 60°F. One pound of carbon bisulfide to each 100 cubic feet of space or 80 bushels of grain, or 2 to 3 gallons per 1,000 bushels of grain, is usually sufficient. It cannot be depended upon to give a satisfactory kill to a depth of more than 6 feet in a bin of grain. The liquid may be applied directly to grain or seeds, but this may injure germination; better results will be obtained by pouring the carbon bisulfide on gunny sacks, rags, or cotton waste. This gives a rapid evaporation which is more effective than when the liquid is exposed in shallow pans. Another method sometimes used is to spray the liquid into the air through a nozzle or to hasten evaporation with hot water conducted through a pan of carbon bisulfide in coils of pipe. While not necessary, it is better to cover the tops of open bins with a tarpaulin or blanket. The room or bin should be kept closed from

36 to 60 hours. Such exposure will not injure the milling qualities of grain or the germination of most seeds and will leave no poisonous residues on feeds. For soil fumigation it is often made into an emulsion (see page 780), and for internal animal parasites it is given in gelatine capsules (see page 952).

The outstanding merit of carbon bisulfide is its superior penetrating powers, which make it effective in masses of grains or fabric where hydrogen cyanide and other light gases cannot reach. It is also simple to use, and because of the foul odor of the commercial form it is not likely to overcome anyone unawares. Seeds can generally be treated without injury to their germination if they are fairly dry when treated, and the use of carbon bisulfide as a soil fumigant stimulates the growth of many kinds of crops. The great limitation of carbon bisulfide is the fire hazard which attends its use. Commercial mixtures of 20 per cent carbon bisulfide, 80 per cent carbon tetrachloride, and a trace of sulfur dioxide, which are said to have a greatly reduced fire hazard, have been placed on the market, but the mixture cannot be made at home. It is recommended as effective at $1\frac{1}{2}$ to 2 pounds per 100 cubic feet of space or 3 to 5 gallons per 1,000 bushels of grain. It is also exceedingly toxic to plants, so that it cannot be used for greenhouse fumigation. There is also some difficulty at moderate temperatures in getting the liquid to evaporate rapidly enough to yield a toxic concentration.

Carbon bisulfide has a high degree of chronic toxicity, and the safe limit for prolonged human exposure is 20 p.p.m. (0.062 milligram per liter).

Sulfur dioxide, SO_2, is used to some extent as an insecticide but is generally inferior to the other fumigants described. The material is a colorless gas with an intensely irritating odor, b.p. $-10°C.$, m.p. $-72.7°C.$, v.p. 2,453 mm. Hg at 20°C., specific gravity of liquid 1.434. The gas has a specific gravity of 1.433 and is noninflammable, but with traces of moisture it forms sulfurous acid, which corrodes metals, ruins paint, and bleaches fabrics and wallpaper. Sulfur dioxide is available as a liquid in pressure cylinders and can be generated by burning sulfur moistened with alcohol, or a mixture of 58 per cent sulfur, 38 per cent potassium nitrate, and 4 per cent potassium chlorate, available as sulfur candles. A concentration of 5 to 10 per cent in air (4 to 8 pounds per 1,000 cubic feet) is effective for most insects. Sulfur dioxide destroys the germinating power of most grains and affects the baking quality of flour. It cannot be used on living plants, and the release of the gas from a large enclosure may kill the surrounding vegetation. The safe limit for prolonged human exposure is 10 p.p.m.

Nicotine, $C_{10}H_{14}N_2$. Nicotine, whose history and chemistry have been discussed under contact insecticides (page 325), is also used extensively as a greenhouse fumigant, where its high degree of safety both to the plants and to the operator have made it a favorite, and to some extent also against orchard and garden pests and poultry lice.

For fumigating, free nicotine and not nicotine sulfate must be used. The nicotine may be volatilized by painting or dropping the liquid over hot steam pipes, by heating it in shallow pans, or by forcing it through heated tubes. The nicotine is also absorbed in known amounts in paper punk, powders, and other combustible materials, from which it is liberated by burning. Such materials should smoulder; if they burn actively, most of the nicotine may be destroyed. Such a preparation is now marketed to be burned in the container.[1] Small holes are punched near the top of the container and the nicotine-impregnated powder is ignited with a "sparkler." The entire house may be fumigated, or the burning pressure fumigators may be attached to poles and carried along rows of plants, thus releasing the toxic gas in close contact with insects to be destroyed. With such a preparation it is possible to do "spot fumigating"; *i.e.*, a few plants or a single bench in a large greenhouse may be fumigated without treating the whole house. The dosage generally required to kill greenhouse aphids is from 0.5 to 1 ounce of 40 per cent free nicotine per 1,000 cubic feet with an overnight exposure. Fumigation of plants is best done under dry conditions at temperatures between 50 and 70°F. and should never be done in bright sunshine. Violets and ferns are among the few kinds of plants likely to be severely injured by nicotine fumigation.

For controlling aphids on peas, beans, cabbage, etc., a device known as a Nicofumer has been used, mounted on a truck, to pump liquid nicotine through heated pipes and to discharge the gas through pipes that are enclosed by the apron of a canvas trailer from 25 to 100 feet long, which holds the fumes about the plants for a brief exposure as it passes over them.

Poultry may be fumigated while roosting by painting the 40 per cent nicotine sulfate upon the perches (page 992). This will free them of lice and certain mites and ticks but will not kill their eggs.

Methyl bromide or bromomethane, CH_3Br, b.p. $-4.5°C$., m.p. $-93°C$., sp. gr. 1.732 at 0°C., v.p. 1,580 mm. Hg at 25°C., is one of the most widely used fumigants. Because of its low boiling point it is a colorless, almost odorless gas at ordinary temperatures, with a specific gravity of 3.29, 1 pound of which occupies about 4 cubic feet at atmospheric pressure. Methyl bromide is slightly soluble in water (1.34 grams per 100 grams at 25°C.) and is readily soluble in organic solvents. It is shipped and stored in sealed tins and is applied by sprinkling or spraying or in mixtures with carbon dioxide. The gas is noninflammable, and mixtures with air cannot be ignited by flame under conditions of ordinary fumigations.

Methyl bromide is the most penetrating of the commonly used fumigants because of its high specific gravity, and this and its low boiling point make it particularly useful for low-temperature fumigation. However, in comparison with hydrogen cyanide, methyl bromide kills insects very slowly, several days sometimes being required. Methyl bromide has the following important uses in insect control: (*a*) for insect pests of stored grains, seeds, dried fruits, and packaged foodstuffs in mills and warehouses, freight cars, and ships. A dosage of 1 to 3 pounds per 1,000 cubic feet is normally used over a 12- to 16-hour exposure. Methyl bromide imparts no objectionable taste or odor to foods and when properly utilized leaves no dangerous residue. Fumigation of dry seeds at normal dosages does not appreciably affect germination. (*b*) Methyl bromide is used for the termites, powder post beetles, and other household and structural pests in buildings, lumber, and household goods. Such structure fumigation is widely practiced using gastight tents of plastic film and a dosage of 2 to 4 pounds per 1,000 cubic feet. (*c*) Fumigation of plants and plant products in greenhouses for pest control at 1 pound per 1,000 cubic feet and of fruits and vegetables, cut flowers, and nursery stock with dosages of 2 pounds per 1,000 cubic feet. This procedure is especially

[1] Such as the Nicofume products.

valuable in connection with plant quarantine regulations against such pests as fruit flies, sweetpotato weevil, Japanese beetle larvae, etc. Some varieties of plants such as azaleas and evergreens are injured by methyl bromide, which also affects the ripening of tomatoes and damages sweet-potatoes. (*d*) Methyl bromide is used for soil fumigation to control insects, nematodes, fungi, and weeds, usually under a plastic tent, using a dosage of about 4.7 milliliters per square foot.

Methyl bromide is a dangerous cumulative poison to warm-blooded animals, affecting the central nervous system and causing disturbances of vision and equilibrium and, in cases of acute poisoning, delirium, convulsions, and death. The appearance of symptoms is sometimes delayed for several months after exposure. The safe limit for prolonged human exposure is 20 p.p.m. (0.1 milligram per liter), and a gas mask should always be worn when working with this fumigant. Methyl bromide is a very active alkylating agent and reacts with sulfhydryl groups (—SH) of enzymes such as succinic dehydrogenase, this reaction accounting for its toxic action to insects and warm-blooded animals. It also reacts with proteins in plant tissues, forming methylated derivatives and leaving residues of inorganic bromide.

Chloropicrin or trichloronitromethane, CCl_3NO_2, is one of the tear gases employed to some extent as an insecticide. At room temperature it is a colorless to yellowish stable liquid, b.p. 112.4°C., m.p. −64°C., sp. gr. 1.651 at 20°C., v.p. 24 mm. Hg at 25°C. The specific gravity of the gas is 5.7, giving it remarkable penetrating power. Chloropicrin is noninflammable and has no bleaching or tarnishing effects but corrodes some metals slightly. It is slightly soluble in water (0.162 grams per 100 milliliters at 25°C.) and is soluble in organic solvents.

Chloropicrin has been employed as a fumigant in flour mills, grain elevators, and bins. It is sprayed into the atmosphere or poured upon sacks laid over materials to be treated. Sacked seeds may be fumigated by injecting a small amount into each sack. Because of its low volatility chloropicrin may be mixed with carbon tetrachloride or ethylene dichloride to promote evaporation and distribution. A dosage of 1 to 3 pounds per 1,000 cubic feet is generally used for most grain and household insects. It is so irritating to insects that it will drive them from holes in nuts or cracks in furniture, to die in the open. However, chloropicrin cannot be used to fumigate growing plants, and it will injure the germination of certain seeds. Chloropicrin is used as a soil fumigant for insects, centipedes, and nematodes and is usually injected into the soil in a solvent such as carbon tetrachloride, ethylene dichloride, or xylene or as a water emulsion at about 5 milliliters per square foot.

Chloropicrin is extremely poisonous to warm-blooded animals, but is so highly lachrymatory and irritant to mucous membranes that dangerous concentrations cannot be tolerated. Tears are produced at concentrations above 2.4 p.p.m. (0.017 milligrams per liter), and 20 p.p.m. (0.12 milligrams per liter) is highly toxic. Chloropicrin, because of its low volatility, is very difficult to remove from premises or materials fumigated, and its powerful irritant effect is therefore likely to cause serious inconvenience. Therefore it is unwise to employ this material in routine fumigations of inhabited structures.

Carbon tetrachloride, CCl_4, is a colorless liquid, b.p. 76.8°C., m.p. −23.0°C., sp. gr. 1.595 at 20°C., v.p. 114.5 mm. Hg at 20°C. The gas has a specific gravity of 5.31 and a pungent chloroform-like odor and is not inflammable or explosive. Carbon tetrachloride is almost insoluble in water (0.08 gram per 100 grams at 25°C.) but is soluble in organic solvents. It is a slow-acting fumigant of comparatively low toxicity to insects and consequently is used alone only where fire hazards are acute or for small-scale fumigation. However, when used with other fumigants such as carbon bisulfide or ethylene dibromide, it greatly decreases the fire hazard. It is also of value as a diluent to increase the volatilization and distribution of other fumigants such as methyl bromide, ethylene dichloride, and chloropicrin.

Carbon tetrachloride is highly toxic to warm-blooded animals, producing severe kidney and liver damage, and can be absorbed through the skin. The safe limit for prolonged exposure is 25 p.p.m. (0.175 milligrams per liter).

Ethylene dichloride or 1,2-dichloroethane, $CH_2Cl \cdot CH_2Cl$, is a colorless liquid, b.p. 83.7°C., m.p. -36°C., sp. gr. 1.257 at 20°C., v.p. 78 mm. Hg at 25°C. It is slightly soluble in water (0.87 gram in 100 grams at 20°C.) and soluble in organic solvents. The specific gravity of the gas is 3.4, and the vapor is inflammable between 6 and 16 per cent in air, with a flash point of about 54°F. Ethylene dichloride is of moderate toxicity as a fumigant and is slow to kill insects, death sometimes occurring from 1 to 3 days after exposure. Emulsions of ethylene dichloride in water are used to control the peach tree borer (page 759) and as a soil fumigant for Japanese beetle larvae, etc. Ethylene dichloride is, however, generally used as a fumigant in a 3-1 mixture by volume with carbon tetrachloride. This is entirely free from fire hazard either as a liquid or a gas. The mixture is stored and transported in ordinary cans and is used by pouring or spraying over cloths or directly upon the material to be fumigated and allowing it to vaporize. In unfurnished rooms it may be sprayed about freely, even upon rugs, clothing, or upholstery, but it ruins paints and varnishes. Ethylene dichloride is rather slow to volatilize, and forced circulation of air over the liquid is advisable. The necessary dosages of the mixture range from 10 to 18 pounds per 1,000 cubic feet in airtight vaults at atmospheric pressure for 24 hours. In trunks, clothes chests, or tightly sealed plastered closets, 3 to 4 gallons per 1,000 cubic feet (3 fluid ounces to 5 cubic feet) are required for a 24-hour exposure. The temperature should be above 75°F. Ethylene dichloride is clean, safe, and pleasant to use and is not injurious to the germination of most seeds. However, it cannot be used on growing plants. Food substances rich in fats absorb appreciable amounts of the gas and require long periods of aeration to remove the taint, and the gas is said to leave a disagreeable odor in tobacco.

Ethylene dichloride is toxic to warm-blooded animals, and prolonged breathing of the vapor should be avoided. It is also absorbed through the skin. Symptoms of poisoning include dizziness, headache, and nausea. The safe limit for prolonged exposure is 100 p.p.m. (0.43 milligram per liter).

Ethylene dibromide or 1,2-dibromoethane, $CH_2Br \cdot CH_2Br$, is a colorless liquid, b.p. 131.6°C., m.p. 9.3°C., sp. gr. 2.172 at 20°C., v.p. 11 mm. Hg at 25°C. The gas has a specific gravity of 6.5 and is slightly soluble in water (0.43 gram per 100 grams at 30°C.) and soluble in most organic solvents. It is noninflammable. Ethylene dibromide is an important fumigant for fresh fruits and vegetables and at 0.5 pound per 1,000 cubic feet destroys the larvae of fruit flies and permits shipment from quarantined areas without affecting the quality of the treated produce. Ethylene dibromide is one of the most important soil fumigants and is used to control wireworms, white grubs, and nematodes by soil injection at 1 to 4 gallons (18 to 72 pounds) per acre in 20 gallons of petroleum naphtha. It is toxic to many plants, and treated soil should be thoroughly aerated before planting.

Ethylene dibromide is highly toxic to warm-blooded animals and can be absorbed directly through the skin, where it may act as a vesicant. The safe limit for prolonged exposure is 25 p.p.m.

β,β'-*Dichloroethyl ether*, $CH_2ClCH_2OCH_2CH_2Cl$, is a colorless liquid, b.p. 178°C., m.p. −50°C., sp. gr. 1.222 at 20°C., v.p. 0.73 mm. Hg at 20°C. It is slightly soluble in water (1.1 grams per 100 grams at 20°C.) and soluble in most organic solvents. The vapor has a specific gravity of 4.9 and is inflammable, with a flash point of 131°F. β,β'-Dichloroethyl ether is used as a soil fumigant for wireworm, sod webworms, and in greenhouses. It is phytotoxic, and treated soil must be thoroughly aerated before planting. The compound is highly toxic to warm-blooded animals and is especially irritant to eyes and mucous membranes in high concentrations. The safe limit for prolonged exposure is 15 p.p.m.

1,3-*Dichloropropene*, $CH_2Cl \cdot CH{=}CHCl$, is widely used as a soil fumigant. The technical mixture[1] is a colorless to brownish liquid, sp. gr. 1.210 at 25°C., v.p. 31.3 mm. Hg at 20°C., is slightly soluble in water (0.1 gram per 100 grams at 20°C.), and is soluble in organic solvents. It consists of about equal parts of *trans*-isomer b.p. 104°C., sp. gr. 1.224 at 20°C., and *cis*-isomer b.p. 112°C., sp. gr. 1.217 at 20°C. The *trans*-isomer is about twice as active as the *cis*-isomer. Another preparation[2] contains in addition about one-third of 1,2-*dichloropropane* or propylene dichloride $CH_2Cl \cdot CHClCH_3$, b.p. 95.4°C., sp. gr. 1.159 at 20°C., v.p. 210 mm. Hg at 19.6°C. It is slightly soluble in water (0.27 gram per 100 grams at 20°C.) and is soluble in organic solvents.

Dichloropropene is used as a soil fumigant for wireworms, centipedes, and nematodes by soil injection at 200 to 400 pounds per acre. It may injure the germination if seeds are planted within 2 to 3 weeks after treatment. It is inflammable, with a flash point of 70°F., and like the other chlorinated hydrocarbon fumigants, is toxic to warm-blooded animals. Dichloropropene is especially irritant to the eyes and mucous membranes and can be absorbed by the skin where it is a vesicant.

Ethylene oxide, $H_2C\overset{\displaystyle O}{\underset{\textstyle}{\diagup\diagdown}}CH_2$, b.p. 10.7°C., m.p. −111.3°C., sp. gr. 0.887 at 7°C., v.p. 1,095 mm. Hg at 20°C., is a colorless gas with a sweetish odor at ordinary temperatures. The gas sp. gr. 1.5, is inflammable and forms explosive mixtures with air between concentrations of 3 and 80 per cent. Ethylene oxide is miscible with water and most organic solvents. It is stored and transported in steel cylinders under pressure.

Ethylene oxide is of moderate toxicity as a fumigant, and the standard dosage in tightly sealed buildings is 3 pounds per 1,000 cubic feet. Because of its low boiling point, it is unusually effective where fumigation must be done at temperatures below 70°F. Ethylene oxide leaves no residual taste or odor on such sensitive products as tobacco, coffee, or nut meats. It has no effect upon the baking or milling qualities of grains

[1] Telone.
[2] D-D.

but may destroy their germinating power. It is not corrosive to metals or injurious to fabrics or finishes. The gas cannot be used to fumigate living plants, and the germination of seeds may be seriously affected. Ethylene oxide is said to kill the eggs of insects before the active stages, so that products freed of visible stages will not be likely to contain viable eggs to reestablish the infestation. Because of its inflammability at concentrations slightly above the effective dosage, ethylene oxide is always used in combination with carbon dioxide. The ethylene oxide from pressure cylinders (1 part) may be poured over crushed, solid carbon dioxide "dry ice" (9 parts), and the mixture shoveled into bins of grain or other enclosures and allowed to vaporize. Three pounds of ethylene oxide and 27 pounds of carbon dioxide are used for each 1,000 bushels of grain. A commercial product[1] consisting of 1 part by weight of ethylene oxide and 9 parts of carbon dioxide as a homogeneous liquid is available in pressure cylinders. Since the two substances have about the same vapor pressures, the gases remain intimately mixed when released from the cylinder. The carbon dioxide makes the material noninflammable, reduces the absorption of the toxicant gas by the materials being fumigated, so that much smaller dosages are effective, and by stimulating the insects to much more rapid respiration, causes them to be killed more quickly. The principal use of this gas mixture is in the sterilizing of packaged-food products, tobaccos, candies, nuts, dried fruits, and furs, in vacuum-fumigating chambers. Ten to thirty pounds of the mixture per 1,000 cubic feet at a 27-inch vacuum effectively destroys all insect life in from 1 to 3 hours. In airtight vaults at atmospheric pressure the same dosage is effective in from 12 to 48 hours.

Ethylene oxide is moderately toxic to warm-blooded animals and is extremely irritating to the eyes and nose in high concentrations. The safe limit for prolonged human exposure is 100 p.p.m. (0.18 milligrams per liter).

Phosphine or hydrogen phosphide, PH_3, b.p. $-87.8°C.$, sp. gr. 1.1829 at $0°C.$, is used to fumigate grain in sacks and bins.[2] The gas is conveniently generated from 3 gram tablets containing aluminum phosphide and ammonium carbamate, which, in the presence of moisture, produce a total of 1 gram of phosphine by the following reaction:

$$AlP + 2NH_4OCNH_2 + 3H_2O \rightarrow PH_3\uparrow + Al(OH)_3 + 4NH_3 + 2CO_2$$

Zinc phosphide, Zn_3P_2, and calcium phosphide, Ca_3P_2, are also used to generate phosphine. By this means of controlled release, phosphine, which is extremely poisonous to warm-blooded animals and often ignites spontaneously in air, can be used safely. The safe limit for prolonged human exposure is 0.05 p.p.m.

Sulfuryl fluoride,[3] SO_2F_2, b.p. $-55.21°C.$, liquid density 1.36 grams per milliliter at $21°C.$, gas density 3.72 grams per milliliter at $0°C.$, is especially useful for the control of dry-wood termites and other structural and household pests. The material is a colorless, odorless, noninflammable, and noncorrosive gas, soluble in water to about 0.1 per cent. It is conveniently released from pressure cylinders and used in structures sealed by plastic tarpaulins, at dosages of 2 to 4 pounds per 1,000 cubic

[1] Carboxide.
[2] LINDGREN, D., L. VINCENT, and R. STRONG, *Jour. Econ. Ento.*, **51**:900, 1958.
[3] STEWART, D., *Jour. Econ. Ento.*, **50**:7, 1957.

feet. Sulfuryl fluoride is of moderate toxicity to warm-blooded animals (safe limit for prolonged exposure is 100 p.p.m.) but is highly phytotoxic to plants.

Para-dichlorobenzene, Cl⟨ ⟩Cl, is a white crystalline material, m.p. 53°C., sp. gr. 1.4581 at 20°C. It vaporizes slowly (b.p. 173.4°C., v.p. 1 mm. Hg at 25°C.) to form a noninflammable gas with an ether-like odor and a specific gravity of 5.1. About 0.7 pound will saturate 1,000 cubic feet of air at 30°C. *Para*-dichlorobenzene is very slightly soluble in water (0.008 gram per 100 grams at 25°C.) and is soluble in organic solvents. It is used as a soil fumigant for the larva of the peach tree borer (page 759) and as a household fumigant and repellent for clothes moths and carpet beetles by scattering the crystals in closets, chests, and through stored clothing. It is also used to protect museum specimens from attack by dermestid beetle larvae. *Para*-dichlorobenzene vapor does not bleach or discolor metals or fabrics. The safe limit for prolonged human exposure is 75 p.p.m. The closely related soil poison, *ortho*-dichlorobenzene, is discussed on page 364.

Naphthalene, [structure], a white crystalline compound, m.p. 80°C., sp. gr. 1.1517 at 15°C., is widely sold as the familiar "moth balls" and moth flakes. It vaporizes slowly (b.p. 218°C., v.p. 0.08 mm. Hg at 25°C.) to form a vapor, sp. gr. 4.4, with a pungent, tarry odor which is inflammable in mixture with air. About 0.06 pound will saturate 1,000 cubic feet of air at 30°C. Naphthalene is very slightly soluble in water (0.003 gram per 100 grams at 20°C.) and is soluble in organic solvents. It is about as effective as *para*-dichlorobenzene for the control of clothes moths and carpet beetles and is used to destroy the gladiolus thrips in flower bulbs (page 865) and to a limited extent as a soil fumigant for the carrot rust fly (page 679) and wireworms. It has also been used as a fumigant in greenhouses to control spider mites and thrips by vaporizing $\frac{1}{8}$ to 2 ounces per 1,000 cubic feet at 70 to 85°F. over an 8- to 15-hour period.

Trichloroethylene, ClCH=CCl$_2$, b.p. 87°C., v.p. 73 mm. Hg at 25°C., sp. gr. 1.470 at 15°C., is a noninflammable fumigant which has been used in mixtures with ethylene dichloride.

β-Methallyl chloride, CH$_2$=C(CH$_3$)CH$_2$Cl, b.p. 72°C., sp. gr. 0.925 at 20°C., and 1,1-*dichloro-1-nitroethane,*[1] CH$_3$CCl$_2$NO$_2$, b.p. 124°C., sp. gr. 1.415 at 20°C., have been used as grain fumigants with 3 to 5 parts of carbon tetrachloride, at 1 to 2 gallons per 1,000 bushels of grain.

Methyl formate, HCOOCH$_3$, b.p. 32°C., v.p. 624 mm. Hg at 25°C., and sp. gr. 0.974 at 20°C., is used for industrial fumigation in vacuum vaults. It can be obtained in cylinders as a mixture of 60 ounces of methyl formate in 50 pounds of liquid carbon dioxide[2] and is used at 15 to 30 pounds of the mixture per 1,000 cubic feet. *Ethyl formate,* HCOOC$_2$H$_5$, b.p. 54°C. sp. gr. 0.906 at 20°C., and *isopropyl formate,* HCOOCH(CH$_3$)$_2$, b.p. 68 to 71°C., sp. gr. 0.873 at 20°C., are used in the package fumigation of dried fruits.

Other compounds have shown considerable promise as fumigants: *acrylonitrile,* CH$_2$=CHCN, b.p. 78°C., sp. gr. 0.801 at 25°C.; *trichloroacetoni-*

[1] Ethide.
[2] Proxate.

trile, CCl_3CN, b.p. 85°C., sp. gr. 1.44 at 25°C.; *cyanogen chloride*, $CNCl$, b.p. 12°C., sp. gr. 1.222 at 0°C.; *tetrachloroethane*, $CHCl_2 \cdot CHCl_2$, b.p. 146°C., sp. gr. 1.600 at 29°C.; *methyl thiocyanate*, CH_3SCN, b.p. 130 to 133°C., sp. gr. 1.069 at 24°C.; *propylene oxide*, $CH_3CH \overset{O}{\overset{\diagup \diagdown}{-\!\!-\!\!-}} CH_2$, b.p. 34°C., sp. gr. 0.8304 at 20°C.; and *ethylene chlorobromide*, $CH_2Cl \cdot CH_2Br$, b.p. 107°C., sp. gr. 1.689 at 19°C.

REPELLENTS

Substances which are only mildly poisonous, or which may not be active poisons but which prevent damage to plants or animals by making the food or living conditions of the insects unattractive or offensive to them, are called *repellents*. These substances are rarely, if ever, repellent to all kinds of insects. Such chemicals can sometimes be employed to advantage where it is impossible to use an insecticide and may afford a greater or less degree of protection to manufactured products, growing plants, or the bodies of animals. Among the many examples may be noted the following: (*a*) *Repellents against crawling insects*. Examples are the creosote lines used as barriers to the migration of chinch bugs (page 485); trichlorobenzene and other chemicals used to protect buildings from termites; ant tapes, usually containing bichloride of mercury, which are placed about table legs and the like to keep ants from crossing; heavy oils at the base of poultry roosts as a barrier to poultry mites; and certain chemical bands about tree trunks. (*b*) *Repellents against the feeding of insects*. Examples are the application of bordeaux, lime, and similar washes to plants to ward off leafhoppers and some chewing insects; the dusting of cucurbits to protect them from cucumber beetles; mosquito repellents and fly sprays to lessen the attacks of bloodsucking flies and mosquitoes; the application of sulfur to the body to keep chiggers from attacking; the use of smoke and smudges to repel biting flies; the chemical treatment of logs to keep beetle borers from destroying log cabins and other rustic work; moth balls, oil of cedar, and mothproofing treatments to protect materials from attack by clothes moths and carpet beetles. (*c*) *Repellents against the egg-laying of insects*. Examples are the use of pine-tar oil and diphenylamine to keep screw-worm flies from laying eggs about wounds of animals (page 970).

Bordeaux Mixture. Bordeaux mixture originated in France as a spray to control the downy mildew disease, *Peronospora viticola*, of grapes, about 1882, and was first used in the United States in 1887. While primarily a fungicide, bordeaux mixture is very repellent to many insects, such as flea beetles, leafhoppers, and potato psyllid, when sprayed over the leaves of plants. By the addition of lead arsenate or DDT, a poisoned bordeaux is made that is widely used as a preventive of, and a remedy for, insects and plant diseases on many crops. It is to some extent an ovicide with a certain residual toxic effect on the sap, killing leafhoppers and psyllids for some days after application.

Bordeaux mixture is made at different strengths for different purposes. The strength is generally indicated by numbers, 4–8–100 or 6–10–100, of which the first number designates the number of pounds of copper sulfate and the second the number of pounds of lime to be mixed with 100 gallons of water. The formula for the widely used 6–10–100 mixture is as follows (any other strength desired can be made by varying the amounts of lime and copper sulfate):

	Field Formula	Garden Formula
Water (cold)............................	100 gal.	3 gal.
Hydrated lime.........................	10 lb.	5 oz.
Pure copper sulfate (bluestone).........	6 lb.	3 oz.

For controlling chewing insects, the above mixture is generally poisoned by adding

Lead arsenate or DDT.................	2 to 4 lb.	1 to 2 oz.

When it can be secured, unslaked stone or rock lime can be used instead of the hydrated lime; 6 pounds to 100 gallons of water, or 1 ounce to 1 gallon, is sufficient. *Air-slaked lime should never be used.* Only wooden, earthenware, or glass containers should be used for mixing and storing bordeaux. For making small amounts, it is best to suspend, the evening before spraying is to be done, the proper amount of copper sulfate in a sack near the top of a tub, barrel, or spray tank containing about three-fourths of the water to be used. If this has not been done, the sulfate can be dissolved quickly in a little hot water and then diluted with cold water. Quick-dissolving powdered copper sulfates can be purchased from most dealers for use instead of the crystal form. Slake the lime, or mix the hydrated lime, in a small quantity of water, dilute it, and then pour the limewater into the copper sulfate solution and stir vigorously. When properly mixed, a beautiful blue, voluminous, colloidal precipitate of tetracupric sulfate, $4CuO \cdot SO_3$, and pentacupric sulfate, $5CuO \cdot SO_3$, is formed, which will remain in suspension for several hours. The precipitated particles are exceedingly thin and, when sprayed upon foliage, dry upon the leaves with remarkable covering power and tenacity. A gallon of good bordeaux has sufficient precipitate to cover 1,000 square feet of leaf surface. Its efficiency is said to depend upon the gradual formation upon the leaf surface of small quantities of soluble copper over a long period of time. For small gardens it is more convenient to buy one of the commercial bordeaux, but these are not generally so effective as the freshly mixed material.

To make good bordeaux observe the following precautions: (a) do not allow the copper sulfate solution or the bordeaux to stand in contact with any metal; (b) use only pure lime and copper sulfate; (c) have one or both of the chemicals diluted with nearly all the water before the other is added; (d) use water as cold as possible; (e) agitate thoroughly after mixing; and (f) use within a few hours after mixing. If spraying is interrupted, the mixture can be preserved for a day or two by adding an ounce of sugar to each 25 gallons and stirring thoroughly. Bordeaux is somewhat incompatible with soaps and must not be used before or shortly after fumigating plants with hydrogen cyanide. It is often combined with an arsenical or other stomach poison, with nicotine sulfate, with DDT, and with oils. It may, however, reduce the efficiency of oil emulsions for scale insects or whiteflies, either by the chemical reaction or because the bordeaux kills the fungi that parasitize the insects.

A number of miscellaneous materials have been used as repellents against the feeding of insects. These include creosote and coal tar used to protect wood from termite attack (page 364) and as repellent barriers against chinch bug migrations (page 485); *tetramethylthiuram disulfide,*

$$(CH_3)_2NC{-}S{-}S{-}CN(CH_3)_2,$$ which has shown considerable repellent
$$\overset{\|}{S} \qquad \overset{\|}{S}$$

properties to plant-feeding insects, especially the Japanese beetle; and *disodium-ethylene-bis-dithiocarbamate,*[1]

$$Na{-}S{-}\overset{\|}{\underset{S}{C}}{-}NH{-}CH_2CH_2{-}NH{-}\overset{\|}{\underset{S}{C}}{-}S{-}Na$$

which is reported to act in a similar manner.

Repellents to Bloodsucking Insects.[2] The use of repellents to prevent the attacks of blood-feeding and disease-transmitting insects upon man and animals represents one of the most spectacular and practical applications of repellency and was the subject of intensive investigations during the Second World War. The limiting criteria for a good repellent are, in order of importance: (a) effective protection of treated area for several hours, on all types of subjects and under all climatic conditions; (b) complete freedom from toxicity and irritation when regularly applied to human or animal skin; (c) cosmetic acceptability, including freedom from unpleasant odor, taste, and touch, and harmlessness to clothing; (d) protection against a wide variety of biting insects; and (e) cheapness and availability. Tests of many thousands of chemical compounds for repellent action to flies, mosquitoes, chiggers, fleas, and ticks have shown that while many possess a significant degree of repellency to the various pests, few meet all the other requirements satisfactorily. The materials which have found practical application as repellents are listed in Table 7.2, together with their properties and uses.

The various repellent compounds exhibit wide differences in their activity against various species of mosquitoes and flies, as well as other biting arthropods, and it has been demonstrated that the over-all protection against various pests is greatly extended by the use of mixtures such as the following:

Parts

(a) Dimethyl phthalate 3
  Indalone 1
  2-Ethyl-1,3-hexanediol 1
(b) Dimethyl phthalate 4
  2-Ethyl-1,3-hexanediol 3
  Dimethyl carbate 3

These repellents can also be incorporated into various creams and lotions, such as the following:

Parts

(c) Dimethyl phthalate 7
  Magnesium stearate 3
(d) Dimethyl phthalate 25
  White wax 19
  Peanut oil 56

[1] Nabam.

[2] *U.S.D.A., Misc. Publ.* 606, 1946; CHRISTOPHERS, R., *Jour. Hygiene,* **45**:176–231, 1947; *Jour. Econ. Ento.,* **39**:627, 1946; *Ann. Rev. Ento.,* **1**:181, 1956; *Advances in Pest Control Res.,* **1**:277, 1957.

TABLE 7.2. REPELLENTS FOR BLOODSUCKING INSECTS, MITES, TICKS

Name	Formula	Properties	Toxicity, LD_{50} to rat, mg./kg.	Uses
Dimethyl phthalate		Clear liquid, b.p. 282°C, sp. gr. 1.189	8,200	General-purpose mosquito repellent
Dibutyl phthalate		Clear liquid, b.p. 340°C, sp. gr. 1.045	21,000	Clothing impregnant for chiggers, mites
Cis-bicyclo-[2.2.1]-5-heptene-2,3-dicarboxylate Dimethyl carbate		White solid, m.p. 40°C.	1,000	Mosquito repellent used in mixtures
2-Ethyl-1,3-hexanediol "6-12"	$HOCH_2CHCHCH_2CH_2CH_3$ with C_2H_5, OH	Clear liquid, b.p. 244°C, sp. gr. 0.94	2,400	General-purpose repellent for mosquitoes, flies, fleas, mites
2-Ethyl-2-butyl-1,3-propanediol	$HOCH_2CCH_2OH$ with C_2H_5, C_4H_9	Solid, m.p. 40-2°C, sp. gr. 0.931	5,040	General-purpose mosquito repellent

Name	Structure	Physical properties		Uses
2-Phenylcyclohexanol		Solid, m.p 41°C.		General-purpose repellent for mosquitoes, flies, mites, ticks
n-Propyl-N,N-diethylsuccinamate	$C_3H_7OCCH_2CH_2CN(C_2H_5)_2$	Sp. gr. 1.01	6,400	Mosquito repellent
o-Chloro-N,N-diethylbenzamide				Mosquito repellent
N,N-Diethyl-m-toluamide Dclphene		Clear liquid, b.p. 111°/1 mm., sp. gr. 0.996	2,000	General-purpose repellent
n-Butyl-6,6-dimethyl-5,6-dihydro-1,4-pyrone-2-carboxylate Indalone		Brownish liquid, b.p. 110–15°/1 mm., sp. gr. 1.06	7,800	Mosquito and fly repellent, used in mixtures
Benzyl benzoate		Oily liquid, b.p. 323°C., m.p. 21°C., sp. gr. 1.12	1,900	Clothing impregnant for chiggers, mites

TABLE 7.2. REPELLENTS FOR BLOODSUCKING INSECTS, MITES, TICKS (Continued)

Name	Formula	Properties	Toxicity, LD$_{50}$ to rat, mg./kg.	Uses
Benzil		Solid, m.p. 95°C.		Clothing impregnant for chiggers, mites
Dibutyl adipate	$C_4H_9OCCH_2CH_2CH_2CH_2COC_4H_9$	Liquid, b.p. 183°/14 mm., sp. gr. 0.965	12,900	Tick repellent
N-Butylacetanilide		Liquid, b.p. 277–81°C., m.p. 22°C., sp. gr. 0.99	2,830	Clothing impregnant for ticks, fleas
Butoxypolypropylene glycol	C_4H_9O—$(CH_2CHO)_n$—CH_2CH—OH with CH_3, CH_3	Liquid, 400 and 800 molecular-weight fractions, d. 0.973–0.990	11,200	Fly repellent for cattle
Di-n-butyl succinate Tabutrex	$C_4H_9OCCH_2CH_2COC_4H_9$	Liquid, b.p. 108°C./4 mm., m.p. −29°C.	8,000	Fly repellent for cattle, repellent for household insects
2,3,4,5-Bis-(Δ²-butenylene)-tetrahydrofurfural MGK 11		Liquid, sp. gr. 1.121	2,500	Cockroach, biting flies on cattle
Di-n-propyl isocinchomeronate MGK 326		Amber liquid, sp. gr. 1.08, b.p. 186°/1 mm.	6,230	Fly repellent for cattle

392

For applications to the skin, a small amount (1 to 2 milliliters) of the repellent is poured into the cupped hands and rubbed over the face and neck, with care to prevent the entrance of the material into the eyes. The viscous compounds such as 2-ethyl-1,3-hexanediol and 2-phenylcyclohexanol are most suitably diluted with alcohol to facilitate application. Human subjects show great variation with regard to the effectiveness of repellents because of such factors as sweatiness, body odors, skin absorption, and spreading rates. However, applications of the materials discussed in Table 7.2 will give from 1 to 6 hours protection against mosquitoes and biting flies.

There is considerable reluctance on the part of many individuals to apply oily materials directly to the skin. Therefore, under many circumstances, the ideal applications of repellents are those made on clothing, gloves, and head nets where the protection time is extended to a week or more. Clothing applications may be made by rubbing or spraying the repellent on the cloth or by dipping the clothing in an emulsion of the repellent. The clothing-impregnation method is especially suited to military operations, and the formulation given in (e) has become the standard clothing impregnant, M-1960, of the U.S. Armed Services for protection from mosquitoes, fleas, ticks, and chiggers:

Parts

(e) Benzyl benzoate........................ 3
    N-Butylacetanilide...................... 3
    2-Ethyl-2-butyl-1,3-propanediol........... 3
    Tween 80............................... 1

Clothing impregnation may be made from acetone solution or preferably from an emulsion of a mixture of 90 per cent repellent and 10 per cent of an emulsifier of the nonionic type. If this is unavailable, soft soap, $\frac{1}{2}$ pint per gallon of water, is effective. Such treatments are applied at the rate of 2 grams per square foot or about 5 per cent of the weight of the cloth and will remain effective after several washings. Most of the repellents are solvents for lacquers and plastics and should not be applied to watch crystals, spectacle frames, synthetic fabrics, paints, or varnishes.

Mothproofing.[1] Fabric pests of which the clothes moths (page 915) and carpet beetles (page 912) are the most important have been estimated to cause from 200 to 500 million dollars damage annually in the United States to clothing, rugs, upholstery, and bookbindings. Therefore there is a great demand for chemical treatment to render these either poisonous or distasteful to fabric pests. Ideally, such mothproofing materials should be fixed in the cloth during dyeing and should be completely effective against the insect pests, give good protection after repeated washings and dry cleanings, and be nonvolatile, stable to light, odorless, colorless, and nontoxic to humans. Several colorless dyestuffs possess these properties and can be applied to woolen goods during the dyeing operation and fixed in the fibers by chemical reactions with the protein. The two most widely used compounds are *Eulan CN* and *Mitin FF*. Both are stated to give protection effective for the lifetime of the article when applied at 1 to 3 per cent of the weight of the cloth.

[1] WATERHOUSE, D., *Advances in Pest Control Res.*, **2**:207–262, 1958.

Eulan CN   Mitin FF

There are several simple processes for mothproofing which may be applied during dry cleaning or by the housekeeper. Sodium fluosilicate alone or with aluminum, magnesium, or ammonium salts, such as sodium aluminum fluosilicate, is an effective mothproofing agent when used at 0.5 to 0.7 per cent in water solution, preferably with 0.25 to 0.5 per cent wetting agent. A popular formulation[1] is stated to contain 0.6 per cent sodium fluosilicate, 0.3 per cent potassium alum, and 0.03 per cent oxalic acid. Such formulations may be applied by spraying or dipping the fabric and cannot be removed by dry cleaning. Fabrics impregnated with 0.25 to 0.75 per cent DDT or 0.05 per cent dieldrin, based on the dry weight of the cloth, are protected from moths and carpet beetles for periods of months to years if not washed or dry-cleaned repeatedly. These applications can be made readily by spraying the cloth with 0.5 per cent DDT or 0.05 per cent dieldrin in a volatile solvent such as petroleum naphtha or by spraying or dipping the fabric in an aqueous emulsion of equivalent strength. The mothproofing treatment can be given during dry cleaning by adding the insecticide to the cleaning bath; it is invisible and has no effect on the properties of the cloth.

Attractants.[2] Many substances are known which serve to attract insects by olfactory stimulation. Such attraction occurs in nature in the response of insects to odors emanating from foods, as sex attractants, and from prey or sites for oviposition. A few successful attempts have been made to use this principle to attract live insects into traps or to poison baits both for control and for the determination of population densities. Japanese beetle adults are trapped by an attractant mixture of 9 parts of *geraniol* and 1 part *eugenol*, which serves as a food lure.

$$(CH_3)_2C{=}CHCH_2CH_2C{=}CHCH_2OH$$
$$CH_3$$

Geraniol   Eugenol

[1] Larvex.

[2] DETHIER, V., "Chemical Insect Attractants and Repellents," McGraw-Hill-Blakiston, 1947; GREEN, N., M. BEROZA, and S. HALL, *Advances in Pest Control Res.*, **3**:129–179, 1960.

Attractants have proved especially suitable for evaluating populations of the fruit flies of the family Trypetidae. *Methyl eugenol* is strongly attractive to males of the oriental fruit fly, *Dacus dorsalis*, luring them from over ½ mile downwind, and this compound has been effectively used in poison baits and traps. *Anisyl acetone* is strongly attractive to the male melon fly, *D. cucurbitae*.

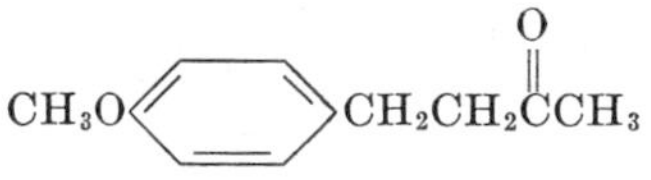

$$CH_3O \langle \underline{\qquad} \rangle CH_2CH_2\overset{\overset{\textstyle O}{\|}}{C}CH_3$$

Anisyl acetone

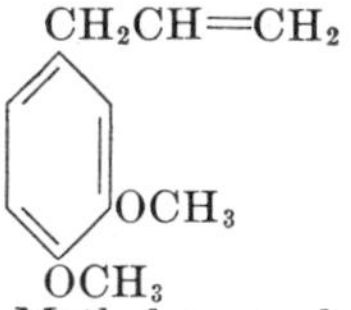

$$CH_2CH{=}CH_2$$
OCH₃
OCH₃
Methyl eugenol

The Mediterranean fruit fly is very strongly attracted to *sec*-butyl-6-methyl-3-cyclohexene-1-carboxylate or *siglure*, and subsequently *sec*-butyl 4-(or 5)-chloro-2-methyl cyclohexane carboxylate or *medlure* proved even more effective.

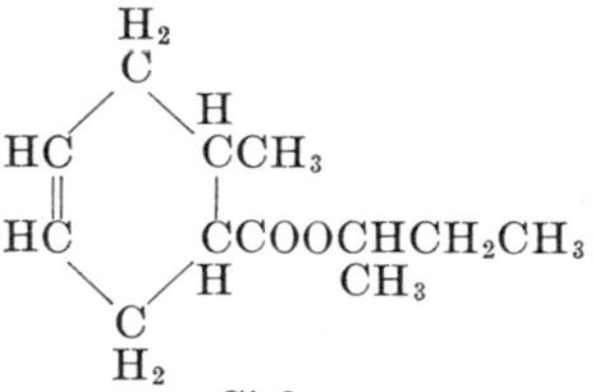

$$
\begin{array}{c}
H_2\\
C\\
HC\diagup\diagdown\!\!\overset{H}{C}CH_3\\
HC\diagdown\diagup\overset{\;}{C}COOCHCH_2CH_3\\
C\;\;\overset{H}{}\;\;CH_3\\
H_2
\end{array}
$$

Siglure

$$
\begin{array}{c}
H_2\\
C\\
HCH\diagup\diagdown\!\!\overset{H}{C}CH_3\\
ClCH\diagdown\diagup\overset{\;}{C}COOCHCH_2CH_3\\
C\;\;\overset{H}{}\;\;CH_3\\
H_2
\end{array}
$$

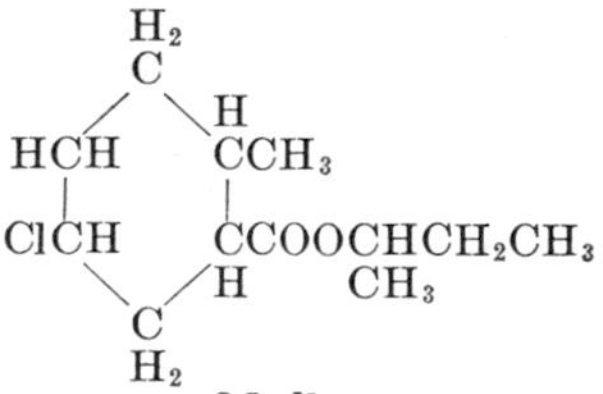

Medlure

The use of fermenting sugars and sirups as attractants for moths and butterflies is well known to every lepidopterist, and these have been supplemented by a variety of essential oils such as *anethol* for the codling moth and *isoamyl salicylate* for the tomato and tobacco hornworm moths.

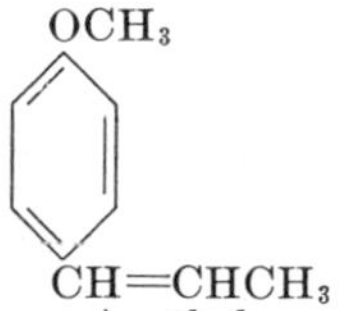

OCH₃

CH=CHCH₃
Anethol

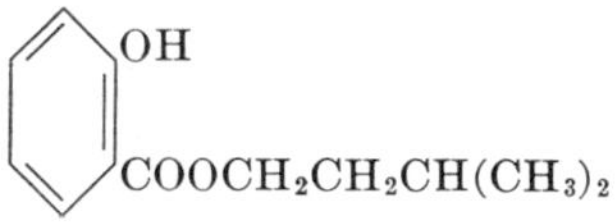

OH
COOCH₂CH₂CH(CH₃)₂

Isoamyl salicylate

The natural sex attractants of female Lepidoptera are known to attract males from distances of several miles, and extracts of female gypsy moth abdomens have been used in bait traps. The natural attractant *gyptol* is an ester, *d*-10-acetoxy-l-hydroxy-*cis*-7-hexadecene (page 115). It is of considerable interest that Butenandt and coworkers, after 20 years of research, have identified the natural sex lure of the silkworm moth as 10,12-hexadecadiene-1-ol (page 115).

Ammonia is strongly attractive to many insects, and its slow liberation from a mixture of glycine, $CH_2(NH_2)COOH$, and sodium hydroxide has been used as a lure for the walnut husk fly. Protein hydrolysates have been found to be active attractants for fruit flies, especially in bait sprays. *Metaldehyde*, $[OCH(CH_3)]_4$-, a crystalline polymer of acetaldehyde, is used at 2 to 3 per cent as an attractant in poison baits for snails and slugs.

Insecticide Residues[1]

The persistence of residues of insecticides on treated produce is a factor of importance not only in determining the degree of residual protection from insect attack, but also in producing possible deleterious effects upon human beings and animals consuming the treated substances. The problems of food contamination are most acute with organic insecticides soluble in the oils, fats, and waxes of plants and animals. The chlorinated hydrocarbon insecticides in particular are stored and sometimes concentrated in the fatty tissues of animals consuming treated forage and may be excreted in butterfat.

The range of persistence of spray residues is extreme, varying from TEPP, which is destroyed by moisture within a few hours, through the plant insecticides pyrethrins, rotenone, sabadilla, and ryania, which are decomposed by light within a few days, to the more stable synthetic organic compounds such as DDT and dieldrin, which may persist for weeks. Most persistent of all are the inorganic substances such as lead arsenate and cryolite, which are removed only by weathering or washing. The systemic insecticides which form toxic residues within plant tissues present a unique case and may persist until decomposed by enzymatic action or hydrolysis (page 359).

Residues of organic insecticides may rapidly penetrate and accumulate in oily tissues of plants and sometimes may reissue to the surface. The residues which may remain on the surface of the plant are exposed to degradation by weathering, light, and volatilization. The penetrated materials are continuously exposed to destructive attack by plant enzymes and acids. The net effect of these attritions is most conveniently expressed as a residue-persistence curve which follows first-order kinetics and is a linear function when plotted as the log of the amount of residue versus time. A curve of this type gives the half-life of the insecticide on a particular plant surface, which is relatively constant, and also indicates most suitably the number of days after application when the residue reaches any practical level.

Limiting values in the permissible amounts of insecticide residues consistent with consumer safety have been expressed as residue tolerances in parts of insecticide per million parts of foodstuff (p.p.m.). In the United States these tolerances are calculated to provide at least a hundredfold factor of safety over the lowest levels producing evidence of tissue damage in 2-year chronic-feeding studies with laboratory animals. Typical values for tolerance ranges, in parts per million of common insecticides are given as follows (see page 422 for legal procedures for their establishment):

Aldrin	0–0.75	Demeton	0.3–1.25
Aramite	0	Diazinon	0.75–1
Calcium arsenate	3.5 (As_2O_3)	Dieldrin	0–0.75
Chlordane	0.3	DNOC	0
Chlorobenzilate	5	Endrin	0
Cryolite	7	EPN	0.5–3
DDT	7	Fluorine	7

[1] Gunther, F., and R. Blinn, "The Analysis of Insecticides and Acaricides," Interscience, 1955; Schechter, M., and I. Hornstein, *Advances in Pest Control Res.*, **1**:353, 1957; Sun, Y., *ibid.*, p. 449.

Guthion	0.5–2	Pyrethrins	Exempt
Heptachlor	0	Rotenone	Exempt
Kelthane	5–10	Ryania	Exempt
Lead arsenate	7	Sabadilla	Exempt
Lindane	10	Sevin	5–25
Malathion	8	Sulphenone	8
Methoxychlor	2–100	TDE	7
Methyl bromide	5–200 (Br)	Tedion	2
Nicotine sulfate	2	TEPP	0
Ovex	3–5	Toxaphene	7
Parathion	1	Thiodan	2
Perthane	15	Trithion	0.8–2
Phosdrin	0.25–1		

Foods containing residues in excess of the tolerances are adulterated and cannot be shipped across state lines. They are subject to seizure by federal and state agents. This can be prevented by following exactly the directions on the manufacturer's label as to use only on crops for which the insecticide is recommended and at specified amounts and times of application.

INSECTICIDE RESISTANCE[1]

The continuous and intensive use of certain insecticides against various insect pests has resulted in the development of races or strains sufficiently resistant to the action of the insecticide as to necessitate a complete change in control measures. Such resistance is believed to arise from the selection of naturally occurring mutants possessing biochemical factors which confer some degree of resistance. Dieldrin resistance in *Anopheles gambiae*, for instance, is attributed to the mutation of a single gene, and this factor has been found to be present in from 0.4 per cent to 6 per cent of unselected wild populations. Several years of selection by dieldrin residual spraying for malaria control resulted in the resistance factor becoming the characteristic of the population, with a frequency of 90 per cent.

The first well-studied case of resistance was that of California red scale, *Aonidiella aurantii*, to hydrogen cyanide fumigation, which was first detected in a single area in 1916 and has slowly spread until fumigation has been abandoned in most of the California citrus area, since from four to five times the original dosage is now required to control the resistant scales. The most spectacular example is that of DDT resistance in the house fly. Within 2 to 3 years after the first widespread applications of this insecticide, strains of flies resistant to several hundred times the normal lethal dosage appeared in many parts of the world. The substitution of chlordane, lindane, and dieldrin resulted in good control for short periods, but resistance to these compounds has also appeared and flies are generally controlled by organophosphorus insecticides. Kearns and coworkers[2] have provided major insight into the biochemical mechanism of resistance by showing that DDT-resistant house flies have vastly increased amounts of an enzyme "DDT-dehydrochlorinase," which rap-

[1] BROWN, A., "Insecticide Resistance in Arthropods," World Health Organization, Geneva, 1958, and *Advances in Pest Control Res.*, **2**:351, 1958; METCALF, R., *Phys. Rev.*, **35**:197, 1955.

[2] *Advances in Pest Control Res.*, **3**:253, 1960.

idly detoxifies the insecticide into the nontoxic ethylene DDE. The acaricide DMC (page 361) and other compounds such as bis-(p-chlorophenyl)-chloromethane, closely related to DDT, act as competitive inhibitors for this enzyme, and their use in combination with DDT has

resulted in practical control of DDT-resistant flies. Such increased biochemical detoxication systems are responsible for most cases of insecticide resistance, especially to organophosphorus compounds and carbamates, although the picture is one of great complexity and other factors such as decreased cuticular absorption may also be involved.

Resistance to acaricides has occurred with red spider mites in greenhouses and on fruit trees, to DDT with the codling moth, *Carpocapsa pomonella*, and to chlorinated hydrocarbons with the cotton boll weevil, *Anthonomus grandis*, and with many other pests. Fortunately, such resistance is generally fairly specific, and the resistant races can be controlled by the use of insecticides with differing biochemical behavior, selected according to the following groups: (*a*) DDT, DDD, methoxychlor, (*b*) aldrin, dieldrin, heptachlor, chlordane, lindane, toxaphene, (*c*) organophosphorus compounds. The cross resistance is often not entirely group-specific, especially for organophosphorus compounds.

The most practical means for dealing with this vast problem consists in maintaining a careful check on the susceptibility of pest populations, using standard susceptibility tests. When it appears that an appreciable amount of resistance has resulted from the use of a given chemical, this should be replaced with another effective insecticide chosen from a different one of the groups listed above.

Insecticide Formulation

The successful employment of any insecticide depends upon its proper formulation. Insecticides are commonly formulated for use as dusts, water dispersions, emulsions, and solutions. The preparation and use of these formulations involves the utilization of accessory agents such as dust carriers, solvents, emulsifiers, wetting and dispersing agents, stickers, and deodorants or masking agents.

Dust Carriers. Insecticides are most simply applied as dusts, in which the concentration of toxicant is low, usually from 0.1 to 20 per cent, to aid in evenness of application. Therefore the properties of the carrier largely determine the quality of the finished dust. Carriers in common use have been classified as follows:[1] (*a*) organic flours such as *walnut-shell, soybean,* and *wood-bark;* (*b*) minerals such as *sulfur;* silicon oxides— *diatomite* and *tripolite;* calcium oxide—*lime;* calcium sulfate—*gypsum,* $CaSO_4 \cdot 2H_2O$; silicates—*talc,* $H_2O \cdot 3MgO \cdot 4SiO_2$; *pyrophyllite,* $H_2O \cdot Al_2O_3 \cdot 4SiO_2$; clays such as *bentonites,* $(OH)_4Al_4Si_8O_{20} \cdot xH_2O$, *kaolins,* $(OH)_8Al_4Si_4O_{10}$, and *attapulgite,* $(OH)_2)_4(OH)_2Mg_5Si_8O_{20} \cdot 4H_2O$; and *volcanic ash.*

[1] *Jour. Econ. Ento.,* **40**:211, 1947; Weidhaas, D., and J. Brann, "Handbook of Insecticide Dust Diluents and Carriers," Dorland, 1955.

Selection may be made on the basis of compatibility with the desired insecticide (including pH, moisture content, and stability), particle size, abrasiveness, absorbability, specific gravity, wettability where the same product may be used as a water dispersion, and cost. The mixture between the toxicant and the diluent may be made by simple operations such as ball-milling, hammer-milling, or ribbon-mixing; by impregnation from volatile or nonvolatile solvents; by fusing and grinding; and by a variety of special operations designed to reduce particle size to the sub-sieve range and to improve homogeneity. Among such processes are the "micronizing" and "reductionizing" methods, in which the material to be ground is suspended on an air cushion and reduced in size by impaction against other particles projected by high-velocity air jets. With some organic insecticides, grinding operations are facilitated by cooling with dry ice.

TABLE 7.3. PROPERTIES OF DUST DILUENTS COMMONLY USED IN
INSECTICIDE FORMULATIONS[1]

Type	Diluent	Surface mean diameter, microns	Specific gravity	Bulking value, g. per cc.	pH slurry
Organic flour..............	Redwood bark	3.2	0.44	0.23	2.8
	Walnut shell	3.6	1.3	0.40	5.3
Clay, attapulgite..........	Attaclay	1.1	2.6	0.30	7.4
Clay, montmorillonite......	Clay Spur Bentonite	3.1	3.1	0.64	9.4
Clay, kaolinite............	Kaolin Type 41	1.3	2.6	0.21	4.6
Pyrophyllite..............	Pyrax ABB	2.2	2.7	0.45	5.6
Talc, natural..............	Emtco	1.8	2.6	0.44	8.3
Synthetic...............	Silene EF	0.6	2.1	0.15	10.0
Calcium oxide............	Hydrated lime	2.2	2.2	0.43	12.0
Calcium sulfate...........	U.S.G. Dusting Gypsum	2.4	2.3	0.62	6.3
Silicon oxide diatomite.....	Celite 209	0.4	....	0.073	6.2
Silicic acid...............	Aerogel 255	3.5	0.20	.....	2.6
Sulfur...................	Dusting sulfur	1.9	2.1	0.38	6.1

[1] Data from original measurements and from manufacturer.

Granular or Pelleted Insecticides. Granulated formulations of insecticides in which the particles range from 30- to 60-mesh have significant advantages for certain uses over ordinary dust formulations. The greater particle weight minimizes drift considerably and prevents undue loss of insecticides and undesirable contamination of areas bordering those being treated. The granular materials adhere less readily to plant surfaces and thus decrease residual deposits. Granular formulations commonly contain 2.5 to 5 per cent toxicant applied by solvent impregnation to highly absorptive clays, bentonites, and diatomaceous earths of the proper particle size. They have proved especially effective for the control of mosquito larvae, soil insects, ants, the Japanese and white-fringed beetles, and the European corn borer. The materials are readily applied by fertilizer spreaders and ground or airplane seeders. The wide variation in number of particles per unit weight of carrier for different-sized granules is shown in the accompanying table.

Mesh	Particle diameter range, microns	No. of particles per pound[1]
20/40	420–840	3,000,000
30/35	500–590	3,850,000
35/40	420–500	5,200,000
40/60	250–420	18,000,000

[1] KRAUSCHE, K., *Agr. Chem.*, April, 1959.

Sorptive Dusts. Certain dust carriers such as montmorillonite clays and silicic acids (Table 7.3) have been found by Ebeling[1] to be rapidly effective in killing termites, cockroaches, fleas, etc. These materials absorb the lipid protective layer of the insect epicuticle (page 80), and the insect dies by desiccation. They are especially effective as carriers for fluoride or organophosphorus toxicants.

Water Suspensions. Agricultural insecticides are very often applied as water suspensions of solid materials. Because most insecticides are difficult to wet and suspend, wettable powders are usually prepared which contain from 15 to 95 per cent of the toxicant blended with a dust carrier, such as attapulgite, which wets and suspends well. One to two per cent of surface-active wetting and dispersing agents are usually added to improve the quality of the product.

Solvents. The increasing use of organic insecticides, insoluble in water, has resulted in the employment of many organic solvents for use in insecticide sprays, aerosols, and emulsions. The important properties of the most commonly used solvents are given in Table 7.4. The choice of a solvent for a particular formulation will depend upon such factors as solvency, toxicity to plants and animals, fire hazard, compatibility, odor, and cost. In many cases these materials possess some insecticidal properties of their own, and intelligent selection may greatly improve the properties of the finished insecticide. Many of the solvents are phytocidal when used at high concentrations but can often be used as mutual solvents or solubilizers to increase the solubility of the insecticide in petroleum oils, which are generally poor solvents for organic compounds.

Emulsification, Wetting, and Spreading.[2] Liquid insecticides, oils, and solutions of insecticides in water-insoluble solvents are generally formulated and applied as water emulsions of the oil-in-water type. This dispersion or emulsification, although produced by mechanical agitation varying from that of the power spray pump to simple hand shaking, is usually not sufficient to produce an emulsion of the desired stability and wetting and spreading characteristics. To obtain these properties, a small amount of a surface-active material must be incorporated into the mixture. Such materials, which are preferentially absorbed at a liquid interface, are of considerable chemical diversity but have in common the

[1] *Jour. Econ. Ento.*, **52**:190, 1959.
[2] *U.S.D.A., Bur. Ento. Plant Quar., Cir.* E-504, 1940, and E-607, 1943; "Emulsion Technology Symposium," Chemical Publishing Co., 1946; SUTHEIM, G. M., "Introduction to Emulsions," Chemical Publishing Co., 1947.

TABLE 7.4. PROPERTIES OF ORGANIC SOLVENTS COMMONLY USED IN INSECTICIDE FORMULATIONS[1]

Solvent	Boiling point, °C.	Specific gravity at 20°C.	Flash point, °F.
Amyl acetate	127–155	0.862	111
Carbon tetrachloride	75–77	1.595	
Cyclohexanone	130–173	0.946	145
Dibutyl phthalate	340	1.047	322
Ethylene dichloride	82–84	1.256	70
Kerosene	147–261	0.82	100–165
Kerosene, refined[2]	176–243	0.798	170
Monochlorobenzene	130–134	1.108	95
Petroleum naphtha or Stoddard's Solvent	153–199	0.780	105
Pine oil	99–226	0.925–0.935	185
Technical di- and trimethylnaphthalenes[3]	240–290	0.98	245
Xylene, 10-degree	135–145	0.880	80

[1] See *Jour. Amer. Oil Chem. Soc.*, **25**:279–295, 1948; "Industrial Petroleum Naphthas," Gulf Oil Co., 1943.

[2] Such as Deobase.

[3] Such as Velsicol AR 60.

ability to reduce the surface tension of water. Their general chemical nature is such that the molecule contains both water- and oil-soluble groups, which enable it to orient itself at the interface between the water and the oil droplets and thus stabilize the emulsion. The specific chemical structure of these surface-active materials largely determines the ultimate stability and behavior of the emulsion. The classes of surface-active agents of importance in insecticide formulation and application are as follows:

1. Alkaline soaps, which are the sodium (hard) or potassium (soft) salts of long-chain fatty acids, as, for example, *sodium oleate*, $C_{17}H_{33}COONa$. These soaps are most useful in alkaline solution but are incompatible with the calcium and magnesium salts of hard water or with lime.

2. Organic amines which form amino soaps with fatty acids, such as *triethanolamine*, $N(CH_2CH_2OH)_3$. These materials have superior emulsifying properties to the soaps and better oil dispersibility but are also incompatible with calcium and magnesium ions. They are especially used to promote oil flocculation of materials such as lead arsenate.

3. Sulfates of long-chain alcohols such as *sodium lauryl sulfate*, $C_{12}H_{25}OSO_3Na$.* These materials are stable in the presence of calcium and magnesium salts and acids and alkalies.

4. Sulfonated aliphatic esters and amides, such as *dioctyl sodium sulfosuccinate*,[1] $NaSO_3CH(CH_2COOC_8H_{17})COOC_8H_{17}$, and *sodium sulfoethylmethyloleylamide*,[2] $CH_3(CH_2)_7CH=CH(CH_2)_7CON(CH_3)CH_2CH_2SO_3Na$. These compounds excel as wetting and spreading agents and are used as dispersing agents for wettable powders. Also included in this class are sulfonated petroleum products of uncertain composition, which are complex mixtures averaging about 16 carbon atoms.[3,4] Such materials

* Gardinol.

[1] Vatsol OT.

[2] Igepon T.

[3] Ultrawet.

[4] Penetrol.

are by-products of petroleum refining and are used in the preparation of miscible oils.

5. Mixed aliphatic-aromatic sulfonates such as *sodium decylbenzene sulfonate,*[1] $C_{10}H_{21}\cdot C_6H_4SO_3Na$.

6. Nonionic types which are ethers, alcohols, and esters of polyhydric alcohols and long-chain fatty acids. Examples are *glycerol monoleate,* $C_{17}H_{33}COOCH_2CHOH$-CH_2OH, and *manitan monolaurate,*[2] $C_{11}H_{23}COOC_6H_8(OH)_5$. Other materials of this class which are widely used are *aliphatic esters of sorbitol*[3] and their *polyoxyalkalene derivatives*[4] and *polyalkalene ether alcohols.*[5] These materials are becoming of increasing importance in insecticidal formulation and in certain cases possess considerable insecticidal activity. They are of varying degrees of oil- and water-solubility and are widely used in the preparation of emulsifiable concentrates.

7. Natural agents such as proteins, gums, lipids, carbohydrates, alginates, and saponins. Such materials as blood albumin, casein, dried milk, gelatin, and lecithin are often used to prepare emulsions, usually of the quick-breaking type. They are inexpensive, can be used with any type of water, and also act as wetters and stickers.

8. Finely divided solids such as bentonite and other clays, flours, etc., which serve primarily as sticking and spreading agents.

The most important property of the insecticidal emulsion is its rate of breaking or separation into its immiscible constituents. This factor can be controlled by the degree of agitation applied and by the type and amount of emulsifier employed. Quick-breaking emulsions are generally favored for agricultural sprays because they produce heavy deposits, but they may be deficient in wetting and spreading properties. On the other hand, emulsions which are very stable produce excessive wetting and wasteful runoff. For many purposes such as residual spraying of dwellings, mosquito larviciding, or clothing impregnation, very stable emulsions are desirable. A water-soluble emulsifier is generally used where the emulsion is to be produced as a tank mix by adding oil, emulsifier, and other ingredients to the water in the spray tank, but where a miscible-type stock solution is to be prepared which can be added to water in any desired amount, oil-soluble emulsifiers are used. Examples are given on pages 347, 369, and 393. Creaming of emulsions or partial separation of the dispersed and continuous phases upon standing is a common occurrence. This is due to differences in specific gravity of the phases and is usually of no consequence, as slight agitation will restore the homogenous state. Creaming can best be prevented by closely matching the specific gravities of the two phases of the emulsion. Phase reversal of emulsions sometimes occurs, in which an oil-in-water emulsion may change to a water-in-oil type, completely altering its properties.

The ability of insecticidal sprays to wet and spread so as to thoroughly cover plant and insect surfaces is an important factor in determining spray performance. Nearly all plants and insect cuticles are provided with exterior lipid layers (page 80) which are strongly hydrophobic and upon which water sprays tend to form discrete droplets. In order to cause such sprays to spread and make intimate contact, it is necessary to reduce markedly the surface tension of the water by adding a small amount of surface-active material, as discussed above.

Stickers. The amount of spray deposit adhering to the treated surface is a function of the wetting and spreading properties of the spray. In

[1] Santomerse D.
[2] NNO.
[3] Spans.
[4] Tweens.
[5] Tritons.

actual practice, where emulsions and wettable powders are applied, the addition of certain supplementary adhesive agents has sometimes proved of value. Oil-flocculated or "dynamite" sprays, in which the suspended solid, initially wet by water, becomes wet by oil just prior to application, have given improved results over ordinary water suspensions. However, they require the addition of special tank-mix wetting agents such as ethanolamine oleates and a high degree of agitation. Adhesive materials such as casein, gelatin, soybean flour, blood albumin, various clays and bentonites, and petroleum and vegetable oils have been used as stickers.

Deodorants or Masking Agents. Because of the unpleasant odors of certain toxicants such as the thiocyanates, pyrethrins, and methylated naphthalenes commonly used in household insecticides, various substances such as pine oil, cedar oil, or various flower scents are often incorporated in the finished insecticides at concentrations of 0.1 to 1 per cent to disguise or mask the odor.

Stabilizing Agents. With the increased use of the relatively unstable organic insecticides, stabilizing agents are often necessary in formulations to retard decomposition during storage. Examples include the use of antioxidants such as mixed isopropyl cresols to prevent the decomposition of the pyrethrins in louse powder, hexamethylene tetramine to stabilize endrin in wettable powders, and acid inhibitors such as epichlorohydrin to prevent the dehydrochlorination of formulations of aldrin and toxaphene. Traces of metal from storage drums sometimes catalyze the decomposition of chlorinated hydrocarbons, and this is prevented by special interior lacquer linings.

Physical and Mechanical Control

Aside from the destruction of insects that may be accomplished by *ordinary* farm practices (see page 407), there are certain *special* physical and mechanical measures that are of value. The more important ones are outlined and defined as follows. These differ from chemical control measures in the nature of their effect upon the insects, which is a physical action not involving a chemical action upon the insect. They may be arbitrarily distinguished from the cultural control measures in that they involve the use of special equipment or operations, which would not be performed at all were it not for the insects, and they generally give immediate, tangible results. On this account they are psychologically good and generally popular. They are in general costly in time and labor, often do not destroy the pest until much damage has been done, and rarely give adequate or commercial control. There are two subdivisions, the *mechanical measures*, which involve the operation of machinery or manual operations; and the *physical measures*, which employ in a destructive way certain physical properties of the environment.

The destruction of insects or their egg masses by hand is sometimes the most practical method to employ in areas where labor is very cheap and where the insects or their eggs are large or conspicuous, not too active, or occur in relatively restricted areas. In other cases it is possible to prevent the invasion of a crop by migrating insects, through the use of physical barriers. The crawling hordes of chinch bugs and armyworms may often be stopped by constructing deep, dusty-sided furrows around the fields toward which they are traveling (page 485).   The same result may be achieved by using barrier lines of certain heavy-bodied oils, poured along on the ground.   Low fences of sheet metal are employed against mormon and coulee crickets (page 602).   Such linear barriers are practicable against the nonflying insects, for crops of limited area, or for a very limited period of attack.   Screening of houses has come to be a regular practice in all civilized countries as a protection from flies and mosquitoes and the diseases they carry.   Sometimes individual plants, storage houses, seedbeds, fields, or vineyards are protected by covering them with screens of thin cloth or wire screen.   Because of the expense involved, this measure can be resorted to only in cases where the crop has a high money value.   Sticky bands around tree trunks are often employed against insects that infest trees by climbing up the trunks; they are of no value for insects that fly into the trees to feed or to lay eggs.   Collars about individual plants to protect from cutworms, bags over clusters of fruits to protect from fruitworms, chips under melons to protect from the melonworm, or the complete wrapping of tree trunks with paper against the flatheaded borers are other examples.   Fly nets, muzzles, and other similar devices are of some value in protecting animals from certain insect parasites.

Mechanical devices such as hopperdozers, hoppercatchers, aphidozers, fly traps, moth traps, maggot traps, light traps, electric traps, and others have been used successfully for catching and killing a variety of insects. Since insects for the most part lack cunning or intelligence, insect traps are often surprisingly simple.   They generally take advantage of some dominating, fixed tropism or instinct which the insect species has been observed to follow rigidly.   Insect traps are sometimes *merely mechanical* such as window traps for flies, the maggot trap for the house fly (page 1031), or boards for squash bugs (page 635); sometimes they employ a *bait*, as in the cone traps for flies or Japanese beetle traps (page 394).   Sometimes they are *stationary*, as the sticky bands about tree trunks or the electrified screens or light traps to which many insects are attracted; and sometimes *moving*, as the hopperdozers (page 470).   The best-known example is the electric fly trap, which employs a potential of 3,500 volts with low amperage connected alternately across closely set wires so that insects passing between produce an arc and are electrocuted.   Such traps properly baited, can destroy as many as 100,000 flies a day.   Rarely, crushing, grinding, suction, or dragging machinery may be successfully employed to destroy some kind of pest.

Physical measures involve especially manipulations or changes in temperature or humidity or employ radiant energy in some way to destroy a pest.   Cranberries are protected from some of their insect enemies by flooding the bogs at the proper time, and some other crops, in areas where irrigation is possible, are protected in the same way.   The draining of swamps, marshes, and other standing water is the most effective method of destroying mosquitoes and horse flies.

The Use of Low Temperatures.   Artificial heating or cooling of stored products, or the mills or factories where such products are processed, is a common method of preventing insect damage.   Nearly all insects become inactive at temperatures between 60 and 40°F.   Few insects are killed at these temperatures unless exposed to them for a considerable length of time.   Insects in hibernation frequently withstand temperatures of -20 to -30°F., or lower.   It is not certain that exposure to such temperatures will kill the eggs of such species as the grain weevils. But practically no damage from insects will occur at temperatures below 40°F. Low temperatures are not so effective as high temperatures in killing insects, but storage of food products or clothing at points below or near freezing will prevent all insect damage. Changes from low to high temperatures and back to cold are more effective in killing insects than constant low temperatures.

Superheating. Abnormally high temperatures are employed against (a) insects in cereals, coffee beans, and other seeds, and their processed derivatives; (b) insects, such as the Mediterranean or Mexican fruit flies in citrus fruits; (c) insects in clothing, bedding, baggage, bales of cotton, and other fibers; (d) insects, mites, and eelworms infesting bulbs; (e) insects infesting soil; and (f) insects infesting logs. Sometimes exposure to the sun's rays is sufficient, especially in tropical regions. For insects and mites infesting bulbs, a temperature of 110 to 111.5°F. for a few hours is used. Careful experiments by a number of entomologists have shown that no insect can long survive when exposed to temperatures of 140 to 150°F. Most insects, including those which attack stored grains, are killed by 3 hours' exposure to temperatures from 125 to 130°F., and this, or a slightly longer exposure, will destroy all stages of these insects. Many mills and large elevators have equipped their buildings with enough heating pipes to enable them to raise the temperature to 125 to 150°F. for several hours during periods of warm weather and thus kill all insects in the buildings at much less expense than by fumigating. In general, the higher the humidity, the more effective superheating will be.

There are a number of heat-treating machines now on the market which raise grain to a high temperature while it is being passed through them. Such machines are fairly effective in cleaning infested grain and other seeds, but the exposure of the grain to high temperature must be of sufficient duration to kill all stages of the insects. It should be understood that the heat treatment of grain will cause a certain shrinkage due to loss of moisture. In applying heat to piles of clothing, bins of grain, or bales of goods, it must be borne in mind that it requires a long time for the heat to penetrate and that the temperature on the surface will have reached the killing point long before the insects within the material have been affected.

Radiation.[1] The use of radiant energy to control insects has been a favorite subject for experimentation. Light has been utilized to attract many strongly phototropic species into traps from which they cannot escape or where they are drowned or poisoned. Herms records that over 85,000,000 adult Clear Lake gnats, *Chaoborus astictopus*, weighing about 85 pounds, were caught in a single night by a light trap fitted with a fan to suck the insects into a porous container. In general, insects appear to be most strongly attracted to radiation in the ultraviolet region, about 3,650 Angstrom units, and ultraviolet lamps are used to trap such night-flying moths as the European corn borer and tobacco and tomato hornworms. The phototropic response declines sharply to the violet region at 4,400 Angstrom units. Minor peak attractions may occur at 4,920 Angstrom units (blue to blue-green), 5,150 Angstrom units (green), or 5,550 Angstrom units (yellow-green), but the longer wavelengths of 5,500

[1] YEOMANS, A. H., *U.S.D.A.*, *Yearbook*, **1952**:411; HASSETT, C., and D. JENKINS, *Nucleonics*, **10**:42, 1952.

to 7,000 Angstrom units (yellow to red) are relatively unattractive. The familiar yellow "bug-repellent" light bulb is an example of the practical application of this knowledge to secure adequate illumination without strongly attracting night-flying insects.

Radiant energy from many sources has been evaluated as a means of destroying insects. However, in general, the high absorption of the plant materials in which the insects are often imbedded, the destructive effects of the radiation on foodstuffs, and the relatively enormous power requirements necessary have prevented practical usage. Ultraviolet radiation is virtually ineffective. High-frequency radiowaves (2,450 megacycles, 12.25 centimeters wavelength, and 940 watts) generated temperatures of 172 to 187°F. in grain and killed granary weevils and confused flour beetles in 15 to 21 seconds.[1] Infrared radiation will heat insects to the death point, but penetration is poor and the substrate is generally heated to the same temperature. Ultrasonic waves of 400 kilocycles were lethal to exposed insects upon exposures of 4 to 30 minutes at 500 watts. However, the lethal effect was largely nullified by shielding from air bubbles or water films, and codling moth larvae were not affected by a 1-hour exposure.

Ionizing radiation or radioactivity consists of alpha particles or helium nuclei, beta particles or high-speed electrons, neutrons, and electromagnetic gamma rays or short-wave X rays. Alpha particles that are too highly charged to penetrate tissues and neutrons which, although extremely penetrating, produce undesirable chemical reactions in most substances have not proved useful for insect control. High-energy beta particles used experimentally at dosages of 70,000 to 350,000 roentgen equivalents have killed powder post beetles in wood, confused flour beetles in flour, and codling moth and potato tuberworm larvae in plant tissues, but penetration is limited. Gamma rays are extremely penetrating, but the ionizing power is low and insects are surprisingly resistant, requiring dosages of the order of 65,000 roentgens for lethality as compared with 1,000 roentgens for laboratory mammals. Such radiation can be produced from Co^{60} or from waste fission products such as Ce^{137}. Much lower dosages are capable of sterilizing insects. Stored-product insects are sterilized by about 16,000 roentgens, and adults of the screwworm fly by 2,500 for males and 5,000 for females. Knipling, Lindquist, and Bushland[2] have utilized this in a very clever way to control this insect. A Co^{60} source of gamma radiation was used to sterilize the pupae, which developed into adults which mated normally. Inasmuch as the screw-worm females mate only once, it proved possible, by infesting the island of Curaçao to saturation with the release of about 400 sterile males per square mile each week for 7 weeks, to cause the production of 100 per cent sterile egg masses, so that eradication was achieved. This technique was applied in 1958 to an area of about 64,000 square miles of Florida, where after the release of nearly 3 billion sterile flies over a 17 months' period, eradication of the screw-worm was achieved at a cost of about $55 per square mile.

This procedure is doubtless applicable to the control of other insects, and the following criteria have been established for its successful employment: (a) a method of mass rearing of the insect must be available; (b) adequate dispersion of the released sterile males must occur; (c) sterili-

[1] *Jour. Econ. Ento.*, **49**:33, 1956.

[2] BUSHLAND, R., *Advances in Pest Control Res.*, **3**:1–25, 1959.

zation must not adversely affect the mating behavior of the males; (*d*) the female must normally mate only once or, if more frequent matings occur, the sperms of sterilized males must compete with those of normal males; and (*e*) the population density of the insect must be inherently low or be reduced by some other means to low levels which will make it economically feasible to release a dominant population of sterile males over an extended period of time.

CULTURAL CONTROL OR THE USE OF FARM PRACTICES

The cultural control measures differ from physical and mechanical control in generally involving the use of *ordinary* farm practices and farming machinery and in being usually preventive, indirect, or intangible, so that the farmer has much difficulty in being sure how effective they are. They must usually be employed far in advance of the time when damage by the pest becomes apparent, and they often do not make a strong appeal to the farmer. However, they are the cheapest of all control measures, once research has revealed an effective and practicable procedure; in fact, they often cost the farmer nothing at all because they are merely variations in the time or manner of performing operations which are necessary in the production of a crop. Often, with crops of great acreage and low unit value, they are the only control measures that can be employed profitably. The opportunity for cultural control of insects usually results from the interplay of the complicated metamorphoses of insects and the change of the seasons. The result is often some particularly *weak point* in the life cycle or adaptation of the insect pest to its environment, at which point it may be attacked by a cultural control measure.

In order to control insects by cultural practices, it is necessary that one understand the life history and habits of the insect with which one is dealing. A control that would be effective against one kind of insect might be useless against a closely related kind because of a difference in habits. These operations, to be effective, must also be used at the proper stage of the development of the insect. It is useless to try to destroy white grubs by late fall or winter plowing, after they have gone down a foot or more below the surface of the soil, or to kill insects by burning their hibernating places, before they have entered them in the fall or after they have left them in the spring.

Crop Rotations. In a state of nature, the plants growing on the land in any of the great agricultural areas of the world are quite different from those which are grown after such lands have been placed under cultivation. There was in most of these areas a predominance of grasses but with a mixture of legumes and plants of many other botanical families. Such plants grew from year to year with little change in the proportion of one over the other. The insects depending on these wild plants were always assured of a food supply sufficient to maintain them, but, with the exception of a few general feeders, the food plants were not abundant enough to permit a great increase of any one species.

Under farming conditions great changes take place in the character of the plants grown on the land. There are no longer a great number of species, generally intermixed, but a few species occupying the land in nearly pure stands of thousands and hundreds of thousands of acres. This affects the insect population of the land in two general ways. Many of those which depend on the plants of one family, or even on one species of plant, find their food supply cut off, except in the small uncultivated areas, and may nearly, or quite, disappear from the region, as certain species of billbugs in drained bottom lands. Others take to the cultivated crop closely related to their wild food plant and find it, perhaps, more palatable. Such insects may, and generally do, increase enormously and become very destructive, as have the chinch bug and the cotton boll weevil.

Among insects that injure cultivated crops, the number of general feeders is very small. Those which feed on the plants of one family are numerous, and there are many that feed on only a few very closely related

TABLE 7.5. A COMPARISON OF THE MOST IMPORTANT INSECTS ATTACKING THREE MAJOR FIELD CROPS IN ILLINOIS

Corn Insects	Wheat Insects	Red Clover Insects
European corn borer	Hessian fly	Clover bud weevil
Northern corn rootworm	Chinch bug[1]	Clover leaf weevil
White grubs[1]	Wheat jointworm	Grasshoppers[2]
Wireworms[1]	Wheat stem maggot	Clover root curculio
Chinch bug[1]	Wireworms[1]	Clover seed chalcid
Corn earworm	Grasshoppers[2]	Pea aphid
Southern corn rootworm[1]	Armyworm[2]	Variegated cutworm[2]
Corn root aphid	Wheat head midge	Clover seed caterpillar
Armyworm[2]	Wheat stem sawfly	Clover seed midge
Grasshoppers[2]	Wheat sawfly	Green cloverworm
Black cutworm	Billbugs[1]	Clover leaf tier
Seed-corn maggot	Frit fly	Armyworm[2]
Common stalk borer	Wheat head armyworm	Clover root borer
Sod webworm	English grain louse	Clover stem borer
Billbugs[1]	Green bug	Leafhoppers
Morning-glory flea beetles	Sorghum webworm	
Corn-seed beetles	White grubs[1]	
Carrot beetle	False wireworms	
Clover rootworm	Southern corn rootworm[1]	
Corn leaf aphid	Variegated cutworm[2]	
Imbricated snout beetle		
Pale-striped flea beetle		
Thief ant		
Green June beetle		
Fall armyworm		
Variegated cutworm[2]		

[1] Species of importance to corn and wheat only.
[2] Species of importance on clover, wheat, and corn.

species. The above lists of the principal insect enemies of corn, wheat, and clover in Illinois include only those insects that are of sufficient importance to be considered as doing commercial damage to these crops. Under each crop, the pests are listed in the order of their destructiveness to this crop in Illinois.

In Table 7.5, it will be noted that 8 of the insects which are listed as pests of corn are also listed as pests of wheat, but that only 3 of the 50 insects are serious pests of all three of these crops. Wheat and corn are grasses, clover is a legume; and it will be seen from a study of these lists that much can be accomplished in preventing insects from becoming seriously abundant in our fields if a good rotation is practiced, where a crop of one plant family follows that of a different family. It is not possible in many cases in grain-farming areas to put such a rotation into effect in all fields each year, but a large part of the increased yields obtained from a rotation where grains follow legumes, and legumes grains, is due to the reduction in insect damage. Crops of the same group, such as corn, oats, and wheat, grown on the same land, year after year, give a condition favorable to the insects that attack the grass crops, and the same is true of a number of years' cropping of ground with plants of any one family.

Crop rotations will be most effective for insects that are restricted feeders, that have limited powers of migration or sluggish habits, and that are slow breeders spending a relatively long time in the feeding stage. Because of the ease with which most insects move about, many of the species which feed on any crop will be found in the fields the first year they are planted to such a crop; these insects may occur in numbers sufficient to cause severe damage. For this reason, rotation of crops cannot be depended upon for combating all insects attacking field crops. Generally, however, infestations in such fields will be later and lighter than in fields continued in the same crop, and crop rotations are by far the best, and in some cases almost the only, means of controlling certain insects. The application of rotations to the control of different species will be discussed in more detail under the insects attacking certain crops.

Tilling or Cultivating the Soil. Insects are greatly affected, directly, by the texture of soils, their chemical composition, the percentage of soil moisture, the temperature, and other soil organisms, and indirectly, by the influence of these things upon their food plants. Consequently various methods of stirring and managing the soil have a profound effect upon many insects. When the exact effects are understood, much can be accomplished in the control of some crop pests by cultivating the soil at a certain time of the year or in some special manner. The best method to employ will depend on the life history and habits of the species to be controlled. Deep, thorough, and frequent cultivation of fields infested by the corn root aphid and its attendant ant is the best method of freeing the soil of these insects. Some species of insects that go through a part of their development in the ground can be easily killed if the soil is cultivated while they are in their pupal cells; the plum curculio and certain wireworms are examples. Others may be killed in the same way in the hibernating shelters in which they pass the winter, *e.g.*, the oriental fruit moth. Some degree of control may be obtained over certain insects by planting infested land to row crops which require frequent cultivation; white grubs and certain flea beetles are examples. Caking of soil usually works a hardship on subterranean insects, and the pale western cutworm and certain thrips never become abundant in caking soil. Tillage may, therefore, favor certain pests and under some conditions should be avoided at particular seasons. Rolling or packing the soil tends to raise the water level and may drive certain subterranean insects above the surface where their natural enemies can get at them.

Infestation may sometimes be entirely prevented if the ground is kept in a state of clean cultivation during the egg-laying period of some of the crop-infesting insects, as the southern corn rootworm, that will not deposit their eggs on the bare soil. Certain other species, as the pale western cutworm, prefer the bare ground, and with these, cultivation should be avoided until after the eggs have been laid. With some of the soil-infesting insects, such as white grubs, plowing at a certain time in the year will destroy large numbers of the larvae, or aid in their destruction, by exposing them to birds and other animals that feed upon them, while plowing at other times will be of no value in reducing their numbers.

Destruction of Crop Residues, Weeds, and Trash. The destruction of crop residues is often of great importance in insect control. In some sections of North America where the European corn borer is well established, it has become necessary to practice rotations and cultural methods that permit the utilizing, plowing under, or destruction by burning during the fall, winter, or early spring of all crop residues and weeds remaining in the fields. In some of the areas infested by the European corn borer such a cleanup of all corn refuse has been made compulsory. Insects often have a much longer season of activity than the annual crops they attack. They are often supported by, and increase their numbers upon, weeds and volunteer plants growing earlier in spring and later in fall than the planted crop. Much can be accomplished in the control of flea beetles, common stalk borer, corn root aphid, the greenbug, hornworms, southern corn rootworm, and many others by eliminating weeds, especially those closely related to the planted crop, from the field and field margins. During the winter many crop insects hide under surface trash such as boards, boxes, sacks, brush heaps, stone piles, dense grass, fallen leaves, and other dead vegetation. Many codling moths winter in such shelter, asparagus beetles often gather in great numbers in cracks of wooden posts, and squash bugs are frequently plentiful in board piles. Neat husbandry, especially over the winter, reduces the insect population that will have to be fought the following season.

Variations in the Time of Planting and Harvesting. The time of planting a crop has a very great influence on the infestation of the crop by some insects. By changing or carefully selecting the time when a crop is planted, we may avoid the egg-laying period of a particular pest; get young plants well established before the attack comes; allow a shorter period of susceptibility during which the insect will attack, as in the case of the seed-corn maggot in a cool, wet spring; or even get a crop matured before a certain pest becomes abundant, as with the cotton boll weevil or an early radish crop and the cabbage root maggot. Early-planted corn will largely escape injury by the corn earworm in most sections of the country. On the other hand, early-planted corn may be heavily infested by the southern corn rootworm or the European corn borer. In places where both these kinds of insects are present, the best time of planting corn will depend to a considerable extent on which of these insects is the more destructive.

There is no better example of the importance of farm practices in insect control than the effect on the infestation by the Hessian fly of early and late seeding of wheat. During most seasons early-sown wheat will be moderately to heavily infested, and medium-late-sown wheat will not be seriously infested. Indeed, there are many years when a difference of a few days in the time of seeding will make the difference between a good crop and a very poor one, all because of the difference in the amount of infestation by the Hessian fly (page 534). An 8-year experiment conducted in eight different localities of Illinois showed that the average per cent infestation by this insect decreased from 39.7 per cent in wheat sown before the normal safe seeding date to 3.8 per cent in wheat sown after the safe date, and the average yield increased from 23.1 bushels per acre in wheat sown before the safe date to 29.8 bushels per acre in wheat sown after the safe date.

With crops of indeterminate growth such as clover, alfalfa, strawberries, etc., it should be possible to do much to destroy populations of such insects as the clover seed midge and chalcid and the clover head caterpillar by clipping or harvesting the crop at carefully chosen times before a particular brood of the pests has completed that part of their development which is dependent upon the growing crop. However, recommendations must be worked out for each locality and with reference to conditions in different years.

The Use of Resistant Varieties.[1] Some strains or varieties of cultivated plants are more or less resistant to certain of the insects that attack them, and the breeding of such resistant varieties of important crop plants has become a major weapon against insect attack. It is only by taking advantage of the resistance of the American grape roots to the grape phylloxera, which is native to North America, that we are able to grow European grapes in this country or in most of the large vineyard areas of Europe. By grafting these varieties upon American rootstocks, the injury is avoided. There is good evidence to show that a marked difference exists in the resistance of several varieties of grain to attacks by insects, and some varieties of corn and sorghums have been found to resist the attacks of chinch bugs. Resistant varieties of alfalfa are important in avoiding major damage from the feeding of the spotted alfalfa aphid.

The factors which are responsible for plant resistance to insect attacks are generally complex and involve interrelations between physiological and biochemical aspects of (a) the preference of the insect for oviposition, food, and shelter, (b) tolerance of the plant to insect damage, and (c) antibiosis or adverse effects of the plant on the insect. Such complex factors are poorly understood, but fortunately it is possible to breed for resistance factors without a detailed knowledge of their modes of action. Sometimes, however, fairly simple processes are involved. Northern Spy and other varieties of apple have so much hard tissue (sclerenchyma) in the circumference of the roots that woolly apple aphids cannot penetrate it with their mouth parts. Roughness of surface or hairiness of some varieties of soybeans, cotton, and red clover gives these plants a high degree of immunity from attack by tiny leafhoppers, which are very destructive to smooth-leaved varieties. The rind of many citrus fruits, while green, contains an oil which kills the young maggots of the Mediterranean fruit fly. Pistillate varieties of strawberries and figs are not attacked by the strawberry weevil and the fig wasp, respectively. Certain kinds of cattle have skins so thick and tough that cattle ticks and horn flies are much less troublesome. Others have a very oily, thin coat of hair unfavorable to lice. Acidity or distastefulness of sap, thickness of husks, vigor, early maturity, and unusual recuperative ability are other qualities which may give plants and animals a valuable degree of tolerance to insect attack. By hybridization, grafting, and pure-line selection, the desirable resistance factors may be combined or intensified in crops or animals. With hybrid corn, in particular, great progress has been made

[1] PAINTER, R. H., "Insect Resistance in Crop Plants," Macmillan, 1951.

in developing strains resistant to certain insect pests such as the European corn borer.

This field has been given renewed emphasis by biochemical studies, showing that chemical factors in plants are responsible for the feeding habits of insects. Thus the mustard oil glycosides of the Cruciferae are responsible for the feeding response of the cabbageworms *Pieris* and *Plutella*, while the specific distribution and structure of the solanidine alkaloids in the Solanaceae determine the food preference of the Colorado potato beetle. The fat-soluble factor (RFA), responsible for limiting the feeding of young corn borers in the pretassel stage of corn, has been identified as 6-methoxybenzoxazolinone, and the presence of this substance is correlated with varietal differences in corn borer resistance.[1]

6-Methoxybenzoxazolinone

Other Cultural Control Measures. Finally, the importance of good husbandry, as defined by agronomists, horticulturists, entomologists, and plant pathologists, working in cooperation, cannot be overstressed. The use of good seed, excellent preparation of seedbeds, conservation and regulation of soil moisture, proper pruning and thinning when necessary, and the judicious use of fertilizers all offer possibilities of stimulating plant growth in such way as to make possible the growing of profitable crops, where the neglect of one or more of these factors may result in loss or disaster.

BIOLOGICAL CONTROL: THE INTRODUCTION AND ENCOURAGEMENT OF NATURAL ENEMIES

Among the many adverse factors which continually affect every insect species in the struggle for existence are the other living things that feed upon it. These are collectively known as its natural enemies. The fact that man has, during the last few centuries, learned something about the habits, ecology, and interrelations of insects now enables him to take sides in the constant warfare that insects are carrying on against each other. The facts at hand concerning the insect population of a given area of the earth show that more than two-thirds of the species present are feeders on plants or plant products or, in other words, are competing with man for the products of the soil. From one-fourth to one-third of the insects present in any given area feed on other insects, and many of these are of great benefit to man in reducing the plant-feeding species. Others feed on those that attack the plant feeders and so become the enemies of man. This complicated relationship of man and insects has already been discussed in Chapter 2.

As long ago as the seventeenth century man first conceived the idea

[1] SMISSMAN, E., J. LAPIDUS, and S. BECK, *Jour. Amer. Chem. Soc.,* **79**:4697, 1957; BECK, S., *Jour. Insect Physiol.,* **1**:158, 1957.

of taking advantage of the food preferences of these natural enemies of insects to destroy or suppress pest species. *Biological control* may be defined as the destruction or suppression of undesirable insects, other animals, or plants by the introduction, encouragement, or artificial increase of their natural enemies.

Among the natural enemies of insects which may be used in this way are:

1. Predaceous and parasitic insects (page 61).
2. Predatory vertebrates.
3. Nematode parasites.
4. Protozoan diseases.
5. Parasitic fungi.
6. Bacterial diseases.
7. Virus diseases.

Of the many possible applications of biological control only two seem to have been used with success, and the entomologist is vitally interested in both of them. *First*, the control in this way of insect pests of growing crops, living animals, or stored products. The plant- or animal-feeding insect is the pest; the attacking organism may be of any one of the groups listed as natural enemies above. *Secondly*, the control of weeds or plant pests by the insects that feed upon them. In this case the insect is the attacking organism, a benefactor, not a pest. Four principal methods have been employed: (*a*) Collecting parasites or predators in places where they have naturally developed or assembled in great numbers and releasing them (perhaps after storing them over winter) in places where they may do the most good: either concentrating the beneficial organism on a small area or dispersing it more widely from a center of great abundance. (*b*) Collecting and storing or handling the host insects in such a way as to kill them but permit any parasites or predators among them to escape. (*c*) Rearing under favorable conditions great numbers of parasites or predators and releasing them, whenever and wherever needed, especially at the time when the normal fluctuations of the pest insect have reached their point of greatest abundance. (*d*) Importing parasites, predators, or diseases from a foreign country. The last-named has been the method most extensively used. It has been used especially in cases where the damage is being caused by a species of insect that was unintentionally imported, whose natural enemies have not been brought with it to that part of the earth where it has become established. The beneficial organism is usually sought in the original home of the pest. There is no hope of *exterminating* insects by this method, but it is sometimes possible so to reduce them that no other control measures need be used, as in the case of the citrophilus mealybug, *Pseudococcus gahani*, and the cottony-cushion scale, *Icerya purchasi*. When any parasite becomes so abundant that it nearly wipes out a pest insect, it too must suffer a marked decline in numbers because of the reduction of its food supply. A scarcity of parasites permits the pest insect again to increase until the parasite once more overtakes it. Thus a single active parasite of a pest species will tend to cause more or less regular periods of abundance and scarcity of the parasite and its host. Consequently it is often of advantage if a parasite can be introduced which will find several hosts in the region where it is liberated. During the periods of the pest abundance, it may be necessary to depend on artificial measures for controlling the plant-feeding insect. However, if biological control results in a pronounced reduction in the abundance of a pest, so as to decrease the amount of damage it would have done or the extent of other control measures required, it should be considered a success. This is the only type of control that is self-perpetuating. Since it may be expected to continue indefinitely, a tremendous initial expense may prove to be a very low total cost. Clausen states that at least 30 pest species have been adequately controlled in one or more countries by the biological method.

Parasitic and Predaceous Insects.[1] The classical example of this type of biological control was the introduction of the Australian lady beetle or vedalia, *Rodolia cardinalis*, into California to destroy the cottony-cushion scale, *Icerya purchasi*. The cottony-cushion scale was unwittingly introduced into California about 1868. It soon became a most serious pest of citrus, spreading rapidly over the state, and by 1890 had killed hundreds of thousands of trees, threatening to wipe out the orange industry of the entire area. It was traced to Australia and New Zealand; in New Zealand it was very destructive also, but in Australia little damage resulted from it. Accordingly, the United States government sent an entomologist to Australia to search for natural enemies that it was believed must be holding it in check there. The lady beetle was discovered, and about 514 of them were carefully shipped to California. They were liberated on screened orange trees and allowed to feed on the cottony-cushion scale. Within $1\frac{1}{2}$ years the progeny of these few beetles had increased to such numbers that they had checked the cottony-cushion scale over the entire state. They have since nearly eliminated the scale as a serious pest of citrus in California. This lady beetle has subsequently been shipped to 40 different countries, becoming established in at least 32, and has never failed to control the cottony-cushion scale wherever it has become established.

Some of the other outstanding examples of success with imported entomophagous insects are the control of the citrophilus mealybug, *Pseudococcus gahani*, by two chalcidoid parasites from Australia, *Coccophagus gurneyi* (page 71) and *Tetracnemus pretiosus*; the long-tailed mealybug, *P. adonidum*, by the internal parasites *Anarhopus sydneyensis* from Australia and *Tetracnemus peregrinus* from Brazil; the citrus mealybug, *P. citri*, by *Leptomastidea abnormis* from Sicily and the lady beetle, *Cryptolaemus montrouzieri*; and the black scale, *Saissetia oleae*, by the chalcid, *Aphycus helvolus* (page 70). In Hawaii, the mealybug, *P. nipae*, which attacks avocado, fig, guava, and cocoanut, has been controlled by the chalcid, *Pseudoaphycus utilis*; the sugarcane leafhopper, *Perkinsiella saccharicida*, by a predaceous bug, *Cyrtorhinus mundulus*, that devours the egg; the New Guinea sugarcane weevil, *Rhabdocnemis obscura*, by a parasitic tachnid fly, *Microceromasia sphenophori*.

In the United States, elaborate programs of biological control which, although not providing complete control, have appreciably reduced the frequency and destructiveness of insect outbreaks have been waged against many pest species. These include the establishment of 13 species of parasites and predators of the gypsy moth, *Porthetria dispar*, and brown-tailed moth, *Nygmia phaeorrhoea*, in New England forests; the parasite *Bathyplectes curculionis*, which destroys many larvae of the alfalfa weevil, *Hypera postica*; the wasp, *Aphelinus mali* (page 70), an internal parasite of the woolly apple aphid, *Eriosoma lanigerum*; the mass

---

[1] Clausen, C. P., *U.S.D.A.*, *Tech. Bul.* 1139, 1956.

rearing and release of the wasp, *Macrocentrus ancylivorus* (page 72), which has effectively controlled the oriental fruit moth, *Grapholitha molesta*, in some areas; and the establishment of 6 species of parasites of the European corn borer, of which the two most effective are *Lydella stabulans grisescens* (page 69) and *Macrocentrus gifuensis*.

In introducing a parasite into a locality where it has not previously been known to occur, great care must be taken. It must be ascertained without question (*a*) that the insect to be introduced is a parasite on the particular insect it is desired to control or on other plant-feeding species, (*b*) that it is never by any chance a plant feeder, and (*c*) that it will not attack some of the other primary parasites already present in the locality and so do more harm than good. The mass rearing of suitable parasites and predators as developed by H. S. Smith and coworkers has permitted the widespread colonization of imported species within a short time after their introduction and has in some instances, as with the Australian lady beetle, *Cryptolaemus montrouzieri*, permitted the use of a species not climatically adapted to the area of its importation. Mass-production techniques depend upon the successful rearing of the host pest insect. In the case referred to, it was discovered that enough mealybugs could be reared upon sprouts from a ton of potatoes to produce more than 125,000 *Cryptolaemus*, and by this technique more than 40,000,000 beetles were distributed annually in California orchards for many years at a cost of about $2.50 per thousand. With *Macrocentrus ancylivorus* potatoes are used to rear an alternate host, the potato tuberworm, *Gnorimoschema operculella*, and about 235,000 parasites are produced for each ton of potatoes, some 29,000,000 being produced in 1946. The minute egg parasite, *Trichogramma minutum* (page 69), has been reared by the hundreds of millions on eggs of the Angoumois grain moth, *Sitotroga cerealella*, and offered for sale by commercial entomologists at a price of 15 to 20 cents per thousand.

Integrated Control and Selective Application of Insecticides.[1] The most efficient way to utilize insecticides for the control of many agricultural pests lies in the use of integrated control, *i.e.*, the combined application of biotic agents and chemicals. It has been realized for a number of years that the widespread application of general-purpose residual insecticides is often highly destructive to populations of beneficial parasites and predators, and this has sometimes resulted in abnormal population increases of pests other than those for which the treatment was made. Familiar examples include the outbreaks of red spider mites following the use of DDT and other chlorinated hydrocarbons and the resurgence of the cottony-cushion scale following the application of DDT to citrus. It is obvious that such occurrences result from the improper application of insecticides and that by recognizing biological and chemical agents as complementary in action and fitting insecticides into the ecosystem of the pest, the entomologist is utilizing his professional capacities to the highest degree. The successful development of integrated control involves (1) knowledge of the biotic environment or ecosystem of the pest, (2) recognition of pest population levels responsible for economic damage, (3) development of selective insecticides, and (4) encouragement or augmentation of natural enemies.

Insecticides with Specific Action. With few exceptions, stomach poisons such as cryolite and lead arsenate are relatively nontoxic to the

[1] STERN, V., R. SMITH, R. VAN DEN BOSCH, and K. HAGEN, *Hilgardia*, **29**:81, 1959.

beneficial insects living in the treated environment. TEPP and Phosdrin have very short residual action and will, at worst, only temporarily reduce the populations of beneficial insects. Generally, such materials will not destroy the eggs of predators or the pupae of predators and parasites. The systemic insecticide schradan is virtually nontoxic to many insects but highly effective against aphids. Disease pathogens or their toxins, such as *Bacillus thuringiensis* toxin, are generally very specific in their action against a few related insects. It is important to realize that the ideal selective action often leaves a few hosts undamaged so as to support beneficial insects in the treated area. Thus 100 per cent mortality in chemical control is not always desirable.

Selective Application of Nonselective Insecticides. Demeton and its derivates are rapidly absorbed into the interior of leaves and fruits and poison the plant juices, thus becoming highly toxic to sucking insects but not damaging to parasites and predators. The selectivity of Di-Syston, phorate, and dimethoate can be assured by applying them to the soil as granular formulations so that root absorption results in selective systemic action. Granular applications of chlorinated hydrocarbons for ant control do not seriously disturb the ecosystem. Alternate strip treatments in which only a portion of the crop is treated at any given time leave a haven for beneficial insect populations to continue development. Proper timing of insecticide applications so that treatment is made when the pest is vulnerable but the parasites or predators are in pupation, hibernation, etc., produces highly selective effects.

The choice and selective usage of insecticides so as to minimize damage to bees, so essential in pollination, and to birds and other wildlife are equally a part of entomological common sense. Very often timing of application is the critical factor, to ensure that insecticides are not applied when orchard crops are in bloom or when bees are actively gathering nectar. It is only through the utilization of the principles of selective action that insect control can be carried out with precision and with a minimum of disturbance to other living creatures.

Predatory Vertebrates. Among the vertebrates that are predatory upon insects, the birds are doubtless most effective, because their ecological activities are nearly identical with those of insects. More than half the food of this most abundant and most mobile group of terrestrial vertebrates is insects. They stand supreme among the vertebrate enemies of insects. Much has been done by man to attract them about his habitations and encourage their abundance, but there have been no spectacular cases of the transportation of birds from one country to another to combat insects (page 424). Fishes of certain species have been employed to destroy mosquito larvae cheaply and effectively. The giant toad of Mexico, Central America, and South America has been introduced into the Hawaiian Islands, the Philippines, and the West Indies and is said to have brought under control the white grubs destructive to sugarcane in Puerto Rico.

Nematode Parasites. Many species of the phylum Nemathelminthes (page 176) are parasitic in the bodies of beetles, grasshoppers, cockroaches, moths, and other insects. Species attacking the striped cucumber beetle and the Japanese beetle are considered important in the control of these pests, and some effort has been made to use the latter, *Neoaplectana glaseri*, in biological control. Much more study of this subject is needed.

The Use of Insect Diseases.[1] Insects, like other animals, suffer from the attacks of diseases. At times, under favorable conditions, a disease may become epidemic on a species of insect and within a few days or weeks reduce the species from a point of great abundance to one of scarcity. Insect diseases may be caused by protozoa, fungi, bacteria, or viruses. All these minute organisms live abundantly on and in the bodies of insects. Among such diseases, which have been employed in biological control, are the brown,[2] red,[3] and yellow[4] fungi that live on the bodies of whiteflies which attack citrus in Florida. The spores of these fungi are sometimes mixed with water and sprayed over trees infested with the whiteflies, much as a chemical spray would be applied. The Empusa diseases of grasshoppers, house flies, and other pests; *Beauveria globulifera* attacking chinch bugs; and *Metarrhizium* and *Cordyceps* on white grubs kill millions of their hosts, in certain areas and certain seasons, but their utilization in biological control has not yet been demonstrated as practicable. The bacterium, *Coccobacillus acridiorum*, has been used against grasshoppers in parts of Africa; *Bacillus thuringiensis* and other species against the European corn borer in Europe and the pink bollworm in Egypt; and the protozoan, *Perezia pyraustae* against the European corn borer.

The outstanding example of the practical utilization of an insect disease is that of the milky disease, *Bacillus popillae,* for the control of the Japanese beetle, *Popillia japonica.*[5] The disease is produced by inoculating healthy beetle larvae with about 1 million spores, and the larvae are kept at 86°F. in boxes of soil containing grass plants as food. After 10 to 12 days, the larvae contain about 2 billion spores and are screened from the dirt, washed, and stored in ice water at 35°F. For use, a brei of the larvae is powered with chalk, dried, and diluted with talc to produce a standardized powder containing 100 million spores per gram, which is stable for years. This dust is applied to the surface of the soil in 2-gram patches from 3 to 10 feet apart, using from 1.75 to 20.6 pounds per acre. Such treatment, usually made with a modified rotary corn planter, results in the slow spread of the disease over a period of several years, so that satisfactory control is obtained.

The polyhedrosis virus disease, *Borrelina campeoles*, has been employed for the commercial control of the alfalfa caterpillar, *Colias philodice eurytheme*, by airplane spraying of a virus suspension, from breis of infected larvae, containing 5 million polyhedra per milliliter, at the rate of 5 gallons per acre.[6] The spore-forming bacterium, *Bacillus thuringiensis*, has also proved highly effective against this insect.[7] Commercial control has been obtained by spraying a suspension of about 1,500,000 spores per milliliter of water at the rate of about 15 gallons per acre.

Biological Control of Weeds. The classical example of biological control of weeds concerns the fight against the cacti in Australia. About 1840 a doctor immigrating to Australia carried with him a single potted plant of the prickly pear, *Opuntia inermis.* This plant was a curiosity

[1] STEINHAUS, E. A., "Principles of Insect Pathology," McGraw-Hill, 1949.
[2] *Aegerita webberi.*
[3] *Aschersonia aleyrodis.*
[4] *Aschersonia goldiana.*
[5] *U.S.D.A., Bur. Ento. Plant Quar., Cir.* E-801, 1950.
[6] THOMPSON, C., and E. STEINHAUS, *Hilgardia,* **19**:411, 1950.
[7] STEINHAUS, E., *Hilgardia,* **20**:359, 1951.

in that country and cuttings from it were spread far and wide. They grew and thrived beyond belief, and within 30 years it was realized that the innocent curio had developed into a terrible weed pest. Whereas in America this cactus grows to heights of 6 to 10 inches, under Australian conditions it reaches a height of 6 to 10 feet, with prickly branches so dense that no one can penetrate through it. It spread rapidly over farms and grazing land, crowding out and smothering all other crops. By 1910 it had claimed 10 million acres. By 1916, 23 million acres were overrun with it, and it was spreading at the rate of about 1 million acres a year. Eventually 50 to 60 million acres were rendered absolutely useless by the prickly pear—a gigantic jungle, made impenetrable by the riotous growth of spiny cactus. Every conceivable method of destruction was tried. Mechanical cutters and rollers and poison sprays and gases proved to be either inefficient or too costly. In 1913 the Australian government undertook to control this weed by the use of its insect enemies, introduced from Texas, Mexico, India, Ceylon, Uruguay, and South Africa where the cactus was native. Of the many insects introduced, the most promising is a moth borer, *Cactoblastis cactorum*. From the original importation in 1925 many thousand millions of caterpillars have developed. With as many as a million per acre eating the interior of the plant and opening it to further destruction by rots, the cactus is soon reduced to dry skin and fiber. Other enemies of the plant including certain kinds of mealybugs, true bugs, and red spider mites have contributed to the good work; and it is hoped that the entire infested area may be eventually reclaimed by the careful manipulation of these insect benefactors. In every case elaborate tests are made to insure that the insects so introduced will feed on nothing but cactus. The pest plant *Hypericum perforatum*, native of Europe and known in Australia and New Zealand as St. John's wort and in the United States as the Klamath weed, has invaded the ranges of these countries and has gradually driven out the valuable native forage plants of many hundreds of thousands of acres. The Australian government has undertaken with considerable success to apply the biological-control method to this pest plant. In California the introduction of the beetle, *Chrysomela gemellata*, which the Australian entomologists discovered in Europe, has controlled the weed over many thousands of acres. Many other attempts have been made in various countries to control plant pests with their insect enemies and with sufficient success to warrant the belief that under certain conditions plant-eating insects may be among man's greatest allies.

LEGISLATION FOR INSECT CONTROL[1]

In the early days of agricultural development in this and other countries, plants and plant products were brought into, or sent out of, the country with little or no thought concerning the insect pests that might be transported along with them. In fact, it is only since the middle of the past century that any serious attempt at legislation to restrict the spread of insect pests has been attempted by any country. The introduction of the grape phylloxera from America to the vineyards of France, some time about 1860, caused such serious destruction to the French vineyards and to those in other European countries to which it was carried that it became apparent something must be done to prevent the unrestricted movement of infested vines to all parts of the world. In 1881

[1] STRONG, LEE A., *U.S.D.A.*, *Bur. Ento. Plant Quar.*, *Cir.* E-455, 1936.

representatives of many of the European countries in which grapes were extensively grown met and agreed on regulations restricting the movement of infested grape stalks. The spread of certain plant pests in the United States in the latter part of the past century, particularly the San Jose scale, stimulated the passage of insect legislation in this country. By the close of the century nearly every state had passed laws restricting the shipment of infested nursery stock. Although an act was passed by the federal government in 1905 "to prohibit importation or interstate transportation of insect pests," it was not until 1912 that the United States had adequate federal laws to control the menace of foreign plant pests.

At present, there are five classes of insect legislation: (a) legislation to prevent the introduction of new pests from foreign countries; (b) legislation to prevent the spread of established pests within the country or within the state; (c) legislation to enforce the application of control measures that have been found effective in preventing damage by established pests; (d) legislation to prevent the adulteration and misbranding of insecticides and to determine their permissible residue tolerances in foodstuffs; and (e) legislation to regulate the activities of pest-control operators and the application of hazardous insecticides.

Quarantine and Inspection Laws. Federal regulatory measures dealing with the introduction and spread in the United States of injurious insect pests are the responsibility of the U.S. Department of Agriculture, Plant Quarantine Division. This organization is responsible for (a) enforcing quarantines against the entry of foreign plants, plant products, and insect pests into the United States, (b) the administration of domestic quarantines against inter- or intrastate movement of such products, and (c) the inspection of domestic nursery stock, fresh fruits, vegetables, seeds, and other plant products exported for propagation and certification that they are apparently free of dangerous insects and diseases. Every state now has in force regulatory measures forbidding the movement of certain plants into or within the state, at least until inspected and fumigated, or both. It is illegal to ship nursery stock anywhere in the United States unless it is accompanied by a certificate stating that it has been found apparently free from certain seriously destructive insects and plant diseases.

The federal government has the power, under the Plant Quarantine Act of 1912, to prevent the introduction or spread of any dangerous insect or plant disease by prohibiting the importation or shipment, interstate, of any class of plants or plant products from any foreign country or locality and from any state or portion of a state in this country. Such specific prohibitions are called *quarantines*. Many of the states also have quarantine laws.[1]

The National Plant Board, which is an organization formed by representatives of the various regional plant boards to bring about greater uniformity and efficiency in the establishment and enforcement of plant quarantines in the various states, has suggested the following fundamental prerequisites for the establishment of a quarantine: (a) the pest concerned must be a threat to substantial interests, (b) the pest must not be susceptible to control by other measures involving less interference with normal activities, (c) the objectives of the quarantine in preventing

[1] See "Plant Regulatory Announcements," *U.S.D.A., Plant Quar. Div.,* and *U S.D.A., Misc. Publ.* 80, 1946.

the introduction or in limiting the spread of the pest must be reasonably possible of accomplishment, and (d) the economic gains from the quarantine must exceed the cost of enforcement. Domestic quarantines are enforced by inspection and examination procedures which include the inspection of field and market for certification and the inspection of physical premises, accounts and records, and vehicles and common carriers passing through the quarantine area.

At present foreign plant quarantines are in force restricting the importation of certain fruits and vegetables; cotton lint and cottonseed products; certain cereals, plants, and plant products for propagation or as packing materials; fruit and vegetable host plants of the various fruit flies (page 812) and the virus of foot-and-mouth disease. To indicate the need for such legislation it should be pointed out that during a 7-year period federal authorities intercepted:

From Germany, 12 infested shipments containing 15 kinds of insect pests.
From England, 154 infested shipments containing 62 kinds of insect pests.
From Japan, 291 infested shipments containing 108 kinds of insect pests.
From France, 347 infested shipments containing 89 kinds of insect pests.
From Holland, 1,051 infested shipments containing 148 kinds of insect pests.
From Belgium, 1,306 infested shipments containing 64 kinds of insect pests.

At present domestic plant quarantines are in force against the unrestricted movement of certain commodities from the known infested areas on account of the gypsy and brown-tail moths, the Japanese beetle, the pink bollworm of cotton, the Mexican fruit fly, the white pine blister rust, and the black stem rust of grains; and certain state quarantines against the European corn borer, the alfalfa weevil, the phony peach disease, and others are in force.

In general, it may be said of quarantine measures that the enforcement of such measures will check the spread of certain insects but cannot be depended upon to stop such spread. A considerable expense is justified if the quarantine checks the spread of a new pest long enough to permit the development of control measures, the introduction of parasites, or changes in agricultural practices best suited to prevent loss by the newly established pest. It should be borne in mind, however, that regulatory measures are always expensive measures and that it is impossible to keep a strong-flying insect out of territory adjacent to that at present occupied by it merely by passing laws. It is certainly desirable that all reasonable restrictions be placed on the importation of new insect pests from foreign countries and that every precaution possible be taken to inspect plants and plant products entering this country, to see that they are free from foreign insects.

The enormous increase in air travel between various foreign ports and the United States has vastly increased the opportunities for the introduction of insect pests dangerous to agriculture and to public health. This is especially true because of the short intervals required for passage between ports, which permit the introduction of many short-lived and delicate insects which could not survive long sea voyages. For example, in 1945, of 45,728 aircraft arriving in the United States from foreign countries and overseas possessions, 7,299 were found to contain prohibited plant material and 2,442 interceptions of insects and plant diseases were made.[1] To safeguard against the admission of such pests into

[1] *Jour. Econ. Ento.*, **40**:129, 1947.

the United States, the U.S. Department of Agriculture has instituted the regular inspection and aerosol treatment of all aircraft entering this country through trans-Pacific routes at Hawaii and the Aleutians. The U.S. Public Health Service also carries out inspection and aerosol treatment of aircraft entering United States ports which may carry insects and other arthropods which are known or suspected vectors of human diseases. As an illustration of the importance of this work, in the 10-year period from 1937 to 1947, 80,716 planes were inspected and 28,752 were found to contain arthropods. From these, 12,852 mosquitoes were collected comprising 10 genera and 73 species, of which 25 were not indigenous to the United States.[1]

The history of an insect in its native country cannot always be relied on as a criterion of what the insect will do when established in some locality in another part of the world, where it is not held in check by the natural enemies found in its native home. For this reason it is best from a legal standpoint to consider that all foreign plant-feeding insects and zoophagous parasites are dangerous, until they have been proved otherwise. At the present time practically all civilized nations have in effect regulatory measures restricting the movement of plant products from other countries, and it is probable that these measures will become more strict in the future, rather than more lenient. Unfortunately, there have been a few attempts to use quarantines as trade barriers. It is hardly necessary to state that this should never be tolerated.

Compulsory Cleanup Measures. While the regulations governing the control of pests in nurseries are, in most states, broad enough to allow for the enforcement of cleanup measures against orchard or field-crop insects as well, it is only in a comparatively few states that such measures are generally enforced. Indeed, it usually requires serious loss of property or personal injury before the public will wholeheartedly back up such measures. Some of the eastern states, notably Massachusetts, have rigidly enforced the control of such pests as the gypsy and brown-tail moths. If premises are infested, in some states, the property owner is given legal notice that he must, before a given date, take measures to control certain insect pests. If such measures have not been taken before this date, the work is done by a force of men employed by the city or town in which the property is located. The cost of this work is assessed against the property in the form of a tax. If unpaid, it constitutes a lien against the property as much as any other unpaid tax and is collected in the same way. This measure has had a thorough test in the courts and has now been in force for a number of years.

Insecticide Laws. The federal government and many of the states have laws regulating the sale and usage of insecticides. The Federal Insecticide, Fungicide, and Rodenticide Act of 1947 provides for the registration with the Pesticides Regulation Branch of the U.S. Department of Agriculture of all economic poisons, including insecticides, fungicides, rodenticides, and herbicides. Labels of such materials are required to state (a) the name and address of the manufacturer, (b) the name, brand, or trademark, (c) the net contents, (d) a statement of the ingredients, including a well-known common name or the chemical name, and the percentage of active and inert ingredients, and (e) adequate directions for use, including recommended rates of application and precautions. Economic poisons highly toxic to man, as defined by the Act, are required

[1] *Pub. Health Rept.*, Suppl. 210, July, 1949.

to carry a poison label with skull and crossbones, a warning or caution statement, and a statement of antidote. Claims for the use of the product must conform to the registration, and the manufacturer must guarantee that the poison conforms to the provisions of the Act. An economic poison is considered adulterated if its strength or purity falls below the professed standard under which it is sold, if any substance has been substituted wholly or in part for the substance named, and if any valuable ingredient has been abstracted. An economic poison is considered misbranded if the label contains false or misleading statements concerning the composition or effectiveness of the product and the amount and nature of its ingredients, its value for purposes other than an economic poison, its comparison with other economic poisons, its safety to plants and animals, or if any statement is made implying that it is recommended or endorsed by the federal government.

The establishment of tolerances for residues of economic poisons on raw agricultural commodities is regulated by the Federal Food, Drug, and Cosmetic Act as amended by the Miller Bill of 1954 (Public Law 518). Under this law, tolerances for insecticides in common use were established and tolerances for new products are determined as follows. The manufacturer or other interested party presents to the U.S. Food and Drug Administration a petition containing (a) data about the chemical composition of the pesticide, (b) the amount, frequency, and timing of applications, (c) complete data on its safety, including the results of 2-year chronic-feeding studies, (d) analytical methods for the determination of residues and complete data on harvest residues, (e) any practical method for removal of excess residues, and (f) proposed tolerances on various commodities. Simultaneously, a petition is filed with the U.S. Department of Agriculture, Pesticide Regulation Branch, containing details of experimental work supporting the effectiveness and usefulness of the pesticide and requesting a certificate of usefulness. If these agencies report favorably on the safety and usefulness of the material, the proposed tolerance is published in the Federal Register and becomes effective in 90 days. Examples of tolerances for insecticides are given on page 396. The Delaney Bill of 1959 amends this law to apply to food additives.

NATURAL CONTROL

Natural control includes control (a) by climatic factors such as rainfall, sunshine, cold, heat, and wind movement; (b) by the physical character of the country, such as large bodies of water, mountain ranges, streams, the character of the stream flow, and the type of soil; (c) by the *natural* presence and abundance of predaceous and parasitic insects, birds, fishes, reptiles, and mammals and by cannibalism; and (d) by the presence of diseases which attack insects and conditions favorable to the spread of such diseases.

Control by Climatic Factors

Under natural control, climatic factors are perhaps the most important. A few species of insects have become adapted to variations in climate to such an extent that they occur throughout the world, in temperate and, in some cases, in temperate and tropical zones. Few, if any, species of insects occur in all three zones—the arctic, temperate, and tropical—except such species as infest stored products, the dwellings of man, or the bodies of animals, which are therefore not subjected to a very marked degree to the climatic changes of any region.

As a general statement, it may be said that the insect life of a region is dependent, directly or indirectly, on the temperature of the region, the soil, and the amount of moisture. There are many species of insects which have become adapted to certain climatic conditions and thrive under these conditions even though they may seem at first to be unfavorable. Great numbers of mosquitoes and certain species of flies occur in the arctic regions during the brief summers. At the height of warm weather, the total number of insects in such regions is very large, but never, on the whole, as large as the numbers found in the tropics. Winter temperatures control the distribution of insects; the harlequin cabbage bug, which cannot survive our northern winters, is an example.

A very warm, moderately humid climate and fertile soil offer conditions favorable for the greatest development of insect life. A poor soil can support only a limited amount of plant growth and, therefore, a limited insect population. A warm, wet climate is unfavorable to many insects. Such a climate creates a condition where insect diseases will flourish and also presents many physical factors unfavorable to insect life. A hot and very dry climate also is unfavorable, only a comparatively few species of insects having become adapted to life under desert conditions.

The amount of sunshine occurring in a given region also is important. Many species of insects are influenced to a marked degree by the rays of the sun. Some apparently seldom fly except during periods when the sun is shining; and, as flight is the chief means of dissemination of most species over any area, this is an important factor influencing the general abundance of insects; the chinch bug is one of the best examples.

Wind movement is also of great importance. Many of the smaller and frailer species of insects which normally fly for considerable distances are unable to leave the ground during strong winds or, if they do take flight, are so buffeted and beaten by the wind that they soon die; the Hessian fly is thus affected. Some species of insects, as certain mosquitoes, habitually fly against the wind, while many others fly with the wind; their dispersal, or the direction of their dispersal, is to an even greater extent dependent upon wind movement at the time of year when they are in their winged or adult stage.

The brown-tail moth has spread very slowly in a westerly direction from the original point of establishment, but its spread to the east and north has been rapid, due chiefly to southwesterly winds during the time when the adults are flying.

CONTROL BY TOPOGRAPHIC FACTORS

Large bodies of water, such as oceans, offer effective barriers to the natural spread of nearly all species of insects. Certain species that do not possess the power of flight are affected in their spread by smaller bodies of water, such as lakes or large streams, and are largely dependent upon man or other animals for the passage of such barriers. The granary weevil is unable to fly but has been carried by man over much of the earth. Mountain ranges also are effective barriers to the spread of insects and offer varying conditions of climate through which many insects cannot pass unaided. The Colorado potato beetle began spreading from Colorado to the East about 1859 and reached the Atlantic Coast in 15 years, while it was more than 50 years in crossing the Rocky Mountains.

The character of the streams and the number of ponds and lakes control to a great extent the insect life of a country. Certain flies, mosquitoes, and beetles live in their immature stages in slowly moving streams or still water, while others, such as black flies and certain caddice-flies, live only in swift-flowing streams.

The character of the soil of any region exerts a marked influence over the insect inhabitants of that region. This is true not only because the soil has a direct influence on the plant growth of the region and thus indirectly affects the insects that live on plants, but also because many insects spend the whole or part of their life in the soil. Certain soils are very favorable to their growth, while they would be unable to live in different soils, or in the same type of soil under different conditions. Certain species of wireworms live only in poorly drained soils, and certain species of tiger beetle larvae, which live in sandy soils, are unable to exist in clay soils.

Control by Natural Enemies

Predaceous and Parasitic Insects. Of the natural factors that tend to reduce the plant-feeding insects, the number of the insects feeding on other insects in a locality is sometimes as important as the climatic factors. It has been demonstrated many times that a plant-feeding insect removed to a part of the world where its insect enemies are not present, and with suitable food plants present, is able to increase to far greater numbers than was the case in its native home. Indeed, it would probably be very difficult to produce crops in most regions of the earth, if the predaceous and parasitic insects were not present to keep down the plant-feeding species.

Weiss, in classifying the insects of New Jersey according to their food habits, found that 28 per cent of the approximately 10,000 kinds of insects known to occur in that state were feeders on other insects.

Cannibalism, the devouring of individuals by their own kind, is an important factor in reducing the numbers of certain insects. A conspicuous example is the larvae of the corn earworm.

Birds. From an insect standpoint a birdless country would be a highly desirable place in which to live, as in such a country insects would be safe from the attacks of many of their most persistent enemies. A proportion of the food of most birds is made up of insects, and the food of many species is largely of insect origin. The actual number of insects eaten in a day by certain birds is surprising, in some cases being almost or quite equal to the weight of the bird itself. This is especially true of the nestlings during the most rapid period of their growth. Thanks to the studies which have been carried on by the United States Bureau of Biological Survey, we now have sufficient data on the food of nearly all the common species of birds to enable us to know their value as insect destroyers.[1] Some of the common species, such as robins and catbirds, eat insects mainly during the summer when this kind of food is most abundant. Others, such as some of the woodpeckers and creepers, subsist largely on an insect diet throughout the year. Even those species that are largely grain feeders, such as the blackbird and English sparrow, will often congregate in large numbers in areas where insect outbreaks are occurring and will feed mainly on insects during the period when they can be easily obtained (page 61).

[1] Henderson, Junius, "The Practical Value of Birds," Macmillan, 1927.

While birds cannot be expected to become sufficiently abundant in any thickly settled farming area for us to depend upon them alone to prevent insect damage, they are of great value. Most birds earn, many times over, the fruit and berries they take from our orchards and gardens. The value of most birds as insect destroyers alone will warrant all the protection we can afford them. This protection is needed, not only against killing by the use of firearms and nest robbing, but also against the cat, which is the birds' worst enemy in many sections. The number of birds in a local area may be increased by providing food during the winter months and water and suitable nesting materials and places during spring and summer, and by planting seed- and fruit-bearing shrubs. The setting aside of tracts of land as game preserves also will aid in increasing the numbers of song birds valuable as insect destroyers.

Mammals and Other Animals That Feed on Insects. Many of the small mammals feed to a great extent on insects. Some of the ground squirrels eat white grubs and other soil-infesting insects, but these make up only a small part of their food. Moles, shrews, and skunks depend largely on insects for their food and destroy very large numbers of the soil-infesting kinds. Some species of snakes, newts, and salamanders also subsist largely on an insect diet. The toad is one of the most useful of our common small animals, its food consisting almost entirely of insects, more than 60 per cent of which are of injurious species. These are eaten in very great numbers, the toad devouring in 24 hours an amount of insect food equal to about four times its stomach's capacity.

Value of a Knowledge of Natural Control. As the factors of natural control cannot be greatly influenced by man, it might seem that a knowledge of such factors would not be of much value. The opposite is true, however. Knowing the climate, soil, and topography of a region, we may, to a certain extent, be able to tell the kinds of insects that will be most common in that region, and, with a knowledge of their food-plant preferences, the crops that will be most subject to injury. This is of the greatest value in estimating the amount of injury that may be expected from foreign insects newly established in a country.

This knowledge of natural control is also of great practical value in enabling one to tell the effect of the weather of a season on insects and from this to predict the relative abundance of injurious species the next year. For example, a winter period of very low temperature with no snow will kill most of the eggs of the gypsy moth. Long periods of dry, hot weather during the summer will prevent the apple maggot from ever becoming a serious pest in areas where such weather is the rule. Heavy rains during the time when the eggs of the chinch bug are hatching will often terminate a period of several years of serious destruction by this insect. In the case of the pale western cutworm, in the prairie provinces of Canada, it has been found possible, according to Seamans, to predict the irregular and very serious outbreaks of this cutworm by noting the number of wet days in the preceding May and June. If there are fewer than 10 days in these 2 months when it is too wet to work in the soil, there will be an increase and probably an outbreak of the cutworm the following spring. If there are more than 15 such wet days in May and June, little trouble may be expected from this insect the following season. The explanation of this correlation between weather and cutworm abundance is as follows: The pale western cutworms work below ground except when the soil is wet. When driven aboveground by heavy rains or by irrigation, they are attacked by several kinds of parasites, which reduce their numbers so that no outbreak is possible the following year. If the soil is dry enough so that they may remain below ground during most of their period of larval activity in May and June, the parasites do not reach them, and the cutworms remain healthy and may increase their numbers to epidemic proportions by the following year.

Taking into account what we know of the effect of various natural factors on insect abundance, it is now possible to warn growers of threatening outbreaks of certain of the more carefully studied species of insects in time to apply measures of control. In Illinois, for example, for many years, fruit growers have been informed, about 14 days in advance, of the time when the eggs of the codling moth will start hatching. Field checks have shown that the predictions have usually been accurate to within 24 hours and that there has never been an error of more than 3 days. The importance of a knowledge of the effect of these natural factors on insect life is just beginning to be realized, and a more thorough study of them is badly needed.

CHAPTER 8

APPLICATION OF INSECTICIDES

The usefulness of any insecticide depends in a very large measure upon its proper application, and this is determined by the properties of the insecticide, the nature of the pest or pest complex to be controlled, and the site to which the application is to be made. The three general methods of applying insecticides are as *sprays*, in which water or oil is used as the carrier for the toxicant, as *dusts*, in which a fine dry powder is the carrier, and as *fumigants*, in which the insecticide is applied as a gas. The application of fumigants is discussed in Chapter 7.

The equipment for the application of insecticides ranges from such simple devices as the puff duster and "flit gun" to complex machines such as the mist blower and spraying helicopter, and this development is entirely a product of the last 100 years. Before that time, liquid insecticides were applied by a bundle of twigs, a feather, or a brush broom, and dusts by a bellows or blowing tube. The progress which has been made is a product of detailed knowledge of physics and engineering coupled with a vast amount of trial and error. Much remains to be learned, and relatively simple devices may yet be produced which will effect veritable revolutions, as for example, the development of the aerosol "bomb," which from its humble beginning in 1942 has become a standard household article, of which 65,900,000 units were sold in 1959 for insect control. The practicing entomologist needs an intelligent understanding of the theoretical background underlying the dispersal of insecticides, and it is to this purpose that this chapter is directed.

GENERAL CONSIDERATIONS AND CAUTIONS IN INSECTICIDE APPLICATION

The application of insecticides must ordinarily be timed accurately in order to obtain the best results possible. Usually insecticides are applied when populations of a pest species reach the economic threshold. In this way economic damage is prevented and needless treatments are avoided. Other factors must often be considered as well, such as the timing of applications so as to coincide with the stage of development of the insect pest which is most easily killed or with the stage in the seasonal development of the plant when it will best withstand the treatment. Timing to observe the safe interval before harvest which is required to attenuate insecticide residues to legal levels and to avoid periods of bloom for protection of bees and other pollinating insects is also extremely important. Since proper timing varies widely with different crops, different insecticides, and in various areas of the country, the grower should follow only approved spray programs of state and federal agencies, observe all label precautions and directions, and whenever in doubt should secure the advice of a trained entomologist who knows local conditions regarding spray programs for different crops and insects. In general, the earlier the insecticide is applied after the insects appear, the easier it is to destroy them, but in most cases applications should be made only when populations reach economic thresholds.

Many insecticides are violent poisons to human beings and livestock, as well as to insects. Therefore they should be plainly labeled and, together with mixing vessels, kept out of reach of children. Animals must be kept away from liquid sprays and not allowed to pasture in treated areas. All discarded insecticide containers should be burned or buried. The spray applicator should familiarize himself with the hazards of various insecticides before use and determine the proper protective measures necessary for safe application, as furnished by the manufacturer. In general, these will include the use of protective clothing, rubber gloves, and goggles when handling spray concentrates and a protective mask if there is any possibility of inhaling toxic dust, mist, or vapor. Prolonged wetting by sprays or other contamination should be avoided, and clothes changed at least twice daily, followed by the liberal application of soap and water to remove any skin contamination. In case of spillage of liquid spray concentrates on the skin or clothing, instant thorough washing is required.

Spraying and dusting should not be carried out in rainy weather, although many sprays will adhere satisfactorily to plants if the spray has time to dry before rain falls. Winter sprays should not be applied when the temperature is below freezing or when the trees are wet with snow or rain. When possible, spraying should also be avoided in very hot weather, above 90°F., because of increased susceptibility of plants to insecticide damage and because of the increased hazard to the operator from skin and inhalation absorption of poisons.

SPRAYS AND SPRAYING

Spraying is the most common means of insecticide application and in recent years, because of the increased effectiveness of the new organic insecticides, concentrate sprays have largely replaced dusts on many field and vegetable crops. Sprays are made up of solutions, emulsions, or suspensions of the toxicant (page 398). The liquid phase is usually water, but light oils are also employed. Insecticidal sprays are conveniently described as *space sprays* or *aerosols*, directed against flying insects, and *residual sprays*, applied to surfaces of plants, animals, or structures frequented by insects.

PHYSICAL PRINCIPLES OF SPRAYING

Importance of Droplet Size. The fundamental property of a spray droplet is its diameter, which is conveniently measured in microns (μ or 0.001 millimeter). In every spray cloud there is a considerable spectrum of droplet sizes and the cloud is most conveniently characterized by a mean value which expresses the normal frequency distribution of the droplets. The most commonly used characteristic is the mass median diameter MMD or D_m, which is that theoretical droplet diameter which divides the volume of the spray into equal parts. The surface median diameter SMD or D_0 is the theoretical droplet diameter which has the same surface to volume ratio as that of the total spray. D_m is a useful measure of the effectiveness of the spray in terms of the amount of active material deposited per unit area, while D_0 defines the amount of exposed liquid surface and thus the coverage which may be obtained. D_m ranges from 1.1 to 1.6 times D_0, depending upon the nature of the atomizing device. However, with the common fan, swirl, and twin-fluid atomizers used in the application of insecticides, Fraser[1] has found that the average ratio D_m/D_0 is relatively constant with a value of 1.28.

D_m is most conveniently determined experimentally by classifying the droplet diameters of a representative sample of the spray cloud and plotting the cumulative percentage of the total volume against droplet diameter on log-probability paper. This gives a straight-line plot where D_m is the intersection with the 50 per cent point. Spray clouds may be sampled by collecting them on filter paper and measuring the droplet diameters after correcting for the spread factor, approximately 0.16. More accurate determinations, especially of small droplets, can be made by waving glass microscope slides through the cloud to impact the droplets and counting at least 200 under an ocular micrometer. Oil droplets can be collected on a very clean glass surface, or more suitably on oleophobic slides prepared by rubbing with aluminum stearate or by dipping in a 1 per cent alcoholic solution of mannitan monolaurate or 2 per cent silicone in carbon tetrachloride. Aqueous sprays are best sampled on slides treated with a thin film of paraffin, polyisobutylene, or of resin in castor oil. With this method, correction must be made for the spread factor relating the apparent diameter of the lenses formed on the slide to that of the original spherical droplet. This factor ranges from 0.3 to 0.6.[2]

Droplet-size Requirements. The droplet-size requirements for various insecticidal spraying operations cover a 300-fold range of droplet diameters and a 27,000,000 range of droplet volumes and extend from the 1- to 30-micron size of true aerosols to the 100- to 500-micron size of hydraulic sprays. With contact sprays, the zone of efficiency for air-blast sprays is about 30 to 80 microns and for hydraulically driven sprays, between 100 and 300 microns. Special problems are encountered in aircraft spraying, and in general the most efficient range is from 100 to 300 microns. Sprays of 500 microns are in the lawn-sprinkling range and have little value in insecticide applications.

Space Sprays or Aerosols. The principle of space spraying is to suspend a cloud of droplets in the space through which the insect pests are flying and thus to force the insect to accumulate a lethal deposit by colliding with the droplets. The successful application of this method is dependent upon physical laws governing the behavior of aerosols, which are defined as colloidal suspensions of matter in air and involve an understanding of the insect flight behavior. It is clear that the insecticide will be effective only as long as the droplets remain suspended in the free-air space.

STOKES' LAW. Aerosol droplets, unlike gas molecules, possess no diffusive properties, and their only motion other than slow settling due to gravity is in response to moving air currents. The settling rate of aerosol droplets is governed by Stokes' law and proceeds very slowly; thus these particles will drift for enormous distances at low wind velocities.

In simplified terms, ignoring the density of air, which is negligible compared with the density of the aerosol droplet, Stokes' law, which is applicable to droplets up to about 100 microns, states:

$$v = \frac{2dgr^2}{9n} \tag{1}$$

[1] FRASER, R., *Advances in Pest Control Res.*, **2**:1–106, 1958.

[2] BROWN, A., "Insect Control by Chemicals," p. 458, Wiley, 1951.

where v = velocity of settling, centimeters per second
    d = density of insecticide formulation
    g = acceleration of gravity, 980 centimeters per second per second
    r = radius of aerosol droplet, centimeters
    n = viscosity of air, 1.8×10^{-4} poise at room temperature

The approximate settling rates and other properties of spray droplets are given in Table 8.1, and it will be noted that droplets above about 25 microns settle so rapidly as to be of little value for space-spraying of insects in enclosures.

TABLE 8.1. PROPERTIES OF AEROSOLS

Droplet diameter, μ	Approximate no. of droplets formed from 1 g. of material, density = 1.0	Approximate rate of settling in still air, ft. per min.	Approximate drift by 1 m.p.h. wind while settling 1 ft., ft.
100	1,923,000	60	1.5
50	15,430,000	15	5.8
25	125,000,000	3.7	23.4
10	1,923,000,000	0.6	146
5	15,430,000,000	0.15	584
1	1,923,000,000,000	0.006	14,600

DEPOSITION OF SPRAYS AND AEROSOLS. The deposition or impactability of aerosol droplets on the flying insect is directly dependent upon the velocity of collision between the droplet and insect. This collision is influenced not only by the flight speed of the insect, which may vary over wide limits, $e.g.$, from 17 centimeters per second in *Aedes aegypti* to 275 centimeters per second in *Locusta migratoria migratoriodes*, but also the wing-tip speeds, which in the two species mentioned are 220 and 360 centimeters per second, respectively. The wings may be a very important factor in collecting spray droplets, which are subsequently transferred to the mouth parts, legs, or other parts of the body in cleaning operations. Thus it has been shown in *Aedes* that the wings collect six times as much insecticide as the remainder of the body, and similar results occur with the house fly.

Wind-tunnel experiments[1] have shown that the deposition of aerosol droplets on flying insects is directly dependent upon the velocity of collision between the droplet and the insect and upon the square of the droplet diameter, up to a point where the mass of each droplet is so large that all impinge and maximum deposition occurs (Table 8.2). With *Aedes aegypti* it was determined that this relationship holds up to the limits of:

$$D^2 v = 1,000 \tag{2}$$

where D = droplet diameter, microns
    v = velocity of collision, miles per hour

Sell[2] has developed a formula to express the efficiency of deposition of aerosol particles upon objects of various shapes placed perpendicular to the drift of the droplets.

$$\delta = \frac{0.00005 D^2 v}{s} \tag{3}$$

where δ = deposition coefficient
    D = diameter of droplet, microns
    v = velocity of collision, miles per hour
    s = maximum width of object, inches

[1] LATTA, R., L. ANDERSON, E. ROGERS, V. LAMER, S. HOCHBERG, H. LAUTERBACH, and I. JOHNSON, *Jour. Wash. Acad. Sci.*, **37**:397, 1947.

[2] See YEOMANS, A., E. ROGERS, and W. BALL, *Jour. Econ. Ento.*, **42**:591, 1949.

When this formula is applied to consideration of a mosquito 0.025 inch in width, flying through an aerosol at 2 miles per hour, the greatest deposition efficiency is predicted for droplets of about 16 microns. This compares with an experimental value of 12 microns obtained from the data in Table 8.2. Similarly for the house fly, the calculated value is about 22 microns at a flight speed of 4 miles per hour.

Apart from other considerations, it is apparent that the most efficient aerosol consists of droplets of a size so that each contains a lethal dosage of the insecticide. For example, with *Aedes aegypti* adults, the approximate median lethal dosage of DDT is 0.03γ. This amount is contained in a single spherical particle of DDT of 34 microns or in a droplet of 10 per cent DDT in oil solution of 83 microns. Such particle sizes, however, are well above the limits for useful aerosol behavior.

An additional factor in influencing the behavior of sprays and aerosols in outdoor treatment is the penetration of the spray cloud through vegetation and, conversely, the deposition of the droplets upon vegetation. Everyone is familiar with the tendency for fine particles such as cigarette smoke (0.1 to 0.3 microns) to follow the streamlines about objects upon which large spray droplets will readily deposit. For this

TABLE 8.2. RELATION BETWEEN DROPLET SIZE AND TOXICITY OF DDT AEROSOLS TO *Aedes aegypti* FEMALES[1]

Droplet diameter, μ	LD_{50}, mg. DDT at collision velocities			
	16 m.p.h.	8 m.p.h.	4 m.p.h.	2 m.p.h.
1.1	18.4	126.0	184.0	80.5
2.5	2.35	5.90	14.2	32.0
5.0	1.45	1.74	3.24	6.38
10.0	0.59	0.70	1.20	1.75
20.4	0.87	0.87	0.87	0.87

[1] LATTA, R., *et al.*, *Jour. Wash. Acad. Sci.*, **37**:397, 1947.

reason, true smokes are of very limited value as dispersal media for insecticides. An appreciation of the quantitative aspects of this phenomenon is of importance in choosing a spray discharge of the proper dimensions for the problem at hand, *i.e.*, for either the penetration through vegetation or deposition upon it, and such knowledge would have prevented many costly failures in insect control, such as the widespread efforts to use thermal aerosol machines to deposit insecticides for codling moth control.

The physics of droplet deposition upon obstructions is an extension of the factors mentioned above for the deposition of droplets upon flying insects, and Sell's equation (3) relating deposition as a direct function of the velocity and the square of the droplet diameter and as an inverse function of the width of the object is directly applicable. The shape of the obstruction is also a factor, and the efficiency of deposition increases in the order sphere < circular disk < cylinder < rectangular plate < concave surface.

The deposition efficiency has also been described by Brooks[1] as the *dynamic catch*, E_m, which is the per cent of approaching particles in an air stream of any given width, impinged by a cylinder of the same width. The dynamic catch is a function of particle size and air-stream velocity as shown in Fig. 8.1 and varies inversely with the diameter of the obstructing cylinder. It will be noted from Fig. 8.1 that a cylinder 0.125 inch in diameter catches only about 3 per cent of 10-micron droplets at 2 miles per hour, but nearly 30 per cent of 20-micron droplets and that at 5 miles per hour the catch of 100-micron droplets is virtually complete. The extension of this information to the practical problems of insect control involved in such diverse operations as space-spraying for adult mosquitoes in dense vegetation or residual spraying of fruit trees with a mist blower is obvious.

[1] BROOKS, F., *Agr. Eng.*, **28**(6):233, 1947.

Surface and Residual Sprays. No agricultural operation requires greater care and thoroughness than residual spraying. When complete film coverage is necessary, as in the application of stomach poisons or contact sprays for the control of sessile scale insects, it is important to cover the entire surface of every leaf, fruit, bud, stem, and branch, and the difference between 95 and 99 per cent coverage is often appreciable in terms of successful insect control. In the application of short-lived contact insecticides such as nicotine, TEPP, or pyrethrins, it must be remembered that only the insects wet by the spray are sure to be killed. For the control of mobile insect populations, random spray coverage is more often used, where the aim is to apply small patches of toxicant in a dispersed pattern so that the insects are certain to contact the poison during their movements about the plant surface.

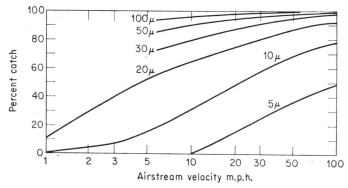

FIG. 8.1. Dynamic catch of spray droplets of various diameters, showing relation of droplet speed to catch by cylinder 0.125 inch in diameter. (*After Brooks*, 1947).

Thorough coverage or *high-volume sprays* are applied to the point of run-off, and the amount of deposit is determined by the concentration of the toxicant in the spray mixture. This may require from 500 to 2,500 gallons per acre with orchard crops. Excessive runoff may waste as much as 95 per cent of the spray liquid. Recent trends in spray applications have been toward the use of *concentrate* or *low-volume spraying*, in which the concentration of the toxicant is increased from 8 to 25 times that used in high-volume spraying, and a corresponding reduction is made in total volume to 2 to 25 gallons per acre. An intermediate range of 2 to 4 times concentration and application at 25 to 100 gallons per acre is called *semi-concentrate spraying*. With concentrate sprays the foliage is never completely wetted and an increased deposit can be obtained by increasing the rate of application, thus making possible a considerable increase in the efficiency of the spray operation.

The use of concentrate sprays has greatly speeded up many spray operations, decreased the amount of insecticide required, simplified transport of water and filling of the spray tank, and decreased the weight of the spraying equipment and its undesirable compaction effect on orchard soils. On the other hand, the use of concentrate sprays requires a much more uniform and finer control of the particle size of the discharge, and successful operations are dependent upon relatively still air conditions, with winds of less than 5 miles per hour.

Satisfactory insect control with concentrate, low-volume sprays depends upon adequate random coverage. It is obvious that if satisfactory deposition could be obtained, small droplets of the 10-micron range would provide perhaps 10 times the coverage of surfaces at limited rates of application, as that obtained with 100-micron droplets. This is shown in Table 8.3, for the ideal circumstances of uniform particle sizes and deposition. However, the question of dynamic catch (Fig. 8.1) sets a definite limitation upon the use of small droplets since the critical factor is the deposition velocity with which the impelled droplets strike the object to be treated. The dynamic catch of 10-micron droplets on a 0.125-inch cylinder is less at 100 miles per hour than that of 100-micron droplets at 5 miles per hour, and this clearly indicates the impracticability of attempting to obtain adequate spray coverage with aerosol droplets. This subject will be considered further under Air Blast Sprayers (page 442).

TABLE 8.3. RELATION OF DROPLET SIZE TO SPRAY COVERAGE AT 1 GALLON
PER ACRE[1]

Droplet diameter, μ	Weight, μg., sp. gr. = 1.0	Droplets per sq. cm. surface	Mean distance apart, cm.	Fraction of surface covered at 90° contact angle
10	0.000525	213,500	0.00216	0.266
20	0.0042	26,700	0.00615	0.133
50	0.0655	1,708	0.0242	0.0532
100	0.525	213.5	0.0685	0.0266
200	4.200	26.7	0.193	0.0133
500	14.175	1.7	0.708	0.00532

[1] Modified from RIPPER, W., *Ann. Appl. Biol.*, **42**:288, 1955.

Atomization of Liquids. The proper application of all types of sprays depends, as we have seen, upon the atomization of liquid into sprays or aerosols. Because of the importance of this process in the combustion of liquid fuels and other industrial applications, a great deal of basic knowledge is available which can be applied to the prediction of the performance of atomizers under a wide variety of conditions. The following discussion and classification of atomizing devices is based on that of Fraser.[1]

Pressure Atomizers. Atomization at a nozzle takes place in two steps: (a) the emission of the liquid from the nozzle orifice as a high-velocity stream or sheet, and (b) the interaction of the liquid sheet with the air to produce filamentation and breakup into droplets. The degree of droplet formation is largely controlled by surface tension, and as the larger drops separate, the connecting filaments between them form smaller drops. The energy requirement for atomization is needed to overcome the surface tension and viscosity of the fluid, which resist deformation into droplets.

Simple pressure atomizers include *swirl spray nozzles,* which may be either hollow-cone or solid-cone types, the *flat-fan nozzle,* and the *impact nozzle.* In the hollow-cone design, which operates at pressures of 40 to 1,000 pounds per square inch, the liquid is made to rotate by a whirl plate with angle holes or by a screw thread in the delivery tube and passes through the disk orifice as a conical sheet of liquid. When a hole is present in the center of the whirl plate, the center of the spray cone is filled, producing a solid-cone spray. The rate of discharge of such a nozzle is a function of the size of the orifice, increasing as the square of the diameter and of the pressure of discharge. The approximate rates of discharge of single-nozzle spray guns at various disk apertures and pressures are given in Table 8.4. The droplet size produced by a hollow-

[1] FRASER, R., *Advances in Pest Control Res.*, **2**:1–106, 1958.

TABLE 8.4. DISCHARGE OF SINGLE-NOZZLE SPRAY GUNS AT VARIOUS APERTURES AND PRESSURES

Size of aperture, in.[1]	Pressure, p.s.i.			
	300	400	500	600
	Rate of discharge, gal. per min.			
$\frac{3}{64}$	1.1	1.2	1.3	1.4
$\frac{5}{64}$	2.4	2.7	3.0	3.2
$\frac{6}{64}$	2.7	3.0	3.3	3.5
$\frac{7}{64}$	4.3	4.8	5.3	5.7
$\frac{8}{64}$	5.6	6.3	7.0	7.7
$\frac{11}{64}$	9.4	10.9	12.3	13.6

[1] Nozzle disks are numbered to represent aperture diameters in 64ths of an inch. Thus a No. 3 disk has an aperture of $\frac{3}{64}$-inch diameter.

cone nozzle is a function of the whirl-plate opening and the size of the swirl chamber and varies inversely with the square root of the pressure (Table 8.5), but is not influenced by the nozzle-aperture size between the common ranges of $\frac{3}{64}$ and $\frac{8}{64}$ in. The common orchard spray gun has a variable eddy chamber whose depth is regulated by turning the handle at the base. A shallow chamber produces a wide-angle cone of fine spray at low output, and deepening the chamber narrows the angle of the cone and increases the droplet size and output. The aperture disks are generally removable and available in various sizes numbered from 2 to 11, corresponding to the diameters of the apertures in 64ths of an inch. These should be replaced when worn since the abrasiveness of the spray may produce irregularities which distort the spray pattern and droplet size.

TABLE 8.5. AVERAGE PARTICLE SIZE OF WATER SPRAYS FROM PRESSURE NOZZLES[1]

Discharge, g.p.m.	Pressure, p.s.i.	Flow number	Surface mean diameter, D_0, μ		
			Fan nozzle	Swirl nozzle	Impact nozzle
0.033	25	0.4	147	110	
	50	0.28	104	78	
	100	0.2	74	55	
0.132	25	1.6	234	175	
	50	1.13	165	123	
	100	0.8	117	88	
0.5	60	3.9		175	465
	120	2.7		123	327
	240	1.94		90	240
2.0	60	15.5		280	740
	120	10.8		195	520
	240	3.9		140	370
10.0	200	42.6		280	695
	400	30.0		220	485
	600	24.4		190	400

[1] After FRASER, R., *Advances in Pest Control Res.*, **2**:1–106, 1958.

In the flat fan nozzle, the oblong orifice lies at the bottom of a semicircular groove or chamber milled on the orifice plate. Such a nozzle produces a relatively coarse fan-shaped spray with an included angle of 65 to 80 degrees. The droplet spectrum is directly proportional to the nozzle capacity and inversely proportional to the pressure. These nozzles operate at the lower pressures of 25 to 125 pounds per square inch. The component parts of typical commercial nozzles are shown in Fig. 8.2.

In the impact nozzle, a high-velocity jet of liquid strikes a smooth surface at a high angle of incidence, for example, 30°, forming a liquid sheet whose angle of spread depends upon the angle of impact. Such a nozzle produces relatively coarser spray than the swirl spray nozzle at equivalent pressure (Table 8.5).

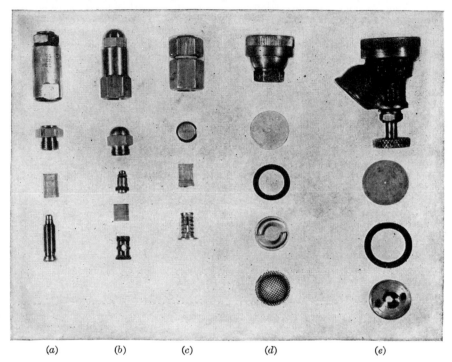

(a)   (b)   (c)   (d)   (e)

Fig. 8.2. Some spray nozzles used in insecticide applications, showing construction. (a) and (b) nozzles for mist or aerosol discharge, (c) flat spray nozzle for applying residual sprays, (d) agricultural nozzle producing hollow-cone discharge, and (e) nozzle with variable discharge. (Original.)

Fraser lists the factors affecting droplet size from swirl spray nozzles as flow number, spray angle, liquid pressure, viscosity, surface tension, and ambient atmosphere. The flow number (FN) is the product of the orifice area and the discharge coefficient (K_Q), and for water sprays

$$FN = 293 K_Q A \tag{4}$$

where A = orifice area, square centimeters

The discharge coefficient depends upon the geometry of the nozzle, and typical values for the flow numbers of fan, swirl, and impact nozzles are given in Table 8.5. The surface mean diameters D_0 for the sprays from pressure nozzles can be estimated by means of an empirical equation,

$$D_0 = C \left(\frac{FN}{P} \right)^{1/3} \tag{5}$$

where P = pressure, pounds per square inch, and C = a constant which has the values of 437 for swirl nozzles for FN up to 5 and pressures to 200 pounds per square inch and a 90-degree spray angle; 585 for fan nozzles to 100 pounds per square inch and 90-degree spray angle; and 1,160 for impact nozzles for FN up to 5, pressure to 100 pounds per square inch, and angle between jet and impact plate of 30 degrees. D_0 values determined by these equations for well-designed pressure nozzles of the three types at typical pressures and rates of discharge are listed in Table 8.5.

The data in Table 8.5 serve not only to indicate the range of droplet sizes which may be obtained from commercial pressure-atomizing nozzles, but also some of the relationships between pressure, capacity, and D_0. It will be noted that under equal conditions the swirl nozzle produces smaller droplets than the fan nozzle and that the discharge from the impact nozzle is relatively very much larger than for the swirl nozzle. The droplet size, D_0, is in all cases approximately inversely proportional to the square root of the pressure; thus, to halve the droplet size, the pressure must be increased 4 times. Increasing the rate of discharge of the nozzle by 4 times increased the D_0 by about 1.5 times. In orchard spraying, the smaller droplets decrease the distance of carry of the spray, as shown in Fig. 8.5. Therefore, to secure increased carry, the discharge should be increased by using a larger orifice rather than by increasing the pressure. The use of increased pressure, up to 600 to 800 pounds per square inch, however, generally results in a better dispersal of the liquid and consequently in better spray coverage.

Rotary Atomizers. These devices effect atomization by means of centrifugal force and at low feed rates produce droplets of more uniform size than pressure atomizers and are especially useful for viscous liquids. In practice, the liquid is fed onto a rotating surface in the form of a disk, cup, bowl, or slotted wheel from 1 to 18 inches in diameter with peripheral velocities of from 50 to 600 feet per second. When fed at levels below the saturation point, the disintegration of the spreading sheet takes place at the rim of the disk and the drops leave tangentially. Such a device may be combined with an air blast to produce a directional flow or to still further break up the droplets and has been employed successfully in aircraft and mist blowers, with capacities up to 250 gallons per hour and D_m values of 40 to 100 microns. The droplet diameter may be predicted from the equation

$$D_m = \frac{360,000}{S} \sqrt{\frac{\gamma}{D\rho_L}} \tag{6}$$

where S = revolutions per minute of disk
γ = surface tension, dynes per centimeter
D = diameter of disk, centimeters
ρ_L = density, gram per cubic centimeter

Flat disks are unsuitable for liquid breakup because of the slip of the fluid, and cupped or saucer shapes or rotors with radial vanes are most practicable.

Twin-fluid Atomizers. In this type of equipment, of which the "flit gun" and paint sprayer are the most familiar examples, atomization results from the impingement of a jet of liquid into a stream of air or gas, and the degree of breakup is an inverse function of the relative velocity between the air stream and liquid stream and is directly proportional to the ratio of the volume of flow of liquid to flow of air. Many designs are possible operating at low or high pressures and air velocities, depending upon whether the gaseous medium is provided by fan, rotary blower, compressor, steam, or exhaust from an internal-combustion engine and whether the gas and liquid meet internally in the body of the atomizer or outside in the atmosphere.

Nukiyama and Tanasawa[1] have devised an empirical equation quantitatively to express the effects of the factors on degree of atomization, which is applicable to any twin-fluid atomizer with sub- or supersonic gas velocity.

$$D_0 = \frac{585}{V} \left(\frac{\gamma}{\rho_L}\right)^{1/2} + 597 \left(\frac{\eta}{\sqrt{\gamma\rho_L}}\right)^{0.45} \left(1,000 \frac{Q_L}{Q_A}\right)^{1.5} \tag{7}$$

[1] See *Ind. Eng. Chem.*, **40**:67–74, 1948.

where γ = surface tension, dynes per centimeter
  η = viscosity, poise
  ρ_L = liquid density, gram per cubic centimeter
  V = relative gas/liquid velocity, meters per second
 Q_L/Q_A = volume liquid/volume free gas
Where water is the liquid, the equation can be simplified to

$$D_0 = \frac{16{,}400}{V} + 39.4 \left(\frac{M_W}{M_A}\right)^{1.5}$$

(8)

where V = relative air/liquid velocity at nozzle, feet per second
M_W/M_A = water/air ratio, pounds, at 65°F.

Twin-fluid atomizers are exceptionally efficient in producing sprays or aerosols of fine droplet size at relatively high rates of discharge (page 442).

SPRAYING EQUIPMENT

The size and type of sprayer to be used will depend upon the amount and kind of work to be done. However, spraying is at best a disagreeable task, and every effort should be made to have equipment of sufficient capacity so that the work will be done thoroughly and efficiently. The best spraying equipment can easily be ruined by failure to give it proper care. Clean water should always be pumped through the sprayer after use, to flush out all corrosive spray materials, and the metal parts should be oiled to prevent rust. It is not advisable to apply insecticides with spraying equipment which has previously been used for the application of herbicides of the plant-growth-regulator type such as 2,4-dichlorophenoxyacetic acid (2,4-D), because very small traces of such materials may seriously injure certain plants.

There are many types of sprayers, and to facilitate this discussion they will be classified as *hydraulic sprayers*, which employ only pressure nozzles for the formation and distribution of the spray, and *air-blast sprayers*, which utilize an air stream for distribution and/or breakup of the spray.

Hydraulic Sprayers. The various types of hydraulic sprayers include (a) the compression sprayer, which is actually a special type in which compressed air is used to force the liquid through a pressure atomizing nozzle. The most familiar type is a 4-gallon cylindrical tank with a hand piston pump which is operated at 40 to 80 pounds per square inch and is very popular with the home gardener. The bucket or stirrup pump sprayer (b) has a single- or double-action plunger pump clamped into a 2-gallon bucket and operates at up to 150 pounds per square inch; (c) the knapsack sprayer is a 2- to 5-gallon tank carried as a knapsack and containing a diaphragm pump and agitator operated at 50 to 80 pounds per square inch by a lever carried under the operator's shoulder. These are especially suitable for small garden operations or treatments in inaccessible spots, as in mosquito larviciding or residual house spraying. The barrel pump (d) is a larger plunger-type spray pump which operates at a pressure of 200 to 300 pounds per square inch and uses a 15- to 45-gallon container as a reservoir. The wheelbarrow sprayer (e) is essentially a portable barrel pump which is suitable for spraying small fields, orchards, or farm buildings. Most modern wheelbarrow sprayers have rubber tires and contain a two-cycle gasoline-engine-driven pump of 1.5 to 3.0 gallons per minute at 200 to 250 pounds per square inch and have a capacity of 15 to 50 gallons. Hydraulic power sprayers (f) range from the power wheelbarrow sprayer through trailer-type field and row-crop sprayers to truck-mounted orchard sprayers. These power sprayers have 50- to 500-gallon capacities, operate up to 400 to 800 pounds per square inch, and have discharge capacities ranging from 5 to 80 gallons per minute. Discharge may be carried through one or more spray hoses (Fig. 8.3) or through a horizontal or vertical spray boom equipped with multiple swirl spray nozzles (Fig. 8.6).

Fig. 8.3. Spraying citrus with modern high-pressure spray equipment. Note the hydraulic tower for spraying the tops of the trees. (*From Hardie Manufacturing Co.*)

Parts of a Sprayer. The important components of a spray outfit are similar for the various types, some of which are as follows:

The *cylinder* is the portion of the pump in which the pressure is developed. It must be of noncorrosive material, and in large power sprayers, which have two to four cylinders, the lining is generally of acid-resistant porcelain.

The *plunger* or *piston* forces the spray liquid through the cylinder and is made to fit tightly inside the cylinder by means of *packing.* Plungers may be fitted with molded rubber and fabric *plunger cups* which act as inside packing, expanding against the cylinder wall on the pressure stroke, or outside packing may be used which is compressed against stainless-steel plungers.

Valves, usually of the ball type, and *valve seats* made of hardened stainless steel, direct the flow of spray liquid. They must be readily accessible for cleaning in case of clogging by particles of dirt or spray materials.

An *air chamber* is used to equalize the pressure and remove excessive strain on the pump. The discharge opening leading to the nozzle is located at the bottom of the air chamber.

The *tank,* in which the spray ingredients are mixed and held, is usually constructed of metal with an enamel lining or of stainless steel.

Since many spray ingredients, such as wettable powders, are not soluble in water, an *agitator* is necessary to keep the finely divided particles in suspension or the emulsion evenly distributed. In power sprayers this is commonly effected by two or more flat paddles or propellers on a rotating shaft near the bottom of the tank.

A *pressure gage* indicates the pressure being maintained, and the *pressure regulator* maintains uniform pressure on the spray nozzles and allows the pump to operate at a greatly reduced load when spray is not being discharged. This is accomplished by a spring-and-ball valve, which can be set to lift at the desired pressure and permit the excess liquid to bypass to the tank.

Strainers, over the intake to the tank and over the suction or intake pipe leading to the cylinders, keep the spray liquid free of troublesome foreign matter.

The *discharge pipe* should be free from abrupt angles that cut down the pressure between pump and nozzles.

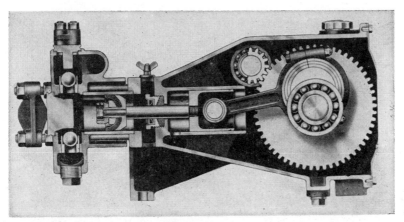

FIG. 8.4. Sectional view of a high-pressure spray pump capable of developing pressures of 600 to 1,000 pounds per square inch. When the plunger moves right, spray solution is drawn through the lower ball valve. Upon the return stroke of the plunger, the lower ball valve is forced shut and the solution under pressure passes through the upper valve into the compression chamber. (*From Food Machinery and Chemical Corp., Bean Division.*)

The *hose* for orchard spraying should be at least 25 feet long, of ½- to ¾-inch inside diameter and four- to seven-ply strength. The friction of the spray liquid against the inside surfaces of the spray hose results in a considerable loss of pressure at the nozzle. For example, with a 50-foot hose ½-inch in diameter and a flow rate of 10 gallons per minute, the friction loss is 70 pounds per square inch.

The *pump* of most power sprayers is of the direct-displacement-plunger type (Fig. 8.4). The recent development of high-capacity, high-pressure boom sprayers for orchard spraying has been possible through the utilization of centrifugal pumps with capacities of up to 125 gallons per minute at pressures of 800 to 1,000 pounds per square inch. Such pumps, however, are subject to severe wear by abrasive wettable powders.

Spray Booms. The boom is a light hollow tube or pipe which is used to carry the spray liquid under pressure from the pump to one or more nozzles. The simplest type is the spray rod or lance, which is a tube from 3 to 8 or even 12 feet in length, with a cutoff valve at one end and a simple swirl spray or flat fan nozzle at the other. With power sprayers, a broom or fog-drive gun is sometimes used for spraying fruit trees. This consists of a spray rod from 3 to 5 feet long, with a cutoff at the base and a group of three to eight nozzles at the tip. Such broom guns produce a fog-like discharge (Fig. 8.5) which does not cause mechanical injury to foliage even when held almost against it. Broom guns are usually equipped with swirl spray nozzles of $\frac{4}{64}$-inch opening and have discharge capacities of about 5 to 10 per cent less per nozzle than single-nozzle guns.

For row-crop and field-crop spraying, horizontal booms of many designs have been developed. These are aluminum, brass, or stainless-steel tubes 1 to 2 inches in diameter and range from a 4- to 14-row coverage and up to 60 feet in length and may have from 8 to 70 nozzles. The larger booms are equipped with a hydraulic mechanism for lifting over obstructions and

FIG. 8.5. Spraying an apple orchard with tractor-drawn sprayer, operated by a power take-off from the tractor. Such an outfit will maintain 500 pounds pressure for 14 nozzles. The caterpillar tractor and large tires permit operations on very soft, wet ground. The spray tower, to enable one operator to cover the tops of the trees, was made from an empty steel drum. (*From John Bean Manufacturing Co.*)

are trailer- or tractor-mounted. The boom lines are equipped with 100- to 150-mesh screens to help prevent clogging of the nozzles. The height of the boom and the placement of the nozzles are dependent upon the crop being sprayed. For small row crops, the boom is usually 16 to 24 in. above the ground and the nozzles are adjusted so that the edges of the fans just touch at ground level. The spacing depends upon the angle of the fan and the height of the boom above the ground and for a 40-degree cone should be 0.7 times, for a 70-degree cone 1.4 times, and for a 100-degree cone 2.4 times the height. For taller row crops such as corn or cotton, vertical downpipes extend from the boom and generally end in two nozzles at an included angle of about 60 degrees, pointing toward the plants on either side (Fig. 8.6). Many other nozzle-placement combinations are used for specific operations. Special "high-clearance" rigs with up to 72-inch vertical clearance may be used for crops such as corn and tobacco.

The pressure used in row- or field-crop spraying depends upon the performance characteristics of the nozzles, but is generally within the range of 40 to 125 pounds per square inch, where small swirl spray nozzles produce D_0 values from 100 to 200 microns (Table 8.5) or well within the optimum range. The nozzle and pump capacity are clearly interdependent and related to the length of the boom and the number of nozzles. With a common eight-row (25-foot) boom with 24 nozzles, a driving speed of 5 miles per hour (440 feet per minute) will cover 11,000 square feet per minute and an acre approximately every 4 minutes. Thus for a semiconcentrate spray of 25 gallons per acre, a pump delivery of 6.2 gallons per minute will be required and the nozzle discharge at 100 pounds per

FIG. 8.6. Horizontal boom for field-crop spraying. (*H. T. Reynolds, Univ. of California.*)

square inch should be about 0.26 gallon per minute. For a full-coverage spray of 100 gallons per acre, the pump delivery should be about 25 gallons per minute and the nozzle discharge should be about 1 gallon per minute, unless the driving speed is reduced materially. In order to control dosages at these low rates of discharge, an accurate slow-speed speedometer is essential.

Vertical booms with alternating nozzles pointing to either side and with mechanical oscillation to improve penetration and coverage are used in orchard spraying. A typical boom of this type for citrus tree spraying, 20 feet long, is shown in Fig. 8.7. The protective cab shelters the operator from a continuous drenching by the high-volume spray. These boom sprayers may operate at 500 to 2,500 gallons per acre and at 2 miles per hour will cover about 0.05 acre per minute. Thus the nozzle discharge should range from about 1 to 5 gallons per minute at 600 pounds per square inch, and a pump capacity of 25 to 125 gallons per minute is necessary. The efficiency of coverages with such equipment is a function of nozzle discharge, oscillation rate, and driving speed.

Residual Spraying Indoors.[1] The spraying of houses and barns with a heavy deposit of a long-lasting residual insecticide is widely practiced for the control of flies, mosquitoes, bed bugs, cockroaches, silverfish, and ectoparasites of animals and is one of the principal weapons for the eradication of malaria. This type of application is most efficiently performed with a small-capacity, portable spray pump such as the knapsack, compression, or bucket sprayer or the wheelbarrow power sprayer. The World Health Organization has specified that residual-spraying operations be conducted with a spray pump having a regulator to produce a constant pressure of 40 pounds per square inch and that a flat fan nozzle discharging at 0.2 gallon per minute with a spray angle of 60 to 65 degrees be used.

Fig. 8.7. High-pressure boom sprayer in operation. (*From Hardie Manufacturing Co.*)

During the spraying operation, the nozzle should be held approximately 18 inches away from the surface being treated, and about 150 to 250 square feet of wall surface should be covered per minute. Adherence to the suggested conditions will avoid wasteful and unsightly discharge and produce a spray of maximum impingement and deposition.

Air-blast Sprayers. Recent trends in spray equipment for certain crops have been toward the development of air-blast sprayers in which a relatively large volume of high-velocity air is used to break up the spray droplets and to carry them to the target, as in the twin-fluid atomizers. For convenience, however, we shall also discuss air-blast equipment in which the spray may be produced by other means but dispersed by means of an air blast.

The common hand atomizer or "flit gun" is perhaps the simplest familiar device of this type. Here a piston-type air pump furnishes the air

[1] "Handbook of DDT Residual Spray Operations," U.S. Public Health Service, March, 1945, *W.H.O. Tech. Rept., Ser.* 1952, p. 46.

blast, which also forces the spray liquid from a reservior up a small-diameter delivery tube, and atomization results from the apposition of the air and liquid streams, according to equation (7).   The compressed air–paint spray gun is a larger example of this type.

In the use of air-blast equipment for the treatment of field or orchard crops the 30- to 80-micron droplet range seems most suitable because of the questions of coverage and dynamic catch (page 431), and thus the matter of air volume and velocity is of great importance.   This is particularly true in orchard spraying, where it is necessary to have sufficient air-moving capacity to agitate the air in all parts of the tree and to displace most of it.   Table 8.6 shows the volumes of air required with various-sized delivery tubes to produce adequate velocities at various distances from the blower.   It is evident that to secure adequate dynamic catch of 30-micron droplets at distances of 25 to 50 feet from the blower, very large volumes of air, ranging from 7,500 to 28,000 cubic feet per minute are required.   It has also been shown that to spray a moderate-sized orchard tree 20 by 20 by 20 feet (8,000 cubic feet) with a mist blower traveling at 0.5 mile per hour, 17,600 cubic feet per minute of spray are required, and of course this requirement is directly proportional to the volume of the tree and the speed of travel of the sprayer.

TABLE 8.6. AIR VELOCITIES AT VARIOUS DISTANCES FROM MIST BLOWERS[1]

Diameter of round outlet, in.	Volume of air, cu. ft./min.	Velocity at indicated distance from blower, m.p.h.				
		Outlet	10 ft.	25 ft.	50 ft.	100 ft.
4	1,300	170	40	8	2	0
8	3,800	125	40	10	3	0
12	7,500	125	60	30	7	2.5
24	28,000	120	75	40	17	7
54	27,000	50	35	30	8	3

[1] After POTTS, S., P. GARMAN, R. FRIEND, and R. SPENCER, *Conn. Agr. Exp. Sta. Cir.* 178, 1950.

Air-blast sprayers are of two standard types, the *spray blower* and the *mist blower*.   In the spray blower, the liquid is broken up by hydraulic nozzles and carried to the target by the air blast, while in the mist blower the spray liquid is finely dispersed by means of air-atomizing nozzles or spinning disks and then carried by the air blast.   Both types of equipment employ axial-flow turbines or centrifugal fans to generate from 4,000 to 60,000 cubic feet of air per minute at velocities ranging from 60 to 250 miles per hour.   The successful application of air-blast sprayers to field and row crops is difficult because of problems in equalizing spray distribution at varying distances from the point of discharge.   Various devices, including manifolds with a number of outlets, fishtails, elongated slots, and rotating or oscillating outlets to sweep the area, have improved the evenness of coverage.   Air blasts of 3,500 to 4,000 cubic feet per minute at 60 miles per hour have given effective results over swaths of 15 to 35 feet.   The commercial units available, however, are generally modifications of orchard sprayers and use air blasts of 20,000 to 40,000 cubic feet per minute at moderate velocities of 70 to 90 miles per hour to avoid crop damage, directed at right angles to the path of travel from either one or both sides. The rows over which the machine passes are simultaneously sprayed with a low-volume boom. Such equipment will treat a swath 40 to 60 feet wide using 50 to 100 gallons per acre.

FIG. 8.8. Speed Sprayer in operation. (*From Food Machinery and Chemical Corp., Bean Division.*)

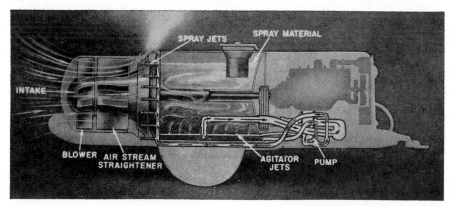

FIG. 8.9. Sectional view of Speed Sprayer. (*From Food Machinery and Chemical Corp., Bean Division.*)

A number of large spray blowers are in extensive commercial use for orchard spraying, and the Speed Sprayer is perhaps the best known (Figs. 8.8 and 8.9). This equipment will apply from 50 to 500 or more gallons per acre and delivers up to 45,000 cubic feet per minute of air, with nozzle pressures of 50 to 70 pounds per square inch and pump capacities of 55 to 140 gallons per minute. The spray is delivered into the air blast from as many as 58 to 264 swirl spray nozzles. The spray duster (Fig. 8.10) employs 40,000 cubic feet per minute of air passing through two vertical fishtail outlets to distribute high-pressure spray at 100 to 1,000 gallons per acre from vertical banks of nozzles. The air flow is periodically changed in direction by the oscillation of a hinged section of the outlet to aid the spray penetration by moving leaves and twigs. The most difficult problem with air-blast sprayers in orchard spraying is equalizing the coverage between the top and bottom of the tree.

FIG. 8.10. Large spray duster showing the spray nozzles and dust orifice on one side. This equipment can be used for applying either sprays or dusts or spray-dust combinations. (*Univ. Calif. Agr. Exp. Sta.*)

Aerosol Generators.[1] Aerosols for insecticidal use have been produced by a number of methods: (*a*) by burning the insecticide; (*b*) by spraying a solution of insecticide onto a heated surface; (*c*) by forcing the insecticide, dissolved in a liquefied, low-boiling gas through a capillary tube; (*d*) by spraying at high pressure through a swirl spray nozzle of low capacity; (*e*) by forcing a solution of insecticide between two closely apposed and rapidly spinning disks; and (*f*) by the use of a high-velocity stream of gas, air, or steam acting upon a stream of liquid in some type of twin-fluid atomizer.

The exhaust aerosol generator uses the hot exhaust gases from an internal-combustion engine to produce atomization. The droplet size is directly proportional to the ratio of flow of the liquid to the flow of air and can be predicted from Eq. (7). The use of the hot exhaust gases favors the production of droplets of aerosol dimensions because of the very high velocities, 600 to 1,400 feet per second, obtainable with the hot gases and

[1] YEOMANS, A., *U.S.D.A.*, *Cir.* EF-258, 1948.

the decrease in the viscosity of the spray liquid which occurs at the gas temperatures, which are about 540°C. To eliminate harmful back pressure on the engine and to efficiently break up the spray, the following features are important in the construction of the exhaust generator: (a) injection of the spray liquid at the periphery or center of the throat of a venturi tube where the gas velocity is highest, (b) venturi tube with ratio of throat diameter to exhaust stack diameter of 0.25 to 0.50, (c) throat section of smooth machined construction and as short as possible, with included angle of entrance cone of 20 to 30 degrees and of exit cone of 7 degrees. Venturi exhaust generators have been used on a variety of internal-combustion engines, and the liquid flow rates have varied from 1 gallon per hour for a small 1.5-horsepower gasoline engine, to 0.3 to 2.0 gallons per minute for truck or tractor engines, and up to 10 to 20 gallons per minute for large aircraft engines.

In the TIFA aerosol generator, air at 150 cubic feet per minute is passed through a combustion chamber heated by a gasoline burner to 800 to 1200°F. The insecticide in oil solution is sprayed into the hot air blast through pressure nozzles, and the particle size, which is controlled by the rate of delivery of the spray solution, ranges from a D_m of 5 microns at 10 gallons per hour to 50 microns at 50 gallons per hour.

In the Beskil aerosol generator, steam is generated by a flash boiler and the insecticide solution is injected into the steam jet issuing at high pressure and velocity. The particle size is controlled by the steam temperature and may be varied from a D_m of 20 microns at 800°F. to 100 microns at 300°F. The output of insecticide is 38 gallons per hour.

In the production of aerosols by heat there is a certain amount of volatilization of the solvent, and some insecticides may partially decompose at high temperatures. However, the various types of aerosol generators have been very successfully used for the control of mosquitoes and black flies by allowing the fog to drift downwind across the infested areas and have also been employed in buildings for the control of household pests. In general, unless operated so as to produce relatively large droplets, 50 to 100 microns, they have not been useful for the application of residual treatments of insecticides.

The Microsol-type generators use the principle of atomization by spinning disks as described on page 436. One large gasoline-driven model utilizes 21 disks, each 8 inches in diameter, spinning at 6,500 revolutions per minute, and will atomize up to 250 gallons per hour of insecticide solution. The aerosol discharge has a minimum D_m of 40 microns and is distributed by an air-blower arrangement producing 4,500 cubic feet per minute. A smaller electrically driven model for indoor use rotates at 17,000 revolutions per minute and produces an aerosol with a D_m of about 10 microns. The spinning-disk equipment, since it employs no destructive heat, is favored for the distribution of heat-sensitive insecticides such as the pyrethrins.

Insecticidal Smokes. Smokes produced by burning or evaporating insecticides are not ideally suited for insect control because the particles, which range from about 0.3 to 2 microns in diameter, are too small to impinge readily upon flying insects and produce little or no surface residue (page 430). Contrary to popular opinion, such smokes, unlike gases, possess no diffusive properties and do not appreciably penetrate into cracks and crevices. Furthermore, nearly all insecticides are appreciably decomposed by the temperatures produced in pyrotechnic mixtures, and even relatively stable compounds such as lindane and DDT may undergo up to 30 per cent decomposition. Nevertheless, the idea and appearance of such smokes are so appealing that marketable products have appeared in many forms.

Nicotine incorporated in a slow-burning powder has been widely used as a greenhouse fumigant for aphids, and it is probable that at least a portion of the discharge is in particulate rather than in vapor form. Azobenzene is similarly used for the control of greenhouse mites and is either burned or volatilized by application to steam pipes as a paste. Benzene hexachloride and DDT have been formulated as slow-burning pellets composed of about 58 to 60 per cent insecticide, 30 to 40 per cent of equal portions of sucrose and potassium chlorate, and 2 to 10 per cent of a retardant such as clay, diatomaceous earth, or magnesium oxide or carbonate. The smokes of these materials condense on surfaces as tiny supercooled droplets, which eventually may crystallize.

Electrical evaporating devices[1] which volatilize purified DDT and lindane at temperatures of 118 to 120°C. have attained considerable popularity for household and commercial pest control. The vapors produced recondense in the cooler room air, forming an aerosol with particles having median diameters between 0.5 and 5 microns. These droplets may remain in the supercooled state for some time before crystallizing on surfaces. Because of the possibility of human health hazards, the vaporizers should be thermostatically controlled to regulate the upper limit of evaporation to 1 gram of insecticide per 15,000 cubic feet of enclosed space per day.

Liquefied Gas Aerosols.[2] The familiar "aerosol bomb" is the most convenient method for the production of relatively small amounts of aerosol and is especially useful indoors. The aerosol is formed by the release of a solution of insecticide in a liquefied gas through a small capillary tube with a diameter of the order of 0.008 to 0.017 inch. The vapor pressure of the gas provides the propellant force, and boiling occurs in the capillary delivery tube, which serves as a mixing nozzle, leaving small droplets of insecticide. The insecticide formulation is contained in a metal container strong enough to withstand the pressure of the gas and fitted with a delivery tube extending from the nozzle at the top to the bottom of the container, in order to ensure the delivery of the solution rather than of the gas at the top. The particle size of the aerosol produced is determined by (*a*) the pressure of the propellant, (*b*) the amount of nonvolatile material present, (*c*) the physical properties of the solution, and (*d*) the size of the nozzle orifice. The propellant gases used should be nontoxic to man and animals and of suitable vapor pressure and should preferably be solvents for the insecticides and carrier used. The properties of suitable propellant gases are given in Table 8.7.

The original bomb as developed by Goodhue and Sullivan contained dichlorodifluoromethane as a propellant and developed a gage pressure of about 70 pounds per square inch at 70°F. and about 120 pounds per square inch at 100°F. These pressures necessitated the use of heavy steel or brass containers. In order to reduce the cost of these applicators, aerosol dispensers of drawn steel of the "beer-can" variety were developed which showed little or no decrease in insecticidal efficiency while operating at pressures of 25 to 38 pounds per square inch gage at 70°F. Effective insecticidal

[1] Such as the Aerovap. SPEAR, P., and H. SWEETMAN, *Jour. Econ. Ento.*, **45**:869, 1952; FULTON, R., W. SULLIVAN, and G. MANGAN, *ibid.*, **46**:639, 1953.

[2] *Ind. Eng. Chem.*, **34**:1456, 1942; *Jour. Econ. Ento.*, **37**:338, 1944; *Bul. Continental Can Co.*, vol. 14, 1947; *Soap and Sanitary Chemicals*, **25**(2):122, 1949; GOODHUE, L., *Ind. Eng. Chem.*, **41**:1523, 1949.

aerosols are not, however, produced at pressures below 25 pounds per square inch gage. These low- or moderate-pressure aerosols require a more critical nozzle design than the simple capillary tube of the high-pressure aerosol.   In order to introduce turbulence and the formation of small bubbles of gas which serve as nuclei for boiling, a mixing chamber is provided into which the liquefied gas flows through a 0.015-inch-diameter constriction and from which the mixture issues through a 0.020-inch orifice.   The dispenser valve consists of a ball of metal or nylon held by a spring against a valve seat of synthetic rubber.

TABLE 8.7. PROPELLANTS FOR LIQUEFIED-GAS AEROSOLS

Propellant	Formula	Boiling point, °C.	Vapor pressure, p.s.i.a. at 30°C.
I. Dichlorodifluoromethane....	CCl_2F_2	−29.8	107.9
II. Trichloromonofluoromethane	CCl_3F	23.7	18.3
III. 1,1-Difluoroethane.........	CH_3CHF_2	−24.7	108.0
IV. 1,1-Difluoro-1-chloroethane .	CH_3CClF_2	− 9.2	59.0
V. Methylene chloride.........	CH_2Cl_2	40.1	11.0
VI. Methyl chloride............	CH_3Cl	−24.1	95.5
VII. Dimethyl ether............	CH_3OCH_3	−23.7	97.3
VIII. Propane...................	C_3H_8	−42.2	155
IX. Carbon dioxide...........	CO_2	−78.5 sublimes	1,039.6

The use of low-pressure aerosols requires higher-boiling propellants or mixtures such as 54 per cent dichlorodifluoromethane and 46 per cent trichlorofluoromethane, which develops 37 pounds per square inch gage at 70°F.   Other useful combinations include the addition of up to 25 per cent by volume of isobutane or n-pentane to the above mixture, II and III, I and IV, and I and V from Table 8.7.   Mixtures such as these form bubbles in the nozzle more readily than single-component propellants and produce better aerosol dispersion at low pressures.

The insecticides most commonly dispersed by the liquefied-gas method are DDT, methoxychlor, synergized pyrethrins or allethrin, lindane, chlordane, and dieldrin.   High purification is necessary to prevent clogging of the capillary nozzle and decomposition of the insecticide and consequent corrosion of the container, although this can be inhibited by including 0.1 per cent propylene oxide.   Auxiliary solvents such as petroleum distillates, polymethylnaphthalenes, or cyclohexanone have been used to solubilize the insecticides in the propellant gas.   A small percentage of nonvolatile material such as 10-W lubricating oil has been found effective in preventing too rapid volatilization of the aerosol droplets. The ultimate droplet size is determined by the concentration of nonvolatile materials, and the optimum for flying insects of 5 to 20 microns (D_m 10 to 15 microns) is obtained with about 20 per cent nonvolatile material in a high-pressure formulation and about 15 per cent in a low-pressure formulation.   Typical formulations of both types are:

High-pressure Aerosol	*Per Cent*
DDT.............................	3
Pyrethrins, 20% extract..........	1
Alkylated naphthalenes...........	16
Dichlorodifluoromethane.........	80

Low-pressure Aerosol	*Per Cent*
Methoxychlor.................	2.0
Pyrethrins....................	0.25
Piperonyl butoxide.............	1.0
Alkylated naphthalenes.........	11.75
Dichlorodifluoromethane........	30.0
Trichloromonofluoromethane....	55.0

Such aerosols should be used at the rate of 2 to 4 grams per 1,000 cubic feet of space, and the discharge rate is approximately 1 gram per second. The most popular containers are the 1-pound size containing 12 to 14 ounces of formulation, but 5- and 10-pound sizes are available for larger enclosures and for greenhouse work. For the latter purpose a formulation of 5 to 10 per cent parathion or TEPP in methyl chloride, 90 to 95 per cent, is widely used in a 10-pound bomb with a 2-foot rod and nozzle. This should be used only when wearing a protective mask (Fig. 8.11).

FIG. 8.11. Applying aerosol for the control of greenhouse insects and mites. Note the protective mask, which should be worn when applying highly toxic materials such as parathion and tetraethyl pyrophosphate. (*From Eston Chemicals, Inc.*)

The liquefied-gas method has been used also to produce bombs delivering residual-type sprays for the control of cockroaches, ants, and clothes moths and carpet beetles, at a low pressure of 20 pounds per square inch gage. For this purpose, the amount of propellant is reduced to 30 to 35 per cent of the total, and this produces a relatively coarse discharge of about 30 to 50 microns, which settles rapidly and has good wetting properties. Typical formulations are:

Roach and Ant Spray	Per Cent	Mothproofing Spray	Per Cent
Dieldrin	0.5	DDT	3.0
Pyrethrins	0.04	Perthane	3.0
Piperonyl butoxide	0.1	Petroleum distillate	59.0
Petroleum distillate	69.36	Propellant	35.0
Propellant	30.0		

DUSTS AND DUSTING

Although the original method of poisoning plants for insect control was by dusting, spraying has been much more extensively practiced during the past 50 years. This is largely due to the following undesirable features of dusting: (a) decreased efficiency due to less efficient deposition on the plant, (b) increased drift problems with poisonous materials, (c) increased cost of dust diluents compared with water, (d) tendency of carrier and toxicant to separate in the air unless the diluent particles are coated with the toxicant, (e) difficulties in incorporating several insecticides or other agricultural chemicals into a single dust, and (f) increased hazard to operator from inhalation of toxic dusts. Despite these disadvantages, dusting is in many cases much easier, lighter, and several times faster than thorough coverage spraying. The equipment is considerably simpler and lighter than the spray rig and can be better used in hilly territory and under muddy conditions. Dusting is obviously the favored means of insect control in regions with a limited water supply and is a useful means of achieving insect control just prior to harvest without exceeding insecticide-residue tolerances.

PHYSICAL PRINCIPLES OF DUSTING

Most insecticidal dusts are very finely divided, and typical surface mean diameters are talc 1.8 microns, pyrophyllite 2.2 microns, lead arsenate 6 to 10 microns, calcium arsenate 1 to 2 microns, and ground rotenone roots 6 microns (Table 7.3). Such dusts pass nearly completely through a 325-mesh screen of 44-micron aperture. Thus, when falling free in air, these dust particles should behave similarly to aerosol droplets (page 429). However, dust particles not only vary greatly in size and shape but also tend to agglomerate during the dusting operation, so that many particles adhere tightly together, and it is very difficult to predict their settling and drifting pattern. Since the settling velocity of these small particles is directly proportional to particle density, dusts of such materials as lead and calcium arsenates settle considerably faster than dusts of botanicals such as pyrethrum or rotenone. The density of the dust diluent and the presence of dust conditioners such as stabilizers and fluffing agents also affect the dusting behavior.

The electrostatic charge on dust particles has often been suggested as a means of importance in causing the dust to adhere to plant surfaces which are negatively charged. A number of devices have been designed to increase the charge on the insecticidal dusts, either by friction or by passing the dust particles through a flow of positive ions from a high-potential electrical discharge. Inorganic dusts such as arsenicals, fluorides, copper, lead, and sulfur assume positive charges, while botanicals assume negative charges. Diluents such as pyrophyllite and gypsum are also positively charged, and diatomite, clays, and talcs are negatively charged. The use of electrostatically charged dusts has been claimed to prevent agglomeration, to increase adherence and even distribution on both sides of leaves, and to increase the dust deposition as much as 4 to 10 times.

Insecticidal dusts are commonly applied to crop plants at rates of 10 to 50 pounds per acre. However, organic insecticides such as DDT give somewhat similar insect control as long as the amount of active ingredient and plant coverage are constant, regardless of whether applied as

3 to 10 per cent dust.   The use of dusts impregnated with oily materials such as mineral oil or polymethyl naphthalenes has sometimes been of advantage in increasing deposits and decreasing drift.

For some insecticidal applications, such as in the control of the European corn borer in the whorls of the corn plant, ants and other soil-inhabiting insects, and mosquito larvae in water under dense vegetation, the use of relatively coarse granular dusts of 30- to 60-mesh (250 to 590 microns) serves to prevent the insecticide from adhering to plant foliage and enables it to reach the soil or water surface.   The use of these granular materials (page 399) also greatly decreases drift problems.

DUSTING EQUIPMENT

A good duster, like a good sprayer, is one that will spread the insecticide—in this case in dry form—in such a manner as to give the most uniform coating possible to the plants being treated.   The simplest dusters are the small *hand dusters* adapted to the home garden in which a plunger discharges an air blast through a chamber containing from 0.5 to 2.0 pounds of dust and the dust cloud is emitted through a small flared nozzle or tip.   Next in size are the *blower dusters*, which consist of an enclosed fan rotated by a hand crank which sends a continuous blast of air through a small chamber into which the dust is fed by an adjustable gate.   An agitator stirs the dust in the hopper and provides for an even feed of the dust into the discharge chamber.   The dust cloud is forced out through a delivery tube extending nearly to the ground, which may end in a Y or fishtailed nozzle.   Such a duster will contain 5 to 10 pounds of dust and can be used to protect 1 to 2 acres of truck crops.   The most recent innovation in blower dusters is a back pack unit containing a small gasoline engine which drives a fan providing a steady stream of dust. The *bellows* or *knapsack duster* of similar capacity uses the extension and compression of a bellows attached to the back of the duster to force air through the discharge chamber.   This gives a discharge of the dust in puffs instead of the continuous stream as in the blower type.   This design is best adapted to somewhat isolated plants such as small trees and shrubbery.

Traction dusters have been widely used for vegetable- and field-crop dusting, especially for small acreages of cotton.   In this type the fan is driven by the wheel or wheels on which the duster runs.   Such machines, containing 50 to 100 pounds of dust, may develop enough power to turn a 12- to 16-inch fan at 2,500 to 3,000 revolutions per minute and may operate with a boom containing up to eight nozzles for a four-row operation.   Because the speed of the machines will vary with the rate at which they are drawn through the field and the wheels may slip in soft ground, it is always difficult to maintain an even discharge from traction-type dusters.

Power dusters operated by small gasoline engines are the most practical for orchard and field-crop work.   The fan is operated at about 3,000 revolutions per minute.   For field and row crops, booms up to 30 feet in length with from 8 to 18 delivery nozzles are used.   These may be individually connected by flexible tubing to a peripheral manifold surrounding the fan, or they may be attached to a tapering hollow-boom manifold to equalize the discharge rates at each nozzle.

The orchard dusters commonly have a single discharge outlet or a double fishtail arrangement for discharging dust from both sides and a centrifugal or squirrel-cage fan to discharge large volumes of air, up to 20,000 to 40,000 cubic feet per minute. This equipment will project dusts for considerable distances even to the tops of forest trees (Fig. 8.12). In these machines the dust is mechanically fed by an agitator or metered by a worm or helix arrangement directly into the fan or blower, assuring a fairly even rate of discharge. However, in practice, the uniformity of discharge of power dusters may vary over a severalfold range because of the changes in the head of dust above the hopper opening and the tendency of dusts to cake and is greatly inferior in this respect to comparable

Fig. 8.12. Dusting forest trees with high-velocity blower. (*From Agricultural Equipment Co.*)

spray equipment. The most recent developments to remedy this deficiency include equipping the hopper with an elevator which lifts the dust to the top, where it is force-fed into the fan, or the use of a rotating hopper which scoops up the same measure of dust at each revolution and feeds it into the fan, and the employment of a conveyer-belt system at the bottom of the hopper which transports an even layer of dust through a variable shutter into the fan. Such designs have reduced the variability of discharge to a few per cent.

The spray duster (Fig. 8.10) was developed to apply dusts simultaneously with a mist spray from nozzles along the edge of the fishtail orifice, which served to wet the foliage and increase the initial deposit of dust. Certain types of mist blowers are also equipped for simultaneous spray-dusting, and results have been secured by the simultaneous emission of 1 gallon of aqueous oil emulsion to each pound of dust, which were equal to or better than results of high-pressure spraying. However, this type of application has been superseded by the use of concentrate sprays.

Many other innovations have been developed for dust applications. For use with volatile materials such as nicotine, a hopper with a mixing attachment in which the dust is mixed immediately before application has been used. The performance of nicotine dusts has also been improved by partially confining them under a gasproof canvas trailer 10 to 30 feet wide dragged behind the duster for a distance of 10 to as much as 100 feet. The trailer serves to confine the vapor of the insecticide given off by the dust and to thus expose the insects to it for a longer period.

AIRCRAFT APPLICATION OF INSECTICIDES[1]

The first practical use of the airplane for insect control was in the application of lead arsenate dusts to control the catalpa sphinx, *Ceratomia catalpae*, in Ohio in 1921. The advantages of aircraft applications in rapidity, cheapness, and ease of treatment were readily apparent, and the uses increased rapidly. Calcium arsenate dust was applied by air in 1923 to control the cotton boll weevil, *Anthonomus grandis*, and the cotton leafworm, *Alabama argillacea*, and paris green dust was applied by air to control anopheline mosquito larvae in the same year. By 1925, forest insects were being dusted by air in Germany, the United States, and Canada. During the next 20 years, most aircraft operations employed dusts because of the difficulties in formulating suitable sprays with the water-insoluble arsenicals.

During the Second World War, malaria control became an essential adjunct to military operations in many areas of the globe and DDT was found to be exceptionally effective for mosquito control. However, its poor dusting qualities resulted in the development of aircraft spraying, and literally hundreds of devices were employed using a range of aircraft such as the Piper Cub L-4 and Stearman PT-17 equipped with booms and nozzles, fighter planes with wing tanks, and heavy multiengined ships such as the B-25 bomber and C-47 transport with large tanks emptying through vertical discharge pipes. Both larviciding and adulticiding operations were proved practicable on the very largest scale and have since been extended to pest mosquito control in Alaska and Canada and to the control of the tsetse flies, *Glossina* spp., in Central Africa and to black flies, *Simulium* spp., in the Adirondack mountains.

Aircraft have been employed extensively for the control of locusts and grasshoppers by a variety of means. Oil solutions of dinitro-*o*-cresol and BHC have been sprayed on swarms of *Locusta migratoria migratorioides* both in flight and at rest on the ground. Aldrin, chlordane, and toxaphene have also been used both as sprays and baits for various species of *Melanoplus*, while arsenical-molasses baits have been applied on a very wide scale. Aircraft applications have been used very successfully for the control of forest insects where other means of application are completely impractical. Notable campaigns have involved the use of calcium

[1] "Handbook on Aerial Application in Agriculture," Texas A. & M. Col., 1956; WEICK, F., and G. ROTH, *Ann. Rev. Ento.*, **2**:297 1957; *U.S.D.A., Farmers' Bul.* 2062, 1954; BROWN, A., "Insect Control by Chemicals," Chap. 8, Wiley, 1951.

arsenate dust and, more recently, of DDT dusts and sprays for the control of the gypsy moth, *Porthetria dispar*, and of DDT sprays for the spruce budworm, *Choristoneura fumiferana*. It has been estimated that in the gypsy moth operations, the payload of a single C-47 airplane treats an area as large as that which could be covered by a truck-mounted spray rig in 4 years.

Today, in the United States, the airplane plays a major role in the application of pesticides. For example, in 1955, 45,316,000 acres were treated by air for insect control and 265,808,000 pounds of dust and 51,274,000 gallons of spray were applied. On California farms in the

Fig. 8.13. Aerial view of airplane duster, power duster, traction duster, one-mule duster, saddle gun, and hand-duster dusting cotton. (*U.S.D.A. photograph.*)

same year, of a total of 5,756,941 acres treated by commercial pest-control operators, 4,853,462 were treated by aircraft.

The aircraft has brought an astonishing increase in efficiency to many insect-control operations (Fig. 8.13). An ordinary dusting plane flying at 100 miles per hour can treat an area at the rate of about 10 to 40 acres per minute, depending upon the effective swath width, varying from 50 to 200 feet; allowing for an operating period of only about 3 hours per day and the time required for loading, from 500 to 2,000 acres are commonly treated. Even greater rates of treatment are obtained in specific operations with larger aircraft. Thus in mosquito-larviciding operations, several square miles may be treated in an hour, and in grasshopper-bait-spreading operations with a C-47, 10,000 acres have been treated per day. In tussock moth, *Hemerocampa leucostigma*, control, 450,000 acres of forest were sprayed within a few weeks.

The aircraft operations have also made insecticidal applications possible in many areas which were previously inaccessible. These include swamps and marshes where mosquitoes and other biting insects breed, jungles and forests, vast plains areas where grasshoppers and locusts are found, and certain cultivated crops such as rice, sugarcane, mature corn, and cotton, where ground treatment is impractical or destructive.

In assessing the place of aircraft in any insect-control operation, it is well to keep in mind that the advantages of aircraft are (a) speed and timeliness of application, (b) freedom from crop injury and soil compaction, (c) no need for farmer preparation of area to be treated, and (d) accessibility of all types of areas. Against these must be balanced the disadvantages: (a) dependency upon optimum weather conditions, (b) lack of uniform coverage, especially under leaves, (c) difficulty in confining application to treatment area which results in severe drift problems, (d) very hazardous operation, with equipment and sometimes human lives lost through accidents, (e) expense of operation and inefficiency on small acreages, and (f) inflexibility after application has begun due to difficulty in communicating with pilot.

PHYSICAL PRINCIPLES OF AIRCRAFT APPLICATION

The use of the airplane or helicopter for the application of insecticides introduces an additional dimension of complexity into the operation, that of the turbulences imparted to the air by the rotation of the propeller and the downdraft resulting from the flight of the aircraft. The influence of these factors on sprays and dusts has been given a great deal of study, and the principles involved are well understood although they are often ignored in practice. A thorough appreciation of these factors is very important in designing and operating aircraft application equipment which will have the maximum effectiveness on the insects to be controlled and reduce hazardous or objectionable drift to a minimum.

Airflow about the Aircraft. The lift of an aircraft in flight is obtained by imparting a downward motion to the air. This downwash or downdraft may amount to about 600 feet per minute for the ordinary biplane flying at 80 to 100 miles per hour and as much as 1,100 feet per minute for the helicopter. The downdraft helps to carry the insecticidal discharge toward the ground and also moves foliage, which aids in penetration and distribution of the insecticide. The lift is secured by the airfoil of the wing and results in a decreased air pressure above the wing and an increased pressure below it. As a result, the high-pressure air under the wing flows out and around the wing tips to the low-pressure area above the wing, producing a rotary movement at each wing tip, the trailing wing-tip vortex, which may persist for several seconds after the passage of the aircraft. Superimposed on these forces is the rotary slipstream of the propeller, which displaces the airflow in a counterclockwise direction. The velocities of the downdraft and the wing-tip vortices increase as the speed of the aircraft decreases, since the downward displacement of the air must equal the lift required to support the aircraft.

With the helicopter, the forces are very similar. Pronounced vortices with outward and upward components occur at the ends of the rotor, while the downdraft is most pronounced under the central section of the rotor (Fig. 8.14). The strength of these forces is greatest when the helicopter is hovering, and they are materially decreased under conditions of normal forward flight (Fig. 8.15).

Effects of Particle Size of Discharge. The air currents following in the wake of an aircraft largely determine the movement of particles released in spraying or dusting. Small particles of aerosol dimensions, with their very slow settling velocities (Table 8.1) follow the airflow almost perfectly and, when released in the region of the wing-tip vortices, may be thrown upward to heights of several thousand feet and will travel vast distances in very low wind currents. Larger particles, such as coarse sprays,

FIG. 8.14. Crop-dusting with the helicopter. (*From Bell Aircraft Corp.*)

FIG. 8.15. Helicopter applying thermal aerosol fog for the control of black flies and mos-quitoes. (*From Bell Aircraft Corp.*)

pelletized or granular insecticides, or baits respond much more directly to gravitational forces and have a high settling velocity. Nevertheless, the distribution of these is also influenced by the flight turbulence. Reed[1] has calculated the theoretical paths of various-sized spray droplets released from the A-1 airplane at 85 miles per hour and 10-foot altitude as shown in Fig. 8.16.

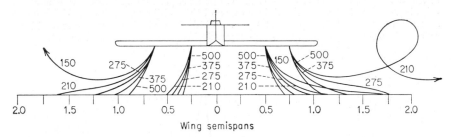

FIG. 8.16. Theoretical path of various-sized spray droplets released by aircraft flying at 85 miles per hour and 10-foot altitude in still air. (*Redrawn after Reed*, 1954, *Nat. Advisory Comm. Aeronautics Tech. Rept.*)

The Swath Cross-section. The practical assessment of the aircraft distribution of insecticides depends upon the dimensions of the swath of deposit. These are determined by such variables as the type, altitude, and speed of the aircraft, the particle-size spectrum of the discharge, the nature of the venturi, spray boom, or other device for producing the discharge, and by the meteorological conditions. The swath cross-section is far from uniform, since the largest particles tend to fall most directly underneath the aircraft while the small ones are more readily deflected by air currents and are dispersed further in the lateral directions, or may escape the treatment area altogether. Typically, the resulting swath cross-section is that of a bell-shaped distribution curve skewed in the direction of the rotation of the propeller or, where a long boom is employed, more of a modified trapezoid. In some cases, with airplane spraying using nozzles near the wing tips, a bimodal or two-peaked swath cross-section results. Typical swath cross-sections obtained with sprays and dusts by various investigators are shown in Fig. 8.17. These curves are generally plotted as the averages of a number of determinations, since the cross-section is somewhat variable, especially because of the corkscrew pattern produced by the rotation of the propeller and the presence of occasional large droplets in sprays. It will be noted that the total recovery of insecticide over a 50- to 200-foot swath, which represents the maximum treatment widths generally obtained in various operations, is only a fraction of the total amount of material discharged. Many of the fine particles drift for relatively enormous distances, as has been graphically demonstrated with 2,4 D herbicidal applications, which have severely damaged cotton fields more than a mile from the intended treatment area (Table 8.1). With dusts, careful recovery studies have shown that only about 27 per cent of a dust with a D_m of about 20 microns was recovered over a 200-foot swath, while increasing the particle size to a D_m of 30 to 40 microns increased the recovery to about 43 per cent. Recoveries with oil-based aerosols and sprays have ranged from 9 per cent for an aerosol of D_m of 35 microns to about 32 per cent for a spray of D_m of 200 microns, based on a 200-foot swath. Even with water-based sprays having a D_m of 300 microns, the recovery over a 100-foot swath was only 50 per cent.

From a consideration of the parameters of the swath cross-section it is clear that for the successful application of any relatively uniform coverage of the treatment area it will be necessary to overlap the edges of the swaths so that the total deposit will everywhere exceed the minimum lethal dosage. Therefore the distances between swath centers (line of flight) must be gaged on the recovery pattern, as shown in Fig. 8.17. Inasmuch as the dynamic catch (Fig. 8.1) of the dust or spray particles on vegetation

[1] REED, W., *Nat. Advisory Comm. Aeronautics, Tech. Rept.* 1196, 1954.

will greatly influence the deposit, this factor must be considered in gaging the rate of discharge. This factor acts in a positive or a negative way, depending upon whether the application is to form a residual deposit on vegetation or to penetrate through vegetation to kill flying insects or insect larvae breeding in water. The importance of the dynamic catch on dosage is shown by the fact that six times as much paris green dust was required in heavy plant cover to produce the same larvicidal effectiveness as occurred under light cover and that while 2 pounds of paris green per acre per 100-foot swath produced 90 per cent kill of anopheline larvae over a width of 115 feet in light plant cover, the same dosage in heavy cover produced no kill.

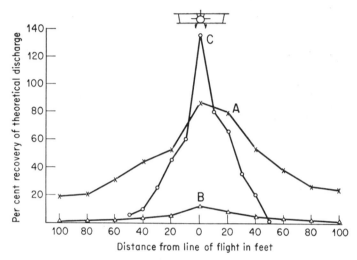

Fig. 8.17. Swath cross-sections from airplane applications of sprays and dusts, showing relation of particle size to per cent recovery. *A*, dust with 88 per cent in 20- to 50-micron range, at 20 feet; *B*, oil aerosol with D_m of 35 microns, at 20 feet; *C*, water spray with D_m of 300 microns, at 5 feet above the ground. (*Original.*)

With aircraft sprays a similar relationship exists: the recovery of a spray of D_m of 110 microns was approximately seven times the mass in the open of that recovered under heavy cover, and the D_m of the material recovered under the heavy vegetation was 50 microns, showing that the larger droplets were filtered out by the vegetation.

The swath dimensions are not materially affected by changes in air speed or height of flight in low-level applications. As has been indicated, the air currents around the plane are strongest at slow speeds, and in practice the swath width in airplane spraying was about 10 feet wider at 50 miles per hour than at 80 miles per hour. An increase in height (to the wheels) from 1 to 10 feet did not appreciably change the swath width, but the pattern was somewhat more regular at the higher altitude.

It should be noted that in most aircraft operations for the control of insect pests of field and vegetable crops, the maximum practical swath width is about 40 feet since the deposits achieved beyond the wing tips are not generally heavy and uniform enough to secure practical insect control. To complete the coverage of a treated field it is essential for the pilot to "dress out" the ends by flying one or two swaths across the ends of the rows at right angles to the regular treatment swaths. For accurate work in large fields, the use of flagmen to indicate the desired line of flight is essential.

EQUIPMENT FOR THE AIRCRAFT APPLICATION OF INSECTICIDES

Aircraft. The aircraft most commonly used for commercial application of insecticides are biplanes originally used as trainers, such as the Stearman PT-17 and the N3N. These originally had 220-horsepower engines and can carry from 600 to 800 pounds payload, but have often been modified by a 450-horsepower engine to carry 1,500 to 1,700 pounds. Several smaller high-wing monoplanes such as the Piper J-3 and PA-18 have also been used, with from 85 to 150 horsepower, and these can carry payloads of 400 to 800 pounds. Several new aircraft especially designed for agricultural operations have recently been developed. They include the Ag-2, with a 450-horsepower engine and a payload of 2,000 pounds, and the Fletcher Utility, of similar size. Both these planes are low-wing monoplanes of especially thick wing section. They provide good pilot visibility together with exceptionally slow flying speed.

Weick has summed up the desirable characteristics of applicator airplanes as (a) ability to carry a spray or dust load of at least 35 to 40 per cent of gross weight, (b) ability to take off and climb to 50 feet within $\frac{1}{4}$ mile with full load, (c) safe operating speed of 60 to 100 miles per hour and minimum safe flying speed of 45 miles per hour, (d) easy-handling controls with high maneuverability, (e) excellent forward and lateral visibility, (f) special design to protect the pilot in a slow-speed crash, (g) facilities for quick and easy loading, and (h) simple rugged construction and simplicity in maintenance and repair.

Dust Distribution. The components for aircraft dusting are as follows: The hopper (a) is sloping-sided, holding from 200 to 2,000 pounds (15 to 50 cubic feet) and should be constructed of aluminum, or better, of stainless steel or plastic reinforced with fiber glass for corrosion resistance. It should contain a large, easily accessible loading door with tight sealing gasket, an air vent, and a small window to serve as a gage for the pilot. The agitator (b) is located directly above the throat of the dust hopper and consists of a rotating reel to sweep the dust into the hopper throat, powered by a wind-driven propeller (c) through a gear box to reduce the speed by about 50:1. The agitator should be equipped with a brake operated from the cockpit. The rate of dust discharge is controlled by a sliding gate (d) at the bottom of the hopper and operating from the cockpit. The gate should be made of aluminum to reduce the fire hazard from sulfur. A venturi-type distributor (e) is commonly used to disperse the dust into the air stream. This employs the principle of the venturi tube to produce increased air velocity and lower static pressure at the hopper throat, which aids the flow of dust. An accurate metering device such as a variable auger or conveyer surface, although not generally used, would be highly desirable to promote the even flow of dust from the hopper.

Spray Distribution. The generalized components for aircraft spraying are as follows: (a) the spray tank which is made of aluminum, stainless steel, or plastic impregnated with fiber glass and may have a capacity of 35 to 200 gallons. It should have a large, conveniently located filler opening with a screen, a large protected vent, a gage, and sump and drain petcock. The spray pump (b) is preferably of the centrifugal type and should be corrosion-resistant, with adequate seals and sufficient capacity to produce agitation by bypass action into the tank. Centrifugal pumps of suitable type operate at a maximum of about 70 pounds per square inch and at 3,000 to 4,000 revolutions per minute. Gear pumps are also satisfactory for emulsions and solutions, but are badly worn by suspensions of wettable powders. The simplest method for driving the pump is the wind-driven propeller (c) mounted in the airplane-propeller slipstream and connected to the pump by a universal joint. The pump assembly should be equipped with a brake. Hydraulically driven spray pumps are also used in some installations. The spray boom (*d*) is usually a ⅝- to 1¼-inch-diameter aluminum or steel tube and is best mounted about a foot below the lower wing or from 4 to 6 inches aft of the trailing edge of the wing. It should be approximately three-quarters as long as the wing span to minimize the amount of spray entering the wing-tip vortices (Fig. 8.16). The type and position of the nozzles is of great importance in determining the particle size of the discharge and the swath cross-section. A typical spray boom for agricultural spraying has from 20 to 30 nozzles. These should be mounted progressively closer together toward the boom tips, with a cluster of two to three at each tip. To compensate for the air currents from the counter-clockwise rotation of the propeller, which displaces much of the spray from a distance of 1 to 4 feet right of center to the left of center, four or five nozzles should be massed close together 2 to 4 feet right of center of the boom and a corresponding reduction made by eliminating the nozzles in the 4 feet of space to left of center.

For insecticide distribution, where sprays with a D_m of 50 to 200 microns are most desirable, hollow-cone or flat spray nozzles are most desirable, operated at 30 to 40 pounds per square inch. The direction of nozzle placement with regard to the slip-stream has an important effect on the degree of liquid breakup, as predicted from equation (7), which indicates that atomization is proportional to the difference in velocities of flow of air stream and liquid stream. Thus in one experiment at 35 pounds per square inch the D_0 was 185 microns, while stationary, decreased to 164 microns in flight at 110 miles per hour with nozzles forward, and increased to 188 microns with the nozzle downward and 290 microns with the nozzle pointed in a trailing direction.

Flight speed has an appreciable effect on the particle size produced by a given nozzle, as would be predicted from equation (7). Thus with water sprays, the D_m decreased from 375 microns at 50 miles per hour to about 300 microns at 90 miles per hour. In another set of experiments, with a heavy aromatic oil, the D_0 decreased from 90 microns at 20 miles per hour to 80 microns at 120 miles per hour, when the nozzles pointed forward. When the nozzles were oriented in a trailing direction, the particle size increased from a D_0 of about 100 microns at 20 miles per hour to 180 microns at 70 miles per hour, and then decreased to 120 microns at 120 miles per hour. The maximum diameter occurred at about the point where liquid and air velocity were equal and poor atomization should result, according to equation (7). Because of the importance of eliminating dripping outside the treatment area, nozzles should be fitted with check valves.

Other means of liquid breakup and distribution have been used in aircraft spraying. The rotating brush disk or drum uses a wind-driven propeller to produce the centrifugal force necessary to break up the liquid, which is fed by gravity (page 436). This equipment will distribute a wide variety of liquids, including those containing a high degree of suspended matter.

The venturi exhaust generator (page 445) has been used to produce aerosols in the 10- to 100-micron range, which have been used successfully for the control of adult and larval mosquitoes.

The most desirable particle sizes for aircraft spraying vary with the type of operation. In the control of adult mosquitoes, black flies, or tsetse flies, where jungle or forest canopy is to be penetrated, an aerosol of 10 to 50 microns is most useful, and this provides swath widths up to 250 feet, but gives very low recoveries on vegetation and is highly sensitive to meteorological effects. For the direct spraying of locust swarms a particle-size range of 50 to 100 microns seems to be most suitable and will give swath widths up to 200 feet. For agricultural spraying for maximum deposit under conditions of low flight a range of 100 to 300 microns combines the best dynamic catch and the least drift away from the treatment area, but the effective swath width is only about 100 feet. Coarse sprays of 200 to 500 microns produce narrow swath widths of about 50 feet.

Most insecticidal dusts will pass about 50 to 75 per cent through 325-mesh or 44-micron diameter. These behave something like aerosols of equivalent D_m and produce swath widths of up to 200 feet with low recoveries. Granular or pelletized materials commonly range from 30- to 60-mesh, 250 to 590 microns, and behave like coarse sprays producing swath widths of 50 feet or less but a high percentage recovery.

Bran baits used for grasshopper control (page 323) generally consist of irregularly shaped particles from 10- to 40-mesh (420 to 2,000 microns), which fall at rates similar to those of oil sprays of 200 to 300 microns. With conventional dusting planes, the swath widths obtainable ranged from about 20 feet at 15-foot altitude to 90 feet at 50-foot altitude. By means of the drift-spraying technique (page 461) with large aircraft, swath widths up to 450 feet have been obtained.

METEOROLOGICAL EFFECTS

Wind and air currents appreciably affect the performance of aircraft in dusting and spraying operations. In general, the most satisfactory results are obtained under conditions of very low wind velocity and when vertical currents or turbulence are at a minimum. These conditions usually occur from the early-morning hours and again until about 1 hour before sunset. At these times, the ground surface and the air immediately above it are cooler than the upper layers of air, and a stable condition of *inversion*, free from rising air currents, exists. When the sun warms the ground so that the lower air level is warmer than the air above, a condition of *lapse* results and the rising air currents or turbulence occurs. Inversion conditions can be measured by the difference of 1 to several degrees in temperature readings on two thermometers, one near ground level and the other at head height, or can be observed by the use of smoke, for everyone has seen the layering of smoke from bonfires or steam engines which commonly occurs at dawn and dusk. Because of the sensitivity of dust and spray clouds to disturbances by turbulence, most aircraft operations of this nature are carried out under inversion conditions.

Surface winds also play an important role, and these are generally minimal under inversion conditions. An 8-year study in California showed that wind velocities in June to August were least from 3 to 7 A.M. and that at sunrise, lulls of less than 2 miles per hour wind velocity occurred on approximately 15 days per month.

Drift Spraying. In this method, the wind is used to disperse the spray discharge from an aircraft flying at an angle across the wind direction. This method is generally suitable only for the treatment of large areas such as for mosquito and locust control or for forest insects, but has also been employed to control the cotton jassid, *Empoasca lybica*, in the Sudan. Here, the spray is emitted from a large discharge pipe and is atomized at air speeds of 150 or more miles per hour into droplets ranging from 10 to 1,000 microns. These droplets will fall at velocities indicated by Stokes' law (page 429) (Table 8.1), which are proportional to the squares of their diameters. Thus the large droplets will be deposited closest to the path of the plane and the small droplets will drift far downwind. The distances to which droplets drift, and hence the swath characteristics, are determined by the product of the velocity of the cross-wind in miles per hour (calculated from the wind velocity × cosine of angle θ, which it makes with a right angle to the flight path) and the height of flight in feet (HU). With a spray with a D_m of about 300 microns, an HU value of 1,500 feet–miles per hour produced satisfactory mosquito control over a 200-yard strip. Applications are generally not successful when the wind is at an angle of less than 15 degrees to the line of flight or when the plane is higher than 500 feet, because of the evaporation of the spray droplets.

CHAPTER 9

INSECTS INJURIOUS TO CORN

Corn is the most important farm crop in the United States, having an average annual value of more than \$3,500,000,000. Insects attack every part of the plant throughout its growth, and year after year destroy about 9 per cent of the crop. Control is difficult, on the whole, because of the extensive areas and the relatively low value of the crop per acre, which make it necessary to depend mostly upon farm practices and other indirect control measures. Nevertheless, many of the insect pests of corn can be very effectively controlled if the best known remedies are properly applied.

FIELD KEY FOR IDENTIFICATION OF INSECTS INJURING CORN

A. Insects chewing leaves or stalk above ground leaving visible holes:
  1. Leaves of plant ragged and sometimes entirely eaten off, the injury beginning about the margins of the fields. Tips of ears gnawed. Fields presenting a ragged appearance, bare stalks standing without leaves.....................
  ....*Grasshoppers*, page 465.
  2. Leaves of young corn with green portion eaten out, presenting a whitened, bleached appearance. Plant growth is retarded and leaves wilt and hang limp. Leaves covered with dull or shining black or greenish-black small jumping beetles from a little smaller than a pinhead to several times as large. Some kinds transmit a bacterial wilt disease causing death of plants..............
  ....*Flea beetles*, page 514.
  3. Unfolding corn leaves with rows of holes running across the leaves. Cavities eaten in the sides of the plant by black, tan, or brown, often mud-covered beetles, with slender snouts nearly ⅓ as long as rest of body. Beetles from ¼ to ⅞ inch long. Injury most severe on reclaimed or sod land..........
  ....*Billbugs*, page 487.
  4. Grayish long-headed beetles, ½ inch long, with faint white stripes along their sides eat the margins of the leaves............*White-fringed beetle*, page 607.
  5. Corn leaves with irregular holes with ragged edges eaten out, giving plant a ragged appearance. Often the plant is entirely stripped of leaves. Dark-green worms up to nearly 2 inches in length with light stripes on the sides and down the middle of the back, feed usually at night and hide under clods or in the heart of the plant during the day. Skin when seen through a lens appears smooth. Worms often crawling into the corn in large numbers from near-by fields of grass or small grains........................*Armyworm*, page 471.
  6. Injury similar to *A*, 5, by worms which often crawl in great numbers from field to field. Holes eaten in leaves have smooth edges. Distinguished from true armyworms, under a lens, by the greater length of the hairs and their more prominent black bases on the smooth skin, by the prominent ⋏ on the front of the head, and by the somewhat different striping..*Fall armyworm*, page 473.
  7. Corn plants under 1 foot in height cut off, mostly at night, below, at, or slightly above the surface of the ground. Fat, well-fed, smooth-appearing worms with 3 pairs of slender legs and 5 pairs of prolegs, of varying sizes up to 2 inches in

462

length and of several shades and markings, hide under the ground or clods near the injured plants. Many kinds curl the body and "play 'possum" when disturbed...*Cutworms*, page 476.

8. Young corn plants up to 8 or 12 inches in height eaten into or cut off near the surface of the ground. A loose silken web containing many bits of dirt, leading to a short, silk-lined tunnel in the ground. Short, dirty-colored, brown-spotted, coarse-haired worms, from ¼ to ½ inch long, usually hidden in these silken tunnels, try to escape when disturbed............*Webworms*, page 513.

B. *Small insects sucking sap from leaves or stems, not leaving visible holes, but causing wilting, spotting, discoloration, or death of the plants:*

1. Leaves of the corn plant covered with masses of soft-bodied bluish-green plant lice or aphids, of the size of pinheads, most abundant in the curl of the plant and on the developing tassel. Infested plants scattered over the field.......
....*Corn leaf aphid*, page 480.

2. Corn wilting, drying out, and falling down, on the side of the field next to small grains. Small, active, reddish or black-and-white, somewhat flattened, sucking bugs clustered behind the lower leaves and over the entire lower part of the stalk; when crushed, giving off a vile odor. Fields invaded by hordes of these insects at the time of small-grain harvest..............*Chinch bug*, page 480.

C. *Insects boring or tunneling in the stalks or stems:*

1. White, chunky, legless, humpbacked grubs, with a distinct, hard, brown or yellow head, tunnel in the pith of the stalk, dwarfing and often killing the plants. Most severe on old corn ground.................................
...................................*Maize billbug* or *curlew bug*, page 488.

2. Flesh-colored caterpillars from ½ to 1 inch long, with inconspicuous, small, round, brown spots scattered over the body, boring in all parts of the stalk, shank of ear, and ear. Tassels frequently broken off from injury at the base. Many caterpillars often found in one stalk......*European corn borer*, page 490.

3. Cornstalks, especially those around margins of fields, bored during June and July by very active dark-brown worms or caterpillars, with two white stripes on each side, which are broken for about ¼ their length near the middle of the body, and a continuous white stripe running down the center of the back. The caterpillars reach a length of 1½ inches........*Common stalk borer*, page 493.

4. Stalks bored throughout their length by white worms, up to 1 inch in length, conspicuously marked with rounded brown spots; not common in the North..
...............*Southern cornstalk borer* and *southwestern corn borer*, page 495.

5. Stalks bored by slender greenish worms with fine longitudinal brownish markings, about ¾ inch long when full-grown. Stalks become much distorted and curled by the early-season injury. Most abundant in the South...........
.....................................*Lesser cornstalk borer*, page 497.

D. *Insects attacking the ear:*

1. Dark-green and coppery-red beetles about ½ inch long by ¼ inch across, feeding on silks and husks, especially at the tip of the ear. Wing covers of the beetles shining. Four prominent white spots on the tip of the abdomen, which projects from under the greenish-brown wing covers..*Japanese beetle*, page 749.

2. Silk fouled with moist masses of excreta, many of the silks eaten off; kernels at the tip of the ear eaten by worms up to 2 inches long, varying in color from very dark green to light green; live beneath the husks. Bodies of the worms sparsely haired, skin rough-appearing under a lens. Worms nearly 1¾ inches long when full-grown and usually only one to the ear. Occasionally the worms enter at the butt of the ear, but usually at the tip. They do not tunnel into the cob...*Corn earworm*, page 498.

3. The ear, its cob and shank tunneled throughout and many of the kernels and the silk eaten by flesh-colored, inconspicuously spotted caterpillars up to 1 inch long. Many worms often found in one ear. Husks perforated with small holes with exuding frass.................*European corn borer*, pages 490 and 500.

4. Ripening kernels of grain sorghum heads and broomcorn eaten out or entirely consumed by sluggish caterpillars, somewhat flattened and thickly clothed with spines and hairs, the body greenish with four, red or brown, longitudinal

stripes above...................................... ....

Sorghum webworm, Celama sorghiella (Riley) (see *Tex. Agr. Exp. Sta. Bul.* 559).

E. *Insects attacking the roots or underground stem:*

1. Plants weak, leaves reddish or yellowish. Bluish-green plant lice or aphids, about the size of pinheads, sucking the sap from the roots; always attended by ants..*Corn root aphid,* page 501.

2. Plant makes slow growth, dies, and sometimes falls over. Often distinct areas in field showing injury. Roots eaten off clean, not tunneled, by white, curved-bodied or U-shaped, six-legged grubs, from ½ to over 1 inch long, with large brown heads and distinct jaws.....................*White grubs,* page 503.

3. Slick, shining, smooth, reddish-brown, very tough-skinned six-legged worms, 1 to 1½ inches long, with the last segment of the abdomen often curiously ornamented, bore through the soil and chew off small roots or tunnel larger roots...*Wireworms,* page 506.

4. Corn on land which has been in this crop for one or more years falls over about the time the tassels appear, frequently after a rain has softened the ground. Roots eaten off or containing many small brown tunnels in which will sometimes be found slender whitish worms not more than ½ inch long with distinct brown heads and 6 short legs..............*Northern corn rootworm,* page 509.

5. Corn falls as in *E,* 4, but the injury is not confined to old corn ground. Underground parts of stalk as well as roots show tunnels. Slender yellowish-white worms having much the same appearance as those described under *E,* 4, but slightly larger and more robust...........*Southern corn rootworm,* page 510.

6. Corn plants following clover sod are stunted, seldom reaching a height of over 8 or 10 inches. Plants wilt during hot, dry days. Small, short, white, fat-bodied grubs, not over ⅛ inch long with light-brown heads, gnaw on the roots. Grubs hold body in a curved position. Cease feeding during June.........
....*Grape colaspis,* page 512.

7. Corn wilts and dies when from 1 to 4 feet high Sometimes falls over. Injury most common to corn in old sod ground. Large, fleshy, legless whitish grubs from ½ to 2 inches long, with the body enlarged just behind the head, not curved, and with strong brown jaws....................................
....*Corn prionus* (see *Eighteenth Rept. Ill. State Ento.,* page 128, 1891–1892).

8. Plants wilt and die from attacks of white, legless, but not curved-bodied grubs, up to ½ inch long, which eat off the underground stem and taproot........
....*White-fringed beetle,* page 607.

9. Roots or stem eaten off. Plump, smooth-appearing grayish or greenish-white worms, up to 2 inches long, unmarked except for brownish head and shield behind head, with 3 pairs of slender legs and 5 pairs of prolegs, remain in the soil day and night; curl the body when dug out or disturbed.............
....*Cutworms,* page 476.

10. Roots and stem eaten by dirt-colored, brown-spotted, coarse-haired caterpillars from ¼ to ½ inch long, which are hidden in a loose-woven dirt-covered silk-lined tunnel in the soil..................*Webworms,* page 513.

11. Plantlet weak, reddish to yellowish in color. Small brown ants make conical mounds of dirt and tunnel the ground along the corn roots, sometimes feeding on the starchy parts of the kernel. Carrying about and caring for small bluish-green aphids on the corn roots..............*Cornfield ant,* page 501.

F. *Insects attacking planted seed, eating into it or devouring the plantlet as it germinates, so that plants often fail to come up:*

1. Seeds fail to sprout or plantlet is weak. White, very slender worms or larvae, slightly over ¼ inch long when full-grown, with yellowish-brown head and 6 short legs, burrowing in the kernels and sometimes in the sprout...........
....*Pale-striped flea beetle,* page 516.

2. Seeds fail to sprout, or plants die when small. Slick, shining, brown to reddish-brown, smooth, hard, six-legged worms, 1 to 1½ inches long, boring through the kernels and young plants......................*Wireworms,* page 506.

3. Seed does not sprout. Brownish or blackish-brown beetles, about ⅓ inch in length, feeding in kernels of corn in the ground.............................

................*Seed-corn beetles:* the darker, brown-striped one, *Agonoderus lecontei;* the uniform chestnut-brown one, *Clivina impressifrons,* page 517.
4. Seed produces a weak, sickly sprout. Starchy parts of kernel eaten out and scattered through the ground by very small orange-colored ants.............
....*Thief ant* or *fire ant,* page 518.
5. Seeds fail to sprout or make a weak sprout. Dirty yellowish-white legless maggots about ¼ inch long, blunt at the posterior end, tapering sharply to the head, may be found burrowing in the kernels of corn in the ground, or in the earth around the kernels......................*Seed-corn maggot,* page 518.

References.   Eighteenth and Twenty-third Repts., Ill. State Ento., 1891, 1892, 1905.

1 (27, 75, 146). GRASSHOPPERS OR LOCUSTS[1]

Importance and Type of Injury.   Few, if any, other species of insects have caused greater direct loss of crops than have grasshoppers. From ancient to modern times they have caused the death through famine of

FIG. 9.1. Differential grasshopper, *Melanoplus differentialis,* laying eggs in soil; enlarged. A part of the soil has been removed to expose the abdomen and the egg mass. (*From U.S.D.A., Farmers' Bul.* 691.)

millions of human beings. In any section of the world when these insects are abundant, man has to make a determined fight to save his crops. Damage is most severe in those parts of the world where the annual rainfall is 25 inches or less. Corn is seldom attacked by grasshoppers until the plant has reached a height of 20 or more inches. Plants that are attacked have the tips of the ears, the tassel, and the leaves eaten, and the stalks present a generally ragged or bare appearance. Grasshopper injury usually starts on the sides of the field, as the insects seldom originate in the cornfield.

Plants Attacked.   Various species of grasshoppers attack nearly all cultivated and wild plants.

Distribution.   Grasshoppers occur over the entire world.

Life History, Appearance, and Habits.   Grasshoppers that attack corn practically all pass the winter in the egg stage. These eggs are laid in packet-like masses nearly 1 inch long and from ½ to 2 inches

[1] Many species of the Order Orthoptera, Family Locustidae.

below the surface of the soil (Fig. 9.1). Each egg mass consists of from 20 to 120 elongate eggs, securely cemented together, the whole mass somewhat egg-shaped and dirt-covered (Fig. 9.2). A single female may deposit from 8 to 25 egg masses. They are mainly deposited in uncultivated ground such as field margins, pasture land, and roadsides. In the middle-western and western states they are frequently laid in considerable numbers in clover, alfalfa, and stubble fields. The pellucid grasshopper lays its eggs chiefly in sod land and almost entirely in heavy soils. The migratory grasshopper and *Melanoplus packardii* oviposit chiefly in fields planted to crops, while the two-striped and the differential grasshoppers deposit their eggs mostly at edges of fields, along roadsides, or in

Fig. 9.2. A clump of grama grass, showing a number of grasshopper egg masses among the stems and roots (indicated by arrows). (*From S.D. Agr. Exp. Sta. Bul.* 172.)

drift soil. A few species winter as partly grown nymphs or as adults. In the latitude of central Illinois, hatching of the more common and typical species begins in mid-May and continues until July. The young hoppers (Fig. 5.9,*E*) differ but little from the adult, except in size and the fact that they lack wings. There are usually five or six nymphal instars that require 40 to 60 days to reach the adult stage. With most of the species which injure corn, growth is completed from the middle of August to the first of September. The adults, however, continue to feed until the first heavy frost. The eggs are mainly deposited during the latter part of September and October. Of the more than 600 species of grasshoppers occurring in the United States, the following are the most destructive. The species destructive in any given year will vary with the locality, altitude, latitude, weather conditions, and kinds of crops that are prevalent there.

The red-legged grasshopper[1] is one of the smallest of the more destructive species, being less than 1 inch long when full-grown. It is very destructive in fields of legumes and is common along roadsides. In the Middle West this and the migratory grasshopper[2] have been very destructive to soybean crops by cutting through the pods and causing the seeds to mold. Its general color is brownish red, the hind tibiae being usually pinkish red with black spines. It can be distinguished from the migratory grasshopper by the shape of the cerci (Fig. 3.2), which in this species are twice as long as wide, widest at the base, and incurved on both upper and lower margins; the tip of the abdomen in the male is not notched.

The migratory grasshopper[2] is one of the most widespread and generally destructive species, able to survive well on dry native grasses and waste land, as well as in nearly all cultivated crops. It shows some preference for light sandy soils. It lays its eggs over a wide area and in diverse situations. This species is very similar in size and

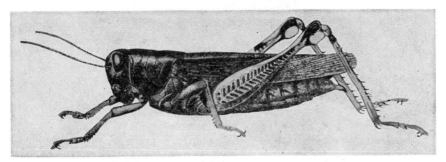

Fig. 9.3. Adult of the differential grasshopper, *Melanoplus differentialis.* Enlarged about ½. (*From Ill. Natural History Surv.*)

appearance to the red-legged grasshopper, being scarcely 1 inch long when full-grown. The hind tibiae are generally not so bright pink as in the red-legged grasshopper. The cerci at the tip of the abdomen (Fig. 3.2) are about two-thirds as broad as long, widest beyond the middle and incurved on the upper margin only; the tip of the abdomen in the males has a distinct median notch or incision. This species is called migratory because the larger nymphs usually migrate from their breeding grounds to more succulent vegetation, and, when abundant, the adult often flies for many miles to new feeding grounds.

The clear-winged grasshopper[3] (Fig. 9.4) is perhaps second in importance only to the migratory grasshopper. It is one of the most common western species, especially at higher elevations and in more northern latitudes, but occurs also in the eastern part of the country. Like the migratory grasshopper it is adapted to survive upon a great diversity of vegetation, and it survives drought well. It prefers to lay its eggs in sod land and in heavy soils. Its hind wings are nearly colorless and transparent.

The differential and the two-striped grasshoppers are better adapted to cultivated crops, such as vast areas of lush grain, and do not survive drought conditions. In dry years they persist only in irrigated districts and along streams. The differential grasshopper[4] (Fig. 9.3) is, next to the Carolina grasshopper, the largest of the destructive species, reaching a length of 1½ to 1¾ inches. It is brownish or olive green in color, with a good deal of yellow on the under parts and with the chevron-like black markings on the hind femora very prominent. The cerci of the male have a short rounded thumb-like projection on the lower margin. This species is especially

[1] *Melanoplus femur-rubrum* (De Geer).

[2] *Melanoplus bilituratus* (Walker) [= *mexicanus* (Saussure)]. The Rocky Mountain grasshopper, *Melanoplus spretus* (Walsh), is considered by some authorities to be an extreme migratory phase of the migratory grasshopper which invaded the Middle West in great swarms during the late nineteenth century.

[3] *Camnula pellucida* (Scudder).

[4] *Melanoplus differentialis* (Thomas).

destructive to corn. The two-striped grasshopper[1] ranges from 1 to 1½ inches in length. It is a robust species, the upper part of the body olive with a yellow stripe on each side, extending from the head to the tip of the wing. There is a dark stripe on the upper half of the hind femur. The species is very common in clovers. It frequently matures by late June, and there may be a partial second generation of both this species and the migratory grasshopper.

The Carolina grasshopper[2] is a very large species, reaching a length of 2 inches. While generally less destructive than the others described, it is so common along roadsides, railroads, and paths that it is one of the most commonly observed species.

Fig. 9.4. Adult of the clear-winged grasshopper, *Camnula pellucida.* About twice natural size. (*From Ill. Natural History Surv.*)

It is brown, mottled with fine specks of gray and red on the wing covers. The hind wings are black with distinct yellow margins. It flies readily when disturbed.

Control Measures. Grasshopper control consists of two distinct measures: (*a*) to destroy the eggs in the fall and winter and (*b*) to combat the grasshoppers at the time they are attacking crops. The first of these measures consists of fall plowing or disking of areas in which the grasshoppers have laid their eggs, thus exposing the eggs to the action of the weather and to birds during the winter and early spring. Plowing or disking to a depth of 5 inches, followed by packing or firming of the soil, is usually sufficient to destroy the eggs. This is often quite effective, especially in the West, where there are areas of uncultivated land interspersed with the cultivated areas, if wind erosion and drying out of the soil are not too serious.

The use of poisoned baits is an effective means of controlling grasshoppers in many areas. Hundreds of different baits have been tried. Practically all of them consist of (*a*) a *base* or *carrier* of bran diluted with either sawdust, cottonseed hulls, ground corncobs, or similar materials; (*b*) a strong inorganic or organic *stomach poison*; and (*c*) *water* or *oil* to make the carrier moist. Some of the baits which have been used most generally and which give the highest kill under average conditions are

[1] *Melanoplus bivittatus* (Say).
[2] *Dissosteira carolina* (Linné).

(A)

Bran... 100 lb.
Sodium arsenite, or white arsenic, or sodium fluosilicate, 4–6 lb., or chlordane, 0.5 lb., or toxaphene, 1 lb., or aldrin, 2 oz., or heptachlor, 4 oz., as
wettable powder or emulsive concentrate
Sufficient water to make a stiff mash, usually...................... 7–12 gal.

(B)

Bran... 100 lb.
Blackstrap molasses... 2 gal.
Insecticide as in (A)
Sufficient water to make a good mash, usually...................... 7–8 gal.

(C)

Bran... 100 lb.
Lubricating oil, 20 to 30 viscosity.................................... 2 gal.
Insecticide as in (A)

In any of the above formulas, hardwood sawdust or corncobs which have been
ground in a hammer mill to the same particle size as bran can be substituted for one-
half the bran. The best results are obtained by spreading the bait at the rate of about
10 pounds per acre. This may be done by hand from pails; two men scattering from
a truck can cover 20 acres per hour. An inexpensive bait-spreading machine has been
developed with which 40 to 50 acres can be treated per hour.[1] Airplanes fitted with
special equipment for bait spreading can cover 100 to 150 acres per hour.[2]

The bait should be applied early in the morning as this is the time of heaviest
feeding of the grasshoppers. They do not feed to any extent at temperatures below
65 or above 90°F. It is of the utmost importance to poison the small grasshoppers
as soon after they hatch as possible. Sometimes the adults can be poisoned as they
congregate for egg-laying. The oil used should be a fresh, lubricating mineral oil
such as is used for automobile crankcases of 20 to 30 S.A.E. rating, the latter being
preferable in very hot dry weather. Advantages claimed for the oil baits are that
they remain moist and attractive to grasshoppers for a week or more, whereas the
standard bait dries out in a few hours and then is not attractive to hoppers. Grasshoppers take water baits only while they are fresh and moist. Water baits applied
during the middle of the day or evening are a complete loss, but in a rush the oil baits
may be applied any time of the day. Water baits ferment, cake, and mold in storage
and cannot be made up in advance or saved for later application; oil baits can be kept
for months if necessary. Water baits containing arsenicals will burn crops if applied
at a rate of more than 10 or 12 pounds per acre and heavy rains follow; oil baits do
not burn plants to which they cling. With the oil baits, the flakes of bran separate
better, which helps to avoid excessive applications and prevents danger to birds and
livestock. The oil bait is not so hard on the hands of men scattering it. One application usually checks damage promptly and kills from 50 to 90 per cent of the hoppers
within 5 days; a second and third application may be needed. For many years organized grasshopper-control campaigns have been carried on, covering the upper Mississippi and Missouri River Valleys. From the period of 1925 to 1934 it is estimated
that grasshoppers destroyed crops to the value of $249,000,000 in the most heavily
infested states. During this same period more than $4,500,000 was spent on grass-
hopper control, and it is believed that every dollar expended resulted in a saving of
about $50 worth of crops, on the average.

Spraying infested areas after the main hatch is completed with dieldrin
at 0.5 to 2 ounces, aldrin or heptachlor at 2 to 4 ounces, chlordane at

[1] PARKER, R. L., "Report of the Fourth International Conference on Anti-locust
Research," Cairo, Egypt, 1936, *Can. Dept. Agr. Publ.* 606, 1938.
[2] *U.S.D.A., Bur. Ento. Plant Quar., Cir.* EC-2, 1948; *Jour. Econ. Ento.,* **41**:656–657,
1948.

0.75 to 1.5 pounds, or toxaphene at 1 to 1.5 pounds per acre has generally proved superior to the use of baits. Dusts of these materials may also be used, but the concentration should be increased by 50 per cent. Sprays are cheaper, last longer, and are much more effective than baits in areas of abundant vegetation such as roadsides, canal banks, field margins, fields of legumes, and orchards. However, aldrin, heptachlor, chlordane, or toxaphene should not be applied to forage for livestock. Where it is necessary to spray pastures or crops intended for animal

Fig. 9.5. A hopper-catching machine or hopperdozer designed to be pushed in front of a tractor, truck, or automobile, over heavily infested fields containing crops that will not be ruined by it.   The upper figure shows the front of the machine, consisting of a back, against which the jumping hoppers strike, and pans below to hold a mixture of oil and water, which kills them as they drop into it.   The lower figure shows the means of attachment to tractor or truck.   (*From Ill. Natural History Surv.*)

forage, or orchards, methoxychlor at 3 pounds per acre has given satisfactory results. For the protection of vegetable crops, malathion at 1 pound per acre has been suggested.

Aircraft spraying with 4,6-dinitro-*o*-cresol at 0.5 to 1 pound per acre as a 10 to 20 per cent kerosene solution has given very effective control of swarms of grasshoppers, particularly of the migratory locusts in Africa.[1]

Where for any reason it is impossible to control grasshoppers by the use of poisons, hoppercatchers or hopperdozers may be used to lessen the hopper damage.   The hopperdozers (Fig. 9.5) are merely devices placed

[1] See *Jour. Econ. Ento.*, **39**:676, 1946; Gunn, D. L., *et al.*, "Aircraft Spraying against the Desert Locust in Kenya, 1945" and "Locust Control by Aircraft in Tanganyika," Anti-locust Research Centre, London, 1948.

on the front end of an automobile, truck, or tractor and pushed slowly over the field at from 5 to 7 miles per hour, or drawn by teams, causing the hoppers to jump or fly up and fall back in the catching device. There are many homemade types of these catchers. The hopperdozer consists of a shallow pan partly filled with water with a little kerosene over the top. On the whole, hopperdozers are much more expensive to operate and less efficient than poisoning. It is rarely possible to catch 50 per cent of the hoppers present. In certain crops such as soybeans they may be used to advantage but cannot be used in corn, small grains, and other crops that would be seriously broken down by the machine. As high as 4 to 8 bushels of hoppers per acre have been caught with these machines. There are about 200,000 grasshoppers in a bushel. It has been estimated that where grasshoppers are present at the rate of 15 to 20 per square yard, they will eat 1 ton of alfalfa, per day, in each 40-acre field.

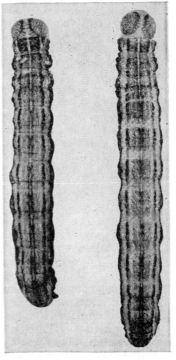

References. *Minn. Agr. Exp. Sta. Tech. Bul.* 141, 1914; *Colo. Agr. Exp. Sta. Bul.* 280, 1923, and *Ext. Cir.*, Ser. 1, 180 *A*, 1921; *S. D. Agr. Exp. Sta. Bul.* 172, 1917; *U.S.D.A., Farmers' Buls.* 1691, 1938, and 2064, 1957; *Jour. Econ. Ento.*, **7**:67, 1914; **10**:524, 1917; **11**:175, 1918; **12**:337, 1919; **13**:232, 237, 1920; and **14**:138, 1921; *U.S.D.A., Tech. Bul.* 190, 1930; *Iowa Agr. Coll. Ext. Bul.* 182, 1932; *Ill. Agr. Exp. Sta. Bul.* 442, 1938; *Jour. Econ. Ento.*, **40**:91, 137, 896, 1947; and **41**:16–19, 945–948, 1948; *U.S.D.A., Bur. Ento. Plant Quar., Cir.* EC-1, 1948; E-771, 1949; and E-774, 1949.

2 (28). ARMYWORM[1]

Importance and Type of Injury. This insect fluctuates greatly in abundance, undergoing cycles which reach destructive peaks at greatly varying periods of years. During epidemics it often destroys much of the vegetation over many hundreds of square miles. Corn under 8 inches in height that is attacked by armyworms will usually have the leaves eaten off entirely. With larger corn the midrib of the leaves will sometimes be left, but the center of the

Fig. 9.6. Full-grown armyworms, *Pseudaletia unipuncta*, the left one showing eggs of a tachinid fly parasite attached to the skin. Twice natural size. (*From Ill. Natural History Surv.*)

young stalk is so eaten out that it dies. The dark-green worms (Fig. 9.6), up to 2 inches in length, with white stripes on the sides and down the middle of the back, will be found hiding under clods and stones or in the center leaves of the plant during the day. The damage usually starts at the sides of the field, where the worms have moved in from some other crop.

Plants Attacked. All grass crops, especially corn, timothy, millet, bluegrass, small grains, and some legumes, and under stress of hunger, many other plants.

[1] *Pseudaletia* (= *Cirphis*) *unipuncta* (Haworth), Order Lepidoptera, Family Noctuidae.

Distribution. United States and Canada, east of the Rockies, and many other parts of the world.

Life History, Appearance, and Habits. The winter is passed mainly in the partly grown larval stage; but the fact that the moths are abroad very early in the spring in the northern states would indicate that some of the insects winter as adults, or as pupae, or that there is a spring flight northward from the southern part of the range of the insect. The partly grown worms shelter in the soil about clumps of grasses, or under litter on the ground. They begin feeding early in the spring, become full-grown by the latter part of April in the latitude of central Illinois, and pupate just below the surface of the soil. The pupae are dark brown, about ¾ inch long, tapering sharply at the tail, and blunt at the head end. They remain in this stage for 2 weeks, or longer if the weather is cool, and then transform to uniform, pale-brown or brownish-gray moths with a wing expanse of about 1½ inches (Fig. 9.7). There is a single, small, but prominent, white dot in the center of each front wing. The moths are strong fliers, but remain hidden during the day, becoming active at night. They are attracted to lights, and strongly so to sweets or decaying fruit. The females lay their greenish-white eggs in long rows or clusters on the lower leaves of grasses to the number of 500 or more. The leaf is generally folded lengthwise, and fastened about the eggs with a sticky secretion. The young worms are pale green in color and have the looping habit of crawling until about half grown. They may often be found by thousands in fields of grass or small grains, and because of their habit of feeding at night, their presence is generally not suspected until the crop is nearly destroyed.

Fig. 9.7. Armyworm. Adult moth, natural size. (*From Ill. Natural History Surv.*)

When the food supply becomes exhausted in the fields where they have hatched, these caterpillars move out in hordes or armies and attack crops in near-by fields. These crawling masses of worms have given them their common name. On becoming full-grown, the worms are nearly 1½ inches long, of a general greenish-brown color, with longitudinal stripes as follows: a narrow broken stripe down the center of the back, bordered by a wide, somewhat darker, mottled one reaching halfway to the side; as seen from the side there are three stripes of about equal width: next to the wide mottled one on the upper side a pale orange, white-bordered stripe, next a dark-brown, light-mottled one just reaching to the spiracles, and just below the spiracles, a pale-orange, unmottled one edged with white. The head is honeycombed with dark lines, and each proleg has a dark band on its outer side and a dark tip on the inner side. They then enter the ground and change to the pupal stage, emerging as moths in from 14 to 20 days. There are from two to three generations each year. The larvae of the first generation do most of the damage, in June in the latitude of central Illinois. The larvae of the last generation are abundant in late August and September.

Control Measures. One of the most effective methods of controlling an outbreak of armyworms is to poison them by scattering a poison-bran mixture in the fields where they are feeding, or across the line of march of the worms when they are leaving fields where food is scarce. Directions for making and applying poisoned bait are given under Grasshoppers, page 469, and on page 323. For armyworms, as for cutworms, the bait should be spread in the late afternoon or early evening. Spraying or dusting with DDT at 1 to 2 pounds per acre or with toxaphene at 2 to 3 pounds per acre is effective in preventing attacks on seedlings and young plants, but treated fodder should not be fed to livestock.

Where the worms are advancing from one field to another, they may be stopped by plowing deep furrows in front of their line of advance and dragging a log or keg of water back and forth in the furrow until a very fine dust mulch has been worked up. The worms tumbling into this furrow will be unable to crawl up the steep dusty side, and may be crushed by the continued passage of the log, or by spraying with kerosene or other contact poison.

Armyworm outbreaks usually originate in fields of small grain or grasses, especially where there is a very rank growth of vegetation, or where the grain has fallen down and lodged. Such situations should be watched, especially during May, and if the young worms are found, the poison-bran bait should be applied immediately.

The armyworm is preyed upon by a number of insects, especially certain parasitic flies,[1] which lay their eggs on the backs of the worms, mostly on the fore part of the body (Figs. 9.6 and 2.16). The young maggots hatching from these eggs bore into the worms and kill them. They are also preyed upon by several ground beetles and certain parasitic wasps. Perhaps the most efficient insect enemy of the armyworm is an extremely small, black, wasp-like insect[2] that deposits its eggs inside the eggs of the armyworm. The other parasites attack the worms when they are partly to nearly fully grown, and thus prevent an excessive increase in the next generation, but do not kill the worms until after most of their feeding has been done. The egg parasite, on the other hand, by preventing the eggs from hatching, stops all damage by these insects.

References. N.Y. (Cornell) Agr. Exp. Sta. Bul. 376, 1916; Jour. Agr. Res., **6**:799, 1916; U.S.D.A., Farmers' Buls. 731, 1916, and 1850, 1951; Ill. Natural History Surv., Ento. Ser., Cir. 7, 1921; Conn. Agr. Exp. Sta. Bul. 408, 1938; Calif. Agr. Ext. Cir. 87, 1944; Jour. Econ. Ento., **39**:669, 1946.

3. FALL ARMYWORM[3]

Importance and Type of Injury. Besides the army cutworms and the true armyworm (page 471), this insect, which is a member of the same family, often develops the marching habit, the caterpillars crawling in great droves, which may be very injurious to field and vegetable crops. They eat the foliage and tender stems of many plants, often taking everything clean as they go, and then disappear suddenly. The larvae often attack the ears of corn in a manner identical with the corn earworm. They are especially bad in the South in seasons following a cold, wet spring.

[1] Winthemia quadripustulata (Fabricius) and other tachinid flies, Order Diptera, Family Tachinidae.

[2] Telenomus minimus Ashmead, Order Hymenoptera, Family Scelionidae.

[3] Laphygma frugiperda (Smith), Order Lepidoptera, Family Noctuidae. A closely related species, Spodoptera (=Laphygma) exigua (Hübner), is known as the beet armyworm (p. 610).

Plants Attacked. Corn, sorghums, and other plants of the grass family are probably the preferred food, and the insect is often called the "grassworm"; but it attacks also alfalfa, beans, peanuts, potato, sweetpotato, turnip, spinach, tomato, cabbage, cucumber, cotton, tobacco, all grain crops, clover, and cowpeas.

Distribution. This insect is a continuous resident of the Gulf states and the tropics of North, Central, and South America and some of the West Indies. It often migrates northward as far as Montana, Michigan, and New Hampshire.

Life History, Appearance, and Habits. This tropical insect is apparently unable to live through the winter in any section where the ground freezes hard. In southern Florida and along the Gulf Coast, several stages may be present and more or less active during the winter months.

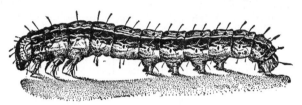

Fig. 9.8. Larva of the fall armyworm. About twice natural size. (*From U.S.D.A., Tech. Bul.* 34.)

In the spring, as they increase in numbers, swarms of moths are produced that fly northward, sometimes covering hundreds of miles before they alight to lay their eggs. About 1,000 eggs are laid by each female, in masses averaging about 150, usually on green plants, and covered with hairs from the moth body. The small larvae feed down near the ground gregariously at first, especially in the heart of the plant, and are not generally noticed until they have reached a length of 1 or 1½ inches, by which time, if abundant, they are consuming so much grass or grain that they create alarm. They do not leave the plant to hide in the soil during the daytime, as do the armyworm and climbing cutworms. The full-grown larvae (Fig. 9.8) vary in color from light tan or green to nearly black. They have three yellowish-white hair lines down the back from head to tail; on the sides next to the yellow lines is a wider dark stripe and next to it an equally wide, somewhat wavy, yellow stripe, splotched with red. These worms are very similar to the true armyworm in appearance but can be distinguished by the more prominent white inverted Y on the front of the head and by the more prominent black tubercles from which the fine scattered hairs on the body arise. The corresponding hairs on the true armyworm are much shorter and the tubercles smaller. The fall armyworm can also be distinguished from the armyworm by the fact that it feeds on cotton, tobacco, legumes, and many vegetables as well as grasses, while the armyworm feeds chiefly on grains and grasses and attacks other plants only when driven by hunger.

When abundant, the caterpillars eat all the food at hand and then start to crawl in great armies into adjoining fields. While these "forced marches" may come in the fall in the North, in the extreme South they occur in midsummer or even in early spring. Gardens may be invaded and consumed in a few nights. Suddenly, when full-grown, all the caterpillars disappear almost as if by magic, having dug into the ground about an inch to pupate.   Within 2 weeks a new swarm of moths emerges from the ground, which generally flies far before laying eggs, and so the entire country may be invaded during the summer.   The adult moth (Fig. 9.9) is similar to many cutworm moths, about 1½ inches across the wings, the hind wings grayish white and the front pair dark gray, mottled with lighter and darker splotches and having a noticeable whitish spot near the extreme tip.   They are active mainly at night and not much noticed.   Only one generation of larvae is usually abundant in any one community in the North, but in the South there may be 5 to 10 generations in the same locality in one year.

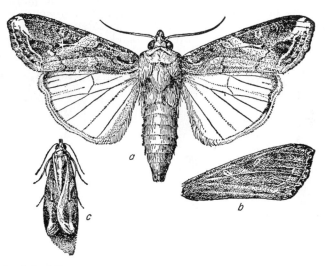

Fig. 9.9. Adult of the fall armyworm.   *a*, Male; *b*, right front wing of female, about twice natural size; *c*, moth in resting position, about natural size.   (*From U.S.D.A., Tech. Bul.* 34.)

Control Measures.   In favorable seasons a number of parasitic enemies[1] keep the fall armyworm caterpillars down to moderate numbers.   As is generally the case, however, cold, wet springs check the parasites more than they do the insects that they feed upon.   In such seasons, especially, watch should be kept of grassy fields for the appearance of the young worms.   The worms may be controlled by the measures suggested for the armyworm (page 472), using baits or DDT or toxaphene sprays.   The migration of the worms may be largely prevented by applying the insecticides along dust furrows.   The larvae may be prevented from feeding in the ears of sweet or seed corn by applications of DDT as described for the corn earworm (page 500).   After the worms have disappeared, fields in which they have been feeding should be disked or otherwise lightly cultivated, if practicable, to break up the pupae and throw them out on the surface, where natural enemies and weather conditions will destroy many of them.   Keeping fields of cotton and corn free of grass will do much to prevent injury by this insect, since the infestations almost always start among grasses.

[1] Including *Winthemia rufopicta* Big., Order Diptera, Family Tachinidae, and *Apanteles marginiventris* (Cresson), Order Hymenoptera, Family Braconidae.

References. *U.S.D.A., Farmers' Bul.* 752, 1916, *Tech. Buls.* 34, 1928, and 138, 1929; *Jour. Econ. Ento.,* **40**:220–228, 1947; **41**:822–823, 928–935, 1948; and **42**:502–506, 1949.

4 (48, 67, 80). Cutworms[1]

Importance and Type of Injury. There are a great many species of cutworms, and they vary greatly in numbers from year to year. They frequently make it necessary to replant corn, and they destroy from 5 to 50 per cent of the stand of other crops. They injure plants in four principal ways: (*a*) The solitary, surface cutworms eat off the plants just above, at, or a short distance below, the surface of the soil and sometimes drag them to their burrows in the soil. Most of the plant is not consumed, merely being eaten enough to cause it to fall over. Consequently these caterpillars have great capacity for doing damage. Among the important surface cutworms are the black cutworm, the bronzed cutworm, the clay-backed cutworm, and the dingy cutworm. (*b*) The

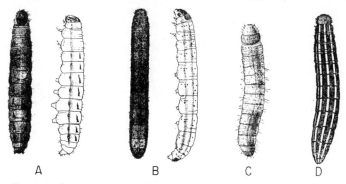

A   B   C   D

Fig. 9.10. Cutworm larva. *A*, spotted cutworm, *Amathes c-nigrum; B*, black cutworm, *Agrotis ypsilon; C*, glassy cutworm, *Crymodes devastator; D*, bronzed cutworm, *Nephelodes emmedonia.* About natural size. (*From Ill. Natural History Surv.*)

climbing cutworms climb the stems of herbaceous plants, vines, shrubs, and trees and eat buds, leaves, and fruits of vegetables and orchard and vineyard crops. The variegated and the spotted cutworms are important species that sometimes assume the climbing as well as the army habit. (*c*) The army cutworms are those which occur in great numbers and, after consuming nearly all the vegetation in one area, crawl along on the ground by the thousands to adjacent fields. They feed largely from the tops of the plants, without cutting them off, but, when abundant, will consume succulent plants clean to the ground. The armyworm (page 471) and the fall armyworm (page 473) are army cutworms that are separately discussed because of their great importance. The army cutworm[2] also has this habit developed to an unusual degree. (*d*) The subterranean cutworms, unlike all the others, remain in the soil to feed upon roots and underground parts of the stems. Consequently they cannot be controlled by the use of poisoned baits. The pale western and the glassy cutworms are important subterranean forms. In most cases the smooth, brownish, greenish, or nearly white well-fed-appearing worms (Fig. 9.10) will be found during the daytime hiding in the soil close to the stems of the plants which they have cut off or fed upon.

[1] Many species of Order Lepidoptera, Family Noctuidae.
[2] *Chorizagrotis auxiliaris* (Grote).

Plants Attacked. Nearly all plants, except those with hard, woody stems are fed upon by cutworms. Some of the crops most seriously injured are corn, beans, cabbage, cotton, tomatoes, tobacco, and clover.

Distribution. Cutworms of various species are of world-wide distribution. Certain species are confined largely to southern, and others to northern, climates. Some prefer dry conditions, while others are most abundant in wet areas or overflowed land.

Life History, Appearance, and Habits. The majority of cutworms pass the winter in the partly grown to fully grown larval stage. Some, however, hibernate as adults, and others as pupae, in the soil. In typical cases, the worms remain as small larvae in cells in the soil, under trash or in clumps of grasses during the winter. They start feeding in the spring and continue growth until early summer, when they change in the soil to a brown pupal stage and later to the adult or moth stage (Fig. 9.11). With most of our common species, there is but one generation a year; a few species have two to four generations; and, in others, the generations are so broken up that adults may be found at almost any time from late spring to midautumn. The eggs of most species are laid on the stems of grasses and weeds or behind the leaf sheath of such plants. Certain of the moths, notably the black cutworm,[1] lay their eggs on low spots in the field, or land that has been subject to overflow. Certain others lay their eggs on the bare ground, or on ground that has been somewhat packed by the passage of vehicles or animals. The egg stage commonly lasts from 2 days to 2 weeks. The larvae in most cases remain below the surface of the ground, under clods, or other shelters during the day and feed at night. The time required to grow from newly hatched caterpillars, about $\frac{1}{25}$ inch long to nearly 2 inches long, varies from 2 weeks to 5 months. They then dig down several inches in the soil where they make cells in which they pupate from 1 to 8 weeks or over winter. The adults upon emerging crawl out of the ground through the tunnel made by the larva going down. So far as their life cycles are concerned, the commoner cutworms fall into two distinct groups. (*a*) Those with a single generation a year are mostly northern in distribution and winter as larvae, regardless of the latitude or length of season where they live. They are held to a single generation by the remarkable fact that, while all other stages are accelerated by higher temperatures, the *prepupal stage* (from the time the full-grown larvae cease feeding until they pupate) is delayed by higher temperatures thus compensating for the longer season in the South and preventing additional generations. (*b*) Those which undergo several generations a year are most abundant in the South and winter as pupae.

The abundance of a given species from year to year is greatly affected by rainfall, which may prevent the moths from laying their eggs or, by flooding the soil, may force the larvae up to the surface during the daytime so that their parasites destroy nearly all of them. From the several dozen destructive species a few of the worst are briefly characterized to show something of the range of appearance and habits in the family. There are many other species of nearly equal importance.

The black cutworm[1] is a cosmopolitan species with a restless, pernicious habit of cutting off many plants while satisfying its appetite. It lays its eggs singly or a few together on the leaves or stems of plants often in low or overflowed land. The winter

[1] *Agrotis ypsilon* (Rottemburg).

is spent in the larval or pupal stage, and there are two generations a year in Canada, and four in Tennessee. The larva is greasy gray to brown above, with faint lighter stripes. The skin has strongly convex, rounded, isolated granules of large and small size.

The dingy cutworms[1] are northern species which winter as partly grown larvae and have but one generation a year. The larvae are dull dingy brown with a broad buff-gray dorsal stripe, subdivided into triangular areas on each segment and margined by a narrow dark stripe on each side. The skin granules are round, coarse, isolated, and slightly convex. These species are said to be very resistant to drought, and they sometimes assume the climbing habit. The eggs are laid singly or a few together.

The bronzed cutworm[2] (Fig. 9.10,D) is one of the most strikingly marked of all species, the larva being dark bronzy brown, striped from head to tail with five, clear-cut pale lines about half as wide as the brown between them. One stripe runs down the middle of the back and one below the spiracles on either side of the body. The skin is granulate. This is a northern species, especially troublesome on corn, grains, and grasses. It winters as partly grown larvae and has a single generation a year.

Fig. 9.11. Adult of the clay-backed cutworm, *Feltia gladiaria*, slightly enlarged. (*From Ill. Natural History Surv.*)

The variegated cutworm[3] is found throughout most of the cultivated parts of the earth. It is said to have destroyed $2,500,000 worth of crops in the United States in a single year. It shows some preference for garden crops and the foliage, buds, and fruits of trees and vines, tobacco, ornamentals, and greenhouse plants. The eggs are laid in bare patches of 60 or more on stems or leaves of low plants, on twigs or branches of trees, or on fences and buildings. It may complete 3 or 4 generations a year, wintering mostly as pupae. The larva has a distinct pale yellow dot on the mid-dorsal line of most of the segments and frequently a dark W on the eighth abdominal segment. The skin is smooth, and the general color ashy or light dirty brown, lightly mottled with darker brown.

The spotted cutworm[4] is a troublesome species of general distribution throughout North America, Europe, and Asia, though scarce in the South. It shows a preference for garden crops. It winters as large larvae and undergoes two or three generations a year. Eggs are laid either singly or in rows or patches of a hundred or more mostly on leaves. On the posterior half of the larva, each segment has a pair of elongate wedge-shaped black dashes on the upper side, which increase in size and lie closer together posteriorly. There is also a dark stripe through the spiracles. The skin is smooth.

The army cutworm[5] is a western species, adapted to arid conditions. It is a surface feeder, burrowing but little, and often assumes the army habit. It is said to have destroyed 100,000 acres of winter wheat in 1 year in Montana alone. It has one generation a year, wintering as half-grown larvae. The eggs are laid singly in or upon the soil. The larvae are pale greenish gray to brown with the back pale-striped and finely splotched with white and brown but without prominent marks. The skin is covered with fine close-set pavement-like granules.

[1] *Feltia subgothica* (Haworth) and *Feltia ducens* Walker.
[2] *Nephelodes emmedonia* (Cramer).
[3] *Peridroma saucia* (Hübner) [= *margaritosa* (Haworth)].
[4] *Amathes* (= *Agrotis*) *c-nigrum* (Linné).
[5] *Chorizagrotis auxiliaris* (Grote).

The pale western cutworm[1] is an underground feeder which has destroyed millions of dollars worth of small grains, beets, and alfalfa in the western half of the United States and Canada. The body is grayish, unmarked by spots or stripes; the skin has fine, flat, pavement-like granules. This species lays its whitish spherical eggs singly or a few together on the soil. The larvae hatch during warm periods in winter or very early spring, and the single generation of larvae have usually completed feeding by the end of June.

The glassy cutworm[2] is widespread except in the more southern states. It is a strictly subterranean species which prefers sod and is most troublesome to crops following sod in low ground. There is one generation a year, the species wintering as small larvae. The body is greenish white like a grubworm, uncolored except for the reddish head and cervical shield, and has a somewhat translucent or glassy appearance. The skin is not granulated.

The yellow-striped armyworm or cotton cutworm[3] is a day-feeding species that has been very destructive to cotton by devouring the young plants as well as by boring into squares and bolls, but it is a general feeder on many crops. The female lays her eggs in masses on foliage, trees, or buildings and covers them with scales from her body. The species winters in the pupal stage, has several generations a year, and is most abundant in the South. The larva has a pair of dorsal, triangular, black spots on most of the segments and commonly a bright orange stripe just outside these spots on each side.

The southern armyworm[4] is a climbing cutworm which is a pest of vegetable crops in the South. The eggs are laid in irregular masses and are covered with whitish hairs from the female moth. The full-grown larva is dark gray to nearly black in color and is marked with median, subdorsal, and lateral yellow stripes. There are four or more generations a year.

Control Measures. The species of cutworms which attack the plant above or at the surface of the ground, including the climbing cutworms, may be controlled very effectively by the use of the poison-bran bait, described on page 323. Spraying or dusting with DDT or toxaphene as suggested for armyworms (page 473) is effective, but these should not be applied to forage intended for livestock. However, in many instances these materials may be applied to the soil before planting. The eggs of many species of cutworms are laid very largely in grasslands. One of the best methods of avoiding damage by these insects is to rotate the crops in such a manner that corn is not planted on sod ground unless such sod has been broken early in the fall or during late summer. Summer plowing, before the eggs are laid, and fallowing until frosts occur are of value against all species which lay their eggs upon low-growing vegetation. Ditches and dusty furrows with postholes are of value in checking the advance of the army species. For the climbing cutworms, bands of tanglefoot about the trunks of trees and grapevines will give protection. Where the underground species are abundant, a special study of conditions will have to be made, as no general recommendations will apply.

Cutworms are subject to attacks by other insects, especially by certain flies which lay their eggs on the backs of the worms, and by ground beetles. They are readily fed upon by many species of birds, and the eggs are attacked by certain small wasp-like parasites.

References. CROSBY, C., and M. LEONARD, "Manual of Vegetable-garden Insects," pp. 260–301, Macmillan, 1918; *Can. Dept. Agr., Div. Ento. Bul.* 3, 1912; *Jour. Agr. Res.,* **46**:517–530, 1933; *U.S.D.A., Tech. Bul.* 88, 1929, and *Cir.* 849, 1950; *Mont.*

[1] *Agrotis orthogonia* Morrison.
[2] *Crymodes* (= *Sidemia*) *devastator* (Brace).
[3] *Prodenia ornithogalli* Guenée.
[4] *Prodenia eridania* (Cramer).

Agr. Exp. Sta. Bul. 225, 1930; *Jour. Econ. Ento.*, **41**:631, 655, 1948; *Calif. Agr. Ext. Cir.* 146, 1948.

5. Corn Leaf Aphid[1]

Importance and Type of Injury. Corn infested by this insect shows numerous greenish or greenish-blue aphids in the curl of the leaves and upper parts of the stalk. Leaves are sometimes entirely covered with these aphids. Winged and wingless individuals will be found during the summer. Infested corn leaves are frequently mottled with yellowish or reddish-yellow patches. It is more destructive in the South. By feeding on the tassel and silk and coating them with honeydew, it may seriously interfere with pollinization of corn. This honeydew may attract great numbers of corn earworm moths to the ears and result in increasing the infestation by that pest. The feeding of the aphids causes a discoloration of the brush of broomcorn. It is a serious pest of fall and winter barley in the southwestern states, sometimes weakening these crops to such an extent that very little grain is produced. It is said to have reduced the yield of grain sorghums in western Kansas by one-third. It is the disseminator of a serious mosaic disease of sugarcorn and one of the vectors of the yellow dwarf virus of barley.

Plants Attacked. The insect has been found on corn, barley, sugarcane, millet, broomcorn, sorghums, Sudan grass, and many other wild and cultivated plants of the grass family. It shows a preference for sorghums. It winters on barley in the Southern states and California.

Distribution. The insect is common throughout the corn-growing areas of the United States and Canada, being more abundant in the South. Its range extends throughout the tropical and temperate regions of the world.

Life History, Appearance, and Habits. Our knowledge of the life history of this insect is incomplete. In the North Central states, it appears in cornfields about midsummer. In the South the insect multiplies rapidly and does its greatest damage in the winter months. Of the females, only the winged and wingless ovoviviparous forms are known. Males have been noted only very rarely and the egg-laying true females have never been found. No observations have been made on the winter stages in the northern states, and it is not known whether this species passes the winter in the egg stage in this section or whether it migrates up from the South during the spring and early summer. The female aphids cluster in large numbers on the plants, sometimes almost entirely covering the leaves. Some winged females are present throughout the summer. The number of ovoviviparous generations produced in a year varies from about 9 in central Illinois to as many as 50 in southern Texas. The insects feed until they are killed by a heavy frost, or the drying up of their food plants.

Control Measures. The damage which this insect causes can be largely prevented by early planting of the crops and by proper tillage and fertilization to hasten their growth and maturing. Pasturing infested winter barley is recommended to free the crop of this pest. See also the general measures suggested for aphid control (page 613).

References. *Twenty-third Rept. Ill. State Ento.*, p. 123, 1905; *U.S.D.A., Tech. Bul.* 306, 1932; *Jour. Econ. Ento.*, **53**:924, 1960.

6 (31). Chinch Bug[2]

Importance and Type of Injury. For more then 150 years chinch bugs have caused serious losses to American agriculture. Webster estimated

---

[1] *Rhopalosiphum* (= *Aphis*) *maidis* (Fitch), Order Homoptera, Family Aphidae.

[2] *Blissus leucopterus* (Say), Order Hemiptera, Family Lygaeidae. The hairy chinch bug, *Blissus leucopterus hirtus* Montandon, predominantly a short-winged form, is sometimes abundant in turf in the northeastern United States, killing the grass in spots. Short cutting, frequent watering, and top dressing help to prevent destruction of the grass, and rotenone, dieldrin, or nicotine dusts or sprays will kill the bugs.

Andre has also described what he believes to be a distinct species, *Blissus iowensis* Andre, which is also predominantly short-winged and shorter haired than *B. leucopterus hirtus,* and requires much longer to develop than *B. leucopterus.*

that the insect caused a total damage in this country, from 1850 to 1900, of $350,000,000.   In 1934 they caused an estimated loss of over $40,000,-000 in Illinois alone.   The first indication of the presence of chinch bugs in the field will often be the wilting and drying out of the infested corn. Usually this occurs on the side of the field next to small grains.   Occasionally injured plants may appear in any part of the field.   Small black-and-white to gray-and-white or red insects will be found behind the sheaths of leaves or in the soil about the base of the plant, often with slender beaks inserted in the plants from which they suck the sap.   At small-grain harvest, hordes of these insects will be found crawling over the ground from cut grain into fields of corn or other growing grass crops. In the East chinch bugs often severely damage lawns and golf greens.

Plants Attacked.   The insect feeds only on the plants belonging to the grass family.   This includes all of our cultivated and wild grasses, corn, and small grains.

Distribution.   The chinch bug has been found throughout the United States, in southern Canada, in Mexico and in Central America.   Its areas of greatest destructiveness are in the Mississippi, Ohio, and Missouri River Valleys.

Life History, Appearance, and Habits.   The chinch bug (Fig. 9.12) hibernates only in the adult stage, the full-grown insect being about $\frac{1}{6}$ to $\frac{1}{5}$ inch in length, with a black body.   The white wing covers are each marked with a triangular black patch at the middle of their outer margins. The legs are reddish to reddish yellow.   The insect gives off a vile odor when crushed, that is somewhat distinctive and always remembered by one who has smelled it.   They hide away in almost any kind of shelter during the winter; but in the Middle West few of them remain in corn-fields, and they are found chiefly along the south side of hedgerows, bushy and grassy fence rows or roadsides, and the south and west edges of wood-lands.   They have been found in large numbers in soybean stubble, where the beans have been harvested with a combine, and they occur in considerable numbers in the underground nests of field mice.   Where the clump-forming native prairie grasses are present, the bugs seem to prefer such clumps for winter quarters.   They have been taken, however, in a great variety of different shelters.   As many as 5,000 bugs may frequently be found on a square foot of surface in favorable hibernating places. The adult chinch bugs remain in their hibernating quarters until the temperature reaches a point above 70°F. for several hours during which the sun is shining.   They may move about on warm days earlier in the spring, and occasionally mating takes place before they leave their hibernating quarters.   When the above-mentioned temperature is reached on sunny days, they crawl up the stems of grasses or other plants about their hibernating quarters and take flight, usually going to fields of small grain.   Here they feed by sucking the sap from wheat, rye, oats, or barley, and the females deposit their eggs behind the "boots" of the lower leaves or, if the ground is loose, upon the roots.   The insects mate repeatedly, laying a few eggs each day for 3 weeks or 1 month, an average of about 200 eggs being laid by each female.   The eggs are nearly cylindrical, are three or four times as long as broad, are yellow, and have four short nipple-like projections on the cap at the head end.   These eggs hatch into small, very active, reddish bugs with a band of white on the back just behind the wing pads.   They become dark as they grow older and, at the last molt, acquire full-sized wings.   The bugs require about 30 to 40 days

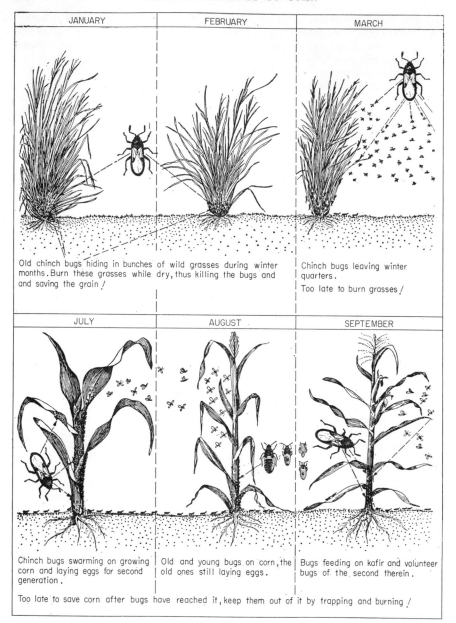

JANUARY	FEBRUARY	MARCH

Old chinch bugs hiding in bunches of wild grasses during winter months. Burn these grasses while dry, thus killing the bugs and and saving the grain !

Chinch bugs leaving winter quarters.
Too late to burn grasses !

JULY	AUGUST	SEPTEMBER

Chinch bugs swarming on growing corn and laying eggs for second generation.

Old and young bugs on corn, the old ones still laying eggs.

Bugs feeding on kafir and volunteer bugs of the second therein.

Too late to save corn after bugs have reached it, keep them out of it by trapping and burning !

Fig. 9.12. Chart showing seasonal history of chinch bug, *Blissus leucopterus*, in the central states. During the winter, from December to February, bugs are hibernating at the edges of woodland, under fallen leaves, and in bunches of grass and other shelters. In February or March, flight to young grain begins and continues until about the middle of May. In June and July, the bugs crawl in great numbers from ripening wheat to young succulent

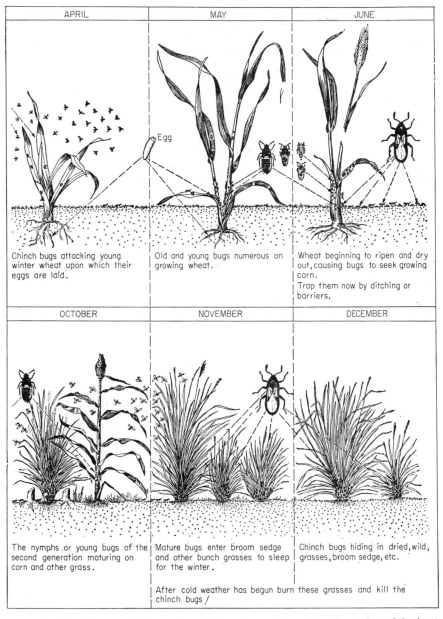

APRIL	MAY	JUNE

Chinch bugs attacking young winter wheat upon which their eggs are laid.

Old and young bugs numerous on growing wheat.

Wheat beginning to ripen and dry out, causing bugs to seek growing corn.
Trap them now by ditching or barriers.

OCTOBER	NOVEMBER	DECEMBER

The nymphs or young bugs of the second generation maturing on corn and other grass.

Mature bugs enter broom sedge and other bunch grasses to sleep for the winter.

After cold weather has begun burn these grasses and kill the chinch bugs.

Chinch bugs hiding in dried, wild, grasses, broom sedge, etc.

corn and at that time may be trapped in ditch or chemical barriers. Becoming adult about August 1, they scatter among the corn plants, where eggs are laid and the second generation is begun. When these become adult from September 23 to the early part of November, they find hibernating quarters, where they remain until spring. (*After U.S.D.A., Farmers' Bul.* 1498.)

to complete their development.   This usually does not occur until after small-grain harvest, especially that of wheat.   They are dependent for their food supply on the sap of growing grass plants.   It is, therefore, necessary for them to leave the dried stubble field when grain is cut.   As they are still wingless, they usually migrate on foot to fields of corn, oats, or grasses, where they complete their growth.   The adult stage is reached during the early part of the summer to midsummer.   The adults may remain in the situations where they are feeding but usually fly for a few days after reaching the adult stage.   Mating again takes place, and the eggs of a second generation are deposited on corn or grasses.   The first-generation adults die by mid-September, and the second-generation nymphs complete their growth by the approach of cold weather.   In parts of the Southwest, as in Oklahoma and Texas, a partial third generation often occurs, the first generation becoming adult about the time small grains mature.   Migration to corn and sorghums, therefore, takes place on the wing and creosote barriers are not practicable.   During the warm sunny afternoons of early fall, they fly from the cornfields to their winter quarters.   As they are seeking warm, sheltered places at this time, most of them will congregate on the south and west sides of the situations which afford them shelter during the winter.

Two forms of the chinch bug occur.   In the more common form the adult has black-and-white well-developed usable wings.   The less common form is sometimes considered a distinct species, the hairy chinch bug.[1]   It occurs more generally in the northeastern states and at more northern latitudes, feeds more on grasses, often becoming a pest in lawns and greens, and does not have the pronounced migration from one food plant to another that occurs in the case of the long-winged form in the large grain-growing sections.

Control Measures.   As the chinch bug feeds only on plants of the grass family, the growing of nongrass crops is of great value during years of chinch bug outbreaks, not only because these crops will not be injured in the least by the bugs, but also because the larger the area in such crops, the less will be that in which the chinch bug will find feeding and breeding places.   Neighbors should cooperate in planning rotations so that corn is not planted adjacent to fields of winter or spring wheat, barley, or oats.

It has also been found possible to reduce the injury by chinch bugs to corn by planting a strong-growing legume crop such as soybeans or cowpeas in the field of corn.   These plants are not of themselves repellent to the chinch bugs, but, by producing a dense shade around the base of the corn plants, they give a condition which is unfavorable to the bugs and which they avoid.   Chinch bugs are primarily sun-loving insects and always seek the thinner parts of fields or poorer stands of any of the crops on which they feed.   Certain varieties of corn and grain sorghums have been found very resistant to the attacks of second-generation chinch bugs, but none chinch bug-proof.

Winter burning over the hibernating quarters has been of value in the areas west of the Mississippi River, where the insects winter mostly in bunch grasses.   This practice is of little value in other areas.   In many cases it should be discouraged as causing more harm than good.   Probably this measure never destroys more than 25 to 50 per cent of the bugs over any large area.   However, it is the last opportunity to prevent dam-

---

[1] *Blissus leucopterus hirtus* Montandon.

age to spring wheat, barley, and oats, and it is often recommended that wasteland, roadsides, ditch banks, and the margins of woodlands in which a dozen or more bugs are found per square foot be burned over with a backfire, if this can be done without endangering property.

One of the most effective methods of combating the chinch bugs is trapping them at the time of small-grain harvest when they are traveling on foot from small-grain fields to fields of corn or other growing grass crops. This may be done by constructing a barrier line around the margin of the small-grain field along which the bugs can be stopped and killed or trapped. One of the effective barriers is made by pouring a narrow line of crude creosote (Fig. 9.13) along the brow of a smooth ridge thrown up with the plow around the margin of the infested small-grain field. The creosote should be poured on the side of the ridge next the small grain so the bugs will be climbing the ridge as they approach it. A strip of creosote making a line 1 inch wide on the soil is sufficient to turn the bugs, as they are strongly repelled by the odor of this chemical. Daily applications of the creosote are necessary for a period of 10 days to 2 weeks; 50 gallons of creosote are usually sufficient to maintain $\frac{1}{4}$ mile of barrier for a season. Postholes 18 inches to 2 feet deep may be dug on the inner or small-grain side of this line, the tops of these holes flared and dusted. The bugs may be caught in such holes by the bushel as they travel along the creosote line seeking a place where they can escape from the field. Dusting of the top of the holes makes it impossible for the bugs to obtain a foothold upon it, and they roll into the holes. A small amount of kerosene or 1 to 2 tablespoonfuls of calcium cyanide flakes or 4,6-dinitro-o-cresol dust poured into the holes will kill all the bugs contacted. Paper barriers have come into use especially in gravelly or gumbo-soil areas. These are constructed by plowing a shallow furrow and digging postholes as for the creosote line. Tarred felt (not asphalt) or red rosin paper strips about 4 or 5 inches wide and thoroughly soaked in creosote are placed against the steep edge of the furrow and the soil banked against the paper so as to hold it erect and projecting 2 or 3 inches above the soil (Fig. 9.14). Additional creosote is applied against the side of the paper strips every 2 or 3 days. Advantages claimed for the paper strips are that they require only about half as much creosote, that bugs are not blown across them by wind as sometimes happens with the creosote line on the soil, and that in certain soils they are much easier to make bugproof than the earlier type of barrier.

The use of toxic-dust barriers has in many cases superseded the creosote types. Such a barrier consists of a shallow groove on smooth, tilled ground between corn and small grain, filled with a continuous band of 2 to 8 per cent 4,6-dinitro-o-cresol dust (DNOC). One-quarter to one-half pound of dust is applied per rod of barrier. The dust barrier must be renewed after rain or winds which destroy its effectiveness. The DNOC dust can also be used in conjunction with the creosote line and treated paper barriers to kill the trapped bugs. The use of the toxic-dust barrier, although more expensive than the creosote barriers, has the decided advantage of killing all the bugs contacting the dust.

Superior control of migrating chinch bugs has been obtained by spraying barrier strips 4 rods (66 feet) wide with dieldrin at 0.5 pound per acre, half on the edge of the small grain field and half on the corn field, with a strip a few rods long and 2 rods (33 feet) wide at right angles on each end. Two or more applications at 1- to 2-week intervals may be necessary.

Fig. 9.13. Barriers for protecting corn from migrating chinch bugs. The large photo shows a barrier of creosote poured on the ground at the brow of a ridge made by throwing a furrow toward the field to be protected. Note that the bugs, which came out of a grain field at the left, had already killed several rows of corn, while the corn on the right has been protected. The insert shows a toxic-dust barrier of 4,6-dinitro-*o*-cresol dust laid in a shallow groove. (*From Ill. Natural History Surv.*)

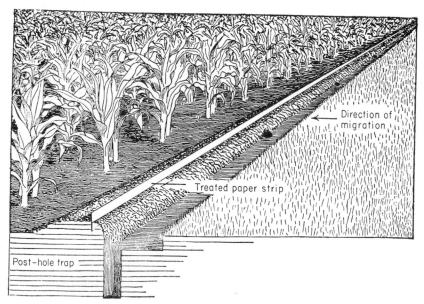

Fig. 9.14. The paper-strip barrier for the control of chinch bugs. Tarred paper, saturated with creosote and cut into strips about 4 inches wide, is set on edge and buried half in the soil at the summit of a furrow thrown toward the corn. This presents a barrier wall in addition to the repellent odor of the creosote and prevents bugs being blown across the barrier. (*From Ill. Natural History Surv.*)

Grain treated in this manner should not be harvested for 7 days, and straw or fodder should not be used for animal forage.

Where large numbers of chinch bugs are concentrated in a few outer rows of corn, spraying or dusting the base of the plants where the bugs congregate with dieldrin at 0.25 pound, toxaphene at 2 pounds, lindane at 0.25 pound, or parathion at 0.25 pound per acre will give practical control. However, forage should not be fed to livestock.

References. Kans. Agr. Exp. Sta. Bul. 191, 1913; U.S.D.A., Farmers' Buls. 1498, 1934, and 1780, 1937; Iowa Agr. Ext. Cir. 199, 1934; Ill. Agr. Exp. Cir. 431, 1935; Ill. Natural History Surv. Bul. 19, Art. 6, 1932; Ill. Agr. Exp. Sta. Cirs. 590, 1945, and 707, 1953; Jour. Econ. Ento., **36**:658–661, 1945; and **38**:283, 391, 713–714, 1945.

7 (40). BILLBUGS[1]

Importance and Type of Injury. Billbugs injure plants in two important ways: (*a*) The adult snout beetles eat small holes in the stalk of the seedling plant at about ground level, feeding on the tender leaf tissue in the center. The center leaves or buds wilt, and excessive suckers or sprouts may form, so that the plant becomes stunted, deformed, and unproductive. Heavy feeding may kill seedling plants, and two or three replantings of corn are sometimes almost totally destroyed. Plants which survive early attack show typical transverse rows of holes across the leaves (Fig. 9.15) as a result of punctures in the developing leaves curled in the heart of the plant. The perforated leaves often fall or twist into a curl so as to interfere with the growth of following leaves. (*b*) The second phase of injury results from the feeding of the larvae or grubs. The larvae of the maize billbug and the southern corn billbug tunnel in the cornstalk, causing further stunting of the plant. A number of species feed upon the

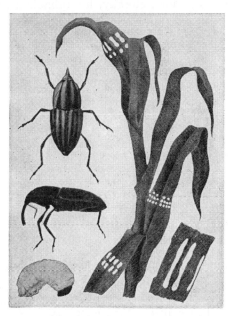

FIG. 9.15. Billbugs; adults, larva, and characteristic injury, from adults feeding on corn. (*From Ill. Natural History Surv.*)

fibrous roots of small grains and cultivated grasses, in timothy corms, and in the stems of small grains, causing the heads to bleach and the straw to fall or lodge.

Plants Attacked. While corn is attacked most conspicuously by most of the adult billbugs, nearly all the cultivated and wild grasses, the small grains, rice, peanuts, reeds, rushes, and cattails are also hosts for the grubs and adults of various species.

Distribution. Taken together, the billbugs may be said to be present throughout the grasslands and cultivated areas of the United States and

[1] A number of species of *Calendra* (= *Sphenophorus*), Order Coleoptera, Family Curculionidae.

Canada.   Their destructiveness has been most apparent from the Great Plains eastward and that of the maize billbug and southern corn billbug in the southern states.

Life History, Appearance, and Habits.   Billbugs usually winter in the adult stage.   The several species range from ⅕ to ¾ inch in length. The head is prolonged into a cylindrical curved snout, about ⅓ to ¼ as long as the rest of the body, at the end of which are small chewing mouth parts.   The body wall and wing covers are very hard.   Corn billbugs seldom fly but may crawl as much as ¼ mile or more in search of food. The beetles "play 'possum" when disturbed and are often so covered with mud that they are practically invisible on the soil, so long as they remain motionless.   Upon coming out of hibernation, the insects feed, mate, and lay about 200 eggs over a 2-month period.   For each whitish kidney-shaped egg laid, a small hole is gouged out in the stem of the host plant with the mouth parts.   The tiny grubs hatch in 4 to 15 days.   They are white, short, chunky, humpbacked grubs, without legs, and with a distinct, harder, brown or yellow head.   The larvae feed and grow for several weeks, eating out the pith of the stem and, if this becomes exhausted, descending to the soil to complete growth by feeding upon the fibrous roots.   The larvae pupate either in the stems of plants or in the soil among the roots.   The adults transform in the fall and may remain within the pupal cell over winter, or they may emerge and feed for some time before entering hibernation.   In general there is one generation a year, but in the warmer parts of the country they winter in various stages.

Among the most important species are the following:

The maize billbug[1] is ⅖ to ⅗ inch long, rather broad-bodied, reddish brown or black, but often so covered with mud that the ground color is invisible.   The raised longitudinal lines on the wing covers run about two-thirds of their length.   The attacks of the adults upon young corn plants cripple them severely, causing excessive suckering and killing small plants outright.   The larvae feed in the pith of the corn stalk for 40 to 50 days from early June to September.   They pupate in August and early September, always in the larval tunnel in the upper part of the taproot where they usually hibernate until the soil temperature reaches 65°F.   Adults that emerge early leave the pupal cells and find winter shelter in coarse grasses or other litter about the fields, but those transforming later winter in the base of the stubble.   The larval injury also stunts and sometimes kills the plants.   Corn is the principal food of both larvae and adults, although the adults feed on almost any of the large swamp grasses.

The southern corn billbug or curlew-bug[2] averages ⅜ inch long and is naturally brown with golden reflections, the more elevated bumps being polished black.   Each wing cover has, besides the longitudinal lines, a prominent dent near the base and an elevation or hump near the tip.   Like the maize billbug, this species feeds in the larval stage in the base of the stalk and the taproot of corn.   It sometimes destroys entire fields of corn and, in certain years, in the South Atlantic states, may do more damage than all other corn insects combined.   Besides corn, it injures rice and peanuts and, after the corn is older, lays its eggs in nut grass or chufa.   When pupating in corn the pupae are so low down in the root that they usually remain in the soil when stalks are pulled out by hand.

The remaining species do not develop as larvae in the stalks of corn but often attack it destructively in the adult stage.   The bluegrass billbug[3] is about ¼ inch long, the body heavily and evenly marked with rounded punctures on the upper surface and tapering strongly to the tip of the abdomen.   The grubs excavate the stems and corms or eat the rootlets of timothy, bluegrass, redtop, and all the small

---

[1] *Calendra maidis* (Chittenden), Order Coleoptera, Family Curculionidae.

[2] *Calendra callosa* (Olivier).

[3] *Calendra parvulus* (Gyllenhal).

grains. Satterthwait considers this the most destructive billbug, because of the reduction it causes in hay and pasture crops and the premature failure of sod lands. It often dwarfs corn crops planted on spring-plowed old sod. The larval period is only a little over 3 weeks, and pupation occupies about 8 days more, in the corms or in the soil. Adults may emerge 45 days after the eggs were laid.

The bluegrass billbug prefers upland, while the timothy billbug[1] works in timothy in the larval stage, in a similar way, but prefers lower ground. The adult is larger than the bluegrass billbug, nearly $\frac{3}{16}$ inch long, and the punctures on the prothorax form a distinct pattern. This species is believed responsible for many of the early failures of stands of grass in meadows. The adults attack corn, causing excessive suckering and failure to form ears.

The clay-colored billbug[2] is more than $\frac{1}{2}$ inch long, buff-colored, with fine punctures on the upper surface. The adults usually kill corn plants when they feed upon them. Sometimes they feed upon the kernels of wheat in the unripened heads, as they do also on millet and wild grasses. The grubs normally develop in the "nuts" at the ends of the roots of rushes and on sedges and reeds.

The cattail billbug[3] is a large species, about $\frac{5}{8}$ inch long, with pale yellow velvet-like hairs in the pits on its back. Normally breeding in cattail flags and bur reed, this species may be disastrous to corn, when it is planted in low infested wasteland. The adults attack the stalks below the surface of the ground, causing the plants to be dwarfed and unproductive.

Control. Crop rotations in which corn does not follow corn are of great help in eradicating the maize billbug and the curlew-bug. Chufas and other sedges must also be eliminated in order to check the curlew-bug. If cornstalks are plowed out and dragged with a spring-tooth harrow to remove the dirt from the taproots, after cold weather has begun, the hibernating beetles of these species will die. Merely raking and burning the stalks is only partly effective, especially for the curlew-bug, since the stalks break off, leaving the taproot containing the beetles in the soil. For the bluegrass, timothy, and clay-colored billbugs, crop rotations in which corn is not planted on ground that has been in sod or on reclaimed swampland where reeds, rushes, or cattails grew, during the preceding season, are of chief importance. Such land should be planted to soybeans, cowpeas, cotton, potatoes, flax, and other crops not known to be attacked by billbugs, the first year after breaking. Proper drainage is generally of great benefit. Fall plowing, followed by clean cultivation to keep the seedbed in perfect condition until the soil is warm and there is sufficient moisture to insure quick germination before planting, will go far to prevent serious losses from these pests.

Billbugs are so well protected during their life cycle that parasites and predators are not important factors in control. However, the fungus *Beauveria bassiana* often causes considerable mortality. Where crop rotation is impractical, chemical control of the adult billbugs by soil application of aldrin or heptachlor at 2 pounds per acre has given adequate protection of the stand of corn and decreased the number of damaged plants. The treatments have been broadcast before planting and disked into the soil or applied as granulars after planting.

References. *Sixteenth Rept. Ill. State Ento.*, 1890; *Twenty-second Rept. Ill. State Ento.*, 1903; *U.S.D.A.*, *Farmers' Bul.* 1003, 1932, and *Bur. Ento. Bul.* 95, Parts II and IV, 1911, 1912; *Kans. Agr. Exp. Sta.*, *Tech. Bul.* 6, 1920; *N.C. Agr. Exp. Sta.*, *Tech.*

[1] *Calendra zeae* (Walsh).
[2] *Calendra aequalis* (Gyllenhal).
[3] *Calendra pertinax* (Olivier).

Bul. 13, 1917; *S.C. Agr. Exp. Sta., Bul.* 452, 1957; *Ga. Agr. Exp. Sta. Mimeo.* 93, 1960; *Jour. Econ. Ento.*, **50**:707, 1957.

8. EUROPEAN CORN BORER[1]

Importance and Type of Injury. The presence of the European corn borer is often indicated by cornstalks with the tassels broken or bending over. Sometimes the stalks are so heavily infested that they break over at various points and collapse. Other indications of its attack are small areas of surface feeding on the leaf blades, with fine sawdust-like castings on the upper sides of the leaves or stalks; small holes in the stalks,

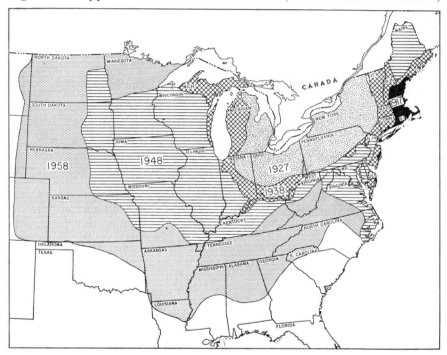

FIG. 9.16. Map showing the known distribution of the European corn borer in the United States. The figures indicate the spread of the borer from its discovery near Boston, Mass., in 1917, through its maximum known distribution in 1927, 1938, 1948, and 1958. (*Original.*)

often with slimy borings protruding from the holes; worms boring through the stem and along the entire length of the ear and cob; numerous flesh-colored, inconspicuously spotted caterpillars, from ½ to 1 inch long, boring in all parts of the stalks. This insect has caused the complete loss of crops of early sweet corn.

Plants Attacked. Nearly all herbaceous plants large enough for the worms to enter. In the single-generation areas of the Middle West and in Ontario the corn borers feed almost entirely on corn, although some of the common weeds, vegetables, flowers, and field crops are often found infested when such plants are grown in close proximity to badly infested cornfields. In the area in New England where the corn borer has two

[1] *Pyrausta nubilalis* (Hübner), Order Lepidoptera, Family Pyralididae. *U.S.D.A., Tech. Bul.* 59, 1928.

generations a year, the insect feeds extensively on potatoes, beans, beets, celery, dahlias, asters, chrysanthemums, gladioli, and many weeds, as well as corn, even when such plants are not associated with corn. It has been found feeding on more than 200 kinds of plants.

Distribution. This insect is distributed over the greater part of Europe and parts of Asia. While the insect has been known to be present in North America only since 1917 and was probably introduced into this country in shipments of broomcorn from Italy or Hungary some time about 1908 or 1909, it has shown itself capable of being one of the most destructive insect pests of corn. By 1952, the insect had spread over practically all the major corn-growing areas of the United States (Fig. 9.16) and in 1949 alone it was estimated to have caused a loss of 313,000,-000 bushels of corn and a total damage in excess of $349,000,000.

Life History, Appearance, and Habits. This insect passes the winter in the form of a full-grown worm or caterpillar (Fig. 9.17), in the stems of

Fig. 9.17. The larva of the European corn borer, greatly enlarged. It is easily confused with the smartweed borer, which is native to the United States. In the European corn borer the two spots on the back of each abdominal segment are usually much smaller than in the smartweed borer, and more widely separated, their distance apart usually exceeding the width of one spot. In the native species, the distance between the spots is usually less than the width of one spot. A faint stripe can usually be seen on the middorsal line of the European corn borer. *(From Ill. Natural History Surv.)*

the food plants on which it has been feeding. These worms are from $\frac{3}{4}$ inch to nearly 1 inch in length. The body is flesh-colored, rather inconspicuously marked with small, round, brown spots. The two spots nearest the middle of the back on any segment are farther apart than their own diameter. They may be found in all parts of the stem and ear; but, especially in cornstalks, are most abundant, in winter, just above the ground surface. In the spring of the year, the caterpillar constructs a flimsy cocoon in its burrow, and in this transforms to a smooth, brown pupal stage. The moths (Fig. 9.18) begin emerging during June and continue to come out in the northern states until August. The adult female moths are a pale yellowish brown with irregular darker bands running in wavy lines across the wings. The male moth is distinctly darker, having the wings heavily marked with olive brown. The moths have a wing expanse of about 1 inch. They are strong fliers but move about mainly at night. The females lay their eggs (Fig. 9.19) in groups of 5 to 50 on the undersides of the leaves of their food plants, especially on the lower leaves of young corn plants. Each female will lay, on the average, from 500 to 600 eggs, sometimes many more. The eggs ordinarily hatch in a week or less, depending upon the temperature; the young larvae feed until nearly half-grown in spaces between closely appressed leaves, in the tassel, beneath the husks, or between the ear and the stalk.

When about half-grown, they begin to eat into the stalk, the ear, or the thicker parts of the leaf stem and become borers. They continue to feed in this way until they are full-grown. Dry summers, extremely cold winters, and heavy rains at the time of hatching are very unfavorable to this insect. In dry summers many of the larvae perish before they can bore into the plant.

The large numbers of them which frequently occur in a single plant often cause the plant stem to collapse. One hundred and sixty-seven borers have been taken in a single plant, and 42 have been taken from a single ear of field corn. Entire fields sometimes average 20 borers per plant.

The infestations in the United States apparently did not originate from a single importation. The corn borers predominating in the eastern

Fig. 9.18. Adults of the European corn borer; female at left, male at right. Slightly enlarged. (*From Mich. Agr. Exp. Sta.*)

Fig. 9.19. Egg masses of European corn borer on underside of corn leaves. About natural size. (*From Spencer and Crawford, Ontario Dept. Agr. Bul. 295.*)

and southern portions of the infested area are of a multiple-generation strain or variety in which the number of generations varies with climatic conditions, varying from one generation in the northern areas to three in eastern Virginia, and as many as five on the island of Guam. Typically, this strain matures adults in June and early July and again in August and September. In the single-generation strain which predominates in the North Central States, the adults emerge in June and July and the larvae from their eggs become practically full-grown by September but remain as larvae in their tunnels over the winter. The multiple-generation corn borers are much more destructive, since they feed on many vegetables and flowers as well as corn. In this strain hibernation takes place mainly in the full-grown larval stage.

Control Measures. The corn borer can be greatly reduced in numbers by the destruction or utilization, during the fall, winter, or early spring, of all crop residues and plant refuse in which the borers may pass the winter. The most effective methods of control are *clean* plowing under of all crop and weed refuse or raking together of this plant refuse and burning. The borers are killed in corn that is cut and shredded or placed in a silo. Disking the fields or *ordinary* plowing under of corn-stalks is of little value. Late planting of corn is of some value in pre-venting injury in the one-generation areas. It is desirable to plant hybrid varieties of corn which are resistant to borer attack.

Other measures which may prove of considerable importance in con-trolling this insect are rotations in which a maximum acreage will be in crops not seriously affected. In general the legume crops have suffered very slightly, although some damage has been done to cowpeas. Soy-beans, red clover, and alfalfa have had almost no damage from this insect. If shock fodder is fed to cattle in feed lots, all the refuse stalks should be cleaned up from the lots and burned by early spring, as it has been shown that many borers may come through the winter in such stalks and spread from them to fields planted in corn in the vicinity the next season. Where corn is cut for silage or shredded for fodder, the stubble should not be left over 2 inches high. Machines for cutting the corn close to the ground have been developed to leave only a 2-inch stubble. As corn is the favorite food plant among the cultivated crops, considerable damage to other crops may be avoided in the two-generation area, if corn is planted at some distance from them. This holds particularly true for beets, beans, spinach, rhubarb, celery, and outdoor flowering plants. All weeds should be kept down around roadsides and field margins or burned during the winter. Valuable or valued plants, such as dahlias and other flowers, may be largely protected by growing them beneath tobacco cloth.

At least 24 species of parasites have been introduced into the United States as enemies of the corn borer, and 6 have become established, of which *Lydella stabulans grisescens* (page 69) and *Macrocentrus gifuensis* have become sufficiently abundant to be of value in field control. The fungus, *Beauveria bassiana*, and the protozoan *Perezia pyraustae* also attack the corn borer.

Corn borer injury to field corn may be controlled profitably by appli-cation of granular DDT at 1 pound, endrin at 0.25 pound, toxaphene at 2 pounds or EPN at 0.2 pound per acre. The granules collect on the leaves and roll down into the whorl. Sprays are somewhat less effective. Applications should be made when three-fourths of the plants have first-generation larvae feeding in the whorl and repeated 1 week later if neces-sary. Treatment for the second generation is most suitable when the egg masses average one per plant. Sweet corn may be protected by applications of DDT or toxaphene granulars as above, repeated four times at 5-day intervals after the borers hatch. Forage treated with DDT, endrin, or toxaphene should not be fed to dairy animals or those being finished for slaughter. Ryania at 15 to 20 pounds per acre has been used for control where it is desirable to avoid this residue problem.

References. Mass. Agr. Exp. Sta. Bul. 189, 1919; *N.Y. (Cornell) Agr. Exp. Sta. Ext. Bul.* 31, 1919; *U.S.D.A., Farmers' Buls.* 1548, 1948, and 2084, 1955, *Dept. Bul.* 1476, 1927, and *Tech. Buls.* 53, 1927, and 77, 1928; *Va. Truck Exp. Sta. Bul.* 102, 1939; *Kans. Agr. Exp. Sta. Cir.* 262, 1950; *Minn. Agr. Ext. Ser. Bul.* 257, 1949; *Ill. Agr. Exp. Sta. Circs.* 637, 1949, and 768, 1955; *Conn. Agr. Exp. Sta. Buls.* 462, 1942, and 495, 1945; *Iowa Agr. Exp. Sta. Bul.* P 60, 1944; *Wis. Agr. Exp. Sta. Cir.* 358, 1945; *Jour. Econ. Ento.*, **40**:395, 401, 1947; **42**:88, 1949; **51**:133, 1958; and **52**:49, 1959.

9. STALK BORER[1]

Importance and Type of Injury. Injury by the common stalk borer is usually confined to the margins of the field for 2 or 3 to 20 rows into the field. Occasionally

---

[1] *Papaipema nebris* (Guenée), Order Lepidoptera, Family Noctuidae.

when fields are very weedy the previous year, the injury may extend over the entire field.   Stalks of corn or other plants attacked by this insect will show irregular rows of holes through the unfolding leaves.   Plants will often show an unnatural growth, twisting or bending over, presenting a stunted appearance, and often not producing ears.   Holes will be found in the sides of the stalks with moist castings being thrown out.   Inside the stalks will be found very active dark brown worms ranging from ¾ inch to nearly 2 inches in length.   All but the larger of these worms (Fig. 9.20,*b*) have a single continuous white stripe down the back, with broken white stripes on the sides extending from the head about one-sixth the length of the body, interrupted for one-fourth the length and then starting again and extending over the posterior half of the body.

Plants Attacked.   This insect is almost a universal plant feeder, attacking and working in stems of any plants large enough to shelter it and soft enough so that it can

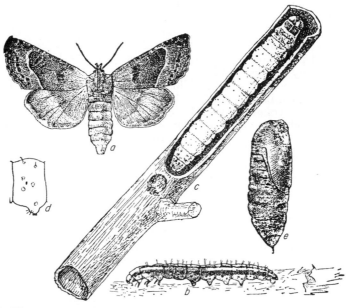

Fig. 9.20. The common stalk borer, *Papaipema nebris.* *a*, Adult; *b*, half-grown larva; *c*, full-grown larva in burrow; *d*, side of one segment of larva; *e*, pupa. All slightly enlarged. (*From Chittenden, U.S.D.A.*)

bore into them.   It apparently prefers giant ragweed and corn in its later stages. It is often a pest in flower and vegetable gardens, where it tunnels in the stalks of dahlias, hollyhocks, tiger lilies, asters, rhubarb, peppers, potatoes, tomatoes, and tobacco.

Distribution.   It is distributed generally throughout the United States east of the Rocky Mountains.

Life History, Appearance, and Habits.   So far as known, the insect hibernates only in the egg stage, the subglobular, whitish to grayish, ridged eggs being laid during the late summer and fall on grasses and weeds, often in the creases of rolled or folded leaves, sometimes over 2,000 from a single female.   They hatch very early in the spring, mostly in May, into the small, brown, white-striped caterpillars which frequently bore first into the stems of grasses, particularly bluegrass, on which the eggs are often laid.   They may enter corn and other plants at the side of the stem and burrow upward, destroying the heart of the plant while the outer leaves remain green, or they may enter near the top of a stem and work downward, causing that part of the plant to wilt and die.   As the caterpillars grow and grass stems become too small for them, they migrate, about July 1, to attack larger-stemmed plants. They are extremely uneasy individuals, never seeming contented with their location, and frequently changing from the stem of one plant to that of another. This habit is responsible for their causing a much greater amount of injury than would be the case if they stayed in one plant. While the giant ragweed seems to be the favorite food of this insect, it will frequently leave these weeds for corn or garden crops, if the two are growing close together. The worms become full-grown during the latter part of July and the first of August, at which time they may lose their striping, and become a plain dirty-grayish color (Fig. 9.20,*c*). They are then about 1 to 1½ inches in length. They transform just below the soil surface or, rarely, inside the stems of their food plants into a brown pupal stage (Fig. 9.20,*e*) and after 2 to 6 weeks, emerge in late August and September as grayish moths (Fig. 9.20,*a*) with a wing expanse of a little over 1 inch. The front wings are usually a dark grayish brown, with a number of small white spots on the disk and along costal and apical margins, or (in the variety *nitela*) without spots and the apical third with diagonal lines or bands of lighter and darker color. The hind wings are a pale gray-brown. There is considerable variation in the appearance of the adults. They mate and the females lay their eggs on grasses and weeds as above mentioned. Eggs have been taken on ragweed, dock, pigweed, burdock, and several grasses. There is only one generation a year.

Control Measures. A thorough cleanup, preferably by burning over the margins of the fields during the fall, and a cleanup and burning of all crop refuse is the best means of fighting this insect. Mowing fence rows about mid-August, just before the adults begin laying eggs, will greatly reduce the number of eggs laid about those fields; but mowing in the early part of the growing season may drive great numbers of the borers into adjacent crops. Individual plants may be saved by splitting the infested stem very carefully lengthwise, removing the borer, and binding the stem together; or by injecting, from an oilcan, a half teaspoonful of carbon bisulfide through the entrance hole and plugging it with clay or gum.

References. *Twenty-third Rept. Ill. State Ento.*, p. 44, 1905; *Rept. N.J. State Ento.* for 1905, p. 584, 1906; *Iowa Agr. Exp. Sta. Res. Bul.* 143, 1931.

10. Southern Cornstalk Borer[1] and Southwestern Corn Borer[2]

Importance and Type of Injury. The southern cornstalk borer is one of the most destructive corn insects in many parts of the South, often responsible for reduction in yields of 15 to 50 per cent; but, because of the insidious method of attack, the damage it does is not generally appreciated. Corn infested by the southern cornstalk borer is usually twisted and stunted, often with an enlargement of the stalk at the surface of the ground. The leaves will sometimes be ragged, broken, and dangling, showing many holes along the leaf which have been eaten out while it was still curled in the heart of the plant. Inside of the stalk, usually well above the ground, will be found dirty grayish-white worms about 1 inch in length when full-grown, conspicuously marked with many dark-brown spots (Fig. 9.21). The southwestern corn borer is equally destructive and often infests 100 per cent of the stalks. It differs from the southern cornstalk borer only in its habit of internally girdling the stalks. It is reported to have caused a loss of $22,000,000 in 1951.

Plants Attacked. The southern cornstalk borer feeds principally on corn. It has also been taken on sorghum and Johnson grass. The southwestern corn borer feeds largely on corn but infestations have been observed in sorghum, sugarcane, broomcorn, Sudan grass, and Johnson grass.

[1] *Diatraea crambidoides* (Grote), Order Lepidoptera, Family Pyralididae.
[2] *Zeadiatraea* (= *Diatraea*) *grandiosella* (Dyar).

Distribution. The southern cornstalk borer is a southern insect, damage being limited mainly to states from Maryland and Kansas on the north to, and including, the southern and southwestern states. The insect is found also in Mexico, and southward to South America. The southwestern corn borer is a native of Mexico and entered the United States about 1913. It is now found in Arizona, New Mexico, Texas, Oklahoma, Colorado, Kansas, Nebraska, Arkansas, and Missouri.

Life History, Appearance, and Habits. The appearance of these two insects is identical, and they can be distinguished only by specialists. The life histories and habits are also so similar that the following discussion will serve for both. The winter is passed in the full-grown larval stage, mostly in the taproots of the old cornstalks. The larvae, which are about 1 inch long, are yellowish, with very pale spots during the

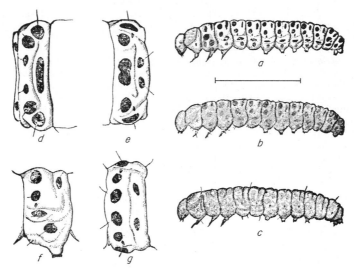

Fig. 9.21. The southern corn stalk borer, *Diatraea crambidoides.* *a–c*, Color varieties of the larvae; *d*, third thoracic segment; *e*, eighth abdominal segment; *f* and *g*, a middle segment in side and dorsal views. The line shows natural size. (*From Howard, U.S.D.A.*)

winter; but during their feeding period in the summer they are conspicuously spotted with eight, rounded, brown or black spots in a transverse row on the front of each body segment and two others behind these. The insect will be found in the lower part of the stalk, just above the roots. It remains in the larval condition until early spring when it changes inside the stalk to a naked brown pupa, the larva having first made a silk-lined exit tunnel to the outside of the stalk, the cover to which is not completely eaten away. The adult moths emerge from the larval burrows in midspring. These moths are of a general light-straw color, with a wing expanse of 1¼ inches. The labial palps extend in front of the head like a short beak. They are active at night only, unless disturbed. They lay their flattened, whitish or yellow, oval eggs, in small groups, overlapping shingle-fashion. Each female commonly lays 300 to 400 eggs on the undersides of the leaves. The worms hatching from these eggs feed at first on the leaves but soon enter the stalk, boring up and down in the pith. They change from one plant to another, as does the common stalk borer. The first generation worms become full-grown a little before midsummer and pupate inside the stalks. Those of the second generation reach maturity in the early fall and may remain as larvae during the winter. There are from one to three generations annually. The shortest recorded development from egg to adult is 36 days.

Control Measures. As the insects hibernate in the stalks of corn and other food plants, a thorough cleaning up and burning of cornstalk refuse and corn stubble, immediately after harvest, is the most effective control measure. Rotation in which corn follows some other crop, at as great a distance as practicable, is also helpful in keeping down these insects. Late-fall and winter plowing followed by thorough harrowing or breaking will destroy most of the hibernating larvae in the stalks.

References. Va. Agr. Exp. Sta., Tech. Bul. 22, 1921; *N.C. Dept. Agr. Bul.* 274, 1920; *S.C. Exp. Sta. Bul.* 294, 1934; *U.S.D.A., Tech. Bul.* 41, 1928; *Kans. Agr. Exp. Sta. Bul.* 317, 1943.

11. LESSER CORNSTALK BORER[1]

Importance and Type of Injury. In the southern part of the United States, corn is sometimes injured by slender greenish worms boring into the lower part of the stalk, usually within 2 inches of the soil surface. Corn under 18 or 20 inches high attacked by this insect becomes much distorted and curled, and frequently fails to produce ears or good stalks. A dirt-covered silken tube usually leads away from the tunnel in the plant. Mostly troublesome on dry sandy soils.

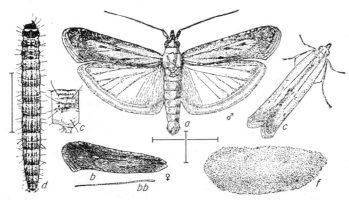

Fig. 9.22. The lesser cornstalk borer, *Elasmopalpus lignosellus*. *a*, Male; *b*, forewing of dark female; *bb*, antenna of female; *c*, male at rest; *d*, larva, dorsal view; *e*, side view of a middle segment; *f*, cocoon. Lines indicate natural size. (*From Chittenden, U.S.D.A.*)

Plants Attacked. Besides corn, the insect feeds on peas, beans, peanut, cowpeas, crab grass, Johnson grass, wheat, turnips, and several other crops.

Distribution. The range of the insect is from Maine to southern California but it is largely confined, so far as injury goes, to the more southern states. It is also found in Mexico, Central America, and South America.

Life History, Appearance, and Habits. The insect hibernates in the larval, pupal, and adult stages, but usually in the southern states as a larva, which transforms to a pupa before spring. The moths (Fig. 9.22,*a,c*) emerge from the pupae, or become active, early in the spring and lay their greenish-white eggs on the leaves or stem of the plants on which the larvae feed. The eggs hatch in about 1 week; the bluish-green brown-striped caterpillars (Fig. 9.22,*d*) feed at first on the leaves or roots but later

[1] *Elasmopalpus lignosellus* (*Zeller*), Order Lepidoptera, Family Pyralididae.

burrow into the stems of corn and other plants. They eat into the heart of the unfolding leaves of the corn, sometimes killing it. The larvae become full-grown in about 2 to 3 weeks. Their presence in the corn is indicated by masses of borings which are pushed out through the holes in the stalk. The insects leave their burrows when full-grown and spin silken cocoons (Fig. 9.22,f) under trash on the surface of the ground, in which they change to brownish pupae about ⅓ inch long. From these the moths emerge in from 2 to 3 weeks. They have a wing expanse of nearly 1 inch. The front wings are brownish yellow with grayish margins with several dark spots. In the female the front wings are nearly black. A second generation is produced in all the southern states where this insect is injurious.

Control Measures. Fall or winter cleanup of the fields and field margin and rotations with crops not attacked by this pest materially reduce the damage. Winter plowing has given control in some cases or is at least helpful in reducing the numbers of the insects in the field. Early planting also is recommended in some sections in the southern states as a control for this pest. The application of granular aldrin at 12 ounces per acre in a band 4 to 5 inches wide over the seed furrow or over seedling plants as they emerge has given effective control.

Reference. *U.S.D.A., Dept. Bul.* 539, 1917.

12 (63, 71, 111, 260). CORN EARWORM[1]

Importance and Type of Injury. This insect has been called the worst pest of corn, considering the United States as a whole. It is claimed that the American farmers grow on the average two million acres of corn each year to feed the corn earworm. Corn attacked by the corn earworm will show the ears with masses of moist castings at the end, and the kernels, especially about the tip of the ear, eaten down to the cob by large brownish to greenish, striped worms, which are nearly 2 inches long when full-grown (Fig. 9.23). In the worst years 70 to 98 per cent of the ears of field corn the country over are attacked and as high as 5 to 7 per cent of the kernels of field corn and 10 to 15 per cent of canning corn may be consumed. Molds may be carried in through diseased ears that may cause death among livestock to which they are fed. The presence of the worms in ears of sweet corn is most repulsive to consumers and very troublesome to commercial canners. These worms vary greatly in color from a light green or pink to brown or nearly black and are lighter on the under-parts. They are marked with alternating light and dark stripes running lengthwise on the body. The stripes are not always the same on different individuals, but there is usually a double middorsal dark line the length of the body. The head is yellow and unspotted, and the legs are dark or nearly black. The skin of the insect is somewhat coarse and, when looked at under a magnifying glass,

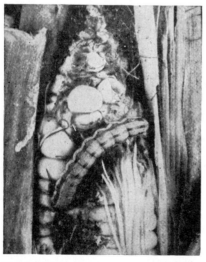

FIG. 9.23. Nearly full-grown larva of corn earworm, feeding in tip of corn. (*From Univ. Calif. Col. Agr.*)

[1] *Heliothis zea* (= *armigera*) (Boddie), Order Lepidoptera, Family Noctuidae.

shows many small thorn-like projections.   Usually the injury starts at
the tip of the ear.   Occasionally the worms enter through the side or at
the butt.

Plants Attacked.   The corn earworm is a very general feeder, attacking
many cultivated crops and weeds.   It is seriously injurious to the tomato,
tobacco, cotton, and vetch, as well as to corn, and has been given the
names of tobacco budworm, tomato fruitworm, cotton bollworm, and
vetchworm.

Distribution.   This insect is of world-wide distribution.   Its damage is
most severe, however, in the South.

Life History, Appearance, and Habits.   In the southern United States,
at least, the insect passes the winter in the form of a brown pupa, which
will be found from 2 to 6 inches below the surface of the soil.   In the
spring and early summer moths (Fig. 9.24) emerge from these pupae and
crawl up exit holes which the larvae prepared before pupating.   The
moths have a wing expanse of about 1½ inches.   They vary in color,
the average having the front wings of a light grayish brown, marked
with dark-gray irregular lines and with a dark area near the tip of the

Fig. 9.24. Adult corn earworm with wings spread.   Somewhat enlarged.   (*From Ill.
Natural History Surv.*)

wing.   The irregular lines often shade into an olive-green.   The hind
wings are white with some dark spots or irregular dark markings.

The moths fly during warm cloudy days, but mainly at dusk of the
evening.   They feed on the nectar of many flowers and, during the warm
evenings, deposit their eggs on the plants in which the larvae feed.   Each
moth will lay from 500 to as many as 3,000 eggs, the average being probably over 1,000.   These eggs are laid singly and are of a hemispherical
shape with ridges along their sides resembling very much a minute sea
urchin.   They are yellowish, about half as large as a common pinhead.
The early generations feed in the curl of young corn plants and on tomatoes, cotton, tobacco, beans, and legumes.   Fresh corn silk is one of the
favorite places for egg-laying of the moths of the later generations.   The
eggs hatch in from 2 to 10 days, and the worms feed at first on the leaves,
or bore directly into the corn silk.   They feed on the silk until it becomes
dry and then on the kernels at the tip of the ear for 2 to 4 weeks, molting
5 times.   On becoming full-grown the larvae crawl down the stalk or drop
to the ground, into which they burrow and excavate a small smooth-
walled cell, commonly 3 to 5 inches deep, where they pupate, coming out
as moths again after another period of 10 to 25 days, although this period

may be prolonged during cold weather. The worms do not always remain in the first ear which they entered but frequently go from one ear to another. They are cannibalistic and usually only one full-grown worm is found in each ear. There are from two to three generations of the insect each season, the number depending on the latitude. While the winter is passed in the pupal stage, it is very doubtful in view of recent work on the life history of this insect if the pupae ever survive the winter north of about 40° north latitude. The infestations in the northern states are, in all probability, caused by a migration of the adults from points farther south. Some moths may also develop from worms in green corn or string beans, shipped to northern markets in the early spring.

Control Measures. No practical control of the corn earworm in field corn has been devised. Ears of sweet or seed corn can be almost completely protected from earworm damage by injecting into the silks 0.5 milliliter of 1 per cent DDT in refined mineral oils (viscosity 120 to 125 seconds). This treatment, which should be applied within a few days after silking, may be made with a small oil can or an applicator designed for the purpose. Hand-spraying the silks when they first appear with a mixture of 1 per cent DDT and 5 per cent mineral oil or dusting with 5 per cent DDT, using a paint brush to treat each silk, gives good control. Fixed boom-spraying with the DDT–mineral oil combination at 1.5 to 2 pounds per acre or ground-dusting with 5 per cent DDT at the same rate will give satisfactory control, but several applications may be needed at about 3-day intervals. Fodder from corn treated with DDT must not be fed to livestock. Spraying or dusting with Sevin at 1.5 to 2 pounds per acre applied as described for DDT is an effective control, and the fodder may be fed to livestock after 7 days following the last application. Pyrethrins applied at 0.2 per cent as the individual ear treatment or as a spray gives less effective control but leaves no objectionable residue.

The time of planting will have a marked effect on injury by this insect but will not always be the same in different years; *i.e.*, in some years early-planted corn will be injured, while in most years the latest corn suffers the worst damage. The moths prefer to lay their eggs on fresh corn silks, so that corn which silks before or after the greatest abundance of moths will largely escape infestation. Fall plowing to disturb the overwintering pupal stage may be of importance in reducing the numbers of moths that emerge in the spring, in areas where the insect winters in the pupal stage. Planting of resistant varieties of sweet corn will greatly reduce earworm damage. See also pages 587 and 658.

References. *Ky. Agr. Exp. Sta. Bul.* 187, 1914; *U.S.D.A., Farmers' Bul.* 1310, 1923, *Tech. Bul.* 561, 1937, and *Leaflet* 411, 1957; *Md. Agr. Exp. Sta. Bul.* 348, 1933; *Ark. Agr. Exp. Sta. Bul.* 320, 1935; *Jour. Econ. Ento.*, **9**:395, 1916; **13**:242, 1920; **31**:459, 1938; **36**:330, 1943; **41**:928, 1948; **52**:1111, 1959; and **53**:22, 1960.

OTHER INSECTS FEEDING ON THE EAR

A large number of different species of beetles are sometimes found feeding on the tip of corn ears where the kernels have been exposed by the feeding of the corn earworm or by birds. These beetles may be regarded as accidental feeders on corn, and practically never attack uninjured ears. The European corn borer described on page 490 is the most serious insect attacking the ears of corn in sections where it is abundant. The Japanese beetle also causes severe damage to corn, in areas where it is abundant by feeding on the husks, foliage, kernels, and silk at the tip of the ear (page 749).

13 (285). CORN ROOT APHID[1] AND CORNFIELD ANT[2]

Importance and Type of Injury. Corn infested by the corn root aphid and its ever-attendant ant germinates normally, and the plants reach a height of from 3 or 4 to 9 or 10 inches, when growth becomes greatly retarded, especially during dry years. The plants often take on a yellowish or reddish tinge to the leaves. An examination of the field will show numerous small anthills around the injured corn plants, and small brownish ants tunneling along the corn roots. Clinging to the corn roots will be found many bluish-green aphids about the size of pinheads when fullgrown—the younger aphids being much smaller.

Plants Attacked. This species of aphid is known to infest the roots of corn, cotton (on which also it is a serious pest), a number of different grasses, and several weeds, particularly smartweed. A very similar species[3] occurs on the roots of aster.

Distribution. The insect is common throughout the corn- and cottongrowing areas east of the Rocky Mountains.

Life History, Appearance, and Habits. The winter is passed, at least in the northern part of the country, only in the egg stage. These eggs (Fig. 9.25,*E*) are collected in the fall by the small brown cornfield ants and stored in their nests over winter. The ants (Fig. 9.25,*F*) pile the eggs in their nests and move them about according to moisture and temperature conditions in the soil. In the early spring, about the time the young smartweed plants begin to appear in the field, the aphid eggs begin hatching. The ants seem instinctively to know that the young aphids must have something to feed on and carry them in their jaws to the roots of the smartweed and some of the grasses on which the aphids feed. Here the young aphids insert their beaks and suck the sap, growing rather rapidly, and in about 2 or 3 weeks all become full-grown females and begin giving birth to living female young. These young become mature females and begin giving birth to others after a period of from 8 days to 2 weeks, reproduction being by parthenogenesis. Only female aphids appear during the summer months. During July and August winged individuals (Fig. 9.25,*B*) frequently make their appearance on the roots and will sometimes crawl to the surface of the ground and fly to other fields, thus spreading the infestation. Local distribution of the aphid in the field is almost entirely dependent on the cornfield ants. This aphid is rarely found on roots except where attended by the ants and, if placed on the surface of the ground, is apparently helpless so far as finding a place to feed is concerned. An ant finding one of these aphids, however, immediately picks it up, carries it underground, and places it on the roots of one of its food plants. The aphids apparently have been dependent on the ants so long that they have entirely lost the faculty of taking care of themselves. This interrelation between the ant and the aphid is one of mutual benefit, as the aphid is protected and kept by the ant where a supply of food is accessible; while the ant, in turn, derives a large part of its food from the sweet sticky exudation known as *honeydew* given off from the anal opening of the aphid. Honeydew is slightly modified sap which the aphids suck from the plant in excess of their needs, so that the ants, in effect, make use of the piercing-sucking mouth parts of the aphids to get

[1] *Anuraphis maidiradicis* (Forbes), Order Homoptera, Family Aphidae.
[2] *Lasius alienus* (Förster), Order Hymenoptera, Family Formicidae.
[3] *Anuraphis middletoni* (Thomas).

their food.   The ants act as though they knew the importance of preserving the eggs of the aphids over winter and that corn is the favorite food of these aphids.   They will carry the aphids for some distance in order to pasture them on the roots of corn.   In one instance under observation, the ants moved 156 feet from a timothy meadow into the third row of a

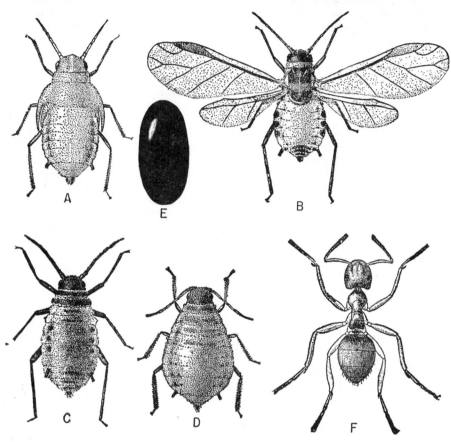

Fig. 9.25. The corn root aphid and the cornfield ant.   A, full-grown nymph of the winged female aphid, × 20; B, winged ovoviviparous female, which produces young without mating, × 14; C, the male aphid, which appears only in autumn, × 25; D, his mate, the oviparous female, which also occurs only in autumn, × 14; E, one of the overwintering eggs produced by the pair, × 25; F, worker of the cornfield ant, which cares for the aphids, × 8. (*From Ill. Natural History Surv.*)

field of corn, carrying with them not only their own young, but also a large number of aphids which they began pasturing on the corn roots.

While only the female aphids are present in the fields during the summer, on the approach of cold weather these females give birth to different forms which consist of wingless true males and females.   These (Fig. 9.25,C,D) mate, and the females, instead of giving birth to young, lay dark-green shiny eggs, dying shortly afterward.   The eggs are gathered up by the ants and stored in their nests during the winter.

Control Measures.   The best and most effective method of combating
these aphids is a measure directed, not against the aphids, but against
their attendant ants.   This consists of thorough deep cultivation of the
soil early in the spring before planting.   The land should be plowed 6½
to 7 inches deep, followed by two or three deep diskings at about 3-day
intervals.   The object of this heavy cultivation is to break up, scatter,
and destroy the ant nests in the soil.   If the field is plowed to a depth of
only 4½ or 5 inches, many of the lower chambers of the ant nests will not
be thrown out in the furrow; 6-inch plowing, however, throws out about
95 per cent of the ant nests, and the disking, following this, breaks up and
scatters the young of the ants and their aphids, so that they are not able
to reestablish their nests in time to cause injury to the young corn.   This
treatment will also drive many of the ants out of the fields, as has been
shown by watching at night the margins of fields where such treatments
have been given.

Extensive experiments have been carried on to find a treatment which
could be applied to seed corn that would repel the cornfield ants and thus
prevent their placing the corn root aphids on the roots of young corn.
Owing to the fact that soil is one of the best deodorants, most of the
chemicals which have been used for treating the seed have not proved
effective.   One of the most effective treatments of this sort has been to
moisten the seed thoroughly with a solution made by stirring 3 to 4 fluid
ounces of oil of tansy into 1 gallon of wood alcohol, and moistening, but
not wetting, the corn seed with this solution before planting.   Some
injury is likely to occur if cool, wet weather follows planting of such
treated seed.   See page 897 for the chemical control of ants.

References.   Ill. Agr. Exp. Sta. Buls. 178, 1915, and 130, 1908; *U.S.D.A., Farmers'*
Bul. 891, 1917; *Forty-sixth Rept. Ill. Agr. Exp. Sta.*, 1933.

14 (39, 87, 216, 273).   WHITE GRUBS OR JUNE BEETLES[1]

Importance and Type of Injury.   White grubs are among the most
destructive and troublesome of soil insects.   When fields infested with
white grubs are planted to corn, the corn usually comes up but the plants
cease growing after reaching a height of from 8 inches to 2 feet.   The corn
will show a patchy growth with varying sized areas in the field where the
plants are dead or dying.   If the injured plants are pulled up, the roots
will be found to have been eaten off, and from 1 to as many as 200 white
curved-bodied grubs, from ½ to over 1 inch in length, will be found in
the soil about the roots (Fig. 9.27).   The grubs (Fig. 9.26, lower figure)
are white with brown heads and six prominent legs.   The hind part of
the body is smooth and shiny, with dark body contents showing through
the skin.   There are two rows of minute hairs on the underside of the
last segment which will distinguish the true white grubs from similar-
looking larvae (Fig. 15.54,*E*).   Injury is usually most severe to crops
following sod.

Plants Attacked.   All grasses and grain crops; potatoes, beans, strawberries, roses, nursery stock, and nearly all cultivated crops.

Distribution.   Throughout North America.

Life History, Appearance, and Habits.   The winter is passed in the
soil both as adults and as larvae of several distinct sizes.   In the spring
after the trees have put forth leaves, the adults (Fig. 9.26, center) become

[1] *Phyllophaga* or *Lachnosterna* spp., Order Coleoptera, Family Scarabaeidae.

active, fly about during the night, and feed on the foliage of trees and the leaves of some other plants. They leave the soil just at dusk and remain on the trees during the night, mating and feeding. At the first streaks of dawn, they return very promptly to the soil, where the females lay their pearly white eggs (Fig. 9.26, left) from one to several inches below the surface. The eggs are generally laid in grasslands or patches of grassy weeds in cultivated fields. Clean stands of clover or alfalfa and clean-cultivated row crops are not likely to be infested by the egg-laying females. The eggs hatch in 2 or 3 weeks, and the young grubs feed on the roots and underground parts of plants until early fall, when they are about ½ inch long. They then work their way down in the soil usually below the frost line and have been taken 5 feet below the surface.

As the soil warms in the spring, they work upward and, by the time plant growth is well started, they are feeding a few inches below the surface. Feeding continues throughout the season and, on the approach of cold weather, they again go deep into the soil, where the second winter is passed, the grubs then being about 1 inch long. The third season they come up near the surface of the ground and feed until late spring or early summer; they then change to the pupal stage (Fig. 9.26, upper figure) in cells in the earth about 6 or 8 inches below the surface. During the latter part of the summer, they change to the adult beetle but do not leave the soil until the following spring. There may be some movement of the beetles downward to below the line of severe freezing. The overwintering population of white grubs, therefore, consists of adults that have not yet taken flight from the soil and of larvae usually of two distinct sizes, the smaller about 9 months, and the larger about 1 year and 9 months old.

Fig. 9.26. Life stages of a white grub, *Phyllophaga rugosa*. Above, pupa; at left, egg; at center, adult; below, larva. (*From Ill. Natural History Surv.*)

The adults are the well-known brown or brownish-black June beetles, May beetles, or "daw bugs." About 200 species are known. They vary somewhat in their life history, some completing their growth in 1 year, while others require as much as 4 years. The 3-year life cycle is by far the most common. There are several other closely related beetles that attack corn, the grubs differing somewhat in structure but having the same general appearance. There are also a number of grubs, very similar in appearance, which occur in the soil and manure, but which eat only decaying vegetable matter and do not feed on the living roots of plants. These lack the double row of spines on the underside of the last body segment which is characteristic of the true white grubs.

Control Measures. The most important control measures for white grubs are based upon three observations regarding the life cycle. (*a*) The larvae prefer to feed upon crops of the grass family, such as corn and other cereals, upon potatoes, or upon strawberries; while legumes such as clover, alfalfa, and soybeans are much less severely damaged. Consequently, land in which numerous white grubs are found while plowing the soil should not be planted to corn, potatoes, or other seriously injured crops, but to soybeans, clovers, or one of the small grains. (*b*) The beetles prefer to lay their eggs in grassy, weedy fields and will not as a rule deposit eggs in fields of clover or alfalfa, unless there is a considerable admixture of grass or weeds in such fields. (*c*) While white grubs are troublesome every year, the most severe injury occurs in regular 3-year cycles. This

FIG. 9.27. A badly injured pasture sod; the bluegrass roots have been entirely eaten off by white grubs, permitting the rolling back of the sod, like a carpet, to expose the grubs. (*From Ill. Natural History Surv.*)

is because most of the insects reach the adult stage in the years 1959, 1962, 1965, and each third year thereafter. Severe damage occurs the year after the adults are abundant and lay their eggs, or the second year of the life cycle. Throughout the central and eastern United States, the years of most severe damage, as shown by Davis, will be 1960, 1963, 1966, and each third year thereafter. During the years when heavy flights of beetles are expected, farmers should make every effort to keep corn and other cultivated fields free of grass and weedy growth during April, May, and June and would do well to have as much of their farms in legume crops as possible. This will reduce the number of eggs laid on their farms. The year following heavy flights of May beetles, it is well to avoid planting corn and potatoes on fields that were in sod or covered with a grassy, weedy growth the preceding spring.

One of the best ways of cleaning grubs out of fields is to pasture the land with hogs during summer and early fall. When hogs are allowed to run on heavily infested land, they will usually root out and eat the grubs, fattening themselves and nearly freeing the land of grubs. If this is done during a summer in which most of the grubs hatch from eggs (1962, 1965, etc.), most of the damage by these pests may be prevented. Hogs intended for breeding stock should not be pastured on grub-infested land, however, as the white grubs are the intermediate host of the giant thorn-headed worm,[1] one of the intestinal parasites of hogs.

Plowing infested fields between mid-July and mid-August, especially in the year when most of the grubs are pupating (1961, 1964, etc.), will kill many of the pupae and newly transformed adults. Plowing until about October first will also crush many larvae and expose them to birds, if done before cool weather when they work their way down below the plow line.

White grubs in the soil are attacked by several insect parasites, especially the larvae of certain wasps[2] which often greatly reduce their numbers, and whose cocoons are often abundant in grub-infested fields. Birds, especially crows and blackbirds, often follow the plow, picking up the grubs as they are turned out in the furrow. On lawns, parks, and golf courses, where the expense is not prohibitive, the use of soil insecticides, such as lead arsenate, DDT, and chlordane, will kill the grubs in the soil. Directions for their use are given on page 522.

References. Ill. Agr. Exp. Sta. Buls. 116, 1907; 186, 1916; and 187, 1916; *U.S.D.A., Farmers' Bul.* 1798, 1938; *Jour. Econ. Ento.,* **24**:450–452, 1931; and **30**:615–618, 1937.

15 (37, 68, 86, 274). WIREWORMS[3]

Importance and Type of Injury. Wireworms are among the most difficult insects to control, the most destructive and most widespread pests of corn, small grains, grasses, potatoes, and other root crops, vegetables, and flowers. Crops that are attacked by wireworms often fail to germinate, as the insects eat the germ of the seeds or hollow them

FIG. 9.28. Larva of the corn wireworm, *Melanotus cribulosus*.   About 3 times natural size. (*From Ill. Natural History Surv.*)

out completely, leaving only the seed coat. The crop may not come up well, or it may start well and later become thin and patchy, because the worms bore into the underground part of the stem causing the plantlet to wither and die, though they do not cut it off completely. Later in the season the worms continue to feed upon the small roots of many plants. The wireworm larvae (Fig. 9.28) are usually hard, dark-brown, smooth, wire-like worms, varying from ½ to 1½ inches in length when grown. Some species are soft, and white or yellowish in color. Their injuries are usually most severe to crops planted on sod ground, or the second year from sod.

[1] *Macrocanthorhynchus hirudinaceus*, Phylum Nemathelminthes.

[2] *Tiphia* spp. and *Elis* spp., Order Hymenoptera, Family Scoliidae.

[3] Many species of the Order Coleoptera, Family Elateridae.

Plants Attacked. Wireworms are especially destructive to corn and grasses, but all the small grains and nearly all cultivated and wild grasses are attacked. Among the garden crops severely damaged are potatoes, beets, sugarbeets, cabbage, lettuce, radish, carrot, beans, peas, onions, asters, phlox, gladioli, and dahlias. Leguminous plants, such as velvet beans, and certain small grains, such as oats, are more resistant to wireworms than other crops, but clovers, alfalfa, peas, and beans may suffer considerable damage.

Distribution. Throughout North America and most of the world.

Life History, Appearance, and Habits. There are many different species of wireworms that attack our cultivated crops, including corn. The winter is passed mainly in the larval and adult stages in the ground. In the early spring the adults (Fig. 9.29) become active and fly about, some species being strongly attracted to sweets; these can be taken in large numbers by placing a few drops of sirup on tops of fence posts or other exposed places out-of-doors. They are "hard-shelled," usually brownish, grayish, or nearly black in color, somewhat elongated, "streamlined" beetles, with the body tapering more or less toward each end. The head and thorax fit closely against the wing covers, which protect the back of the abdomen. The joint just in front of the wing covers is loose and flexible, and, when beetles are placed or fall on their backs, they flip the middle part of the body against the ground in such a manner as to throw themselves several inches into the air. The chances of their alighting on their feet seem to be about fifty-fifty; but they will generally keep trying until they come down right side up, when they make use of their legs to escape. This habit has afforded amusement to most country boys and girls and has given the insects such names as click-beetles, snapping beetles, and skipjacks. The females of the species that are most injurious to corn burrow into the soil and lay their eggs mainly around the roots of grasses. The adults live 10 to 12 months, most of which time, and all that of the other stages, is spent in the soil. The egg stage requires a few days to a few weeks. The larvae hatching from these eggs spend from 2 to 6 years in the soil feeding on the roots of grasses and other plants. As the soil becomes hot and dry, the larvae migrate downward so that it is often hard to find them in dry summer weather, even in severely infested fields. The last segment of the larva is usually characteristically ornamented and serves to distinguish different species during this stage (Fig. 9.30). Most species change to a naked, soft pupa, and a few weeks later to the adult stage, in cells in the ground, during the late summer or fall of the year in which they become full-grown. The adults, which are commonly about ½ inch long, remain in the soil until the following spring. There is much overlapping of the genera-

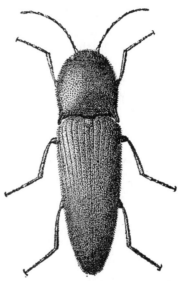

FIG. 9.29. A click beetle, adult of the wireworm, *Melanotus fissilis*. About 5 times natural size. (*From Ill. Natural History Surv.*)

tions so that all stages and nearly all sizes of larvae may be found in the soil at one time.   The larvae move only a few yards, at most, during their long lifetime, and the adults often remain and lay their eggs near where they developed so that marked differences in infestation often occur in near-by fields.

Control Measures.   On large acreages devoted to field crops, chief dependence for avoiding wireworm damage will have to be placed upon cultural control measures.   The nature of these measures will vary with the species of wireworm present, and much more research is needed to develop the best control for the various sections of the country.   In the northwestern United States and western Canada, clean summer fallowing or resting of the land every second or third year; shallow tillage to prevent the growth of all vegetation, especially large weeds, in the early part of the summer; and the avoidance of deep plowing which allows the wireworms to penetrate into the soil more successfully are recommended.   On the other hand, plowing to a depth of 9 inches about the

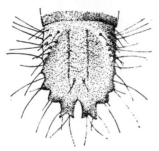

FIG. 9.30. Terminal segment of larva of a wireworm, *Drasterius elegans*. Much enlarged. (*From Ill. Natural History Surv.*)

first of August and allowing the dry lumpy soil to lie undisturbed for a few weeks is said to kill great numbers of the pupae and adults, by breaking up their cells in the soil.   Hawkins claims that from 1 to 4 years of clean cultivation of infested fields will reduce the wireworm population to harmless numbers, in Maine, and that new generations do not start to any extent in cleanly cultivated fields.   In parts of the West, however, the heaviest infestations of wireworms are in fields that have been under continuous cultivation.   The growing of small grains or legumes on second-year sod helps in lessening the damage by these insects, as the percentage of such plants killed by them is not large enough seriously to affect the yield of these crops.   Hay crops or long-standing pasture or grain crops are favorable to wireworms and should be omitted as much as possible from the rotation.   Certain species of wireworms are abundant only in poorly drained soils.   The proper draining of such soils will entirely prevent damage by these species.   In irrigated districts all stages of wireworms can be killed by flooding the land so that the water stands a few inches deep for a week during warm weather when the temperature of the soil at a depth of 6 inches averages 70°F. or higher. Allowing the top 18 inches of soil to become very dry for several weeks during the summer at least once in 6 years is also recommended in the Pacific Northwest.   The most effective rotations, methods of tillage, planting dates, and other farm practices will have to be worked out for each farm area, with especial reference to the kinds of crops desired and the particular species of wireworms present.   Applications at time of planting using aldrin or heptachlor at 1.5 pounds per acre, applied to row or hill or at 2 pounds per acre broadcast and worked into the top 3 or 4 inches of soil, have given effective control.   Other chemical control measures, including seed treatments, are discussed on page 618.

References.   U.S.D.A., Farmers' Buls. 725, 1916, and 733, 1916; *Can. Dept. Agr., Pamphlet* 33 (n.s.), 1923; *Maine Agr. Exp. Sta. Bul.* 381, 1936; *Clemson, S.C., Agr. Coll. Cir.* 163, 1938; *Pa. State Coll. Exp. Sta. Bul.* 259, 1930; *Can. Dept. Agr., Ent. Branch, Saskatoon Leaflet* 35, 1933.

16. Northern Corn Rootworm[1,2]

Importance and Type of Injury. This is one of the most important corn pests in the upper Mississippi Valley. Corn makes a slow growth, this checking of growth being most noticeable about the time the tassel appears. Plants are undersized and frequently fall over after a heavy rain. Small roots are eaten off and larger roots tunneled by thread-like white worms about ½ inch long (Fig. 9.31) with yellowish-brown heads

FIG. 9.31. Northern corn rootworm, *Diabrotica longicornus*, larva. Five times natural size. (*From Ill. Natural History Surv.*)

and six small legs on the fore part of the body. The skin of the body is somewhat wrinkled. The larvae of this insect are also able to transmit bacterial wilt of corn (page 514). The insects are so numerous that fields are practically sure to be heavily infested by the end of the second year in corn.

Plants Attacked. The larvae attack only corn, so far as known. The adults feed on a large number of plants that flower in summer and early fall.

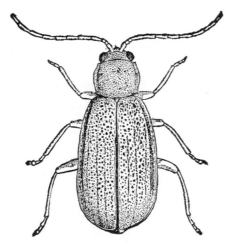

FIG. 9.32. Adult of northern corn rootworm, *Diabrotica longicornis*. About 10 times natural size. (*From Ill. Natural History Surv.*)

Distribution. The insect occurs in greatest abundance in the north part of the Mississippi Valley. It is not injurious in the southern states, and occurs in small numbers east of New York and west of Kansas and Nebraska.

Life History, Appearance, and Habits. The winter is passed only in the egg state. The eggs are deposited in the fall in the ground around the roots of corn, and in no other known situation. They hatch rather late

[1] *Diabrotica longicornis* (Say), Order Coleoptera, Family Chrysomelidae.

[2] The closely related western corn rootworm, *Diabrotica virgifera* LeConte, has a similar life history and control.

in the spring, and the larvae work through the ground until they encounter the roots of corn.   Although the larvae feed to a slight extent on some of the native grasses, most of them will die if corn is not planted in the field where the eggs were laid.   Larvae have been found in the roots of corn growing in ground following 1 year in oats.   The worms burrow through the roots, making small brown tunnels.   They become full-grown during July, leave the roots, and pupate in cells in the soil, the pupa being pure white and very soft.   The adult stage (Fig. 9.32) is reached during the latter part of July and August.   The beetles leave the soil and feed on the silk of corn and the pollen of this and many other plants.   The eggs are deposited in cornfields during September and October, and nearly all the beetles die at the time of the first heavy frost.   They are about $\frac{1}{6}$ to $\frac{1}{4}$ inch long, of a uniform greenish to yellowish green, and are very active, tumbling off the flowers or out of the corn silk when disturbed.

   Control Measures.   As the eggs are laid only in cornfields and the larvae feed mainly on the roots of this plant and cannot migrate to other fields, a rotation that will put any crop other than corn on the land for a period of 1 year, between corn crops, will effectively prevent damage. Soil treatments with aldrin, heptachlor, or chlordane, as described on page 511, have given effective control.

   References.   *Eighteenth Rept. Ill. State Ento.*, p. 135, 1891–1892; *Ill. Agr. Exp. Sta. Bul.* 44, 1896; *Jour. Econ. Ento.*, **25**:196–199, 1932; and **41**:392, 1948.

17 (95). SPOTTED CUCUMBER BEETLE OR SOUTHERN CORN ROOTWORM[1]

Importance and Type of Injury.   Cornfields infested with this insect start growth in a normal way, and the plant begins to show the effect of the infestation when from 8 to 20 inches tall in the North or much

FIG. 9.33. Larva of the spotted cucumber beetle or southern corn rootworm, *Diabrotica undecimpunctata howardi*.   About 5 times natural size.   (*From Ill. Natural History Surv.*)

earlier in the South.   From then on, the plant makes a very poor growth, or none at all, and frequently dies.   Sometimes the heart of the plant is killed by the larvae, the lower leaves remaining green.   As when attacked by the northern corn rootworm, the larger plants will fall after heavy rains.   Examination of the plants will show the roots tunneled and eaten off by larvae about $\frac{1}{2}$ to $\frac{3}{4}$ inch in length, with yellowish-white, somewhat wrinkled bodies, six very small legs, and brownish heads (Fig. 9.33).   The last segment of the abdomen has a nearly circular margin and is brownish in color.   In addition to the injury to the roots, the lower part of the stalk will usually be bored through by the grubs.   This insect has also been shown to have a part in the dissemination of bacterial wilt of corn (see page 514).   As a garden pest, both larvae and adults do a large total damage, which is seldom fully appreciated because distributed over so many different crops (page 631).

   [1] *Diabrotica undecimpunctata howardi* Barber, Order Coleoptera, Family Chrysomelidae.   The western spotted cucumber beetle, *D. undecimpunctata undecimpunctata* Mannerheim, is common in Colorado, Arizona, Oregon, and California, where it causes similar damage and has a nearly identical appearance and life history.

Plants Attacked. This insect has been taken from a very large number of plants, including more than 200 of the common weeds, grasses, and cultivated crops. It goes by several names, according to the food plants, and is perhaps best known as the twelve-spotted cucumber beetle. It is also frequently called the overflow worm, and the budworm.

Distribution. The insect is widely distributed, occurring over the greater part of the United States east of the Rocky Mountains, in Southern Canada, and in Mexico. It is more abundant and destructive in the southern part of its range. The variety *tenella* extends into New Mexico, Arizona, and California.

Life History, Appearance, and Habits. This insect passes the winter in the form of a yellowish or yellowish-green beetle about ¼ inch long, with 12 conspicuous black spots on the wing covers (Fig. 9.34). The head is black, and the antennae, which are about one-half to two-thirds as long as the body, are dark or nearly black. The beetles hibernate in nearly any kind of shelter but seem to prefer the bases of plants which are not entirely killed down by the frost. They become active very early in the spring, flying about during the first days when the temperature reaches 70°F. or above. There is some evidence that the adults migrate northward in the spring. The females deposit their eggs in the ground around the bases of plants. The young larvae, on hatching, bore in the roots of plants and the underground parts of the stem. They become full-grown during July. The insect has two generations in the southern part of its range and at least a partial second generation is produced in the North.

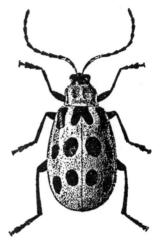

Fig. 9.34. Adult, spotted cucumber beetle, *Diabrotica undecimpunctata howardi.* About 6 times natural size. (*From Ill. Natural History Surv.*)

Control Measures. It is extremely difficult to prevent damage to corn by these insects, as the eggs are frequently laid in the fields after the corn is up and there is no method by which infested soil can be cleaned of the larvae. About the most effective cultural method is late planting on land which has been plowed early in the spring or in the fall and cultivated frequently before planting, so that all vegetation has been kept down. In certain seasons when the beetles have been very abundant, fields handled in this way have been practically the only ones to escape injury. Rotation of crops is of no value in controlling this species. The injury is usually most severe during wet years or the first season following wet years, and is also often serious on land that has been overflowed. Damage is often most severe on land of high fertility that produces a heavy early growth of vegetation. This may be due to preference of the female beetles for such soil in which to lay their eggs or to the fact that they are attracted to the rank vegetation which generally follows an overflow. Soil treatment with aldrin or heptachlor at 0.5 to 1 pound or chlordane at 1 to 1.5 pounds per acre as spray, dust, or fertilizer mixture has given effective control. The lower dosage is applied to rows or hills at time of planting, and the higher is used as a broadcast treatment which is cultivated into the top 3 to 4 inches of soil prior to planting.

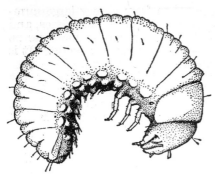

FIG. 9.35. Larva of the grape colaspis, *Colaspis flavida*. About 15 times natural size. (*From Ill. Natural History Surv.*)

References. *U.S.D.A., Farmers' Bul.* 950, 1918; *S.C. Agr. Exp. Sta. Bul.* 161, 1912; *Eighteenth Rept. Ill. State Ento.*, p. 129, 1891–1892; *Ark. Exp. Sta. Bul.* 232, 1929; *Ala. Agr. Exp. Sta. Bul.* 230, 1929, and *Cir.* 65, 1934; *Jour. Econ. Ento.*, **39**:781, 1946, and **42**:558, 1949.

18 (58). GRAPE COLASPIS[1]

Importance and Type of Injury. Corn that has been planted on clover sod will sometimes wilt when the plants are about 6 to 10 inches high. The plants may die, or merely be greatly retarded in growth. An examination of the roots and soil about them will show numerous curved, fat-bodied, very short-legged grubs from $\frac{1}{8}$ to $\frac{1}{6}$ inch in length (Fig. 9.35). Injury is most severe during late May and June, and usually on corn following clover sod.

Plants Attacked. Adults of this insect have been found on a number of crops including timothy, June grass, grapes, strawberries, beans, clover, buckwheat, potatoes, cowpeas, muskmelons, and apples. Its habit of feeding on grape has given

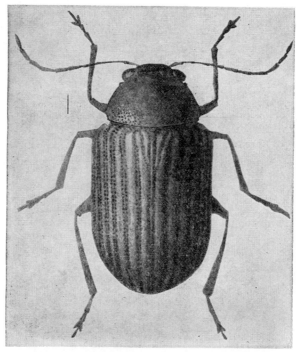

FIG. 9.36. Adult of the grape colaspis, *Colaspis flavida*. About 15 times natural size. Line shows natural size. (*From Ill. Natural History Surv.*)

it the common name of grape colaspis. It is also known as the clover rootworm.

Distribution. This beetle ranges throughout eastern North America and into Arizona and New Mexico.

[1] *Colaspis flavida* (Say), Order Coleoptera, Family Chrysomelidae.

Life History, Appearance, and Habits.   So far as known, the winter is passed in the young larval stage.   The larvae are active early in the spring and generally become full-grown during the first part of the summer; in central Illinois, by about June 15. They pupate in the earthen cells in the soil and emerge during July as pale brown elliptical beetles (Figs. 9.36, 11.19).   The body of the beetle is about ⅙ inch long and covered with rows of evenly spaced punctures.   The adults fly about freely in the field and, as above stated, are very general feeders.   Mating takes place and eggs are deposited in midsummer about the roots of several of the above mentioned food plants but particularly on those of timothy, grape, and clover.   There is only one generation of the insect each year.

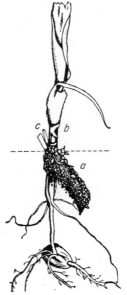

Control Measures.   Injury by this insect frequently occurs on spring-plowed red clover sod or spring-plowed timothy, but seldom where such ground is broken in the fall. Injury to corn has rarely been recorded following crops other than clover or timothy.   A rotation that will avoid putting corn on spring-plowed clover or timothy sod will nearly always prevent injury by this insect.

References.   Twenty-third Rept. Ill. State Ento., p. 129, 1891; *Trans. Ill. State Acad. Sci.,* **24**:235, 1931.

19 (47, 69, 79). WEBWORMS[1]

Importance and Type of Injury.   Corn on spring-plowed sod land will sometimes be cut off near the surface of the ground in much the same manner as where attacked by cutworms.   An examination of the surface of the ground will show a loose silken web containing bits of dirt leading to a short silk-lined tunnel in the ground, usually at the base of the plant (Fig. 9.37).   Short, rather thick-bodied, usually spotted and coarsely haired, active worms, from ¼ to ¾ inch long, will be found in these silk-lined tunnels (Fig. 9.38).   Frequently the cutoff corn plants are dragged to the tunnel.   In years when any one of several species is abundant they may destroy from 10 to 80 per cent of a stand of corn.   Several species of webworms cause serious and widespread injury to golf greens and lawns, causing a ragged, patchy appearance or sometimes killing large areas completely.   The presence of many blackbirds, robins, flickers, and other birds pecking holes in the turf is an almost sure sign of the presence of these or other insect pests.

FIG. 9.37. Corn plant injured by a sod webworm, *Crambus luteolellus. a,* Silk-lined tunnel in the soil; *b,* gnawed surface of stalk; *c,* tip of severed leaf drawn into mouth of nest.   *(From Ill. Natural History Surv.)*

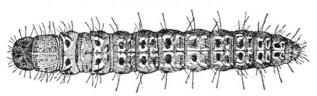

FIG. 9.38. Larva of a sod webworm, *Crambus mutabilis,* dorsal view.   About 4 times natural size.   *(From Ill. Natural History Surv.)*

Plants Attacked.   The webworms feed on the grass plants, including corn, more particularly on bluegrass, timothy, and other pasture and field grasses, and on tobacco and certain weeds.   Certain closely related species attack clovers.

[1] Species of the subfamily Crambidae, Order Lepidoptera, Family Pyralididae. Among the dozen or more species of importance are the corn root webworm, *Crambus caliginosellus* Clemens, the bluegrass webworm, *Crambus teterrellus* (Zincken), and the striped or black-headed webworm, *Crambus mutabilis* Clemens.

Distribution. There are a number of species of webworms occurring in different parts of the United States and Canada.

Life History, Appearance, and Habits. The webworms, which are injurious to corn, pass the winter in the larval stage in silk-lined nests in grass and sod lands. They become active early in the spring, feed in much the same way as cutworms, although eating the leaves to a greater extent, and become full-grown during late June and July. Pupation takes place in a silken cocoon a short distance underground. The moths (Fig. 9.39) are, in most cases, pale brown in color, and have a pronounced projection from the front of the head. They vary in size from ½ to nearly 1 inch in length. The projection from the head, which is formed by the labial palps held close together, has given them the name of snout moths. One will frequently stir them up when walking across grasslands. They have a very quick, jerky, zigzag flight, usually going only a rod or two, when they alight, roll their wings close about the body, and hide by crawling down into the grass. The moths lay their eggs around the lower parts of grass stems or drop them at random as they fly about over the grass, soon after emerging, and the young larvae hatch and feed for a short time before going into hibernation. With most species, there are two or three generations each season. The corn webworm has only one generation a year, so far as known.

Fig. 9.39. Adult of a webworm, *Crambus mutabilis.* About twice natural size. (*From Ill. Natural History Surv.*)

Control Measures. The most effective control measure for these insects is early fall plowing of sod land which is to be used for corn the next season. Adults of the webworms will not lay their eggs in plowed ground, so that this measure will effectively prevent damage. If it is impossible to plow early, the land should be plowed in the fall and harrowed to expose the hibernating worms to the weather and their natural enemies. If fields are damaged early in the spring to such an extent that they should be replanted, the replanting should be made between the rows and the first planting of corn left as long as possible, undisturbed, to serve as food for the worms until the second planting becomes too large for them to cut off. For the control of these pests on lawns and golf greens see page 522.

References. *U.S.D.A., Farmers' Bul.* 1258, 1922; *Twenty-third Rept. Ill. State Ento.*, p. 36, 1905; *Jour. Agr. Res.*, **24**:399, 1913; *Rept. Ento. Soc. Ontario* for 1934, pp. 98–107; *Tenn. Agr. Exp. Sta. Bul.* 186, p. 37, 1943.

20 (21, 65, 76, 108). FLEA BEETLES[1]

Importance and Type of Injury. The injury by these beetles is most severe in cold or wet years. They may reduce the stand or yield from 5 to 25 per cent. The adults eat very small holes in the green portion of the leaves, giving the whole plant a bleached appearance; growth is retarded, and the leaves wilt even during wet weather. Small to very small, shining, roundish, black, brown, or grayish-black beetles will be found feeding on the leaves. These beetles jump readily when approached. The hind legs are distinctly enlarged and thickened. Occasionally the injury will be caused by somewhat elongated dark-green beetles with white stripes on the back (Fig. 9.43). The most serious phase of injury by these insects is the dissemination of a bacterial wilt of corn, known as Stewart's disease[2] which causes great damage, frequently amounting to the loss of practically the entire crop, especially in the early-maturing varieties of sweet and field corn in the middle and southern states. The adults of the brassy or corn flea beetle[3] not only are responsible for most of the spread of bacterial wilt throughout the summer, by directly infecting the leaves as the beetles feed, but are largely responsible for the successful wintering of the bacteria which

[1] Order Coleoptera, Family Chrysomelidae.

[2] Caused by *Bacterium* (= *Aplanobacter*) *stewarti* Smith. See *U.S.D.A., Tech. Bul.* 362, 1937.

[3] *Chaetocnema pulicaria* Melsheimer.

occur in enormous numbers in the alimentary canal of these beetles in hibernation. The toothed flea beetle[1] has also been shown to be capable of spreading the disease, and the larvae of the northern and southern corn rootworms (pages 509 and 510), start infections at the base of the plants as they bore through the roots and crown.

Plants Attacked. Nearly all kinds of plants are attacked by adult flea beetles. The species that are most injurious to corn also feed on millet, sorghum, broomcorn, sweetpotato, sugarbeet, oats, morning-glory, bull nettle, and cabbage.

Distribution. Flea beetles of various species are of world-wide distribution. The corn flea beetle and sweetpotato flea beetle,[2] perhaps the two most injurious species on corn, occur generally over the eastern part of the United States. Their place is taken in the West by the western black flea beetle.[3]

Life History, Appearance, and Habits. Practically all the flea beetles injurious to corn, with the probable exception of the pale-striped flea beetle (see page 516), pass the winter in the full-grown beetle stage. They shelter largely along bushy fence rows, roadsides, or the edges of woodlands. Some of the species prefer shelters afforded under trees. In the spring the insects become active as soon as vegetation is well started. The beetles, after mating, lay their eggs on the leaves of plants or in the ground about the roots or underground stems. The larval habits of most species

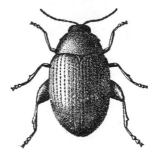

Fig. 9.40. Adult of the corn flea beetle, *Chaetocnema pulicaria.* About 20 times natural size. (*From Ill. Natural History Surv.*)

Fig. 9.41. Adult of the sweetpotato flea beetle, *Chaetocnema confinis.* About 25 times natural size. (*From Ill. Natural History Surv.*)

are not well known. Nearly all the damage to corn is caused by the adults feeding on the leaves and disseminating wilt disease. This occurs during the first 2 or 3 weeks after the corn has come up, and the injury is usually most severe during cold seasons, when the growth of the corn plant is slow, thus giving the flea beetles a long period during which they are attacking the small plants. In seasons when the weather is favorable for growth, the plant usually outgrows the attack of these insects. Bacterial wilt disease, on the other hand, is serious only after mild winters, because cold winters are deadly to the flea beetles which harbor the bacteria in their bodies over the winter. Several of the species attacking corn have a single generation, while others produce two generations each season. Only the overwintering adults are of importance, however, so far as their injury to corn goes. The species which most commonly cause injury to corn are the western black flea beetle, the corn flea beetle (Fig. 9.40), the toothed flea beetle, the sweetpotato flea beetle (Fig. 9.41), and the smartweed flea beetle.[4]

Control Measures. In combating flea beetles on corn, keeping the fields free from weeds is probably of first importance. Larvae of some of the species injurious to corn feed on the roots of weeds and grasses, and the adults on the leaves of these plants as well as on corn. Fields that have been kept clean the previous season, both in the

[1] *Chaetocnema denticulata* (Illiger).
[2] *Chaetocnema confinis* Crotch.
[3] *Phyllotreta pusilla* Horn.
[4] *Systena hudsonias* Förster.

field and around the margins, are very seldom injured by flea beetles. The next important control measure is planting sufficiently late so that the corn will make a quick growth. If the beetles are working on the corn, frequent cultivation will drive them temporarily from the plants and prevent continued feeding, thus lessening damage. In cases of severe infestation it is possible to catch large numbers of the beetles by attaching vertical sheets covered with sticky tanglefoot to the cultivator frames in such a manner that the sheets pass between the corn rows and just above the surface of the ground. Such measures will seldom be necessary if frequent cultivation can be given. Flea beetles attacking sweet corn may be controlled by spraying or dusting with 1.5 to 2 pounds of DDT per acre, but the crop refuse should not be used for animal forage (page 396).

References. *Twenty-third Rept. Ill. State Ento.*, pp. 109–111, 1905; *U.S.D.A., Dept. Bul.* 436, 1917, and *Farmers' Bul.* 1371, 1934, and *Tech. Bul.* 362, 1937; *Phytopathology*, **25**:32, 1935; *Jour. Agr. Res.*, **52**:585–608, 1936; *Jour. Econ. Ento.*, **38**:197, 1945.

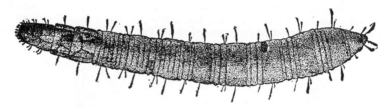

FIG. 9.42. Larva of the pale-striped flea beetle, *Systena blanda*, dorsal view. About 15 times natural size. (*From Ill. Natural History Surv.*)

21. PALE-STRIPED FLEA BEETLE[1]

Importance and Type of Injury. Corn seed attacked by the larva of this flea beetle often fails to sprout or produces a pale, weak plant. If the seed is examined, it will be found to be injured by very slender white worms, a little over $\frac{1}{4}$ inch long, with light-brown heads, six very short legs, and the body tapering slightly toward the head (Fig. 9.42). Injury is usually most serious during periods of cool weather, which retards the germination of the seed after planting.

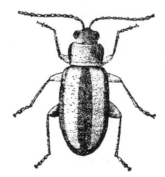

FIG. 9.43. Adult of the pale-striped flea beetle, *Systena blanda.* About 10 times natural size. (*From Ill. Natural History Surv.*)

Plants Attacked. The adults of this species have been found feeding on a great variety of cultivated plants and weeds, including watermelon, pumpkin, pea, bean, eggplant, potato, sweetpotato, mint pigweed, lamb's-quarters, purslane, ragweed, cocklebur, wild sunflower, alfalfa, and many others. Lamb's-quarters and shepherd's-purse seem to be preferred by the larvae.

[1] *Systena blanda* Melsheimer, Order Coleoptera, Family Chrysomelidae.

Distribution. This species is distributed generally over temperate North America. It is probably a native species.

Life History, Appearance, and Habits. The winter stage is not known, but the insect probably hibernates as an adult or pupa. The adult beetles (Fig. 9.43) are about ⅙ inch in length, with the margins of each wing cover pale brown to nearly black and with a broad, median, white stripe. The legs are dull red. The beetles appear by May 1 in the southern part of the cotton belt. Larvae will often be found on the earliets planted corn seed. The larvae bore through the kernel, often destroying the germ and thus preventing growth. The larval period has not been definitely worked out but is, judging from the time of appearance of the first adults in the summer, about 1 month. There is probably one complete generation a year in the latitude of central Illinois.

Control Measures. The damage by this insect has been most severe in fields which were weedy the previous season. Keeping down weeds will help in preventing damage the next season. Early plowing and late planting of cornfields are also of value. These measures will starve out many of the larvae in the soil before the corn is planted. The most effective measure of control is planting good seed sufficiently late so that the corn will make a quick, strong growth, and the seed will not lie in the soil long enough to be seriously damaged by the larvae. See page 605 for the chemical control of flea beetles.

References. Twenty-third Rept. Ill. State Ento., p. 107, 1905; *N.Y. (Cornell) Agr Exp. Sta. Memoir 55,* 1922; *Mich. State Board Agr. Rept.* for 1933–1934.

22. SEED-CORN BEETLES[1,2]

Importance and Type of Injury. Sometimes when corn seed fails to sprout, an examination of the kernels will show dark-brown, striped,[1] or nearly chestnut-brown[2] beetles about ¼ to ⅓ inch long eating out the contents (Fig. 9.44). This injury seldom occurs except where seed of low vitality has been used or when cold weather has greatly delayed germination.

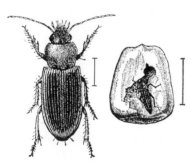

Fig. 9.44. Adult of the seed-corn beetle, *Agonoderus lecontei,* and injury to kernel. Line indicates natural size. *(From Ill. Natural History Surv.)*

Fig. 9.45. Worker of the thief ant, *Solenopsis molesta.* About 25 times natural size. *(From Ill. Natural History Surv.)*

Plants Attacked. These two beetles are mainly feeders on insects, or insect remains, and only rarely attack seeds.

Distribution. The greater part of the United States and Canada.

Life History, Appearance, and Habits. The life history of these beetles is not known. They probably pass the winter in the pupal or adult stage, as they are abroad very early in the spring. They may often be seen in large numbers at electric lights.

Control Measures. Planting sufficiently late to insure quick germination of the corn seed, and using seed of good vitality is the best method of overcoming the damage caused by these insects.

[1] *Agonoderus lecontei* Chaud., Order Coleoptera, Family Carabidae.
[2] *Clivina impressifrons* LeConte, Order Coleoptera, Family Carabidae.

References. *Eighteenth Rept. Ill. State Ento.*, p. 11, 1891–1892; *Trans. Ill. State Acad. Sci.* for 1934, pp. 138, 139.

23 (285). THIEF ANT[1]

Importance and Type of Injury. An examination of a corn kernel that has failed to sprout or that has produced a weak plant, will sometimes show many little starch grains scattered through the soil about the kernels and the entire inside of the seed hollowed out. Frequently very small orange-red ants will be found actively working in the kernels (Fig. 9.45).

Plants Attacked. Seeds of corn, sorghum, millet, and probably other plants.

Distribution. General throughout North America.

Life History, Appearance, and Habits. The winter is passed in all stages of development in nests in the soil. These nests are often made in the walls of the nests of larger species of ants. The workers of this small species obtain a part of their food by preying upon the helpless larvae and pupae of the larger species. This has given them their common name of thief ant. The workers of this species are about $\frac{1}{20}$ inch long, of an orange-yellow color. The males are slightly larger and black. The females, or queens, are brown and very much larger, being about $\frac{1}{4}$ inch long. Certain other species of ants rarely cause similar injury.

Control Measures. The seed treatments that have been used for preventing injuries by these ants have not proved of much value. The best method of preventing their injuries is thorough cultivation of the soil before planting. This will break up their nests, scatter the young, and greatly reduce their numbers in the field. Surface planting aids in control. The chemical control of ants is discussed on page 897.

References. Ill. Agr. Exp. Sta. Bul. 44, p. 214, 1896; *Ill. Agr. Exp. Sta. Cir.* 456, 1936.

24. SEED-CORN MAGGOT[2]

Importance and Type of Injury. The seed attacked by the seed-corn maggot usually fails to sprout, or, if it does sprout, the plant is weak and sickly. The pale or dirty-colored yellowish-white maggots (Fig. 9.46)

FIG. 9.46. Larva of the seed-corn maggot, *Hylemya cilicrura*, side view, 8 times natural size. (*From Ill. Natural History Surv.*)

will be found burrowing in the seed. Injury is usually most severe in wet, cold seasons and on land rich in organic matter.

Plants Attacked. Corn, beans, peas, cabbage, turnip, beets, radish, seed potatoes, and several others.

Distribution. This species is widely distributed in Europe. It was first found in this country in 1856, in New York. It has now spread over nearly the entire United States and southern Canada.

Life History, Appearance, and Habits. The winter is probably passed in the soil of infested fields in the maggot stage inside of a dark-brown capsule-like puparium about $\frac{1}{5}$ inch long, or as free maggots in manure or about the roots of clovers. The flies (Fig. 9.47), which are grayish brown in color and about $\frac{1}{5}$ inch long, are abroad in the fields early in May in the latitude of central Illinois. They deposit their eggs in the soil where there is an abundance of decaying vegetable matter or on the

[1] *Solenopsis molesta* (Say), Order Hymenoptera, Family Formicidae.
[2] *Hylemya cilicrura* (Rondani), Order Diptera, Family Anthomyiidae.

seed or plantlet.   The eggs hatch readily at temperatures as low as 50°F., and the larvae and pupae may develop at any temperature from 52 to 92°F.   The maggots burrow in the seed, often destroying the germ. When full-grown, they are of a yellowish-white color, about ¼ inch long, sharply pointed at the head end, legless, and very tough-skinned.   They change to the pupal stage inside the brown puparium in the soil and in from 12 to 15 days emerge as adults.   Since the insect has been carried through its entire life cycle in 3 weeks, there are probably from 3 to 5 generations each year throughout the corn belt.

Control Measures.   Shallow planting in a well-prepared seedbed, sufficiently late to get a quick germination of the seed, is probably the best means of preventing injury.   Land that is heavily manured, or where a cover crop is turned under, should be plowed early in the fall if

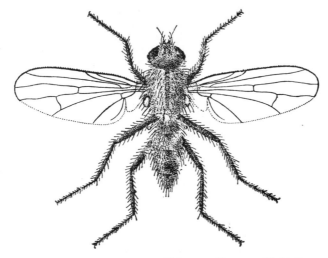

Fig. 9.47. Adult of the seed-corn maggot, *Hylemya cilicrura*.   Eight times natural size. (*From Ill. Natural History Surv.*)

possible, so it will be less attractive to the egg-laying flies the following spring.   Prompt resetting or replanting of the damaged crops will usually give a stand.   This insect is particularly susceptible to control by seed treatments with lindane, aldrin, dieldrin, heptachlor, or endrin, as described for wireworms, page 618.

References.   U.S.D.A., *Div. Ento., Bul.* 33 (n.s.), p. 84, 1902; *N.Y.* (*Cornell*) *Agr. Exp. Sta. Memoir* 55, 1922; *Jour. Econ. Ento.*, **42**:77, 1949.

INSECTS THAT ATTACK SORGHUM, GRAIN SORGHUMS, BROOMCORN, AND SUDAN GRASS

The insects attacking these grasses and forage crops are largely the same as those attacking corn.   The corn leaf aphid occasionally causes considerable damage to broomcorn because of the discoloration of the broomcorn head.   Chinch bugs are very serious pests of millet and Sudan grass, making it almost impossible to grow either of these crops in areas where the bugs are very abundant.   They also find Sudan grass an especially favorable place in which to pass the winter, and many of them hibernate in this grass in areas where it is grown.   In general the sorgo and Kafir varieties of grain sorghums are more resistant to chinch bug injury than the milos and feteritas.[1] Millet is one of the favorite places for second-generation army-worm moths to lay eggs. For the general corn insects, control measures are the same as given in preceding pages.

The ripening kernels of grain sorghum heads and broomcorn are often eaten out or entirely consumed by sluggish caterpillars, somewhat flattened and thickly clothed with spines and hairs, the body greenish, with four red or brown longitudinal stripes on the upper part, which are the larval stage of the sorghum webworm.[2] Early planting and other measures to hasten maturity of the crop and a thorough cleanup of the crop residue are suggested for control.

24A. SORGHUM MIDGE[3]

Importance, Type of Injury, and Life History. Grain sorghums, sweet sorghum, Sudan grass, and broomcorn in the southeastern quarter of the United States are severely damaged by tiny grayish to red headless maggots which feed on the juices of the developing seeds, blasting or blighting the grain spikelets. An infestation of one larva per spikelet is sufficient to cause a loss of the grain. The adult midge is a minute orange fly which, during its lifetime of about 1 day, lays 30 to 100 eggs in the spikelets or seed husks. The eggs hatch in 2 days, and the maggots are fully grown in 9 to 11 days and pupate in the spikelets. A generation requires from 14 to 16 days, and there are as many as 13 generations a year in Texas. The midges winter as larvae within cocoons in the spikelets of the host plant and usually emerge as adults the following spring, although some carry over until the second or third season.

Control Measures. It is important to plant only pure seed of a uniformly blooming strain and to produce a uniform crop located as far as possible to the leeward from sources of infestation such as Johnson grass and old fields of sorghum, broomcorn, and Sudan grass. Johnson grass should be cut and prevented from producing heads near sorghum fields and cultivated or burned early in the spring to destroy hibernating midges. Crop refuse from sorghum fields should be plowed under or burned during the winter.

INSECTS THAT ATTACK SUGARCANE

Most of the insects that attack corn also attack sugarcane, but there are some species especially destructive to this crop.

25. SUGARCANE BORER[4]

Importance and Type of Injury. This insect has been estimated to cause an annual loss of sugarcane ranging from 4 to 30 per cent with an average value of $10,000,000. The larvae bore into the interior of the sugarcane stalk, and in young plants this produces a yellowed, dying condition of the inner whorls of leaves known as "dead hearts." In older canes the tunneling of the borers causes the tops to die and weakens the supporting tissues, so that the stalks break off in strong winds. Borer injury also decreases the amount and purity of juice that can be extracted from the cane and lowers the sucrose content of the juice as much as 10 to 20 per cent. Borer attacks in seed cane increase the susceptibility to destructive rots.

Plants Attacked. This insect attacks sugarcane, corn, sorghums, rice, and wild grasses.

Distribution. The sugarcane borer was introduced into the United States from the West Indies in about 1855 and is found in Florida, Mississippi, Louisiana, and Texas.

Life History, Appearance, and Habits. This insect, which is a close relative of the southern cornstalk borer (page 495), overwinters as the larva in a tunnel in the host plant. It becomes active early in the spring and bores out toward the surface of the

---

[1] *U.S.D.A., Tech. Bul.* 585, 1937.

[2] *Celama sorghiella* (Riley), Order Lepidoptera, Family Noctuidae (Nolidae).

[3] *Contarinia sorghicola* (Coquillet), Order Diptera, Family Cecidomyiidae (Itonididae). See *U.S.D.A., Tech. Bul.* 788, 1941, and *Farmers' Bul.* 1566, 1959.

[4] *Diatraea saccharalis* (Fabricius), Order Lepidoptera, Family Pyralididae (Crambidae).

stalk, where it transforms to the pupa and the adult emerges. The moth is straw-colored, with the forewings marked with black dots in a V-shaped pattern, and has a wingspread of about 1 inch. The female lays its elliptical eggs in clusters of 2 to 100 or more (average about 25) on the leaves of the host plant, and these hatch in 4 to 9 days. The first-instar larvae feed for a short time on the leaves and then bore into the stalk, where they tunnel up and down, occasionally emerging and reentering at another place. After 20 to 30 days at summer temperatures, the larvae pupate and the moths emerge 6 to 7 days later. The full-grown larva is about 1 inch long, yellow-ish white with brown spots. There are four to five generations a year. Hibernation in cane trash and stubble, seed cane, wild grasses, and rice stubble begins in September to November, and the moths emerge in March to April, depending upon the latitude.

Control Measures. The egg parasite, *Trichogramma minutum* (page 69), destroys many eggs, especially in late fall, but its artificial propagation and release at rates as high as 45,000 adults per acre have had no appreciable effect in reducing borer injury under Louisiana conditions. The imported tachinid fly, *Lixophaga diatraeae,* and the braconid wasp, *Agathis stigmaterus,* have proved effective in reducing the percentage of injured canes in southern Florida. Thorough cleanup and burning of cane trash will greatly reduce the number of overwintering larvae. Where stubble is to be retained for next year's crop, the trash should be plowed under not later than February and left covered until the end of May to prevent the adults from emerging. Borer survival in stubble can also be decreased by cutting the stalks close to the ground. Soaking infested seed cane in water for 3 days will kill the majority of the borers. Where infestations are heavy, cutting out and burning dead hearts of cane are of value in controlling the first-generation borers. Resistant varieties of cane are of considerable importance in decreasing borer injury.

Dusting the cane with cryolite or 40 per cent ryania at 8 pounds per acre by ground equipment or 10 pounds by airplane has given good control of first- and second-generation larvae and considerably increased sugar yields. The applications should be made four times at weekly intervals, beginning with the general hatch of the borers.

References. Jour. Econ. Ento., **41**:914, 1948; *U.S.D.A.,* Cir. 878, 1951.

INSECTS THAT ATTACK LAWNS, GOLF GREENS, PASTURES, AND HAY GRASSES

Among the most serious pests of turf are the webworms (page 513), cutworms (page 476), white grubs (page 503), annual white grub (page 523), wireworms (page 506), green June beetle (page 751), billbugs (page 487), Japanese beetle (page 749), Asiatic garden beetle (page 606), ants (page 803), chinch bugs (page 480), and earth worms. Occasionally, large, dark-colored, thick-skinned maggots will be found around the roots of grasses in pastures. These maggots are the young of the crane flies.[1] They are often called "leatherjackets." In most cases these insects are entirely harmless, feeding only on the decaying vegetable matter in the soil. One species, the range crane fly,[2] has occasionally been destructive in the West. These species all work more or less in the soil. Grasses are often attacked above ground by the armyworm (page 471), the fall armyworm (page 473), chinch bugs (page 480), grasshoppers (page 465), crickets (page 600), and leafhoppers.[3] In some of the western States the range caterpillar[4] has caused serious losses of range grasses. On hay and pasture crops about the only control measures the expense of which

[1] Many genera and species, Order Diptera, Family Tipulidae (see *U.S.D.A., Bur. Ento., Bul.* 85, 1910).

[2] *Tipula simplex* Doane, Order Diptera, Family Tipulidae (see *U.S.D.A., Dept. Cir.* 172, 1921).

[3] See *U.S.D.A., Bur. Ento. Bul.* 108, 1912.

[4] *Hemileuca oliviae* Cockerell, Order Lepidoptera, Family Saturniidae (see *U.S.D.A., Bur. Ento. Bul.* 85, part V, p. 59, 1910).

can be justified are rotations with crops not of the grass family, and pasturing with hogs as recommended on page 506. In some cases sowing pastures with mixtures of grass and legume seeds has reduced the damage by these pests.

Diagnosis of Infestation. In order to apply the most effective insecticidal treatment, it is important to identify the pest concerned before extensive damage occurs. The *pyrethrum test*, which is made by sprinkling areas where insect damage is suspected with pyrethrum extract at the rate of 1 tablespoonful (15 cubic centimeters) per gallon of water per square yard, irritates webworms, cutworms, and other caterpillars and causes them to come to the surface within 10 minutes. Extensive insecticidal treatments are usually not justified unless more than 5 cutworms or 15 webworms are found per square yard. The pyrethrum test does not bring white grubs or the grubs of billbugs to the surface, and for these it is necessary to examine the soil around the grass roots. These are often completely cut away in heavy infestations so that the turf can be rolled back (Fig. 9.27). Treatment should be applied if 3 to 4 grubs per square yard are found.

Treatment with Soil Insecticides. Control measures for insects attacking golf greens, lawns, parkways, etc., include the application of sprays, dusts, and granular insecticides, often in combination with fertilizers. The larvae of the Japanese beetle, the oriental beetle, the Asiatic garden beetle, billbugs, and white grubs may be controlled by applications of DDT at 0.5 to 1 pound, chlordane at 4 to 8 ounces, and aldrin, heptachlor, or dieldrin at 2 to 4 ounces per 1,000 square feet, applied to the upper 3 inches of topsoil in constructing a new lawn or to established sod, followed by heavy watering to wash the insecticide to the roots. As much as 3 to 6 weeks may be required for the insecticide to penetrate and achieve control. These materials may remain effective for several years.

Cutworms, sod webworms or lawn moth larvae, and skipper larvae may be controlled by sprays of DDT or toxaphene at 2 ounces per 1,000 square feet. Aldrin, dieldrin, and heptachlor sprays are effective against sod webworms at 1 ounce, and chlordane at 2 ounces per 1,000 square feet, using about 20 gallons of water. The lawn should be mowed and well watered before treatment, and the application should be made when the lawn is dry, without watering again until necessary.

The cornfield ant or American lawn ant[1] may be eradicated from turf by applications of aldrin, heptachlor, dieldrin, or chlordane, as described above, or by treatment of localized infestations by spraying the hills with chlordane at $\frac{1}{8}$ ounce per gallon of water, or by washing about $\frac{1}{8}$ teaspoonful of wettable chlordane into each hole.

Frequent cutting and watering and top-dressing help to prevent destruction of lawns by chinch bugs (page 480), and dusting or spraying with chlordane, aldrin, dieldrin, or heptachlor, as described above, will kill the bugs. Leafhoppers and various leaf bugs (Miridae) may be controlled by spraying or dusting with malathion at 4 ounces per 1,000 square feet.

Earth worms sometimes become a problem in golf and bowling greens and may be controlled by sprinkling with mercuric chloride, 3 ounces per 50 gallons of water per 1,000 square feet, or by spraying with lead arsenate, 2 to 5 pounds, or chlordane, 4 ounces per 1,000 square feet.

---

[1] *Lasius alienus* (Förster).

References. *Mich. State Coll. Ext. Bul.* 125, 1932; *U.S.D.A.*, *Cir.* 238, 1932; *Jour. Econ. Ento.*, **41**:48, 1948; **42**:499, 1949; and **52**:966, 1959; *Hilgardia*, **17**:267, 1947.

26. ORIENTAL BEETLE,[1] EUROPEAN CHAFER,[2] AND ANNUAL WHITE GRUB[3]

Importance and Type of Injury. The larvae of these three species of beetles are white grubs with an annual life cycle and feed on the roots of grasses, often killing areas of lawn or turf by eating off the roots close to the soil surface. The larvae of the oriental beetle also attack nursery stock and sugarcane in Hawaii. The larvae of the European chafer also feed on the roots of winter grains, clover, alfalfa, strawberries, gladiolus, and other plants. The adult oriental beetles occasionally cause some damage by chewing the blossoms of flowers, and the adult annual white grubs may chew green peaches and other fruits and the leaves of walnuts and roses. They have also occasioned human injury by burrowing in the external ear of sleeping persons.

Distribution. The oriental beetle is a native of Japan and is also found in Hawaii. It was discovered in Connecticut in 1920 and is now present also in New York and New Jersey. The European chafer was introduced from Europe into New York and

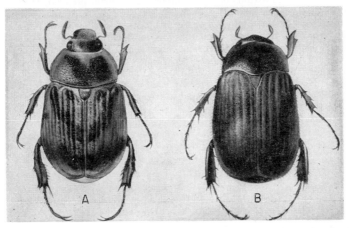

FIG. 9.48. The oriental beetle, *Anomala orientalis*, *A*, and the Asiatic garden beetle, *Autoserica castanea*, *B*. Each enlarged about 3 times. (*From N.J. Dept. Agr.*)

was discovered in 1940. It now infests about 700,000 acres in New York, Connecticut, and West Virginia. The annual white grub is generally distributed over the United States.

Life History, Appearance, and Habits. All three species have very similar life histories, wintering as nearly full-grown larvae below the frost line in the soil. In the spring the larvae resume feeding in late April or May and pupate in earthen cells at a depth of about 6 inches. The adult oriental beetles emerge from June to August and fly from 8 A.M. to 4 P.M. for a few hundred yards. They may swarm in large numbers on a few hot days. The adults are broad-bodied, spiny-legged, convex-backed beetles (Fig. 9.48) about ⅝ inch long, varying in color from yellow to black, but mostly straw-colored with variable dark markings. The females lay an average of about 25 nearly spherical white eggs, singly in the upper 5 inches of soil. The newly hatched larvae feed until cold weather and are of various sizes when they go into hibernation. Most of them complete their life cycle in 1 year, but some pass two winters as larvae. They can best be distinguished from common native white grubs by the transverse anal opening and typical rastral pattern (Fig. 15.54). As many as 550 larvae have been found per square foot of turf.

[1] *Anomala orientalis* Waterhouse, Order Coleoptera, Family Scarabaeidae.
[2] *Amphimallon majalis* (Razoumowsky), Family Scarabaeidae.
[3] *Ochrosidia villosa* Burmeister, Family Scarabaeidae.

The adults of the European chafer emerge from early June until late July and fly in swarms about sunset, mate in trees, and return to the soil at sunrise, burrowing down to about 4 inches. The adults are light-brown or tan oval beetles about ½ inch long with shallow grooves in the elytra. The females may travel up to 2 miles in mating flights and then lay 20 to 40 oval white eggs, singly, in the upper 6 inches of soil. These hatch after 16 to 19 days, and the first larval instar requires 35 days, the second 16 days, and the third about 8 months. The larva has a Y-shaped anal slit with two nearly parallel rows of spines in front (Fig. 15.54).

The adult annual white grubs emerge in late June and in some years may be present in enormous numbers until August. They fly at night and are strongly attracted to lights. The beetle is about ½ inch long, pale yellow to dull brown, and sparsely clothed with fine hairs. The female lays from 10 to 20 oval white eggs, singly, in the soil, and the average total developmental time is about 350 days. As many as 50 larvae have been found per square foot of soil. They have transverse anal slits without the double row of small spines (Fig. 15.54).

Control Measures. The larvae of these beetles feeding in lawns and turf may be controlled by soil applications of DDT, chlordane, aldrin, dieldrin, or heptachlor, as described under Japanese beetle (page 522). Natural enemies such as birds, skunks, moles, and shrews destroy many of the grubs, but will not prevent population build-ups. The importation of natural enemies has not been successful. Federal and state quarantines limit the spread of the oriental beetle and European chafer.

References. Oriental beetle: *Conn. Agr. Exp. Sta. Bul.* 304, 1929; *U.S.D.A., Cirs.* 117, 1930, and 238, 1932; *Jour. Econ. Ento.*, **41**:905, 1948. European chafer: *N.Y. (Geneva) Agr. Exp. Sta. Bul.* 703, 1943; *Jour. Econ. Ento.*, **35**:289, 1942; **36**:345, 1943, **39**:168, 1946, and **45**:313, 1952. Annual white grub: *Jour. Econ. Ento.*, **11**:136, 1918; and **31**:340, 1938.

CHAPTER 10

INSECTS INJURIOUS TO SMALL GRAINS

FIELD KEY FOR THE IDENTIFICATION OF INSECTS INJURING WHEAT AND OTHER SMALL GRAINS

A. Insects chewing leaves, heads, or stalks aboveground, leaving visible holes:
  1. Early-sown wheat in the fall sometimes eaten to the ground. In the spring, the heads and leaves are eaten off...................*Grasshoppers*, page 526.
  2. Fields of wheat stripped of leaves and the awns of the heads eaten off by large green and dark-green striped worms up to 2 inches in length, with 3 pairs of legs and 5 pairs of prolegs. Worms feed mainly at night and remain at base of plants during the day..............................*Armyworm*, page 527.
  3. Part of wheat head eaten off by grayish or greenish-gray worms, with clear-cut brown and yellow stripes on the body; worms hide about the base of the plant during the day. Body slenderer and head relatively bigger than in the armyworm, and with two straight dark bands over top of head...................
  Wheat-head armyworm, Faronta diffusa (Walker) [= *Protoleucania* (= *Neleucania*) *albilinea* (Hübner)] (see *Iowa Agr. Exp. Sta. Bul.* 122, 1911).
  4. Pale-green larvae up to ⅔ inch long, with distinct heads and at least 10 pairs of legs and prolegs, feeding on the edges of the wheat leaves; often hold the hind part of the body curled against the leaf............*Wheat sawflies*, page 527.
B. Small insects sucking sap from leaves, stems, or developing grain, not leaving visible holes, but causing wilting, discoloration, falling over, or death of the plants:
  1. Areas of dead and whitened plants appearing in fields in early spring, especially during periods of cool weather. Plants in and about such areas covered with very small, winged and wingless, green, sucking aphids....*Greenbug*, page 527.
  2. Injury much the same as in *B*, 1, but with the insects more generally distributed over the fields. Later large numbers of the aphids are found in the heads....
  ....*English grain aphid*, page 529.
  3. Patches of wheat dying in the late spring in parts of the field where the stand is the poorest, or on thin ground. Great numbers of small, red, brown, or black-and-white bugs, the largest only ⅕ inch long, clustered on the lower parts of the stems and on the lower leaves...................*Chinchbug*, page 530.
  4. Plants in the fall with stiff, bluish-green leaves; center shoot often missing; or plants dead. Whitish, legless and headless maggots up to 3/16 inch long; or brown, capsule-like cases of about the same length, behind the lower leaf sheaths. In the spring, many of the straws fall when the heads are beginning to fill; brown seed-like cases about ⅛ inch long found behind the lower leaf sheaths but not inside the straw......................*Hessian fly*, page 531.
  5. Heads of wheat full of very small pink or reddish maggots, 1/12 inch long or less, that lie among the bracts and feed on the kernels. Kernels shriveled........
  Wheat midge, Sitodiplosis mosellana (Gehin) (= *Contarinia tritici* Kirby) (see *Purdue Agr. Exp. Sta. Cir.* 82, 1918).
C. Insects boring or tunneling in the stem or straw:
  1. Stalks damaged by a brown-and-white-striped caterpillar working inside the stem. The stem always with a hole in the side from which castings are pushed out.......................................*Common stalk borer*, page 493.

525

2. Plants in fall having much the appearance of *B*, 4, but with very slender, greenish, footless maggots in the enlarged part at the base of the stem.   In the spring, scattered wheat heads are blasted, turn white just after forming, but do not fall over.   The stalk is eaten off inside the leaf sheath at first or second joint below the head, by a pale-green very slender maggot, about ¼ inch long ....*Wheat stem maggot*, page 540.

3. Straw falls as the heads fill, as in *B*, 4, but the inside of the straw is eaten out down to the ground by a pale-yellow, wrinkled-bodied larva about ½ inch long ....*Wheat stem sawfly*, page 535.

4. Wheat in early spring with crown of plant eaten out by small, yellowish, maggot-like larva less than ¼ inch long, with distinct brown jaws, stunting or killing the plant.   Later in the spring, after the plant has jointed, it may be injured at the joint by a similar worm eating inside the stem, and the straw weakened so that the head falls.............................*Wheat strawworm*, page 538.

5. Straw fallen as in *B*, 4, the weakening of the straw caused by small, yellow, legless, brown-jawed larvae, working in hard, knotty galls, usually just above one of the lower joints.   Many bits of the galls and hardened straws coming through the thresher in the grain...............*Wheat jointworm*, page 536.

6. Stems neatly severed with a concave cut from the inside, at or slightly above the surface of the ground, the adjacent stem packed with frass.   A light-yellow larva, up to ½ inch long found inside the straw near the base.............. *Black grain-stem sawfly, Cephus* (= *Trachelus*) *tabidus* (Fabricius) (see *Jour. Econ. Ento.*, **28** :457, 1935).

D. *Insects that attack the plant underground:*

1. Tender portions of grain plants eaten off just below the ground, or in wet weather aboveground, especially in the spring............................... ....*Pale western and other cutworms*, page 476.

2. Young wheat plants killed out in the fall and sometimes in the spring, over irregular areas in the field.   Slender, shining, smooth, tough, brown or yellow-ish-brown larvae, the largest 1 to 2 inches long, with 6 slender legs, found around the wheat roots.............................*Wireworms*, page 541.

3. Germinating seeds in the ground and young plants eaten by slender, pale-brown or yellowish-brown or nearly black, 6-legged larvae, from ½ to 1 inch in length, the body prominently jointed.   Legs and antennae longer than in true wire-worms.   Injury occurs in drier parts of wheat belt.   *False wireworms*, page 542.

4. Wheat plants, especially in early-sown fields, killed out in the fall over irregular areas in the field.   Plants eaten off just above the roots by slender, white, 6-legged, brown-headed larvae which are less than half an inch long and very difficult to find.........................*Spotted cucumber beetle*, page 510.

5. Roots of plants eaten off in the late spring and sometimes in the fall in early-sown fields, by white, curved-bodied grubs, from ½ to over 1 inch in length, with reddish-brown heads and 6 long legs.   Grubs feed just below the surface of the ground......................................*White grubs*, page 543.

6. Wheat stalks die at the ground just before harvest.   Lower part of stalk hollowed out by white, short, very fat, legless larvae about ¼ inch in length. Stalks sometimes fall.................................*Billbugs*, page 544.

A. INSECTS THAT ATTACK WHEAT

27 (1, 75, 146). GRASSHOPPERS

Grasshoppers, when abundant, especially in the spring-wheat-growing sections, frequently cause severe injury to the wheat by eating off the bracts or sometimes cutting off newly formed heads.   As a rule, severe damage by grasshoppers to wheat occurs only in the western states.   Early-sown wheat is sometimes killed by grass-hoppers feeding upon it in the fall.   Grasshoppers may be controlled on wheat by the methods given for controlling these insects on corn.

For a full description of grasshoppers, see Corn Insects, page 465.

28 (2, 48). Armyworm[1]

Importance and Type of Injury. Armyworms are most apt to be destructive to wheat and other small grains when wet weather has caused a rank growth in the fields. The female moths lay their eggs in large numbers among the rank or lodged grain in the fields, and the worms hatching from these eggs may suddenly appear in such numbers as completely to strip the leaves from the grain; they then crawl out into other near-by fields. For full description of this insect, see Corn Insects, page 471.

Control Measures. In wheat and other small-grain fields the only method of controlling the armyworm is to watch the fields closely in years when the moths are known to be abundant and, if the worms are found in the fields, to apply poison bran at once (page 469).

Sawflies That Feed on the Wheat Leaves[2]

There are several species of sawflies that feed on the leaves of wheat. In all these the larvae may be recognized by their wrinkled bodies, in most cases of a pale green color, and by the number of abdominal prolegs of which there are always six pairs or more. They occur throughout the wheat-growing areas of the country but are seldom if ever of economic importance. Occasionally, some of these species will become abundant enough partially to strip the plants of leaves, but, as this stripping does not occur until the wheat head is partially filled, it has but little effect on the yield of the grain. They are, in general, heavily parasitized, and no special control measures are necessary to keep down their numbers.

29. Greenbug[3]

Importance and Type of Injury. This grain aphid, because of its general distribution and great prolificacy, causes a loss of from 1 to possibly 3 per cent of the wheat crop of the entire world. In some years the damage has amounted to 25 per cent of the wheat crop in the southwestern part of the United States. Wheat or other small-grain fields infested by the greenbug usually show small deadened areas appearing in the field during the late winter or early spring. An examination of such areas will show the plants swarming with numbers of tiny green aphids or plant lice (Fig. 10.1,*A*), which are sucking the sap. While feeding, the aphid injects a toxic saliva which causes discoloration and tissue destruction. During periods of spring weather, favorable to the aphids, these deadened spots may spread rapidly and the entire field be killed out.

Plants Attacked. The greenbug feeds on all the small grains and many of the wild and cultivated grasses. It has also been found on rice, corn, sorghum, and several other cultivated plants.

Distribution. The insect is of European origin and was first recorded in the United States in 1882 in Virginia. It has now spread so it is generally distributed over practically all the United States, its range extending on the north into Canada. The greenbug is not common in the New England states. Its greatest damage has been done in the large grain-growing states west of the Mississippi, including Oklahoma, Texas, Kansas, and Nebraska.

Life History, Appearance, and Habits. In the southern states this insect passes the winter in the active nymphal and adult stages, feeding

---

[1] *Pseudaletia* (= *Cirphis*) *unipuncta* (Haworth), Order Lepidoptera, Family Noctuidae.

[2] *Dolerus arvensis* Say, *Dolerus collaris* Say, and *Pachynematus extensicornis* (Norton), Order Hymenoptera, Family Tenthredinidae.

[3] *Toxoptera graminum* (Rondani), Order Homoptera, Family Aphidae.

on the stems of plants and giving birth to living young during the warmer periods of weather.   In the more northern states the winter is passed in the form of black shiny eggs, deposited on the leaves of plants on which the insect feeds.   These eggs hatch during the winter or early spring, producing numbers of pale-green, wingless female insects, about $\frac{1}{16}$ inch in length when full-grown, and having a dark-green stripe down the back (Fig. 10.1,*A*).   In from 7 to 18 days after hatching, these females begin giving birth to living young.   These may become either winged or wing-less.   The winged individuals differ slightly in appearance from those hatching from the winter eggs, being slightly larger and with filmy wings having an expanse of about $\frac{1}{4}$ inch.   The head is brownish yellow and there are blackish lobes on the back of the thorax.   In about 15 days these females, in turn, begin giving birth to living young, and the insects continue thus, generation following generation.   Each female begins reproducing when from 7 to 18 days old and continues to reproduce for about

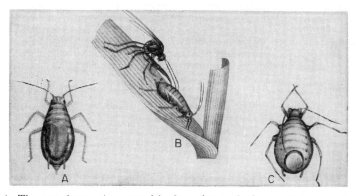

Fig. 10.1. The greenbug.   *A*, a parasitized specimen, the larva of the parasite showing through its body wall; *B*, a female of the parasite, *Aphidius testaceipes*, ovipositing in a greenbug; *C*, the dead "shell" of a greenbug showing the circular hole through which an adult parasite has emerged.   All much enlarged; actual size about the same as the letters in this figure.   (*Modified from Kans. Univ. Bul., vol. 9, no. 2.*)

20 to 30 days, giving birth during this time to an average of from 50 to 60 young.   It will be seen from these figures, that the rate of reproduction is enormous and almost beyond comprehension.   If unattacked by natural enemies, in the course of a single season the mass of aphids which could be produced would be so great as to destroy all vegetation upon which they could feed.   On the approach of cold weather, the female aphids give rise to winged males and females.   These mate, and the mated females produce eggs, in which stage the insect passes the winter in the colder parts of the country.   A single female may at times both lay eggs and give birth to living young.   There are from 5 to 14 generations each season, all except the last being composed entirely of females.

Control Measures.   The abundance of this insect and resultant injury are very largely dependent on weather conditions and can be but little influenced by man.   No serious outbreak of the greenbug has occurred except when the preceding summer was comparatively cool and moist. The worst outbreaks have always occurred in seasons of mild winters followed by cool, late springs.   The reason for this is that the greenbug will start reproducing at temperatures a little above 40°F. and can reproduce at a fairly rapid rate at temperatures between 55 and 65°F. It is preyed upon by a number of insect enemies, but particularly by a little wasp[1] (Fig. 10.1,*B*). This tiny wasp stings the aphids and deposits its eggs within their bodies. These eggs hatch into little maggots which eat out the body substance of the aphid, and finally emerge as adults, through a hole cut in the back of the aphid (Fig. 10.1,*C*). These wasps are practically always present in areas where the greenbug is abundant. When the temperature is below 65°F., the wasp will reproduce very slowly or hardly at all. Long periods of cool wet weather thus permit the greenbug to increase in enormous numbers, while its most effective natural enemy can increase only very slowly. This relationship between the two insects, and the effect of the weather upon them, is apparently responsible for the abundance of the aphids during the years of mild winters and cool springs. In normal years the parasites reproduce at a sufficiently rapid rate to hold down excessive abundance of the aphids.

Cultural control measures which may reduce the damage caused by this insect include the destruction of volunteer grain, especially oats, which is particularly attractive to the aphids in the early spring and serves as centers from which the insects can spread to adjacent fields. Crop rotations which increase the nitrogen levels in the soil produce vigorous, more rapidly growing plants, which are better able to withstand aphid attack. Progress is being made in the development of resistant varieties of small grains.

The most satisfactory chemical control measure has been the spraying of infested fields with parathion or methyl parathion at 0.25 pound per acre. More than 2,500,000 acres were treated by this method during the outbreak of 1950, with very acceptable results. Spraying should not be begun until the aphid population reaches the threshold of economic damage, about 11 to 50 aphids per row-foot, and should be carried out only above 50°F.

References. *Kans. Univ. Bul.* 9, no. 2, 1909; *U.S.D.A., Farmers' Bul.* 1217, 1921, *Bur. Ento. Bul.* 110, 1912, and *Tech. Bul.* 901, 1915; *Okla. Agr. Exp. Sta. Tech. Bul.* T-40, 1951, and T-55, 1955; *Texas Agr. Exp. Sta. Bul.* 845, 1956; *Jour. Econ. Ento.*, **41**:520, 1948; **44**:347, 1951; **45**:82, 1953; and **49**:567, 600, 1956.

30. ENGLISH GRAIN APHID[2]

Importance and Type of Injury. Grain infested by the English grain aphid will show somewhat the same appearance as that infested by the greenbug. However, the infestation is not usually confined to small spots in the field. After the wheat or small grain begins to head, very large numbers of these aphids will often be found clustered in the bracts of the wheat heads, or in the heads of other grain (Fig. 10.2). Their feeding may shrivel the growing wheat kernels and in the early spring cause the death of the wheat plants.

Plants Attacked. This insect feeds on all the small grains and many of the wild and cultivated grasses. It has been found in small numbers on corn but is not an important pest of this plant.

Distribution. The English grain aphid is generally distributed throughout the United States and southern Canada wherever small grains are grown.

Life History, Appearance, and Habits. This insect passes the winter mainly in the fully or partly grown stages. A few individuals go through the winter in the egg stage. It will be found in the heaviest growth of grain and especially in clumps of volunteer

[1] *Aphidius* (= *Lysiphlebus*) *testaceipes* (Cresson), Order Hymenoptera, Family Braconidae.

[2] *Macrosiphum granarium* (Kirby), Order Homoptera, Family Aphidae.

oats, rye, or wheat that has made a rank growth. The overwintering forms are all females, and, as the weather becomes warm in the spring, they begin giving birth to living young which may become winged or wingless. They feed during the early spring on the growing grains, sucking the sap from the leaves and stems. As the heads begin to form, many of the aphids will gather in the heads, causing shriveling and shrinking of the newly formed grain. After the harvest of small grains the insects migrate to wild or cultivated grasses where they spend the summer. In the fall, after the winter grains are planted, they go back to them or, as above stated, gather in large numbers in clumps of volunteer grain. The males appear during the fall and early winter and mate with the true females, which lay eggs on the grains where they have been feeding. Only a comparatively small number of eggs are laid, the average number being about eight. The wingless females are of a pale-green color with long black antennae and have a long black cornicle extending backward from each side of the abdomen. The winged individuals are about the same size and of the same general color. The lobes on the thorax, however, are brown or blackish. The wing expanse is a little over $\frac{1}{4}$ inch.

Fig. 10.2. The English grain aphid clustered about the bracts of a wheat head, enlarged about $3\frac{1}{2}$ times. (*From Ill. Natural History Surv.*)

Control Measures. As in the case with the greenbug, we are largely dependent upon natural enemies for the control of this insect. It is usually held in check by several parasites, by lady beetles, and other aphid eaters. During cool springs, it may become sufficiently abundant to cause damage. Destruction of volunteer grain in the fall is of value in controlling this species. The use of insecticides for aphid control is discussed on page 613. The minimum economic level justifying insecticidal treatment has been placed at 25 to 30 aphids per head.

Reference. Jour. Agr. Res., **7**:463, 1916.

31 (6). CHINCH BUG[1]

Importance and Type of Injury. The chinch bug is as destructive to small grains as it is to corn. Wheat fields infested with the chinch bug show a deadening and drying out of the plants early in the spring. As the wheat begins to head, deadened areas will appear over the fields. Usually these areas are in the spots where the soil is the poorest or where the stand of wheat has been partly killed out by the winter or by heavy rains early in the spring. Thus in dry years these spots will usually appear in the higher parts of the field, while following wet springs they will usually be noticed in the low, wet spots. For full description of the chinch bug, see Corn Insects, page 480.

Control Measures. The control of chinch bugs by insecticides is discussed on page 485. Such measures are not, in general, practical to apply to wheat fields which have already been infested by the bugs. The sowing of wheat on fertile soil, which will promote a heavy growth, is of value, as chinch bugs avoid shade and dampness and will not be found in large numbers in fields having good stands of wheat with a strong uniform growth. Sowing clovers in wheat will also help to keep down the number of bugs in the field, as the growth of these plants shades the ground. Very

[1] *Blissus leucopterus* (Say), Order Hemiptera, Family Lygaeidae.

few chinch bugs spend the winter in wheat stubble, so that burning over infested stubble during the latter part of the summer is practically of no value for the control of these pests.

References. U.S.D.A., Farmers' Bul. 1498, 1936; *Kans. Agr. Exp. Sta. Bul.* 191, 1913; *Ill. Agr. Exp. Sta. Buls.* 243, 1923, and 249, 1924; *Ill. Agr. Exp. Sta. Circs.* 268, 1923, and 431, 1935.

32. HESSIAN FLY[1]

Importance and Type of Injury. The type of injury caused by the Hessian fly is not conspicuous. Wheat infested in the fall is stunted in growth; the leaves of the plants take on a dark bluish-green color, become distinctly thickened, and stand more erect and stiff than those of uninfested plants. The central growing shoot is often lacking. Small white or greenish-white, shiny, legless and headless maggots about $3/16$ inch in length; or brown, elongated, capsule-like cases (*puparia*), about $1/8$ inch long, containing white maggots, will be found behind the sheaths of the lower leaves of the plant, usually below the surface of the ground (Fig. 10.3, *December to March*). The injury is caused entirely by the larvae, which withdraw the sap from the lower parts of the stem. Heavily infested plants generally die during the winter. In the early spring the appearance of the injured plants is much the same as that in the fall. Later in the spring the capsule-like "flaxseeds" will be found behind the leaf sheath above the surface of the ground, sometimes as high as the second or third joints. Infested straws usually break over when the heads begin to fill (Fig. 10.3, *June*). Heavily infested fields will frequently have 50 to 75 per cent or more of the straws fallen. The yield of infested grains is seriously reduced.

Plants Attacked. The principal food plants of the Hessian fly are wheat, barley, and rye and are preferred in about the order named, wheat being by far the favorite food of this insect. Oats are never injured by this insect. It has been taken rarely and in very small numbers on certain species of wild grasses. It has also been taken in very small numbers on emmer and spelt. Eggs have been found on oats, foxtail, and einkorn, but no larvae have ever been known to develop on these plants.[2]

Distribution. The original home of the Hessian fly was possibly in the southern Caucasus region of Russia. It was probably introduced into North America in straw bedding used by the Hessian troops during the Revolutionary War, as it was first noted on Long Island about 1779. The insect has now spread to all the principal wheat-growing areas of the world. It is not found in a few of the arid wheat-growing sections in the plains states, but occurs on the Pacific coast.

Life History, Appearance, and Habits. The Hessian fly passes the winter in the full-grown maggot stage. Occasionally a partly grown maggot will survive. In most cases, however, the maggot is within the brown puparium, or what is commonly called the "flaxseed." These overwintering stages will be found hidden away behind the leaves of the volunteer or early-sown wheat (Fig. 10.3, *November to March*), between leaf sheath and stem, or in some cases among the stubble of the previous season's crop. The insect is inactive during the winter, all those in the flaxseed stage having finished feeding. In the spring, shortly after the

[1] *Phytophaga destructor* (Say), Order Diptera, Family Cecidomyiidae (Itonididae).
[2] *Kans. Agr. Exp. Sta. Tech. Bul.* 11, 1923.

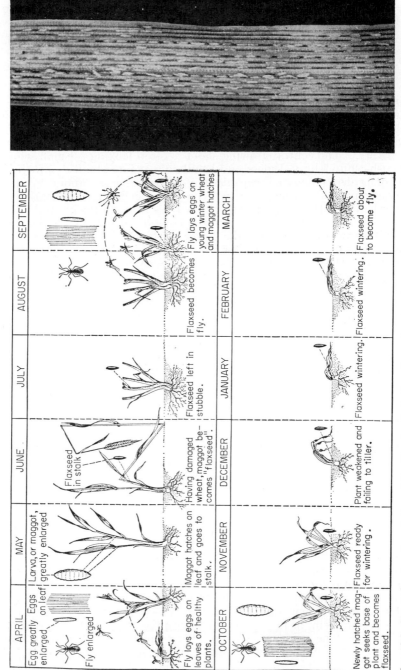

Fig. 10.3. Chart showing the relation of the Hessian fly to the wheat plant, each month of the year. (*From Lochhead, "Economic Entomology," McGraw-Hill-Blakiston, after U.S.D.A.*) At the right, photograph of Hessian fly eggs on leaf of wheat, enlarged about 4 diameters, showing how the eggs are laid in the natural grooves of the wheat leaf on the upper side. (*Original.*)

wheat plant starts its active period of growth, the maggots change inside the puparia to the pupal stage and in a week or two emerge as small two-winged flies a little less than ⅛ inch in length. These flies are all a sooty-black color, much smaller than the common house mosquito, and are very frail creatures, never feeding so far as known. During windy weather they remain clinging to the leaves or clustering about the base of the wheat plants. On warm days they fly about over the fields and mate, and the females, whose abdomens are of an orange-red color, lay eggs of the same reddish color in the grooves on the upper sides of the wheat leaves. The eggs (Fig. 10.3, *April and October*) are very small and can just be seen without the use of a magnifier. They are slender, often laid 2 to 15 in a string, end to end, and, when viewed through a lens, they have somewhat the appearance of a string of "wieners." They are likely to be overlooked by any but the most careful observers. The adult flies probably never live more than 4 days and most records show that they usually die in 3 days or less. While incapable of flying long distances by their own efforts, they may be carried by a moderate wind for a distance of several miles. Adult Hessian flies may be found fairly abundant at a height of 25 feet above the ground, and they may occur in considerable numbers at least 2 miles from any of their known food plants. The female flies lay from 250 to 300 eggs, the average number of the fall generation being 285, and for the spring generation slightly less. Many eggs are often deposited on a single plant, as many as 319 having been counted in the fall on one plant. The eggs hatch in from 3 to 10 days, depending on the temperature. The young maggots, which are reddish when they first emerge from the egg, soon turn white. They work their way down the grooves of the leaves as far as they can go behind the leaf sheath, without cutting through the sheath or the stem. Here they start feeding by rasping on the straw and sucking up the sap which oozes out from the irritated surface. They do not move about after feeding has begun. The maggots never enter the straw of the plant as do those of some other insects which are sometimes mistaken for the Hessian fly. With favorable weather the maggots will become full-grown in about 2 weeks (Fig. 10.3, *May and October*). The outer skin then loosens from an inner skin and forms the brown protective case known as the flaxseed or puparium. Most of the maggots developing from eggs laid in the spring reach the flaxseed stage some time before the wheat begins to head. Normally, there is one spring generation. Under certain weather conditions, particularly those of an early wet spring, a second or supplementary spring generation may be produced. Flies of the supplementary generation emerge and lay their eggs on the late tillers, normally causing but little injury. Nearly all the flies will be in the flaxseed stage 2 weeks or more before wheat harvest. They remain in this stage in the dry stubble during the summer (Fig. 10.3, *July, August*) and emerge again as flies as soon as sufficient rain has fallen to cause a growth of volunteer wheat in the fields. The stimulus that causes the estivating larvae in the puparia to pupate is the soaking of the straw and the puparia by late summer rains with the average temperature above 45°F. The same conditions start the growth of volunteer wheat. Having pupated, the time before the flies emerge is dependent upon temperature. At 40°F. the pupal period averaged 30 days; at 50°F., 15 days; at 60°F., about 11 days; while at 66°F., it averaged a little over 7 days. It is said that above 75°F. the larvae will not pupate. During wet seasons, adult flies may begin coming out by the middle of July and a nearly full, summer generation may be produced in volunteer wheat. In normal years the flies do not start emerging before late summer or early fall and lay their eggs in such volunteer wheat as may be present in the fields or in early-sown wheat. In central Illinois, emergence starts usually about September 1. If no green wheat, rye, or barley is available, the flies will die without depositing many of their eggs. Emergence ceases on the approach of cold weather. The maggots developing from the fall generation of flies will nearly all become full-grown before the first hard frost or at least before the ground freezes in the fall. Maggots that are less than half-grown are likely to be killed by freezing weather.

It will be seen from this description that there are normally two full generations of flies each year, but under exceptionally favorable weather conditions there may be three, four, or even five. In California there is generally a single generation, the insects remaining in the puparia from

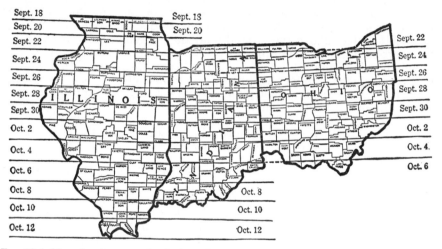

Fig. 10.4. Map showing normal safe dates for sowing wheat in several North Central states to escape injury by the Hessian fly. (*Original.*)

June to January, and infestation of the new crop comes mostly from the stubble of the preceding year.

Control Measures. The three measures which have been found of most value in preventing damage by this insect, throughout most of the wheat belt, are:

1. Sowing sufficiently late in the fall so that the wheat will not come up until after the adult flies have emerged, laid their eggs, and died. The accompanying map (Fig. 10.4) shows the normal safe dates for sowing wheat to escape injury by the Hessian fly and to make the largest yields, in some of the North Central states.

2. Keeping down all growth of volunteer wheat in the fields, or about stacks, on which the fall generation of flies might deposit their eggs, and thus carry the insect through until the following spring.

3. Plowing under infested wheat stubble as soon as possible after harvest. The adult Hessian fly is such a weak insect that it cannot work its way up through the soil, and where wheat stubble is thoroughly

plowed under and firmed, this will prevent the Hessian fly from emerging. This is a most effective control measure but frequently cannot be practiced because wheat is often used as a nurse crop for grasses or clovers which remain on the ground for a year or more after the wheat is harvested.

The proper date of seeding wheat to escape infestation by the fall generation of Hessian fly has now been worked out by the entomologists in the experiment stations of all the principal wheat-growing states. There will be some variation in the time of emergence of the fly during different seasons, as its development is so largely dependent on weather conditions. However, long-continued experimental seedings in many states have shown that it is nearly always possible to sow late enough to avoid any but a very light infestation by the Hessian fly and still to secure a sufficient plant growth to permit the wheat to withstand the cold weather of the winter. The results secured from experimental seedings conducted for 8 years, in Illinois, are summarized on page 410. Under California conditions, where the adults emerge from February to April, *early* planting and the stimulation of rapid growth are suggested.

Many experiments have been carried on to develop other methods of control. Those sometimes suggested, but which have been found of practically no value, are pasturing wheat with cattle or sheep, rolling wheat to crush the maggots or puparia, mowing wheat in the spring after the spring generation of flies has emerged, and early planting of strips of wheat as traps. Under certain conditions rotation of crops is of some value in preventing injury. Where no grass or clover is sown in the wheat, and the weather is sufficiently dry, summer burning of stubble may be of slight value. In most cases, however, the flaxseeds are sufficiently low in the wheat so that they are not killed by the fire. Some varieties of wheat have been found resistant to Hessian fly, but up to the present time no varieties have been found sufficiently resistant to attacks of this insect, and having other desirable qualities, to warrant recommending them generally in any of the large wheat-growing areas. Maintaining the wheat ground in a good state of fertility is of considerable value, as a wheat plant on rich ground will overcome the attack of one or two maggots of the Hessian fly and still produce a fairly good yield, whereas on poor ground such a plant would be killed. Where late seeding is practiced, the seedbed should be put in the best condition possible so that a vigorous growing plant may be obtained. No method has yet been developed for controlling Hessian fly by direct application of insecticides.

References. Ohio Agr. Exp. Sta. Bul. 177, 1906; *Jour. Agr. Res.*, **12**:519–527, 1918; *U.S.D.A., Tech. Bul.* 81, 1928; *Mo. Agr. Exp. Sta. Cir.* 212, 1941, *Kans. Agr. Exp. Sta. Tech. Buls.* 11, 1923, and 59, 1945.

33. Wheat Stem Sawfly[1]

Importance and Type of Injury. This native grass-feeding sawfly has acquired an appetite for small grains and become a menace to the small-grain crops in the more northern wheat belt. In some years it destroys 50 per cent of the crop. Wheat infested by this sawfly will show fallen straw in much the same manner as fields infested by Hessian fly or joint-

---

[1] *Cephus cinctus* Norton, Order Hymenoptera, Family Cephidae.

worm. Examination of the straw will show the inside filled with fine sawdust-like cuttings, among which will be found a wrinkled-bodied, nearly legless, brown-headed larva $\frac{1}{3}$ to $\frac{1}{2}$ inch long, of a pale-yellow color and with a short, pointed projection at the "tail" end.

Plants Attacked. Wheat, spring rye, barley, spelt, timothy, quack grass, and some of the native grasses.

Distribution. The insect is native to North America and is most destructive in the northern wheat-growing states west of the Mississippi River. A closely related species, the European wheat stem sawfly,[1] has been found in New York and Pennsylvania.

Life History, Appearance, and Habits. The winter is passed as a mature larva in the base of the wheat straw near the surface of the soil. In the spring the larva transforms to a pupa inside the wheat straw and the adult emerges during June. The adult is wasp-like in appearance, black, with yellow rings on the abdominal segments. The females lay their eggs by thrusting them into the plant tissues on the upper parts of the wheat stem. The larva feeds within the stem, boring down through the joints, and, by late summer, has reached the lower parts of the plant close to the surface of the ground. Here it cuts a V-shaped groove entirely around and inside the stem, which causes the stem to break off, and plugs itself in the base of the plant with its frass, thus forming a chamber in which it hibernates and later pupates.

Control Measures. Plowing under infested stubble in the fall, making sure that it is thoroughly turned under to a depth of 5 or 6 inches is the best method of control. The overwintering larvae remain so close to the surface of the soil that it is impossible to kill many of them by burning. Cutting grains as early as possible without seriously affecting the yield or grade of wheat does much to reduce damage by this insect. Rotation of crops which will put some immune crops, as corn, winter rye, flax, oats, alfalfa, or sweet clover, on the wheat-stubble land is also a help in combating this insect. Some varieties of wheat, especially the solid-stemmed varieties, are resistant.

References. U.S.D.A., *Tech. Bul.* 157, 1928; *Sci. Agr.*, **15**:30, 1934; *Bul. Ento. Res.*, **22**:547, 1931.

34. WHEAT JOINTWORM[2]

Importance and Type of Injury. In the large wheat-growing areas east of the Mississippi the wheat jointworm is probably second in importance only to the Hessian fly as an insect pest of wheat. Infested fields just before harvesttime will show many of the straws broken off and bent over in a manner similar to fields infested by the Hessian fly. An examination of the fallen straws will show numerous, hard, gall-like swellings filling the entire straw, often for an inch or more, usually just above a joint (Fig. 10.5). Inside these swellings, in oval cavities, are small, yellowish maggots about $\frac{1}{7}$ to $\frac{1}{6}$ inch in length. When infested fields are threshed, many small bits of broken straw containing these galls will come through into the grain or be thrown out in large numbers around the separator.

Plants Attacked. Wheat is the only host known in the East, but the insect is reported from grass in California.

---

[1] *Cephus pygmaeus* (Linné).

[2] *Harmolita tritici* (Fitch), Order Hymenoptera, Family Chalcididae (Eurytomidae).

Distribution. The wheat jointworm is a native insect and is generally distributed in the states east of the Mississippi. Its range extends, however, into some of the states west of the Mississippi, but it has caused little damage in this area except in Missouri. The absence of the insect west of the Mississippi appears to be due to the harder stems of the wheat in that area.

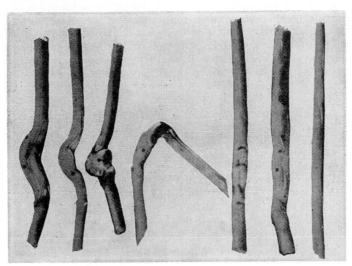

FIG. 10.5. Characteristic galls in wheat straws caused by the wheat jointworm, showing exit holes of adults or parasites. About natural size. (*From U.S.D.A., Dept. Bul.* 808.)

Life History, Appearance, and Habits. The winter is passed in the gall-like hardened swellings inside the wheat straw, the insect being in the pupal or larval stage, mainly in the former. Those that have passed the winter in the larval stage change to pupae early in the spring, and

FIG. 10.6. Adult female jointworm with her ovipositor thrust through the leaf sheath into the straw. Enlarged about 6 times. (*From Ohio Agr. Exp. Sta.*)

all emerge as adults about the time the active growth period of the wheat starts, or the plants are beginning to form joints. The adult insects are about $\frac{1}{10}$ to $\frac{1}{8}$ inch in length and are jet black with the exception of the joints of the legs and two spots on the shoulders which are yellow. The females after mating insert their eggs just above the wheat joints and inside the straw (Fig. 10.6). They drill a tiny hole into the wall of the straw by means of a stiff hair-like ovipositor attached to the under-side of the abdomen.   Usually a number of eggs are laid in one place, sometimes as many as 25.   Occasionally, only one egg will be deposited in a plant.   The larvae feed within the walls of the straw, and the irrita-tion set up by this feeding causes the straw to thicken, each larva being separated from the others in a little cavity of its own.   Often the swellings around the larvae cause the straw to twist or the formation of the galls makes the straw so brittle that it is broken over in the field.   The larvae complete their growth about the time the wheat matures but remain in the larval condition inside the straw until fall, when most of them pupate. The height of the jointworms inside the straw will vary in different seasons.   Sometimes the galls will be just above the first joint, within 8 or 10 inches of the surface of the ground, and at other times the galls may be as high as the third joint and will, therefore, be cut off with the straw when the wheat is harvested.   The variation in the height of the galls is due to the difference in the development of the plant at the time the eggs are deposited, the females tending to lay their eggs in the upper-most parts of the plant.

Control Measures.   If wheat stubble can be thoroughly burned, prac-tically all the overwintering jointworms remaining in the stubble may be killed.   Plowing the infested stubble under shortly after harvest, being sure that all stubble is turned under to a depth of 5 or 6 inches, is a good control measure.   These measures are especially advisable where combine harvesters and threshers are used.   In seasons when the larvae are well up in the straw, it is possible by cutting the wheat low to remove nearly all the insects in the straw.   Such straw may then be baled and sold for use in cities or, if it cannot be disposed of in this way, it should be burned.   Rotation of crops, putting any other crop than wheat on the land, also may be of some help in reducing the numbers of this insect.   In ordinary years, most of the infestation originates from not burning, or not plowing under, the stubble of the previous season, so that every effort should be made to burn or turn under infested stubble as soon after harvest as possible.

References.   *U.S.D.A.*, *Farmers' Bul.* 1006, 1918; *Ohio Agr. Exp. Sta. Bul.* 226, 1911; *U.S.D.A.*, *Dept. Bul.* 808, 1920; *Utah Agr. Exp. Sta. Bul.* 243, 1933.

35. WHEAT STRAWWORM[1]

Importance and Type of Injury.   Injury by this insect is of two dis-tinct kinds: that caused by the first-generation larvae in young plants in early spring and the later injury by the second generation in the matur-ing straw.   In the spring plants attacked by the wheat strawworm show a stunted appearance; the crown of the plant is usually eaten out, includ-ing the developing head; and the plant killed or so injured that no head is produced.   The later injury, after the plant has started to form joints, has a somewhat stunting effect, weakening the straw, although a head may be produced.

Plants Attacked.   Wheat.   A closely related species[2] attacks rye.

Distribution.   The insect is generally distributed in the wheat-growing regions west of the Mississippi River and is found in small numbers, but is rarely destructive, in the states east of the Mississippi River.

[1] *Harmolita grandis* (Riley), Order Hymenoptera, Family Chalcididae (Eurytomi-dae).

[2] *Harmolita websteri* (Howard).

Life History, Appearance, and Habits. The insect passes the winter in the larval and pupal stages in the stubble or in stacks of straw. In the early spring the adult insects gnaw small round holes in the straw, through which they emerge. Usually emergence begins during April or, in early spring, in March. These adults from the overwintering pupae are wingless and about ⅙ inch in length. The general color of most of the insects is brownish. They have much the appearance of ants. Upon examination with a lens their bodies will be seen to be quite hairy. They deposit their eggs in the stem walls at the base of the young wheat plants a little above the ground. The larvae, on hatching, work their way into the stem, eating off the developing head and preventing the formation of any tillers on the plant. The larvae, which are yellowish, legless, with small

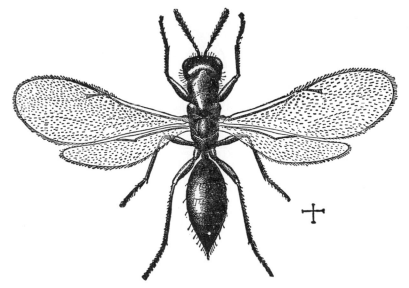

Fig. 10.7. Adult female of the summer generation of wheat strawworm, greatly enlarged. Lines show natural size. (*From U.S.D.A., Dept. Bul.* 808.)

heads, and only about ⅕ inch long, become full-grown the latter part of April or first of May. They pupate within the plant and emerge as adults the latter part of May. These adults (Fig. 10.7) are larger than the early spring adults, less hairy, somewhat black in color, and are practically all winged. The females deposit their eggs inside the straw during the late spring, usually about the time the wheat is heading. In general, only one egg is laid in a wheat stem. The larva feeds within the stem and remains in this stage during the summer, either in the cut straw or, if it was working in the lower part of the stem, in the standing stubble. They change to the pupal stage in midautumn and pass the winter in this stage. There are thus two generations of the insect a year, the adult females of the two generations being strikingly different both in size and in appearance.

Control Measures. As the majority of the adults of the spring generation are wingless, and those of the summer generation are not strong fliers, one of the best methods of controlling this insect is by rotation of crops. Wheat should not be sown within 300 feet of any straw of the previous season. Volunteer wheat should be destroyed by early spring. In dry-farm areas, allowing the land to lie fallow every other year is recommended. Baling and shipping infested straw to cities or towns is also a help in reducing the numbers of this pest. Many of the overwintering pupae may be destroyed where the stubble can be closely burned.

References. U.S.D.A., Dept. Buls. 808, 1920, and 1137, 1923; and *Farmers' Bul.* 1323, 1923; *Utah Agr. Exp. Sta. Bul.* 243, 1933; *Jour. Econ. Ento.,* **24**:414, 1931.

36. Wheat Stem Maggot[1]

Importance and Type of Injury. Wheat attacked by this insect in the fall of the year appears much the same as that infested by the Hessian fly, the plants taking on a darker appearance and remaining stunted, with

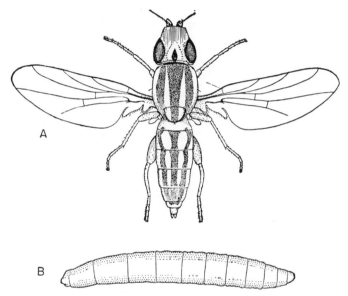

Fig. 10.8. The wheat stem maggot, *Meromyza americana. A,* adult, *B,* larva; 11 times natural size. (*From Ill. Natural History Surv.*)

stiff, somewhat thickened leaves. An examination of these plants will reveal slender, pale-green maggots (Fig. 10.8,*B*) working inside the lower part of the stem or crown of the plant. These maggots are about ¼ inch long.

The summer type of injury differs from that of the winter. The first indication of the presence of the insects is usually the dying out and whitening of the wheat heads and upper parts of the straw shortly after the head begins to fill while the lower stem and leaves are still green. The maggots will be found at this time of the year inside the straw just above the last or next to the last joint. The whitened heads are very conspicuous in the green fields of wheat, often giving an exaggerated idea of the importance of the insect. Its injury rarely amounts to 1 to 2 per cent, but occasionally may be much higher.

[1] *Meromyza americana* Fitch, Order Diptera, Family Chloropidae (Oscinidae).

Plants Attacked. The principal food plants among the cultivated crops are wheat, rye, barley, and oats. It also feeds on bluegrass, timothy, and a number of the wild and introduced grasses.

Distribution. The insect is a native species, occurring over practically the entire United States, in Mexico, and in the principal agricultural regions of Canada.

Life History, Appearance, and Habits. The insect, so far as known, passes the winter only in the larval stage, the maggot being hidden away inside the lower parts of the stem of the wheat or the other plants on which it feeds. In the spring these larvae change inside a green puparium to the pupal stage and emerge mostly in June as yellowish-white flies, about $\frac{1}{5}$ inch long, with three conspicuous black stripes on the thorax and abdomen and with conspicuous bright-green eyes (Fig. 10.8,*A*). The females, after mating, deposit their eggs on the leaves or stems of wheat and grasses on which the larvae feed. The young maggots crawl down behind the leaf sheaths to the tender soft part of the stems and tunnel into them—in the case of wheat, feeding along the stem for a distance of 2 or 3 inches. The injured stem is partly severed and the head turns white and dies. When the larva becomes full-grown, the outer skin loosens from an inner skin (the integument of the last instar larva) and forms a pale-green slender puparium in which the maggot changes to the pupal stage and later to the adult stage. These adults emerge about midsummer and lay their eggs on wild grasses or volunteer grain. The larvae of this summer generation become full-grown by the last of August or during September and transform to adults that emerge and lay eggs for the fall generation, which develops on winter wheat as above described.

Control Measures. Rotations with other crops and the destruction of volunteer grains will tend to reduce the numbers of this pest. Destruction of straw that is heavily infested by these maggots, or the baling and selling of straw off the farm, if done soon after wheat harvest, will help somewhat in reducing the number of insects.

References. U.S.D.A., Bur. Ento. Bul. 42, p. 43, 1903, and *Dept. Bul.* 1137, 1923; *S.D. Agr. Exp. Sta. Bul.* 217, 1925.

37 (15, 68). WIREWORMS[1]

Importance and Type of Injury. The different species of wireworms doing the greatest amount of injury to wheat are much the same as those which injure corn. So far as is known, all the wheat wireworms (Fig. 10.9) have a 4-year, or longer, life cycle. The adult beetles (Fig. 10.9) have the same appearance and character as those attacking corn. For full description of these insects, see Corn Insects, page 506.

Control Measures. In general, the control measures for combating these insects given under Corn Insects apply to the species attacking wheat. In grain-growing sections of the Northwest, the following recommendations have been made for reducing the great losses caused by these pests: (*a*) summer fallowing of all fields where wireworms have recently caused damage; (*b*) shallow tillage, especially in spring and early summer, leaving the soil below 3 inches as firm as possible and destroying absolutely all vegetation; (*c*) seeding only when there is sufficient soil moisture to produce prompt germination; (*d*) the use of 5 to 10 pounds of extra

[1] *Agriotes mancus* (Say), and others, Order Coleoptera, Family Elateridae.

seed per acre to allow for thinning by wireworms; (*e*) the application of commercial fertilizers.   The chemical control of wireworms is discussed on page 618.

References.   U.S.D.A., Farmers' Bul. 1657, 1931; *Can. Ento. Branch, Saskatoon Leaflet* 35, 1933; *Jour. Econ. Ento.,* **27**:308, 1934; and **18**:90, 1925; *Idaho Agr. Exp. Sta. Res. Bul.* 6, 1926.

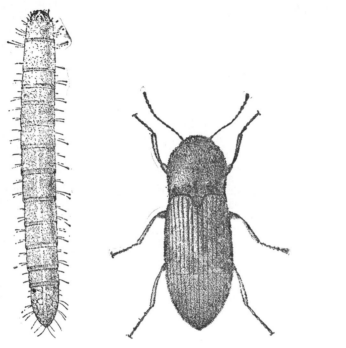

Fig. 10.9. The wheat wireworm, *Agriotes mancus*.   At left, larva, 5 times natural size; right, adult, 7 times natural size. (*From Ill. Natural History Surv.*)

Fig. 10.10. A false wireworm, *Eleodes letcheri vandykei*, dorsal view of larva.   Much enlarged. (*From U.S.D.A., Bur. Ento. Bul.* 95.)

38. FALSE WIREWORMS[1]

Importance and Type of Injury.   In the drier wheat-growing sections of the West, wheat is often seriously injured by the larvae of this and several closely related species of beetles.   The germinating seed in the ground and the young plants are fed upon in the fall, and the small plants are often destroyed in the spring.   The seed grain in the ground is nibbled, and the germ often eaten out.   The young wheat plants are killed over irregular areas in the field, frequently in the vicinity of straw stacks or weed patches.   Injury is most severe during dry years.

Plants Attacked.   Native, dry prairie grass, oats, millet, corn, kafir, alfalfa, cotton, beans, sugarbeets, garden crops, and other plants.   Wheat is much the preferred food plant.

[1] *Eleodes opaca* (Say), *E. suturalis* (Say), and others, Order Coleoptera, Family Tenebrionidae.

Distribution. States between the Mississippi River and the Pacific Coast, ranging northward into British Columbia and southward to Texas.

Life History, Appearance, and Habits. These insects pass the winter in the form of partly grown larvae or as adults. The adults (Fig. 10.11) are dark to black beetles a little under 1 inch in length. The wing covers of some species are distinctly ridged, others are smooth or granulate, and they are grown fast together, making it impossible for the insects to fly. When disturbed, they have a peculiar habit of placing their heads on the ground and elevating the hind part of their bodies as though standing on their heads. The larvae (Fig. 10.10) closely resemble wireworms in appearance, but have longer legs and antennae. Their bodies are brown or yellowish brown in color and prominently jointed. Some species of larvae are nearly black. The adult insects become active early in the spring and lay their eggs in the soil, from 10 to as many as 60 being deposited in a place. The larvae feed on the seeds, roots, and underground parts of the stems of the plants which they attack and under favorable conditions complete their growth in from 110 to 130 days. They pupate in earthen cells in the soil, the pupal period lasting from 10 to 25 days. Most of the damage is caused by the larvae. The adult

Fig. 10.11. Adults of a false wireworm in characteristic attitudes. Somewhat enlarged. (*From U.S.D.A., Bur. Ento. Bul.* 95.)

beetles are general feeders. They are very long-lived, in some cases having been known to live for as much as 3 years in the adult stage. A second period of egg-laying usually occurs in the late summer or early fall, the larvae from these eggs hibernating in the soil in a partly grown condition.

Control Measures. The most effective control measure for these insects is rotation of crops that will bring corn or other cultivated crops on the ground for at least 2 years between crops of wheat. General, thorough cleanup of straw and wheat refuse in the fields is also of much help in preventing damage. A dust furrow with post holes, as for armyworms, has been used to check the crawling adults as they migrate into new fields. Seed treatment as described under Wireworms, page 619, has given satisfactory control.

References. Jour. Agr. Res., vol. 22, no. 6, 1921, and vol. 26, no. 11, 1923; *U.S. D.A., Bur. Ento. Bul.* 95, Part V, pp. 73–87, 1912; *Tex. Agr. Exp. Sta. Ann. Rept.*, **56**:52–53, 1933.

39 (14, 87, 216, 273). WHITE GRUBS[1]

Importance and Type of Injury. Where wheat is sown early on ground heavily infested with white grubs, the young plants may be damaged in the fall by having their

[1] *Phyllophaga* or *Lachnosterna* spp., Order Coleoptera, Family Scarabaeidae.

roots eaten off by these insects. The most serious damage to wheat occurs in the spring from grubs feeding on the roots. Heavily infested fields may be nearly destroyed. Feeding usually starts during the latter part of May or the first of June, and continues until wheat harvest. The brown-headed, curved-bodied, six-legged grubs, 1 inch or more in length, will be found just below the surface of the soil about the wheat roots. For a full description of white grubs see Corn Insects, page 503.

Control Measures. The control measures given for preventing damage by this insect on corn apply equally well to wheat. In addition, most of the fall damage may be prevented by seeding sufficiently late so that the grubs will not feed on the wheat roots before they start their winter migration to below the frost line. Seeding late enough to avoid damage by Hessian fly will prevent most of the fall damage by grubs. Wheat is not so severely damaged as corn but, where possible, should not be planted on ground known to be infested by white grubs. Soil treatments for white grub control are discussed on page 522.

40 (7). BILLBUGS[1]

Wheat infested by billbugs will often show deadened stalks shortly before harvest, or the stalks may fall in much the same manner as where wheat is heavily infested with Hessian fly or jointworm. An examination of the lower parts of the plant will show short, white, curved-bodied, legless grubs with brown heads, feeding in the lower part of the stem and crown of the plant. The burrows made by the insect in the plant are filled with fine, sawdust-like castings. The grubs are rather small, seldom attaining $\frac{1}{4}$ inch in length. Damage is usually more severe on sod ground, in lowlands, or about the margins of fields. Damage by billbugs to wheat is not usually severe, but has apparently been increasing somewhat during the last few years in the midwestern states. For a full description of billbugs and their control, see Corn Insects, page 487.

B. INSECTS THAT ATTACK RYE

Rye is not so subject to injury by insects as is wheat. It is fed upon to a limited extent by the Hessian fly, but the injury to rye is not nearly so severe. In one case where comparisons could be made of the number of eggs laid on wheat and rye, in adjoining strips alternating through a field, it was found that only about one-sixth as many eggs of the Hessian fly were laid on the rye as on the wheat. The Hessian fly maggots seem to have greater difficulty in developing on rye than on wheat. The same control measures apply as on wheat.

Rye is especially subject to attack by chinch bugs, unless the stand is heavy so that the bugs are repelled by the shady condition. It is also subject to the attack of the sawflies and jointworm. None of these insects are serious pests on rye.

C. INSECTS THAT ATTACK OATS

Oats are comparatively free from serious insect injury. They suffer more severely perhaps than other small grains during outbreaks of armyworms. Oats are never attacked by the Hessian fly. In most years oats have not made sufficient growth to be attractive to chinch bugs when they leave their winter quarters and, therefore, escape injury, except such as may occur from migrating bugs at wheat harvest. In years when the English grain aphid is abundant, it sometimes severely injures oats.

D. INSECTS THAT ATTACK BARLEY

Barley is somewhat more subject to insect injury than oats. It is one of the favorite foods of the chinch bug, and fields of barley growing

[1] *Calendra* spp., Order Coleoptera, Family Curculionidae.

in areas generally infested by the chinch bug are usually more severely injured and contain greater numbers of these insects than adjoining or near-by fields of rye, oats, or wheat. Barley is also fed upon to some extent by the Hessian fly, which, however, does not damage barley to nearly the extent that it does wheat. Armyworms, grasshoppers, stem maggots, and aphids also attack barley. For the control of these insects, see the discussions under Wheat.

E. INSECTS THAT ATTACK RICE

40A. RICE STINK BUG.[1]

Importance, Type of Injury, and Life History. Straw-colored shield-shaped bugs up to ½ inch long suck the milky contents from developing rice grains, leaving an empty seed coat or a discolored spot on the kernel and lowering the quality. The bugs winter as adults in wild grasses growing in and around rice fields and lay as many as 150 eggs there in the spring. The nymphs pass through five instars over 2 to 4 weeks, and there may be four to five generations annually on grass and two or three on rice.

Control Measures. Burning or plowing under of wild grasses helps to decrease the number of overwintering adults. Spraying about 1 week after heads appear with dieldrin at 0.25 pound, aldrin at 0.5 pound, or toxaphene at 2 pounds per acre is effective, but the treated straw should not be fed to animals.

The other principal insect pests of rice are the rice water weevil,[2] the legless, small-headed white grubs of which at first tunnel and later chew at the roots from the outside; the larvae of the sugarcane borer (page 520); the rice stalk borer,[3] the caterpillars of which burrow through the rice stalks, causing them to break over and reducing the yield of grain; the sugarcane beetle,[4] the adult of which gnaws the stem between the surface of the ground and the roots of the plant; the rice leaf miner,[5] which attacks the floating leaves of seedling plants; the southern corn rootworm (page 510); the fall armyworm (page 473); and the chinch bug (page 480).

General control measures include the keeping down of wild grasses near rice; winter plowing, burning over, or pasturing; rotating rice with crops not attacked by these insects; and submersion of fields when young rice is attacked by the sugarcane beetle, rootworms, or the chinch bug.

References. U.S.D.A., Farmers' Bul. 1543, 1927, and *Cir.* 632, 1957.

[1] *Oebalus* (= *Solubea*) *pugnax* (Fabricius), Order Hemiptera, Family Pentatomidae.
[2] *Lissorhoptrus oryzophilus* (Kuschel), Order Coleoptera, Family Curculionidae.
[3] *Chilo plejadellus* Zincken, Order Lepidoptera, Family Pyralididae.
[4] *Euetheola rugiceps* (LeConte), Order Coleoptera, Family Scarabaeidae.
[5] *Hydrellia scapularis* Loew, Order Diptera, Family Ephydridae.

CHAPTER 11

INSECTS INJURIOUS TO LEGUMES

FIELD KEY FOR THE IDENTIFICATION OF INSECTS INJURING CLOVERS, ALFALFA, SWEET CLOVER, COWPEAS, AND SOYBEANS

A. *Chewing insects that eat away portions of leaves, buds, or stems aboveground:*

1. Irregular holes, usually extending from the margin of the leaf inward, made by brownish or grayish jumping insects up to $1\frac{1}{2}$ inches long................ ....*Grasshoppers*, page 465.

2. Leaves of clover and alfalfa eaten in early spring by green, legless, curled, narrow-headed grubs with a whitish stripe down the middle of the back, up to $\frac{1}{2}$ inch in length. The grubs hide about crown of plant during day and feed at night. Robust, brown, gray-mottled, oval-bodied beetles, $\frac{1}{3}$ inch long, with the thorax and head narrowed to a short snout, feed on leaves in late spring and early fall..........................*Clover leaf weevil*, page 548.

3. Alfalfa plants have upper leaves shredded and growing tips eaten off in spring by dark-green larvae about $\frac{1}{3}$ inch long. Badly infested plants have a whitened, bleached appearance. Small, dark-brown, grayish-mottled, oval, snout beetles about $\frac{3}{16}$ inch long feed on the leaves....*Alfalfa weevil*, page 550.

4. Red clover buds dying and the growth of the plant stunted. Small, pale, brown or green, legless grubs, up to $\frac{1}{4}$ inch long, feed in the heads and inside the lateral buds. Green or bluish-green snout beetles, about $\frac{1}{8}$ inch long, with black heads and snouts, found on leaves and stems. Injury occurs in spring months only and is most severe during dry seasons................ ....*Lesser clover leaf weevil*, page 553.

5. Long gray, black, or striped beetles, from $\frac{1}{2}$ to $\frac{3}{4}$ inch long, feed on leaves of alfalfa during late summer and early fall. The growth of the plants is stunted and the leaves present a ragged appearance.........*Blister beetles*, page 554.

6. Oval beetles about $\frac{1}{5}$ inch long, green or greenish yellow in color and with 6 conspicuous black spots on each wing cover, often abundant on cowpeas and soybeans eating the leaves................*Spotted cucumber beetle*, page 510.

7. Reddish or yellowish beetles, about $\frac{1}{6}$ inch long, with 3 black spots near the inner edge of each wing cover, eat holes in the leaves of cowpeas. Slender, white larvae feed on the roots and root nodules....*Bean leaf beetle*, page 624.

8. Soybeans and cowpeas grown near garden beans in certain sections have the leaves skeletonized by coppery-brown beetles, $\frac{1}{4}$ inch long, with 8 small black spots on each wing cover; and by oval, yellow, very spiny larvae up to $\frac{1}{3}$ inch long which feed from underside of leaves.......*Mexican bean beetle*, page 622.

9. Grayish long-headed beetles, $\frac{1}{2}$ inch long, with white stripes along their sides, eat the margins of the leaves.................*White-fringed beetle*, page 607.

10. Short-snouted, dark-gray or brownish beetles, about $\frac{1}{6}$ inch long, eat off small soybean plants on spring-broken clover sod as fast as they appear through the ground....................................*Clover root curculio*, page 572.

11. Alfalfa and, more rarely, clover leaves are eaten by dark-green caterpillars, about 1 inch long, with a light stripe containing a crimson hair line along each side of the body. Caterpillars with 5 pairs of prolegs and 3 pairs of thoracic legs. Sulfur-yellow butterflies with black borders on the wings fly about over the fields..............................*Alfalfa caterpillar*, page 555.

12. Leaves of clover, alfalfa, soybeans, and cowpeas eaten off by light-green worms, about 1¼ inch long, with a narrow white stripe and a second faint white line on each side. These caterpillars have only 4 pairs of prolegs in addition to the 6 slender thoracic legs. They drop off the plants when disturbed. Injury most common in southern part of the United States........................
....*Green cloverworm*, page 556.

13. Light webs of silk cover alfalfa plants or surface of ground about the base of the plants in newly sown fields in early fall. Yellowish-green worms, up to 1 inch long, with scattered hairs and conspicuous black spots, feed on the leaves and new growth within these webs...................................
................*Garden, sugarbeet, and alfalfa webworms*, pages 513 and 557.

14. Leaves and stems of cowpeas, alfalfa, and vetch, and pods of cowpeas and soybeans eaten by greenish, white-striped caterpillars, up to 1¾ inches long, sparsely haired, and very variable in color; the skin rough-appearing under a lens...*Corn earworm*, page 498.

15. Leaves of clover or alfalfa eaten, or plant stripped of foliage and tender shoots, by dark-green worms up to 2 inches in length with light stripes on the sides and down the middle of the back. These caterpillars, which feed at night and hide under clods, stones, or in heart of the plant during the day, have a skin that appears smooth under a lens. Worms often crawl over the soil in great armies.
....*Armyworm*, page 559.

16. Plants of clover, alfalfa, and other legumes are cut off at the surface of the ground, or leaves eaten, by plump, cylindrical worms of several shades and markings up to 1½ or 2 inches long, with 6 short slender legs near the head and 5 pairs of prolegs..*Cutworms*, page 559.

B. *Piercing-sucking insects that take the sap only, causing wilting, whitening, browning, reddening, and dying of the leaves and stems:*

1. Stems and leaves of clover and alfalfa covered with small, green plant lice or aphids. Plants wilt and die; leaves and stems coated with sticky fluid from the aphids...*Pea aphid*, page 626.

2. Attack similar to *B*, 1, on alfalfa, by small yellowish or grayish, very active aphids....................................*Spotted alfalfa aphid*, page 561.

3. Attack similar to *B*, 1, on cowpeas.................*Cowpea aphid*, page 564.

4. In the Rocky Mountain states, plants are stunted and seeds stuck together in pellets when threshed, by an insect similar to *B*, 1......*Clover aphid*, page 563.

5. Leaves of clover and alfalfa have a somewhat mottled, whitened appearance due to many very fine, white spots, or become dwarfed and yellowish or reddish in color. Numerous, elongate, active, wedge-shaped bugs, mostly less than ¼ inch long, feeding on underside of leaves. Fields swarming with these small, variously colored, flying and jumping insects.........*Leafhoppers*, page 564.

6. Greenish to yellowish-brown, flat bugs, about ¼ inch long, cause blasted buds, flower drop, and shriveled seeds, by puncturing the terminal growth of alfalfa..
....*Lygus bugs*, page 565.

7. Large, flattened, shield-shaped, bright-green, stinking bugs about ⅔ inch long, and various sized nymphs with reddish markings, suck sap from and poison soybeans and cowpeas, causing pods to drop or to form hardened, knotty areas or to produce stunted and distorted seeds.....................................
Green stink bug, Acrosternum (=Nezara) hilare (Say) and *southern green stink bug, Nezara viridula* (Linné) (see *Va. Agr. Exp. Sta. Bul.* 294, 1934).

8. Frothy masses on alfalfa and clover conceal small frog-like insects, which suck juices and may cause plants to wilt and die......*Meadow spittlebug*, page 560.

C. *Insects that bore in the stems:*

1. Stems of red and sweet clover are swollen or cracked open, with the pith eaten out. Stems sometimes break off. Yellowish, smooth-sided, cylindrical worms about ½ inch long with two curved hooks at end of body, feeding on the pith in these tunnels. Parent beetle a smooth, narrow, hard-shelled insect ⅛ inch long with blue wing covers and bright red head and prothorax...................
Clover stem borer, Languria mozardi Latreille (see *U.S.D.A., Dept. Bul.* 889, 1920).

D. *Insects that eat into, suck sap from, or live within the flowers, heads or seeds:*

1. Cowpeas in the pod, or in storage, contain white footless grubs or short, chunky, brownish beetles, about $\frac{1}{10}$ inch long. Both beetles and grubs feed inside seed, later leaving the seed through a small round hole..............................
........................*Cowpea weevil or four-spotted bean weevil*, page 576.

2. Legless, whitish grubs, up to $\frac{1}{4}$ inch long, with small yellowish heads, bearing a white Y on the front, feed inside the seeds within the growing pods of cowpeas. Bronzy-black snout beetles, $\frac{1}{5}$ inch long, the thorax and convex elytra having very conspicuous round punctures, feed on the plants and deposit eggs in holes eaten through the pods. They rarely fly and "play possum" at the slightest disturbance......................*Cowpea curculio* or *pod weevil*, page 576.

3. Red clover heads with much the same appearance as in D, 5, but with a small, somewhat hairy caterpillar, about $\frac{1}{4}$ inch long and with distinct head and legs, feeding on the developing seeds and destroying many of the florets at the base..
....*Clover head caterpillar*, page 567.

4. Seeds of red and other clovers and alfalfa each completely occupied and later broken and cracked open, by a very small, fat, white, legless, maggot-like larva that reaches full growth inside the seeds. Infested seeds are often dull-colored. Eggs laid by very small, black, four-winged, wasp-like insects, $\frac{1}{16}$ inch long, that fly about fields and crawl over the heads....*Clover seed chalcid*, page 568.

5. Red clover heads fail to develop evenly, only a part of the pink florets opening, the rest of the head remaining green. Very small, pinkish, legless maggots, $\frac{1}{12}$ inch long, feeding on the outside of the green seeds causing them to shrivel and dry up................................*Clover seed midge*, page 570.

E. *Insects that attack the plant underground:*

1. Plants wilt and die during periods of dry weather. Plants are scored along the roots and often girdled near the crown, by short-snouted, dark-gray or brownish beetles or grayish-white, legless, brown-headed grubs, about $\frac{1}{6}$ inch long. Injury most severe in late spring and early fall..*Clover root curculio*, page 572.

2. Clover plants turn brown, wilt, and die. No feeding is apparent aboveground but roots are found scored on the surface and tunneled through, by very small, black or dark-brown, cylindrical beetles or very small, legless, curved, brown-headed grubs about $\frac{1}{10}$ inch long. Injury is most common in old stands of clover...*Clover root borer*, page 573.

3. White, curved-bodied grubs, about $\frac{1}{8}$ inch long, with 6 slender legs and brown head and prothoracic shield, gnaw at the roots; chunky light-brown beetles, $\frac{1}{10}$ inch long, feed on the leaves...................*Grape colaspis*, page 575.

4. Plants wilt and die from the attacks of white, legless, but not curved-bodied grubs, up to $\frac{1}{2}$ inch long, which eat into the underground stem and taproot but do not eat the smaller roots................*White-fringed beetle*, page 607.

A. CLOVER AND ALFALFA INSECTS

41. Clover Leaf Weevil[1]

Importance and Type of Injury. The damage by this insect is most apparent in clover fields during the early spring. In late, cool, dry springs, red clover and alfalfa plants are frequently totally destroyed. The insect is never of great importance in wet springs. At this time of the year, the leaves of the clover or alfalfa plants will be found with smooth-edged notches eaten out of their sides or occasionally with whole leaves eaten off. Small, green, fat-bodied, legless larvae will be found during the daytime hidden away around the base of the clover plants, occasionally feeding on the leaves. Their bodies are nearly always curved so that the head and tail nearly touch. There is a pale yellowish-white

[1] *Hypera punctata* (Fabricius), Order Coleoptera, Family Curculionidae.

stripe, edged with red, down the center of the back. The full-grown larvae are approximately ½ inch in length (Fig. 11.1).

Plants Attacked. Red clover, white clover, sweet clover, and alfalfa are the main plants fed upon by the larvae of this beetle. Beans have been very little injured. The adult insects feed on a great variety of flowers and have been observed in large numbers on goldenrod and also on wheat, corn, soybeans, as well as many of the common weeds and flowering plants.

Distribution. This insect is probably a native of southern Europe. It was not known in this country until 1880, when it was first reported as injuring clover in New York. It has now spread over the clover- and alfalfa-growing areas of the United States and into Canada.

Life History, Appearance, and Habits. This insect passes the winter mainly in the form of partly grown larvae, around the crowns of the plant; to a lesser extent in the egg stage; and to a still lesser extent in the full-grown or adult beetle stage. All eggs have hatched by early spring and the young larvae feed mainly at night on the leaves. During the day, they remain well hidden in trash on the surface of the ground, or about the crown of the plants. The larvae become full-grown during late spring and then spin coarse brown or greenish-brown cocoons just beneath the ground

Fig. 11.1. Larva of the clover leaf weevil, *Hypera punctata.* About 3 times natural size. (*From Ill. Natural History Surv.*)

surface, about the crowns of the plants, or occasionally on the stems of the leaves. These cocoons are thin, about ⅓ inch in length, and have the appearance of a coarse network of somewhat stiff threads. The pupal stage is passed in this cocoon, the adult beetle coming out in early summer. The beetles are dark brown, flecked with black on the back and paler brown underneath. A strong, robust snout projects from the head (Fig. 11.2). The beetles feed very actively for a short time after emerging from the cocoon, and then become rather sluggish and feed but little until fall; they again become active in the fall, mating takes place, and the females lay their eggs mainly during September and October. The eggs are pale yellow in color and are deposited in the stems of the leaves, on the stalks, or about the crown of the food plant. Some of these eggs hatch during the fall, while others do not hatch until the spring of the following year.

Control Measures. In most seasons, the insects are held in check by a fungus disease, *Entomophthora sphaerosperma,* which attacks the larvae. The larvae infested with this disease turn yellowish and later brownish in color, and usually remain curled around the tips of the leaves during the daytime, not hiding away as do the healthy larvae. Where numbers of the insect are found in this condition, one may be sure that little damage to the crop will result, although the weevils may be very abundant in the field. Clover fields that are found very heavily infested early in the spring, so that growth has practically stopped and the plants are being killed, should be plowed up and planted to some of the grass crops, or where possible, to small grain. Spraying infested fields in the spring when the clover or alfalfa is 2 to 6 inches high, with methoxychlor at 1 pound per acre, has given good control.

Fig. 11.2. Adult of the clover leaf weevil, *Hypera punctata.* About 6 times natural size. (*From Ill. Natural History Surv.*)

References. Ill. Agr. Exp. Sta. Bul. 134, 1909; *U.S.D.A., Dept. Bul.* 922, 1920, and *Farmers' Bul.* 1484, 1956; *N.Y. (Cornell) Agr. Exp. Sta. Bul.* 411, 1922; *Okla. Agr. Exp. Sta. Rept.* 1932–1934, pp. 268, 269, 1935; *Jour. Econ. Ento.,* **40**:751, 1947; and **51**:195, 1958.

42. Alfalfa Weevil[1]

Importance and Type of Injury. This is the most important insect enemy of alfalfa. It may, and many times it does, destroy at least one cutting of hay during the season. The plants attacked show a skeletonizing or shredding of the tips of the new growth, this injury increasing from early spring until shortly before the time of the first cutting of alfalfa. In heavily infested fields, the growing tips are eaten off, the growth of the plants stunted, and the green part of the leaves eaten out to such an extent that the fields appear to be suffering from severe frost injury, presenting a bleached-out appearance. The plants are covered with green larvae about ¼ inch long when full-grown. They are plump-bodied, legless, but with well-developed ridges on the underside of the body which take the place of legs (Fig. 11.3,*b*). The adult snout beetles (*d*) will also be found in numbers during the spring and early-summer months in the injured fields.

[1] *Hypera postica* (Gyllenhal) [= *Phytonomus variabilis* (Herbst)], Order Coleoptera, Family Curculionidae. The Egyptian alfalfa weevil, *Hypera brunneipennis* (Boheman), was introduced near Yuma, Ariz., about 1939 and is an important pest of alfalfa in the Southwest. It is very similar in appearance, life history, and control to the alfalfa weevil, but the adults estivate (*Jour. Econ. Ento.,* **48**:297, 1955).

Plants Attacked. Alfalfa is the most important food plant. Bur clover, yellow sweet clover, and rarely vetch and a few of the true clovers are also attacked.

Distribution. The insect was probably imported from southern Europe some time about 1900. The first noticeable injury occurred in the vicinity of Salt Lake City, Utah, in 1904. Since that time the insect has spread through most of the alfalfa-growing areas of the West, including California, Oregon, Washington, Arizona, Nevada, Utah, Idaho, Montana, New Mexico, Colorado, Wyoming, Kansas, and North and South Dakota, and is also widely distributed in the East, in Georgia,

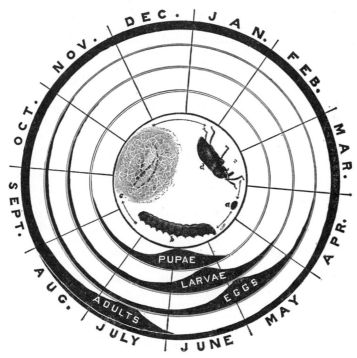

Fig. 11.3. Diagram of seasonal history of the alfalfa weevil in Colorado. *a*, Egg; *b*, larva; *c*, pupa in cocoon; *d*, adult. Enlarged 3 times. (*From Eleventh Ann. Rept. Colo. State Ento.*, 1919.)

North and South Carolina, Virginia, West Virginia, Maryland, Delaware, New Jersey, Pennsylvania, New York, Connecticut, Rhode Island, and Massachusetts. It has spread at a rate of 20 or more miles a year, depending upon the type of farming and natural factors in the areas where infestations occur.

Life History, Appearance, and Habits. This weevil (Fig. 11.3,*d*) generally winters only in the adult stage although the eggs overwinter in some areas. The beetles are grayish brown, or nearly black, with short grayish hairs giving them a somewhat spotted appearance. They are about ⅛ to ¼ inch long with a medium-sized beak about one-half the length of the thorax, projecting downward from the front of the head. On leaving their hibernating quarters about the crowns of the alfalfa plants, or under leaves and rubbish, the beetles feed for a few days, and mate, and the females lay their shining, oval, yellowish eggs (*a*) in the stems of the alfalfa.   They first make cavities in the stems with their beaks, and in these insert from 1 to as many as 40 eggs.   Each female beetle lays from 600 to 800 eggs during the spring.   The weevils fly actively on still days at temperatures above 70°F., and flight is an important means of dissemination.   The larvae (*b*) on hatching are nearly white but soon become green with a prominent, white, middorsal stripe. They feed in the interior of the stalk for 3 or 4 days then make their way to the opening leaf buds at the tips of the plants, where they feed for some time concealed.   There are three to four larval instars over a period of 29 to 58 days.   The feeding stunts the plant and produces a new growth below the tip, which in turn is eaten by the weevil larvae, leaving nothing but woody fibers.   In heavily infested fields, practically the entire first cutting may be ruined and the growth of the second cutting delayed. Most of the larvae become full-grown about the time of cutting of the first crop and go to the soil, where they spin a net-like, nearly spherical cocoon (*c*) in which they pupate.   After about 10 days, they emerge as adults, which feed on alfalfa the remainder of the summer and go into winter quarters early in the fall.   The adults may live from 10 to 14 months, and there is one complete and sometimes a partial second generation a year.

Control Measures.   Careful timing of the cutting of the first and second crops is effective in controlling this insect and will often avoid the necessity of using insecticides.   The most suitable schedule for any particular area should be learned from the authorities in each state.   However, whenever the feeding of the weevils has retarded the growth of the first crop and most of the eggs have been laid, the alfalfa should be cut and the hay removed promptly to allow the hot sun to kill the larvae in the stubble.   Proper farm practices which maintain a vigorous stand of alfalfa are also important.   The imported ichneumonid parasite, *Bathyplectes curculionis*, is an important factor in control and has been reported to parasitize more than 90 per cent of the weevils attacking the first crop. However, it cannot be depended upon to entirely prevent damage by the weevil.   Quarantines against the importation of hay, straw, alfalfa meal, and some other farm products from the infested areas have been established to prevent transportation of the weevil to other areas.

Where economic damage is regularly sustained, insecticidal treatments should be applied.   Alfalfa hay may be protected by spraying with methoxychlor at 1.5 pounds, parathion or methyl parathion at 4 ounces, or malathion at 1 pound per acre, applied when the buds and shoots first appear ragged from weevil feeding.   Seed alfalfa may be protected by spraying with DDT at 1.5 to 2 pounds, or dieldrin at 4 ounces per acre, as soon as the buds appear.   In the West, effective control of the overwintering adults has been obtained by spraying the alfalfa when it is less than 2 inches high with dieldrin at 4 ounces per acre.

References.   *U.S.D.A.*, *Farmers' Bul.* 1528, 1927, *Bur. Ento. Plant Quar.*, *Cir.* E-697, 1946, and *Tech. Bul.* 975, 1949; *Calif. Agr. Exp. Sta. Bul.* 567, 1933, and *Dept. Agr. Spcl. Publ.* 166, 1939; *Colo. Agr. Exp. Sta. Bul.* 567, 1933; *Wyo. Agr. Exp. Sta. Bul.* 167, 1929; *Utah Agr. Exp. Sta. Ext. Cir.* 213, 1955; *Jour. Econ. Ento.*, **25**:681, 1932; **27**:960, 1934; **42**:554, 1949; **48**:283, 1955; **50**:559, 810, 1957; and **52**:663, 942, 1959.

43. Lesser Clover Leaf Weevil[1]

Importance and Type of Injury. The lesser clover leaf weevil since about 1910 has become one of the most important and destructive pests of red clover in the middle-western states. The injury is most severe during dry seasons and frequently amounts to an infestation of 90 to 98 per cent of all plants in the field. Infested clover plants show a deadening of the leaves and a general checking of the growth, which is particularly noticeable during dry seasons. If such plants are examined, small slits will be found cut in the stem, usually just above an axil of the stem or at the lateral buds, and the buds eaten into both at the terminal and on the sides of the stems. The heads of the plant become stunted and misshapen. Examinations made during May will show small pale-green larvae which are feeding in the stems, the newly forming buds, or the florets in the head.

Plants Attacked. The insect seems to prefer red clover, but feeds on all the common species of clovers, and also on sweet clover and alfalfa. On the last two plants it has been of little importance.

Distribution. The insect was probably introduced into this country in the eastern part of Canada some time about 1875 or 1880. It has now spread over the entire eastern half of the United States and into the Pacific Northwest. In the midwestern states it has increased very rapidly in abundance since about 1915.

Life History, Appearance, and Habits. The winter is passed only in the adult stage, the overwintering beetles being about ⅛ inch long, of a beautiful deep-green or blue-green color, with small black heads and a glossy-black, slender beak approximately as long as the thorax

Fig. 11.4. Adult lesser clover leaf weevil. About 5 times natural size. (*From Ill. Natural History Surv.*)

(Fig. 11.4). They shelter to some extent around the crowns of the clover plants in the field but in Illinois have been found in greatest abundance in woodland areas. A number of hibernating adults have been taken from around the base of a single tree or stump in the center of large areas of heavy woodlands. They are also found along bushy hedges, fences, roadsides, and other areas where trash occurs to give protection from the winter weather. They fly from their hibernating quarters about the time clover growth starts in the spring. The adults feed for a few days on the clover leaves; the females then begin laying their eggs. The eggs are deposited in small slits, cut in the stem of the clover plant or in the bud at the axil of the leaf or in the terminal bud of the plant. Usually but one egg is laid in a place, although two or three have occasionally been found. Egg-laying extends over nearly a month, from early April to early May,

[1] *Hypera* (= *Phytonomus*) *nigrirostris* (Fabricius), Order Coleoptera, Family Curculionidae.

each female laying a total of from 200 to 300 eggs. The eggs hatch in from 2 to 3 weeks, and the young larvae, which are at first white, but later change to a brownish white, begin feeding on the plant tissue. Where feeding starts at the buds, these may be entirely killed. If they start feeding in the head, it is destroyed entirely or in part. They occasionally tunnel in the stems, causing the stem above the point where they are working to wilt and die. The larval period extends from 20 to 25 days, or possibly somewhat longer, depending on the weather conditions of the season. Full-grown larvae are legless whitish grubs, with a black head and a dark line across the body just behind the head; they usually lie in a curve. They may spin their cocoons on the ground around the base of the plant, but more commonly in the part of the plant where they have been feeding. These cocoons are elliptical or nearly round, about ⅟₇ inch across. They are transparent, of a somewhat whitish or yellowish-white color. The insects remain in the pupal stage for from 5 to 12 days, emerging as full-grown beetles about 10 days before the time for cutting the first crop of red clover. Newly emerged beetles are brown in color but begin to show green by the end of 1 or 2 days, and by the third or fourth day after emergence are a pronounced grass green. They feed in the clover field for 2 to 3 weeks, gradually becoming less abundant as they fly from the field to seek shelter in which to pass the winter. In Illinois most of the beetles have left the fields by mid-July. There is probably only one generation of the insect each season.

Control Measures. Granular applications of aldrin, dieldrin, or heptachlor at 2.5 pounds per acre have given good control of the weevils on seed clover and have increased the yields of seed. Clover on fertile soil during a moderately wet or wet season will not suffer severely from the attacks of this insect; but on poor soil in a normal year, or in a dry season, it may be severely damaged. Pasturing clover lightly until about May first may avoid much of the damage to the seed crop, by delaying the maturing of the crop.

References. *Ohio State Univ. Ext. Bul.*, vol. 16, no. 10, 1920–1921; *Okla. Agr. Exp· Sta. Rept.* for 1932–1934, pp. 268, 269, 1935; *Jour. Econ. Ento.*, **49**:542, 1956; **50**:224, 1957; and **51**:459, 1958.

44 (102). BLISTER BEETLES[1]

Importance and Type of Injury. Clover, alfalfa, and soybeans are rarely seriously damaged by these insects. Frequently, however, they are present in such numbers in alfalfa fields as to cause some alarm and a little damage. The long, black, gray, spotted or striped black-and-yellow beetles, with very conspicuous heads and necks, long legs, and rather soft wing covers which do not completely cover the tip of the abdomen (Fig. 14.25), will be found clustered on the tips of alfalfa and clover plants, feeding on the flowers and leaves. Growth of the plants is somewhat stunted, and the field presents a ragged appearance where the beetles are numerous.

Plants Attacked. Blister beetles are general feeders attacking many flowering plants, field and garden crops, clover, alfalfa, soybeans, and weeds.

Distribution. Throughout the United States and arable parts of Canada.

Life History, Appearance, and Habits. The blister beetles, whose life histories are known, pass the winter in the larval stage. The larvae of some species feed on the eggs of grasshoppers and others in the cells of certain burrowing bees. They have a very interesting and complicated life history. Those which feed on the eggs of grasshoppers are, of course, beneficial in their larval stage. Some of the blister beetles which feed on clover and alfalfa, notably the gray blister beetle, have two

[1] *Epicauta* spp., *Macrobasis* spp., Order Coleoptera, Family Meloidae.

generations a year.   Nearly all the other species have only one.   The adult beetles vary in appearance, the gray blister beetle and the black blister beetle being of a uniform color and from ½ inch to nearly 1 inch in length.   Several other species are spotted, striped, or marked with different colors (see also page 642).

Control Measures.   In most cases it is not necessary to take any active steps to combat these insects on clover and alfalfa.   Passing once over the field with a hopperdozer as described for catching grasshoppers will probably clean out serious infestations.   The chemical control of blister beetles is described on page 643.

Reference.   U.S.D.A., Dept. Bul. 967, 1921.

45. ALFALFA CATERPILLAR[1]

Importance and Type of Injury.   This insect is particularly a pest of alfalfa in the southwestern states.   Infested fields show part of the leaves eaten out or entirely consumed by a green worm, which in the southern states appears during late February or March.   Previous to

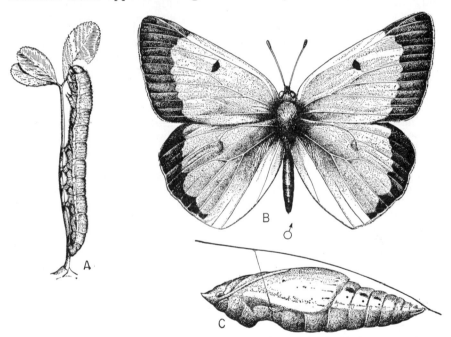

Fig. 11.5. The alfalfa caterpillar.   *A,* full-grown larva feeding on leaf; *B,* adult male butterfly; *C,* chrysalid or pupa, showing how it is suspended from the plant.   Slightly enlarged.   (*From U.S.D.A., Dept. Bul.* 124.)

the appearance of the worms, sulfur-yellow butterflies with black margins on the upper surface of the wings will be found in large numbers hovering over the alfalfa plants, on which they are depositing their eggs.

Plants Attacked.   The insects feed mainly on alfalfa, although they are occasionally taken on clover and several other legumes.

Distribution.   The alfalfa caterpillar is found throughout the United States and southern Canada, with the possible exception of the extreme southeastern United States.   Its heaviest damage, however, occurs in the Southwest, and the alfalfa-growing regions along the Pacific slope.

[1] *Colias* (= *Eurymus*) *eurytheme* Boisduval, Order Lepidoptera, Family Pieridae.

Life History, Appearance, and Habits. Throughout most of its range the insect passes the winter in the pupal stage (Fig. 11.5,*C*) on the plants. In the extreme southern part of its range, larvae may be found during the winter months, and, occasionally, adults will be seen on the wing in these sections; so it probably hibernates in all stages. In the spring the overwintering pupae change to yellow butterflies (Fig. 11.5,*B*) having a wing expanse of about 2 inches. The entire undersurface of the wings is solid sulfur yellow. The upper surface of the wings is bordered with black. The butterflies, upon emerging, lay their eggs in numbers of from 200 to 500, singly on the undersides of the alfalfa leaves. These eggs hatch in a few days into small dark-brown worms which soon change to a green color. The worms grow very rapidly, becoming full-grown in from 12 to 15 days, at which time they are nearly 1½ inches in length, of a dark grass-green color, with a fine white stripe on each side of the body through which runs a very fine red line (Fig. 11.5,*A*). These worms change to the pupal stage without spinning a cocoon. They attach the narrow tail end of the pupa to the alfalfa stalk and throw a loop of silk about their bodies a little above the middle, which holds the head upright. The pupal stage lasts from 5 to 7 days before the emergence of the adult butterflies. In the extreme southwestern states, there are from five to seven generations a year, while probably at least two generations always occur in the northern part of the insect's range.

Control Measures. The most effective cultural practice consists in cutting the infested alfalfa as low as possible and removing the hay from the field so that the food supply of the young larvae is removed and they are also exposed to the sun and to their many insect enemies, of which the braconid wasps, *Apanteles medicaginis* and *A. flaviconchae*, are the most effective. The caterpillars are highly susceptible to diseases caused by the polyhedral virus, *Borrelina campeoles*, and by *Bacillus thuringiensis*, and these have been used in applied microbial control (page 417). When larval populations in fields of normal growth average at least 10 unparasitized larvae per sweep of net and cutting is impractical, spraying with methoxychlor at 1 pound, Phosdrin at 2 ounces, or Dylox at 5 ounces per acre has given good control. Alfalfa growers should learn to recognize the yellow butterflies (Fig. 11.5) which are the parents of this caterpillar and, when swarms of these are noted about the field, should keep close watch of the growing alfalfa crop so that proper control measures can be applied if necessary, before the caterpillars have caused serious damage.

References. U.S.D.A., *Farmers' Bul.* 1094, 1920, and *Dept. Bul.* 124, 1914; *Bul. Brooklyn Ento. Soc.*, **28**:108, 1933; *Jour. Econ. Ento.*, **42**:487, 859, 1949.

46. Green Cloverworm[1]

Importance and Type of Injury. The green cloverworm is only occasionally of importance as a pest of clover and alfalfa. It is nearly always present in fields of these crops grown in the eastern United States. When it is abundant, the infested fields present a ragged appearance, many of the leaves having been eaten off by green worms with two narrow white stripes down each side of the body (Fig. 11.6). The insect is more important in the southern than in the northern states.

Plants Attacked. Alfalfa, clover, garden and field beans, soybeans, cowpeas, vetch, strawberry, raspberry, and many of the common weeds and other legumes.

Distribution. Eastern United States to the plains states and southeastern Canada.

[1] *Plathypena scabra* (Fabricius), Order Lepidoptera, Family Noctuidae.

Life History, Appearance, and Habits. The green cloverworm passes the winter in the pupal or adult stage. The adults are dark-brown, black-spotted or mottled moths with a wing expanse of about 1¼ inches (Fig. 11.7). When at rest, the wings are held in such a position as to give the insect a triangular appearance. The moths shelter around barns, haystacks, and other protected places. They become active about the time clover growth is well started, and, after mating, the females lay their eggs singly on the undersides of the leaves of the plants which they attack. The eggs hatch into small worms which are nearly the shade of the alfalfa leaf. They feed on the leaves, completing their growth under normal conditions in about 4 weeks. They then crawl down the plant and work their way under litter or just below the surface of the soil, where they change, inside a light silken cocoon, to a brown pupal stage. They remain in this stage for from 10 days to 3 weeks and then emerge as moths. There are from three to four generations in the southern part of the United States, and probably two in the northern part of the insect's range.

Control Measures. In most years these insects are not sufficiently abundant to necessitate taking active measures of control. When fields are being injured, the alfalfa or clover

Fig. 11.6. Larva of the green cloverworm, *Plathypena scabra*, and its injury to clover. About natural size. (*From Ill. Natural History Surv.*)

Fig. 11.7. Adult of the green cloverworm. Enlarged about ½. (*From Ill. Natural History Surv.*)

should be cut as soon as possible and removed from the field. This exposes the worms to the bright sunlight and to their insect and bird enemies and is usually the only control measure necessary. Hopperdozers have been used with a fair degree of success in combating this insect.

References. U.S.D.A., *Farmers' Bul.* 982, 1918, and *Dept. Bul.* 1336, 1925; *Calif. Mon. Bul.*, **22**, 156–60, 1933; *Rept. Ont. Ento. Soc.*, **62**: 75–82, 1931.

47 (19, 69). GARDEN WEBWORM[1] AND ALFALFA WEBWORM[2]

Importance and Type of Injury. These insects are of some importance on clover but are more particularly pests of alfalfa and some other crops. During the spring and summer months, fields that are infested will show light webs over the leaves, in which will be found greenish to yellowish-green, somewhat hairy worms with black dots over their bodies (Fig. 11.8). Under protection of the webs, the larvae may consume all the green tissue of the leaves. When full-grown, they are an inch or a little over in length, greenish to nearly black, with a light stripe extending down the middle of the back and with three dark spots on the side of each segment, from each of which projects one to three bristle-like hairs. When disturbed, the worms drop to the ground or crawl down into tubular webs which they have spun. Fields that are heavily infested will show

[1] *Loxostege similalis* (Guenee), Order Lepidoptera, Family Pyralididae.
[2] *Loxostege commixtalis* (Walker).

a considerable amount of webbing over the alfalfa, and the leaves inside these webs will be nearly all eaten off. In the fall, the worms work in somewhat the same manner as cutworms on the newly sown alfalfa. At this time, they hide in silken-lined burrows on or in the ground to which they retreat when disturbed.

Plants Attacked. The garden webworm is a general feeder. Among the plants attacked are alfalfa, clover, corn, beans, soybeans, cowpeas, sugarbeets, peas, strawberries, wild sunflower, thistles, pigweed, ragweed, sweet clover, lamb's-quarters, and a number of others. The alfalfa webworm will eat almost any succulent growth, except small grains and grasses. It has been recorded as a very serious pest of alfalfa and sugarbeets, especially in seasons following a dry year.

Distribution. The insects are native to both North and South America and occur generally in the farming areas of the United States, Canada, and Mexico.

Life History, Appearance, and Habits. The winter is passed in the pupal stage in the soil about the plants on which the fall generation of larvae fed. In the extreme South it is possible that some of the insects live through the winter in the larval stage. Adult moths emerge from

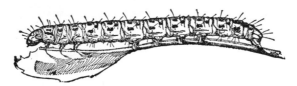

Fig. 11.8. The garden webworm, *Loxostege similalis*, caterpillar. About twice natural size. (*From U.S.D.A., Farmers' Bul.* 944.)

the pupae early in the spring and deposit their eggs on the leaves of their food plants. The eggs are laid in masses of from 2 or 3 to 20 or 50. The garden webworm moths are about ¾ inch across the wings, buff-colored with shadings and irregular markings of light and dark gray; the alfalfa webworm is similar in appearance, but measuring 1 to 1¼ inches from tip to tip of wings. They are rather weak insects, probably living but a short time. When one is going through infested fields, they will frequently fly up, going a short distance, but usually alighting within a rod or two of the point where they were first disturbed. They are not active during the daytime, unless disturbed, but often assemble about lights on warm nights. The larvae hatch from the eggs in 3 days to 1 week. They begin feeding on the underside of leaves, protecting themselves inside the light webs above described. It usually requires about 3 to 5 weeks for the worms to reach the full-grown stage and go into the ground for pupation. When short of food, the alfalfa webworm caterpillars sometimes migrate like armyworms. The pupal period is ordinarily from 7 days to 3 weeks. There are several generations each season. There are said to be five full generations each year in Oklahoma. In the northern part of the range, there are probably two or three generations each season.

Control Measures. Fields which become infested during the summer months can usually be cleaned of the webworms by cutting the alfalfa. This cuts off the food supply of the worms, as they are unable to feed on the dried alfalfa hay, and also by destroying their webs exposes many of them to bird and insect enemies.   On alfalfa hay, webworms may be controlled by spraying with Dylox at 4 ounces per acre, 14 days before cutting.   On seed alfalfa, spraying or dusting with DDT at 1 to 1.5 pounds or toxaphene at 2 to 3 pounds per acre is also effective.   Webworms attacking sugarbeets or corn may be controlled by spraying with Dylox at 0.5 pound, parathion or endrin at 6 ounces, or toxaphene at 3 pounds per acre.   However, plant products treated with DDT, toxaphene, or endrin should not be used for animal forage.   To be most effective sprays should be applied as soon as the eggs hatch and before there is much webbing over of the plants.   The migrating alfalfa webworms may be stopped by dusty or vertical-sided furrows, constructed across the path of their migrations.   Maintaining field margins closely cut to keep down all growth of the weeds on which this insect feeds is also of considerable help in preventing damage.

References.   Okla. Agr. Exp. Sta. Bul. 109, 1916; *U.S.D.A., Farmers' Bul.* 944, 1918; *Calif. Monthly Bul.*, **23**:236, 1934; *Colo. Agr. Coll. State Ento Cir.* 58, 1933.

48 (2, 4, 67, 80).   CUTWORMS AND ARMYWORMS

The true armyworm and the army cutworm[1] prefer grass crops and normally cause but little damage to clover and alfalfa.   In the middle-western states the variegated cutworm[2] (Fig. 11.9) is the one most generally abundant in clover and alfalfa during the spring and summer months.   The dingy cutworm,[3] clay-backed cutworms,[4] and several others are also common in fields of clover and alfalfa at this time of year.

FIG. 11.9. Larva of the variegated cutworm, dorsal view.   Enlarged about ½.   *(From Ill. Natural History Surv.)*

In the fall, newly sown alfalfa fields are often seriously damaged by the yellow-striped armyworm,[5] the western yellow-striped armyworm,[6] the fall armyworm,[7] and the beet armyworm.[8]   The larvae feed on alfalfa in the general manner of cutworms, sometimes destroying the entire field.

Control Measures.   Properly timed cutting of alfalfa fields will often prevent severe damage by the various cutworms and armyworms.   Heavy infestations attacking seed alfalfa may be controlled by spraying or dusting with DDT at 1.5 pounds or toxaphene at 2 pounds per acre.   Alfalfa hay crops may be treated with Dylox at 0.5 to 1 pound per acre, applied at least 2 weeks before cutting, or with malathion at 1 pound per acre, applied at least 1 week before cutting.   Fields may be protected from migrating larvae of the western yellow-striped armyworm by the use of barrier strips of 10 per cent DDT or toxaphene dust, 8 to 12 inches wide.   From 4 to 6 pounds of dust should be applied to 100 feet of barrier, preferably located in the bottom of a furrow.

References.   U.S.D.A., Farmers' Bul. 739, 1916; *Can. Dept. Agr. Bul.* 22, 1923.

[1] *Chorizagrotis auxilaris* (Grote), Order Lepidoptera, Family Noctuidae.
[2] *Peridroma saucia* (Hübner) [= *margaritosa* (Haworth)].
[3] *Feltia subgothica* (Haworth).
[4] *Feltia gladaria* Morrison.
[5] *Prodenia ornithogalli* Guenée.
[6] *Prodenia praefica* Grote.
[7] *Laphygma frugiperda* (J. E. Smith).
[8] *Spodoptera* (= *Laphygma*) *exigua* (Hübner).

49. Meadow Spittlebug[1]

Importance and Type of Injury. This insect hides under frothy masses of spittle (Fig. 6.26) and sucks plant juices with its piercing-sucking mouth parts. In severe infestations, which may number several hundred nymphs per plant, yields of leguminous hays may be reduced by as much as one-third and alfalfa seed crops almost entirely lost. Strawberry yields have been reduced by 0.5 to 1 ton per acre.

Plants Attacked. The meadow spittlebug is a general feeder and has been recorded as feeding on nearly 400 species of plants. It is especially destructive to alfalfa and clover hays and to strawberries.

Distribution. The meadow spittlebug is present over most of the United States east of the Mississippi and along the Pacific Coast, but is a serious pest only in regions of high humidity in the northeastern states and in Oregon.

Life History, Appearance, and Habits. The insect winters in the egg stage, which is ovoid and about 1 millimeter long, and laid in masses of 1 to 30 held together with a hardened frothy cement. These are inserted between the sheath and stem of grain stubble at an angle of about 45 degrees to the axis of the stem and from 3 to 6 inches above the ground. The eggs hatch in late March to early June, at temperatures of 55 to 70°F. After hatching, the first-instar nymphs make their way to suitable host plants, seeking protected sites of high humidity. Almost immediately after beginning feeding, they commence the excretion of the frothy spittle, which gives the insect its name. This is formed from excess plant fluids discharged from the alimentary canal, apparently with the addition of viscous secretions of the Malpighian tubes and special glands, and the excretion is stirred into a stable froth by contractions of the abdomen, which forces bubbles of air from a specialized "air canal" formed by extensions of the tergites of the terminal abdominal segments. The froth completely covers the insect and protects it from desiccation and perhaps from attacks by natural enemies (Fig. 6.26). There are five nymphal instars, which require from about 31 days at 75°F. to 100 days at 50°F. The adults, which are about 6 millimeters long, are pale-straw-colored to dark brown and have short blunt heads with prominent eyes, giving them a frog-like appearance (Fig. 6.25). The elytra are held tent-shaped over the abdomen and are marked with darker spots, stripes, or bands. They appear in late May and early June and live until fall, with only a single generation a year. The adults are somewhat sluggish, but have been observed to fly as much as 100 feet at a time as they disperse in search of succulent foliage, moving from harvested meadows and pastures into legumes, corn, wheat, and oats, where they deposit their eggs in late August and early September.

Control Measures. The development of high nymphal populations in alfalfa or other legumes in the spring may be reduced by crop-management practices which either remove a cutting just prior to the oviposition period in the late fall or else allow the crop to mature at this time and thus reduce its succulence and consequent attractiveness to the adult spittlebugs. Certain varieties of alfalfa are somewhat resistant to spittlebug attack. No important natural enemies have been found. The young nymphs are best controlled by spraying within a week of hatching with methoxychlor at 1 pound per acre. Where residues are not a prob-

[1] *Philaenus leucophthalmus* (Linné), Order Homoptera, Family Cercopidae.

lem as on seedling plants, spraying with endrin at 0.12 pound or hepta-chlor at 0.25 pound per acre also gives good control. Fall treatment of the adults with methoxychlor is effective in preventing oviposition.

References. Ohio Agr. Exp. Sta. Res. Bul. 741, 1954; *Jour. Econ. Ento.,* **48**:204, 592, 1955; **51**:218, 1958; and **52**:240, 904, 1959.

50 (92). Spotted Alfalfa Aphid[1] and Other Aphids

Importance and Type of Injury. The spotted alfalfa aphid was intro-duced, probably as a single female, into central New Mexico early in 1954 and spread through alfalfa-growing areas of the United States so rapidly that by 1957 it was present from coast to coast in a broad area extending from central Wisconsin to Mexico (Fig. 11.10). This extremely destructive aphid injects a toxic salivary secretion into the plants while feeding, which causes chlorosis at the feeding site and yellowing of the veins as it is translocated in the phloem. In heavy infestations, young seedlings may be killed and the growth of mature plants stunted. In warm weather, the aphids congregate on the bottoms of the lower leaves and, as these are killed and dry up, progressively move up the stems to healthy leaves until the plant is largely defoliated. In uncontrolled infestations, which may number more than 1,000,000,000 individuals per acre, the yield and protein and carotene content of the alfalfa may be reduced by half. The aphids also secrete a profuse, sticky honeydew which often becomes infested with sooty mold fungus, and this combina-tion makes harvesting and baling of the hay very difficult and greatly reduces its quality.

Plants Attacked. This aphid prefers plants of the genus *Medicago,* especially alfalfa, bur clover, and medic. It also will infest some species of *Trifolium* and *Melilotus,* including crimson clover, sour clover, and berseem.

Distribution. The spotted alfalfa aphid is a native of the Middle East and is now distributed generally throughout the alfalfa-growing areas in at least 30 states. It is especially destructive in the Southwest.

Life History, Appearance, and Habits. This aphid reproduces parthe-nogenetically the year around in warm climates, both apterous and alate females giving birth to living young. These are produced at rates of from 1 to 2 per day in cool weather up to 8 per day in hot weather, and a single female may give birth to 100 offspring. The nymphal period of four instars averages 18.7 days at 59°F. and 5.5 days at 86°F. at about 70 per cent relative humidity. Reproduction is inhibited at 95°F., and the lower-threshold temperature is about 45°F. The time required to grow from birth to maturity ranges from about 1 month in winter to less than 1 week in summer, and as many as 20 generations may be produced each year. In cooler areas, oviparous (sexual) females and males may appear in the fall and produce eggs. The winged aphids are capable of flying distances up to 70 miles with the aid of the wind.

The spotted alfalfa aphid (Fig. 11.11) is a small, rapidly moving pale-yellow or gray aphid, with four to six conspicuous rows of dark spots with spines on the upper abdomen and with smoky-veined wings.

Control Measures. Proper cultural practices can minimize damage by this insect. Timing of cutting in the one-tenth bloom stage when there is little green-leaf residue will kill many aphids by exposure to the hot

[1] *Therioaphis maculata* (Buckton), Order Homoptera, Family Aphidae.

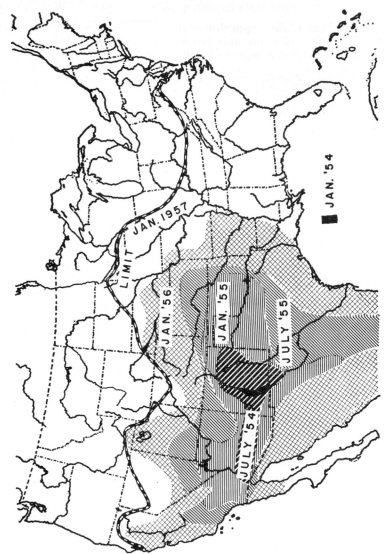

FIG. 11.10. Map showing the spread of the spotted alfalfa aphid, *Therioaphis maculata*, across the United States from its original infestation in New Mexico. (*Drawn by R. C. Dickson, Univ. Calif., Coll. Agr.*)

sun. Establishing and maintaining a vigorous stand of alfalfa through proper water management, fertilization, and weed control is important. A number of beneficial insects when present in sufficient numbers can check increasing aphid populations. These include predators such as lady beetles, syrphid fly larvae, lace-winged fly larvae, and the hemipterans *Nabis* and *Geocoris*. Three species of imported parasites, *Praon pollitans*, *Trioxys utilis*, and *Aphelinus semiflavus*, have been established and aid in control. Two species of parasitic fungi of the genus *Entomophthora* (= *Empusa*) have given effective control of the aphid under favorable conditions. Where the aphid populations increase to damaging numbers, about ½ to 1 aphid per seedling or 20 to 40 per stem for mature plants, insecticidal control will be necessary under most conditions since the reproductive potential of the aphid is far greater than that of its natural enemies. The most effective results have been obtained with integrated control measures, spraying with 1 to 2 ounces of demeton or Phosdrin per acre, as these do not seriously affect the populations of

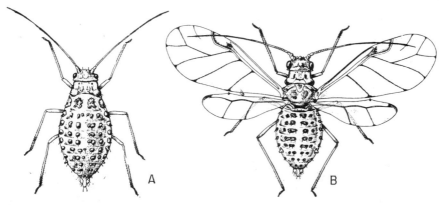

Fig. 11.11. Parthenogenetic forms of the spotted alfalfa aphid, *Therioaphis maculata*. *A*, apterous female; *B*, alate female. About 12 times natural size. (*From R. F. Smith, Hilgardia, vol. 28, no. 21.*)

natural enemies. General-purpose aphicides such as parathion at 4 ounces or malathion at 10 ounces per acre may be used under emergency situations, but usually severely damage the populations of natural enemies. Seed treatment with phorate (Thimet) or Di-Syston at 1.5 pounds per 100 pounds of seed as charcoal powder, or these materials applied as granulars at time of planting, effectively protects young seedlings from attack. Certain varieties of alfalfa such as Lahontan have shown considerable resistance to aphid attack.

References. Hilgardia, **24**:93, 1955; and **28**:647, 1959; *Jour. Econ. Ento.,* **48**:668, 1955; **50**:352, 805, 817, 1957; and **52**:136, 249, 368, 714, 1959.

The pea aphid (page 626) is often a serious pest of clover and alfalfa, which are its winter hosts. In the northwestern United States the clover aphid[1] has in some instances caused the total loss of seed crops of red and alsike clovers, especially because of the formation of honeydew on the seeds, which causes them to lump or cake in storage. The winter hosts

[1] *Anuraphis bakeri* (Cowen).

of this aphid are pome fruit trees.	The cowpea aphid[1] attacks cowpeas much as the pea aphid injures peas.	The yellow clover aphid[2] is a minor pest of red and Ladino clovers in the northeast, middle-west, and southern states, but has not been a pest of alfalfa.	These aphids may be controlled as discussed under Spotted Alfalfa Aphid (page 563) and Pea Aphid (page 627).	Cutting clover and alfalfa as early as possible, or the use of a brush drag on hot days, after the first crop has been cut for hay, has given good results.	Where these crops are being grown for seed and become infested, closely pasturing with sheep, or clipping as close to the ground as possible, promptly removing the hay, and allowing the second crop to produce seed, will often prevent a heavy loss of the crop.

Reference.	*Idaho Agr. Exp. Sta. Res. Bul.* 3, 1923.

51. Leafhoppers[3]

Importance and Type of Injury.	Where leafhoppers are abundant in any crop, the plants show a lack of vigor, growth is retarded, and in most cases the leaves have a somewhat whitened, mottled appearance, or turn yellow, red, or brown, due to sucking out of the sap by the hoppers,

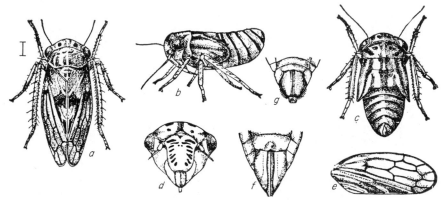

Fig. 11.12. The clover leafhopper, *Aceratagallia sanguinolenta.*	*a*, Adult; *b*, nymph, side view; *c*, nymph, dorsal view; *d*, face; *e*, front wing; *f* and *g*, genitalia and last segment of female and male.	Line shows natural size.	(*From Osborn and Ball.*)

which feed mainly on the undersides of the leaves.	In walking through infested fields, large numbers of tiny mottled or speckled, green, yellow, brown-gray, or various-colored insects (Fig. 11.12) will hop or fly for short distances ahead of one.	With certain species of leafhoppers the feeding produces a burning effect on the plants, and causes the tips to wither and die somewhat as though scorched by bright sunshine or injured by drought.	Leafhoppers are especially destructive to clovers during the seedling stage and just after a cutting has been made.	Some species are carriers of plant diseases.	See Potato Leafhopper (page 643) and Beet Leafhopper (page 676).

[1] *Aphis medicaginis* Koch.

[2] *Therioaphis trifolii* (Monell).

[3] Many species of the Order Homoptera, Family Cicadellidae, including the potato leafhopper, *Empoasca fabae* (Harris), the clover leafhopper, *Aceratagallia sanguinolenta* (Provancher), and the six-spotted leafhopper, *Macrosteles fascifrons* (Stål).

Plants Attacked. Nearly all cultivated and wild plants are attacked by various species of leafhoppers. All the clovers, alfalfa, and many of the grasses and small grains, as well as orchards, vineyards, and forest and shade trees are infested and damaged to some extent.

Distribution. Leafhoppers occur throughout the world.

Life History, Appearance, and Habits. The winter is passed in various stages, according to the different species. Some go through the winter in the egg stage in the stems of various plants, a large number of species pass the winter in the form of full-grown insects, which hide away in shelters around and in the field crops which they attack, while a few pass the winter in the partly developed, or nymphal stages. In the Gulf states, the females may continue to reproduce through the winter and the winged forms migrate northward in spring. In most cases that are known, the insects lay their eggs in the stems, buds, or leaves of their food plants. These hatch into wingless but very active nymphs. The nymphs feed by sucking the sap and sometimes inject a substance which is distinctly poisonous to the plant tissue, killing the areas around their feeding punctures. They develop from small nymphs to adults, molting their skins several times during this process, but not passing through any distinct pupal stage and never spinning cocoons or forming a chrysalis. The adults vary in size from $\frac{1}{20}$ to $\frac{1}{4}$, and rarely $\frac{1}{2}$, inch in length. They are all good jumpers or hoppers, as their common name implies. The adults are winged but use their legs to a large extent in jumping from one part of the plant to another. The general outline of their bodies is long and slender, and, as above stated, they vary greatly both in color and in shape.

Control Measures. Very effective control of leafhoppers on alfalfa has been obtained by spraying with methoxychlor at 1 pound or parathion at 0.5 pound per acre. DDT at 0.5 to 2 pounds per acre is used on seed alfalfa, but should not be applied to alfalfa intended for animal forage, because of the residue hazard. Cutting and removing the crops from heavily infested clover and alfalfa fields, especially just after the majority of the eggs have been laid, removes the eggs, drives out some of the leafhoppers, and starves some of the young nymphs. Immediately after cutting, large numbers of leafhoppers can be caught by running a hopperdozer over the infested field, or this machine may be used on the growing crops. It is most effective when the back and sides of the machine have been thoroughly coated with some sticky material such as tree tanglefoot that will catch the hoppers as they fly against it. Certain pubescent native strains of clover are noticeably resistant to leafhopper attack and should be used when they can be obtained.

References. Jour. Agr. Res., **17**(no. 6): 399–404, 1927; *Ann. Ento. Soc. Amer.,* **16**:363, 1923; *U.S.D.A., Farmers' Bul.* 737, 1916, and *Bur. Ento. Bul.* 108, 1912; *Ky. Agr. Exp. Sta. Cir.* 44, 1936; *Jour. Econ. Ento.,* **42**:496, 1949; and **50**:270, 1957.

52. Bugs of the Genus Lygus Affecting Alfalfa[1]

Importance and Type of Injury. Several species of Lygus injure the tender growing or fruiting parts of alfalfa by puncturing the tissues with their piercing-sucking mouth parts to obtain food. In addition to the physical injury a toxic reaction upon the plant cells near the puncture results from their feeding. In alfalfa, the feeding of Lygus bugs causes

---

[1] *Lygus hesperus* Knight, *L. elisus* Van Duzee, and *L. lineolaris* (Palisot de Beauvois), Order Hemiptera, Family Miridae.

"blasted" buds, excessive flower fall, and brown, shriveled, worthless seeds. In many western areas the species *Lygus hesperus* constitutes approximately 80 per cent of the Lygus population on alfalfa producing seed. Alfalfa seed fields often fail to produce a profitable crop due to the damage caused by these bugs.

Plants Attacked. These species of Lygus are general feeders and are found on many herbaceous plants and also on trees. The cultivated crops damaged in the western and southwestern states are alfalfa, cotton, beans, carrots, and sugarbeets grown for seed. Weed hosts include winter mustard, lamb's-quarter, mare's-tail, and slim aster.

Distribution. The species of Lygus considered here occur throughout the Rocky Mountain and Pacific states.

Life History, Appearance, and Habits. In most regions these species of Lygus pass the winter as adults in hibernation, but in warmer climates the adults can be swept from plants all winter. Eggs and nymphs have been found in southern Arizona in all months except December. At an average mean temperature of 85.5°F. the egg stage requires about 8 days, the five nymphal instars about 11 days. A generation requires from 20 to 30 days in the summer under the climatic conditions of Arizona and 6 to 7 weeks in Utah. The adults (Fig. 14.9) are about $\frac{1}{4}$ inch long by $\frac{3}{32}$ inch broad, flattened, oval, and show a variation in color, from pale greenish to yellowish brown. Eggs are found in the flowers, buds, bracts, nodes, and internodes. The egg is slightly curved, elongated, and bears a lid on its truncated apex. When the nymphs first hatch they are very pale green and have an orange spot on the middle of the abdomen. Shortly after feeding begins, they become darker green in color and the third, fourth, and fifth instars have four noticeable black spots on the thorax.

Control Measures. Excellent control of Lygus bugs and as much as ten-fold increases in the yield of alfalfa seed have resulted from spraying this crop with DDT or toxaphene at 1.5 to 2 pounds per acre or dusting at an equivalent rate. Fifty per cent sulfur should be included in the dust treatments to control spider mites. Where the bugs are resistant to DDT or toxaphene, a combination of 1 pound DDT and 1 pound Dylox per acre is effective. Treatments should not be made until necessary because of the danger of poisoning bees, which are essential to alfalfa seed production. Toxaphene is preferred if treatment is to be made during the bloom stage, and it should be applied at night when the bees are not foraging. Alfalfa treated with DDT or toxaphene should not be used for animal forage. To determine the necessity for treatment, Lygus bug counts should be made in the early morning, using standard 180-degree sweeps of an insect net and averaging the results from 10 to 20 locations about the field. In the early bloom stage, treatment should be made when the bugs average 1 per sweep. During the period of seed set, treatment should be made when the count averages 12 per sweep and in maturing fields when the count averages 20 per sweep. In all these counts, nymphs should be counted as two individuals and adults as one. Cultural practices such as clean cutting of the hay crop preceding the seed crop will result in a reduction of the nymphal population.

References. *Idaho Agr. Exp. Sta. Res. Bul.* 11, 1933; *Utah Agr. Exp. Sta. Ext. Bul.* 221, 1951; *U.S.D.A., Tech. Bul.* 5, 1927, and 741, 1940; *Jour. Agr. Res.*, **61**:791, 1940; *Hilgardia*, **17**:165, 1946; *Jour. Econ. Ento.*, **29**:454, 1936; **39**:638, 1946; **48**:148, 509, 1955; **49**:94, 689, 1956; and **52**:461, 1959.

53. Clover Head Caterpillar[1]

Importance and Type of Injury. Clover heads infested by the clover head caterpillar present somewhat the same appearance as those infested by the clover seed midge. The head is usually irregular in appearance and will have the flowers opening on only one side of the head. Such heads are often pink on one side and green on the other. An examination of the infested heads will show a small, somewhat hairy caterpillar feeding on the seeds at the base of the clover florets, preventing them from opening (Fig. 11.13,*A*). These caterpillars are very small, being only about ¼ inch long when full-grown.

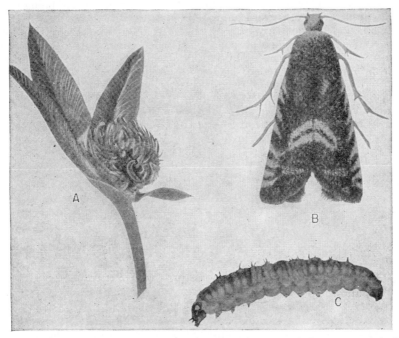

Fig. 11.13. Clover head caterpillar, *Grapholitha interstinctana.* *A*, larva at work in head of red clover, slightly enlarged; *B*, adult, with wings folded, about 8 times natural size; *C*, larva, side view, about 8 times natural size. (*From Ill. Natural History Surv.*)

Plants Attacked. The insect feeds mainly in the heads of red clover but has also been found on white clover, alsike clover, and mammoth clover. It also attacks the leaves of these plants when no heads are present.

Distribution. The clover head caterpillar is distributed generally over the eastern part of the United States and southern Canada.

Life History, Appearance, and Habits. The insect passes the winter in both the larval and pupal stages, both being found under trash and around clover fields. Pupation occurs inside of small silken cocoons about ⅓ inch in length. These are very hard to find because of the dirt which is usually attached to the outside of them. The overwintering larvae

[1] *Grapholitha* (= *Laspeyresia*) *interstinctana* (Clemens), Order Lepidoptera, Family Olethreutidae.

pupate in the spring, and, from these and the overwintering pupae, the moths begin emerging just about the time that red clover is starting to come into bloom. They are small dark-brown moths about ¼ inch long. The wing margins are marked with six or seven short dashes, silvery white in color. The marks on the inner margins form a double crescent when the wings are folded (Fig. 11.13,*B*). They lay their eggs on the leaves, stems, and heads of clover. The young larvae hatching from these eggs feed on the clover leaves, or more usually work their way into the green clover heads, feeding at the base of the florets and destroying from half to all the florets in a head, so that seed is not produced. The larvae are said not to destroy the hardened seed, but eat the green, soft, newly formed seeds. They complete their growth in 4 or 5 weeks, being about ¼ inch long and of a greenish to greenish-white color (Fig. 11.13,*C*). They spin their cocoons either in the head or about the base of the clover plants and change inside the cocoons to the brown pupal stage, which lasts about 15 to 20 days. The moths of the second generation emerge about midsummer in time to lay eggs upon the normal second crop of blooms. A third generation occurs throughout the entire southern part of the insect's range, the larvae feeding in the crowns of the plant.

Control Measures. The only measures that have been found at all practical for reducing the numbers of these insects are the same as those given for the control of the clover seed midge. Fall clipping will help to destroy the overwintering stages. The first crop may best be used for seed when this insect is abundant. Rotation of crops is also of some benefit in keeping down the numbers of this insect.

References. N.Y. (Cornell) Agr. Exp. Sta. Bul. 428, 1923; Ill. Agr. Exp. Sta. Bul. 134, 1909.

54. Clover Seed Chalcid[1]

Importance and Type of Injury. This is one of the most important insect pests of alfalfa and clover seed but has no effect on the production of hay from these crops. The worst damage from this insect occurs when both first and second crops of alfalfa are used to produce seed in the same community. Infested plants have little to distinguish them from those that are uninfested. A close examination of the seeds, however, will show many of them broken or cracked open. Threshed seed will show many empty shells of the seed, or parts of such shells. Very small, white, maggot-like larvae develop in the infested seeds, eating the contents (Fig. 11.14,*e*). In some sections this insect has made the growing of clover and alfalfa seed unprofitable. Examinations of alfalfa seed in Utah over a 4-year period showed that from 10 to 60 per cent of the seed was destroyed by this insect.

Plants Attacked. The insect attacks nearly all the clovers and alfalfa.

Distribution. General over the United States and southern Canada.

Life History, Appearance, and Habits. The insect passes the winter in the full-grown larval stage (Fig. 11.14,*b*) inside the infested seeds on the surface of the ground. Occasionally the insect may pupate in the fall and remain inside the seed in this stage. The adult insects begin emerging during the late spring, usually the latter part of May in the latitude of central Illinois. Adults of the first generation occur in great numbers

[1] *Bruchophagus gibbus* (Boheman) (= *B. funebris* Howard), Order Hymenoptera, Family Chalcididae (Eurytomidae).

during the first part of June. The adult insect (Fig. 11.14,*a*) is a very active little wasp-like creature, jet metallic black in color, with legs of a dark-brownish color, and tarsal claws of a light yellowish brown. They are so small, being only about $\frac{1}{15}$ inch in length, that one has great difficulty in seeing them in the field and will never suspect their numbers unless they are collected in a fine-mesh net. The female insects lay their eggs in clover in which the seed has formed but has not yet hardened. The egg is pushed inside the soft seed and hatches into a white, footless, maggot-like larva that consumes the entire inside of the seed, leaving only a thin outer shell. During threshing, most of the light, infested seeds are blown out with the straw. The larvae complete their growth in 2 weeks or more, depending on the weather. As the adult insects continue to

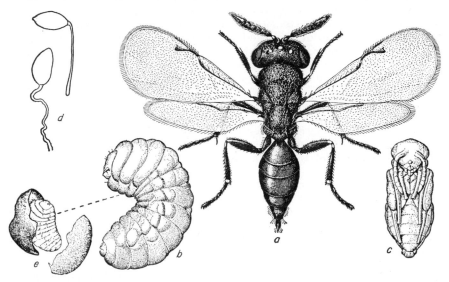

Fig. 11.14. Clover seed chalcid. *a*, Adult; *b*, larva; *c*, pupa; *d*, egg; *e*, larva in broken seed. All greatly enlarged. (*a–c, from U.S.D.A., Farmers' Bul.* 636; *d, from Ill. Natural History Surv.; e, from Ohio State Univ.*)

emerge over a long period, the first-generation larvae occur over a correspondingly long period. Upon becoming full-grown, they pupate inside the seed, either in the head or in seed which has dropped to the surface of the ground, and emerge about midsummer as adults. These adults lay the eggs for a second generation, and in the South a third generation occurs late in the season. The generations overlap, however, so that the insects are present in all stages in the field practically at all times from the first of June until September.

Control Measures. This insect has been found extremely difficult to control. In certain years the use of the first crop of clover, instead of the second crop, for seed will greatly increase seed yields. When the seed crop is known to be heavily infested, as indicated by the presence of many adult wasps about the heads, many of the insects can be killed by early cutting of the crop and removing the hay from the field before the seed has had time to mature. If this is generally practiced in a community and volunteer red clover, alfalfa, and bur clover are kept down, the numbers of the insect may be somewhat reduced. Destruction of chaff and screenings left in the hulling operation is important in preventing the emergence of adults the following spring. Fall or spring cultivation to bury chalcid-infested seeds is also beneficial. However, at present, we are nearly dependent on the natural enemies of the insect to hold it in check, as no really effective control measure has been developed.

References. U.S.D.A., Farmers' Bul. 1642, 1931, and Dept. Bul. 812, 1920; Ill. Agr. Exp. Sta. Bul. 134, 1909; Proc. Utah Acad. Sci., 11:241–244, 1934.

55. CLOVER SEED MIDGE[1]

Importance and Type of Injury. The clover seed midge, if very abundant, is capable of practically destroying the red clover seed crop. Its presence is indicated by the failure of many of the clover florets to open (Fig. 11.15, *May and June*). Where many maggots are present in the head, it will present a stunted appearance with irregular bloom, most of the head remaining green and the florets never opening. Infested seed presents a shriveled appearance or scarcely forms within the base of the clover head. The insect has no effect on the production of clover for a hay crop.

Plants Attacked. Red clover seed is apparently the favorite food of this insect. It has been found in small numbers in alsike, mammoth, crimson, white, and sweet clover but is of no importance on these crops.

Distribution. The insect occurs generally throughout the United States from the East to the Pacific Coast; also in the southern part of Canada.

Life History, Appearance, and Habits. The clover seed midge passes the winter in the larval stage inside a frail silken cocoon which is spun on, or shortly below, the surface of the ground (Fig. 11.15, *September*). Occasionally the larvae will crawl under trash and not spin a cocoon. In this stage the insect is about $\frac{1}{10}$ inch in length and of a reddish-pink color. In early spring the larvae change inside the cocoons to the pupal stage and emerge during April and May as very small, fragile, delicate-winged flies. These flies are shaped like mosquitoes but much smaller, dusty gray to black on the body, with a bright-red abdomen which is especially noticeable in the female. The female is equipped with a long ovipositor equal in length to her body. With this, she deposits her eggs in the young clover heads just as they are appearing from the buds (Fig. 11.15, *May*). The pale-yellow eggs are attached singly or in clusters to the hairs about the calyx of the clover blossoms, each female laying about 100 eggs. The eggs hatch in from 3 to 5 days, and the young maggots work their way to the top of the flowers, and down inside the unopened petals. Here they feed by sucking the sap and thus destroy the ovules and prevent the formation of seed. Upon becoming full-grown, in about a month, the larvae drop to the ground, generally during periods of rain (Fig. 11.15, *June*). They work their way below the surface, and there change inside their cocoons to the pupal stage. The first of the summer-generation adults usually appear during the first part of July. They lay their eggs in the second-crop clover, and from these come the overwintering maggots. In the southern part of the country, a third generation is produced.

[1] *Dasyneura leguminicola* (Lintner), Order Diptera, Family Cecidomyiidae. A very closely related species, *D. gentneri* Pritchard, attacks Ladino and alsike clover in Oregon (*Jour. Econ. Ento.*, 47:141, 1954).

Control Measures. When this insect alone is causing serious damage to clover seed, most of the injury may be prevented by cutting the clover a little before the uninjured heads have come into full bloom and removing the hay from the field as soon as it is dry. This is a most effective control

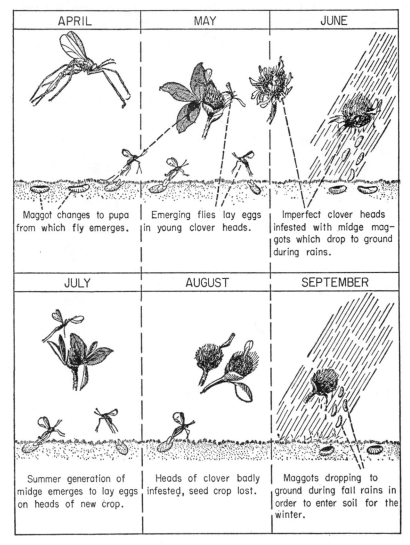

APRIL	MAY	JUNE
Maggot changes to pupa from which fly emerges.	Emerging flies lay eggs in young clover heads.	Imperfect clover heads infested with midge maggots which drop to ground during rains.

JULY	AUGUST	SEPTEMBER
Summer generation of midge emerges to lay eggs on heads of new crop.	Heads of clover badly infested, seed crop lost.	Maggots dropping to ground during fall rains in order to enter soil for the winter.

Fig. 11.15. Diagram showing seasonal history of the clover seed midge, *Dasyneura leguminicola.* (*From U.S.D.A., Farmers' Bul. 971.*)

measure, as the cutting at this time will kill practically all the young midge larvae within the clover heads. The insect may also be controlled to some extent by clipping clover about 2 weeks before any of the heads show the bloom, and again about 1 month later. This brings the first-crop clover into bloom before the summer-generation midges have emerged.   These measures apply mainly to the control of the clover seed midge and also to the clover head caterpillar.   Where several other clover insects are present, however, this practice may not be found to give the best production of seed.

References.   Ore. Agr. Exp. Sta. Bul. 203, 1917; *Ill. Agr. Exp. Sta. Bul.* 134, 1909; *U.S.D.A., Farmers' Bul.* 971, 1932; *Rept. Ento. Soc. Ont.,* **65**:22, 1935.

56. Clover Root Curculio[1] and Sweet Clover Weevil[2]

There are several species of clover root curculios, or clover Sitonas, that attack red, sweet, and alsike clover and alfalfa.   Our knowledge of the different species is incomplete.   Probably the most important is *Sitona hispidula* (Fabricius), but there are two other common species.[3]

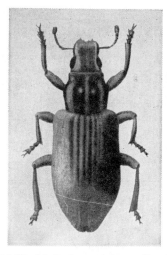

Fig. 11.16. Work of clover root curculio on alfalfa root.   Enlarged about 3 times. (*From Ill. Natural History Surv.*)

Fig. 11.17. Adult clover root curculio, *Sitona flavescens*, enlarged about 12 times.   (*From Ill. Natural History Surv.*)

The sweet clover weevil is a very closely related insect of considerable economic importance.

Importance and Type of Injury.   These insects work in such a way that much of their injury is not noticed, although the larval populations may number more than 10,000,000 per acre.   They may destroy from 60 to 80 per cent of the plants in young stands of alfalfa.   In most years they are not of much importance.   Clover or alfalfa plants infested by these insects wilt and often die, especially during periods of dry weather.   If the plant is dug and examined, the roots will be found scored and furrowed on the outside with numerous burrows, oftentimes nearly girdled (Fig. 11.16).   Small, grayish-white, footless, brown-headed grubs, about ⅙ inch long, will be found on the roots.   Small, grayish or brownish beetles, of about the same length, with blunt short snouts (Fig. 11.17), may be found eating rounded areas from the leaves or gnawing stems and leaf buds during the day, or hidden away among the trash on the ground

[1] *Sitona hispidula* (Fabricius), Order Coleoptera, Family Curculionidae.
[2] *Sitona cylindricollis* Fåhraeus.
[3] *Sitona flavescens* Marsham, and *S. crinitus* Gyllenhal.

or around the crown of the plant. The adult sweet clover weevils are voracious feeders on sweet clover, sometimes defoliating the plants. The larvae feed on the roots but do not cause severe damage.

Plants Attacked. The insects feed on all the common clovers, alfalfa, soybeans, and cowpeas, and doubtless on some other legumes. White clover seems to be slightly preferred by the insects, although they are often very abundant on the other varieties of clover as well as on alfalfa.

Distribution. These insects are quite generally distributed over the United States and southern Canada. These species are probably all of European origin but have been known in North America since 1876, with the exception of the sweet clover weevil, which was introduced into Canada about 1924 and has since spread across the United States from the East Coast to the Rocky Mountains.

Life History, Appearance, and Habits. The winter is passed in the egg, adult, and larval stages. Most of the insects pass the winter as young larvae. In the spring these larvae develop by feeding on the clover roots and crown of the plant, pupate during late March and April, and emerge during May and June as beetles. These beetles feed actively for about 1 month or 6 weeks, often when abundant doing considerable damage to clover fields. They become less active during the middle of the summer, and, although they remain about the clover fields, they feed but little. In the early fall they again become active, feed, and mate, and the female clover root curculio deposits as many as 500 eggs about the crowns of the plants. Egg-laying occurs during October and November and during warm periods of the winter, but approximately 75 per cent of the eggs are laid the following April and May. Eggs laid in the winter hatch in 150 to 200 days, while those laid in the spring hatch in 23 to 26 days. There is a single generation a year.

Control Measures. Crop rotation which will put infested fields in a grass or cultivated crop will drive out the beetles. If the land is plowed late in the fall or early in the spring, many of the insects will be destroyed. However, if the land is plowed early in the fall, many of the adult beetles will migrate out of the field and crawl considerable distances to other fields containing crops on which they feed and survive. The fungus *Beauvaria globulifera* greatly reduces the populations of these insects in some years. Granular applications of heptachlor at 2.5 pounds or dieldrin at 2 pounds per acre worked into the soil just before planting in the fall or spring have given excellent control of the clover root curculios and have increased hay yields as much as 44 per cent. The sweet clover weevil should be controlled when 50 per cent of new seedling plants show the characteristic marginal notches resulting from adult feeding. Dusting or spraying with DDT or toxaphene at 1.5 pounds per acre has given effective control, but these should not be applied to clover intended for animal forage. Parathion at 0.5 pound per acre is also effective.

References. Ill. Agr. Exp. Sta. Bul. 134, 1909; *U.S.D.A., Farmers' Bul.* 649, 1915; *Ky. Agr. Exp. Sta. Cir.* 42, 1934; *Jour. Econ. Ento.,* **23**:334, 1930; **27**:807, 1934; **42**:318, 1949; **44**:792, 1951; **48**:184, 1955; **50**:645, 1957; and **52**:1155, 1959.

57. CLOVER ROOT BORER[1]

Importance and Type of Injury. This insect is apparently of much less importance than formerly. This is due, in part at least, to better systems of crop rotation. Infested clover plants turn brown, wilt, and die, generally having the appearance of

[1] *Hylastinus obscurus* (Marsham), Order Coleoptera, Family Scolytidae.

suffering from the attack of some disease.   An examination of the roots, however, will show numerous burrows running through them and also grooves on the surface of the roots (Fig. 11.18).   These burrows cut off the circulation of the plant and frequently kill it.   Small, white, brown-headed, footless grubs, about $\frac{1}{10}$ inch long, will be found boring in the roots.   Dull-black or dark-brown, somewhat hairy, cylindrical, hard-bodied beetles, about $\frac{1}{12}$ to $\frac{1}{10}$ inch in length, may be found in the burrows in the roots or around the crown of the plant.

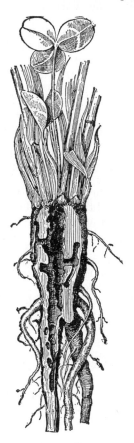

Fig. 11.18. Tunnels of the clover root borer, *Hylastinus obscurus*, in clover root.   About natural size.   (*From U.S.D.A.*)

Plants Attacked.   This insect prefers red and mammoth clover.   It has also been taken on white and sweet clover and alfalfa, peas, and vetch but is apparently of little importance on these crops.

Distribution.   The clover root borer is distributed over the northern part of the United States and eastern Canada, with the possible exception of the Great Plains. It is of European origin and was probably brought into this country about 1870.

Life History, Appearance, and Habits.   The insect passes the winter in the larval and adult stages, nearly all the insects being in the adult stage.   The winter is passed in the ground in the clover roots.   In the spring the female beetles deposit a few eggs in cavities eaten out in the crown of the clover plant, on the sides of the roots, or in burrows inside the clover roots.   Egg-laying extends over the spring months.   The larvae, upon hatching, tunnel through the roots, making irregular branched burrows.

Most of them become full-grown during late summer and transform to pupae during midfall.   All stages of the insect can be found during the late summer and fall months. There is but one generation each year.

Control Measures.   The clover root borer is seldom seriously destructive except in fields where clover has been allowed to stand more than two seasons.   Planting clover seed directly over bands of aldrin, heptachlor, or lindane at 0.75 pound per acre in fertilizer mixture or similar granular applications have given approximately 90 per cent control and have increased hay yields by about 25 per cent.

References.   U.S.D.A., Bur. Ento. Cir. 67, 1905, and *Cir.* E-811, 1950; *U.S.D.A., Dept. Bul.* 1426, 1926; *Jour. Econ. Ento.,* **42**:315, 1949; **43**:82, 1950; **47**:327, 1954; **48**:190, 1955; **50**:255, 1957; **51**: 491, 1958; and **52**:1166, 1959; *Ohio Agr. Exp. Sta. Bul.* 827, 1959.

58 (18). Grape Colaspis

This insect has been fully covered as a pest of corn, which is the crop most seriously damaged by it (page 512).   It sometimes causes considerable injury to clover plants by the feeding of the grubs on the clover

Fig. 11.19. Injury to clover foliage by beetles of the grape colaspis, and an adult, natural size on leaf.   (*From Ill. Natural History Surv.*)

roots.   The adults eat the leaves (Fig. 11.19).   No method of control on the clover plant has been developed.

B. SOYBEAN AND COWPEA INSECTS[1]

Soybeans and cowpeas have been, on the whole, more free from serious insect injury than most of the field crops.   Grasshoppers are probably the worst insect pests of these crops and, in years when they are abundant, may seriously damage, or nearly destroy, soybean fields.   They may be controlled by the same methods as those given for the control of these pests on corn; see page 468.   The poison-bran bait, however, should be applied to adjoining fields to kill the hoppers before they enter the soy-

[1] See *Ohio Agr. Exp. Sta. Bul.* 366, 1923.

beans, as they seldom originate in these fields, but migrate into them from adjoining fields. If the poison bait is applied in the soybean fields, it falls between the plants where the ground is heavily shaded, and very little of it is eaten.

The green cloverworm (page 556) has, during certain years, proved to be a very destructive pest of soybeans and cowpeas in the southern states. It may be controlled on these crops by dusting with calcium arsenate, at the rate of 5 to 7 pounds per acre, applying the material with a cotton duster or a field-crop duster. In most years, the insect is of no importance on these crops (see *North Carolina Agricultural Extension Circular* 105, 1920).

The clover root curculio (page 572) may completely destroy soybeans if they are planted on spring-broken clover sod. The adult beetles of the curculio attack the soybeans as they first come up and eat off the plants as fast as they appear through the ground. No injury has been reported to these crops when planted on clover-sod ground, plowed in the fall or early winter. Where soybeans are to follow spring-broken clover sod, it should be well harrowed and planting delayed as late as possible.

In the southern states the cowpea is sometimes badly damaged by the caterpillars of the corn earworm or cotton bollworm. There is no effective method of controlling this insect on cowpeas. Damage occurs only in scattered seasons during periods of their greatest abundance.

Throughout the Gulf states cowpeas, string beans, Lima beans, and to some extent, cotton and strawberry plants are severely injured by the cowpea curculio,[1] which causes wormy peas and beans within the growing pods. Spraying or dusting with methoxychlor at 1.75 pounds per acre and the frequent picking of the ripe pods and storing on a clean, tight, dry floor are helpful in control. The cowpea weevil[2] and some other weevils often are very destructive to stored cowpea seed. The insects deposit their eggs in the cowpeas in the field, the infestation originating from infested seed or from scattered cowpeas remaining in the field from the previous year. (The control of these insects is given under Stored Grain Insects, page 932.) (See U.S. Department of Agriculture, Technical Bulletin 599, 1938.)

The striped blister beetle (page 642) is sometimes a serious pest in soybeans (see *U.S. Department of Agriculture, Leaflet* 12, 1927). The spotted cucumber beetle is sometimes abundant on soybeans and cowpeas but never sufficiently injurious to warrant taking special control measures. Several aphids also attack these crops but are not of special importance. Leafhoppers are often abundant, especially in fields of soybeans, and occasionally cause some injury to the leaves.

The Mexican bean beetle[3] (see Garden Insects, page 622) has caused some injury to soybeans and cowpeas where these crops were grown in close proximity to heavily infested garden beans and is becoming an increasingly important pest of soybeans in the South. The bean leaf beetle[4] (page 624), the Japanese beetle (page 749), and the white-fringed beetle (page 607) are pests of soybeans in some areas.

---

[1] *Chalcodermus aeneus* Boheman, Order Coleoptera, Family Curculionidae (see *Ala. Poly. Agr. Exp. Sta. Bul.* 246, 1938).

[2] *Callosobruchus maculatus* (Fabricius), Order Coleoptera, Family Mylabridae.

[3] *Epilachna varivestis* Mulsant, Order Coleoptera, Family Coccinellidae.

[4] *Cerotoma trifurcata* (Förster), Order Coleoptera, Family Chrysomelidae.

CHAPTER 12

COTTON INSECTS

FIELD KEY FOR THE IDENTIFICATION OF INSECTS INJURING COTTON

A. Insects that chew the foliage, often stripping the plants of leaves, or cutting off and devouring young seedling plants:

1. Large and small, lubberly grasshoppers often invade cotton from near-by waste lands and defoliate the plants...
..........*Lubber grasshoppers, differential grasshoppers,* and others, page 465.

2. Seedlings cut off, gnawed, or rasped, about 1 inch above the ground, or on larger plants just below the apical bud, by brownish to black, jumping insects, with long antennae...
Field cricket, Acheta assimilis Fabricius, page 600 (see also *U.S.D.A., Cir.* 75, 1929).

3. Cotton in newly broken fields, or fields that were very grassy the year preceding, attacked while small by swarms of robust, long-legged, grayish or brown beetles, ½ to 1 inch long, that lack the underwings and cannot fly. They feed in late afternoon and at dusk, and may eat off the plants over a large field..
..*Wingless May beetles* or *June bugs* (see *U.S.D.A., Farmers' Bul.* 223, 1905).

4. Dark-brown to black, humpbacked beetles, nearly ¼ inch long, with a slender snout ⅓ as long as the body, bent down between the legs, and prominent, round punctures on the back, appear during early summer in fields that were in cowpeas the previous year, and eat small holes in leaves or tender parts of stem.........*Cowpea pod weevil* (see *N.C. Dept. Agr. Bul.,* vol. 29, no. 6).

5. Cotton sometimes seriously defoliated, especially in wet seasons, by slender, green, black-and-white-striped caterpillars, with four black dots on each segment above, and ranging up to 1½ inches long. They crawl with a slightly looping movement and feed only on cotton.......*Cotton leafworm,* page 579.

6. Cotton, along with most other plants, defoliated by tan or green to nearly black caterpillars with three yellow hairlines down the back and a wider one on the side, that is splotched with red. Body covered with fine scattered hairs arising from black tubercles, and a prominent inverted Y on the front of the head......................................*Fall armyworm,* page 473.

7. Young plants cut in two at or near the surface of the soil, or the leaves eaten, at night. Plump dirty-looking caterpillars, of various shades and markings of green, brown, and black, up to 1½ inches long, found in daytime in the soil. Often curl the body when disturbed..................*Cutworms,* page 476.

8. Foliage devoured by very hairy or woolly caterpillars, up to 2 inches long, black-bodied and covered with long black and red hairs....................
...*Salt-marsh caterpillar,* page 611 (see *U.S.D.A., Farmers' Bul.* 223, 1905).

9. Pieces of the leaves cut off and carried to their nests in the soil by reddish-brown ants varying from 1/12 to ⅓ inch in length...........................
.......*Leaf-cutting ant,* page 897 (see *U.S.D.A., Bur. Ento. Cir.* 148, 1912).

10. Many-legged hard-shelled crayfish, with a pair of pincer legs in front, crawl out of burrows in wet soil and devour the plants. Not insects but Crustacea
....*Crayfish* or *crawfish.*

B. Insects that suck the sap from leaves, stem, squares, or bolls:

1. Very slender, almost microscopic, yellowish, active nymphs and brown-bodied, bristly winged adults, only 1/25 inch long, suck sap from the leaves and buds,

577

causing retarded growth, defoliation, and malformed seedlings..............
Thrips, especially *the tobacco thrips, Frankliniella fusca* (Hinds), *the flower thrips, F. tritici* (Fitch), and *the bean thrips, Hercothrips fasciatus* (Pergande) (see *S.C. Agr. Exp. Sta. Bul.* 306, 1936; *Calif. Agr. Exp. Sta. Bul.* 609, 1937).

2. Leaves of cotton curled or dwarfed; the undersides thickly dotted with small, winged or wingless, tan, brown, green, or black plant lice of various sizes, which suck the sap........................*Melon aphid* or *Cotton aphid*, page 581.

3. The lint, especially of long-staple cotton, is stained an indelible yellow color by flattened, rather narrow, long-legged bugs of various sizes up to $\frac{3}{5}$ inch long, with head and prothorax bright-red and the rest of the body dark-brown, crossed with light-yellow lines. The bugs puncture the seeds in the developing bolls and cause a juice to exude that stains the lint........................
Cotton stainer, Dysdercus suturellus (Herrich-Schaeffer) and other species (see *U.S.D.A., Bur. Ento. Cir.* 149, 1912; and *Jour. Agr. Res.*, **31**:1137–1147, 1925).

4. Squares and young bolls drop, shrivel, or decay; small, round, blackened spots appearing on their surface, and stems often split lengthwise where they have been punctured by sap-sucking bugs.....................................
Cotton-leaf bug, stink bugs, and leaf-footed plant bugs (see *U.S.D.A., Farmers' Bul.* 223, 1905, and *Tech. Bul.* 296, 1932).

5. Shield-shaped, bright-green, bad-smelling, flattened bugs, the winged adults about $\frac{5}{8}$ inch long, suck sap from the buds and bolls, causing them to shed...
Green stink bug, Acrosternum hilare (Say) (see *Va. Agr. Exp. Sta. Bul.* 294, 1934).

6. Small squares shed, buds blasted, and growth stunted; resulting in shortening of internodes and development of an excessive growth of branches close together, with deformed leaves, on young plants; and an excessively tall, whip-like growth of the main stem and a suppression of fruiting branches on older cotton. Small greenish bugs from $\frac{1}{25}$ to $\frac{1}{8}$ inch in length, with scarlet eyes, cluster on the growing tips..............................*Cotton fleahopper*, page 581.

7. During hot weather the leaves of cotton become blotched with large irregular areas of reddish brown so that entire patches of varying size in the fields look red. Undersides of such leaves, when examined under a lens, are seen to be swarming with minute, pale-greenish to reddish, six- or eight-legged mites....
....*Spider mites*, page 616.

C. *Insects that bore or tunnel in the stalks or puncture and split them:*

1. Caterpillars, up to $1\frac{1}{2}$ inches long, prominently striped with brown and white in front and behind, but with a large grayish-brown "bruised-looking" area in front of the middle, burrow through the heart of the stem, killing the terminals. Especially bad around weedy margins of fields in early summer. Stripes disappear when caterpillars become full grown.....*Common stalk borer*, page 493.

D. *Insects that eat into, or feed inside of, the squares and bolls:*

1. Squares flaring, and either falling to the ground, or hanging withered and dry on the plant. Squares and bolls with wart-like scars, covering punctures about the diameter of a pin. Inside of bolls decayed and soiled. White, curved-bodied, brown-headed, legless grubs, up to $\frac{1}{2}$ inch long, feed in the squares and bolls destroying the unopened buds, lint, and seed. Hard-shelled, grayish to brown, long-legged beetles, averaging $\frac{1}{4}$ inch long, with a slender snout half as long as the body, puncture squares and bolls to feed and lay eggs in them..
....*Boll weevil*, page 582.

2. Flowers fail to open, squares fall, and bolls are eaten out but not decayed-looking. A pinkish-white, brown-headed caterpillar up to $\frac{1}{2}$ inch long, found inside the fruits or seeds; easily distinguished from *D*, 1 by having 8 pairs of legs and prolegs. Larvae often bore inside of the seeds, and web 2 seeds together, so that "double seeds" are found when the cotton is ginned. Distinguished from other caterpillars by having 4 teeth on the mandibles and the crochets on the prolegs horseshoe-shaped....................*Pink bollworm*, page 587.

3. Light- to dark-green, pinkish or nearly black, more or less striped caterpillars, up to $1\frac{3}{4}$ inches long, bore into squares and bolls and eat out the interior. The worms have 3 pairs of slender legs behind the head, 5 pairs of fleshy prolegs, and the skin is rough and thorny appearing under a lens........................
...............................*Bollworm* or *corn earworm*, page 587.

4. Squares eaten out as in *D*,3 by bright-green, oval, distinctly flattened caterpillars, covered with short velvety hairs, and with the head drawn under the front of the body; not exceeding ¾ inch in length...........................
   Cotton square borer, Strymon (= *Uranotes*) *melinus* (Hübner) (see *Tex. Agr. Exp. Sta. Bul.* 401, 1929).

E. *Insects that attack the roots, below ground:*

1. Seeds fail to sprout or plant dies when small. Slick, whitish, slender, cylindrical worms up to 1 inch long, with 6 small legs behind the head, are found in the soil or eating into the seeds and roots.................................
   Sand or *corn* and *cotton wireworm* (*Horistonotus uhlerii* Horn) (see *S.C. Agr. Exp. Sta. Bul.* 180, 1914).

2. Seedlings weak, yellowish in color. Many small brown ants tunneling in the soil about the roots and caring for bluish-green, wingless plant lice or aphids about the size of pinheads, that cluster on the roots and suck the sap........
   ....*Corn root aphid* and *cornfield ant*, page 501.

F. *Insects that lay eggs in the stalks of the plant:*

1. Cotton stalks, in the fall of the year, are punctured and may be split by rows of round holes of the diameter of a pin and with an elongate curved white egg within. No injury to cotton......................*Tree crickets*, page 788.

The following important cotton insects are not discussed in this chapter, but accounts of their damage, habits, and control are given in connection with other crops, which they also attack:

Cutworms, page 476
European corn borer, page 490
Fall armyworm, page 473
Garden webworm, page 609
Grasshoppers, page 465
Lygus bugs, page 565
Salt-marsh caterpillar, page 611

Seed-corn maggot, page 518
Spider mites, page 616
Thrips, pages 742 and 800
Tobacco budworm, page 596
White-fringed beetle, page 607
Wireworms, page 506
Yellow-striped armyworm, page 479

General References. U.S.D.A., Ento. Res. Div., "Annual Conference Report on Cotton Insect Research and Control"; GAINES, J., *Ann. Rev. Ento.,* **2**:319, 1957; LITTLE, V., and D. MARTIN, "Cotton Insects of the United States," Burgess, 1942.

59. COTTON LEAFWORM[1]

Importance and Type of Injury. Slender, greenish, looping caterpillars with black-and-white stripes and a number of small rounded black spots scattered over the body (Fig. 12.1,*B*) rag or strip the leaves of cotton, especially late in the season, and when food is scarce, may eat the squares or bolls. The adult moth sometimes causes injury to ripe peaches, grapes, and other fruits by lacerating them with spines on the end of its tongue. In this respect it is a striking exception to the rule that adults of Lepidoptera are not injurious.

Plants Attacked. The larva feeds only on cotton or wild cotton. The adults sometimes feed destructively on ripe peaches, grapes, and other fruits.

Distribution. Eastern United States from the Great Lakes southward into the tropics. It is more destructive in the Gulf states than further north.

Life History, Appearance, and Habits. This insect is a native of the tropics, and the adults migrate into the United States each year from Central America, flying northward in great numbers as far as the Great Lakes and Canada. All stages die during the winter in the United States, the species surviving only in the tropics. The adult female deposits from

[1] *Alabama argillacea* (Hübner), Order Lepidoptera, Family Noctuidae.

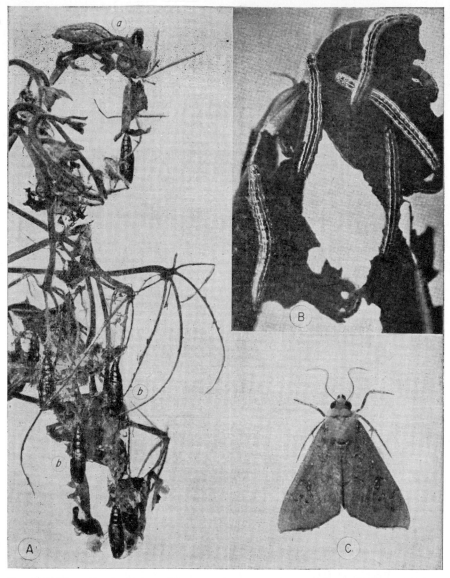

Fig. 12.1. The cotton leafworm, *Alabama argillacea*. *A*, larva, pupae, and injury; note how the leaves have been stripped, the larva at *a*, and the pupae attached by their posterior ends at *b,b* (*from U.S.D.A.*). *B*, larvae natural size, devouring foliage. *C*, adult female moth, enlarged ½, in resting position. (*From Ala. Poly. Inst.*)

400 to 600 eggs on the underside of the larger cotton leaves over a period of about 10 days. The larva passes through six instars over a period of 2 to 3 weeks and when full-grown is about 1½ inches long, with four black spots in a square on the back of each segment and the last pair of prolegs standing out conspicuously behind the body. The larva pupates in a flimsy, silken cocoon inside a folded leaf or fastened by its tail to the

foliage (Fig. 12.1,*A*). A generation is completed in about 4 weeks, and there are from three to seven generations in the Cotton Belt, the moths from each flying farther and farther north. The moths (Fig. 12.1,*C*), which measure about 1¼ inches across the wings, are olive-tan in color, with three more or less prominent, wavy, transverse bars on each front wing.

Control Measures. The cotton leafworm is susceptible to the insecticides normally used to control the boll weevil and is not normally a problem where the latter is controlled. If it appears early in the season, it may be controlled by dusting with calcium arsenate at 7 to 10 pounds; spraying or dusting with parathion at 2 to 4 ounces; endrin, Guthion, malathion, or methyl parathion at 4 to 8 ounces; Sevin at 0.5 to 1 pound; toxaphene or Strobane at 2 to 3 pounds; or by spraying with Dylox at 0.25 to 1 pound per acre, applied to the foliage as needed. Late in the season the feeding of the leafworm is considered a benefit in boll weevil territory because it destroys the top growth, which could not make cotton and which would sustain the weevil late in the fall. When the adults are troublesome in orchards, it has been found possible to prevent much injury by dusting with sulfur. In small orchards or vineyards many of the moths may be killed by placing pans of crushed fruits poisoned with sodium arsenite or other stomach poison around the trees.

60 (13, 96). APHIDS[1-3]

Almost as soon as cotton has put out leaves, small, soft-bodied, pale-green plant lice fly to them and start to reproduce. In cool, wet seasons, when their natural enemies cannot work against them so well, they may become abundant enough to stunt and deform the plants. Often, when the hot weather of summer arrives, they practically disappear.

The most important species, feeding above ground, is the cotton or melon aphid.[1] The cowpea aphid[2] often migrates into cotton from fields of soybeans and cowpeas. Aphids may become abundant after the use of DDT or calcium arsenate for boll weevil control, since these kill the parasites and predators without controlling the aphids. However, this increase does not usually occur after the use of endrin, Sevin, toxaphene, or Strobane. Heavy infestations of aphids may be controlled promptly by spraying or dusting with parathion at 2 to 4 ounces, methyl parathion at 4 to 8 ounces, Trithion at 0.25 to 1 pound, Ethion at 0.5 to 1 pound, or malathion at 1 to 2 pounds per acre, or by spraying with demeton at 2 to 6 ounces, Diazinon at 4 ounces, or Dylox at 0.25 to 1 pound per acre. The corn root aphid[3] and its attendant ant are sometimes pests of cotton, attacking the roots just as they do those of corn. The life history, descriptions, and control are discussed under Corn Insects (page 501).

References. *Jour. Econ. Ento.*, **40**:374, 536, 1947; **41**:851, 1948.

61. COTTON FLEAHOPPER[4]

Importance and Type of Injury. In scattered areas throughout the cotton belt this small green sucking bug has sporadically caused serious losses by sucking the sap from the very small squares and other terminal

[1] *Aphis gossypii* Glover, Order Homoptera, Family Aphidae.
[2] *Aphis medicaginis* Koch, Order Homoptera, Family Aphidae.
[3] *Anuraphis maidi-radicis* (Forbes,) Order Homoptera, Family Aphidae.
[4] *Psallus seriatus* (Reuter), Order Hemiptera, Family Miridae.

growth, resulting in excessive shedding and an abnormal whip-like growth of the plant as described in the key.

Plants Attacked. In addition to cotton, croton, evening primrose, goatweed, sage weed, horsemint, orach, wild sunflower and at least 30 other species of weeds.

Distribution. General throughout the cotton belt and over much of the United States.

Life History, Appearance, and Habits. The cotton fleahopper hibernates in the egg stage on croton, sage weed, goatweed, Atriplex, and other weeds. In southern Texas it appeared on horsemint early in March, migrated to cotton late in April, and deserted the cotton by the end of July, feeding for the remainder of the season on croton, evening primrose, snap beans, and potatoes. The eggs, which are yellowish-white, about $\frac{1}{30}$ inch long by a fourth as wide, are inserted beneath the bark, especially just below the growing tips. In a little over a week they hatch, and the greenish nymphs begin sucking the sap from the terminal bud cluster, including young leaves, stems, and squares. The extent of the injury suggests that a toxin may be inserted in the plant as the insect feeds. The nymphs molt five times, and within 10 to 30 days are mature bugs. The adult is about $\frac{1}{8}$ inch long, flattened, elongate ovate in outline, with prominent antennae, the body pale yellowish-green in color, with minute black hairs and black specks over the upper surface. The generations are not distinctly separated, as reproduction is continuous throughout the warm season.

Control Measures. The eradication of weeds and the destruction of cotton stalks during fall and winter will reduce the spring populations of this insect. It is generally controlled by spraying or dusting with (a) γ-benzene hexachloride or endrin at 0.1 pound, dieldrin at 0.15 pound, aldrin or heptachlor at 0.25 pound, DDT at 0.5 pound, toxaphene or Strobane at 1 pound, Sevin at 0.5 to 1 pound, or (b) Guthion at 0.25 pound or malathion at 0.25 to 1 pound per acre. Livestock should not be pastured on cotton treated with late dosages of the materials in (a).

References. *Tex. Agr. Exp. Sta. Bul.* 339, 1926, and 380, 1928; *S.C. Agr. Exp. Sta. Bul.* 235, 1927; *U.S.D.A., Dept. Cir.* 361, 1926, and *Cir.* E-430, 1938; *Jour. Agr. Res.*, **40**:485, 1930; *U.S.D.A., Tech. Bul.* 296, 1932; *Jour. Econ. Ento.*, **30**:125, 1937; **41**:548, 735, 1948; and **51**:489, 1958.

62. BOLL WEEVIL[1]

Importance and Type of Injury. No insect pest in all the world has gained greater notoriety than the cotton boll weevil, which has been estimated to have caused a direct loss in cotton and cottonseed of more than $200,000,000 annually since 1909. In addition to this direct loss, there have been enormous financial losses due to depreciated land values, the abandonment of cotton growing, the closing down of cotton gins and oil mills, the interruption of railroad, bank, and mercantile business, and many other economic disturbances in territory where the boll weevil has appeared. It should not be supposed that the cotton farmer sustains all the loss. On the contrary, it is borne chiefly by the users of cotton goods.

The injury is caused by the adult weevils and their young or grubs. The adults (Fig. 12.2,c) puncture the squares and bolls by chewing into them with their long slender bills to feed on the tissues inside and to lay

[1] *Anthonomus grandis* Boheman, Order Coleoptera, Family Curculionidae.

their eggs in the holes. This causes the squares to flare (*b*) and either drop off or hang withered and dry (*a*). The grubs that hatch from the eggs feed inside the squares and bolls, destroying the developing flower so that it fails to bloom or else develops seeds with little fiber. A good indication of the importance of this feeding is the fact that where it is prevented by the application of insecticides, gains of from 500 to 1,000

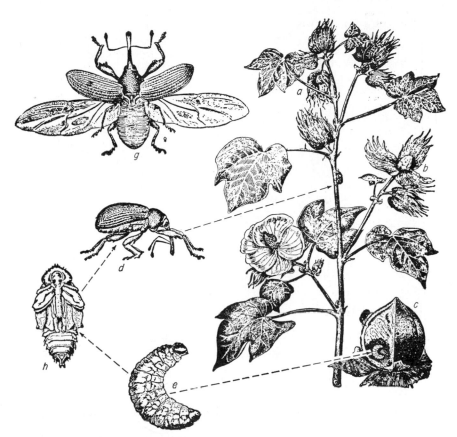

Fig. 12.2. The cotton boll weevil, *Anthonomus grandis*. On the right a cotton plant attacked by the boll weevil, showing at *a*, a hanging dry infested square; *b*, a flared square with weevil punctures; *c*, a cotton boll sectioned to show attacking weevil and larva in its cell; *g*, adult female with wings spread as in flight; *d*, adult; *h*, pupa; *e*, larva. (*Rearranged from U.S.D.A.*)

pounds of seed cotton per acre are often secured, the average gain being 300 to 400 pounds per acre.

Plants Attacked. This pest feeds and reproduces almost exclusively upon the cotton plant, although it will also feed upon other malvaceous plants such as okra, hollyhock, *Hibiscus syriacus*, and *Thespesia populnea*. A variety, the thuberia weevil,[1] feeds on wild cotton, *Thuberia thespesioides*, in the Southwest and also attacks cultivated cotton.

[1] *Anthonomus grandis thurberiae* Pierce, Order Coleoptera, Family Curculionidae.

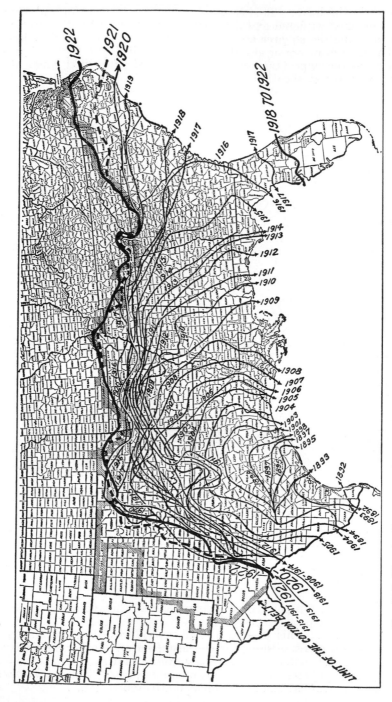

Fig. 12.3. Map showing spread of cotton boll weevil over the southern states from 1892 to 1922. (*From U.S.D.A.*)

Distribution. The cotton boll weevil is not a native of the United States. Its original home was in Mexico or Central America. Little was known about it previous to 1892, when it first appeared in southernmost Texas. From that date it spread to the north and east, an average distance of about 60 miles a year (Fig. 12.3). An average of more than 20,000 square miles of new territory was infested by the weevil in each of the first 35 years after it had crossed the Rio Grande. It now occupies practically all the important cotton-growing area in the United States, except Arizona and California. The weevil does not seem to be able to maintain itself in the extreme northern edge of the cotton-growing area. It also ranges southward through Mexico to Guatemala and Costa Rica and is known to occur in Cuba.

Life History, Appearance, and Habits. The boll weevil winters chiefly in the adult stage in all kinds of shelter; under dead leaves and piled cotton stalks, among Spanish moss on trees, under loose bark, about gins and barns, in woods, along fences, and in many other protected places. The adult beetle (Fig. 12.2,*d,g*) is a small, hard-shelled weevil, averaging about $\frac{1}{4}$ inch long, of a yellowish, grayish, or brownish color, becoming nearly black with age. It has a slender snout, half as long as the body, and close-fitting smooth wing covers with fine parallel lines and covered with short gray down or fuzz. The most characteristic feature about the boll weevil is the two spurs or teeth near the end of the front femur, the inner one being much longer than the other, and a single tooth on the middle femur.

From about the time early cotton is up (March), until early or even late June, the adults straggle out of hibernation, feeding at first on the tender terminal growth of the plants. They concentrate at first on the older or earlier planted fields. The weevils prefer to attack the blossom buds or squares after they are about 6 days old. They eat cavities into them and lay a single egg in each hole and usually one egg to a square. Each female may lay from 100 to 300 eggs. Later in the season eggs are laid in the bolls, but the squares are always preferred. The egg hatches in an average of 3 to 4 days, and the larva feeds inside the square or boll and undergoes two or three molts over an average period of about 8 to 9 days. The mature larva is a white, legless, curved-bodied grub about $\frac{1}{2}$ inch long and much wrinkled, with a brown head and mouth parts (Fig. 12.2,*e*). It transforms into the pupal stage inside the hollowed-out cavity in which it has fed. The pupal stage (*h*) lasts an average of about 5 days, when the adult emerges and eats its way out of the square or boll. Mating occurs and eggs are laid within 7 to 8 days. The average time required for a generation from egg to egg is about 25 days, and there are from 2 to 3 to 8 to 10 generations per year. However, because of the irregular emergence of the adults from winter hibernation, the generations are not distinct. As the cotton becomes mature (mid-August to early September), the weevils spread extensively, taking flights by short stages that may aggregate 20 to 50 miles or more, and these account for the chief spread of the insect. At this time many of the adults accumulate reserves of body-fat over a period of about 3 weeks and then enter diapause, a condition marked by atrophy of testes and ovaries and a decrease in water content and respiratory rate. The onset of diapause apparently provides the stimulus for the insect to leave the cotton plant and seek winter shelter, and only diapausing weevils overwinter. These emerge from this condition continuously over the growing season and may be found in every month. They live an average of about 21 days after diapause, and although the males resume spermatogenesis immediately, the females do not begin oögenesis until after feeding on young cotton.

Control Measures.   Because all the developing stages of the weevil are spent inside the squares and bolls of the cotton plant and the adults insert their slender snouts below the surface to feed, this insect is one of the most difficult of all known pests to control.   Generally speaking the weevils will be least destructive following low winter temperatures and when hot dry weather prevails during June, July, and August.   The bolls or fruits of the crop are produced over a period of about 2 months and protection from weevils during this heavy fruiting period is essential to making a crop.   Successful control can be achieved only by a combination of measures, of which the following are the most important:

FALL DESTRUCTION OF PLANTS IN THE FIELD.   When examination shows that nearly all the squares in the field are being punctured by weevils, the cotton should be harvested as soon as possible and the stalks cut or plowed out, immediately collected together, and burned or plowed under deeply.   No more cotton can be made from punctured squares, and the prompt destruction of the plants prevents the maturing of thousands of late weevils, which are the chief ones to winter over.   The destruction of the plants before the first killing frost, and fall plowing, will remove hibernating places and make possible the early planting of the next crop.

MAKING AN EARLY CROP.   Since the weevils prefer squares in which to lay their eggs and such squares are doomed to make no cotton, it is of the greatest importance to get the blossoms past the square stage to the boll stage before weevils become abundant enough to damage many of them.   The use of early-maturing varieties; planting as early as the seedbed can be thoroughly prepared; enriching the soil by the use of legumes and commercial fertilizers; and frequent shallow cultivation not too close to the plants are important in making the early crop.

INSECTICIDE TREATMENTS.   When the crop has been produced by the best farming possible, it should be protected and further increased by spraying or dusting the foliage with one of the following: (*a*) dieldrin or endrin at 0.2 to 0.5 pound, (*b*) aldrin or heptachlor 0.25 to 0.75 pound, (*c*) γ-benzene hexachloride at 0.3 to 0.45 pound, (*d*) Guthion at 0.25 to 0.5 pound, (*e*) methyl parathion at 0.25 to 0.75 pound, (*f*) malathion or Sevin at 1 to 2 pounds, (*g*) toxaphene or Strobane at 2 to 4 pounds, or (*h*) calcium arsenate dust at 7 to 15 pounds per acre.   The choice will depend upon the relative susceptibility of the weevils in the area, the presence of other pests, and economics.   DDT is usually added to prevent the build-up of bollworms and an acaricide or 40 per cent sulfur to control spider mites.   Early-season applications may be made to large populations of overwintering weevils at about 7-day intervals, and mid- or late-season applications every 3 to 5 days when control is necessary.   Where proper insecticidal control is practiced, an average gain of 300 to 500 pounds of seed cotton per acre may be obtained.   Dairy animals or those being finished for slaughter should not be pastured upon cotton or fed on gin waste where late applications of the materials in (*a*), (*b*), (*c*), or (*g*) were made, and heptachlor, Sevin, or Strobane should not be applied after the bolls open.

No successful introduced parasites or predators of the boll weevil are known.   However, more than 50 native insects attack the weevil and may destroy an average of about 16 per cent of the larvae.

References. U.S.D.A., *Ento. Buls.* 45, 1904, and 114, 1912, *Farmers' Bul.* 1329, 1923, *Tech. Buls.* 112, 1929, and 487, 1935, and *Cirs.* E-430 and 431, 1938; *Ark. Agr. Exp. Sta. Bul.* 271, 1932; *Jour. Econ. Ento.*, **25**: 772, 1932; **40**:374, 508, 600, 644, 1947; **41**:22, 543, 548, 851, 986, 1948; **49**:696, 1956; and **52**:403, 453, 603, 1042, 1138, 1959.

63 (12, 71, 111, 260). BOLLWORM OR CORN EARWORM[1]

This caterpillar, which also attacks corn (page 498), tomatoes (page 657), tobacco, beans, vetch, alfalfa, and many garden plants and flowers, destroys the squares and bolls by eating into them. The tobacco budworm[2] causes very similar damage and is controlled in the same way. The eggs of the bollworm are laid on the leaves and outsides of the squares, and the caterpillars do not remain in a single boll, but bore out again and crawl from square to square, a single bollworm often destroying all the fruits on a branch of the plant. The pupal stage is passed in the soil at a depth of 1 to 4 inches. The first two generations usually occur on corn, tobacco, and other plants, and it is not until the third generation (August or later) that the bollworm becomes destructive to cotton. Therefore it is important to produce an early crop. Deep plowing and cultivating the soil in fall and winter are effective in turning up the pupae and exposing them to natural enemies and weather. The bollworm has been controlled by dusting or spraying the foliage every 5 to 7 days as needed with DDT 0.5 to 2 pounds, endrin 0.2 to 0.5 pounds, Sevin 1 to 2 pounds, or toxaphene or Strobane 2 to 4 pounds per acre (see cautions under Boll Weevil). Treatment is commonly made when 4 or more small worms are found per 100 plants.

References. U.S.D.A., *Bur. Ento. Bul.* 50, 1905, *Farmers' Bul.* 1595, 1929, and *Leaflet* 462, 1960; *Ark. Agr. Exp. Sta. Bul.* 320, 1935; *La. Agr. Exp. Sta. Tech. Bul.* 482, 1953; *Jour. Econ. Ento.*, **40**:374, 1947; **41**:406, 548, 583, 986, 1948; **51**:782, 1958; and **52**:975, 1959.

64. PINK BOLLWORM[3]

Importance and Type of Injury. The pink bollworm is one of the world's most destructive insects and, where uncontrolled, regularly causes damage to 20 to 40 per cent of the cotton bolls and may cause a total crop loss in limited areas. Many millions of dollars have been spent in controlling this pest in the United States and in preventing its spread through quarantine and eradication programs. The larvae bore into the squares, causing them to fall or the blossoms to fail to open. As the plant matures, the bolls are attacked and the caterpillars tunnel through the immature lint and into the seed, where they eat out the contents. As a result of this attack, the lint may be unpickable or so stunted as to greatly lower the yield and grade, and the yield of oil from the seeds is greatly decreased. Double seeds (Fig. 12.5) are found when the cotton is ginned, two partly eaten seeds being fastened together.

Plants Attacked. Cotton and rarely okra, hollyhock, hibiscus, and other malvaceous plants.

Distribution. The pest is believed to be a native of India and has been transported to most of the other cotton-producing countries of the world. It is a well-established and very destructive pest in nearly all the cotton-

---

[1] *Heliothis zea* (Boddie) [(= *armigera*) (Hübner)], Order Lepidoptera, Family Noctuidae.

[2] *Heliothis virescens* (Fabricius).

[3] *Pectinophora gossypiella* (Saunders), Order Lepidoptera, Family Gelechiidae.

growing countries of Asia and adjacent islands, including Hawaii and the Philippines; in North, East, and West Africa; in Australia; in Brazil; in the West Indies; and in Mexico. It was brought to Mexico in 1911, in seed imported from Egypt, and in 1915 or 1916 was carried across the border into Texas in cotton lint and seed. Since its discovery there in 1917, it has spread through most of Texas and Oklahoma into southern New Mexico and Arizona and into Arkansas and Louisiana. It is also present in southern Florida, where it attacks wild cotton. A number of outlying infestations have been eradicated at great cost.

Fig. 12.4. Larva of pink bollworm. Enlarged about 6 times. (*From U.S.D.A., Dept. Bul.* 1397.)

Fig. 12.5. Cotton seeds containing pink bollworms opened to show the cells. Both single- and double-seeded cells are shown, the double-seeded ones being broken apart. (*From U.S.D.A., Dept. Bul.* 1397.)

Life History, Appearance, and Habits. The adult pink bollworm is a small dark-brown moth measuring about ¾ inch across the wings, which are narrow with a wide fringe and are peculiarly pointed at the tips. The first segment of the antenna has five or six long stiff hairs, and the palps are long and curved. The moths are seldom seen since they hide in the daytime but are active fliers at night. They have been taken at altitudes of 3,000 feet and can migrate long distances with favorable winds. The female begins egg-laying within a day or so after emergence and deposits about 200 to 400 white, oval eggs, usually singly or in groups of 5 to 10 all over the cotton plant, but concentrated on the terminals in early season and later on the developing bolls. The eggs hatch in 3 to 5 days at optimum temperature, and the larvae promptly bore into the squares and eat the developing flowers, or into the bolls, where they consume both lint and seeds. In the summer, larval development is complete in an average of about 13 days and requires four instars. The mature larva is about ½ inch long with a prominent pinkish coloration (Fig. 12.4), and in the summer eats a hole in the boll and drops to the ground, where it pupates in the upper 2 inches of soil, in surface trash, or in seed or boll. It emerges as an adult after about 9 days, and under optimum conditions, the entire life cycle requires about 31 days from egg to egg, and there are four to six generations a year in the cotton-growing areas of the United States. Under conditions of cool, dry weather and decreased food supply a large percentage of the larvae enter diapause, and these "resting larvae" pass the winter curled up in a small cocoon in partially opened bolls, in cotton lint, in stored seed, or in the soil. The diapausing larva normally emerges in the spring, but under extremely dry conditions may remain quiescent for as long as 2½ years.

Control Measures. Cultural practices are very effective in reducing the overwintering population of the pink bollworm, which hibernates only in the fields which it has attacked, unless it is removed in harvesting the crop. These include (*a*) early maturity of the cotton by planting high-quality seed of an early variety and using farm practices which ensure a thick stand and protection of early fruit and hasten opening of the bolls, and (*b*) destruction of old stalks and bolls of cotton by shredding (the flail-type shredder will destroy about 85 per cent of the pink bollworm larvae left in the field after harvest) and plowing under to a depth of 6 inches as early as possible in the fall and not later than 10 days before frost. Okra stalks should also be shredded and plowed under, since this plant serves as a secondary host. In arid areas, the fields may be irrigated after plowing under the crop debris, to increase the mortality of the pink bollworms. In cold arid areas, where winter irrigation is not feasible, the stalks may be left standing until the lowest winter temperatures occur, in order to kill as many larvae as possible before plowing under.

Federal and state quarantines to prevent the spread of the pink bollworm regulate the movement of cotton and cotton products and require the treatment of seeds and lint before shipment. Cotton seed may be freed from pink bollworm infestation by heating to 150°F. for 30 seconds or more or by fumigation with methyl bromide or vacuum fumigation with hydrogen cyanide, without injury to germination or the quality of cottonseed oil. Cotton lint may be treated by passing it between rollers set to crush the larvae, by compression of the cotton into bales, or by vacuum fumigation with hydrogen cyanide. Proper sanitation of gins, mills, and warehouses by burning trash and waste and by fumigation is also required.

Crop losses may be materially reduced by spraying or dusting cotton at 7-day intervals with DDT or Sevin at 2 to 3 pounds per acre. These treatments kill not only the larvae, but also the female moths before oviposition.

References. *U.S.D.A., Dept. Buls.* 918, 1921, and 1374, 1926; *Jour. Agr. Res.,* **9**:343, 1917; *Jour. Econ. Ento.,* **41**:120, 1948; **43**:491, 1950; **48**:767, 1955; **50**:122, 487, 609, 642, 1957; **51**:567, 1958; and **52**:301, 385, 1959; *U.S.D.A., Yearbook,* **1952**:505.

CHAPTER 13

TOBACCO INSECTS

FIELD KEY FOR THE IDENTIFICATION OF INSECTS INJURING TOBACCO

A. *Insects chewing leaves or stalk aboveground, leaving visible holes:*

1. Very young plants in seedbeds partially to completely defoliated by tiny dark-purple yellow-spotted bugs, only $\frac{1}{25}$ inch long, with a spherical soft body and very distinct head, that jump actively by the use of a forked tail-like appendage...
   Garden springtail, Bourletiella (= *Sminthurus*) *hortensis* (Fitch) (see *Conn. Agr. Exp. Sta. Bul.* 379, 1935).
2. Large, grayish to brownish, actively jumping hoppers often invade tobacco fields in great numbers from surrounding grass and clover, when these crops are harvested, and rag the leaves severely...............*Grasshoppers*, page 476.
3. Very small, oval, active, jumping beetles, about $\frac{1}{16}$ inch long, dark-brown in color, with more or less black across the wing covers, or entirely black eat tiny "shot holes" in the leaves, especially of newly set plants or plants in seed beds..................*Tobacco flea beetle* or *potato flea beetle*, page 592.
4. Large green caterpillars, up to 4 inches long, with diagonal white bars on the sides and a slender horn at the tip of the body, cling to the vines and rapidly strip off the foliage, including the veins of the leaves. Numerous pellets of excrement on the ground...........................*Hornworms*, page 594.
5. Large holes, not including the veins, are eaten in the leaves or whole seedlings devoured by plump, smooth, greenish or brownish, more or less striped or spotted caterpillars, up to $1\frac{1}{2}$ inches long, which usually lie curled up in the soil during the daytime.............................*Cutworms*, page 594.

B. *Insects sucking sap from leaves or stems, not leaving visible holes, but causing wilting, yellowing, spotting, or death of the leaves:*

1. Very slender, almost microscopic, yellowish, active nymphs and the brown-bodied bristly-winged adults, only $\frac{1}{25}$ inch long, suck sap from the leaves, causing the veins to become outlined with silver and the leaves peppered with minute black spots..
   Tobacco thrips, Frankliniella fusca (Hinds) (see *Conn. Agr. Exp. Sta. Bul.* 379, 1935; *S.C. Agr. Exp. Sta. Bul.* 271, 1931).
2. Small plants or the leaves or tops of older plants wilt suddenly. Broad, flat, shield-shaped bugs, about $\frac{1}{2}$ inch long and of a buff to greenish color, suck at stems, petioles, or veins...
   Brown stink bug, Euschistus servus (Say) and *E. variolarius* (Beauv.) (*N.C. Dept. Agr. Spcl. Bul.* (Supplement, October, 1909), pp. 58–59).
3. Second-crop and late tobacco leaves yellow and later split and become ragged-looking, due to the feeding of a greenish-black slender bug, the largest only $\frac{1}{8}$ inch long and about $\frac{1}{4}$ as broad, with long, slender legs and antennae.....
   ......*Suckfly, Cyrtopeltis notatus* Distant [= *Dicyphus minimus* (Uhler)] (see *Fla. Agr. Exp. Sta. Bul.* 48, 1898).
4. Flattened, coppery-brown, somewhat mottled, winged bugs, $\frac{3}{8}$ inch long, and their greenish soft-bodied nymphs suck sap from the buds before the leaves unfold, causing leaves to be distorted and curly. Winged ones fly actively when disturbed............................*Tarnished plant bug*, page 614.

590

C. Insects boring in stalk or stem:

1. Smooth, tough, yellowish to reddish-brown, six-legged slender worms, up to 1½ inches long, burrow into the plant at the surface of the soil and hollow out the stalk or taproot above and below the entrance hole, causing the plants to wilt. The worms may also eat off the roots.............................. *Wireworms*, or *"pithworms"* (see *N.C. Dept. Agr. Spcl. Bul.* (Supplement, October, 1909), pp. 46–53).

2. Yellow to pinkish-white, swift-crawling caterpillars, up to ⅗ inch long, live in silken tubes covered with grains of soil, at the surface of the ground, near the plant. They burrow into the plant below ground and then tunnel upward, causing the plant to wilt and sometimes cutting off the terminal bud........ ....*Corn root* or *tobacco webworm*, page 596.

3. Young plants are tunneled up and down the stem for an inch or so from a small entrance hole just below the ground by small dirty-white maggots, up to ¼ inch long, broad at the posterior end and tapering to a pointed head, causing them to wilt and die...............................*Seed corn maggot*, page 518.

D. Insects eating into the buds or seed pods:

1. Greenish caterpillars with pale longitudinal stripes, up to 1½ inches long, tunnel holes into the leaf buds as the plants begin to top, causing the leaves to be misshapen and ragged with large holes. Later the caterpillars bore into the seed pods.....................................*Tobacco budworm*, page 596.

2. Green, tan, or blackish caterpillars, nearly 1¾ inches long when full-grown, attack the plant as in *D,1*, but mostly later in the season. Bodies of the worms sparsely haired, the skin rough and thorny-looking under a lens, often heavily striped and with more or less reddish markings........................... ....*Corn earworm* or *false budworm*, page 597.

E. Insects mining between upper and lower surface of the leaves:

1. Grayish, irregular, blotch mines (later turning brown), especially in the older leaves; grayish-white caterpillars, up to ⅓ or ½ inch long, with a pinkish or greenish tinge and brown at each end, may be found in the mines between upper and under leaf surface.. ....*Tobacco leaf miner* or *splitworm*, or *potato tuberworm*, page 597.

F. Insects attacking the plant underground, or uprooting and eating off the young plants near the surface of the ground:

1. Garden crops, peanuts, tobacco, and others grown in moist, light soils have the seedlings uprooted, the roots cut off, and pits eaten into the underground parts by a weird-looking brown cricket, up to 1½ inches long, covered with fine velvety hairs; with long hind wings, half-length front wings, large hind legs, and broad, rake- and shovel-like front legs, which are used to make small, mole-like burrows just beneath the soil................................... *Mole crickets, Scapteriscus vicinus* Scudder, *S. acletus* R. & H., and *Gryllotalpa hexadactyla* Perty, Order Orthoptera, Family Gryllidae (Gryllotalpidae) (see *U.S.D.A., Farmers' Bul.* 1561, 1928).

2. Seedbeds in use for more than 1 year become infested with large, white, six-legged grubs, that crawl on their backs over the surface of the ground at night and burrow and work up the soil to such an extent that the small plants are covered or uprooted and seriously damaged........*Green June beetle*, page 751.

G. Insects attacking tobacco and tobacco products in storage:

1. Cigars, cigarettes, and package tobaccos, infested with small, white, curved, very hairy grubs and very small light-brown beetles (1⁄16 inch long) which eat holes through the wrappers or feed within the packages. The beetles have smooth wing covers and antennae not enlarged at end...................... ....*Cigarette beetle*, page 910.

2. Small, white, curved grubs, similar to *G,1*, but not hairy; and small, rather narrow, oval, reddish-brown beetles, with parallel lines on the wing covers and last 3 segments of the antennae enlarged, tunnel through tobacco and many other substances...........................*Drug-store beetle*, page 911.

3. Very active, whitish caterpillars, peppered with many small scattered brown spots and sometimes tinged yellow, brown, or pink, up to ⅝ inch long, eat

holes much larger than those made by the cigarette beetle in tender parts of leaves and contaminate stored tobacco and other dried vegetable products with silk and excrement..........................*Tobacco moth*, page 911.

65 (76). TOBACCO FLEA BEETLES[1]

Importance and Type of Injury.   This is considered the most injurious insect of tobacco, at least in some sections.   The adult flea beetles chew small rounded holes into or through the leaves, especially from the underside (Fig. 13.1).   They attack the young plants in the seedbeds almost as soon as they come up, and often ruin entire beds.   After the plants are

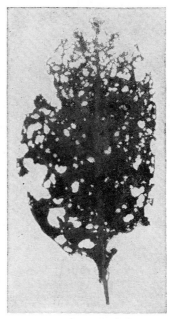

FIG. 13.1. Tobacco leaf badly damaged by tobacco flea beetle.   (*From Ky. Agr. Exp. Sta. Bul.* 266.)

transplanted, they weaken or kill them in the field.   Damage continues until the crop is harvested, the mature leaves often being spotted with holes which greatly lessen the quantity and the quality, especially of cigar-wrapper tobacco.   It has been estimated that a flea beetle eats ten times its own weight in a day.   The larvae feed on the roots of tobacco and other plants of the same family, cutting off the small roots and sometimes tunneling into the stalk.   As a result of the feeding of the beetles, fungus diseases are encouraged and spread among the tobacco.   The tobacco flea beetle is most destructive in the southern tobacco areas, while in Connecticut the adults of the closely related potato flea beetle[2] (page 603), by migrating into tobacco from potato and tomato fields, cause more damage than the tobacco flea beetle.   Morgan estimated the loss in Kentucky and Tennessee in 1 year (1907) as close to $2,000,000.

[1] *Epitrix hirtipennis* (Melsheimer), Order Coleoptera, Family Chrysomelidae.
[2] *Epitrix cucumeris* (Harris).

Plants Attacked. Tobacco, tomato, potato, eggplant, pepper, night-shade, Jimson weed, ground cherry, and horse nettle are attacked regularly, and many other plants to a less extent.

Distribution. Probably in all tobacco-growing sections of the United States and Canada.

Life History, Appearance, and Habits. The brownish black-clouded hard-shelled little beetles, only $\frac{1}{16}$ inch long (Fig. 13.2), hide during the winter in great numbers under leaves, grass, and trash on the ground about tobacco fields, especially along the margins of woods. Here they hibernate or remain more or less active, depending on the temperature. As soon as the plants come up in the seedbeds the beetles attack them, increasing in numbers from then until harvesttime, when they begin to disappear.

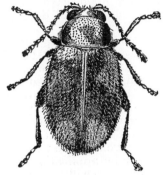

Fig. 13.2. Tobacco flea beetle. Adult, enlarged about 20 times. (*From U.S.D.A., Farmers' Bul.* 1425.)

The eggs are laid mostly on the surface of the soil under the plant. They hatch in a week, and the delicate, slender, white larvae burrow into the soil and feed on the fibrous rootlets mostly within a few inches of the surface. They become full-grown in about 2 weeks. They are then about $\frac{1}{6}$ inch long by $\frac{1}{60}$ inch thick and dirty white except for the brownish mouth parts. After 4 or 5 days spent as a whitish pupa in the soil, the new adult emerges. There are probably three or four generations a year in the South.

Control Measures. The ground surrounding the seedbed should be burned over in late winter. Less injury will occur if the seedbeds are located as far from good hibernating quarters as possible and at a considerable distance from the new fields. Seedbeds should be boarded or planked up on the sides and covered with tobacco cloth having at least 25 strands to the inch, making the covering beetle-tight. Surrounding the seedbed should be a 2- or 3-foot strip of early-seeded tobacco, protected with poles and cloth and kept dusted with one of the materials suggested below or with lead arsenate. The topsoil of seedbeds should be sterilized with steam or other heat before planting. Old seedbeds should be plowed and harrowed after transplanting. If the beetles get into the seedbeds or attack the plants later in the season, they may be controlled by dusting or spraying with DDT at 1 pound, dieldrin at 0.2 pound, or endrin at 0.2 to 0.4 pound per acre. After the crop is harvested, much can be done to check flea beetles by destroying the stalks and the "suckers" that grow up in the fields.

References. N.C. Agr. Exp. Sta. Bul. 239, 1919; U.S.D.A., Farmers' Buls. 1352, 1923, and 1425, 1924; Conn. Agr. Exp. Sta. Bul. 364, 1934; Ky. Agr. Exp. Sta. Res. Bul. 266, 1926; U.S.D.A., Bur. Ent. Plant Quar. Cir. E-373, 1937; Jour. Econ. Ento., 25:1187, 1932; 26:233, 1933; and 37:13 and 224, 1944.

66 (110). TOBACCO HORNWORMS[1]

The best known of tobacco insects, and among the most injurious, are the large green tobaccoworms with white bars on the sides and a slender horn at the end of the body and the parent "tobacco flies," or hawk moths, that lay the eggs of the hornworm. These worms infest the plants all summer long and are such ravenous feeders as to ruin many leaves and, where abundant, defoliate the plants. Because of their concealing coloration, they can often be located most easily by the pellets of excrement which they drop upon leaves or the ground, above which they are feeding. The life history and descriptions of these pests are given on page 655.

Control Measures. Where labor is cheap and plentiful, hand-picking of these worms is fairly effective and laborers should always be instructed to destroy the worms wherever they are encountered while working among the plants. These worms may be controlled by spraying or dusting the plants with TDE (or DDD) at 1 to 2 pounds or endrin at 0.2 to 0.4 pound per acre. Dusting with DDT at 1.5 to 3 pounds per acre is effective against *Protoparce quinquemaculata* but not against *P. sexta,* although thorough applications of DDT spray will control this latter insect. The first application should be given as soon as the young worms become numerous on the plants and later applications should be given as needed. Tobacco grown under shade cloth will be largely free of these worms, if the cloth is kept intact throughout the growing season. Fall and winter plowing to destroy the pupae, where it is not otherwise objectionable, reduces the population of hornworms the following season. Stalks and stubble should be destroyed immediately after harvest. It has been found possible to poison the adult moths by using isoamyl salicylate as an attractant with a stomach poison (page 395), exposed in special bait feeders (see *Jour. Econ Ento.,* 26:227, 1933).

References. U.S.D.A., Farmers' Bul. 1356, 1923; Conn. Agr. Exp. Sta. Bul. 379, 1935; Tenn. Agr. Exp. Sta. Bul. 120, 1918; N.C. Agr. Exp. Sta. Ext. Cirs. 174, 1929, and 207, 1936; Tenn. Agr. Exp. Sta. Bul. 186, 1943; Jour. Econ. Ento., 39:610, 1946.

67 (4, 48, 80). CUTWORMS

Importance and Type of Injury. Cutworms of a score of species often cut off newly transplanted tobacco at the surface of the ground in the manner discussed under Corn Insects (page 476) and garden crops (page 610) (Fig. 13.3). Some climbing cutworms attack the leaves of older plants.

Life History, Appearance, and Habits. The cutworms injurious to tobacco are divided by Crumb (*l.c.*) into two groups. One group, including the variegated cutworm[2] and the black cutworm,[3] pass the winter as naked brown pupae in the soil, transform to moths very early in spring,

---

[1] *Protoparce sexta* (Johanssen) and *P. quinquemaculata* (Haworth), Order Lepidoptera, Family Sphingidae.

[2] *Peridroma saucia* (Hübner) [= *Lycophotia margaritosa* (Haworth)], Order Lepidoptera, Family Noctuidae.

[3] *Agrotis ypsilon* (Rottemburg), Order Lepidoptera, Family Noctuidae.

and produce during the season three or four generations. The species of this group are destructive throughout the tobacco-growing regions of the United States and Canada. The other group consists of species that lay their eggs chiefly in weedy or grassy fields in late summer or fall. The eggs hatch and the larvae spend the winter partly grown and feed destructively upon the newly set plants in the spring. These species include the dingy cutworm,[1] the clay-backed cutworm,[2] the dark-sided cutworm,[3] and the spotted cutworm.[4] Species of this group are rarely destructive south of Virginia and Tennessee and all but the last-named have a single generation a year.

Fig. 13.3. Tobacco plant ruined by a cutworm with the larva in feeding position. (*From U.S.D.A., Farmers' Bul.* 1494.)

Control Measures. If cutworms are present, poisoned-bran bait consisting of:

Wheat bran.................................	100 lb.
Aldrin or dieldrin, 2 oz., heptrachlor, 8 oz., or chlordane, 1 lb., active ingredient as wettable powder	
Water......................................	Enough to moisten

should be applied several days before the tobacco plants are set. To determine whether the bait should be applied, Crumb suggests placing rather large compact bunches of freshly cut clover, dock, or chickweed on well-plowed soil. If cutworms are present in the soil they will collect under such vegetation, and an examination in 2 or 3 days will indicate whether the bait should be applied. Poison-bran bait may be broadcast late in the afternoon at the rate of 10 to 20 pounds dry weight per acre, or 4 pounds to 100 square yards of seedbed. Or a small handful may be placed about each hill as the plants are set. It should not be scattered

[1] *Feltia subgothica* (Haworth) and *F. ducens* Walker.
[2] *Feltia gladiaria* Morrison.
[3] *Euxoa messoria* (Harris).
[4] *Amathes* (= *Agrotis*) *c-nigrum* (Linné).

upon the leaves.   Spraying or dusting plants and adjacent soil with DDT at 1 to 1.5 pounds per acre is also effective.   Fall or late spring plowing is recommended to reduce the cutworms.

References.   U.S.D.A., Farmers' Bul. 1494, 1926; *Fla. Agr. Exp. Sta. Bul.* 48, 1898; *Tenn. Agr. Exp. Sta. Bul.* 159, 1936.

68 (15, 37). WIREWORMS OR PITHWORMS[1]

Wireworms are among the most destructive tobacco insects.   Newly transplanted tobacco often wilts and dies within a few days.   If the stems are split open, the larvae are found tunneling near the ground level.   Other larvae will be found in the adjacent soil, where they may feed throughout the season, by chewing off the smaller roots.   These pests are discussed more fully on page 617.

69 (19, 47). CORN ROOT OR TOBACCO WEBWORM[2]

Importance and Type of Injury.   This small caterpillar, also known as the corn webworm, causes much damage to tobacco in Virginia and adjoining states by cutting into the young plants at the surface of the ground and then boring up or down in the stem.   Usually the attack comes after the plants have become established, and it may necessitate one or more replantings.
Plants Attacked.   Tobacco, corn, and certain weeds such as buckhorn, wild carrot, plantain, asters, and daisies.
Distribution.   Throughout the eastern states, especially destructive in Virginia.
Life History, Appearance, and Habits.   Similar to that given for sod webworms (page 513).
Control Measures.   See page 558.
Reference.   *U.S.D.A., Dept. Bul.* 78, 1914.

70. TOBACCO BUDWORM[3]

Importance and Type of Injury.   Tiny, rust-colored to pale-green, striped caterpillars eat into the buds or unfolded leaves of tobacco as the plants begin to top.   If the holes are made in the tips of the buds, the leaves that expand from the buds are often ragged and distorted.   If the tiny holes penetrate the unfolded leaves, these leaves will have large unsightly holes when they are fully expanded.   Only a single larva is commonly found on a plant, and the larvae migrate but little.   The attack on the buds renders the leaves unfit for cigar wrappers and greatly cuts the price.   The larvae of the second generation often feed on the "suckers" and eat into the seed pods.
Plants Attacked.   Tobacco, cotton, ground cherry, and other solanaceous plants and also geranium and ageratum.
Distribution.   From Missouri, Ohio, and Connecticut, southward. Most injurious in the Gulf states; rarely in Kentucky and Tennessee.
Life History, Appearance, and Habits.   The budworm winters in the soil as a mahogany-colored spindle-shaped pupa, about ¾ inch long. The moths emerge in the spring and deposit their eggs singly on the underside of leaves.   They are about 1½ inches across the spread wings. The front wings are light-green and crossed by four oblique light bands,

[1] Order Coleoptera, Family Elateridae.
[2] *Crambus caliginosellus* Clemens, Order Lepidoptera, Family Pyralididae (Crambinae).
[3] *Heliothis* (= *Chloridea*) *virescens* (Fabricius), Order Lepidoptera, Family Noctuidae.

the inner three of which are edged with black. Eggs are laid singly on the underside of the leaves. The tiny larvae that hatch from the eggs make their way to the buds and crawl down among the unfolded leaves, where they may cause great damage before they are detected. The full-grown larva is 1½ inches long, of a pale-green color, and marked with several longitudinal pale stripes. Pupation occurs in the soil, and there are two generations of the insect each season.

Control Measures. A small quantity of a mixture of 1 pound of DDT in 75 pounds of corn meal should be applied directly into the bud of each plant by hand or with a sifter can, using 6 to 10 pounds per acre once or twice a week. DDT and TDE 10 per cent dusts may also be applied directly into the buds at 6 to 10 pounds per acre, or the upper parts of the plants may be dusted or sprayed with DDT or TDE at 0.5 to 1 pound or endrin at 0.2 to 0.4 pounds per acre. Much of the damage may be prevented if all worms seen while working among the tobacco are destroyed. As soon as the plants are harvested, the remnants should be cut and burned and the soil plowed. Infestation the following year may be reduced if the leftover plants in seedbeds are destroyed as soon as planting is finished and if suckers are removed and destroyed. The seedbeds should be carefully covered as suggested for the tobacco flea beetle.

References. *U.S.D.A., Farmers' Bul.* 819, 1917, and 1531, 1927; *Jour. Econ. Ento.,* **29**:282, 1936.

71 (12, 63, 111, 260). Corn Earworm or False Budworm[1]

This very destructive pest of corn and cotton (pages 498 and 587) attacks tobacco in a manner very similar to the tobacco budworm. It is usually not so destructive to tobacco as the true tobacco budworm. Very early in the season in the extreme south, and after corn has become mature in most sections, the moths, of which there are several generations, may lay their eggs on tobacco. The caterpillars may be destroyed by the same measures recommended for the tobacco budworm.

72 (106). Potato Tuberworm or Tobacco Splitworm[2]

This insect is discussed as the potato tuberworm on page 648. When attacking the tobacco, the pinkish-white caterpillars, about ⅓ inch long, mine between the upper and lower surface, especially of older leaves, causing unsightly gray to brown blotches that render the product unfit for cigar wrappers.

Control Measures. It is advisable to transplant as early as possible and make every effort to mature an early crop. Early leaves that are badly infested should be pruned off and the stubble and sucker growth destroyed. Cleaning up trash about the field and barns will help to prevent injury. Potatoes should not be followed with tobacco or grown near tobacco fields. See also page 649.

Reference. *U.S.D.A., Dept. Bul.* 59, 1914.

[1] *Heliothis zea* (Boddie) [= *armigera* (Hübner)] Order Lepidoptera, Family Noctuidae.
[2] *Gnorimoschema* (= *Phthorimoea*) *operculella* (Zeller), Order Lepidoptera, Family Gelechiidae.

CHAPTER 14

INSECTS INJURIOUS TO VEGETABLE GARDENS AND TRUCK CROPS

An abundance of fresh vegetables in the diet is known to be very important to human health. The use of vegetables, now negligible in many families, would be much more extensive if the control of the many insects that discourage the home gardener were better understood. The quality of commercially grown vegetables would be better and the price could be lowered, also, if insects were not permitted to add so much to the expense of production. A very slight amount of feeding upon vegetables such as cucumbers, melons, tomatoes, and lettuce renders them unsalable. This makes the insect losses to truck crops high in proportion to the amount of the plant consumed.

A knowledge of the habits, life cycles, and control measures effective against these pests will enable any grower to reduce very much the extensive damage he suffers from garden insects.

The more important garden insects are discussed under the following groups:

A. General Garden Pests, page 598.
B. Insects Injurious to Peas and Beans, page 620.
C. Insects Injurious to Cucurbits, page 628.
D. Insects Injurious to Potatoes, page 639.
E. Insects Injurious to Sweetpotatoes, page 649.
F. Insects Injurious to Tomatoes, Peppers, and Eggplant, page 654.
G. Insects Injurious to Onions, page 658.
H. Insects Injurious to Cabbage and Related Vegetables, page 662.
I. Insects Injurious to Beets, Spinach, Lettuce, Carrots, Parsnips, Celery, and Related Vegetables, page 672.
J. Insects Injurious to Asparagus, page 680.
K. Insects Injurious to Sweet Corn, page 683.

General References. CROSBY, C. R., and M. D. LEONARD, "Manual of Vegetable-garden Insects," Macmillan, 1918; L. M. PEAIRS and R. H. DAVIDSON, "Insect Pests of Farm, Garden, and Orchard," 5th ed., Wiley, 1956; *U.S.D.A., Farmers' Bul.* 1371, 1938; *Can. Dept. Agr. Bul.* 161 (n.s.), 1932; *Cornell Ext. Bul.* 206, 1931; *Mich. Agr. Exp. Sta. Spec. Bul.* 183, 1929; *Mich. State Coll. Ext. Bul.* 180, 1937; *Purdue Univ. Ext. Bul.* 186, 1932; *N.D. Agr. Exp. Sta. Cir.* 42, 1933; *Fla. Agr. Exp. Sta. Bul.* 370, 1942; *Calif. Agr. Ext. Ser. Cir.* 87, 1944 and 146, 1948; *Tenn. Agr. Exp. Sta. Bul.* 186, 1943; *N.J. Agr. Exp. Sta. Bul.* 740, 1948; *Ill. Agr. Exp. Sta. Cir.* 514, 1941; *Ga. Agr. Exp. Sta. Bul.* 254, 1947; *Ohio Agr. Ext. Ser. Bul.* 76, 1947; and *U.S.D.A., Misc. Publ.* 605, 1946.

A. GENERAL GARDEN PESTS

While many garden insects feed on only one kind of crop or a few closely related ones, there are a number of others that feed on a variety of plants. Some of these are discussed in this section before the pests of special crops are taken up. The following species are general garden pests. They are also included in the field keys under the crops to which they are most destructive.

FIELD KEY FOR THE IDENTIFICATION OF SOME GENERAL GARDEN PESTS ATTACKING MANY CROPS

A. *Insects chewing leaves, buds, or stems of plants:*
1. Black, jumping, somewhat square-backed insects, with antennae longer than the body and the females with a needle-like ovipositor almost as long, eat seedling plants or the fruits of vegetables mostly at night and hide among surface trash or in soil cracks in the daytime........*Field cricket*, page 600.
2. Insects somewhat similar to *A*,1, but heavier and clumsier, with very small wings and long, sword-shaped ovipositors. Migrate from waste land often attacking crops in great migrating swarms, feeding during warm sunny days. Only in the Rocky Mountain states..*Mormon cricket* or *coulee cricket*, page 601.
3. Brownish, grayish, jumping insects up to 1½ inches long, often migrate into gardens from surrounding fields and strip leaves, eat tender stems and devour fruits of vegetables of all kinds...................*Grasshoppers*, page 602.
4. Many minute holes eaten in leaves by very small, hard-shelled black beetles from ⅟₁₆ to ⅛ inch long, which rest and feed chiefly on the underside of leaves and jump readily when disturbed..................*Flea beetles*, page 603.
5. Cinnamon-brown velvety-looking beetles, something like small May beetles, ⅜ inch long, eat irregular holes from margins of leaves of nearly all garden and flower plants at night. Insects congregate about artificial lights. Only along Atlantic seaboard........................*Asiatic garden beetle*, page 606.
6. Grayish long-headed beetles, ½ inch long, with faint white stripes along their sides, eat the margins of the leaves..........*White-fringed beetle*, page 607.
7. Brownish snout beetles with a V-shaped marking on the wing covers, feed on foliage of vegetables..........................*Vegetable weevil*, page 608.
8. Yellowish-green caterpillars, up to 1 inch long, with scattered fine hairs and conspicuous small dark spots, skeletonize and devour the leaves, spinning considerable silk over the foliage, webbing the leaves together..............
....*Webworms*, page 609.
9. Plant stems eaten off at night and left lying on the ground, or leaves devoured, or fruits eaten into by plump, smooth, greasy-looking, greenish, brownish, or grayish, spotted or striped caterpillars, up to 1½ inches long, which dig into the soil to hide during the daytime..................*Cutworms*, page 610.
10. Very hairy, yellowish, brown, or reddish-brown- and black-banded caterpillars up to 2 inches long, eat large holes in leaves..........*Woollybears*, page 611.
11. Large smooth-edged holes eaten in leaves, or plants stripped of leaves, by caterpillars up to 1½ inches long, striped in several colors, with a white λ on front of head, often migrating in "armies.".......*Fall armyworm*, page 473.
12. Injury as in *A*,11 by green caterpillars with dark, lateral stripes, up to 1¼ inches long, which often lightly web the foliage....*Beet armyworm*, page 610.
B. *Insects sucking sap from leaves, buds, fruits, or stems, not leaving visible holes but causing wilting, rolling of leaves, spotting and discoloration of foliage, and discoloration and distortion of fruits:*
1. Small, soft-bodied, sluggish, green or blackish bugs of various sizes, the largest scarcely larger than a pinhead, either winged or wingless, cluster on underside of leaves and terminal shoots. Leaves often sticky with honeydew..........
....*Aphids*, page 612.
2. Olive-green bugs, flecked with yellow and dark-brown, with flattened, oval bodies, somewhat triangular in front, up to ¼ inch long by half as wide, cause spotted and deformed leaves, discolored stems, and dwarfed or dying buds, by sucking sap and poisoning tissues. Bugs run or fly readily when disturbed..
....*Tarnished plant bug*, page 614.

3. Blackish, rather soft-winged, jumping bugs, only $\frac{1}{10}$ inch long, and their greenish nymphs cause spotting and dying of leaves. Resemble flea beetles but have piercing-sucking mouth parts.........*Garden fleahopper*, page 615.
4. Leaves turn yellow, become blotched with reddish-brown or white spots, and die. Underside of leaf with inconspicuous white webs among which are many microscopic, 8-legged, reddish or greenish mites and their round pearly eggs, looking, to the naked eye, like fine white dust................*Spider mites*, page 616.

C. *Insects living in the soil and feeding on the roots underground:*
1. Young plants may be rooted out and the roots of others eaten off or scarred by weird-looking brown crickets, up to $1\frac{1}{2}$ inches long, and covered with fine velvety hairs. The rake- and shovel-like front legs are used to make tunnels beneath the soil in which they live..............................
.....................*Mole crickets* (see *U.S.D.A., Farmers' Bul.* 1561, 1928).
2. Slender, tough-skinned, brownish, polished-looking worms, up to $1\frac{1}{4}$ inches long, with chewing mouth parts and 6 short legs, eat into planted seeds or eat off rootlets of plants....................................*Wireworms*, page 617.
3. White thick U-shaped grubs with brownish heads and 6 slender legs, up to $1\frac{1}{2}$ inches long, eat off roots or gnaw into tubers.......*White grubs*, page 619.
4. Plants wilt and die from the attacks of white legless, not curved-bodied grubs, up to $\frac{1}{2}$ inch long, which eat into underground stems and larger roots. Only in Gulf states.................................*White-fringed beetle*, page 607.
5. Roots of vegetables chewed off by a somewhat slender white grub with a Y-shaped anal opening and a transverse row of short setae in front of it......
....*Asiatic garden beetle*, page 606.
6. Very delicate, slender, cylindrical, white larvae, not over $\frac{1}{8}$ inch long, with distinct heads and 6 very short legs, eat tiny rootlets....*Flea beetles*, page 603.
7. Caterpillars similar to *A*,8, but usually paler, feed on roots or stems below ground..*Cutworms*, page 610.
8. Cylindrical, many-segmented worms, with 2 pairs of legs on each segment, ranging up to 1 or 2 inches long, eat roots, tunnel into tubers and attack leaves and fruits lying on the soil.........................*Millipedes*, page 620.
9. Rootlets eaten off and underground parts of plant scarred by small, nearly white, very active creatures with a pair of legs on each principal segment of the body, of all sizes up to $\frac{1}{4}$ inch long.............*Garden centipede*, page 884.

73. FIELD CRICKET[1]

Importance, Type of Injury, and Plants Attacked. Everywhere present in small numbers, this familiar insect (Fig. 6.13) sometimes becomes abundant enough to cause serious losses. In California and some Gulf states they have damaged cotton by cutting off seedling plants just above the ground or just below the apical bud and cutting off many leaves that drop to the ground. In the Great Plains, they destroy the seeds of alfalfa and all the cereals by devouring them or cutting off the heads especially in the "dough" stage, but also after the grain is shocked or stacked. The fruits of tomatoes, all the cucurbits, peas, beans, strawberries, and others may be seriously damaged or entirely ruined especially in the Gulf states. These pests also frequently eat holes in paper and rubber and in cotton, linen, woolen, or fur garments, either out-of-doors or indoors, especially when soiled with perspiration or foods.

Distribution. Throughout the United States, southern Canada, and much of South America.

Life History, Appearance, and Habits. The life history varies greatly throughout the range of the insect, but in the northern states the majority hibernate in the egg stage and complete a single generation in a year, while in the Imperial Valley of California and the Gulf Coast they winter

[1] *Acheta* (= *Gryllus*) *assimilis* Fabricius, Order Orthoptera, Family Gryllidae.

as nymphs or remain more or less active the year round and may produce as many as three generations. Hatching of overwintered eggs takes place during late May and June. There are 8 to 12 nymphal instars, in all but the first of which the nymphs feed like the adults. The total developing period occupies from 9 to 14 weeks. The body of the adults varies from $\frac{3}{5}$ to 1 inch in length, the antennae are a half longer than the body, the hind legs are very heavy for jumping, the cerci are about $\frac{1}{3}$ as long as the body, and the ovipositor of the female is $\frac{2}{3}$ to nearly 1 inch long. The adults vary so much in color, length of wings, and other structural features that no fewer than 45 different scientific names have been proposed. Throughout most of the northern states, adults are present from late July until heavy frosts and lay their eggs mostly in August and early September. The chirping of the males during this part of the year is one of the most familiar out-of-doors sounds. During sunshiny days crickets remain most of the time under shelter of vegetation, under surface trash, or in cracks or excavations that they make in the soil. Feeding, mating, and egg-laying take place from late afternoon until late morning. Damp soil is preferred in which to lay the eggs: this may be found along ditches, beneath cracks in baked soil, or by digging pits the size of the end of a finger with the four front legs. The long ovipositor is then forced down into the soil and a small number of eggs laid in each hole beneath the soil but not enclosed in a pod-like secretion as grasshopper eggs are. The eggs are banana-shaped, $\frac{1}{12}$ inch long, and straw- to cream-colored. Females commonly produce about 300 eggs.

Control Measures. The use of poisoned-bran baits, dusts, and sprays as for grasshoppers (page 468) is effective against this pest. Deep fall plowing to bury the eggs, the maintenance of a fine dust mulch to kill and drive away the active stages, rotations in which cotton does not follow small grains, and, in houses, placing bran bait (page 323) on papers on the basement floor, out of the way of pets and children, or dusting or spraying with chlordane, pyrethrins, lindane, or malathion have been found useful control measures.

References. S.D. Agr. Exp. Sta. Bul. 295, 1935; U.S.D.A., Tech. Bul. 642, 1939; N.D. Agr. Exp. Sta. Bimonthly Bul. **11**:11, 1948.

74. MORMON CRICKET[1] AND COULEE CRICKET[2]

Importance and Type of Injury. These insects, which are not true crickets but are related to the katydids and longhorned grasshoppers, have somewhat the appearance and habits of the field cricket. For nearly a century the mormon cricket has periodically overwhelmed settlers and homesteaders in the Rocky Mountain area by migrating from its native breeding grounds in the high rugged hills into the arable lands of the valleys and eating garden crops, small fruits, legumes, and especially the heads of grain in the milk or dough stage. A very serious outbreak of these pests in the Great Salt Lake basin in 1848 was terminated so spectacularly by great flocks of gulls that the grateful settlers erected an imposing monument in commemoration.

Distribution. The mormon cricket occurs in the states west of the Missouri River and is most destructive in Oregon, Washington, Nevada,

[1] *Anabrus simplex* Haldeman, Order Orthoptera, Family Tettigoniidae.
[2] *Peranabrus scabricollis* (Thomas), Order Orthoptera, Family Tettigoniidae.

Utah, Idaho, Montana, Wyoming, and Colorado. The coulee cricket has been very destructive in Montana and Washington.

Plants Attacked. Nearly all field and garden crops, and many fruits, shrubs, trees, and weeds, including more than 250 species of range plants, especially bitterroot and wild mustards. Severe damage often occurs to wheat and alfalfa. The coulee cricket is especially fond of sagebrush, *Artemisia rigida*, and also eats dung and dead animals. Both species are very cannibalistic.

Life History, Appearance, and Habits. The winter is passed in the egg stage, $\frac{1}{4}$ to 1 inch deep in the soil, the mormon cricket favoring barren, sandy soil in sunny locations, the coulee cricket choosing the base of tiny grass stools. The elongate eggs are deposited singly by thrusts of the long ovipositor, and the female mormon cricket lays an average of about 150 eggs over a period of a week or more. These eggs hatch in the first warm days of spring, Feburary to April, usually a full month earlier than grasshoppers. The crickets pass through seven nymphal stages in about 75 to 100 days and become adults from early June to mid-July. Coulee crickets nearly all disappear by July, but the mormon cricket continues egg-laying all summer long, and the embryos become well developed before winter but do not hatch until the following spring. The adults are about 1 inch long, heavy-bodied, and clumsy. The wings are very small and useless, except that the male uses them to produce his mating chirp. The tarsi are four-segmented, and the antennae as long as the body, and the female has a sword-shaped ovipositor also as long as the body. Unlike the field cricket, the mormon and coulee crickets are active only during the warm sunny part of the day and seek shelter at night and in cloudy or rainy weather. From the time they are about half-grown, they begin migrating from their rangeland breeding grounds, moving from $\frac{1}{8}$ to over 1 mile per day, going in a straight line in no predictable direction, over all kinds of obstacles, often in bands covering a square mile in area and composed of as many as 100 to 500 individuals per square foot. The migrations occur at air temperatures of 65 to 95°F. and when the wind velocity is less than 25 miles per hour.

Control Measures. Since these insects cannot fly at any stage of their lives, linear barriers of 10-inch strips of 28- to 30-gage galvanized iron, held on edge with stakes driven into the ground, will stop migrating swarms. Vertical-walled ditches have also been used for the coulee cricket, and soil pits or water traps may be made at intervals to catch the crickets halted by the barrier. For irrigated crops, cheap and effective protection is given by dripping a low-grade petroleum distillate on the water surface of the irrigation ditches so that a continuous film of oil is maintained to kill the migrating crickets as they plunge into the water. Bran baits, preferably of the wet type, containing aldrin, heptachlor, chlordane, or toxaphene (see formulas on page 468), spread at 10 to 20 pounds per acre, provide the most effective control for the mormon cricket.

References. U.S.D.A., *Tech. Bul.* 161, 1929; *Wash. Agr. Exp. Sta. Bul.* 137, 1937; *Univ. Idaho Ext. Bul.* 100, 1936; U.S.D.A., *Farmers' Bul.* 2081, 1959.

75 (1, 27, 146). GRASSHOPPERS[1]

In years when grasshoppers are abundant, they may become serious on garden crops, especially where the gardens are surrounded by fields of clover or by waste

[1] Many species of Order Orthoptera, Family Locustidae.

lands in which the grasshoppers have not been controlled.   The ravenous grass-hoppers strip the leaves, devour fruits, and eat the tender stems from almost all kinds of vegetable crops.   These insects and their control are fully treated under Corn Insects, page 465.

76 (20, 65, 108, 121). FLEA BEETLES[1]

Importance and Type of Injury.   The name *flea beetle* is applied to a variety of small beetles which have the hind legs enlarged and jump vigorously when disturbed.   When flea beetles are abundant, the foliage of garden plants may be so badly eaten that it can no longer function and the plant dies.   Since they are small, and also rather active, they do not take much food in one spot; their injury consists in very small, rounded or irregular holes eaten through or into the leaf, so that leaves look as though they had been peppered with fine shot (Fig. 4.3,*I*).   These small holes give an opportunity for the entrance of destructive plant diseases, and the beetles may carry the disease organisms from one plant to another and spread them as they feed.   The potato flea beetle is a means of spreading early potato blight in this way, and the corn flea beetle spreads the bacterial wilt of corn.   Besides the injury by the adults, the larvae of some flea beetles commonly feed on the roots of the same plants, riddling them with tunnels or eating off small rootlets. The larvae of other kinds feed, along with the adults, on the foliage, or mine in the leaves or tunnel in the stems.

Plants Attacked.   Some flea beetles are rather general feeders, but perhaps the majority attack only one plant or the closely related crops of a single plant family.   Among the most destructive garden species are the potato flea beetle[2] and western potato flea beetle,[3] eggplant flea beetle,[4] spinach flea beetle,[5] horse-radish flea beetle,[6] sinuate-striped flea beetle,[7] striped flea beetle,[8] and western striped flea beetle,[9] western black flea beetle,[10] sweetpotato flea beetle,[11] and the mint flea beetle.[12]   Toma-toes, peppers, and cucumbers are also subject to severe injury.

Life History, Appearance, and Habits.   The life history varies greatly with the different species.   Usually the winter is passed in the adult stage, the beetles hibernating under leaves, grass, or trash about the margins of fields, along ditch banks, fence rows, margins of woods, and similar protected places.   The potato flea beetle (Fig. 14.1) and eggplant flea beetle are about $\frac{1}{16}$ inch long and nearly a uniform black in color. The equally small, tobacco flea beetle[13] is yellowish-brown with a dark cloud across the wings, and the sweetpotato flea beetle[11] (Fig. 14.31) is of about the same size but with a bronzy reflection.   The striped flea beetle (Fig. 14.2) and the sinuate-striped flea beetle are approximately $\frac{1}{12}$

[1] Order Coleoptera, Family Chrysomelidae.
[2] *Epitrix cucumeris* (Harris).
[3] *Epitrix subcrinita* (LeConte).
[4] *Epitrix fuscula* Crotch.
[5] *Disonycha xanthomelas* (Dalman).
[6] *Phyllotreta armoraciae* (Koch).
[7] *Phyllotreta zimmermanni* (Crotch).
[8] *Phyllotreta striolata* (= *vittata*) (Fabricius).
[9] *Phyllotreta ramosa* (Crotch).
[10] *Phyllotreta pusilla* Horn.
[11] *Chaetocnema confinis* Crotch.
[12] *Longitarsus menthaphagus* Gentner (see *Mich. Agr. Exp. Sta. Spcl. Bul.* 155, 1926).
[13] *Epitrix hirtipennis* (Melsheimer).

inch long, with a curious, crooked, yellowish stripe on each wing cover. The pale-striped[1] and horseradish flea beetles[2] are ⅛ inch long, with a broader, nearly straight, yellowish stripe on each wing cover; the smart-weed flea beetle,[3] of the same size, is bluish-black, without stripes, and very straight-sided.   Among the largest of our common species is the spinach flea beetle[4] (Fig. 14.5), which is fully ⅕ inch long, with greenish-black wing covers, a yellow prothorax, and a dark head.   Nearly all these species are elongate oval in outline, with narrowed prothorax and

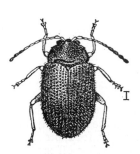

FIG. 14.1. The potato flea beetle, *Epitrix cucumeris*, adult.   Line shows natural size. (*From U.S.D.A.*)

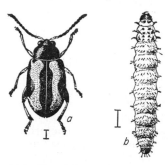

FIG. 14.2. The striped flea beetle, *Phyllotreta striolata*. *a*, Adult; *b*, larva.   Lines show natural size.   (*From U.S.D.A.*)

narrower head.   The antennae are one-half to two-thirds as long as the body, and the hind femurs are distinctly thickened, enabling the beetles to jump away quickly when they are disturbed.   Many of the species emerge from hibernation, beginning in late May, and feed on weeds and the foliage of trees until the garden plants are available, when they migrate to them.   They are frequently serious pests in seedbeds and on newly transplanted vegetables.   The eggs, which are so small as never to be seen by the grower, are scattered in the soil about the plants by the potato flea beetles and the tobacco flea beetle and laid in clusters upon

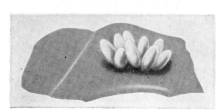

FIG. 14.3. Eggs of the spinach flea beetle, *Disonycha xanthomelas*, enlarged about 20 times.   (*From Ill. Natural History Surv.*)

the leaves by the spinach flea beetle (Fig. 14.3).   They require about 10 days to hatch.   The striped flea beetle deposits her eggs in tiny cavities which she gnaws in the stem of the plant, the horseradish flea beetle deposits them in clusters on the leaf petioles, and the sinuate-striped flea beetle lays her eggs singly on the leaves of cabbage, turnip, and radish.   The larvae of

flea beetles are mostly whitish, slender, delicate, cylindrical worms from ⅛ to ⅓ inch long when full-grown, with tiny legs and brownish heads.   They feed on the roots, underground stems, or tubers of vegetables or of weeds, for 3 or 4 weeks.   The

[1] *Systena blanda* Melsheimer.
[2] *Phyllotreta armoraciae* (Koch).
[3] *Systena hudsonias* Förster.
[4] *Disonycha xanthomelas* (Dalman).
[5] *Epitrix tuberis* Gentner.

larva of the tuber flea beetle[5] tunnels into the potato tuber; it is an important pest in the West.   The larva of the sinuate-striped species, however, mines in the leaves, the horseradish flea beetle burrows in the leaf petiole, and the spinach flea beetle feeds exposed to the underside of the leaves.   The last-named species (Fig. 14.4) is a short leaden-gray wrinkled grub about ¼ inch long in the full-grown larval stage.   Pupation usually occurs in the soil for 7 to 10 days.   There are generally one or two generations a year.

Control Measures.   Flea beetles may be controlled by spraying or dusting with DDT, methoxychlor, or malathion at 1 to 1.5 pounds or Thiodan at 0.5 to 1 pound per acre, observing the proper interval between the

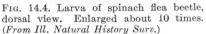

Fig. 14.4. Larva of spinach flea beetle, dorsal view.   Enlarged about 10 times. (*From Ill. Natural History Surv.*)

Fig. 14.5. Adult of spinach flea beetle.   Enlarged about 10 times.   (*From Ill. Natural History Surv.*)

last application and harvest.   On potatoes the adult flea beetles are controlled with DDT, often applied in bordeaux mixture at 1 pound per 100 gallons; with Thiodan or with dieldrin or endrin at 0.5 pound per acre. The larvae attacking potato tubers may be controlled by soil treatments of aldrin at 3 pounds or dieldrin at 2 pounds per acre, thoroughly disked before planting.

Seedbeds may be covered with strips of gauze or tobacco cloth, 25 strands to the inch, to exclude these pests from delicate young plants. Sticky shields or specially made boxes having the inner walls covered with tree tanglefoot may be passed over infested plants and catch thousands of the beetles as they jump off the disturbed plants.   Keeping down weeds in and around the garden is often the most important method of holding these pests in check, since the adults often feed on weeds in early spring and late in fall, and the larvae may develop in great numbers on the roots of certain weeds. In the case of the mint flea beetle care should be taken not to introduce the eggs in soil about the roots used for planting new fields.

References. CROSBY and LEONARD, "Manual of Vegetable-garden Insects," Macmillian 1918; *U.S.D.A., Dept. Buls.* 535, 1917, and 902, 1920; *N.Y. (Geneva) Agr. Exp. Sta. Bul.* 442, 1917; *N.Y. (Cornell) Agr. Exp. Sta. Memoirs,* No. 55, 1922; *Can. Dept. Agr. Ento. Cir.* 2, and *Pamphlet* 80 (n.s.), 1927; *U.S.D.A., Farmers' Bul.* 1371, 1934; *Ohio Agr. Exp. Sta. Bul.* 595, 1938; *Jour. Econ. Ento.,* **29**:586, 1936; **40**:651, 1947; and **41**:275, 1948.

77. ASIATIC GARDEN BEETLE[1]

Importance and Type of Injury. Irregular holes are eaten, especially from the margins of the leaves on warm nights, by beetles which may be found during the day in loose soil at the base of the injured plants. The adults occur in such numbers around bright lights that they are a serious nuisance about residences, and patrons are driven away from amusement parks, swimming pools, and restaurants. Their whitish grubs feed especially on the roots of grasses but do not cut them off so destructively as do the Japanese beetle larvae. The adults have caused personal injury by burrowing into the external ear of sleeping persons.[2]

Plants Attacked. More than 100 different plants, including vegetables such as beets, carrots, corn, eggplant, kohlrabi, parsnips, peppers, and turnips; ornamentals such as asters, dahlias, chrysanthemums, roses, zinnias, delphiniums, azaleas, rhododendrons, and viburnum; and peach, cherry, and strawberry.

Distribution. First found in America in New Jersey, in 1922, this insect is now widely distributed along the eastern seaboard in Connecticut, Massachusetts, New York, New Jersey, Pennsylvania, Delaware, Virginia, West Virginia, and North and South Carolina. It is believed that this pest cannot thrive in areas of low summer rainfall.

Life History, Appearance, and Habits. The winter is passed as small grubs, 6 to 10 inches deep in the soil. They resume feeding near the soil surface in early spring, pupate in late May and June, and emerge as adults in June and July. The adults (Fig. 9.48) are shaped like May beetles, but are only ⅜ inch long. They are cinnamon-brown in color, with fine longitudinal striae on the wing covers and a fine pubescence which gives them a velvety appearance. They are troublesome all summer. Eggs are laid in greatest numbers in grassy areas, overgrown with such weeds as orange hawkweed, goldenrod, sorrel, and asters, which keep the soil moist and cool. There is a single generation a year. The larvae may be distinguished from related white grubs, by their more slender bodies, the Y-shaped anal opening, and the transverse row of short setae in front of it, as shown in Fig. 15.54.

Control Measures. In lawns and golf courses, complete protection from the larvae has been secured by treating the soil with lead arsenate, DDT, or chlordane, as described on page 522. Strawberry plots have been protected by mulching with hay. When infested overgrown land is plowed up, it may be planted the first year to some other crop that is not injured. Spraying the preferred food plants with DDT, methoxychlor, Sevin, or malathion is effective in killing the beetles, but in areas

---

[1] *Autoserica castanea* Arrow, Order Coleoptera, Family Scarabaeidae.
[2] *Jour. Econ. Ento.,* **51**:546, 1958.

of great abundance successive hordes of beetles may attack the plants each warm night and destroy them in spite of the sprays. Funnel-shaped light traps have captured as many as 2,000 beetles per hour per trap.

References. U.S.D.A., *Cirs.* 238, and 246, 1932; *Jour. Econ. Ento.*, **27**:476–481, 1934; and **29**:348–356, 1936.

78. THE WHITE-FRINGED BEETLE[1]

Importance and Type of Injury. As many as 200 to 300 grayish long-headed beetles, ½ inch long, with faint white stripes on their sides, are found per plant feeding upon the margins of the leaves or crawling upon the ground. The lower part of the stem and taproot or the planted seeds of many field and garden crops are attacked in the spring by vast numbers of small white legless grubs, up to ½ inch long, which cause the plants to yellow, wilt, and die. Although the insect cannot fly, the eggs are deposited upon many parts of plants that are moved in commerce, the larvae may be transported in small quantities of soil, and the adults, which have a pronounced tendency to climb upward, readily cling to objects being transported. Entomologists who have studied the insect feel that it may become a serious pest in many regions of the United States.

Plants Attacked. These insects are very general feeders. The adults have been observed feeding upon more than 170 different kinds of plants, ranging from herbaceous weeds and crops to vines and trees. The adults are not voracious feeders and most of the damage is done by the larvae, which feed on the roots of at least 385 species of plants, including the following crops: peanuts, corn, sugarcane, cotton, cowpea, velvet beans, cabbages, collards, sweetpotatoes, chufa, lucerne, Mexican clover, and blackberries.

Distribution. The white-fringed beetle is a native of South America, where it is widely distributed over the southern part of the continent. It was first found in Florida in 1936 and now is scattered over 500,000 acres in Arkansas, Louisiana, Mississippi, Alabama, Georgia, Florida, and North and South Carolina.

Life History, Appearance, and Habits. The insect usually passes the winter months as large larvae, mostly in the upper 9 to 12 inches of the soil, but may also overwinter as the egg. The whitish naked pupae have been found mostly in the ground 3 to 6 inches deep, from late May to the end of July. From early May until mid-August the adult beetles (Fig. 14.6) emerge and crawl to a near-by food plant. The elytra of the beetles are grown together so that they cannot fly. They are dark-gray beetles, slightly less than ½ inch long, less than ⅓ as broad over the basal half of the wing, whence the body narrows gradually to the end of the short, broad snout and tapers rapidly to the tips of the wing covers. The margins of the elytra are banded with white, and there are two pale lines extending along the sides of the head and prothorax, one above the eye and one below it. The body is covered with dense, short, pale hairs, longer toward the tip of the elytra. The adults feed for some days upon the margins of older leaves toward their base but eat relatively little and

---

[1] *Graphognathus* (= *Pantomorus*) *leucoloma* (Boheman), Order Coleoptera, Family Curculionidae. Recent studies indicate that the white-fringed beetles represent several closely related species or races.

are active chiefly in the afternoon.   No males of this species have been found.   The females, which reproduce parthenogenetically, begin to lay eggs in 10 to 12 days after emerging from the pupal stage.   A female may lay many hundreds of eggs, the maximum observed being 3,258.   The average lifetime of the females is 2 to 5 months, during which time they frequently crawl from ¼ to ¾ mile.   The whitish oval eggs are deposited in gelatinous masses, as many as 60 eggs per cluster.   They adhere to sticks, stones, the base of plant stems, and other objects on or near the ground.   The soil adheres to the egg masses, making them very difficult to see.   In midsummer the eggs hatch in about 2 weeks, but in cooler weather, the egg stage may last 1 to 3 months, hatching resulting from proper conditions of temperature and moisture.   The larvae feed entirely below ground down to a depth of 6 inches or more, beginning in late July and continuing until cold weather. They chew away the lower part of the stem and taproot of many kinds of plants.   When full-grown the larvae measure about ½ inch long. They are yellowish-white, the back evenly rounded, upward; they are legless, and sparsely covered with short hairs.   From late May to the end of July, the larvae pupate at depths of 3 to 6 inches in the soil.   There is normally one generation a year.

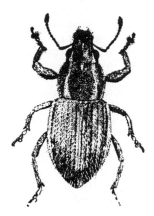

FIG. 14.6. White-fringed beetle, *Graphognathus leucoloma*. About twice natural size. (*From U.S.D.A., Bur. Ento. and P.Q.*)

Control Measures.   The known infested areas have been placed under rigid federal and state quarantines to restrict the movement of all products likely to convey the beetles in any stage. Since the insect does not fly, ditches about a foot wide and a foot deep, with well-packed vertical sides, are very effective in stopping the migrations of the insects, and they may be trapped in postholes in the bottoms of the ditches and destroyed with kerosene. Larval populations may be controlled by soil applications of the following materials, broadcast as dusts, granulars, or sprays, and cultivated into the upper 3 inches of soil before planting, or applied as fertilizer mixtures to rows at time of planting: DDT 10 pounds (2 to 3 pounds to rows), chlordane 5 pounds (1 to 2 pounds to rows), dieldrin 1.5 pounds (0.5 to 0.75 pound to rows), aldrin or heptachlor 2 pounds (0.75 to 1 pound to rows).

References.   U.S.D.A., Bur. Ent. Plant Quar., Cir. E-420, 1938, E-422, 1938; E-750, 1948, and E-779, 1949, and *Leaflet* 401, 1959.

78A. Vegetable Weevil[1]

Importance, Life History, Appearance, and Habits.   This insect, which is a native of Brazil, was first reported from Mississippi in 1922 and since that time has spread throughout the Gulf states and the coastal areas of California.   The cream-colored larvae feed extensively on truck crops such as carrots, turnips, and spinach in the fall and winter, while the parthenogenetic adults damage potatoes and tomatoes in the late spring and early summer.   The grayish snout beetles are about ⅓ inch long and have a V-shaped marking across the wing covers.   The eggs are laid

[1] *Listroderes costirostris obliquus* (Klug), Order Coleoptera, Family Curculionidae.

in the crowns of plants, on leaf petioles, or in soil near the base of the plants during September to March, each female depositing from 300 to 1,500.

Control Measures.   The adults and larvae may be controlled by spraying or dusting with DDT at 1 to 2 pounds or parathion at 1 pound per acre.   Rotenone at 0.25 pound per acre may be used in the home garden. The adults may also be controlled by applications of baits as suggested for the strawberry root weevil, page 695.

References.   Calif. Agr. Exp. Sta. Bul. 546, 1932; *U.S.D.A., Cir.* 530, 1939.

79 (19, 47, 69). BEET WEBWORM,[1] ALFALFA WEBWORM,[2] AND GARDEN WEBWORM[3]

Importance and Type of Injury.   These small webbing caterpillars, by devouring the foliage, have caused the abandonment of thousands of acres of sugarbeets, often destroy fields of alfalfa, and take from 10 to 100 per cent of truck crops.   When they have eaten everything in a field, they often migrate like armyworms.

Plants Attacked.   The larvae are general feeders and destructive to almost any succulent garden or field crop, but grains and grasses are not preferred.   Among the garden crops seriously attacked are cabbage, carrots, beets, beans, peas, potatoes, spinach, and cucurbits.

Distribution.   From the Mississippi Valley westward to the Continental Divide.

Life History, Appearance, and Habits.   These pests winter as larvae in silk-lined cells or tubes in the soil, within which they pupate in late spring.   The adults from this overwintering generation are present in the fields from late March to late June.   They are night-active moths, with a wing expanse of 1 to $1\frac{1}{4}$ inches.   The alfalfa webworm is buff-colored, irregularly marked with light and dark gray and with a row of spots on the underside of the hind wings near the apical margin.   The adult beet webworm is smoky brown, mottled with dusky and straw-colored spots and lines.   It can be distinguished from the alfalfa webworm by the dark markings on the underside of the hind wings, being in the form of a continuous dark line near the margin.   The whitish to yellow or green, oval eggs are laid in groups of 2 to 20, mostly on the underside of the leaves. The beet webworm eggs are laid in a single row, end to end, while the alfalfa webworm eggs are in overlapping groups.   As the larvae skeletonize and devour the leaves, they spin a web, drawing leaves together, and also forming a tube several inches long which leads to a hiding place, as under a clod of earth, into which they retreat when disturbed.   The fullgrown larvae are yellowish or greenish to nearly black, 1 to $1\frac{1}{4}$ inches long.   The beet webworm larva has a black stripe down the middle of the back, while the alfalfa webworm has a broad light-colored stripe down the middle of the back.   On each side of each segment there are three small dark spots, from which arise one to three setae.   The insects pupate 2 to 3 weeks in earthen silk-lined cells, an inch or so underground.   The pupa of the beet webworm has eight, bristle-like appendages at the end of the abdomen, while in the alfalfa webworm these appendages are spoonshaped.   There are three partial generations over much of the range. The garden webworm is discussed on page 557.

[1] *Loxostege sticticalis* (Linné), Order Lepidoptera, Family Pyralididae.
[2] *Loxostege commixtalis* (Walker).
[3] *Loxostege similalis* (Guenée).

Control Measures. On vegetable crops the control measures are the same as for cabbageworms (page 664). Control measures for sugarbeets and alfalfa are given on page 558. Dusting with 0.2 per cent pyrethrins at 20 to 30 pounds per acre or spraying with 0.5 ounce per acre is effective in the home garden. Growers should watch their fields and gardens for the appearance of the moths, which fly up as the plants are disturbed, and for the egg masses on the underside of the leaves and should begin spraying just as soon as the eggs start hatching. The migrating caterpillars can be stopped by trap furrows as recommended for armyworms and the white-fringed beetle (pages 475, 608). Moths lay fewer eggs in fields kept free from weeds, especially pigweed and lamb's-quarters. Late-seeded alfalfa may escape serious damage.

References. *Colo. State Ento. Cir.* 58, 1933; *U.S.D.A., Bur. Ent. Bul.* 109, 1912.

79A. BEET ARMYWORM[1]

Importance and Type of Injury. This insect, which is a close relative of the fall armyworm (page 473), is a general feeder and attacks the foliage, stems, and sometimes the roots of field and vegetable crops. It sometimes appears in enormous numbers and has completely defoliated hundreds of acres of sugarbeets. In Florida it is an important pest of asparagus, chewing the buds and younger leaves.

Plants Attacked. Sugarbeets are the favorite host, but table beets, asparagus, cotton, alfalfa, corn, lettuce, tomatoes, potatoes, onions, peas, citrus, wild sunflower, saltbush, mallow, plantain, and even wild grasses may be attacked.

Distribution. The beet armyworm is a native of the Orient and was first recorded from California in 1876. It is distributed throughout the Gulf states, westward to the Pacific Coast, and northward to Kansas and Nebraska.

Life History, Appearance, and Habits. The insect overwinters as the pupa and in warmer areas as the adult moth, which has a wingspread of approximately $1\frac{1}{4}$ inches. The forewings are grayish brown with a pale spot in the mid-front margin, and the hind wings are white with a dark anterior margin. The female begins to lay eggs early in the spring, depositing these in irregular masses of about 80 eggs covered with hairs and scales from her body. She may deposit an average of 500 to 600 eggs over a 4- to 10-day period. These hatch in 2 to 5 days, and the larvae feed for about 3 weeks, spinning slight webs over the foliage and passing through five instars. The mature larva is green, with prominent, dark, lateral stripes, and is about $1\frac{1}{4}$ inches long. It usually pupates in the upper $\frac{1}{4}$ inch of soil in a cell formed by gluing together soil particles and trash with a sticky secretion. The entire life cycle from egg to adult required 36 days at 80°F. and about 24 days in midsummer in Florida. There are normally four generations a year in warmer areas.

Control Measures. This insect may be controlled on sugarbeets by spraying or dusting with DDT 1.5 pounds and toxaphene 3 pounds, endrin 0.5 pound, or Dylox at 1 pound per acre. Tops should not be fed to livestock where treated with DDT, toxaphene, or endrin. For additional control measures, see page 475. The braconid parasites, *Chelonus texanus* and *Meteorus autographae*, and the chalcid, *Apanteles marginiventrus*, may cause severe mortality under favorable conditions.

References. *U.S.D.A., Bur. Ento. Bul.* (n.s.) 33, 1902, and 57, 1906; *Colo. Agr. Exp. Sta. Bul.* 98, 1905; *Fla. Agr. Exp. Sta. Tech. Bul.* 271, 1934; *Jour. Econ. Ento.* **53**:616, 1960.

80 (4, 48, 67, 259). CUTWORMS[2]

Importance and Type of Injury. Recently set or young seedling plants are often cut off during the night at the surface of the soil (Fig. 13.3) and

---

[1] *Spodoptera* (= *Laphygma*) *exigua* (Hübner), Order Lepidoptera, Family Noctuidae. It has also been called the asparagus fern caterpillar.

[2] Various species of Order Lepidoptera, Family Noctuidae.

left lying to wilt on the ground near by.   Plump, soft-skinned, greasy-looking caterpillars, varying in length up to 1½ inches or more, which generally roll the body tightly when disturbed and "play 'possum," are found in shallow holes in the soil about the base of the plants.   Some species climb up the plants and eat at the leaves or chew their way into fruits, such as tomatoes, while still others feed entirely below ground. Several species, when they become abundant, crawl from the fields where they developed, in great numbers like marching armies, and may invade gardens and rapidly devour all kinds of garden crops.

Plants Attacked.   Nearly all garden vegetables, as well as flowers, field crops, and fruit trees are attacked by cutworms.

The distribution, life history, and control of cutworms are discussed under Corn Insects, page 476, and Tobacco Insects, page 594.   Sod or weedy land intended for vegetables should be plowed in the late summer and kept fallow until late in the fall, since it is during this period that the cutworm moths lay their eggs, chiefly on rough, grassy, or weedy land.   Some species of cutworms, however, deposit their eggs on fence posts and the stems of large weeds along fences and ditches.   In small gardens certain measures are practicable that cannot be used on field crops.   The worms may be concentrated by laying small boards about the garden or they may be dug out and destroyed by hand.   Some gardeners place cylinders of tin (such as tin cans with bottoms cut out) partly sunken in the soil about choice plants when they are transplanted, to keep the cutworms away.   The poisoned baits and insecticide treatments discussed under Corn Insects are effective for many species.

81. Banded Woollybear,[1] Yellow Woollybear,[2] and Salt-marsh Caterpillars[3]

Importance and Type of Injury.   Very hairy or woolly, yellowish and brown caterpillars (Fig. 14.7), ranging up to 2 inches in length, riddle the foliage of many

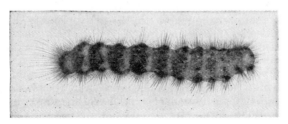

Fig. 14.7. The salt-marsh caterpillar, *Estigmene acrea*, larva.   *(From Ill. Natural History Surv.)*

garden crops in summer and autumn.   The smaller worms are usually found feeding together on the underside of leaves.   The larger ones feed exposed and, when full-grown, are often seen in the autumn scurrying over the surface of the ground, apparently in great haste to get somewhere.

Plants Attacked.   One or more of these species attack practically all the garden and field crops.

Distribution.   General throughout the United States.

Life History, Appearance, and Habits.   The banded woollybear winters as a larva in some protected place and feeds briefly in the spring before changing to a pupa.

[1] *Isia isabella* (Smith and Abbott), Order Lepidoptera, Family Arctiidae.

[2] *Diacrisia virginica* (Fabricius).

[3] *Estigmene acrea* (Drury).

The other two species hibernate in the pupal stage inside thin silken cocoons heavily covered with interwoven hairs from the body of the caterpillar. The adults are snow-white or yellowish-winged moths, from 1½ to 2 inches across the wings, with yellowish black-spotted abdomens. The yellow woollybear adult is nearly pure white except for the abdomen, each wing having a few small black spots. The salt-marsh caterpillar adult has the wings peppered with a number of small black spots, white above and yellow below, in the female. In the male (Fig. 14.8) the hind wings are yellow both above and below. The adult of the banded woollybear is entirely yellowish and with a few black spots on the wings. The adults appear in spring and lay their spherical eggs in patches on the leaves. The larvae attain full size in a month or two. The banded woollybear larva is black at each end, with a median band of brown of variable extent, and has an evenly clipped appearance. The other two species have hairs of different lengths, yellowish, tawny, or grayish in color, and not so dense as completely to hide the skin. There are generally two generations a year in the North.

FIG. 14.8. The salt-marsh caterpillar, male moth. Natural size. (*From Ill. Natural History Surv.*)

Control Measures. Fields may be protected from migrating caterpillars by an upright barrier made from aluminum foil or heavy paper, 4 inches wide, set on edge in the soil. The caterpillars may be controlled by spraying or dusting with toxaphene at 4 pounds, alone or in mixture with DDT at 2 pounds per acre, where residue restrictions permit; or with parathion or Dilan at 1 pound or Phosdrin at 0.5 pound per acre. Two dipterous parasites, *Exorista larvarum* and *Rileymyia adusta*, are of importance in control.

References. *Can. Dept. Agr. Bul.* 99 (n.s.), 1934; *Jour. Econ. Ento.*, **50**:279, 1957.

82 (50, 60, 96, 117, 126, 267). APHIDS OR PLANT LICE[1]

Importance and Type of Injury. Aphids may attack and severely damage or destroy every vegetable crop. These insects are soft-bodied "plant lice," generally about ⅟₁₆ to ⅛ inch in length, and are usually green, although some species are brown, yellow, pink, or black. They all feed by thrusting sharp, hollow stylets from their beaks in among the plant cells and sucking out the sap, and during the feeding process they may inject a toxic saliva into the plant. The result is a blighting of buds, dimpling of fruits, curling of leaves, and the appearance of discolored spots on the foliage. Where large numbers of aphids are present, the plants may gradually wilt, turn yellowish or brown, and die. Aphids are the most important agents in the dissemination of plant virus diseases (Table 1.1), and a brief period of feeding by a single infected aphid may infect and eventually kill the plant.

The presence of aphids makes vegetables unattractive, detracts from their flavor and market value, and occasions much more work in preparing the vegetables for use. The aphids often secrete a sticky, sugary honeydew, which gums up the plants and serves as a medium for the growth of sooty mold fungus, which further spoils vegetables.

Plants Attacked. All vegetables, including those of the cabbage family, cucumbers, melons, beans, peas, potatoes, tomatoes, lettuce, turnips, spinach, and other garden crops, have serious aphid pests.

Distribution. As a group, world-wide, and many species are nearly cosmopolitan.

[1] Many species of the Order Homoptera, Family Aphidae.

Life History, Appearance, and Habits. Aphids typically winter as fertilized eggs on some perennial plant; some winter on the dead remnants of annual vegetables; while the overwintering condition of some species is not known. The eggs are small, ovate, blackish objects, glued on their sides generally to the stems of plants or in crevices about the buds (Fig. 15.23). When the weather becomes warm enough, small nymphs hatch from the eggs, which grow quickly to full size but never get wings. Since each of these is the start for a great colony of aphids that may be produced during the season, they are called *stem-mothers*. They are all females which have the remarkable ability to reproduce young like themselves, without mating. These young are born ovoviviparously, *i.e.*, already hatched from the egg, and differ from their stem-mothers, in having only one parent and in not passing through an exposed egg state. They are like the stem-mothers in being wingless and in producing young ovoviviparously, beginning when they themselves are only a week or so old and producing from a dozen to 50 or 100 active nymphs within the next week or two. In this way a succession of generations is produced, the young clustering about their mothers until patches on the plant may be crowded with them. At some time during this period, either all or a part of certain generations of these females may develop wings. These may fly to other plants of the same kind, or in some species they habitually fly to a different kind of plant (usually an annual), known as the *summer host*. Such winged ones are called *spring migrants*. They settle down on the new host plant and start a succession of generations there, all produced, as before, from unfertilized eggs that hatch in the body of the mother.

As shortening days forecast the end of the season, and before the summer-host plant dies, a generation is usually produced that is all winged but is often of two kinds. Some of them are winged males, the first appearance of males in the aphid colonies being at the approach of cold weather. The others are winged females which are called *fall migrants* and which may serve to return the species to the kind of perennial plant from which their distant ancestors flew away in the spring. These fall migrants give birth to nymphs in the normal manner, but the nymphs, when grown, are wingless true females that cannot reproduce unless they mate with the males which are of the preceding generation. After mating, the true female lays from one to four or more large fertilized eggs in a sheltered place about the plant, and dies; or, sometimes, simply dries up about the single egg she is capable of maturing. From these eggs arise the stem-mothers of the next spring, which differ from all the hosts of other aphids produced during the year in having both male and female parents. In some species the males and true females have no mouth parts.

This is a kind of standard life cycle. Variations from it will be noted in the discussion of particular species, but the essential features of this life cycle should be fixed in mind because the details are not repeated for the aphids discussed under particular crops.

Control Measures. Aphids may be satisfactorily controlled by spraying with demeton (Systox) at 0.25 to 0.5 pound, spraying or dusting with malathion at 1 to 1.5 pounds, parathion at 0.4 to 0.5 pound, Phosdrin at 0.25 to 0.5 pound, or TEPP at 0.4 pound per acre, taking care to observe residue restrictions and safe intervals before harvest. For the home garden, spraying with 0.05 to 0.06 per cent nicotine sulfate in soapy water is effective. Heavy infestations of aphids will usually require several applications of these materials at weekly intervals, for control. For certain low-growing dense crops such as melons, turnips, and the like, dusting is superior to spraying because the dust will circulate among the plants and even penetrate into curled leaves and reach the aphids in positions where a spray could not be driven.

References. "The Life-cycle of Aphids," *Ann. Ento. Soc. Amer.*, **13**:156, 1920; *Ann. Appl. Biol.*, **22**:578, 1935; *Jour. Econ. Ento.*, **31**:60, 1938.

83 (52, 182). TARNISHED PLANT BUG[1]

Importance and Type of Injury. This small, brownish, flattened bug is provided with piercing-sucking mouth parts with which it takes the sap

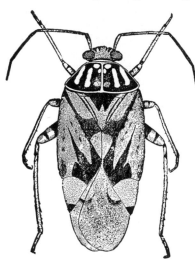

of a great variety of plants. As it feeds it introduces a toxic saliva into the plant. Its feeding causes various sorts of injuries; the leaves may become deformed, as in beets and chard; the stems or leaf petioles scarred and discolored, as in the "black joint" of celery; or the buds and developing fruit dwarfed and pitted, as in the case of beans, strawberries, peaches (Fig. 15.58), and pears.

Plants Attacked. Beet, chard, celery, bean, potato, cabbage, cauliflower, turnip, salsify, cucumber, cotton, tobacco, alfalfa, many flowering plants, and most deciduous and small fruits— more than 50 economic plants, besides many weeds and grasses.

Distribution. Throughout the United States, and in many other parts of the world.

FIG. 14.9. Adult of the tarnished plant bug, *Lygus lineolaris.* Enlarged about 8 times. (*From Ill. Natural History Surv.*)

Life History, Appearance, and Habits. The adult bugs, and probably also the nymphs, hibernate under leaf mold, stones, bark of trees, among the leaves of such plants as clover, alfalfa, and mullein, and in many other protected places. However, in warmer climates the bugs are active throughout the winter. They are about ¼ inch long by less than half as broad, flattened, oval in outline, with the small head projecting in front, and of a general brown color much mottled with small, irregular splotches of white, yellow, reddish-brown, and black. Along the side of the body at the posterior third is a clear-yellow triangle (the cuneus) tipped with a small, triangular, intensely black dot (Fig. 14.9). They become active very early in spring, when they attack the buds of fruit trees, causing serious injury to the terminal shoots and fruits (page 754). They do not appear to lay their eggs on these plants to any great extent, but migrate to various herbaceous weeds, vegetables, and flowers, where the eggs either are inserted full length into the stems, petioles, or midribs of leaves or into buds, or are tucked in among the florets of the flower head.

[1] *Lygus* (= *oblineatus*) *lineolaris* (Palisot de Beauvois), Order Hemiptera, Family Miridae.

The egg is elongate, slightly curved, and the outer end is cut off squarely, the lid which covers this end being usually flush with the stem.   After about 10 days, a small yellowish-green nymph $\frac{1}{25}$ inch long, oval in outline and provided with long legs, antennae, and piercing-sucking mouth parts, emerges from the egg and begins feeding on the sap.   It grows rapidly, molting five times, the larger nymphs gradually taking on the appearance of

the adult and being marked with four rounded black dots on the thorax and one on the base of the abdomen (Fig. 14.10).   The life cycle is completed in 3 or 4 weeks, so that probably three to five generations occur each season.   By late summer they occur everywhere in profusion but, because of their obscure and protective coloration and shy and hiding habits, are not much noticed.

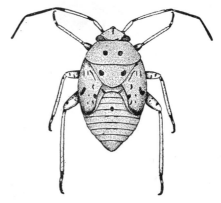

Control Measures.   Satisfactory control of the tarnished plant bug may be obtained by spraying or dusting with DDT at 1 to 2 pounds or toxaphene at 2 to 4 pounds per acre, or a combination of these, on crops such as lima beans, where they do not cause a residue problem.

Fig. 14.10. Tarnished plant bug, nymph, last instar.   Enlarged about 5 times. *(From Ill. Natural History Surv.)*

Parathion at 0.3 to 0.4 pound and malathion at 1 to 2 pounds are also effective.   Cleaning up of weeds and destruction of favorable hibernating places may help to keep its numbers down.

References. Mo. Agr. Exp. Sta. Res. Bul. 29, 1918; *N.Y. (Cornell) Agr. Exp. Sta. Bul.* 346, 1914; *Jour. Econ. Ento.,* **25**:671, 1932; **26**:148, 1933; **38**:389, 1945; and **41**:403, 1948; *Contrib. Boyce Thompson Inst.,* **16**:279, 429, 1951; and **17**:347, 1954.

84. Garden Fleahopper[1]

Importance and Type of Injury.   These small bugs look somewhat like black aphids, and they suck the sap from leaves and stems in a manner like aphids.   Small pale spots often appear on the leaves where the fleahoppers have sucked out the sap, and badly infested leaves are killed.   The insect occurs sporadically.

Plants Attacked.   Bean, beet, cabbage, celery, corn, cowpea, cucumber, eggplant, lettuce, pea, pepper, potato, pumpkin, squash, sweetpotato, tomato, various legumes, ornamentals, and many weeds.

Distribution.   General throughout the United States, except in the western part.

Life History, Appearance, and Habits.   The insect generally passes the winter as the winter or diapausing egg.   In the spring the females insert their eggs in the leaves or stems of the plants on which they feed, the eggs being laid in punctures made by the mouth parts.   The greenish nymphs of various sizes appear on the underside of the leaves in early spring and grow rapidly to blackish adults, passing through five nymphal instars.   The average time of development from egg to adult ranged from about 41 days at 55°F. to 11 days at 75°F.   The females laid an average of about 105 summer eggs over a period of about 50 days.   The females (Fig. 14.11) are of two kinds: a long-winged form (*b*), which is $\frac{1}{12}$ inch long by about one-third as wide, nearly straight-sided, and with the overlapping tips of the wings transparent; and a short-winged form (*a*), about $\frac{1}{10}$ inch long, oval-bodied, and more than half as wide as long. The latter form lacks the transparent tips of the wings, and this, together with the

[1] *Halticus bracteatus* (Say), Order Hemiptera, Family Miridae.

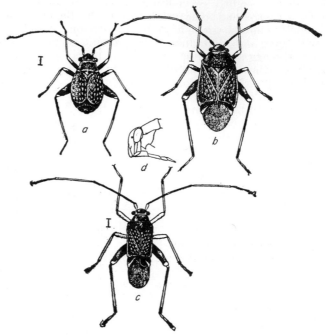

Fig. 14.11. The garden fleahopper, *Halticus bracteatus*. *a*, Short-winged female; *b*, long-winged female; *c*, male; *d*, side view of head showing piercing-sucking mouth parts.   Lines indicate natural size.   (*From U.S.D.A.*)

jumping habit of the species, gives it a strong resemblance to a flea beetle.   The legs are long and the antennae are longer than the body and very slender.   Five generations have been recorded in South Carolina.

Control Measures.   This insect may be controlled by spraying or dusting with DDT or malathion at 0.5 pound per acre.   Destruction of the weeds on which the insect lives will do much to keep it from reaching serious numbers.

References.   *U.S.D.A., Dept. Bul.* 964, 1921; *Conn. Agr. Exp. Sta. Bul.* 344, pp. 160, 161, 1933; *Va. Agr. Exp. Sta. Tech. Bul.* 101, 1946; 107, 1947.

85 (158, 269). Spider Mites[1]

Importance and Type of Injury.   In periods of dry hot weather the leaves of beans and other plants become blotched with pale yellow and reddish-brown, in spots ranging from small specks to large areas, on both upper and undersurfaces; the leaves have a pale sickly appearance and gradually die and drop.   The undersurfaces of such leaves look as though they had been very lightly dusted with fine white powder.   When examined under a lens, the fine white specks are seen to consist of empty wrinkled skins and minute spherical eggs and to be suspended upon almost invisible strands of silk.   Upon this silk, and beneath it on the surface of the leaf, are resting or running about numerous minute, whitish, greenish or reddish, eight-legged mites of several sizes, up to about $\frac{1}{60}$ inch long (Fig. 14.12).   These mites live on the sap of the plant, which is drawn by piercing the leaf with two sharp slender lances attached to the mouth.   They generally overwinter as orange hibernating females in

[1] A number of species of the Class Arachnida, Order Acarina, Family Tetranychidae.

protected locations.   A number of species attack vegetable and field crops which can be distinguished with certainty only by the specialist.

The two-spotted spider mite, *Tetranychus telarius* (Linné), is the most common and widely distributed species and is described on page 875. The strawberry or Atlantic spider mite, *T. atlanticus* McGregor, feeds in definite colonies on the undersides of the leaves and causes heavy leaf drop.   The Pacific spider mite, *T. pacificus* McGregor, the Schoene spider mite, *T. schoenei* McGregor, and the four-spotted spider mite, *T. canadensis* (McGregor), are described on page 717.

Plants Attacked.   The vegetable crops most seriously injured are beans, corn, tomato, eggplant, celery, and onion. These pests also do serious damage to cotton, fruit trees, and ornamental and greenhouse plants.

Distribution.   The two-spotted spider mite is found throughout the United States, and the other species are widely distributed in various sections.

Control Measures.   The overwintering mites may be reduced in numbers by the destruction of weeds such as pokeweed, Jerusalem oak, Jimson weed, wild blackberry, and wild geranium about the garden and by the destruction or spraying of violets, berry bushes, and other host plants, wherever they retain green foliage over the winter.   The mites are commonly controlled on vegetable crops by dusting with finely divided sulfur at 20 to 40 pounds per acre or by spraying with wettable sulfur.   Several applications at weekly intervals may be required.   The addition of 25 to 50 per cent sulfur to dusts of DDT, toxaphene, and other chlorinated hydrocarbons will largely prevent the destructive increases in spider mite populations which sometimes result

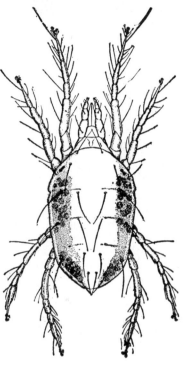

FIG. 14.12. Two-spotted spider mite, *Tetranychus telarius*, adult female, highly magnified.   (*From U.S.D.A., Farmers' Bul.* 831.)

from their use.   Spider mites may also be controlled by spraying or dusting with the following, subject to residue restrictions: malathion 1 to 2 pounds, Trithion 1 pound, parathion 0.5 to 1 pound, Kelthane 0.5 to 1 pound, and demeton (Systox) spray at 0.25 pound, per acre.   Protection to within 3 to 4 days of harvest is given by spraying or dusting with TEPP or Phosdrin at 0.5 pound per acre.

References.   U.S.D.A., Dept. Bul. 416, 1917; *Ore. Agr. Exp. Sta. Bul.* 121, 1914; *Va. Agr. Exp. Sta. Tech. Bul.* 113, 1949.

86 (15, 37, 68).   WIREWORMS[1]

Importance and Type of Injury.   When garden or truck crops are planted on land that has been in sod for several years, the wireworms

[1] Several species of the Order Coleoptera, Family Elateridae.

that were attracted to the grass plants remain in the soil until their life cycle is completed and may damage the vegetables on such plots for from 1 to 5 years. Other species breed and thrive in the intensely cultivated truck-crop areas and are not bound up with sod conditions in any way. They feed below ground, chewing off small roots and scoring the surface or tunneling through larger roots, tubers, and underground parts of stems. Another serious injury is their habit of feeding on newly planted seeds, which prevents the plants from coming up.

Plants Attacked. The seeds of corn, beans, peas; the tubers of potatoes; and the roots of turnips, sweetpotatoes, carrots, rutabagas, radish, sweet corn, cabbage, cucumber, tomato, onion, watermelon, many other vegetables, and nursery stock suffer from the attacks of these pests.

Distribution. See Corn Insects, page 506.

Life History, Appearance, and Habits. These points are discussed under Corn Insects (page 506). Some of the species injurious in gardens have a 2-year life cycle, while others require at least 6 years.

Control Measures. Cultural control practices are difficult to apply for wireworm control because of the relatively long life cycles and the necessity for adapting them to each region and crop. In colder climates, summer plowing of infested grasslands followed by winter cultivation will destroy the exposed pupae and adults. For some species, crop rotations planting rape, buckwheat, or clover which are not seriously attacked for 2 years following the plowing of grasslands are effective. However, certain western species prefer to lay their eggs in cultivated land, and crop rotations are valueless. In small vegetable gardens, where the soil has been plowed in the summer or fall and kept free of vegetation, wireworms have been trapped at the rate of 6,000 to 80,000 per acre by planting baits, 2 to 4 inches deep, at intervals of 3 to 10 feet, consisting of germinating peas, beans, or corn; cull potatoes; or stiff flour dough. These are covered with boards or tiles and dug up after a week and the worms destroyed.

Chemical control measures are much more effective, relatively inexpensive, and simple to apply, and do not necessitate keeping land out of crop production. Soil fumigation with ethylene dibromide at 2 to 3 gallons (36 to 54 pounds) per acre, applied as a 40 or 85 per cent solution in petroleum naphtha, has given good control of wireworms. The lower dosage is used in coarse, sandy soils, and the higher in fine, clay soils. Special application equipment or a simple gravity-feed applicator arranged to discharge the fumigant directly ahead of the plow sole is used to inject the fumigant to a depth of 8 inches, with 12-inch spacing. The soil should be rolled, dragged, or harrowed immediately after treatment to compact the surface and prevent the rapid escape of the fumigant. Small plots can be treated by pouring about 2 milliliters ($\frac{1}{2}$ teaspoonful) of 10 per cent ethylene dibromide solution into 8-inch-deep holes punched into the soil at 1-foot intervals and covered immediately after treatment. Ethylene dibromide fumigation should not be applied close to growing plants, and crops should not be planted in the treated soil for 3 weeks.

Economical wireworm control has been obtained by applications of aldrin, dieldrin, or heptachlor at 2 to 3 pounds or chlordane at 4 to 6 pounds per acre, made to the soil surface just prior to planting, as a spray, dust, or granular, and disked into the soil in two directions. The proper dosage will vary with soil type and moisture content. Certain of these materials may accumulate in the soil and affect the yield, flavor, and amount of pesticide residue in root crops.   Therefore care should be taken not to treat more frequently than necessary for wireworm control.   Such applications are often made in combination with fertilizer treatments.[1]

Seed treatments with lindane, heptachlor, aldrin, dieldrin, or endrin have been shown to produce 70 to 95 per cent mortalities of wireworms attracted to germinating seeds and thus ensure the production of a satisfactory stand of young plants.   The cost is minimal, and the amount of insecticide, usually 0.25 ounce or less per acre, is so small as to preclude difficulties with off flavors, soil accumulations, or plant residues.   Seed treatments are best made with slurries of 75 per cent wettables, except with seeds such as beets or spinach, where spray application is more effective.   Methyl cellulose is usually added to the water at 2.5 to 3 per cent to act as a sticker.   Treatments of seeds with insecticides have often been found to predispose them to attack by soil fungi such as *Pythium*, and a fungicide such as thiuram, captan, chloranil, dichlone, or organic mercurial is generally included.[2]   Seed-treatment preparations as well as pretreated seeds containing the proper selection and dosage of insecticide and fungicide are commercially available.   The following rates of application of insecticides to various seeds have been used in California:[3]

Seed	Lindane	Aldrin, heptachlor, dieldrin, or endrin
	Ounces 75% wettable per 100 lb. seed	
Alfalfa, clover..........................	0.66	
Beans, broccoli, cabbage, turnip, oats, rye, sorghum, soybean.................	0.66	0.66
Lima bean, wheat.......................	0.33	0.66
Carrot, melons, cucumber, squash, corn, grasses, lettuce, peas...................	1.33	0.66
Barley..................................	1.33	1.00–2.00
Pepper.................................	1.33	1.33
Onion, spinach.........................	1.33	2.00
Tomato, cotton........................	2.66	2.00
Beets, chard...........................	5.33	5.33

References.   U.S.D.A., Bur. Ento. Bul. 123, 1914, and *Farmers' Buls.* 725 and 733, 1916, and 1866, 1959, and *Dept. Bul.* 156, 1915; *Can. Dept. Agr. Pamphlet* 33 (n.s.), 1923; *N.Y. (Cornell) Agr. Exp. Sta. Buls.* 33, 1891, and 107, 1896; *Can. Dept. Agr. Leaflet* 35, 1935; *Jour. Econ. Ento.,* **17**:562, 1924; **19**:636, 1926; **38**:643, 1945; **40**:724, 727, 1947; and **49**:111, 1956.

87 (14, 39, 216, 273). WHITE GRUBS[4]

Like wireworms, white grubs develop from eggs laid chiefly in sod land.   When such soil is broken and planted to vegetables, the grubs

[1] LILLY, J., *Ann. Rev. Ento.,* **1**:203–219, 1956.

[2] LEUKEL, R., *U.S.D.A., Yearbook,* 1953: 134–45.

[3] REYNOLDS, H., *Advances in Pest Control Res.,* **2**:135–182, 1958; LANG, W. H., *Ann. Rev. Ento.,* **4**:363–388, 1959.

[4] *Phyllophaga* spp., Order Coleoptera, Family Scarabaeidae.

seriously injure the latter and remain in the same fields 2 to 3 years before they are full-grown.   If gardens are allowed to grow up to grass and weeds, the parent May beetles may lay their eggs directly in the garden.   All kinds of roots may be eaten, and tubers and fleshy roots like potatoes and beets have the surface scarred and gouged with broad shallow cavities $\frac{1}{4}$ to $\frac{1}{2}$ inch deep.

Control Measures.   Vegetables should not be planted in soil that has been covered with a grass sod for years, nor in fields where white grubs are abundant.   Legumes may be grown in such soil for a year or two until the grubs have disappeared.   In general, garden crops should not be planted in soil which was covered with grasses or small grains during a spring in which a heavy flight of May beetles occurred, since such soil is nearly certain to be infested with eggs and grubs for the next 3 years. Clover sod, or fields bearing clean-cultivated row crops during beetle flights, should be given preference for trucking during the years that immediately follow.   Gardens that are surrounded by such trees as poplars, oaks, lindens, willows, ash, maple, and walnut are usually infested with white grubs every year, because the May beetles are attracted to such trees to feed and lay their eggs in the soil near by.

The distribution, life cycle, additional control measures, and references for white grubs are given under Corn Insects, page 503.   See also control for Wireworms, page 618.

88 (280).   MILLIPEDES OR THOUSAND-LEGGED-WORMS[1]

Importance and Type of Injury.   Although not insects, these pests, which are often wrongly called wireworms, injure plants in much the same way as wireworms and white grubs.   They eat the roots of various plants, tunnel into the root vegetables and tubers, eat planted seeds, and also devour leaves and bore into fruits that lie in contact with the soil.

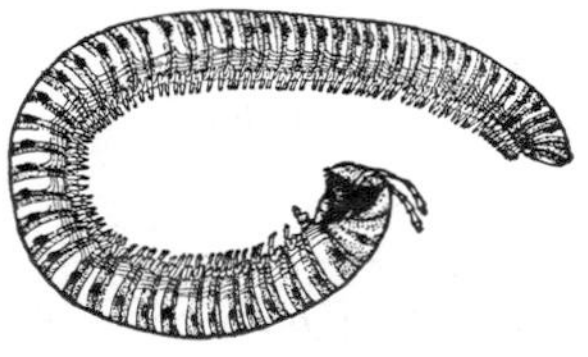

FIG. 14.13. A typical milli-pede, *Julus impressus*, en-larged. (*From Ill. Natural History Surv.*)

Plants Attacked.   Corn, potatoes, parsnips, carrots, beets, turnips, radishes, lettuce, cabbage, cauliflower, beans, peas, tomatoes, muskmelons, cucumbers, squash, and others.

Distribution.   Although these pests are widely distributed, chief complaints of injury in the field have been from New York, Ohio, and neighboring states and the Pacific Coast states.

Life History, Appearance, and Habits.   The life cycles of millipedes are not well known.   They have a simple development.   The eggs are laid in the soil, sometimes in masses, and hatch into small worms differing from the full-grown ones in having fewer segments and legs.   The adults (Fig. 14.13) are 1 to 2 inches long and are at once distinguished from insect larvae by the large number of legs that they possess; two pairs to each apparent body ring, the total number often being more than 100.

Control Measures.   Millipedes are readily controlled by soil treatments, as described for wireworms (page 618).   The other control measures suggested for wireworms may also be effective.

Reference.   Calif. Agr. Exp. Sta. Bul. 713, 1949.

B. INSECTS INJURIOUS TO PEAS AND BEANS

FIELD KEY FOR THE DETERMINATION OF INSECTS INJURING PEAS AND BEANS

A. *Insects eating holes in the leaves:*
    1. Irregular holes, usually extending from the margin of leaf inward, eaten by robust, brown or grayish, elongate wedge-shaped insects up to $1\frac{1}{2}$ inches long,

---

[1] *Julus impressus, J. hortensis* Wood, *J. hesperus* Cham., Class Diplopoda.

with short or long wings and the third pair of legs very long.   Jump vigorously
when disturbed..................................*Grasshoppers*, page 465.

2. A coppery-brown oval beetle, $\frac{1}{4}$ inch long, with 8 small black spots on each
wing cover, eats irregular holes from underside of leaves; or oval, yellow, very
spiny larvae up to $\frac{1}{3}$ inch long eat rectangular holes from underside of leaves,
the holes separated by slender parallel strips of leaf that are untouched.   Both
the adult and the larva leave the transparent upper epidermis uneaten.......
....*Mexican bean beetle*, page 622.

3. Reddish to yellowish beetles, $\frac{1}{6}$ inch long, usually with three black spots in a
row along inner edge of each wing cover and costal margin also edged with black,
eat rounded holes through the leaf and chew at the stems..................
....*Bean leaf beetle*, page 624.

4. A greenish or yellowish beetle $\frac{1}{4}$ inch long, with 6 black spots on each wing
cover, eats irregular holes in the leaves; especially around their margins......
....................................*Spotted cucumber beetle*, page 625.

5. A uniformly tan-colored beetle, about $\frac{1}{6}$ inch long, the wing covers with regular
rows of small punctures, eats irregular holes in the leaves..................
....*Grape colaspis*, page 512.

6. Cinnamon-brown velvety-looking beetles, about $\frac{3}{8}$ inch long, eat irregular
holes in the leaves, especially at the margin.   Much attracted to lights.   Found
in the eastern states........................*Asiatic garden beetle*, page 606.

7. Very hairy, yellowish, brown or reddish-brown, and black-banded caterpillars
eat the foliage....................................*Woollybears*, page 611.

8. Very large holes or entire leaves eaten at night by fleshy grayish or brownish,
slimy worm-like creatures, without legs or segmentation, up to 3 inches long.
Hide under trash about base of plants during the day......................
....................................*Garden slugs* (not insects), page 885.

B. *Insects sucking sap from leaves and stems:*
1. Foliage of beans, peas, and many other crops is silvered, bleached, wilted,
and covered with tiny black dots of excrement by very slender, reddish-yellow,
active nymphs and grayish-black adult bugs, only $\frac{1}{25}$ inch long, with 2 white
bars across the bristly front wings.......................................
Beans thrips, Hercothrips fasciatus (Pergande) (see *Calif. Agr. Exp. Sta. Bul.* 609,
1937).

2. Small, greenish, long-legged, winged or wingless aphids up to $\frac{1}{6}$ inch long with
2 slender tubes projecting from body near tip of abdomen, cluster on underside
of leaves and along stems of peas and suck sap, causing plants to wither and
bronzy-looking patches to appear in the field...........*Pea aphid*, page 626.

3. Similar to the above on terminal growth of beans, but not over half so large,
and nearly black in color.   Plants often become covered with a black "soot."
Bean aphid, Aphis fabae Scop. (see *Ann. Appl. Biol.*, **8**:51; **9**:135; and **10**:35).

4. Shield-shaped bright-green bad-smelling flattened bugs, the winged adults
about $\frac{5}{8}$ inch long, suck sap from pods of beans, causing pods to shed and
the seeds to be distorted...
Green stink bug, Acrosternum hilare (Say) (see *Va. Agr. Exp. Sta. Bul.* 294, 1934).

5. Similar to *B,3*, but $\frac{1}{10}$ inch long and lacking the tubes on abdomen.   Jumps
readily when disturbed.   Wings thick, at least at base....................
....*Garden fleahopper*, page 615.

6. The foliage of beans is dwarfed, crinkled, and curled, and rosettes form, or
small triangular brown areas appear at tips of leaves, gradually spreading
around entire margin.   Few pods and few beans per pod are produced.   One
or more winged or wingless, elongate, wedge-shaped green bugs up to $\frac{1}{8}$ inch
long by $\frac{1}{4}$ as broad, found on lower side of leaves; run or jump quickly when
disturbed.....................................*Potato leafhopper*, page 643.

7. Leaves become blotched with red and yellow, and die.   Underside of leaf with
inconspicuous white webs among which are many microscopic 8-legged reddish
or greenish mites and their round pearly eggs. . . . . . *Spider mites*, page 616.

C. *Insects attacking the seeds within the growing pods in the field or living inside of the
seeds in the field or in storage:*

1. Interior of peas in the field and in storage devoured by short, white, footless grubs and chunky, brownish, white-flecked beetles, ⅕ inch long, with 2 small black spots at tip of abdomen, beyond wing covers, and a single sharp tooth near apex of hind femur. Only 1 insect to a seed............*Pea weevil*, page 935.
2. Interior of beans and peas eaten out in the field and in storage by grubs similar to *C*,1, and by smaller brownish beetles, ⅛ inch long, with 1 large and 2 small teeth near apex of hind femur. Often several insects in 1 seed..............
   ....*Bean weevil*, page 936.
3. Interior of cowpeas eaten out as in *C*,1 and 2, by a more cylindrical, larger-headed grub, which is found in seeds in the field only. A black humpbacked beetle, nearly ¼ inch long, with a slender snout ⅓ as long as the body, and prominent round punctures on the back, makes punctures in the pods and lays eggs in the seeds...
   Cowpea curculio, Chalcodermus aeneus Boheman (see *Fla. Agr. Exp. Sta. Bul.* 232, 1931).
4. Peas within the pod partly eaten, showing irregular superficial cavities, web-covered, and surrounded by granular pellets of excrement, among which is a white caterpillar, up to ½ inch long....................*Pea moth*, page 627.
5. Seeds of peas, beans, vetch, and locust trees attacked in the green pods by a caterpillar up to 1 inch in length, and varying from white or greenish to reddish in color...
   Lima-bean pod borer, Etiella zinckenella (Treit.) (see *U.S.D.A., Bur. Ento. Bul.* 95, Part VI, p. 82, 1912).

D. Insects cut off plants near surface of ground or feed on underground parts:
1. Plants chewed just below surface of soil, or roots eaten off by a slender white worm, up to ⅓ inch long, with 6 short legs, and dark head and tip of body. Plants wilt and die......................Larvae of *bean leaf beetle*, page 624.
2. Plants chewed off near the surface of the ground and sometimes dragged to burrows in the soil. Dull-green, brown, gray, or black, cylindrical, greasy-looking worms, variously striped or spotted, with distinct head, 6 short slender legs on the thorax, and 5 pairs of prolegs, up to 1½ to 2 inches long, are found in the soil about plants during the daytime..............*Cutworms*, page 610.
3. Elongate, brownish, cylindrical or somewhat flattened worms, with many legs, up to 1 or 1½ inches long, eat the roots and the foliage that rests on the soil. Distinguished from wireworms by having 1 or 2 pairs of legs on each body segment..........*Millipedes* or *thousand-legged-worms* (not insects), page 620.

E. Plants fail to come up from seed:
1. Seeds have been eaten into. Slender tough-skinned brownish polished-looking worms about 1 inch long, with chewing mouth parts and 6 short legs, are found in seeds or in soil near by...........................*Wireworms*, page 617.
2. The germ of the softened kernel has been eaten out by one or more yellowish-white footless maggots, pointed at the head end; ¼ inch long...............
   ....*Seed corn maggot*, page 518.

89. MEXICAN BEAN BEETLE[1]

Importance and Type of Injury. Where it occurs, this insect is a serious enemy of all kinds of snap beans and Lima beans. It varies greatly in the same area according to weather and other factors affecting it. It may cause heavy damage one season and be very difficult to find the next year. Both larvae and adults feed on the leaves, usually on the undersurface, leaving the upper surface more or less intact except as it breaks through upon drying out. The larvae eat out somewhat regular areas, leaving slender parallel strips of untouched leaf between them, giving the plants a characteristic lace-like skeletonized appearance. When abundant, the insects also attack the pods and stems, and the plants

[1] *Epilachna varivestis* Mulsant (= *E. corrupta* Mulsant), Order Coleoptera, Family Coccinellidae.

may be shredded and dried out so that they die within a month after the attack begins, often before any crop is matured.

Distribution.   For three-quarters of a century the Mexican bean beetle, also known as the bean lady beetle, was a more or less serious pest only in the western part of the country, from Colorado southward.   In 1920 it was discovered in northern Alabama, having been shipped there, it is believed, in alfalfa hay a year or two earlier.   It is now present throughout the United States except in the Pacific Coast states.

Plants Attacked.   All kinds of garden beans and cowpeas, soybeans, and beggar-tick.   It has been most injurious to garden and field beans and occasionally to soybeans and cowpeas.   In cases of extreme infestation, alfalfa, clovers, vetch, and some grasses and weeds are fed upon.

Fig. 14.14. Mexican bean beetle, showing eggs, larvae, pupa (above), and adults (right) on bean leaves.   About natural size.   (*From Tenn. State Board of Entomology.*)

Life History, Appearance, and Habits.   This insect passes the winter only in the adult stage, occurring on the ground among leaves and other rubbish, sometimes in groups, especially in woodland and hedgerows, near the fields where beans were grown.   The beetles (Fig. 14.14) are from ¼ to ⅓ inch long, very convex, short ovate in outline, and from yellow to coppery-brown in color.   Each wing cover has eight small black spots that form three rows across the body when the wings are at rest. The general appearance is similar to that of our common beneficial lady beetles (page 66), but it is larger than most of them.   Some of the beetles appear in gardens and fields of beans about the time the earliest garden beans are coming up from the seed, while others continue to straggle out of winter quarters for nearly 2 months.   After feeding a week or two on the beans, the adults deposit eggs on the underside of the leaves.   These are about 1⁄20 inch long, orange-yellow in color, and fastened on end in close groups of 40 or 50 or more.   The eggs hatch in 5 to 14 days, depending on the temperature, and the larvae feed as already described from 2 to 5 weeks.   When full-grown, they are ⅓ inch or more in length by half as broad, are oval, are yellow, and have their backs protected by six rows of long branching black-tipped spines (Fig. 14.14). When growth is completed, the larvae cement the hinder part of their bodies to the underside of uninjured leaves of beans or other plants, often gathering in groups.   The pupa (Fig. 14.14) pushes out of the larval skin crowding it back to the tip of the abdomen, which remains covered with this spiny wrinkled skin.   The exposed part of the pupa is nearly bare, smooth, orange-yellow, and rounded in front.   The adult emerges in about 10 days and may lay eggs for the second generation within 2 weeks more.   From egg to adult occupies a month, on the average, and in the southeastern states there are three or even four partial generations. In the western and northern states there is one generation, with a partial second.   Most injury occurs in July and August.   Dispersal takes place chiefly in late summer by the flight of the adults and during the autumn months, when the beetles gradually leave the plants and seek hibernating places.

Control Measures.   Control may be obtained by thoroughly spraying or dusting the foliage of beans *so as to cover the underside of the leaves* with rotenone at 0.2 to 0.3 pound, parathion at 0.5 pound, Sevin at 0.5 to 1 pound, or malathion or methoxychlor at 1.5 pounds per acre.   The application should be made as soon as the beetles or their eggs are found on the plants.

Where the beetles are abundant, it is suggested that bush beans be grown instead of pole beans, that no larger acreage be planted than can be properly sprayed or dusted, that pods be picked promptly, and that the remnants of crops of green beans be plowed under as soon as the crop is harvested.   Since damage occurs mostly during July and August, very early and late summer, quick-maturing varieties of green beans may be grown with less injury.

References.   U.S.D.A., Farmers' Bul. 1624, 1960, and *Dept. Bul.* 1243, 1924; *Jour. Econ. Ento.,* **21**:178, 1928; **38**:101, 1945; and **40**:103, 1947.

90. BEAN LEAF BEETLE[1]

Importance and Type of Injury.   Injury by the bean leaf beetle is twofold: the reddish to yellowish, dark-spotted adult beetles feed on the underside of the leaves, eating rounded holes in them and upon the stems of seedling plants at or below the soil level.   The slender white larvae chew the roots and nodules, and feed on the stem just below the surface of the soil, more or less completely girdling the plant.   This insect, which is said to disseminate cowpea mosaic, may cause a loss of from 10 to 50 per cent of the crop.

Plants Attacked.   Beans, peas, cowpeas, soybeans, corn, beggar-tick, bush clover, tick trefoil, quail pea, hog peanut, and other weeds.

Distribution.   Abundant in the southeastern states and ranging north to Kansas, Minnesota, Canada, and New York, and westward to Texas and New Mexico.

Life History, Appearance, and Habits.   Winter is passed in the adult stage, in or near bean fields of the preceding season, and the beetles are ready to attack the plants as soon as they appear aboveground.   The adults (Fig. 14.15,*a*) vary much in color and markings, but are typically reddish to yellowish in color, about ⅛ to ¼ inch long, with three or four black spots in a row along the inner edge of each wing cover and a black band all around near the outer margin of the wing covers.   They are found on the underside of the leaves and when disturbed generally drop.   The females descend to the ground to lay their eggs in the soil about the bases of the plants.   The lemon-shaped orange-colored eggs are found in small clusters of a dozen or two and each female may lay 40 or more such clusters during a period of about a month.   The eggs hatch in from 1 to 3 weeks, depending upon the season, and the slender white larvae find their way to the base of the stem or roots and feed, as described above, for 3 to 6 weeks.   The larvae (Fig. 14.15,*c*) are whitish, dark brown at both ends, con-

[1] *Cerotoma trifurcata* (Förster), Order Coleoptera, Family Chrysomelidae.

spicuously segmented, and have six very small legs near the head.   When full-grown, the larva forms an earthen cell within which the white soft-bodied pupal stage (*b*) is completed in about a week.   In the North these adults constitute the overwintering population.   In the South there are one or two partial generations in addition.

Control Measures.   When this insect has been destructive, applications of rotenone at 0.25 pound or DDT at 1 pound per acre should be made to the vines as soon as the beetles appear on them, to control the females before they have laid their eggs. Special care should be taken to cover the undersides of the leaves.   The destruction of the food plants, by plowing so as to expose the roots as soon as possible after harvest,

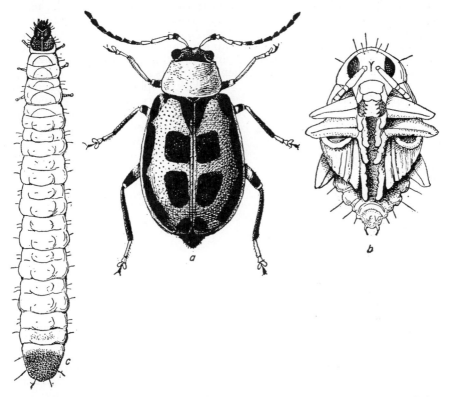

FIG. 14.15. Bean leaf beetle.   *a*, Adult; *b*, pupa; *c*, larva.   About 8 times natural size. (*From U.S.D.A., Farmers' Bul.* 856.)

and of the wild host plants, such as beggar-tick, hog peanut, tick trefoil, and bush clover, will tend to prevent abundance of the pests.   Timing the planting of the crops so that they germinate between the times of appearance of the generations of adults is recommended.

References.   U.S.D.A., Bur. Ento. Bul. 9, 1897; Jour. Econ. Ento., **8**:261, 1915; S.C. Agr. Exp. Sta. Bul., 265, 1930; Ark. Agr. Exp. Sta. Bul. 248, 1930.

91 (17, 95). SPOTTED CUCUMBER BEETLE

This general pest often eats large irregular holes in the foliage of beans and cuts off the growing tips.   When the leaves are parted, the yellowish-green, black-spotted adults (Fig. 9.34) drop or fly away.   The stems of the plants are often girdled by the feeding of many beetles at or near the surface of the ground so that the plant gradually dies.   The insect is further discussed under Corn Insects, as the southern corn rootworm (page 510), and under Cucurbit Crops (page 631). A thorough application of the dusts or sprays suggested for the Mexican bean beetle or bean leaf beetle will control this insect on beans.

92 (50). PEA APHID[1]

Importance and Type of Injury. This insect is not of special importance in small gardens, but it is most feared of all the pea insects by the large growers and canners. Infested peas wilt and bronzy patches appear in the fields. Upon brushing or shaking of the plants, small green insects, the largest of them only ⅙ inch long, fall to the ground. If the plants are examined, the terminal shoot and leaves and the stems may be found crowded or lined with myriads of these green, long-legged, either winged or wingless aphids, which suck the sap and possibly also poison the plant. When the aphids are abundant, their shed skins give the plants and the ground a whitish appearance. When the insects are not abundant enough to kill the plants or even to cut the yield greatly, the quality of the peas is often affected. The pea aphid is also an efficient vector of pea enation mosaic virus, which infects peas, vetch, sweet clovers, and other legumes, and of the yellow bean mosaic of peas and alfalfa.

FIG. 14.16. Winged, ovoviviparous female adult of the pea aphid. (*From Ill. Natural History Surv.*)

Distribution. Throughout the pea- and alfalfa-growing sections of the United States and Canada.

Plants Attacked. Garden and field peas, sweet peas, clover, sweet clover, alfalfa, and weeds of the legume family.

Life History, Appearance, and Habits. This aphid winters on alfalfa, clovers, especially red and crimson clover, and other perennial plants. It may go through the winter either in the egg stage or as ovoviviparous females in the northern states and, farther south, may continue to breed all winter on these and other plants. In the spring it increases on the winter-host plants and in the latitude of central Illinois winged migrants begin to spread to other plants, including peas, about May 1. These winged females (Fig. 14.16) start colonies on new plants by giving birth to active young nymphs which molt four times, reach the adult stage, and begin reproducing in about 12 days. Each female commonly produces 6 or 7 young a day until from 50 to 100 or more have been born. There are from 7 to 20 or more such generations of females in the course of a year. When unchecked by natural enemies their increase is phenomenal, but they are much subject to fluctuations due to weather conditions, the attacks of many predators and parasites, and fungus diseases. They usually become more abundant and injurious to peas during June, in the latitude of Illinois. Most of the adults are wingless, but when they become crowded on the plant some winged ones appear, and these spread the insect widely. During midsummer they are found chiefly on other leguminous plants but in early fall again become abundant on peas wherever these are available.

[1] *Macrosiphum* (= *Illinoia*) *pisi* (Harris), Order Homoptera, Family Aphidae.

As the amount of daylight diminishes in the fall and the temperature drops, the ovoviviparous females give birth to young, some of which become sexually mature, egg-laying, wingless females and others winged males. The eggs are laid chiefly on the leaves and stems of alfalfa and red clover and are about $\frac{1}{30}$ inch long, light green in color when newly laid but turning a shiny black. These fertilized eggs live over winter and give rise in the following spring to the ovoviviparous stem-mothers which start the next season's infestations.

Control Measures. The pea aphid may be controlled on peas by spraying or dusting with parathion or TEPP at 0.4 pound, Phosdrin at 0.25 pound, or malathion at 1 to 1.25 pounds per acre (see page 561 for control on alfalfa or clover).

References. U.S.D.A., Dept. Bul. 276, 1915, and *Farmers' Bul.* 1945, 1956; *Md. Agr. Exp. Sta. Bul.* 261, 1924; *Maine Agr. Exp. Sta. Bul.* 190, 1927; *Utah Agr. Exp. Sta. Ext. Cir.* 221, 1955; *Jour. Econ. Ento.*, **40**:101, 190, 199, 1947; **49**:878, 1956; **50**:770, 1957; **52**:541, 758, 1959; and **53**:881, 1960.

93. PEA MOTH[1]

Importance and Type of Injury. This insect is a serious pest in the more northern areas of canning peas, dried peas, and seed peas industries. Growing peas within the pod have irregular cavities eaten out of the side, seldom exceeding half of their substance, by a yellowish-white caterpillar up to $\frac{1}{2}$ inch long with both extremities of the body darker, and small dark spots and pale short hairs scattered over the body. The presence of the worms is not easily detected except by opening the pods, when the seeds are seen to be spoiled and the pod partially filled with the pellets of excrement and silk of the caterpillars (Fig. 14.17). Affected pods yellow, or ripen prematurely.

Plants Attacked. This insect attacks all varieties of field and garden peas, sweet peas, and vetch.

Distribution. The insect has been present in America only since about 1900, and has been recorded from New York, eastern Canada, Michigan, Wisconsin, Washington, and British Columbia.

Life History, Appearance, and Habits. Winter is passed as inactive larvae enclosed by strong cocoons of fine silk, about $\frac{3}{8}$ inch long and covered with soil particles. These are found a short distance below the surface of the soil in the fields. Others winter in similar cocoons in cracks and crevices about the barns where the peas were stored before threshing. They change to brownish pupae in late spring.

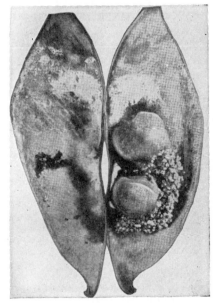

FIG. 14.17. Work of the pea moth. (*From Wis. Agr. Exp. Sta. Bul.* 310.)

About the time peas come into blossom (latter half of July in northern Wisconsin) the adults appear from the cocoons. They are frail, very small, day-active moths, about $\frac{1}{2}$ inch across the spread wings, of a general brown color with short black-and-white oblique lines along the front margin of the forewings. They may be found zigzagging about the plants in late afternoon, mating and laying eggs, but not themselves

[1] *Laspeyresia* (= *Grapholitha*) *nigricana* (Stephens), Order Lepidoptera, Family Olethreutidae.

injuring the peas.   The eggs are white, flattened, and a little smaller than an ordinary pinhead.   They are deposited one in a place upon the pods, leaves, flowers, or stems of the peas, and on other plants growing in the pea fields.   Promptly upon hatching, the minute caterpillars drill into the pods, casting out a little frass, which however soon blows or wears away, leaving almost no indication of their entrance.   The larvae devour parts of several seeds, which are further contaminated with their excrement and silk. Within 2 to 4 weeks they are full-grown when they eat out of the pods and enter the soil or other protected places mostly during August.   Here they form cocoons in which some of the larvae may transform to moths and emerge 2 to 12 weeks later; but the majority of them remain in the cocoons as larvae during the next 10 or 11 months.

Control Measures.   Seed and dried peas or vetch should not be grown in the same areas where canning peas are an important industry.   Other control measures include threshing promptly within a day or two of harvesting, so that larvae still in the pods might be killed by the thresher; burning all remnants of the crop immediately; thorough disking of the soil after harvesting; deep plowing in the fall to destroy the larvae in their winter nests; and the use of early varieties, maturing before mid-July, in Wisconsin.

References.   *Wis. Agr. Exp. Sta. Bul.* 310, 1920; *Proc. Ento. Soc. Nova Scotia for 1919*, p. 11, 1920; *Mich. Agr. Exp. Sta. Rept.* 44, p. 266, 1931; *Wash. Agr. Exp. Sta. Bul.* 327, 1936.

C. INSECTS INJURIOUS TO CUCURBITS

FIELD KEY FOR THE DETERMINATION OF INSECTS INJURING CUCUMBER, SQUASH, MELONS, AND OTHER CUCURBITS

A. *Insects eating holes in the leaves or chewing at the stems:*
  1. Small beetles, about ⅕ inch long, with three black stripes down the back, separated by wider stripes of bright-yellow, eat irregular holes in the leaves and feed especially about the base of the stem, often girdling the plant at or near the surface of the ground.................*Striped cucumber beetle,* page 629.
  2. Beetles of the same general shape but slightly larger than A,1, green or greenish-yellow in color, and with 6 conspicuous black spots on each wing cover, attack especially the leaves and flowers of the plants.............................
  ....*Spotted cucumber beetle,* page 631.
  3. Leaves have small rounded holes eaten into them so that they look as though peppered with fine shot.   Very small black beetles about 1/16 inch long, or larger ones ⅛ inch long, with a broad yellow longitudinal stripe on each wing cover, are found mostly on underside of leaves, jumping vigorously when disturbed..*Flea beetles,* page 603.
  4. Stems of young plants are chewed or cut off near or below the ground, by dull-greenish, brownish, or grayish, cylindrical, "greasy-looking" worms, variously marked with spots, stripes, or oblique bands; the largest ones 1½ inches long; found in the soil during the day; often coil up when disturbed..............
  ....*Cutworms,* page 610.
B. *Insects sucking sap from the leaves or stems, causing wilting, curling, and dying of leaves:*
  1. Elongate, flattened, oval, blackish-brown bugs, ⅔ inch long, and powdery-white black-legged nymphs, from 3/16 to ½ inch long, hiding under wilted leaves, clods, etc., or shying about the vines, suck the sap and poison the plants. Have a stinking odor when crushed..................*Squash bug,* page 632.
  2. Developing leaves at tips of vines are curled, wilted, and shriveled, the undersides thickly dotted with small, winged or wingless, tan, brown, green, or black aphids of various sizes, which suck out the sap.......*Melon aphid,* page 631.
C. *Insects boring through the center of the vines causing extensive wilting and rotting of the stem:*
  1. Yellowish sawdust-like excrement accumulates in small masses from holes in the vines.   Within the vine are found whitish grub-like caterpillars up to 1 inch long and ¼ inch in diameter, with brown heads and short legs.........
  ....*Squash vine borer,* page 635.

    2. Injury similar to *C*,1 especially late in the season, caused by a smaller and slenderer green worm, not over ¾ inch long, the smaller ones with a row of black spots across each segment................*Pickleworm*, page 637.

D. Insects boring into the fruits:

    1. Greenish to yellowish, brown-headed caterpillars, up to ¾ inch long, those under ½ inch with conspicuous cross rows of blackish spots, bore into fruits of muskmelon, cucumber, and squash, pushing out a small mass of excrement behind them and causing the fruits to rot and sour.....*Pickleworm*, page 637.

    2. Injury very similar to *D*,1 is caused by a larger, more slender worm, ranging up to 1¼ inches long, greenish-yellow in color, with light-brown head and 2 long, whitish well-separated stripes down the back. They wriggle very actively when disturbed.................................*Melonworm*, page 638.

    3. Plain green or greenish black-spotted, or yellow and black-striped beetles, about ⅕ inch long, gnaw into the rind of squashes, pumpkins, melons, and related fruits in the fall of the year, often working together in great numbers. They disfigure the fruits but seldom penetrate deeply enough to cause rotting or fermentation...
Striped and spotted cucumber beetles and southern corn rootworm, pages 629 and 510.

E. Insects attacking the roots and boring in the underground stem, or devouring the planted seeds:

    1. Very slender whitish worms, the largest ⅓ inch long, with the two extremities brown, with 6 short legs behind the head and a pair of blunt prolegs on the last segment, devour the smaller roots and tunnel through the underground stems and larger roots................*Larvae of striped cucumber beetle*, page 629.

    2. Thick white soft-bodied grubs, up to an inch or more long, with reddish-brown heads, 6 slender legs, and body somewhat slenderer near the middle, chew the roots. Lie curled in a "U" or semicircle.............*White grubs*, page 619.

    3. Slender, tough, smooth, brown worms up to 1 inch long, with chewing jaws and 6 small legs, eat into seed before or as it germinates and later chew at the roots and underground parts of the stem..............*Wireworms*, page 617.

94. Striped Cucumber Beetle[1]

Importance and Type of Injury. This is the most serious pest of cucurbits throughout America east of the Rocky Mountains. On the Pacific Coast, a distinct but very similar species, the western striped cucumber beetle,[2] has similar habits, life history, and control. The beetles feed on the plants from the moment they appear aboveground in the spring until the last remnants of the crop are removed or destroyed by frost. They work down to meet the germinating plants before they reach the surface of the soil. They chew the leaves and tender shoots, and especially the stem near or below the surface, partially or completely girdling it. They feed in the blossoms and, in autumn especially, gnaw holes in the rind of the fruits. They are known carriers of the bacterial wilt of cucurbits, the bacillus[3] causing this disease living over winter in the intestines of this and the related spotted cucumber beetle. In the spring the beetles inoculate the disease into the interior tissues of the new plants as they feed, and spread it from plant to plant and field to field wherever infective beetles go and feed. This insect is also one of the most important agencies in the spread of cucumber mosaic. Furthermore, the larvae of the beetle injure the vines during the summer by devouring the roots and tunneling through the underground parts of the stems. Many plants

---

[1] *Acalymma* (= *Diabrotica*) *vittata* (Fabricius), Order Coleoptera, Family Chrysomelidae.

[2] *Acalymma* (= *Diabrotica*) *trivittata* (Mannerheim).

[3] *Erwinia tracheiphila* (see *U.S.D.A., Dept. Bul.* 828, 1920).

are killed early in the season by this beetle and the wilt disease it spreads, and such vines as survive the first attack have the yield greatly reduced by the work of the adults and larvae.

Plants Attacked.   Cucumbers, muskmelons, winter squashes, pumpkins, gourds, summer squashes, and watermelons appear to be injured about in the order named.   So far as known the larvae can develop only on these and related cucurbits.   The beetles, however, also feed on beans, peas, and corn and the blossoms of many other cultivated and wild plants.   This pest is not troublesome in sandy soils.

Distribution.   This native insect ranges from Mexico into Canada, east of the Rocky Mountains.

Life History, Appearance, and Habits.   Only the unmated adults live through the winter.   They are ordinarily not found in old cucurbit

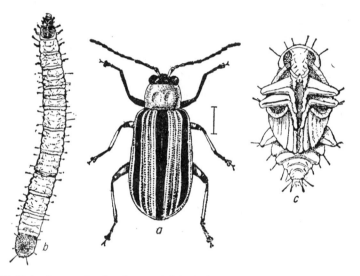

Fig. 14.18. Striped cucumber beetle.   *a*, Adult; *b*, larva; *c*, pupa.   About 7 times natural size.   (*From U.S.D.A.*)

patches, but in neighboring woodlands, under fallen leaves, strips of bark or rotten logs, or under protecting trash in lowlands, hedgerows, or weedy fence rows, practically always in direct contact with the soil, near the previous crop or about their wild food plants, such as goldenrod and asters.   The beetles emerge from hibernation from early April on, in central Illinois, becoming active at temperatures above 55°F., but not taking flight below 60°F.   Before cucurbits are available for them to feed upon, they subsist on the pollen, petals, and leaves of buckeye, willow, wild plum, hawthorn, thorn apple, apple, elm, syringa, and related plants for several weeks.   As soon as cucumber, squash, or melon vines appear above ground, they come to these plants, settling upon the young vines and devouring the seed leaves and the stems.   The beetles (Fig. 14.18,*a*) are familiar to nearly every grower.   They measure about ⅕ inch long by ⅒ inch wide.   The upper surface is about equally black and yellow, the folded wing covers forming three longitudinal black stripes. Mating and egg-laying take place soon after the beetles migrate to the vine crops.   The orange-yellow eggs are laid about the base of these plants, often below the surface of the soil or in cracks in the ground.   The larvae (*b*) that hatch from them work their way into the soil and feed for from 2 to 6 weeks on the roots and underground parts of the stem, frequently entirely destroying the root system.   When full-grown, they are about ⅓ inch long by one-tenth as broad.   The last segment of the abdomen is more flattened than in the related northern and southern corn rootworms.   The whitish pupal stage (*c*) is also found in the soil and lasts for about a week.   Adults from this first generation appear in midsummer over a period of 6 weeks or more and feed extensively on stems, leaves, and blossoms of cucurbits and also on legumes.   In the warmer latitudes these beetles soon mate and produce another generation during late summer and early fall, and the adults from these and even from partial third or fourth generations are the ones found in late fall on goldenrod, sunflower, and aster blossoms and on the fruits of cucurbits.   In the northern part of its range there is only one generation.

Control Measures.   The Diabrotica beetles are readily controlled by dusting or spraying with methoxychlor at 1 to 2 pounds, parathion at 0.25 to 0.5 pound, or malathion at 1 to 1.5 pounds per acre.   Rotenone at 0.1 to 0.2 pound per acre is suitable in the home garden.   Repeated applications may be necessary to control beetles flying in from surrounding areas.

Where the acreage is small enough to make it practicable, the plants may be covered, from the moment they appear through the ground, with wire- or cloth-screen protectors made in the form of cones or hemispheres. Such protectors keep the beetles off until the plants get a good start, when they must be removed.   It is well to plant an excess of seed, and thin out after the plants are started.

References.   Ohio Agr. Exp. Sta. Bul. 388, 1925; *U.S.D.A., Farmers' Bul.* 1322, 1923; *Wis. Agr. Exp. Sta. Bul.* 355, 1923.

95 (17, 91). SPOTTED CUCUMBER BEETLE[1]

This species, which is closely related to the striped cucumber beetle and northern corn rootworm, is a more general feeder than either one of them.   While the total damage done by it, because it also injures field crops, is probably greater than either of the others, its injury to cucurbits is much less noticeable.   This injury is similar to that of the striped cucumber beetle.   The larva of this insect is the well-known southern corn rootworm.   In addition, the adult is a constant pest of string and Lima beans, peas, potato, beet, asparagus, eggplant, tomato, cabbage, and many other garden plants, and the larvae develop on the roots of corn, beans, small grains, alfalfa, and many wild grasses.   A description of the insect is given under Southern Corn Rootworm, page 510.   On cucurbits it may be controlled by the same methods as suggested for the striped cucumber beetle.

96 (60). MELON APHID[2]

Importance and Type of Injury.   After the vines of cucurbits begin to "run," a single hill here and there will often be found to have the

---

[1] *Diabrotica undecimpunctata* (= *duodecimpunctata*) *howardi* Barber, Order Coleoptera, Family Chrysomelidae.   The western spotted cucumber beetle, *D. undecimpunctata* Mannerheim, replaces this species west of the Rocky Mountains.

[2] *Aphis gossypii* Glover, Order Homoptera, Family Aphidae.

edges of the leaves curled downward or some of them wilted, shriveled, and browning. The undersides of such leaves, inside the curl, are generally crowded with very small, yellow, green, and black aphids, some winged, others wingless, and of several sizes. These insects suck the sap from the leaves, weakening the plants and reducing both the quantity and quality of the fruit. In years of abundance they kill the plants and ruin the crops over extensive areas. In the North they are especially destructive in hot dry summers, following cool wet springs which have reduced the efficiency of their natural enemies. This has been considered the most destructive aphid occurring in this country.

Plants Attacked. This insect feeds on a wide variety of plants but is of economic importance chiefly on the cucurbits, cotton (on which it is a very serious pest), okra, and citrus fruits. It also feeds on strawberry, bean, beet, spinach, eggplant, asparagus, a number of ornamental plants, and many weeds, especially shepherd's-purse, peppergrass, pigweed, and dock. It also occurs on certain of these plants in the greenhouse.

Distribution. Throughout the United States, southward to Central and South America; most destructive in the South and Southwest.

Life History, Appearance, and Habits. In the North the insect winters on live-forever, in the egg stage, where this plant is found. These eggs are fertilized in the fall by males that developed on the same plant. In the extreme South, at least, the insect continues breeding ovoviviparously throughout the winter. It appears on cucurbits in late spring or early summer, and, if the weather is favorable and it is not checked by sprays or natural enemies, increases and spreads with astonishing rapidity. Fifty-one generations were recorded in 12 months in Texas, the average young per female being more than 80. Many winged individuals are produced, and wherever they fly or are blown by the wind a new colony is soon started.

Control Measures. Since the attack on cucurbits commonly begins in small scattered spots over the field, such spots should be watched for and the aphids destroyed upon them before they become generally established over the whole crop. Besides the wilting and curled leaves, attention may be attracted to the aphids early by the visits of ants, bees, wasps, and flies to the colonies to get the honeydew and by the white cast skins of the aphids sticking to the leaves. Every effort should be made to apply one of the control measures suggested on page 613, before the leaves on the terminal shoots curl too badly.

References. U.S.D.A., *Farmers' Bul.* 1499, 1926; *Tex. Agr. Exp. Sta. Bul.* 257, 1919; *Ill. Agr. Exp. Sta. Cir.*, 297, 1928; *U.S.D.A., Farmers' Bul.* 1282, 1922; *Maine Agr. Exp. Sta. Bul.* 326, 1925.

97. SQUASH BUG[1]

Importance and Type of Injury. There is no more vexatious pest of the garden than the squash bug. Leaves of plants attacked by it rapidly wilt as though the sap flow had been cut off or poisoned, and soon become blackened, crisp, and dead. Small plants may be killed entirely; larger plants usually show certain leaves or runners that are affected. In many localities it is practically impossible to grow certain varieties of squashes, as the plants are killed by these bugs before any fruits are matured. The bugs possess a remarkable vitality and tenacity of life, to the great

[1] *Anasa tristis* (De Geer), Order Hemiptera, Family Coreidae.

annoyance of the gardener.   Plants in farm or city gardens suffer most severely although large commercial plantings also are badly injured.

The attack of this pest may be identified by finding the brownish-black, flat-backed adult bugs, about ⅝ inch long (Fig. 14.20,*a*), and their numerous, whitish, black-legged nymphs (*c,d,e*), which range in size from 3⁄16 to ½ inch long.   They are usually more or less hidden about the base of the plant under the deadened leaves or under clods, and, when approached, they shy around the vines or walk rapidly to cover.

Plants Attacked.   All the cucurbits or vine crops are attacked, but the bugs show a marked preference for squashes and pumpkins.   Among

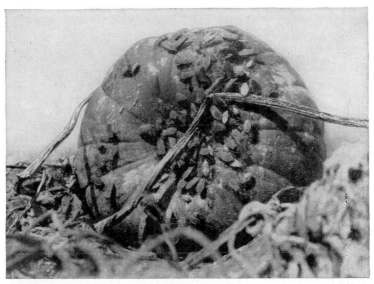

Fig. 14.19. Squash bug adults and nymphs massed upon cucurbit fruit in fall after foliage has been killed by frost.   About ⅓ natural size.   (*Original.*)

the squashes, the winter varieties, such as hubbards and marrows, suffer most severely.

Distribution.   Throughout the whole United States from Central America to Canada.

Life History, Appearance, and Habits.   Only the unmated adult bugs pass the winter.   They hibernate in all kinds of shelter under the protection of dead vines, leaves, clods, stones, piles of boards, outbuildings, and dwellings which they enter in the fall.   They are rather slow to appear in spring and feed only on cucurbits.   By the time vines begin to "run," the adults fly into the fields and gardens and apparently locate their food by the odor of the plants.   Mating occurs in the spring and egg-laying begins soon afterward.   The egg clusters (Fig. 14.20,*b*) will usually be found on the underside of leaves in the angle formed by the veins.   The eggs are yellowish-brown to very dark bronzy-brown according to their age, elliptical in outline, and about 1⁄16 inch long.   They are laid in groups commonly numbering a few dozen.   The eggs lie on their sides, sometimes close together, sometimes separated by more than their own diameter.   They are usually placed in rows in two directions, the rows meeting each other at an acute angle.   The overwintering adults live and continue laying eggs until about midsummer.

A week or two after the eggs are laid, the small nymphs hatch.   They are at first strikingly colored, the abdomen being green, the head, thorax, antennae, and legs crimson, soon darkening to reddish-brown.   The older nymphs are grayish-white in color, with nearly black legs and antennae. There are five nymphal instars.   When the insects are crushed, a disagreeable odor is given off from two oval spots on the middle of the upper side of the abdomen; or, in the adults, from near the base of the legs.   This gives rise to the name "stink bug," often wrongly applied to these insects.[1]

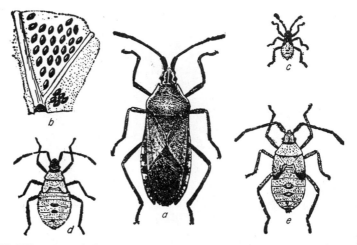

Fig. 14.20. Life stages of the squash bug.   a, Adult; b, eggs (about natural size); c–e, nymphs in different stages.   a, c–e twice natural size.   (From U.S.D.A.)

Nymphs and adults all feed by sucking the sap of the plant.   The youngest nymphs are gregarious, those from one egg cluster feeding close together.

From 1½ to 2 months after the eggs were laid, the new bugs begin to transform to adults, there being usually a period of scarcity of mature bugs following the disappearance of the parents, before the young have become adult.   The new adults do not mate or lay eggs until the following spring, at least in the North, and there is probably only one generation a year throughout the country.   These adults and the nymphs from later laid eggs are usually present in great numbers when the first frosts kill the leaves and vines.   The bugs then collect in dense groups upon the protected or sunny faces of unripe fruits, where they continue to suck the sap (Fig. 14.19).   They gradually fly and crawl away in search of winter shelter, and such nymphs as do not succeed in reaching the adult condition die during the early winter.

Control Measures.   The squash bug is one of the most difficult insects to control satisfactorily.   Spraying or dusting with parathion at 0.5 pound per acre is effective.   Spraying with 40 per cent nicotine sulfate at 1.5 pints per acre will kill the young nymphs but not the mature bugs. In the home garden, dusting with 20 per cent sabadilla at 20 to 40 pounds

---

[1] The true stink bugs are the shield-shaped bugs of the related family Pentatomidae.

per acre will give satisfactory control.   These treatments may have to be repeated at intervals.   One of the most effective controls is to collect the bugs by hand and to crush the egg masses as fast as they appear on the young plants in the spring.   Pieces of boards, shingles, and similar flat objects are often placed out among the plants.   The bugs collect under them at night, and if they are examined each morning, many may be destroyed.   In the fall great numbers of the bugs can be caught and killed on the fruits after the vines have died.   As soon as the crop is harvested, the vines should be removed from the field and burned or worked into a compost heap.   The margins of fields and the grounds surrounding the garden should be as free as possible of rubbish, piles of leaves, boards, and other shelter for the bugs during the winter.

References.   N.H. Agr. Exp. Sta. Bul. 89, 1902; *Thirty-fourth Rept. Conn. State Ento. Bul.* 368, p. 224, 1935; *Okla. Agr. Exp. Sta. Bul.* 34, p. 269, 1935; *Jour. Econ. Ento.*, **13**:416, 1919; **16**:73, 1923; **38**:389, 391, 661, 1945; **39**:255, 1946.

98. Squash Vine Borer[1]

Importance and Type of Injury.   This pernicious borer is in many localities a very destructive pest of squashes and pumpkins, sometimes destroying 25 per cent or more of the crop.   Attention will often first be called to it by the sudden wilting of a long runner or an entire plant. Examination of the wilted vine will reveal masses of coarse greenish-yellow excrement which the borer has pushed out of the vine through holes in the side.   If such a vine is split open, it will be found hollowed out and partly filled with moist slimy frass similar to that cast out on the ground, In the midst of this material is a thick, white, wrinkled, brown-headed caterpillar, the largest ones 1 inch long and almost $\frac{1}{4}$ inch thick (Fig. 14.21,*d*).   Plants injured to the point of wilting usually contain a number of borers.   Their excrement should be watched for, especially about the base of the plant close to the roots.   Infested vines are often completely girdled and usually become rotten and die beyond the point of attack.

Plants Attacked.   Squashes, such as hubbards, marrows, cymblings, and other late varieties, summer squash, pumpkins, gourds, cucumbers, and muskmelons are injured, about in the order named.   The wild cucumber also is infested.

Distribution.   East of the Rocky Mountains from Canada to South America.

Life History, Appearance, and Habits.   The insect winters an inch or two below the surface of the soil inside of a tough dirt-covered silk-lined black cocoon (Fig. 14,21,*e*) about $\frac{3}{4}$ inch long, in either the larval or pupal stage.   If as a larva, the change to the mahogany-brown pupa (*f*) occurs in the spring.   Two or three weeks later, about the time the vine crops are beginning to run, the pupa tears open the end of its cocoon and wriggles up through the soil to the surface.   Its skin then splits down the back and reveals a beautiful wasplike moth, 1 to $1\frac{1}{2}$ inches across the wings (Fig. 14.21,*a*,*b*).   The front wings are covered with metallic-shining olive-brown scales, but the hind wings are transparent.   The abdomen is ringed with red, black, and copper; and, like its relative the peach borer (page 757), the moth flies swiftly and noisily about the plants during the daytime, seeming more like a wasp than a moth.   Small, oval, somewhat flattened, brownish eggs, $\frac{1}{25}$ inch long, are glued one in a place on the

[1] *Melittia cucurbitae* (Harris), Order Lepidoptera, Family Aegeriidae.

stems and leaf stalks, especially toward the base of the plant.   The small
borers enter the stem a week or two later and tunnel along, eating out the
inner tissues for about a month.   They have a brownish head, six short
slender legs on the thorax, and five pairs of short prolegs.   Each proleg
bears two transverse rows of crochets.   Small borers may often be found
in the leaf stems, but most of them occur toward the base of the plant.
Later in the season they are often found throughout the stem, and even
in the fruits.   The larvae are full fed in 4 to 6 weeks.   They then desert
their burrows and make cocoons in the soil.   In the more northern states
the larvae remain in these cocoons until spring before pupating, there
being only one generation a year.   Farther south, at least a part of the

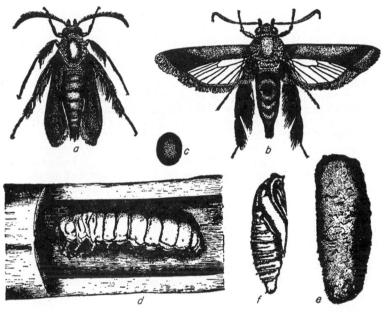

Fig. 14.21. The squash vine borer.   *a*, Male; *b*, female; *c*, egg; *d*, larva in stem; *e*, earth-
covered cocoon; *f*, pupa.   About twice natural size.   (*From Iowa Agr. Exp. Sta. Cir.* 90.)

larvae pupate promptly after leaving the plants, and some of these change
to adults, giving rise to a second generation of larvae during August and
September.   In the Gulf states, two generations are believed to be the
rule.

Control Measures.   Like most borers, this insect presents great difficulties in control, since no insecticides can reach the point of feeding.
The following dust or spray treatments applied once or twice a week to
the stems and vines at the base of the plant are effective in killing many
of the young insects before they bore into the plant: malathion at 1.75
pounds, parathion at 0.5 pound, or lindane at 0.2 to 0.3 pound per acre.
Rotenone at 0.2 to 0.4 pound per acre is effective in the home garden.
The best thing to do with infested plants is to slit the stems lengthwise
at the point of attack, crushing or removing the borers and immediately
covering the stems with moist earth.   Since girdling is most likely to
occur at the base of the plant, it is well to cover the vines with soil a few

feet from the base so that they may form supplementary roots and save the vine in case it is cut off at its base.   To reduce injury the following year, all vines should be raked together and burned as soon as the crop can be harvested.   The soil should be harrowed in the fall and turned under deeply in the spring to prevent the emergence of adults from the cocoons.

References.   U.S.D.A., Farmers' Bul. 668, 1915; *Mass. Agr. Exp. Sta. Bul.* 218, pp. 70–80, 1923; *Conn. Agr. Exp. Sta. Bul.* 328, 1931; *Jour. Econ. Ento.,* **28**:229, 1935; **40**:716, 1947; and **41**:352, 1948.

99. PICKLEWORM[1]

Importance and Type of Injury.   Throughout the Gulf states, ripening fruits of cantaloupes, cucumbers, and squashes are bored into, from the side next the ground, by a whitish to greenish caterpillar, up to ¾ inch long, brownish at the head end and the smaller ones with a transverse row of black spots on each segment.   The larvae push out small masses of green sawdust-like excrement from their holes in the fruit. The fruits soon rot, sour, and mold after the interior has been exposed to the air by these burrows.   Earlier in the season larvae work in the stems, terminal buds, and especially in the blossoms of squash. Late maturing crops are frequently almost totally destroyed by these worms.

Plants Attacked.   Cantaloupe, cucumber, and squash are seriously injured; watermelon rarely; and pumpkin not at all.

Distribution.   Especially destructive in the southern states but occasionally so as far north as Missouri, Illinois, Michigan, and New York.   The insect is found from Canada to South America and as far west as Texas, Kansas, and Nebraska.

Life History, Appearance, and Habits. The insect hibernates in the pupal stage

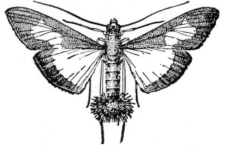

FIG. 14.22. Adult of the pickleworm.   Twice natural size.   (*From Crosby and Leonard, "Manual of Vegetable-garden Insects,"* copyright, 1918, Macmillan.   Reprinted by permission.)

surrounded by a thin cocoon of silk and usually in a roll of leaf from the food plant. The cocoons generally lie on the ground in or near the old food plants but are sometimes suspended on weeds and other plants near by.   These overwinter only in semitropical areas such as southern Florida, and the adults spread northward each spring. They are striking-looking moths a little more than 1 inch wide across the spread wings (Fig. 14.22).   The two pairs of wings are margined with a band of yellowish-brown, about ⅛ inch wide, and the body above is of the same color, with purplish reflections in certain lights; while a median spot on the front wings and the basal two-thirds of the hind wings are transparent yellowish-white.   The tip of the abdomen has a prominent rounded brush of long hair-like scales.   The moths fly late at night and eggs are deposited in clusters of two to seven on the tender buds, new leaves, underside of fruits, or stems.   The larvae work at first in the buds, blossoms, and tender terminals, and some of them complete their growth in the vegetative part of the plant.   Many, however, begin to wander about when they are partly grown and find their way to the fruits.   Several fruits, especially of muskmelon, may be entered and ruined by a single caterpillar before its growth is completed.   When full-grown they are greenish or coppery all over, except for the brown head and brown area just behind the head, although younger larvae are conspicuously marked with about 100 black spots evenly scattered over the body.   After feeding for about 2 weeks and passing through five instars, they desert their burrows, roll a leaf about themselves and spend the next week

[1] *Diaphania nitidalis* (Stoll), Order Lepidoptera, Family Pyralididae.

or 10 days in the pupal stage inside of a thin cocoon. The life cycle requires 22 to 28 days at mean temperatures between 75 and 77°F. The first generation is few in numbers, but the moths emerging from these pupae in early July lay sufficient eggs so that the second generation may do some damage to early cantaloupes. It is, however, during August and September that the really severe injury by this insect occurs, probably by the third- or fourth-generation larvae. Cantaloupes harvested before early July or mid-July are rarely injured. Four or five weeks are required for a generation, and there are four full generations and a partial fifth in North Carolina.

Control Measures. Spraying or dusting the foliage when the worms appear in the blossoms with parathion at 0.25 to 0.5 pound, malathion at 1.25 pounds, or lindane at 0.2 to 0.4 pound per acre is effective. In the home garden, rotenone at 0.2 to 0.4 pound per acre is satisfactory. Several applications may be necessary following blossoming. Carrying out of the following protective measures during the preceding fall and spring will aid in reducing the damage from this pest. As soon as a crop is harvested, the vines and unused fruits and adjoining weeds and trash should be collected together and burned or converted into compost, to destroy the worms in them. Early in the fall the fields should be plowed in order to bury the pupae that have fallen to the ground. Every effort should be made to get the crop matured early, since the early cantaloupes and cucumbers almost always escape. To protect muskmelons and cucumbers, early squash may be used as a trap crop. Four to eight rows of squash to the acre should be planted at 2-week intervals, the first when the main crop is planted, in order that there may be an abundance of fresh squash blossoms throughout the season. The moths prefer to lay their eggs on squash, and the worms feeding in squash do not wander from fruit to fruit as those on other crops. Before the worms become full-grown in the squash blossoms, *either the infested blossoms must be picked off and destroyed or else the entire vines removed and destroyed as the later planted ones come into bloom.* Cantaloupes may be completely protected by this means if the trapped worms in the squash plants are destroyed

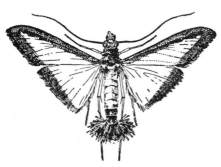

regularly. The flavor of the fruits is not affected by growing squash among them, but the seeds from such fruit cannot be used.

References. N.C. Agr. Exp. Sta. Bul. 214, 1911; *Ga. Agr. Exp. Sta. Bul.* 54, 1901, and 5 (n.s.), 1955; *N.C. Agr. Exp. Sta. Tech. Bul.* 85, 1947; *U.S.D.A., Bur. Ento., Cir.* E-856, 1953; *Mo. Agr. Exp. Sta. Cir.* 122, 1924; *Jour. Econ. Ento.,* **40**:220, 1947; **41**:334, 1948; and **49**:870, 1956.

FIG. 14.23. Adult of the melonworm. Twice natural size. (*From Crosby and Leonard, "Manual of Vegetable-garden Insects," copyright, 1918, Macmillan. Reprinted by permission.*)

100. MELONWORM[1]

The melonworm is a close relative of the pickleworm, and its life cycle and habits are very similar, except that this species feeds much more extensively on the foliage than does the pickleworm, rarely enters the vine or leaf petioles, appears a little later, and attacks pumpkins as well as the other cucurbits, but it is rare in watermelons. It is rarely injurious north of the Gulf states, although the adults at least are found from South America to Canada. The adult of the melonworm (Fig. 14.23) is a beautiful moth with a wing expanse of about 1¾ inch. The wings are pearly white, with a narrow dark-brown band about ⅟₁₆ inch wide all around the front and outer margins. The body in front of the wings is dark-brown, while the hinder part of the thorax and abdomen is silvery white with a large bushy tuft of darker hair-like scales at the tip of the body. The greenish caterpillar may be distinguished from the pickleworm in all but the smallest and largest stages by having two white well-separated slender stripes

[1] *Diaphania hyalinata* (Linné), Order Lepidoptera, Family Pyralididae.

the full length of the body on the upper side and by lacking the dark spots. They are somewhat more slender than pickleworms, and more active in their movements.

Control Measures. Since the larvae of this species feed a good deal on the foliage, they are readily controlled by the measures suggested for the pickleworm. The application should be made while the worms are still small, and should be directed against the undersides of the leaves. Dusting is best because the cloud of dust will cover the undersides of the low-growing leaves more successfully.

Reference. *N.C. Agr. Exp. Sta. Bul.* 214, 1911.

D. INSECTS INJURIOUS TO POTATOES

FIELD KEY FOR THE IDENTIFICATION OF INSECTS INJURING THE POTATO

A. *Insects chewing holes in the leaves:*
    1. Very convex, nearly hemispherical beetles, about $\frac{3}{8}$ inch long, yellow in color, with black spots on the prothorax and 5 black stripes on each wing cover; and brick-red humpbacked soft-bodied larvae of various sizes up to $\frac{3}{5}$ inch in length, with 2 rows of black spots along each side of the body; eat the leaves and tender shoots..........................*Colorado potato beetle*, page 640.
    2. Elongate, nearly parallel-sided, not very hard, active, long-legged beetles, either black, grayish, or black with narrow gray or yellow stripes or margins on the wing covers, swarm upon the plants, devouring blossoms and leaves......
    ....*Blister beetles*, page 642.
    3. Very small, elongate oval, black beetles from $\frac{1}{16}$ to $\frac{1}{8}$ inch in length rest on the leaves and jump readily when disturbed. Eat small rounded holes into the leaves. Leaves look as though peppered with fine shot. Very delicate slender white larvae, up to $\frac{1}{5}$ inch long, with brown heads and 6 very short legs near the head, sometimes found feeding on roots or tubers...*Flea beetles*, page 603.
B. *Insects sucking sap from leaves or stems:*
    1. Small, very active, greenish, slender, wedge-shaped jumping bugs, from $\frac{1}{8}$ inch down in length, suck sap from underside of leaves, causing tips of leaves to turn brown, followed by the browning and curling of the entire margin, the tissue along the midrib dying last of all..............*Potato leafhopper*, page 643.
    2. Small, soft-bodied, green, or sometimes pink aphids, the largest from $\frac{1}{6}$ to $\frac{1}{8}$ inch long, some winged and others wingless, cluster on underside of leaves and terminal shoots causing them to wilt, curl, and die. Vines become covered with sticky "honeydew."..........................*Potato aphid*, page 646.
    3. Tiny scale-like flat nymphs, margined with a white fringe, suck sap from shaded parts of foliage, producing rolled or cupped, yellow or reddish leaves ("purple-top" or "psyllid yellows"), killing or stunting the plants, and causing tiny malformed unmarketable potatoes. The adults are small "jumping plant lice." *Potato psyllid, Paratrioza cockerelli* (Sulc) (see *Jour. Econ. Ento.*, **30**: 377, 891, 1937).
    4. Large, slender, shield-shaped, flat-backed bugs suck sap, causing tops to wilt; one, an inch long with hind femora and tibiae greatly swollen...............
    ....*Big-footed plant bug, Acanthocephala femorata* (F.). Another, with the hind tibiae expanded like a leaf, $\frac{3}{4}$ inch long and with a transverse yellow line across the wing covers.................................... *Leaf-footed bug, Leptoglossus phyllopus* (L.) (see *Fla. Agr. Exp. Sta. Bul.* 232, p. 80, 1931).
C. *Insects boring in the stem and leaves:*
    1. Legless white grubs, about $\frac{1}{4}$ inch long, with brown heads, bore up and down the stalks, causing the leaves and stem to wilt and die. Small dark-gray straight-sided snout beetles, about $\frac{1}{5}$ inch long, with 3 small black spots at the junction of prothorax and wing covers, feed by gouging out slender deep holes in the stems....................................*Potato stalk borer*, page 647.
    2. Caterpillars, up to $1\frac{1}{2}$ inches long, prominently striped with brown and white in front and behind, but with a large grayish-brown "bruised-looking" area about the middle, mine in the leaves or burrow through the heart of the stems,

killing the terminals.   Especially troublesome ar(and weedy margins of fields in early summer.   Stripes become fainter and disappear when caterpillars are full-grown. . . . .*Common stalk borer*, page 493.

3. White caterpillars, up to ½ inch long, with a pinkish or greenish tinge, and brown at each end, form blotch mines in the leaves or bore through petioles and terminal stems, causing the shoots to wilt and die. .*Potato tuberworm*, page 648.

D. *Insects attacking the tubers underground:*

1. Tubers tunneled with deep, more or less cylindrical burrows, about the diameter of a match, by shining, slick, reddish-brown, tough, 6-legged worms up to 1½ inches long by ⅛ inch thick. . . . .*Wireworms*, page 617.

2. Tubers gnawed and eaten away in irregular, broad, scabby areas over the surface, by white curved-bodied grubs with reddish-brown heads and 6 long slender legs. . . . .*White grubs*, page 619.

3. White caterpillars, up to ½ inch long, with a pinkish or greenish tinge, and brown at both ends, tunnel through tubers in the field or in storage. . . . .
. . . .*Potato tuberworm*, page 648.

4. Tubers in low ground or when stored in damp places occasionally attacked by a slender white black-headed maggot, only ⅙ inch long when full-grown, that bores through the flesh, causing superficial wounds that resemble potato scab. . *Potato scab gnat, Pnyxia scabiei* (Hopk.) (see *Ohio Agr. Exp. Sta. Bul.* 524, 1933).

5. Yellowish-white legless maggots, about ¼ inch long, burrowing over the surface and through the tubers of seed potatoes. . . . .*Seed corn maggot*, page 518.

101. COLORADO POTATO BEETLE[1]

The common yellow- and black-striped "potato bug" is perhaps the best-known beetle in all America.   When first known to man, it occupied the eastern slopes of the Rocky Mountains from Canada to Texas, and its food was the weed known as buffalo bur or sand bur.[2]   It was described and named by Thomas Say, one of the earliest American entomologist-explorers, in 1824, and for 30 years longer it continued to live as an obscure beetle of no importance to man.   The pioneer settlers, pushing westward across the continent, finally brought to this insect a new food, the potato.   The insect soon largely deserted the weeds for the cultivated plant and began spreading eastward from potato patch to potato patch, often destroying the entire crop wherever it appeared.   It was recorded in Nebraska in 1859, in Illinois in 1864, in Ohio in 1869, and reached the Atlantic coast in 1874.   Its average annual spread was about 85 miles a year.   Nothing was known about spraying in those days, and the insect multiplied and spread almost unchecked until about 1865, when it was discovered that paris green could be used to poison it.

Importance and Type of Injury.   Both the yellow- and black-striped hard-shelled beetles and their brick-red black-spotted soft-skinned young or larvae feed by chewing the leaves and terminal growth of the potato.   Unless killed by stomach poisons, they soon devour so much of the vines that the plants die and the development of tubers is prevented or the yield greatly reduced.

Distribution.   Throughout the United States except parts of Florida, Nevada, California, and in eastern Canada.

Plants Attacked.   Potato is the favorite food.   When this cannot be found, the insect may survive on tomato, eggplant, tobacco, pepper, ground cherry, thorn apple, Jimson weed, henbane, horse nettle, bella-donna, petunia, cabbage, thistle, mullein, and perhaps other plants.   It is a pest chiefly of potato.

[1] *Leptinotarsa decemlineata* (Say), Order Coleoptera, Family Chrysomelidae.
[2] *Solanum rostratum.*

Life History, Appearance, and Habits.   The adult stage goes through the winter buried in the soil to a depth of several inches, seldom more than 8 or 10.   This is the only stage that survives the winter.   The beetles come out of the ground in spring in time to meet the first shoots of volunteer or early-planted potatoes.   These adults (Fig. 14.24,*a*) are familiar to nearly everyone and may be recognized by the alternate black and yellow stripes that run lengthwise of the wing covers, five of each color on each wing cover.   They are about ⅜ inch long by ¼ inch wide and very convex above.   The orange-yellow eggs are deposited on the underside of the leaves, in close-standing groups averaging a couple of dozen each.   A number of batches of eggs are matured by each female until an average of about 500 are deposited in the course of 4 or 5 weeks.   The overwintering adults then die, and, from 4 to 9 days after the eggs were laid, small, humpbacked, reddish, chewing larvae hatch and likewise attack the leaves.   They grow very fast, passing through four instars, similar except for size, and become full-grown in 2 or 3 weeks.

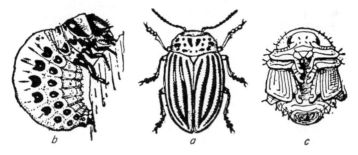

Fig. 14.24. Colorado potato beetle, *Leptinotarsa decemlineata.   a*, Adult; *b*, larva; *c*, pupa. About twice natural size.   (*From U.S.D.A., Farmers, Bul.* 1349.)

The largest (Fig. 14.24,*b*) are a little more than ½ inch long, the back arched in almost a semicircle, with a swollen head and two rows of black spots on each side of the body.   They usually feed in groups, completely consuming the leaves.   These slug-like, reddish larvae are frequently supposed to be a different kind of potato pest from their parents.

If the full-grown larvae are watched, however, they will be found to descend into the soil, make a spherical cell, and transform to a yellowish, motionless, pupal stage (*c*) which lasts 5 to 10 days.   Then the adult beetles appear from the pupae, crawl up out of the ground and, after feeding for some days, may lay eggs for a second generation.   Two generations appear to be the rule, although there may be only one in the North and there is a partial third in the more southern part of their range.

Control Measures.   Both adults and larvae are controlled by spraying or dusting the foliage with DDT at 1 to 1.5 pounds, dieldrin at 0.25 to 0.5 pound, or Thiodan at 0.5 to 1 pound per acre.   These treatments should be made at any time that beetles or larvae appear on the vines. When potatoes are being sprayed with bordeaux mixture for flea beetles, leafhoppers, or potato blights, the DDT should be added to the bordeaux when needed to control the potato beetle, thus saving the expense of a separate application.   For small patches of potatoes, the insecticide may be dusted lightly over the foliage from an open-meshed bag, sifter can, or hand duster.

References. Iowa Agr. Exp. Sta. Bul. 155, 1915; *U.S. Bur. Ento. Cir.* 87, 1907; *Jour. Agr. Res.*, **5**:917, 1916; *Jour. Econ. Ento.*, **26**:1068, 1933; **40**:640, 1947; **51**:828, 1958; and **52**:564, 1959; *Boyce Thompson Inst. Contrib.*, **3**:1, 1931; *Purdue Univ. Agr. Exp. Sta. Cir.* 338, 1948.

102 (44). BLISTER BEETLES[1]

Importance and Type of Injury. These insects were once called "old-fashioned potato bugs," because they were much more noticeable on potato before the Colorado potato beetle invaded the central and eastern states. They are slender beetles, about four times as long as wide, rather soft, with the head distinctly set off from the prothorax and the tip of the abdomen exposed beyond the tip of the wing covers (Fig. 14.25). They are black- or grayish-colored or black with narrow yellowish or gray stripes or margins on the wings. Only the adults feed on the foliage, but they are very ravenous and may destroy many plants.

Plants Attacked. The several species of blister beetles attack potato, tomato, egg-plant, sweetpotato, bean, pea, soybean, cowpea, melons, pumpkin, onion, spinach, beets, carrot, peppers, Swiss chard, radish, cabbage, corn, oats, barley, alfalfa, clover, cotton, clematis, aster, chrysanthemum, zinnia, and other crops and weeds.

Distribution. Blister beetles are found in all parts of the United States and Canada, their prevalence being controlled by that of the grasshoppers, whose eggs they prey upon.

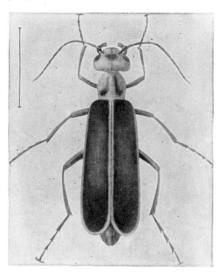

FIG. 14.25. Adult of the margined blister beetle, *Epicauta pestifera*. Line shows natural length. (*From Ill. Natural History Surv.*)

Life History, Appearance, and Habits. The life histories of the destructive species are quite similar and that of the striped blister beetle is as follows: The insect generally winters as a full-fed but not fully transformed sixth-instar coarctate larva, known as the pseudopupa (Fig. 14.26,*g*), in an earthen cell in the soil, but under favorable conditions the fifth-instar larva may overwinter. The pseudopupa is about ⅖ inch long, yellow, tough-skinned, and with much reduced legs and mouth parts. It is highly resistant to desiccation and to high and low temperatures. Under some conditions the pseudopupa may survive for 2 years or more without changing to the sixth-instar larva, but the average duration of this stage is about 232 days. The pseudopupa generally molts in the late spring to the sixth-instar larva, acquiring functional legs and moving about for approximately 14 days before it molts again to the true pupal stage, which lasts for an average of 12 to 13 days. The adult beetles, which may live for 4 to 6 weeks, emerge rather suddenly in great numbers in June and July. They are very restless, active beetles that tend to feed together in swarms. Their bodies contain an oil known as cantharidin (page 55), which will blister tender skin if the beetles are crushed on it. The females lay their eggs in clusters of 100 to 200 in holes which they make in the soil. The eggs are elongate, cylindrical yellow objects which,

[1] Striped blister beetle, *Epicauta vittata* (Fabricius), margined blister beetle, *E. pestifera* Werner, ash-gray blister beetle, *E. fabricii* (LeConte), spotted blister beetle, *E. maculata* (Say), and gray blister beetle, *E. cinerea* (Förster), Order Coleoptera, Family Meloidae.

in an average of about 12 days, develop into very active, strong-jawed, little larvae (*c*) that burrow through the soil until they find an egg mass of a grasshopper, where they gnaw into the egg pod and feed on the eggs.   During the next 25 to 28 days the larva molts four times, undergoing a remarkable series of changes in form and appearance known as *hypermetamorphosis*, during which its legs, mouth parts, and other appendages grow progressively smaller (Fig. 14.26,*c,d,f,g*).   Under normal summer temperatures, the average times required for the development of the various instars are first, 5.9 days; second, 2.4; third, 2.7; fourth, 1.9; and fifth, 13.6.   During warm temperatures in midsummer, the fifth instar may pass directly into the pupal stage, and thus there are one and a partial second generation in the latitude of Arkansas.

The margined blister beetle and the ash-gray blister beetle have somewhat similar life histories, both feeding on the eggs of the differential and two-striped grasshoppers. These two species regularly pass through seven larval instars, overwintering as the coarctate larva or pseudopupa.   The margined blister beetle deposits from 150 to 350 eggs, which require about 26 days to hatch.   The first five larval instars require approximately 31 days, and the sixth instar requires about 260 days for development.   The

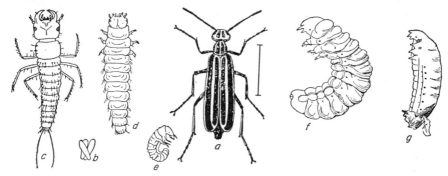

Fig. 14.26. Life stages of the striped blister beetle, *Epicauta vittata*. *a*, Adult; *b*, eggs; *c*, first larval (*triungulin*) stage; *d*, second (*caraboid*) stage; *e*, same as *f*, as doubled up in pod; *f*, third (*scarabaeoid*) stage; *g*, pseudopupa or *coarctate* larval stage.   All but *e* enlarged. (*From U.S.D.A.*)

larva of this species consumes about 43 grasshopper eggs in order to reach maturity. The ash-gray blister beetle deposits from 50 to 335 eggs and has a longer prehatching period, requiring about 77 days.   The first five larval instars require 25 days, and the sixth instar about 230 days, for development.   Both species have one generation a year.   Although the larvae of these species may be considered as beneficial, even where very abundant they do not destroy more than 25 per cent of the grasshopper egg masses.

Control Measures.   A highly efficient program of grasshopper control will do much to reduce the populations of blister beetles.   Where the beetles are very abundant, they may be controlled by spraying or dusting with DDT at 1 to 1.5 pounds, toxaphene at 1.5 to 2.5 pounds, or parathion at 0.5 pound per acre.   On potatoes, DDT may be combined with bordeaux mixture, which is repellent to the beetles, and bordeaux used alone will give fair protection to the vines.   Individual plants may be protected by knocking the beetles into a pan of kerosene, or if very valuable, they may be covered with mosquito netting.

References.   Iowa Agr. Exp. Sta. Bul. 155, 377, 1915; *S.D. Agr. Exp. Sta. Bul.* 340, 1940; *Ark. Agr. Exp. Sta. Bul.* 436, 1943; *U.S.D.A., Dept. Bul.* 967, 1921; *Jour. Econ. Ento.*, **25**:71, 1932; and **27**:73, 1934.

103 (155). POTATO LEAFHOPPER[1]

Importance and Type of Injury.   This little wedge-shaped green leafhopper, only ⅛ inch long, is the most injurious pest of potatoes in the

[1] *Empoasca fabae* (Harris), Order Homoptera, Family Cicadellidae.

eastern half of the United States.   The insects feed on the underside of the leaves, sucking out the sap from the veins and, in some manner not fully explained, cause the trouble known as tipburn or hopperburn.   It appears that only two species of leafhoppers produce hopperburn and only upon certain plants, such as potato, eggplant, rhubarb, dahlia, and horsebean.   The first symptom of this trouble is the appearance of a triangular brown spot at the tip of the leaf.   Similar triangles may appear at the end of each lateral veinlet, or the entire margin may roll upward and turn brown at one time, as though scorched by fire or drought. These brown margins increase in width until only a narrow strip of the leaf along the midrib remains green, the rest is shriveled and dead, with the leaf veins much distorted.   Older leaves below the growing tips usually burn first, but in cases of heavy infestations every leaf rapidly succumbs and the vines die long before the normal development of the tubers has been completed, thus greatly cutting the yield by producing mostly tubers so small that they are often not worth harvesting.   Many other agencies cause a browning of the leaf area of potatoes but the typical relation of the spots to the veins, just described, is characteristic of this particular trouble.   The diagnosis may be confirmed by finding the leafhoppers on the underside of the leaf.   The number of hoppers to an acre of potatoes may run between five and six million.[1]   Upon bean and apple, stunting, dwarfing, crinkling, and tight curling of the leaves are characteristic symptoms.   Alfalfa leaves become yellowed and clover leaves reddened when attacked, with a great reduction of the vitamin A content.   It has been shown that this species feeds on the phloem cells of the veins, which become torn and distorted and the xylem tubes plugged, so that food substances in the leaves are not properly translocated.

Several other species, so similar to *E. fabae* that only microscopic examination of the internal genitalia can be depended upon to distinguish them, have been found occurring as major pests of truck crops in various parts of the United States.   The intermountain leafhopper, *E. filamenta* DeLong, occurs in arid regions at high elevations in the Rocky Mountains, where the annual rainfall is under 10 inches, as a pest on potatoes, sugarbeets, and beans.   The western potato leafhopper, *E. abrupta* DeLong, occurs at low altitudes and low relative humidity from Texas to Oregon, along the Mexican border and Pacific Coast.   It causes a speckled whitestippled appearance of leaves of potato and bean, which is caused by withdrawing of sap from the mesophyll tissues.   The arid leafhopper, *E. arida* DeLong, also occurs in low altitude and low relative humidity areas in California, Arizona, and Utah.   The southern garden leafhopper, *E. solana* DeLong, is generally distributed throughout the United States and causes serious damage to potato, cotton, lettuce, and beans.   Like *E. fabae*, it is a phloem feeder and produces hopperburn.

Plants Attacked.   For many years this insect has been known as the apple leafhopper because its most extensive known injury was to apple nursery stock.   It feeds on more than 100 cultivated and wild plants, including beans, potatoes, eggplant, rhubarb, celery, dahlia, alfalfa, soybeans, clovers, and sweet clover.   On peanuts it causes a diseased condition known as "peanut pouts."

Distribution.   The potato leafhopper occurs over the eastern half of the United States, westward to Colorado and Wyoming, at elevations

[1] *Jour. Econ. Ento.*, **14**:62, 1921.

below 4,600 feet and where the average annual rainfall is 25 inches or more, with resultant high average humidity.

Life History, Appearance, and Habits.   Although many kinds of leaf-hoppers may be found under surface vegetation during the winter in the northern states, it has been impossible to find the potato leafhopper in hibernation, or to keep them alive through the winter when placed in promising hibernating quarters.   No wild host plant has been found in the North, upon which it breeds in early spring, before cultivated crops are up.   The adults apparently migrate into the northern states from areas of milder climate, instead of wintering over in the North.   In Florida and other Gulf states it breeds on alfalfa and other legumes or on castor bean and other weeds during the entire winter.   The adults are about ⅛ inch long, by one-fourth as broad, of a general greenish color and somewhat wedge-shaped (Fig. 14.27). They are broadest at the head end, which is rounded in outline, and taper evenly to the tips of the wings.   There are a number of faint white spots on the head and thorax, and one of the characteristic marks of this species is a row of six rounded white spots along the anterior margin of the prothorax, which can be seen with a hand lens.   The hind legs are long and enable the insect to jump a considerable distance.

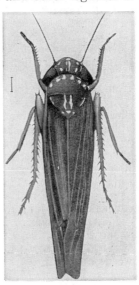

FIG. 14.27. Adult of the potato leafhopper, *Empoasca fabae*. Greatly enlarged. Line at left indicates natural size. (*From Ill. Natural History Surv.*)

Large numbers of flying adults often appear suddenly in fields of beans as soon as these plants come up in early June.   Except in cool wet seasons they are not attracted to potatoes until the plants are considerably larger.   The possible reason is that bean plants are much higher in sugar content when they first break through the ground than potato plants are.   Potatoes become much sweeter as they grow older, and it is then that the leafhoppers swarm upon them. Beginning from 3 to 10 days after mating, the very small, whitish, elongate eggs, only 1/24 inch long, are thrust into the main veins or petioles of the leaves on the underside, by use of the female's sharp ovipositor.   An average of two or three eggs are laid daily, and the females live about a month or more.   The eggs hatch in about 10 days, and the nymphs become full-grown in about 2 weeks.   The nymphs are similar in shape to the adults but lack the wings and are very small and pale colored so they are really hard to see on a leaf.   They usually complete their growth on the leaf where they hatched, feeding from the underside and increasing in size, greenness, and activity as they shed their skins; at the fifth molt they appear as adults.   Both nymphs and adults are very active, the adults flying or jumping when disturbed, while the nymphs more characteristically run sideways over the edge of the leaf to the side which is turned downward.   Two nearly complete generations and a partial third and fourth are produced at the latitude of central Illinois, so that there are individuals to infest both early and late potatoes and other truck crops.

Control Measures.   This leafhopper may be controlled by spraying or dusting with DDT at 1 to 1.5 pounds or parathion at 0.2 to 0.4 pound per acre or by spraying with malathion at 0.5 to 1 pound or methoxychlor at 2.25 pounds per acre.   Applications should be started as soon as the plants are 4 to 8 inches high and continued at 10- to 14-day intervals as long as the vines can be kept green.   From five to eight applications may be required.

References.   U.S.D.A., Farmers' Bul. 1225, 1921; *Fla. Agr. Exp. Sta. Bul.* 164, 1922; *Jour. Econ. Ento.,* **21**:183, 261, 1928, **24**:361, 475, 1931, **27**:525, 1934, **28**:442, 1935; **41**:275, 1948; and **52**:908, 1959; *Iowa Agr. Exp. Sta. Res. Bul.* 78, 1923; *U.S.D.A. Tech. Bul.* 231, 1931, and 850, 1943; *Jour. Agr. Res.* **43**:267, 1931; *U.S.D.A., Tech. Bul.* 618, 1938; *Purdue Univ. Agr. Exp. Sta. Cir.* 338, 1948.

104. POTATO APHID[1]

Importance and Type of Injury.   The epidemics of this insect are extremely sporadic.   Severe outbreaks result in the complete browning and killing of the vines of the potato by curling and distorting the leaves from the top of the plant downward (Fig. 1.4).   On tomato the most noticeable injury is the devitalizing of the blossom clusters so that the blossoms fall and no tomatoes set.   Green or pinkish, winged or wingless aphids cluster in shaded places on the leaves, stems, and blossoms. Winged migrants spread from field to field, so that the epidemic may sweep over a district in an alarming manner.   Following such an outbreak, the insect may not visit a community in conspicuous numbers again for many years.   The transmission of tomato and potato diseases, such as mosaics, leaf roll, and spindling tuber, by the feeding of these aphids, causes more injury to the plants than sucking the sap.

Plants Attacked.   Potato, tomato, eggplant, pepper, ground cherry, sunflower, pea, bean, apple, turnip, buckwheat, aster, gladiolus, iris, corn, sweetpotato, ragweed, lamb's-quarters, shepherd's-purse, and many other weeds and crops.

Distribution.   Maine to California and Florida to Canada; probably in every state.

Life History, Appearance, and Habits.   The winter eggs (page 613) are deposited chiefly on rose, and the aphids may be found regularly on the succulent parts of rose bushes in the spring.   During the first half of July, in Maine, winged aphids develop that fly to potatoes, and some of the others crawl to their summer host.   A generation may be developed on potato every 2 or 3 weeks, and each unmated female may give birth to 50 or more active nymphs within 2 weeks' time.   Thus the vines rapidly become covered with aphids, which blight the stems and wither the leaves.   On a single large tomato plant, 24,688 aphids have been counted.   By the middle of September, in Maine, and usually by early October, in Ohio, the aphids have all deserted potatoes and dispersed to other plants, of which the rose is the favorite.   Here wingless, egg-laying females develop as the last generation of the season, and, after mating with males that fly over from the summer-host plants, the winter eggs are laid on stems and leaves of the rose.

The potato aphid when full-grown is nearly ⅛ inch long, of a clear green or pink, glistening color and with long slender cornicles (Fig. 4.5,*B*). The wingless ones tend to drop from the plant when disturbed.

[1] *Macrosiphum* (= *Illinoia) euphorbiae* (= *solanifolii*) (Thomas), Order Homoptera, Family Aphidae.

Control Measures.   This aphid may be controlled by spraying or dusting with DDT at 0.66 to 1 pound, Thiodan at 0.5 to 1 pound, malathion at 0.5 to 1 pound, Diazinon at 0.4 to 0.5 pound, parathion at 0.2 to 0.45 pound, or endrin at 0.25 to 0.5 pound per acre, applied as needed. Because of the many host plants on which this aphid may develop, clean cultivation is important in preventing outbreaks.   Rosebushes should not be permitted to grow in abundance near potato fields, since they afford a place for an abundance of overwintering eggs.

References.   *Maine Agr. Exp. Sta. Bul.* 242, 1915, and 323, 1925; *Ohio Agr. Exp. Sta. Bul.* 317, 1917; *Jour. Econ. Ento.*, **25**:634, 1932, **39**:189, 205, 1946, **40**:220, 1947.

105. POTATO STALK BORER[1]

Importance and Type of Injury.   In some sections, this stalk borer is abundant enough in certain years to destroy entire fields of potatoes.   Throughout much of its

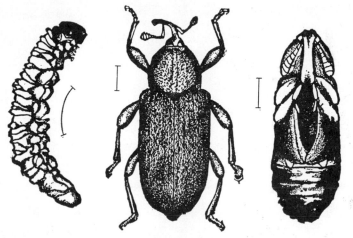

FIG. 14.28. Potato stalk borer; larva, adult, and pupa; the lines show natural size.   (*From Sanderson and Peairs, "Insect Pests of Farm, Garden, and Orchard," copyright, Wiley.*)

range it is of little importance.   Chief injury is due to the larvae eating out the interior of the stalks, causing the entire plant to wilt and die.   The adults eat slender deep holes in the stems.

Plants Attacked.   Potato, eggplant, and related weeds, such as Jimson weed, horse nettle, and ground cherry.   Most injurious to early potatoes.

Distribution.   Over the United States, except the northernmost states.

Life History, Appearance, and Habits.   The insect goes through the winter in or among the old vines in the adult stage.   These adults are blackish snout beetles, about $\frac{1}{5}$ inch long, covered with flattened gray hairs or scales that give them a frosted appearance.   There are three distinct black dots at the base of the wing covers (Fig. 14.28).   In late spring they emerge and feed by eating deep holes in the stems of the new plants.   The eggs are placed singly in similar cavities in the stem or leaf petioles and hatch in a week or 10 days.   The larvae eat up and down in the stems, completely hollowing them out for several inches.   They are yellowish-white, legless, wrinkled grubs with brownish heads, and range up to about $\frac{1}{3}$ inch long (Fig. 14.28).

[1] *Trichobaris trinotata* (Say), Order Coleoptera, Family Curculionidae.   The tobacco stalk borer, *T. mucorea* (LeConte), is very similar in appearance and habits and attacks potatoes in Southern California and Arizona.

Before pupating, the larva packs its burrow with excelsior-like scrapings from the stem and chews an exit hole nearly through the stem for the escape of the adult.   A week or two later the pupae transform to adults which may be found in the stalks, from late July on, in the northern states.   They do not come out of the larval burrows until the following spring unless the stalks are broken open.

Control Measures.   The only practical control is to collect and burn all potato vines in infested fields, as soon as the crop is harvested.   As with most potato insects, the destruction of Jimson weed, horse nettle, and ground cherry will help to keep down their numbers.

References.   *U.S.D.A., Bur. Ento. Bul.* 33, 1902; *Kans. Agr. Exp. Sta. Bul.* 82, 1899.

106 (72). Potato Tuberworm[1]

Importance and Type of Injury.   The tubers of potatoes in the field and in storage are riddled with slender, dirty-looking, silk-lined burrows

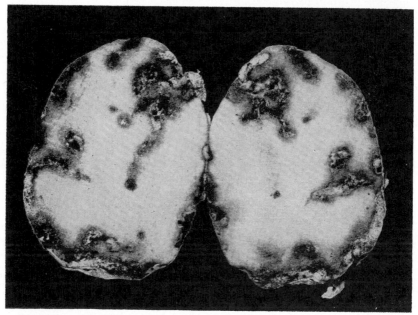

Fig. 14.29. Potato cut open to show injury by larvae of potato tuberworm, *Gnorimoschema operculella.*   (*From U.S.D.A., Dept. Bul.* 427.)

of pinkish-white or greenish caterpillars (Fig. 14.29), that range up to ¾ inch in length, with dark-brown heads.   Some of them burrow in the stems and petioles or mine in the leaves.   When working on tobacco, the insect is known as the split worm.   Brown blotches caused by this worm, between the upper and lower epidermis of tobacco leaves, make the leaves unfit for wrappers.   It is very destructive to potatoes in warm dry regions where it occurs.

Plants Attacked.   Potato, tobacco, tomato, eggplant, and weeds of the same family.

Distribution.   Southern United States from California to Florida and northward to Washington, Colorado, Virginia, and Maryland.

[1] *Gnorimoschema* (= *Phthorimoea*) *operculella* (Zeller), Order Lepidoptera, Family Gelechiidae.

Life History, Appearance, and Habits.   In warm storage the insect may continue to reproduce as long as the potatoes contain enough food to mature the larvae.   The moths escape from storehouses in early spring, or winter in all stages in sheltered culls out-of-doors.   They lay 150 to 200 eggs, one in a place, chiefly on the underside of the leaves, or upon exposed tubers.   Most of the larvae first produce blotch mines on the leaves but subsequently work down into the stems.   They become mature in 2 or 3 weeks under summer conditions, pupate in a grayish, silken, dirt-covered cocoon, about ½ inch in length, entangled in the dead leaves or among trash on the ground, and emerge as adults in a week or 10 days. The adults are very small narrow-winged night-active moths, ½ inch from tip to tip of the wings, and grayish-brown, mottled with darker brown. An entire generation may develop in a month of warm weather, and five or six in a year.   The damage is most severe in years of low rainfall and high temperatures.   The later generations infest the tubers in the field by working down through cracks in the soil to lay eggs upon them, or the larvae may migrate from the stems to the tubers, or, especially at digging time, the eggs may be laid upon the exposed tubers.   The caterpillars at first work just under the skin of the potato but later tunnel through the flesh.   Before pupating, they come out of the potatoes and spin up among them or in cracks about the storehouse.

Control Measures.   Potatoes should be kept well cultivated and deeply hilled during their growth.   At harvesttime, to prevent the caterpillars migrating from the wilting vines to the tubers, infested vines should be cut and burned or removed from the field a few days before digging and never piled over dug potatoes.   The tubers should not be left exposed to the egg-laying moths during late afternoon or overnight. Harvest should be very thorough and all culls should be destroyed. Tubers infested in storage may be saved by fumigating when the temperature is above 70°F. with carbon bisulfide at 5 pounds to 1,000 cubic feet for 48 hours, or with methyl bromide at 2.5 pounds to 1,000 cubic feet for at least 3 hours.   Potatoes and potato bags should be fumigated before being transported.   Bagged potatoes may be protected from infestation by storing in sacks treated with a 1 per cent solution of DDT in xylene. Seed potatoes, not to be used for food or forage, may be treated at time of storage with 5 per cent DDT or methoxychlor at 1.5 to 2 ounces per 100 pounds.   Infestations in growing potatoes may be controlled by spraying or dusting the foliage when the worms begin to web the leaves with DDT at 1 to 1.75 pounds or Thiodan at 0.5 to 1 pound or by spraying with endrin at 0.4 pound per acre.   The application should be repeated after 10 days.   A thorough cleanup of potato storage sheds will help prevent reinfestation.

References.   U.S.D.A., *Dept. Bul.* 427, 1917; *Va. Truck Exp. Sta. Bul.* 61, 1927, and 111, 1949; *Calif. Agr. Ext. Ser. Cir.* 99, 1936; *Jour. Econ. Ento.*, **25**:625, 1932; **37**:539, 1944, **41**:198, 1948, and **53**:868, 1960.

E. INSECTS INJURIOUS TO SWEETPOTATOES

FIELD KEY FOR THE IDENTIFICATION OF INSECTS INJURING SWEETPOTATOES

A. *Insects chewing holes in the leaves or vines:*

   1. Oval beetles, about ¼ inch long, that look like a drop of molten gold when alive, nearly hemispherical, with the margins of the body extended so as to

hide the head and most of the legs; and elongate oval, brown larvae, up to $\frac{3}{8}$ inch long, the margin of their bodies surrounded by about 30 thorny spines, but their backs completely covered with a dirty mass of excrement and shed skins that conceal the body, eat rounded holes in the leaves or devour them completely...*Golden tortoise beetle*, page 652.

2. Similar to *A*,1, but the beetle is dull-yellow in color, with five longitudinal black stripes on the wing covers. The larva with shorter marginal spines and yellowish-white in color with a median gray line. A thick tail-like projection, carried at an angle of 45 degrees from the leaf, bears the shed skins, but no excrement.............................*Striped tortoise beetle*, page 652.

3. Similar to *A*,1, but $\frac{5}{16}$ inch long, golden yellow, with three small black spots on each wing cover, arranged in a triangle. The larva straw-yellow with two dark spots behind the head; the marginal spines not very long, black at their tips; the back covered with excrement carried on two long spines and drawn out sideways into long shreds..........*Black-legged tortoise beetle*, page 652.

4. Beetle similar to the above, golden around the margin, the disk mottled with black and yellow, the black extending out as a slender tooth to each shoulder. The larva is dull-green, bluish along the back, and covered with broad branching masses of excrement supported on the anal spines....................
....*Mottled tortoise beetle*, page 652.

5. Beetle larger than the above, $\frac{1}{3}$ inch long, yellow to brick-red in color, with 15 to 20 small rounded black spots on the forward two-thirds of the back; very convex, and the margins of the body not extended. Larva light-yellow, with many small brown spots, $\frac{1}{2}$ inch long, marginal spines long, black-tipped. An irregular mass of excrement held slanting back from the body or vertically over it.....................................*Argus tortoise beetle*, page 651.

6. Very small, black jumping beetles only $\frac{1}{16}$ inch long, with a bronzy reflection eat long narrow channels in the leaves, especially on the upper surface, parallel with the veins, during May and early June...*Sweetpotato flea beetle*, page 652.

7. Metallic, bluish-green, oblong-oval beetles, a little over $\frac{1}{4}$ inch long by two-thirds as wide, eat the tender leaves about the crown of the plant, completely devouring them from the margin inward (see also *C*,2)....................
Sweetpotato leaf beetle, *Typophorus nigritus viridicyaneus* (Crotch) (see *U.S.D.A.*, *Cir.* 495, 1938).

8. Leaves and vines eaten by shiny, slender, ant-like snout beetles, $\frac{1}{4}$ inch long, with blue-black head, wing covers, and abdomen, but the middle region of the body and the legs bright-red (see also *C*,1, below)....................
....*Sweetpotato weevil*, page 653.

9. Holes eaten in leaves or leaves skeletonized by bluish-green caterpillars with yellow heads, up to 1 inch in length, which feed inside of folded leaves held together by silk. Injurious in the Gulf states............................
Sweetpotato leaf roller, *Pilocrocis tripunctata* (Fabricius) (see *U.S.D.A. Dept. Bul.* 609, 1917).

10. Plants or separate leaves cut off and the heart of the plant eaten out during the night. Plump, greenish-gray or blackish, more or less mottled or striped caterpillars found during the day just below the surface of the soil.........
....*Cutworms*, page 610.

B. *Insects sucking sap from the plants:*
  1. Greenish, motionless, oval flat nymphs with white waxy spines radiating from their bodies, the largest only $\frac{1}{16}$ inch long, suck sap from the underside of the leaves. The small but conspicuous four-winged white bugs, like tiny moths, fly up when the plants are disturbed....................................
Sweetpotato whitefly, *Bemisia tabaci* (*Gennadius*) [= *inconspicua* (Quaintance)] (see *Fla. Agr. Exp. Sta. Bul.* 134, 1917).

C. *Insects burrowing in the fleshy roots or "tubers" in the field or in storage:*
  1. Legless, white, fat grubs with pale-brown heads, up to $\frac{1}{3}$ inch long, make winding, excrement-filled tunnels through the tubers, causing them to decay and become bitter and unfit for use. The red-and-blue adult beetles also found in the tunnels (see also *A*,8, above)............*Sweetpotato weevil*, page 653.

2. Pale-yellow, rather plump larvae, up to nearly ½ inch long, with 6 small legs and a distinct brownish head, at first burrow through the vine under ground and later tunnel through the fleshy roots during the growing season.   Usually lie curled when in the soil (see also *A*,7, above)..............................
Sweetpotato leaf beetle, Typophorus nigritus viridicyaneus (Crotch) (see *U.S.D.A., Cir.* 495, 1938).

107. Sweetpotato or Tortoise Beetles or "Gold Bugs"[1]

Importance and Type of Injury.   The foliage of sweetpotatoes is very commonly cut full of holes, or entire leaves may be eaten by beautiful oval beetles a little squared

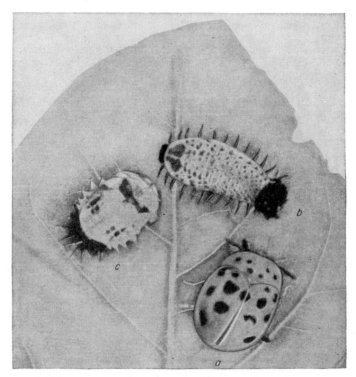

Fig. 14.30. Argus tortoise beetle.   Sweetpotato leaf showing: *a*, adult; *b*, larva; and *c*, pupa. About 3 times natural size.   (*From Jour. Agr. Res., vol.* 27, *no.* 1.)

at the shoulders, of a golden color, sometimes with black stripes or spots, and about ¼ inch long.   Similar injury is performed by the spiny dirt-laden larvae found on the underside of the leaves.   When they attack newly set plants the injury may be severe.

Plants Attacked.   These beetles restrict their feeding largely to the plants of a single family, the morning-glory family, of which the best known kinds are sweet-potato, morning-glory, and bindweed.

Distribution.   As a group, the tortoise beetles occur over nearly all the United States and arable Canada.

Life History, Appearance, and Habits.   The tortoise beetles live through the winter in the beetle stage in dry sheltered places, under bark or trash.   They come out of hibernation rather late and are found feeding and mating on the plants during May and June.   They are turtle-shaped, flat below, and the sides of the prothorax and

[1] Order Coleoptera, Family Chrysomelidae.

wings are extended beyond the sides of the body so as to hide the head and much of the legs.   The recognition marks of the more important species[1] are given in the key on pages 649 and 650.

The females of the striped tortoise beetle[2] lay their eggs one in a place, on the leaf stems or veins of the underside of the leaf, covering each white egg with a little daub of black pitchy material that hides and protects it.   The argus tortoise beetle[3] (Fig. 14.30,a) lays her eggs in clusters of 15 to 30, each egg attached to the leaf by a slender pedicel.   In about a week or 10 days the eggs hatch, and during June and July the curious larvae of these beetles (Fig. 14.30,b) are found feeding mostly on the underside of the leaves.   They are about ⅜ inch long, provided with conspicuous thorny spines all around the margin, two of which at the posterior end are nearly as long as the body.   On these long spines, which may be turned up over the back like a squirrel's tail, the larva packs all its excrement and the skin it sheds at each molt, tying the dirty mass together with silk.   The larvae thus come to look like moving bits of dirt or excrement and are often called "peddlers."   When growth is completed, the larvae fasten themselves to leaves.   The somewhat different-looking pupae (c) are exposed by molting the skin, and a week or so later the new adults emerge. They feed a little during August and then go into winter quarters.

Control Measures.   These insects may be controlled by spraying or dusting with DDT at 1 to 1.5 pounds or rotenone at 0.2 to 0.4 pound per acre.   These treatments may be applied in the seedbed or in the field.

References.   N.J. Agr. Exp. Sta. Bul. 229, 1910; Jour. Agr. Res., vol. 27, no. 1, 1924; U.S.D.A., Farmers' Bul. 1371, 1934.

108 (76). SWEETPOTATO FLEA BEETLE[4]

Importance and Type of Injury.   Long narrow grooves are eaten in the leaves, especially on the upper surface, along the veins, during May and early June by very small, chunky, black, jumping beetles about $\frac{1}{16}$ inch long, with bronzy reflections and reddish-yellow appendages (Fig. 14.31).   When these channels are numerous, the leaf may wilt and turn brown and the plant be killed or badly stunted.

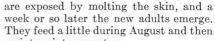

FIG. 14.31. Sweetpotato flea beetle, adult, about 24 times natural size.   (*From Ill. Natural History Surv.*)

Plants Attacked.   Besides sweetpotato, bindweed, and morning-glory, corn, wheat, oats, rye and other grasses, red clover, sugarbeets, raspberry, and boxelder.

Distribution.   General east of the Rocky Mountains.

Life History, Appearance, and Habits.   The small beetles winter under protecting trash in fence rows, the margins of wood lots, and other sheltered places.   They come out of hibernation and attack the plants about the time they are set out from the seedbeds.   By the end of June, all have usually left the sweetpotato, migrating especially to bindweed, about which they lay their eggs, and then die.   Their white larvae feed on the small roots of bindweed, becoming full-grown and producing the new generation of beetles in late July and August.   These beetles make their char-

---

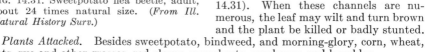

[1] Golden tortoise beetle, *Metriona bicolor* (Fabricius).   Striped tortoise beetle, *Cassida bivittata* Say.   Black-legged tortoise beetle, *Jonthonota nigripes* (Olivier). Mottled tortoise beetle, *Deloyala* (= *Chirida*) *guttata* (Olivier).   Argus tortoise beetle, *Chelymorpha cassidea* (Fabricius).

[2] *Cassida bivittata* Say.

[3] *Chelymorpha cassidea* (Fabricius).

[4] *Chaetocnema confinis* Crotch, Order Coleoptera, Family Chrysomelidae.

acteristic feeding channels on bindweed and morning-glory in the fall, but rarely attack sweetpotato until they come out of hibernation the following spring.   They sometimes injure small grains in the fall.

Control Measures.   Plants in the seedbed may be protected by spraying or dusting with DDT, methoxychlor, or malathion at 1 to 1.5 pounds per acre (page 605). Plants set out late, after the beetles have migrated to weeds, are less liable to injury.

Reference.   *N.J. Agr. Exp. Sta. Bul.* 229, 1910.

109. Sweetpotato Weevil[1]

Importance and Type of Injury.   Also known as the sweetpotato root borer, this insect is most injurious to the roots or "tubers" which may be

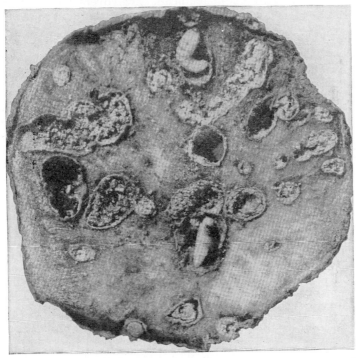

Fig. 14.32. Sweetpotato cut open to show injury by sweetpotato weevil.   Larva in burrow at top; pupa below.   3 times natural size.   (*From U.S.D.A., Farmers' Bul.* 1020.)

honeycombed by numerous, fat, legless, white grubs, with pale-brown heads, ranging up to ⅓ inch in length.   Their tunnels are tortuous and filled with excrement, and when badly infested (Fig. 14.32), the roots are unfit even for stock feed.   From 25 to 75 per cent of the crop is often destroyed.

Plants Attacked.   Sweetpotato, morning-glory, and other plants of the same family.

Distribution.   Confined in this country to South Carolina, Georgia, Florida, Alabama, Mississippi, Louisiana, and Texas.   Probably imported from Asia.

---

[1] *Cylas formicarius elegantulus* (Summers), Order Coleoptera, Family Curculionidae.

Life History, Appearance, and Habits.   Breeding is continuous throughout the winter months, especially in potatoes in storage; and all stages may be found practically every week of the year.   The eggs are deposited singly, in small cavities eaten out of the stem, or by preference in the "tuber."   They hatch in less than a week and the grubs eat down through the stem or into the potato, feeding for 2 or 3 weeks and causing the potato to develop a bad odor and a bitter taste.   The larvae may also be found in the slips, draws, vines, or smaller roots, before the potato develops.   About a week or more is spent in the pupal stage in a cavity in the tuber, and then the beetle eats its way out.   The adult is the only stage generally seen.   It is a shiny, slender-bodied, ant-like snout beetle, about ¼ inch long, with blue-black head, wing covers, and abdomen, but the middle region of the body (prothorax) and the legs are bright-red. The adults feed on the stems and leaves, and soon deposit eggs for another generation.   The generations require 1 month to 6 weeks each and follow each other as long as growing plants or stored potatoes are available. The adults may live for 8 months and fly over 1 mile in search of food. Spread to new fields or territory usually results from the planting of infested slips or draws.

Control Measures.   Sweetpotatoes or slips from infested territory should never be used for planting.   Other measures that will prevent serious damage by this weevil are cleaning up the vines and all infested tubers from the fields promptly after harvest and feeding, burning, or burying them deeply; turning hogs into the fields after the potatoes have been dug; observing quarantine and certification regulations designed to protect the growers; destroying volunteer sweetpotatoes and related weeds; putting sweetpotato plantings as far away from those of the preceding year as possible; dusting soil at the base of the plants with dieldrin in two bands 6 to 8 inches wide, at 1.5 pounds per acre, at least 21 days before harvest.   Cultivating or mulching the soil to prevent exposure of the roots by cracking of the soil is of value.   Treatment of storage houses, bags, and potatoes with residual applications of DDT spray or dust has given good control of the weevil.   Federal and state quarantines prohibit the movement from a known infested area of plant material which is likely to carry the insect.

References.   U.S.D.A., Farmers' Bul. 1371, 1934; *Jour. Econ. Ento.,* **41**:563, 1948, and **48**:644, 1955.   *U.S.D.A., Bur. Ento. Plant Quar., Cir.* E-691, 1946; *La. Agr. Exp. Sta. Tech. Bul.* 483, 1954; *U.S.D.A., Leaflet* 431, 1960.

F. INSECTS INJURIOUS TO TOMATOES, PEPPERS, AND EGGPLANT

FIELD KEY FOR THE IDENTIFICATION OF INSECTS INJURING THE TOMATO, PEPPERS, OR EGGPLANT

A. *Insects cutting off the newly set plants close to the ground:*
  1. Plants chewed off at night and left lying on the soil.   Plump, greasy-looking, green, tan, or blackish caterpillars up to 2 inches in length, some of them spotted or striped, found in the soil about plants during the daytime. . . . .
  . . . . *Cutworms,* page 610.

B. *Insects devouring or eating holes in the foliage:*
  1. Very small, oval, black, brassy or pale-striped beetles, from ¹⁄₁₆ to ⅛ inch long, eat small round "shot holes" in the leaves, and jump vigorously when disturbed.
  . . . . *Flea beetles,* page 603.
  2. Elongate, nearly parallel-sided, not very hard, long-legged beetles, black, brown, grayish, or black with light margins on the wing covers, swarm actively over the plants, devouring leaves and blossoms. . . . . . *Blister beetles,* page 642.

3. Large green caterpillars, up to 4 inches long, with diagonal white bars on the sides and a slender horn at the tip of the body, cling to the vines and strip off the foliage.........................*Tomato* and *tobacco hornworms*, page 655.

C. *Insects that suck sap from the stems, buds, or leaves:*

1. Small, soft-bodied, green or pinkish plant lice, the largest ⅙ inch long, winged or wingless, cluster on underside of leaves and on terminal shoots, sucking their sap and causing them to wilt, curl, and die.   Vines become covered with sticky honeydew..*Potato aphid*, page 646.
2. Slender, wedge-shaped, greenish or yellowish leafhoppers, under ⅛ inch long, suck sap of tomatoes and transmit a virus which causes tomato yellows or curly top; the leaves droop, thicken, become crisp, and later develop a yellow color with purple veins or, in the greenhouse, transparent veinlets............ ....*Beet leafhopper*, page 676.
3. Tiny scale-like flat nymphs, margined with a white fringe, suck sap from shaded parts of foliage, producing rolled or cupped, yellow or reddish leaves ("psyllid yellows"), killing or stunting the plants, and causing tiny malformed unmarketable fruits.. *Potato psyllid, Paratrioza cockerelli* (Sulc) (see *Jour. Econ. Ento.*, **30**:337, 891, 1937).
4. Stems and leaves develop a white fuzzy appearance almost like a white mold, owing to the development of hair-like outgrowths under which many minute mites live.. *Tomato erinose, Eriophyes cladophthirus* Nalepa (see *Fla. Agr. Exp. Sta. Bul.* 76).

D. *Insects that bore into the fruit or buds, or mine into the leaves:*

1. Fruits tunneled, soured, and decayed.   Plump striped light-green or tan to nearly black worms, 1¾ inches long when full-grown, found partly or wholly buried in the fruits which they are eating, or hiding about the ground in daytime......*Corn earworm* or *tomato fruitworm* and *climbing cutworms*, page 657.
2. Leaves of tomato, in the greenhouse or out-of-doors, have serpentine mines or large white blotches adjacent to folded leaves, which are held together by light webs.   Developing buds and ripening fruits have pinholes bored into them, especially near the stem, by tiny, yellowish, gray, or green purple-spotted brown-headed caterpillars, only ¼ inch long............................. *Tomato pinworm, Keiferia* (= *Gnorimoschema*) *lycopersicella* (Busck), Order Lepidoptera, Family Gelechiidae (see *U.S.D.A., Cir.* 440, 1937; *Mon. Bul Calif. Dept. Agr.,* **24**:301, 1935; *Jour. Econ. Ento.* **26**:137, 1933).
3. Damage similar to *D,2,* especially on the fruits, is caused by a whitish caterpillar, up to ½ inch long, with a pinkish or greenish tinge and brown head... ....*Potato tuberworm*, page 648.
4. Legless white brown-headed grubs, less than ¼ inch long, are found tunneling in the seed mass in the center of the pods of peppers.   Shining, brownish-black to grayish snout beetles, averaging ⅛ inch in length, with a single stout spine at the middle of the front femur, lay eggs in buds or fruits.................. *Pepper weevil, Anthonomus eugenii* Cano (= *A. aenotinctus* Champ) (see *Fla. Agr. Exp. Sta. Bul.* 310, 1937).
5. Fruits of peppers and eggplant are stung by barred-winged flies laying eggs in them.   Legless, nearly headless maggots, up to ½ inch long, devour the core causing the fruits to drop or decay.. *Pepper maggot, Zonosemata* (= *Spilographa*) *electa* (Say) (see *N.J. Agr. Exp. Sta. Bul.* 373, 1923).

General References.   *Cornell Univ. Ext. Bul.* 206, 1931; *Calif. Agr. Ext. Serv. Cir.* 99, 1936; 384, 1948; *Calif. Agr. Exp. Sta. Bul.* 707, 1948.

110 (66). Tomato and Tobacco Hornworms

Importance and Type of Injury.   The best-known tomato insects are the large, green, white-barred worms, up to 3 or 4 inches long, with a slender horn projecting from near the rear end (Fig. 14.33).   They eat the foliage ravenously.   They are more seriously injurious to tobacco and are known among tobacco farmers as tobaccoworms and "tobacco flies." Some people suppose that they can sting with their horns, but they are entirely unable to hurt a person in any way.

Plants Attacked. Tomato, tobacco, eggplant, pepper, potato, and related weeds.

Distribution. Both species occur throughout most of the United States, often in the same garden. The southern or tobacco hornworm[1] ranges from the northern states southward far into South America. The northern or tomato hornworm[2] ranges from the southernmost United States into Canada.

Life History, Appearance, and Habits. The winter stage of the hornworms is very often spaded up or plowed out in the spring. It is a

Fig. 14.33. Adult female moth and larva of the tobacco hornworm, *Protoparce sexta* (top), and the tomato hornworm, *P. quinquemaculata* (bottom). About one-half natural size. (*From H. T. Reynolds.*)

mahogany-brown, hard-shelled, spindle-shaped pupa about 2 inches long, with a slender tongue case projecting from the front and bent around like a pitcher handle. From these cases appear in May or June large swift-flying hawk moths or hummingbird moths, 4 or 5 inches from tip to tip of wings. They fly at dusk and hover about beds of petunias or patches of Jimson weed and other flowers with deep tubular corollas, sipping the nectar with their very long tongues. When not in use, the tongue is coiled up like a watch spring under the head. The moths are grayish or brownish in color, with white and dark mottlings. The adult of the tomato hornworm[2] may be distinguished from the adult of the other species by the two clear-cut, narrow, zigzag, dark stripes that extend diagonally across each hind wing; these stripes are indefinite and obscured in the tobacco hornworm (Fig. 14.33). The tomato hornworm, adult,

[1] *Protoparce sexta* (Johanssen), Order Lepidoptera, Family Sphingidae.
[2] *Protoparce quinquemaculata* (Haworth), Order Lepidoptera, Family Sphingidae.

has five pairs of orange-yellow spots on the abdomen, while the tobacco species has six pairs of such spots.

The moths themselves do no injury; but deposit spherical greenish-yellow eggs, one in a place, on the lower side of the leaves.   The small larvae, which hatch from the eggs in about a week, feed ravenously for about 3 or 4 weeks, during which they shed their skins five times and increase to a length of 3 or 4 inches.   The two species are similar, but may easily be distinguished by the diagonal white stripes on each side of the body.   In the tomato hornworm larvae the horn is black and there are eight stripes, each hooking backward from its lower end, forming an L or V; while in the tobacco hornworm (Fig. 14.33) the horn is red and there are seven oblique stripes which do not turn backward.

When full-grown, the larvae, by using their mouth parts and legs, dig into the soil 3 or 4 inches and change to pupae, which may go through the following winter before they transform to moths.   In the southern half of their range, however, the pupal stage lasts only about 3 weeks, and at least a part of the adults emerge and produce a second generation late in the season.

Control Measures.   Since the large caterpillars are hard to poison and the small ones hard to see, gardeners will do well to dust their tomato plants with TDE (DDD) or toxaphene at 2 to 3 pounds per acre as a routine practice.   The addition of 50 to 75 per cent sulfur to these dusts will control the tomato russet mite,[1] which sometimes causes severe defoliation.   Spraying with TDE at the same dosage will also give good control.   In small gardens the worms may often be destroyed by hand more easily than by spraying the plants, locating the well-concealed worms by their droppings.   Fall plowing destroys many of the pupae. A 5 per cent solution of tartar emetic or an organic stomach poison in isoamyl salicylate as an attractant, put out in exposed pans or feeders, one per acre, attracts and kills many of the adult moths.

These caterpillars would be much more destructive if it were not for their natural enemies.   The most commonly noticed is a braconid wasp.[2] The caterpillars are often found with small white objects covering their backs (Fig. 2.20), which are generally thought to be eggs.   They are, however, cocoons enclosing the pupal stage of the parasite.   The eggs of the parasite had been previously thrust through the skin of the horn-worm, and the larvae, after feeding within the worm body, ate out through the skin and spun the cocoons.   The adult wasps later cut out circular lids and escape from the cocoons to attack other worms.   Worms with cocoons on their backs should not be destroyed.

References.   Tenn. Agr. Exp. Sta. Bul. 93, 1911; *Ky. Agr. Exp. Sta. Bul.* 225, 1920; *U.S.D.A., Farmers' Bul.* 1356, 1923; *Jour. Econ. Ento.,* **26**:227, 1933.

111 (12, 63, 71, 260).   TOMATO FRUITWORMS

Importance and Type of Injury.   Among the most serious enemies of the tomato are the plump, greasy, greenish or brownish, striped caterpillars that eat into the fruits from the time they are formed until they are ripe (Fig. 14.34).   These insects frequently make it necessary to discard from 5 to 25 per cent of the tomatoes in commercial tomato-

---

[1] *Vasates lycopersici* (Massee) (= *Phyllocoptes destructor*), Order Acarina, Family Eriophyidae.

[2] *Apanteles congregatus* (Say), Order Hymenoptera, Family Braconidae.

growing areas.   Slightly damaged fruits, containing larvae or their shed skins, often contaminate the products from tomato canneries, causing great loss of prestige.   The worms are rather restless, and shift from one fruit to another so that a single caterpillar may spoil many fruits without eating the equivalent of a single one.   Fifty to eighty per cent of the fruits are sometimes destroyed by these caterpillars.   The most serious of these tomato fruitworms is the corn earworm[1] or tobacco false bud-worm, which has been discussed on page 498.   It also attacks the fruits of beans, peppers, okra, and eggplant but is not injurious to any of the cucurbit or cruciferous vegetables.   Certain cutworms[2] attack the fruit

Fig. 14.34. Tomatoes with larvae of the tomato fruitworm feeding in them.   About ½ natural size.   (*From Tenn. Agr. Exp. Sta. Bul.* 133.)

in the same way (page 478).   One of the worst infestations by cutworms that the writers have seen was brought about by the owner's cutting a quantity of wild grasses and spreading over the tomato fields as a mulch.

Control Measures.   Tomatoes can be protected from injury by these fruitworms by dusting or spraying with DDT or TDE (DDD) at 3 pounds per acre or toxaphene at 5 pounds per acre.   If the tomato russet mite is a problem, 50 to 75 per cent sulfur should be included.   Two to three applications are usually required, beginning when the first fruit sets and repeating at 2-week to monthly intervals.   The infested fruits should be picked and destroyed by burning or burying a foot or more deep.

References. Tenn. Agr. Exp. Sta. Bul. 133, 1925; *Ill. Agr. Exp. Sta. Cir.* 428, 1935; *Calif. Agr. Ext. Serv. Cir.* 99, 1936.

G. INSECTS INJURIOUS TO ONIONS

There are two very serious pests of the onion widely distributed in America: the onion maggot, which feeds in the bulbs; and the onion thrips, a very small, slender bug that draws the sap from the leaves.

---

[1] *Heliothis zea* (Boddie) (= *armigera*), Order Lepidoptera, Family Noctuidae.

[2] Especially the variegated cutworm, *Peridroma saucia* (Hübner), Order Lepidoptera, Family Noctuidae.

112. Onion Thrips[1]

Importance and Type of Injury.   The onion thrips is a very minute insect that punctures the leaves or stems and sucks up the exuding sap, causing the appearance of whitish blotches and dashes on the leaves.   As the attack increases in severity, the tips of the leaves first become blasted and distorted and later whole plants may wither, brown, and fall over on the ground.   The insects may be found in greatest numbers between the leaf sheaths and the stem.   The bulbs become distorted and remain undersized.   Entire fields are often destroyed by this pest, especially in dry seasons.

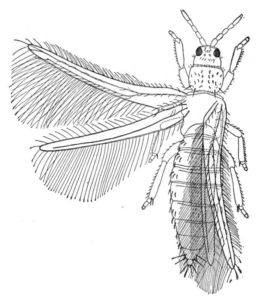

Fig. 14.35. The onion thrips, *Thrips tabaci*, adult.   Enlarged about 50 times, left wings spread to show the fringe of bristles.   Note also tarsi without claws.   (*Drawn by Kathryn Sommerman.*)

Plants Attacked.   Nearly all garden plants, many weeds, and some field crops.   Seriously injurious to onion, cauliflower, cabbage, bean, cucumber, squash, melon, tomato, turnip, beet, sweet clover, and other plants.

Distribution.   In all onion-growing sections of the United States and Canada.

Life History, Appearance, and Habits.   The adults and nymphs both winter on plants or rubbish in the fields or about weedy margins.   They are slender, yellow, active bugs, pointed at both ends, the largest of them only $\frac{1}{25}$ inch long (Fig. 14.35).   The males are wingless and very scarce—the females regularly reproducing without mating.   The females have four extremely slender wings which could hardly serve for flight except for the fringe of very long hairs on their hinder margins.   The feet

[1] *Thrips tabaci* Lindeman, Order Thysanoptera, Family Thripidae.   The flower thrips, *Frankliniella tritici* (Fitch), causes similar injury in the Western states.

also are remarkable in these insects, the tarsus ending in a small bladder, without claws. These bugs squirm in between the leaves and feed, mostly out of reach of insecticides. They rasp and puncture the surface of the leaf with their stabber-like mouth parts (Fig. 4.12) and swallow the sap, together with bits of leaf tissue. White bean-shaped eggs are thrust into the leaves or stems nearly full length and hatch in 5 to 10 days. The nymphs are very similar to the adults, but paler in color. They become full-grown in from 15 to 30 days, passing through four instars, two of which are passed in the soil and without taking food. After the fourth molt the adult females return to the plants and soon lay eggs for another generation. Eggs, nymphs, and adults are found together throughout the summer. There are usually five to eight generations a year.

Control Measures. The onion thrips may be controlled on both green and dry onions by spraying or dusting with Diazinon at 0.5 to 1 pound, malathion at 0.75 pound, parathion at 0.5 pound, or Phosdrin at 0.5 pound per acre. Two to three applications at 1- to 2-week intervals are usually required, beginning when the thrips become numerous enough to cause scarring of the leaves. Treatments applicable only to dry onions include spraying or dusting with DDT at 1.5 to 3 pounds, dieldrin at 0.5 pound, or toxaphene at 2 to 3 pounds per acre. Set onions should not be grown near seed onions. After the crop is harvested, the tops should be raked together and burned. The margins of fields should be burned over where practicable to destroy the weeds on which the thrips develop. Certain varieties of sweet Spanish onions possess considerable resistance to injury.

References. U.S.D.A., *Farmers' Bul.* 1007, 1919; *Jour. Econ. Ento.*, **27**:109, 1934; **29**:335, 1936; **30**:332, 1937; **40**:603, 1947; and **41**:378, 694, 1948.

113. ONION MAGGOT[1]

Importance and Type of Injury. In dry years the onion maggot is of little or no importance, but during the second, third, or later years of a series of wet springs, it may destroy 80 to 90 per cent of the crop. Small white maggots up to $\frac{1}{3}$ inch in length (Fig. 14.36,*d*) bore through the underground stem and into the bulbs, causing the plants to become flabby and turn yellow. They mine out the small bulbs completely, leaving only the outer sheath and causing a thin stand that is often blamed to poor seed. Larger bulbs are attacked, often by several maggots that eat out cavities which, if not completely destructive to the bulbs, cause subsequent rotting in storage.

Plants Attacked. This insect is of no importance to any crop except the onion, rarely, if ever, attacking other plants.

Distribution. Like its close relative the cabbage maggot, the onion maggot is a pest in the northern part of the United States and in Canada, rarely injurious in the South.

Life History, Appearance, and Habits. The insect winters mostly as larvae or pupae in chestnut-brown puparia (Fig. 14.36,*h*) which resemble grains of wheat, often buried several inches in the soil. They are frequently found very abundantly in piles of cull onions. Some adults may survive under the protection of sheds and trash. Those in puparia trans-

[1] *Hylemya antiqua* (Meigen), Order Diptera, Family Anthomyiidae.

form to adults and emerge from the soil over a period of several months in late spring.   The adults are slender, grayish-bodied, large-winged, rather bristly flies, only about ¼ inch long (*a, b*).   The females lay elongate white eggs about the base of the plant or in cracks in the soil.   The eggs hatch in 2 to 7 days, varying with temperature and humidity.   The maggots crawl down the plant, mostly behind the leaf sheaths, and enter the bulbs, consuming and spoiling them as stated above.   They feed for 2 or 3 weeks.   When full-grown, they are about ⅓ inch long and can be distinguished from the closely related cabbage maggot by the middle lower pair of tubercles at the rear end, which are single- and not double-pointed as in the cabbage maggot.   Pupation occurs in the soil about

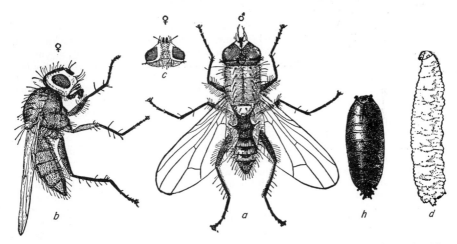

Fig. 14.36. The onion maggot.   *a*, Adult male from above; *b*, Adult female from the side; *c*, head of female; *d*, larva; *h*, puparium.   About 5 times natural size.   (*From Ill. Natural History Surv.*)

the plant and after 2 or 3 weeks the adults emerge and lay eggs for another generation.   A third generation often attacks the onions shortly before harvest, causing them to rot very badly in storage.

Control Measures.   The onion maggot may be controlled by spraying the foliage when the flies appear with Diazinon at 0.5 pound (repeat every 2 weeks) or malathion at 1.5 pound (repeat every 4 days) per acre. Other treatments include the application of Trithion at 1.5 pounds per acre in the furrow with seeds at planting and spraying seedlings with dieldrin at 1 to 2 pounds per acre.   Cull onions should be burned, buried, or hauled far away from onion land, immediately after harvest.   The first-generation adults can be attracted to rows of cull onions, planted around the margins of the field and at intervals through the field.   These come up earlier and grow faster than the seeded onions and are attractive to the egg-laying flies, thus giving partial protection to the main crop. The maggots in the culls must later be destroyed by spraying with oil, using about 1 gallon to 25 feet of row.

References.   Jour. Econ. Ento., **18**:111, 1925; **11**:82, 1918; **39**:631, 1946; **43**:950, 1950; **47**:494, 852, 1954; **50**:577, 1957; and **52**:851, 1959; *Penn. Agr. Exp. Sta. Bul.* 171, 1922; *Ill. Agr. Exp. Sta. Cir.* 337, 1936.

H. INSECTS INJURIOUS TO CABBAGE AND RELATED VEGETABLES
FIELD KEY FOR THE IDENTIFICATION OF INSECTS INJURING CRUCIFERAE

A. Insects that eat holes in the leaves and into the heads:
   1. Young plants showing many small holes with yellow margins made by the feeding of very small, jumping, black beetles with a crooked yellow stripe on each wing cover..............................*Cabbage flea beetle*, page 603.
   2. Chunky, ash-gray weevils, about ⅛ inch long, with short snouts, gouge out leaves and stems, and their white brown-headed grubs hollow out the stems.. *Cabbage curculio (Ceutorhynchus rapae* Gyllenhal) (see *U.S.D.A., Div. Ento. Bul.* 23, pp. 39, 1900).
   3. Velvety green caterpillars with a very slender orange stripe down the middle of the back; of all sizes up to 1¼ inches long, and with 8 pairs of legs and prolegs, rag the leaves and eat their way beneath the outer leaves, leaving accumulations of dirty pellets in the leaf axils.   White butterflies, nearly 2 inches from tip to tip of the wings, each of which has a few black spots on it, are usually found about the patch....................*Imported cabbageworm*, page 662.
   4. Caterpillars of similar size and habits to *A*,3, but more striped, the body smooth, and with only 6 pairs of legs and prolegs, loop over the plants, humping the back high at each "step."   Eggs laid at night by somber-brown moths with a silvery spot at the middle of each front wing...*Cabbage looper*, page 664.
   5. Small pale-green caterpillars, not over ⅓ inch long, eat small rounded holes in the leaves from the underside.   Wriggle actively when disturbed............ ....*Diamond-back moth*, page 666.
B. Insects that suck the sap from leaves and stems:
   1. Very small, whitish-green bugs or aphids rest in groups, or great soggy patches, in the heart of the plant or on the undersurface of leaves, causing leaves to cup and curl, wilt, and turn yellow.........*Cabbage and turnip aphids*, page 667.
   2. Showy, red- and black-spotted, stinking bugs, flat and shield-shaped, up to ⅜ inch long, cause the plants to wilt and die.   Not injurious north of fortieth parallel..*Harlequin bug*, page 668.
C. Insects that eat out the interior of the leaf, forming blotches but not holes through the leaf:
   1. Winding white trails or broad whitish spots appear on leaves, made by small white maggots feeding between the two surfaces of the leaf................. ....*Leaf miners*, page 670.
D. Insects that attack the roots:
   1. Plants become stunted and frequently wilt down suddenly during the day and die without apparent external cause.   Roots scarred and tunneled by white maggots up to ⅓ inch long, without legs or distinct head................. ....*Cabbage maggot*, page 670.
General Reference.   Ill. Agr. Exp. Sta. Cir. 454, 1936.

114. IMPORTED CABBAGEWORM[1]

Importance and Type of Injury.   The first-formed outer leaves of cabbage, cauliflower, and related plants, unless sprayed or dusted, are usually riddled with large holes of irregular shape and size, and the outer layers of the cabbage heads are eaten into by velvety-green worms of all sizes up to 1¼ inches long (Fig. 14.37).   If the leaves are parted, masses of greenish to brown pellets (the excrement of the worms) are found caught in the angles of the leaves.   So much of the leaf tissue is generally devoured by these worms that the growth of the plants is seriously interfered with (Fig. 1.3), the heads of cabbage and cauliflower are stunted

---

[1] *Pieris* (= *Ascia*) *rapae* (Linné), Order Lepidoptera, Family Pieridae.

or do not form at all; other leafy vegetables are rendered unfit for consumption.

Plants Attacked.   All the vegetables of the cabbage or mustard family are attacked by these worms—cabbage, cauliflower, kale, collards, kohl-rabi, Brussels sprouts, mustard, radish, turnip, horseradish, and many related weeds.   The worms also feed on nasturtium, sweet alyssum, mignonette, and lettuce.

Distribution.   This very common pest was unknown in the New World previous to 1860, when the butterflies were first taken at Quebec, Canada. Within 20 years it had spread over all the United States east of the Missis-

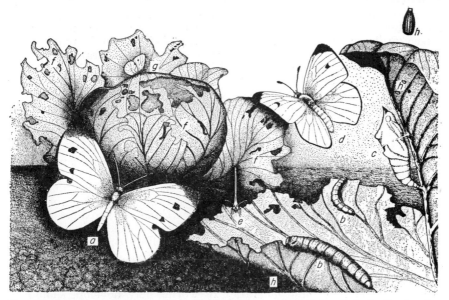

Fig. 14.37. The imported cabbageworm, *Pieris rapae.* *a,* Adult female butterfly; *b,* cater-pillars or larvae; *c,* the pupa or chrysalis, showing how it is suspended by a girdle of silk about the middle of the body and a pad at the tip of the abdomen; *d,* adult male; *e,* front view of an adult with wings folded, in the manner characteristic of butterflies; *f,* larvae and their destructive work on cabbage plant; *g,* adult with folded wings, side view; *h,* eggs, nat-ural size on the leaves and much enlarged at upper right.   (*Drawn by Kathryn Sommerman.*)

sippi River.   It now occurs as a pest throughout the United States and most of Canada.

Life History, Appearance, and Habits.   Neither the greenish worms nor their well-known white butterfly parents persist through the winter, but only the pupal stage.   This is a naked, grayish, greenish, or tan-colored chrysalid with some sharp, angular projections over its back and in front.   It is suspended from some part of the plant or on a building or other object near the cabbage patch.   The tail end of the pupa is fastened with a button of silk and kept from hanging head downward by a single loop of silk that encircles the body near the middle, like a girdle (compare Fig. 11.5,*C*).   Early in spring the familiar white butterflies (Fig. 14.37), with three or four black spots on the wings, split out of the chrysalids and fly about the gardens, alighting frequently to glue an egg to the underside of a leaf of cabbage or related plant.   In all, several hundred eggs are laid by a female.   They are just big enough to be seen, are shaped like a short, very thick bullet, are deep yellow in color, and have ridges running both lengthwise and crosswise. Each egg gives rise in about a week to a very small greenish caterpillar which feeds voraciously on the leaves and reaches a length of an inch or a little more in about 2 weeks. These caterpillars (Fig. 14.37) are an intense leaf green, except for a very slender orange stripe down the middle of the back, and another broken stripe along each side of the body, which is formed by a pair of elongate yellowish spots near each spiracle. The worm has a velvety appearance, due to numerous, close-set, short, white and black hairs that form a kind of white bloom over the body. The crawling of these caterpillars is slow and even, the body being supported by three pairs of slender legs and five pairs of fleshy prolegs. When full-grown, they frequently crawl some distance away, fasten their tails with silk to some support, spin a silken girdle about the middle of the body, and change to the pupal stage. In summer this stage lasts 1 to 2 weeks and other generations succeed until from three to six are completed.

Control Measures. The imported cabbageworm may be controlled by spraying or dusting with one of the following treatments: (*a*) Dibrom at 1 to 2 pounds, (*b*) malathion at 1.25 pounds, (*c*) parathion at 0.5 to 1 pound, (*d*) Phosdrin at 0.5 pound per acre. The following treatments should be used only before the formation of the edible portions of cabbage, cauliflower, broccoli, Brussels sprouts, and collards: (*e*) DDT at 1.5 to 2 pounds, (*f*) toxaphene at 2 to 4 pounds, or (*g*) endrin at 0.5 pound per acre. (*h*) Methoxychlor at 1.25 pound or (*i*) rotenone at 0.25 pound per acre may be used in the home garden. Mixtures of DDT and toxaphene or toxaphene and parathion may be used where the worms are difficult to control. Dusting with *Bacillus thuringiensis* toxin is an effective and selective control for cabbageworms. Weekly treatments of these materials may be required, and every effort should be made to destroy the worms while they are small.

After the crop is harvested, the old stalks should be destroyed and the field plowed. Weeds such as wild mustard, peppergrass, and shepherd's-purse, on which the first generation of worms may develop, should be destroyed. A number of natural enemies prey on the cabbageworm and in certain seasons and sections greatly reduce its numbers. The most important are the braconid wasp, *Apanteles glomeratus*, and the chalcid wasp, *Pteromalus puparum* (page 71).

References. U.S.D.A., *Farmers' Bul.* 766, 1916, and 2099, 1957; *Wis. Agr. Exp. Sta. Res. Bul.* 45, 1919; *N.Y.* (Geneva) *Agr. Exp. Sta. Bul.* 640, 1934; *Jour. Econ. Ento.,* **27**:440, 1934; **39**:184, 1946; and **41**:948, 1948.

115. CABBAGE LOOPER[1]

Importance and Type of Injury. This species attacks the plant in the same manner as the imported cabbageworm, and the two species are commonly found on the same plant. In certain seasons or sections the cabbage looper is more destructive than the imported cabbageworm.

Plants Attacked. In addition to all the plants of the cabbage family, this species also attacks lettuce, spinach, beet, pea, celery, parsley, potato, tomato, carnation, nasturtium, and mignonette.

[1] *Trichoplusia ni* (Hübner) (= *Autographa brassicae* Riley), Order Lepidoptera, Family Noctuidae.

Distribution. Throughout the United States, from Canada into Mexico; a native species.

Life History, Appearance, and Habits. The cabbage looper winters as a greenish to brownish pupa nearly ¾ inch long, wrapped in a delicate cocoon of white tangled threads attached by one side usually to a leaf of the plant on which the larva fed. The cocoon is so thin that the outline of the pupa can be seen inside. These pupae transform in the spring to moths of a general grayish-brown color, about an inch long, with a wingspread of nearly 1½ inches (Fig. 14.38). The mottled brownish front wings have a small silvery spot near the middle, somewhat resembling the figure 8; the hind wings are paler brown to bronzy. They are nocturnal and much less conspicuous about the fields than the cabbage butterflies, but nevertheless manage to lay 275 to 350 round greenish-white eggs, singly, on the upper surface of the leaves.

Fig. 14.38. Moth of the cabbage looper. Enlarged by ⅓. (*From Crosby and Leonard, "Manual of Vegetable-garden Insects," copyright, 1918, Macmillan. Reprinted by permission.*)

All injury is by the greenish larvae (Fig. 14.39) which are similar in size and habits to the imported cabbage-worms. The body tapers to the head. There is a thin but conspicuous white line along each side of the body just above the spiracles and two others near the middle line of the back. The larva has three pairs of slender legs near the head and three pairs of thicker club-shaped prolegs behind the middle. The median half of the body is without legs, and this region is generally humped up when the insect rests or moves. From this looping habit the common name is derived. Two to four weeks of feeding bring the small looper to full size. It then spins a cocoon similar to that in which the winter is passed, and, in the summer months, appears as an adult again within 2 weeks. There may be three, four, or more generations in a year, the number of worms usually increasing with each generation.

Fig. 14.39. Full-grown cabbage looper. About 1½ times natural size. (*From Crosby and Leonard, "Manual of Vegetable-garden Insects," copyright, 1918, Macmillan. Reprinted by permission.*)

Control Measures. The same measures are suggested as for the imported cabbageworm, but very thorough dusting or spraying must be done, because the worms crawl very actively and will migrate to parts of a plant that have not been covered by the insecticide. It is very important to kill these worms while they are small, as the larger ones are more difficult to kill. The looper caterpillars are often almost completely destroyed, usually late in the season, by a wilt disease which causes their bodies to rot.

References. U.S.D.A., Bur. Ento. Bul. 33, 1902; Ill. Agr. Exp. Sta. Cir. 437, 1939.

116. DIAMOND-BACK MOTH[1]

Importance and Type of Injury. This is one of the minor cabbageworms, seldom devouring more than a small percentage of the leaves. The very small caterpillars work on the underside of the leaves, eating many small holes, giving a shot-hole effect all over the leaves. In dry seasons they sometimes become abundant enough to cause appreciable injury to young cabbage. On the vegetables, of which the outer leaves are eaten, and on greenhouse plants they are more serious.

Plants Attacked. In addition to practically all the Cruciferae, the diamond-back moth attacks some ornamental and greenhouse plants such as sweet alyssum, stocks, candytuft, and wallflower.

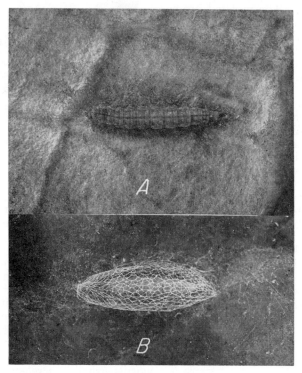

FIG. 14.40. Diamond-back moth: larva and silken cocoon. 4 times natural size. (*From Mont. Agr. Exp. Sta. Cir.* 28.)

Distribution. Introduced into the United States from Europe sometime before the middle of the nineteenth century, it now occurs wherever its host plants are grown.

Life History, Appearance, and Habits. The small grayish moths spend the winter hidden under the remnants of the cabbage crop left in the field. They are about ⅓ inch long, the folded wings flaring outward and upward toward their tips, and, in the male, forming a row of three diamond-shaped yellow spots where they meet down the middle of the back. The hind wings have a fringe of long hairs. The minute yellowish-white eggs are glued to the leaves, one, two, or three in a place, and in a few days the very small greenish larvae are at work on the underside of the leaves. They become full-grown in from 10 days to a month. They rarely exceed ⅓ inch in length, are pale-yellowish-green in color with fine scattered erect black hairs over the body (Fig. 14.40,*A*), and can be distinguished from small cabbageworms of other kinds by

[1] *Plutella maculipennis* (Curtis), Order Lepidoptera, Family Plutellidae.

their nervous habit of wriggling actively when disturbed or dropping on a silken thread. The cocoon (*B*), within which the full-grown caterpillar changes to the moth, is a beautiful gauzy sack ⅜ inch long, but so thin and loosely spun that it hardly conceals the pupa.   It is usually fastened to the underside of a leaf.   The little moth emerges from it within a week or two and promptly starts another generation, of which there may be from two to six or more a year in temperate regions.

Control Measures.   The same as for the imported cabbageworm.

References.   Jour. Econ. Ento., **30**:443, 1937; *Ill. Agr. Exp. Sta. Cir.* 437, 1939.

OTHER CABBAGEWORMS

There are several other kinds of caterpillars that often feed on cabbage and related plants and some of them may be locally more abundant than the ones described above. The potherb butterfly,[1] the southern cabbageworm,[2] and the Gulf white butterfly[3] are very closely related to the imported cabbageworm; while the cross-striped cabbageworm[4] has numerous black transverse bands across the body and is the young of a small yellowish-brown moth, and the zebra caterpillar[5] is light-yellow in color with three broad, longitudinal, black stripes, across which run many fine, yellow, transverse lines.   The cabbage webworm[6] and the purple-backed cabbageworm[7] usually feed beneath a protecting silken web or burrow into the leaves.   All these worms may be controlled by the same methods given for the imported cabbageworm, but for the last two named the applications must be made early, before the young worms gain protection under their webs or in their tunnels.

Reference.   CROSBY, C., and M. LEONARD, "Manual of Vegetable-garden Insects," Macmillian, 1918.

117 (82).   CABBAGE[8] AND TURNIP[9] APHIDS

Importance and Type of Injury.   These two plant lice, while easily distinguished by specialists, are very similar in general appearance, and indeed, were not recognized as separate species until 1914.   The nature of attack is similar, and they may be considered together.   Plants in seedbeds and at all subsequent stages of their growth are frequently covered with dense clusters of whitish-green plant lice about the size of the smallest bird shot, which suck the sap from the leaf.   The affected leaves curl and crinkle or form cups, completely lined with the aphids, and, in severe infestations, wilt and die.   The plants, if not killed, are dwarfed, grow slowly, and form small light heads not suitable for marketing.   In cases of bad infestation the entire plants become covered with a disgusting mass of the small, soggy lice, and the dying leaves and plants rapidly decay.

Plants Attacked.   The cabbage aphid is recorded from cabbage, cauliflower, Brussels sprouts, kohlrabi, collards, kale, turnip, and radish; the turnip aphid from cabbage, collards, kale, rape, mustard, rutabaga, lettuce, wild mustard, and shepherd's-purse.   Doubtless both occur on other plants of this family.

Distribution.   Both species probably occur throughout North America wherever their host plants grow.

---

[1] *Pieris oleracea* Harris, Order Lepidoptera, Family Pieridae.

[2] *Pieris protodice* Boisduval and LeConte, Order Lepidoptera, Family Pieridae.

[3] *Pieris monuste* Linné.

[4] *Evergestis rimosalis* (Guenée), Order Lepidoptera, Family Pyralididae.

[6] *Ceramica* (= *Mamestra*) *picta* (Harris), Order Lepidoptera, Family Noctuidae.

[6] *Hellula rogatalis* (= *undalis*) (Hulst), Order Lepidoptera, Family Pyralididae.

[7] *Evergestis pallidata* (= *straminalis*) (Hufnagel), Order Lepidoptera, Family Pyralididae.

[8] *Brevicoryne brassicae* (Linné), Order Homoptera, Family Aphidae.

[9] *Rhopalosiphum pseudobrassicae* (Davis), Order Homoptera, Family Aphidae.

Life History, Appearance, and Habits. The life cycle is, in general, like that given as typical for aphids (page 613). The cabbage aphid winters in the northern states as small black fertilized eggs laid in depressions upon the petioles and underside of leaves of cabbage. The turnip aphid probably winters in a similar way, although the sexual individuals and eggs of this species have not been described. Farther south the species continues to reproduce ovoviviparously throughout the winter. In cage experiments Paddock carried the turnip aphid through 25 generations in 12 months in Texas, while 16 generations of the cabbage aphid have been observed from April to October. When their food becomes unsatisfactory from any cause, winged females are developed which spread the species from plant to plant and start new families wherever they alight. Each female commonly produces from 80 to 100 young during her lifetime of about a month.

Control Measures. The control measures recommended for aphids on page 613 are effective for these species. Dusts containing malathion, parathion, Phosdrin, or TEPP will penetrate into and under the leaves very effectively. These dusts should be applied when the plants are dry and the temperature above 70°F. On account of the waxy powder that covers the bodies of the aphids, and the tendency of the leaves to form pockets or cups in which the aphids are protected, it is essential, when sprays are used, to add a good spreader, to use a spray pressure of 200 pounds or more, to apply 100 to 125 gallons per acre, and to repeat 2 or 3 days later. The destruction of the old stalks of cabbage and other crops as soon as the crop is harvested will help to prevent destructive outbreaks of these aphids.

References. U.S.D.A., Dept. Cir. 154, 1921; *Purdue Agr. Exp. Sta. Bul.* 185, 1916; *Tex. Agr. Exp. Sta. Bul.* 180, 1915.

118. HARLEQUIN BUG[1]

Importance and Type of Injury. The harlequin bug, "fire bug," or "calico back" is the most important insect enemy of cabbage and related crops in the southern half of the United States. It often destroys the entire crop where it is not controlled. It sucks the sap of the plants, taking its food entirely from beneath the surface, sapping them so that they wilt, brown, and die. The gaudy, red-and-black-spotted, stinking bugs (Fig. 14.41), about ⅜ inch long, flat and shield-shaped, and the smaller similar-looking nymphs have a very characteristic pattern. They may be found in all stages of development from early spring to winter, and dozens to the plant in severe cases.

Plants Attacked. Horseradish, cabbage, cauliflower, collards, mustard, Brussels sprouts, turnip, kohlrabi, radish, and, in the absence of these favorite foods, tomato, potato, eggplant, okra, bean, asparagus, beet, and many other garden crops, weeds, fruit trees, and field crops.

Distribution. This is a southern insect ranging from the Atlantic to the Pacific, and rarely if ever injurious north of about the fortieth parallel. It first spread over the South from Mexico, shortly after the Civil War, and it was believed by many that it was brought in by the Yankee troops.

Life History, Appearance, and Habits. Throughout most of its range the insect continues to feed and breed during the entire year. Farther north the approach of winter drives the bugs into the shelter of recumbent

[1] *Murgantia histrionica* (Hahn), Order Hemiptera, Family Pentatomidae.

cabbage stalks, bunches of grass, and other rubbish, and only the adults survive severe winter weather.   The first warm days of spring tempt them out of hiding, and they begin feeding on weeds, being ready to lay eggs by the time the earliest garden plants are set out.   The eggs are laid mostly on the underside of the leaves.   They are like tiny white kegs standing on end in double rows, about a dozen glued together, each "keg" bound with two broad black hoops and with "round black spots set in the proper place for bungholes" (Fig. 14.41,b).   The eggs hatch in from 4 to 29 days, the time varying with the temperature, and the very young

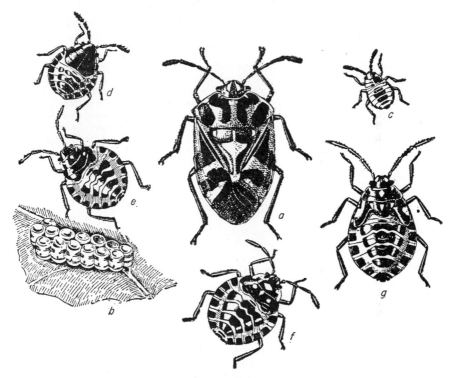

FIG. 14.41. Harlequin cabbage bug, *Murgantia histrionica*.   a, Adult; b, egg mass; c, first-stage nymph; d, second-stage nymph; e, third-stage nymph; f, fourth-stage nymph; g, fifth-stage nymph.   All enlarged about 3 times.   (*From U.S.D.A., Farmers' Bul. 1061.*)

bugs begin the business of destroying the plants.   They feed and grow for 4 to 9 weeks, passing through five distinct instars before they are capable of mating and laying eggs for a second generation.   Three generations, and a partial fourth, may succeed each other before cold weather puts a stop to their rapid increase.

Control Measures.   This bug may be satisfactorily controlled by spraying or dusting with DDT at 1.25 pounds or Thiodan at 0.75 pound per acre.   After the plants have begun to head, the bugs may be controlled by dusts or sprays of rotenone or by dusting with 20 per cent sabadilla. Hand destruction of the adult bugs in the fall and spring as they come out of hibernation and before they have begun egg-laying is an effective control.   This may be facilitated by the use of trap crops of mustard, kale, turnip, or radish, planted very early in spring or late in fall after the main crop is harvested.   When the bugs have concentrated on these small patches, they should be killed by spraying with kerosene, or by covering the trap crop with straw and burning.   Trap crops should never be used unless they can be given careful attention to destroy the bugs attracted to them.   Weeds such as wild mustard, Amaranthus, and others of the mustard family should be kept down.   After the bugs have been reduced by the diligent use of the above methods, the stragglers that remain must be cleaned up by hand-picking early in the day and destroying the bugs and their egg masses wherever found.

References.   U.S.D.A., Farmers' Bul. 1712, 1933; *Jour. Econ. Ento.,* **41**:808, 1948.

119. LEAF MINERS

This group of plants is often disfigured and damaged by several species of small flies that live in the maggot stage by eating the tissue of the leaves, between the upper and lower surfaces.   Their feeding causes the production of large whitish blotches or blasted areas, or, in the case of the serpentine leaf miner,[1] slender, white, winding trails through the interior of the leaf.   The leaves are greatly weakened and the mines serve as points where disease and decay may start, but the chief loss is to those vegetables of which the green leaves are eaten and which are rendered unattractive and unsalable by these flies.   The leaf miners may be controlled by spraying or dusting with parathion at 0.4 pound or Diazinon at 0.5 pound per acre.

References.   Jour. Agr. Res., **1**:59, 1913; *Jour. Econ. Ento.,* **40**:496, 1947; **41**:653, 1948; and **51**:357, 1958; *Fla. Agr. Exp. Sta. Press Bul.* 639, 1947.

120. CABBAGE MAGGOT[2]

Importance and Type of Injury.   Plants attacked by the cabbage maggot appear sickly, off color, and runty, and, if the attack is severe, they wilt suddenly during the heat of the day and die.   Roots of cabbage, cauliflower, rape, and the fleshy parts of turnips and radishes show brownish grooves over their surface and slimy winding channels running through the flesh, while many of the small fibrous roots are eaten off.   Legless white maggots, from $\frac{1}{4}$ to $\frac{1}{3}$ inch long, blunt at the rear end and pointed in front, are often found in these burrows.

In most sections early cabbage after transplanting, late cabbage while still in the seedbed, early turnips, and late spring radishes are most severely injured.   Very early radish and late cabbage, after being set out, usually escape injury.   The pest fluctuates much in abundance in different sections and years, but frequently 40 to 80 per cent of the plants are destroyed, resulting in the loss of thousands of dollars annually in many of the states where it occurs.

Plants Attacked.   This fly is chiefly injurious to plants of the mustard family or Cruciferae, such as cabbage, cauliflower, broccoli, Brussels sprouts, radish, and turnip, but also attacks beet, cress, celery, and some other vegetables to a slight extent.

Distribution.   Introduced from Europe early in the nineteenth century, it has spread widely over North America.   It is a serious pest in Canada and the northern part of the United States but is seldom injurious south of the fortieth degree north latitude.

---

[1] *Liriomyza brassicae* (Riley) (= *Agromyza pusilla*), Order Diptera, Family Agromyzidae.

[2] *Hylemya brassicae* (Bouché), Order Diptera, Family Anthomyiidae.

Life History, Appearance, and Habits. This insect goes through the winter chiefly as a pupa in a hard brown egg-shaped puparium, about ¼ inch long, and buried from 1 to 5 inches in the soil. In spring, about the time early cabbage plants are set out (mid-May in the latitude of Chicago) the end of the puparium is broken open and a small gray fly emerges and crawls out of the soil. These flies (Fig. 14.42) are similar in general appearance to the common house fly but only about half as long (¼ inch long), dark ashy gray with black stripes on the thorax and many black bristles over the body. The cells of the wing that open nearest to its tip are both wide open at the margin. They fly about close to the ground and deposit their small, white, finely ridged eggs on the plants near where the stem meets the ground or in cracks and crevices in the soil.

FIG. 14.42. Female of the cabbage maggot, *Hylemya brassicae.* About 4 times natural size. *(From Can. Dept. Agr.)*

FIG. 14.43. Root of cabbage showing cabbage maggots and their destructive work. *(From Can. Dept. Agr.)*

Three to seven days later, the eggs hatch and the very small maggots promptly seek the roots and eat into them. Each larva feeds for 3 to 4 weeks, and the roots often become riddled with their tunnels (Fig. 14.43). The larva has at the blunt rear end 12 short, pointed, fleshy processes arranged in a circle around the two button-like spiracles. The two processes nearest the middle line below are double-pointed.

When the maggots are abundant, the underground parts of the plants soon become honeycombed and rotten. Over 125 maggots have been taken from the roots of a single plant. Upon completing its growth the larva may pupate in its burrow, but more generally crawls away from the root into the soil a short distance and there forms its puparium. Two or three weeks later, on the average, the adults break out of the puparium and may push up through the soil from a depth of 6 inches or more. Undoubtedly some of these puparia of the first generation remain until the following spring, but most of them transform to adults in late June and July, and lay eggs upon late cabbage and other plants. In most sections the injury from this second generation during dry midsummer weather is not severe, since the insect requires cool moist weather and succulent plants in which to thrive. Enough transform, however, to produce a partial third generation in autumn when they are sometimes very destructive to fall radishes and turnips. In some sections a partial fourth generation has been reported.

Control Measures. Satisfactory control of this insect may be obtained by thoroughly dusting the plants before transplanting with 2.5 per cent aldrin, 1.5 per cent dieldrin, or 5 per cent chlordane applied to the lower stem and roots. Immediately after transplanting, the rows may be treated by spraying or dusting the soil at the base of the plants with aldrin at 1.5 pounds or chlordane at 3 pounds per acre. The application should be repeated weekly as needed. Seedbeds may be treated before planting by broadcasting chlordane at 5 pounds, aldrin at 2.5 pounds, or dieldrin at 1.5 pounds per acre as spray or dust and mixing thoroughly into the upper 2 inches of soil.

Seedbeds may be protected from these flies and from flea beetles and other pests by covering them with thin cloth, such as hospital gauze, having 20 to 30 threads to the inch, securely tacked to the framework around the bed and supported by wires across the beds at intervals of 5 or 6 feet.

References. N.Y. (Geneva) Agr. Exp. Sta. Bul. 442, 1917, and 419, 1916; *Proc. Ento. Soc. Nova Scotia,* 1919, p. 41, 1920; *Can. Agr. Dept., Ento. Bul.* 12, p. 9, 1916; *Conn. Agr. Exp. Sta. Bul.* 338, 1932; *Jour. Econ. Ento.,* **25**:709, 1932; **41**:98, 362, 865, 1948; **43**:899, 1950; and **49**:354, 1956.

I. INSECTS INJURIOUS TO BEET, SPINACH, LETTUCE, CARROT, PARSNIP, CELERY, AND RELATED VEGETABLES

FIELD KEY FOR IDENTIFICATION

A. *Insects chewing the foliage, tender stem, or seed heads:*
  a. *Insects feeding exposed upon the foliage:*
    1. Small, greenish-black, hopping beetles, ¼ inch long, with a broad yellow collar behind the head, and grayish to purple, short, warty grubs (up to ⅓ inch long) eat small holes in the leaves or skeletonize them...........
    ....*Spinach flea beetle,* page 674.
    2. Elongate, nearly cylindrical, rather soft-bodied long-legged beetles, with head and prothorax well marked off and tip of abdomen exposed beyond the wing covers, and either black, gray, or striped yellow and black, swarm upon beets and rapidly devour the tops................*Blister beetles,* page 642.
    3. Yellowish-brown hopping beetles, ⅛ inch long, with a yellow stripe lengthwise of each wing cover, which divide the back into 5 stripes of equal width, eat tiny "shot holes" in the leaves of beets and beans from the underside..
    ....*Pale-striped flea beetle,* page 603.
    4. Small cinnamon-brown velvety May beetle-like pests, only ⅜ inch long, eat irregular holes from the margins of leaves on warm nights...............
    ....*Asiatic garden beetle,* page 606.
    5. Large green caterpillars, up to 2 inches long, with a black crossband and 6 yellow spots on each segment, eat the leaves of celery, parsnip, and carrot
    ....*Black swallowtail butterfly,* page 675.
    6. Lettuce, celery, and beets are frequently attacked by a pale-green caterpillar, up to 1¼ inches long, narrowing gradually to the head, and with light and dark stripes. They have only 3 pairs of prolegs and consequently crawl with a looping movement..
    Celery looper, Anagrapha (= *Autographa*) *falcifera* (Kirby). Control as for

celery leaf tier (see *Eleventh Rept. Ill. State Ento.*, pp. 38–43, 1882; *Fla. Agr. Exp. Sta. Bul.* 250, 1932).

b. *Caterpillars working under the protection of a silken web:*

1. Lettuce, beet, cabbage, cucurbits, pea, beans, potato, and tomato are attacked by yellowish-green worms, up to 1 inch long, with scattered hairs and conspicuous black spots, feeding within light webs of silk that they spin over the plants, especially near the ground or in the soil................ ....*Garden webworm*, page 557.

2. Sugarbeets, garden beets, carrots, cabbage, peas, spinach, beans, potatoes, squash, and other foliage are ragged and webbed by pale-green caterpillars, up to an inch in length, with one middorsal and two lateral dark stripes. Worm retreats into a tubular, silken tunnel when disturbed.............. ....*Beet webworm*, page 609.

3. Caterpillars similar to *b*, 2, but with a light, middorsal stripe.............. ....*Alfalfa webworm*, page 609.

4. The flower heads of parsnip and celery are webbed together and devoured by small yellowish to grayish-green black-spotted caterpillars, the largest about $\frac{3}{5}$ inch long, which also mine in the stems.....*Parsnip webworm*, page 674.

5. Celery, beet, and spinach foliage is ragged by greenish watery-looking white-striped caterpillars, the largest $\frac{3}{4}$ inch long, which work on the underside of leaves or fold and web the leaves together and feed within. Squirm actively when disturbed...................*Greenhouse* or *celery leaf tier*, page 675.

B. *Insects sucking the sap from leaves and stems:*

1. Slender wedge-shaped greenish or yellowish leafhoppers, under $\frac{1}{8}$ inch long, suck sap of beets and poison them, causing the rolling and shriveling of leaves and the appearance of warts along the veins.......*Beet leafhopper*, page 676.

2. Pale-green or pinkish, winged or wingless aphids, only $\frac{1}{12}$ inch long, suck the sap of spinach, beets, celery, and about 100 other plants, causing stunting, wilting, and unmarketable condition of the plants............................... ....*Green peach aphid* or *spinach aphid*, page 676.

3. Broad, oval, black bugs, up to $\frac{1}{10}$ inch long, convex above, suck the sap of celery and other plants causing the leaves to wilt and die. They give off a vile odor when crushed................................*Negro bug*, page 677.

4. Very active, shy, flat-backed, sucking bugs, less than $\frac{1}{4}$ inch long, cause wilted leaves or large brown dead spots on the stem of celery and related vegetables, known as "black joint".....................*Tarnished plant bug*, page 614.

C. *Insects mining in the leaves:*

1. Blister-like or blasted spots on the leaves of spinach, chard, beets and related plants are made by small maggots, not over $\frac{1}{3}$ inch long, eating the interior of the leaf without consuming either surface.......*Spinach leaf miner*, page 677.

D. *Insects attacking the roots:*

1. Aphids cluster on the roots of beets and other vegetables, sucking the sap and forming moldy white-looking clumps.......................... ....*Sugarbeet root aphid* and others, page 678.

2. Stout, broad, reddish-brown, spiny-legged beetles, $\frac{1}{2}$ inch long by $\frac{1}{4}$ inch broad, gouge out unsightly holes in the roots of celery, carrots, parsnips, sugarbeets, sunflowers, and other vegetables and field crops..*Carrot beetle*, page 678.

3. Irregular zigzag dark grooves over the surface, or burrows through the roots of carrots and tunnels in the stalks and heart of celery, parsley, and dill are made by fat, white, legless grubs....................................... *Carrot weevil, Listronotus oregonensis* (LeConte) (see *Jour. Econ. Ento.*, **20 :** 814, 1927, and **31 :**262, 1938; *Cornell Ext. Bul.* 206, 1931).

4. Roots of plants chewed off by a somewhat slender, whitish grub, with a Y-shaped anal opening and a transverse row of short setae in front of it ....*Asiatic garden beetle*, page 606.

5. Yellowish-white maggots, about $\frac{1}{3}$ inch long, chew off the small roots of celery and the bottom of the taproot of carrots and parsnips causing plants to yellow and make stunted growth. Roots may be riddled with rust-red burrows and surface scars...................................*Carrot rust fly*, page 679.

121 (76). SPINACH FLEA BEETLE[1]

Importance and Type of Injury.   Small holes are eaten in the leaves, or the leaves are skeletonized from beneath, by small, jumping, greenish-black beetles, with a yellow collar behind the head, and by grayish to purple, warty, short, cylindrical worms, all under $\frac{1}{4}$ to $\frac{1}{3}$ inch long.

Plants Attacked.   Spinach, beet, and pigweed, chickweed, lamb's-quarters, and other weeds.

Distribution.   General east of the Rocky Mountains.

Life History, Appearance, and Habits.   The insect winters as a greenish-black oval beetle, $\frac{1}{5}$ to $\frac{1}{4}$ inch long, with a yellow prothorax (Fig. 14.5).   In April and May the beetles appear on the plants and lay small clusters of orange eggs (Fig. 14.3) placed on end at the base of the plant or on the soil near by.   The dirty-gray to purplish young or larvae (Fig. 14.4) feed on the underside of the leaves, becoming $\frac{1}{4}$ to $\frac{1}{3}$ inch long within 2 to 4 weeks.   They are very warty, cylindrical grubs, each wart terminating in a short black hair.   When disturbed, the larvae and beetles "play 'possum" and drop to the ground.   A pupal stage of a week or 10 days is passed in the soil, and the new adults appear in July, lay eggs over a period of nearly 2 months, and the second generation matures before winter sets in.

Control Measures.   Control measures are given on page 605.

Reference.   *U.S.D.A., Farmers' Bul.* 1371, pp. 12, 13, 1934.

122. PARSNIP WEBWORM[2]

Importance and Type of Injury.   The flower heads of parsnip and celery are webbed together with silk and devoured by small, yellow, greenish, or grayish caterpillars covered with small black spots and short hairs. They interfere seriously with the production of celery and parsnip seed. After consuming the unripe seed the caterpillars mine in the stems, and when full-grown are about $\frac{3}{5}$ inch long.

Plants Attacked.   Parsnip, celery, wild parsnip, wild carrot, and related weeds.

Distribution.   Southern Canada and the northern states east of the Mississippi.

Life History, Appearance, and Habits.   The grayish moth, an inch across the wings, winters under loose bark and in other protection and lays its eggs in late spring on the developing flower head and other parts of the plant.   After destroying the flower buds and seeds, the caterpillars pupate in their mines, emerging as adults in late summer, when they seek hibernating places.   The insect is especially abundant on the heads of wild parsnips.

Control Measures.   The suggestions given under webworms (page 610) are effective, but much of the damage has usually been done before the caterpillars are noticed.   Injured flower heads can be cut and burned in August, before the moths emerge.   Wild host plants should be destroyed about the farm.

Reference.   *Ont. Dept. Agr. Bul.* 359, 1931.

---

[1] *Disonycha xanthomelas* (Dalman), Order Coleoptera, Family Chrysomelidae.
[2] *Depressaria heracliana* (Linné), Order Lepidoptera, Family Oecophoridae.

123 (256). CELERY OR GREENHOUSE LEAF TIER[1]

Importance and Type of Injury. This insect, which is more fully discussed as the greenhouse leaf tier (page 861), is a major pest of celery culture and also destructive to spinach, beans, and beets grown in the field, especially following mild winters. It usually appears in injurious numbers within a few weeks of harvesting the crop, feeding upon the tenderest growth just above the heart of the plant, covering the leaves with its webs and excrement and greatly lowering the value of the crop. For descriptions of the insect see page 861. There are three or four generations a year in the South.

Control Measures. On plants less than 2 months old, this insect may be controlled by spraying or dusting with DDT at 1.25 pounds per acre. On more mature plants pyrethrum dust containing 0.1 to 0.2 per cent pyrethrins or the equivalent of pyrethrins plus activator such as piperonyl cyclonene, applied at 25 to 35 pounds per acre, is effective. Spraying with pyrethrum, 0.5 ounces per acre, is also effective. This application should be directed into the celery hearts and will kill many of the larvae and drive the others from their webs. A second application about ½ hour later will destroy these exposed worms. Early harvesting of infested crops and plowing under of all crop refuse immediately after harvest will aid in control.

Reference. Fla. Agr. Exp. Sta. Buls. 250 and 251, 1932.

124. BLACK SWALLOWTAIL BUTTERFLY, CELERYWORM, OR PARSLEYWORM[2]

Importance and Type of Injury. Large green caterpillars, with a black crossband on each segment, which is indented by six yellow spots on its front margin, ranging up to 2 inches in length, eat the foliage of celery and related plants, stripping the leaves clean as they go (Fig. 6.58,*a,b*). They sometimes seriously injure young plants but are usually more noticeable because of their gaudy appearance than for the injury they do.

Plants Attacked. Celery, dill, parsnip, carrot, parsley, caraway, and many other plants of the same family.

Distribution. Throughout North America east of the Rocky Mountains. In the West it is replaced by the western parsley caterpillar.[3]

Life History, Appearance, and Habits. In the northern states the winter is passed as a dirty tan-colored chrysalid suspended from the host plants and other objects by a silk button and girdle, as described for the imported cabbageworm (page 663). The adults are large black swallowtail butterflies, expanding nearly 4 inches, with numerous yellow spots on the outer part of the wings and also a row of blue patches on the hind wing (Fig. 6.58,*c*). The eggs are scattered about on the leaves, in May and June, and hatch into the curious caterpillars. When disturbed, the caterpillars protrude from the head end two, soft, orange-colored horns (known as *osmeteria*), that give off a sickening sweet odor which is probably a protection from some enemies. The larval stage lasts about a month, and the pupal stage 9 to 15 days in summer. There are two or three generations a year.

Control Measures. The insecticidal treatments described for the imported cabbageworm (page 664), will readily destroy the worms. DDT, toxaphene, or endrin should not be applied after bunching or when plant refuse is to be fed to livestock.

Reference. SCUDDER, "Butterflies of the Eastern United States," p. 1353, 1889.

[1] *Udea* (= *Phlyctaenia*) *rubigalis* (Guenée) (= *P. ferrugalis* Hübner), Order Lepidoptera, Family Pyralididae.

[2] *Papilio polyxenes asterius* Stoll (= *ajax* Linné), Order Lepidoptera, Family Papilionidae.

[3] *Papilio zelicaon* Lucas, Order Lepidoptera, Family Papilionidae.

125. Beet Leafhopper[1]

Importance and Type of Injury. Beets are attacked by a disease known as "curly top" or blight, which is caused by the feeding of a small leafhopper. This stunts the plants, kills them, or greatly reduces the sugar content of the beets and the crop of seed. The leaf veins become warty, the veinlets transparent, the petioles kinked, and the leaves rolled upward at the edges, brittle, and shriveled. In many localities the growing of sugarbeets has been abandoned because of this pest. The attack is generally sporadic, and there is no way of predicting when a destructive outbreak will occur. This insect is also the carrier of tomato yellows.

Plants Attacked. Sugarbeets, table beets, mangels, tomatoes, and certain weeds, a total of more than 100 kinds of plants.

Distribution. Western United States from Canada to Mexico and east to Missouri, Illinois, and Texas.

Life History, Appearance, and Habits. The cause of the trouble known as curly top is a small wedge-shaped leafhopper, of a pale-greenish or yellow color, often with some darker blotches, about ⅛ inch long, with long slender hind legs that enable it to jump quickly into the air. It also flies readily, and when flying looks like a tiny white fly. The adult females winter on and about their wild food plants such as salt bush, Russian thistle, greasewood, wild mustard, filaree, and sea blite. Two generations usually develop on sugarbeets. Egg-laying in these desert host plants begins by early or mid-March, and the first generation normally matures on these plants. From early May to early June the first-generation adults may fly for hundreds of miles in great swarms, alighting in beet fields wherever the crop is up. The adults, and later the nymphs, feed by inserting the slender mouth parts into the plant and introducing the virus that causes the curly top condition to develop. The eggs are inserted full length into the veins, leaf petioles, or stems, and hatch in 2 weeks to tiny, pale-colored wingless nymphs that settle in the center of the plant. In from 3 weeks to 2 months the bugs are full-grown. There may be from one to three or more generations.

Control Measures. The feeding of this insect causes the disease, curly top, and although the leafhopper is readily killed by insecticides (page 646), it is very difficult to prevent disease transmission. Considerable success has been obtained in California by removal of the winter host plants such as Russian thistle and by spraying the overwintering leafhoppers on these plants with 1 to 1½ pounds of DDT per acre in 3 gallons of kerosene. Early planting and thorough cultivation are recommended to produce a crop in spite of the presence of the leafhoppers. The insect is generally less destructive in seasons following severe winters and in areas of high rainfall. Some progess has been made in developing resistant varieties of sugarbeets.

References. Utah Agr. Exp. Sta. Bul. 234, 1932; U.S.D.A., Tech. Bul., 855, 1943; Hilgardia, **7**:281–350, 1933; Jour. Econ. Ento., **27**:945, 1934; Bul. Calif. Dept. Agr. **37**:216, 1948.

126 (82, 181). Green Peach Aphid (Spinach Aphid)[2]

This insect, which is further discussed as the green peach aphid, has been very destructive to spinach in the large trucking sections of the Atlantic Coast. It has

[1] *Circulifer* (= *Eutettix*) *tenellus* (Baker), Order Homoptera, Family Cicadellidae.
[2] *Myzus persicae* (Sulzer), Order Homoptera, Family Aphidae.

been estimated that $750,000 worth of damage was done to the spinach crop in Virginia in a single year (1907). It also attacks celery, lettuce, beets, tomato, eggplant, potato, the Cruciferae, cucurbits, and other vegetables. Its life cycle on the peach is discussed on page 754. Control measures for spinach and other vegetables are given on page 613.

127. NEGRO BUG[1]

Importance and Type of Injury. Short, oval, black bugs (Fig. 14.44), about $\frac{1}{10}$ inch long, together with small black and reddish nymphs, sometimes congregate on celery, corn, wheat, and other crops and suck out the sap, stunting the plants and causing the leaves to wilt and die. Outbreaks of the bug in Michigan and Ohio have destroyed thousands of dollars worth of celery, but the pest occurs only sporadically. Secretions of the bugs give a foul taste to raspberries and blackberries over which they crawl.

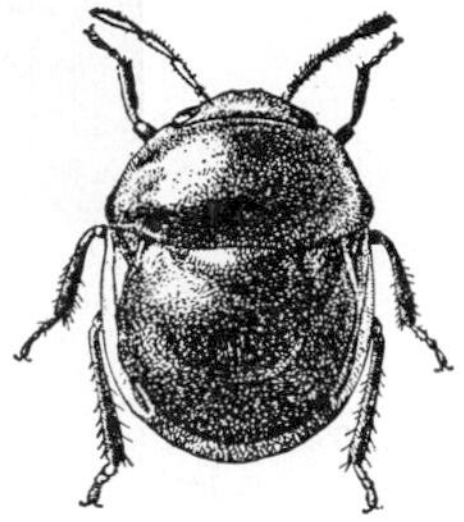

Plants Attacked. Celery, corn, wheat, other grasses, some ornamental flowers, and many weeds, such as beggar-ticks, Lobelia, and Veronica.

Distribution. Throughout the United States and Canada, east of the Rocky Mountains.

Life History, Appearance, and Habits. The insects winter as adult bugs which are often mistaken for beetles because of the hard shell over the back. This is not formed of the two wing covers but is a greatly enlarged thoracic shield, under the edges of which the wings slip when the bug comes to rest. The eggs are laid singly on the leaves and hatch in about 2 weeks into reddish nymphs which gradually grow into the form of the adult as they feed on the plants. They become adult by midsummer, and, after feeding for a few weeks, seek hibernating places long before cold weather.

Fig. 14.44. The negro bug, *Corimelaena pulicaria*, adult. About 10 times natural size. *(From Ill. Natural History Surv.)*

Control Measures. Weeds on which the bugs feed should be destroyed. When they attack cultivated crops a spray of nicotine sulfate, 1 quart, water 100 gallons, and soap, 8 pounds, will kill all the bugs hit by it.

Reference. Mich. Agr. Exp. Sta. Bul. 102, 1893.

128. SPINACH LEAF MINER[2]

Importance and Type of Injury. This insect is of less importance in the United States and Canada than in Europe. Blasted spots or blister-like blotches appear on the leaves of spinach, chard, and related plants, where small maggots have eaten out the tissue of the leaf between upper and lower surfaces. The leaf vegetables are rendered unfit for greens, and the development of seeds and roots, such as beets, is decreased by the partial defoliation.

Plants Attacked. Spinach, beet, sugarbeet, chard, mango, and many weeds, including chickweed, lamb's-quarters, and nightshade.

Distribution. Probably introduced from Europe previous to 1880, it is now generally distributed over the United States and Canada.

Life History, Appearance, and Habits. The winter is probably passed mostly in puparia (Fig. 14.45,*3*) in the soil. In April and May the slender-bodied, grayish, black-haired, two-winged flies (*5*), about $\frac{1}{4}$ inch long, appear in the fields, and the females deposit small white eggs (Fig. 14.45,*1*), one to five side by side, on the underside of the leaves. Upon hatching, the tiny maggot at first eats a slender winding mine in the leaf, but as it increases in size the mine is widened to form a blotch that often joins the mines of other maggots in the same leaf. The maggots (*2*) may migrate from leaf to leaf, and become full-grown in 1 to 3 weeks. Pupation takes place chiefly in the upper 2 or 3 inches of the soil, but some transform among trash on the ground or even in the larval mines. Within 2 to 4 weeks the adults appear from the pupae and start a new generation. Three or four generations may be completed during the season.

[1] *Corimelaena* (= *Allocoris*) *pulicaria* (Germar), Order Hemiptera, Family Cydnidae.

[2] *Pegomya hyoscyami* (Panzer), Order Diptera, Family Anthomyiidae.

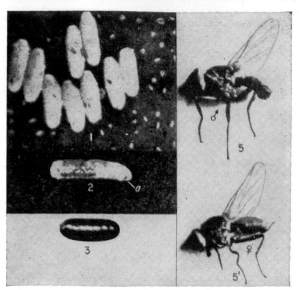

FIG. 14.45. Spinach leaf miner, *Pegomya hyoscyami*. *1*, Egg on leaf, greatly magnified; *2*, maggot or larva, the mouth hooks at *a*; *3*, puparium; *5*, male adult; *5'*, female adult, about 4 times natural size. (*From N.Y.* (*Geneva*) *Agr. Exp. Sta. Bul.* 99.)

Control Measures. The increase of the flies may be checked by destroying their host weeds. Screening the beds with cheesecloth or tobacco cloth, as suggested on page 593, will keep the plants free of this pest. The screen should be removed a week before harvesting. Spinach grown very late in fall, in winter, or very early in spring may escape injury. Spraying or dusting with parathion at 0.4 pound or Diazinon at 0.5 pound per acre is effective.

References. *N.Y.* (*Geneva*) *Agr. Exp. Sta. Bul.* 99, 1896; *Ann. Appl. Biol.*, **1**:43–76, 1914.

129. SUGARBEET ROOT APHID[1]

Aphids are frequently found on the roots of the vegetables in this group. One of the most destructive is the sugarbeet root aphid, which is found in the western half of the United States on the roots of sugarbeets, beets, mangels, and many weeds, such as lamb's-quarters, yarrow, dock, goldenrod, and grasses. It reduces both the size and the quality of the beets by sucking the sap from the roots.

The insects winter in part as fertilized eggs on the bark of poplar trees, and in part as wingless females on the roots of herbaceous plants. The insects are yellow in color and have a mass of fine cottony-looking waxy threads toward the end of the body, so as to appear like white mold on the roots. A migration of winged aphids from the poplars to beets takes place in July, and a return migration to poplars in September and October.

Control Measures. Where beets are grown on irrigated land, aphids can be kept in control and the yields increased by giving five or more irrigations at 10-day intervals, during July and August.

Reference. *Utah Agr. Exp. Sta. Leaflet* 41, 1934.

130. CARROT BEETLE[2]

Importance and Type of Injury. The roots of celery, carrot, and parsnip are gouged by the feeding of broad reddish-brown stout-legged beetles, about ½ inch long, and slightly over half as wide (Fig. 14.46).

[1] *Pemphigus betae* Doane, Order Homoptera, Family Aphidae.
[2] *Bothynus* (= *Ligyrus*) *gibbosus* (De Geer), Order Coleoptera, Family Scarabaeidae.

Plants Attacked.   Carrot, parsnip, celery, beet, potato, cabbage, corn, cotton, sunflower, dahlia, and other crops and weeds, especially Amaranthus.

Distribution.   Over much of the United States except the most northern states.

Life History, Appearance, and Habits.   The adult beetles winter in the soil to a depth of 4 feet, emerge in spring, and lay eggs at night in the soil.   The eggs increase greatly in size before hatching.   The larvae resemble the common white grubs, being curved and white with a bluish cast and red-brown heads.   They feed largely on grasses and decaying vegetation in the soil but often attack the roots of crops, a dozen or more beetles sometimes being found around a single plant.   A generation is completed in a year, the adults being present and injurious from late April to August.

Control Measures.   Row treatment with dieldrin at 3 pounds per acre has given satisfactory control.

References   *Jour. Econ. Ento.*, **10**:253, 1917; and **50**:369, 1957; *Ill. Agr. Exp. Sta. Cir.* 437, 1939.

131. CARROT RUST FLY[1]

Importance and Type of Injury.   Celery plants, after getting a good start, wilt, and the outer leaves turn yellow, as a result of the eating off of most of the fibrous roots by a very slender yellowish-white legless maggot, about ⅓ inch long when full-grown.   Carrots and parsnips become stunted, and the lower end of the taproot

FIG. 14.46. The carrot beetle, *Bothynus gibbosus*, adult.   Line indicates natural size.   (*From Ill. Natural History Surv.*)

is found to be eaten off.   In severe attacks the entire root becomes scarred and riddled with the burrows of the larvae, the burrows taking on a rust-red color.   Injury may continue in stored carrots if the temperature is favorable.

Plants Attacked.   Carrots, parsnips, celery, parsley, celeriac, wild carrots, coriander, caraway, dill, and fennel.

Distribution.   Starting near Ottawa in 1885, as an importation from Europe, the insect has spread over much of eastern Canada and northern United States as far west as Oregon and Washington.

Life History, Appearance, and Habits.   The winter is passed in the slender brown puparia, about ⅕ inch long, buried in the soil, or as maggots in the roots.   The shining-green, yellow-headed flies are abroad in May and deposit eggs about the base of the plants.   These hatch in from 3 to 17 days, and the maggots work down in the soil to attack the smaller tender tips of the roots.   As they increase in size and numbers, nearly the entire root system may be destroyed during the month or more while the larvae are growing.   The pupal stage is passed in the soil near the roots, and a second generation of flies emerges in August and a partial third generation in September to attack late carrots and celery.

[1] *Psila rosae* (Fabricius), Order Diptera, Family Psilidae.

Control Measures.   This insect has been generally controlled by treating the soil before planting with aldrin or chlordane at 2 pounds per acre, broadcast as a spray or dust and cultivated into the top 6 inches.   Higher dosages are required in heavy soils, and where the chlorinated hydrocarbons have not been satisfactory, furrow treatment with Diazinon at 1 to 2 pounds per acre has given good results.   Screening the plants as described for the cabbage maggot (page 672) is also effective.

References.   Mass. Agr. Exp. Sta. Bul. 352, 1938; *Wash. Agr. Exp. Sta. Bul.* 405, 1941; *Ore. Agr. Ext. Cir.* 304, 1943; *Jour. Econ. Ento.,* **24**:189, 1931; **51**:165, 1958; and **52**:963, 1959.

J. INSECTS INJURIOUS TO ASPARAGUS

FIELD KEY FOR THE IDENTIFICATION OF INSECTS INJURING ASPARAGUS

1. Young shoots are gouged and scarred and, later in the season, the foliage is devoured and the stems scarred, by hordes of brilliant, blue-, red-, and yellow-spotted beetles, about ¼ inch long, and by their slug-like dull-gray larvae. Eggs, ¹⁄₁₆ inch long and dark-brown, are found standing on end in rows like the teeth of a comb, along the stems and leaves........*Asparagus beetle,* page 680.
2. New shoots, foliage, and stems scarred and chewed by a light-brown to reddish-orange, rather straight-sided beetle, a little over ¼ inch long, with 6 small black spots scattered over each wing cover.   Orange to brownish larvae ⅓ inch long feed mostly in the berries.   Eggs laid flat on their sides.....................
   ....*Spotted asparagus beetle,* page 682.
3. Whitish legless and headless maggots, up to ⅕ inch long, mine beneath the epidermis of the stems, sometimes girdling the plants and causing them to yellow and die.   Adults are small, black, shiny two-winged flies, ⅙ inch long........
   ....*Asparagus miner,* page 682.

General Reference.   Iowa Agr. Exp. Sta. Cir. 134, 1932.

132. ASPARAGUS BEETLE[1]

Importance and Type of Injury.   Wherever asparagus is grown, in the area infested by this insect, the voracious beetles (Fig. 14.47) make their appearance as soon as the shoots push above the soil in the spring.   They gnaw out the tender buds at the tips and cause the tips to be scarred and browned.   The eggs are also deposited in great numbers on the tips and make the plants unfit for sale.   After the leaves come out, the beetles and their grayish slug-like larvae gnaw the surface of the stems and devour the leaves, thus robbing the root system of the food materials needed to form a good crop of shoots the following season.   The larvae also excrete a black fluid that stains the plants.   The injury is especially serious in new beds.

Plants Attacked.   Asparagus only, so far as known.

Distribution.   The northeastern fourth of the United States, from Missouri eastward and from Tennessee and North Carolina northward into Canada; also throughout much of California and Colorado and Oregon.   The insect is an importation from Europe, first found in America on Long Island in 1860.   In Illinois it first became destructive from 1910 to 1915.

Life History, Appearance, and Habits.   The beetles winter in sheltered places such as decayed or split fence posts, under loose bark of trees, or in the hollow stems of old asparagus plants.   The adult is a brilliantly colored, active beetle, about ¼ inch long, with bluish-black straight-sided wing covers, each with three large, yellowish, squarish spots along each side and reddish margins.   The reddish prothorax and the head are much narrower.   Egg-laying begins soon after they appear in the field in April

[1] *Crioceris asparagi* (Linné), Order Coleoptera, Family Chrysomelidae.

or May.   Shoots of asparagus are often seen which are literally blackened by hundreds of eggs, all standing on end in rows of from three to eight (Fig. 14.47).   The eggs hatch within a week, and the very small grubs migrate to the tips of the leaves and begin feeding upon them.   The later stages of the larvae are similar to the young except for a gradual increase in size up to ⅓ inch long.   The color is dull-gray, with black head and legs.   The body is plump, soft, wrinkled, and not hairy.   After feeding for 10 days to 2 weeks, the larvae disappear into the soil and form a yellowish pupa.   A week or two later the new adults emerge from the soil and promptly start another generation.   Each generation requires from

Fig. 14.47. Asparagus beetle, *Crioceris asparagi*.   At left, eggs on asparagus shoots, natural size; at right, adult and larva, about 4 times natural size.   (*From Conn Agr. Exp. Sta.*)

3 to 7 or 8 weeks, and there are from two to five generations in the course of a year.   Cold weather kills the eggs and larvae and drives the beetles into hibernation.   The insect appears to be much less destructive in wet seasons.

Control Measures.   During the cutting season, the tips should be protected by spraying or dusting with rotenone at 0.1 to 0.3 pound, or malathion at 1.25 pounds per acre.   After the cutting season, spraying or dusting with 1.5 pounds DDT per acre is effective.   These treatments may be applied to newly set beds as soon as the beetles appear and again when the foliage is fully formed.   Keeping down volunteer plants and cutting the shoots very clean and deep every day or two will tend to remove the eggs before the larvae can establish themselves in the patch. A few plants here and there may be left to grow and attract beetles and then should be treated as suggested above.

References. *U.S.D.A., Farmers' Bul.* 837, 1917; *N.Y. Agr. Exp. Sta. Bul.* 278, pp. 70–71, 1934.

133. SPOTTED ASPARAGUS BEETLE[1]

Importance and Type of Injury. Injury by the adult is the same as by the asparagus beetle, and the gnawing is especially noticeable on the young shoots. The larvae however, do little damage, since they feed almost entirely in the fruits or berries. In some sections this species is more abundant and destructive than the former.

Plants Attacked. Asparagus only.

Distribution. Imported from Europe some time previous to 1881, when it was reported in Maryland, the insect has now spread over much the same territory as the asparagus beetle, ranging as far west as Utah and Idaho. It was first taken in Illinois in 1925.

Life History, Appearance, and Habits. The life history is similar to that of its close relative, described above. The adults, which are reddish-orange or tan, with six prominent black spots scattered over each wing cover, appear a little later in the spring and do not lay eggs until shortly before the berries form. The greenish eggs are glued, singly, on their sides to the leaves and hatch in 1 to 2 weeks. The orange-colored larvae, upon hatching, crawl to a developing berry, bore into it, and eat out the pulp.

FIG. 14.48. Spotted asparagus beetle, *Crioceris duodecimpunctata.* 4 times natural size. (*From Conn. Agr. Exp. Sta.*)

Before they become full-grown, they usually destroy three or four berries, by eating out the seeds. Pupation takes place in the soil. The new adults (Fig. 14.48) appear in late July and produce another generation in early September.

Control Measures. The same as for the asparagus beetle. In small gardens, gathering and burning the berries will give control.

References. *U.S.D.A., Farmers' Bul.* 837, 1917, *N.Y.* (*Geneva*) *Agr. Exp. Sta. Bul.* 331, 1913; *Ill. Agr. Exp. Sta. Cir.* 437, p. 45, 1939.

134. ASPARAGUS MINER[2]

Importance and Type of Injury. The maggots of this small fly mine up and down the stems of asparagus just beneath the surface, near the base of the plant. Where abundant, they may girdle the plants, causing the foliage to yellow and die prematurely.

Plants Attacked. Asparagus.

Distribution. Northeastern United States and southern Canada and also California.

[1] *Crioceris duodecimpunctata* (Linné), Order Coleoptera, Family Chrysomelidae.

[2] *Melanagromyza* (= *Agromyza*) *simplex* (Loew), Order Diptera, Family Agromyzidae.

Life History, Appearance, and Habits. The insect winters in puparia in the larval tunnels under the epidermis of the stems and from 1 to 6 inches below the surface of the soil. The flies appear in the fields the latter half of May and thrust their eggs beneath the epidermis of the stem near or below the surface of the soil. The egg stage lasts 2 or 3 weeks, and the larvae feed for about the same period before pupating. In 3 weeks more, chiefly July, the new adults are abroad, and the second-generation larvae mine in the stems during late summer. There are two generations a year.

Control Measures. The rust-resistant strains of asparagus are said to be less injured by this insect. If old stalks are pulled up and burned, the puparia in the stems will largely be destroyed. In irrigated districts, winter flooding is recommended. Insecticidal treatments for leaf miners are given on page 678.

References. U.S.D.A., Bur. Ento. Cir. 135, 1911; N.Y. (Cornell) Agr. Exp. Sta. Bul. 331, 1913.

K. INSECTS INJURIOUS TO SWEET CORN

The insects of sweet corn are, without exception, the same as those of field corn, already discussed in Chapter 9. One of the most destructive to sweet corn is the corn earworm, which eats the kernels at the end of the ears (page 498). In seasons when this insect is abundant, nearly every ear will be attacked by the worms, rendering the ears repulsive to the consumer and causing great additional expense to canners. If corn can be produced to be past the fresh-silk stage before the moths become abundant, they will pass it by and lay eggs elsewhere. In the latitude of northern Illinois, early-maturing or early-planted corn is injured least, while in the latitude of central and southern Illinois medium plantings suffer least.

In the area where it is abundant, the European corn borer is by far the most destructive pest of sweet corn. Sweet corn is more severely injured than field corn because of the smaller stalks. The control of this pest is given on page 493.

The soil-infesting larvae, such as white grubs, wireworms, and cutworms, are especially destructive to gardens when planted on sod land. This practice should never be followed. Sod land that is to be planted with vegetables should be seeded to clover, alfalfa, or small grains for at least 1 year. Gardens surrounded by the favored trees of the May beetles are especially liable to injury by white grubs every year. Since corn insects of many kinds, notably the corn rootworms and corn root aphid, accumulate and increase in numbers when corn is planted year after year on the same soil, sweet corn should not be planted in fields that have been in corn for several years. Sweet corn will never be injured severely by the northern corn rootworm unless grown in soil that has been in corn for one or more years previously. For the corn root aphid, the cultivation of the soil suggested on page 503 will be most effective.

Armyworms and grasshoppers are periodically very destructive but can be brought under control by the use of poisoned-bran bait or sprays and dusts (page 469). It may be necessary, in the case of small gardens, to apply these treatments to surrounding fields. Keeping down the growth of weeds about gardens will help in the control of many pests of corn, especially the corn root aphid and the common stalk borer.

For further information regarding the special control of insects on sweet corn, see pages 462 to 519.

CHAPTER 15

INSECTS INJURIOUS TO DECIDUOUS FRUITS AND BUSH FRUITS

A large number of insects find their favorite or only food supply on the roots, trunk, branches, leaves, or fruit of our deciduous tree and bush fruits. Many of the insects which formerly fed on the fruits or foliage of uncultivated plants have found the abundant food supply furnished by large orchards or plantations of the bush fruits much to their liking, and now feed almost exclusively on these cultivated crops. The world-wide commerce in fruits and fruit-producing plants has led to the importation and general spread of many serious fruit pests in this and other countries. In most parts of this country it is now impossible to produce marketable fruit, regardless of the fertility of the soil, favorable climate, or varieties grown, unless the insects are controlled. In other words, insects have come to be one of the limiting factors in fruit production.

In the United States and Canada fruit trees are subject to attack by many insects, among which are several that, in a single season, may ruin the crop or even destroy the trees of a well-established mature orchard. The most important of these fruit insects are discussed in the following order:

General References. SLINGERLAND, M. V., and C. R. CROSBY, "Manual of Fruit Insects," Macmillan, 1915; *N.Y. (Cornell) Agr. Exp. Sta. Ext. Bul.* 711, 1947; *Ill. Agr. Exp. Sta. Cir.* 634, 1949; *Ohio (Wooster) Agr. Exp. Sta. Bul.* 670 *(Rev.)*, 1949.

A. APPLE INSECTS

FIELD KEY FOR THE IDENTIFICATION OF INSECTS INJURING THE APPLE

A. External chewing insects that eat holes in leaves, buds, bark, or fruit:
    1. Foliage of apple skeletonized and eaten off by grayish or brownish measuring-worms about 1 inch in length, with 3 pairs of legs near the head and 2 pairs of prolegs near the end of the body. Damage occurs in the spring about the

time the trees have come into full foliage.   Injured trees have the appearance
of having been scorched by fire.   The caterpillars causing the injury spin
down from the foliage on silken threads when the trees are jarred or the insects
are disturbed.   Grayish, wingless, soft-bodied, spider-like moths, crawling up
the trunks of the trees in the spring and mating with grayish-brown males,
with wings about 1½ inches from tip to tip. . . . *Spring cankerworm*, page 689.

2. Injury and appearance of insects the same as in *A*,1, except the moths appear
on trunks of trees during the fall and the caterpillars have 3 pairs of prolegs
instead of 2.   Injury to the trees occurs at the same time in the spring. . . . .
. . . .*Fall cankerworm*, page 691.

3. Loosely woven white webs, enclosing the leaves of several branches, within
which very hairy, pale-yellow, black-spotted caterpillars may be found feeding
on the protected foliage.   Caterpillars are about 1¼ inches long when full-
grown.   The webs contain black pellets of excrement, giving them an unsightly
appearance.   Injury most conspicuous in the late summer and early fall. . . .
. . . .*Fall webworm*, page 692.

4. Large, thick, white webs in the forks or crotches of trees in the early spring;
leaves stripped from the branches within a considerable distance of these nests.
Brownish-black, hairy caterpillars, ½ inch to 2 inches in length, with a light
stripe down the back, sheltering within the webs during the daytime or feeding
on the new foliage.   Silken roadways or paths lead from the web to other parts
of the tree.   Small shiny varnished-appearing black egg masses, about ½ inch
long, encircle small twigs during the winter. . *Eastern tent caterpillar*, page 693.

5. Apple leaves skeletonized by yellow-striped, hairy caterpillars, up to 1½
inches in length, with 2 pencil-like tufts of long black hairs projecting in front,
and a single similar tuft projecting backward from near the tail.   Frothy,
white, egg masses, about ½ inch across, attached to grayish, hairy, empty
cocoons on folded dead leaves or bark will be found on the twigs and branches
of infested trees during the winter. . . . . . *White-marked tussock moth*, page 828.

6. During the winter season several leaves, tightly webbed together and firmly
attached to a twig near the tip of a branch, enclose from 25 to several hundred
small dark-brown, very hairy caterpillars about ¼ inch long.   If caterpillars
or their hairs come in contact with the skin, a distinct burning or itching
results, which may persist for several days.   White moths, with brown
abdomens, flying in great numbers at night, during July.   The moths are
strongly attracted to lights. . . . .*Brown-tail moth*, page 832.

7. Pale-yellow masses of eggs about 1 inch across, with a felt-like covering of
tawny hairs, attached to the bark of trees, or on buildings, walls, and other
objects during the winter.   Somewhat flattened, bluish, stiff-haired caterpillars, feed on the foliage of all kinds of deciduous trees and also upon evergreens.   Caterpillars with 5 pairs of conspicuous bluish tubercles on the back,
behind the head, followed by 6 pairs of somewhat larger, red tubercles on the
following segments.   Foliage often completely stripped from trees in infested
areas. . . . .*Gypsy moth*, page 830.

8. Yellow- and black-striped caterpillars, with a distinct ridged yellow ring
about the neck, or with a red hump just back of the head with a row of black
spines projecting from it, feed, many together, on the foliage during the summer and early fall.   Caterpillars, when full-grown, are 1½ to 2 inches in
length.   They rear head and tail when disturbed.   Small trees, or single
branches on large trees, completely stripped of leaves by these caterpillars. . .
. . . . *Yellow-necked* and *red-humped caterpillars*, page 694.

9. Leaves, especially those at tips of new growth near the top of the tree, skeletonized or eaten out in part, by dark-green active caterpillars with four shining
black tubercles on the back just behind the head.   Caterpillars web 2 or 3
leaves together and feed within the enclosed area.   Injury is most severe
during the late summer and early fall. . . . . . . *Apple leaf skeletonizer*, page 696.

10. Very tough, blackish, conspicuously curled cases, about ¾ inch long, usually
partly enclosed in 1 or 2 dead leaves, attached to the ends of twigs and in the
forks of small twigs during the winter.   Cases enclose dark-brown hairy

caterpillars, about ⅓ inch long, which eat out the opening buds in the spring. Fruit sometimes specked with small feeding punctures during August.......
....*Leaf crumpler*, page 697.

11. Tiny brownish-gray tough silken cases, about ⅙ inch long, bent at the top, so that they roughly resemble a pistol in outline, standing out at right angles to twigs and branches during the winter. Brown caterpillars, about ⅛ inch long, sheltering in these cases, feed on the newly opening buds in the spring and on the fruit and foliage during the late summer.......................
....*Pistol casebearer*, page 698.

12. Cases attached at right angles on the twigs and branches as in *A*,11, but cases not bent over at the top. In the spring the insect moves about over the leaves eating or mining out small areas.................*Cigar casebearer*, page 698.

13. The newly opening buds and leaves of the apple are fed upon by a small brown caterpillar, about ¼ inch long, with a black head; leaves and buds webbed with silk where the caterpillar is feeding. In the winter small silken cases containing brown caterpillars are attached to the axil of the bud or twig. Fruit on infested twigs often shows small areas of feeding, somewhat resembling injury by late codling moth larvae....*Budworms* or *bud moths*, page 699.

14. Large, irregular cavities eaten in the sides of small apples, from 1 to 3 weeks after the fruit has set. Green caterpillars about ¾ inch long, sometimes with fine stripes on the sides, are found on the twig or branch adjacent to the injured apple or actively feeding on the apple. Apples at picking time deformed or misshapen, with brown, healed-over cavities in the sides.......
....*Green fruitworms*, page 700.

15. Injury resembling *A*,14, but injured areas much more shallow, often merely the surface of the apple eaten away. Injury most common where the leaves rest against the fruit. Very active, translucent, greenish caterpillars, with 4 white stripes on the back, the outer stripes broader than the inner ones, feeding on the fruit, or skeletonizing the leaves during early summer. Caterpillars about ½ inch long when full-grown................................
Palmerworm, Dichomeris ligulella Hübner [see *N.Y. (Cornell) Agr. Exp. Sta. Bul.* 187, 1901].

16. Injury similar to *A*,14, but usually including nearly all the apples on a branch. Foliage ragged and buds eaten off by small, greenish-brown, very active worms about ¾ inch long when full-grown. Injury occurs from the time the buds separate until about 4 weeks thereafter. Injured leaves rolled, folded, or drawn together about the fruit with light silken threads. Apples at picking time with brown, scarred areas in the sides....*Fruit tree leaf roller*, page 701.

17. Fruits of apple scarred by shallow feeding, most common where a leaf touches the fruit. Skin of the fruit is eaten. Injury occurs late in the season......
....*Red-banded leaf roller*, page 701.

18. Brownish or grayish, winged, vigorously jumping insects, from 1 to 2 inches in length, with conspicuous heads and eyes, feed on the foliage of apple during the late summer, sometimes stripping the trees and scoring the bark............
....*Grasshoppers*, page 704.

19. Very small, dull-black, snout beetles, about ⅒ inch long, which hop vigorously when disturbed, eat small holes in the underside of leaves from the time when the foliage first starts until midsummer. Foliage of heavily infested trees has the appearance of having been riddled with fine bird shot. Yellowish-brown mines start near the center of the leaf and run to conspicuous blister-like cells in the outer leaf margin.........................*Apple flea weevil*, page 704.

20. Twigs and small branches of apple, up to 2 feet in length, are pruned off, especially on young trees. Injury occurs during May and the first of June. At this time large grayish snout beetles nearly ¾ inch long, with bodies spotted with black, may be found resting on the trees, or cutting off the twigs and buds. Injury occurs only in the vicinity of woodlands or on newly cleared ground...
New York weevil, Ithycerus noveboracensis (Förster) (see Slingerland and Crosby, "Manual of Fruit Insects," page 210).

21. Buds of apple and the newly set fruit cut off by grayish plump-bodied snout beetles, slightly smaller than *A*,20. Beetles are about ½ inch in length, of a greenish-gray color, with 2 irregular light bands across the wing covers. The wing covers are bent down at the end and terminate in an acute angle. Body covered with overlapping scales............*Imbricated snout beetle*, page 705.

B. Insects that suck sap from leaves, buds, twigs, branches, trunks, or fruits:

1. Small gray or brownish-gray spots about ¹⁄₁₆ inch across, each with a central nipple, on twigs and fruit, often surrounded by reddish or pinkish, inflamed-appearing areas. Twigs covered with a gray coating, having the appearance of ashes. If the larger grayish dots on the bark are lifted, a lemon-yellow, soft-bodied creature, nearly the size of a pinhead, will be found beneath. Trees where the grayish covering occurs on the trunk and branches have thin, yellow, spotted foliage, often in a dying condition........................ ....*San Jose scale*, page 705, or *Forbes' scale*, page 768.

2. Grayish-white, pear-shaped, flat scales, about ⅛ inch long, adhering tightly to the bark of branches and trunk. Smaller three-ridged, straight-sided scales scattered among the larger pear-shaped individuals. In the winter, the larger grayish-white scales will be found to cover a number of small reddish-purple eggs. In the summer, a yellowish, soft-bodied sucking insect will be found under the scale....................................*Scurfy scale*, page 708.

3. Brownish to grayish-brown, hard, polished scales, very closely resembling a half oyster shell in appearance, and about ⅛ inch in length, by one-third as wide, adhering tightly to the bark of the apple, so thick as to overlap. In the winter, from 40 to 60 pearly white eggs will be found under most of the larger scales. Scales often in patches on the bark. Heavily infested trees lacking in vigor or dying. Fruit sometimes specked with scales, but usually lacking in reddened areas such as occur around San Jose scale...................... ....*Oyster-shell scale*, page 709.

4. Greenish-black shiny eggs about the tips and buds of twigs. Many small soft-bodied sucking aphids, clustered on the buds or curling the newly formed leaves of apple trees in the spring. The feeding causes the dwarfing of the fruit, or practically stops growth, although the apples remain on the tree. Winged and wingless aphids will be found after the time of bloom. Bodies of the insects with a distinct waxy coating, some of them with a pinkish tinge. These aphids migrate from the tree by early summer....................... ....*Rosy apple aphid*, page 710.

5. Dark, greenish-black, shiny eggs about the buds and in crevices in the bark during the winter. Soft green-bodied aphids appear on the buds as soon as the green begins to show in the spring. These aphids continue to feed on the buds, new growth, and water sprouts throughout the summer, often causing a curling of the new growth. Some dwarfing of apples, as in *B*,4............. ....*Apple aphid*, page 712.

6. Shiny greenish-black eggs on twigs and bark as in *B*,5. Large numbers of green-bodied aphids clustered on the opening buds. They nearly all leave the apple before the time of bloom. Injury by this species is very slight........ ....*Oat aphid* or *apple-grain aphid*, page 712.

7. White cottony masses of wax-like material covering the backs of dark purplish-brown aphids, about ¹⁄₂₀ inch long, which cluster on wounds along the trunk and branches of apple and other trees (see also *E*,1)........................... ....*Woolly apple aphid*, page 733.

8. Small greenish, active, slender, winged insects, about ⅛ inch long, accompanied by pale, greenish-white, wingless, active nymphs, sucking the sap from the underside of leaves of apple. Foliage of infested trees pale in color, with small white dots showing on the leaf; new foliage curled, leaf margins slightly burned; and very small black specks of excrement on the fruit............. ....*Apple leafhoppers*, page 713.

9. Fruit with irregular russeted spots, misshapen, pitted, or dimpled in appearance, but without cavities, and dwarfed in size; flesh often hard or woody under the injury. Elongate-oval reddish bugs, about ¼ inch long, somewhat

plump of body, with head of orange color, feeding on the new foliage, and later on the fruit, by sucking the sap through their slender beaks..........
....*Apple red bugs*, page 714.

10. Very small, yellowish-green or reddish, 8-legged mites, about $\frac{1}{50}$ inch long, feeding on the underside of foliage of apple and other trees. Infested foliage has a pale sickly appearance. Numerous delicate webs over the foliage and other parts of the tree...
....*Two-spotted spider mite*, page 875; *Pacific mite* or *Schoene mite*, page 717.

11. Large numbers of reddish, minute, rounded eggs on the ends of twigs of apple during the winter. Leaves of apple, plum, peach, and other fruits show a pale, sickly appearance, or with brownish-red specks. At a distance, foliage has the appearance of being dusty. Very light webs over the surface of the leaves, upon which are crawling minute, reddish 8-legged mites...................
....*European red mite*, page 715.

12. Similar to *B*,11, but eggs somewhat larger and more reddish in color, often occuring in such numbers as to give a reddish tinge to the twigs. Infested trees with pale foliage which sometimes drops off during late summer. Small 8-legged reddish mites about $\frac{1}{30}$ inch long, with very long front legs, working on the underside of the foliage......................*Brown mite*, page 717.

13. Very small, elongated, 4-legged whitish mites, about $\frac{1}{125}$ inch in length, under bud scales during the winter. Dark-brown, blister-like, scabby areas on the leaves and occasionally on the fruit. Badly infested trees with distinct yellowish tinge to the foliage...............*Pear leaf blister mite*, page 745.

C. *Insects that bore into the trunk, branches, or twigs:*

1. Shallow, irregular burrows in the inner bark, mostly on the sunny sides of apple and other trees. During the early summer months, dark olive-gray flattened beetles, about $\frac{1}{2}$ inch long, blunt at the head end and tapering to a rounded point at the tail, and with metallic-colored, roughened wing covers will be found on the sunny parts of the trunk and branches. White, flattened, legless grubs from $\frac{1}{2}$ to 1 inch long, with the body greatly enlarged just behind the head, are found in the shallow burrows under the bark..................
....*Flatheaded apple tree borer*, page 718.

2. Nearly round holes, a little larger than a lead pencil, in the lower parts of the trunks of apple trees. Bits of yellowish-brown frass or sawdust forced from small holes in the bark at the base of apple trees. Yellowish-white legless grubs up to 1 inch long, with strong brown jaws and body enlarged but not flattened just behind the head, bore in the inner bark and wood of trees about the base. Velvety-brown cylindrical beetles, about 1 inch long, with two conspicuous white stripes on the back, crawling over the trunk and feeding on the foliage of apple from June to September............................
....*Roundheaded apple tree borer*, page 720.

3. Many small round holes, about the size of a No. 6 shot, in the bark of the twigs and branches of apple; seldom occur except on trees lacking in vigor. Small black beetles about $\frac{1}{10}$ inch long, the body blunt at either end, crawling over the bark and excavating small holes, usually starting at the base of a bud or twig. Underneath the bark are numerous, fine, sawdust-filled burrows up to 4 inches in length, radiating from a short parent gallery, and often containing small white grubs..............................*Shot-hole borer*, page 722.

4. Small holes through the bark of apple somewhat smaller than *C*,3, but extending directly into the wood, where they branch several times. Burrows in the wood usually stained a dark color. Brownish or brownish-black blunt-ended minute beetles, working in these holes and occasionally crawling over the bark of the tree...............................*Pinhole borers*, page 846.

D. *Caterpillars, worms, maggots, or weevil grubs burrowing into, or feeding inside, the fruits:*

1. Newly set apples with crescent-shaped punctures in the side of the fruit, often with a small hole cut on the inside of the apex of the crescent. Early-injured apples often containing small white fat-bodied footless grubs. Small round cavities or pits eaten in the surface of the apple from midsummer until midfall.

Dark-brown snout beetles, about ⅛ inch in length, with grayish-white patches on the back and four slightly raised humps on the wing covers, feeding on the surface of the fruit...........................*Plum curculio*, page 724.

2. Injury similar to *D*,1, but lacking the crescent-shaped cuts. The cavities are eaten deeper into the fruit and are often close together, so that the side of the fruit may be peppered with small holes about ⅛ inch deep. Apples knotty, misshapen, and undersized. Dead areas of skin often present around the cavities. A brown to light-brown beetle, with slender snout ½ the length of the body, and with 4 very distinct humps on the wing covers, but lacking the white markings of *D*,1, may be found feeding on the fruit. Injured fruit seldom drops unless heavily fed upon early in the season...................
....*Apple curculio*, page 726.

3. Apples with holes eaten in the flesh of the fruit, the burrow generally running to the core, and seeds more or less injured. Masses of dark castings often protruding from these holes especially at the calyx. Pinkish-white, brown-headed worms, about ¾ inch long, feeding inside the apple, or resting in tough cocoons of white silk, spun under the bark on the trunk or in other shelters about the tree. Minute, flat, white, shiny eggs, ¾ the size of a pinhead, on the leaves adjacent to the growing apples or on the skin of the apple...........
....*Codling moth*, page 727.

4. Injury similar to *D*,3, but holes in the apple about half the size, seldom reaching the core, usually going in from the side of the fruit. Areas of varying size where the larva has eaten under the skin, without feeding deeply on the flesh.
....*Lesser appleworm*, page 771.

5. Numerous, brown, twisting mines, containing yellowish-white maggots, up to ⅜ inch in length, run through the flesh of the fruit. These mines are most abundant in early varieties of apples, particularly in sweet apples. Mines in the fruit indicated by dark lines on the surface of the apple. Flies a little smaller than the house fly and of a general black color, marked with white bands on the abdomen and black bands on the wings, rest on the surface of the fruit, and deposit their eggs in the flesh.......*Apple maggot*, page 732.

E. *Insects that attack the roots of the tree:*

1. Trees stunted or unthrifty. Large knots on the roots and base of the trunk, often with many short, fibrous roots extending from the knots or tumors. Numerous water sprouts about the crown of the trunk. Purplish soft-bodied aphids on the affected parts, more or less heavily covered with a white powdery or cottony mass of secretion..................*Woolly apple aphid*, page 733.

F. *Insects that injure twigs and branches by laying their eggs beneath the bark:*

1. Twigs and small branches of apple and many other trees split and scored where bark of twigs has been pushed back and the wood at short intervals raised in small bundles of splinters, with double row of egg punctures beneath. During May and June wedge-shaped, transparent-winged, black-bodied insects, about 1½ inches long, blunt at the head end and with conspicuous eyes at each corner of the head, resting on the foliage, or flying about the trees. Many of them utter a shrill high-pitched song.......................................
.......................*Periodical cicada*, or *seventeen-year locust*, page 735.

2. Double rows of punctures or slits, about ¼ inch long, in the bark of twigs and branches, usually most abundant in trees standing in sod or surrounded by weeds and grasses. Slits are crescent-shaped and farthest apart at the middle. The scars remain on the injured twigs for several years. In the late summer and early fall, small, triangular, green, very active, jumping insects, about the size and shape of a beechnut, make these punctures in the twigs, in which their eggs are deposited...........................*Buffalo treehopper*, page 739.

135. Spring Cankerworm[1]

Importance and Type of Injury. For more than two centuries out-breaks of cankerworms have periodically defoliated shade and fruit trees

[1] *Paleacrita vernata* (Peck), Order Lepidoptera, Family Geometridae.

in the eastern United States. In unsprayed or poorly sprayed orchards they may cause complete defoliation and loss of the crop; but they are of no importance in well-sprayed orchards. The foliage of the trees is eaten and skeletonized by measuring-worms. The injury occurs just about the time the trees have come into full foliage. Silken threads are spun from branch to branch on the tree and from the branches to the ground. Brown and brownish-green measuring-worms about 1 inch long spin down from the tree when it is jarred or shaken. Heavily infested orchards have much the appearance of having been scorched by fire.

Trees Attacked. Apple, elm, and many other fruit and shade trees.

Distribution. General, east of the Rocky Mountains; southeastern Canada; also in California and Colorado.

Life History, Appearance, and Habits. The winter is passed in the form of naked brown pupae about ½ inch long by ⅛ inch thick. These pupae are found in the soil from 1 to 4 inches below the surface, and in greatest numbers close to the base of the trees. The moths begin emerging during warm periods in February and continue coming out until the end of April. The male moth is strongly winged and is of a dull-gray appearance, being much the color of a well-weathered piece of board. These moths may be seen flitting about from tree to tree at dusk and after dark in the spring evenings. The female moth is wingless, with a gray spidery body. She differs from the fall cankerworm female (Fig. 15.2) by having a dark stripe down the middle of the back and two transverse rows of small reddish spines across each abdominal segment on the upper side. On emerging from the ground, the female crawls to a tree and up the trunk,

FIG. 15.1. Larvae of the spring cankerworm, *Paleacrita vernata*, showing the two pairs of prolegs near tip of abdomen. About twice natural size. (*From U.S.D.A., Farmers' Bul. 1270.*)

or onto the branches, where she mates with the male and deposits her oval dark-brown eggs in irregular masses under the loose scales of bark. These eggs hatch in about a month into small greenish or brownish measuring-worms, which at once begin to feed on the foliage. These worms (Fig. 15.1) can be distinguished from the fall cankerworm by having only two pairs of prolegs, near the end of the body. They vary from light-brown to nearly black and usually have a yellowish stripe below the spiracles and the under parts partly black. When not feeding, the larvae tend to rest upon the twigs more than upon the leaves. They feed for 3 weeks to 1 month and, if abundant, may completely strip the foliage from the trees. At the end of the feeding period they crawl or spin down to the ground where they excavate the small cells in which they change to the pupal stage and pass the remainder of the summer and the following winter.

Control Measures. As the wingless female moths crawl up the tree to deposit their eggs, they may be stopped and trapped by placing a band of sticky material, such as tanglefoot, around the trunk of the tree at from 2 to 4 feet from the ground. This should be put on before the first warm weather in February or March. When the moths are abundant, the bands should be watched very closely from February to May, as the trapped moths will often become so numerous as to bridge the bands with their bodies, allowing those emerging later to cross without being caught. All trees within 200 feet of others should be banded, since the moths and small larvae may be blown into banded trees by strong winds, from infested trees near by.   As the caterpillars are easily poisoned and do their feeding when the heaviest spray applications of the year are being made, they are never abundant in well-sprayed orchards.   They may be controlled by spraying in the pink-bud stage with DDT at 0.5 to 1 pound, endrin at 0.25 pound, as wettable powders, or with lead arsenate 3 pounds and hydrated lime 3 pounds per 100 gallons of spray.   In unsprayed orchards late spring and summer cultivation, with care to work close to the trunks of the trees, will help to some extent in reducing their injuries, by destroying the pupae.

References. Ohio Agr. Exp. Sta. Cir. 65, 1907; *U.S.D.A., Farmers' Bul.* 1270, 1922, and *Dept. Bul.* 1238, 1924.

136. Fall Cankerworm[1]

This insect has practically the same life history as that of the spring cankerworm and is controlled by the same measures.   It differs from the other species in that the moths emerge late in the fall, some time after

 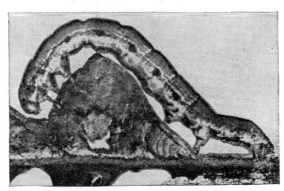

Fig. 15.2. At left, female moths of fall cankerworm, *Alsophila pometaria*, about twice natural size.   The female of the spring cankerworm is similar but has a black stripe down the middle of the back and transverse rows of reddish spines on the upper side of the abdomen.   At right, larva of the fall cankerworm, showing the three pairs of prolegs near the tip of abdomen.   About twice natural size.   (*From U.S.D.A., Farmers' Bul.* 1270.)

the ground has been frozen.   The insect passes the winter in the egg stage instead of in the pupal stage, as is the case with the spring species.   The ashy-gray eggs are laid in single-layer groups, on branches and trunks of trees, each egg having a flowerpot-like shape.   The worms hatch and feed on the leaves in the same manner and at the same time as the spring cankerworm.   In this species the larva has three pairs of prolegs near the tip of the body (Fig. 15.2).   They are brownish above, with three narrow whitish stripes along the sides above the spiracles and a yellow stripe below the spiracles, and the underparts light apple-green.   They remain more upon the leaves and spin cocoons in the soil before pupating.   The females may be distinguished from those of the spring cankerworm by

[1] *Alsophila pometaria* (Harris), Order Lepidoptera, Family Geometridae.

their lack of the dark middorsal stripe, and the double rows of reddish spines across the upper side of the abdominal segments.

Trees Attacked. Apple, elm, and many other fruit and shade trees.

Distribution. Much of the northern United States and southern Canada; also Colorado, New Mexico, and California.

Control Measures. The control measures are the same as for the spring cankerworm, except that the trees should be banded in the late fall and not in the spring. Cultivation is of little value for this species, since it forms a tough cocoon.

137. FALL WEBWORM[1]

Importance and Type of Injury. This insect is of no importance in orchards that are sprayed regularly for the control of codling moth and other pests, but in neglected orchards and upon shade trees and shrubs, it often makes very unsightly webs. Its presence is indicated by loosely woven, dirty white webs (Fig. 15.3) enclosing the foliage on the ends of the branches. Several branches are sometimes covered by one of these webs. The webs enclose many pale-yellow, black-spotted, very hairy caterpillars, about 1 inch long, which feed upon the surface of the leaves. These webs contain a quantity of black pellets of excrement from the worms, making them very unsightly.

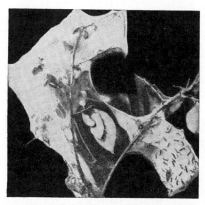

FIG. 15.3. Nest or web of the fall webworm showing worms inside the web. Much reduced. (*From U.S.D.A., Farmers' Bul. 1270.*)

Trees Attacked. The fall webworm has been found feeding on more than 100 fruit, shade, and woodland trees. It does not attack evergreens.

Distribution. General over the United States and southern Canada.

Life History, Appearance, and Habits. This insect passes the winter in the form of brown pupae, enclosed in lightly woven, silken cocoons. These cocoons will be found under trash on the ground or sometimes under the bark of trees. The moths begin emerging during the spring and continue to come out over a long period. Both sexes are winged, satiny white, sometimes with brown or black spots. They lay their eggs on the leaves, in masses, partly covered with white hairs, and the caterpillars hatching from these eggs construct webs over the leaves inside of which they feed. They continue feeding for about 1 month to 6 weeks, and upon becoming full-grown, crawl down the tree and construct the cocoons in which they pupate. The adults emerge late in the summer and lay eggs for a second generation of the worms in early fall, which, upon becoming full-grown, spin the cocoons in which they pass the winter as pupae.

Control Measures. Sprays for the second and third generations of codling moth will usually control the webworm. DDT at 1 pound, parathion at 0.3 pound, or lead arsenate at 2 pounds with 2 pounds of hydrated lime per 100 gallons of spray have been suggested for specific control.

[1] *Hyphantria cunea* (Drury), Order Lepidoptera, Family Arctiidae.

In young orchards and on shade trees and shrubs the webs may easily be removed by hand at less expense than will be incurred for spraying.

References.   Ann. Rept. Smithsonian Inst. for 1921, pp. 395–414, 1923, Can. Dept. Agr. Bul. 3(n.s.), 1922; U.S.D.A., Farmers' Bul. 1270, 1931.

138. EASTERN TENT CATERPILLAR[1]

Importance and Type of Injury.   This insect, which has been observed in America since 1646, periodically at intervals of 10 years or so, becomes so abundant as to defoliate unsprayed orchards and many kinds of shade trees.   Large thick webs, containing many hairy brown caterpillars, are constructed in the forks and crotches of trees (Fig. 15.5) in the spring, The leaves may be stripped from all the branches within a yard or more of these nests.   These caterpillars do not feed within their webs, but congregate there during the night and in rainy weather.

Trees Attacked.   Wild cherry, apple, peach, plum, and more rarely witch hazel, rose, beech, birch, barberry, oak, willow, and poplar.

Distribution.   This species occurs throughout the eastern United States and Canada, westward to the Rocky Mountains, and with closely related forms (page 828) covers the United States.

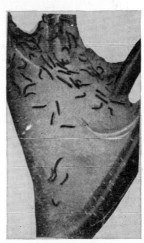

FIG. 15.4.   Eastern tent caterpillar, egg masses on twig. About natural size. (*From U.S.D.A., Farmers' Bul.* 1270.)

FIG. 15.5. Eastern tent caterpillar.   Larvae and nest in crotch of wild cherry tree. Greatly reduced.   (*From U.S.D.A., Farmers' Bul.* 1270.)

Life History, Appearance, and Habits.   This insect passes the winter as a dark-brown collar-like mass of eggs securely attached to, and often encircling, small twigs.   These egg masses (Fig. 15.4) are about ¾ inch long by ½ inch in diameter and contain several hundred eggs.   They have a shiny, varnished appearance.   The eggs hatch early in the spring as soon as the apple leaves begin to unfold or a little earlier.   The caterpillars (Fig. 15.6) gather in a near fork of the limbs, a colony often being made up of all the caterpillars hatching from several egg masses.   Here they construct their webs and sally forth to attack the newly opening leaves.   They spin a fine thread of silk wherever they crawl and, in the

[1] *Malacosoma americanum* (Fabricius), Order Lepidoptera, Family Lasiocampidae.

course of a few days, well-defined silken pathways lead from the nest to the favored feeding spots on the tree. As the caterpillars grow, their nest webs are enlarged. They become full-grown in 4 to 6 weeks. They are then about 2 inches long, thinly covered with long soft light-brown hairs. The general color is black. There is a white stripe down the back bordered with reddish-brown, and along each side there is a row of oval blue spots and brown and yellow lines. They now scatter to some distance from the nest and spin white cocoons, usually on the tree trunk or some near-by object, in which they change to brown pupae and later emerge as light reddish-brown moths with two whitish stripes running obliquely across each forewing. The females deposit their eggs on the twigs for the next season's caterpillars early in the summer. There is only one generation each year, about 9 months being spent in the egg stage.

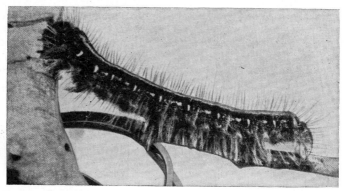

Fɪɢ. 15.6. Larva of eastern tent caterpillar, *Malacosoma americanum.* Slightly enlarged. (*From U.S.D.A., Farmers' Bul.* 1270.)

Control Measures. Spraying with DDT, methoxychlor, chlordane, or malathion at 1 pound, toxaphene at 1.2 pounds, as wettable powders, or with lead arsenate 3 pounds with 3 pounds hydrated lime to 100 gallons of spray will give satisfactory control. Spraying for the control of codling moth will control this pest. Wild cherries and plums should not be allowed to grow within a quarter-mile of an apple orchard. The removal of such trees will aid greatly in keeping down several other orchard pests. The winter egg clusters may be readily seen and should be pruned out and burned in winter. The nests may be removed by winding them upon the end of a pole bearing a conical brush, or having nails driven into the end of it. They should then be burned.

Reference. Conn. Agr. Exp. Sta. Bul. 378, 1935.

139. Yᴇʟʟᴏᴡ-ɴᴇᴄᴋᴇᴅ[1] ᴀɴᴅ Rᴇᴅ-ʜᴜᴍᴘᴇᴅ Cᴀᴛᴇʀᴘɪʟʟᴀʀs[2]

Importance and Type of Injury. Black- and yellow-striped caterpillars, up to 2 inches long, with yellow rings around their necks, or, in the case of the red-humped caterpillar, with a pronounced red hump just back of the head, with a row of spines projecting from it. These caterpillars will be found feeding in colonies on the leaves of the apple, pear, and some forest trees during July and August, completely defoliating small trees, or single branches on large trees.

Trees Attacked. Apple, pear, cherry, quince, and many shade and forest trees.

[1] *Datana ministra* (Drury), Order Lepidoptera, Family Notodontidae.
[2] *Schizura concinna* (Smith and Abbott), Order Lepidoptera, Family Notodontidae.

Distribution. General over the United States and Canada.

Life History, Appearance, and Habits. The yellow-necked caterpillar passes the winter in the form of a brown naked pupa 2 or 3 inches below the surface of the ground.

The red-humped caterpillar passes the winter as a full-grown larva in a cocoon on the ground, pupating early in the summer. Both emerge as brown moths, about 2 inches across the wings, which fly to the apple and related trees, where the females lay their eggs on the undersides of the leaves in masses of 50 to 100 (Fig. 15.7). The young caterpillars hatching from the eggs feed at first on a single leaf with their heads all pointing toward the outer edge of the leaf. At first they skeletonize the leaf, but within a few days they increase in size and of necessity spread over a number of leaves, sometimes all the leaves on a single twig or small branch, and begin consuming the entire leaf. During the course of their feeding, they sometimes migrate from one part of the tree to another. When disturbed, these caterpillars raise both ends of their bodies in the air, clinging to the plant with the prolegs near the middle (Fig. 15.8). They become full-grown in about 3 weeks, at which time the yellow-necked caterpillars enter the ground and change to the pupal stage, and the red-humped caterpillars spin their cocoons on the ground.

Control Measures. In young orchards where the trees are 3 years old or under, a careful inspection of the trees during late July or early August will show where these

Fig. 15.7. Egg mass of yellow-necked caterpillar. Slightly enlarged. (*From U.S.D.A., Farmers' Bul.* 1270.)

insects are starting to feed, and they may be easily removed by hand and killed. They are normally controlled by the regular spray schedule. Spraying with malathion at 0.5 pound or parathion at 0.15 pound has been suggested for specific control.

Fig. 15.8. Cluster of larvae of yellow-necked caterpillar, *Datana ministra*, showing position assumed when alarmed. About natural size. (*From U.S.D.A., Farmers' Bul.* 1270.)

Trees Attacked.   Apple, plum, crab apple, quince, cherry, wild cherry, wild plum, and pear.

Distribution.   This insect, like the preceding one, is abundant in the upper Mississippi Valley and other northern states.

Life History, Appearance, and Habits.   The winter is always passed as a dark-brown, somewhat hairy caterpillar, about ⅓ to ½ inch long, enclosed in the tough, curled, grayish cases above mentioned.   In the spring of the year, about the time that the apple buds open, these worms become active, loosen their cases from the points where they have been attached to the tree during the winter and, carrying the cases with them, begin feeding on the newly opening buds, later fastening several leaves together with silken threads.   The latter part of May and during June, they change to the pupal stage, and later emerge as moths having a wing expanse of about ¾ inch. The wings are brownish in color, with white mottlings.   The moths deposit their eggs on the new apple leaves, the young caterpillars of the next generation appearing in about 2 to 3 weeks.   These begin feeding at once on the shoots and leaves, and they construct the silken, curved cornucopia-shaped cases in which they feed for the remainder of the summer.   In the early fall they attach these cases securely to the twigs and there pass the winter.   This insect is never destructively abundant in well-sprayed orchards.

Control Measures.   In orchards where this insect is causing injury, particular attention should be given to applying DDT or lead arsenate in the cluster-bud, calyx, and 3-weeks' applications, as suggested under Codling Moth (page 730).   No special sprays are required in commercial orchards other than those regularly given for the control of codling moth, curculio, and other leaf-feeding insects.

References.   U.S.D.A., *Farmers' Bul.* 1270, 1931; *Fourth Rept. Ill. State Ento.,* pp. 65–74, 1889.

Fig. 15.12. Case bearers.   *a,* Cigar case-bearers and their work on apple leaves; *b,* pistol casebearer and its work on young fruits.   Natural size.   (*From Lochhead, "Economic Entomology," McGraw-Hill-Blakiston.*)

142. Pistol Casebearer[1] and Cigar Casebearer[2]

Importance and Type of Injury.   Tiny, brownish-gray, tough, silken cases, about ¼ inch long, either slightly curved[2] or curled or bent over at the top, so that they roughly resemble a pistol in outline[1] are attached to the leaves, twigs, fruits, and branches, and stand at right angles to them (Fig. 15.12,*b*).   Small brown worms, enclosed in these cases, which they drag about with them, feed on the leaves, buds, and fruit and eat numerous small holes over the surface.   (See also Fig. 5.17.)

Trees Attacked.   Apple, pear, quince, plum, cherry, and haw.   The Jonathan apple is apparently immune.

Distribution.   General in apple-growing sections, Virginia to Kansas and northward into Canada.   The cigar casebearer ranges westward to New Mexico, Montana, and British Columbia.

Life History, Appearance, and Habits.   The winter is passed in the larval or worm stage inside the little pistol- or cigar-shaped cases, the cases being about ⅛ inch long

---

[1] *Coleophora malivorella* Riley, Order Lepidoptera, Family Coleophoridae.

[2] *Coleophora occidentis* (= *fletcherella*) Zeller, Order Lepidoptera, Family Coleophoridae.

and firmly attached to the bark of twigs or branches.   The partly grown larvae within these cases are light-brown with dark-brown heads.   The insects begin feeding about the time the buds unfold in the spring, and reach the full-grown stage about June 1.   The cigar casebearer, by protruding the head end from the case, eats in between leaf surfaces, making small blotch mines (Fig. 15.12,*a*).   This species also makes complete new cases in the spring when the larvae start to feed.   About June 1, they change to the pupal stage, and a little later to small mottled gray moths, with lanceolate wings fringed with long hairs and expanding about ½ inch, which deposit their eggs on the underside of the leaves.   The larvae of the next generation appear during the late summer.   These worms feed on the leaves until early fall, when they migrate to the twigs and branches, where they pass the winter.   These little insects are usually not abundant enough to cause commercial loss.

Control Measures.   The ordinary spray schedule for bearing trees will usually control these insects.   Where this schedule is not followed and the casebearers are abundant, spraying in the cluster-bud stage with DDT 1 pound to 100 gallons of spray is effective

References.   *N.Y. (Cornell) Agr. Exp. Sta. Buls.* 93, 1895, and 124, 1897; *U.S.D.A., Farmers' Bul.* 1270, 1922; *W. Va. Agr. Exp. Sta. Bul.* 246, 1931.

143. Bud Moths[1]

There are several species of bud moths which cause injury to the apple. They vary somewhat in their life history, but the means of control are the same for all species.

Importance and Type of Injury. The larvae of these insects eat out the newly opening buds (Fig. 15.13). The worms form small silken cases on the leaves or twigs, webbing together bits of the foliage. Small pits are eaten in the fruit, usually where a leaf lies in contact with it.

Trees Attacked. Apple, pear, cherry, wild plum, haw, and possibly others.

Distribution. Most important in the northern apple-growing sections.

Life History, Appearance, and Habits. The bud moths or budworms pass the winter as small brown worms with black heads, about ¼ inch long, in small silken cases attached to the axil of a twig or bud. These cases are inconspicuous and not easy to find. In the spring when the leaves

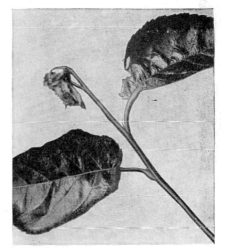

Fig. 15.13. Spring foliage injury by bud moth on apple. Reduced ½. (*From U.S.D.A., Farmers' Bul.* 1273.)

begin to open, the little dark-brown caterpillars emerge from their hibernating cases and attack the buds and leaves.   They eat out the buds or cut off the stem of a leaf and, folding the edges together, attach it by silken threads to other leaves on the tip of the twig.   They live in these cases for from 5 to 7 weeks, feeding at night.   In early summer the worms, which are then about ½ inch long, change within their silken homes into a brown pupal stage, and later emerge as grayish or brownish moths. The moths deposit eggs for the next generation, which appears from mid-

[1] The eye-spotted bud moth, *Spilonota ocellana* (Denis and Schiffermüller), and others, Order Lepidoptera, Family Olethreutidae.

June to mid-July.   These little worms feed in much the same manner as those of the spring generation, although one of the species, the lesser bud moth,[1] mines the leaves to some extent.

Control Measures.   These insects may be controlled by spraying with malathion or parathion at 0.5 pound to 100 gallons of spray, as wettable powder, applied as a delayed dormant to first-cover application.   Lead arsenate is also effective at 3 pounds to 100 gallons of spray in the pink-bud stage.

References.   U.S.D.A., Dept. Bul. 1273, 1924; *Can. Ento.,* **65**:160–168, 1933.

Fig. 15.14. Green fruitworm hollowing out a small apple.   About natural size.   (*From Ill. Natural History Surv.*)

144. Green Fruitworms[2]

Importance and Type of Injury.   It frequently happens that an orchardist will find large holes eaten in the young apples, from 3 to 4 weeks after the petals fall.   This

[1] *Recurvaria nanella* (Hübner), Order Lepidoptera, Family Gelechiidae.

[2] *Orthosia hibisci* Gn. (= *Graptolitha insciens*); *Lithophane* (= *Graptolitha*) *bethunei* G. and R.: *L. laticinerea* (= *cinerosa, grotei*) Grt.: *L. antennata* (Walker), and others, Order Lepidoptera, Family Noctuidae.

injury is often caused by the larvae of several species of moths which are known as green fruitworms.   When the apples are about the size of marbles, one may find the entire side or end of the fruit eaten out (Fig. 15.14).   A close examination will often show that the majority of the young apples on a single branch or twig have been injured in this way.   Usually one will be able to find a greenish or greenish-white worm, from $\frac{1}{4}$ to $1\frac{1}{4}$ inches long, somewhere upon the twig or branch, possibly feeding actively on the side of the apple.   The injury is sometimes severe enough entirely to destroy the apple; more often the fruit will continue to grow but will be worthless. These worms are seldom very abundant in an orchard.

Plants Attacked.   Nearly all the deciduous forest trees, all common tree fruits and some field crops.

Distribution.   General over the eastern United States and Canada.

Life History, Appearance, and Habits.   The winter is passed in two stages, but most commonly as a grayish moth with a wing expanse of a little over 1 inch.   These moths hibernate in woodlands or sheltered places about orchards.   A small percentage of the insects pass the winter in the pupal stage.   This stage will be found in a closely woven, silken cocoon 2 or 3 inches beneath the surface of the soil.   Those that have gone through the winter as pupae emerge as moths early in the spring.   The moths fly to the orchards as soon as growth starts, and deposit their eggs, one in a place, on the twigs and branches of trees, the moths dying after they have completed egg-laying. There are at least three species, and probably more, which cause this injury.   All these larvae are of a pale grass-green color, some marked with whitish stripes down each side of the body and a narrower stripe down the middle of the back.   When full-grown, they are fat squatty-appearing worms from 1 to $1\frac{1}{4}$ inches in length. The full-grown stage is usually reached during the first of June; then the caterpillar descends from the tree, burrows into the ground, and there constructs the cocoon in which it changes to the pupal stage.   Aside from feeding on apples, the insects have occasionally become abundant enough to cause defoliation of woodland trees, and have been known to destroy a 20-acre field of corn in Illinois.

Control Measures.   The fruitworms may be controlled by spraying with DDT at 1 pound or lead arsenate at 3 pounds with 3 pounds of hydrated lime to 100 gallons of spray, in the petal fall or first-cover applications.

References.   *U.S.D.A., Farmers' Bul. 1270, 1931; N.Y. (Geneva) Agr. Exp. Sta. Bul. 423, 1916; Can. Dept. Agr. Ento. Branch, Tech. Bul. 17, 1919; U.S.D.A., Bur. Ento. Plant Quar. Cir. 270, pp. 31, 32, 1933.*

145. Fruit Tree Leaf Roller[1] and Red-banded Leaf Roller[2]

Importance and Type of Injury.   The fruit tree leaf roller varies greatly in numbers from year to year.   In years of abundance it may ruin 80 to 90 per cent of the apple crop if uncontrolled.   From shortly after the buds open to about 3 weeks after the petals fall, small greenish to greenish-brown caterpillars feed on the leaves, buds, and small fruits of the apple.   In most cases, a light web is spun about several leaves, and these are rolled or drawn together, often enclosing a cluster of newly formed apples. Small apples have cavities eaten out of the side or center, somewhat like those made by the green fruitworm.   The trees may be partially to completely defoliated, with numerous fine, white, silken webs over the bark and trunk.   At picking time the apples have deep, russeted, elongate scars in the side (Fig. 15.15).   The red-banded leaf roller has become a serious pest in many orchards sprayed with DDT for codling moth control, probably by the destruction of its natural enemies.   The larvae skeletonize the leaves close to the midribs from the underside, folding and webbing them together.   When abundant they feed on the fruits, eating shallow irregular channels especially where leaves touch the fruits.

[1] *Archips* (= *Cacoecia*) *argyrospila* (Walker), Order Lepidoptera, Family Tortricidae.
[2] *Argyrotaenia velutinana* (Walker), Order Lepidoptera, Family Tortricidae.

Plants Attacked.   Nearly all kinds of deciduous fruits, many forest trees, some bush fruits, and some herbaceous plants.

Distribution.   The fruit tree leaf roller is generally distributed in the apple-growing sections of the United States and Canada, while the red-banded leaf roller is found only in the eastern areas.

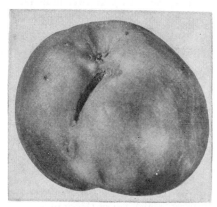

FIG. 15.15. Mature apple showing injury caused by fruit tree leaf roller feeding on the young fruit. (*From Ill. Natural History Surv.*)

Life History, Appearance, and Habits.   The fruit tree leaf roller passes the winter in the egg stage.   These eggs are laid in masses of 30 to 100, and are closely plastered on the twigs, branches, and occasionally on the trunk of the tree.   They are covered with a smooth coating of dull-brown or gray varnish-like material, which protects them from the weather and prevents the individual eggs from showing in the cluster (Fig. 15.16). The egg masses blend almost perfectly with the bark on which they are deposited, making it extremely difficult to see them.   The eggs begin hatching about the time the apple fruit buds are beginning to separate in the spring.   The young worms crawl to the leaves, and feed for about 1 month.   They vary somewhat in color but in general are pale-green with

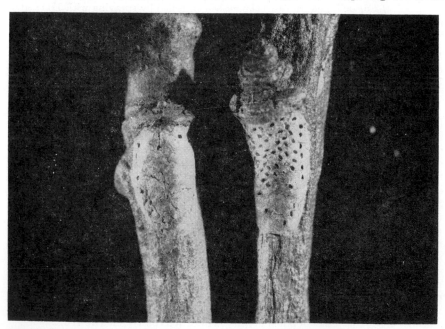

FIG. 15.16. Egg masses of the fruit tree leaf roller on twigs of apple.   Egg mass on the right shows exit holes of newly hatched larvae.   About twice natural size.   (*From Ill. Natural History Surv.*)

a brown head and a brown plate just back of the head.   When about ¾ inch in length, the larvae pupate within the folded or rolled leaves, or crawl to the trunk or branches of the tree and there construct a somewhat flimsy cocoon.   From this cocoon the moths, which have brownish front wings, expanding ¾ to 1 inch, mottled with pale gold and usually forming two large patches on the front margin (Fig. 15.17), emerge during late June or July.   Within a few days they mate and lay their eggs in the situations above described.   The insect remains in the egg stage until the following spring.   There is but one generation each year.   The red-banded leaf roller overwinters in the pupal stage beneath leaves and debris on the ground, and the adult moths emerge as the buds open to deposit egg masses on the undersides of the larger limbs.   The egg masses, which contain about 50 eggs, resemble those of the fruit tree leaf roller but are more irregular and flattened.   The eggs hatch about petal fall into pale-green larvae with yellowish heads which feed on leaves and fruits through June.   These pupate and a second generation appears in August;

Fɪɢ. 15.17. Adult female of the fruit tree leaf roller.   About twice natural size.   (*From Ill. Natural History Surv.*)

the larvae feed until apple harvest in northern areas.   A third generation takes place in southern areas.

Control Measures.   The fruit tree leaf roller may be controlled by spraying the trees, during the dormant stage, with 3 to 4 per cent mineral oil emulsion.   Commercial miscible oils or oil emulsions should be used as recommended by the manufacturer.   Very thorough application is necessary to wet each egg mass since these are often laid in the smaller twigs and in the axils of the twigs.   Spraying when the buds begin to separate with wettable powders of DDT at 1 pound or lead arsenate at 4 pounds with 4 pounds of hydrated lime or other safener to 100 gallons of spray, or at petal fall and again 10 days later with parathion at 0.3 pound to 100 gallons of spray, is also effective.

The red-banded leaf roller may be controlled by spraying, at the indicated times, with one of the following materials in 100 gallons of spray: (*a*) endrin at 0.25 pound (petal fall), (*b*) Guthion at 0.3 pound, parathion at 0.25 pound, or Phosdrin at 0.25 pound (petal fall, first cover, and fifth or sixth cover), (*c*) TDE (DDD) at 0.75 pound (petal fall and fifth or sixth cover), (*d*) lead arsenate at 3 pounds with 3 pounds of hydrated lime or other safener (petal fall and first cover).   Sevin at 1 pound per

100 gallons of spray may be used in the fifth or sixth cover.   All materials except Phosdrin are used as wettable powders.   The granulosis virus, *Bergoldia clistorhabdion*, readily infects the larvae and may have possibilities for control.

References.   Colo. State Ento. Cir. 5, 1912; *Mont. Agr. Exp. Sta. Bul.* 154, 1923; *Idaho Agr. Exp. Sta. Bul.* 157, 1928; *Ill. State Natural History Surv., Ento. Ser. Cir.* 9, 1926; *N.Y. (Geneva) Agr. Exp. Sta. Buls.* 583, 1930; 733, 1948; and 755, 1952; *Jour. Econ. Ento.*, **42**:29, 354, 1949; and **51**:378, 454, 1958.

146 (1, 27, 75).   GRASSHOPPERS[1]

Grasshoppers are not generally considered as orchard pests, but occasionally, during years of abundance, when they have eaten most of the green growth in fields, or during periods of drought, they will migrate to orchards and sometimes completely strip the foliage from trees during July and August.   Sometimes the bark is scarred and roughened by the feeding of these insects.   Outbreaks of grasshoppers in the vicinity of orchards should be controlled before the insects migrate to the trees, by spraying ground cover and fence rows with aldrin or dieldrin at 4 ounces, chlordane at 1 pound, parathion at 6 ounces, or toxaphene at 2 pounds per acre.   Poison baits (page 468) may also be effective.   If the grasshoppers have already invaded the orchard, spraying the orchard floor and lower portions of the trees with methoxychlor at 3 pounds per acre has been suggested.

References.   Mont. Agr. Exp. Sta. Bul. 148, 1922; *Iowa Agr. Ext. Bul.* 182, 1932; *Ill. Agr. Exp. Sta. Cir.* 634, 1949.

147. APPLE FLEA WEEVIL[2]

Importance and Type of Injury.   This insect has been more destructive in Ohio and Illinois than in other states.   Very small dull-black snout beetles about $\frac{1}{10}$ inch long puncture the newly opening leaves and buds of apple trees early in the spring. In the summer the leaves have numerous small holes eaten out from the underside, in cases of severe injury appearing as if riddled by very fine bird shot.   Yellowish-brown larval mines start from near the center of the leaf and run to small blister-like cells at the margin of the leaf.

Trees Attacked.   Apple, haw, winged elm, hazelnut, quince, wild crab, blackberry.

Distribution.   From Missouri and Illinois to eastern New York and southward to the Ohio River.

Life History, Appearance, and Habits.   The adults pass the winter in trash, grass, and leaves just at the surface of the ground, usually under apple trees.   They become active as soon as the buds begin to swell in the spring, some crawling up the trunk of the trees, while others mount the stems of weeds and grasses and fly to the newly opening buds.   They feed for 1 or 2 weeks on the newly forming leaves, dropping to the ground during periods of cold weather and storms.   When the leaves are about two-thirds grown, the females begin depositing their eggs along the midribs.   These eggs hatch into small grubs, which feed between the upper and under surface of the leaf, eating out a little mine which is extended to the margin of the leaf.   Here they hollow out a cell about $\frac{1}{4}$ inch across and, in this, complete their growth and change to pupae, emerging as full-grown beetles during the latter part of May and June. The newly emerged beetles feed for from 2 weeks to 1 month on the leaves and then seek shelter in the trash about the base of the trees, remaining there until the following spring.   There is but one generation each year.

Control Measures.   This insect may be controlled by spraying in the prepink or pink stage of bloom development with wettable DDT at 1 pound or parathion at 0.15 pound per 100 gallons of spray.   Care should be taken to apply the spray to the undersides of the leaves.   This spray is to be given only when the spray schedule for the control of other apple insects(page 730) is not effective in controlling this insect.

[1] Various species of Order Lepidoptera, Family Locustidae.

[2] *Rhynchaenus* (= *Orchestes*) *pallicornis* (Say), Order Coleoptera, Family Curculionidae.

References.   *Ill. State Natural History Surv. Bul.*, vol. 15, art. 1, 1924; *Ohio Agr. Exp. Sta. Bul.* 372, 1923; *Jour. Econ. Ento.*, **29** :381, 1936, and **43** :947, 1950.

148. IMBRICATED SNOUT BEETLE[1]

Importance and Type of Injury.   Occasionally, buds of apple and newly forming fruits are injured by a grayish snout beetle, about ½ inch long.   This insect has a body which is very plump and well rounded (Fig. 15.18,*a,b*).   It injures the apple by eating out the buds, or cutting off the young fruit and leaves.   It is a very general feeder and is found on many other plants, causing great damage to strawberries. It has seldom been abundant enough to cause serious injury in well-cared-for orchards.

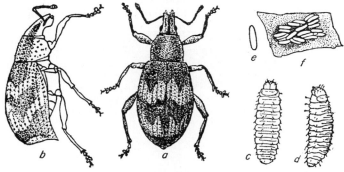

FIG. 15.18. Imbricated snout beetle, *Epicaerus imbricatus.*   *a, b,* Adult, dorsal and side views, 3 times natural size; *c, d,* larva, dorsal and side views, enlarged; *e,* egg more enlarged; *f,* eggs on leaf, twice natural size.   *(From Ill. Natural History Surv.)*

Life History, Appearance, and Habits.   The eggs of this insect (Fig. 15.18,*e,f*) are laid on the leaves of many plants, and the immature stages (*c,d*) are passed in the stems or on the roots of legumes or some other field crops.   The adult insects make their appearance on the trees during the latter part of May and June, and continue feeding for about 1 month.   There is one generation a year.

Control Measures.   If the beetles are causing damage to young trees, many of them may be caught by jarring them upon sheets spread under the trees.   They will normally be controlled by the spray schedule for the control of other apple insects (page 730).

149 (172, 179, 189). SAN JOSE SCALE[2]

Importance and Type of Injury.   If infestation by this insect is allowed to develop unchecked, it will result in the death of the trees in home or commercial orchards.   In 1922, more than 1,000 acres of mature apple trees were killed in southern Illinois by the San Jose scale.   Trees lightly infested show small grayish specks on the surface of the bark which are disk-shaped and just discernible to the eye (Fig. 15.19).   Under the hand lens these specks show a raised nipple-shaped spot at the center.   Frequently the bark is reddened for a short distance around each of the scales, this being especially true on young trees and on new growth of old trees. On heavily infested trees the entire surface of the bark is covered with a gray layer of overlapping scales, appearing as if the twig or branch had been sprinkled with wood ashes when wet.   Infested trees show a general decrease in vigor and thin foliage, which is usually more or less yellowed and spotted because of the presence of scales upon it.   Terminal twigs characteristically die first.   Sometimes there is an abundance of water

[1] *Epicaerus imbricatus* (Say), Order Coleoptera, Family Curculionidae.
[2] *Aspidiotus perniciosus* Comstock, Order Homoptera, Family Coccidae.

sprouts along the larger branches.   The fruit on infested trees is also attacked by the scale, the insects being most abundant around the blossom and stem ends, often forming a gray patch about the calyx of the apple.   The fruit on infested trees often has a spotted or mottled appearance because of a small red inflamed area surrounding each of the scales (Fig. 4.5,*E*).

Trees Attacked.   Apple, pear, quince, peach, plum, prune, apricot, nectarine, sweet cherry, currant, gooseberry, and osage orange, besides many bush fruits and shrubs and some shade and forest trees.

Distribution.   Throughout the commercial fruit-growing sections of the United States and Canada and in many other parts of the world.

Fig. 15.19. Twig of apple encrusted with San Jose scale, *Aspidiotus perniciosus.*   Many of the scales show exit holes of a hymenopterous parasite.   Both male and female scales can be recognized.   Enlarged several times.   (*From Ill. Natural History Surv.*)

Life History, Appearance, and Habits.   The insect passes the winter in a partly grown condition.   Nearly all the scales surviving a temperature of 10°F. above zero are in the second nymphal instar, often known as the sooty-black stage, and are about one-third grown.   The insect remains dormant, tightly fastened to the bark of the tree until the sap starts flowing in the spring.   It then begins to grow and usually becomes full-grown about the time the trees come into bloom.   At this time two forms of scales will be noted, one nearly round about $\frac{1}{12}$ inch across, with a raised nipple in the center, and the other oval, about $\frac{1}{25}$ inch long by half as broad, with a raised dot nearer the larger end of the scale.   The latter scales are the waxy covering of the males, which emerge as small, yellow, two-winged insects (Fig. 15.20) at about this time.   The larger round scales cover the bodies of the females, which remain under the scales throughout their lives, and, after mating, begin to give birth to living young.   They continue to reproduce for a month or more, depending on the temperature.   These young have the appearance of very small yellow mites or lice (Fig. 4.5,*F*).   They have six well-developed legs and two antennae; they crawl about over the surface of the bark for a short time.   On finding a place which is attractive to them, they insert their slender, thread-like mouth parts through the bark and begin sucking the sap.   Very shortly after this they molt or shed their skins and with the old skin lose their legs and feelers, becoming mere flattened yellow sacks attached to the bark of the tree by their sucking mouth parts. As the insect grows, a waxy secretion is given off from the body and this hardens into the protective scale under which the insect lives.   There

are from two to possibly six generations of the scale each year; the smaller number occurring in the northern part of the country, and the larger in the extreme southern. The insect increases most rapidly in dry hot seasons, it having been determined that the progeny of a single female insect could be well over 30,000,000 in a single year.

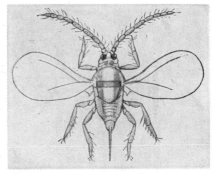

FIG. 15.20. San Jose scale, adult male, greatly enlarged. (*From Ill. Natural History Surv.*)

The insect is carried accidentally from orchard to orchard on the bodies of birds and larger insects and also to a greater extent by being blown through the air by the wind. It has been spread throughout the entire country on shipments of infested nursery stock, the original introduction having resulted from an importation of Chinese plants which were set out on the grounds of a large estate in San Jose, Calif., about 1870.

Control Measures.   This insect may be controlled by spraying the trees during the dormant stage with 2 to 3 per cent mineral-oil emulsions. Commercial miscible oils or oil emulsions should be used as recommended by the manufacturer, superior oils (page 368) at 2 gallons to 100 gallons of spray, and others at 3 gallons.   In some areas the oil spray has been supplemented with 0.6 pound of DNOC or 0.4 pound of DNOCHP, but care should be used to avoid plant injury.   Oil-spray applications may be made at any time the tree is in a dormant condition and the temperature is above freezing, although injury has sometimes resulted when applications were made just before a drop in temperature to 20°F. below zero or lower.   In order to secure satisfactory control it is necessary to do a most thorough and careful spraying, since the tiny insects are generally distributed all over the bark, twigs, branches, and trunk and only those hit with the spray will be killed.

During the growing season, the scale may be controlled by spraying when the crawlers are present, about the second, sixth, and seventh cover sprays, with wettable parathion at 0.15 to 0.25 pound, malathion at 0.5 pound, Guthion at 0.3 pound, or Sevin at 0.5 pound to 100 gallons of spray.   Parathion should not be applied to McIntosh and related varieties.

References.   Ill. Agr. Exp. Sta. Cir. 180, 1915; *Bul. Ill. State Nat. Hist. Surv.,* vol. 17, art. V., 1929; *U.S.D.A., Farmers' Bul.* 650, 1915, and *Bur. Ento. Bul.* 62, 1906.

150 (237). SCURFY SCALE[1]

Importance and Type of Injury. This insect is found on the bark of the tree in the form of a grayish-white scale, pear-shaped in outline, about ⅛ inch long, rounded at one end and tapering to a rather sharp point at the other (Fig. 15.21). The scales are usually more abundant on the shaded parts of the tree and in orchards where trees have not been properly pruned or the foliage is too dense. The fruit is sometimes spotted by

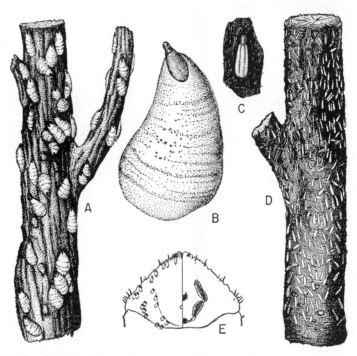

FIG. 15.21. Scurfy scale, *Chionaspis furfura*. *A*, female scales on twig, about twice natural size; *B*, a single female shell about 16 times natural size, showing exuviae of first and second nymphal stages at the pointed end; *C*, a single male shell about 16 times natural size, showing the three longitudinal ridges; *D*, twig bearing male scales, which are usually assembled on a separate twig from the females; *E*, pygidium or terminal abdominal segment from body of female, showing microscopic features used in classifying Coccidae; the dorsal surface is shown on the left, the ventral surface on the right. *(Original.)*

this insect. However, it is not of much importance in vigorous well-sprayed orchards.

Plants Attacked. Pear, apple, gooseberry, currant, black raspberry, Japan quince, mountain ash, and other common deciduous trees and bush fruits.

Distribution. This is a native insect which is found in the United States, from Idaho and Utah eastward.

Life History, Appearance, and Habits. The scurfy scale passes the winter as reddish-purple eggs, averaging about 40 under each female scale. The eggs hatch rather late in the spring (late May or June) after the apple

[1] *Chionaspis furfura* (Fitch), Order Homoptera, Family Coccidae.

trees have come into full leaf.   The minute, purplish, young nymphs crawl about for a few hours and then settle down on the bark for the remainder of their lives.   The female scales are, when full-grown, about ⅛ inch in length; the males, which are usually concentrated on certain twigs by themselves, are only about one-fourth as long, and are covered by a narrow, straight-sided, intensely white scale with three distinct ridges along the back of it.   From these scales the minute two-winged males emerge and fly about to seek the females.   The males die soon after mating and the females after they have laid their eggs.   There are two generations of the insect each year throughout most of the country, the eggs of the summer generation being deposited in July and hatching the latter half of that month. Some of these nymphs usually settle upon the fruit. Eggs of the overwintering generation are usually laid in late August and September; but only one generation occurs in the North.

Control Measures.   This insect may be controlled by spraying when the eggs hatch with wettables of DDT at 1 pound or parathion at 0.3 pound to 100 gallons of spray.   In some areas, dormant oil sprays or oil plus dinitro compounds have been used, as described on page 707.

References.   *N.Y. (Geneva) Agr. Exp. Sta., Farm Res.,* January, 1938; *Jour. Econ. Ento.,* **26**:912, 1933.

151 (238). OYSTER-SHELL SCALE[1]

Importance and Type of Injury.   Small dark-brown scales are found adhering closely to the bark of the tree, appearing very much like half of a minute oyster shell (Fig. 15.22).   The scales are about ⅛ inch long by one-third as wide.   They usually are more or less clustered on the bark and a heavily infested tree may have the bark entirely covered. The bark of the injured tree usually becomes cracked and scaly.   The trees lose vigor, the foliage is undersized and specked with yellow, and in severe infestations the death of the tree results.

Fig. 15.22. Twig infested with apple oyster-shell scale, *Lepidosaphes ulmi.* About 3 times natural size. (*From Ill. Natural History Surv.*)

Plants Attacked.   Apple, pear, mountain ash, quince, plum, raspberry, currant, almonds, apricots, grapes, figs, Persian walnuts, and many shade trees and shrubs (page 835).

Distribution.   This scale is largely confined to the northern two-thirds of the United States and southern Canada.

Life History, Appearance, and Habits.   The insect passes the winter in the form of grayish-white minute eggs, tightly enclosed under the wax of the parent scale.   From 40 to 150 or more of these eggs will be found under each female scale.   The eggs hatch late in the spring after the apple trees have bloomed.   The young nymphs, which are very small and whitish in color, crawl about over the bark for from a few hours to 1 or 2 days.   Once having inserted their beaks into the bark, they begin the formation of a waxy scale coating, which covers their bodies, and soon

[1] *Lepidosaphes ulmi* (Linné), Order Homoptera, Family Coccidae.   A very similar species, *Lepidosaphes ficus* (Signoret), is a pest of figs in California.

shed their skins and their antennae and legs along with them.   The scale is white at first but later changes to polished brown.   The insects become full-grown about the middle of July; the males then emerge as winged insects, and the females, after mating, deposit their eggs under their scales, the body gradually shrinking toward the pointed end of the scale as the eggs are deposited.   The females die shortly after the last eggs are laid. These eggs hatch in about 2 weeks, and the second generation of scales becomes full-grown during the early fall, there being two generations of this insect each season over much of its range.   In the northern part of its range, probably only one generation occurs.

Control Measures.   The oyster-shell scale does not yield readily to dormant treatments of ordinary oil emulsion or lime-sulfur.   Dormant applications of DNOC have been used in some areas (page 707).   The most generally satisfactory control has resulted from spraying just at the time of hatching of the young scales, with wettables of DDT at 1 pound or parathion at 0.3 pound to 100 gallons of spray.   No fixed date can be given for applying this spray and it will be necessary, in order to accomplish the best results, to watch the trees carefully and spray as soon as the crawling young appear in numbers.   In the latitude of central Illinois hatching occurs about the first of June.

References.   Ohio Agr. Exp. Sta. Cir. 143, 1914; *U.S.D.A., Farmers' Bul.* 1270, 1922; *Proc. N.Y. Hort. Soc.,* **79**:15, 1934; *N.Y. (Cornell) Agr. Exp. Sta. Memoir* 93, 1925.

Apple Aphids

Importance and Type of Injury.   Aphids of three different species are common on the foliage, fruit, and twigs of apple trees throughout the country.   These are known as the rosy apple aphid, the apple aphid, and the apple-grain aphid.   The last named is of little importance, although usually the most abundant early in the season.   The relative abundance varies greatly with different seasons, sometimes one species being extremely abundant and the others scarce, while in other years all three may be very numerous.   In many years the injury from aphids is slight.   When they are abundant, their feeding causes the leaves to curl and the stems and twigs to become stunted and unhealthy in appearance. The new twig growth sometimes assumes a curled and twisted appearance, even forming a loop.   In many cases, particularly when attacked by the apple and rosy aphids, the apples remain very small.   Many of these fruits are also somewhat misshapen, hard, and knotty in texture and characteristically puckered around the calyx end.   The rosy apple aphid is the most injurious species of aphid occurring on the foliage of the apple. A fourth species, the woolly apple aphid, does not injure the fruits but is common on trunks, branches, and roots.

152. Rosy Apple Aphid[1]

Plants Attacked.   Apple is the favorite host, although the insect feeds also on pear, thorn, and Sorbus and, during the summer, on the narrow-leaved plantain.

Distribution.   General throughout the United States and apple-growing sections of Canada.

Life History, Appearance, and Habits.   The dark-green shiny ovate eggs, in which stage the insect passes the winter, are attached to the bark

[1] *Anuraphis roseus* Baker, Order Homoptera, Family Aphidae.

of the twigs and branches on all parts of the tree and usually hidden away in crevices in the bark or the depressions and wrinkles formed around the buds, twigs, and old wounds (Fig. 15.23). The eggs begin to hatch when the buds start opening in the spring. With this species, the eggs do not all hatch at once but continue for 2 weeks or sometimes longer. The young aphids (Fig. 15.24) make their way to the newly opening buds and feed on the outside of the leaf-bud and fruit-bud clusters, until the leaves have begun to unfold. They then work their way down the inside of the clusters and begin sucking the sap from the stems and newly formed fruits. Their feeding causes the leaves to curl, and this affords the aphids protection from some of their natural enemies and from sprays or dusts applied to the tree for their control. The aphids which hatch from the eggs are all females and are called the stem-mothers, as they are the mothers of the season's brood. In about 2 weeks or a little longer, depending on the weather, these stem-mothers, without mating, begin giving birth to young, and these young in turn begin reproducing in about 2 weeks. The body of the aphid has a somewhat waxy coating and usually a slight purplish or rosy tinge. They continue on the apple during May, and in smaller numbers through June and July. During the early summer they migrate to stems and stalks of the narrow-leaved plantain, where they feed and reproduce until fall. In the fall winged ovoviviparous females fly back to apple trees and give birth to the true egg-laying females. The males develop a little

FIG. 15.23. Eggs of the three common species of apple aphids. Enlarged. (*From N.Y. (Cornell) Agr. Exp. Sta. Memoir* 24.)

later and fly to mate with the true females, which then deposit their eggs in the situations above mentioned. The eggs hatch the following spring.

Control Measures. Damage by this and the other species of aphids that winter in the egg stage on apple can be prevented by dormant spraying with DNOC at 0.6 pound or DNOCHP at 0.4 pound per 100 gallons of spray; by spraying during prepink- or pink-bud stage with lindane,

FIG. 15.24. Newly hatched nymphs of the three common apple aphids. *1*, The apple aphid; *2*, the rosy apple aphid; and *3*, the apple-grain aphid, much enlarged. (*From N.Y. (Geneva) Agr. Exp. Sta. Bul.* 431.)

Trithion, or demeton (Systox) at 0.25 pound per 100 gallons of spray; or by spraying when the aphids become abundant with demeton or Diazinon at 0.25 pound, Guthion or parathion at 0.15 pound, malathion at 0.5 pound, or TEPP at 0.12 pound per 100 gallons of spray. Other materials which have been used in certain areas are delayed dormant spraying with the triethanolamine salt of 4,6-dinitro-*o-sec*-butylphenol at 1 pound, or spraying as needed with nicotine sulfate at 0.4 pound with 1 pound of hydrated lime to 100 gallons of spray.

It is only during an occasional season that these insects become destructively abundant. They are preyed upon by many natural enemies, including the lady beetles, syrphid flies, and aphid-lions. In seasons when the spring is warm, the natural enemies usually become sufficiently abundant to control the aphids, and it is rarely necessary in such years to resort to artificial measures of control. When the spring is cold and backward, the aphids usually increase more rapidly than their enemies, and it is in such seasons that the greatest damage occurs.

References. Maine Agr. Exp. Sta. Bul. 233, 1914; *N.Y.* (*Cornell*) *Agr. Exp. Sta. Memoirs* 24, 1919; *N.Y.* (*Geneva*) *Agr. Exp. Sta. Bul.* 636, 1933; *U.S.D.A., Farmers' Bul.* 1128, 1920.

153. APPLE APHID[1]

The apple aphid has much the same life history as the rosy apple aphid. It passes the winter in the same manner in the egg stage on the bark of wood of the previous season's growth. Unlike the rosy aphid, it remains on the apple trees during the entire summer but does not cause curling of the leaves. It is difficult for the orchardist to distinguish the nymphs and eggs of this species from those of the rosy aphid and apple-grain aphid (Figs. 15.23 and 15.24). The adults are yellowish-green, with the head, tips of antennae, legs, and cornicles dark.

Trees Attacked. Apple, pear, wild crab, hawthorn, and possibly others.

Distribution. General in the apple-growing sections of North America.

Control Measures. The control measures are the same as for the rosy aphid, but somewhat more effective because of the fact that the eggs of this species nearly all hatch within a few days and the young will all be clustered on the buds at the time when the most effective spraying can be done. This species is also subject to wide fluctuations in abundance, being controlled by the same natural enemies as those attacking the more injurious rosy aphid.

References. See under Rosy Apple Aphid.

154. APPLE-GRAIN APHID[2]

This aphid spends practically its entire feeding period on various grains and grasses, being particularly abundant on the small grains commonly grown in the United States. It is of very little importance as an apple pest, as it does not remain on the apple tree long enough to cause serious deforming of the fruit, twigs, or foliage. Its eggs are laid on the apple twigs and cannot readily be distinguished from the two species previously discussed. In some states its eggs are by far the most abundant of the species found on apple, and for this reason one cannot tell from the number of aphid eggs found on twigs during the winter whether or not

[1] *Aphis pomi* De Geer, Order Homoptera, Family Aphidae.

[2] *Rhopalosiphum fitchii* (Sanderson), Order Homoptera, Family Aphidae.

serious injury from these insects is likely to occur the following spring. The appearance of great numbers of these aphids on the buds in early spring may cause alarm, but they soon disappear from the apple and do not curl the leaves.   The adults are yellowish-green, with darker bands across the abdomen.   The antennae, legs, and cornicles are yellow except at the extreme tips.

Plants Attacked.   Apple, pear, wild crabs, and haws; and grains and grasses during the summer.

Distribution.   General throughout the country.

Control Measures.   This species is not generally a serious pest on apple but may be controlled, if necessary, by the measures suggested under Rosy Apple Aphid.

References.   See under Rosy Apple Aphid.

155 (103). APPLE LEAFHOPPERS[1]

Importance and Type of Injury.   The importance of these insects varies greatly in different years.   Damage results from both the devitalizing effects of their feeding upon the tree and the spotting of the fruit by their excrement.   During late summer and fall apple foliage becomes pale in color, with little specks of greenish-white showing through from the under surface of the leaves.   The new foliage becomes lightly curled, the margins of leaves sometimes burned, and the fruits speckled with minute deposits of excrement.   Many flimsy, white, shed skins of the leafhoppers are left on the undersides of the leaves.   Severely injured leaves fall from the trees.   One of the species[2] is also a pest of first importance on potatoes (page 643).

Plants Attacked.   Apple, rose, currant, gooseberry, raspberry, potato, sugarbeet, bean, celery, many other trees, shrubs, and some herbaceous flowers, grains, grasses, and weeds.

Distribution.   General in the United States.

Life History, Appearance, and Habits.   The apple and potato leafhopper[3] apparently does not winter in the northern states (page 645). The rose leafhopper,[4] the white apple leafhopper,[5] and the common apple leafhopper[6] pass the winter in the egg stage in the bark of apple, rose, and other plants.   Those hibernating as adults become active very early in the spring, the common species flying about and mating before the buds of the trees have begun to show green.   When the leaves appear, the insects begin laying their eggs, which are pushed into the midrib or the larger veins and stems of the leaves.   The first-generation nymphs appear about the time the leaves become full-grown, or shortly thereafter. They, as well as the adults, feed by sucking the sap from the underside of the leaves.   The nymphs are a pale-green to greenish-white in color, are wingless, but are very active, running forward, backward, or sidewise with equal ease.   They reach maturity about midsummer, change to the adult stage, and deposit eggs for the second generation, which becomes full-grown during the early fall.   Some of the species found in orchards

---

[1] Several species of the Order Homoptera, Family Cicadellidae.
[2] *Empoasca fabae* (Harris).
[3] *Empoasca fabae* (Harris).
[4] *Edwardsiana* (= *Typhlocyba*) *rosae* (Linné).
[5] *Typhlocyba pomaria* McAtee.
[6] *Empoasca maligna* (Walsh).

differ somewhat in their life history from the above, but not to an extent to be of any significance in the application of control measures.

Control Measures. Where the leafhopper nymphs average 50 or more per 100 leaves in the spring (May in central Illinois), control is usually profitable. The leafhoppers are readily controlled by thorough spraying, especially of the undersides of the leaves, with DDT at 0.5 pound, Guthion at 0.3 pound, malathion at 0.5 pound, or parathion at 0.15 pound per 100 gallons of spray. Nicotine sulfate at 0.4 pound per 100 gallons with soap or hydrated lime has also been used. The codling moth spray schedule will control the first generation, but a special application may be necessary for the control of the second-generation nymphs.

References. U.S.D.A., Dept. Bul. 805, 1919; N.Y. (Geneva) Agr. Exp. Sta. Bul. 541, 1918; Jour. Econ. Ento., **17**:594, 1924; Conn. Agr. Exp. Sta. Cir. 111, 1936.

156. APPLE REDBUGS

Importance and Type of Injury. The apple-growing states in the East often suffer considerable losses from certain sucking insects known as apple redbugs. Infested fruit (Fig. 15.25) has a pitted or dimpled appearance, is dwarfed in size, is somewhat hard or woody in texture, and is sometimes russeted in spots. The general appearance is somewhat like that of the injury by the rosy apple aphid except for the dimpling or pitting and russeting, which do not usually occur in the case of injury by aphids.

Trees Attacked. Apple, pear, haw, and probably others.

Distribution. States east of the Mississippi River. Most destructive in New York, New England, and southeastern Canada.

FIG. 15.25. Apples deformed by apple red-bug. Note the dimpled appearance. (*From U.S.D.A., Farmers' Bul.* 1270.)

FIG. 15.26. Adult apple redbug, *Lygidea mendax.* Enlarged about 3½ times. (*From U.S.D.A., Farmers' Bul.* 1270.)

Life History, Appearance, and Habits. Redbugs pass the winter in the egg stage. These eggs are laid in the bark of branches of the trees in the case of the dark apple redbug,[1] and in the bark lenticels in the case of the apple redbug.[2] They hatch early in the spring. The young nymphs feed at first on the foliage and later on the young apples. They are piercing-sucking insects and feed entirely on the sap of the leaves or juice of the fruit. Wherever they insert their beaks in the apple, the surrounding tissue becomes hardened and ceases to grow, and the entire fruit is stunted. These two species have essentially the same habits. Upon becoming full-grown, the adults of the dark apple redbug are about ¼ inch long, reddish-black in color, and covered on the upper surface with white flattened hairs. In the apple redbug (Fig. 15.26) the

[1] *Heterocordylus malinus* Reuter, Order Hemiptera, Family Miridae.
[2] *Lygidea mendax* Reuter, Order Hemiptera, Family Miridae.

head and front part of the body are of an orange color.   Development from egg to adult through 5 nymphal instars requires 5 to 6 weeks.   There is but one generation annually, nearly all the injury being caused by the young or nymphs, during the first month after the fall of the petals.

Control Measures.   Light to moderate infestations of redbugs may be controlled by dormant sprays of 3 to 4 per cent oil.   Where the bugs are abundant, they may be controlled by spraying at petal fall with DDT at 1 pound or parathion at 0.3 pound to 100 gallons of spray.   Spraying with nicotine sulfate at 0.4 pound with 1 pound of soap applied in the cluster-bud and calyx stages is also effective.   The spray should be applied on warm days by two men working from opposite sides of the tree, as the redbugs are so active that they may escape being hit by a spray applied to one side of the tree at a time.   On cool days the insects will not be feeding in the trees.

References.   *N.Y. (Cornell) Agr. Exp. Sta. Buls.* 291, 1911, and 396, 1918; *N.Y. (Geneva) Agr. Exp. Sta. Bul.* 490, 1921, and *Tech. Bul.* 52, p. 66, 1933; *N.Y. (Geneva) Agr. Exp. Sta. Bul.* 716, 1946.

157. European Red Mite[1]

Importance and Type of Injury.   This mite has become one of the most important fruit pests of the United States and Canada.   The injured trees, if the infestation is slight, show specking of the foliage; if the infestation is heavy, the foliage is pallid and sickly or bronzed in appearance, and from a little distance it has the appearance of being covered with dust.   Many of the injured leaves drop.   The fruit is undersized and of poor quality and color, and fruit buds are greatly weakened or prevented from forming.

Trees Attacked.   The mite occurs on many deciduous fruit and shade trees and shrubs but is most injurious to plum, prune, apple, and pear.

Distribution.   This pest, which is widely distributed over continental Europe, was first found in the United States in 1911 and has rapidly appeared in many localities in the northeastern and northwestern quarters of the United States, north of latitude 37°N.

Life History, Appearance, and Habits.   The mite passes the winter as somewhat spherical eggs (Fig. 15.27,*A*), of a bright-red to orange color, on twigs and smaller branches of the trees.   The egg has a distinct style or stalk about as long as the diameter of the egg and without guy lines. These winter or diapausing eggs require from 150 to 200 days at temperatures below 50°F. before hatching occurs in the spring just before the time of apple bloom.   The young six-legged larvae crawl to the unfolding leaves and suck the sap, passing through the protonymph and deutonymph stages (Fig. 15.27) before becoming adults.   The life cycle from hatching to adult ranges from an average of 20 days at 55°F. to 4 days at 77°F.   The adult female lives for an average of about 19 days and deposits an average of about 20 summer eggs, which develop without interruption.   Eggs from unmated females develop into males, those from fertilized females into a mixture of about 63 per cent females and 37 per cent males.   The adult females are bright to brownish-red and unspotted, and the body quite elliptical in outline, about $\frac{1}{75}$ inch in length.   There are four rows of long curved spines down the back, each borne on a whitish tubercle.   The production of diapausing winter eggs is determined by photoperiod, temperature, and by the nutritional quality of the leaves.   Thus females feeding on senescing leaves or those bronzed by heavy mite feeding may produce winter eggs.   Winter eggs

[1] *Panonychus ulmi* (Koch) (= *Paratetranychus pilosus*), Order Acarina, Family Tetranychidae.

are also produced if the photoperiod is 6 to 13 hours at a moderate temperature around 60°F. or if the temperature falls below 50°F.   These factors are somewhat independent in that low temperatures produce some diapausing eggs even with long photoperiod and high temperatures above 77°F. tend to prevent the formation of diapausing eggs even with short photoperiod.   There are normally from four to nine generations of mites a year.

Control Measures.   Thorough spraying with a dormant-oil emulsion at 2 per cent for superior oils, or 3 per cent for regular oils, will kill the overwintering eggs and greatly reduce the population of mites during the growing season.   Spraying with 2 gallons of dormant oil containing 3 per cent Trithion or 2 per cent Ethion to 100 gallons of spray is an effective

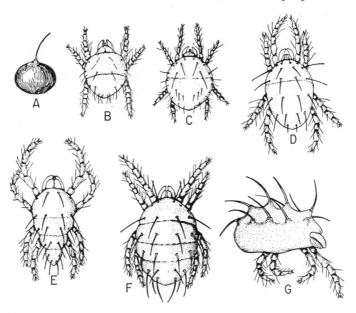

Fig. 15.27. European red mite.   *A*, egg; *B*, first instar or six-legged stage; *C*, protonymph; *D*, deutonymph; *E*, adult male; *F*, adult female; and *G*, adult female in side view.   All enlarged about 75 times.   (*From U.S.D.A.*)

ovicidal treatment.   These oil applications should be made during the delayed dormant period.   When four to six mites are to be found per leaf in June, July, or August, summer treatment is necessary to prevent injury.   This mite has developed resistant strains to a number of acaricides, and control will depend upon the careful choice of one or more of the following materials, applied at the indicated time: (*a*) chlorbenside at 0.4 pound, Genite at 0.75 pound, or Tedion at 0.36 pound (prepink or pink only); (*b*) ovex at 2 to 4 ounces, demeton (Systox) at 0.25 pound, Chlorobenzilate at 0.36 pound (as needed); (*c*) EPN at 0.12 to 0.19 pound, Guthion at 0.3 pound, Kelthane at 6 ounces, malathion at 0.5 to 0.6 pound, parathion at 0.15 pound, Sulphenone at 1 pound, or TEPP at 0.13 pound (2 cover sprays 7 to 10 days apart) per 100 gallons of spray.

References.   *Conn. Agr. Exp. Sta. Bul.* 252, 1923; *Can. Dept. Agr. Ento. Branch, Cir.* 39, 1925; *U.S.D.A., Tech. Bul.* 25, 1927; and 89, 1929; *Jour. Agr. Res.,* vol. 36, no. 2,

1928; *Can. Jour. Res.*, **13**:19, 1935; *Mass. Agr. Exp. Sta. Bul.* 305, p. 28, 1934; *Va. Agr. Exp. Sta. Tech. Bul.* 98, 1946; *Ann. Appl. Biol.*, **40**:449, 487, 1953.

158. PACIFIC MITE[1] AND SCHOENE MITE[2]

Importance and Type of Injury. These mites are destructive pests of deciduous fruit orchards. The leaves of infested trees become bronzed and heavily webbed. In heavy infestations which may exceed 1,000 mites per leaf, defoliation occurs, fruits fail to color properly, and there is usually a heavy fruit drop just before harvest.

Plants Attacked. Apple, pear, plum, prune, almond, walnut, grape, blackberry, cotton, beans, alfalfa, clover, and ornamentals.

Distribution. The Pacific mite is found along the Pacific Coast from California northward to British Columbia. The Schoene mite occurs throughout the southeastern states.

Life History, Appearance, and Habits. These mites closely resemble the two-spotted mite and can be distinguished only by the specialist. They are both about $\frac{1}{60}$ inch long, and the adult summer females are greenish with dark spots. Both species overwinter as bright-orange females, under bark of trees or leaves and trash on the ground. In the spring they migrate up the tree trunks about the green-tip stage and feed on the new growth. Shortly after migration, each female lays about 50 globular white eggs. All the unfertilized eggs produce males, and the fertilized eggs about 80 per cent females. The newly hatched mites pass through the typical six-legged, larval, protonymph, and deutonymph stages (Fig. 15.27) over a period of about 2 weeks in hot, dry weather. The life cycle of the Schoene mite from hatching to adult ranges from 28 days at 55°F. to 5 days at 80°F. There is a preoviposition period of 1 to 5 days, and the life span of the female is 38 to 40 days. These mites produce very heavy silk webbing and are spread by being blown about on strands of silk. After as many as nine generations a year, diapausing eggs are deposited in late July to August, from which the overwintering females develop.

Control Measures. These mites may be controlled by spraying with the summer acaricides as indicated under European red mite (page 716). They have developed resistant strains to a number of acaricides, and control will depend upon the careful choice of materials. Moderate infestations have been checked by dusting with sulfur or spraying with 1.5 per cent summer-oil emulsion.

References. *U.S.D.A., Cir.* 157, 1931; *Wash. Agr. Exp. Sta. Cir.* 64, Rev. 1948; *Hilgardia*, **21**:253, 1952; *Va. Agr. Exp. Sta. Tech. Bul.* 87, 1943; *Jour. Econ. Ento.*, **34**:111, 1941.

159. BROWN MITE[3]

Importance and Type of Injury. In dry seasons this mite sometimes does considerable harm to apples by sucking the sap from the buds and leaves. Infested foliage becomes stippled and takes on a yellowish appearance, and during prolonged drought,

---

[1] *Tetranychus pacificus* McGregor, Order Acarina, Family Tetranychidae. The McDaniel mite, *T. mcdanieli* McGregor, very closely resembles the Pacific mite, of which it may be a variety, and is a pest of deciduous fruits from Michigan and North Dakota west.

[2] *Tetranychus schoenei* McGregor. The four-spotted mite, *T. canadensis* (McGregor), is very closely related to the Schoene mite and is a pest of deciduous fruits in the northeastern states.

[3] *Bryobia arborea* (Morgan and Anderson), Order Acarina, Family Tetranychidae.

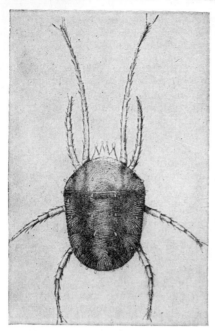

FIG. 15.28. Adult brown mite, *Bryobia arborea*. Greatly enlarged. (*From U.S.-D.A., Farmers' Bul.* 1270.)

many of the leaves fall. Occasionally, during the dormant stage of the tree, one will find the small spherical reddish eggs attached to the surface of the bark around the buds and tips of the twigs. They are sometimes so numerous that the twigs have a reddish cast. Until 1957, this species was confused with the clover mite,[1] which attacks only herbaceous plants.

Plants Attacked. Apple, peach, prune, plum, pear, cherry, almond, walnut, and raspberry.

Distribution. Northern and southwestern United States and Canada.

Life History, Appearance, and Habits. The brown mite overwinters, as the egg, on leaf buds and roughened areas of the bark. The eggs hatch in the early spring, and the mites feed on the foliage at night or under low light intensity and congregate on the wood during the day. They may feed on both surfaces of the leaf but prefer the upper. On apple, pear, and plum there are a number of generations during the growing season. These deposit summer eggs on twigs and the midrib of the leaf until September, when they lay overwintering eggs. On almond, there are three generations and overwintering eggs are laid in June. The adult mite (Fig. 15.28) is brown to reddish, slightly smaller than a pinhead, and with the front pair of legs much longer than the other three pairs.

Control Measures. This mite is controlled by the measures suggested for the European red mite (page 716).

References. *Colo. Agr. Exp. Sta. Bul.* 152, 1909; *Conn. Agr. Exp. Sta. Bul.* 327, p. 574, 1931; *Hilgardia,* **21**:253, 1952; *Can. Ento.,* **75**:41, 1943; and **89**:485, 1957.

160. FLATHEADED APPLE TREE BORER[2] AND PACIFIC FLATHEADED BORER[3]

Importance and Type of Injury. These borers are among the worst enemies of deciduous trees and shrubs and kill many trees and shrubs in the nursery and many more after they have been set in orchards, parks, city streets and lots, and along highways. They are especially destructive during the first 2 or 3 years after the trees are planted, or in very dry seasons, or where parts of trees, that have been shaded, are exposed to the sun by pruning. The presence of this insect is indicated by shallow, broad, irregular mines or burrows on the main trunk or large branches, just under the bark and in the wood, the larger ones going into the wood a distance of an inch or two. Above these burrows are dark-colored dead areas of bark, often with sap exuding. These burrows are packed tightly with fine sawdust, except where they go into the wood; here they are usually packed with coarse, excelsior-like fibers. They are nearly always on the sunny side of the tree but may extend completely around the tree.

[1] *Bryobia praetiosa* Koch.
[2] *Chrysobothris femorata* (Olivier), Order Coleoptera, Family Buprestidae.
[3] *Chrysobothris mali* Horn.

Injuries usually result in killing large areas of bark and sometimes in girdling and killing the tree or infested branches. There are several other closely related species that also sometimes attack apple.

Trees Attacked. Nearly all fruit, woodland, and shade trees.

Distribution. Generally distributed throughout the United States and the fruit-growing sections of Canada. The Pacific flatheaded borer occurs throughout western North America from Canada to Arizona and Texas.

Life History, Appearance, and Habits. The winter is passed in the grub or borer stage. These are of different sizes, from ½ to 1 inch in length. The larger, nearly full-grown borers will be found from 1 to 2 inches deep in the wood of the tree, usually to lesser depth in the southern states. In the spring they change to yellow pupae (Fig. 15.29,*d*) and later to beetles. The full-grown grub of the flatheaded apple tree borer is about 1¼ inches in length, legless, of a yellow to yellowish-white color with a broad flat enlargement of the body just back of the head (Fig. 15.29,*a*). It usually lies with the body curved to one side. The adult beetle (*b*) is about ½ inch long by ⅕ inch wide. It is of a dark olive-gray to brown color with a metallic luster on the irregularly corrugated wing covers. The body is very blunt at the head, and tapers to a rounded point at the posterior end. They are decidedly sun-loving insects, and will be found in greatest numbers on the sunny sides of trees or logs. The female beetle lays her yellow disk-like wrinkled eggs in cracks in the bark of trees, nearly always selecting a tree that is unhealthy, or a spot on a healthy tree where the bark has been injured, as by sunscald or a bruise. The eggs are laid from May to August, and most of the borers complete their growth by fall, the life cycle occupying only 1 year.

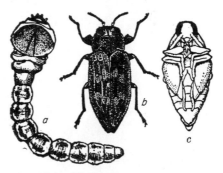

Fig. 15.29. Flatheaded apple tree borer, *Chrysobothris femorata.* *a*, Larva; *b*, adult beetle; *c*, pupa, about twice natural size. (*From Chittenden, U.S.D.A.*)

The adult Pacific flatheaded borer is dark-reddish bronze and from ¼ to ½ inch long. The mature larva is slightly over ½ inch in length. This species has an almost identical life cycle with the flatheaded apple tree borer.

Control Measures. Nearly all damage by this insect may be prevented by wrapping the trunks of the trees the first year they are set. Wrappings should be placed before the middle of May and should extend from the ground level to the lower branches. Any good grade of paper, even several thicknesses of old newspapers, may be used. The paper should be held in place with twine and should remain on the trees through the second year. Whitewashing the trunk up to the lower branches or spraying with parathion at 1 pound or with DDT at 2 to 5 pounds to 100 gallons of spray has given good control. The only effective remedy, once a tree is infested, is to cut out the grubs with a sharp-pointed knife. This should be done during the late summer or early fall. The presence of the insects is indicated by darkened areas of bark and fine bits of sawdust protruding through the bark. Tree borer paints, such as *para*-dichlorobenzene, 2 pounds, in cottonseed oil or dormant oil emulsion, 1 gallon, applied when growth is not rapid, will kill the small larvae but are of little value. Care should be taken to avoid planting of nursery stock that is infested and to keep trees in a healthy, vigorous condition by proper planting, watering, pruning, cultivation, and avoidance of wounds. Wounds should be promptly covered with a good tree paint. The shading of the trunks of young trees by pruning to head the trees low, or by wide stakes, tends to keep the adults away.

References. *U.S.D.A., Farmers' Buls.* 1065, 1919, and 1270, 1931, and *Tech. Bul.* 83, 1929; *Jour. Econ. Ento.,* **52**:255, 1959.

161. ROUNDHEADED APPLE TREE BORER[1]

Importance and Type of Injury. This insect may cause a loss of 10 to 20 per cent of the apple trees in an orchard in years when it is abundant.

This damage is caused by the borer, or grub, which feeds on the inner bark and sapwood of the tree. The burrows of this insect are usually made in the base of the trunk, from 1 or 2 inches below the surface of the ground to a foot or more aboveground. They extend through the sapwood and heartwood, often seriously weakening young trees, and sometimes girdle the tree. The presence of the insect is indicated by coils or piles of sawdust-like particles adhering to the bark, or on the ground about the base of the trunk, and by darkened areas in the bark about the base of the tree. To make a thorough examination for the presence of this insect, it is necessary to remove the earth about the base of the trunk to a depth of 2 or 3 inches and examine the bark carefully.

FIG. 15.30. Larva of roundheaded apple tree borer, nearly full grown. About natural size. (*From U.S.-D.A., Farmers' Bul.* 1270.)

Trees Attacked. Apple, pear, quince, haw, mountain ash, serviceberry, and occasionally some others.

Distribution. United States and Canada, from the Dakotas and Texas, eastward; also New Mexico and British Columbia.

Life History, Appearance, and Habits. The winter is passed only in the larval or borer stage. The borers are of two sizes; those which have hatched from eggs laid in the past season, and those from eggs laid a year earlier, the latter being full-grown and an inch or over in length. They are of a creamy-yellow color, with a brown head and a rounded thickening of the body just behind the head (Fig. 15.30). The two-year-old larvae are usually found in the tree to a depth of 1 to 2 inches. In the spring of the year these larger larvae change to the pupal stage and, after 2 weeks' to a month's time, emerge as robust velvety-brown cylindrical beetles with a conspicuous white stripe on each side of the body above, and the underside of body, legs, and head, except the eyes, also white (Fig. 15.31). The beetles are very striking in appearance. They are abroad from June to the first of September. The adults crawl over the surface of the tree and feed to some extent on the foliage and on the

[1] *Saperda candida* Fabricius, Order Coleoptera, Family Cerambycidae.

new twig growth. They have well-developed wings, but are rather sluggish creatures and usually fly only short distances. They lay their eggs during the summer months, these being deposited in the bark of the trunks of apple trees from just below the surface of the ground to 18 or 20 inches up on the trunk, or occasionally even higher than this. They are usually laid in cracks of the bark, which are sometimes enlarged by the beetles with their strong jaws. The borers, upon hatching 2 or 3 weeks later, feed at first on the outer bark and, as they increase in size, work into the wood of the tree. The older, or nearly full-grown ones are found several inches within the wood and are usually a foot or more above the surface of the ground, while the young borers may work several inches

Fig. 15.31. Adult male and female roundheaded apple tree borer, *Saperda candida*. Male on right, female on left. About twice natural size. (*From U.S.D.A., Farmers' Bul.* 847.)

below the surface. When full-grown, the larva hollows out a cell in the wood in which to pupate. In most cases, the insect requires 2 years in which to complete its growth but may require 3 under unfavorable conditions.

Control Measures. Hand worming in late August or early September and again in late April is advised. The base of the tree should be carefully examined, 1 or 2 inches of the soil around the trunk removed, and the young borers dug out or probed out. So far as possible, workmen should avoid cutting across the grain of the bark. A flexible wire is used to reach the borers deep in the wood. The larger borers may be killed by injecting into the burrows a solution of 1 gram of *para*-dichlorobenzene in 1 milliliter of carbon bisulfide, pyrethrum extract in alcohol, or a 5 per cent rotenone extract in acetone, using an oil can or similar device.   The ordinary spray schedule for controlling other orchard insects aids to some extent in controlling the roundheaded apple tree borer, as the adults will frequently get enough insecticide to kill them when feeding on the new bark or the leaves.   All growth of crabs, haws, and particularly serviceberry and mountain ash should be kept down within ¼ mile of the apple orchard.

References.   *U.S.D.A., Farmers' Bul.* 1270, 1931, and *Dept. Bul.* 847, 1920; *Ark. Agr. Exp. Sta. Bul.* 146, 1918; *Can. Dept. Agr. Ent. Branch Cir.* 73, 1930.

162 (183).   SHOT-HOLE BORER[1]

Importance and Type of Injury.   Small holes, about the size of a pencil lead, are often eaten through the bark on the twigs of healthy fruit trees, especially above a bud or other projection.   The holes are sometimes indicated by a small amount of sawdust or borings on the bark of the tree.   On peach, cherry, and other stone fruits these holes are usually covered and sealed in by dried droplets of gum, which hang from the twigs like tear drops.   When the insects become very abundant, wilting and yellowing of the foliage occur and are usually followed by the death of the tree.   Bark of twigs, branches, and trunks of weakened, infested trees is perforated with numerous small shot-hole-like openings, from which the beetle gets the name "shot-hole borer."   The removal of the bark exposes many, small, winding, sawdust-filled, gradually enlarging galleries leading out from a shorter central gallery (Fig. 15.32,*C*).   In nearly all cases, the attack of the borers results in the death of the tree or branches where they are numerous, but these beetles are very rarely the *primary* cause of the death of the tree.

Trees Attacked.   Apple, peach, pear, plum, cherry, quince, serviceberry, chokeberry, and many other trees.

Distribution.   This insect is a native of Europe but is now generally distributed over the United States.

Life History, Appearance, and Habits.   During the winter, this insect is in the grub or larval stage in the inner bark.   At this time, the legless grubs are about ⅛ inch long, of a pinkish-white color, with a slight enlargement of the body just behind the head (Fig. 15.32,*A*).   In the early spring they change to the pupae and emerge as full-grown insects during June and July.   The adult insects (*B*) are about ¹⁄₁₀ inch long, by nearly half as wide, black in color, the body very blunt at either end. They have well developed wings and are capable of flying considerable distances.   The insects mate, and females seek out trees that are in a somewhat unhealthy condition.   They enter the bark at a point along the branch or twig, usually just above a slight projection, or at a lenticel, and excavate a gallery about 1¼ to 2 inches long, usually running in the same direction as the length of the trunk or branch.   They deposit their white spherical eggs at short intervals on either side of this parent gallery. The female usually dies with the tip of her body blocking the entrance to her egg gallery.   The young grubs hatching from these eggs start burrowing in the inner bark, in general at a sharp angle from the parent gallery.   They continue their burrows until they become full-grown, about 6 to 8 weeks.   The larval burrows are from 2 to 4 inches long, enlarging slightly throughout their length and diverging gradually outward from the parent gallery.   They are packed with frass, while the

[1] *Scolytus rugulosus* (Ratzeburg), Order Coleoptera, Family Scolytidae.

parent gallery is clean.   When full-grown, the larva changes to a pupa and later to an adult beetle at the end of the larval burrow beneath the bark.   The beetles emerge from holes bored directly outward through the bark.   There are from one to three generations of the insect each year, the larger number occurring in the South.

Control Measures.   Probably the best method of preventing and overcoming the attack of this insect is to provide adequate water and food for the trees.   Trees that are in a backward, somewhat sickly, condition should be given a heavy treatment in the spring with some strong nitrogenous fertilizer, applying the material to the surface of the soil above the roots of the tree.   The dosage should vary with the size of the tree.

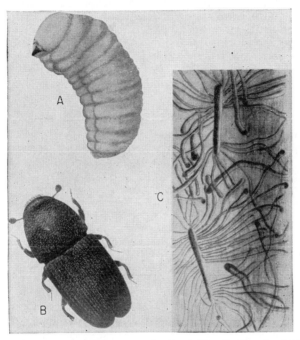

Fig. 15.32. Shot-hole borer.   *A*, larva or grub; *B*, adult or beetle, about 15 times natural size; *C*, parent and larval galleries in sapwood of injured twig from which the bark has been removed, about natural size.   (*From Ill. Natural History Surv.*)

Painting with a light distillate type of oil in which flake naphthalene has been dissolved, ¾ pound per gallon, will kill a large proportion of the insects.   Spraying the infested trees with DDT at 1 pound or parathion at 0.3 pound to 100 gallons of spray, applied when the adults are active in the spring and fall, is of value in controlling this insect.

During the winter all badly diseased trees or branches in or near the orchard should be cut out and all prunings promptly burned.   Infested firewood should be burned before the spring following cutting.   This applies not only to the apple, but to the other favored food plants of this insect (pages 755 and 846).

References.   U.S.D.A., Farmers' Bul. 1270, 1922; *Ohio Agr. Exp. Sta. Bul.* 264, 1913; *Rept. Neb. State Ento.*, 1909; *Univ. Colo. Agr. Ext. Serv. Cir.* 64, 1932; *U.S.D.A., Farmers' Bul.* 1666, p. 131, 1931; *Ore. Agr. Exp. Sta. Cir.* 162, *Rev.* 1948.

163 (187, 191, 194). PLUM CURCULIO[1]

Importance and Type of Injury.   This snout beetle, which is primarily a pest of peach, plum, cherry, and other stone fruits, is sufficiently fond of the apple to make it second to the codling moth in importance as a pest of this fruit.   Injury on apple is shown by small crescent-shaped cuts in the skin of small fruits, some of them with a little round hole opposite the concave side of the crescent, into which an egg is usually deposited. Later these injuries develop into swellings or knots, protruding from the surface of the fruits, each with a small puncture in the skin at its apex. Apples will sometimes show depressions instead of swellings, with the curculio injury at the center of the depression.   An examination of such apples will sometimes reveal a grayish-white curved worm inside.   Many of the infested fruits drop during late May and June.   During late sum-mer, numerous round feeding holes or punctures are made through the skin of the apple and other fruits, and

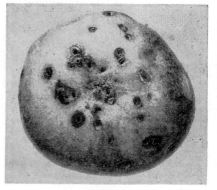

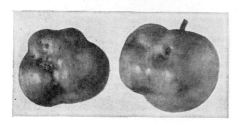

FIG. 15.33. Apples deformed by the plum cur-culio.   (*From U.S.D.A., Farmers' Bul.* 1270.)

FIG. 15.34. Late-summer feeding punctures of the plum and apple curculios, on apple. (*From Ill. Natural History Surv.*)

the flesh is eaten out beneath these punctures (Fig. 15.34).   The infested apples are often hard, knotty, and misshapen (Fig. 15.33).   In peaches, plums, and cherries serious losses result from the feeding of the grubs in the fruit and their presence in ripened marketed fruits (pages 763, 766, and 769).

Trees Attacked.   Plum, pear, apple, peach, cherry, apricot, prune, nec-tarine, quince, and other cultivated and wild fruits.

Distribution.   East of the Rocky Mountains in the United States and Canada.

Life History, Appearance, and Habits.   This insect passes the winter as a dark-brown snout beetle, about ¼ inch long, with grayish or whitish patches on its back and four humps on the wing covers (Fig. 15.35).   A strong curved snout, about one-third the length of the body, projects forward and downward from the head of the insect.   These beetles seek protection in and around orchards or near-by woodlands, where they find shelter during the winter, under fallen leaves, in piles of stone, and about rock outcroppings and fences.   They become active about the time the apples and peaches bloom, or possibly in some years a little earlier than this.   They fly to the trees, feed on the newly forming apples or buds, petals, shucks, and newly set fruits.   They then mate, and the females

[1] *Conotrachelus nenuphar* (Herbst), Order Coleoptera, Family Curculionidae.

begin laying their eggs.   In this operation the female first eats a small round hole into the skin of the fruit, then turns around and deposits a shiny white egg in the hole, and finally cuts a crescent-shaped slit beneath the egg with her mouth parts so as to leave the egg in a gradually dying flap of the fruit.   Over 1,000 eggs have been deposited by single females, but the average under field conditions ranges from 145 to 200.   Upon hatching, 2 to 12 days later, the young grub eats into the flesh of the apple to the core and seeds.   The curculio larva is grayish-white, legless, curved-bodied, with a small brown head (Fig. 15.35).   It is about ⅓ inch long when full-grown.   Infested apples nearly always fall to the ground before the curculio has completed its growth.   When the apple remains on

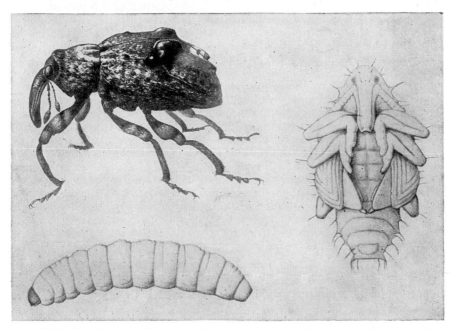

Fig. 15.35. The plum curculio, *Conotrachelus nenuphar*.   Adult above at left in side view; larva below at left, side view; pupa at right, ventral view.   About 8 times natural size. (*From Ill. Natural History Surv.*)

the tree, the larva does not develop, but peaches infested after the stone starts to harden, and nearly all cherries, remain on the tree until ripe. Two or three weeks are generally spent in the fruit in the larval stage. Upon becoming full-grown the insect leaves the apple and works its way into the ground, an inch or two, excavating a little cavity, in which, after about 2 weeks, it changes to the pupal stage (Fig. 15.35).   About 1 month after the larva enters the soil, it changes to the adult insect, and the summer generation of beetles begins to appear in the orchard.   At 80°F. the life cycle from egg to ovipositing adult required an average of 57 days and the females lived for an average of 138 days.   In the latitude of central Illinois, most of the beetles come out during July.   In the southern states, there is, in some seasons, a partial second and possibly sometimes a partial third generation.   North of the 39° north latitude there is probably only one generation annually.   The adult beetles on emerging fly to the fruit and, during the remainder of the summer, feed on the apples, making small holes through the skin and, with the aid of their curved snouts, eating out a cavity in the flesh beneath these holes. Some of them begin seeking winter quarters in August, while others remain on the trees as late as mid-October or, in the South, the first of November.

Control Measures. To control curculio in badly infested areas, it may be necessary to combat the insect by spraying, jarring, and the removal of infested dropped and cull fruits. Sprays applied during the maximum feeding and egg-laying period of the beetles are the quickest method of control. These should be applied at petal fall and once or twice more at 7- to 10-day intervals. Materials which may be used include Guthion or parathion at 0.3 pound, dieldrin at 0.5 pound, methoxychlor at 1.5 pounds, lead arsenate at 3 pounds with 3 pounds of hydrated lime, or endrin at 0.25 pound (petal fall only) to 100 gallons of spray. Treatment of orchard soil with aldrin, dieldrin, or heptachlor at 2 to 4 pounds per acre has given good control for three to four years after application. Jarring the beetles from the trees in the early morning, upon sheets placed on the ground, especially under trees bordering woodlands, ditches, and hedgerows, enables the grower to collect and destroy great numbers of the beetles and prevent them migrating into the interior of the orchard. If jarring is not practiced, it is well to apply an extra spray to the margins of orchards near favorable hibernating places.

In badly infested orchards it will be of advantage to pick up and destroy the dropped fruits during June and July, if this can be done at small expense. These may be disposed of by soaking them with waste oil or by enclosing in paper-lined sacks and placing in the hot sun for several days or burying. Once well established in an orchard, the curculio may become such a serious pest that it will warrant taking the most vigorous measures possible to secure effective control (page 763).

References. U.S.D.A., Dept. Bul. 1205, 1924; Del. Agr. Exp. Sta. Buls. 175, 1932, and 193, 1935; Conn. Agr. Exp. Sta. Bul. 301, 1929, and Cir. 99, 1934; N.Y. (Geneva) Agr. Exp. Sta. Bul. 606, 1937, and 684, 1938; Va. Agr. Exp. Sta. Bul. 297, 1935; U.S.D.A., Tech. Bul. 188, 1930; Va. Agr. Exp. Sta. Bul. 453, 1952; Jour. Econ. Ento., **50**:187, 457, 516, 1957; and **51**:131, 330, 1958.

164. APPLE CURCULIOS[1-4]

Importance and Type of Injury. There are several forms of these insects: (a) the apple curculio;[1] (b) its variety, the western or larger apple curculio;[2] (c) a distinct species,[3] which also attacks apple; and (d) a variety of the latter, the cherry curculio.[4] Any one can cause very serious damage, amounting to the loss of 50 per cent or more of a crop. This insect is somewhat like the plum curculio in general appearance but differs from it in its habits and manner of injury to the apple. It is not so important as an apple pest. The apples attacked are misshapen, knotty, and undersized. Small holes are eaten in the sides or ends of the apple, many holes often being made close together causing a deadened area on the skin of the apple (Fig. 15.34). There are sunken pits or sharp-pointed protuberances on the apple marked at their center by a small puncture in the skin. Infested apples sometimes drop as is the case with the plum curculio. The injury may be distinguished, on mature fruits, from that of the plum curculio by the larger number of punctures close together through the skin, by the larger deadened areas on the fruit surface as above mentioned, and by the absence of crescents.

[1] *Tachypterellus* (= *Anthonomus*) *quadrigibbus* (Say), Order Coleoptera, Family Curculionidae.

[2] *Tachypterellus quadrigibbus magnus* List.

[3] *Tachypterellus consors* Dietz.

[4] *Tachypterellus consors cerasi* List.

Trees Attacked. Apple, haw, wild crab, quince, pear, shadbush, cherry, and some others.

Distribution. The apple curculio is found east of the Continental Divide; the larger apple curculio from Illinois and Missouri to Nebraska and Texas; *Tachypterellus consors* from the Rocky Mountains to the Pacific Coast; and the cherry curculio is an enemy of cherries in Colorado and New Mexico.

Life History, Appearance, and Habits. This insect, like the plum curculio, passes the winter in the full-grown or beetle stage in leaves and rubbish in dry places on the ground, especially under trees on which they developed. It is of a brown to light-brown color, with four very distinct humps on the back (Fig. 15.36). The snout is longer and more slender than that of the plum curculio, being nearly as long as the insect body. The head is small, and the body enlarges toward the base of the abdomen, giving the insect a distinct triangular outline when viewed from above. Before fruit sets, the adults feed on buds, fruit spurs, and terminal twigs, blighting the tender shoots. The fruit may be attacked as soon as it is set. The females eat similar cavities in the fruit, in which they lay their eggs, but do not make the crescent-shaped slits characteristic of the egg-laying scars of the plum curculio. The eggs are deposited

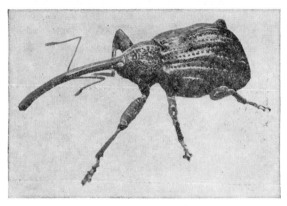

Fig. 15.36. Adult apple curculio, *Tachypterellus quadrigibbus.* Twelve times natural size. (*From N.Y. Agr. Exp. Sta.*)

from May to mid-July, each insect laying up to 125, with an average of several dozen, one in a place. Many of the larvae develop in "June drops," and many in mummied apples on the trees. This insect also differs in its habits from its near relative, the plum curculio, in that the larvae pupate within the apple. A total of 5 to 6 weeks are spent in the fruit as egg, larva, and pupa. The new adults emerge from the fruits from mid-July to early September and feed until they go into hibernation, sometime in September, upon the maturing fruits. So far as we know, there is only one generation of this insect annually.

Control Measures. General measures of orchard sanitation are fully as important in the control of this insect as for the plum curculio. These include keeping the orchard and vicinity free from grass, bushes, or other places offering hibernating quarters for the adult insects. Suitable chemical control measures are given under plum curculio (page 726).

References. N.Y. (Geneva) Agr. Exp. Sta. Tech. Bul. 240, 1936; *Jour. Agr. Res.,* **36**:3–249, 1928; *Jour. Econ. Ento.,* **26**:420, 1933; **29**:697, 1936; *Colo. Agr. Exp. Sta. Bul.* 385, 1932.

165 (175). CODLING MOTH[1]

Importance and Type of Injury. This is the most persistent, destructive, and difficult to control of all the insect pests of the fruit of the apple, and if left to itself, it will usually infest from 20 to 95 per cent of the apples

---

[1] *Carpocapsa pomonella* (Linné), Order Lepidoptera, Family Tortricidae.

in an orchard.   It has forced the abandonment of apple growing as an industry in certain large sections of the country.   Apples attacked by this insect have holes eaten into the side, or from the blossom end, to the core.   The seeds and core are tunneled and eaten by pinkish-white brown-headed worms about ¾ inch long when full-grown.   Dark masses of frass or castings often protrude, especially at the blossom end, from the holes eaten in the apples.   Even poisoned larvae may eat far enough into the fruits before they die to lower the grade of the fruit by their "stings."   These small holes of the size of a pinprick, and less than ¼ inch deep, with a little dead tissue around them, lower the grade of the fruit.

   Plants Attacked.   Apple, pear, quince, wild haw, crab, English walnut, and several other fruits.

   Distribution.   Throughout the apple-growing sections of the world.

   Life History, Appearance, and Habits.   The codling moth passes the winter as the full-grown larva, in diapause, in a thick silken cocoon (Fig. 15.37).   The larvae are pinkish-white caterpillars with brown heads and are about ¾ inch long.   These cocoons are generally spun under loose

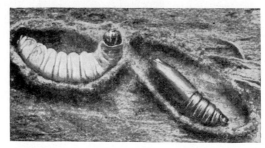

Fig. 15.37. Codling moth larva and pupa within cocoons from beneath bark of apple tree. About twice natural size.   (*From U.S.D.A., Farmers' Bul. 1270.*)

scales of the bark on the trunks of apple trees, under other shelters about the base of the trees, or on the ground near by.   Many of the larvae winter in or around packing sheds.   They remain dormant and are able to withstand low temperatures.   A drop in temperature to −25°F. or below, however, will kill many of the larvae.   During the winter, birds, especially woodpeckers, find and eat large numbers of the larvae.   In mid-spring the worms change inside their cocoons to a brownish pupal stage (Fig. 15.37), and, after a period of from 2 to 4 weeks or more, they emerge from the cocoons, beginning in early May, as grayish moths with somewhat iridescent, chocolate-brown patches on the back part or tip of the front wings and faint wavy crossbands of brown on the rest of the wings. The moths (Fig. 15.39) have a wing expanse of from ½ to ¾ inch.   During the day the moths remain quiet, usually resting on the branches or trunk of the tree.   The coloring of the wings is such that it blends with that of the bark, making the insect very inconspicuous.   About dusk of the evening, if the temperature is above 55 to 60°F., they become active, mate, and the females lay their eggs.   If the temperature is low, they remain quiet and few eggs will be deposited.   Consequently, if the temperature is high and the weather dry during the period of egg-laying and hatching, the codling moth is likely to be very destructive that year. Each female usually deposits more than 50 eggs during her lifetime.   The eggs are white, flattened, pancake-shaped, and about ⅟₂₅ inch in diameter (Fig 15.38).   The eggs of the first generation are laid, one in a place, almost entirely on the upper side of the leaves, the twigs, and the fruit spurs, usually a short distance from a cluster of apples.   Most of the eggs are laid 2 to 6 weeks after the apples have bloomed and hatch in from 6 to 20 days, depending on the temperature and to some extent on the rainfall.   The young larvae feed slightly on the leaves but in a few hours crawl to the young apples and chew their way into the fruit, usually entering by way of the calyx cup at the blossom end.   After entering the fruit, they work their way into the core, often feeding on the seeds (Fig. 15.40).   Some of the infested fruits drop from the tree and the larvae complete their growth on the ground.   Upon becoming full-grown, in 3 to 5 weeks, they burrow to the outside of the apple and either crawl to, or down, the trunk of the tree, or drop to the ground and crawl back to the trunk or to some other object.   Under loose bits of bark or other shelter on the trees

Fig. 15.38. Eggs of codling moth on sections of apple leaves.   About 3 times natural size. From eggs laid in cages.   In the orchard eggs are usually laid singly.   (*From Ill. Natural History Surv.*)

or on the ground, such as discarded sacks, tree prunings, weed stems, and other litter, they spin their cocoons, and change as before to the pupal, and later to the adult, stage.   At a constant temperature of 84°C., the larval feeding period requires 18 days and the complete life cycle about 28 days.

Fig. 15.39. Adults of the codling moth, *Carpocapsa pomonella*.   About 5 times natural size. (*From Ill. Natural History Surv.*)

In the latitude of southern Illinois, there is nearly a full first, nearly a full second, and a partial third generation of this insect each season. In the latitude of northern Illinois, there is nearly a full first generation, a partial second, but no third generation. The emergence of the moths of the second generation extends over about 6 weeks, and eggs of this generation may be deposited in the northern parts of the United States from early June to mid-September. In the South, eggs may be laid as late as October. The larvae enter diapause in response to a decrease in the photoperiod below 15 hours, and the hibernating larvae may consist of individuals from all generations. Development of the codling moth is largely dependent on the temperature and is nearly at a standstill below 50°F. and is retarded above 86°F. Temperatures between these extremes have been called effective temperatures, and it has been found that approximately 550 day-degrees of effective temperature are required to bring about hatching of the earliest larvae of the first generation and 1,000 additional day-degrees for hatching of the larvae of each succeed-

Fig. 15.40. Apple injured by codling moth, showing larva in fruit. Slightly enlarged. (*From U.S.D.A., Farmers' Bul.* 1270.)

ing generation. This method has been used to forecast the time of appearance of the larvae of each generation so that proper spray measures may be applied.

Control Measures. In areas where the climate is favorable to the development of the codling moth, the proper control of this insect is an exacting task. The measures necessary vary considerably, depending upon the section of the country, the crop and variety grown, the age of the trees, the proximity of other orchards and woodland, the presence of other fruit pests, and the fauna of natural enemies. Information concerning the most suitable spray schedules for use in each locality should be obtained from the local experiment station. In using the insecticides suggested, the manufacturers' directions should be followed and proper timing observed to prevent excessive residues from persisting onto the harvested fruit. Spraying with wettable powders of DDT at 1 pound, Guthion at 0.3 to 0.5 pound, or Sevin at 0.75 pound to 100 gallons of spray provides the most generally satisfactory control. Lead arsenate at 3 pounds plus 3 pounds of hydrated lime or other safener is satisfactory where there are light infestations, but if more than two applications are made, it may be necessary to wash the harvested fruit in dilute nitric acid to reduce the residue of lead below the legal tolerance. Ryania at 5 to 6 pounds provides adequate codling moth control with minimum disturbance of beneficial insects and is used in integrated control programs and for application close to harvest to prevent undesirable residues. Other insecticides which may be employed are Diazinon at 0.5 pound, parathion at 0.25 pound, or methoxychlor at 1.5 pounds, or mixtures of DDT at 1 pound plus parathion or Guthion at 0.25 to 0.5 pound, or malathion at 0.5 pound to 100 gallons of spray. The codling moth spray is generally applied in combination with various acaricides and fungicides.

It is very important that sprays for the codling moth be applied with proper timing. In the older lead arsenate spray schedule, the first and most important spray is the *petal-fall* or *calyx spray*. This is applied when about three-fourths of the petals have fallen and should thoroughly cover the calyx cup of the fruit since more than half of the young larvae enter the fruit through this area. Sprays should not be applied when the trees are in full bloom to avoid poisoning bees. Following the petal-fall spray, several *cover sprays* are generally applied at 10- to 14-day intervals to cover the fruit just before the times when newly hatched larvae are entering.

The widespread use of DDT for codling moth control has greatly reduced populations of the insect so that the calyx spray is generally unnecessary. In the eastern states, three to four cover sprays are generally applied at 10- to 14-day intervals and similarly one to three times for second- and third-brood larvae. In the Pacific Coast states, two to four cover sprays are generally applied at 21- to 25-day intervals. Where there is resistance to one or more insecticides, alternation of materials or the use of combinations of insecticides may be employed. The proper timing of the cover sprays may be aided by using fermenting-bait traps to make a daily count of the adult moths. These traps consist of small pails containing 1 part of good-grade molasses to 10 to 15 parts of water and are suspended, four to five in an orchard, on poles near the tops of the trees so that they may be lowered for inspection. When increasing numbers of moths are trapped on several consecutive nights at temperatures above 60°F., cover sprays should be completed within 8 to 10 days.

Aside from spraying, there are several other measures which help in controlling the codling moth. These consist in a thorough cleanup of the orchard, scraping the loose bark from old trees, and removing rubbish from the ground. In cases of abundance, the trees should be banded during the summer with chemically treated bands, which will kill the codling moth larvae that seek shelter under them to spin their cocoons. The chemical most generally used for this purpose is β-naphthol in an oil carrier. Bands can be purchased treated ready to apply or may be prepared by the fruit grower. Directions for making them can be obtained from the state experiment stations or the United States Bureau of Entomology and Plant Quarantine. The bands should be in place not later than 5 weeks after petal fall: June 1 in the latitude of southern Illinois and June 15 in the latitude of northern Illinois. Removing cull apples from the orchard and a thorough cleanup of refuse and rubbish around the packing shed also will help in keeping down the numbers of this insect. The codling moth is preyed upon by many insect enemies, but these are never sufficient to reduce its numbers so that artificial control measures may be omitted.

References. *Ill. State Natural History Surv., Bul.,* vol. 14, art. VII, 1922; *N.M. Agr. Exp. Sta. Tech. Bull.* 127, 1921; *U.S.D.A., Dept. Bul.* 932, 1921, and *Farmers' Bul.* 1326, 1931; *Ill. State Hort. Soc. Trans.,* **67**:184, 1934; *Jour. Econ. Ento.,* **30**:404, 1937; **45**:66, 1952; **46**:414, 1953; **47**:1093, 1954; **50**:756, 1957; **51**:422, 1958; and **52**:1103, 1959; publications of nearly all experiment stations.

166. APPLE MAGGOT[1]

Importance and Type of Injury. Apples in the colder sections of the United States are often badly injured by the maggots of medium-sized black, white, and yellow flies. These maggots bore through the flesh of the apple and are known as the apple maggot, or more commonly as the "railroad-worm." Where this insect is abundant, it is one of the most serious pests of apples, especially early varieties. Infested apples have brown winding galleries running through the superficial part of the flesh and minute egg punctures and distorted, pitted areas on the surface.

FIG. 15.41. On left, apple maggot, *Rhagoletis pomonella;* adult on fruit. At right, section through an apple infested with the apple maggot, showing a full-grown larva. Natural size. (*From Ont. Dept. Agr. Bul.* 271.)

Heavily infested early varieties of fruit will be reduced to a brown rotten mass filled with yellowish legless maggots, about ¼ inch in length and tapering toward the head. In later varieties the injury consists of corky streaks. When the fruit is slightly infested, there is no external indication of the presence of the maggots, but, when the fruit becomes ripe, the burrows show as dark lines under the skin (Fig. 15.41). There is a marked difference in the susceptibility to attack by different varieties, the thin-skinned early maturing varieties being most severely injured.

Plants Attacked. The apple maggot is a native insect, probably feeding originally on hawthorn or *Crataegus* spp. It has been found in wild crabapples, is a serious pest of blueberries, and feeds to some extent in huckleberries, European plums, pears, and cherries.

Distribution. The apple maggot is a northern insect occurring as far west as North and South Dakota, southward and eastward to Arkansas, Ohio, and Georgia, and throughout the northeastern states and southeastern Canada. A small variety of this species, breeding in snowberry, has been taken in the western states.

[1] *Rhagoletis pomonella* (Walsh), Order Diptera, Family Trypetidae.

Life History, Appearance, and Habits. The winter is passed in the pupal stage within a brown puparium about ¼ inch long. These puparia are buried in the soil to a depth of from 1 to 6 inches or more. The adult flies emerge over a period of a month or two in summer. They are black in color, with white bands on the abdomen, four on the female and three on the male, and are a little smaller than the house fly. The wings are conspicuously marked with four oblique black bands (Fig. 15.41). They drink drops of water that have accumulated on the fruit and leaves. The females lay their eggs singly in punctures in the skin of the apple, made by a sharp ovipositor attached to the tip of the abdomen. Egg-laying does not usually take place until 8 to 10 days after the flies have emerged. The eggs hatch in from 5 to 10 days, and the maggots develop slowly in the green fruit and do not usually complete their growth until the infested apples have dropped from the tree. After the fruit has fallen, growth is rapidly completed, and the larvae leave the apple and enter the ground, where the puparia are formed within which they pupate. In the southern part of the range of the insect there is a partial second generation, the adults emerging in the early fall. Some of the insects remain in the puparium for a year or two before emerging. These long-term pupae emerge later than the others and complicate control.

Control Measures. An effective control for the apple maggot is to spray all the trees in the orchard at the time the adults make their appearance in midsummer, using lead arsenate at 3 pounds plus 3 pounds of hydrated lime to 100 gallons of spray. Throughout most of the insect's range, this spray should first be given during the last week in June and should be followed by a second, and possibly a third application, at intervals of 2 or 3 weeks. Spraying with DDT at 1 pound to 100 gallons of spray in the second, third, fourth, and fifth cover sprays is also an effective control. The date of emergence of the flies can be determined by placing a screen or cheesecloth cage over an area at least a yard square, where maggots pupated in the soil the previous year or by trapping the adults using glycine–sodium hydroxide or ammonium carbonate lures. The flies feeding on the surface of the fruit and foliage are poisoned by the spray. The insects in picked fruits may be destroyed by placing the fruits in cold storage for 4 or 5 weeks. Picking up all early-dropped fruit every few days and feeding it to hogs will destroy many of the larvae before they have left the apples.

References. Jour. Agr. Res., vol. 28, Apr. 5, 1924; *Nova Scotia Dept. Agr. Bul.* 9, 1917; *N.H. Agr. Exp. Sta. Bul.* 171, 1914; *U.S.D.A., Tech. Bul.* 66, 1928; *N.Y. (Geneva) Agr. Exp. Sta. Buls.* 644, 1934, and 789, 1960; *Vt. Agr. Exp. Sta. Bul.* 43, pp. 8–14, 1935; *Jour. Econ. Ento.*, **29**:542, 1936; **38**:330, 1945; **40**:183, 1947; and **41**:61, 1948.

167 (243). Woolly Apple Aphid[1]

Importance and Type of Injury. White cottony masses cover purplish aphids, clustered in wounds on the trunk and branches of apple, quince, elm, pear, and mountain ash, or on large knots on the roots and underground parts of the trunk (Fig. 15.42). Infested trees often have many short fibrous roots. These injuries sometimes cause the death of the tree, stunting, or serious retardation of growth. The injury on elm causes the formation of close clusters of stunted leaves or rosettes, at the tips of the twigs, the leaves being lined with purplish masses of aphids, covered with white powdery secretion.

[1] *Eriosoma lanigerum* (Hausmann), Order Homoptera, Family Aphidae.

Trees Attacked.   Apple, pear, hawthorn, mountain ash, elm.
Distribution.   World-wide.

Life History, Appearance, and Habits.   In the North the winter is passed in the two forms, the eggs and the immature nymphs.   The nymphs hibernate underground on the roots of apple.   In the warmer

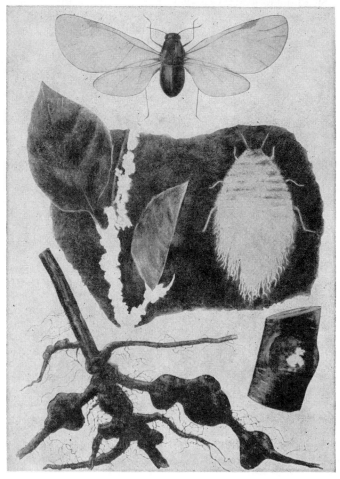

Fig. 15.42. Woolly apple aphid, *Eriosoma lanigerum*. Winged female above, greatly magnified.   At center, cluster of wax-covered aphids on twig, a single wingless female much enlarged at right.   Roots showing characteristic galls produced by root-infesting form, below.   A cluster of aphids in pruning scar at lower right.   The last two reduced in size.   (*From Ill. Natural History Surv.*)

parts of the country the egg-laying females may winter over on apple trunks and branches.   Wherever apple and elms are grown in the same community, eggs are normally deposited in cracks or protected places on the bark of elm trees in the fall.   The eggs hatch early in the spring.   The aphids that emerge from these eggs are wingless and feed on the elm buds and leaves for two generations during May and June.   They then produce a winged form which migrates to the apple, hawthorn, and mountain ash.   They feed to some extent in wounds on the trunk and branches, and many work their way down the trunk below the surface of the ground. Their most severe injury is caused by feeding on the roots.   During the summer the aphids reproduce by giving birth to living young.   In the fall the wingless males appear and mate with wingless females, each female laying a single egg in the situations above described.   Some winged females are present during the entire summer.   The body of this aphid is really of a reddish or purplish color but is nearly hidden under masses of bluish-white cottony wax that is exuded by the insect.

Control Measures.   Infested nursery stock should never be planted. Some varieties of apples, particularly the Northern Spy, show resistance to attack.   The most important insect enemy of this aphid, a wasp-like parasite, *Aphelinus mali*, has been transported to 28 different foreign countries in attempts to control this pest.   In the Pacific Northwest and some other areas it has been so successful that no other control measures are necessary.   The forms of the woolly apple aphid living on the trunk and branches of the tree may be killed by thorough spraying with one or two applications of parathion at 0.15 pound, demeton (Systox) at 0.2 pound, Diazinon at 0.25 pound, Guthion at 0.45 pound, or malathion at 0.5 pound to 100 gallons of spray.   The spray will have to be applied with strong pressure in order to hit the bodies of the aphids, which are well protected by their waxy covering.   The root-infesting forms may be killed on nursery stock by dipping the roots in a strong nicotine solution. In California *para*-dichlorobenzene, applied from September to November, has been successful against the root-infesting forms, using ¾ to 1 ounce for 4- to 6-year-old trees and removing the residue after 2 weeks. Various other methods of soil fumigation and applications of liquids have been tried against these insects but, up to the present time, without much success.   Cultivation and fertilization that will keep the trees in a vigorous growing condition will help in lessening the damage by these insects, and resistant rootstocks offer promise of decreasing injury.

References.   Va. Agr. Exp. Sta. Tech. Bul. 57, 1935; *U.S.D.A., Farmers' Bul.* 1270, 1922; *Maine Agr. Exp. Sta. Buls.* 217 and 220, 1913, and 256, 1916; *Jour. Econ. Ento.,* **43**:463, 1950; and **50**:402, 1957.

168. Periodical Cicada or Seventeen-year Locust[1]

This insect (Figs. 15.45 and 15.46) is so well known that a description is hardly needed.   The body is wedge-shaped, nearly black, and from 1 inch to 1½ inches long, including the wings.   It is, however, frequently confused with the common "harvestmen" or dog-day cicadas,[2] which are believed to have 2-year life cycles.   There are two races of the periodical cicada which cause damage to fruit trees.   One of these has a life cycle of 13 years and is abundant only in the southeastern part of the United States (Fig. 15.43).   The other race, which is the true seventeen-year cicada, or seventeen-year "locust," is more abundant in the northeastern part of the United States and has a 17-year life cycle (Fig. 15.44).

[1] *Magicicada septendecim* (Linné), Order Homoptera, Family Cicadidae.   The race with the 13-year life cycle is *M. septendecim tridecim* Riley.   Two forms of the race with the 17-year life cycle are known.   *M. septendecim cassinii* (Fish), which averages about 1 inch long, has a distinctly different song and mating and resting behavior from *M. septendecim septendecim,* which averages about 1.14 inch long.

[2] *Tibicen linnei* (Smith and Grossbeck), and other species, Order Homoptera, Family Cicadidae.

There are a number of broods which overlap, the adults appearing in different years. Adults of both races are smaller than the dog-day cicadas and appear during May, June, and very early July, while the dog-day cicadas are present every year during July, August, and September. The dog-day cicadas never cause extensive injury. In both races of the periodical cicada, the body is brownish-black, unspotted above, the margins of the wings have a distinct reddish tinge, and a black "W" is present near the lower margin of the front wing. The dog-day cicadas are much larger, have a greenish margin to the wings, and numerous lighter markings on the thorax and abdomen.

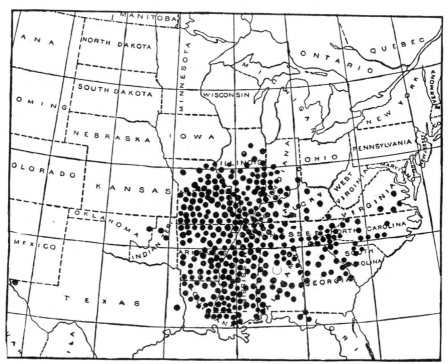

Fig. 15.43. Periodical cicada. Map showing distribution of the combined broods of the 13-year race, broods of which appeared in some part of the dotted area during 1950, 1958, 1959, and 1961, and will appear in 1963 and each thirteenth year thereafter. (*From U.S.D.A., Bur. Ento. Bul. 71.*)

Importance and Type of Injury. Roughened punctures are found in the twigs and small branches of apple and many other trees, usually from 1 to 4 inches in length. The bark is pushed from the wood and the wood cut and raised so that a series of small bundles of splinters protrudes from the surface (Fig. 1.7,*A*). Unprotected fruit trees have had as many as 95 per cent of the terminals destroyed. This injury is caused, not by the feeding of the insect, but by the female cicada depositing her eggs in the twigs (Fig. 15.46). More than 80 plant species have been reported as oviposition hosts.

Distribution. The thirteen-year form ranges from Virginia and southern Iowa and Oklahoma on the north and west, to the Atlantic Coast and Gulf of Mexico. The seventeen-year form ranges from Massachusetts, Vermont, Michigan, and Wisconsin on the north, Kansas on the west, to Texas, northern Alabama, and northern Georgia on the south. Both are most abundant east of the Mississippi River. This insect is not known to occur outside of eastern North America.

Life History, Appearance, and Habits. This species has the longest developmental period of any known insect. The eggs are laid during late May, June, and early July, in the above-described punctures, in the twigs and small branches of trees. The females lay from 400 to 600 eggs, depositing from 12 to 20 in each puncture beneath the bark. Many egg punctures may occur in a single line, as many as 50 having been found

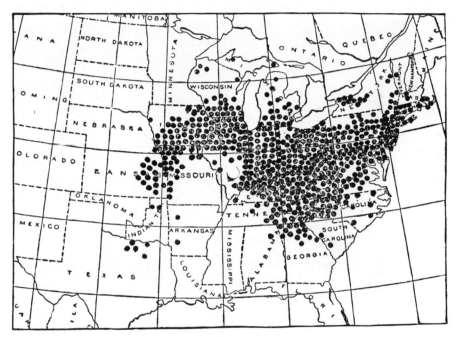

FIG. 15.44. Periodical cicada. Map showing distribution of the combined broods of the 17-year race, broods of which appeared in some part of the dotted area in 1950 to 1952, 1953, 1956, 1957, and 1961, and will appear in 1962 to 1967 and each seventeenth year after each of these years. (*From U.S.D.A., Bur. Ento. Bul.* 71.)

along one branch. The eggs hatch in about 6 or 7 weeks (Fig. 5.6,*F* to *J*). The somewhat ant-like young drop to the ground and enter the soil through cracks, or at the base of plants. Here they excavate a small cell about a tree rootlet from which they suck the sap. They grow very slowly, and their feeding usually has no noticeable effect on the trees, on the roots of which thousands of the young cicadas occur. After 13 or 17 years of this underground existence, the nymphs have become full-grown. The insects are now about an inch in length and somewhat resemble a small crayfish (Fig. 5.9,*L*). The full-grown nymphs burrow to the surface of the soil and emerge through small holes about ½ inch in diameter. They sometimes construct mud cones or "chimneys" about these holes to a height of 2 to 3 inches, with the opening near the ground.

They appear in large numbers at about the same time, emergence usually starting soon after sunset. They crawl upon the trunk of some tree or the stem of a weed, which they grasp firmly. The skin splits down the middle of the back, and the adult insect gradually works its way out (Fig. 15.45). They remain on the support until their wings harden and the bodies dry, but by the following day they are ready to take flight, the empty skins often remaining clinging to tree trunks and other supports for months. As many as 20,000 to 40,000 may emerge from the ground under one large tree. They fly about during the day, mate, and feed by sucking the sap from the twigs of trees. The injury caused in this way is very slight. Four or five days after emergence the males start "singing." This song is a very high-pitched, shrill call, produced by

Fig. 15.45. Adult periodical cicada beginning to issue from nymphal shell. About twice natural size. (*From U.S.D.A., Bur. Ento. Bul.* 71.)

two drum-like membranes on the sides of the first abdominal segment, and not by rubbing wings or legs together, as many people wrongly believe. The adults live from 30 to 40 days and have all disappeared by the second week in July, in Illinois.

Control Measures. Young trees, or particulary valuable fruit trees, may be covered with mosquito bar or other cheap cloth, during the period when the adult cicadas are on the wing. It has been found possible to reduce their numbers greatly, in valuable groves or parks, by banding the trees with sticky "tanglefoot" and raking off the trapped cicada nymphs each morning. Sprays of Sevin at 1 pound or TEPP at 0.5 pint in 100 gallons of water, applied at intervals during egg-laying, will greatly reduce the damage to fruit trees. Orchards set on soil where trees were standing during the last appearance of the adults will be more heavily infested than those set on prairie soil, or at some distance from woodland. Young fruit trees should not be heavily pruned the winter or spring before the appearance of the adult cicadas.   After the cicadas have disappeared, the seriously weakened twigs should be pruned off and burned.   The years in which adult cicadas will appear in any locality can be obtained from the entomologists of the state experiment stations.

References.   U.S.D.A., Bur. Ento. Bul. 71, 1907; *Ohio Agr. Exp. Sta. Bul.* 311, 1917, and *Cir.* 142, 1914; *Mo. Agr. Exp. Sta. Bul.* 137, 1915; *Jour. Econ. Ento.,* **29**:190, 1936; **30**:281, 1937; **41**:722, 1948; **42**:359, 1949; **50**:713, 1957; and **53**:961, 1960.

FIG. 15.46. Periodical cicada.   Female depositing eggs in apple tree and characteristic egg punctures.   (*From Wellhouse, "How Insects Live," copyright, 1926, Macmillan.   Reprinted by permission.*)

169. BUFFALO TREEHOPPER[1]

Importance and Type of Injury.   Double rows of curved slits are found in the bark of small branches and twigs.   If these slits (Figs. 1.7,*C* and 15.47) are carefully cut open, a row of from 6 to 12, small, elongated, yellowish eggs will be found embedded in the inner bark just under each slit.   The bark of infested trees presents a roughened, somewhat scaly and cracked appearance and never makes a very vigorous growth.

Plants Attacked.   Apple, pear, peach, quince, cherry, elm, locust, cottonwood, and many other trees.   The nymphs feed on weeds, grasses, corn, and legumes.

Distribution.   General throughout the United States and southern Canada.

Life History, Appearance, and Habits.   The insect remains in the egg stage during the winter, hatching rather late in the spring into tiny, pale-green, very spiny nymphs (Fig. 5.9,*B*), which drop from the tree and feed on the sap of various grasses and weeds, reaching the adult stage during August.   The new adults then deposit their eggs in the bark for the next year's generation, after which the adults die.   The full-grown buffalo treehopper (Fig. 15.48) is a peculiarly shaped little insect of a light-green color, triangular in outline, very blunt at the head end with a short horn at each upper corner, and pointed behind.   They are about ¼ inch long

---

[1] *Stictocephala* (= *Ceresa*) *bubalus* (Fabricius), Order Homoptera, Family Membracidae.

by two-thirds as wide at the head end. The female treehopper has a very sharp, knife-like ovipositor with which she cuts the slits in the twigs and through these forces her eggs into the inner bark. A single slit is made through the bark, eggs are thrust to the right and left beneath the bark and the scar later separates to form the characteristic double crescent.

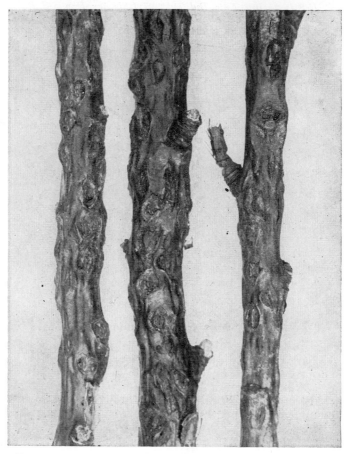

FIG. 15.47. Twigs of apple showing injury by the egg punctures of the buffalo treehopper. Slightly enlarged. *(From Ill. Natural History Surv.)*

Several other species of treehoppers[1] attack apple and other fruit trees, most of them having practically the same life history as that of the buffalo treehopper. Some species differ in their manner of laying eggs, and produce a single row of slits in the bark of the tree; others lay their eggs in gummy covered masses on the surface of the bark.

Control Measures. The best method of controlling this insect is clean cultivation of the orchard, keeping down all weeds or grassy growth and avoiding cover crops of alfalfa. When the insect is abundant, the growing of summer cover crops, such as clover or cowpeas, should be discontinued

[1] The green-clover treehopper, *Stictocephala inermis* Fabricius; the dark-colored treehopper, *S. basalis* Walker; *S. gillettei* Godg.; and others.

for one or two seasons, and the orchard kept clean of all vegetation until the latter part of July.   Dormant sprays of 4 to 6 per cent of oil have been found to kill 70 to 100 per cent of the overwintering eggs.

References.   U.S.D.A., Farmers' Bul. 1270, 1922; *N.Y. (Geneva) Agr. Exp. Sta. Tech. Bul.* 17, 1907; *U.S.D.A., Cir.* 106, 1930, and *Tech. Bul.* 402, p. 30, 1934.

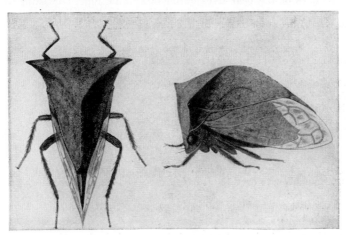

Fɪɢ. 15.48. Buffalo treehopper, adults in dorsal and side views.   About 6 times natural size. (*From Ill. Natural History Surv.*)

B. PEAR INSECTS

FIELD KEY FOR THE IDENTIFICATION OF INSECTS INJURING THE PEAR

A. External chewing insects that eat holes in leaves, buds, bark, or fruit:
  1. Soft, fleshy, dark-green to orange, slimy, slug-like larvae, up to ½ inch in length, feed on the surface of pear and cherry leaves, skeletonizing the leaves. Most abundant during late spring and again in late summer.   In late spring black and yellow wasp-like insects, about ⅕ inch in length, lay their eggs on the leaves..*Pear-slug,* page 767.
B. Insects that suck sap from leaves, buds, twigs, branches, trunk, or fruit:
  1. Buds and young flowers of pear, prune, and some other fruits fail to open, and become brown and blasted in appearance.   Very small, black, slender-winged insects feed within the buds or opening flowers.   Young fruits shrivel and drop..*Pear thrips,* page 742.
  2. Pear trees with foliage showing a brownish color, blackening or drying up about midsummer.   Very small, shining, cicada-like insects, about 1/10 inch long, under the bark of infested trees during the winter.   Minute orange-yellow eggs on bark at base of fruit spurs in spring.   Translucent, yellow, olive, or black, very small, but broad, wingless nymphs, sucking the sap from the stems of leaves, and sometimes from the undersurface of the leaf.   Sticky drops of nearly colorless liquid on the leaves and fruit, sometimes covered by a black growth of soot-like fungus........................*Pear psylla,* page 743.
  3. Fruit and branches specked with small blackish or grayish-brown scales, circular in outline and from very small to a little larger than a pinhead, with a raised, dark-gray, nipple-shaped area in the center.   Bright lemon-yellow, soft-bodied insects lying under the protecting scales.   In heavy infestations, bark of twigs and branches completely coated with a gray covering of scales............. ....*San Jose scale,* page 745.
  4. Very minute, elongated, nearly white-bodied, four-legged mites, about 1/125 inch long, sheltering under the bud scales of trees in winter and forming reddish-brown blister-like galls on the undersides of the leaves during the growing

season. Galls often so thick as entirely to coat the undersurface of the leaves giving them much the appearance of being infected with some fungus........
....*Pear leaf blister mite*, page 745.

C. *Insects that bore into the trunk and branches of the tree:*

1. Slender, bronzy, shining beetles, about ⅓ inch in length, on the bark of the sunny sides of the pear trees during May and June. Twisting or winding, brown burrows running through the inner bark, indicated by swelling or cracking of the outer bark. Where burrows are numerous, trees are sometimes girdled and killed.......................*Sinuate pear tree borer*, page 746.

D. *Caterpillars, worms, or maggots that burrow into or feed inside the fruits:*

1. Dark masses of wet frass protrude from the sides of the fruits or from the blossom end, being forced out from holes which extend through the flesh of the fruit usually to the core. Pinkish-white brown-headed worms feeding in these holes. Flat, shining, white eggs on the leaves in the vicinity of fruits or on the skin of the fruit...............................*Codling moth*, page 747.

2. Caterpillars similar to *D*,1, but more pinkish, less conspicuous, and not over ½ inch long, boring inside the fruit, often without external evidence of their presence...............................*Oriental fruit moth*, page 763.

3. Small fruits misshapen, bloated, lopsided, and with dark blotches; usually drop a few weeks after setting. White to orange maggots, not over ¼ inch long, from a few to over 100 per fruit, may consume entire interior of the small fruits...
Pear midge, Contarinia pyrivora (Riley) (see *N.Y. (Geneva) Agr. Exp. Sta. Tech. Bul.* 247, 1937).

Many other apple insects may be found injuring the pear. See the Key on pages 684–689.

170. PEAR THRIPS[1] AND BEAN THRIPS[2]

Importance and Type of Injury. Pear thrips attack the buds of fruit trees very early in the spring, before the buds open, causing them to shrivel and turn brown. They also attack the developing fruit, causing scabbing or russeting of the fruit surface. The female thrips also injure young fruits by depositing their eggs in the stems of the blossoms. These egg punctures cause the fruit to drop. The bean thrips feeds on both foliage and fruit, producing a silvery appearance. In severe infestations the leaves may die and the fruit will fail to mature.

Distribution. The pear thrips is an imported species and was first found in California in 1904. It now occurs along the Pacific Coast, in California, Oregon, and British Columbia. In the eastern part of the country, it is recorded from New York, Pennsylvania, and Maryland. The bean thrips is distributed throughout the country.

Plants Attacked. The pear thrips is primarily a pest of pear, but it attacks also apple, apricot, cherry, grape, peach, plum, prune, and several other fruit trees, as well as poplar, maple, shadberry, willow, currant, and several shrubs and weeds. The bean thrips is a very general feeder.

Life History, Appearance, and Habits. The pear thrips pass the winter as the newly formed adults in small cells 5 to 7 inches deep in the soil. The thrips remain in these cells until early spring, appearing on the trees in New York about the first of April and in California in February or March. The adults are very active, working their slender bodies in between the bud scales and feeding upon the developing buds. They are about 1/20 inch long. The females soon begin laying their eggs in the

[1] *Taeniothrips inconsequens* (Uzel), Order Thysanoptera, Family Thripidae.
[2] *Hercothrips fasciatus* (Pergande).

fruit stems, midribs, and stems of leaves.   The egg-laying period extends over about 3 weeks.   The young nymphs begin hatching in 2 weeks and feed in large numbers within the opening fruit buds.   The young are full-grown in about 4 weeks, and, still in the nymphal stage, they drop to the ground and form the cells in which they pass the summer and hibernate during the winter (Fig. 6.24).

The bean thrips also overwinters as the adult but develops through two or more generations upon weed hosts during the spring, and the adults fly to pear foliage during early summer.   There they insert their eggs in the foliage and pass through three or more additional generations.

Control Measures.   These insects are most readily controlled by spraying with 1 pound of DDT to 100 gallons of spray or dusting with 5 per cent DDT, applied when about one-half the buds are in the green-tip stage.   Serious infestations can be greatly reduced by thorough, deep cultivation of the orchard in the late summer and early fall, if this can be done without injury to the trees, or the forms in the soil may be killed by irrigation.

References.   U.S.D.A., Bur. Ento. Bul. 68, pp. 1–16, 1909; N.Y. (Geneva) Agr. Exp. Sta. Bul. 484, 1921; Jour. Econ. Ento., 27:879, 1934; Calif. Agr. Exp. Sta. Bul. 687, 1944, and Cir. 478, 1959.

171. Pear Psylla[1]

Importance and Type of Injury.   In sections where this insect has become established, it is one of the most important pests of the pear. The leaves on heavily infested trees turn brown and often drop, the fruit drops prematurely, or is undersized and of poor quality.   Dark, reddish-brown, four-winged, cicada-like insects, about $\frac{1}{10}$ inch in length (Fig. 15.49,4), will be found under the bark during the winter; and much smaller, very broad, active, yellow nymphs (Fig. 15.49,1,2) will be found on the fruit and leaves during the growing season.   The leaves and fruit of badly infested trees will be covered with honeydew, which in turn is generally coated with a black fungus, later in the season.

Trees Attacked.   Pear and occasionally quince.

Distribution.   This insect is of European origin and was brought to Connecticut about 1832 and has spread over the eastern states.   It appeared on the Pacific Coast in 1953 and has spread to most of the pear-growing areas.

Life History, Appearance, and Habits.   The pear psylla passes the winter as the adult under the bark of trees or in other sheltered places about the orchard.   The adults come out of hibernation early in the spring, and the females deposit their pear-shaped, orange-yellow eggs (Fig. 15.49,3) in cracks in the bark or about the buds.   The eggs are attached by a short stalk and have a thread-like filament projecting from the unattached end.   The eggs hatch in 2 weeks to a month, and by the time the trees are in full bloom, many very small, yellow, wingless nymphs, about $\frac{1}{80}$ inch long, may be found on the stems and undersides of the leaves, from which they are sucking the sap.   These nymphs pass through five instars and complete their growth in 1 month or slightly less.   There are from three to five generations each season, the eggs of the later generations being laid on the leaves and stems.

Control Measures.   Delayed dormant spraying as the buds swell, with 2 per cent superior-oil or 3 per cent regular-oil emulsion, will repel the

[1] *Psylla* (= *Psyllia*) *pyricola* Förster, Order Homoptera, Family Chermidae

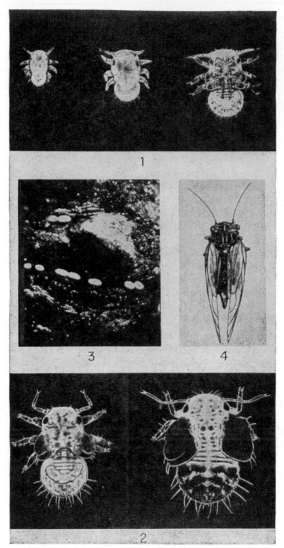

FIG. 15.49. The pear psylla, *Psylla pyricola*. *1*, First-, second-, and third-instar nymphs; *2*, fourth- and fifth-instar nymphs; *3*, eggs; *4*, winter adult. All much enlarged. (*From N.Y. (Geneva) Agr. Exp. Sta. Bul.* 527.)

adults and kill many of the newly hatched nymphs. The oil spray may be fortified with 0.25 pound of parathion, Trithion, Diazinon, or Ethion to 100 gallons of spray. When the nymphs become abundant during the growing season, spraying at petal fall and when necessary during the summer with parathion at 0.15 pound, EPN at 0.25 pound, Diazinon, Guthion, or malathion at 0.5 pound, or Sevin or toxaphene at 1 pound to 100 gallons of spray (as wettable powders) will give satisfactory control. Rotenone at 0.1 pound in 2 quarts of summer oil to 100 gallons of spray is another effective summer treatment. Timing of these sprays is of great importance since the young nymphs are much more susceptible than the last-instar nymphs or the adults.

References. N.Y. (Geneva) Agr. Exp. Sta. Bul. 387, 1914, and 527, 1925, and *Cir.* 129, 1932; Jour. Econ. Ento., **40**:234, 1947; **41**:244, 1948; and **42**:338, 1949.

172 (149, 179, 189, 204). San Jose Scale

Varieties of pears, such as Duchess, Seckel, Bartlett, and Bosc, are subject to attack by the San Jose scale and should be given the same dormant treatment as the apple. Mature Kieffer and Garber pears are

seldom seriously injured by San Jose scale and rarely need to be sprayed for the control of this insect (page 707).

173. Pear Leaf Blister Mite[1]

Importance and Type of Injury. This insect is of moderate importance in most pear-growing sections, but control measures are generally necessary in commercial orchards. Brownish blisters appear on the undersides of the pear and apple leaves. The blisters are commonly ⅛ inch across, or massed together in such a way as nearly to cover the underside of the leaf surface (Fig. 15.50). Upon examination with a lens, these blisters will be found swarming with very small, whitish or pinkish mites, with an elongate, tapering, ringed abdomen and only two pairs of legs located near the head (Fig. 15.51). Fruit buds turn brown and flare

Fig. 15.50. Pear leaves infested with the pear blister mite. (*From Univ. Calif., Coll. Agr.*)

open during the winter, or produce weak flowers, and russeted globular or misshapen fruits, due to the work of this mite under the fruit bud scales.

Trees Attacked. Pear, apple, and related wild trees.

Distribution. General in fruit-growing sections of North America. It was introduced into this country about 1870.

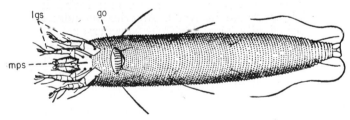

Fig. 15.51. Pear leaf blister mite, *Eriophyes pyri*, ventral view, magnified about 450 diameters. *mps*, Mouth parts; *lgs*, the two pairs of legs; *go*, opening of the reproductive system. (*From U.S.D.A.*)

Life History, Appearance, and Habits. The adult mites, which are only about ½₂₅ inch long (Fig. 15.51), enter the bud scales in August and September where they spend the winter. In warmer regions eggs are deposited within the buds, where they hatch and develop during the winter, destroying the bud tissues as already described. In the Northwest the mites winter in the egg stage and hatch early in the spring.

[1] *Eriophyes pyri* (Pagenstecher), Order Acarina, Family Eriophyidae.

As soon as the foliage has started to come out in the spring, they become active and start feeding on the undersides of the leaves, where they cause the formation of blisters or galls in which the eggs are laid. The colonies of young mites feed on the plant tissue, entirely within the blister, and are almost completely protected, although they may move in and out through a small central opening. A number of generations develop on the leaves, and although activity may decrease during the hot summer months, fresh blisters are often produced on the fall flush of new growth. The mites also attack the fruit, where they cause depressed russeted spots and, in severe infestations, dwarfing and malformation.

Control Measures. The mites may be controlled by dormant sprays as the buds are swelling, using emulsifiable mineral oil, 3 gallons; lime sulfur, 8 gallons; or a mixture of mineral oil, 3 gallons, and lime sulfur, 2 gallons to 100 gallons of spray. Spraying in the early fall after harvest with lime sulfur, 4 gallons, plus 2 pounds of wettable sulfur or with Diazinon at 1 pound to 100 gallons of spray is also effective and will prevent the mites from migrating to the buds. If the dormant treatment was omitted or ineffective, spraying during the spring or summer with demeton (Systox) at 0.125 to 0.25 pound to 100 gallons of spray will give good control.

References. *N.Y. (Geneva) Agr. Exp. Sta. Bul.* 306, 1908; *Calif. Agr. Exp. Sta. Cir.* 324, 1932; *U.S.D.A., Farmers' Bul.* 1666, p. 64, 1931; *Jour. Econ. Ento.*, **21**:676, 1928; **25**:985, 1932; **47**:1020, 1954; **50**:695, 1957; and **52**:183, 1959.

BORERS

Pears are subject to attack by the flatheaded apple tree borer and Pacific flatheaded borer and some species are also injured by the roundheaded apple tree borer. The control methods for these insects are the same as those given for their control on apple (pages 718–722).

174. SINUATE PEAR TREE BORER[1]

Importance and Type of Injury. Trees infested by this insect have narrow winding burrows in the inner bark and sapwood and discolored canker-like areas from base of trunk to small branches, the presence of these burrows being indicated by splitting and dark, dead lines in the outer bark. If the burrows are numerous, the bark will be killed. Entire plantings of young trees are sometimes girdled and killed where the insects are abundant.

Trees Attacked. Pear, hawthorn, mountain ash, and cotoneaster.

Distribution. This is a European insect first found in New Jersey in 1894. Its known range in North America includes several of the eastern states, principally New York and New Jersey.

Life History, Appearance, and Habits. The winter is passed in the larval stage in burrows in the tree. The larvae, which are very slender, are of two sizes in winter: the smaller, about ½ inch long, in the inner bark, and the larger, about 1½ inches long, in small cells in the sapwood which are plugged at each end with coarse sawdust. The larger larvae pupate in the early spring and emerge as beetles during the last of May and June. These beetles are very slender, about ⅓ inch long and about one-fifth as wide, and of a purplish-bronze color. They are somewhat flattened and boat-shaped in outline, tapering to a blunt point at the tail end. The beetles feed on the foliage of the pear and related wild and cultivated fruit trees; the females, after mating, lay their eggs in cracks in the bark. The grubs on hatching burrow in the bark as above described. The larvae grow rather slowly, changing to pupae the second spring after the eggs are laid. There is probably one generation each two years.

[1] *Agrilus sinuatus* (Olivier), Order Coleoptera, Family Buprestidae.

Control Measures. The dead or dying trees, or branches that are very heavily infested, should be cut off and burned during the winter. If the foliage of the pear trees is sprayed heavily with lead arsenate at 6 pounds or DDT at 1 pound to 100 gallons of spray, just about the time the adult beetles emerge and again 2 weeks later, most of them will be killed when they feed on the poisoned leaves. This is the most effective control.

References. *Conn. Agr. Exp. Sta. Bul.* 266, 1921; *N.Y. Agr. Exp. Sta. Bul.* 648, 1934.

175 (165). CODLING MOTH

The fruit of the pear is subject to attack by the codling moth, but not to so great an extent as is the apple. The control measures are the same as for the apple (page 730), but fewer sprays are required.

C. QUINCE INSECTS

FIELD KEY FOR THE IDENTIFICATION OF INSECTS INJURING THE QUINCE

1. Grayish scales adhering to the bark and fruit (see Field Key for the Identification of Insects Injuring the Apple, *B*,1)............*San Jose scale*, page 705.
2. Worms feeding in, and boring holes through, the fruit (see Field Key for the Identification of Insects Injuring the Apple, *D*,3)......*Codling moth*, page 727.
3. Pinkish to creamy-white, short-legged worms with brown heads, up to ½ inch long, boring in the twigs, causing them to wilt, and also in the fruit. Entrance holes of worms often through the stem, not showing on the outside of the fruit...
....*Oriental fruit moth*, page 763.
4. Irregular cavities eaten in the flesh of the quince, with very small openings through the skin. White legless grubs feeding in the fruit of the quince during early summer but seldom causing the fruit to drop. Grayish-brown, rather broad-shouldered snout beetles, the snout about one-third as long as the body, feeding on the flesh of the fruit. Injured fruit becomes deformed and knotty...
................*Quince curculio*, page 747.

APPLE INSECTS THAT INJURE THE QUINCE

The quince is often infested with San Jose scale (page 705). It is particularly subject to injury by the roundheaded apple tree borer (page 720), which seems to prefer the quince above any of the other fruit trees. The oriental fruit moth (page 763) prefers the quince to most other fruits. The fruit is attacked by the codling moth (page 727), and the leaves by several of the insects common on apple.

176. QUINCE CURCULIO[1]

FIG. 15.52. The quince curculio, *Conotrachelus crataegi*, adult. Natural size and enlarged. (*From N.Y. (Cornell) Agr. Exp. Sta. Bul.* 148.)

The most important insect which confines its attack to the quince is the quince curculio. This curculio differs slightly from the apple and plum curculios in that it goes through the winter in the grub stage in the soil. The adult (Fig. 15.52) is a broad-shouldered snout beetle, grayish-brown in color, without humps on the back.

[1] *Conotrachelus crataegi* Walsh, Order Coleoptera, Family Curculionidae.

Control Measures. Summer cultivation of the orchard is of no value in combating this insect, since it does not enter the soil until late summer. As the grubs leave the fruit before it drops, picking up the dropped quinces cannot be practiced as a control measure.

This insect may be controlled by the spray program suggested for the control of the plum curculio on apple (page 726).

References. N.Y. (Cornell) Agr. Exp. Sta. Bul. 148, 1898; *N.Y. State Dept. Agr. Bul.* 116, 1919; *Conn. Agr. Exp. Sta. Bul.* 344, pp. 174, 175, 1933.

D. PEACH INSECTS

FIELD KEY FOR THE IDENTIFICATION OF INSECTS INJURING THE PEACH

A. External chewing insects that eat holes in the foliage or fruit; or moths sucking juices from the ripening fruits:
  1. Robust metallic-green to greenish-bronze, or copper-colored beetles, from ⅓ to ⅝ inch long, feeding on the fruit of the peach. The abdomen of the insect projects back from the wing covers and is marked by two conspicuous white spots. The injured fruit appears to have been gouged or partly peeled; sometimes where the beetles are numerous, the entire skin is eaten off..........
  ....*Japanese beetle,* page 749.
  2. Large, somewhat flattened, dark-green beetles up to 1 inch in length, with a brownish-yellow tinge to the wing covers and thorax, feeding on the fruit of the peach, and also on the foliage, during July and August. Fruit scored somewhat as in *A,*1. Beetles drop to the ground when the trees are shaken in the daytime. Feed mainly at night...............*Green June beetle,* page 751.
  3. Tan-colored moths, with a purplish tinge and a small, oval, dark spot near the middle of each front wing, distinctly triangular in outline and about ⅞ inch long, suck the juice from cracks or from small punctures that they make in ripening fruits. Feed on peach, grape, apple, and other fruits.............
  ....Adult of *cotton leafworm,* page 579.
B. Insects that suck sap from buds, leaves, fruits, and terminal twigs, sometimes killing the latter:
  1. Black shiny eggs on the tips of twigs and in cracks in the bark of larger branches of peach during the winter. Pale-greenish or black aphids sucking the sap of the newly formed fruit, leaves, and young twigs. Leaves on infested trees curl and turn yellow. Insects sometimes so numerous as to cover completely the twigs and foliage............*Green peach aphid,* or *black peach aphid,* page 754.
  2. Young peach trees in the orchards, and particularly trees in the nursery row, with the terminal growth and some of the laterals dying back, but not containing borers. Coppery-brown bugs, with oval bodies, somewhat triangular in front and about ¼ inch long by half as wide, feeding on the terminals, twigs, foliage, and peach fruits. Body of the bug is somewhat spotted and flecked with yellow and dark brown. Bugs fly readily when approached............
  ....*Tarnished plant bug,* page 754.
C. Motionless insects found on the bark of trunk and branches and on leaves, sucking the sap:
  1. Round, flattened, grayish scale insects, about 1⁄16 inch across, closely adhering to the bark of trees, often with reddened areas of bark about the point where the insects are attached. Lemon-yellow, sack-like, sucking insects sheltered under the waxy scales...............................*San Jose scale,* page 753.
  2. Peach twigs, particularly on the undersides, nearly covered with sharply convex, brownish scales, about 1⁄12 inch in diameter, yellow-brown to dark brown in color. In summer, fruit and leaves of infested branches coated with honeydew or masses of sooty, black fungus..............................
  ....*Terrapin scale* or *brown apricot scale,* page 753.
  3. Brown flattened scales, less than ⅛ inch long, adhere tightly to the bark on the underside of the branches in winter. Whitish masses of cottony material, nearly ⅓ inch across, beneath and behind these scales from May to July....
  ....*Cottony maple scale,* page 837.

D. Insects that burrow in the trunk, branches, or twigs:

1. Small bits of gum exuding from many points along the trunk and branches of peach. Galleries about the size of a pencil lead and a little over 1 inch long, with sawdust-filled burrows radiating from them, in the inner bark of the tree. Brownish, or brownish-black, blunt-ended beetles, $\frac{1}{16}$ inch long, are found boring the galleries in the inner bark, or crawling over the bark of the tree... ....*Shot-hole borer* and *peach bark beetle*, page 755.

2. Masses of gum, mixed with brown sawdust-like bits of frass and bark, exuding from around the base of the trunk. White caterpillars, with brown heads, up to 1 inch long, burrowing in the bark from 8 or 10 inches above, or 3 or 4 inches below, the surface of the soil. Brown tough-skinned pupal cases protruding from the gummy masses about the base of the trunk during the latter part of the summer and early fall. Black and yellow wasp-like moths flying rapidly, in daytime, about the base of the trunk, in late summer or early fall........ ....*Peach tree borer*, page 757.

3. Masses of gum containing brown sawdust, exuding from the upper part of the trunk and branches of the peach, most frequently from forks of limbs. White caterpillars with brown heads, up to a little over $\frac{3}{4}$ inch in length, working in the bark under masses of gum. Metallic, blue-black, yellow-marked, wasp-like moths flying rapidly about the trees after midsummer................. ....*Lesser peach tree borer*, page 760.

4. Very small, inconspicuous, silk cocoons closely attached to the larger branches and the trunk of the tree during the winter. Brown worms with black heads, from $\frac{1}{16}$ to $\frac{1}{8}$ inch in length, inside these cocoons. Small cavities eaten into the ends of the new growth of peach shoots in the spring, causing them to wilt and die back. Small masses of gum exude from injured twigs. Occasionally the small brown worms will be found in the peach fruits................... ....*Peach twig borer*, page 761.

5. Pinkish to creamy-white, short-legged worms with brown heads, the largest $\frac{1}{2}$ inch long, bore in the twigs of peach, causing them to wilt. Injury is similar to D,4...............................*Oriental fruit moth*, page 763.

E. Worms—caterpillars or weevil grubs—burrowing into and feeding inside the fruits, or weevils scarring the surface of the fruits:

1. Crescent-shaped punctures, or slits, with a small hole often cut at the inside of apex of the crescent, on the young peach fruits for the first 2 or 3 weeks after the blooming period. Fat-bodied, brown-headed, white, footless grubs feeding in the flesh of the peach, causing irregular greenish areas on the surface of the fruit and black cavities within the fruit. Many of the grub-infested fruits drop when less than one-fourth grown. Dark-brown snout beetles about $\frac{1}{3}$ inch long, with grayish-white patches on the back of the wing covers, feeding and laying eggs on peach fruits for a few weeks after the fruit has set and feeding again during the latter part of the summer.........*Plum curculio*, page 769.

2. Pinkish to creamy-white short-legged worms, with brown heads, up to $\frac{1}{2}$ inch long, bore in the fruits of the peach. Entrance holes of worms often through the stem, not showing on the outside of the fruit. Gray moths with chocolate-brown markings on the wings, about $\frac{1}{4}$ inch long, flying about the trees at dusk or resting on the trunk... ....*Oriental fruit moth*, page 763.

177 (274). JAPANESE BEETLE[1]

Importance and Type of Injury. Metallic green or greenish-bronze beetles, $\frac{1}{3}$ to $\frac{5}{8}$ inch long, with reddish wing covers and with two prominent, and several smaller, white spots near the tip of the abdomen and along the sides, feed on the surface of the fruit and leaves of deciduous fruits. The fruit may be partly peeled and gouged in irregular shallow patches, or nearly devoured. The leaves are skeletonized on many trees and plants. Grass is sometimes killed by the feeding of the larvae on the roots. Where the beetle has become established in the eastern United

[1] *Popillia japonica* Newman, Order Coleoptera, Family Scarabaeidae.

States, it is so abundant as to cause serious injury to tree fruits and also to many field crops and some truck crops on which it feeds.

Plants Attacked. The adult beetles feed on the foliage and fruits of about 250 kinds of plants, including nearly all the deciduous fruits and small fruits, shade trees, shrubs, corn, soybeans, garden flowers, vegetables, and weeds. The larvae are serious pests of lawns and grass roots, vegetables and nursery stock.

Distribution. This insect was imported into New Jersey from Japan, on the roots of nursery stock, about 1916. It is present in all the states east of the Mississippi River except Florida and Wisconsin and is well established in the Atlantic Coast states.

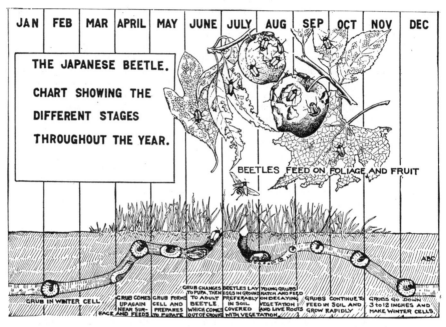

Fig. 15.53. Diagram of the life cycle of the Japanese beetle, *Popillia japonica.* (*From Pa. Agr. Bul.* 390.)

Life History, Appearance, and Habits. The winter is passed as a grub, about ½ to ¾ inch long, buried in the soil (Figs. 15.53 and 15.54,*B*). Growth is completed during June, and the adults emerge in greatest numbers in July. The female beetles lay their white spherical eggs, in groups, 2 to 6 inches deep in the soil. The grubs feed mainly on decaying vegetation, at first, but later on the fine roots of grasses and other plants. They frequently cause serious damage to lawns and golf greens. The grub resembles the common white grub but is only ¾ to 1 inch long, when full-grown, and can be distinguished from its near relatives by the appearance of the last ventral segment, as shown in Fig. 15.54,*B*. The nature of the life cycle is shown by Fig. 15.53. The larvae pass through three instars, which require an average of 136 days at 25°C. The adults are most abundant during July and August, flying and feeding actively on warm sunny days. They prefer to feed on parts of the plant exposed to the sun.   As a rule there is one generation annually; but the grubs may take 2 years to develop in wet cold soils.

Control Measures.   Fruits and foliage may be protected from damage by the adult beetle by spraying when the beetles appear with DDT at 1 pound, methoxychlor at 1.5 pounds, malathion at 0.5 pound, parathion at 0.3 pound, or Sevin at 1 pound to 100 gallons of spray.   The treatments may be repeated after 10 to 14 days if necessary.   Spraying with rotenone at 2 ounces to 100 gallons of spray, repeated at 7- to 10-day intervals if necessary, is especially suitable for the home garden and may be applied to mature fruit without a residue problem.   Thousands of beetles can be caught by jarring them from trees and shrubs upon sheets or canvas in the early morning.   Home grounds, parks, and estates can be protected to some extent in the less heavily infested districts by the use of properly constructed traps, baited with geraniol and eugenol, suspended about

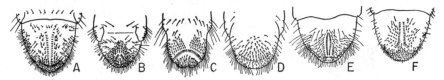

Fig. 15.54. Terminal abdominal segments of the larvae of six species of "white grubs," ventral surface, showing the features by which the different species may be distinguished from each other. *A*, the oriental beetle, *Anomala orientalis; B*, the Japanese beetle, *Popillia japonica; C*, the Asiatic garden beetle, *Autoserica castanea; D*, the annual white grub, *Ochrosidia villosa; E*, one of the native white grubs with a 3-year cycle, *Phyllophaga hirticula;* and *F*, the European chafer, *Amphimallon majalis.*   All enlarged about three diameters.   (*Redrawn from Conn. and N.J. Agr. Exp. Sta. publications.*)

every 50 feet in sunny spots near trees or shrubs throughout the summer (page 394).   Directions for securing and using such traps should be secured from the U.S. Department of Agriculture.   The larvae may be controlled by treating permanent sod and grass pastures within ½ mile of fruit plantings with spores of the milky disease,[1] which should be applied in teaspoonfuls (2 grams) every 10 feet in rows 10 feet apart. The sod of lawns, golf courses, etc., may be protected from larval feeding by applying DDT or the cyclodiene insecticides, as described on page 522.   Such treatments will remain effective for several years.

References   *N.J. Dept. Agr. Bur. Statistics and Inspection, Cir.* 46, 1922; *Pa. Dept. Agr. Bul.* 390 (vol. 7, no. 11), June, 1924; *U.S.D.A., Dept. Bul.* 1154, 1923; *Jour. Econ. Ento.,* **19**:786, 1926; *Conn. Agr. Exp. Sta. Bul.* 411, 1938; *U.S.D.A., Cirs.* 237, 1932, and 332, 1934, and *Tech. Bul.* 478, 1935; *Conn. Agr. Exp. Sta. Buls.* 491, 1945, and 505, 1947; and *Jour. Econ. Ento.,* **41**:905, 1948.

178. Green June Beetle[2]

Importance and Type of Injury.   Serious damage is caused by this insect to lawns, golf courses, vegetable gardens, and ripening fruits. Large, somewhat flattened, green beetles, with the margins of the body bronze to yellow, nearly 1 inch long and half as broad (Fig. 15.55), feed on the foliage of the peach and also on the fruit just before ripening.   Thick-

[1] *Bacillus popillae.*

[2] *Cotinis nitida* (Linné), Order Coleoptera, Family Scarabaeidae.   *C. texana* Casey, the fig beetle of southwestern United States, is very similar in appearance and habits, the adults feeding on ripe peaches, nectarines, apricots, pears, apples, figs, grapes, melons, and tomatoes.

bodied dirty-white grubs, always crawling on their backs (Fig. 15.55), feed on grass roots, tobacco plants, and decaying vegetation in the soil.

Plants Attacked. The adult beetles feed on the foliage of a number of different trees and plants; also occasionally attacking ears of corn, fruits, and vegetables of the garden. The larvae also do considerable injury to the roots of grasses in lawns and golf courses, to a number vegetables of and ornamental plants, and in the tobacco seedbeds.

Distribution. This insect is confined to the southern United States, extending north into Long Island and southern Illinois.

Life History, Appearance, and Habits. The winter is passed as a grub deep in the soil. In the spring the grubs burrow close to the surface and feed mainly on decaying vegetable matter. After a heavy rain they occasionally come out of the soil and crawl on the surface of the ground. When so crawling, they generally lie on their backs and work themselves forward by a peculiar movement of the body ridges. The grubs become full-grown by midspring, change in an earthen cell in the ground to the

pupal stage, emerge as beetles, and feed on fruits or foliage during July and August. Their eggs are laid in soil rich with decaying vegetable matter, on which the grubs feed until cold weather. There is one generation each year.

Fig. 15.55. Left, adult of green June beetle, *Cotinis nitida*. Right, full-grown larva of green June beetle in natural position when crawling on its back. Twice natural size. (*From U.S.D.A., Dept. Bul.* 891.)

Control Measures. Piles of grass clippings or manure should not be left near lawns or orchards, since they attract the beetles as a site for egg-laying. Thorough wetting of lawns in the evening will bring the grubs to the surface, so that they can be destroyed. V-shaped troughs or flowerpots sunk in the ground have proved effective in trapping the larvae. Soil treatments with granular dieldrin at 0.4 pound, aldrin or heptachlor at 1 pound, or DDT or toxaphene at 2 to 3.5 pounds per acre are effective against the younger larvae, where these materials can be applied without a residue problem. In tobacco seedbeds, the larvae may be controlled by applying methyl parathion at 0.2 pound, parathion at 0.15 pound, or Dylox at 1 pound in a drench of 100 gallons of water to 100 square yards of plant bed. The adults are difficult to control, although promising results have been obtained with lindane at 0.125 to 0.25 pound and aldrin or heptachlor at 0.5 pound to 100 gallons of spray. Large numbers of adult beetles have been trapped in peach orchards by bait pails containing fermenting malt extract, molasses and geraniol, or caproic acid.

References. N.C. Agr. Exp. Sta. Bul. 242, 1921; U.S.D.A., Dept. Bul. 891, 1922, and *Farmers' Bul.* 1489, 1926; Va. Agr. Exp. Sta. Bul. 454, 1952; Ky. Agr. Exp. Sta. Bul. 559, 1951; Jour. Econ. Ento., **37**:855, 1944; **45**:736, 1952; **48**:621, 1955; **49**:828, 869, 1956; and **50**:96, 1957.

179 (149, 172, 189, 204). SAN JOSE SCALE

The San Jose scale is fully as serious a pest of the peach as of the apple. The life history on the peach is practically the same as on the apple, and the control measures are similar (page 707).

180 (241). TERRAPIN SCALE[1] AND BROWN APRICOT SCALE OR EUROPEAN FRUIT LECANIUM[2]

Importance and Type of Injury. During the winter the undersides of peach and apricot twigs are sometimes found to be nearly covered with shiny, convex, brownish scales about $\frac{1}{12}$ inch in diameter (Fig. 15.56). In the summer the fruit will be covered with masses of honeydew on which a sooty black fungus grows.

Plants Attacked. All the common fruit trees and many shade trees and shrubs.

Distribution. The terrapin scale is found in eastern and southeastern United States, and the brown apricot scale is generally distributed through most of the fruit-growing areas, including the Pacific Coast.

Life History, Appearance, and Habits. The terrapin scale passes the winter as fertilized females closely adhering to the bark of the twigs (Fig. 15.57). These vary somewhat in color but are generally dark brown, with some black lines marking off squarish areas about the margins of the scale. The scale margins are also generally somewhat crinkled. They resume feeding early in the spring, and by the last of May or early June the female, which is then about $\frac{1}{4}$ inch in diameter, begins to produce hundreds of oval white eggs. These accumulate in a dome-shaped cavity underneath her body, and a day or two after birth the tiny nymphs, which are oval, white, and with well-developed legs and antennae,

FIG. 15.56. Mature female of the terrapin scale, *Lecanium nigrofasciatum.* *a,* Ventral, *b,* dorsal, and *c,* lateral views. About 8 times natural size. (*From U.S.D.A., Dept. Bul.* 351.)

crawl from under the dead female's body and out to the undersides of the leaves. Here they insert their beaks along the larger veins and feed by sucking the sap. After about a month on the leaves, the females migrate

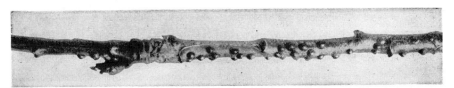

FIG. 15.57. Terrapin scale on peach twigs during winter. About natural size. (*From U.S.D.A., Dept. Bul.* 351.)

back to the twigs, and a week or two later the winged males find them and mating occurs. The males die in midsummer, but the females continue to feed until cold weather, hibernate through the winter, feed again

---

[1] *Lecanium nigrofasciatum* Pergande, Order Homoptera, Family Coccidae.

[2] *Lecanium corni* Bouché.

in spring as stated above, and finally die during the following summer. There is a single generation a year.

The life history of the brown apricot scale is very similar. It overwinters as immature scales on the twigs, and these grow rapidly in the spring, reaching maturity about April or May. The adult female scale is nearly hemispherical, shiny brown, and ⅛ to ⅜ inch long. The eggs hatch under the female in June and early July, and the crawlers feed on the leaves until October and November, when they move back to the twigs to overwinter. There is a single generation a year.

Control Measures. These scales may be controlled by spraying at the completion of hatching of the crawlers with malathion at 0.5 pound or EPN or parathion at 0.3 pound to 100 gallons of spray. They are also susceptible to dormant sprays of 3.5 to 4 per cent oil emulsions or 2 per cent oil emulsions fortified with 0.5 pound of parathion, or Diazinon 0.25 pound, or Ethion 0.32 pound.

References. *U.S.D.A., Dept. Bul.* 351, 1916, and *Farmers' Bul.* 1557, p. 32, 1931; *Calif. Agr. Exp. Sta. Cir.* 478, 1959.

181 (126). GREEN PEACH APHID[1]

Importance and Type of Injury. Greenish smooth-looking aphids or plant lice, suck the sap from the new fruits and twigs of peach and a number of other trees and herbaceous plants.

Plants Attacked. The food plants of this insect include peach, plum, apricot, cherry, many ornamental shrubs, and garden and flowering plants (page 676).

Distribution. This insect is a native of Europe but is now generally distributed over North America.

Life History, Appearance, and Habits. The winter is passed as black shining eggs on the bark of the peach, plum, apricot, and cherry. The young aphids, which are pale yellowish-green in color with three dark lines on the back of the abdomen, begin to hatch about the time the peach blooms. On becoming full-grown, they begin giving birth to living young, generally remaining on the peach for two or three generations, after which most of the individuals acquire wings and migrate to garden plants in late spring. On the approach of cold weather in the fall, the females fly to the peach, where they give birth to the true sexual females. These mate with males that fly over from the summer host plants, and the fertilized females deposit their eggs in the above-mentioned situations.

Control Measures. Spraying as the buds swell with 3 per cent (superior) or 4 per cent (regular) oil emulsion will kill most of the winter eggs. The aphids may be controlled as they appear in the spring, with nicotine sulfate at 0.3 to 0.4 pound (with soap or hydrated lime), parathion 0.15 pound, Guthion 0.3 pound, or malathion at 0.5 pound to 100 gallons of spray.

References. *Pa. Agr. Exp. Sta. Buls.* 185 and 186, 1924; *Colo. Agr. Exp. Sta. Bul.* 133, 1908; *N.J. Agr. Exp. Sta. Cir.* 107, p. 5, 1919; *Va. Agr. Truck Exp. Sta. Bul.* 71, pp. 812–815, 1930; *Jour. Econ. Ento.*, **39**:661, 1946; and **40**:915, 1947.

182 (83). TARNISHED PLANT BUG[2]

Importance and Type of Injury. This insect, which feeds on many different kinds of plants, is particularly injurious to the peach. In the nursery rows peach trees will be found with the main terminal and a number of laterals wilting and dying back, but not containing borers. Young trees thus injured become scrubby or brushy. The insect also causes sunken areas on the sides of the fruit which are free from down,

---

[1] *Myzus persicae* (Sulzer), Order Homoptera, Family Aphidae.

[2] *Lygus lineolaris* (= *oblineatus*) Palisot de Beauvois, Order Hemiptera, Family Miridae.

looking as though the peach had been gouged or partly peeled, when small. This injury is commonly called "cat-facing" (Fig. 15.58).

Plants Attacked. Many kinds of trees and herbaceous plants; a very general feeder.

Distribution. Throughout the world.

Life History, Appearance, and Habits. The tarnished plant bug passes the winter as a full-grown insect in many different kinds of shelter but seems to prefer plants which do not entirely die down. Large numbers of them have been found in winter between the leaves of mullein, wild parsnip, alfalfa, and clover. It is of a coppery-brown color, flecked with darker brown and yellow. The body is oval in shape, somewhat triangular in front, about ¼ inch long, and about half as wide (Fig. 14.9). The insects become active during the first warm days of spring. They fly to trees, feed by sucking the sap and, while feeding, inject some substance that is highly injurious to the plant tissue. The feeding causes the peach to die back, resulting in a condition usually referred to by nurserymen as "stopback." There are several generations of the insect each season, the eggs being laid mainly in the stems of herbaceous plants, including some cultivated plants and weeds. A few eggs are laid in the peach terminals and also occasionally in the peach fruits (page 614).

Control Measures. This insect may be controlled by spraying with DDT at 1 pound, dieldrin, EPN, or parathion at 0.25 pound, or Sevin at 1 pound to 100 gallons of spray, applied just before bloom and at petal

FIG. 15.58. Typical "cat-facing" of peach, caused by feeding of the tarnished plant bug. (*From Ill. Natural History Surv.*)

fall. Keeping down weeds about the orchard is of some value in reducing its numbers, but the weeds should not be allowed to grow in early spring and cut later, as this practice may drive the bugs to cultivated plants.

References. Mo. Agr. Exp. Sta. Res. Bul. 29, 1918; *N.Y. (Cornell) Agr. Exp. Sta. Bul.* 346, 1914; *Ill. State Natural History Surv. Bul.,* vol. 17, art. VI, 1928; *Jour. Econ. Ento.,* **41**:403, 1948; **42**:335, 1949.

183 (162). Shot-hole Borer[1] and Peach Bark Beetle[2]

Importance and Type of Injury. These two important species of bark beetles attack the trunk and branches of the peach but are of little importance on vigorously growing trees. They are so similar in their habits that a separate description will not be given. The shot-hole borer, or fruit tree bark beetle, is further discussed on page 722. Small beads of gum will be found exuding from points on the bark of the trunk and branches of the peach, also from small holes in the bark, about the size of a pencil lead. Galleries of borers are found in the inner bark radiating from a main or parent gallery and terminating in small round exit holes through the bark.

[1] *Scolytus rugulosus* (Ratzeburg), Order Coleoptera, Family Scolytidae.
[2] *Phloeotribus liminaris* (Harris), Order Coleoptera, Family Scolytidae.

Trees Attacked. The shot-hole borer attacks peach, cherry, plum, apple, and other fruit and shade trees. The peach bark beetle attacks cherry and other stone fruits, but *not* the pome fruits.

Distribution. The peach bark beetle is most abundant in the eastern half of the United States. The shot-hole borer occurs throughout the country.

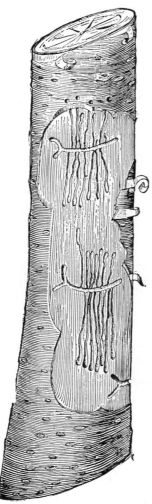

Fig. 15.59. Work of the peach bark beetle, *Phloeotribus liminaris*, in wood of peach tree, showing parent galleries, larval galleries, and exit holes. (*From U.S.D.A., Farmers' Bul.* 763.)

Life History, Appearance, and Habits. These insects are small brownish or blackish beetles about $\frac{1}{10}$ inch or less in length. The peach bark beetle winters in the adult stage, either in its pupal cells in dead or dying wood or in special hibernating cells cut in the bark of healthy trees. The shot-hole borer passes the winter in the form of a small pinkish-white legless grub in the inner bark of the tree. These grubs transform to beetles in their burrows and emerge in the late spring. Both species are good fliers and are seemingly attracted to trees that are in a poor condition or those having dying branches. The parent beetles excavate a burrow in the inner bark of such trees, and along the sides of this the female lays her eggs. The parent gallery of the shot-hole borer (Fig. 15.32) generally runs the long way of the branch. That of the peach bark beetle (Fig. 15.59) generally runs crosswise and forks to form a Y at one end of the gallery. The grubs hatching from these eggs work out through the inner bark for a distance of 2 to 3 inches, gradually enlarging their burrows as they go. The larval burrows of the peach bark beetle follow mostly the grain of the wood, while the larval burrows of the shot-hole borer run chiefly across the grain. Upon becoming full-grown, they change to beetles and emerge through small holes which they cut in the bark. There are at least two generations of these insects in the latitude of central Illinois, and a partial third farther south.

Control Measures. As the insects are attracted mainly to unhealthy or injured trees, the best control measure is to keep the trees in a vigorous condition during the summer, either by cultivation, or by fertilizing with nitrogenous fertilizers. All peach prunings and dying or diseased trees in or around the orchard should be removed and burned during the winter. Spraying the trunk and branches of infested trees with DDT at 1 pound or parathion at 0.3 pound, or with a mixture of these, to 100 gallons of spray, when the beetles are active in late spring and in September after the crop is harvested, is of value in controlling these insects.

References. U.S.D.A., *Farmers' Bul.* 763, 1916; *Ohio Agr. Exp. Sta. Bul.* 264, 1913; *Mich. Agr. Exp. Sta. Bul.* 284, pp. 30, 32, 1933; *Ore. Agr. Exp. Sta. Cir.* 162, *Rev.* 1948.

184. Peach Tree Borer[1]

Importance and Type of Injury.   The most important insect enemy of the peach is the peach tree borer.   Nearly everyone who has tried to grow peaches, in a commercial orchard, a farm orchard, or a back yard, is familiar with the work of this insect.   Masses of gum exude from around the base of the trunk from about 1 foot above, to 2 or 3 inches below the surface of the soil, with bits of brownish frass or sawdust, mixed with the gum.   White worms with brown heads will be found burrowing in the bark of the peach trunk from 2 or 3 inches below the surface of the ground to 10 inches above.   Deadened areas in the bark, where these worms have eaten out the living tissue, in many cases cause the death of the tree.

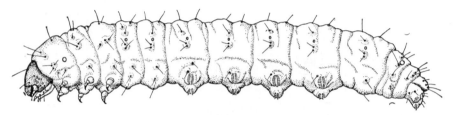

Fig. 15.60. The peach tree borer, *Sanninoidea exitiosa.*   Upper figure, the larva about 4 times natural size, slightly from beneath to show the prolegs, each with the characteristic double transverse row of crochets.   (*Drawn by Kathryn Sommerman.*)   Lower figure, the frass-covered silken cocoon of the pupal stage, from which the pupa has projected its body before the adult moth emerged, leaving the empty exuviae clinging to the end of the cocoon. (*From Ill. Natural History Surv.*)

Trees Attacked.   Peach, wild and cultivated cherry, plum, prune, nectarine, apricot, and certain ornamental shrubs of the genus *Prunus.*

Distribution.   All sections of the United States and Canada.

Life History, Appearance, and Habits.   This insect always passes the winter in the worm or larval stage.   These larvae (Fig. 15.60) vary greatly in size, some being over ½ inch in length, while others are very small, in some cases not over ⅛ inch long.   The difference in the size of the larvae comes from the fact that the eggs are laid over a considerable period of time so that the developmental periods of the six larval instars may overlap.   In the spring of the year the worm becomes active as soon as the soil is sufficiently warm.   In the latitude of southern Illinois the

[1] *Sanninoidea* (= *Conopia*) *exitiosa* (Say), Order Lepidoptera, Family Aegeriidae. The western peach tree borer, *S. exitiosa graefi* (Hy. Edwards) is a variety.

larger ones will complete their growth by the middle or latter part of May and will be found under the bark close to the ground.   They are then about 1 inch in length, whitish, with a dark-brown head and plate behind the head.   They change, in closely spun, silken, dirt-and-gum-covered cocoons on the surface of their burrows or in the soil, to the brown pupal stage.   Just before the moth emerges, the pupa forces its way out of the cocoon, and the empty pupal skin generally remains protruding from the cocoon.   The first adults begin to appear in July and emergence continues until the latter part of September or possibly in a few cases into October, the greatest number coming out during August.   The female insect is a blue-black moth having clear hind wings, with an orange crossband on the abdomen (Fig. 15.61).   When in flight the moths closely

Fig. 15.61. Peach tree borer, *Sanninoidea exitiosa*, adults, natural size.   The ones at center and lower left are males, the other two females.   (*From Sanderson and Jackson, "Elementary Entomology."*)

resemble some of the larger wasps, for which they are frequently mistaken. Unlike most moths they are active in the daytime.   The male has both wings nearly clear and several narrow yellow bands across the abdomen. The females lay from 200 to 800 eggs mostly on the trunks of the trees or in cracks in the soil within a few inches of the trunk.   They are seemingly attracted to trees which previously have been infested by the borer or to those to which some mechanical injury has occurred.   Upon hatching from the eggs, in about 10 days, the worms work their way into the bark of the tree and feed during the late summer or early fall in the outer layers of the bark, penetrating deeper into the bark as they become larger.   Where the eggs are laid over a period of several months, it is easy to understand the great variation in the size of the larvae found in the bark in the fall.   There is, as a rule, one generation each year.

Control Measures.   While this insect, if not combated, is almost sure to kill peach trees within a few seasons, we are fortunate in having a control measure which, if properly applied, is nearly 100 per cent effective. This control method consists of placing a small amount of the crystals of *para*-dichlorobenzene, generally called "PDB," on the surface of the soil around the trunk of the tree during the fall (Fig. 15.62).   The amount of PDB to be applied will vary with the size of the tree.   The following rules should be observed in using this chemical.   For trees under 3 years of age, use ½ ounce per tree, for trees from 3 to 6 years old, ¾ ounce, and for older trees from 1 to 1½ ounces, depending on the diameter of the trunk at the surface of the ground.   Apply the crystals in a ring completely encircling the trunk, not closer to the bark than 1 inch, nor at a greater distance than 3 inches.   Cover the crystals with several shovelfuls of earth to confine the PDB gas.   Do not apply the treatment during the summer, as the borers are not then in the tree, nor should it be applied in the late fall, when the soil temperature is likely to be below 60°F. for the

Fig. 15.62. Control of the peach tree borer.   At left, a ring of *para*-dichlorobenzene placed about the trunk of a peach tree to kill the borers.   The ground has been cleared of trash before applying the chemical.   Right, soil is placed over the ring to confine the fumes. (*From Ill. Natural History Surv.*)

first 2 weeks after the material is applied   The crystals applied in warm soil quickly volatilize into a heavy gas which penetrates through the soil crevices and into the burrows of the borers and kills them.   The best dates for applying *para*-dichlorobenzene can be obtained from entomologists of the state experiment stations.   In the latitude of southern Georgia the best dates for making these applications will usually be between Oct. 15 and 20; in central Georgia, between Oct. 10 and 15; in northern Georgia, between Sept. 25 and Oct. 5; in the latitude of southern New Jersey, Oct. 1 to 10; in northern New Jersey, Sept. 20 to Oct. 1; in southern Illinois, Sept. 25 to Oct. 15; and in northern Illinois, Sept. 20 to Oct. 20.

An alternate treatment consists of the application in ethylene dichloride emulsion in a ring around the trunk, using ⅛ pint of 15 per cent emulsion for 1-year-old trees, ¼ to ½ pint of 15 per cent emulsion for 2- to 3-year-old trees, and ½ pint of 20 per cent emulsion for 4-year and older trees.   If there are signs of borer injury above the soil line, the soil level should be raised, and after treatment the area around the trunk should be covered with a little soil, to delay the escape of the vapor.

The ethylene dichloride emulsion should not be sprayed or poured on the tree trunk.   This treatment is most satisfactory in the fall, although if the fall treatment is not made, it is sometimes advisable to treat during the first warm period in the spring.   The treatment will not be effective if given before the soil temperature has reached an average of about 60°F.; by this time the young borers have been actively at work in the bark for several weeks and will have caused serious damage.   The mounds of earth may be removed from around the trees 4 to 6 weeks after treatment and should be leveled before July in order not to force the moths to lay their eggs high up on the trunk.

Effective control of the borers has been obtained by spraying the tree trunks, from the main limbs to the soil level, with DDT at 3 to 4 pounds, parathion at 0.3 pound, or Guthion or EPN at 0.5 pound to 100 gallons of spray.   The applications should be made just after the moths emerge and again at the peak of emergence.   In the latitude of Virginia and northward, two applications, in early July and again 3 to 4 weeks later, are suggested.   From Virginia southward, three applications, in the middle of July and at 3- to 4-week intervals, are required.   In the latitude of southern Georgia, four applications at monthly intervals, from the first of August, are required.   The organophosphorus insecticides, at least, have considerable ovicidal action.

The borers may be dug out by the use of a sharp knife, or probed out with a flexible wire, first removing the soil from around the base of the tree to a depth of about 3 inches.   This method is slow and laborious, and it is not possible in most cases to get all the borers.

References.   N.J. Agr. Exp. Sta. Bul. 391, 1922; *U.S.D.A., Dept. Buls.* 796, 1919, and 1169, 1923, and *Tech. Buls.* 58, 1928, and 854, 1943; *Ohio Agr. Exp. Sta. Bul.* 329, 1928; *N.Y. (Geneva) Agr. Exp. Sta. Cir.* 172, 1937; *N.C. Agr. Ext. Cir.* 277, 1944; *Wash. Agr. Exp. Sta. Cir.* 77, Rev. 1955; *Jour. Econ. Ento.,* **25**:786, 1932; **29**:754, 1936; **41**:240, 1948; **42**:343, 1949; **45**:611, 1952; **46**:704, 1953; **47**:359, 1954; **49**:397, 574, 1956; **51**:557, 1948; and **52**:804, 1959.

185. LESSER PEACH TREE BORER[1]

Importance and Type of Injury.   This, too, is one of the most destructive insect enemies of stone fruits, causing injury similar to, and only slightly less important than, that of the peach tree borer.   Masses of gum exude from the trunk and branches of the peach tree where injury has occurred.   Frequently these masses of exuding gum will be found in the forks, especially where the forks have split, and around wounds in the bark (Fig. 15.63).   White worms with brown heads will be found working in the inner bark.   Bits of brownish castings and sawdust-like material are mixed with the masses of gum.   The larvae very closely resemble those of the peach tree borer, but they may usually be distinguished in the orchard by the fact that they attack the upper part of the trunk and the larger branches.

Trees Attacked.   Peach, plum, cherry, wild plum, wild cherry, Juneberry, and possibly some others.

Distribution.   Peach-growing sections of the United States, except the western states; most abundant in the South.

Life History, Appearance, and Habits.   The lesser peach tree borer passes the winter in the inner bark of the tree in the form of partly grown larvae.   These larvae vary in size from $\frac{1}{4}$ inch to nearly 1 inch in length

[1] *Synanthedon pictipes* (Grote and Robinson), Order Lepidoptera, Family Aegeriidae.

and may be in the second to sixth larval instar. They start to feed very early in the spring, the larger individuals completing their growth by late April and the smaller ones by the last of June. They pupate in cocoons inside the burrows, but always approach the surface of the bark, leaving the exit of the burrow covered by a thin silken web. Just before the moths emerge, the pupae push themselves out of the cocoon and partly out of their burrows. Emergence of the moths occurs in May and June. They are clear-winged, of a metallic blue-black color, with pale-yellow markings on the abdomen. The moths are extremely active and dart about with great rapidity on warm sunny days. Their general appearance is much like that of some of the larger wasps. The female moth lays her tiny reddish-brown eggs in cracks and crevices in the bark, usually around the crotch of the tree, and is especially attracted to wounds in the bark. The eggs hatch in 8 to 20 days, depending on the temperature, and the first-instar larvae work their way into the bark, generally through wounds or abrasions. The larval period of six instars requires about 50 days for the first brood and about 240 days for the second brood, which overwinters. The pupal period requires from 15 to 28 days. A single generation occurs in South Dakota, a partial second generation over most of the North Central States, and a full second generation in the South.

Control Measures. This insect may be controlled by spraying the trunk and larger branches with parathion, EPN, or Guthion at 0.5 pound, or with malathion at 1 pound to 100 gallons of spray, four times at intervals of 3 weeks, beginning with the emergence of the adults in June. In the latitude of Georgia, the appli-

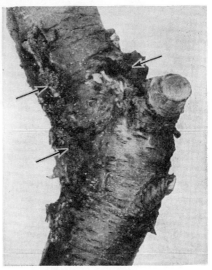

Fig. 15.63. Damage caused by the lesser peach tree borer. Arrows indicate masses of exuding gum and frass around wounds made by the borer. (*From Ill. Natural History Surv.*)

cations should be made at monthly intervals, for the first brood in April and May and for the second brood in August and September.

References. U.S.D.A., Bur. Ento. Bul. 68, 1907; Ohio Agr. Exp. Sta. Bul. 307, 1917; S.D. Agr. Exp. Sta. Bul. 288, 1934; U.S.D.A., Cir. 172, 1931; Ill. Natural History Surv. Cir. 31, 1938; Jour. Econ. Ento., **29**:754, 1936; **44**:685, 1951; **45**:607, 611, 1952; **49**:397, 1956; and **52**:634, 1959.

186. Peach Twig Borer[1]

Importance and Type of Injury. In unsprayed or neglected orchards, the peach twig borer is of considerable importance. Small brown worms may be found working inside the twigs and new growth, as well as in the fruits. The twigs die back and small masses of gum exude from them.

Trees Attacked. Peach, plum, apricot, and almond.

Distribution. General in peach-growing sections, most serious on the Pacific Coast.

[1] *Anarsia lineatella* Zeller, Order Lepidoptera, Family Gelechiidae.

Life History, Appearance, and Habits. This insect passes the winter as a partly grown, chocolate-brown caterpillar, with a black head, from ⅛ to 1/16 inch in length, hidden in a silken cocoon closely attached to the bark of trunk and branches (Fig. 15.64a,b,c). About the time the peach leaves appear, the larvae leave their winter nests and start boring in the ends of the tender new growth, causing a wilting and dying back of the twigs (g) in which they are working. A larva may feed in more than one twig before it completes its growth. Upon becoming full-grown, the larvae are about ½ inch in length. They then spin cocoons on the larger branches or trunk. Here they change into small moths, gray in color, with a wingspread of about ⅓ to ½ inch (i). The moths mate, and the female lays her eggs on the twigs. Another generation of the worms soon hatches and feeds almost entirely in the fruits. There

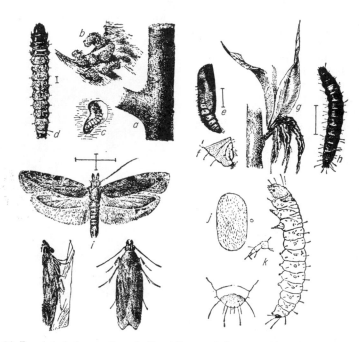

FIG. 15.64. Peach twig borer, *Anarsia lineatella.* *a–d,* Larva and its winter nests; *e, f,* pupa with detail of posterior end; *g,* new shoot of peach withering from attack of larvae; *h,* larva enlarged; *i,* adult moths; *j,* egg greatly enlarged; *k,* larva, side view and details of structure. Lines show natural size. (*From U.S.D.A.*)

are from one to four generations of the insect in the United States each season, the larger number occurring in the South.

Control Measures. Damage by this insect is largely confined to unsprayed orchards. The dormant-oil or lime-sulfur sprays regularly applied for the control of the San Jose scale kill many of the overwintering larvae in cocoons and largely prevent serious damage in the central states. In western areas, however, regular treatments are often necessary and the insect may be controlled by spraying in the prepink or petal-fall period and again 10 days later with DDT or Sevin at 1 pound, parathion or Diazinon at 0.5 pound to 100 gallons of spray, or with basic lead arsenate at 4 pounds to 100 gallons of spray, in the pink-bud stage, and again at petal fall. The local state experiment station should be consulted for the exact timing.

References. *Colo. Agr. Exp. Sta. Bul.* 119, 1907; *Calif. Agr. Exp. Sta. Buls.* 355, 1923, and 708, 1948; *Jour. Econ. Ento.,* **15**:395, 1922; **29**:156, 1936; **42**:22, 1949; **44**:935, 1951; and **52**:340, 637, 1959.

187 (163, 191, 194). PLUM CURCULIO[1]

The plum curculio, which has been described under apple insects (page 724), is one of the most serious insect pests of the fruit of the peach. The adults damage the fruits by their feeding and egg-laying punctures and the larvae tunnel through the fruit, feeding on the pulp. Infested fruits generally drop. The plum curculio is also one of the main agencies in spreading the brown rot of the peach.

Control Measures. The method of control is, in general, the same as that given for the insect on the apple. In peach orchards which have been badly infested, it is often advisable to gather the peach "drops" during May and the first part of June. The orchard should be gone over every few days, if possible, and the dropped peaches gathered and burned or sacked and buried to a depth of at least 2 feet. For the control of this insect, general orchard sanitation is as important in peach orchards as in apple orchards. Cultivating the soil during the late spring and early summer will destroy the larvae and pupae of the curculio in their cells in the earth. Jarring the tree early in the morning and catching the beetles on sheets spread under the tree is of help in control.

Effective control may be obtained by spraying with Guthion or parathion at 0.3 pound, EPN at 0.36 pound, malathion at 0.5 to 0.75 pound, or methoxychlor at 1.5 pound to 100 gallons of spray. In northern areas, three to four applications should be made at 7- to 10-day intervals, beginning as soon as the brown skin or shuck is pushed off by the forming fruit. In southern areas, three applications should be made at 10-day intervals, beginning at petal fall, and single applications at 4 to 5 weeks and again at 2 weeks before harvest. Dieldrin at 0.25 to 0.5 pound may be used through the second-cover spray, and lead arsenate at 2 pounds plus 8 pounds of hydrated lime, up to 4 weeks before harvest.

Treatment of orchard soil with aldrin, dieldrin, or heptachlor at 2 to 4 pounds per acre has given good control for 3 to 4 years after application.

References. U.S.D.A., Tech. Bul. 188, 1930; Del. Agr. Exp. Sta. Buls. 175, 1932, and 193, 1935; Ill. Natural History Surv. Cir. 37, 1939; Jour. Econ. Ento., **41**:217, 1948; **42**:7, 330, 1949; **50**:457, 1957; and **51**:330, 1958.

188. ORIENTAL FRUIT MOTH[2]

Importance and Type of Injury. This is, in many parts of the eastern United States, the most important insect pest of the fruits of the peach. The earliest indication of injury by this insect is similar to that of the peach twig borer and consists of a dying back of the new growth of twigs in the spring. The worms found burrowing in the twigs, however, are not brown like the peach twig borer, but pinkish or creamy white, with brown heads. The fruit shows injury similar to that of the codling moth in apples. A pinkish-white, short-legged larva, ½ inch long, with a five-toothed, comb-shaped plate on the last segment of the body, may be found inside the fruit (Fig. 15.66). In many cases the fruit will not show any blemish through the skin, the young worms having entered through the stem. The fruit may look perfect at the time of picking but breaks down shortly after packing and shows numerous feeding burrows of the larvae. Their attacks cause an increase in the amount of brown rot.

---

[1] *Conotrachelus nenuphar* (Herbst), Order Coleoptera, Family Curculionidae.

[2] *Grapholitha* (= *Laspeyresia*) *molesta* (Busck), Order Lepidoptera, Family Tortricidae.

Trees Attacked.   Peach, quince, apple, pear, apricot, plum, and several other fruits.

Distribution.   The oriental fruit moth is an imported insect, having been brought in from the Orient previous to 1915, on imported nursery

Fig. 15.65. Adult of oriental fruit moth, *Grapholitha molesta.*   At left, with wings folded; at right, with wings spread.   About 5½ times natural size.   *(From U.S.D.A.)*

stock.   It is now well established over the eastern United States in areas where peaches are grown and has also been found in the West.

Life History, Appearance, and Habits.   The winter is passed as full-grown larvae about ½ inch in length, closely resembling the codling moth. The larvae enclose themselves in silken cocoons which they spin on the

Fig. 15.66. Peach injured by oriental moth, showing larva and result of its feeding. About natural size.   *(From Purdue Agr. Exp. Sta. Cir. 122.)*

bark of the trunk or in many cases in rubbish, weeds, and mummied fruits on the ground around the peach orchard.   In the spring the insect changes to the pupal, and then to the moth stage.   The moths (Fig. 15.65) are gray, with chocolate-brown markings on the wings. They very closely resemble the codling moth but are somewhat smaller in size, being about ¼ inch long, with a wingspread of only ½ inch. The females lay their flat whitish eggs on the leaves, and, in some cases, on the twigs, generally shortly after the peaches bloom.   The first-generation worms, upon hatching, bore in the tender twigs and, upon becoming full-grown, spin cocoons in which they transform to moths.

There are from one to seven generations of the insect each year.   The later generations attack the fruit in much the same manner as the codling moth attacks the apple, except for the fact that the larvae do practically no external feeding upon the leaves and in working their way through the skin of the peach apparently swallow little or none of the particles which they bite off.   As already mentioned, many of the larvae enter the fruit through the stems.

Control Measures. In many areas early-maturing varieties may be planted which are often picked before this pest attacks the fruits, even where the insect is abundant. The mass colonization and release of the braconid parasite, *Macrocentrus ancylivorus*, at rates of about 500 adults per acre, has effected a reduction of 50 to 80 per cent in fruit injury in some areas, but has not been as successful in others. Such parasite releases combined with a single preharvest spray have given very satisfactory control. Where a complete insecticidal program is desirable, spraying with DDT or Sevin at 1 pound, or parathion or Guthion at 0.3 pound, or EPN at 0.37 pound to 100 gallons of spray has given excellent control. Several applications at about 10-day intervals, beginning when the shucks split, may be required. Cultivation of the soil of infested orchards to a depth of 4 inches, 1 to 3 weeks before blooming time, will kill many of the overwintering larvae in the soil. Prompt destruction of the cull fruits, the screening of packing sheds, and the cleaning of used baskets are advised in northern peach areas.

References. *Va. Agr. Exp. Sta. Tech. Bul.* 21, 1921; *Pa. Dept. Agr. Bul.*, vol. 8, no. 9, June, 1925; *Purdue Agr. Exp. Sta. Cir.* 122, 1925; *U.S.D.A, Tech. Buls.* 152, 183, and 215, 1930; *Conn. Agr. Exp. Sta. Bul.* 313, 1930; *Ill. Nat. History Survey Cir.* 26, 1934; *Ohio Agr. Exp. Sta. Bul.* 569, 1936; *Jour. Econ. Ento.*, **42**:348, 351, 1949; **44**:418, 1951; **45**:462, 1952; and **47**:147, 1954.

E. PLUM INSECTS

The insects that attack the plum are very much the same as those that attack the peach; in fact, there are only one or two insects of special importance on the plum.

FIELD KEY FOR THE IDENTIFICATION OF INSECTS INJURING THE PLUM

1. About the time the plum trees come into full foliage, silken webs are found enclosing the leaves at the ends of branches. Feeding on the leaves within the webs are smooth-bodied, grayish-yellow, many-legged caterpillars, up to ¾ inch long. Black wasp-like insects deposit eggs on the midribs of the plum leaves just previous to the appearance of the webs.....................................
   Plum web-spinning sawfly, Neurotoma inconspicua (Norton) (see *S.D. Agr. Exp. Sta. Bul.* 190, 1920).
2. Tips of new growth on plum trees literally covered with brownish, blackish, or light-green aphids. Leaves badly curled, growth stunted, fruit sometimes misshapen and shriveled............*Rusty plum aphid* or *hop aphid*, page 766.
3. Plums of the Japanese and European varieties have the branches coated with grayish scales, having much the color of wet ashes. Mature trees of the Americana varieties are seldom injured (see *C*,1 under Key for Insects Injuring the Peach).................................*San Jose scale*, page 765.
4. Leaves of plum and other fruits show a pale sickly appearance as though covered with dust or brownish-red specks. Very light webs over the surface of the leaves, upon which are crawling minute, reddish, eight-legged mites..........
   ....*European red mite*, page 715.
5. Partly grown plums are "stung" and drop from the tree (see *E*,1 under Key for Insects that Attack the Fruit of the Peach)........*Plum curculio*, page 766.

189 (149, 172, 179, 204). SAN JOSE SCALE[1]

This scale is very destructive to plums belonging to the Japanese or European varieties, *i.e.*, such varieties as Lombard, Green Gage, Fallenberg, Burbank, and many others. The scale is not, however, of great importance on mature American plum trees, such as the Wild Goose or

[1] *Aspidiotus perniciosus* Comstock, Order Homoptera, Family Coccidae.

other varieties of Americanas. The San Jose scale may cause injury to young plum trees of any variety (page 707). Control is the same as on peach.

190. PLUM APHIDS

There are several species of aphids which attack the plum. They often appear in very large numbers on the tips of the plum twigs (Fig. 15.67) and by their feeding cause splitting of the fruits, curling of the leaves, stunting of the growth, and injury of the tree and fruit by the discharge of honeydew. The hop aphid,[1] the rusty plum aphid,[2] and the mealy plum aphids[3] are common species.

FIG. 15.67. Rusty plum aphids, *Hysteroneura setariae*, clustered on plum. About natural size. (*From U.S.D.A., Farmers' Bul.* 908.)

Control Measures. While several species of aphids are involved, the control measures are essentially the same. These are discussed under the green peach aphid (page 754) and include dormant spraying with mineral-oil emulsion to kill the overwintering eggs and spraying with parathion, malathion, or nicotine when the aphids appear.

References. Pa. Agr. Exp. Sta. Bul. 182, 1923; *U.S.D.A., Dept. Bul.* 774, 1913; *Okla. Agr. Exp. Sta. Bul.* 88, 1910; *Calif. Agr. Exp. Sta. Bul.* 606, 1937.

191 (163, 187, 194). PLUM CURCULIO[4]

The plum curculio has already been discussed under both apple and peach (pages 724 and 763). It is perhaps more destructive to plums than to any other fruit. The curculio is one of the main agencies in spreading brown rot. If this very destructive disease of the fruit is to be controlled, it is necessary to control the plum curculio.

Control Measures. The methods of control which are given for this insect on the peach will apply equally well for the plum. Particular attention should be given to clean cultivation of the plum orchard and to picking up the early-dropped plums. Spraying with parathion at 0.3 pound, EPN at 0.36 pound, or methoxychlor at 1.5 pounds to 100 gallons of spray is effective when applied three to four times at 10-day intervals after petal fall. Dieldrin at 0.25 pound may be used through the first-cover spray.

F. CHERRY INSECTS

The cherry, like the plum, is attacked by many of the insects which affect the peach. Sour cherries are not subject to severe injury by San Jose scale, but it is a serious pest of sweet cherries. Cherries are sometimes attacked by the peach tree borer, lesser peach tree borer, and bark beetles; the same control measures may be used as on peach. Some of the

---

[1] *Phorodon humuli* (Schrank), Order Homoptera, Family Aphidae.

[2] *Hysteroneura setariae* (Thomas), Order Homoptera, Family Aphidae.

[3] *Hyalopterus pruni* (Geoffroy), and *H. arundinis* (Fabricius).

[4] *Conotrachelus nenuphar* (Herbst), Order Coleoptera, Family Curculionidae.

aphids that attack the plum attack also the cherry, and there are several other species which confine their work largely to this tree.   The methods of control are the same as for these insects on plum.

The sweet cherry is attacked by practically the same insects as those feeding on the sour cherry and is also subject to injury by the San Jose scale.   Wherever sweet cherries are grown, especially while the trees are young, they should receive a dormant treatment for San Jose scale, the same as that given the apple or peach.

FIELD KEY FOR THE IDENTIFICATION OF INSECTS INJURING THE CHERRY

1. Fawn-colored, very long-legged, slender-bodied beetles, about ½ inch long, attack ripening fruits often in swarms.   Most abundant in sandy areas and disappear within a month............................. *Rose chafer*, page 772.
2. Soft, fleshy, dark-green to orange, slimy slug-like larvae, up to ½ inch in length, feed on the surface of cherry leaves and skeletonize the leaf.   Most abundant during late spring and again in late summer.   In the late spring black-and-yellow wasp-like insects, about ⅕ inch long, lay their eggs on the leaves of cherry................................... *Pear-* or *cherry slug*, page 767.
3. Grayish, very thin, flaky scales, up to ⅛ inch across, and with a raised reddish nipple-like area in the center of the scale are found clustered or massed on the bark of cherry.   The scales covering the reddish-yellow bodies of the insects are thinner and more translucent than those of the San Jose scale...............
   .... *Forbes' scale*, page 768.
4. Grayish scales on the bark of sweet cherry, rarely on sour cherry (see *C*,1, under Key for Insects Injuring the Peach)................. *San Jose scale*, page 707.
5. Cherries are "stung" but seldom drop.   White footless grubs, up to ⅓ inch long, feed inside the fruit (see *E*,1, under Key to Insects that Attack the Fruit of the Peach)................................... *Plum curculio*, page 769.
6. Injury similar to *5*, but the egg punctures in the fruits lack the crescent.   Blossoms as well as fruits are eaten into by a brown snout beetle with four humps on the wing-covers..
   ....*Cherry curculio* (page 726, also *Colo. Agr. Exp. Sta. Bul.* 385, 1932).
7. Cherries misshapen, undersized, ripening prematurely, or the fruit partly decayed or shrunken and wrinkled on one side.   Yellowish-white maggots, up to ¼ inch long, burrow in the injured fruit.   Black flies, smaller than a house fly, with the abdomen marked with white crossbands and the thorax margined with yellow, and with four black bands on the wings, are found resting on the foliage or on the fruit.......................... *Cherry fruit flies*, page 769.
8. Cherries deformed and discolored, often dropping.   Pinkish larvae up to ⅓ inch long, with three pairs of legs and five pairs of prolegs, tunnel in the injured fruit.   Small gray moths lay eggs on the fruit or leaves, in late spring........
   .... *Cherry fruitworm* or *lesser appleworm*, page 771.

192. Pear-slug[1]

Importance and Type of Injury.   Small, fleshy, dark-green to orange, slug-like, slime-covered larvae, up to ½ inch in length, with the front part of the body enlarged, feed on the surface of the cherry leaves and skeletonize the leaves, leaving only a framework of veins (Fig. 15.68).

Trees Attacked.   Cherry, pear, and plum.

Distribution.   Throughout the United States.

Life History, Appearance, and Habits.   This insect, like many other sawflies, passes the winter in a cocoon formed in an earthen cell 2 or 3 inches below the surface of the ground.   In the late spring, shortly after

[1] *Caliroa cerasi* (Linné) (= *Eriocampoides limacina* Retzius), Order Hymenoptera, Family Tenthredinidae.

the cherries have come into full leaf, black-and-yellow sawflies, $\frac{1}{5}$ inch long emerge from these cocoons. The insects are a little larger than the common house fly. They have four wings. The female sawfly inserts her eggs in the leaves, and after a few days these eggs hatch into the soft-bodied worms or slugs which feed on the leaf as above described. The feeding period varies from 2 to 3 weeks. As the slugs grow in size, they become somewhat lighter in color, until when full-grown they are nearly orange-yellow. They then crawl or drop to the ground, into which they burrow, and there change to the pupal stage. Adults emerge during late July and August and lay eggs for the second generation of slugs. It is this generation that usually causes the greatest amount of injury, especially on young trees, which they may completely defoliate. When this second generation becomes full-grown, they go into the ground and remain there during the winter.

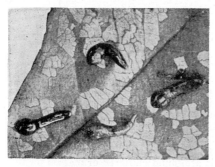

FIG. 15.68. Larvae of the pear-slug, *Caliroa cerasi*, and characteristic injury on leaf. About twice natural size. (*From Slingerland and Crosby, "Manual of Fruit Insects," copyright, 1915, Macmillan. Reprinted by permission.*)

Control Measures. The standard spray program for the control of curculio and fruit flies will control the pear-slug. Where special treatment is necessary, it may be controlled readily by spraying the foliage, 2 to 3 weeks after bloom, with DDT at 1 pound, parathion or EPN at 0.25 pound, or lead arsenate at 2 pounds with 2 pounds of hydrated lime to 100 gallons of spray. The slugs may be washed from the foliage of backyard trees with a strong stream of water.

References. State Ento. S.D. Cir. 27, 1918; Mich. Agr. Exp. Sta. Spcl. Bul. 244, 1933.

193. FORBES' SCALE[1]

Importance and Type of Injury. As already stated, the San Jose scale is of little or no importance on sour cherries. The Forbes' scale, which resembles the San Jose scale in general appearance, is often very abundant, although not seriously destructive to the tree. Grayish, thin, flaky scales are massed on the bark of the trunk and branches of the tree, sometimes completely covering the bark (Fig. 15.69). Examination with a microscope will show a raised reddish area in the center of each scale which will distinguish it from the San Jose scale. On apples, the crawlers may infest the fruits, causing a high percentage of culls.

Plants Attacked. Cherry, apple, apricot, pear, plum, quince, and currant.

Distribution. Throughout the United States east of the Rocky Mountains.

Life History, Appearance, and Habits. The winter is passed as a partly grown scale, somewhat resembling the San Jose scale, but the raised area in the center of the scale has a reddish or orange color. In Illinois, the first young appear in May, and are produced from eggs and by the birth of nymphs. There are from one to three generations each season.

Control Measures. This insect may be controlled by dormant or delayed dormant spraying with 2 per cent (superior) or 3 per cent (regular) mineral-oil emulsions or by spraying when the crawlers are active with parathion at 0.15 to 0.25 pound, malathion at 0.5 pound, or Guthion at 0.3 pound to 100 gallons of spray, as given for the control of the San Jose scale (page 707).

References. Twenty-second Rept. Ill. State Ento., p. 115, 1902; Mich. Agr. Exp. Sta. Spcl. Bul. 239, 1933; Jour. Econ. Ento. **46**:494 and 968, 1953.

[1] *Aspidiotus forbesi* Johnson, Order Homoptera, Family Coccidae.

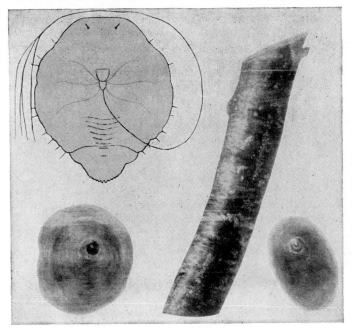

FIG. 15.69. Forbes' scale, *Aspidiotus forbesi*. Mature female removed from scale, upper left; female scale, lower left; male scale, lower right. All much enlarged. The infested twig, somewhat reduced. (*From Ill. Natural History Surv.*)

194 (163, 187, 191). PLUM CURCULIO

Most of the wormy cherries in many parts of the country are caused by the plum curculio. This insect is especially abundant on cherry trees grown in small orchards or in cities. The control measures for this insect are the same as those given for its control on the plum (page 766).

195. CHERRY FRUIT FLIES[1,2]

Importance and Type of Injury. There are two closely related flies which attack the cherry and cause wormy fruits. Cherries are somewhat misshapen and undersized, turning red ahead of the main crop, oftentimes with one side of the fruit partly decayed and shrunken or wrinkled and closely attached to the pit. Yellowish-white footless maggots, up to $\frac{1}{4}$ inch long and pointed at the head end, will be found in the flesh of the fruit (Fig. 15.70). The small ones are very hard to detect in raw fruits but become more visible and tend to sink to the bottom when broken fruits are boiled for a minute. Broken burrows extend through the fruit.

Trees Attacked. Cherry, pear, plum, and a number of species of wild cherries. The black cherry fruit fly prefers sour to sweet cherries.

Distribution. Northern United States and Canada.

Life History, Appearance, and Habits. The winter is passed in brown, capsule-like puparia in the soil. The pupae are in diapause, and approxi-

---

[1] The cherry fruit fly, *Rhagoletis cingulata* (Loew), Order Diptera, Family Trypetidae. Two distinct subspecies are known, *R. cingulata cingulata* (Loew), in eastern United States, and *R. cingulata indifferens* Curran, of the Pacific Northwest.

[2] The black cherry fruit fly, *Rhagoletis fausta* (Osten Sacken).

mately 150 days at 32 to 40°F. are required to give maximum emergence of the adults. The adult cherry fruit flies emerge from the soil about the middle of May to the middle of July, in the latitude of Yakima, Wash. They fly to cherry trees, where the females feed on exudates

from puncturing the leaves and fruits with the ovipositor. After mating, there is a preoviposition period of 5 to 6 days; then egg-laying takes place over a period of about 25 days, during which an average of 386 eggs may be laid. These are inserted through small slits cut with the ovipositor into the flesh of the fruit. The adult female (Fig. 15.71) is a little smaller than the house fly, of a general black color, with yellow margins on the thorax and with four white crossbands on the abdomen. The wings have blackish bands. After oviposition, the eggs hatch in 5 to 8 days and the young larvae mine directly into the surface of the cherry seed. They pass through three instars over an average of 11 days at 77°F., feeding on the flesh of the fruit. The last-instar larva forms one or two breathing holes to the surface of the fruit, and about 3 days later emerges and drops to the ground, where it seeks a pupation site in the upper 3 inches of soil. About 1 per cent of the flies emerge in August and September as a second generation, but the remainder over-winter as the pupa.

FIG. 15.70. Section of ripe cherry showing maggot of the cherry fruit fly at work. About 3 times natural size. (*From N.Y. (Cornell) Agr. Exp. Sta. Bul. 325.*)

The black cherry fruit fly has an almost identical life history, emerging about 1 week earlier than the cherry fruit fly. This fly does not have the white abdominal bands of the cherry fruit fly, and the blackish bands on the wings are more intense.

Control Measures. The cherry fruit flies may be controlled by spraying the fruit at the time the first adults are trapped, using a 6- by 8-inch yellow sticky-board baited with 7 grams of ammonium carbonate suspended below it in a small jar with perforated top. Two or three applications of malathion at 0.5 pound, parathion at 0.15 to 0.3 pound, Diazinon at 0.25 pound, methoxychlor or Perthane at 1 pound, or rotenone at 1.25 to 2.2 pounds to 100 gallons of spray are usually required at 10-day intervals, starting just after the blossom buds open.

FIG. 15.71. Adult of the cherry fruit fly, *Rhagoletis cingulata*. About 4 times natural size. (*From N.Y. (Cornell) Agr. Exp. Sta. Bul. 325.*)

Dusting with parathion at 1 pound, Perthane at 5 pounds, or rotenone at 3 pounds per acre has also been used. These applications protect the fruit by killing the adult flies. Cull cherries around canneries and picking stands should be destroyed immediately by heat or buried deeply.

The cherry fruit fly is attacked by several parasites, of which the braconid, *Opius ferrugineus*, is most important. Parasitism may reach 50 per cent of the flies breeding in the native wild cherries but is less than 3 per cent in cultivated cherries, where the larvae are much better protected. The black cherry fruit fly is parasitized by the ichneumonid, *Phygadeuon epochrae*.

References.   N.Y. (Cornell) Agr. Exp. Sta. Bul. 325, 1912; *Wash. Agr. Exp. Sta. Tech. Bul.* 13, 1954; *Calif. Dept. Agr. Bul.* 44, p. 77, 1955; *Ont. Dept. Agr. Bul.* 227, 1915; *Mich. Agr. Exp. Sta. Cir.* 131, 1930; *Jour. Econ. Ento.*, **26**:431, 1933; **28**:205, 1935; **45**:262, 1952; **46**:896, 1953; and **50**:584, 1957.

195A.   CHERRY FRUITWORM[1] AND LESSER APPLEWORM[2]

Importance and Type of Injury.   The larvae of these two closely related insects bore into cherry fruits, causing deformation and discoloration, fruit drop, and as many as 6 to 8 per cent cull fruits.   The lesser appleworm is also a pest of prunes and apples, where it eats out a blotched mine under the skin, which may weather away, causing a scabby area that disfigures the fruit.

Plants Attacked.   The cherry fruitworm attacks cherry and blueberry and wild chokecherry.   The lesser appleworm attacks cherry, apple, prune, and plum.

Distribution.   Throughout the deciduous fruit-growing areas of the United States.

Life History, Appearance, and Habits.   The cherry fruitworm winters as the full-grown larva in a silken cocoon, tunneled in the pruned stub of a dead twig, under bark, or debris on the ground.   It pupates in May, and the adult emerges after an average of 29 days.   The flattened circular eggs are laid on the cherry fruit or stems and hatch after about 7 days, when the young larvae bore into the fruit.   They tunnel about the stone for approximately 3 weeks, until full-grown, and then leave through large exit holes, to pupate.   The mature larva is pinkish and about $\frac{1}{3}$ inch long, and the adult moth is dark gray and about $\frac{1}{3}$ inch in wingspread.   There is one and sometimes two generations a year.   The life history of the lesser appleworm is very similar, the insect wintering as the full-grown larvae in a cocoon under debris on the ground.   Pupation requires 15 to 20 days, and the adult emerges in early May, in Oregon.   The adult moth has a wing spread of $\frac{7}{16}$ inch and closely resembles the codling moth.   The eggs are laid on the leaves of the cherry and hatch in 5 to 10 days, the larvae immediately boring into the fruit.   They mature in 18 to 24 days and drop to the ground to pupate.   There are two and sometimes three generations a year.

Control Measures.   These fruitworms may be controlled by spraying with methoxychlor at 1 pound to 100 gallons of spray, applied at 4 weeks after petal fall.   The lesser appleworm is normally controlled by the codling moth spray schedule on apples.

References.   Cherry fruitworm: *Can. Dept. Agr. Ento. Branch* C79, 1931; *Jour. Econ. Ento.*, **45**:800, 1952.   Lesser Appleworm: *N.Y. (Cornell) Agr. Exp. Sta. Bul.* 410, 1922; *Jour. Econ. Ento.*, **46**:163, 1953.

G. APRICOT INSECTS

The apricot is injured by practically the same insects as the plum, and the same control measures will apply on this tree.   Plum curculio is often very destructive to apricot, and, where this fruit is grown, special attention should be given to this insect.   San Jose scale is also often very serious, especially on the younger trees.

H. GRAPE INSECTS

FIELD KEY FOR THE IDENTIFICATION OF INSECTS INJURING THE GRAPE

A. *External chewing insects that eat holes in the leaves, buds, or fruit:*
   1. Fawn-colored, long-legged, slender-bodied beetles, about $\frac{1}{2}$ inch long, feed on the blossoms, newly set grapes, and leaves.   Most abundant in sandy areas and during the first 2 weeks after blooming................*Rose chafer*, page 772.

[1] *Grapholitha packardi* Zeller, Order Lepidoptera, Family Olethreutidae.
[2] *Grapholitha* (= *Laspeyresia*) *prunivora* Walsh.

2. Small, active, jumping beetles of a metallic greenish-blue color, feed on the unfolding leaves of grape in the spring.   Light-brown black-spotted grubs, the largest about ⅓ inch in length, feed in company with the beetles and, later, on the opened leaves..*Grape flea beetle*, page 774.

3. Small, chain-like holes are eaten in the leaves by brown to grayish beetles about ⅓ inch long; especially abundant about 2 weeks after grapes bloom..........
....*Grape rootworm*, page 780.

4. Leaves skeletonized; all the green eaten and buds and fruits devoured by metallic-green to coppery-red beetles, ⅓ to ⅝ inch long, with white spots at the tip of the abdomen beyond the wing covers............*Japanese beetle*, page 749.

5. Greenish-gray to purplish caterpillars up to ½ inch in length, with brown heads, feed on or in the grape berries, under protecting webs, and often attach bits of leaves to the berries............................*Grape berry moth*, page 775.

6. Buds and new growth, especially of vines growing in or near grass sod, eaten by plump, nearly bare, caterpillars, with 6 slender legs and 5 pairs of prolegs, somewhat ornamented with spots or stripes, which climb into the vines to feed at night, and hide on or in the soil during the day..........................
Climbing cutworms, Lampra alternata Grt., and others (see *Mich. Agr. Exp. Sta. Spcl. Bul. 239,* 1933).

B. *Insects that suck sap from the leaves:*

1. Vines lacking in vigor, sickly in appearance, the foliage pale, with many tiny whitish spots over the leaves.   Very slender yellowish bugs, with red marks on the wings, about ⅛ inch long by a fourth as wide, and their active light-colored nymphs, suck the sap from the underside of leaves.   The nymphs run sidewise as readily as forward.............................*Grape leafhopper*, page 777.

C. *Insects that produce small galls or swellings on the leaves:*

1. Vines lack vigor, often dying.   Rounded, irregular galls, about half the size of a pea, on the leaves, sometimes covering them.   The galls open on the underside of the leaves, and the inside is lined with many, small, wingless, pale-yellow aphids (see also *D*,1)............................*Grape phylloxera*, page 779.

D. *Subterranean insects that attack the roots of the vines:*

1. Vines lack vigor.   Knot- or gall-like swellings on the roots are covered with aphids similar to *C*,1, causing the roots to rot off...*Grape phylloxera*, page 779.

2. Vines show a lack of vigor, the foliage yellowing, and little growth being made. White grubs, about ½ inch long, with brown heads, eat off the small, fibrous roots and score the bark of the larger roots during late summer (see also *A*,3)
....*Grape rootworm*, page 780.

3. Whitish caterpillars with brown heads, up to 1¾ inches long with crochets on the prolegs in 2 transverse rows, burrow in the roots eating out the wood and inner bark, most of them a foot or more from the base of the vine, killing the roots and disastrously weakening the vines................................
Grapevine rootborer, Memythrus polistiformis Harris (see *W. Va. Agr. Exp. Sta. Bul.* 110, 1907).

Among the other pests which sometimes attack grapes are the Pacific mite (page 717), bean thrips (page 742), citrus thrips and flower thrips (page 800), grape mealybug (page 807), cottony maple scale (page 837), two-spotted mite (page 616), red-banded leaf roller (page 701), and European fruit lecanium (page 753).

196. ROSE CHAFER[1]

Importance and Type of Injury. The leaves of grape, and especially the blossoms, are eaten by gray or fawn-colored, long-legged, slender beetles about ½ inch long (Fig. 15.72,*a*).   Newly set grapes are eaten and bunches of grapes nearly ruined.   The insects are most abundant on the grape for the first 2 or 3 weeks after bloom, but will be found in smaller

[1] *Macrodactylus subspinosus* (Fabricius), Order Coleoptera, Family Scarabaeidae. The western rose chafer, *Macrodactylus uniformis* Horn, is very similar in appearance and habits and causes similar injury in Arizona and New Mexico.

numbers for the next week or 10 days.   The beetles also eat ripening cherries and riddle the buds and leaves of roses.   Poultry eating these beetles may be poisoned and killed.

Plants Attacked.   Grape, apple, peach, cherry, pear, strawberry, rose, hydrangea, peony, blackberry, raspberry, Virginia creeper, corn, bean, beet, pepper, cabbage, poppy, hollyhock, mullein, clover, small grains and grasses, and a great variety of other plants, trees, and shrubs.

Distribution.   The insect is generally distributed over the eastern United States and southeastern Canada, and westward to Colorado and Texas.   Its injuries are more severe in sandy areas.

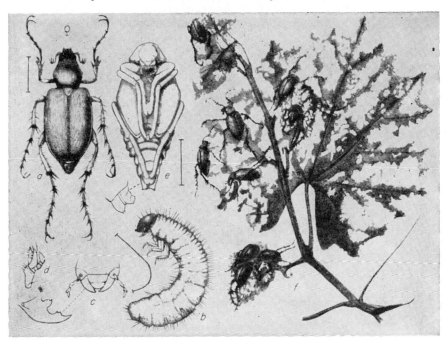

Fig. 15.72. The rose chafer, *Macrodactylus subspinosus.*   *a,* Adult; *b,* larva; *c* and *d,* mouth parts; *e,* pupa, *f,* injury to leaves and blossoms of grape, with beetles at work.   *a, b, e,* Enlarged, the lines showing natural size; *c* and *d,* more enlarged; *f,* slightly reduced. (*From U.S.D.A., Farmers' Bul.* 721.)

Life History, Appearance, and Habits.   The winter is passed in the larval or grub stage.   The larva (Fig. 15.72,*b*) closely resembles that of the common white grub but is somewhat more slender and much smaller. When full-grown, it is about ¾ inch in length.   The larvae are found in uncultivated land, especially in sandy areas, to a depth of 10 to 16 inches. In the spring the nearly full-grown larvae work their way toward the surface of the ground and feed for a short time on the roots of grasses, grains, weeds, and other plants.   They pupate during May, and remain in this stage (*e*) for about 3 weeks, emerging as adults at about the time when the grapes come into bloom.   The adult (*a,f*) is a very ungainly beetle, nearly ½ inch in length, with reddish-brown head and thorax and the undersurface of the body blackish in color.   The whole body is covered with small yellow hairs, which give the beetle a fawn-colored appearance.   These beetles feed chiefly on the surface of the plants, and the females, after mating, deposit their eggs in groups of from 6 to 25, at a depth of about 6 inches in the soil.   While the eggs are grouped, each egg is laid in a separate pocket in the soil.   The eggs hatch in 1 to 2 weeks, and the young larvae feed on the roots of grasses and other plants for the remainder of the summer, going down in the soil on the approach of cold weather.

Control Measures.   The most important method of controlling this insect is thoroughly to cultivate the areas about the vineyard where the eggs may have been deposited.   Very few grubs of the rose chafer have been found in land where cultivated crops, such as potatoes and corn, are being grown.   Cultivation is most effective in killing the insects if carried on during May and early June, when the rose chafer is in the pupal stage.   In order to secure the maximum protection from cultivation, it is necessary that all grape growers unite in keeping the areas, surrounding vineyards, under cultivation.   Grapes may be protected by spraying with DDT or methoxychlor at 1 pound to 100 gallons of spray.   It is important to control the rose chafer when it first appears, because after it becomes established in an orchard or vineyard it is very difficult to drive away or kill.   Choice flowers and fruits may be protected by covering with cheesecloth during the 4 or 5 weeks when the adults are present.

References.   Fifty-third Rept. Ento. Soc. Ont., p. 60, 1922; *U.S.D.A., Bur. Ento. Bul.* 97, Part III, 1911; *Boyce Thompson Inst. Contrib.* 4, 1932; *U.S.D.A., Farmers' Bul.* 1666, 1931.

FIG. 15.73. Grape flea beetle, *Altica chalybea.*   Natural size and enlarged. (*From N.Y. (Cornell) Agr. Exp. Sta. Bul.* 157.)

197. GRAPE FLEA BEETLE[1]

Importance and Type of Injury.   Small jumping beetles, about ⅕ inch long, having a dark metallic greenish-blue color (Fig. 15.73), feed on the buds of the grape just as they are starting to unfold in the spring. The buds are eaten off, and the newly opening foliage presents a ragged, tattered appearance.   Light-brown, black-spotted grubs about ⅓ inch long will be found feeding on the newly opening leaves, along with the beetles.

Plants Attacked.   Grape, plum, apple, quince, beech, elm, and Virginia creeper.

Distribution.   Eastern two-thirds of the United States.

Life History, Appearance, and Habits.   The adult grape flea beetles pass the winter in hibernation in or near the grape vineyards under any shelter which they can find.

[1] *Altica chalybea* (Illiger), Order Coleoptera, Family Chrysomelidae.

They become active fairly early in the spring and at once start feeding on the opening grape leaves.   The females soon start laying their eggs, which are deposited in masses underneath the loose bark of canes, or occasionally on the upper surface of the leaves. The eggs are light yellow in appearance and are fairly conspicuous.   Small dark-brown to blackish, spotted grubs hatch from these eggs and feed on the surface of the leaves and on the clusters of the blossom buds.   On becoming full-grown, the larvae, which are then lighter brown, with regular rows of blackish spots (Fig. 15.74), drop or crawl to the ground, enter the soil, and there change to the beetle stage.   The adult beetles of the summer generation make their appearance during July and August, and feed on the grape leaves until the approach of cold weather.   They rarely cause much injury during this part of the season.

FIG. 15.74. Larvae of the grape flea beetle, dorsal and lateral views.   About 4 times natural size.   (*From N.Y. (Cornell) Agr. Exp. Sta. Bul.* 157.)

Control Measures.   Spraying with DDT at 1 pound to 100 gallons of spray, applied as soon as the beetles make their appearance on the vines in the spring, is somewhat effective as a control measure, but because of the limited amount of foliage which can be covered with insecticide at this time, it is often hard to protect the vines from injury by this insect.   When the grubs are present on the leaves, the same spray will control them effectively.   On a few backyard vines, the insect may be controlled by spreading a piece of cloth, dipped in oil, under the vines and jarring the beetles upon it.   Fall cleanup of the vineyards, as recommended for the control of the grape berry moth, will also help in keeping down the numbers of this insect.

References.   U.S.D.A., *Dept. Bul.* 901, 1920; N.Y. *(Geneva) Agr. Exp. Sta. Bul.* 331, 1910; *Mich. Agr. Exp. Sta. Spcl. Bul.* 139, 1933.

198. GRAPE BERRY MOTH[1]

Importance and Type of Injury.   This insect is almost universally present wherever grapes are grown, either as a few vines in a back yard or in extensive vineyards.   The grape berries are webbed together, turning dark purple in color, and drop from the stems when the grapes are about the size of garden peas.   Small holes are eaten in the nearly ripened grapes, the sides of which are attached by a light web to a bit of leaf, or to adjoining berries   Small silken cocoons lie in small, semicircular flaps cut in the grape leaves, folded over, and held together by a light web. When this insect is abundant, it will often destroy as much as 60 to 90 per cent of the fruit in unsprayed vineyards.

Plants Attacked.   Cultivated and wild grapes.

Distribution.   Troublesome chiefly in the northeastern fourth of the United States and southeastern Canada, but extending westward to Wisconsin and Nebraska, and southward to Louisiana and Alabama.

Life History, Appearance, and Habits.   This insect passes the winter in grayish silken cocoons, nearly always folded in fallen grape leaves. Occasionally the cocoons may be attached to the loose scales of bark or in rubbish on the ground.   In the late spring, shortly before or just after the grape has bloomed, a grayish or grayish-purple moth, about ½ inch across the wings (Fig. 15.75), emerges from the cocoon, and the female lays an average of 20 flattened, circular, cream-colored eggs at dusk, upon

[1] *Paralobesia* (= *Polychrosis*) *viteana* (Clemens), Order Lepidoptera, Family Olethreutidae.

the fruit, stems, flower clusters, or newly forming grape berries. The little worms hatching from these eggs spin a silken web wherever they go and thus web together the fruit clusters, or parts of clusters (Fig. 15.77). They feed on the grape berries and each worm of this generation will

FIG. 15.75. Grape berry moth, *Paralobesia viteana*, adult. Natural size and enlarged. (*From N.Y. (Cornell) Agr. Exp. Sta. Bul.* 223.)

usually destroy a number of grapes. On becoming full-grown, the worms are ⅓ to ½ inch long, greenish gray with brown heads (Fig. 15.76). Each cuts out and folds over a little flap of leaf and within this spins a cocoon. The moths of the second generation emerge during July and deposit their eggs on the grape berries as did those of the first generation. While a single worm may find sufficient food in one berry to complete its development, it often goes from one grape to another, especially where the berries are touching, and thus may destroy three or four grapes before it completes its growth. Development, from the deposition of the eggs to the emergence of the adults, commonly averages about 5 weeks. When full-grown, they again go to the leaves where they spin their cocoons, pupate, and remain in this stage during the winter. There are two generations normally in the northern states and three farther south.

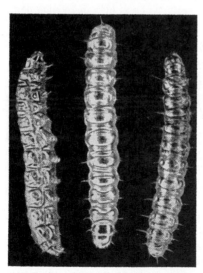

FIG. 15.76. Grape berry moth. Larvae or caterpillars. About 4 times natural size. (*From N.Y. (Cornell) Agr. Exp. Sta. Bul.* 223.)

Control Measures. This insect may be controlled by spraying at petal fall and 1 to 2 weeks later, when the eggs of the second brood are found, with DDT at 0.75 pound, methoxychlor at 1 pound, parathion at 0.15 pound, EPN at 0.25 pound, or Diazinon at 0.5 pound to 100 gallons of spray. To cover the grapes thoroughly, it is necessary to use a high-pressure sprayer that will break the spray into very fine particles. In addition to the spraying, a partial control of this insect may be accomplished by thoroughly cleaning up around the grape vineyards and raking and burning the fallen leaves during the fall or winter or by plowing the vineyards and adjacent land as soon as the frost is out of the ground in the spring.

References. N.Y. (Cornell) Agr. Exp. Sta. Bul. 223, 1904; U.S.D.A., Dept. Bul. 550, 1917; Del. Agr. Exp. Sta. Buls. 176, 1932, and 198, 1936; Mich. Agr. Exp. Sta. Spcl. Bul. 239, 1933; Jour. Econ. Ento., 40:845, 1947; 43:76, 1950; 44:256, 1953; 45:101, 1952; 46:77, 85, 1953; and 50:455, 1957.

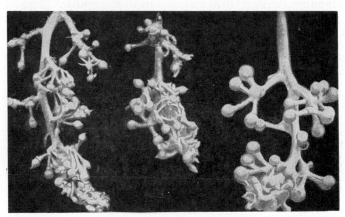

Fig. 15.77. Work of spring generation of grape berry moth larvae among blossoms and young fruits in June.   Natural size.   *(From N.Y. (Cornell) Agr. Exp. Sta. Bul. 223.)*

199. Grape Leafhoppers[1]

Importance and Type of Injury.   There is no insect which attacks the grape that is so universally present, year in and year out, as the grape leafhopper.   The quantity of fruit harvested may be reduced as much as 30 per cent and the quality greatly lowered.   Very small whitish spots appear over the grape leaves (Fig. 1.4,*B*), and they later become brown and shriveled.   The entire leaf becomes pale greenish yellow, the vines show very little vigor, and the foliage presents a general sickly appearance.   Numerous small, pale, red-flecked, very active insects are found sucking sap from the undersides of the leaves.

Plants Attacked.   Grape, Virginia creeper, apple, and many other plants.

Distribution.   Throughout the United States.

Life History, Appearance, and Habits.   The adult grape leafhoppers (Fig. 15.78, right) are of a pale yellowish color, with red markings on the wings.   They pass the winter among the fallen grape leaves, grasses, or other shelters in the vicinity of the vineyard.   The hoppers are about ⅛ inch or less in length by one-fourth as wide.   They become active about the time the grape leaves are half-grown, and fly to the leaves, on which they feed by sucking the sap from the underside.   After feeding 2 or 3 weeks, the female hoppers begin to lay their eggs.   These eggs, which are very small, are pushed into the tissue of the leaf.   Each female lays about

---

[1] *Erythroneura comes* (Say), Order Homoptera, Family Cicadellidae.   In northwestern Ohio and southwestern Michigan, at least, the three-banded leafhopper, *Erythroneura tricincta* Fitch, is more injurious than *E. comes*.   *Erythroneura elegantula* (Osborn) in central California and *E. variabilis* Beamer in southern California are the most injurious western species.

100 eggs, and these hatch after 8 to 20 days, depending upon the temperature, into pale-green or greenish-white nymphs (Fig. 15.78, left), which are wingless, but extremely active, and which feed by sucking the sap, remaining almost entirely on the undersides of the leaves.   They become full-grown in from 18 days to 5 weeks, depending largely on the temperature of the season.   There are probably two generations of the insect in most of its range, with a third generation in the South.   The nymphs of the last generation become full-grown during September, and the adults seek the shelters above described about the time of the first frost.

Control Measures.   The grape leafhoppers may be controlled by pre- and post-bloom spraying with DDT at 1 pound, Sevin at 0.5 pound, malathion at 0.5 pound, Diazinon at 0.125 pound, Trithion at 0.25 pound,

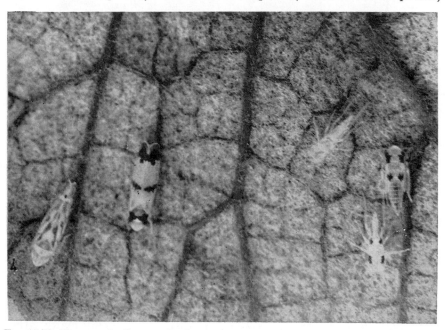

Fig. 15.78. The grape leafhopper, *Erythroneura comes*.   About 7 times natural size.   (*From U.S.D.A., Dept. Bul.* 19.)

or demeton (Systox) at 0.25 pound per acre.   Dusting with DDT at 2.5 pounds, Sevin at 1.25 pounds, malathion at 1 pound, or Trithion at 0.75 pound per acre is also effective and may be preferred in postbloom applications to avoid spotting the fruit.   The fungus disease, *Entomophthora sphaerosperma*, is the most important natural enemy.   As the adult leafhoppers, like many of the other grape insects, pass the winter in shelters afforded by trash, weeds, or grasses, thoroughly cleaning up around the vineyards is of value in controlling these insects and should be done as a general practice.

References.   Calif. Agr. Ext. Serv. Cir. 72, 1933; *Ariz. Agr. Exp. Sta. Cir.* 146, 1924; *Rept. Ento. Soc. Ont. for* 1922, p. 48, 1922; *N.Y. (Geneva) Agr. Exp. Sta. Buls.* 344, 1922, and 738, 1949; *U.S.D.A., Dept. Bul.* 19, 1914; *Del. Agr. Exp. Sta. Bul.* 198, 1936; *Calif. Agr. Exp. Sta. Cir.* 365, 1946; *Jour. Econ. Ento.*, **40**:195, 487, 1947; **47**:238, 1954; and **50**:411, 1957.

200. Grape Phylloxera[1]

Importance and Type of Injury. This aphid is the most destructive grape pest known in the western United States and Europe. Fortunately for the grape growers of the eastern United States, it practically never causes serious damage in this section of the country. Within 25 years after this insect was introduced into France from America, about 1860, it had destroyed nearly one-third of the vineyards in that country— more than 2,500,000 acres. Small galls about the size of half a pea form on the leaf surface (Fig. 15.79), sometimes so numerous as practically to cover the entire leaf. The galls are open on the underside of the leaf. They contain many small, wingless, yellowish aphids. This form rarely occurs in the west. Numerous knots or galls form on the grape roots and rotting of the roots, yellowing of the grape foliage, and general decrease in

Fig. 15.79. Phylloxera galls on wild grape leaf. About natural size. (*From Slingerland and Crosby, "Manual of Fruit Insects," copyright, 1915, Macmillan. Reprinted by permission.*)

vigor, or the death of the vines, result from injury by this insect. In western vineyards, infestation by the root-infesting form usually results in death of the vines within 3 to 10 years.

Plants Attacked. Grape.

Distribution. General in North America where grapes are grown.

Life History, Appearance, and Habits. The life history of this insect is extremely complicated, as there are four distinct forms of adults, besides the immature stages. Only a brief outline of its life history can be given. The winter is passed both as eggs attached to the canes of the grape plants and in the form of yellowish aphids on the nodules or galls on the grape roots (Fig. 15.80). The root-infesting forms become active, feeding on the roots as soon as growth starts in the spring. The eggs on the canes hatch in the spring after the foliage of the grape has come out, and the

---

[1] *Phylloxera vitifoliae* (Fitch), Order Homoptera, Family Phylloxeridae.

yellow aphids developing from these eggs migrate to the leaves where they begin feeding. This injury to the leaf causes the formation of galls. As soon as the aphids have become full-grown, they give birth to living young inside the galls, and these young shortly begin forming other galls, several generations being passed on the leaf. Some of these leaf-inhabiting aphids drop to the ground and burrow beneath the soil to the roots, where they cause the formation of the root galls and where they can live for a number of generations. Toward the fall of the year, winged forms are produced on the grape roots which leave the ground and lay eggs on the vines. These eggs hatch into males and true females, mating takes place, and each fertilized female lays a single egg, which remains on the cane during the winter.

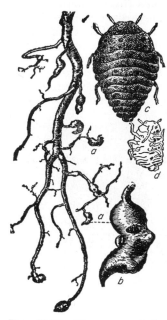

FIG. 15.80. The grape phylloxera, *Phylloxera vitifoliae*. *a*, Galls on grape roots caused by feeding of the insects; *b*, gall, much enlarged, with aphids feeding; *c*, adult aphid, greatly enlarged; *d*, shed skin of the same. (*From Herrick, "Manual of Injurious Insects."*)

Control Measures. In the eastern United States, the form of phylloxera which causes the galls on the leaves is very abundant, especially on some of the native grapes. The root-infesting form, while present, is very rare in the East and does not cause any serious damage but is the only form found in the Pacific Coast states. The insect is a native of eastern United States, and the grapes growing in that section have acquired practical immunity to its attack. The best-known and most effective remedy for combating this insect is the grafting of European grapes on resistant rootstocks native to eastern United States. This practically does away with any injury by the insect. Nearly all the grapes sold in nurseries are grafted on native rootstocks. In certain parts of California and in Europe, where the European rootstocks are used, the insect is controlled by flooding the vineyards at certain times during the season, also by soil fumigation with carbon bisulfide. Quarantine measures are important in limiting the spread of the pest. In Europe, to kill all the phylloxera in the soil of an infested vineyard, before setting new vines, chemicals are sometimes injected into the vines, killing vines and insects together.

References. U.S.D.A., Dept. Bul. 903, 1921; *U.S.D.A., Tech. Bul.* 20, 1928; ESSIG, "Insects of Western North America," pp. 225–227, Macmillan, 1926.

201. GRAPE ROOTWORM[1]

Importance and Type of Injury. Vines show a lack of vigor, the leaves turn yellow, and little new growth is made. An examination of the roots

[1] *Fidia viticida* Walsh, Order Coleoptera, Family Chrysomelidae. A very similar species, the western grape rootworm, *Adoxus obscurus* (Linné), occurs in the western states.

will sometimes show small whitish grubs eating off the small feeding roots and gouging out numerous channels in the bark of the larger roots.   Grayish-tan beetles feed upon the leaves, throughout the summer.   The entire leaf is not eaten, but a series of small holes are made through the leaf in chain-like rows.

Plants Attacked.   Grape and related wild plants.

Distribution.   Eastern United States, except extreme north and south.

Life History, Appearance, and Habits.   The grape rootworm passes the winter as small curved-bodied brown-headed grubs of various sizes, up to ½ inch long.   During the winter these worms make their way deep into the soil and will be found 2 feet or more below the surface of the ground.   As the soil warms in the spring, the grubs migrate back to within 1 or 2 inches of the surface and excavate small cells, in which they change to very soft white pupae in late May and June. About 2 weeks after the grapes bloom, the insects emerge from these cells as small brown or grayish beetles.   These beetles are about ¼ inch long and are rather chunky in appearance (Fig. 15.81). They feed on the upper surface of the grape leaf, producing the effect on the foliage above described.   Shortly after commencing to feed, the females deposit their yellowish eggs, in masses averaging about 20 to 30 in a cluster.   The eggs are attached to the grape canes, usually under the loose bark scales, although sometimes not so protected.   The beetles do not all come out at once, and the period of their feeding and egg-laying may extend over a month or 6 weeks. The eggs hatch after a short time, and the young grubs drop to the ground, where they burrow into the soil until they encounter grape roots.   They at once start feeding on these roots, cutting off the slender feeding roots and channeling or gouging the bark of the larger roots.   On the approach of cold weather they work their way deeper into the soil, where they remain during the winter.   There is one generation each year.

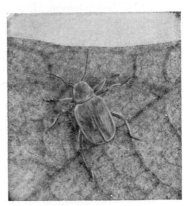

Fɪɢ. 15.81. Adult of the grape rootworm, *Fidia viticida*, on leaf.   About 3 times natural size.   (*From N.Y.* (*Geneva*) *Agr. Exp. Sta. Bul.* 453.)

Control Measures.   As the beetles of the grape rootworm feed extensively on the upper surface of the leaves, they may be controlled by spraying with 0.75 pound of DDT to 100 gallons of spray, applied as soon as any of the feeding punctures are noticed on the leaves.   Thorough spraying of all grape foliage is essential in the control of this insect.   Bordeaux mixture alone has no marked repellent effect on this beetle, and vineyards sprayed with bordeaux, without the addition of the DDT, have in a number of cases been very seriously damaged by the grape rootworm.   Sprays for the grape berry moth will also destroy these adults.   Intensive shallow cultivation of the vineyard soil, up to the time of emergence of adults in late June, will destroy many of the pupae.

References.   U.S.D.A., *Bur. Ento. Bul.* 68, Part VI, 1908; *N.Y.* (*Geneva*) *Agr. Exp. Sta. Bul.* 453, 1918; *N.Y.* (*Geneva*) *Agr. Exp. Sta. Bul.* 519, 1924; *Pa. Agr. Exp. Sta. Bul.* 433, 1926; *Mich. Agr. Exp. Sta. Spcl. Bul.* 239, 1933.

I. CURRANT INSECTS

FIELD KEY FOR THE IDENTIFICATION OF INSECTS INJURING CURRANTS AND GOOSEBERRIES

1. Greenish many-legged worms with black-spotted bodies, up to 1 inch in length, feed on the edges of the currant leaves. Injury usually starts about the time the plants come into full foliage and is first noticeable in the center of the bushes. Black-bodied, wasp-like sawflies, about ¼ inch in length, deposit pearly-white elongated eggs, in rows, on the vines and midrib on the underside of the currant leaves................................*Imported currantworm*, page 782.

2. Currant leaves show bright-red, cupped, or wrinkled areas. Numerous, small greenish-yellow, somewhat flat-bodied aphids on the undersides of the curled and distorted leaves. Where injury is severe the leaves drop from the plants.. ....*Currant aphid*, page 783.

3. Scales up to ⅙ inch across, rounded in outline, with a grayish, nipple-shaped projection in the center, are clustered or distributed singly over the bark of the currant. Lemon-yellow, sacklike, sucking insects on the bark under the scale. Infested plants often so heavily covered as to entirely coat the bark and kill the plants......................................*San Jose scale*, page 784.

4. Small whitish areas on the upper sides of the currant leaves. Leaves turn brown and drop. New growth of the currant wilts and dies. During late spring, bright-red wingless bugs, with black dots on the thorax and yellow stripes on the sides, suck the sap from the leaf and new twig growth. Yellowish adult bugs, about ⅓ inch long by one-half as wide, with 2 distinct black stripes on each wing cover, feed on the currant leaves and new growth during late summer......................................*Four-lined plant bug*, page 784.

5. Currant canes put forth undersized foliage in the spring, the tips of the canes often being dead and sometimes broken off. Burrows running the entire length of the cane, mainly in the pith, and containing yellowish grub-like larvae, about ½ inch long. Small round holes bored through the sides of the cane and in the early spring covered with a silken web. Yellow clear-winged moths fly very actively about the currant bushes in the early summer................... ....*Currant borer*, page 785.

6. Fruits of currants and gooseberries turn red and drop from the plants. A small white maggot, up to ¼ inch long, feeds inside of the berry................... *Currant fruit fly, Epochra canadensis* (Loew) (see *Wash. Agr. Exp. Sta. Bul.* 155, 1938).

202. IMPORTED CURRANTWORM[1]

Importance and Type of Injury. This insect is present on practically every currant bush each season, and control measures should be a part of the regular routine of currant growing. Many-legged, smooth, greenish worms, with numerous black spots over their bodies, feed on the edges of the currant leaf (Fig. 15.82). When disturbed, the worms raise the front and hind parts of their bodies from the leaf. When they are abundant, the leaves are stripped from the currant, the injury usually starting in the thick foliage near the center of the plant.

Plants Attacked. Currant and gooseberry.

Distribution. General over the United States and southern Canada. Imported from Europe about 1857.

Life History, Appearance, and Habits. The winter is passed in a small capsule-like cocoon on or near the surface of the ground in the larval or pupal stage. The adult sawflies (Fig. 15.82) are black, about ⅓ inch long, with the abdomen marked with light yellow. They deposit their

[1] *Nematus* (= *Pteronidea*) *ribesii* (Scopoli), Order Hymenoptera, Family Tenthredinidae.

white, flattened, shining eggs in rows on the veins and midribs of the underside of currant leaves shortly before the plant comes into full foliage (Fig. 15.82, *upper leaf*). The worms hatch just about the time that the currant leaves are full-grown and feed along the margins of the leaves, consuming the entire leaf as they go. They feed for from 2 to 3 weeks, when they go to the ground and transform within their cocoons to the pupal stage. The second generation of the insect is usually not so numerous as the first and appears on the vines during late June or July. A partial third generation occurs in the South. The larvae of the later generations construct cocoons in the soil in which they pass the winter.

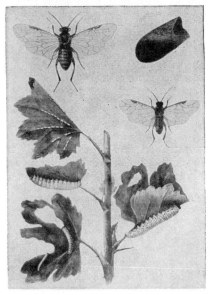

FIG. 15.82. Imported currantworm, *Nematus ribesii*. Adult female at upper left, empty cocoon at upper right, male below it.   Characteristic strings of eggs on veins on underside of upper leaf. Larvae of several instars feeding below. (*From Ill. Natural History Surv.*)

Control Measures. This insect is very susceptible to insecticides and has traditionally been controlled with sprays or dusts of hellebore (page 336). When the worms are present on the plants before the fruit is set they may be controlled by 3 to 4 pounds of lead arsenate to 100 gallons of spray. However, this material should not be applied to the fruits, and control during the latter part of the season may be obtained by spraying with malathion at 2 pounds or rotenone at 0.25 to 0.3 pound per acre, or by a 1 per cent rotenone dust.

References. *S.D. Tenth Rept. Ento.*, p. 26, 1919; *Can. Ento.* **52**:106, 1920; *Cornell Univ. Ext. Bul.* 306, 1934.

203. CURRANT APHID[1]

Importance and Type of Injury. Currant leaves become crinkled or cupped. The leaf surface becomes bright red in color, just above the points where the cupping occurs. Numerous, small, greenish-yellow, somewhat flat-bodied aphids are found on the undersides of the curled and distorted leaves, most numerous inside the cups.

Plants Attacked. Currants and sometimes gooseberry.

Distribution. Throughout the United States and Canada.

Life History, Appearance, and Habits. The currant aphid goes through the winter in the form of shining black eggs, which are found on the canes of the currant plants, particularly on the new growth. The young aphids hatch from these eggs soon after the leaves appear in the spring. They crawl to the leaves and suck the sap from the underside. Their feeding soon causes the distortion of the leaf described above. If very numerous,

[1] *Capitophorus* (= *Myzus*) *ribis* (Linné), Order Homoptera, Family Aphidae. *Aphis varians* Patch also attacks currants.

the infested leaves will drop from the plants. There are a number of generations each season, all those appearing during the summer consisting of females which give birth to living young as soon as they have become full-grown. Winged females migrate to certain weeds and reproduce during the summer, their offspring returning to the currants in fall to join those wingless strains which remained upon the currant all summer. In the fall of the year these females give birth to males and females, which mate and deposit their eggs on the canes where the insect passes the winter.

Control Measures. The aphids may be controlled by thoroughly spraying the undersides of the leaves early in the season with 40 per cent nicotine sulfate at 0.4 quart or malathion at 2 pounds per acre. Dusting with a 2 per cent nicotine dust also is very effective for the control of this insect. In spraying or dusting, care must be used to get the material on the underside of the leaves and up into the ridges and cup-shaped depressions caused by the feeding of the aphids.

References. N.Y. (*Geneva*) Agr. Exp. Sta. Bul. 517, 1924; N.Y. (*Geneva*) Agr. Exp. Sta. Bul. 139, 1897.

204 (149, 172, 179, 189). SAN JOSE SCALE

The San Jose scale is a serious pest of currants and will soon kill out currant bushes if a plantation becomes infested and no treatment is given. It may be controlled by dormant applications of lime-sulfur at 24 gallons or mineral-oil emulsion at 4 gallons per acre.

205. FOUR-LINED PLANT BUG[1]

Importance and Type of Injury. Where this insect is feeding on the leaves of currants, distinct white or dark-colored spots, $\frac{1}{16}$ to $\frac{1}{8}$ inch in diameter, looking much like fungus disease spots, appear on the upper sides of the leaves. If the insect is abundant, these areas come together and the leaf turns brown and usually drops. New growth of the currant will sometimes wilt where this bug has fed upon it.

Plants Attacked. General; gooseberry, rose, and many other shrubs, garden flowers, and herbaceous plants.

Distribution. General east of the Rocky Mountains.

Life History, Appearance, and Habits. This insect passes the winter in the form of slender white eggs, about $\frac{1}{6}$ inch long, which are inserted in slits in the canes of currants and other plants. Several eggs are generally laid at one place. They are forced into the cane at right angles to the long axis of growth, the tips of the eggs usually protruding from the cane.

FIG. 15.83. Four-lined plant bug, *Poecilocapsus lineatus*. Adult, natural size and enlarged. (*From Sanderson and Peairs, "Insect Pests of Farm, Garden and Orchard," Wiley.*)

The eggs hatch from May to the latter part of June. The bright-red to orange nymphs have black dots on the thorax and, in the last stage, a yellow stripe on each side of the wing pads. They feed by inserting their beaks in the leaves and new twig growth and sucking the sap. They become full-grown in about a month to 6 weeks. The adult insects (Fig. 15.83) are $\frac{1}{4}$ to $\frac{1}{3}$ inch long, of a general greenish-yellow color, with four distinct black stripes down the wing covers on the back. In the fall the female insects

[1] *Poecilocapsus lineatus* (Fabricius), Order Hemiptera, Family Miridae.

lay their eggs in the canes as above described. There is but one generation of the insect each year.

Control Measures. As these insects are exceedingly active, they are somewhat hard to control. The best method is to spray with a contact poison such as 40 per cent nicotine sulfate, ½ pint, soap, 1 pound, in 25 gallons water, or derris or pyrethrum with a good wetting agent, in May or early June before the adults appear, or to dust with a 4 per cent nicotine or 1 per cent rotenone dust.

References. N.Y. (Cornell) Agr. Exp. Sta. Bul. 58, 1893; *Mo. Agr. Exp. Sta. Bul.* 342, pp. 19, 20, 1934.

206. CURRANT BORER[1]

Importance and Type of Injury. This insect is of European origin but is now well distributed in America. It is almost sure to cause trouble in both commercial and garden currant plantations. Canes have yellowish undersized foliage appearing on them in the spring. Such canes usually die within 2 or 3 weeks. Dead canes usually have the tips broken off showing a burrow running nearly the entire length of the cane, partly in the pith and partly in the wood.

Plants Attacked. Gooseberry, black elder, sumac, and currant. The insect is more destructive to the black than to the red currant.

Distribution. Throughout North America.

Life History, Appearance, and Habits. This insect passes the winter as nearly full-grown yellowish borers, or larvae, about ½ inch long, which will be found inside the canes usually a short distance above the ground (Fig. 15.84). The larvae feed a little in the spring and eat an exit hole through the side of the cane which they cover with a silken web. They transform to the pupal stage inside the burrow a short distance from this exit hole. The adult insects emerge during June or July, in central Illinois. They are black-and-yellow clear-winged moths, about ½ inch long, which one would readily mistake for small wasps. They are extremely active, flying with great rapidity. They deposit their eggs on the bark of the currant canes, and the grubs hatching from these eggs bore into the canes, feeding on the pith and the wood. They are nearly full-grown by mid-fall and remain in the canes during the winter as above stated.

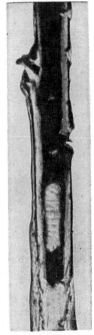

FIG. 15.84. The currant borer, *Ramosia tipuliformis*, shown in its burrow ready to pupate. Enlarged ½. (*From Slingerland and Crosby, "Manual of Fruit Insects,"* copyright, 1915, Macmillan. Reprinted by permission.)

Control Measures. As practically no external feeding is done by this insect and as it cannot be reached with contact sprays, the only effective method of control is to cut out and burn the infested canes. This can best be done shortly after the currant leaves appear in the spring, as it will then be very easy to distinguish the infested canes because of their weak, sickly appearance. The canes should be cut close to the ground and removed and burned before the last week in May, in the latitude of southern Illinois.

References. Wash. Agr. Exp. Sta. Bul. 36, 1898; *U.S.D.A., Farmers' Bul.* 1398, 1933.

[1] *Ramosia* (= *Synanthedon*) *tipuliformis* (Clerck), Order Lepidoptera, Family Aegeriidae.

J. GOOSEBERRY INSECTS

The gooseberry is injured by much the same insects as the currant. San Jose scale is often very destructive on gooseberry but is not nearly so conspicuous on this plant as on the currant, the heaviest infestation often occurring close to the ground.   If the scale is present in the vicinity of gooseberry plantations, they should be thoroughly sprayed, even though a casual examination does not show much scale on the canes. The imported currant sawfly is fully as destructive to the gooseberry as to the currant and can be controlled by the same measures.   The gooseberry fruitworms[1] feed on the pulp of the fruits, causing browning at the blossom end, premature coloring, and often the drying up of the berries.   They may be controlled by spraying or dusting with rotenone at 0.25 to 0.33 pound per acre, when webbing appears and 7 to 10 days later.

K. RASPBERRY INSECTS

FIELD KEY FOR THE IDENTIFICATION OF INSECTS INJURING THE RASPBERRY AND BLACKBERRY

1. Raspberry leaves with pale-green, spiny, many-legged worms feeding on the edges of the leaf.   Insects more abundant about the time raspberry comes into full foliage, sometimes nearly stripping the plants of leaves.................... *Raspberry sawflies, Monophadnoides geniculatus* (Htg.) and *Priophorus rubivorus* Rohwer (see *N. Y. (Geneva) Agr. Exp. Sta. Bul. 150*, 1898).

2. Bases of raspberry plants, particularly on the more heavily shaded parts of the stem, covered with white, scurfy-appearing scales.   Reddened areas on the bark around the points where the scale is attached.   These scales are about 1/5 inch across when full-grown, rounded in outline, and flattened.   In the winter many reddish eggs will be found beneath the protecting scale...................... *Rose scale, Aulacaspis rosae* (Bouché) (see Essig, "Insects of Western North America," p. 307, 1926).

3. Leaves of raspberry and blackberry become spotted with white or brownish flecks, due to feeding on the undersurface of minute 8-legged reddish or greenish mites, causing the leaves to fall and the fruits to dry up....................... ....*Two-spotted spider mite*, page 875.

4. Fleshy enlargements or swellings of the raspberry canes, up to several inches long, and 1/2 inch or more in diameter.   Slender, white-bodied grubs, up to 1/2 inch in length, feed on the pith of the canes in, or near, the swollen areas. Dull, bluish-black beetles, 1/3 inch in length by 1/4 as wide, with a dull coppery-red area on the back just behind the head, are found on the canes during late spring and summer.........................*Red-necked cane borer*, page 787.

5. Buds and blossoms of the raspberry and loganberry with numerous holes eaten in them; leaves partly skeletonized; whitish grubs about 1/4 inch long feed inside the ripening fruit.   Small light-brown beetles about 1/8 inch in length feeding on the tender leaves and newly opening buds......................... ....*Raspberry fruitworms*, page 787.

6. Raspberries and blackberries make very little growth.   Many of the canes wither and die.   Fruit, if any, is small.   Roots and lower parts of the canes girdled and burrowed by whitish, brown-headed caterpillars, up to 2/3 inch long, which enter the canes near the ground and tunnel into crown of plant, just beneath the bark... *Raspberry crown borer, Bembecia marginata* (Harris), Order Lepidoptera, Family Aegeriidae (see *Wash. Agr. Exp. Sta. Bul. 155*, 1938).

7. Tips of shoots of young canes wilted, with a purple discoloration at the base of the wilted part, or broken off clean as though cut with a knife; caused by the

[1] *Zophodia convolutella* (Hübner) and *Zophodia franconiella* (Hulst), Order Lepidoptera, Family Pyralididae.

feeding of a small white maggot which tunnels the pith and girdles the canes from the inside, then tunnels on down below the break.....................
Raspberry cane maggot, Pegomya (= *Hylemya*) *rubivora* (Coquillet), Order Diptera, Family Anthomyiidae (see *Ill. Agr. Exp. Sta. Cir.* 427, 1935).   Control by cutting and burning infested canes.

8. Irregular rows of minute round punctures, each about the diameter of a pin and extending into the pith of the raspberry.   The raspberry cane frequently dies above these punctures, or splits and breaks off during the winter.   Cricket-like pale-green insects, about 1 inch in length, with very long feelers and long hind legs crawl about over the plants and insert their eggs in the canes during the late summer... *Tree crickets*, page 788.

General Reference.   N. Y. (Cornell) Agr. Ext. Bul. 719, 1947.

207. RED-NECKED CANE BORER[1]

Importance and Type of Injury.   The injury caused by this beetle is shown by enlargements or swellings of the raspberry or blackberry canes (Fig. 15.85).   The canes frequently die and sometimes break off at the point where the swelling occurs.

Plants Attacked.   Raspberry, blackberry, and dewberry.

Distribution.   Eastern United States.

Life History, Appearance, and Habits.   The winter is passed in the form of a slender white grub inside the pith of the raspberry cane.   These grubs are about ½ inch in length.   In the spring they complete their growth, change inside the stem to the pupal stage, and emerge during May and June as dull bluish-black beetles, with a metallic luster, about ⅓ inch in length by one-fourth as wide, with a coppery-red or brassy prothorax.   These beetles lay their eggs in the bark of the cane, usually near the base of the leaf.   The young larvae burrow upward in the sapwood and also around the cane, sometimes making several girdles, and causing the formation of the galls or swellings.   There is but one generation of the insect each year.

Control Measures.   Partial control of this insect may be obtained by cutting out and burning the infested canes during the winter.   Spraying with DDT at 1.5 pounds or rotenone at 0.25 pound per acre, applied just before the plants come into bloom, will control the adult insects.   For severe infestations, a second spray of rotenone should be applied when the last petals have fallen.   DDT or rotenone dusts are also effective.

Fig. 15.85. Section of blackberry cane showing gall or swelling caused by larva of red-necked cane borer. (*From   U.S.D.A.,   Farmers' Bul.* 1286.)

References.   U.S.D.A., Farmers' Bul. 1286, 1922; *Mich. Agr. Exp. Sta. Quar. Bul.* 14, pp. 267–269, 1932.

208. RASPBERRY FRUITWORMS[2]

Importance, Type of Injury, and Life History.   These are among the most destructive pests of raspberry and loganberry in the northern part of

[1] *Agrilus ruficollis* (Fabricius), Order Coleoptera, Family Buprestidae.   *A. rubicola* Abeille, the rose stem girdler or bronze cane borer, has an identical life history and control.

[2] *Byturus bakeri* Barber, the western raspberry fruitworm, and *Byturus rubi* Barber, the eastern raspberry fruitworm, Order Coleoptera, Family Dermestidae.

the United States.   Light-brown beetles, $\frac{1}{8}$ to $\frac{1}{6}$ inch long, feed on buds, blossoms, and tender leaves and lay eggs on blossoms and young fruits from which slender white grubs, up to $\frac{1}{3}$ inch long, with a brown area on the upper side of each segment and prominent prolegs at the tip of the abdomen, bore into the fruit, making it unfit for food; the grubs feed especially in the receptacle.   The grubs become full-grown as the fruit ripens, drop to the ground, and pupate in the soil.   They winter in the adult stage in the soil, emerge about mid-April, and begin mating and egg-laying as the berries blossom.

Control Measures.   Control may be obtained by spraying with rotenone at 0.25 pound per acre or by dusting with 1 per cent rotenone.   The first application should be made just after the blossom buds appear and the second a few days before blooms appear.   A special effort should be made to obtain thorough coverage and to penetrate the blossom clusters. Thorough cultivation during the late summer will destroy many of the insects in the soil.

References.   *Conn. Agr. Exp. Sta. Bul.* 251, 1923; *U.S.D.A., Bur. Ento. Cir.* E-433, 1938; *Wash. Agr. Exp. Sta. Bul.* 155, 1938, and *Bul.* 497, 1947; *Jour. Econ. Ento.,* **41**:436, 1948.

209. TREE CRICKETS[1]

Importance and Type of Injury.   The insects known as tree crickets (Fig. 15.87) sometimes cause injury to tree and bush fruits.   Small round holes about the diameter of a pin (Fig. 15.86) are drilled singly in the twigs or brambles.   In each hole, which extends into the cambium or sapwood, the insect deposits a single pale-yellow egg, about $\frac{1}{8}$ inch long.   The punctures are usually made in a single row along one side of the cane or stem, sometimes as many as 50 to 75 in a row, and about 25 to the inch. These egg punctures often serve as the entrance points for tree and bramble diseases.   Canes frequently die above these punctures, and, where they are numerous, the canes split and break off.   The adult insects are sometimes very injurious by eating holes in ripe fruits.

Plants Attacked.   Apple, prune, plum, peach, cherry, raspberry, loganberry, blackberry, and others.

Life History, Appearance, and Habits.   The insects winter in the egg stage in twigs or brambles, the young hatching in the spring.   The pale-green slender nymphs (Fig. 5.10) feed on the foliage of various plants or on small, sluggish insects, fungi, pollen, or ripe fruits.   They grow rather slowly, reaching the adult stage in late summer.   The adult has somewhat the appearance of a cricket but is pale green in color and has a longer, more slender body and smaller head.   The antennae or feelers are much longer than the body.   The males have stiff veins in the flat wings, which are adapted to making sounds.   The songs of the males in late summer are described as a series of short, clear, musical whistling notes, indefinitely repeated, often synchronized, and varying in frequency with the temperature.   The females deposit their eggs in the fall, and there is only one generation each year.

Control Measures.   Spraying or dusting with DDT or TDE (DDD) at 1 to 2 pounds per acre in early summer, while the crickets are young,

---

[1] *Oecanthus niveus* (De Geer), the snowy tree cricket; *O. nigricornis nigricornis* Walker, the black-horned tree cricket; and *O. nigricornis quadripunctatus* Beutenmüller, the four-spotted tree cricket, Order Orthoptera, Family Gryllidae.

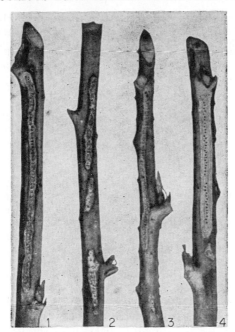

Fig. 15.86. Egg punctures of a tree cricket in raspberry canes.   (*From N.Y. (Geneva) Agr. Exp. Sta. Bul.* 388.)

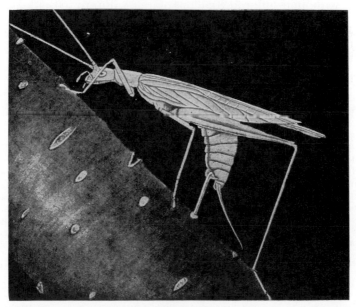

Fig. 15.87. The snowy tree cricket, *Oecanthus niveus;* characteristic posture of female in act of laying eggs   Reduced ½   (*From N.Y. (Geneva) Agr. Exp. Sta. Bul.* 388.)

has given good control. Pruning out and burning the parts of canes containing the eggs, after the last crop has been removed and before spring, aids in control.

References. N.Y. (Geneva) Agr. Exp. Sta. Bul. 388, 1914; Ore. Agr. Exp. Sta. Bul. 223, 1926; U.S.D.A., Cir. 270, pp. 56–58, 1933.

L. BLACKBERRY INSECTS

The blackberry is comparatively free from insect injuries. It is infested to some extent by the rose scale, reference to which has been made under Raspberry Insects, page 786. The canes are occasionally punctured by tree crickets (page 788). The raspberry sawfly also attacks the blackberry and can be controlled by the same methods as recommended for other sawflies on page 783. The canes of the blackberry are occasionally injured by the larvae of the red-necked cane borer (page 787), which causes irregular swellings or galls along the canes. The insect is rarely numerous enough to be considered of any economic importance. Leafhoppers (page 564) and mites (page 616) sometimes become thick enough on the foliage to cause a whitening of the leaves. The blackberry psyllid[1] curls and stunts the new foliage in some sections. Certain leafhoppers suck the sap of the leaves, causing white mottling. Whitish, sawfly larvae,[2] up to $\frac{1}{3}$ or $\frac{1}{2}$ inch long with brown heads and plates on the thorax, mine between upper and lower epidermis of the leaves, causing large blotch mines so that the plants appear as though singed by fire. On the Pacific Coast the redberry mite[3] feeds near the base of the drupelets around the core preventing the fruits from ripening in part or in whole— the so-called "redberry disease." The affected fruit becomes very bright colored, in parts varying from a single drupelet to an entire berry, hard, and clings to the canes until winter. Control may be effected by dormant spraying with lime-sulfur at 8 gallons of lime-sulfur to 100 gallons of spray just before the leaf-buds open or 5 gallons when the new growth is about $1\frac{1}{2}$ inches long. A preharvest spray at 2.5 gallons may be necessary just before the blossom buds open.

M. STRAWBERRY INSECTS

FIELD KEY FOR THE DETERMINATION OF INSECTS INJURING THE STRAWBERRY

A. *External chewing insects that eat holes in the leaves or chew off the lower surface of the leaves:*
   1. Very small, green or metallic-blue, jumping beetles, less than $\frac{1}{6}$ inch across, feeding on the strawberry plants early in the spring. Leaves riddled with small round holes, often drying up and turning brown............................ ....*Strawberry flea beetle*, page 791.
   2. Grayish plump-bodied snout beetles, about $\frac{1}{2}$ inch long, the wing covers crossed by 2 light bands, bent down at the end and terminating in an acute angle, and most of the body covered with small flattened scales, sometimes defoliate the plants......................*Imbricated snout beetle*, page 705.
   3. Small greenish or bronze caterpillars folding or rolling together the strawberry leaves and feeding within the rolled portion. Heavily infested plants have a

[1] *Trioza tripunctata* Fitch, Order Homoptera, Family Chermidae (*N.J. Agr. Exp. Sta. Bul.* 378, 1923).
[2] *Metallus rubi* Forbes, Order Hymenoptera, Family Tenthredinidae (see *N.Y. (Geneva) Agr. Exp. Sta. Tech. Bul.* 133, 1928).
[3] *Aceria* (= *Eriophyes*) *essigi* (Hassan), Order Acarina, Family Eriophyidae (see *Wash. Agr. Exp. Sta. Bul.* 279, 1933; *Calif. Agr. Exp. Sta. Bul.* 399, 1925).

whitish appearing foliage. Injury occurs from early spring until early fall. Small grayish-brown moths, about ¼ inch long, the wings marked with wavy bands of light brown, fly about strawberry beds from April to September......
....*Strawberry leaf rollers*, page 792.

B. *Insects that injure the leaves by sucking the sap:*

  1. Almost microscopic, white to light-brown mites, feed on young unfolded leaves, in crown of plant, stunting and dwarfing them and causing a rosette of leaves due to failure of stems to elongate..
  Strawberry crown mite, or *cyclamen mite*, page 876 (see also *Wash. Agr. Exp. Sta. Bul.* 155, 1938; *Calif. Agr. Exp. Sta. Ext. Cir.* 484, 1959).

  2. Leaves curled, spotted, and dying, with minute white and dark specks and some silk on underside, among which almost invisible, 6- and 8-legged greenish or reddish mites crawl and feed..............*Two-spotted spider mite*, page 875.

C. *Insects injuring the crown of the plant:*

  1. White, thick-bodied grubs, about ⅕ inch in length, boring in the crowns of the strawberry plant. Crown of the plant eaten out so that it dies or is so weakened that very few runners or new growth is produced. Reddish-brown snout beetles, about ⅙ inch in length, shelter under litter or about the plants in strawberry fields during the winter.........*Strawberry crown borer*, page 796.

  2. Strawberry plants eaten off close to the ground by curved-bodied, light brown-headed grubs, about ⅕ inch long. Grubs most abundant during late spring and again during the late summer. Black beetles, ⅙ inch long, with a short blunt snout protruding from the front of the head, clustered about the bases of strawberry plants and sometimes feeding on the leaves..................
  ....*Strawberry root weevil*, page 795.

D. *Insects injuring the roots:*

  1. Strawberry plants lacking in vigor, foliage of a pale color, the fruit drying up or failing to mature properly. Roots of the injured plants covered with dark bluish-green aphids, their slender beaks inserted in the roots, from which they suck the sap. Usually attended by small brown ants. Aphids of the same description feeding on the leaves during early spring......................
  ....*Strawberry root aphid*, page 793.

  2. Large white grubs, up to 1 inch in length, with brown heads, usually holding the body in a curved position, and with distinct, well-developed, slender legs, feed on the roots of strawberries from early spring to early fall..............
  ....*White grubs*, page 796.

  3. Small, white, brown-spotted grubs, about ⅛ inch in length, feeding on the roots of strawberries during May and June. Infested plants are weakened and have poorly colored foliage. Brownish or coppery-colored beetles, about ⅛ inch long, feeding on the foliage of strawberries during the early fall......
  ....*Strawberry rootworms*, page 794.

E. *Insects injuring the buds and fruits:*

  1. Buds and newly formed fruit of the strawberry dried up on the partly severed stems, or entirely eaten off. This injury is caused by dark reddish-brown snout beetles, from 1/12 to ⅛ inch long, which make punctures in the buds, in which they insert eggs. Very small, legless, white soft-bodied grubs, feed within the strawberry buds.............................*Strawberry weevil*, page 797.

210. STRAWBERRY FLEA BEETLE[1]

Importance and Type of Injury. Very small metallic-blue beetles feed on the foliage of the strawberry plants. The leaves are riddled with large numbers of small round holes, often drying up and browning around these holes.

Plants Attacked. The beetles lay their eggs on strawberry, evening primrose, and other plants of the same family. They also feed on some greenhouse plants.

Distribution. Throughout the United States and Canada.

Life History, Appearance, and Habits. These beetles, in common with many other kinds of flea beetles, hibernate as adults. They emerge early in the spring, and their

---

[1] *Altica ignita* Illiger, Order Coleoptera, Family Chrysomelidae.

principal damage is done before the strawberry plants bloom.   There are from one to two generations a year, the larger number occurring in the South.

Control Measures.   Thorough spraying or dusting of the plants with DDT or methoxychlor at 1 to 2 pounds per acre, applied whenever the insects are abundant, is very effective.   Usually this application should be made about a week before the strawberries bloom.

Reference.   Conn. Agr. Exp. Sta. Bul. 344, p. 165, 1933.

211. STRAWBERRY LEAF ROLLERS[1]

Importance and Type of Injury.   Small greenish or bronze caterpillars, up to ½ inch long, fold or roll the strawberry leaves (Fig. 15.89), fastening them together and feeding within.   The leaves have a brown appearance and much of the foliage is killed.   Heavily infested beds have a

FIG. 15.88. Larva of the obsolete-banded strawberry leaf roller, *Archips obsoletana*, on leaf. 3 times natural size.   (*From Slingerland and Crosby,* "*Manual of Fruit Insects,*" *copyright, 1915, Macmillan.   Reprinted by permission.*)

whitened or grayish appearance instead of the usual green of healthy strawberry plants, and the fruits are withered and deformed.

Plants Attacked.   Strawberry, blackberry, dewberry, and raspberry.

Distribution.   Northern United States, Louisiana, Arkansas, and southern Canada.

Life History, Appearance, and Habits.   The winter is passed in the larval and pupal stages, the pupae inside of silken cocoons in the folded leaves and the larvae in silken shelters under trash on the surface of the ground.   In the spring of the year the larvae transform to pupae and a little later emerge as small grayish or brownish moths, with a wing expanse of about ⅖ inch.   The wings are marked with wavy bands of light and dark color.   In southern Illinois the moths are abroad in April, and in the northern part of the state by the middle of May or a little earlier. They lay their flattened, transparent eggs singly, 20 to 120, on the undersides of the strawberry leaves.   These eggs hatch in about 1 week and the caterpillars coming from them (Fig. 15.88) feed at first on the undersurface of the leaves, under a silken cover, migrating when about half-grown to the upper side of the leaves, where they fold a leaf about themselves in the characteristic manner, holding it with fine silken

[1] *Ancylis comptana fragariae* (Walsh & Riley), Order Lepidoptera, Family Olethreutidae, and *Archips obsoletana* Walker, Order Lepidoptera, Family Tortricidae.

threads.   They feed for 35 to 50 days and then transform into brown pupae inside the folded leaves.   There are two, and a partial third, generations throughout most of the range of the insect.   The hatching of the second generation begins about the middle of July.

Control Measures.   These insects may be controlled by spraying or dusting with malathion at 1 to 1.5 pounds, parathion at 0.4 pound, or TDE (DDD) at 3 pounds per acre.   The treatment may need to be repeated for the second generation.   The applications should be made,

Fig. 15.89. Leaves folded by larvae of the strawberry leaf roller, *Ancylis comptana fragariae.* (*From Iowa Agr. Exp. Sta. Bul.* 179.)

however, before the leaves have become folded to any extent, as the insecticides will not be as effective against the leaf rollers after they have constructed their silken cases.   The number of insects may be kept down if the strawberry beds are mowed close to the ground and burned over shortly after the fruit is picked.

References.   Neb. State Ento. Cir. 7, 1908; *Iowa Agr. Exp. Sta. Bul.* 179, 1918; *Iowa Agr. Exp. Sta. Cir.* 110, 1928; *Jour. Agr. Res.,* **44**:541–558, 1932; *Jour. Econ. Ento.,* **28**:388, 1935; and **44**:424, 1951; *Wash. Agr. Bul.* 246, Rev. 1948; *Ohio Agr. Exp. Sta. Bul.* 651, 1944.

212. Strawberry Root Aphid[1]

Importance and Type of Injury.   Strawberry plants are lacking in vigor, the foliage of a pale color.   The fruit dries up or fails to mature properly.

Plants Attacked.   Strawberry.

Distribution.   United States east of the Rocky Mountains.

Life History, Appearance, and Habits.   These insects usually pass the winter in the form of black shining eggs which are attached to the leaves and stems of the strawberry plant, although in mild climates they may reproduce parthenogenetically throughout the winter.   The eggs hatch early in the spring into dark-bluish-green aphids (Fig. 15.90), which feed on the new leaves of the strawberry.   Wherever these aphids become abundant, they are soon found by colonies of the brown cornfield ant and

[1] *Aphis forbesi* Weed, Order Homoptera, Family Aphidae.

are carried by these ants to the strawberry roots, where they feed by sucking the sap from the roots.   There are a number of generations of the insect during the year.   All those occurring in the field during the summer are females and reproduce by giving birth parthenogenetically to living young.   At the approach of cold weather winged females are produced, and these make their way to the leaves of the plant and there give birth to the sexual males and females.   After mating, these true females lay the winter eggs described above.

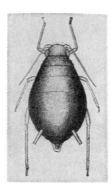

Fig. 15.90. Wingless adult female of the strawberry root aphid, *Aphis forbesi*.   Greatly magnified.   (*From Ill. Natural History Surv.*)

Control Measures.   In setting new beds, uninfested plants should be selected and the ground should be given a thorough and deep cultivation in the early spring, to break up and drive out the ants which may be present in the soil.

The leaf-infesting forms may be controlled by spraying or dusting with demeton (Systox), parathion, or TEPP at 0.4 pound, Diazinon, Phosdrin, Thiodan, or malathion at 1 pound, or nicotine sulfate at 2 pints of 40 per cent solution per acre.   Drenching the bed, when there is no fruit on the plants, with chlordane at 0.125 pound to 100 gallons of water to 1,000 square feet will control the ants and prevent the establishment of the aphids on the roots.

References.   *N.J. Agr. Exp. Sta. Bul.* 225, 1909; *Trans. Ky. State Hort. Soc.*, **77**:120, 1932; *Calif. Agr. Exp. Sta. Ext. Cir.* 484, 1959.

213 (255). STRAWBERRY ROOTWORMS[1]

Importance and Type of Injury.   There are several species of small beetles, the young of which are injurious to strawberries.   Small, white, brown-spotted grubs feed on the roots of the strawberries during May, June, and July.   These grubs (Fig. 15.91) are about ⅛ inch in length and are very much smaller than the common white grub.   The foliage of strawberries is destroyed by small bronze-brown or copper-colored adult beetles, about ⅛ inch long (pages 512 and 860).

Plants Attacked.   Strawberry, raspberry, grape, rose, and many other plants.

Distribution.   Over most of the United States.

Life History, Appearance, and Habits.   The strawberry rootworms pass the winter as full-grown beetles (Fig. 15.92).   They come out of their winter shelters in early spring.   The females deposit an average of 125 eggs in the ground near the plants.

[1] *Paria* (= *Typophorus*) *canella* (Fabricius), *Colaspis flavida* (Say), and *Graphops pubescens* Melsheimer, Order Coleoptera, Family Chrysomelidae.

The young grubs hatching from them feed on the roots for about 3 months, passing through four instars, becoming full-grown during August and transforming to a pupal stage and later to the adult beetle. The total life cycle requires 82 days at 70°F. and 48.5 days at 80°F. There are two generations annually in California. The beetles do considerable feeding on the foliage of the strawberry during the early fall.

Control Measures. These beetles may be controlled by spraying or dusting the plants as the beetles leave hibernation

Fig. 15.91. Strawberry rootworm, *Paria canella*, larva or grub. About 15 times natural size. (*From U.S.D.A., Farmers' Bul.* 1344.)

Fig. 15.92. Adult of the strawberry rootworm, *Paria canella*. About 10 times natural size. (*From U.S.D.A., Farmers' Bul.* 1344.)

with DDT at 1 to 2 pounds per acre. The application is usually made about the time the blossoms appear. July plowing of infested beds will destroy many larvae, and rotation of beds is also of value.

References. U.S.D.A., Farmers' Buls. 1344, 1923, and 1362, p. 59, 1922; *Thirteenth Rept. Ill. State Ento.*, p. 159, 1883. *U.S.D.A., Dept. Bul.* 1357, 1926; *Mich. Agr. Exp. Sta. Spcl. Bul.* 214, pp. 94, 95, 1931; *Hilgardia*, **19**:25, 1949; *Jour. Econ. Ento.*, **46**:1102, 1953.

214. STRAWBERRY ROOT WEEVIL[1] OR CROWN GIRDLER

Importance and Type of Injury. Strawberry plants stunted, the leaves closely bunched together and very deep colored, or dying, the fine roots and the crown having been eaten off close to the ground, by small, fat, curved grubs with light-brown heads.

Plants Attacked. Strawberry, the related small fruits, and many other plants. Nursery evergreens have been seriously attacked by the larvae.

Distribution. Northern United States and Canada.

Life History, Appearance, and Habits. The insect passes the winter in both the larval and adult stages, under surface trash and in old strawberry crowns. The adults are nearly black beetles, about ¼ inch long or a little less, having short blunt snouts protruding from the front of their heads. They feed upon the foliage and berries. No males have ever been found and the females reproduce parthenogenetically. The grubs, or larvae, are white, legless, up to ⅜ inch long, with light-brown heads, and hold their bodies in more or less of a curved position. The grubs begin feeding as soon as the weather becomes warm in the spring and about the same

[1] *Brachyrhinus ovatus* (Linné), Order Coleoptera, Family Curculionidae. The black vine weevil, *B. sulcatus* (Fabricius), has very similar habits. The adults are ⅓ to ⅖ inch long and have patches of golden hairs upon the wings when freshly emerged (see *Wash. Agr. Exp. Sta. Bul.* 199, 1926; *Ore. Agr. Exp. Sta. Cir.* 79, 1923; *U.S.D.A., Tech. Bul.* 325, 1932). The rough strawberry weevil, *B. rugostriatus* Goeze, is a very similar species, destructive in the West. *B. meridionalis* (Gyllenhal) and *B. cribricollis* (Gyllenhal) sometimes attack strawberries in California (*Calif. Agr. Exp. Sta. Ext. Cir.* 484, 1959).

time the beetles leave the shelters in which they have passed the winter and gather in the strawberry beds. The wing covers of the beetles are tightly grown together and the insect is unable to fly. In southern Illinois the insect is most abundant in the adult stage during May and June, and again in late July and August. The female beetles lay small, white, spherical eggs among the roots or in the crown of the strawberry plants, these eggs hatching into the grubs above described. There are probably two generations a year in the latitude of Illinois.

Control Measures. This insect may be controlled by applications of aldrin or dieldrin at 5 pounds or chlordane at 10 pounds per acre, made to the soil surface as dust, granular, or spray, and cultivated into the upper 6 to 8 inches before planting. The foliage may be protected by dusting or spraying with malathion at 1 to 1.5 pounds or parathion at 0.4 pound per acre. Destroying old beds promptly after the last picking and rotating new beds to fields that have been in cultivation but have not had strawberries on or near them for at least 1 year are of value. This insect may be kept out of the strawberry fields by constructing a barrier of tarred boards around the new fields. Various poison baits of bran and molasses—or chopped raisins, dried apples, or other fruits and shorts poisoned with 5 per cent sodium fluosilicate—have given good control. About a tablespoonful is dropped on the ground about each plant, just before the overwintering weevils attack them and again just before the new adults appear in midsummer. The following formula has given good control.

Ingredients	For 1½ acres	For 30 acres
Bran..	5 lb.	100 lb.
Molasses......................................	1 pt.	2½ gal.
or brown sugar.................................	1 lb.	20 lb.
Calcium arsenate or sodium fluosilicate..............	¼–½ lb.	5–10 lb.
Water...	2 qt.	10 gal.

References. Maine Agr. Exp. Sta. Bul. 123, 1905; *Can. Dept. Agr., Pamphlet* 5 (n.s.), 1931; *Wash. Agr. Exp. Sta. Bul.* 199, 1926; *Ore. Agr. Exp. Sta. Bul.* 330, 1934; *Wash. Agr. Ext. Bul.* 246, *Rev.* 1948.

215. STRAWBERRY CROWN BORER[1]

This pest damages strawberries in much the same manner as the strawberry root weevil. It lives upon cultivated and wild strawberries and cinquefoil. It has been especially destructive in the bluegrass region of Kentucky and Tennessee but ranges also to Illinois, Missouri, and Arkansas. Most of the insects winter as reddish-brown snout beetles, about ⅙ inch long, under trash in strawberry patches. Most of the eggs of the first generation are laid after mid-April in small holes eaten in the leaf bases of the plants. The white thick-bodied grubs, up to ⅕ inch long, tunnel through the strawberry crowns, killing or stunting the plants and greatly reducing the yield. The control measures are much the same as for the root weevil and consist (*a*) of setting only plants or crowns dug before the first beetles are active in the spring (Mar. 1, in Kentucky) and certified as free from infestation; (*b*) the destruction of old beds promptly after the last picking; (*c*) setting new beds on soil in cultivation for at least 1 year, and at least 1,000 feet from a source of infestation; and (*d*) the use of barriers of boards set on end with an L-shaped iron turned outward at the top.

References. Tenn. Agr. Exp. Sta. Bul. 128, 1923; *Jour. Econ. Ento.,* **31**:385–387, 1938; *Ohio Agr. Exp. Sta. Bul.* 651, 1944.

216 (14, 39, 87, 273). WHITE GRUBS[2]

White grubs or grubworms are among the insects most commonly injurious to strawberries. Large white grubs with brown heads feed

[1] *Tyloderma fragariae* (Riley), Order Coleoptera, Family Curculionidae.
[2] Many species of *Phyllophaga*, Order Coleoptera, Family Scarabaeidae.

on the roots of the strawberry plants, causing the plants to die over areas of varying size.   It is practically impossible to clean a strawberry bed, once infested with white grubs, without plowing up the bed.   In setting new patches, ground should be selected that has been in some clean-cultivated crop for 1 or 2 years previously; as the June beetles prefer to lay their eggs in grassland or land with a heavy growth of grassy weeds.   If possible, strawberry beds should be located at some distance from the trees on which the beetles feed and near which they usually lay their eggs.   Thorough cultivation of the soil in the spring before the plants are set will be of some benefit but will not clean all the grubs out of the soil (see further discussion of white grubs on page 503).   The application of 1 handful (about 1.5 ounces) of a 1-20 mixture of lead arsenate and sand to the hole made for setting out the strawberry plant will control the grubs.   The most satisfactory control may be obtained by spraying or dusting the soil with aldrin or dieldrin at 5 pounds or chlordane at 10 pounds per acre and harrowing the insecticide into the upper 4 to 6 inches.

References. N.Y. (Cornell) Agr. Exp. Sta. Bul. 770, 1941; *Ohio Agr. Exp. Sta. Bul.* 651, 1944.

217. Strawberry Weevil[1]

Importance and Type of Injury.   The work of this beetle kills buds and fruits, leaving them hanging on partly severed stems (Fig. 1.7,*E*).

Plants Attacked.   Strawberry, wild blackberry, raspberry, dewberry, and cinquefoil.

Distribution.   Eastern United States.

Life History, Appearance, and Habits.   This insect passes the winter in the form of a dark reddish-brown snout beetle, from $\frac{1}{12}$ to $\frac{1}{8}$ inch long (Fig. 15.93), sheltered under trash.   These adults become active early

Fig. 15.93. Strawberry weevil, *Anthonomus signatus*, on stem of strawberry.   About 4 times natural size.   *(From Ark. Agr. Exp. Sta. Bul.* 185.)

in the spring, about the time the strawberries are coming into bloom. The adult beetle makes a puncture in the strawberry bud with her long beak, and in this inserts an egg.   She then crawls down and girdles the stem of the bud.   The young grubs, which are legless, white, and soft-bodied, feed within the buds and, after about 4 weeks, change within the bud to a pupal stage.   The adults emerge a little before midsummer. They feed for a short time and then go into hibernating quarters about midsummer, remaining there until the following spring.

Control Measures.   This insect may be controlled by dusting or spraying the foliage as needed with DDT or methoxychlor at 1 to 2 pounds per acre.   Clean cultivation and resistant varieties help in control.   The

[1] *Anthonomus signatus* Say, Order Coleoptera, Family Curculionidae.

planting of one row of staminate variety to each five rows of pistillate varieties has also given good results.

References. Tenn. State Bd. Ento. Bul. 30, 1919; *N.J. Agr. Exp. Sta. Bul.* 324, 1918; *Ark. Agr. Exp. Bul.* 185, 1923; *U.S.D.A., Cir.* E-346, 1935; *Rept. Ento. Ark. Agr. Exp. Sta.* 1930–1931, pp. 45–49; *Ore. Agr. Exp. Sta. Bul.* 330, 1934; *Jour. Econ. Ento.,* **30**:437, 1937; **38**:678–682, 1945, and **42**:559, 1949; *N.C. Agr. Exp. Sta. Cir.* 336 (C), 1949.

N. BLUEBERRY AND HUCKLEBERRY INSECTS

The most injurious insect attacking this group of fruits is the blueberry maggot, a variety of the apple maggot which is discussed on page 732. Eggs are laid in the ripe berries and the maggots eat the pulp of the fruits, causing many fruits to drop, spoiling the sale of others, and occasioning great difficulties in sorting the fruits for canning. Dusting blueberry land with malathion at 1 pound or rotenone at 0.5 pound per acre at 7- to 10-day intervals after the berries turn blue has given excellent control by killing the adult flies. The cherry fruitworm (page 771) attacks blueberries and may be controlled by dusting or spraying with malathion or methoxychlor at 1.5 to 2 pounds per acre, applied when the berries are about ¼ inch in diameter and 7 to 10 days later. Many other pests of these fruits are recorded in *Maine Agricultural Experiment Station Bulletin* 356, 1930. The blueberry thrips and certain caterpillars and sawflies are recorded as of economic importance.

References. U.S.D.A., Cir. 196, 1931; *Jour. Econ. Ento.,* **30**:294, 1937.

CHAPTER 16

CITRUS INSECTS

Only the more important pests of citrus are discussed here. For more complete information about these species and for the other minor species, the reader is referred to the various publications issued by the states where citrus is extensively grown.

General References. *U.S.D.A., Dept. Bul.* 907, 1920; *Calif. Agr. Exp. Sta. Bul.* 542, 1932; *Fla. Agr. Ext. Ser. Bul.* 88, 1937; *Calif. Agr. Exp. Sta. Cir.* 123, 1941; *Calif. Fruit Growers' Exchange Monthly Pest Control Circulars;* ESSIG, E. O., "Insects of Western North America," pp. 269–322, Macmillan, 1926; QUAYLE, H. J., "Insects of Citrus and Other Subtropical Fruits," Comstock, 1938; BATCHELOR, L. D., and H. J. WEBBER "The Citrus Industry," vol. II, Chap. 12, H. S. SMITH, and Chap. 14, A. M. BOYCE, Univ. of California Press, 1948; EBELING, W., "Subtropical Fruit Pests," Univ. of California Press, 1959.

FIELD KEY FOR THE DETERMINATION OF INSECTS INJURING CITRUS FRUITS

A. *Motionless insects, covered by a firm waxy shell, sucking sap from fruits, leaves, or twigs:*
  1. Foliage of trees showing yellow leaves or brown areas. Fruits uneven in color, with elongate, oyster-shell-shaped scales attached to the skin of the fruit. Many dead twigs......................*Glover scale* or *purple scale*, page 802.
  2. Leaves entirely yellow or spotted with yellow. No honeydew on the leaves or fruit:
    a. Many yellow spots on the fruits. Entire branches of trees covered with a thin coating of circular yellowish or reddish scales, about $\frac{1}{12}$ inch in diameter................................*California red scale*, page 803.
    b. A sprinkling of purplish-red round scales on leaves and fruit only..........
    ....*Florida red scale*, page 803.
    c. Scale predominately on fruit and leaves............*Yellow scale*, page 803.
  3. Foliage of trees somewhat discolored. Fruits and leaves covered with honeydew or a black, sooty growth of mold. Bodies of insects not covered with a separable shell:
    a. Nearly round scales, about $\frac{1}{5}$ inch across, of a brown or blackish color, often with a raised H on the back, are found over the twigs, leaves, and fruits..*Black scale*, page 805.
    b. Oval, brownish or grayish scales on twigs and leaves, up to $\frac{1}{6}$ inch long and very flat.................*Soft scale*, page 869, or *citricola scale*, page 806.
    c. Large white, cottony, fluted objects, nearly $\frac{1}{2}$ inch long, on the twigs. Smaller oval yellowish-brown scales on the leaves......................
    ....*Cottony-cushion scale*, page 414.
B. *Insects usually more or less active; not covered with a separable shell; sucking sap from fruits, leaves, or twigs:*
  1. Leaves of trees showing pale spots and with light webs on their undersurface. Minute red or yellow mites on the leaves or fruits. Fruits with a grayish or silvery sheen to the skin........................*Red spider mites*, page 809.

2. Buds show retarded growth and the new growth distorted. Many tiny, slender, dark or yellowish insects among buds and flowers. Skins of fruits show erect shallow scabs or smooth scars...........................*Thrips*, page 800.
3. Masses of cottony-white material covering flat, oval, purplish, soft-bodied bugs on leaves or at the angles where fruits touch. Fruits coated with very sticky honeydew or black sooty mold. Fruit drops..........*Mealybugs*, page 807.
4. Foliage of trees covered with honeydew or blackened with a heavy growth of sooty mold. Fruits undersized or of poor color. At certain seasons, very small white or black insects fly from the tender growth of trees when disturbed. Small pale-green short-oval flat scale-like nymphs fixed to the underside of the leaves. Brown, bright-red, or yellow fungi often growing on the bodies of the nymphs...*Whiteflies*, page 808.
5. Foliage more or less tightly curled and more or less covered beneath with soft-bodied plant lice...
............*Melon aphid* (page 631), *green citrus aphid*, and others, page 814.
6. Fruits russet-brown or, on grapefruit, chamois color, on lemons, silvery, but smooth...*Rust mite*, page 811.
7. Fruits, buds, twigs, and leaves distorted and deformed....*Bud mite*, page 811.
C. *Legless maggots tunneling in pulp of fruits:*
....*Fruit flies*, page 812.

Occasional damage to citrus may result from the feeding of a number of species of chewing insects. These include grasshoppers and katydids (page 465), cutworms (page 476), beet armyworm (page 610), fruit-tree leaf roller (page 701), orange tortrix, *Argyrotaenia citrana* (Fernald), black orangeworm, *Holcocera iceryaeella* (Riley), pink scavenger caterpillar, *Pyroderces rileyi* (Walsingham), garden tortrix, *Clepsis peritana* (Clemens), and cucumber beetles (page 629).

218. THRIPS

Importance and Type of Injury. Thrips injure the citrus fruits by attacking the buds and new growth. The growth of the young trees is commonly retarded and new growth distorted by their feeding on the foliage. The fruit also is attacked and the skin is scarred, often with a definite ring around the stem end where the thrips feed beneath the sepals while the fruit is small. In heavy infestations as much as 90 per cent of the fruit may become culls (Fig 16.1).

Plants Attacked. All citrus, many deciduous fruits, and many other plants.

Distribution. Thrips are world-wide in their distribution. The citrus thrips[1] is the species causing the greatest amount of injury in the warmer, more arid regions of California and Arizona; it does not occur in Florida. The flower thrips[2] and the orchid thrips[3] are of some importance in Florida. The greenhouse thrips (page 863) does severe damage in limited areas.

Life History, Appearance, and Habits. The citrus thrips pass the winter in the egg stage on the stems and leaves of the infested trees. The eggs are inserted into the tissue in the new foliage, or in the stems bearing leaves or fruits. The female may lay as many as 250 eggs. The young thrips are yellowish, very small, slender, active creatures. The nymphs, during the first two instars, feed by rasping the plant surface and sucking the sap that flows from these injured spots. These two stages last from 4 to 14 days, depending upon the temperature, and the nymphs usually drop to the ground and seek shelter among trash on the surface of the

---

[1] *Scirtothrips citri* (Moulton), Order Thysanoptera, Family Thripidae.
[2] *Frankliniella tritici* (Fitch), Order Thysanoptera, Family Thripidae.
[3] *Chaetoanaphothrips orchidii* (Moulton).

soil or in soil crevices. Here they pass through two nonfeeding nymphal instars, sometimes termed the prepseudopupa and the pseudopupa, which require from 4 to 20 days. From the last of these, the adults emerge, to begin a new generation every 2 to 3 weeks in midsummer. There are from 10 to 12 generations a year.

Control Measures. The thrips may be controlled by dusting with superfine sulfur at 100 pounds per acre, using three applications at about monthly intervals in the spring or spraying at petal fall with lime sulfur 2 gallons and wettable sulfur 4 pounds to 100 gallons of spray. Other treatments may be necessary to control this insect, of which several resistant races are known. These include spraying at petal fall with DDT at 2 pounds, dieldrin at 0.5 pound, parathion at 1.5 pounds, or sabadilla at 1 to 1.5 gallons of 0.5 per cent plus 10 pounds of sugar per acre in 100 to 200 gallons of spray. A second application may be required in summer to protect new growth. Tartar emetic at 1.5 pounds plus 1.5 pounds of sugar per acre in 25 to 50 gallons of spray is effective in some areas. Dusting with 2 per cent DDT in sulfur is also effective when

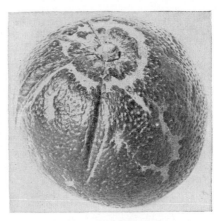

Fig. 16.1. Orange injured by citrus thrips, *Scirtothrips citri*, showing characteristic scars. (*From Univ. Calif. Coll. Agr.*)

applied in the spring, but DDT applications should be used with caution since they may cause disturbing upsets of the cottony-cushion scale because of the killing of the lady beetle predator.

CITRUS SCALE INSECTS

The scale insects of citrus are probably the most destructive of any group of insects which attack these trees. As with the greenhouse scales, the citrus scales can be divided into three classes: armored scales, unarmored scales, and mealybugs.

ARMORED SCALES

In this class of scales, a protective covering of wax is secreted from the body of the insect to form two protective scales, one above and the other beneath the body of the scale. The upper covering is thick and hard; the lower plate, or scale, is very thin and delicate, fitting closely to the surface of the plant where the insect is feeding. In this class of scales the eggs are laid under the protective scale, or in some cases the young are born alive. In either case the young scales move about for a short time, select a favorable location on their food plant, and there insert their beaks; in the case of the females they do not move for the remainder of their lives. On starting to feed, they begin to secrete fibers of wax, which form the covering or scale over the body. The insects molt or shed their skins very shortly after beginning to suck the sap, and at this first molt lose their legs. The females later molt a second time and become adults, while the males, after a further metamorphosis, develop into small two-winged adults, which are incapable of feeding.

They mate with the females and die very shortly thereafter.   The female after being fertilized increases very rapidly in size and produces her eggs or begins giving birth to living young.   The scales that attack citrus develop more slowly during the colder periods of the year, but all stages of these insects can usually be found on trees at any season.   Some of the most destructive of the armored scales are the red scales, the purple scale, the Glover scale, and the yellow scale.

219. PURPLE SCALE[1]

Importance and Type of Injury.   This is one of the most widely distributed and destructive of citrus pests.   The foliage of trees infested by

FIG. 16.2. The purple scale, *Lepidosaphes beckii*, male and female scales on orange leaf. About 7 times natural size.   (*From Univ. Calif. Coll. Agr.*)

the purple scale turns yellow about the areas where the scales are feeding. Fruit attacked by this scale is stunted, ripening is delayed, the coloring of the fruit is very uneven, and flavor is affected.   The scales are difficult to remove from the fruit before marketing, requiring a vigorous scrubbing to detach.   Their feeding also permits the entrance of various fungi. Severe infestations may kill the wood.

Distribution.   This scale insect is the most important pest of orange groves in Florida and other Gulf states; and, although mainly confined to coastal regions, it is rated as third in importance in California, being outranked only by the California red scale and the black scale.

---

[1] *Lepidosaphes beckii* (Newman), Order Homoptera, Family Coccidae.   The Glover scale, *L. gloverii* (Packard), is very similar to the purple scale but is much narrower and straighter.   It is a pest of citrus in Florida and to a limited extent in California and is controlled in the same manner as purple scale.

Plants Attacked. The purple scale is primarily a pest of all citrus fruit and particularly of the orange and grapefruit. It also occurs on avocado, croton, eucalyptus, fig, olive, yew, and other plants.

Life History, Appearance, and Habits. The female insect deposits from 40 to 80 eggs beneath her scale, dying soon after her full quota of eggs is laid. These eggs hatch in from 2 weeks to 2 months. The young, very pale scales crawl about the tree for a short time. They are said to be strongly repelled by light and to seek the moderately shaded part of the tree and fruit as a place for inserting their beaks before starting to feed. Shortly after settling, the insects secrete two long, coarse threads, which become a protective fuzzy coating for the body while the permanent scale covering is being formed. This grows by elongation, the insect moving its body from side to side as it adds threads of wax to the posterior. The female molts twice at 3- to 4-week intervals, becoming thicker and of a reddish- or purplish-brown color. The full-grown scale is about ⅛ inch long and is shaped like an oyster shell (Fig. 16.2). The male molts four times, and after about 2 months emerges as a two-winged insect which flies about and mates. The entire life cycle requires an average of about 77 days, and there are three main generations a year.

Control Measures. The purple scale is attacked by a number of natural enemies, including the lady beetles, *Chilocorus* spp. and *Lindorus lophantae*, and the hymenopterous parasites, *Aspidiotiphagus citrinus* and *Aphytis lepidosaphes.* It is also attacked, in Florida, by the fungi, *Hirsutella besseyi* and *Myiophagus* sp. Where chemical control is necessary, purple scale may be controlled by thorough coverage spraying in the fall with light-medium- or medium-grade mineral-oil emulsions at 1.75 per cent (emulsive) or 2 per cent (emulsion) or by thorough coverage spraying immediately after bloom with wettable formulations of parathion at 0.375 to 0.625 pound or malathion at 0.675 to 0.875 pound to 100 gallons of spray. Mixtures of oil emulsion and one-third the dosage of parathion or malathion have also been used.

References. Calif. Agr. Exp. Sta. Bul. 226, 1912, and Cir. 123, 1941.

220 (263). CALIFORNIA[1] AND FLORIDA[2] RED SCALES AND YELLOW SCALE[3]

Importance and Type of Injury. The California red scale[1] is the most important pest of citrus in California. This scale infests all parts of the tree, including leaves, fruits, and twigs, and the injury results solely from the feeding of the insect and is apparently caused by the injection of a toxic substance into the tree through the mouth parts. Infested trees have the leaves spotted with yellow or entire leaves turning yellow and yellow spots on the fruit, but not the marked discolorations that often appear with other scales. The entire bark of twigs and branches may be covered by round, or nearly round, distinctly reddish scales up to about ¹⁄₁₂ inch in diameter, with central exuviae (Fig. 16.3). The yellow scale[3] has only recently been recognized as a separate species and infests primarily the leaves and fruits. Florida red scale[2] is one of the most important citrus pests in Florida and the Gulf states, where it attacks primarily the leaves and fruits (Fig. 18.4). All these scales are difficult to detach

---

[1] *Aonidiella aurantii* (Maskell), Order Homoptera, Family Coccidae.
[2] *Chrysomphalus aonidum* (Linné).
[3] *Aonidiella citrina* (Coquillet).

from the fruits before marketing and lower the grade of the fruit and in severe infestations may seriously injure the health of the tree.

Plants Attacked. These scales primarily infest citrus but also attack acacia, eucalyptus, fig, grape, privet, quince, rose, English walnut, willow, and many other plants. Florida red scale is known to infest 191 species of plants.

Distribution. California red scale is found in California, Arizona, and Texas; yellow scale in California, Texas, and Florida; and Florida red scale in Florida and the Gulf states. The latter is found in greenhouses in California. These scales occur in many other parts of the world where citrus is grown.

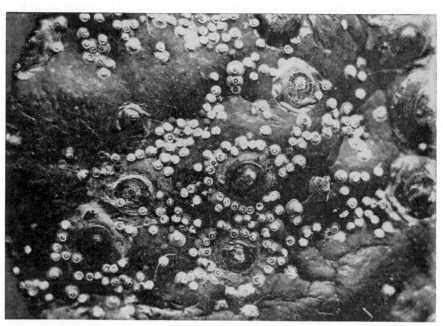

Fig. 16.3. Adults and immature stages of California red scale, *Aonidiella aurantii*, on lemon. About 10 times natural size. (*From Univ. Calif. Coll. Agr.*)

Life History, Appearance, and Habits. The young of California red scale are born alive at the average rate of two to three a day for a period of about 2 months during the summer. Small numbers of young are produced during warmer periods of the winter, but few, if any, are produced during the coldest weather. The first-instar "crawlers" have well-developed legs and antennae and move about for an hour or so before settling, when they begin to cover themselves with a white, waxy covering. The female molts twice at 10- to 20-day intervals, loses its legs and antennae, and incorporates its cast skins into the waxy covering, which becomes circular, depressed, and reddish in color. The female rotates while forming the scale covering, so that the waxy threads secreted from the pygidium are added uniformly to the periphery of the armor. The female reaches maturity in 2½ to 3½ months and lives for several months longer while reproducing. The male scale becomes elongate after the first molt; the third instar is a "prepupal" stage, and the fourth instar a "pupal" stage. The winged adult emerges in from 1 to 2 months, fertilizes the female, and dies. There may be as many as four generations a year in California. The life history and appearance of the yellow scale are virtually identical.

The Florida red scale produces lemon-yellow eggs, which are deposited under the scale armor and hatch within a few hours. The females produce an average of about 150 eggs, and the life cycle requires about 1½ months under summer temperatures. Otherwise the development is very similar to that of California red scale. There are commonly four generations a year.

Control Measures. California red scale and yellow scale may be controlled by thorough coverage spraying with mineral-oil emulsions or with parathion or malathion as discussed under purple scale (page 802). Biological control has not been generally practical, although the aphelinid parasites, *Aphytis lingnanensis* and *A. chrysomphali*, are effective natural enemies in some localities. The encyrtid parasite, *Comperiella bifasciata*, is an effective natural enemy of yellow scale. The encyrtid parasite, *Pseudohomalopoda prima*, and the eulophids, *Aspidiotiphagus lounsburyi* and *Prospaltella aurantii*, may provide effective biological control of Florida red scale under some conditions. Where chemical control is needed, summer spraying with 1.3 to 1.5 per cent oil emulsion or early-fall spraying with parathion may be used.

References. Calif. Agr. Exp. Sta. Cir. 123, 1941; *Jour. Econ. Ento.*, **43**:610, 1950; **48**:432, 444, and 584, 1955; **49**:103, 534, 1956; **51**:194, 1958; and **52**:577, 857, 1959.

Unarmored Scales

The unarmored scales are often called also soft scales and are, as a class, larger than the armored scales. No true scale of wax is formed separate from the body wall. The protective covering is the body wall, which is heavily sclerotized. Both sexes move about during the early part of their lives, but the females cannot crawl after the eggs have formed. These scales discharge quantities of honeydew from their bodies, and the sooty fungus that grows in the accumulations of this honeydew often causes nearly as much injury as the feeding of the scales. The black scale, the soft scale (page 869), and the citricola scale are typical examples of this class.

221 (264). Black Scale[1]

Importance and Type of Injury. This scale is considered one of the destructive pests of citrus in California. The principal damage is caused, in addition to the feeding of the insect, by the sooty mold fungus[2] which grows on the honeydew given off by this scale. This fungus interferes with the physiological functions of the leaves.

Trees Attacked. Orange, grapefruit, lemon, plum, almond, apple, pear, apricot, olive, beech, fig, grape, pepper tree, oleander (the most common host in Florida), rose, English walnut, and a number of other plants.

Distribution. All the principal citrus-growing regions of the world; and in greenhouses in colder areas. It is not important in the Gulf states.

Life History, Appearance, and Habits. In California there are two broods, a single-brooded form in the interior and a double-brooded form

[1] *Saissetia oleae* (Bernard), Order Homoptera, Family Coccidae.
[2] *Meliola camelliae* Catt.

in coastal areas. The full-grown females are nearly hemispherical in shape, being about $\frac{1}{5}$ inch across and from $\frac{1}{25}$ to $\frac{1}{8}$ inch thick, dark

brown to black in color, with a median longitudinal ridge and two transverse elevations on the back forming a letter H (Fig. 16.4). They deposit an average of 2,000 eggs. The eggs, which are about $\frac{1}{80}$ inch in length, are white at first, later changing to orange. The eggs hatch in about 20 days. The young remain beneath the parent scale for some hours and then emerge and crawl about, but always start feeding within 3 days. Most of the young settle on the leaves or new growth. In the fall or winter, usually after the second molt, most of the scales migrate to the twigs and branches. From 8 to 10 months are required for the single-brooded scales to complete their growth. The males pass through prepupal and pupal stages and in the adult stage are active, two-winged insects. They are, however, very rare, and reproduction is generally by parthenogenesis.

FIG. 16.4. The black scale, *Saissetia oleae*, on twig. About natural size. (*From Univ. Calif. Coll. Agr.*)

Control Measures. The encyrtid parasite, *Aphycus helvolus* (page 70), has given effective biological control in some areas but is very susceptible to cold winters. Chemical control may be obtained by thorough coverage spraying with light-medium- or medium-grade mineral-oil emulsions at 1.75 per cent (emulsive) or 2 per cent (emulsion) or with wettables of malathion at 0.645 pound, parathion at 0.375 pound, or DDT at 0.75 pound to 100 gallons of spray. The oil sprays may be fortified with rotenone or with malathion, parathion, or DDT at about one-third the dosage mentioned above. Applications should be made as soon as possible after the major hatch of the scales.

References. Calif. Agr. Exp. Sta. Cir. 123, 1941; *Jour. Econ. Ento.*, **50**:593, 1957; and **52**:596, 1959.

222. CITRICOLA SCALE[1]

Importance and Type of Injury. This species is an important citrus pest in interior California. It sucks the juices from the tree and produces large quantities of honeydew, which may become infested with sooty mold fungus.

Plants Attacked. Primarily citrus.

Distribution. World-wide, but of economic importance in the United States only in California.

Life History, Appearance, and Habits. The adult female is grayish, elongate, and about $\frac{1}{4}$ inch long, closely resembling the soft scale (page 869). The female deposits from 1,000 to 1,500 eggs during a period of 1 to 2 months in the spring. The young nymphs are very flat and transparent and feed principally upon the undersides of the leaves during the summer, passing through two instars over a period of about 2 months.

[1] *Coccus pseudomagnoliarum* (Kuwana), Order Homoptera, Family Coccidae.

In the winter and early spring they migrate to the twigs and smaller branches and become adult. Reproduction is by parthenogenesis, and males are not known. There is one generation a year.

Control Measures. This insect may be controlled by thorough coverage spraying with lime sulfur at 5 gallons plus wettable sulfur 4 pounds to 100 gallons of spray, in January or February; with mineral-oil emulsion at 1.75 per cent (emulsive) or 2 per cent (emulsion), in late summer or early fall; or with wettable parathion at 0.375 pound, or malathion at 0.625 pound to 100 gallons of spray, at any time other than bloom. Other treatments which are used include oil and DDT, or dusting with sulfur, or 2 per cent DDT in sulfur at 100 pounds per acre, at petal fall and 2 to 4 weeks afterward. The encyrtid parasite, *Aphycus luteolus*, is the most effective natural enemy.

223 (265). MEALYBUGS[1-4]

Importance and Type of Injury. Citrus trees infested with mealybugs will have masses of white cottony-appearing insects clustered on the leaves and twigs and at the angles where fruits touch (Fig. 16.5). Infested fruit is generally coated with a very sticky honeydew. Heavy infestations cause ripe fruit to drop and affect future crops.

Plants Attacked. All citrus and many ornamental and greenhouse plants.

Distribution. World-wide. In this country, the insects are more destructive in California, although they are found in the other citrus-growing states.

Life History, Appearance, and Habits. The insects can be found in all stages of development throughout the year. The adult long-tailed mealybug has four waxy filaments at the tip of the abdomen which are nearly as long as the rest of the body; the citrophilus and grape mealybugs have these filaments about one-third as long as the body; whereas in the citrus mealybug they are scarcely longer at the tail end than elsewhere. The

FIG. 16.5. A colony of female citrophilus mealybugs, *Pseudococcus gahani*, on orange twigs. About twice natural size. (*From Univ. Calif. Coll. Agr.*)

citrophilus mealybug has a much darker body fluid than the other species. On the citrus mealybug the white powder over the back is very dense. Except in the case of the long-tailed mealybug, which is ovoviviparous, the mature females, which are about ⅛ to ⅓ inch long (Fig. 18.10), deposit from 300 to 600 eggs in a cottony mass of wax secreted from their

[1] Long-tailed mealybug, *Pseudococcus adonidum* (Linné), Order Homoptera, Family Coccidae.
[2] Citrophilus mealybug, *Pseudococcus gahani* Green.
[3] Grape mealybug, *Pseudococcus maritimus* (Ehrhorn).
[4] Citrus mealybug, *Pseudococcus citri* (Risso).

bodies. The eggs hatch in from 6 to 20 days, and the young mealybugs feed by sucking the sap or juices from the leaves or fruit. They move about but little and require from 1 to 4 months to complete their growth. The female passes through three nymphal instars before transforming to the adult, while the male after three nymphal instars passes through a pupal stage in a flimsy cocoon before becoming an adult. The adult males have two wings, the females are wingless throughout life. There are generally two to four generations a year.

Control Measures. Biological control methods have successfully controlled these mealybugs under most conditions. The annual mass rearing and release of millions of individuals of the coccinellid beetle, *Cryptolaemus montrouzieri*, has been an effective factor in the control of mealybugs in California citrus orchards. The imported hymenopterous parasites, *Coccophagus gurneyi* and *Tetracnemus pretiosus*, of the citrophilus mealybug; *Leptomastidea abnormis*, of the citrus mealybug; and *Anarhopus sydneyensis* and *T. peregrinus*, of the long-tailed mealybug, have given outstanding control of these pests. Where chemical control is required, the mealybugs may be controlled by thorough coverage spraying with parathion at 0.375 pound to 100 gallons of spray.

Reference. U.S.D.A., Farmers' Bul. 1309, 1923.

224. WHITEFLIES[1-4]

Appearance and Type of Injury. Infested trees have a blackened appearance due to a sooty mold which grows in the honeydew given off by the whitefly nymphs. This sweetish, sticky honeydew is discharged in large quantities from the alimentary tract. Trees are stunted, through loss of sap, and the fruit is undersized and of poor color.

Distribution. Whiteflies are widely distributed throughout the world. They are serious pests of citrus in Florida and the Gulf states but are, at present, of no importance in California. The common citrus whitefly[1] (Fig. 16.6) formerly found in California has been eradicated. The cloudy-winged whitefly[2] and the woolly whitefly[3] are also pests in Florida. The citrus blackfly,[4] which occurs in Mexico and in the West Indies, in the nymphal stages has a black body with a white fringe; the adult has a dark-brown body with the wings dark bluish.

Food Plants. All species of citrus and many other plants, especially camellia. In Florida these insects breed in large numbers on chinaberry.

Life History, Appearance, and Habits. The life histories of all species of whiteflies which are of importance on citrus are very much alike. All stages of the insects may be found throughout the year, but little breeding takes place during cold periods. The oval eggs, less than $\frac{1}{100}$ inch in length, are attached to the undersides of the leaves by a short stalk. The eggs of the citrus whitefly are pale yellow, those of the cloudy-winged whitefly are black, and those of the woolly whitefly are brown and curved like miniature, fat sausages. They hatch in from 4 to 12 days into active pale-yellow flattened six-legged "crawlers," or nymphs. They move about for a short time, mainly on the lower sides of the leaves, as they avoid strong light. These crawlers soon insert their beaks into the leaves and begin sucking the sap. They soon molt, losing their legs in the

[1] *Dialeurodes citri* (Ashmead), Order Homoptera, Family Aleyrodidae.
[2] *Dialeurodes citrifolii* (Morgan).
[3] *Aleurothrixus floccosus* (Maskell).
[4] *Aleurocanthus woglumi* Ashby (see *U.S.D.A., Dept. Bul.* 885, 1920).

process, and then have the appearance of minute, flattened, oval bodies, attached to the undersides of the leaves by their sucking beaks (Fig. 18.13), and with a marginal fringe of short, white, waxy filaments somewhat like a mealybug. After two more molts the adults (Fig. 16.6) emerge. They are small four-winged insects about $\frac{1}{12}$ inch long. Both sexes are winged, and both feed by sucking sap. They have a white appearance because of the fine white powder which completely covers the wings and body. In Florida there are usually three generations a year of the common citrus whitefly.

Fig. 16 6. Citrus whitefly, *Dialeurodes citri*, adults and eggs on leaf. Slightly enlarged. (*From Florida Agr. Exp. Sta. Bul.* 183.)

Control Measures. Spraying with oil emulsions or parathion as discussed under purple scale (page 802) will give control. Keeping down water sprouts is of some value in controlling these insects. Entomogenous fungi of the genera *Aegerita* and *Aschersonia* may be of value in control under ideal weather conditions.

References. Fla. Agr. Ext. Serv. Bul. 67, 1932; *U.S.D.A., Bul.* 885, 1920; *Jour. Econ. Ento.,* **40**:499, 1947.

225. RED SPIDER MITES[1-6]

Importance and Type of Injury. These mites are among the most destructive of all pests of citrus fruits. The citrus red mite[2] produces a silvery, speckled effect on the leaves by extracting the chlorophyll, and those heavily infested turn silver or brown and drop. The fruits also become gray or yellow. The feeding of the six-spotted mite[3] is restricted to limited areas on the undersides of the leaves, where the mites form colonies. These areas become depressed, yellow, and webbed beneath and swollen, shiny, and yellow on the upper side, and the leaves often drop. The Lewis spider mite[4] attacks the fruit, where it causes a russeting of oranges and a silvering of lemons. The flat mite[5] causes a pitting

[1] Order Acarina, Family Tetranychidae.
[2] *Panonychus* (= *Metatetranychus*) *citri* (McGregor).
[3] *Eotetranychus* (= *Tetranychus*) *sexmaculatus* (Riley).
[4] *Eotetranychus lewisi* (McGregor).
[5] *Brevipalpus lewisi* McGregor.
[6] *Eutetranychus banksi* Pritchard and Baker (= *Anychus clarki*).

or spotting of citrus fruits.   The Texas citrus mite[1] causes leaf injury
similar to the citrus red mite.

Plants Attacked.   These red spider mites do little damage to plants
other than citrus.

Distribution.   The citrus red mite is injurious in California and Florida.
The six-spotted mite is important throughout Florida and the Gulf states
and in limited coastal areas of California.   The Lewis mite and flat mite
are pests in California, and the Texas citrus mite in Texas and Florida.

Life History, Appearance, and Habits.   In most of the citrus-growing
area of the United States all stages of the mites are found throughout the
year, although their numbers are generally greatly reduced during the
winter and often in midsummer.   The eggs are fastened to the leaves or

Fig. 16.7. The citrus red mite, *Panonychus citri*, mites and one of their eggs showing the
silken threads, like guy lines, that stretch from the leaf to the top of the egg stalk.   (*From
Quayle, Univ. of California.*)

to the silk spun by the mites.   The female citrus red mite lays from 20
to 50 eggs, at the rate of 2 to 3 a day, on the leaves, twigs, or fruit.   The
bright-red eggs have a vertical stalk like a mast, from the top of which
about a dozen threads extend like guy lines, to the leaf.   The mite passes
through the six-legged larval stage and the eight-legged protonymph and
deutonymph stages, which closely resemble those of the European red
mite (Fig. 15.27) before becoming adult.   The entire life cycle requires
from 3 to 5 weeks, depending upon the temperature, and there may be
from 12 to 15 generations a year.   The adult citrus red mite is velvety red
or purplish in color, with about 20 prominent bristles over the body, each
arising from a conspicuous tubercle (Fig. 16.7).   The life history of the
six-spotted mite is very similar.   The colorless to greenish-yellow eggs
are deposited among the webbing of the colony and bear a stalk but no

[1] *Eutetranychus banski* Pritchard and Baker (= *Anychus clarki*).

guys. The adult is pinkish, greenish, or yellowish, with six dark spots. The bristles do not arise from conspicuous tubercles. The Lewis mite is very similar.

Control Measures. The citrus red mite may be controlled by thorough coverage spraying with light-medium- or medium-grade mineral-oil sprays at 1.75 per cent (emulsive) or 2 per cent (emulsion) applied in the fall or by mist-spraying with ovex (Ovotran) at 4 pounds, Kelthane at 3 pounds, Tedion at 3 pounds, Delnav at 4 pounds, or DNOCHP at 2 pounds per acre as the mites appear. These treatments also control the other species of red spider mites attacking citrus. The six-spotted mite and the Texas citrus mite may be controlled by spraying with wettable sulfur at 4 pounds to 100 gallons of spray or by dusting with sulfur at 80 pounds per acre.

References. *Calif. Agr. Exp. Sta. Bul.* 234, 483, 1912; *Jour. Agr. Res.,* vol. 36, no. 2, 1928; *Jour. Econ. Ento.,* **29**:125, 1936; **39**:813, 1946; **43**:807, 1950; **46**:10, 1953; **50**:293–307, 1957; and **51**:232, 1958.

226. Citrus Rust Mite[1] and Citrus Bud Mite[2]

Importance and Type of Injury. The rust mite is second only to the purple scale as a citrus pest in the Gulf states. By rupturing the cells

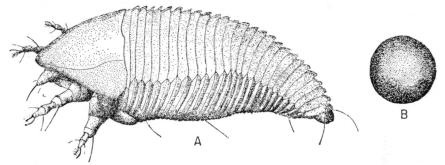

FIG. 16.8. Citrus rust mite, *Phyllocoptruta oleivora.* *A,* adult, side view, × 700; *B,* egg, × 825. *(From U.S.D.A.)*

and sucking sap from the skin of the fruit and leaves it causes a serious russeting of oranges and a silvering of lemons, lowering the quality and attractiveness of the fruit. The bud mite has only recently become a serious pest of lemons in California, where it occurs in the buds and blossoms and beneath the calyx, causing curious deformities of the fruits, leaves, blossoms, buds, and twigs.

Plants Attacked. The rust mite injures oranges, grapefruit, lemons, and limes, in the order named. All other citrus fruits are attacked. The bud mite is restricted to citrus and attacks principally lemons.

Distribution. The rust mite is most important in the Gulf states but also occurs in limited areas of California. The bud mite is found only in California.

Life History, Appearance, and Habits. The rust mites occur on the trees throughout the year but are least abundant in January and February. The females deposit up to 30 pale-yellow, smooth, spherical eggs

[1] *Phyllocoptruta oleivora* (Ashmead), Order Acarina, Family Eriophyidae.
[2] *Aceria sheldoni* (Ewing), Order Acarina, Family Eriophyidae.

in depressions on the fruits and leaves.   After 2 to 8 days the nymphs
hatch.   They feed like the adults, molt twice at from 1- to 6-day intervals,
and are then slender, elongated females which are only $\frac{1}{200}$ inch long,
with two pairs of legs at the head end and a tapering, ringed abdomen
(Fig. 16.8).   Generations succeed each other at intervals as short as 1 or
2 weeks, and the mites become most abundant about July 1.   The bud
mite deposits white spherical eggs in protected places on the tree, and
these hatch in 2 to 6 days.   The mites pass through three molts in 10 to
20 days and become tiny cream-colored to pinkish adults which are about
$\frac{1}{150}$ inch long and worm-like in appearance.

   Control Measures.   For the rust mite, dusting with sulfur at 40 to 80
pounds' per acre or spraying with wettable sulfur at 4 to 10 pounds to
100 gallons of spray is effective.   Two to four applications are usually
required in Texas and Florida.   Lime-sulfur sprays are also effective,
although oil sprays provide but partial control.   Spraying with zineb at
0.65 pound or Chlorobenzilate at 0.25 pound to 100 gallons of spray when
needed is very effective.   The bud mite may be controlled by spraying
with light-medium-grade oil emulsion at 1.75 per cent (emulsive) or 2
per cent (emulsion) in the fall or spring or with Chlorobenzilate at 1.5
pounds per acre.

   References.   U.S.D.A., Tech. Bul. 176, 1930; *Jour. Econ. Ento.*, **34**:745, 1941;
45:271, 1952; **48**:375, 1955; and **51**:657, 1958.

227. FRUIT FLIES[1-3]

   Importance and Type of Injury.   These pests are capable of preventing
successful growing of a number of important fruits in areas of mild winters
where there is a year-round succession of cultivated or wild fruits upon
which they can develop.   The maggots develop in the pulp of the fruit,
devouring it and favoring the development of bacterial and fungus diseases; the egg punctures made by the adults may affect shipping qualities
of the fruit.

   Plants Attacked.   The Mediterranean fruit fly[1] attacks many citrus and
deciduous fruits, including especially oranges, grapefruit, peaches, nectarines, plums, apples, pears, quinces, coffee, and nearly 100 other cultivated and wild fruits.   The oriental fruit fly[2] attacks all citrus and most
deciduous fruits, bananas, pineapples, avocados, tomatoes, melons, and
at least 150 other plants.   The Mexican fruit fly[3] is primarily a pest of
citrus and mangoes.

   Distribution.   The Mediterranean fruit fly is not now known to be present in the continental United States.   It was discovered in Florida in
1929, in scattered locations over an area of about 10,000,000 acres.   It
was never found in wild host plants away from cultivated land.   A most
vigorous and remarkably successful campaign of eradication, involving
the expenditure of over $7,000,000 of state and federal funds, resulted in
its complete extermination.   The fly again invaded Florida in 1956 and
spread into 28 counties before it was eradicated at a cost of over $10,000,-
000.   It occurs in Bermuda and Hawaii and in nearly all subtropical
countries except North America.   Constant vigilance is required to pre-

---

[1] *Ceratitis capitata* (Wiedeman), Order Diptera, Family Trypetidae.
[2] *Dacus dorsalis* Hendel, Order Diptera, Family Trypetidae.
[3] *Anastrepha ludens* (Loew), Order Diptera, Family Trypetidae.   The melon fly,
Dacus cucurbitae Coq., and the walnut husk fly, *Rhagoletis completa* Cresson, are
important fruit fly pests of other crops.

vent its reintroduction to this country in shipments of horticultural products or in the baggage of travelers. The oriental fruit fly was introduced into Hawaii in 1946 from the Marianas Islands. It is also found in Formosa, the Philippines, Burma, and India. The Mexican fruit fly is a serious pest in Mexico and Central America and has several times appeared in the Rio Grande valley of Texas. Several specimens of both the Mexican and oriental fruit flies have been trapped in California orchards, but no general infestations exist.

Life History, Appearance, and Habits. In cool regions the Mediterranean fruit fly winters as pupa or adult, while in warmer regions, where fruits are available, reproductive activity may be continuous throughout the year. From 2 to 10 eggs are deposited through a small hole, the size of

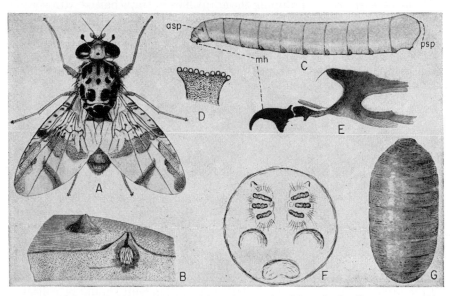

Fig. 16.9. Mediterranean fruit fly, *Ceratitis capitata*. *A*, adult male fly; *B*, two egg cavities in rind of grapefruit, showing conical elevation left by withering of the rind, the right one in section to show mass of eggs and a single tunnel made by a newly hatched larva; *C*, a larva in side view, *asp*, anterior spiracle, *psp*, posterior spiracles, *mh*, mouth hooks; *D*, anterior spiracle, more enlarged; *E*, mouth hooks and cephalopharyngeal skeleton of larva; *F*, posterior view of last segment of larva, showing posterior spiracles of full-grown larva; *G*, puparium. (*Rearranged from U.S.D.A.*)

a pinprick, made in the rind of the fruit by the ovipositor, but many additional eggs may be laid in the same hole by other females. A single female may produce 800 eggs. The eggs may hatch in 2 to 20 days, and the larvae burrow in the pulp for 10 days to 6 weeks before completing growth, when they are a little more than $\frac{1}{4}$ inch long. They then desert the fruit and form a puparium in the soil, within an inch or two of the surface or under other protection, or even exposed in boxes or wrappers. Pupation may be completed in 10 to 50 days. The adult (Fig. 16.9) is about the size of the house fly. The thorax is glistening black with a characteristic mosaic pattern of yellowish-white lines. The abdomen is yellowish with two silvery crossbands, and the wings are banded and blotched with yellow, brown, and black. The adults have sponging mouth parts and take only liquid foods. The females may oviposit successfully after as much as 10 months of inactivity. Under most favorable conditions the life cycle may be passed in 17 days, commonly in 3 months; and there may be from 1 to 12 or more generations a year in various parts of the world. The habits of the oriental and Mexican fruit flies are very similar. The oriental fruit fly deposits an average of 1,200 to 1,500 eggs, which are inserted under the skin of the host fruit in groups of up to 10. The larva passes through three instars in the fruit over a period of 6 to 35 days, and then drops to the soil to pupate. The entire life cycle from egg to adult requires about 16 days at summer temperatures, and there is a preoviposition period of 8 to 12 days. The adult fly is somewhat larger than the house fly and has clear wings, and the body is bright yellow with dark markings on the thorax and abdomen.

Control Measures. Trapping the adult flies in traps baited with specific attractants such as methyl eugenol for the oriental fruit fly or medlure for the Mediterranean fruit fly (page 395) is the most effective way to determine the extent of infestation. These may be poisoned with 0.5 per cent DDVP or 1 per cent Dibrom. Area control, as in the Florida eradication campaign, is most satisfactorily carried out by the biweekly application of bait sprays such as:

Yeast protein	1 lb.
Malathion, 25% wettable powder	3 lb.
Water	40–150 gal.

Conventional spraying with parathion, DDT, Dilan, or methoxychlor at 3 to 4 pounds per acre or with malathion at 5 to 6 pounds per acre has protected fruits from attack. Plant quarantines against the transportation of infested commodities are important in preventing the reinfestation of mainland areas of the United States. Fruit may be sterilized for interstate shipment by treatment with moist heat to raise the temperature to 110°F. for 10 to 14 hours, by exposure to 33°F. for 15 days, or by fumigation with ethylene dibromide at 8 to 10 ounces per 1,000 cubic feet for 2 hours. In Hawaii, the introduced parasites *Opius longicaudata*, *O. vandenboschi*, and *O. oophilus* have substantially reduced the numbers of the oriental fruit fly. In infested areas, fallen fruit should be collected daily and buried to a depth of several feet or burned or cooked to destroy the larvae.

References. *U.S.D.A., Dept. Buls.* 536, 1918, and 640, 1918; *Misc. Publ.* 531, 1944; *State Calif. Spcl. Bul.,* "Oriental Fruit Fly," 1949; *Jour. Econ. Ento.,* **40**:483, 1947; **45**:241, 388, 838, 1952; **48**:310, 1955; **49**:176, 1956; **50**:16, 1957; and **51**:679, 1958.

227A (82, 96, 181, 267). APHIDS[1-5]

Importance and Type of Injury. The important species of aphids attacking citrus in California and Florida, in order of abundance, are the green citrus or spirea aphid,[2] the melon aphid[3] (page 631), the green peach aphid[4] (page 754), and the black citrus aphid.[5] These aphids cause severe curling and stunting of young leaves and twigs and stimulate the growth of the sooty mold fungus by secreting honeydew. Sometimes

[1] Order Homoptera, Family Aphidae.
[2] *Aphis spiraecola* Patch.
[3] *Aphis gossypii* Glover.
[4] *Myzus persicae* (Sulzer).
[5] *Toxoptera aurantii* (Fonscolombe).

they attack the blossoms, causing them to drop, and young fruits, causing stunting. The most serious injury is caused by the melon aphid, and to a lesser extent by the green citrus aphid, which are vectors of the tristeza or quick-decline virus, *Corium viatorum*. This disease is invariably fatal to infected trees with susceptible sour rootstock.

Life History, Appearance, and Habits. Reproduction of these aphids on citrus is entirely parthenogenetic. The squat greenish nymphs of the green citrus aphid appear in large numbers in early spring. These become reproductive females in 4 to 16 days and may give birth to as many as 100 young during a lifetime of 2 to 4 weeks. Whenever the colony becomes crowded or the young spring growth begins to harden, darker-winged forms appear and spread to more favorable locations. The melon aphid has a very similar life history, with a nymphal period of 3 to 20 days and a life span of 9 to 64 days, during which an average of 67 young are produced. There may be as many as 47 generations a year in Florida.

Control Measures. Natural enemies such as lady beetles, syrphid larvae, lace-winged flies, and hymenopterous parasites destroy large numbers of aphids and generally keep them under control during unfavorable seasons of the year. However, during the spring the reproductive potentials of the aphids are so great that damaging numbers often develop. The aphids may be controlled by mist-spraying with nicotine at 0.75 pint of 40 per cent, rotenone at 1 pint of 2.5 per cent, malathion at 0.375 pound, demeton (Systox) at 0.125 pound, or TEPP at 0.2 pound to 100 gallons of spray. Dusts of nicotine and TEPP are also effective. Applications should be made before large populations develop.

References. Fla. Agr. Exp. Sta. Buls. 203, 1929, and 252, 1932; *Jour. Econ. Ento.,* **44**:172, 1951.

CHAPTER 17

INSECTS ATTACKING SHADE TREES AND SHRUBS

Nearly all our forest trees, shade trees, and shrubs are attacked by a large number of insects. No attempt is made in this book to treat all the insects of forest trees. More than a thousand different species are known to feed upon the oaks, and correspondingly large numbers attack various other shade trees. A few shade trees are relatively free from injury. Shrubs are subject to attack by scale insects, especially the oyster-shell scale and San Jose scale. Poplar, linden, dogwood, willow, rose, and lilac are some of the trees and shrubs most subject to insect attack.

TABLE 17.1 RELATIVE RESISTANCE OF TREES TO INJURY BY INSECT PESTS
(TREES HAVING THE HIGHER NUMBERS SHOULD BE GIVEN PREFERENCE
FOR PLANTING)

Sweet gum	6	European linden	3
Tree of Heaven	6	Horse chestnut	3
Ginkgo	5	Buckeye	3
Red oak	5	American elm	3
Scarlet oak	5	Hackberry	3
Oriental plane	5	Water or red elm	3
Tulip, or tulip poplar	5	Soft or silver maple	2
Sycamore	5	European elm	2
Sugar maple	5	Scotch elm	2
Norway maple	5	American linden	2
White oak	5	Cottonwood	1
Burr oak	5	Carolina poplar	1
American plane	4	Lombardy poplar	1
Red maple	4	Balm of Gilead	1
Honey locust	4	Black locust	1
Spruces	4	Boxelder	1
Blue ash	4	American mountain ash	1
European mountain ash	4	Green ash	1
White pine	3	Black ash	1
Catalpa	3	White ash	1

This table follows a plan originated by Dr. E. P. Felt, formerly State Entomologist of New York.

Where an especially destructive insect pest is well established, one should avoid planting trees or shrubs particularly likely to be attacked by this pest. In most cases such trees or shrubs can be grown if frequently sprayed, but other species which are not subject to attack by the same insect and which are nearly as attractive in appearance could well be substituted. As an example, one should avoid planting white, green, black, or red ash in any of the cities of central Indiana, Illinois, Ohio, and neighboring states, as such trees are certain to be infested with the oyster-shell scale. The blue ash, sycamore, various maples, and many other trees can be planted in these localities, and will never be injured in the least by this scale.

Table 17.1 shows the relative resistance to insect attack of the different varieties of shade trees commonly grown in the latitude of central Ohio and Illinois. The figure 6 has been placed opposite trees which are practically immune to insect injury; 5 indicates some damage; trees having one somewhat serious enemy are rated at 4; and those having at least one notorious insect pest at 3. Greater likelihood of injuries is indicated by 2, and still more by 1. The species of trees are arranged according to the comparative resistance to insect attack.

General References. DOANE, VAN DYKE, CHAMBERLIN, and BURKE, "Forest Insects," McGraw-Hill, 1936; HERRICK, "Insect Enemies of Shade Trees," Comstock 1935; FELT, "A Manual of Tree and Shrub Insects," Macmillan, 1924; GRAHAM, S., "Forest Entomology," McGraw-Hill, 1952; WESTCOTT, C., "The Gardener's Bug Book," Doubleday, 1946; *U.S.D.A., Misc. Publ.* 273, 1938; *N.Y. State Museum Memoir* 8, Vols. 1 and 2, 1905; *Ohio Agr. Exp. Sta. Bul.* 332, 1918; *U.S.D.A., Farmers' Bul.* 1169, 1921; *Univ. Md. Agr. Ext. Ser. Bul.* 84, 1939; *U.S.D.A., Misc. Publ.* 626, 1948; *Ore. Agr. Exp. Sta. Bul.* 449, 1948; ANDERSON, R. F., "Forest and Shade Tree Entomology," Wiley, 1960.

FIELD KEY FOR THE IDENTIFICATION OF INSECTS INJURING SHADE TREES AND SHRUBS

A. *External chewing insects that eat holes in leaves, buds, bark, or fruit:*
1. Yellow-and-black beetles and grubs, about ¼ inch long, eating out the green parts of elm leaves. Large numbers of grubs collect on the bark about the base of the trunk or on the ground near by.........*Elm leaf beetle*, page 821.
2. Leaves of willows, poplars, and cottonwood partly or completely eaten. Convex, oval, yellow or reddish beetles, spotted or striped with black, about ¼ to ½ inch long, feed, along with their soft-bodied, more elongate, blackish young or larvae, on underside of leaves................................
Cottonwood, poplar, aspen, and *willow leaf beetles, Chrysomela scripta* Fabricius, *Chrysomela lapponica* Linné, *Chrysomela tremulae* Fabricius, *Chrysomela interrupta* Fabricius, and others.
3. Spruce or balsam trees appearing as though scorched by fire and slowly dying. Thick dark-brown caterpillars, ¾ inch or less in length, covered with pale-yellow warts, eat off the needles and spin silken threads over the terminal shoots, webbing them together, in early spring....*Spruce budworm*, page 822.
4. Shade trees, particularly evergreens, stripped of their leaves by brownish, rather fat-bodied worms, which live within tough silken bags. Bags up to 2 inches in length hanging in large numbers from the leaves and twigs of infested trees. Bags remaining on the trees during the winter months......
....*Bagworm*, page 823.
5. Catalpa trees completely stripped of their leaves by dark-colored caterpillars marked with varying amounts of green and yellow. Caterpillars up to 3 inches in length, with a curved black horn projecting from the last segment of their bodies.......................................*Catalpa sphinx*, page 825.
6. Large colonies of black caterpillars stripping the leaves from hickory, walnut, and related trees. All foliage completely stripped from single branches. Caterpillars coming down the trunks of the trees in large numbers and hanging in masses on the trunks while shedding their skins......................
....*Walnut caterpillar*, page 826.
7. Loosely woven, flimsy, white webs of silk, enclosing the leaves of branches of varying size up to several feet across, contain very hairy pale-yellow black-

spotted caterpillars, up to 1¼ inches long, which eat the foliage without leaving their tent. Injury most noticeable in late summer or fall in the North......
....*Fall webworm*, page 692.

8. Tent-like webs of silk in the forks of wild cherry, wild plum, crab apple, and other trees in spring. Brownish, very hairy caterpillars, 2 inches or less in length, with a light stripe down the back, hide in these tents during the day. At night they make silken pathways along the branches to feed upon the foliage.................................*Eastern tent caterpillar*, page 693.

9. Pale-blue hairy caterpillars, with a row of keyhole-shaped white spots down the middle of the back and pale-yellow stripes on the sides, defoliate poplars, oaks, maples, and many other trees, without spinning tents; and often swarm into buildings...........................*Forest tent caterpillar*, page 828.

10. Leaves of trees skeletonized or stripped by yellowish caterpillars with 2 long tufts or pencils of dark hairs protruding from near the head, and one tuft from near the tail. Trunk and branches of trees during winter months with dark-gray cocoons on the bark, cocoons often having frothy-white masses of eggs attached to them................*White-marked tussock moth*, page 828.

11. Buff-colored masses of eggs, up to 1 inch across, covered with felt-like hairs, on the trunks of shade trees and surrounding objects during the winter months. Large dark-gray, somewhat flattened caterpillars, with pairs of red and blue spots down the back, strip the leaves from trees during June and early July..
....*Gypsy moth*, page 830.

12. During the winter season several leaves tightly webbed together and firmly attached to a twig near the tip of the branch, enclose 25 to several hundred small dark-brown very hairy caterpillars, about ¼ inch long. If the caterpillars or their hairs come in contact with the skin, a distinct burning or itching results, which may persist for several days. White moths, with brown abdomens, flying in great numbers at night during July, strongly attracted to lights.......................................*Brown-tail moth*, page 832.

13. Poplar and willow leaves eaten off during midsummer by rather large black-bodied caterpillars. When full-grown, these caterpillars are about 2 inches long, with irregular whitish colorings on the sides of the back and a nearly square white patch of hairs on the middle of the back of each segment. The adult insect is a white moth with a wing spread of nearly 2 inches..........
...*Satin moth*, *Stilpnotia salicis* (Linné) (see *U.S.D.A.*, *Dept. Bul.* 1469, 1927).

14. Smooth greenish or brownish measuring-worms or looping caterpillars, which lack prolegs near the middle of the body, eat foliage of elm, hackberry, oak, and other deciduous trees, dropping on silk threads when disturbed........
.....................................*Cankerworms*, pages 689 and 691.

15. Greenish, yellowish, whitish, or bluish worms, 1 inch or less in length, often spotted or striped with black or yellow, with dark heads and 9 to 11 pairs of legs and prolegs, defoliate elm, birch, poplar, willow, pine, larch, spruce, balsam, or other trees during the summer. Worms usually hold their bodies or tails coiled or curved over edge of leaves...............................
...................*Sawflies* (Order Hymenoptera, Family Tenthredinidae).

B. *Motionless scale insects on twigs, trunks, or leaves, sucking sap:*

1. The bark of shade trees, particularly elm and willow, with small grayish-white flat scales, often nearly covering the bark. From September to late spring, minute purplish eggs to the number of 50 or more will be found under each of these scales......................................*Scurfy scales*, page 834.

2. Surface of the bark of ash, poplar, and many other shade trees covered with brownish-gray scales, which, upon close examination, will be found to resemble the half of a minute oyster shell. During the winter months, numerous pearly white, very small eggs to the number of 75 to 100 will be found packed beneath the larger of these scales........................*Oyster-shell scale*, page 835.

3. Small whitish scales on the needles or leaves of pine, spruce, hemlock, and various other evergreens. Often present in such numbers as to give the needles a grayish appearance. Very small, purplish eggs beneath these scales during the winter months..............................*Pine needle scale*, page 835.

4. Underside of limbs and trunks of elm trees more or less completely covered with oval reddish-brown scales, up to ⅜ inch long, surrounded with a white fringe. Sticky honeydew over the trees and objects beneath, and many flies and wasps buzzing about..................... *European elm scale*, page 836.

5. Whitish cottony-appearing masses of wax, nearly ⅓ inch across, on the undersides of the branches of maple, linden, and other shade trees. Such masses most conspicuous during June and July. Small, flat, brownish-appearing scales, adhering tightly to the bark of shade trees during the winter months...
.... *Cottony maple scale*, page 837.

6. Plump, rounded, reddish-brown scales, about ⅛ inch across, sharply convex and oval in outline, clustered on the twigs and branches of various shade trees during the winter months. Scales often so thick as completely to cover the bark... *Terrapin scale*, page 837.

7. Very plump, almost hemispherical, brownish-gray hard scales up to ¼ inch or more in diameter, on the leaves and terminal twigs, especially of oaks of the burr oak group. Some of the twigs with leaves dead at the tips............
.... *Oak Kermes* (see general references, page 817).

C. *Soft-bodied, more or less active, sap-sucking insects, on the leaves, branches, twigs, or trunks:*

1. Leaves of elm curled and bunched together in the spring, but no galls formed upon them. Numerous brownish aphids, with a purplish-white woolly wax covering their bodies, feeding within the curled leaves. Copious discharge of honeydew from such leaves.................... *Woolly apple aphid*, page 839.

2. The trunks and undersides of the branches of pines and balsam become covered with white cottony flecks. The flocculent tufts of wax cover dark-brown aphids, that suck the sap and cause a sickly condition of the trees...........
.... *Pine bark aphid*, *Pineus strobi* (Htg.) (= *Chermes pinicorticis* Fitch).

3. Large numbers of bright-red bugs, ½ inch or less in length, the larger ones with dark wings bordered with red and with 3 red lines on the thorax, suck sap from leaves and new growth of boxelder or ash. In the fall the bugs cluster in masses on the trunks or wander in great numbers up and down the tree trunks or crawl over walls and porches and enter houses...... *Boxelder bug*, page 839.

D. *Borers working in the trunks and branches:*

1. Shallow mines in the bark and sapwood of the trunk and larger branches of shade trees, these mines generally occurring on the south and southwest sides of the trees. Grayish-white grubs, with a pronounced flattened enlargement of the body just back of the head, working in these burrows. Adults are flattened beetles with the body tapering back from the shoulders, the antennae short, and their backs often irregularly roughened and metallic-colored.......
.... *Flatheaded borers* or *metallic wood borers*, page 840.

a. Birch trees with the leaves on the upper branches dying. Brown spots on the bark, and the inner bark of branches and trunk with many zigzag sawdust-filled burrows running through it. White slender grubs, up to 1 inch in length, in these burrows. Entire trees dying after 1 or 2 seasons.......
.... *Bronze birch borer*, page 840.

2. Injury similar to D,1, but mines more often extend through the solid wood of the tree, and the enlargement of the front end of the body is not flattened. Adults are generally more or less cylindrical beetles, with very long antennae, and often beautifully colored...
.... *Long-horned borers* or *roundheaded borers*, page 841.

a. Large amounts of excelsior-like sawdust accumulating about the base of poplar, particularly Carolina poplars, aspen, and cottonwood. Ragged, usually dark-colored holes in the trunk and branches, from which dark-colored sap is seeping and coarse sawdust is being forced out. Branches breaking, disclosing large burrows running through the wood, and greatly weakening the structure. Injury most severe on Carolina poplar, Lombardy poplar, and cottonwood.................... *Poplar borer*, page 842.

b. Black locust trees with swollen areas on the trunk and larger branches, the bark on these areas often cracking open, or trees breaking over at the point

of injury. Wood of the trees honeycombed with burrows of rather large
borers...*Locust borer*, page 843.

c. Elms in a weakened, sickly condition with areas of the bark along the trunk
becoming loose and easily detached from the tree. Such bark containing
numerous burrows filled with brown sawdust. Yellowish-white grubs, up to
a little over 1 inch in length, will be found working in these burrows. Injury
occurring only on weakened or sickly trees...........*Elm borer*, page 845.

3. Terminals, or leaders, of pines, fir, or spruce turn brown and die, from midsummer on, as a result of the work of fat, white, legless grubs that tunnel
through the wood and make "shot holes" through the bark of the twigs, which
exude small drops of resin. Eggs laid by a reddish-brown snout beetle, ¼ to
⅓ inch long, irregularly blotched with white on the back....................
White pine weevil, *Pissodes strobi* (Peck) (see *N.Y. (Cornell) Agr. Exp. Sta. Bul.*
449, 1926; *U.S.D.A., Cir.* 221, 1932).

4. Branches and limbs of willows or poplars in northeastern United States are
bored with tunnels, deformed with knotty swellings, and splitting and breaking,
with much sawdust and sap oozing at points of attack, and foliage wilted. Fat,
white, legless grubs, ½ inch long or less, may be found tunneling in the trees
from fall to midsummer. Eggs laid by blackish, white-flecked, chunky, snout
beetles, about ⅓ inch long, with the rear third of the wing covers, the sides of
the thorax, and parts of the legs pink.....................................
Poplar and willow borer, *Cryptorhynchus lapathi* (Linné) (see *N.Y. (Cornell)*
Agr. Exp. Sta. Bul. 388, 1917).

5. Fine sawdust deposited in small amounts over the trunk and branches of
deciduous trees. Small, round holes in the bark, a little larger than a pinhead.
Very small, blunt-headed black beetles crawling over the bark or boring through
the bark. Numerous galleries in the inner bark and outer sapwood, branching
out from a short parent gallery...
Elm bark beetles, page 848; *fruit tree bark beetle*, page 722; *hickory bark beetle*,
page 846; *peach bark beetle*, page 755; and others.

6. Swarms of very small, chunky, cylindrical, brown or black beetles, usually
⅛ to ¼ inch long, attack the trunks of various conifers or evergreen trees,
eating many holes into the bark and making galleries in the inner bark and
outer sapwood. Small white grubs make similar tunnels radiating from the
parent gallery. Sawdust sifts down, and pitch or resin accumulates, at the
entrance holes. Affected trees show reddish tops and soon die..............
............................*Bark beetles* and *ambrosia beetles*, page 846.

7. Large dark-colored burrows in the wood of locust, willow, chestnut, maple, and
some other shade trees, but particularly abundant in black and red oaks. Dark-
colored sap oozing from such burrows and discoloring the bark for some distance below the burrow. Large, pinkish, brown-headed borers, up to 3 inches
in length, with well-developed prolegs, working in these burrows............
....*Carpenterworm*, page 850.

8. Shade trees along the eastern coast of the United States, especially elm and
maple, with many dead branches in the tops of the trees, or with small branches
broken over and hanging partly severed. Numerous holes along the infested
branches from which sawdust is being forced out. Whitish brown-spotted
caterpillars, up to 3 inches long, in these burrows.....*Leopard moth*, page 851.

9. The canes of lilacs dying or breaking over. Slightly enlarged, swollen areas
on the canes, just above the surface of the ground. Bark cracked from infested
canes, showing the presence of numerous burrows through the wood.........
....*Lilac borer*, page 852.

E. *Insects causing the development of galls on leaves and twigs:*
1. Leaves of elm with raised, much wrinkled, greenish or pinkish galls on the
upper surface. Such galls when broken open, found packed almost solid with
greenish or brownish aphids. Small slits on the undersides of the leaves below
the galls......................................*Elm cockscomb gall*, page 853.

2. See also Felt, "Key to American Insect Galls," *N.Y. State Mus. Bul.* 200,
1917.

F. *Twigs girdled, severed, or splintered so that they die and hang broken over or fall to the ground:*

1. Many twigs, ⅓ to ¾ inch in diameter and several feet long, neatly cut off and lying on the ground beneath hickory, persimmon, oak, poplar, sour gum, honey locust, and other deciduous shade, fruit, and nut trees. Severed end of twig is convex and no borers in the new fallen twigs. Grayish cylindrical hard-shelled beetles, nearly 1 inch long, with antennae longer than the body, girdle the twigs by cutting round and round them from the bark inward....................
.... *Twig girdler, Oncideres cingulata* (Say), and other species (Fig. 1.7,*D*).
2. Twigs severed as in *F*,1, from oak, maple, hickory, chestnut, locust, and other shade and fruit trees. Severed end of twig is concave and with a central burrow leading from it up the twig, which is plugged with shavings and contains a white, cylindrical, conspicuously segmented grub, which cut off the twig from the inside.... *Twig pruner, Ellaphidion* (= *Hypermallus*) *villosum* (Fabricius).
3. Tips of small branches breaking off or hanging with dead leaves during the early summer, wood split at point of break, with tufts of splinters sticking up at short intervals along the twig...................... *Periodical cicada*, page 735.

228. ELM LEAF BEETLE[1]

Importance and Type of Injury. This beetle is one of the most destructive pests of the elm tree throughout the United States. Infested trees have a general yellow appearance of the foliage, with many leaves skeletonized. Yellowish to dull-green beetles, about ¼ inch long, with an indistinct black stripe along each side (Fig. 17.1,*5,6*), or small yellow to black larvae (*2,3*) will be found skeletonizing the leaves or crawling about on the bark of the trunk, sometimes clustered in great numbers about the base of the trunk.

Trees Attacked. The different species of elm, the English elm and camperdown elm being most subject to attack.

Distribution. The elm leaf beetle is of European origin and was probably brought into this country some time about 1834. It is now established throughout the United States.

Life History, Appearance, and Habits. The adult beetles go through the winter hidden away in sheltered places which will afford them some protection from the weather, often in buildings, where they may be a nuisance in the fall or during warm winter weather. They are about ¼ inch long, of a light-yellow to brownish-green color, with several black spots on the head and thorax, and a somewhat indefinite, black or slate-colored stripe on the outer margin of each wing cover. The beetles fly to the elm trees shortly after they come into foliage in the spring and deposit double rows of yellowish eggs resembling minute lemons (Fig. 17.1,*1*) on the undersides of the elm leaves, usually about 25 in a place. The slug-like larvae hatching from these eggs are yellow in color, spotted and striped with black. They feed for about 3 weeks and, when full-grown, are ½ inch in length. They then crawl down the trunk of the tree, gathering in large masses about the base of the tree or in any shelter near by. Here they pupate, emerging as beetles in from 1 to 2 weeks. There are two, three, or more generations a year, depending on the locality.

Control Measures. The elm leaf beetle may be held in check so that no damage from it will occur, if the trees are thoroughly sprayed with DDT at 1 pound or lead arsenate at 6 pounds to 100 gallons of spray,

[1] *Galerucella luteola* (= *xanthomelaena*) (Müller), Order Coleoptera, Family Chrysomelidae.

with casein, lime, or soybean flour as a sticker. The spray should be applied when the leaves are nearly full-grown or as soon as feeding is noticed. The first spray should be applied about a week after the elms have come into full foliage. If only a part of the trees in the neighborhood have been sprayed, it will be necessary to make another application for the second generation of beetles, putting this application on about

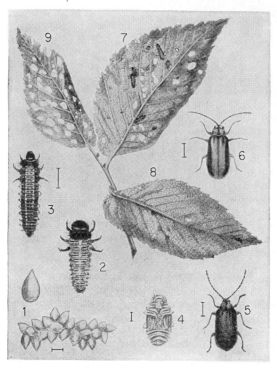

FIG. 17.1. The elm leaf beetle, *Galerucella luteola*. 1, Egg mass, 2, young larva, 3, full-grown larva, 4, pupa, 5 and 6, adult beetles, about twice natural size; 7–9, injured leaves reduced. (*From Felt, "Manual of Tree and Shrub Insects," copyright, 1924, Macmillan. Reprinted by permission.*)

midsummer. Many of the larvae may be killed by spraying with DDT about the base of the tree, when they have come down to pupate.

References. Conn. Agr. Exp. Sta. Bul. 155, 1907; *N.Y. (Cornell) State Museum Bul.* 156, 1912; *Ore. Agr. Exp. Sta. Cir.* 92, 1920; *N.Y. (Cornell) Agr. Exp. Sta. Bul.* 333, 1933; *Proc. Nat. Shade Tree Conf.,* **7**:66, 1931; *Jour. Econ. Ento.,* **41**:770, 1948.

229. SPRUCE BUDWORM[1]

Importance and Type of Injury. In a list of the 20 most destructive insects in the United States, compiled by a group of federal entomologists a few years ago, this insect was given third place, being outranked only by the cotton boll weevil and the corn earworm. It is said to have destroyed 200,000,000 cords of balsam fir and red spruce in the forests of

[1] *Choristoneura* (= *Archips, Cacoecia*) *fumiferana* (Clemens), Order Lepidoptera, Family Tortricidae.

the eastern United States and Canada between 1910 and 1925. The damage is all done by the caterpillars feeding on the foliage of the terminal shoots which are webbed together to form shelters in which the larvae pupate and hibernate. The tops of the trees first appear as though scorched by fire and, if heavily infested, the trees may die over large areas.

Trees Attacked. Firs, spruces, larch, hemlock, and pines.

Distribution. The insect occurs over the coniferous forests of the northern United States and southern Canada.

Life History, Appearance, and Habits. Winter is passed as tiny caterpillars in very small silken cases attached in crevices on the twigs near the buds. As the balsam buds burst in the spring the caterpillars emerge and feed for 3 to 5 weeks on the tender needles, becoming full-grown in June or July. The mature caterpillars are about 1 inch long, dark brown, with a yellowish stripe along each side and covered with yellowish tubercles. They pupate for 1 week or 10 days within loose cocoons of silk among the damaged foliage and then appear as moths from early July to early August. The moths, which are reddish brown to yellowish gray, with the forewings splotched with a number of golden-brown spots and expanding about 1 inch, lay their greenish scale-like eggs in overlapping masses of about a dozen, along the underside of the balsam and spruce needles. After 10 days as eggs, the caterpillars hatch but feed only a short time before going into hibernation. There is only one generation a year.

Control. The insect can be controlled on shade trees and ornamentals by spraying with DDT at 1 pound or lead arsenate at 3 pounds to 100 gallons of spray, just after the buds open. Under forest conditions airplane spraying with DDT in fuel oil at about 1 pound per gallon per acre is effective. Maximum kill is obtained when the larvae are in the fourth and fifth instars. It is recommended that the oldest stands of fir be cut first, since fast-growing trees are less severely injured.

References. Can. Dept. Agr. Div. Forest Insects, Sp. Cir., 1931; *Maine Agr. Exp. Sta. Bul.* 210, 1913; *Can. Dept. Agr. Tech. Bul.* 37 (n.s.), 1924; *U.S.D.A., Bur. Ento. Plant Quar., Cir.* E-684, 1946, and *Yearbook,* **1952**:683; *Jour. Econ. Ento.,* **43**:774, 1952; **49**:338, 1956; and **52**:212, 1959.

230. BAGWORM[1]

Importance and Type of Injury. Infested trees have the foliage stripped or very much ragged. The stripped trees usually die. Numerous spindle-shaped sacks or bags (Fig. 17.2) from ¼ to 1½ inches in length hang down from the twigs, leaves, branches, and sometimes on the bark of the trunk. During the summer these bags contain dark-brown shiny-bodied worms.

Trees Attacked. This insect is a very general feeder, attacking practically all deciduous and evergreen trees.

Distribution. The bagworm is found from the Atlantic states to the Mississippi Valley. Related species extend westward and southward to Texas.

Life History, Appearance, and Habits. The winter is passed as pale, whitish eggs, enclosed by the pupal skin inside the bag in which the female worm lived during the summer. These eggs hatch rather late in the spring, after the trees have come into full foliage. In the latitude of central Illinois they hatch about June 10 to 15. The young worms, on

[1] *Thyridopteryx ephemeraeformis* (Haworth), Order Lepidoptera, Family Psychidae.

hatching, almost immediately spin a silken sack or bag about themselves and then begin feeding on the foliage.   As they feed, they attach to the bag bits of the leaves on which they are feeding.   The bag is carried about by the insect wherever it goes, the larva merely protruding the front end of its body from the bag.   It is almost impossible to draw the larva out of the bag without crushing its body.   This bag offers almost complete protection from birds, but the worms are parasitized by several species of flies and wasp-like parasites.   The worms are of a brown color over the head and thorax, the parts enclosed by the bag being lighter and softer. When full-grown, the insect measures about 1 to $1\frac{1}{4}$ inches in length, the bags at that time being from $1\frac{1}{2}$ to $2\frac{1}{2}$ inches in length.   They pupate within the bags during early September, attaching the bag to the twigs with numerous threads of silk, forming a strong thick loop (Fig. 17.2).

Fig. 17.2. Cases of the bagworm, *Thyridopteryx ephemeraeformis*, fastened to twigs of cedar, as found in winter.   Natural size.   (*From Ill. Natural History Surv.*)

The male changes to a black-winged moth, with a wing expanse of about 1 inch.   These moths emerge from the bags and fly actively.   The female moths, however, are wingless and merely protrude their abdomens from the tip of the bag during mating and then retreat within the bag, where they deposit their eggs and shortly afterward wriggle out of the bag, drop to the ground, and die.

Control Measures.   On small trees, or those that can be readily reached, a thorough cleanup of the bags during the winter by hand-picking and burning will effectively control the insects.   This method is one of the best to use when infestations are first starting, but for a general infestation over shade trees in a neighborhood, the most effective control is spraying with toxaphene, dieldrin, or malathion at 1 pound, Dibrom at 0.5 pound, or lead arsenate at 4 to 6 pounds plus sticker to 100 gallons of spray. These are most effective, applied when the caterpillars are not over half-grown, during the latter half of June in the latitude of central Illinois. Where the bags are collected, care should be taken that they are removed from the trees and burned for, if they are merely thrown about the ground, the eggs will hatch, and many of the worms will find their way to the trees.

References. Mo. Agr. Exp. Sta. Bul. 104, 1912; *N.J. Dept. Agr. Cir.* 243, 1934; *Jour. Econ. Ento.,* **41**:264, 1948; **51**:367, 1958; and **52**:353, 1959.

231. Catalpa Sphinx[1]

Importance and Type of Injury.   This is one of the most serious insect pests of the catalpa tree.   Infested trees will have the leaves eaten off by

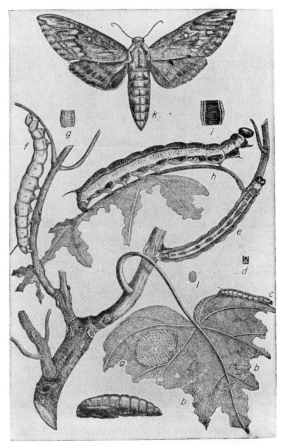

Fig. 17.3. Catalpa sphinx, *Ceratomia catalpae.*   Above at *k,* adult female moth; *e, f, h,* three caterpillars stripping leaves; *a,* an egg mass on leaf; *j,* the pupa; *b* and *c,* partly grown larvae; *l,* a single egg enlarged; *d, g,* and *i* show the variation in color pattern of single segments of larvae from above.   Slightly reduced.   *(From U.S.D.A.)*

dark or black caterpillars from 1 to 3 inches in length with dark-green markings on their bodies, and a sharp horn at the tip of the abdomen (Fig. 17.3).   The markings vary greatly in different individuals.

[1] *Ceratomia catalpae* (Boisduval), Order Lepidoptera, Family Sphingidae.

Trees Attacked. Catalpa.

Distribution. This insect is widely distributed in the United States and is of greatest importance in a zone from New York to Colorado.

Life History, Appearance, and Habits. The winter is passed as brown naked pupae in the soil, 2 or 3 inches below the surface. These pupae (Fig. 17.3,*j*) will be found under, and in close proximity to, catalpa trees. The moths (*k*) emerge shortly after the catalpas have come into full leaf, and deposit their white eggs in masses on the undersides of the leaves. Sometimes as many as 1,000 eggs have been found in a single mass. The moths are of a general gray color with a wing expanse of from 2½ to 3 inches. They fly mainly at night and are seldom seen. The eggs hatch in 10 days to 2 weeks, and the young caterpillars at once begin feeding on the foliage. At first, they feed in groups (*b,b*) but later separately. The full-grown caterpillars are about 3 inches in length, with a moderately large black horn protruding from the tip of the body. The backs of the worms are often almost completely covered with the whitish cocoons of a wasp-like parasite.[1] Upon becoming full-grown, they go down the trunk of the tree, enter the ground, and there change to the pupal stage. There are two generations of the insect each season, at 40° north latitude, the second generation of worms appearing on the trees during late August and early September.

Control Measures. While these worms are large and very ravenous feeders, they may be easily controlled by spraying or dusting. Spraying with lead arsenate, at the rate of 4 pounds or DDT at 1 pound to 100 gallons of spray, applying the spray as soon as any of the caterpillars are noticed on the leaves, will afford almost complete protection. The insects are heavily parasitized, and severe outbreaks usually last for only one or two seasons, with a break of about the same period before the worms will again become numerous enough to strip the trees completely. If discovered while small, on small trees, hundreds of them can be killed by burning single leaves.

Reference. *Ohio Agr. Exp. Sta. Bul.* 332, pp. 238–241, 1918.

232. WALNUT CATERPILLAR[2]

Importance and Type of Injury. Walnut, hickory, and other trees have the branches or entire trees stripped of their leaves during July and August. Masses of large dark-bodied white-haired caterpillars cluster on the trunk of the tree (Fig. 17.5) or feed together on the leaves.

Trees Attacked. Walnut, including English and Japanese walnuts, butternut, pecan, hickory, and occasionally peach, willow, honey locust, beech, sumac, apple, and oak.

Distribution. The insect is common throughout the eastern and southern United States and westward to Kansas.

Life History, Appearance, and Habits. The winter is passed in the form of a brown naked pupa about 1 inch or a little over in length. The pupae will be found from 2 to 6 inches beneath the surface of the soil in the vicinity of the trees on which the insect feeds. The adult moths emerge from these pupae during late June and July. They measure 1½ to nearly 2 inches across the expanded wings. The wings are of a general

[1] *Apanteles congregatus* Say, Order Hymenoptera, Family Braconidae.
[2] *Datana integerrima* Grote and Robinson, Order Lepidoptera, Family Notodontidae.

light brown, with dark-brown wavy lines running across them.   The hind wings are lighter brown without the crosslines.   A dark-brown tuft of hair covers the back of the thorax.   The moths are strong fliers and deposit their eggs in masses of 200 to 300 (Fig. 17.4) on the underside of the leaves of their food plants.   These hatch in about 2 weeks into small reddish white-striped worms with black heads.   As they grow, the color of the body changes to brown and later, in the full-grown stage, to black.   They are covered with rather soft, long, frowzy white hairs. The full-grown caterpillar is 2 inches or a little over in length.   They feed together, several hundred in a place, and a single colony will often strip one or two branches on the tree. As the worms grow, they have a peculiar habit of coming down the tree to change their skins.   The entire colony crawls down the trunk at the same time, so that one will often find masses of worms, from 4 to 8 inches across, clinging to the trunk of the tree (Fig. 17.5).   The shed skins remain on the tree trunk, having much the appearance of dead worms.   On becoming full-grown, the caterpillars leave the tree, crawl away for a short dis-

Fig. 17.4. Walnut caterpillar, egg mass. About twice natural size.   (*From Ohio Agr. Exp. Sta. Bul.* 332.)

tance, and enter the ground, where they later change into the brown pupal stage.   There is but a single generation over most of the range of the insect, although possibly two in the southern areas where it occurs.

Fig. 17.5. A cluster of walnut caterpillars, *Datana integerrima*, molting on the trunk of a walnut tree.   (*From Ohio Agr. Exp. Sta. Bul.* 332.)

Control Measures. The most effective method of controlling this insect is to spray the infested trees with DDT at 1 pound or lead arsenate at 4 pounds to 100 gallons of spray. This spray should be applied as soon as the caterpillars are seen feeding upon the foliage. On small trees the colonies may be removed before much of the foliage has been destroyed, using a pole pruner, and cutting off and burning the few leaves on which the caterpillars are starting to feed. If the trees are watched every day, it is possible to crush many of the caterpillars when they come down the trunk to shed their skins.

Spraying, however, is the only really effective remedy that can be depended upon to protect trees from all damage by these insects. Banding trees with sticky material is of no benefit, as the moths do not crawl up the trunk to deposit their eggs.

References. U.S.D.A., Farmers' Bul. 1169, 1921; *N.Y. State Museum Memoir* 8, **1**:303, 1905; *Ark. Agr. Exp. Sta. Bul.* 224, 1928.

233. FOREST TENT CATERPILLAR[1]

This "forest armyworm" is a serious defoliator of many shade and forest trees, and sometimes, as it migrates by millions to new food plants, it swarms over and into cabins and other buildings to the annoyance and dismay of farmers and tourists, who may be driven away from the woods. It appears to prefer aspen or poplar, especially where they are growing in pure stands, but it is seriously destructive to oaks and maples and feeds on basswood, ash, elm, birch, conifers, and, during its migrations, upon field and vegetable crops.

It is a close relative of the tent caterpillar (page 693) and has a similar life history, but it makes no tent. Eggs are laid in a similar manner and carry the insect over winter, hatching the first half of May. The pale-blue larvae, which have a row of keyhole-shaped, white spots down the middle of the back and pale-yellow stripes on the sides, are troublesome from mid-May to the end of June. After 10 to 12 days in cocoons, the adults appear in swarms about mid-July. Spraying with DDT at 1 pound or lead arsenate at 3 pounds to 100 gallons of spray, when the larvae are about $\frac{1}{2}$ inch long, has given good control over areas accessible to power sprayers. Farm premises and summer resorts may be protected by spraying protective zones about 400 feet wide around them. In heavily forested areas, aircraft spraying with 0.5 pound of DDT in 1 gallon of fuel oil has been widely used. Similar sprays of toxaphene at 1.5 pound or dieldrin or endrin at 0.15 pound per acre are also effective. The parasitic fly, *Sarcophaga aldrichi*, often aids in terminating outbreaks of this insect.

References. Minn. Agr. Exp. Sta. Tech. Bul. 148, 1941; *Jour. Econ. Ento.*, **32**:396, 1939; *U.S.D.A., Forest Pest Leaflet* 9, 1956.

234. WHITE-MARKED TUSSOCK MOTH[2]

Importance and Type of Injury. This insect is usually considered more of a shade tree than orchard pest, but sometimes, especially in the

[1] *Malacosoma disstria* Hübner, Order Lepidoptera, Family Lasiocampidae. The California tent caterpillar, *M. californicum* (Packard), the prairie tent caterpillar, *M. lutescens* (Neumoegen & Dyar), and the western tent caterpillar, *M. pluviale* (Dyar), cause similar damage.

[2] *Hemerocampa leucostigma* (Smith and Abbott), Order Lepidoptera, Family Lymantriidae.

Plants Attacked. Apple, pear, cherry, oak, willow, and other deciduous trees and shrubs; rare on buckeye, ash, hickory, chestnut, never on evergreens.

Distribution. The insect is a native of Europe and was introduced into eastern Massachusetts on imported nursery stock about 1897. It is

4
Male pupa

5
Female pupa

2
Winter nest

6
Male moth

3

Full grown
caterpillar

1

Egg mass and
moth laying eggs

7
Female moth

Fig. 17.10. The brown-tail moth, *Nygmia phaeorrhoea*, showing the various stages of the insect and a winter nest. About natural size. (*From Mass. State Forester.*)

found in Maine, Massachusetts, New Hampshire, Vermont, and Rhode Island and in Nova Scotia and New Brunswick.

Life History, Appearance, and Habits. The insect passes the winter in the form of very tiny caterpillars, living together in colonies of 25 to 500. Each colony webs together several leaves and attaches them firmly to twigs by threads of very tough silk. These winter nests (Fig. 17.10,*2*) are rather conspicuous on the trees. The caterpillars are so well protected within these nests that they can withstand very low temperatures. As soon as the leaves start coming out in the spring, the little caterpillars become active and crawl out from the nest to feed on the tender foliage. They go back within the nest at night, for a time, but as they become larger remain on the foliage. In the latitude of Boston they become full-grown about the last of June. The full-grown caterpillars (Fig. 17.10,*3*) are about 1½ inches in length, dark brown in color, with a broken white stripe on each side of the body and a bright-red tubercle on the back of the eleventh and of the twelfth body segment. They seek some sheltered place where they transform to the pupal stage (*4,5*) and remain in this stage for about 2 weeks; they then emerge as medium-sized moths during July. These moths (Fig. 17.10,*6,7*) have a wing expanse of a little over 1½ inches. The wings and thorax are pure white. The abdomen is mostly brown, with a very conspicuous tuft of chestnut-brown hairs at the tip. These moths are strong fliers, flying in large swarms at night. They are attracted to strong lights, and sometimes in localities where they are abundant, they literally cover the sides of electric light and telephone poles and buildings, giving them the appearance of being covered with snow. The female moth deposits from 200 to 400 eggs on the underside of a leaf of the food plant, mostly in July. These eggs are laid in masses (Fig. 17.10,*1*) and are covered with brown hairs, giving them a dark chestnut-brown color. The eggs hatch during August or early September, and the young caterpillars from each egg mass or from several near-by egg masses feed together for a short time on the terminal leaves of twigs and branches and then spin the web shelter in which they pass the winter. The insect has been shipped to many parts of the world on nursery stock in these winter webs. There is one generation a year.

Control Measures. The most effective method for controlling the brown-tail moth is to spray infested trees during the early spring when the caterpillars have started to feed or in early August as the new caterpillars begin hatching, using DDT or lead arsenate as suggested for the gypsy moth, page 832. In light infestations much can be done in keeping down the numbers of the insect by cutting off and burning the winter webs. After a little practice these webs can be easily detected on the tips of the bare branches, and with the aid of a long pole pruner and a ladder, nearly all the webs can be removed from the trees and burned. With federal aid, about 24 million webs were cut and burned in the winter of 1933–1934 and 2 to 4 million in succeeding winters. Since about 1915 natural enemies, especially a fungus disease[1] which kills the caterpillars, low winter temperatures, and applied control measures have brought about a remarkable decrease in the abundance of this insect.

References. Can. Dept. Agr. Bul. 63 (n.s.), 1926; *U.S.D.A., Cir.* 464, 1938, and *Farmers' Bul.* 1335, 1923, and *Bur. Ento. Bul.* 87, 1910.

237 (150). SCURFY SCALES[2]

Importance and Type of Injury. These scales are often very injurious to smaller elm, willow, and dogwood. They are rarely a serious pest on large trees. Infested

---

[1] *Entomophthora aulicae* Reich.

[2] Order Homoptera, Family Coccidae. The elm scurfy scale is *Chionaspis americana* Johnson; the one on dogwood is *C. corni* Cooley; the black-willow scale is *Chionaspis salicis-nigrae* (Walsh); and the one on hickory is *C. caryae* Cooley.

branches show small flattened dirty-white scales, about $\frac{1}{10}$ of an inch long, lying nearly flat on the bark. In the winter many reddish-purple eggs, just discernible with the naked eye, will be found beneath these scales. On heavily infested trees, the entire bark may be coated with the grayish scales.

The life history and control of these scales are similar to those of the apple scurfy scale given on page 708.

238 (151). OYSTER-SHELL SCALE[1]

Importance and Type of Injury. Branches of trees or entire trees are dying, the bark cracking and having much the appearance of drying up on the branches. The bark is covered with small brownish-gray scales about $\frac{1}{8}$ inch long by $\frac{1}{16}$ inch wide, usually curved and closely resembling a miniature oyster shell (Fig. 15.22). The bark may be completely covered with these scales.

Trees Attacked. All species of ash (with the exception of the blue ash), poplar, dogwood, elm, soft maple, linden, horse chestnut, lilac (with the exception of the white lilac), many species of rose, peonies, and many other shade trees and shrubs. The very closely related form attacking apple is described on page 709.

Distribution. General throughout the United States.

Life History, Appearance, and Habits. The winter is passed in the egg stage under the female scale. These eggs are elliptical, nearly white in color, and from 50 to 60 will be found under each female scale. They hatch late in the spring after the trees have come into full foliage. In the latitude of central Illinois, this is about June 1. The white six-legged young, just discernible to the naked eye, crawl about over the tree for a few hours and then insert their beaks into the bark and begin sucking the sap. They soon molt, and the females remain in this position for the rest of their lives. They grow rather rapidly, secreting the wax which forms into the brown protective scale over their bodies. About midsummer, or a little later, the males become full-grown and change, under their scales, to minute yellowish-white two-winged insects, which fly about for a short time, mate with the females, and die. After mating, the female deposits her eggs, her body gradually shrinking to the small end of the scale, where she finally dies. There is but one generation of this variety of oyster-shell scale each year. Another kind of oyster-shell scale, somewhat closely resembling this one, occurs on certain species of dogwood. The dogwood form has two generations each year.

Control Measures. These scale insects have several natural enemies that keep down their numbers. Certain mites feed on the eggs and may prevent damage by this scale for a number of years. Dormant oil sprays of 3 to 5 per cent oil usually give fair control. Spraying when the crawlers appear with 1 pound of DDT or 0.3 pound of parathion to 100 gallons of spray is very effective. The planting of resistant shade trees, such as the blue ash, hackberry, hard maples, and oaks, will keep down this scale, as it does not breed on these trees.

References. Ohio Agr. Exp. Sta. Cir. 143, 1914; *Jour. Econ. Ento.,* **30**:651, 1937; **41**:978, 1948.

239. PINE NEEDLE SCALE[2]

Importance and Type of Injury. Infested trees have the foliage somewhat yellowed, with rather elongated, whitish scales up to $\frac{1}{8}$ inch in length, attached to the leaves.

[1] *Lepidosaphes ulmi* (Linné), Order Homoptera, Family Coccidae.
[2] *Phenacaspis* (= *Chionaspis*) *pinifoliae* (Fitch), Order Homoptera, Family Coccidae.

These white scales (Fig. 17.11), on the green leaves or needles of pine or other evergreens, often first attract notice to the presence of the insects.

Trees Attacked. Pines, spruces, firs, cedars, and hemlocks.

Distribution. Throughout the northern United States and southern Canada.

Life History, Appearance, and Habits. The winter is passed in the form of very minute, purplish eggs underneath the gray parent scale. From 20 to 30 of these eggs will be found under each scale. The eggs hatch in midspring into crawling young, which move about for a short time and then settle down and secrete a scale about their bodies. They become full-grown by late summer, and a second generation is produced from eggs laid during August. These, in turn, become full-grown, and the females deposit eggs, by fall. In the latitude of Illinois and Ohio there are two generations annually, although probably only one farther north.

Control Measures. Spraying just after the eggs hatch (about May 25 in central Illinois) with malathion at 1 pound to 100 gallons of spray has given good control. Summer oil emulsion, as recommended by the manufacturer, but not exceeding 2 per cent oil, has given satisfactory results. This spray should be applied when the temperature is below 80°F. and the humidity high.

References. Ohio Agr. Exp. Sta. Bul. 332, p. 291, 1918; *Jour. Econ. Ento.,* **24**:115, 1931; *Conn. Agr. Exp. Sta. Bul.* 315, p. 578, 1930.

FIG. 17.11. The pine needle scale, *Phenacaspis pinifoliae.* One leaf shows the small male scales and the larger female scales. About 3 times natural size. (*From Ohio Agr. Exp. Sta.*)

240. EUROPEAN ELM SCALE[1]

Importance and Type of Injury. This unarmored scale insect is often found in enormous numbers covering the underside of limbs and the trunks, weakening and killing the trees by sucking the sap, secreting sticky honeydew that gums and soots the leaves or smuts walks, benches, and grass, and attracting thousands of flies and wasps, making shade trees very unattractive.

Trees Attacked. All kinds of elms.

Life History, Appearance, and Habits. The winter is passed as second-instar nymphs, motionless, about the color of the bark, and hidden in protected crevices on trunk or branches. The males form conspicuous white cocoons in very early spring in which they transform to minute, winged or wingless, reddish "gnats" in late April or May. The females also become adult early in May. They are from ⅙ to ⅜ inch long, oval in outline, reddish brown, and surrounded by a white cottony fringe (Fig. 17.12). From late May through June and July the females produce eggs which are deposited beneath them, like a hen sitting upon a nest. These eggs hatch within an hour and the lemon-yellow crawling nymphs swarm over twigs to the leaves, upon the underside of which they settle and feed until fall. Most

[1] *Gossyparia spuria* (Modeer), Order Homoptera, Family Coccidae.

of them migrate back to the limbs or trunk before the leaves fall. There is only one generation a year.

Control Measures. Dormant sprays of 5 per cent miscible oil or 5 per cent oil and 5 per cent lime-sulfur combined are effective. Spraying with DDT at 1 pound to 100 gallons of spray 1 month after the first eggs hatch has given excellent results. Where spraying equipment cannot be procured, powerful streams of water from fire hydrants directed

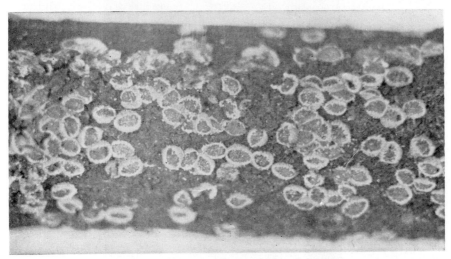

Fig. 17.12. European elm scale, *Gossyparia spuria*, scales of the female on twigs, 3 times natural size. Note the white fringe of secretion that surrounds the body of the insect. (*Original.*)

against the branches about the time the leaf buds are opening will dislodge great numbers of the females.

References. U.S.D.A., Dept. Bul. 1223, 1924; *Nev. Agr. Exp. Sta. Bul.* 65, 1908; *Proc. 24th. Nat. Shade Tree Conf.*, 1948, pp. 93–94.

241 (180). TERRAPIN SCALE[1]

Nearly hemispherical reddish-brown scales, about ⅛ inch across and very convex in outline, cluster on the bark of twigs and branches (Fig. 15.57). The scale is generally somewhat mottled and streaked with black. The bark is often entirely covered for considerable distances along the branches.

Maple, elm, and sycamore are most severely injured. It also attacks osage orange, peach, plum, pear, quince, and some other shade trees. The distribution, life history, and control of this insect are discussed on page 753.

242. COTTONY MAPLE SCALE[2]

Importance and Type of Injury. This is one of the most destructive scales on soft maple and also injures some other trees. Attention is usually attracted to infested trees by the cottony-appearing masses of scale along the underside of the twigs and branches during May and

---

[1] *Lecanium* (= *Eulecanium*) *nigrofasciatum* Pergande, Order Homoptera, Family Coccidae.

[2] *Pulvinaria innumerabilis* (= *vitis*) (Rathvon), Order Homoptera, Family Coccidae.

June (Fig. 17.13).   Branches of heavily infested trees die, and the foliage of the entire tree turns a sickly yellow.   The reduction of the vigor of the tree by the scale often leads to attacks by bark beetles or other borers.

Plants Attacked.   Soft maple, linden, Norway maple, apple, pear, willow, poplar, grape, hackberry, sycamore, honey locust, beech, elm,

Fig. 17.13. The cottony maple scale, *Pulvinaria innumerabilis*.   The whitish masses looking something like popped corn are tufts of wax-like material secreted by females and containing their eggs to the number of more than 1,500 per tuft.   The brownish bodies of the females stand somewhat upright at one end of the tufts they have secreted.   (*Original.*)

plum, peach, gooseberry, Virginia creeper, currant, sumac, and some others.

Distribution.   The insect is distributed throughout the United States and Canada.   It is most destructive in the northern part of the United States.

Life History, Appearance, and Habits.   The cottony maple scale passes the winter as a small, brown, flattened scale, a little less than ⅛ inch long, attached to the bark of twigs and small branches.   These scales are all females.   In the spring, as soon as the sap starts to flow, they grow very rapidly, and soon begin depositing their eggs, which are secreted in cotton-like masses of wax under the scale.   This wax is secreted in such abundance that it forms a mass several times the size of the overwintering insect, and the body of the scale often becomes elevated at an angle from the twig.   From 1,500 to 3,000 eggs are laid by each female.   The eggs hatch during late June and July, in the latitude of central Illinois, and the young scales crawl from the twigs to the undersides of the leaves, where they suck the sap along the midrib or the veins.   They become mature during August and September, mating takes place, and the males die, the females crawling back to the twigs and small branches, where they pass the winter.

Control Measures.   Spraying during the early spring, just before the leaves emerge on the maples, with any of the dependable miscible oils, diluted to give 2 per cent actual oil, or according to the manufacturer's directions, will give very good control.   Care must be taken in spraying maples that the trees are dormant, since these trees, especially hard maples, are very sensitive to oil injury.   Summer spraying at the completion of hatching with malathion at 0.5 pound to 100 gallons of spray is very effective.

References.   Twenty-sixth Rept. Ill. State Ento., p. 62, 1911; *Ohio Agr. Exp. Sta. Bul.* 332, 1918; *Conn. Agr. Exp. Sta. Bul.* 344, p. 127, 1934.

243 (167). Woolly Apple Aphid

This insect is described and the means of control are given under Apple Insects, page 733.   It frequently becomes very abundant on elm where its feeding on the leaves in the spring causes them to curl and bunch together.   No true galls are formed by this insect, so that it can be readily distinguished from the work of the cockscomb gall aphid.

Other Aphids of Shade Trees and Shrubs

Nearly all shade trees and ornamental shrubs suffer to some extent from the attacks of aphids, especially *Spiraea vanhouttei*, roses of all varieties, snowball, poplar, sycamore, and maple.   In general, the life history and control of these aphids are very similar to those described on pages 612 and 872.

244. Boxelder Bug[1]

Importance and Type of Injury.   These strikingly marked, red-and-black sucking bugs (Fig. 17.14) often feed destructively upon the flowers, fruits, foliage, and tender twigs especially of boxelder and ash trees. They are, however, more important as a nuisance during fall and in warm days in winter when they swarm into houses or

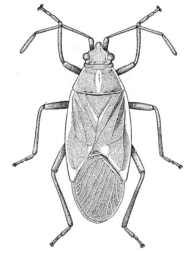

Fig. 17.14. Boxelder bug; the light portions of the upper surface in the drawing are bright red.   Enlarged about 4 times.   (*Drawn by Kathryn Sommerman.*)

congregate in great numbers upon trunks of trees, porches, walls, and walks, often causing great alarm.   They are not capable of biting or

[1] *Leptocoris trivittatus* (Say), Order Hemiptera, Family Coreidae.

harming foods, clothing, or other household articles. The insects winter in buildings and other dry sheltered places in the adult stage. They are flat-backed rather narrow bugs, about ½ inch long, brownish black, with three longitudinal red stripes on the thorax and red veins on the wings. Nymphs of all sizes are bright red. There are two generations a year in the warmer areas.

Plants Attacked. Boxelder, especially the flowers and fruit, is the favorite food, although ash and maple are attacked. When abundant, the bugs may attack the fruits of peach, pear, apricot, almond, raspberry, strawberry, and the foliage of potato.

Control Measures. This insect may be controlled by spraying with DDT or malathion. When the adults invade houses, pyrethrins, DDT, chlordane, or malathion household sprays are effective. It has recently been shown that this insect is an obligate feeder on the pistillate or female boxelder. Therefore only the staminate or male tree should be planted as an ornamental.

Flatheaded Borers[1]

These borers are the most important pests of newly set shade trees in the South and Middle West. There are several species of borers belonging to this group, which are very common, attacking practically all kinds of trees. The typical injury consists of rather shallow, long, winding, oval galleries packed with frass, beneath the bark, usually on the south or southwest side of the trees. Areas of the bark become entirely undermined. Some species kill trees by mining beneath the bark, while others are very destructive to lumber by mining into sapwood and heartwood. In these galleries are to be found medium to rather large, yellowish-white legless grubs, with a pronounced flattened enlargement of the body (thorax) just back of the head, which bears a horny plate on both the upper and the undersides. The adults are often beautifully colored or metallic, boat-shaped beetles, ⅓ to 1 inch long, with the wing covers usually curiously roughened, like bark. A typical life history of these insects is given under Apple Insects, page 718.

The control of the flatheaded borers on shade trees is the same as on apple. Trees that have been taken from the nursery and are set in situations exposed to the sun should be protected at least during the first season by complete wrappings of paper from the ground to the lowest limbs. Under forest conditions, the only practical measure is prompt felling, peeling, and burning as for bark beetles (page 848).

References. Jour. Econ. Ento., **11**:334, 1918; *N.Y. State Museum Memoir* 8, **2**:653–658, 1905.

245. Bronze Birch Borer[2]

Importance and Type of Injury. This flatheaded borer is a very serious pest of birches and in northern states has destroyed nearly every white or paper birch in many localities. Infestation is first indicated by a browning of the tips of the upper branches, followed by the death of the entire tree. Infested branches will often appear somewhat swollen and brown, with ridges around the smaller ones. Small, slightly oval holes,

[1] *Chrysobothris femorata* (Olivier) and many other species, Order Coleoptera, Family Buprestidae.

[2] *Agrilus anxius* Gory, Order Coleoptera, Family Buprestidae.

about ⅛ inch in diameter, will be found in the bark.   An examination of the inner bark will show numerous burrows, tightly packed with sawdust, running in every direction, often extending into the wood, and containing slender white grubs about ¾ inch long, with a slight brownish enlargement of the body just back of the head.

Trees Attacked.   The white or paper birches are most severely injured. The insect has been found also in several other species of birch, poplar, cottonwood, and willow.

Distribution.   It is a native American insect, generally distributed throughout the northern United States and southern Canada, westward to Idaho, Colorado, and Utah.

Life History, Appearance, and Habits.   The winter is passed as full-grown larvae in cells just within the sapwood.   These larvae are from ½ to ⅗ inch in length, white in color, very slender, with a slight enlargement of the thorax, and with two rather slender, brownish projections from the last segment of the body.   In the spring the larvae pupate within their cells in the wood and emerge during May, June, and July, as greenish-bronze beetles ¼ to ½ inch long, with rather blunt heads and slender, pointed bodies (Fig. 17.15).   The female beetles deposit their eggs singly in cracks in the bark or in crevices made by their jaws.   These hatch within 10 to 11 days into tiny borers that work their way into the inner bark, becoming full-grown by fall and excavating cells in the sapwood.   The burrows of the insect may be 4 or 5 feet long, are very crooked, are filled with frass, and cross and recross, often completely cutting off the circulation of the sap.   There is but one generation of the insect each year.

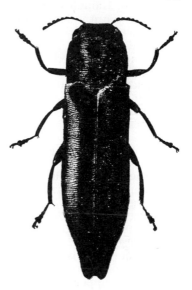

FIG. 17.15.   Bronze birch borer, *Agrilus anxius*, adult.   Enlarged about 7 times.   *(From Ill. Natural History Surv.)*

Control Measures.   The insect may be checked by keeping the trees growing as vigorously as possible, by wrapping the trunks during the period when the adults are present, and by cutting out all infested parts and burning them not later than the first of April.   Spraying the trunks with parathion at 1 pound or with DDT at 4 pounds to 100 gallons of spray has given satisfactory control.

References.   *Jour. Forestry,* **25**:68, 1927 and **29**:1134, 1931; *Jour. Econ. Ento.,* **52**:255, 1959; *Can. Ento.,* **81**:245, 1949.

ROUNDHEADED BORERS[1]

The family of beetles known as long-horned or roundheaded borers[1] are even more numerous and destructive than the flatheaded borers. The larvae, or grubs, work beneath the bark and also tunnel through the heartwood, often riddling the trunks of trees with holes that are as large

[1] Order Coleoptera, Family Cerambycidae.

as a pencil or much larger.   These grubs (Fig. 17.20) also have an enlargement of the body behind the head, but it is not so much flattened as in the flatheaded borers and has a horny plate on the upper side only.   They are generally legless.   The adults are usually cylindrical or straight-sided hard-shelled beetles, with antennae nearly as long as, or longer than, the body and often gorgeously banded, spotted, or striped with contrasting colors.   A typical life history is given under Apple Insects, page 720.

Besides the locust borer, elm borer, and poplar borer, discussed below, there are a number of other species which attack various species of shade trees.   A few of these attack healthy trees, while others are attracted only to trees in a weakened or sickly condition.   The wood may be rendered worthless for lumber.   A minor annoyance often results when wood containing the larvae is stored in dwellings for firewood.   The adults which transform, though harmless, are alarming to persons who do not know what they are.   The maintenance of healthy vigorous trees, and the avoidance or painting of wounds will go far to prevent attack.   The only practical methods of control known are to use carbon bisulfide in the burrows as described for the poplar borer, or to cut and burn the badly infested trees during the winter when the insects are all within the trunk or bark in the larval stage.   Linden, poplar, soft maple, oak, young hickory, and willow are particularly subject to attack by some of these borers.

246. Poplar Borer[1]

Importance and Type of Injury.   Damage by this insect frequently makes it impossible to grow certain species of poplars in some localities. It is the most destructive borer attacking poplar trees.   Infested trees have many large burrows, with openings through the bark, on the trunk and large branches.   The entrance to these burrows is usually packed with coarse, excelsior-like wood fibers.   An accumulation of these fibers or sawdust-like material will often be noticed around the base of the trunk. There is generally a discharge of sap from the opening to the burrow, which wets and discolors the bark of the tree for some distance below it. The wood of the tree is weakened, so that the branches or the main trunk breaks during periods of high wind.

Trees Attacked.   Carolina poplar, cottonwood, aspen, and Lombardy poplar are the most seriously injured.   It has also been found in some other species of poplars and in willow.

Distribution.   This is a native insect, distributed generally over the northern United States and Canada and southward to Texas.

Life History, Appearance, and Habits.   The winter is passed as a yellowish, rather round-bodied grub, from 1 to nearly 1½ inches in length. These grubs will be found in large burrows in the wood of the tree.   They start feeding as soon as the weather becomes warm in the spring and continue throughout the season.   They bore through the wood of the tree, cutting out large galleries, sometimes as much as an inch in diameter. Some of the borers become full-grown by late spring.   They are then about 2 inches long.   They pupate in cells in the wood, and emerge as beetles from July to September.   The adult beetles (Fig. 17.16) are from 1 to 1½ inches in length, with long antennae slightly darker than the body.   The beetle is of a general light-gray color, with irregular, somewhat elongated, small yellowish spots, and the whole body is sprinkled

---

[1] *Saperda calcarata* Say, Order Coleoptera, Family Cerambycidae.

with minute black dots.   The female deposits her eggs in cracks in the bark.   These eggs hatch in a few days into small grubs, which feed at first in the outer bark, but work their way rather quickly into the inner bark and sapwood.   At least 2 years are required for the grubs to complete their growth.   Under unfavorable conditions this period may extend over 3 years.

Control Measures.   Trees that are especially valued for shade may be freed of most of the borers by carefully going over the tree and injecting a small amount of carbon bisulfide into all openings leading into the burrows.   A machine oilcan is best suited for injecting the carbon bisulfide.   The hole should be closed immediately after injecting the chemical, using a wad of wet clay or putty for this purpose.   An examination of the infested trees two or three times from the first of June to the first of October, treating each time all burrows from which fresh sawdust is being forced out, will kill practically all insects in the trees.   Experiments have shown that all grubs within 6 to 10 inches of the point where the carbon bisulfide is injected will be killed. The edges of the egg scars in burrow wounds may be painted with wood-preserving creosote during October.   This will kill the young borers which are frequently found in these wounds.   Badly infested trees should be cut during the winter or early spring and all the large branches and trunk burned to prevent emergence of the adult beetles.   Cutting and piling the wood will not be effective, as the larger grubs will complete their growth and emerge as beetles to reinfest neighboring trees.

Fig.   17.16.   Poplar borer, *Saperda calcarata,* adult female.   Slightly enlarged.   *(From U.S. D.A.,   Farmers'   Bul.* 1154.)

References.   Ohio Agr. Exp. Sta. Bul. 332, p. 319, 1918; U.S.D.A., Farmers' Bul. 1154, 1920; *N.Y. State Museum Memoir* 8, **1**:98, 1905.

247. Locust Borer[1]

Importance and Type of Injury.   Black locust trees have swollen areas on the trunk, often with the bark cracked open, exposing burrows in the tree sometimes ½ inch in diameter.   The wood is nearly always discolored and blackened in and around these burrows.   Young locust trees break over during the late summer, by reason of numerous burrows through the trees which have weakened them (Fig. 17.18).   This insect is so numerous throughout most of the eastern United States as to prevent profitable growing of black locust.   If this insect could be controlled, the black locust would become of much greater value as a farm wood-lot tree.

Trees Attacked.   Black locust.

Distribution.   The locust borer is found throughout the United States, westward to Arizona, New Mexico, Colorado, and Washington, and in southern Canada.

Life History, Appearance, and Habits.   The winter is always passed as a very small grub or larva within the inner bark or barely to the sap-

[1] *Megacyllene* (= *Cyllene*) *robiniae* (Förster), Order Coleoptera, Family Cerambycidae.

wood of the infested trees.   Early in the spring these grubs begin feeding, and burrow their way into the wood of the tree, going through both the sapwood and the heartwood.   These burrows are very irregular in direction and are frequently so numerous as to cause the death of the tree. The tree may break over during periods of high winds.   The larvae become full-grown about midsummer.   At this time they are $\frac{3}{4}$ to nearly 1 inch in length, legless, and tapering from the thorax backward, giving them a club-shaped appearance.   These larvae excavate cells in the wood and here change to the pupal stage.   The adult beetles begin coming out about the first of September in the latitude of Illinois.   They are of a general black color, marked with bright-yellow crosslines (Fig. 17.17).   The legs and antennae are dull red, the underside of the body jet

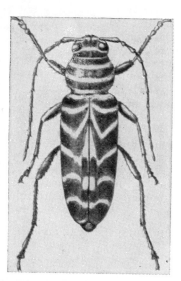

Fig. 17.17. Locust borer, *Megacyllene robiniae*, adult female.   About twice natural size.   (*From Ill. Natural History Surv.*)

Fig. 17.18. Young locust tree broken at point of injury by larvae of the locust borer. (*From Ohio Agr. Exp. Sta. Bul.* 332.)

black.   They are extremely active, flying readily from tree to tree and scuttling about over the bark.   They feed on the flowers of goldenrod and a few other allied plants.   Feeding in the adult stage is not necessary, as the females will lay their eggs without having fed.   The elongate white eggs are tucked into crevices and cracks in the bark.   These hatch in about 2 weeks, and the young work their way through the outer bark and into the inner bark, or just to the surface of the sapwood, before the approach of cold weather.   There is one generation each year.

Control Measures.   Where black locust is interplanted with other trees, there is less damage from this borer.   Dense stands are less injured than open ones.   Rapidly growing trees escape severe injury.   To accelerate growth, mulching with leaves to a depth of 6 inches has been recommended.   If a plantation is growing very slowly or in cases where locust is planted for erosion control, the trees may be cut back very severely, after a few years' growth, thus causing the production of numerous root sprouts, greatly accelerated growth, and a shady condition unfavorable to the insects.   Shade trees may be protected from this borer by spraying the trunks with a 2 per cent DDT-xylene emulsion. This should be applied to the trunks when new growth starts in the spring.

References.   U.S.D.A., Bur. Ento. Bul. 58, Parts I and III, 1906; *Rept. Ky. State Forester,* 1915; *Ohio Agr. Exp. Sta. Bul.* 332, 1918; *Jour. Econ. Ento.,* **25**:713, 1932; *Proc.* 21st. *Nat. Shade Tree Conf.,* 1945, p. 16.

248. ELM BORER[1]

Importance and Type of Injury.   Branches of elm dying, or entire trees with foliage undersized and of a general yellow color, is the symptom of attack by this round-headed borer.   Numerous galleries run through the inner bark and sapwood.   The outer bark is often darkened and loosened from the tree.   These galleries are tightly packed with frass or brownish sawdust.

Trees Attacked.   White elm and slippery elm.   English and Scotch elms are not attacked.

Distribution.   Throughout the northeastern United States.

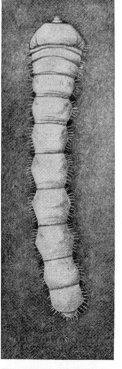

FIG. 17.19. Elm borer, *Saperda tridentata*, adult, about 3 times natural size.   Ground color is dark gray; light stripes on sides of body and bands across elytra are bright red.   (*From Ill. Natural History Surv.*)

FIG. 17.20. Elm borer, larva.   About 3 times natural size.   (*From Ill. Natural History Surv.*)

Life History, Appearance, and Habits.   The winter is passed as partly grown larvae in the bark and sapwood of the tree.   In the spring these larvae begin feeding on the inner bark and wood, running their irregular galleries in all directions.   They become full-grown late in the spring and change to a pupal stage in cells in the sapwood. They emerge during late spring and early summer as gray long-horned beetles, about $\frac{1}{2}$ inch in length (Fig. 17.19).   The wing covers are bordered on the outer margin with a narrow line of red, and there are three fine extensions of this red line across each wing cover.   The eggs are laid in cracks in the bark, and the grubs, on hatching, work their way into the inner bark and sapwood.   The galleries running through the inner bark cut off much of the sap flow.   Most of the grubs (Fig. 17.20) are about two-thirds

[1] *Saperda tridentata* Olivier, Order Coleoptera, Family Cerambycidae.

grown by fall. There is probably one generation of the insect each year, except in cases where conditions are unfavorable to the growth of the grubs, when they may require two seasons to complete their growth.

Control Measures. The most effective control for this beetle, as well as for many other shade-tree insects, is to keep the trees in as vigorous condition as possible. The elm borer seems to be attracted to trees which are in a sickly condition, or lacking in vigor because of insufficient plant food or moisture. Keeping the trees well supplied with water and well fertilized will do much to prevent injury by this insect. Sickly trees, or large branches which are dead or dying, should be cut out and burned during the winter months. Such trees or branches act as breeding places for the beetle and as centers of infestation for all elms in the neighborhood.

References. Ill. Agr. Exp. Sta. Bul. 151, 1911; *N. Y. State Museum Memoir* 8, p. 67, 1905; *Jour. Econ. Ento.*, **32** :848, 1939.

249. Bark Beetles and Ambrosia Beetles[1]

These short, cylindrical, reddish-brown to black beetles, ranging from $\frac{1}{25}$ to over $\frac{1}{3}$ inch long, are the most destructive group of insect pests attacking coniferous trees. Many species also attack deciduous trees.

Destruction of standing timber is estimated to reach 4.5 billion board feet a year in the United States. The adults usually have a very large prothorax overhanging the head, the antennae are clubbed, the tibiae serrate, and the elytra often cut off obliquely at the end and provided with spines. The larvae are whitish, legless, stout, cylindrical, curved grubs. There are three main types of attack upon the trees: (1) The twig and cone beetles bore into the cones of trees or the pith of twigs and eat the wood. (2) The ambrosia or timber beetles tunnel into the sapwood and heartwood of unseasoned lumber, log cabins, and rustic work, making "pinholes" upon which they propagate fungi as food, often rendering lumber practically worthless. These adults do not eat the wood and their larvae do no damage. (3) The most destructive and

Fig. 17.21. Hickory bark beetle, *Scolytus quadrispinosus*, adult. About 7 times natural size. *(From Ill. Natural History Surv.)*

best-known species, called bark beetles, mine just beneath the bark of standing or felled trees, feeding upon the cambium and adjacent tissues, and leaving characteristic engravings upon both the inner surface of the bark and the wood. Reddish, boring dust is pushed out of the tunnels and clings to the bark or accumulates about the base of the trees and often sap exudes and dries to form hard pitch or resin tubes about the entrance holes. These engravings can always be analyzed into (*a*) *feeding holes;* (*b*) *egg tunnels* or *parent galleries*, made by the female who lays her eggs in little niches along the sides; and (*c*) *larval tunnels* radiating from the parent gallery, in which the larvae feed and complete their growth, and at the end of which they pupate and transform to adults (Fig. (17.22). In monogamic species the female starts the burrows, but there are polygamic species in which the male first eats out a nuptial chamber from which his several females each excavates an egg gallery, while the male remains throughout his life in the nuptial chamber plugging the entrance

[1] Order Coleoptera, Family Scolytidae.

hole with his body.   Some attack perfectly healthy trees, others restrict themselves to trees already weakened from some other cause.   Their feeding destroys the cambium, loosens the bark, and rapidly kills the tree usually from the top downward.   Many foresters believe that these insects are largely responsible for forest fires by producing areas of dead and highly inflammable material.   Among the most important groups are the genus *Dendroctonus*, including the mountain pine beetle (*D. monticolae* Hopkins), the southern pine beetle (*D. frontalis* Zimmerman), the western pine beetle (*D. brevicomis* Leconte), the Black Hills beetle (*D. ponderosae* Hopkins), the Engel-

mann spruce beetle (*D. engelmanni* Hopkins), and many others, which are especially destructive to mature trees.   Species of the genus *Ips*, known as pine engravers, are primarily enemies of young trees and the tops of older ones.   Species of the genus *Phloeosinus* attack cypress and the giant redwoods or sequoia.   In this family also are included the hickory bark beetle[1] (Fig. 17.21), which has killed a large per cent of the hickory trees in the states of the Mississippi Valley and caused losses of about 15 million dollars a year, the shothole borer (page 722), the peach bark beetle (page 755), the elm bark beetles (page 848), and dozens of other species of only slightly less importance.

17.22. Work of the hickory bark beetle, *Scolytus quadrispinosus*, in a 12-inch hickory tree.   The dead bark has been removed. Note the straight, vertical egg galleries made by the females and, radiating from each egg gallery, the numerous tunnels made by the growing larvae.   (*From N.Y. State Coll. Forestry, Tech. Bul.* 17.)

The western pine beetle, which attacks only ponderosa and Coulter pine, probably causes more damage than any other species of bark beetle, and its life history is briefly as follows.   The insect overwinters as both larva and adult in galleries in the bark.   The female deposits an average of 50 eggs, singly along the egg gallery in the inner bark, at an average of 4 eggs per inch of gallery.   Oviposition continues for about 23 days.   The eggs hatch after about 7 days, and the first-instar larvae tunnel in the cambium.   The second-instar larvae return, after about $\frac{1}{2}$ inch of tunnel, to the bark and then to the junction of the inner and outer bark, where the third and fourth instars are found.   The duration of the larval period is about 30 days for the first generation, 60 for the second, and 155 for the third or overwintering generation.   The larva then forms a pupal cell in the outer bark, and after a prepupal period of 2 to 7 days and a pupal period of 6 to 51 days, depending upon the temperature, the adult emerges and spends 7 to 18 days in the outer bark before boring to the outside.   The life cycle from egg to adult requires an average of about 46 days for the first generation, 88 for the second, and 243 for the overwinter generation.   The adult beetles may live as long as 244 days.   They become inactive at temperatures below 50°F., but fly during the day at 70 to 85°F. and may travel up to 2 miles.

[1] *Scolytus quadrispinosus* Say.

Control Measures. No practical methods have been discovered to prevent the attack of bark beetles or to save trees over large forest areas. Modern silvicultural practices which greatly reduce the infestations include early removal of trees especially susceptible to attack because of age, size, injury, or lack of vigor and felling of infested trees during the summer period of May to October when they dry out rapidly. To prevent further losses, the infested trees may be converted to lumber during the winter and early spring, burning all slab and slash before the beetles emerge or submerging the logs for at least 6 weeks to drown the insects. If the milling is done 20 to 50 miles from the forest, the slabs need not be burned. Other measures include felling the trees, removing infested bark and burning to destroy the insects, or felling and exposing the bark to the sun so that lethal temperatures of 115 to 120°F. are developed. A cheaper method consists in spraying felled or standing timber with (a) *ortho*-dichlorobenzene 1 gallon to 6 gallons of fuel oil, (b) ethylene dibromide 1 gallon in 5 gallons of fuel oil, or (c) an emulsion of ethylene dibromide 2 pounds, emulsifier 0.5 pound, fuel oil to make 1 gallon, and 4 gallons of water. Spraying infested trees with a mist of 5 per cent DDT in fuel oil will kill most of the emerging beetles. For shade trees and farm wood lots, effective means of control are to keep the trees in vigorous condition by supplying adequate water and fertilizer and to keep all dying and injured wood pruned out. Pruning should be done during the winter or early spring, and all such wood should be burned to prevent it from serving as foci to infest growing trees in the vicinity.

References. *U.S.D.A., Cirs.* 664, 1943; 817, 1949; 864, 1951; and 944, 1954, *Forest Pest Leaflet* 12, 1957, and *Misc. Publ.* 800, 1960; *Can. Dept. Agr. Tech. Bul.* 14, Parts I and II, 1917–1918; *Miss. Agr. Exp. Sta. Tech. Bul.* 11, 1922; *Ore. Agr. Exp. Sta. Bul.* 147, 1918, and 172, 1920, and *Cir.* 162, 1948; *U.S.D.A., Farmers' Bul.* 1188, 1921, and *Bureau Ento. Buls.* 56, 1905, 58, 1906–1909, and 83, 1909; *Jour. Econ. Ento.,* **46**:601, 1953.

250. ELM BARK BEETLES

Importance and Type of Injury. The native elm bark beetle[1] and the smaller European elm bark beetle[2] have taken on especial importance with the discovery that they are carriers of a virulent and destructive parasitic fungus[3] imported from Europe, and first recognized in America in 1930. The disease and the European bark beetle were both imported on Carpathian elm logs used for furniture veneer. The disease now occurs from the East Coast to the Rocky Mountains and from Tennessee north to Ontario and Quebec in Canada. The American elm is especially susceptible, and in many areas nearly all these trees have died and have been cut down as a result of the attack of this fungus. Infested trees show wilting leaves on certain branches, which shrivel, discolor, and fall, except at the end of the branch; the ends of the twigs curl in a peculiar manner; the larger limbs put out numerous trunk suckers, very evident in winter; and the dying twigs when examined in cross-section show a brown discolored ring beneath the cambium. Beetles flying from diseased trees bear spores in or upon their bodies, and when they make feeding or breeding cavities in healthy trees, inoculate the latter with the deadly disease.

[1] *Hylurgopinus rufipes* (Eichhoff), Order Coleoptera, Family Scolytidae.
[2] *Scolytus multistriatus* (Marsham).
[3] *Ceratostomella* (= *Graphium*) *ulmi* Schwarz.

Trees Attacked.   Elms are the favorite host of the beetles but the native elm bark beetle also attacks basswood and ash.

Distribution.   The native elm bark beetle probably occurs wherever the American elm is found throughout the United States and Canada.   The smaller European elm bark beetle has been present in this country at least since 1909.   It is known to occur throughout most of the eastern United States and as far west as Colorado and Wyoming.   Another carrier in Europe, the larger European bark beetle[1], has been intercepted by American quarantine officials but so far is not known to be established in America.   The disease spreads most rapidly where the bark beetle carriers are breeding, although it is readily disseminated also by transportation of diseased wood.

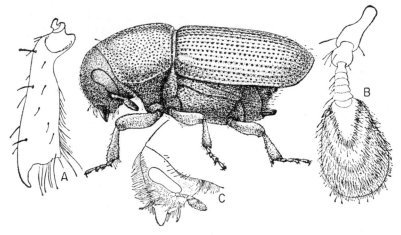

Fig. 17.23. The smaller European elm bark beetle, *Scolytus multistriatus*, about 20 times natural size.   *A*, front tibia; *B*, antenna; *C*, outline of the head from the side, more enlarged.

Life History, Appearance, and Habits.   The breeding habits of these insects are essentially like those described for bark beetles in general.   The smaller European elm bark beetle larvae winter at the ends of their tunnels, pupate in the outer bark, and the adults begin to appear in May. They are stout reddish-brown beetles, about ⅛ inch long, of the appearance shown in Fig. 17.23.   They feed on the twigs, often in the crotches, of healthy elm branches especially in trees which contain, or are very near to other trees that contain, weak, injured, or sickly portions.   The females construct their brood chambers, however, only in the cambium of weakened, dying, or recently dead wood.   The parent galleries or brood chambers run lengthwise of the branches and the larval galleries across the grain.   The parent gallery is 1 to 2 inches long and from 80 to 140 eggs are laid in it.   The larvae develop in a few weeks, and a second or summer generation of adults appears and lays eggs in August and early September.   There are two generations a year, but egg-laying extends over a considerable time, and adults may be continuously present from May to September.   The parent galleries of the small, black, native species are cut across the grain of the wood, and the larval tunnels run more or less lengthwise of the twigs.

[1] *Scolytus scolytus* Fabricius.

Control Measures.   Control consists in keeping trees healthy, vigorous, and free from injuries and in the prompt removal and burning of diseased or infested trees or parts of trees likely to attract the beetles to feed or to lay their eggs.   The rigorous practice of such methods will reduce the number of bark beetles in an area but will not provide satisfactory control, because the smaller European elm bark beetle can fly and carry the fungus for at least 3 miles.   Dormant spraying of the trees with heavy concentrations of DDT or methoxychlor will kill the beetles before they can feed on the elms during the spring and summer when they are most susceptible to the disease.   Special emulsifiable concentrates have been developed which leave a suitable residue in the bark and do not cause tree injury: (*a*) for hydraulic spraying: DDT, 32 per cent; xylene, 66 per cent; emulsifier (Triton X-100), 2 per cent; used at a final concentration of 2 per cent DDT; and (*b*) for mist blower: DDT, 26 per cent; xylene, 48 per cent; mineral oil (horticultural grade), 24 per cent; emulsifier, 2 per cent; used at a final concentration of 12 per cent DDT.   All bark surfaces should be completely covered before the beetles become active in the spring, and for a 50-foot tree this requires about 20 to 30 gallons for hydraulic spraying and 2 to 3 gallons for a mist blower.

References.   Cornell Agr. Exp. Sta. Serv. Buls. 290, 1934, and 841, 1947; *Jour. Econ Ento.,* **41**:327, 1948; *Proc. 24th. Nat. Shade Tree Conf.,* 1948, pp. 113; *U.S.D.A., Information Bul.* 193, 1959.

251. CARPENTERWORM[1]

Importance and Type of Injury.   This insect is very common on many shade trees, especially on oak.   Infested trees have large burrows running through the wood, with occasional openings through the bark of the trunk, from which sawdust may be forced out or from which a discharge of dark-colored sap will be oozing and discoloring the trunk.   The burrows of the insect in the trunk will sometimes be as much as 1 to 1½ inches in diameter.   These galleries occasionally are so numerous as to weaken trees, and cause them to be broken off during high winds.   The injury also greatly lessens the value of lumber or posts from infested trees.

Trees Attacked.   Oaks of the black and red oak groups, elm, locust, poplar, willow, maple, ash, chestnut, and others.

Distribution.   The insect is generally distributed throughout the United States and southern Canada.

Life History, Appearance, and Habits.   The winter is passed in the larval stage in the burrows in the trunk and large branches of the tree.   The borers vary in size, according to their age, from 1 inch up to 2 inches or a little over.   The presence of prolegs distinguishes them from beetle grubs.   The general color is white tinged with pink, with a very dark-brown head and numerous dark-brown tubercles over the body. The insects pupate in large cells excavated in the wood of the tree and, just before emerging, the pupae work part way out of the burrow.   The adult moth on emerging generally leaves the empty pupal skin protruding from the tree.   The moths are abroad at night during the summer months.   In this stage the insect is of very striking appearance (Fig. 17.24), having a wing expanse of as much as 3 inches, in the female. The moths are of a general gray color, mottled with black and lighter shadings, the hind wings being faintly tinged with yellow.   In the male the hind wings have a distinct orange margin.   The eggs are laid in crevices in the bark and the young borers on hatching immediately work their way into the wood of the tree.   Three years are probably required for completion of the larval growth.

[1] *Prionoxystus robiniae* (Peck), Order Lepidoptera, Family Cossidae.   A very similar species, the little carpenterworm, *Prionoxystus macmurtrei* (Guérin-Méneville), occurs in eastern United States and Canada.

Control Measures. These boring caterpillars may be killed by injecting carbon bisulfide or other fumigant into their burrows and closing the openings with mud or putty. This is practical only on valuable shade trees.

Fig. 17.24. The carpenterworm, *Prionoxystus robiniae*, adult female.   Natural size.   (*From Ohio Agr. Exp. Sta. Bul. 332.*)

References. Ohio Agr. Exp. Sta. Bul. 332, p. 329, 1918; *N.D. Agr. Exp. Sta. Bul.* 278, 1934.

252. LEOPARD MOTH[1]

Importance and Type of Injury. Branches, especially in the tops of trees, are dead and smaller branches broken over or hanging partly cut off, during the latter part of the summer. The leaves wilt suddenly on small branches. There are holes along such branches, from which damp sawdust is being pushed out (Fig. 17.25).

Trees Attacked. Elms and maples are the favored food plants, but the insect attacks also many other deciduous shade and fruit trees.

Distribution. This moth, a native of Europe, was first found in the United States in New York, in 1879. It is distributed along the Atlantic Coast from Maine to Delaware.

Life History, Appearance, and Habits. The winter is passed as a partly grown larva, from 1 to 1½ inches in length, in burrows in the heartwood of infested trees. The worms (Fig. 17.25) are of a pinkish-white color, with many dark-brown spots distributed over the body, and have prolegs. In the spring the worms start feeding and boring through the wood, the smaller ones continuing to feed throughout the season, while the larger ones become full-grown in the late spring and change to a brown pupal stage within their burrows. When the moth is ready to emerge, the pupa forces itself partly out of the burrow, and the skin splits down the back, permitting the adult moth to escape. Emergence takes place from late May to early fall. The adults (Fig. 17.25) are very striking in appearance, being of a general white color, blotched and spotted with blue and black. They have a wing expanse of from 2 to 3 inches. The salmon-colored oval eggs, to the number of 400 to 800, are laid in crevices of the bark. They hatch in about 10 days, and the young borers rapidly work their way into the heartwood of the branch. It requires from 2 to 3 years for the insect to complete its growth.

Control Measures. Where this insect is abundant, it is very important that all infested branches and badly infested trees be cut and burned during the fall and winter months. The presence of the insect is indicated by wilting of the leaves and by

[1] *Zeuzera pyrina* (Linné), Order Lepidoptera, Family Cossidae.

masses of wet frass thrown out from the burrows. In valuable shade or fruit trees the insect may be killed by digging or probing out the worms or by injecting commercial borer paste or carbon bisulfide into the burrows. The worms will attempt to clean out the burrows and so come in contact with the paste and be killed.

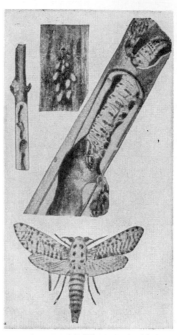

Fig. 17.25. The leopard moth, *Zeuzera pyrina;* work in small twig, borings hanging from bark, nearly full grown larva, and adult moth. ½ natural size. (*From Felt, "Manual of Tree and Shrub Insects," copyright, 1924, Macmillan. Reprinted by permission.*)

References. *Conn. Agr. Exp. Sta. Bul.* 169, 1911; *Jour. Econ. Ento.*, **27**:196, 197, 1934; *N.J. Agr. Exp. Sta. Rept.*, **50**:146, 147, 1929; *U.S.D.A., Farmers' Bul.* 1169, 1925.

253. Lilac Borer[1]

Importance and Type of Injury. Lilac canes are dying, the base of the canes exhibiting swollen areas where the bark is cracked and broken away from the wood. There are numerous holes through the bark and wood. Canes suddenly wilt and show fine, sawdust-like borings forced out from holes in the bark. Figure 17.26 shows typical injury on ash.

Plants Attacked. Lilac and green, white, English, and mountain ash are most seriously infested, although the insect is occasionally taken in some other trees.

Distribution. Eastern United States, north through southern Canada, and westward to Colorado.

Life History, Appearance, and Habits. The winter is passed as a partly grown larva in the stems of lilac, usually near the surface of the

[1] *Podosesia syringae syringae* (Harris), Order Lepidoptera, Family Aegeriidae. The ash borer, *P. syringae fraxini* (Lugger), attacks ash and mountain ash in a similar manner.

soil, and in infested trees.   The insect starts feeding in the spring and completes its growth by early summer.   At this time it is a nearly pure-white worm, about 1½ inches long, with a brown head.   It transforms in the burrow to a brown pupal stage and emerges in about 3 weeks, in June and July, as a clear-winged moth of a somewhat wasp-like appearance.   The forewings are of a general brown or chocolate color, the hind wings are clear, marked with a dark border.   The body is mainly brown and the legs are marked with brown and yellow.   The insect is about 1 inch in length with a wing expanse of 1½ inches.   The moths are very active fliers.   The females deposit their eggs on the bark about the base of lilac canes, or on ash.   The worms, hatching from them, become about half-grown by cold weather.   There is one generation a year.

Fig. 17.26. Ash tree injured by lilac borer, *Podosesia syringae*.   (*From Ill. Natural History Surv.*)

Control Measures.   Painting infested areas on ash and lilac with a solution of *para*-dichlorobenzene, 1 pound, in raw cottonseed oil, 2 quarts; or with *para*-dichlorobenzene, 1 pound, in dormant miscible oil, 2 quarts, and water, 2 quarts, has given good control.   Injecting carbon bisulfide into the burrows is effective if carefully done.   Badly injured canes of lilac should be cut and burned during the winter.

References.   *Twenty-sixth Rept. Ill. State Ento.*, 1911; *Mich. State Bd. Agr. Rept. for* 1934.

254. ELM COCKSCOMB GALL[1]

Importance and Type of Injury.   This aphid seldom causes the death of the elm trees but often becomes annoying because of the fact that its galls make the trees unsightly.   Leaves of elms infested with this aphid have crinkled reddish galls ½ to ¾ inch long on their upper surfaces.   The shape of these galls resembles a miniature rooster's comb (Fig. 17.27).   During the early part of the growing season, the galls are green or reddish in color.   During the late summer, they dry and become brown.

[1] *Colopha ulmicola* (Fitch), Order Homoptera, Family Aphidae.

An examination of the green galls will show that they are filled with small, rather smooth, greenish or brownish aphids. A narrow slit on the underside of the leaves serves as an entrance to the gall, although during the early part of the season this slit is so nearly closed that the aphids cannot get through it.

Trees Attacked. Elm, especially red elm.

Distribution. General over the United States.

Life History, Appearance, and Habits. The winter is passed as dark-brown shiny eggs in cracks on the bark of elm trees. These eggs hatch in the spring when the leaves are partly grown. The aphids crawl to the leaves and begin feeding by sucking the sap. Where they start to feed, the peculiar gall formations start growing, enclosing the aphids. These aphids, which are all females, remain inside these galls, giving

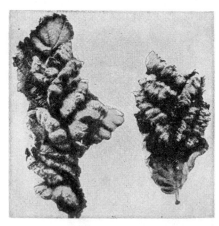

Fig. 17.27. The elm cockscomb gall, *Colopha ulmicola.* A number of galls on leaves of white elm. About natural size. (*From Washburn, "Injurious Insects and Useful Birds,"* Lippincott.)

birth to living young, so that if the gall is broken open during the summer, it will be found swarming with aphids. As the aphids feed, they give off quantities of honeydew, which drops from the gall, as it begins to crack open late in the season, often nearly coating the surfaces of walks and benches under the trees. This honeydew attracts flies and other insects, which add to the unsightly appearance of infested trees. During the summer, the galls crack open and the aphids make their way out. Winged migrants carry the species to the roots of various grasses, where the aphids live during the summer, returning to elms in the fall.

Control Measures. There is no practical control measure for this insect other than cutting off the galls early in the season, as soon as they start showing on the leaves. This, of course, can be done only on small trees or those that are especially valuable.

References. Ohio Agr. Exp. Sta. Bul. 332, p. 311, 1918; *Maine Agr. Exp. Sta. Bul.* 181, 1910; *Wis. Dept. Agr. Bul. 123,* 1931.

CHAPTER 18

INSECT PESTS OF GREENHOUSE AND HOUSE PLANTS AND THE FLOWER GARDEN

Few persons realize the importance of the annual crop of flowers and ornamentals and other hothouse products in this country. Approximately 5,300 acres of ground are enclosed in glass and devoted to the production of flowers and vegetables in the United States and the annual crop is valued at more than $300,000,000. In addition to the commercial production there is an enormous interest in the growing of house plants and garden flowers by individuals who attach to their products a personal, but very real value, far beyond their market price.

The insects that attack greenhouse crops are the same as those that attack related crops in the field. Some of these insects are field pests only in the South but have adapted themselves to the semitropical conditions found in greenhouses in the North. It would seem at first thought that insects could be easily controlled in greenhouses, where climatic factors can be largely regulated by man; but this is far from being the case. Conditions must be maintained in the greenhouse to give a maximum rate of growth to the crop, and such conditions are frequently very favorable for the insects attacking the crops. The cost of producing crops under glass is high, and the loss of 10 to 25 per cent of the crop from attacks by insects generally means that the entire operation of producing and marketing the crop has been carried on at a loss to the grower. Crops grown under glass are more easily injured by heavy applications of insecticides, or other insect-control measures that affect the plants, than the same crops grown in the open. For this reason much greater care must be exercised in the control methods employed.

The best way to prevent damage by insects to greenhouse crops is to keep the insects out of the houses. This may be done by careful inspection of all plants brought into the houses and by thoroughly cleaning up the houses by heat or heavy fumigations in the intervals between crops. Open range houses are, on the whole, more difficult to keep free of insects than those built in sections, as some part of the open range house is nearly always occupied by a crop in some stage of growth. Under such conditions, strong fumigations which will kill nearly all plant and animal life cannot be used for cleaning up the house.

It will help greatly to avoid most insect injury, if the following precautions are taken: (1) Before a crop is put in, or the soil placed in the benches, the houses should be cleaned out as thoroughly as possible, and, if they have recently been infested with insects, a thorough fumigation with hydrogen cyanide gas should be given at a dosage equal to 1 ounce sodium cyanide to 100 cubic feet of space (page 376). Such a fumigation will kill practically all insect life in any stage that may be present in the house.   (2) A careful inspection should be made of the soil being brought into the benches, to make sure that this soil does not contain wireworms, white grubs, eelworms, cutworms, or other insects that may be injurious to greenhouse crops.   If the soil is infested, it should be sterilized with steam.   (2) All plants brought into the greenhouse should be inspected to make sure that they are free from insect infestation.   Many florists have suffered serious losses by bringing a few infested plants into their houses and thus starting an infestation which, before it was noticed, had spread throughout the entire range of their greenhouses.

Many of the best accounts of the life histories, habits, and control of greenhouse insects are contained in general publications dealing with this class of insect pests. Some of these publications are here listed and should be used as general references for further information on the insects treated in this chapter.

References.   Can. Dept. Agr., Bul. 7 (n.s.), 1922, and *Bul.* 99 (n.s), 1934; *Cornell Univ. Ext. Bul.* 371, 1937; *Mich. Agr. Exp. Sta. Spcl. Bul.* 214, 1931; *Ill. State Natural History Surv. Cir.* 12, 1930; *N.J. Agr. Exp. Sta. Bul.* 296, 1916; *Fifth Rept. S.D. State Ento. for* 1924; *U.S.D.A., Farmers' Bul.* 1362, 1923; *Twenty-seventh Rept. Ill. State Ento. for* 1912; *U.S.D.A., Misc. Publ.* 626, 1948; *Calif. Agr. Ext. Ser. Cir.* 87, 1944; WESTCOTT, C., "The Gardener's Bug Book," Doubleday, 1946; *Calif. Agr. Exp. Sta. Bul.* 713, 1949.

KEY FOR THE IDENTIFICATION OF INSECTS INJURIOUS TO GREENHOUSE PLANTS AND THE FLOWER GARDEN

A. *External chewing insects that eat holes in leaves, buds, or stems:*

1. Brown, green, or gray, mottled, jumping insects of all sizes up to 1½ inches long, migrate into gardens from surrounding areas and strip leaves and eat tender stems of many kinds of flowers.............*Grasshoppers,* page 465.

2. Foliage of many kinds of flowers is eaten at night by dark reddish-brown beetle-like insects, up to ⅘ inch long, with a pair of sharp pincers or forceps at the tip of the abdomen, ¼ as long as rest of body.   Insects hide in soil during day...
European earwig, Forficula auricularia Linné, Order Dermaptera, Family Forficulidae.

3. Many, minute holes eaten in leaves by very small hard-shelled black beetles, from ¹⁄₁₆ to ⅛ inch long, which feed on the underside of leaves and jump readily when disturbed............................*Flea beetles,* page 514.

4. Fawn-colored long-legged cylindrical beetles, about ½ inch long, attack blossoms and buds of roses and other flowers and shrubs, during daytime, especially in regions of sandy soil, in June............*Rose chafer,* page 772.

5. Cinnamon-brown velvety beetles, shaped like a May beetle but only ⅜ inch long, eat irregular holes from margins of leaves of many garden flowers at night.   Attracted to lights.   Along Atlantic seaboard....................
....*Asiatic garden beetle,* page 606.

6. Elongate, nearly cylindrical, not very hard, active long-legged beetles, either black, gray, or brown, sometimes with narrow, longitudinal stripes or with small spots, swarm upon plants in gardens, devouring blossoms and leaves. Cause blisters on skin when crushed...............*Blister beetles,* page 642.

7. Bright-red snout beetles, with the snout and undersurface black, about ¼ inch long, eat numerous holes in the buds of roses, so that blooms fail to develop. The whitish legless grubs feed in the "hips" or fruit........................
Rose curculio, Rhynchites bicolor (Fabricius), Order Coleoptera, Family Curculionidae.

8. Undersurface of the leaves eaten off by pale-green, very active caterpillars up to ¾ inch long.   Such leaves are often covered with a light web enclosing several leaves or drawing parts of a single leaf together....................
....*Greenhouse leaf tier,* page 861.

9. Injury appearing much the same as in *A*,8.   Small active greenish worms feeding at first as leaf miners but later eating off the surface of the underside of the leaves.   Leaves rolled or drawn together with light webs............. ....*Oblique-banded leaf roller*, page 862.

10. Ferns with the leaves partly stripped, presenting a very ragged appearance. Newly unfolding leaves gnawed or eaten away.   Dark-green velvety caterpillars, with two wavy white lines down each side of the body, hiding in the soil about the base of the ferns, feeding very actively at night............. ....*Florida fern caterpillar*, page 862.

11. Many kinds of plants cut off at the surface of the ground, or, in some cases, stripped of leaves, by fat, sleek caterpillars or worms of varying sizes and colors...*Cutworms*, page 863.

12. Leaves eaten from the new growth, or buds eaten into or partly eaten out, by greenish, dark-brown, or yellowish worms with slightly hairy bodies, differing from cutworms in their habit of feeding on the upper part of the plants. Injury common during the fall months.............*Corn earworm*, page 863.

13. Roses with the leaves riddled with small holes, bark of the new growth eaten off, and buds eaten out.   Injured plants with most of the feeding roots eaten off by small curved-bodied whitish grubs.   Brownish or brownish black, very active beetles, about ⅛ inch long, feeding on the leaves................. ....*Strawberry rootworm*, page 860.

14. Velvety-green caterpillars, with a very faint, orange stripe down the middle of the back, up to 1¼ inches long, rag the leaves of mignonette, alyssum, and nasturtium.   White butterflies about 2 inches across the wings lay the eggs.. ....*Imported cabbageworm*, page 662.

15. Pale-green, white-striped, looping caterpillars, with only 3 pairs of prolegs, which hump the back high as they crawl, eat holes in leaves of mignonette, geranium, chrysanthemum, and carnation..........*Cabbage looper*, page 664.

16. Pale-green caterpillars, not over ⅓ inch long, very active, eat small rounded holes in the leaves of stocks and wallflowers in midsummer.   Wriggle actively when disturbed.........................*Diamond-back moth*, page 666.

17. Yellowish, brown, or reddish brown-and-black-banded caterpillars, up to 2 to 2½ inches long, very densely covered with short stiff hairs of even length, eat large holes in leaves of dahlias, verbenas, hydrangeas, and many other flowers.   Smaller caterpillars often feed close together.................... ....*Woollybears*, page 611.

18. Velvety-black caterpillars, with 2 conspicuous bright-yellow stripes on each side of the body, between which cross many narrow hair-lines of the same color, eat foliage of sweet peas, gladioli, lilies, and other flowers in late summer; full-grown worms about 2 inches long..................................... *Zebra caterpillar*, *Ceramica picta* (Harris), Order Lepidoptera, Family Noctuidae.

19. Greenish or yellowish-green false caterpillars, up to about ½ inch long, some smooth, others bristly, skeletonize the leaves of roses, by eating the upper surface, leaving the lower epidermis and the veins to dry out and die.   The worms, which have more than 5 pairs of prolegs on the abdomen, are sluggish, and leave a slimy secretion on the leaves which dries to form a glaze........ ....*Rose slugs*, Order Hymenoptera, Family Tenthredinidae.

20. Clean-cut round or oval pieces of rose leaves, from ½ to 1 inch across, are cut off and carried away to a nesting site by large bumblebee-like insects.... *Leaf-cutter bees*, *Megachile* spp., Order Hymenoptera, Family Megachilidae.

21. Slender, active, wingless, very slender-waisted, hard-bodied insects crawl over plants infested with aphids or over the buds of peonies and other flowers to get sweet secretions of the aphids or the flowers.   Others make mounds of soil thrown up from underground nests, especially about aster plants ....*Cornfield ant* and other *ants*, pages 501 and 893.

B. Insects that suck sap from leaves, buds, or stems:

1. Surface of the leaves whitened with many small flecks of light green or yellow. Tips of the leaves curling up and dying.   Flower buds producing distorted

blossoms which open only on one side. Undersides of the leaves with numerous black spots. General vigor of the plants greatly reduced by the feeding of many very small yellow or black, very slender, active insects............
....*Greenhouse thrips*, page 863.

2. Greenhouse plants show a sickly appearance, the foliage turning yellow. Numerous brown or brownish-gray, flattened, motionless scales adhere to the undersides of the leaves or along the stems especially at the leaf axils. Scales usually of sufficient size to be easily seen, some up to ⅛ inch in diameter. The scale covers the body of the insect but is easily lifted from it...........
....*Armored scales*, page 866.

3. Canes of roses specked or nearly covered with snowy-white, thin, flat, rounded scales, with a nipple at the center, about $\frac{1}{12}$ inch in diameter; or smaller, slender scales with the exuviae at one end. Most abundant near the ground level. Beneath the scales, small legless insects suck the sap from the plants.
...*Rose scale, Aulacaspis rosae* (Bouché) Order Homoptera, Family Coccidae.

4. Foliage plants, such as oleander, fern, palm, Ficus, and Vinca having much the same appearance as in *B*,2. Scales somewhat larger and adhering more tightly to the plants. Scales much more conspicuous than in *B*,2, usually very convex. Scale covering cannot be lifted from the body of the insect, but forms a distinct part of the body wall............*Tortoise scales*, page 868.

5. Plants, particularly foliage plants, such as coleus, orchids, poinsettias, ivy, and many others, with whitish clusters of soft-bodied insects, up to ¼ inch long, at the axils of the stems and leaves and along the stems. These insects have many whitish, waxy filaments protruding from the body. These filaments are so thick as to give the body a distinct bluish-white appearance. Insects seldom move unless disturbed.................*Mealybugs*, page 869.

6. Plants infested in much the same way as in *B*,5, by insects with rather small, dark-colored bodies, with rows of short waxy filaments radiating around the outer margin of the body, and forming a waxy tube extending back from the body several times its length.................*Greenhouse orthezia*, page 872.

7. Greenhouse and garden plants having a weakened appearance; frequently with the leaves curled, the young growth distorted, and often coated to some extent with sticky honeydew. Small winged or wingless aphids, or plant lice, of various colors, sucking the sap from the more tender parts of the plants, buds, and blossoms.....................*Greenhouse* and *garden aphids*, page 872.

8. Undersurface of the leaves, with small, oval, flat, pale-green motionless insects, less than $\frac{1}{30}$ inch in length, adhering to them. Many tiny four-winged snow-white flies on the underside of the leaves. These flies leave the plants in swarms when disturbed. Plants more or less covered with a coating of glazed, sticky material on which a sooty black fungus is frequently growing..
....*Greenhouse whitefly*, page 874.

9. Leaves of rose, dahlia, asters, and many other flowers are mottled with very fine whitish spots, or somewhat curled and withered. Tiny, elongate, active, somewhat wedge-shaped bugs, green or varicolored, suck sap from underside of leaves. Bugs jump, fly, or run sideways when disturbed...............
....*Leafhoppers*, page 564.

10. Olive-green bugs, flecked with yellow and dark brown, with flattened, oval bodies, somewhat triangular in front, up to ¼ inch long by half as wide, by piercing and sucking the sap from buds of zinnias and dahlias, dwarf or deform them, and cause spotting and deforming of leaves and dead areas on stems of various flowers, sometimes in greenhouses.....*Tarnished plant bug*, page 614.

11. Bright greenish-yellow bugs, of same shape as *B*,10, but up to ⅓ inch long, with 2 triangular, black spots on the prothorax, 2 black stripes on the fore part of each wing, and a rounded black spot near the posterior end of the stripes, pierce leaves and buds of garden snapdragon, weigelia, zinnia, and dahlia, and sometimes in greenhouses, causing whitish to brown spots on upper side of leaves.................................*Four-lined plant bug*, page 784.

12. Blackish, rather soft-winged, flattened or nearly globular, jumping bugs, only $\frac{1}{10}$ inch long, somewhat resembling flea beetles, and their greenish

nymphs of various sizes, suck sap from primrose, ageratum, and other flowers in the garden, and sometimes in greenhouses, causing spotting and dying of leaves...................................*Garden fleahopper*, page 615.

13. Plants with pale blotches or spots showing through the leaves, or with the entire leaf having a light color, often drying up or turning a reddish brown about the margins. Fine silk threads are spun on the undersides of the leaves, or formed into webs which entirely cover the surface of the plant. Minute, 6- or 8-legged, greenish or yellowish mites crawling about on the webs or underside of the leaves..................*Two-spotted spider mite*, page 875.

14. Cyclamen with the leaves much distorted, buds gnarled and knotty in appearance, usually failing to open; flowers on infested plants when open are streaked and blotched in appearance and quickly die. Infested foliage shows pockets or depressions in the leaf, or dark-purplish areas, often covered with small cracks, on the leaf surface. Tiny white or pale-brown mites, with 3 or 4 pairs of legs, working in the infested flowers, on the leaves, and about the base of the plants..................................*Cyclamen mite*, page 876.

C. *Insects that bore in the stems of the flowers:*

1. Stems of dahlia, hollyhock, bleeding heart, larkspur, golden glow, and other flowers have the heart eaten out or filled with frass from a dark-brown caterpillar with two white stripes on each side of the body, which are interrupted for a space near the middle of the body, giving the appearance of a bruise....
....*Common stalk borer*, page 493.

2. Similar to *C*,1, but the white stripes not interrupted.....................
Burdock borer, Papaipema cataphracta (Grote), Order Lepidoptera, Family Noctuidae.

3. Stems of iris bored by a smooth, cylindrical caterpillar, up to 1½ inches long, flesh-colored to pinkish, which tunnels from stems downward toward base of stalk..
Iris borer, Macronoctua onusta Grote, Order Lepidoptera, Family Noctuidae.

D. *Insects that make galls or mine in the leaves:*

1. Chrysanthemums with small blister-like cone-shaped galls on the upper surface, and occasionally on the undersurface, of the leaves. Leaves curled and stems of heavily infested plants crooked and distorted. Flowers opening imperfectly if at all. Small gnat-like flies, about 1/14 inch in length, and of a reddish-orange color, crawling about over the leaves. Minute yellowish-white maggots in the galls on the leaves and stems........*Chrysanthemum midge*, page 877.

2. Flower buds and leaves of roses much distorted; terminal growth and bud dying and turning brown. An examination of such buds will show many small, white-to orange-colored maggots, up to 1/12 inch long, feeding on the plant tissue within the buds. Small silken cocoons just under the surface of the soil about the plants,......................................*Rose midge*, page 878.

3. Leaves of chrysanthemum or marguerite with irregular light-colored mines extending over their surface. When badly infested, the leaves may dry up, but usually remain attached to the plants. Heavily infested plants are stunted and produce small inferior flowers..
Chrysanthemum leaf miner or *marguerite fly, Phytomyza atricornis* Meigen (see *U.S.D.A., Farmers' Bul.* 1362, 1922).

4. Columbine has conspicuous, slender, white, serpentine lines on the leaves due to the mining of a tiny maggot between upper and lower surface of the leaves. Larkspur leaves are similarly tunneled, the mines often coalescing to make yellowish blotches...
Columbine leaf miner, Phytomyza minuscula Gour., or *larkspur leaf miner, P. delphiniae* Frost, Order Diptera, Family Agromyzidae.

E. *Insects that attack the roots or bulbs:*

1. Greenhouse or garden plants sometimes cut off below the surface of the ground, or wilting and dying, without visible injury to the part of the plant above the ground. Examination of the roots will show large curved-bodied, brown-headed white grubs with six rather conspicuous legs, feeding on the underground parts of the plant..........................*White grubs*, page 879.

2. Appearance and injury much the same as in *E*,1, but with underground parts of the plant eaten off, or bored through, by shining, hard-bodied, brown, slender worms up to 2 inches in length........................*Wireworms*, page 617.

3. Plants presenting a somewhat sickly appearance, with no visible injury to the part above the surface of the ground. Roots showing minute brownish scars or tunnels along their surface. Very small, somewhat thread-like, active, white maggots, not over ¼ inch long, embedded in the root tissues, or working in the soil about the roots................*Fungus gnats* or *Sciara maggots*, page 880.

4. Narcissus and related bulbs with scars on the base of the bulbs, the bulbs soft and frequently rotting. An examination of the bulbs shows whitish or yellow-ish-white fat maggots inside the bulbs, eating out the plant tissue. Maggots, when full-grown, are from ½ to ¾ inch in length..*Narcissus bulb fly*, page 880.

5. Bulbs appearing as in *E*,4, but containing smaller maggots of a grayish-yellow color, with bodies markedly wrinkled. Many maggots often occurring within one bulb. The maggots vary in size up to ½ inch in length................ ....*Lesser bulb fly*, page 881.

6. Plants of various species that are grown from bulbs turn a sickly yellow, failing to produce flowers; or, if producing flowers, with the flowers much distorted. Leaves stunted and plants generally of a very unhealthy appearance. Examination of the bulbs shows numerous pale-white 6- or 8-legged mites, sheltering and feeding behind the bulb scales. Infested bulbs with reddish-brown spots on the bulb scales....................................*Bulb mite*, page 882.

F. *Various small creatures, not true insects, attacking mostly the roots or leaves on the soil surface:*

1. Light-gray or slate-colored, flat-bodied, distinctly segmented creatures, usually about ½ inch long, with 7 pairs of legs, hiding under clods or bits of plant refuse on the surface of the benches, and feeding mainly at night on the roots and tender portions of nearly all greenhouse plants. When disturbed, they usually roll themselves into small tight balls.. ....*Sowbugs* or *pillbugs* (not true insects), page 883.

2. Hard-shelled, very active, many-legged creatures, up to 2 inches in length, usually with 2 pairs of legs on each body segment, crawling over the surface of greenhouse benches or hiding under any shelter on the surface of the soil. Bodies usually have a brown or pinkish-brown color. Creatures scuttle about very actively when disturbed. Are most abundant in damp parts of the bench...................*Greenhouse millipedes* (not true insects), page 884.

3. Roots and underground parts of stems scarred or eaten off by small, nearly white, very active, many-legged creatures, ¼ inch long or less. They are very strongly repelled by light.. ....*Garden centipede, Scutigerella immaculata* (Newport), page 884.

4. Soft, gray or gray-and-brown-spotted, slimy, legless creatures, from ½ inch to as much as 4 inches in length, crawling about on the surface of the soil or on the plants. A sticky, viscid secretion is given off from the body and, on drying, forms a shiny trail where the creatures have crawled. Usually abundant only in the damper parts of the greenhouse, where they will be found under decaying wood, flowerpots, and other shelters. They feed at night on the plant tissue of many greenhouse crops........*Slugs* or *snails* (not true insects), page 885.

5. Plants presenting a sickly appearance, often somewhat distorted, but with no visible injury to the parts aboveground. An examination of the roots will show numerous knots or galls, or the roots distinctly swollen, enlarged, and of a gouty appearance. The cause of the injury cannot be seen, except when the roots are examined under high magnification, when numerous, very minute, nearly transparent, worm-like creatures will be found in the tissue of the swollen and distorted roots............*Nematodes* or *eelworms* (not true insects), page 886.

255 (213). STRAWBERRY ROOTWORM[1]

Importance and Type of Injury. Roses attacked by this beetle have the leaves riddled with small holes, often to such an extent that they appear to have been

[1] *Paria canella* (Fabricius), Order Coleoptera, Family Chrysomelidae.

peppered with shot. The bark is eaten from the new growth, and the eyes of the buds are frequently eaten out. An examination of the roots will show small curved-bodied whitish grubs eating off the small feeding roots and scoring the bark of the larger roots (see also page 794 and Fig. 15.91).

Plants Attacked. Rose, raspberry, blackberry, grape, oats, rye, peach, apple, walnut, butternut, wild crab, mountain ash, and several others.

Life History, Appearance, and Habits. The females of the strawberry rootworm are about ⅛ inch long and two-thirds as wide (Fig. 15.92). Most of the beetles are brownish, but some are brownish-black, usually with four black spots on the wing covers. They lay their eggs in bunches of from 4 to 15 on dead leaves on the surface of the benches. The eggs hatch in from 10 days to 2 weeks, and the small grubs work their way into the soil, and begin feeding on the rootlets. They become full-grown in from 35 to 60 days. They then change in the soil into the soft white pupal stage, in which they remain for about 2 weeks, emerging as adults at the end of this period. There are several generations a year under greenhouse conditions. While this insect has been known for a number of years as a pest of strawberries, its injury to greenhouse plants has been comparatively recent. Both sexes of the insect are found out-of-doors, but only females have been found in the greenhouse.

Control Measures. Spraying the plants thoroughly, immediately after they have been cut back, with a mixture of 8 pounds lead arsenate and 6 pounds fish-oil soap in 100 gallons of water will give good control. Dusting with sulfur, 85 parts, and lead arsenate, 15 parts, will help in keeping down the beetles on the new growth. It is also advisable in badly infested houses to cover the soil at this time with a heavy application of finely ground tobacco dust. As the soil will not be watered during the drying-out period of the plants, this layer of dust will remain effective for some time and will kill many of the beetles or young larvae hatching from the eggs. In greenhouses, the parathion aerosol treatment described on page 876 is an effective control. All dried leaves on the benches should be collected and burned at short intervals.

Reference. U.S.D.A., Dept. Bul. 1357, 1926.

256 (123). GREENHOUSE LEAF TIER[1]

Importance and Type of Injury. The undersurfaces of the leaves are eaten by slender, pale-green, active caterpillars, which leave the upper epidermis intact. The leaves are sometimes covered with a light web, enclosing several leaves or drawing the parts of a single leaf together.

Plants Attacked. Chrysanthemum, snapdragon, cineraria, aster, sweet pea, ageratum, ivy, rose, and many other *soft-leaved* greenhouse plants, as well as many out-of-door weeds and cultivated crops such as celery, beets, spinach, beans.

Life History, Appearance, and Habits. The adult moths (Fig. 18.1) are of a brownish color, with the front wings crossed by dark wavy lines of a distinctive pattern, as shown in the figure. The wings expand to about ¾ inch. They remain quiet about the greenhouse most of the day, flying actively at night. When disturbed during the day, they fly with a jerky motion, seeking shelter on the underside of the leaves or other objects after a very short flight. The female moths lay their flattened, scale-like, watery-looking, shagreened eggs singly or in overlapping groups on the undersides of the leaves, especially close to the soil. These eggs hatch in about 5 to 12 days into small slender caterpillars which soon take on a pale-green color. When full-grown, the caterpillars

FIG. 18.1. Adult of the greenhouse leaf tier, *Udea rubigalis*, in resting position. About twice natural size. (*From Can. Dept. Agr. Bul. 7, n.s.*)

[1] *Udea* (= *Phlyctaenia*) *rubigalis* (Guenée), Order Lepidoptera, Family Pyralididae.

are slightly less than ¾ inch long, of a pale-yellow color, with a broad white stripe running lengthwise over the back, and a dark-green band in the center of this white stripe.   The younger ones have black heads.   The larvae are almost invisible against a leaf.   When ready to pupate, the caterpillars usually form a shelter by rolling over the edge of the leaf and fastening it together with threads of silk.   Inside this shelter, they spin thin silken cocoons, and then change to the pupal stage.   After 10 or 12 days, the adult moths emerge from these cocoons.   The length of the entire life cycle is about 40 days, and there may be seven or eight generations a year in greenhouses.   The insects appear to be adversely affected by temperatures above 80 or 85°F.

Control Measures.   The larvae of this insect may be controlled by spraying the plants with DDT, TDE (DDD), malathion, or Sevin at 1 pound to 100 gallons of spray.   These sprays may discolor mature plants, and spraying with 0.004 to 0.01 per cent pyrethrins, or the equivalent of pyrethrins and activator, is often preferable, using two applications, ½ hour apart.   Dusting with pyrethrum dust containing 0.1 to 0.2 per cent pyrethrins or with 5 per cent DDT or 4 per cent malathion is also effective.   Both the adults and the larvae may be controlled by applications of the parathion aerosol (page 876) or the DDT aerosol (page 448). Weekly fumigation with hydrogen cyanide or with nicotine will destroy the adult insects.   Good greenhouse sanitation is effective in preventing attacks of this pest.

References. Jour. Agr. Res., **29**:137, 1924, and **30**:777, 1925; *Ill. Nat. History Surv. Cir.* 12, p. 39, 1930.

257. Oblique-banded Leaf Roller[1]

Importance and Type of Injury.   This leaf roller is common in greenhouses and also attacks the foliage of a great many shade trees, vegetables, and ornamental shrubs and flowers.   The type of injury is much the same as that of the greenhouse leaf tier.   The pale-green black-headed young larvae, after hatching from the eggs, live for a time as leaf miners, and then feed for the remainder of their life on the undersides of the leaves.   The adult moth is a little over 1 inch across the wings, of a reddish-brown color, with the front wings crossed by three distinct bands of dark brown.   The control measures are the same as those described for the greenhouse leaf tier.

Reference. Jour. Econ. Ento., **2**:391, 1909.

258. Florida Fern Caterpillar[2]

Importance and Type of Injury.   Ferns attacked by this insect have the leaflets stripped from the old growth, and the new growth entirely eaten away.   Large ferns may be completely stripped of leaves in 1 or 2 days.   Ferns are disfigured to such an extent that they are useless for decorative purposes and cannot be sold.

Life History, Appearance, and Habits.   The adult moth is of a general brown color with a rather dark V-shaped patch near the center of the wings.   The front wings, as a whole, are variegated.   The wings expand about 1 inch.   The eggs are laid on the undersides of the leaves and hatch in from 5 to 7 days into uniformly pale-green larvae. These larvae later change to a slightly darker green or to velvety black.   There are two wavy white lines down each side of the body, and a central stripe of somewhat darker color.   The full-grown larvae are about 1½ inch in length.   These caterpillars feed very actively at night, cutting off and devouring the fern leaves.   On becoming full-grown, they work their way into the soil and there spin a cocoon in which they change

[1] *Archips* (= *Cacoecia*) *rosaceana* (Harris), Order Lepidoptera, Family Tortricidae.
[2] *Callopistria floridensis* (Guenée), Order Lepidoptera, Family Noctuidae.

to reddish-brown pupae. The pupal stage lasts about 2 weeks. There may be a complete generation of this insect every 7 or 8 weeks. Control measures are given below.

259 (4, 67, 80). Cutworms

The same cutworms that attack garden and field crops occur also in greenhouses. Greenhouse vents may be screened with $\frac{1}{4}$-inch hardware cloth to prevent entry of the moths. Control measures are given on page 479 and below.

260 (12, 63, 71, 111). Corn Earworm and Other Caterpillars

The corn earworm moths (page 498) sometimes enter greenhouses in large numbers and lay their eggs on chrysanthemums and other plants. The injury caused by young earworms feeding on chrysanthemum buds is often very severe. A number of other caterpillars also attack greenhouse plants. These include the European corn borer (page 490), the cabbage looper (page 664), the beet armyworm (page 610), and the salt-marsh caterpillar (page 611). These insects may be controlled by spraying with DDT, toxaphene (or a mixture of these), malathion, or Sevin at 1 pound or parathion at 0.25 pound to 100 gallons of spray. Dusting with these materials is also effective.

261. Greenhouse Thrips[1]

Importance and Type of Injury. Thrips occur very generally in greenhouses, and several species of the insect are found feeding on various plants. The surface of the leaves becomes whitened and somewhat flecked in appearance. The tips of the leaves wither, curl up, and die. Buds fail to open normally. The underside of the leaves will be found spotted with small black specks. Where such spots are numerous, the appearance of the foliage becomes marred and, in many cases, plants are rendered so unsightly that they cannot be used for decorative purposes.

Plants Attacked. Practically all plants found in greenhouses are attacked by thrips. Some of those which suffer most severely are roses, carnations, cucumbers, fuchsias, chrysanthemums, crotons, and cinerarias.

Life History, Appearance, and Habits. There are slight variations in the life history of the species found in greenhouses. A typical life history is about as follows: The female insect deposits her eggs in slits in the leaf, inserting the minute white eggs in the leaf tissues. These eggs hatch in from 2 to 7 days, into active, very pale, white nymphs. The nymphs feed on the tissues of the leaf, rasping the leaf tissue with their mouth stylets and sucking up the sap which flows from the injured area. They pass through four instars in the course of their growth and, in the last two instars, they are inactive for a few days before changing to the adult. The adults of different species vary in color, some being yellowish, others nearly black, and others dark brown. They are less than $\frac{1}{10}$ inch long, slender-bodied, and are possessed of three pairs of legs, and four very slender wings with a fringe of long hairs around their margins (Fig. 18.2). The greenhouse thrips is dark brown with the appendages light-colored and $\frac{1}{20}$ to $\frac{1}{24}$ inch long; the antennae are eight-segmented; and the body surface is reticulated. Under greenhouse conditions generations follow each other throughout the year, the total time required for each generation being from 20 to 35 days, depending on the climatic conditions and the species of thrips.

[1] *Heliothrips haemorrhoidalis* (Bouché), Order Thysanoptera, Family Thripidae. The banded greenhouse thrips, *Hercinothrips femoralis* (O. M. Reuter), the onion thrips (page 659), the flower thrips (page 800), and others frequently do severe damage in greenhouses also.

Control Measures. These insects may be controlled by spraying with DDT or malathion at 1 pound or with lindane, heptachlor, or dieldrin at 0.25 pound to 100 gallons of spray, or by dusting with these materials. Weekly applications may be necessary to protect flowers. Thrips have

FIG. 18.2. The greenhouse thrips, *Heliothrips haemorrhoidalis.* From left to right, egg immature stages, and adult female. About 40 times natural size. (*From Univ. Calif Coll. Agr.*)

also been controlled by repeated applications of DDT or parathion aerosols or by fumigation with nicotine (page 380) or hydrogen cyanide (page 375). The thrips may also be controlled by fumigation with naphthalene volatilized at 1 to 1½ ounces per 1,000 cubic feet.

References. Jour. Econ. Ento., **38**:173, 1945; **41**:624, 1948.

262. GLADIOLUS THRIPS[1]

Importance and Type of Injury. Since 1929 this pest has spread with great rapidity over practically the entire United States, resulting in great commercial losses and incalculable disappointment to home gardeners and gladiolus fanciers. The injury is of three distinct types: (*a*) The foliage becomes blasted, with a characteristic silvery appearance, eventually browning and dying, where the minute slender insects have punctured the surface cells and sucked up the exuding sap; (*b*) the flowers become flecked, spotted, and greatly deformed, and many spikes fail to bloom at all; (*c*) the corms in storage are rendered sticky by sap exuding from punctured cells, become darker in color, and their surfaces corky, russeted, and greatly roughened. Infested corms may fail to germinate or may develop a weakened root system and plants that produce only small flowers or none at all.

Plants Attacked. Gladioli, iris, lilies.

[1] *Taeniothrips simplex* (Morison), Order Thysanoptera, Family Thripidae. Very similar injury to growing iris is caused by the iris thrips, *Iridothrips* (= *Bregmatothrips*) *iridis* Watson (see *U.S.D.A., Cir.* 445, 1937).

Distribution.   Probably over the entire United States where gladioli are grown.

Life History, Appearance, and Habits.   Apparently the insect does not winter out-of-doors except in very warm climates, but at digging time many of the thrips leave the foliage, collect on the corms, and hibernate where they are stored, or, if infested corms are stored at temperatures above 60°F., breeding and development may continue in storage.   When infested corms are planted, the insects work along the developing shoot to the leaves and flowers.   Although one of the largest of the thrips, the adults are only $\frac{1}{16}$ inch long and very slender.   The basal third of the slender bristly wings and the third segment of the antennae are nearly white, in sharp contrast to the general brownish-black color of the body. The females may produce offspring either with or without mating, but the unfertilized eggs always develop into males.   The tiny kidney-shaped eggs are thrust into the tissues of the growing plants or the corms.   At favorable temperatures the eggs hatch in about a week, and the minute, yellowish, first- and second-instar nymphs begin feeding in a manner like the adults and become nearly as large in about a month.   There follow two additional nymphal instars, during which the wing pads appear on the body and the nymphs are quiescent and take no food.   A complete generation may require from 2 weeks (at 80°F.) to about a month (at 60°F.), and there may be six generations from June to September in the greenhouse.   Effective temperatures for development are from 50 to 90°F.

Control Measures.   There is evidence that certain varieties are resistant to injury by this pest.   The most important control is to avoid planting infested corms, and in warmer areas to avoid planting on soil infested the preceding year.   The corms can be freed of the thrips by placing them in paper bags or covered boxes or trays and dusting with 5 per cent DDT at the rate of $\frac{1}{2}$ ounce to each bushel of corms or with flake naphthalene at the rate of 1 ounce to 100 corms.   After 3 weeks, or at temperatures of 70°F. after 10 days, the excess naphthalene should be removed and the bulbs aired.   Mercuric chloride or corrosive sublimate (poison), dissolved at the rate of 1 ounce in $7\frac{1}{2}$ gallons of water, may be used by submerging the bulbs completely for 8 to 12 hours and then planting.   The thrips can be very effectively destroyed by submerging the corms in a hot-water bath between 112 and 120°F. for 20 to 30 minutes.

After the plants start growing, they should be watched carefully and if the thrips or the characteristic injury appears upon them, the plants should be sprayed with one of the materials suggested for thrips control on page 864, or with the following formula, three to six times at intervals of 2 to 6 days.

	Small quantity for home garden	Large quantity for $\frac{1}{10}$ acre
Tartar emetic........	2 oz.	4 lb.
Brown sugar........	$\frac{1}{2}$ lb.	16 lb.
Water.............	3 gal.	100 gal.

References.   Ohio Agr. Exp. Sta. Bul. 537, 1934; *Colo. State Ento. Cir.* 64, 1935; *Jour. Econ. Ento.*, **40**:220, 1947; **41**:955, 1948.

SCALE INSECTS

Many greenhouse plants are attacked by some 20 to 25 species of scale insects which occur commonly in greenhouses. It is impossible to treat, in this book, all these different species separately. These species are figured and described by Dietz and Morrison in "The Coccidae or Scale Insects of Indiana," a bulletin from the office of the State Entomologist of Indiana, 1916.

The scale insects found in greenhouses may be divided into two large groups: (*a*) those having a distinct, hard, separable shell or scale over their delicate bodies, known as *the armored scales* or *Diaspidinae*, and (*b*) those in which the hard shell is not separable from the body, *the tortoise scales, soft scales*, or *Lecaniinae*. The tortoise scales have their bodies rounded and resemble somewhat the back of a turtle.

263. ARMORED SCALE INSECTS[1]

In the first class, the armored scales, reproduction takes place by means of eggs in most cases, although in a few of these species the young are born alive. In cases where the eggs are laid, these eggs are protected by the scale of the mother insect until they hatch. In whichever manner the young are produced, they crawl from beneath the scale of the parent and move about actively for a short time, until, upon finding a location on the plants which seems favorable to them, they insert their thread-like mouth parts through the epidermis of the leaf or bark and begin feeding by sucking the sap. After feeding a short time, they molt and in this process lose their legs and antennae. The cast skin is incorporated into the scale, which now forms over the body of the insect and which is composed of fine threads of wax which have exuded from the body wall of the scale and have run together. The female scales molt twice during their life but always remain under the scale for their entire life. The males, after their second molt, have a more elongated body and, after passing through the "prepupal" and "pupal" stages, assume the adult form. In this stage they are very minute two-winged yellowish insects, with antennae, eyes, three pairs of legs, and a rather prominent long appendage projecting from the tip of the abdomen. They move about actively, seek out the female scales, and mate with them, but do not feed in this stage. After the female has mated, she continues to feed for some time and produces her eggs, or in the case of a few species as above mentioned, brings forth living young. The life history of all the armored scales is essentially the same.

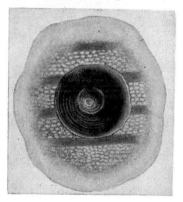

FIG. 18.3. Ivy or oleander scale, *Aspidiotus hederae*, on a bit of leaf. About 9 times natural size. (*From Ill. Natural History Surv.*)

A few of these scales which are most common in greenhouses in the United States are:

[1] Order Homoptera, Family Coccidae.

Oleander or ivy scale, *Aspidiotus hederae* (Vallot) (Fig. 18.3).
Latania scale, *Aspidiotus lataniae* Signoret.
Greedy scale, *Aspidiotus camelliae* Signoret.
Florida red scale, *Chrysomphalus aonidum* (Linné) (Fig. 18.4).
California red scale, *Aonidiella* (= *Chrysomphalus*) *aurantii* (Maskell) (Fig. 16.3).
Dictyospermum or palm scale, *Chrysomphalus dictyospermi* (Morgan).
Boisduval's scale, *Diaspis boisduvalii* Signoret.
Cactus scale, *Diaspis echinocacti* (Bouché).
Rose scale, *Aulacaspis rosae* (Bouché).
Cyanophyllum scale, *Aspidiotus cyanophylli* Signoret.
Mining scale, *Howardia biclavis* (Comstock).
Camellia scale, *Lepidosaphes camelliae* Hoke.
Fern scale, *Pinnaspis* (= *Hemichionaspis*) *aspidistrae* (Signoret) (**Fig. 18.5**).
Black thread scale, *Ischnaspis longirostris* (Signoret) (Fig. 18.6).
Chaff scale, *Parlatoria pergandii* Comstock.

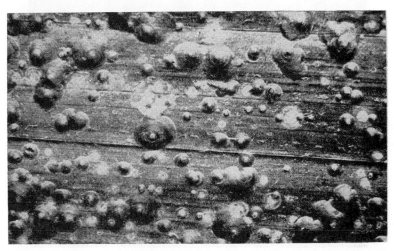

Fig. 18.4. The Florida red scale, *Chrysomphalus aonidum;* scales of adults and nymphs on palm leaf. About 6 times natural size. (*From Mich. Agr. Exp. Sta.*)

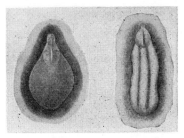

Fig. 18.5. The fern scale, *Pinnaspis aspidistrae.* Female on a bit of leaf, about 20 times natural size. Male, about 60 times natural size, on right. (*From Ill. Natural History Surv.*)

This class of scales attacks a wide variety of plants, including palms, Ficus, lantana, citrus plants, oleander, ivy, hibiscus, rose, and several other greenhouse plants.

Control Measures. These scales may be controlled by spraying with malathion at 1 pound, Diazinon at 0.5 pound, Trithion at 0.375 pound, or parathion at 0.25 pound to 100 gallons of spray. Several applications at 3- to 4-week intervals may be necessary, and care should be taken to avoid plant injury. Sulfotepp as an aerosol or smoke is effective with repeated applications. Fumigation with hydrogen cyanide at ⅛ to ¼

Fig. 18.6. Black thread scale, *Ischnaspis longirostris*, on a bit of palm leaf. About 20 times natural size. *(From Mich. Agr. Exp. Sta.)*

ounce to 1,000 cubic feet is effective in cleaning up infestations of this class of scales. All incoming plants should be inspected and, if infested, should be freed from scale insects before they are placed in the greenhouse.

264. Tortoise Scale Insects[1]

This group of scales has a life history somewhat similar to that of the armored scales. The protective shell is, however, not formed of wax and the cast skins, but of the chitinous body wall, very little wax being secreted. The body is generally smooth in outline, often very convex, and brown, black, or mottled in color. The young female scales may move about for a time after they have begun to feed. They retain their legs and antennae through adult life (Fig. 15.56,*a*). They reproduce by means of eggs and living young. The males in this class of scales, also, are winged and similar to those described above. Among the more common scales of this class are:

[1] Order Homoptera, Family Coccidae.

Soft scale, *Coccus hesperidum* Linné (Fig. 18.7).
Tessellated scale, *Eucalymnatus tessellatus* Signoret.
Hemispherical scale, *Saissetia hemisphaerica* (Targioni) (Fig. 18.8).
Black scale, *Saissetia oleae* (Bernard) (Fig. 16.4).
Long scale, *Coccus elongatus* (Signoret) (Fig. 18.9).

The soft scale[1] (Fig. 18.7) differs from most of the tortoise scales in being very flat.
It is oval in outline and greenish to brownish in color,
often resembling very closely the part of the plant on
which it is resting. It is also peculiar in being ovoviviparous, one or two young being born daily for a
month or two. The young are sluggish, usually settle
down near the female, and grow gradually to the adult
female stage in about 2 months. It is a very general
feeder in the greenhouse and also attacks tropical and
subtropical fruits out-of-doors.

The hemispherical scale[2] (Fig. 18.8) is very convex,
elliptical in outline, smooth, and brown. They rarely
move from the point where the young nymphs settle
and insert their mouth parts. Growth is slow, there
being only about two generations a year. Males are
rare and reproduction is generally by parthenogenesis.
From 500 to 1,000 eggs are laid beneath the female's
body, the ventral wall shrinking dorsad to make a neat
cup to protect them. It is a serious greenhouse pest
and also attacks subtropical fruits.

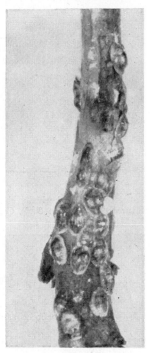

FIG. 18.7. The soft scale, *Coccus hesperidum*. About twice natural size. (*From Univ. Calif. Coll. Agr.*)

The injury by these scales is very similar
to that by the armored scales. The plants
attacked include oleander, bay, Vinca, croton,
cyclamen, fern, palm, Ficus, abutilon, and several other plants.

Control Measures. These scales may be controlled by spraying with malathion at 1 pound,
Diazinon at 0.5 pound, or Trithion at 0.375
pound to 100 gallons of spray. Several applications at 4- to 6-week intervals may be required, and care should be exercised to avoid
plant damage. Sulfotepp as an aerosol or
smoke is effective with repeated applications.

265 (223). MEALYBUGS[3-5]

Mealybugs are closely related to the scale insects; in fact they belong
to the same family. There are several species which occur commonly in
greenhouses and on flowering plants in houses. They are all very much
alike in their life history, and differ but slightly in appearance (Fig 18.10).
Mealybugs may be placed in two groups: (*a*) *The short-tailed mealybugs,*
which reproduce by laying eggs; including the common or citrus mealybug,[3] the cocoanut mealybug,[4] and the Mexican mealybug.[5] In all these
the filaments that surround the body are of about equal length and none
of them more than one-fourth the length of the body. (*b*) *The long-tailed*

[1] *Coccus hesperidum* Linné.
[2] *Saissetia hemisphaerica* (Targioni).
[3] *Pseudococcus citri* (Risso), Order Homoptera, Family Coccidae.
[4] *Pseudococcus nipae* (Maskell).
[5] *Phenacoccus gossypii* Townsend & Cockerell.

Fig. 18.8. Hemispherical scale, *Saissetia hemisphaerica*, on avocado twigs.   About 3 times natural size.   (*From Univ. Calif. Coll. Agr.*)

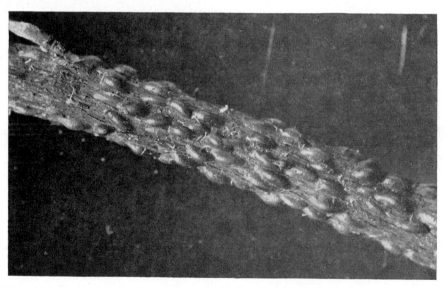

Fig. 18.9. The long scale, *Coccus elongatus*, adults on stem.   About 3 times natural size. (*From Mich. Agr. Exp. Sta.*)

mealybug[1] gets its name from the four filaments near the tip of the abdomen, which may be as long as the body.   This species does not form an egg sac but gives birth to nymphs.

In general, the life history is as follows: The adult mealybugs deposit their eggs in a compact, cottony, waxy sack, beneath the rear end of the

[1] *Pseudococcus adonidum* (Linné) [= *P. longispinus* (Targioni)].

body, to the number of 300 to 600.   Egg-laying continues for 1 or 2 weeks, and, as soon as it is completed, the female insect dies.   These sacks containing eggs will be found chiefly at the axils of branching stems or leaves but occasionally on other parts of the plant.   Under greenhouse conditions the eggs hatch in about 10 days.   The young mealybugs remain in the egg case for a short time and then crawl over the plants.   They are flattened, oval, light-yellow, six-legged bugs, with smooth bodies.   They feed by inserting their slender mouth parts into the plant tissues and sucking the sap.   Shortly after they begin feeding, a white waxy material begins exuding from their bodies and forms a covering over the insect and about 36 prominent leg-like filaments, which radiate from the margin of the body on all sides.   They do not remain fixed but move about to some extent over the plant, although they are always very sluggish.   The female nymphs change but little in their appearance, except to increase in size, being about $\frac{1}{6}$ to $\frac{1}{4}$ inch long when full-grown.   The males, when nearly grown, form a white case about themselves and, inside of this, transform into tiny, active, two-winged, fly-like insects.   On emerging from the case, the males fly about actively and mate with the females, but very soon die.   The male is incapable of feeding in the adult stage.   It takes about 1 month for the completion of a generation under greenhouse conditions.

FIG. 18.10. Four common species of mealybugs.   From top to bottom: the long-tailed mealybug, *Pseudococcus adonidum;* the citrophilus mealybug, *P. gahani;* the grape mealybug, *P. maritimus;* and the citrus mealybug, *P. citri.*   About 5 times natural size.   (*From Univ. Calif. Coll. Agr.*)

Plants Attacked.   Many kinds of greenhouse plants are attacked by mealybugs.   They are especially troublesome on the soft-stemmed foliage plants, such as coleus, fuchsia, cactus, croton, fern, gardenia, and begonia.   Among the many other plants attacked are citrus, heliotrope, geranium, oleander, orchids, poinsettias, umbrella plant, ivy, dracaena, and chrysanthemums.

Control Measures.   Fairly satisfactory control may be obtained by syringing the plants with water applied with as much force as the plants will stand without injury to the leaves.   The most effective results may be obtained by spraying with malathion at 1 pound, Diazinon at 0.5 pound, Trithion at 0.375 pound, or parathion at 0.25 pound to 100 gallons of spray.   The application should be repeated in 3 to 4 weeks.   Repeated applications of parathion or sulfotepp aerosol are also effective (page 876).   Fumigation with hydrogen cyanide is very effective in controlling the Mexican mealybug and will kill many of the younger stages of other species.   Relatively high concentrations of hydrogen cyanide ($\frac{1}{4}$ to $\frac{1}{2}$

ounce per 1,000 cubic feet, or more if the plants will stand it) for short periods of exposure are effective against all stages of the Mexican mealybug and are less likely to injure plants than lower dosages for longer periods of time.　Fumigations should be made three to six times at 5- to 7-day intervals.

266. GREENHOUSE ORTHEZIA[1]

This insect also is a close relative of the scales and mealybugs.　The small dark-green wingless bodies of the nymphs are about the size of pinheads and have rows of minute waxy plates extending back over their bodies.　In the case of the female insect (Fig. 18.11) a white, waxy, fluted egg sac is attached to the body and extends backward for a distance of two or more times the diameter of the body, the total length being about ⅓ inch.　These insects resemble the mealybugs quite closely in their habits and can be controlled by the same methods.

267 (82). APHIDS OR PLANT LICE[2]

Aphids have already been described as pests of many of the outdoor crops (pages 612, 626, 814).　Everyone who has grown plants for a year or

FIG. 18.11. Greenhouse orthezia, *Orthezia insignis,* nymph and adult females; the one on the right has the egg sac broken open to show the eggs and newly hatched nymphs. About 5 times natural size. (*From Mich. Agr. Exp. Sta.*)

FIG. 18.12. Rose aphids, *Macrosiphum rosae,* on a rosebud. Enlarged. (*From Calif. Agr. Ext. Serv. Cir.* 87.)

more in the greenhouse or flower garden is well aware that these little soft-bodied insects are also pests of ornamental crops.　A number of different species of aphids occur in greenhouses, and one or more of them attacks almost every species of plant that is grown under glass.　The aphids differ somewhat in size, appearance, and the color of their bodies.

[1] *Orthezia insignis* Browne, Order Homoptera, Family Coccidae.
[2] Order Homoptera, Family Aphidae.

Some are green, others brown, reddish, or black in color.   Their life histories under greenhouse conditions are very similar, except for some variation in the length of time which it takes the different species to develop. So far as known, all species continue to go on, generation after generation, the year round, producing only the female forms; and these females, on becoming full-grown, give birth to living young without being mated. Under greenhouse conditions, the true sexes do not appear, nor do the insects reproduce by means of eggs.   They feed by sucking the sap from the tender plants, often causing the plants to become deformed, the leaves curled and shriveled; and, in some cases, galls are formed on the leaves. The insects are further injurious because of the fact that some species eject a sweetish honeydew, which is attractive to ants and on which certain sooty-appearing fungi grow.   This may render the plants so unsightly as to hinder their sale for any purpose.   Aphids are also the carriers of certain diseases of greenhouse plants.   Some of the more common species occurring in the greenhouse are as follows:

Small green chrysanthemum aphid, *Myzus rosarum* Kaltenbach.
Chrysanthemum aphid, *Macrosiphoniella sanborni* (Gillette).
Pea aphid, *Macrosiphum pisi* (Harris) (Fig. 14.16).
Rose aphid, *Macrosiphum rosae* (Linné) (Fig. 18.12).
Corn root aphid, *Anuraphis maidi-radicis* (Forbes) (Fig. 9.25).
Violet aphid, *Micromyzus violae* (Pergande).
Green peach aphid, *Myzus persicae* (Sulzer).
Melon aphid, *Aphis gossypii* Glover.

Species of importance in the flower garden include:

Bean aphid, *Aphis fabae* Scopoli.
Green peach aphid, *Myzus persicae* (Sulzer).
Columbine aphid, *Hyalopterus trirhoda* Walker.
Melon aphid, *Aphis gossypii* Glover.

Control Measures.   Aphids may be readily controlled under greenhouse conditions by spraying with malathion at 1 pound, lindane at 0.25 pound, Trithion at 0.375 pound, Thiodan or Diazinon at 0.5 pound to 100 gallons of spray, or by dusting with these materials.   Spraying or soil drenching with schradan at 1 pound to 100 gallons has given effective control for 3 to 5 weeks.   Light applications of hydrogen cyanide gas, generated from calcium cyanide dust scattered upon the walks in the evening, are very effective.   For most plants the dosage used should be ¼ ounce per 1,000 cubic feet of space.   One of the most effective and safest methods of destroying aphids is by the use of pressure fumigation with nicotine.   A combination of nicotine and a combustible powder is prepared commercially.   This mixture is sealed in cans of several sizes. To release the nicotine, two holes are punched in the can and the powder ignited with a slow match or sparkler.   The pressure formed in the can forces out the nicotine vapor.   Such pressure fumigators can be used for treating an entire house, or for "spot fumigation" of a part of a single bench.   This product should be used according to the manufacturer's directions.   The application of parathion, sulfotepp, or TEPP aerosols as described for the two-spotted spider mite (page 876) is very effective for the control of greenhouse aphids.   Nicotine should not be used in any form on violets, ferns, or blooming orchids.   Diazinon, Thiodan, and Trithion should be used with caution until it is certain that no plant damage will result.

268. Greenhouse Whitefly[1]

Importance and Type of Injury. Plants covered, especially on the undersides, with small, snow-white, four-winged flies and very small, oval, flat, pale-green nymphs, less than $\frac{1}{30}$ inch in length, which suck the sap (Fig. 18.13). Infested plants are lacking in vigor, wilt, turn yellow, and die. The leaves are covered with a coating of glazed, sticky material on which a sooty-colored fungus often grows, completely covering the foliage.

FIG. 18.13. Greenhouse whitefly, *Trialeurodes vaporariorum*, adults and nymphs on underside of leaf. 4 times natural size. (*From Can. Dept. Agr. Bul.* 7, *n.s.*)

Plants Attacked. Cucumber, tomato, lettuce, geranium, fuchsia, ageratum, hibiscus, coleus, begonia, solanum, and many other plants.

Life History, Appearance, and Habits. The female whitefly deposits more than 100 minute yellowish eggs. These are attached to the underside of the leaves by a short stalk and are often laid in a small ring, as the female circles about with her mouth parts inserted in the leaf. On hatching, the nymphs are flat and nearly transparent. They settle upon the leaf, near the point where they hatch, and remain in this situation until they become adults. They suck the sap from the leaves, feeding greedily on the plant juices for about 4 weeks. In the course of this time, they pass through four instars. All the nymphs have fine, long and short, white, waxy threads radiating from their greenish bodies. The average duration of the nymphal periods is about 28 to 30 days. The adult whitefly is about $\frac{1}{16}$ inch in length, very active, four-winged, with a yellowish body, and has the appearance of having been thoroughly dusted with some very fine white material (Fig. 16.6). Both males and females fly, and they feed, like the nymphs, on the underside of the leaves, living from 30 to 40 days. Under greenhouse conditions the generations overlap, and all stages of the insect may be found on infested plants at any time.

Control Measures. Spraying with malathion at 1 pound or parathion at 0.25 pound to 100 gallons of spray, or aerosol treatment with parathion or sulfotepp as described on page 876, or fumigation with hydrogen cyanide are the most satisfactory methods of controlling the whitefly (page 375). The dosage of cyanide used will depend upon the plants infested and the tightness of the greenhouse. One-eighth to one-fourth ounce of sodium cyanide per thousand cubic feet, or the equivalent in calcium cyanide, should be used. Three or four applications, at weekly intervals, will usually be necessary before the greenhouse is cleaned of these pests, because some of the nymphs and eggs will be likely to escape the first several treatments.

Reference. Twenty-seventh Rept. Ill. State Ento., p. 130, 1912.

[1] *Trialeurodes vaporariorum* (Westwood), and others, Order Homoptera, Family Aleyrodidae.

269 (85). Two-spotted Spider Mite[1]

Importance and Type of Injury. Leaves of plants infested by spider mites present a peculiar appearance. Those lightly infested have pale blotches or spots showing through the leaf. In heavy infestations the entire leaf appears light in color, dries up, often turning reddish-brown in blotches or around the edge. Plants generally lose their vigor and die. The undersurface of lightly infested leaves will show silken threads spun across them. In heavy infestations these threads may form a web over the entire plant, upon which the mites crawl and to which they fasten their eggs. The underside of leaves, on close examination, will be found covered with minute eight-legged mites, showing as tiny, reddish, greenish, yellowish, or blackish, moving dots on the leaves. The color appears to vary in part with the kind of food.

Plants Attacked. There are very few plants grown in greenhouses which are not subject to injury by these spider mites. Some of the smooth hard-leaved plants, such as certain palms, are only moderately injured. Some of the worst injured are: cucumbers, tomatoes, carnations, chrysanthemums, melons, sweet peas, snapdragons, violets, roses, *Asparagus plumosus*, and fuchsia. Out-of-doors arborvitae, cedars and other evergreens, hydrangeas, roses, and clematis suffer severely. Of all the plants infested, control is probably most difficult on roses.

Life History, Appearance, and Habits. The adult female mite is eight-legged and about $\frac{1}{60}$ inch in length (Fig. 14.12), ranging in color from pale yellow through green to brown to orange. The male is smaller, about $\frac{1}{80}$ inch long, with a narrower body and pointed abdomen. Two dark spots, composed of the food contents, show through the transparent body wall. The body is oval in outline and sparsely covered with spines. The mite feeds through sucking mouth parts, with which it pierces the epidermis of the leaf. After mating, the female begins to lay eggs at the rate of two to six a day, producing an average of 100 over an average lifetime of 68 days. The eggs are spherical, shiny, and straw-colored and are attached to the undersides of the leaves, usually to the web which the mite spins wherever it goes over the plant. After an incubation period, ranging from 19 days at 50°F. to 3 days at 75°F., the unfertilized eggs hatch to produce males and the fertilized eggs, predominantly females. The newly hatched mites pass through the typical six-legged larval, protonymph, and deutonymph stages (Fig. 15.27), the life cycle from hatching to adult ranging from 19 days at 55°F. to 5 days at 75°F.

Control Measures. Spider mites cause the most damage in greenhouses, where plants are grown at high temperature and low humidity. Frequent syringing or spraying of the plants with a stream of clear water, applied with sufficient force to tear apart the webs of the mites and knock them off the plants, will aid in keeping them under control. This treatment cannot be applied to many species of plants, which are injured by frequent watering or become diseased if grown under moist conditions. Frequent dusting of the beds and leaves of plants with superfine sulfur is also of value in control. The most generally satisfactory control may be obtained by spraying with Tedion, Kelthane, Chlorobenzilate, or Aramite at 0.25 pound, Trithion at 0.375 pound, or Ethion at 0.5 pound to 100 gallons of spray, making two applications at an interval of 7 to 10 days. The mite has developed a number of strains resistant to one or

---

[1] *Tetranychus telarius* (Linné), Order Acarina, Family Tetranychidae.

more of the pesticides mentioned, and satisfactory control will depend upon careful selection of materials and their use in alternation or combination. Azobenzene at 1 pound to 100 gallons may be used in combination with the materials listed above, and this chemical has also given effective results when applied as a fumigant by applying a 30 per cent mixture in diatomaceous silica[1] to steam pipes at about 1 pound of mixture to 40,000 cubic feet, or by burning in a candle-type dispenser or pressure fumigator. Schradan applied as a spray or soil drench at 1 pound to 100 gallons or demeton (Systox) as a spray at 0.4 pound to 100 gallons is effective systemic treatment. Aerosols drifted above the plants several times at 3- to 4-day intervals, using sulfotepp, TEPP, or parathion at 5 to 10 per cent in methyl chloride, applied at about 1 gram to 1,000 cubic feet, are also effective. The operator should wear a respirator and gloves. Sodium selenate applied to the soil as a dust or by sprinkling in water at the rate of about 1 ounce to 100 square feet of bench space has been used to control spider mites on carnations, chrysanthemums, and a few other plants. A single treatment will remain effective for many months, but the material may severely injure many plants, especially in overdosages.

The destruction of rank growths of weeds in the immediate vicinity of the greenhouse is of value in preventing the mites from gaining access to the house.

References. *U.S.D.A., Farmers' Bul.* 1306, 1923; *Fifteenth Rept. S. D. State Ento.,* 1925; *Mass. Agr. Exp. Sta. Bul.* 179, 1917; *Contrib. Boyce Thompson Inst.,* **2**:512, 1930; *Va. Agr. Exp. Sta. Tech. Bul.* 113, 1949; *Jour. Econ. Ento.,* **31**:211, 1938; **39**:78, 1946; **40**:733, 1947; and **41**:356, 1948.

270. CYCLAMEN MITE[2]

Importance and Type of Injury. Infested plants have the leaves distorted, with the buds failing to open or with small distorted flowers, presenting a streaked and blotchy appearance. The foliage shows purplish areas. Small, white, green, or pale-brown mites, with six or eight legs, work about the base of the plants or in the buds or the injured areas on the leaves.

Plants Attacked. Cyclamen is injured more than any other plant. This mite is also recorded as a pest of snapdragon, geranium, chrysanthemum, larkspur, begonia, fuchsia, and petunia; and of strawberry, out-of-doors.

Distribution. It is an imported species. It was first noticed in New York in 1898 and in Canada in 1908. It is now generally distributed in greenhouses throughout the country.

Life History, Appearance, and Habits. The adult mites are orange-pink and shiny, with eight legs. They are about $\frac{1}{100}$ inch long and cannot be seen without magnification. The hind pair of legs in the female are thread-like, and those of the male are pincer-like (Fig. 18.14). The eggs are laid about the base of the cyclamen plant and in injured areas of the leaves and on strawberries along the midribs of the unfolding leaves. Each female deposits about 90 eggs, of which some 80 per cent develop into females. The mites develop through a six-legged larval stage and

---

[1] Celite 209.

[2] *Steneotarsonemus* (= *Tarsonemus*) *pallidus* (Banks), Order Acarina, Family Tarsonemidae.

a quiescent nymphal stage with eight legs. The life cycle from egg to adult requires about 2 weeks, and all stages of the mite are found about the foliage of infested plants. Out-of-doors, the adult female overwinters in protected locations around the crown of the strawberry plant.

Control Measures. Great care should be taken to avoid introducing this pest into the greenhouse on plants, hands, or clothing. Infested plants may be treated by immersion in water to 110°F. for 15 to 30 minutes, or by fumigation with 2 pounds of methyl bromide to 1,000 cubic feet at about 70°F. Spraying the plants with Kelthane or endrin at 0.4 pound or Thiodan at 0.5 pound to 100 gallons of spray gives effective control, especially after pruning back the growth. Three applications should be made at 2- to 3-week intervals. The effectiveness of these sprays

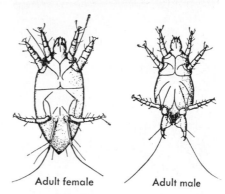

Adult female   Adult male

FIG. 18.14. The cyclamen mite, *Steneotarsonemus pallidus.* Much enlarged. (*From W. W. Allen, Calif. Agr. Exp. Sta. Cir.* 448.)

can be increased by adding 4 ounces of wetting agent[1] to 100 gallons of spray. Stocks of susceptible plants should be sprayed before cuttings are taken, and infested plants should be destroyed immediately. On strawberries the mite predators, *Typhlodromus bellinus* and *T. reticulatus,* often give effective biological control.

References. Proc. Ento. Soc. Wash., **39**:267, 1935; *Calif. Agr. Exp. Sta. Cir.,* 484, 1959; *Jour. Econ. Ento.,* **28**:91, 1935; **36**:286, 1943; and **41**:624, 1948.

271. CHRYSANTHEMUM GALL MIDGE[2]

Importance and Type of Injury. Infested plants have the leaves misshapen, in cases of light infestations, with small, somewhat blister-like cone-shaped galls on the upper surfaces. In severe infestations, the leaves are curled, the flowers are distorted, with crooked stems, and with numerous galls along the stems and leaves (Fig. 18.15).

Plants Attacked. Chrysanthemums of all varieties.

Distribution. This insect is a late importation from Europe, having first been found in this country in 1915. It is now generally distributed in greenhouses over the United States and Southern Canada.

Life History, Appearance, and Habits. The adult insect is a very frail little long-legged orange-colored gnat, about $\frac{1}{14}$ inch in length. The female flies lay about 100 very minute orange-colored eggs on the surface and tips of the new growth of the chrysanthemums. The maggots hatching from these eggs, in 3 to 16 days, bore their way into the plant tissues. The irritation resulting from their feeding causes the growth of small cone-shaped galls. On the leaves these galls are usually on the upper side, but they may occur also along the stems, many galls developing so close together that they form masses or knots on the stems. Inside

[1] Triton X-100.

[2] *Diarthronomyia chrysanthemi* (= *hypogaea*) Ahlberg, Order Diptera, Family Cecidomyiidae.

these galls the maggots develop, reach their full growth, and pupate. On emerging, the empty pupal skin will usually be left protruding from the

FIG. 18.15. Injury caused by the chrysanthemum gall midge, *Diarthronomyia chrysanthemi.* About natural size. (*From Mich. Agr. Exp. Sta. Spcl. Bul.* 134.)

gall. In nearly all cases, the adult flies emerge after midnight and before morning. The life cycle requires, on the average, about 35 days. There may be five or six generations a year under greenhouse conditions.

Control Measures. Careful inspection of all plants brought into the greenhouse, with the destruction of those found infested, is the best method of avoiding trouble from this insect. Galls are most likely to be found on varieties with light-green foliage. All infested foliage should be picked off and burned. Spraying with DDT at 1 pound or lindane at 0.25 pound to 100 gallons of spray, in the early evening, is effective. When a house is generally infested, nightly fumigation with a weak charge of hydrogen cyanide, applied between midnight and daylight, until all the pests have disappeared, or aerosol treatment with DDT or parathion (page 876), are effective methods of control. These insecticides kill the adults, which emerge at night; consequently nightly treatments over a considerable period of time are necessary to eradicate the pest.

Reference.   *U.S.D.A., Dept. Bul.* 833, 1920.

272. Rose Midge[1]

Importance and Type of Injury. Flower buds are distorted, turning brown and dying. Tender growth is sometimes curled and brown, the buds and young shoots failing to develop. An examination of buds will show whitish maggots clustered inside, mainly at the base, or on the upper sides of the tender leaves and leaf petioles. The maggots are about $\frac{1}{20}$ inch in length when full-grown. They are at that time somewhat tinged with red.

Plants Attacked. Roses.

Life History, Appearance, and Habits. The adult midge is a very small two-winged fly, about $\frac{1}{20}$ inch in length, and of a reddish or yellowish-brown color (Fig. 18.16). They are usually most abundant in greenhouses during the summer and early fall. The females deposit their very minute yellow eggs, inserting them into the buds, usually just behind the sepals of the flower buds, or in the unfolding leaves. The whitish maggots issuing from these eggs feed on the tender tissue of the new growth and inside the buds, becoming mature in 5 or 6 days. They then drop to

[1] *Dasyneura rhodophaga* (Coquillett), Order Diptera, Family Cecidomyiidae.

the ground, where they spin a cocoon in which the pupal stage is passed. The length of the life cycle varies with the temperature of the house, but, under favorable conditions, a complete generation may appear every 20 days. Usually the winter is passed in cocoons in the soil.

Control Measures. Almost complete control of this insect has been obtained by spraying rose plants and the soil around them with DDT at 1 pound to 100 gallons of spray. In the greenhouse this treatment

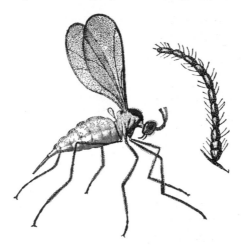

Fig. 18.16. Rose midge, *Dasyneura rhodophaga*, adult female, greatly enlarged, and antenna in detail. (*From Ill. Natural History Surv.*)

should be applied three times at 7- to 10-day intervals, while in the field, three applications should be made at weekly intervals starting in late May or early June and the treatments continued at 2-week intervals after that. When the rose midge has become established in houses, a vigorous campaign should be waged against it, as it is possible for the insect to destroy the entire crop in a short time. Picking off and burning the infested buds will aid in control.

273 (14, 39, 87, 216). White Grubs

Importance and Type of Injury. White grubs of the same species as those attacking the roots of plants outdoors (see Corn Insects, page 503) occasionally occur in numbers in greenhouses. They may be the grubs of the ordinary June beetles, or those of several other beetles, particularly of the genus *Ochrosidia* or the Japanese beet e (page 749) or black vine or cyclamen weevil.[1] In most cases, the insects are brought into the greenhouse with infested soi . In a few cases, such as the annual white grub (page 523), the beetles may gain access to the greenhouses at night, during the period of the year when the insects are abundant, and lay their eggs in the soil of pots or benches. The length of the life cycle under greenhouse conditions has never been carefully worked out. With the last-mentioned species, however, it does not occupy more than 1 year.

Control Measures. The best method of controlling white grubs in greenhouse soils is to sterilize the soil with steam, hot water, or carbon bisulfide (see under Nematodes, page 887). A careful inspection should be made of all soil brought into the greenhouse, to make sure that it does not contain large numbers of these insects. The adult beetles may be kept out of the greenhouses by screening.

[1] *Brachyrhinus sulcatus* (Fabricius), Order Coleoptera, Family Curculionidae.

274. OTHER GREENHOUSE BEETLES

Wireworms (page 617) gain access to greenhouse soils in the same way as white grubs. The treatment for these insects under greenhouse conditions is the same. The black vine or cyclamen weevil[1] is sometimes a pest and may be controlled by two to three applications of dieldrin at 0.25 pound or chlordane at 1 pound to 100 gallons of spray, applied to the surface of the beds at monthly intervals. The Japanese beetle (page 749) is also an occasional greenhouse pest.

275. FUNGUS GNATS OR MUSHROOM FLIES[2]

Importance and Type of Injury. Plants are lacking in vigor and leaves turning yellow, without visible injury to any part of the plant above the ground. Roots with small brown scars on the surface, or with small feeding roots and root hairs eaten off. Very small, thread-like, active, white maggots embedded in the root tissue, or working through the soil about the plant. Often injurious to potted plants. Root rots often follow the attacks of these insects. In mushroom houses the maggots destroy the spawn and tunnel through caps and stems of the mushrooms. These flies also transport the hypopi of mites (page 883).

Plants Attacked. Many species of plants grown in greenhouses.

Life History, Appearance, and Habits. There are several species of so-called fungus gnats which cause injury to the roots and underground parts of the stems of plants in greenhouses. The adults are all very small, sooty-gray or nearly black, long-legged, slender flies or gnats, measuring from $\frac{1}{8}$ to $\frac{1}{10}$ inch in length. These flies deposit their eggs in clusters of 2 to 30 or more in the soil, each female laying from 100 to 300 eggs. The eggs are only about $\frac{1}{100}$ inch in length, almost too small to be seen, although the egg clusters may be readily found. After about 4 to 6 days, small legless maggots, with black heads and nearly transparent bodies, hatch and begin working their way through the soil. The maggots feed for 5 to 14 days and, on becoming full-grown, are about $\frac{1}{4}$ inch in length. They form a flimsy cocoon in or on the ground and there pupate. In 5 or 6 days the adult flies emerge from the cocoons and work their way to the surface of the soil. They live for about 1 week, while they mate and lay eggs for another generation.

Control Measures. Thorough spraying of soil surface, benches, and ground under the benches with malathion or chlordane at 1 pound, lindane or dieldrin at 0.25 pound, Trithion at 0.375 pound, or Diazinon at 0.5 pound, applied several times at 7- to 10-day intervals, will give control. Dusting the soil surface with these materials is also effective. Allowing the soil to dry to as great a degree as possible without injury to the plants also is effective in killing many of the maggots; but it is seldom that all can be cleaned out by this method.

276. NARCISSUS BULB FLY[3]

Importance and Type of Injury. Bulbs of narcissus and other plants fail to grow. The bulbs become soft and the outer scales of the bulbs often have brown scars upon them. An examination of the bulbs will show a large whitish or yellowish-white maggot inside the bulb, feeding on the plant tissue (Fig. 18.17).

Plants Attacked. Narcissus, hyacinth, amaryllis, galtonia, and several others.

Distribution. The insect is of European origin and is now well established throughout the western part of Canada. It has frequently been found in shipments of foreign bulbs and has been reported from many points in the United States.

Life History, Appearance, and Habits. The adult of the narcissus bulb fly is a shiny yellow-and-black hairy fly, about the size of a small bumblebee, which it somewhat resembles in appearance. The eggs are laid in the base of the leaves or in the necks of bulbs. The young maggots hatching from them bore into the bulb and rasp or tear apart the plant tissues by means of their strong, somewhat hooked mouth parts. The maggots are whitish to yellowish-white in color, thick and fat in appearance, and reach

---

[1] *Brachyrhinus sulcatus* (Fabricius), Order Coleoptera, Family Curculionidae.

[2] *Sciara* spp. and others, Order Diptera, Family Mycetophilidae.

[3] *Lampetia* (= *Merodon*) *equestris* (Fabricius), Order Diptera, Family Syrphidae.

a length of ¾ inch. The puparia are formed in the bulb or in the soil. There is probably not more than one generation a year.

Control Measures. Vacuum fumigation of thoroughly cured and clean bulbs with 25 to 30 pounds of carbon bisulfide or 2 to 3 pounds of methyl bromide per 1,000 cubic feet for from 1¾ to 2 hours at a temperature between 70 and 80°F. is very effective. Treating bulbs with hot water has been found one of the most effective methods of controlling this insect. The bulbs should be submerged in water held at a temperature of 110 to 111.5°F. for 2½ hours. Treating in a chamber where the bulbs may be

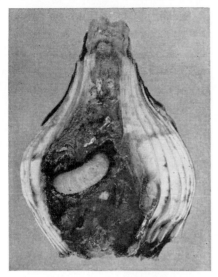

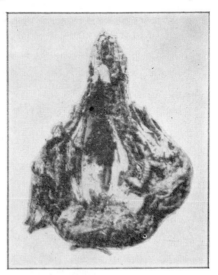

Fig. 18.17. Narcissus bulb infested with larva of the narcissus bulb fly, *Lampetia equestris.* (*From Essig, "Insects of Western North America," copyright, 1926, Macmillan. Reprinted by permission.*)

Fig. 18.18. The lesser bulb fly, *Eumerus tuberculatus.* Bulb showing larvae, somewhat reduced. (*From U.S.D.A., Farmers' Bul. 1362.*)

held at a temperature of 110 to 111°F. for a period of 2 hours has some advantages over the hot-water treatment. The air in the chamber should be held at near saturation by being drawn in through steam or water spray. When bulbs are taken up from the field, the infested ones can usually be sorted out by their lighter weight, softer condition, shrunken and corky basal plates, or partly decayed appearance. Bulbs showing indications of heavy infestation should be destroyed by burning and the remainder of the lot sterilized in hot water.

References. Va. Truck Exp. Sta. Bul. 60, 1927; *Jour. Econ. Ento.,* **21**:342, 1928; and **25**:1020, 1932.

277. LESSER BULB FLY[1]

Importance and Type of Injury. The injury closely resembles that of the narcissus bulb fly. The maggots are grayish or yellowish gray, and the body is markedly wrinkled (Fig. 18.18). Many maggots will often be found in a single bulb. They reach a length of about ½ inch.

Plants Attacked. Narcissus, hyacinth, amaryllis, onion, iris, shallot, and several other plants.

Distribution. The distribution of this insect is about the same as that of the narcissus bulb fly. It is probably already established out-of-doors in some points in the United States.

[1] *Eumerus tuberculatus* Rond., Order Diptera, Family Syrphidae.

Life History, Appearance, and Habits. The adult fly is of a blackish-green color, ⅓ inch long, with the body nearly bare of hairs but with several white lunate markings on the sides of the abdomen. It somewhat resembles a small wasp. The general life history is the same as that of the narcissus bulb fly, but there are probably two generations a year.

Control Measures. The control measures are the same as those suggested for the narcissus bulb fly.

278. Bulb Mite[1]

Importance and Type of Injury. Bulbs of various species of plants rot and fail to produce growth (Fig. 18.19). Plants grown from bulbs turn

Fig. 18.19. The bulb mite, *Rhizoglyphus echinopus*. Mites working in a rotten bulb and their eggs. Enlarged about 8 times. (*From Conn. Agr. Exp. Sta. Bul. 225.*)

yellow and present a general sickly appearance. The leaves of such plants are stunted and distorted, and the plants will generally fail to produce flowers or will produce only misshapen ones. Very small whitish mites, with six or eight brownish or pinkish legs, may be found in large numbers sheltering behind, or boring into, the bud scales.

Plants Attacked. These mites attack practically all classes of bulbs, some of the more commonly infested being narcissus, lilac, hyacinth, crocus, gladiolus, amaryllis, and lily. They may also infest cereals.

Distribution. These creatures are of European origin but have been now generally distributed throughout the United States and Canada in shipments of bulbs sent from Europe, Japan, and the Bermuda Islands.

Life History, Appearance, and Habits. Infested bulbs will contain practically all stages of the bulb mite. These mites are ⅟₅₀ to ⅟₂₅ inch long, whitish, barely visible to the naked eye (Fig. 18.19). The eggs are laid behind the bud scales and soon hatch into the six-legged nymphs. After molting, these nymphs change to the eight-legged form and during

[1] *Rhizoglyphus echinopus* (Fumouze & Robin), Order Acarina, Family Tyroglyphidae.

this second instar of their life are most destructive to the bulbs. After two or more molts they become adults and begin to reproduce. Under certain conditions, probably unfavorable to the species, a remarkable, heavily chitinized, nonfeeding, but very active stage, known as the *hypopus*, may intervene between the six-legged and eight-legged nymphal stages, lasting for 1 or 2 weeks before molting to the eight-legged nymph. The hypopi have a group of suckers on the ventral side, behind the anus. They readily attach to insects, mice, and other creatures and may be distributed in this manner to new and more favorable breeding places. The mites apparently prefer healthy bulbs and migrate through the soil from the decaying bulbs to the more attractive food. They are readily transported from place to place in bulb shipments. The life of the female bulb mite is about 1 or 2 months, while that of the male is usually less. Each female deposits from 50 to over 100 eggs.

Control Measures. All infested bulbs, which can be recognized by their soft, mushy condition, should be destroyed by burning at the time of digging or when the bulbs are planted. The bulbs may be treated by dipping for 10 minutes in a solution of 40 per cent nicotine sulfate, using 1 part of nicotine to 400 parts of water heated to 122°F. before dipping; or in a 2 to 4 per cent lime-sulfur solution at 125°F. for 1 minute. Practically all the mites will be killed by submerging the bulbs for 3 hours in water at 110 to 111.5°F. Sound, healthy bulbs in which the root growth has not started will not be injured by this treatment. Storing the bulbs in tight containers with 2 per cent nicotine dust is also effective.

Reference. *Conn. Agr. Exp. Sta. Bul.* 402, 1937.

279. Sowbugs[1] and Pillbugs[2]

Importance and Type of Injury. These creatures are not insects but are more closely related to crayfish. They frequently cause some damage in greenhouses. The plants infested show feeding about the roots and on the tender portions of the stems near the ground. Light-gray to slate-colored fat-bodied and distinctly segmented creatures, with seven pairs of legs, and up to ½ inch long, will be found hiding about the bases of plants or under clods or bits of manure on the surface of the benches (Fig. 18.20).

Plants Attacked. Roots and tender growth of nearly all greenhouse plants and mushrooms are attacked.

Life History, Appearance, and Habits. These creatures reproduce by means of eggs which are retained in the marsupium of the female for about 2 months. Young sowbugs, on hatching, do not leave the marsupium of the female for some time; 25 to 75 constitute a brood. The young are similar to adults except in size. About 1 year is required for the young bugs to reach full growth. All stages of the bugs will be found in infested greenhouses at the same time.

Fig. 18.20. Sowbugs feeding in manure. About natural size. (*From Can. Dept. Bul.* 7, *n.s.*)

Control Measures. These pests may be eliminated by spraying soil surface, benches, and the soil under the benches with malathion or Sevin at 1 pound, Trithion at 0.375

[1] *Porcellio laevis* Koch, and others, Class Crustacea, Order Isopoda, Family Oniscidae.

[2] *Armadillidium vulgare* (Latrielle), Class Crustacea, Order Isopoda, Family Armadillididae.

pound, or Diazinon at 0.5 pound to 100 gallons of spray. Dusting these areas with 5 per cent DDT or 2 per cent lindane or chlordane is also effective.

280 (88). MILLIPEDES[1]

Importance and Type of Injury. Young shoots of plants are chewed, and the tender parts of the stem have areas eaten out of the bark. Seedlings are sometimes cut off. Infested houses have many long, slender, cylindrical, hard-shelled, very active, crawling creatures (Fig. 14.13), up to about 1 inch in length, with two pairs of legs on nearly every segment, hiding under clods on the benches, or scuttling about among the plants.

Plants Attacked. Many different species of greenhouse plants. However, the main food of these creatures is decaying vegetable matter, particularly manure, used for fertilizing in greenhouse benches.

Life History, Appearance, and Habits. The eggs are laid either in the soil or on the surface. They are deposited in clusters containing 20 to 100 or more. Each female deposits about 300. The eggs are covered with a somewhat sticky material and are nearly translucent when first laid. They hatch in about 3 weeks, giving rise to young millipedes, which differ from the adult only in size, in number of segments on the body, and in having at first only three pairs of legs. They grow slowly, gradually assuming the adult form. There is probably one generation each year.

Control Measures. Millipedes may be controlled by the measures suggested for sowbugs (page 883).

281. GARDEN CENTIPEDE[2] OR SYMPHYLID

Importance and Type of Injury. Tiny, whitish, hundred-legged-worms live in the soil of gardens and ground beds in greenhouses, eating off the fine roots and root hairs to such an extent that entire crops are sometimes ruined. Plants become stunted, grow slowly, or die, and the root system will be found to have been severely pruned.

Plants Attacked. Lettuce, radishes, tomatoes, asparagus, and cucumbers are often severely injured, and many kinds of ornamental plants are also attacked.

Life History, Appearance, and Habits. The worms are found in winter about the roots of plants and in cracks in the soil near the surface. The full-grown ones are only $\frac{1}{4}$ to $\frac{3}{8}$ inch long by $\frac{1}{25}$ inch thick, with a head and 14 body segments. The head bears a pair of antennae, more than one-third as long as the body, and delicate chewing mouth parts. The body has 12 pairs of short legs. The creatures spin silk, from a pair of short cerci at the tip of the abdomen, as a lining for their burrows, which follow earth worm tunnels, soil cracks, and the site of decayed roots. After the winter crops are harvested and the greenhouse soil dries out, the centipedes make their way through cracks and earth worm tunnels down to moist subsoil where they feed upon decaying vegetable matter and lay their white spherical eggs at a depth of a foot or more. Most of the eggs are laid from April to September in small clusters of 5 to 20, from which the very tiny young hatch, a week or 10 days later. At first they have only 10 body segments, six pairs of legs, and very short antennae. As they grow, an additional pair of legs is acquired each time the skin is molted, body segments are added, and the antennae elongate. As the soil is wet down and crops started in the fall, the centipedes work their way upward to feed on the plant roots. A generation may be completed in a few months. The pests are rarely seen on the surface of the soil

---

[1] *Orthomorpha gracilis* Koch, *Julus* spp., and others, Class Diplopoda.
[2] *Scutigerella immaculata* (Newport), Class Symphyla.

and thrive best in deep soil that is undisturbed throughout the season. They are destructive only in ground benches, since the soil of raised benches is too shallow and too dry during the summer months for them to breed.

Control Measures. Fumigating the infested soil with ethylene dibromide or a mixture of 1,2-dichloropropane and 1,3-dichloropropylene,[1] as described on page 887, or with carbon bisulfide at the rate of 1 ounce to each square foot of surface is suggested. This is best done before the first fall planting, at a time when the soil temperature is above 65°F., and some days after the soil has been wet down to attract the centipedes to the surface. Lindane or parathion applied at 1.6 ounces to 1,000 square feet of soil as dust or wettable powder, and worked in, gives good control. All soil and manure brought into greenhouses should be steam-sterilized. Ground beds should be deeply cultivated as often as possible, taking care to disturb the retreats of the pests near foundation supports, walls, walks, and curbings.

References. Ohio Agr. Exp. Sta. Bul. 486, 1931; *Purdue Agr. Exp. Sta. Bul.* 331, 1929.

282. SLUGS[2]

Importance and Type of Injury. The foliage of plants, particularly in the damper parts of greenhouses, is frequently fed upon and injured by grayish, or grayish-brown slimy, legless, soft-bodied creatures, from ½ to 4 inches in length (Fig. 18.21). A shiny

FIG. 18.21. A slug. Enlarged. (*From Can. Dept. Agr. Bul.* 7, *n.s.*)

trail, composed of a sticky, viscid secretion given off from the body of the slug, will mark the course of its travels over benches and plants. Flowerpots, flats, and other supplies standing on or beneath greenhouse benches harbor these pests and should be stored elsewhere.

Plants Attacked. Many kinds of plants are attacked, particularly coleus, cineraria, geranium, marigold, and snapdragon.

Life History, Appearance, and Habits. The eggs are laid in masses, in damp places about the greenhouse benches, underneath boards and flowerpots, or in the soil. They are held together by a sticky secretion which turns yellow before the eggs hatch. In about 1 month the eggs give rise to very small young, which closely resemble the adult slugs except in size. They develop slowly and probably live for a year or more.

Control Measures. Trapping and hand-picking are fairly effective under greenhouse conditions. For this purpose bits of boards or several handfuls of plant refuse may be piled about the benches in the damper parts of the greenhouses where the slugs congregate in greatest numbers. If these piles are lifted each morning and the slugs picked up and killed, they can be kept down to a point where little damage will occur except in the cases of very heavy infestations. The most satisfactory control of slugs and snails under both greenhouse and outdoor conditions is obtained by broad-

---

[1] DD mixture.

[2] The spotted garden slug, *Limax maximus* Linné; the gray garden slug, *Deroceras reticulatum* (Müller); the gray field slug, *D. laeve* (Müller); the greenhouse slug, *Milax gagates* (Linné), and others, Class Gastropoda, Order Pulmonata, Family Limacidae.

casting the following poison bait in the evening at about 20 pounds per acre:

	Small quantity	Large quantity
Bran......................	1 lb.	16 lb.
Calcium arsenate............	1 oz.	1 lb.
Metaldehyde...............	½ oz.	½ lb.
Blackstrap molasses.........	2 tbsp.	1 pt.
Water....................	1 pt.	2 gal.

References.   Jour. Econ. Ento., **34**:321–322, 1941; *Calif. Agr. Ext. Ser. Cir.* 87, 1944.

283. NEMATODES OR EELWORMS[1]

Importance and Type of Injury.   Plants infested by nematodes will present a weakened, sickly appearance, without visible injury to the stem or any part of the plant above ground.   An examination of the roots will show numerous knots or galls; or, in cases of severe infestation, practically all roots will have a swollen, gouty appearance (Fig. 18.22).

FIG. 18.22. Roots of a tomato plant, showing enlargements caused by a heavy infestation of nematodes.   (*From Ill. Natural History Surv.*)

Plants Attacked.   Four or five hundred different kinds of plants are known to be attacked by nematodes, including practically all plants grown in greenhouses in the United States.   They are particularly a pest of tomatoes and cucumbers, and also very bad on cyclamen.

Life History, Appearance, and Habits.   The nematodes that develop in the gall are practically invisible, but the pear-shaped females may be distinguished without a microscope.   Either the adults or the larvae may

---

[1] The garden or root-knot nematodes, *Meloidogyne* spp.; the bulb or stem nematode, *Ditylenchus dipsaci* (Kühn); the leaf and bud nematode, *Aphelenchoides fragariae* (Ritzema-Bos), and others, Class Nematoda, Order Tylenchida.

winter in the root galls or the soil.   After mating, the females lay several hundred eggs in the galls.   The minute worms that hatch from the eggs may work through the soil for a considerable time, but, when they encounter roots, they penetrate them and start new galls.   A life cycle may be completed in about a month but is often much longer.

Control Measures.   The best method of controlling these creatures is to use soil which is known to be free from infestation.   If nematodes have gained access to greenhouse benches, it will be necessary to sterilize the soil, between crops, in order to keep down injury.   Probably the most effective method of soil sterilization is by steam.   Gibson gives the following directions for the use of this method:

"The Pipe Method.   In this method the steam is applied by means of a set of 1¼ inch or 1½ inch pipes, perforated with holes ⅛ to ¼ inch in diameter, and about 1 foot apart.   The number and length of the pipes should conform to the boiler capacity and length of beds.   According to Selby and Humbert,[1] the perforated pipes should not be more than 40 feet in length, nor exceed seven or eight in number. With medium boiler capacity, say 50 to 60 horespower, pipes 30 feet in length are most serviceable.   The pipes should be about 16 inches apart and should be connected with a 2-inch crosshead by means of T-connections.

"In using this system, the pipes are buried at a depth of about 6 inches, care being taken to see that they lie level.   The soil is then levelled and covered with canvas or sacking to check the escape of steam."

Potatoes may be used to determine the duration of the treatment. The potatoes should be placed near the surface and treatment stopped when they are well cooked.

"Steam sterilization not only eradicates nematodes, but it also rids the soil of all insect life; of pests such as sowbugs; of injurious fungi such as those which cause damping-off, lettuce drop, and Rhizoctonia; and of weed seeds.   It also has the advantage of improving the soil conditions, and for this reason it is said that sterilized soil requires less fertilizer than untreated soil."

Hot-water sterilization can be used and is nearly as successful as steam; it is easier to use in most greenhouses.   Soil fumigation with ethylene dibromide or a mixture of 1,2-dichloropropane and 1,3-dichloropropylene[2] is a very effective means of controlling nematodes in greenhouses or gardens.   These materials may be applied by pouring into holes 10 inches apart and 4 to 8 inches deep in the soil.   Ten per cent ethylene dibromide solution and the dichloropropane–dichloropropylene mixture should be used at 2 to 3 cubic centimeters per hole.   This treatment should be made at temperatures above 60°F., and the soil should not be sown for 2 to 3 weeks.   Chloropicrin and 10 per cent methyl bromide solution may also be used, but they are considerably more toxic and require a water seal to retain the vapors, and chloropicrin is highly toxic to growing plants.

Reference.   *Cornell Univ. Agr. Exp. Sta. Bul.* 850, 1948.

284. Miscellaneous Greenhouse Pests

Ants (page 893), earth worms, and termites (page 898) sometimes infest the soil of greenhouses, and the latter are often destructive to the woodwork.

[1] *Ohio Agr. Exp. Sta., Cir.* 151, 1915.
[2] DD mixture.

CHAPTER 19

HOUSEHOLD INSECTS, AND PESTS OF STORED GRAINS, SEEDS, AND CEREAL PRODUCTS

No other insects cause so much personal annoyance and embarrassment as those that frequent our dwellings. They attempt, often successfully, to appropriate our food and clothing to their use, and sometimes even make themselves at home on our persons. Most of these insects may be cheaply and easily controlled, if the right method is employed.

A large number of insects, including many species of beetles, moths, and mites, attack grain and grain products in farmers' bins, elevators, mills, warehouses, retail stores, and the home. The damage done in this way is estimated to exceed 500 million dollars yearly. It is impossible to treat all these insects in this book. The more important insects are mentioned, and some of the points of their life history are given.

As the method of control is much the same for all the insects of stored grains and seeds, no attempt is made to give a separate control for each species, but the general measures for controlling these insects are given near the end of the chapter, pages 932 to 935.

Many of the best publications on these insects deal in a general way with the many species attacking this class of products, and, as some of these publications contain references or short bibliographies, no special references are given for some of the species. Some of the publications dealing with this class of insects are given below:

General References. Kans. Agr. Exp. Sta. Bul. 189, 1913; *Ill. Agr. Exp. Sta. Bul.* 156, 1912; *Minn. Agr. Exp. Sta. Bul.* 198, 1921; *U.S.D.A., Dept. Buls.* 872, 1920, and 1393 and 1428, 1926, and *Farmers' Buls.* 1029, 1919, and 1260 and 1275, 1922, and 1483, 1926; HERRICK, G. W., "Insects Injurious to the Household," Macmillan, 1921; *U.S.D.A., Cir.* 390, 1936; MALLIS, A., "Handbook of Pest Control," McNair-Dorland, 3d ed., 1960; COTTON, R. T., "Pests of Stored Grain and Grain Products," Burgess, 2d ed., 1956.

KEY TO INSECTS INFESTING THE HOUSEHOLD AND STORED GRAIN, SEEDS, AND GRAIN PRODUCTS

A. Ant-like insects, of various sizes from $\frac{1}{16}$ to $\frac{1}{2}$ inch long, and yellow, reddish, brown, or black in color, with a slender waist (pedicel) between thorax and abdomen which always bears 1 or 2 enlargements (petiole). They generally crawl hurriedly about indoors or out and get into foods as they forage out from populous nests in the ground or in wood:

The most important American species may be determined by the following key prepared by Dr. M. R. Smith of the Bureau of Entomology and Plant Quarantine.

1a, With a 1-segmented pedicel...2.

1b—With a 2-segmented pedicel...5.

2a, Petiole poorly developed, hidden beneath base of abdomen (gaster). Tip of
abdomen without a circlet of hairs. Body of freshly-crushed worker with a
sweetish, nauseating, rotten cocoanut-like odor....*Odorous house ant*, page 895.

2b—Petiole well developed and erect. Body of freshly crushed worker with a
different odor..3.

3a, Tip of abdomen without a circlet of hairs; practically uniform-sized ants, about
$\frac{1}{10}$ inch long; color brown. Freshly crushed worker with a musty or greasy
odor. An imported species occurring in southern United States and in California, where they swarm in buildings and over trees, especially citrus.........
....*Argentine ant*, page 894.

3b—Tip of abdomen with a circlet of hairs. Freshly crushed worker with a distinct
acid (formic) odor..4.

4a, Large, black, variable sized ants, from $\frac{1}{4}$ to $\frac{1}{2}$ inch long. Colonies nest entirely
in wood, sometimes of old buildings............*Black carpenter ant*, page 896.

4b—Small, brown, robust, almost uniform-sized ants, approximately $\frac{1}{10}$ inch long.
....*Cornfield ant*, page 895.

5a, Head and thorax with numerous spines; body rusty brown; size variable, from
$\frac{1}{16}$ to $\frac{1}{2}$ inch long. Strictly a leaf-cutting species, known only from Texas and
Louisiana.................................*Texas leaf-cutting ant*, page 897.

5b—Head and thorax either spineless or with only 1 pair of spines on the posterior
portion of the thorax...6.

6a, Thorax and head covered with numerous, longitudinal, parallel ridges or lines.
Posterior portion of thorax with a single pair of spines. Imported species most
abundant along Atlantic seaboard..................*Pavement ant*, page 896.

6b—Thorax and head without parallel ridges or lines as in 6a. Posterior portion
of thorax without spines...7.

7a, With a 2-segmented antennal club......................................8.

7b—With a 3-segmented antennal club......................................9.

8a, Exceedingly small ants, from $\frac{1}{14}$ to $\frac{1}{10}$ inch long, approximately uniform in
size, yellowish. Eyes extremely small...................*Thief ant*, page 895.

8b—Moderately large ants, highly variable in size, from $\frac{1}{16}$ to $\frac{1}{4}$ inch long. Color
varying from yellowish red to black. Eyes prominent. Sting well developed.
Common in Gulf Coast states and southwestern United States............10.

9a, Small shining black ants, approximately $\frac{1}{10}$ inch long.....................
....*Little black ant*, page 895.

9b—Small yellowish-red ants, $\frac{1}{12}$ to $\frac{1}{10}$ inch long. Head and thorax with dense,
pitted impressions or punctulations. Nests in buildings..*Pharaoh ant*, page 895.

10a, Petiole with anteroventral tooth or keel; mandible with 3 biting teeth......
....*Southern fire ant*, page 896.

10b—Petiole without anteroventral tooth or keel; mandible with 4 biting teeth...
....*Imported fire ant*, page 896.

B. *Insects working in the interior of wood and wood products:*

    1. Large numbers of slender dark-brown winged insects, without a slender
waist, emerge from cracks in floors or walls or from holes eaten through the
woodwork of buildings. Small, whitish, brown-headed, soft-bodied, wingless insects, ant-like but with thick waists, feed and tunnel in the woodwork
of buildings, particularly on timbers that come in contact with the ground.
The tunnels are kept free of frass. Injured wood is almost completely
eaten, except an outer shell. Sometimes cement-like tubes are constructed
over brick or stone foundations, leading to the woodwork..............
....*Termites* or *white ants*, page 898.

    2. Wood with holes from $\frac{1}{16}$ to $\frac{1}{4}$ inch in diameter leading to the interior of
timbers, the interior of which may be completely reduced to fine yellowish
powder, packed in the tunnels made by small white grubs with large heads
and 6 small legs. Adults which made the holes to the outside are hard-
shelled beetles, $\frac{1}{12}$ to $\frac{1}{5}$ inch long, brownish, elongate and cylindrical, **or**

short and stubby, with varied sculpturing on body and wings...........
....*Powder post beetles*, page 903.

C. *Insects running very swiftly about the house in dark places or at night: rarely found in foods:*

1. Small wingless grayish-white to brownish carrot-shaped glistening insects, up to ½ inch in length, with slender feelers projecting from the head and from the tail, run rapidly over books, walls, and papers and feed on the binding of books, wall paper, or foodstuffs..........*Silverfish*, page 905.

2. Dark-brown, light-brown, or black, shiny, flat-bodied, long-legged insects, the smaller ones without wings, the larger ones winged, feeding on the binding of books, the sizing of papers, and foods of all kinds. The insects are seldom seen in the daytime. A noticeable, sweetish, sickening odor where these insects are present in numbers............. ....*Cockroaches*, page 906.

3. Slender pale-brown creatures, up to 2 inches in length, with many extremely long legs, scurry rapidly over walls or floors when lights are suddenly turned on. Do not feed on stored foods or fabrics.....*House centipede*, page 909.

4. Rather hard-shelled cylindrical brownish creatures, from 1 to 2 inches long, with many legs, usually 2 pairs to each principal segment of the body. Generally found in basements or under boards or stones around foundation walls where it is damp.........................*Millipedes*, page 884.

D. *Insects feeding in tobacco, drugs, the silk upholstery of furniture, leather, etc.:*

1. Small, white, very hairy grubs and very small light-brown beetles (1/16 inch long), with hairy but not striated wings, eat holes through the tobacco or feed in numbers within packages of cigars, cigarettes, and package tobaccos. Upholstered furniture with numerous holes, eaten through the covering by these grubs or beetles........................*Cigarette beetle*, page 910.

2. Small, white, curved grubs, similar to *D*,1, but not hairy; and small, rather narrow-oval, reddish-brown beetles with striated wings, mine and tunnel through packages of drugs, vegetable poisons, foods, leather, and a great variety of other substances, which they eat....*Drug store beetle*, page 911.

3. Stored tobacco, nuts, chocolate, and other dried vegetable products infested with yellowish-white to pinkish-brown very active caterpillars, up to ⅝ inch long, which eat holes through the products and contaminate them with their silk and excrement. Moths, ½ inch from tip to tip of wings and with base and tip of wings darker than the middle, flutter about the storage........
....*Tobacco moth*, page 911.

E. *Insects destroying clothing, fabrics, furs, and other animal products:*

1. Rugs or carpets with holes eaten in them, or stuffed animals, birds, and insects being fed upon by small dark-brown, very hairy larvae, about ¼ inch in length. Small brown-and-gray, or black, beetles, sometimes flecked with red, crawling over infested materials. Beetles oval in outline, about ⅛ inch long.....*Carpet beetles*, or *buffalo beetles*, and *museum pests*, page 912.

2. Wool or silk clothing, feathers, and other animal products with many silken cases or webs on the surface, containing small whitish worms. Fabrics with numerous holes eaten through them. Small buff-colored moths, about ⅛ inch long, running rapidly over the surface of fabrics when exposed to light, or flying aimlessly about in closets...............*Clothes moths*, page 915.

F. *Insects eating in meats, cheese, and other foods:*

1. Very hairy, brown larvae, somewhat resembling *E*,1, but with two, short, curved hooks on the last segment, feed on meats, particularly dried or smoked meats, other food products, feathers, skins, etc. Dark-brown beetles, about ⅓ inch long, with a pale-yellow band across the middle of the body, crawling about on the surface of food materials where larvae are feeding.......................................*Larder beetle*, page 917.

2. Elongate, slender, purplish larvae, not very hairy, cylindrical but tapering toward the head, with 6 short legs, eat through hams and bacons. Brilliant greenish-blue beetles with reddish legs, up to ¼ inch long, run rapidly from the food substances when disturbed.......*Red-legged ham beetle*, page 918.

3. Insects similar to *F*,1, feeding in meats, hides, tallow, and other animal matter. The beetles are black above, whitish beneath, and from ¼ to ⅓ inch long. . . . . *Hide beetle,* page 917.
4. Cheese and smoked meats with small yellowish maggots crawling over the surface or through the product. These maggots sometimes jump or hop for some distance, by suddenly bending and then straightening their bodies. . . . . *Cheese skipper,* page 919.
5. Cheese, hams, cereals, and other food products become musty and are found on close examination to be swarming with tiny, whitish, 8-legged mites, not over 1/32 inch long. . . . . . *Cheese mites, ham mites,* and *grain mites,* page 919.

G. *Insects attacking stored grains and grain products, dried fruits, and some other products:*

a. *The parents are hard-shelled beetles: the larvae do not have prolegs and spin no silk.*

1. Dark-brown snout beetles, up to ⅙ inch in length, crawling over and through grain products held in storage. Grain heats and becomes moist, sometimes matted together because of sprouting. Individual grains contain small fat-bodied, legless, white grubs. . . . *Granary weevil* and *rice weevil,* page 920.
2. Small dark-red narrow-bodied beetles, about ⅙ inch in length, running rapidly over the surface of grains; often found in flour, meal, breakfast foods, and other food products. Small brownish-white 6-legged larvae, about ⅙ inch long, feeding in and on the grain. . . . . *Confused flour beetle* and *red flour beetle,* page 925.
3. Insects similar to *G*,2, but more slender, flatter, and somewhat darker in color, feeding on dried fruits, grain and grain products, nuts, seeds, and many other foods. Each side of the thorax shows 6 fine saw-tooth-like projections when examined under a lens. Larvae with bodies very slender, nearly bare of hairs, somewhat depressed. . . . . *Saw-toothed grain beetle,* page 926.
4. Very small dark-brown beetles with rather long, prominent antennae, and slight knobs at front angles of thorax, found in same situations as *G*,3, and particularly in bins of oats and barley. . . . . *Foreign grain beetle, Cathartus advena* Waltl.
5. Very small, brown to black, nearly cylindrical beetles, with the head completely hidden by the thorax, only ⅛ inch long. Curved-bodied, grub-like, whitish larvae, with the swollen anterior end bearing a small head and 6 small legs, are found inside grains of wheat and other seeds, which they completely hollow out. Holes often eaten in wooden or paper boxes in which seeds are stored. . . . . *Lesser grain borer,* page 927.
6. Very similar to *G*,5, but the adults a little larger, thicker, smoother, and shinier; eat holes chiefly in corn. . . . . *Larger grain borer,* page 927.
7. Smooth shiny-bodied, uniformly brown to yellow worms, closely resembling wireworms, feeding in the lower parts of, or beneath, grain bins and where the grain is moist. Rather robust, somewhat flattened, black to brown beetles, up to nearly 1 inch in length, in the same situations. . . . . *Yellow mealworm* and *dark mealworm,* page 922.
8. Black beetles, somewhat resembling *G*,7, but smaller in size, about ½ inch in length, with the head and prothorax distinctly separated from the rest of the body by a narrow waist. Whitish to grayish-white soft-bodied larvae, with black heads and 2 short black projections from last segment of abdomen, work among grains, particularly where undisturbed for some time. . . . . *Cadelle,* page 923.
9. Numerous yellow-brown larvae with yellow undersides and yellow areas between the segments and clothed in long brown hairs, up to ¼ inch long, crawling over and chewing holes in grain, seeds, and other stored products and contaminating these with shed skins. . . . . *Khapra beetle,* page 925.
10. Active strong-flying beetles, ⅛ inch long and half as wide, with wings much shorter than the abdomen, each bearing an amber-brown spot at the tip and a smaller one near the base; and their white shortlegged grubs, with brown

head and tail end, work their way to the inside of partially dried figs, other drying fruits, fermenting watermelons, and the like, causing them to rot and sour..*Dried fruit beetle,* page 927.

b. *The parents are delicate winged moths; the larvae have prolegs and often spin much silk among the grain.*

  1. Wheat and corn in storage have small round holes, a little larger than a common pinhead, eaten into the grains. Whitish worms, up to nearly ½ inch in length, living and feeding *inside* the kernels. Brownish-gray moths, a little less than ½ inch in length, with a long fringe of hairs on the wings, fly about the bins or cribs, or crawl over the surface of the grain........ ....*Angoumois grain moth,* page 929.

  2. Flour in mills, and sometimes in storehouses, webbed and matted together in masses, by small white to pinkish worms up to ⅗ inch in length, which are usually concealed in their silk tubes. Masses of webbed flour frequently becoming so tightly packed in mill machinery as to prevent the operation of the machines. Gray, black-marked moths, resting on the surface of grain or flour, in infested mills...............*Mediterranean flour moth,* page 929.

  3. Flour, grain, and grain products and especially dried fruits, nuts, nut candies, and other food products, infested by small whitish to greenish-white worms, up to ½ inch in length. More or less webbing of infested material, but not so much as in *G,b,2.* Small moths with the basal half of the front wings much lighter in color than the tip, resting on infested material, or flying about over it...........................*Indian-meal moth,* page 931.

  4. Grain stored while damp or in damp places, infested with whitish caterpillars, having black heads and prothorax, the body tinged with orange at each end, which rest in tough silken tubes and feed from the open end. Often cut through sacks. Adult moths 1 inch from tip to tip of wings, the forewings dark-brown on basal and distal thirds, light brown across the middle, the light- and dark-brown areas separated by transverse wavy white lines... *Meal moth, Pyralis farinalis* (Linné), Order Lepidoptera, Family Pyralididae.

H. *Insects attacking stored beans and peas, also the developing seeds within pods in the field:*

  1. Peas in the pod contain white, nearly footless grubs or short, chunky, brownish beetles (never more than 1 to a seed), which feed exclusively inside and show no entrance hole except a small dark dot. Stored peas often show a small round hole where the beetle has emerged. Beetles ⅕ inch long with 2 black spots on the extreme tip of the abdomen that is exposed beyond the wing covers. Hind femur with a sharp tooth near apex...... ....*Pea weevil,* page 935.

  2. Similar to the preceding but occurring chiefly in broad beans in addition to peas. The adult beetle smaller, without black spots on that part of abdomen exposed beyond the wing covers, and the tooth near apex of femur broader and more blunt than in the pea weevil........................ ....*Broad bean weevil,* page 936.

  3. Growing beans and peas attacked in the field and in storage as by the pea weevil, by a smaller, brownish beetle, about ⅛ inch long, and by footless white grubs that feed exclusively inside the seeds. Stored beans often show many round holes and fine, powdered frass. Hind femur of beetle has 1 large and 2 small teeth near the tip.............*Bean weevil,* page 936.

  4. Insect and injury similar to the preceding, but prefers cowpeas. Adult with a large dark spot on apex of each wing cover and another pair of spots near the middle of costal margin of wing covers......*Cowpea weevil,* page 936.

  5. Insect and injury similar to *H,3,* and *H,4,* but smaller and with two prominent ivory-white spots at the middle of hind margin of prothorax. Wing covers with a narrow dark crossband at middle. Male with comb-like antennae...........................*Southern cowpea weevil,* page 937.

A. HOUSEHOLD INSECTS

285 (13, 23). Ants[1]

Importance and Type of Injury. There are few insects which have proved themselves more persistently exasperating to the housekeeper than ants. Everyone is familiar with the fact that when ants have invaded houses, the workers will be found crawling over any food that is to their liking, bits of which they cut off and carry to their nest (Fig. 6.87). Some species also cause injury by establishing their nests in the sills and

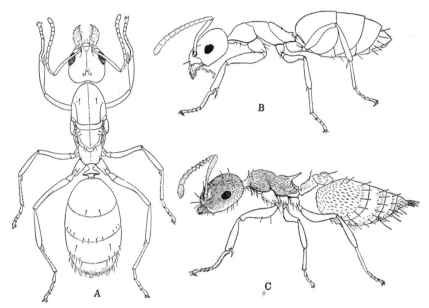

FIG. 19.1. Three common and troublesome American ants. *A*, the queen Argentine ant, *Iridomyrmex humilis,* after discarding her wings. Actual size of queen, about ⅕ inch long; of workers, from 1/16 to ⅛ inch long. *B*, worker of the odorous house ant, *Tapinoma sessile;* length 1/12 to ⅛ inch; black; with a pronounced "rotten pineapple" odor when crushed. Note how the top of the abdomen overhangs the nodus of the petiole. *C*, worker of the acrobat ant, *Crematogaster lineolata,* a shining black species, about ⅛ inch long, that often carries its pointed abdomen turned up over the rest of the body. (*From Woodworth, Calif. Agr. Exp. Sta. Bul. 207.*)

woodwork of old houses; but as a rule, they do not attack perfectly sound wood. Other species throw up mounds of earth about the entrance to their nests, disfiguring lawns and walks. Some species obtain a part of their food from the sweet, sticky honeydew given off from the bodies of aphids, scale insects, and mealybugs, and these species may establish and protect colonies of these insects on plants out-of-doors or in greenhouses or residences.

Food. The food of ants is even more varied than that of man. Various species eat particles of human foods, sweets, honeydew, seeds, fats,

[1] Many species of Order Hymenoptera, Family Formicidae. The authors make grateful acknowledgment for valuable aid in the preparation of the section on ants to Dr. M. R. Smith.

meats, dead insects, roots of vegetables, and fungi, which they grow in their nests.   They are fond of nearly every kind of human food, and in some parts of the country they become a pest from their habit of gathering and storing quantities of grains and seeds in their nests.

Distribution.   As a group, world-wide; certain species are more abundant and annoying in some localities than in others.

Life History, Appearance, and Habits.   The wingless workers (Fig. 6.87,*e*), which are the form of ants generally noticed, are all sterile females and are only one of the forms that occur in the ant colony.   The winged males and females (Fig. 6.87,*a,d*), which swarm out of the nests at certain seasons, are also very evident.   But the wingless females or "queens" (*c*) and the helpless, maggot-like young (*f, b*) remain exclusively in the underground nests and are rarely seen by the layman.   In some colonies even the worker caste is divided into two or three forms.   Some of these forms, as in the honey ants, have abdomens which can be greatly distended and are used as a sort of jug in which to store honey for the colony. Others, as in some of the seed-gathering ants, have certain castes with enormously enlarged heads and very strong jaws which serve as seed crackers.   Ants, as a group, can be distinguished from other insects by having one or two wart-like elevations on the slender pedicel that separates the thorax from the large part of the abdomen (Fig. 19.1).

Individual worker ants cannot exist alone.   A typical colony consists of from one to several females or queens (whose duty it is to lay eggs), and thousands of workers.   The workers are incompletely developed females and seldom lay eggs.   The queens are cared for by the workers and usually remain in the inner chambers of the nest, seldom leaving the nest except at times when the colony may migrate to a new location.   From the eggs laid by the females (Fig. 6.87,*g*), larvae or maggots (Fig. 6.87,*f*) hatch, the majority of which develop into workers after going through the pupal stage (*b*), either in a cocoon, or without such protection.   A certain percentage, however, produce winged males and females, the "kings" and "queens" of the ant colonies.   At certain times of the year, varying with the species, these winged males and females leave the nest or swarm. Usually this occurs over a wide area on the same day for any one species. The males mate and soon die.   The female flies to a situation that is attractive to her as a nesting site, alights, tears off her wings, encloses herself in a small cave which she excavates in the soil and in which she lays a few eggs.   After the eggs have hatched, the female feeds and cares for the young larvae until they have become full-grown.   The larvae of ants are whitish, helpless, maggot-like grubs, without legs and with very small heads.   These larvae develop into workers; and, as soon as the adult workers appear, they start taking care of the queen and her young.   From this time forward, the queen confines her activities entirely to egg-laying. The queen may live from 1 or 2 to 12 to 15 years, during which time she will produce many thousands of eggs and may develop many thousands of individuals.

Ant pests of the first magnitude include the following:

1. Argentine ant, *Iridomyrmex humilis* (Mayr) (Fig. 19.1,*A*).   Although first observed in the United States in 1891, and restricted in its distribution to California, Arizona, Texas, Louisiana, Arkansas, Mississippi, Georgia, North and South Carolina, Florida, and Tennessee, this small ant is the most annoying and destructive household insect wherever it is established.   Locally in cities and towns it drives out all other

kinds of ants; swarms into houses and stores; gets into all kinds of foods; is indirectly very injurious to citrus and other fruits and to shade trees, by attacking the blossoms and by distributing, establishing, and protecting aphids, mealybugs, and scale insects; and is very troublesome in bee and poultry yards. It prevents people from sleeping and often destroys the value of real estate by rendering neighborhoods undesirable. It is believed to have been brought into New Orleans in shipments of coffee from Brazil and has subsequently been widely spread by commerce. The workers are slender, $\frac{1}{12}$ to $\frac{1}{8}$ inch long, brown with lighter appendages. There is a single segment in the pedicel, the petiole is erect, and the mandibles have about a dozen denticles on their inner margins. They have a very slight, musty, greasy odor when crushed and always travel in definite trails. Nests are established in any dark, somewhat moist situation in orchards, lawns, or gardens. The winter is passed in large centralized colonies in particularly favorable nesting sites, such as fairly dry, warm piles of manure or other decaying vegetable matter. All stages may be present throughout the winter but activity is greatly accelerated in warm weather. Mating takes place in the nest and the nuptial flights are less evident than in most ant species. There may be several to many queens in one colony. Eggs hatch in 12 to 50 days. The larvae require about 1 month for their development, and the pupae 15 days. The time from egg to adult averages 75 to 80 days. Widespread and intensive community campaigns of eradication should be initiated wherever this pest appears.

References. U.S.D.A., Bur. Ento. Bul. 122, 1913, and Cir. 387, 1936.

2. Pharaoh ant, *Monomorium pharaonis* (Linné). This tiny, slender, yellowish-red ant, $\frac{1}{12}$ to $\frac{1}{10}$ inch long, generally nests in inaccessible places about the foundations and in the walls of buildings, from which it forages indoors the year round in search of food. It is very difficult to eradicate. It is an imported species, found mostly in the larger towns and cities, but usually sporadically distributed. It is practically omnivorous, showing great fondness for proteins as well as sweet foods, and is especially pestiferous about apartment houses, groceries, cafes, and hotels. The workers of this and the following species have two segments in the pedicel and three segments in the antennal club. There are no spines on the posterior part of the thorax. Most of the body is covered with minute pitted impressions.

3. Little black ant, *Monomorium minimum* (Buckley). The workers of this species (Fig. 6.87) are from $\frac{1}{12}$ to $\frac{1}{10}$ inch long, slender and shining, and generally jet black in color. They typically nest out-of-doors, in the soil, where they make tiny craters, or in stumps; but sometimes indoors in the woodwork or masonry of buildings. This is a native species that is a common house pest in both town and country. It is very general in food habits, taking either sweets, honeydew, fruits, vegetables, or meats.

4. Odorous house ant, *Tapinoma sessile* (Say). The workers of this species (Fig. 19.1,B) are from $\frac{1}{10}$ to $\frac{1}{8}$ inch long, soft-bodied, the pedicel one-segmented, the petiole somewhat horizontal and difficult to see because hidden by the overhanging abdomen. They are darker in color than the Argentine ants, deep brown to black, have a broader abdomen, and are less agile in their movements. When crushed, they have a characteristic, nauseating, sweetish, rotten cocoanut-like odor. This is a native species, widely distributed over the United States. It generally colonizes out-of-doors in shallow nests under diverse conditions, but sometimes in foundations of buildings. It is practically omnivorous, but especially fond of sweets and is a common attendant upon honeydew-excreting insects.

5. Thief ant, *Solenopsis molesta* (Say). The workers of this species (Fig. 9.45) are light yellow to bronze, $\frac{1}{15}$ to $\frac{1}{10}$ inch long. They are similar to, but still smaller than, Pharaoh, ant, lighter in color, and have two segments in the antennal club, a two-segmented pedicel, and extremely small eyes. This native species usually nests out-of-doors, in soil or wood, and is troublesome only in warm weather. It is a common house pest, showing a preference for meats, butter, cheese, seeds, nuts, oils, and other protein foods, rather than sweets. It sometimes does damage to germinating seeds of grains (page 517).

6. Cornfield ant, *Lasius alienus* (Förster). This species nests mostly in open places, in the soil or in rotten wood. The workers are fond of sweets, often attending and fostering honeydew-excreting insects such as aphids (page 501). They are $\frac{1}{12}$ to $\frac{1}{10}$ inch long, robust, soft-bodied, light to dark brown. The anal

opening is terminal, circular, and surrounded by a fringe of hairs. There is no antennal club, and the pedicel is one-segmented. The body when crushed has an acid (formic) odor. It is a native species, distributed over most of the United States. It sometimes makes objectionable nests in lawns.

7. Pavement ant, *Tetramorium caespitum* (Linné). This imported ant is a rather hairy, robust, hard-bodied species, from $\frac{1}{12}$ to $\frac{1}{7}$ inch long, with two segments in the pedicel and no antennal club. The abdomen is black, the appendages pale. The head and thorax are grooved with fine parallel lines, and there are two small spines on the posterior part of the thorax. It is common in lawns, especially in or near towns or cities along the Atlantic seaboard. It nests under stones or the edges of pavements, from which it forages into houses and greenhouses. It is often a pest in gardens, where it gnaws at the roots of vegetables or steals planted seeds.

8. Larger yellow ant, *Lasius interjectus* Mayr. The workers of this species are $\frac{1}{10}$ to $\frac{1}{8}$ inch long and yellow. When crushed they have a pronounced and characteristic lemon-verbena odor. Although a common soil-inhabiting species, which fosters mealybugs and aphids on the roots of plants, it not infrequently nests in walls or beneath the floors of basements, from which the winged, sexual individuals emerge inside or outside the house and cause concern. But it does not attack human foods.

9. Black carpenter ant, *Camponotus pennsylvanicus* (De Geer). This ant is among the largest of our common ants, the workers ranging from $\frac{1}{4}$ to $\frac{1}{2}$ inch long. The body is dark brown to black. The pedicel is one-segmented, there is no antennal club, and the tip of the abdomen has a circlet of hairs. When crushed the ants have a distinct acid (formic) odor. It is a native species generally distributed over the eastern half of the United States. It nests entirely in wood, hollowing the wood out, but not eating it. It is a very common house pest, especially fond of sweet foods. It will also bite people, though it does not sting. In addition to nesting in stumps and tree trunks, it does appreciable injury to telephone and telegraph poles, and the sills and other timbers of older buildings.

10. Southern fire ant, *Solenopsis xyloni* MacCook. The workers are variable in size, from $\frac{1}{16}$ to $\frac{1}{4}$ inch long, shining, the largest ones yellowish red, the smaller ones much darker. The pedicel is two-segmented, and there is a two-segmented antennal club. The eyes are prominent, the body hard, and these ants have a sting like a bee. It is a native species common in the Gulf Coast states, less common eastward, and with subspecies ranging into the southwestern states. It nests generally in open, exposed places, often forming loose mounds or numerous scattered craters. They steal planted seeds, infest houses, and sting severely, often killing young poultry and quail just as they are hatching. While omnivorous, they show some preference for protein foods.

11. Imported fire ant, *Solenopsis saevissima richteri* Forel. This ant was first found in Alabama in 1929 but had apparently been present for a decade or more, as an imported species from South America. It possesses a painful sting, and the wound soon fills with pus as a result of the injection of a powerful necrotizing poison. Small birds and mammals are often killed by the stings. The imported fire ant is often a serious agricultural pest since it feeds on the tender stems of young plants just below the ground, damaging cabbage, collards, eggplant, potato, and germinating seeds. Its mounds, which may number 25 to 100 per acre, are from 15 inches to 3 feet high and covered with a hard crust, so that they interfere with farm operations and damage equipment. The workers are polymorphic, existing as minors $\frac{1}{8}$ inch long and majors $\frac{1}{4}$ inch long. They range from dark brown to tan, with a broad orange band on the dorsum of the first gastric segment. The pedicel is two-segmented, and there is a two-segmented antennal club. In the colonies along the Gulf Coast, sexually mature winged ants are present all year, overwintering in the colonies and making nuptial flights in the spring. The newly mated queen digs a small brood chamber and lays from 75 to 125 eggs, which hatch in 8 to 10 days. The queen rears the first brood of workers, and later broods are reared entirely by the workers. The larval period requires from 6 to 12 days, and the pupal period 9 to 16 days. Colonies are frequently abandoned, and the ants may move 3 to 30 feet by tunneling. This species now infests over 20,000,000 acres in Texas, Arkansas, Louisiana, Mississippi, Georgia, Alabama, Florida, and North and South Carolina.

References. U.S.D.A., *ARS*-33-49, 1958; *Jour. Econ. Ento.*, **52**:1, 455, 1959; and **53**:188, 646, 1960.

12. **Texas leaf-cutting ant,** *Atta texana* (Buckley). This native species is confined to southern and eastern Texas and western Louisiana, where there is well-drained soil of a loamy nature. The workers are variable in size, from $\frac{1}{16}$ to $\frac{1}{2}$ inch long, light brown, hard-bodied, with many spines on head and thorax, and a two-segmented pedicel. Nests are often enormous—10 to 20 feet deep but not very high, with numerous craters, and containing many thousands of individuals. They invade houses and steal seeds and other farinaceous food. The workers cut leaves from both wild and crop plants, macerate them, and grow a fungus for food upon the macerated leaves in their nests. These ants are serious pests of garden and field crops, and also in reforested areas, where they destroy young pine seedlings.

Control Measures. The simplest thing to do when ants appear in dwellings is to scatter a dust in their paths or to spray the infested area with a deodorized oil spray. This treatment is particularly effective when applied entirely around the foundation or under a house. Chlordane used as a 2 to 5 per cent dust or as a 2 to 5 per cent solution in deodorized kerosene is the most effective material for the purpose, as it does not repel the ants and remains effective for a month or more if properly applied. Similar applications of 1 to 2 per cent dieldrin or heptachlor are equally effective. Spraying or dusting with chlordane at 1 to $1\frac{1}{2}$ pounds per acre will destroy free-living ants (page 522 for control of turf-infesting ants). The application of heptachlor or dieldrin at 2 pounds per acre as a spray or granular has given area-wide control of the imported fire ant for 3 to 5 years. A bait of kepone at 0.125 per cent in peanut butter, placed in soda straws and distributed at 4 to 6 pounds per acre, has given excellent control of this species without injury to wildlife. The secret of thorough ant control, however, is the destruction of the queens in the nest, following which all other forms will perish. The most effective method is to find the mounds or nest of the ants in the ground. This can usually be done by following the line of foraging workers back to the place from which they are coming. If the nest is located it may be destroyed by washing about $\frac{1}{8}$ teaspoonful of 50 per cent wettable chlordane into each hole or by spraying the hills with an emulsion concentrate of chlordane used at 1 ounce to 1 gallon of water. Another effective treatment consists in punching several holes into the nest to a depth of about 6 to 15 inches, depending upon the volume of the nest, then pouring in 1 to 2 tablespoonfuls of carbon bisulfide or of calcium cyanide dust through a funnel and closing the holes with mud. Carbon bisulfide must be kept at a distance from any fire, and in using calcium cyanide extreme care must be used not to breathe the fumes or get the poison into the mouth or into sores.

If the nests cannot be located, a poison bait should be used which is attractive to the worker ants and slow acting so that they will carry it to their nests and feed it to all the developing young and the queens before they themselves are killed or suspicious of it. The bait must be carefully selected with reference to the species of ant. The best times to poison are in spring and autumn and, in the south, in winter. Among the best poisons are the following:

THE ARGENTINE ANT BAIT

Boil together the following materials for 30 minutes:

Granulated sugar	$1\frac{1}{4}$ lb.
Water	$1\frac{1}{4}$ pt.

Tartaric acid (crystallized)............ 1 g.
Benzoate of soda.................... 1 g.
Dissolve sodium arsenite in hot water in the following proportions:
Sodium arsenite (C.P.).............. ⅛ oz.
Hot water........................ 1 fl. oz.

When the above solutions have cooled, add the second to the first and stir well.   Then add ⅔ pound of strained honey to the resulting sirup and mix thoroughly.

This bait is successful for the Argentine ant and many other species that show a preference for sweet foods, but not for the Pharaoh ant, the tiny thief ant, or the southern fire ant.   For the Argentine ant, poisoning should be undertaken on a community- or city-wide basis.   The poison is exposed in small aluminum or covered paperoid cups.   Small holes must be provided near the top of the container for the ants to enter, and the cups should have securely fastened lids or be placed high out of the reach of children.   The cups should be placed about 20 to 25 feet apart.   Ten to 20 such containers are needed about an isolated house or 100 to 200 for an average city block.

Thallium sulfate is a very deadly poison and must be handled with great care.   It is, however, a more effective poison for certain kinds of ants than sodium arsenite.   Ready-prepared baits containing this poison, in either a sweet or a greasy base, may be purchased or can be made as follows.

THALLIUM SULFATE BAIT FOR SWEET-LOVING ANTS

Dissolve 2 grams of thallous sulfate, Tl_2SO_4, carefully weighed, in ½ pint of luke-warm, not hot, water.   In a separate container mix ½ pint of water, 1 pound of granulated sugar, 3 ounces of strained honey, and 45 cubic centimeters of glycerin. Bring this mixture to a boil and remove from the fire.   Cool for about 5 minutes, add the solution of thallium sulfate, and stir thoroughly.   Label "Poison" and store in a cool place.

THALLIUM SULFATE BAIT FOR PROTEIN-LOVING ANTS

Thallous sulfate, Tl_2SO_4.............. 0.5 g.
Peanut butter....................... 75 g.
German sweet chocolate.............. 25 g.
Mix together very thoroughly.

The poison baits may be put out in metal bottle caps or in short sections of soda fountain straws.

References.   WHEELER, W., "Ants: Their Structure, Development, and Behavior," Columbia Univ. Press, 1910; *U.S.D.A., Cir.* 387, 1936, and *Leaflet* 147, 1937; *Calif. Agr. Exp. Sta. Cir.* 342, 1937; *Jour. Econ. Ento.,* **39**:663, 1946; and **41**:48, 1948.

286. TERMITES[1]

Importance and Type of Injury.   The presence of termites in dwellings or other buildings is often first indicated by swarms of winged, black, ant-

[1] Several families, a number of genera, and about 1,800 species of the order Isoptera, of which approximately 45 are found in the United States.

like insects (Fig. 19.2) suddenly appearing inside the buildings, often emerging from holes in the walls, floor, or other parts of the woodwork. All termites differ from true ants in not having a slender waist.   Badly infested buildings may suddenly sag, due to the weakening of the frame by the feeding of the termites on the main timbers.   Occasionally their presence is indicated by the breaking through of floors or the cutting through of the rugs or floor covering.   Brick or cement foundations will have small tunnels constructed over their surfaces, of a clay or cement-like substance.   Growing plants may be eaten out.

Food.   The food of termites con-sists of wood, paper and other wood products, fungi, dried plant and animal products, and also partly digested food material passed from the mouth or anus and an oily exu-date sweated through the skin,

FIG. 19.2. Termite queen with wings spread. About 6 times natural size. (*From Ill. Natural History Surv.*)

which are freely circulated from caste to caste and individual to individual throughout the colony (page 207).

Distribution.   Termites of various species occur in nearly all the warmer parts of the world, extending in a general way around the earth between the mean annual isotherms of 50°N. and 50°S., but are much more numerous in the tropics.

Life History, Appearance, and Habits.   As to habits, termites have been classified as follows:

A. *Soil-inhabiting termites*, in which the colony is always partly in the ground or the mating pairs enter the soil or wood in the earth, following swarming.   These are further subdivided into (*a*) subterranean termites, (*b*) desert termites, (*c*) mound-building termites, and (*d*) carton-building termites.   The subterranean termites of the family Rhinotermitidae are those most commonly injuring structures in the United States.   They are small, the winged forms averaging about ¼ inch long and the workers and soldiers about ½ inch.   Subterranean termites make tubes from the soil to wood and do not pack their galleries with fecal pellets.   The following list describes the most important species:

1. The eastern subterranean termite, *Reticulitermes flavipes* (Kollar), is common throughout the region east of the Mississippi River (Fig. 19.4).

2. The arid-land subterranean termite, *Reticulitermes tibialis* Banks, ranges from the Pacific Coast to the Mississippi River and from Montana to Mexico.

3. The western subterranean termite, *Reticulitermes hesperus* Banks, ranges along the Pacific Coast from British Columbia to Mexico and eastward to Idaho and Nevada.

4. The desert subterranean termite, *Heterotermes aureus* Snyder, is an important pest in Southern California and Arizona.

B. *Wood-inhabiting termites*, in which the colony is confined entirely to wood, or swarming pairs enter wood above ground.   These are subdivided into (*a*) damp-wood termites and (*b*) dry-wood termites.   The termites of this group do not make tubes and can be distinguished by their characteristic well-formed fecal pellets, which are found in great quantities within the galleries.   The dry-wood termites of the family Kalotermitidae are larger than the subterranean termites, the winged forms and workers averaging about ½ inch long.

5. The common dry-wood termite, *Kalotermes minor* Hagen, causes severe damage along the Pacific Coast, where it swarms during the middle of the day in brilliant sunlight and at temperatures of 80 to 100°F.   After mating, the winged pair quickly break off their wings and bore into sound wood.   The life cycle from egg to adult

comprises seven instars and requires approximately a year. A mature colony may contain about 3,000 individuals.

6. The southern dry-wood termite, *Kalotermes hubbardi* Banks, extends from the Gulf of California into California and Arizona. The adults swarm at night, and the species is especially adapted to conditions of low moisture.

7. The southeastern dry-wood termite, *Kalotermes snyderi* Light, is found from Texas east to South Carolina. It swarms at night, and the adults are attracted to light.

8. The common damp-wood termite, *Zootermopsis angusticollis* (Hagen), is found along the Pacific Coast from British Columbia to Mexico, where it attacks poles, pilings, bridge timbers, and other structures near water. The soldiers and winged forms may range from ¾ to 1 inch in length.

The most important species in the United States belong to the soil-inhabiting, subterranean group. The kings and queens are always found in the ground, no external mounds are erected, and, although the workers may carry on a large part of their activities in wood aboveground, they

Fig. 19.3. Termites in their galleries in sill of a house. Slightly enlarged. (*From Ill. Natural History Surv.*)

always maintain a connection with the ground nest, through either burrows or covered runways. The colonies of termites, or white ants, somewhat resemble those of the true ants in their organization. In the termite colony, however, the male helps to start the nest and remains with the female or queen throughout life. The latter is usually the mother of the entire colony. In some species more than one royal pair will be found in the colony. There are only three stages in the development of the termites; first, the egg; then the immature or nymphal forms; and, finally, the adults, which are always separated into several different castes, such as workers, soldiers, kings, and queens. They do not have larval and pupal stages, as the true ants do. The so-called workers and soldiers in the termites are always wingless and consist each of *both* males and females, although they never or rarely produce any young. The workers and the nymphs are the destructive forms which are active in foraging and feeding on wood (Fig. 19.3). The winged true males and females develop within the colony and at certain times of the year leave the nest in swarms. Several such swarms may come out from the same nest, usually within short intervals of time.   Most of the termite species swarm in the spring.   These males and females differ greatly from the soft-bodied white, blind and wingless workers and soldiers.   They (Fig. 19.2) are black, rather narrow-bodied, and possessed of rather long, very brittle wings, which are broken off near the base when the kings and queens enter the ground after a single pairing flight.   After leaving the nest, they separate in pairs and select sites for starting new colonies. They then break off their wings, and the females become much enlarged, although, in our species, still capable of moving about.   They are said to lay eggs at the rate of a dozen or more a day for years.   In the absence of the winged (*macropterous*) males and females, reproduction may be accomplished by short-winged (*brachypterous*) or wingless (*apterous*) males and females, which cannot, however, produce the winged caste and so

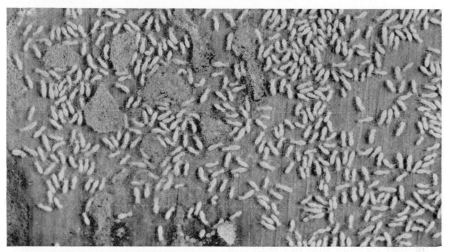

FIG. 19.4. Worker termites, *Reticulitermes flavipes*, as exposed by tearing away a portion of the infested timbers.   Natural size.   (*Original.*)

cannot give rise to any new colonies, though they may start subdivisions of the parent colony in adjacent burrows.   As in the case of ants, a colony may endure for many years.   Except for the winged males and females, which are positive to light and air for a time and escape from the nest to start new colonies, all other castes and all developmental stages live in a completely closed system of intercommunicating tunnels in soil and wood.   It was long supposed that they were negatively phototropic, but it is now known that their seclusion is due to an aversion to air currents and reaction to a closely controlled humidity, which is maintained by carefully sealing themselves into their tunnels.

Control Measures.   To prevent infestation by subterranean termites, great care should be taken to avoid having any woodwork of the buildings within 18 inches of contact with the ground, as such points of contact are nearly always the cause of buildings becoming infested with termites.   The insertion of a thin sheet of metal between the foundation and timbers of the house, running all the way around and projecting about an inch on each side, will prevent termite infestations.   Similar metal strips should be inserted between porch steps and adjacent wooden timbers, in all joints between concrete floors and side walls, and beneath wooden "sleepers." These shields must be carefully fitted to avoid leaving small openings through which termites may pass. They should always be placed so that it will be possible to inspect them from both the inside and outside of the building. Foundation brick- or stonework should be laid with good-grade cement mortar, well capped, and all cracks and crevices should be eliminated. All wood that must have contact with the soil should be termiteproofed by pressure impregnation with coal-tar creosote, zinc chloride, mercuric chloride, sodium fluosilicate, pentachlorophenol and its sodium and copper salts, DDT, or other tested preservative, after the wood is cut. When building, no bit of waste wood, such as grade stakes and concrete forms, should be left buried in the soil near the structure. Stumps and waste wood should be removed from the premises.

Where buildings are infested with subterranean termites, the most satisfactory treatment consists in soil treatment with one of the following: (a) 0.5 per cent aldrin or dieldrin as an oil solution or water emulsion, (b) 0.8 per cent lindane as an oil solution or water emulsion, (c) 1 per cent chlordane as an oil solution or water emulsion, (d) 8 per cent DDT as an oil solution, (e) 25 per cent trichlorobenzene in fuel oil, (f) 5 per cent pentachlorophenol (2.5 pounds dissolved in 1 gallon of pine oil or 2 gallons of methylated naphthalenes and diluted to 7 gallons with fuel oil, or as the sodium salt in water), (g) 10 per cent sodium arsenite in water. Treatments (a) to (d) are approved by the Federal Housing Administration as giving at least 5 years' protection. The newer organic insecticides are free from unpleasant and persistent odors and can be diluted with water rather than with oil, which decreases the fire hazard and danger of damaging plants. These treatments should be applied as follows:

1. Under concrete slab floors, porch floors, and entrance platforms at 1 gallon to 10 square feet.

2. To critical areas along both sides of foundation walls at 1 gallon to 2.5 linear feet per foot of depth.

3. To voids in masonry foundation walls and piers at 1 gallon to 5 linear feet.

These treatments are best applied along foundations by digging a narrow trench about the walls from the ground level to the footings and pouring in about one-half the total volume of material required. The soil is then replaced and tamped, and the remainder of the material added in several applications. A similar treatment can be made by pumping the material under pressure into holes made with a soil auger. These treatments may also be injected into the center space of brick walls and beneath concrete floors and footings by drilling ½-inch holes and pouring 1 to 2 quarts of material into each.

For the treatment of infestations of dry-wood termites, ¼- to ½-inch holes are drilled into the galleries in timbers and adjacent wood at 1- to 2-foot intervals. A dust gun with a delivery tube fitting snugly into the holes is then used to inject free-flowing dusts of sodium fluosilicate, sodium fluoride, calcium arsenate, or paris green at about 1 ounce to 15 to 30 holes. The openings are then sealed with putty, plastic wood, dowling, or cork. Pressure injection, at 40 to 75 pounds per square inch, of 5 per cent oil solution of DDT, dieldrin, chlordane, lindane, pentachlorophenol, or *ortho*-dichlorobenzene, will also give good results. Where large portions of structures are infested, fumigation with methyl bromide at 2.5 pounds, hydrogen cyanide at 2 pounds, or sulfuryl fluoride at 2 to 4 pounds per 1,000 cubic feet, under a plastic film to confine the gas (page 374), is frequently used to destroy the termites. The application of sorptive dusts (page 400) to attics and under houses at 1 to 2 pounds per 1,000 cubic feet appears to be effective in preventing infestation by dry-wood termites. Stored lumber may be protected by spraying thoroughly with sodium fluosilicate at 1 pound to 10 gallons of water, and infestations may be eradicated by methyl bromide fumigation or kiln-drying for several hours at 125 to 150°F.

The control of termites, once they have become established in a building, often means the expenditure of considerable money and labor. Passageways that lead over the foundation walls of buildings should be broken, and the surface of the wall over which these passages have been running should be heavily treated with one of the materials mentioned above for soil treatment. All woodwork of the building, including timbers or parts of the framework which are badly eaten, should be replaced with metal or cement, or if this is not advisable, with wood thoroughly termite-proofed.

References. U.S.D.A., Farmers' Buls. 1472, 1937, and 1993, 1948; *Ill. State Natural History Survey Cirs.* 30, 1938, and 41, 1947; *Conn. Agr. Exp. Sta. Bul.* 382, 1936; *Purdue Agr. Exp. Sta. Ext. Bul.* 225, 1946; *Jour. Econ. Ento.*, **40**:124, 1947; and **42**:273, 1949; KOFOID, C. A., *et al.*, "Termites and Termite Control," Univ. of California Press, 1934; SNYDER, T. E., "Our Enemy the Termite," Comstock, 1948; *Soap*, **33**(8):73,109, 1957; and *Pest Control*, **26**(2):9, 1958; *Calif. Agr. Exp. Sta. Cir.* 469, 1958.

287. POWDER POST BEETLES[1]

Importance and Type of Injury. Second only to termites as destroyers of seasoned wood, these species do not require that the wood be in contact with the soil. The larvae eat the hard, dry wood, tunneling through and through timbers in successive generations until the interior is completely reduced to fine, packed powder and the surface shell is perforated by many small "shot holes" (Fig. 19.5). Some of the species (*Lyctus* spp.) attack only the sapwood of the hardwoods, but others work through pine and fir and the heartwood as well. They may completely destroy the timbers of buildings, log cabins, rustic work, ship and airplane lumber, furniture, tool handles, wheel spokes, oars, casks, and other lumber. Their attack is an insidious one, because they may live beneath the surface for months and bore out through the surface in great numbers after the lumber has been made into furniture, implements, flooring, girders, or interior finish.

Life History, Appearance, and Habits. The winter, in unheated places, is generally passed in the larval stage, pupation occurs in spring, and the adults emerge in spring or early summer. The adults of most of the species are small, ranging from $\frac{1}{12}$ to $\frac{1}{5}$ inch in length, hard-shelled, brownish, elongate, and cylindrical or short and stubby, and with varied sculpturing on body and wings. They are not often seen in the adult stage. Eggs are laid in the pores of the wood. The larvae resemble small white grubs ($\frac{1}{8}$ to $\frac{1}{3}$ inch long) with an unusually large head end. They eat the wood and pack their burrows with exceedingly fine flour-like

[1] Various species of the Order Coleoptera, Families Lyctidae, Ptinidae, Anobiidae, and Bostrichidae. The best-known are species of the genus *Lyctus* and the furniture beetle, *Anobium punctatum* De Geer.

frass. The holes are from $\frac{1}{16}$ to $\frac{1}{4}$ inch in diameter. Generation after generation of the insects may develop in the dry wood with little external evidence of damage until structural timbers or vehicle stock collapses, or furniture and finish are completely ruined.

Control Measures. The sapwood of green trees can be impregnated with solutions of copper sulfate or zinc chloride, $\frac{3}{4}$ pound in $\frac{1}{2}$ gallon of water for each cubic foot of sapwood, in the spring as leaf buds begin to swell, by sawing the tree off and setting the butt in a container of the solution while the top rests against an adjacent tree or by various methods of banding, capping, or collaring the trunk, as described in the references below. Wood so impregnated resists insects and decay for many years. For the *Lyctus* species the complete removal of sapwood from the heartwood when lumbering, and eliminating it entirely from storage where

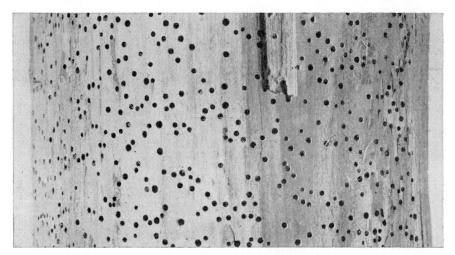

Fig. 19.5. A piece of timber ruined by powder post beetles. The holes are made by the adult beetles emerging to find mates and lay eggs for another generation. (*From Ill. Natural History Surv.*)

lumber and lumber products are kept, is said to be completely effective in preventing attacks. Wood infested with powder post beetles may be treated by kiln-drying at 180°F. for $\frac{1}{2}$ hour, by fumigation with methyl bromide in a vault or under a tent, or by spraying the wood thoroughly with one of the following treatments: (*a*) 5 per cent DDT or toxaphene, 2 per cent chlordane, or 0.5 per cent lindane in refined kerosene, (*b*) 5 per cent pentachlorophenol in oil, (*c*) lime-sulfur, 2 to 3 gallons to 100 gallons of water, or (*d*) borax at 10 pounds to 100 gallons of water containing 4 ounces sodium lauryl sulfate. The oil solutions of DDT or lindane are most satisfactory for treatment of furniture or hardwood floors. For the protection of furniture, finishing all surfaces with varnish or paint will prevent egg-laying but will not destroy any insects present.

References. DOANE *et al.*, "Forest Insects," p. 216, McGraw-Hill, 1936; *U.S.D.A., Dept. Bul.* 1490, 1927, and *Farmers' Buls.* 778, 1917, 1477, 1926, and 1582, 1929, and *Bur. Ento. Plant Quar. Cir.* E-409, 1937; *Mich. Agr. Exp. Sta. Quar. Bul.*, **29**:327, 1947; *Jour. Econ. Ento.*, **49**:664, 1956; *Pest Control*, **26**(1):39, 1958.

288. SILVERFISH[1] AND FIREBRATS[2]

Importance and Type of Injury. The bindings of books, papers, cards, boxes, and the like have the surface eaten off in irregular patches; the paper on walls is gnawed or the paste largely eaten off where the paper is attached to the walls. Small, whitish, grayish-brown, or greenish, glistening, carrot-shaped, entirely wingless insects, up to $\frac{1}{2}$ inch long (Fig. 5.8), scurry rapidly about over shelves and walls on exposure to light.

Food. These insects feed on a large variety of materials, such as starched clothes, rayon fabrics, bindings of books, book labels, the sizing of paper, or any papers on which paste or glue has been used.

Distribution. General throughout the United States and Canada.

Life History, Appearance, and Habits. The insects reproduce by means of eggs, the young closely resembling the adults except in size. The full-grown insect is wingless, about $\frac{1}{3}$ to $\frac{1}{2}$ inch in length, varying with the species; of a silvery, greenish-gray, or brownish color sometimes faintly spotted. The body tapers very markedly from head to tail, and is covered with thin scales which give it a silvery shiny appearance. There are two long antennae on the head and three similar appendages at the tail end, each nearly as long as the body. The preferred food of these insects is vegetable matter high in carbohydrates, such as flour and oatmeal. Most damage, however, is done by their attacks upon wall paper, card files, book bindings, rayon fabrics, starched clothing, and stocks of paper on which paste or glue has been used as a sizing. The silverfish or slicker,[1] which is uniform, silvery or greenish gray, prefers damp places next to the soil about basement rooms and porches and is less abundant in upper stories of houses. The firebrat,[2] which is mottled with patches of whitish and blackish scales, will not lay eggs at ordinary room temperatures, but at 98 to 102°F. and 70 to 80 per cent humidity it breeds rapidly. It is most abundant in very hot rooms, such as furnace rooms, leading people to suppose that it has been brought to them in deliveries of coal. The whitish oval eggs are deposited loosely in secluded places and may hatch in one to many weeks, depending upon the temperature. The young are whitish miniatures of their parents. They grow slowly, with a large and indefinite number of molts, reaching the adult stage in from 3 to 24 months depending upon temperature and humidity.

Control Measures. These insects may be controlled by spraying with 5 per cent DDT, 2.5 per cent chlordane, 1 per cent lindane, 0.5 per cent dieldrin, or 1 per cent pyrethrins. The organophosphorus insecticides such as 2 per cent malathion, 0.5 per cent Diazinon, 2 per cent ronnel, or 1 per cent dicapthon are also effective. These sprays, or equivalent amounts of dusts, should be applied where the insects are most abundant (see cockroach control, page 908).

Silverfish may also be controlled by the use of poison baits such as the following:

Oatmeal finely ground...........	200 parts by weight or about $\frac{7}{8}$ pt.
Sodium fluoride or white arsenic..	16 parts or about $\frac{1}{4}$ tsp.
Powdered confectioners' sugar....	10 parts or about $\frac{1}{2}$ tsp.
Common salt, powdered........	5 parts or about $\frac{1}{4}$ tsp.

[1] *Lepisma saccharina* Linné, Order Thysanura, Family Lepismatidae.
[2] *Thermobia domestica* (Packard).

If white arsenic is used as the poison, enough water should be added to moisten and bind the ingredients together. When thoroughly dried, the bait should be broken into small particles and scattered lightly in out-of-the-way places where it need not be swept up for a long time, or scattered among loosely crumpled paper in uncovered boxes; or the moist bait may be spread upon squares of cardboard and tacked in the secluded harbors of the pests.

References. Idaho Agr. Exp. Sta. Bul. 185, 1931; *Jour. N.Y. Ento. Soc.*, **41**:557, 1933; *U.S.D.A., Leaflet* 149, 1937.

289. Cockroaches

Importance and Type of Injury. These brown, brownish-black, or tan, shiny, flat-bodied, foul-smelling insects are well known to almost everyone. They are mainly active at night or in dark basements. Their filthy habits, repulsive appearance, bad odor, and the possibility that they may spread diseases, such as tuberculosis, cholera, leprosy, dysentery, and typhoid, make them very objectionable.

Food. They feed on many kinds of material, often becoming annoying in houses by eating the binding or leaves of books or magazines, the paper covering of boxes, various food products in pantries, kitchens, bakeries, restaurants, and like places, and by fouling with their excreta the material over which they run.

Distribution. World-wide; especially abundant in the warmer parts of the world.

Life History, Appearance, and Habits. There are 5 or 6 species of cockroaches which are common in houses and other buildings in the United States.

1. The small tan German cockroach[1] (Fig. 19.6, lower figures), about ½ inch long, with two dark stripes on the upper side of the prothorax, is one of the worst species. The females of this species carry their egg capsules protruding from the abdomen for about 2 weeks until they are nearly ready to hatch. There are commonly 25 to 30 eggs in each capsule, and a female produces 1 to 7 or more capsules during her lifetime. The nymphs pass through seven molts in 6 to 8 weeks, and the total life span is 2 to 5 months, with 2 or 3 generations a year in the average house. This species is most common in kitchens and bathrooms. It is very active, but rarely flies. The last ventral segment is entire in both sexes.

2. The large brown American cockroach[2] (Fig. 19.6, upper figures) reaches a length of 1½ inches or more. The female drops the egg capsule, or glues it in a sheltered place by secretions from her mouth, the day after it is formed, and may produce a capsule once a week until 15 to 90 capsules, averaging about 14 to 16 eggs each, have been formed. The nymphs hatch in 35 to 100 days and require 10 to 16 months and 13 molts before becoming adult, the total life span sometimes being as much as 2½ years. This species often becomes abundant in city dumps and is most common in basements, restaurants, bakeries, packing houses, and groceries. The pronotum has a yellowish posterior border and the last ventral segment in the female is notched.

3. The large black Oriental cockroach[3] is about 1 inch long, uniformly black, the females nearly wingless, and the males with wings much shorter than the abdomen. The last ventral segment in the female is long and notched. The life cycle is about 13 months. The female produces an average of 14 or 15 capsules averaging 12 to 16

[1] *Blattella germanica* (Linné), Order Orthoptera, Family Blattidae (see *Jour. Kans. Ento. Soc.*, **11**:94–96, 1938).
[2] *Periplaneta americana* (Linné) (see *Ann. Ento. Soc. Amer.*, **31**:489, 1938).
[3] *Blatta orientalis* Linné.

eggs each, and the incubation period is about 44 days. This species is most prevalent in damp basements and along sewer lines and is considered the filthiest of all roaches.

4. The Australian cockroach[1] is very similar to the American roach, but only about 1¼ inches long, and has a prominent yellow stripe about one-third the length of the front wing, along the costal margin, and a distinct dark spot at the center of the pronotum. There is an average of 24 eggs per capsule, and the incubation period is 40 days. The life cycle is about 9 months.

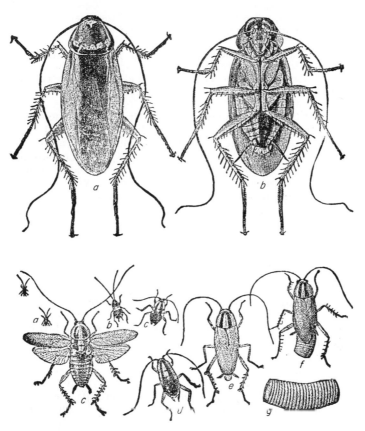

FIG. 19.6. Cockroaches. Above, the American cockroach, *Periplaneta americana: a,* dorsal view; *b,* ventral view. Below, the German cockroach, *Blattella germanica,* adults and young; *f,* female with egg capsule protruding from tip of abdomen; *g,* egg capsule. (*From U.S.D.A.*)

5. The smoky-brown cockroach[2] is a subtropical species that lives in houses and greenhouses in the North, prefers high temperature and humidity, and completes a cycle in about a year. Its egg capsules contain up to 24 eggs each.

6. The brown-banded cockroach[3] has recently spread over many of the Gulf Coast and northern states. It is smaller than the German cockroach, being usually under ½ inch in length. There is a crossband of light yellow at the base of the wings and another about 1⁄16 inch farther back, and the dark stripes on the thorax are much

[1] *Periplaneta australasiae* (Fabricius).
[2] *Periplaneta fuliginosa* (Serville).
[3] *Supella supellectilium* (Serville).

broader than in the German roach. The female is very broad-bodied, the male more slender and lighter-colored. This species seems to prefer living rooms of dwellings and apartments where it hides in cracks in the woodwork or furniture and is particularly hard to control. The female produces an average of 10 capsules over a lifetime of 115 days, and these contain an average of 15 eggs each. The incubation period averages 40 to 58 days, and the seven nymphal instars require an average of 56 to 90 days. There may be two generations a year.

7. The Surinam cockroach[1] is a large dark-brown lubberly insect, about the size of the American cockroach, but with wings only one-third the length of the body which is distinctly oval in outline. They are often common in greenhouses, where they feed upon growing plants.

8. The long-winged males of the wood cockroach,[2] $\frac{2}{3}$ to 1 inch long, may fly for long distances, and are sometimes abundant in houses; and the short-winged females also invade dwellings near woods. The life cycle is about 1 year. The egg capsules contain a maximum of 32 eggs.

The life histories of all species are much the same. The eggs are laid in pod-like or bean-like capsules called *oötheca* (Fig. 19.6,*g*). This pod-like case may be seen protruding from the abdomen of the female (*f*), as she moves about, especially in the case of the German cockroach. The small roaches hatching from the eggs have much the same appearance as the adults, except that they lack wings. They develop rather slowly, requiring 2 to 18 months or more to become full-grown. Cockroaches hide in the cracks of buildings during the daytime, and their abundance is much greater than ordinarily supposed. The household species are active throughout the year in heated buildings. They prefer high temperatures and humidity and are easily killed by cold, although wood roaches live through the winter under the bark of trees and in similar situations.

Control Measures. The first steps in control should be scrupulous cleanliness and guarding against reinfestation from reservoirs of infection, such as neighboring buildings, sewers, covers of wells and cisterns, and stores from which deliveries of groceries, laundry, and furniture are being received. All pipe lines should be tightly sealed where they enter through floors and basement walls, and cracks in which the pests hide should be filled with a crack filler. It would be well to trade only at sanitarily built stores. To eradicate cockroaches from a building the following methods will be found effective: (*a*) Perhaps the most efficient method for controlling cockroaches is to dust or spray protected situations where the insects congregate with chlordane, used as a 2.5 per cent emulsion or oil solution, or dieldrin as a 0.5 per cent oil solution or 1 per cent dust. Races of the German cockroach resistant to these materials are widely distributed and may be controlled with malathion as a 1 to 5 per cent spray or dust, Diazinon as a 0.5 to 1 per cent spray or 5 per cent dust, ronnel as a 2 per cent spray, or dicapthon as a 1 per cent spray. A combination of 1 part malathion and 2 parts Perthane at 4 per cent is also effective. These materials should be applied in dark corners of closets, at the base of the walls in basements, under sinks, around drainpipes, behind baseboards, upon shelves, and in any cracks in the walls, where the cockroaches are likely to hide. Hand sprayers or dusters producing fine streams of insecticide are most desirable for this application. One or two treatments of these materials will usually clean up

[1] *Pycnoscelus surinamensis* (Linné).
[2] *Parcoblatta pennsylvanica* (De Geer).

even severe infestations. (*b*) Other treatments which have been widely used but are generally less efficient include dusting with sodium fluoride or sodium fluosilicate, or with 1 per cent pyrethrins, or spraying with a household formulation containing pyrethrins and a residual insecticide. Care should be taken in applying any of these materials not to contaminate foods. (*c*) Commercial pastes of yellow phosphorus are one of the best methods of controlling the American, oriental, and Surinam cockroaches. These usually contain about 2 per cent phosphorus colloidally dispersed in sirup or other attractant. The paste may be spread upon cardboard, rolled into cylinders with the poison on the inside, and placed or tacked in out-of-the-way places. This material is poisonous and should be kept away from children and pets. A bait consisting of 10 per cent boric acid and 90 per cent powdered sugar has proved effective in controlling the German cockroach. These measures are satisfactory for use when the cockroaches are not too numerous and to supplement other control measures. (*d*) Fumigation of the house with hydrogen cyanide may be used to give immediate cleanup of heavy infestations. Even though all roaches have been cleaned out of a house, the premises will not remain roach-free long if other houses in the immediate vicinity are infested, as these insects move about very freely and actual migrations of the insects from house to house have been noted.

References. *U.S.D.A., Leaflet* 144, 1937; *Proc. Ind. Acad. Sci.*, **47**:281–284, 1938; *Purdue Univ. Agr. Exp. Sta. Bul.* 451, 1943; *Jour. Econ. Ento.*, **33**:193, 1940; **38**:407, 1945; **41**:516, 652, 1948; **46**:1059, 1953; **48**:572, 1955; and **51**:608, 1958; *Ann. Ento. Soc. Amer.*, **51**:53, 1958.

290. HOUSE CENTIPEDE[1]

This creature, which is frequently encountered about houses, is really beneficial, causing no injury whatever to stored products. It becomes annoying to some people because of its appearance and habit of rushing over the walls or floors, occasionally toward a person, when it is suddenly disturbed by the turning on of lights or by other causes. There are a few cases on record where the centipede has inflicted a painful bite when handled.

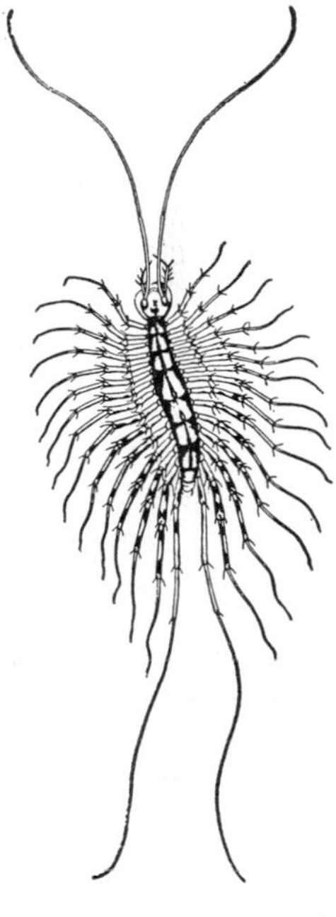

FIG. 19.7. House centipede, *Scutigera cleoptrata*, adult. Natural size. (*From U.S.D.A., Bur. Ento. Cir.* 48.)

In appearance these creatures are of a grayish-tan color, with long antennae and many extremely long legs extending out all around the body (Fig. 19.7). They move so rapidly that it is often difficult to get an accurate idea of their appearance. When full-grown, they are from 2 to 3 inches in length. Their food consists of small insects, such as roaches, clothes moths, house flies, and others which they may encounter about houses.

Unless they become extremely abundant, they should not be killed. In cases where they do become so numerous as to prove annoying, the liberal use of pyrethrum or DDT powder about water pipes, or other damper parts of the house frequented by the centipedes, will be found effective in controlling them. They are killed by fumigation, with either hydrogen cyanide or carbon bisulfide, but it is rarely, if ever, necessary to go to this expense for the control of these creatures.

[1] *Scutigera cleoptrata* (Linné), Class Chilopoda, Family Scutigeridae.

291. Cigarette Beetle or Towbug[1]

Importance and Type of Injury. This is the most important insect pest of tobacco in factories and cigar stores and also causes considerable damage to other products. Package and chewing tobaccos, cigars, and cigarettes have holes eaten through the tobacco, and contain many small, light-colored, brown-headed grubs or brown beetles working in the tobacco. Furniture upholstered in flax, tow, or straw is often grossly infested by these beetles and sometimes has holes eaten through the covers by the larvae or beetles.

Food. Tobacco products, especially cigarette tobaccos with high sugar content, and dried leaves; upholstered furniture, materials used in stuffing furniture, seeds and other dried plant products, especially those used as drugs, black and red pepper, and many others.

Distribution. Throughout the United States and southern Canada.

Life History, Appearance, and Habits. The adult beetle is rounded in outline, of a very light-brown color, and only about $\frac{1}{16}$ to $\frac{1}{10}$ inch in length, with the head and prothorax bent downward so as to give the insect a strongly humped appearance. The wing covers are not striated,

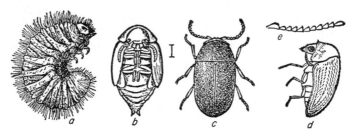

Fig. 19.8. Cigarette beetle, *Lasioderma serricorne*. *a*, Larva; *b*, pupa; *c*, adult, dorsal view; *d*, adult, side view; *e*, antenna. About 6 times natural size. Line indicates natural size. *(From U.S.D.A., Farmers' Bul. 846.)*

and the antennae are of the same thickness from base to tip (Fig.19.8,*c–e*). The female beetle produces about 100 oval whitish eggs, $\frac{1}{50}$ inch long. These are laid in and about the substances on which the insects feed and hatch in 6 to 10 days at summer temperatures. The larvae are yellowish-white, curved-bodied, very hairy little grubs, with light-brown heads. They may become full-grown in 30 to 50 days when they are nearly $\frac{1}{6}$ inch long. They pupate for 8 to 10 days or more in silken cocoons covered with bits of their food material. The entire life cycle may be passed in from 45 to 50 days, and there are commonly three to six generations a year.

Control Measures. Infested tobacco factories, warehouses, or other premises should be thoroughly cleaned before a new crop is stored, removing all refuse material on which the beetles may be feeding and treating the cracks in walls and floors with a strong disinfectant. Infested tobacco in closed storage warehouses may be freed of the insects by fumigating with 16 to 32 ounces of liquid hydrogen cyanide per 1,000 cubic feet. This should be applied just after the peak of adult abundance of any generation has passed and the lowest level has almost been reached, as shown by the catch at light traps, so as to avoid most of the large larvae

[1] *Lasioderma serricorne* (Fabricius), Order Coleoptera, Family Anobiidae.

and pupae which are hardest to kill and likely to be deepest in the tobacco. Kearns suggested that the given dosage be divided and applied in four "shots" at 2-hour intervals, for best results. Fumigation with a 50:50 mixture of acrylonitrile and carbon tetrachloride used at 20 ounces to 1,000 cubic feet has given satisfactory results. Ethylene oxide and carbon dioxide (page 385) and other industrial fumigants are used for imported and manufactured, packaged tobaccos in vacuum vaults to insure freedom from this pest. Space-spraying infested warehouses with 5 per cent DDT or 1 per cent pyrethrins in deodorized kerosene at 100 milliliters to 1,000 cubic feet is effective in controlling the beetles. Warehouses freed of the insects should have all openings covered with 24-mesh screens to prevent reinfestation by the entrance of adults from infested surroundings. Suction light traps should be operated throughout the season in warehouses, beginning in spring when the temperature reaches 65°F., one trap to each 75,000 cubic feet of space. All other lights must be kept off. Heating to 130 to 135°F. is very effective. The heat must be applied long enough to penetrate all parts of the infested materials, and the temperature should be maintained for at least 6 hours. All stages of the insect can be killed in tobacco products stored for a week at 25°F. or 16 days at 36°F., but a week or two more may be required for the cold to penetrate to the center of tobacco hogsheads. In cooler areas this may be accomplished by opening doors and ventilators during the winter. The control of this insect as a furniture pest is much the same as for carpet beetles.

References. U.S.D.A., Dept. Bul. 737, 1919, and *Farmers' Bul.* 846, 1917, and *Cirs.* 356, 1925, and 462, 1938; *Furniture Manufacturer,* **34**:3–7, 1927; *Ill. Agr. Exp. Sta. Cir.* 473, 1938; *U.S.D.A., Bur. Ento. Plant Quar., Cir.* E–717, 1947; *Jour. Econ. Ento.,* **37**:147, 1944, **38**:449, 1945, **41**:13, 1948; and **51**:471, 1958.

292. Drug Store Beetle[1]

This insect closely resembles the cigarette beetle in its life history and habits. Its food is even more varied than that of the cigarette beetle and includes practically all dry plant and animal products. The adult is about $\frac{1}{10}$ inch long, of a reddish-brown color, and densely covered with very short light hairs. The wing covers are plainly striated, and the antennae are enlarged at the end (Fig. 19.9,*c–e*). The larva (*a*) differs from the cigarette beetle in being nearly bare.

Control Measures. The same as for the cigarette beetle.

293. Tobacco Moth[2]

Importance and Type of Injury. Since about 1930 this relative of the Mediterranean flour moth has become an alarming pest of stored tobacco, especially the brighter grades of flue-cured tobacco. The larvae eat the tender parts of leaves, making holes much larger than the cigarette beetle does, and pollute what they do not eat, with their silk and excrement.

Food. Cured tobacco, nuts, chocolate, and many other dried vegetable products.

Life History, Appearance, and Habits. The elliptical, granular, grayish-white to brownish eggs are laid singly or in small clusters, usually upon the tobacco. They hatch in 4 to 7 days. The larvae are whitish, tinged with yellow, brown, or pink and peppered with small brownish, scattered spots. They measure $\frac{3}{8}$ to $\frac{5}{8}$ inch long, when full-grown,

[1] *Stegobium* (= *Sitodrepa*) *paniceum* (Linné), Order Coleoptera, Family Anobiidae.
[2] *Ephestia elutella* (Hübner), Order Lepidoptera, Family Pyralididae.

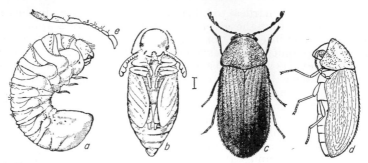

Fig. 19.9. Drug store beetle, *Stegobium paniceum*. *a*, Larva; *b*, pupa; *c*, adult, dorsal view; *d*, adult, side view; *e*, antenna. Enlarged about 12 times. Line indicates natural size. (*From U.S.D.A., Bur. Ento. Bul. 4.*)

and are very active. The larval or feeding period lasts 40 to 45 days. They pupate in almost any sheltering crevice, commonly for 7 to 19 days. The adults are grayish to brownish moths, ½ inch from tip to tip of the wings. The middle half of the front wings is lighter in color than the base or tip and is separated from these darker extremities by a narrow whitish crossband. They are most active at night.

Control Measures. The controls suggested for the cigarette beetle are effective for this species also.

References. U.S.D.A., Circs. 269, 1933, E-325, 1936, E-422, 1937, E-717, 1947; *Tobacco,* **105**:21, 22, 1937; *Jour. Econ. Ento.,* **37**:147, 1944; **38**:449, 1945; and **41**:13, 1948.

294. Carpet Beetles or Buffalo Beetles[1-3]

Importance and Type of Injury. Holes are eaten in fabrics, such as clothing, rugs, the upholstery covers and interior padding of furniture, curtains, especially those containing wool, fur, feathers, or hair, and brushes made of animal bristles, by very hairy or bristly, brownish larvae of all sizes up to more than ¼ inch long (Fig. 19.10). Some of the species[2] have a long tail of black hairs (Fig. 19.11). There is no webbing together of the fabric nor silken threads spun over the surface as in the case with clothes moths. These larvae live a large part of the time in secluded places about the rooms, out of reach of house-cleaning operations, so that they are particularly insidious and are responsible for very many holes, ravelings, and defects in clothing and other fabrics, which are not even suspected of being insect injury. Small, blackish, hard-shelled beetles, sometimes flecked with white or reddish scales, are often seen in the infested materials or about windows. Most severe injury occurs in materials that have lain undisturbed for some time.

Food. Woolen goods of all kinds; sometimes cotton goods; leather, bristles, feathers, hair, silk; dried meat, milk or insect specimens; stuffed animals, fur; grains, flour, and many other animal and plant products.

Distribution. Two of the most destructive species of carpet beetles have been imported from Europe, these being the common carpet beetle[1] and the black carpet beetle.[2] These and several other species[3] occur generally in this country.

---

[1] *Anthrenus scrophulariae* (Linné), Order Coleoptera, Family Dermestidae.

[2] *Attagenus piceus* (Olivier), Order Coleoptera, Family Dermestidae.

[3] *Anthrenus flavipes* (= *vorax*) LeConte and *Anthrenus verbasci* (Linné).

Life History, Appearance, and Habits. Although breeding may take place throughout the year in uniformly heated buildings, most carpet beetles winter as larvae. They pupate from April to June, for 1 to 4 weeks, as a white soft stage inside the split larval skin (Fig. 19.10,*b*) but without the protection of a cocoon, and the adults are most abundant during hot weather from early May to mid-June. After molting to the

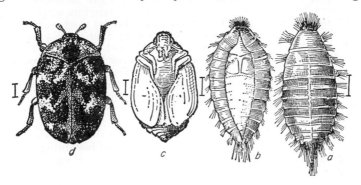

FIG. 19.10. The common carpet beetle, *Anthrenus scrophulariae.* *a*, Larva; *b*, pupa within larval skin; *c*, pupa removed from larval skin, ventral view; *d*, adult. All enlarged about 7 times. Lines indicate natural size. (*From Riley, U.S.D.A.*)

adult stage, the insects may further remain within the larval skin from a few days to a few weeks. The adults may live from 2 weeks to several months but never damage household goods in this stage. They are active, attracted to light, and are often found about windows and out-of-doors upon flowers, especially spiraea, eating the pollen. There is some evidence, however, that the females usually lay their eggs before they leave the building in which they developed. They lay their soft white eggs upon the food materials of the larvae or in dark secluded places, as upon clothing, the angles of upholstered furniture, cracks about baseboards, in warm-air furnace pipes, and other places where dust and lint accumulate and are undisturbed. The eggs hatch in a week or two and the young larvae immediately attack such food as they can find. Unlike their parents, they avoid the light and feed voraciously

FIG. 19.11. The black carpet beetle, *Atta-genus piceus*, larva. About 4 times natural size. (*From Herrick, "Insects Injurious to the Household," copyright, 1914, Macmillan. Reprinted by permission.*)

in dark secluded crevices or folds of their food. Unlike the clothes moths they may crawl away from the food material to molt in cracks and crevices inaccessible to insecticides. They molt from 5 to 11 times in the course of their growth, and the cast skins closely resemble, and often are mistaken for, living larvae (*a*). It requires from less than 1 year to as much as 3 years for the larva to complete its growth, the length of time depending on climatic conditions and food.

The black carpet beetle is believed to be the most important pest of fabrics in storage in many parts of the United States. The adults are uniformly dull black, with legs and antennae brownish, elliptical in outline, $\frac{1}{8}$ to $\frac{5}{32}$ inch long by half as

wide.   Females produce an average of about 50 eggs which hatch in a week or 10 days.
The larvae of this species, which may reach a length of nearly ½ inch, are very different from the others, being elongate, carrot-shaped, golden to chocolate brown, and with
a tuft of very long brown hairs at the tail end of the body.

The larvae of all the other species are short, rarely over ¼ inch long, chubby, and
covered with erect brown or black bristles all over the body and with three tufts of
bristles at each side of the posterior third of the body which lie closely against the
body like folded wings.   The carpet beetle[1] and the furniture carpet beetle[2] are
covered with black bristles while the varied carpet beetle[3] has tawny or brownish
bristles.   The adults of the common carpet beetle are elliptical, $\frac{5}{32}$ inch long and
three-fourths longer than wide, with a band of brick-red scales down the center of
the back and red-and-white scales forming several irregular bands across the body
(Fig. 19.10).   The furniture carpet beetle and the varied carpet beetle are scarcely a
third longer than broad, are irregularly mottled with flecks of white, yellow, and black,
and are white on the underside.   The furniture carpet beetle is very destructive to
upholstery of hairs and feathers.

Control Measures.   The control of these pests is of two distinct phases:
(*a*) measures which will tend to prevent them from becoming established
in the home and (*b*) measures necessary to eradicate them after damage
has begun.   Good construction which does not provide hiding places for
the insects in cracks of floors, baseboards, and quarter round, or the elimination of such crevices with a good crack filler, is an important step in
prevention.   Good housekeeping methods, which promptly eliminate
dust and lint, do not permit floor coverings, draperies, blankets, clothing,
and the like to remain long undisturbed but subject them to frequent
vacuum sweeping and at least semiannual sunning, beating, and brushing, give the insects little opportunity to establish themselves.   The
interior of furnace pipes and pianos is especially likely to be neglected.
Once the insects are found, the nature and extent of the infestation must
determine the wisest procedure.   If the infestation is widespread and
intense, thorough fumigation of the entire house or of infested rooms
with hydrogen cyanide or carbon bisulfide or by superheating, as described
on pages 375 and 405, is the best and quickest method of stamping them
out.   If this is not practicable, the next best method is the use of DDT,
chlordane, lindane, dieldrin, or malathion household spray, applied to
cracks in floors and baseboards with a power atomizer or hand sprayer
(page 908).   Individual pieces of badly infested furniture should be sent
to companies equipped to do vacuum fumigation or treated with one of
the moth sprays applied with nozzles in the form of perforated needles,
for reaching the interior of the upholstery.   There are a number of good
mothproofing materials which may be applied by the manufacturers.
Other processes are widely available through dry-cleaning companies,
and solutions of DDT, dieldrin, sodium fluosilicate, sodium aluminum
silicofluoride, and other poisons may be applied by the housekeeper herself as discussed on page 394.   Many companies offer the facilities of
cold storage for furs and other valuable winter clothing during the summer.   Even though present, the insects cannot injure the goods at temperatures below 40°F.   Articles not to be used for some time may best
be stored in very tight trunks or boxes in the following manner: At various levels among the clothing or blankets place flake naphthalene or

[1] *Anthrenus scrophulariae* (Linné), Order Coleoptera, Family Dermestidae.
[2] *Anthrenus flavipes* (= *vorax*) LeConte.
[3] *Anthrenus verbasci* (Linné).

para-dichlorobenzene between thin sheets of paper, using a pound for 20 to 100 cubic feet. On top of the clothing place a shallow pan or dish and pour carbon bisulfide upon it, in a warm place away from any flame, using at least 1 pound to 100 cubic feet of space. Quickly close the box and seal tightly with strips of gummed paper. The carbon bisulfide will destroy any stages of the insects that may be present, and the other chemical will serve as a repellent against infestation for months.

Cedar chests will not kill these insects. Heating to 135°F. for 6 hours will kill all stages of the insects, but care must be taken in using this method that the desired temperature is maintained in the center of the articles treated (page 405).

References. U.S.D.A., Div. Ento. Buls. 4, 1896, and 8, 1897, and *Cir.* E-395, 1936, and *Leaflet* 150, 1938; *Ill. Agr. Exp. Sta. Cir.* 473, 1938; *Jour. Econ. Ento.*, **31**:280, 1938; and **42**:127, 1949.

295. Clothes Moths[1-3]

Importance and Type of Injury. There are several species of clothes moths that are responsible for damage in this country, the most common being the casemaking clothes moth[1] and the webbing clothes moth.[2] The carpet moth[3] is much less common in this country. Fabrics injured by clothes moths have holes eaten through them by small, white caterpillars, and in most cases the presence of the insect is indicated by silken cases or lines of silken threads over the surface of the materials. Materials left undisturbed for some time or stored in dark places are most severely injured by these insects. Small, buff-colored moths, not over $\frac{1}{2}$ inch across the wings, will be found running over the surface of infested goods when such goods are exposed to light, or flying somewhat aimlessly about in houses or closets. The clothes moths are not attracted to lights.

Food. The clothes moth larvae feed on wool, hair, feathers, furs, upholstered furniture, occasionally on dead insects, dry dead animals, animal and fish meals, milk powders such as casein, and nearly all animal products, such as bristles, dried hair, and leather. The adults take no food.

Distribution. Clothes moths are distributed generally over the world.

Life History, Appearance, and Habits. The adult "millers" or moths (Fig. 19.12,*a*) are entirely harmless and probably take no food of any sort. They lay their eggs singly on the products in which the larvae feed, each female laying from 100 to 150 eggs. Occasionally, with some species, the eggs may be laid in groups. They are small, being about one-tenth the size of a pinhead ($\frac{1}{50}$ inch long), of a white color, which makes them rather conspicuous when deposited on black material. They hatch in about 5 days. The larvae (Fig. 19.12,*b,c*) which hatch from these eggs are the only stage of the insect causing damage. They are white, and vary in size from about $\frac{1}{16}$ inch long, when first hatched, up to about $\frac{1}{3}$ inch when full-grown. The length of the larval period varies greatly according to the conditions and food supply. The complete development of this stage may take from 6 weeks to nearly 4 years. Development is greatly influenced by humidity, the life cycle being shortest, in average room temperature, at about 75 per cent relative humidity. The larvae of some

[1] *Tinea pellionella* (Linné), Order Lepidoptera, Family Tineidae.
[2] *Tineola bisselliella* (Hummel), Order Lepidoptera, Family Tineidae.
[3] *Trichophaga tapetzella* (Linné), Order Lepidoptera, Family Tineidae.

species live in silken cases (*b*) which are dragged about with them and enlarged as they grow. Upon completing its growth, the larva adds to the silk of the case in which it has been living, forming a rather tough cocoon. Within this case (*d*) it changes to a white pupa, about ⅙ inch in length. The pupa later turns brown and in 1 to 4 weeks the adult moth emerges. The complete life cycle may be passed in about 2 months or may be prolonged over 4 years, including long periods of larval inactivity or dormancy. In heated buildings, the adults may be found at any time of the year but are most abundant during the summer months.

Fig. 19.12. Casemaking clothes moth, *Tinea pellionella*. *a*, Adult; *b*, larva in case; *c*, larva removed from its case; *d*, cocoon enclosing pupa. About 7 times natural size. (*From Ohio Agr. Exp. Sta. Bul.* 253.)

Control Measures. Control measures are essentially the same as for the carpet beetles. Frequent dusting, brushing, and vacuum cleaning are very important control measures and should be extended to the most remote cracks, warm and cold air passages, and similar hiding places. Furniture upholstered in leather, silk, rayon, linen, or cotton will not be damaged by clothes moths. Layers of cotton will not be damaged by clothes moths. Layers of cotton batting properly installed beneath the covers of furniture or furniture springs will nearly always prevent moths from attacking mohair or woolen upholstery from the inside. Vault fumigation and mothproofing (page 393) are highly effective, when properly done, and furniture should always be so treated before it is delivered to the buyer. When houses are to be vacant for several weeks, as during a vacation period, a simple method of fumigation consists in closing rooms very tightly, spreading papers over the floor, and scattering *para*-dichlorobenzene crystals over the papers at the rate of 8 to 10 pounds to each 1,000 cubic feet. Clothing which is in daily use is practically never infested by the clothes moths. It is highly important that clothing, or other fabrics placed in storage, should be free from moths.

To insure this, such clothing should be thoroughly brushed and shaken and hung out-of-doors in the bright sun for several hours before being packed away. If placed in trunks or boxes, clothing may be protected from infestation as described for carpet beetles.

Tight-fitting cedar chests will protect clothing from attacks of the clothes moths and will kill larvae hatching from the eggs within the chest. However, partly grown larvae placed in such chests may complete their development and cause considerable damage. Heating and fumigating as suggested for the carpet beetle will kill all stages of the clothes moths. No damage to clothing will occur if kept in cold storage at 45°F., or lower.

References. U.S.D.A., Farmers' Buls. 1353, 1923, and 1655, 1931, and *Dept. Bul.* 1051, 1922; *Furniture Manufacturer,* vol. 35, nos. 1 (n.s.) and 5, 1928; *Mon. Rev. Nat. Retail Furniture Assoc.,* vol. 3, nos. 6 and 8, 1929.

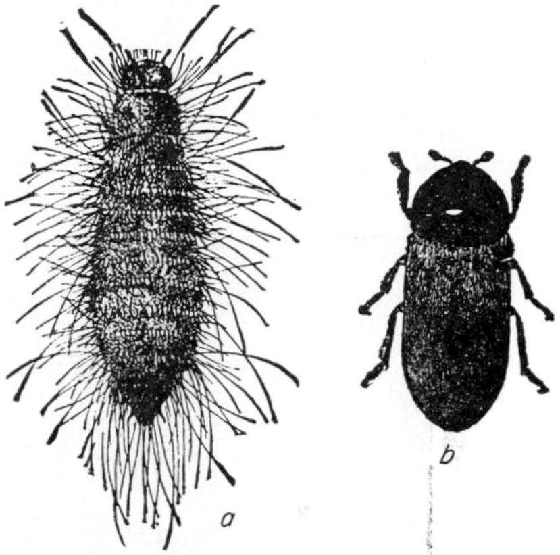

296. LARDER BEETLE[1]

Importance and Type of Injury. Very hairy, brown larvae (Fig. 19.13), tapering toward both ends of the body, feed on meats and animal products of nearly all kinds.

FIG. 19.13. The larder beetle, *Dermestes lardarius.* a, Larva; b, adult. About 4 times natural size. (*From Herrick, "Insects Injurious to the Household,"* copyright, 1914, *Macmillan. Reprinted by permission.*)

Food. Feathers, horn, skins, hair, beeswax, ham, bacon, dried beef, and like products.

Distribution. World-wide.

Life History, Appearance, and Habits. The adult beetles are about ⅓ inch long, of a very dark-brown color, and with a moderately wide yellowish band across the front part of the wing covers. There are six black dots in this band, three on each wing cover, usually arranged in a triangle (Fig. 19.13). The eggs are laid on the food or in sheltered places near by. The larvae, on hatching, increase rapidly in size. They feed chiefly near the surface of the infested materials and become full-grown (a little over ⅓ inch long) in 40 to 50 days. Pupation takes place in the larval skin. The exact number of generations and some details regarding the length of the stages are not known. The adult beetles will occasionally be found in numbers out-of-doors on flowers, where they feed on pollen.

Control Measures. Smoked meats kept in farm storehouses should be carefully sacked or wrapped in paper, muslin, or other cloth immediately after smoking, care being taken that the entire piece of meat is covered and no openings or cracks left through which the beetles may gain access to the meat. Smoked meats held in cold storage will not become infested. If meat storerooms or larders become infested, they should be heated to 130 to 135°F. for 3 hours, or fumigated with carbon bisulfide or hydrogen cyanide gas. All meat in the storeroom should be removed before treatment is applied, and the infested parts of such meat should be carefully trimmed off and burned. Skins and hides may be protected from damage by these beetles by dusting with 10 per cent DDT or 1 per cent lindane or by spraying with these materials.

[1] *Dermestes lardarius* Linné, Order Coleoptera, Family Dermestidae. The hide beetle, *D. maculatus* De Geer, and *D. cadaverinus* Fabricius are very similar in appearance and habits.

References. U.S.D.A., *Div. Ento. Bul.* 4 (n.s.), p. 107, 1902; *Can. Ento.*, **38**:68, 1910; N.Y. (*Geneva*) *Agr. Exp. Sta. Bul.* 202, 1931; *Jour. Econ. Ento.*, **39**:283, 1946.

297. RED-LEGGED HAM BEETLE[1]

Importance and Type of Injury. This is the most destructive pest attacking well-cured and dried smoked meats, occasionally ruining large quantities of hams and bacons. The larvae do most of the damage by burrowing through meat, especially in fatty portions; but larvae and adults both feed, and are found, in the infested products.

Food. Cured pork is most injured, but the larvae or adults feed also in fish, guano, bone meal, cheese, dried egg, hides, copra, dried fruits and nuts, other insects, and insect eggs including their own kind.

Distribution. Nearly cosmopolitan, especially troublesome in the Middle Atlantic states.

Fɪɢ. 19.14. The cheese skipper, *Piophila casei;* eggs, larvae or "skippers," pupae, and adult flies. About 1½ times natural size. (*From E. O. Essig, Univ. of California.*)

Life History, Appearance, and Habits. Winter in unheated buildings is probably passed chiefly as large larvae, the adults becoming evident in May and June. The adults are brilliant greenish-blue, convex, straight-sided, noticeably "punctured" beetles, about ⅐ to ¼ inch long, and less than half as wide, with the legs and base of antennae reddish-brown. The adults disperse mostly by rapid running, but they also fly. The female deposits from 400 to 2,000 elongate, cylindrical, translucent eggs in clusters in dry recesses of the food substances, and these hatch in 4 or 5 days in warm weather. The larvae are elongate, slender, tapering toward the head, purplish in color, with six short legs and short hairs; about ⅖ inch long when full-grown. They are strongly repelled by light. They migrate from greasy foods to dry dark spots to pupate, lining a crevice with silk. The entire life cycle occupies from 36 to 150 days or longer, depending upon the food. About two-thirds of that time is spent in the growing, larval stage.

Control Measures. Very thorough and prompt wrapping of hams, shoulders, and bacons is effective in preventing attack. Screening with wire cloth, 30 meshes per

[1] *Necrobia rufipes* (De Geer), Order Coleoptera, Family Cleridae. Also called copra bug.

inch, will exclude the adults from storage rooms. Fumigation of badly infested premises with hydrogen cyanide is advised.

Reference. Jour. Agr. Res., **30**:845, 1925.

298. CHEESE SKIPPER[1]

Importance and Type of Injury. Small, naked, yellowish maggots crawling in and over cheese or meat (Fig. 19.14), or jumping for short distances by bending their bodies nearly double and then suddenly straightening them.

Food. Cheese and smoked, cured meats.

Distribution. World-wide; probably imported into this country.

Life History, Appearance, and Habits. The adult insects are small, rather shiny two-winged flies, a little less than ⅙ inch in length (Fig. 19.14). They lay as many as 500 eggs singly or in clusters up to about 50. The eggs hatch in 24 to 36 hours into small fleshy maggots which are legless and taper toward the head end. They are yellowish-white in color, and about ⅓ inch long when full-grown. These larvae may complete their growth in from 1 week to several months. They then change to a very light-brown puparium and remain in this stage for about 1 week to 10 days. The entire life cycle averages about 18 days.

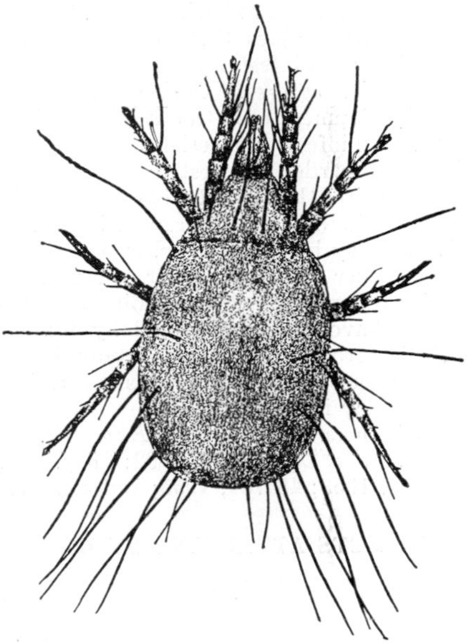

Control Measures. Infested portions of cheese or ham should be cut away and burned. Where cheese is stored, care should be taken that the grubs are kept out of the storeroom by tightly closing all openings or by using 30-mesh screen or cloth to protect windows open for ventilation. Frequent examinations should be made of cheese in storage to see that no infestation has started. Bandages or coatings of paraffin used for covering cheeses should be kept as firm and smooth over the cheese as possible. Where serious infestations have started in storerooms, they should be fumigated with hydrogen cyanide gas, at 20 ounces per 1,000 cubic feet, or thoroughly cleaned out and the walls and entire room washed with very strong soapsuds, using the soap at the rate of 3 to 4 ounces to the gallon of water. The water should be as hot as possible while the washing is done. Synergized pyrethrins aerosols will control the adults. Heating to 125°F. for 1 hour will kill the larvae and pupae. Meats and cheese held in storage at temperatures below 43°F. will not be infested by the cheese skipper.

FIG. 19.15. Cheese or ham mite, *Tyroglyphus* sp. Greatly magnified. (*From Minn. Agr. Exp. Sta. Bul.* 198.)

References. U.S.D.A., Dept. Bul. 1453, 1927; *Calif. Agr. Exp. Sta. Bul.* 343, 1922; *Jour. Exp. Biol.,* **12**:384, 1935; *Jour. Econ. Ento.,* **35**:289, 1942.

299 (365). CHEESE, HAM, AND GRAIN MITES[2]

There are a number of species of mites that infest ham, cheese, flour, stored grains, and other food products. In mushroom houses they often eat holes in the caps and

---

[1] *Piophila casei* (Linné), Order Diptera, Family Piophilidae.

[2] The grain mite, *Acarus siro* Linné (= *Tyroglyphus farinae* De Geer), the cheese mite, *Tyrophagus castellanii* Hirst (= *Tyroglyphus longior* Gervais), the mushroom mite, *T. lintneri* (Osborn), and others, Order Acarina, Family Tyroglyphidae.

stems or consume the spawn.   They are all whitish in color and so small as to be barely discernible to the naked eye (Fig. 19.15).   They vary in size, the largest being only about $\frac{1}{32}$ inch long.   When they are abundant, a quantity of fine brownish powder accumulates over and through the materials, and a musty, sweetish odor is given off by the infested products, which is quite characteristic.   Occasionally a rather intense irritation of the skin is caused by the mites where one has been handling food products infested by them.   This has received the descriptive name of grocer's itch (page 1025).

The mites reproduce by means of eggs, which are laid promiscuously over the food materials.   The young mites grow rapidly, being six-legged at first, but eight-legged on becoming adults.   When conditions are unfavorable, the mites pass into a very active, nonfeeding, but very resistant stage, known as the *hypopus*.   In this stage the body wall hardens and suckers are formed on the underside with which they may attach to other insects or mice and be carried to new locations.   The mites may remain in this condition for a number of months without food, but, when conditions become favorable to their growth, they molt and again become active.   This peculiar adaptation enables them to survive for considerable periods, and for this reason premises once infested are often difficult to clean up.

Control Measures.   Curing cheeses slowly at temperatures of 30 to 36°F. prevents the development of both mites and skippers.   Complete coating with paraffin is effective.   All infested material should be carefully gone over and that which shows serious injury discarded.   Infested parts of hams or cheeses should be cut off and burned. As the mites thrive best where the humidity is rather high, keeping storerooms as dry as possible will tend to prevent infestation.   It has been found that the mites cannot live for any length of time, except in the hypopus stage, where the humidity is less than 11 per cent.   However, when mites are working on food, they give off moisture which often raises the humidity to a point favorable to their development.   Storerooms have been disinfested by fumigation with sulfur at 2 to 3 pounds or with β,β-dichlorodiethyl ether at 1 pound per 1,000 cubic feet.   The most practical treatment to use around foodstuffs is a residual spray of synergized pyrethrins applied at 3- to 4-day intervals.

References.   *Minn. Agr. Exp. Sta. Bul.* 198, 1921; *U.S.D.A., Div. Ento. Bul.* 4 (n.s.), *Insect Life,* **4** :170, 1892; *Calif. Agr. Exp. Sta. Bul.* 343, 1922; *Jour. Econ. Ento.,* **46** :844, 1953; **48** :754, 1955; and **52** :237, 514, 1959.

B. INSECTS ATTACKING STORED GRAIN, SEEDS, AND GRAIN PRODUCTS

300. GRANARY WEEVIL[1] AND RICE WEEVIL[2]

Importance and Type of Injury.   These two true weevils (Fig. 19.16) are perhaps the most destructive grain insects in the world.   They frequently cause almost complete destruction of grain in elevators, in farmers' bins, or on ships where conditions are favorable to their growth and the grain is undisturbed for some length of time.   Infested grain will usually be found to be heating at the surface, and it may be damp, sometimes to such an extent that sprouting occurs.   Many small brown beetles, with distinct snouts projecting from their heads, will be found working over and in the grain.   The kernels of grain are eaten out or contain small, legless, fat-bodied, white grubs feeding on the interior. South of about 39°N. latitude, corn may be attacked by the rice weevil in the field, especially following damage by the corn earworm.   Thence the pest is sure to be carried into storage.

Food.   Wheat, corn, macaroni, oats, barley, sorghum, kaffir seed, buckwheat, and other grain and grain products.   Adults feed upon whole seeds of flours, but the larvae develop only in seeds or pieces of seeds or cereal

[1] *Sitophilus granarius* (Linné), Order Coleoptera, Family Curculionidae.
[2] *Sitophilus oryza* (Linné), Order Coleoptera, Family Curculionidae.

products large enough for them to live within and not in flours unless they have become caked.

Distribution. The granary weevil is probably not native to North America. The rice weevil is supposedly a native of India. Both species have been distributed all over the world in shipments of grain and are very frequently found together. The granary weevil is less prevalent, at least in tropical and semitropical climates.

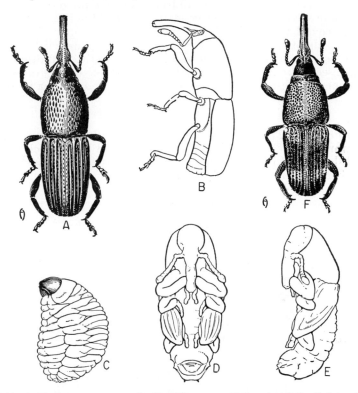

Fig. 19.16 *A–E,* the granary weevil, *Sitophilus granarius. A,* adult; *B,* lateral view of adult; *C,* larva; *D,* ventral view of pupa; *E,* lateral view of pupa; *F,* adult rice weevil, *S. oryza.* All about 13.5 times natural size. (*From Calif. Agr. Exp. Sta. Bul.* 676.)

Life History, Appearance, and Habits. In unheated storages winter may be passed as larvae or adults, the latter stage surviving zero weather for several hours. The adult granary weevil is a somewhat cylindrical beetle, about ⅙ inch in length. It is dark brown or nearly black in color, with ridged wing-covers, and a prolonged snout extending downward from the front of the head for a distance of about one-fourth the length of the body. The rice weevil has much the same general appearance, although it is somewhat smaller. There is usually a patch of somewhat lighter, yellowish color on the front and back of each wing cover. A distinguishing mark is in the shape of the small shallow pits on the prothorax of the beetles: in the rice weevil these are round (Fig. 19.16,*F*), in the granary weevil (*A*) they are oval.

The female weevil chews slight cavities in the kernels of grain or in other foods and there deposits from 300 to 400 small white eggs, one in a cavity, sealing it in with a plug of gluey secretion. The eggs hatch in a few days into soft, white, legless, fleshy grubs which feed on the interior of the grain, hollowing it out; on becoming full-grown, they are about ⅛ inch in length. They change to naked white pupae and later emerge as adult beetles. The entire life cycle may be passed under favorable conditions in from 4 to 7 weeks. Adults can withstand starvation for 2 or 3 weeks, often live 7 or 8 months, and may survive for over 2 years. In Kansas there are from four to five generations of the insect each season. The rice weevil has well developed wings, and frequently flies, especially during periods of high temperature. The granary weevil has the wing covers somewhat grown together and is unable to fly. In other respects the two insects closely resemble each other in their life histories and will very frequently be found associated and working together in the same bins.

301. MEALWORMS[1]

Importance and Type of Injury. Yellowish to brown, smooth, shiny-bodied worms, closely resembling wireworms, and large black beetles up to 1 inch long, feed in and around grain bins, particularly in dark, moist places where the grain has not been disturbed for some time.

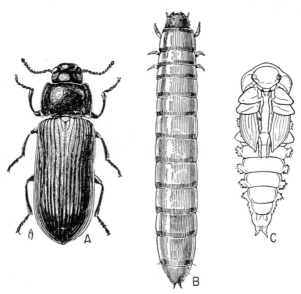

FIG. 19.17. The yellow mealworm, *Tenebrio molitor.* *A*, adult; *B*, larva; and *C*, pupa. About 4 times natural size. (*From Calif. Agr. Exp. Sta. Bul.* 676.)

Food. Cereals and cereal products, including meal, bran, some other food products, meat scraps, feathers, and dead insects.

Distribution. The insects are natives of Europe. They are now distributed throughout the world. The yellow mealworm is less common in the southern states.

[1] *Tenebrio molitor* Linné, the yellow mealworm, and *T. obscurus* Fabricius, the dark mealworm, Order Coleoptera, Family Tenebrionidae.

Life History, Appearance, and Habits. The adults of the two meal-worms rather closely resemble each other. They are black to nearly black beetles, robust, flattened, somewhat shining, from ½ to nearly 1 inch in length (Fig. 19.17,*A*). The female deposits her whitish oval eggs, to the number of 250 to 1,000, singly or in clusters in the food materials. The eggs hatch in from 4 to 18 days into white larvae which become yellow in color as they grow. When full-grown, they are 1 to 1½ inches in length and very closely resemble wireworms in appearance (*B*). The larval period usually requires from 6 to 9 months, and they usually winter in the larval stage, although a complete generation may range from 4 months to nearly 2 years. The pupal stage is white and is passed without any cocoon or protective covering. The two species of mealworms can be separated only by a careful examination and have practically the same life history and habits. They are always most abundant in damp grains, or grain products that have remained undisturbed for some time.

Reference. U.S.D.A., Tech. Bul. 95, 1929.

302. CADELLE[1]

Importance and Type of Injury. This insect is sometimes very important as a pest of stored grains; but it is also, to some extent, a feeder upon other insects. When abundant, however, it becomes seriously

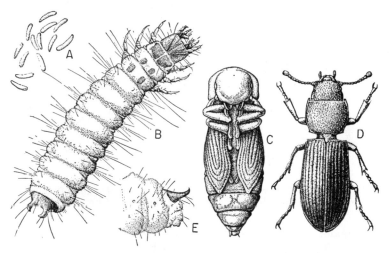

Fig. 19.18. The cadelle, *Tenebroides mauritanicus.* About 4 times natural size. *A*, eggs; *B*, mature larva, dorsal view, showing characteristic spots on thoracic segments and the curious rigid hooks on tip of abdomen; *C*, pupa, ventral view; *D*, adult, showing the strong constriction between prothorax and mesothorax; *E*, last three abdominal segments of larva in side view, showing terminal hooks and spiracles. (*Original.*)

destructive in grain bins. It also causes damage in flour mills by cutting holes in flour sacks, silk bolting cloth, and other silk cloth used in the machinery, and it eats holes in cartons of packaged foods.

Food. The insect is a general feeder on stored grains and seeds. It usually attacks the embryo, eating out only the softer parts of the grain. While the adults will often kill and feed upon other insects with which

[1] *Tenebroides mauritanicus* (Linné), Order Coleoptera, Family Ostomidae.

they come in contact, their general habits are not those of predaceous insects.

Life History, Appearance, and Habits. This insect is, next to the meal-worms, the largest of those attacking stored grains. Both adults and larvae, but not the eggs or pupae, may pass the winter. The adult beetle (Fig. 19.18,*D*) is black or nearly black, ⅓ to ½ inch in length, with the head and prothorax distinctly separated from the rest of the body, to which it is attached by a rather loose, prominent joint.

The eggs are laid on or near the food, by preference in cracks, under flaps of cartons, or in some such protected situation, commonly being deposited in groups of 10 to 60. Over 1,300 eggs have been secured from

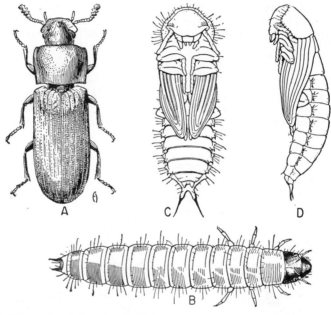

Fig. 19.19. The confused flour beetle, *Tribolium confusum.* *A*, adult; *B*, larva; *C*, pupa, ventral view; *D*, pupa, lateral view. About 13.5 times natural size. (*From Calif. Agr. Exp. Sta. Bul.* 676.)

a single female. The larvae (Fig. 19.18,*B*) that hatch from these eggs, 1 or 2 weeks later, are rather soft-bodied, white to grayish-white in color, with prominent black heads, black spots on the three thoracic segments, just behind the head, and two short dark hooks at the rear end of the body. They are about ⅔ inch long when full-grown. Under favorable conditions the larvae may complete their development in from 70 to 90 days; others require from 7 to 14 months, and they have been kept alive for nearly 3½ years. Full-grown larvae and adults have the habit of boring into wood, adjoining the grain on which they have been feeding, in granaries, storerooms, and grain ships, where they may remain hiding for months, ready to attack the next consignment of foods. The pupal stage is passed in such cavities or in other secluded places. Female adults commonly live for a year.

Reference. U.S.D.A., Dept. Bul. 1428, 1926.

302A. Khapra Beetle[1]

Importance and Type of Injury. This is one of the worst stored-products pests in the world, preferring dried vegetable products but also attacking animal products. The larvae feed on grain having moisture contents as low as 2 per cent, tolerate temperatures to 111°F., and can live for as long as 3 years without food. They congregate in cracks and crevices and are very difficult to remove by ordinary sanitation.

Food. Grains, seeds, flour and cereal products, hay and straw, dried fruits and nuts, dried blood and milk, and fish meal.

Distribution. This insect, which is a native of India, was first found in California in 1953, although it had apparently been present since 1946. It spread into Arizona and New Mexico and is now under federal quarantine, and an eradication program is being carried out.

Life History, Appearance, and Habits. The adult beetle is brownish black, oval, and about ⅛ inch long. The larva (Fig. 1.15) is yellow-brown with yellow intersegmental rings, covered with long brown hairs, and when mature is about ¼ inch long. Under optimum conditions, the female lays an average of about 90 white oval eggs, and these hatch after 3 to 14 days. The life cycle from egg to adult requires an average of 220 days at 70°F. and 26 days at 95°F. The adult beetles do not feed and seldom fly, and distribution is largely through the shipment of infested products.

References. *Hilgardia*, **24**:1, 1955; *Jour. Econ. Ento.*, **52**:312, 1959.

303. Confused Flour Beetle[2] and Red Flour Beetle[3]

Importance and Type of Injury. This insect is one of the most common occurring in situations where grain products are stored. It is one of the most annoying pests in retail grocery stores and warehouses and extremely serious in flour mills. Infested material will show many elongate reddish-brown beetles, about ⅐ inch long (Fig. 19.19), crawling over the material when it is disturbed, and brownish-white somewhat flattened six-legged larvae feeding on the inside of the grain kernels and crawling over the infested seeds. They are generally known among millers as "bran bugs."

Food. This insect feeds on a great variety of products, including all kinds of grains, flour, starchy materials, beans, peas, baking powder, ginger, dried plant roots, dried fruits, insect collections, nuts, chocolate, drugs, snuff, cayenne pepper, and many other foods.

Distribution. The confused flour beetle was first noted in this country in 1893. Both species occur throughout the world, but the confused flour beetle is most abundant in the northern part of the United States, while the red flour beetle is not commonly found north of the forty-first parallel.

Life History, Appearance, and Habits. The adult beetles are very active, moving rapidly when disturbed. They may survive moderately

---

[1] *Trogoderma granarium* Everts, Order Coleoptera, Family Dermestidae.

[2] *Tribolium confusum* Duval, Order Coleoptera, Family Tenebrionidae. The red flour beetle is very similar to the confused flour beetle. The two may be distinguished in the adult stage by the following differences: as seen from the underside of the head, the eyes of the confused flour beetle are separated by about three times the width of either eye, whereas the width of each eye seen from below in the red flour beetle is about equal to the distance between them. The confused flour beetle has the antennae gradually enlarged toward the tip, the red flour beetle suddenly enlarged at the tip; the margin of the head is notched at the eyes in the confused flour beetle, and not so notched in the other species. The red flour beetle seems to be less common.

[3] *Tribolium castaneum* (Herbst) (= *T. ferrugineum* Fabricius).

cold winters in unheated buildings and often live 2 years or more in the adult stage, during which period the female may produce nearly 1,000 eggs. The very small clear-white sticky eggs are laid on sacks, in cracks, or directly on the food material. These hatch in 5 to 12 days into small brownish-white worms, which become full-grown in 1 to 4 months, depending upon temperature and kind of food, and are then about ⅙ inch in length. They change to white naked pupae, remaining in this stage for 1 to 2 weeks. A complete generation is passed in 3 to 4 months when the temperature is high. Under Kansas conditions four or five generations occur annually in heated storehouses or mills. All stages of the insect may be found at any time of the year in such buildings.

References. Jour. Agr. Res., **46**:327, 1933; *Rept. Minn. State Ento.,* **17**:73, 1918.

304. SAW-TOOTHED GRAIN BEETLE[1]

Importance and Type of Injury. Infested material will show many very slender, much-flattened, small, dark-red beetles hurrying over the surface of the food. Its flattened shape enables this species to work into

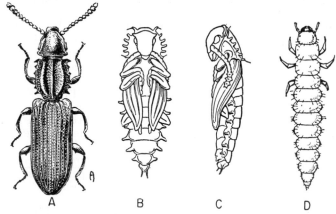

FIG. 19.20. Saw-toothed grain beetle, *Oxyzaephilus surinamensis.* *A*, adult; *B*, pupa, ventral view; *C*, pupa, lateral view; *D*, larva. About 15 times natural size. (*From Calif. Agr. Exp. Sta. Bul.* 676.)

packages of food that are apparently tightly sealed. Feeding consists of scarring and roughening of the surface of the food. Its attack upon stored seeds usually follows that of other insects, since it is not able to attack sound seeds.

Food. As with the confused flour beetle, this insect feeds on a great variety of products including practically all grains and grain products, dried fruits, breakfast foods, nuts, seeds, yeast, sugar, candy, tobacco, snuff, dried meats, and, in fact, almost all plant products used for human food.

Distribution. Throughout the world.

Life History, Appearance, and Habits. In unheated buildings only the adult stage (Fig. 19.20,*A*) winters. They are dark-brown, much flattened, slender insects, about 1/10 inch long. When examined under a

[1] *Oryzaephilus surinamensis* (Linné), Order Coleoptera, Family Cucujidae.

lens, six saw-tooth-like projections will be seen on each side of the thorax. The adults almost never fly. They have been kept alive for over 3 years. The female lays from about 50 to 300 eggs on and near the food, and these hatch in from 3 to 17 days. The larvae (D) are brown-headed, elongate, white, six-legged grubs with abdomens tapering to the tip. They become full-grown in 2 to 10 weeks when they are about 1/8 inch in length. They then form a protective covering by sticking together small bits of the food material. They remain in the pupal stage (B) for from 6 to 21 days, emerging at the end of this time as adult beetles. There are from four to six generations annually throughout most of the United States. It is possible, under very favorable conditions, for the entire life cycle to be passed in from 24 to 30 days.

Reference. Jour. Agr. Res., **33** :435, 1926.

305. LESSER GRAIN BORER[1]

This insect, which is also known as the Australian wheat weevil, is now well distributed through the south and midwest, but not in the more northern areas. It feeds in both larval and adult stages in the interior of nearly all grains and some other substances, such as seeds, drugs, dry roots, and cork, and eats into wood and paper boxes. It is most common in wheat and one of the most destructive wheat insects. The female lays about 300 to 500 eggs in the grain, and under favorable conditions a generation may develop in a month. The adults are brown to black, nearly cylindrical, about 1/8 inch long by 1/4 as wide. The large head is bent under the thorax, and the rear end of the body is blunt. The antenna has a large three-segmented serrate club. The larvae are grub-like, lie curved, and become about 1/10 inch long. The anterior end is much swollen, bearing a small brown head and six short legs.

Reference. U.S.D.A., Bur. Ento. Bul. 96, Part III, 1911.

306. LARGER GRAIN BORER[2]

This species is very similar in appearance and habits to the lesser grain borer. It is larger (1/6 inch long), cylindrical, nearly one-third as wide as long, and the wing covers are smoother and more polished. It is of importance in the United States at only a few places in the south. It feeds chiefly on corn and also bores into wood.

307. DRIED FRUIT BEETLE[3]

Importance and Type of Injury. This small beetle has been reported chiefly as a pest of the fig and date industries in California. The adults work their way to the inside of ripening and partially dried figs and dates, carrying upon their bodies bacteria, yeasts, and fungus spores which cause smut and souring of the fruit.

Food. Fermenting tree fruits, raisins, and watermelons.

Distribution. Widely distributed in California and the Gulf states and many other warm countries.

Life History, Appearance, and Habits. Pupae winter in the soil and the adults in all kinds of cull and fermenting fruits and fruit refuse and in stored fruits, the immature stages becoming adults by March or April. The adults deposit on the average over 1,000 small white eggs on the pulp of ripe figs on the tree, piles of cull oranges, raisins, stone fruits, watermelons, and other fermenting fruits. The eggs hatch on the average in 2 days. The whitish short-legged grubs, with brown head and posterior end, grow in a week or two to 1/4 inch in length, then pupate for about a week in the soil, so that there may be a generation of the insects about every 3 weeks in warm weather. The adults are very active, strong fliers, which readily find fermenting or

---

[1] *Rhyzopertha dominica* (Fabricius), Order Coleoptera, Family Bostrichidae.

[2] *Dinoderus truncatus* Horn, Order Coleoptera, Family Bostrichidae.

[3] *Carpophilus hemipterus* (Linné), Order Coleoptera, Family Nitidulidae. The corn sap beetle, *C. dimidiatus* (Fabricius), is also an important pest of dates.

ripe fruits and in the case of figs, enter promptly through the "eye" of the fruit. They are only ⅛ inch long by half as wide. The short elytra leave the terminal third of the abdomen exposed. Each wing cover has a prominent amber-brown spot at the tip and a smaller one near the base, on the side. The appendages are also amber or reddish, the rest of the body black.

Control Measures. Great numbers of beetles have been caught in traps baited with cull dried peaches moistened with water, 1 pint to 1 pound of fruit. The fermenting bait remained attractive to the beetles for about a month. Such bait was especially

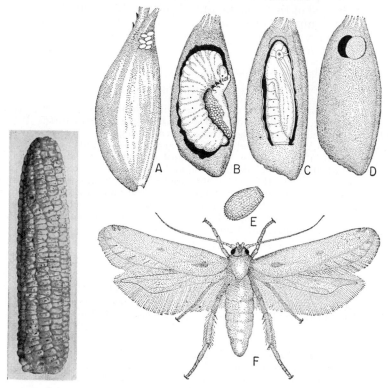

FIG. 19.21. Ear of corn heavily infested by the Angoumois grain moth, *Sitotroga cerealella*, showing emergence holes of the moths. (*From Ill. Natural History Surv.*)

FIG. 19.22. Angoumois grain moth, *Sitotroga cerealella*. *A*, glume or chaff covering a grain of wheat, with a cluster of eggs deposited near outer end; *B*, the larva or caterpillar inside the grain, most of the contents of which it has eaten; *C*, the pupa within a grain of wheat; *D*, hole made by moth through the thin seed coat, the round "lid" having been nearly cut through by the larva before it pupated; *E*, a single egg greatly magnified; *F*, the adult moth: note characteristic shape of tip of hind wing and the long fringe of hairs on both wings. (*Redrawn after King, Pa. Bur. Plant Ind.*)

attractive after it became infested by larvae of the beetle. The larvae were prevented from escaping to the soil to pupate by smearing the top 3 inches of the inside of the container with cup grease. Prompt collection of fallen fruits from beneath the trees, destruction of all culls by burning, drying thoroughly, or feeding to livestock, or spraying culls with benzene hexachloride in oil are suggested. Packages of dried fruits may be fumigated with ethyl or isopropyl formate, ethylene oxide, methyl bromide, or ethylene dibromide. Dusting with 5 per cent malathion is also effective in controlling the beetles.

References. U.S.D.A., Cir. 157, 1931; *Jour. Econ. Ento.*, **28**:396, 1935; *Date Growers' Inst. Ann. Rept.*, 1946, p. 34; 1948, p. 12; *Hilgardia*, **22**:97, 1953.

308. Angoumois Grain Moth[1]

Importance and Type of Injury. Grain in bins or ear corn in storage has small buff moths flying about the bins or crawling rapidly over the surface of the grain when it is disturbed. One or two small round holes are eaten in the kernels of infested corn (Fig. 19.21) or in other grain. This insect is the most destructive grain moth occurring in this country. It is extremely important in the South, causing great damage to corn in cribs and also destroying ripening grain, especially wheat, in the field.

Food. Wheat, corn, and other grains.

Distribution. The insect received its name from having been first reported as injurious in the Province of Angoumois, France, about 1736. It was first reported in the United States, from North Carolina, and is now distributed throughout the United States.

Life History, Appearance, and Habits. The adult insect is a rather delicate moth of a buff color with a wingspread of $\frac{1}{2}$ to $\frac{2}{3}$ inch. This is the only stage commonly observed, since the eggs are almost microscopic and the larvae and pupae live entirely inside of seeds. The hind wings are uniformly light gray, with a heavy fringe of hairs which are longer than the width of the wing membrane. The wing membrane is prolonged at the apical angle like a thumb or finger (Fig. 19.22,*F*). This easily distinguishes it from the clothes moths, for which it is often mistaken. The female moths lay several hundred, very minute, white to reddish eggs, singly, or in clusters of as many as 20, on the grains where the larvae feed. Where the insects are working out-of-doors, the eggs are attached to wheat heads in the field or to the grain in the shock. They hatch in from 4 days to 4 weeks, and the worm-like larvae at once burrow into the kernels or berries of the grain, feeding upon the starchy parts of the kernel. In wheat usually only one larva is found in a kernel, although several may find sufficient food in one kernel of corn. When full-grown the larva is about $\frac{1}{5}$ inch long, white in color, with a yellowish head, six true legs, and four pairs of prolegs. It then eats an exit tunnel for the adult, leaving a very thin transparent film of the seed coat covering the emergence hole. Next it spins a thin silken cocoon within the grain and there changes to the pupal stage. The adult moth emerges through a small round hole in the seed coat. The larval stage normally lasts from 20 to 24 days, and the entire life cycle may be passed in about 5 weeks. However, the larvae may lie dormant over winter. In unheated buildings, there are probably two generations a year in the latitude of Kansas and central Illinois. In the warmer parts of the South, as many as six generations occur annually. The insect is killed by cold winter weather, but in heated buildings, such as storehouses or mills, breeding is continuous throughout the year.

References. Pa. Bur. Plant Ind. Cir. 1, 1920; *U.S.D.A., Farmers' Bul.* 1260, 1931.

309. Mediterranean Flour Moth[2]

Importance and Type of Injury. The Mediterranean flour moth was at one time the most troublesome insect occurring in flour mills in this

---

[1] *Sitotroga cerealella* (Olivier), Order Lepidoptera, Family Gelechiidae.

[2] *Ephestia* (= *Anagasta*) *kühniella* (Zeller), Order Lepidoptera, Family Pyralididae. The closely related almond or fig moth, *Ephestia cautella* (Walker), has very similar habits. The adult is more slender and less conspicuously marked and the larva somewhat tinged with brown and faintly striped with darker dots. It is a serious pest of harvested figs but breeds in nearly all kinds of stored vegetable products.

country. However, modern fumigation methods have reduced its importance. Where this insect is present, masses of flour in chutes and elevators will be found webbed together and containing small pinkish-white caterpillars. The chutes may become entirely clogged and the machinery stopped by these webbed masses of flour. Small gray moths will also be noted flying about the infested buildings.

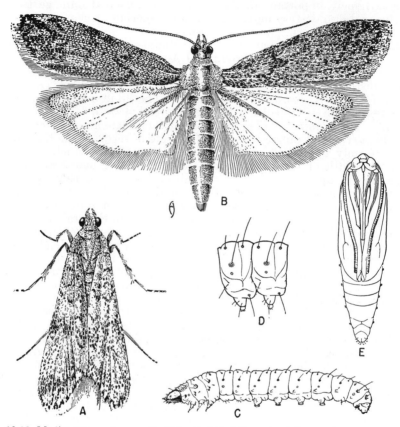

FIG. 19.23. Mediterranean flour moth, *Ephestia kühniella*. *A*, adult moth in normal resting position; *B*, adult with wings spread; *C*, larva; *D*, enlarged drawing of two larval segments in the region of the prolegs; *E*, pupa. About 8 times natural size. (*From Calif. Agr. Exp. Sta. Bul.* 676.)

Food. Flour is the favorite food of the insect, but it will attack also the whole grain of wheat, bran, breakfast foods, corn and other grains, and pollen in beehives.

Distribution. The insect was first reported in North America in 1889, when it was found in Canada. It has now spread throughout the United States and Canada and probably throughout most of the world.

Life History, Appearance, and Habits. The adult moth (Fig. 19.23, *A,B*) is of a pale-gray color and is from ¼ to ½ inch long. The head and tail are slightly raised when the insect is resting, this being very characteristic. The wings are marked with two zigzag lines of black which are not prominent. The eggs are laid in accumulations of flour or other foods, in cracks about buildings, or on the cloth in spouts or bolters in mills. They hatch in from 3 to 6 days, depending on the temperature. The caterpillars immediately begin spinning silken threads which form into little tubes in which they live and feed. It is this web-spinning that causes the greatest amount of damage by the insect. The caterpillars when full-grown are about ⅗ inch long, of a general whitish to pinkish color (Fig. 19.23,C), and have three lengths of crochets in a circle on the

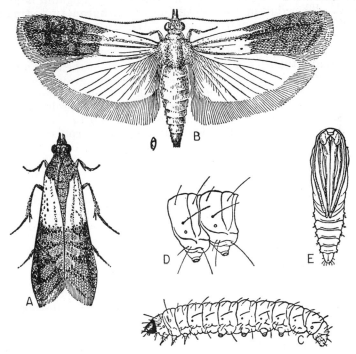

Fig. 19.24. Indian-meal moth, *Plodia interpunctella*. *A*, adult moth in normal resting position; *B*, adult with wings spread; *C*, larva; *D*, enlarged drawing of two larval segments in the region of the prolegs; *E*, pupa. About 6 times natural size. (*From Calif. Agr. Exp. Sta. Bul.* 676.)

prolegs. They pupate in silken cocoons, this stage lasting from 8 to 12 days. Under conditions encountered in mills, the entire life cycle is usually passed in from 9 to 10 weeks.

Reference. Jour. Agr. Res. **32**:895, 1926.

310. INDIAN-MEAL MOTH[1]

Importance and Type of Injury. This is one of the most general and troublesome of the moths infesting stored products. Infested material will be more or less webbed together and often fouled with dirty silken masses containing the excreta of the larvae.

[1] *Plodia interpunctella* (Hübner), Order Lepidoptera, Family Pyralididae.

Food. This insect attacks all kinds of grains, meal, breakfast foods, soybeans, dried fruits, nuts, seeds, dried roots, herbs, dead insects, museum specimens, powdered milk, the excrement, exuviae, and pollen in beehives, and many other substances. It is frequently a pest in candy factories where nut candies are made and has been bred from milk chocolate not containing nuts.

Distribution. The insect is of European origin but is now generally distributed throughout the United States.

Life History, Appearance, and Habits. In unheated buildings, the insects winter as larvae, but in warm situations feeding and breeding are continued. The adult moth is about $\frac{1}{2}$ to $\frac{3}{4}$ inch from tip to tip of wings and is active at night or in dark places. When at rest (Fig. 19.24,*A*) the wings are folded closely together along the line of the body and the antennae flat on the wings. The front or base of the forewings is a grayish-white color, and the tip half or two-thirds, as well as the head and thorax, a contrasting reddish brown; the underwings are grayish white. The palps form a characteristic cone-like beak in front of the head. The minute whitish ovate eggs, some 40 to 350 or more, are deposited singly or in clusters of from 12 to 30 on the larval food or adjacent objects. They hatch in from 2 days to 2 weeks into small caterpillars, of a general white color, but often with a distinct greenish or pinkish tinge, with a light-brown head, prothoracic shield, and anal plate. On becoming full-grown, in from 2 weeks to 2 years, these caterpillars are about $\frac{1}{3}$ to $\frac{1}{2}$ inch in length. They crawl up to the surface of the food material and pupate within a thin silken cocoon, from which the adult moths emerge in 4 to 30 days. The entire life cycle of the insect will require from 4 to 6 weeks under conditions usually encountered in heated buildings, there being from four to six generations of the insect each year.

Reference. U.S.D.A., *Tech. Bul.* 242, 1931.

METHODS OF CONTROLLING STORED-GRAIN PESTS

While the methods used for controlling insects infesting farmers' grain bins and elevators differ somewhat from those used for the control of the same insects in mills, warehouses, retail stores, and dwellings, the most effective means of preventing damage is the same. In all cases this consists in keeping the premises as clean as possible. This means the gathering up and removal of all refuse grain, seeds, grain products, flour, meal, or other material in which the insects breed; thoroughly cleaning the bins, storerooms, mills, or warehouses at periods of the year when they can be emptied; the carrying over from year to year of as little material as possible; and the careful inspection of all material when brought in for storage or processing. When bins, mills, seed houses, or warehouses are built, the insect problem should be given due consideration, and the construction should be such that the stored products will be kept dry, that cracks and crevices are reduced to the minimum, and that the most effective control measures may be applied at the minimum expenditure of effort and expense. Provision should be made in mills for heat-treating, and bins and storehouses should be made sufficiently tight for effective fumigation. Any building where grains or grain products or seeds are held in storage for some time is practically sure to become infested sooner or later and to require treatment. Inaccessible dead spaces in mills and storerooms, where wastes collect, may be treated with miscible oil, 1 gallon, lye, ½ pound, and water, 9 gallons. Screenings and waste grain of all kinds should be kept away from mills and storehouses. Since insect damage is greater the higher the moisture content of the grain, storage in damp situations should be avoided. Storage of grain in hermetically sealed underground vaults eventually kills stored-products pests through oxygen starvation.

Destruction by Heat and Cold. The heat-treatment has given good results for treating mills, retail stores, or storerooms in dwellings, as it can usually be applied so that practically all insects are killed; and the expense of treatment is lower than that of any other method. The heat-treatment should be used during the period when the outside temperature is high. This treatment can seldom be used in elevators, farmers' bins, or other small storages. In most buildings the heating system used in the winter will be sufficient to maintain temperatures high enough to kill the insect life in the mills during the summer, when outside temperatures are from 90 to 100°F. and little wind is blowing. Repeated tests with this method have shown that, in mills of fairly tight construction under the above conditions, it is possible to maintain temperatures from 120 to 150°F. for several hours and that such temperatures are fatal to all insects exposed to them. Owing to the construction of the buildings or the lack of heating facilities, this method cannot always be applied effectively. In most mills, however, the expense of installing sufficient heating units will not be so great as that of annually fumigating with hydrogen cyanide gas or some other fumigant. Heat is not satisfactory for the treatment of nuts and dried fruits. Heat may be used for seeds intended for planting, if the exposure is not over 6 hours, the temperature not higher than 135°F., and the seeds reasonably dry (containing not more than 12 per cent moisture).

If there are large quantities of seeds to be treated, it will be found difficult to raise the temperature sufficiently in the interior of the mass without exposing the outer layers too long or to too high temperatures. Under such conditions it is better to use a heat-treating machine. There are a number of heat-treating machines through which grain may be passed as it is moved from one bin to another and be subjected during its passage through the machine to temperatures of 130 to 140°F. for 30 minutes. Such machines are frequently very valuable, particularly for use in elevators or mills, where conditions are such that frequent infestations are sure to occur from the waste grain about the building and from grain being brought in. Where outside winter temperatures go as low as 20° below zero, opening the building to this temperature will kill most grain-infesting insects. In elevators and mill storage bins it is often possible to keep down insect damage by frequently moving the grain from one bin to another, as the insects do not thrive and multiply rapidly in grain that is frequently disturbed.

Destruction by Fumigation. Stored grains in tight bins may be fumigated by applying the fumigant as a coarse spray over the top of the grain. The temperature of the grain should be above 65°F., and after application of the fumigant, the bin should be covered with a tarpaulin or plastic sheeting and the exposure continued for at least 3 days. The operator should make the application from outside the bin and should be protected by a respirator. The following fumigants and dosages have been suggested for tight steel bins. The dosage should be doubled in a wooden bin.

	Small grain	Corn
	Gallons fumigant per 1,000 bu. (1,250 cu. ft.)	
Carbon bisulfide.............................	1.5	2
Carbon tetrachloride.........................	3	5
Carbon tetrachloride–carbon bisulfide (4:1)	2	5
Ethylene dichloride–carbon tetrachloride (3:1)	4	6
Carbon tetrachloride–ethylene dibromide (19:1) ...	2	5

The carbon bisulfide is highly inflammable, but the others are non-inflammable. Products in paper-lined bags may be fumigated by injecting the proper dosage into the interior of the bag through a pointed-nozzle extension rod. The general methods of fumigating with hydrogen cyanide have already been described on pages 375 to 379. The same methods should be followed in the treatment of warehouses or storerooms where grain insects are causing injury. Hydrogen cyanide cannot be relied upon to penetrate more than a few inches into a mass of seeds, flour, or stored or sacked grain. Modern mills are often piped with copper tubing fitted with nozzles through which liquid hydrogen cyanide can be discharged over the entire mill or directly into the milling machinery. The dosage used is generally 0.5 pound per 1,000 cubic feet although this should be doubled where the mill is not carefully sealed before fumigation. Methyl bromide at 1 pound to 1,000 cubic feet is highly effective in modern tightly sealed mills, where it is best applied through a piping system, and the exposure continued for 16 to 24 hours. This gas, because of its penetrating properties, is the most efficient fumigant for the treatment of bagged commodities in warehouses, where a dosage of 1.5 pounds per 1,000 cubic feet is generally used. Small 1-pound cans of methyl bromide may be used for fumigating small bins or rooms. Chloropicrin is also effective for mill fumigation when applied at 1 pound per 1,000 cubic feet. It may be distributed through a piping system, propelled by methyl chloride, or applied to machinery by sprinkling or evaporation from burlap sacks. Other fumigants which have been used for spot fumigations are mixtures of carbon tetrachloride and acrylonitrile (1-1) or mixtures of carbon tetrachloride, ethylene dichloride, and ethylene dibromide (65-35-5). The efficiency of fumigation of large grain bins, elevators, or mills can be greatly improved by recirculation of the fumigant-air mixture.

Vault fumigation is especially suitable for the treatment of packaged or bagged products. The use of vacuum fumigation aids in the penetration of the fumigant and decreases the exposure necessary to 1 to 3 hours. The following dosages have been suggested for the fumigation of stored products: liquid hydrogen cyanide, 2 pounds; methyl bromide, 4 pounds; ethylene oxide, 1 part, carbon dioxide, 9 parts, at 60 pounds per 1,000 cubic feet. Satisfactory treatments may be made under polyethylene or polyvinyl plastic sheeting (page 374).

Destruction by Insecticides. The application of residual sprays of 2.5 per cent methoxychlor or premium-grade malathion or of 0.5 per cent pyrethrins or allethrin to the walls and woodwork of warehouses, flour mills, and empty grain bins, at 2 gallons per 1,000 square feet, is effective in controlling most stored-product insects. Treatment of stored grains by spraying with synergized pyrethrins (0.15 per cent pyrethrins and 1.5 per cent piperonyl butoxide) or with 1.5 per cent premium-grade malathion, at 5 gallons to 1,000 bushels, will protect against infestation for an entire season. Treatment of stored seeds not intended for animal consumption with DDT at 1:10,000 to 1:50,000 parts by weight of the seeds, or with lindane at 1:100,000 to 1:1,000,000 will protect against insect attack. Dusting at 1:1,000 with magnesium oxide or silica gel with particle sizes of 1 micron or less is also effective. Flour bags treated with pyrethrins or a 10-1 mixture of piperonyl butoxide and pyrethrins at the rate of 10 milligrams of pyrethrins per square foot of cloth will protect the contents against infestation by flour beetles, the cadelle, the lesser grain borer, and the Mediterranean flour moth for many months.

References. *U.S.D.A., Cirs.* 390, 1937; 720, 1945; E-419 and E-429, 1938; E-677, 1945; and E-783, 1949; and *Farmers' Bul.* 1811, 1938, and 1906, 1947; *Jour. Econ. Ento.,* **39**:59, 1946; **40**:136, 1947; **41**:715, 1948; **51**:843, 1958; and **53**:341, 1960.

C. INSECTS ATTACKING STORED PEAS AND BEANS

311. PEA WEEVIL[1]

Importance and Type of Injury. Most housekeepers and seedsmen are familiar with "buggy" peas. The insides of the seeds in storage are eaten out by short chunky beetles (Fig. 19.25,a), about $\frac{1}{5}$ inch long, of a general brownish color, flecked with white, black, and grayish patches; and by the larvae (b) which are white all over except for the small brown head and mouth parts. In the spring and summer many of the old peas are found with neat circular holes, about $\frac{1}{10}$ inch in diameter, leading into the cavity where the insect developed. Heavily infested peas are often reduced to mere shells. Green peas are infested with the minute larvae, but there are only the dot-like

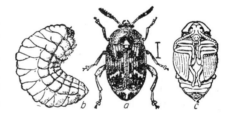

FIG. 19.25. Pea weevil, *Bruchus pisorum.* a, Adult beetle; b, larva or grub; c, pupa. Enlarged about 4 times. The line shows the natural length of the beetle. (*From Fernald, "Applied Entomology," after U.S.D.A.*)

entrance holes to show that they are not sound at this stage, and they are generally overlooked and eaten. If buggy peas are planted, a poor stand of weak, unproductive plants results.

Plants Attacked. Peas.

Distribution. Throughout the country.

Life History, Appearance, and Habits. The winter is passed in the adult stage and, under northern conditions, chiefly in the peas either in the field or in storage. Some, however, and in the South many, of the beetles leave the seeds in the fall and hibernate in protected places, up off the ground, out-of-doors. Some of the overwintering beetles may survive in their hibernating places until the second summer after they became

[1] *Bruchus* (= *Mylabris*) *pisorum* (Linné), Order Coleoptera, Family Bruchidae (= Mylabridae or Lariidae).

adults. If infested seed is not treated, the beetles may be planted with it. They wait until the plants are in blossom, then join others that wintered out-of-doors, and start feeding upon the pollen, the petals, the leaves, or the pods of the plants. The females glue their elongate yellowish eggs to the outside of the pods, from 1 to 12 or more on a pod. The tiny yellowish larva that hatches from the egg, 5 to 18 days later, is well adapted with spines and very short legs to burrow through the pod until it reaches one of the developing seeds, which it enters. Having found a place where it no longer needs to search for food, the larva loses its spines at the first molt, and the legs become very short. The grub grows slowly, consuming a third or more of the contents of the seed and finally eating an exit passage to the surface of the pea, leaving only a thin circular lid (the outer seed coat) intact to protect its tunnel. The larva requires about 4 to 6 weeks to become full-grown. The feeding of the weevils within the peas causes them to heat. This rise in temperature aids in the development of the weevil larva. It then paints the walls of its burrow with a gluey secretion from the mouth and in this snug chamber passes the pupal stage (Fig. 19.25,c), which occupies about 2 weeks in late summer.

There is usually only one generation a year and only one weevil matures in a pea. Eggs are never laid on dried peas, and there is no increase in numbers in storage. The adults must get to the growing plants in the spring or perish without laying eggs.

Control Measures. This insect may be controlled by spraying or dusting with DDT, malathion, or methoxychlor at 1 pound or with parathion at 0.4 pound per acre, applied to the foliage during the period of early bloom before the eggs are laid. To destroy the weevils wintering in seeds left in the field, plowing under the pea straw and chaff to a depth of 8 inches, converting it into silage, or burning it as soon as possible after the last picking destroys most of the insects that would winter out-of-doors. It is unwise to allow peas to ripen for seed in districts where green peas and canning peas are important crops. Infested peas may be fumigated as described on page 933. This is conveniently done by bagging them at harvest, allowing them to dry, and fumigating in a tight barrel, metal garbage can, or in a tight granary, by pouring the proper amount of fumigant over several gunny sacks placed on top of the bags. The weevils may be killed by suspending the seeds in a bag of cold water and heating it to 140°F., then pouring the peas out on a surface where they will dry quickly; or the seeds may be heated dry at a temperature of 135°F. for 3 to 4 hours, thus killing all stages of the beetle without injuring germination. Treatment of seed peas with DDT or lindane as described on page 935 will protect them against attack. Badly infested seeds should not be planted, even after the weevils have been killed.

References. *U.S.D.A.*, *Farmers' Buls.* 1275, 1925, and 1971, 1946, *Tech. Bul.* 599, 1938, and *Cirs.* E-435, 1938, and E-740, 1948; *Jour. Econ. Ento.*, **41**:832, 1948.

312. Bean Weevils and Cowpea Weevils

The broadbean weevil[1] is almost identical in appearance and habits with the pea weevil, but only about two-thirds as large. It can be distinguished by the points given in the key. It prefers the European broad bean as food but attacks also peas and vetches. The only other important difference appears to be that several individuals occur in a

[1] *Bruchus* (= *Mylabris*) *rufimanus* Boheman, Order Coleoptera, Family Bruchidae (= Mylabridae or Lariidae).

single seed in contrast with the invariable one of the pea weevil. It occurs in California.

The best-known species attacking beans is the common bean weevil,[1] which is thought to be native to the American continent. The cowpea weevil[2] is a closely related species that has very similar habits. It prefers cowpeas in which to develop, but attacks various kinds of beans and peas, at least in storage. The southern cowpea weevil[3] is also of economic importance. The two latter species can be distinguished from the bean weevil by the characters given in the key.

Importance and Type of Injury. All kinds of beans and peas stored for seed or food, unless they are protected, are almost sure to be devoured and rendered useless by these hungry weevils. In the field the beans may

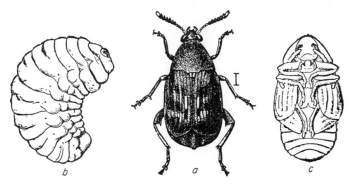

FIG. 19.26. Common bean weevil, *Acanthoscelides obtectus.* *a*, Adult; *b*, larva or grub; *c*, pupa. About 10 times natural size. Line indicates actual length of the beetle. (*From U.S.D.A., Farmers' Bul.* 1275.)

be stunted and deformed so as to be worthless, but often the infestation in green beans is not detected and the infested beans are eaten or stored by the unsuspecting. These weevils are so destructive in some sections as practically to prevent the commerical production of beans either for seed or for food.

Plants Attacked. Kidney beans, Lima beans, and cowpeas, in the field; and all varieties of beans, peas, lentils, and some other kinds of seeds in storage.

Distribution. The bean weevil occurs throughout the United States and in many other countries. The broad bean weevil is confined to the western coast, while the cowpea weevil is more abundant in the southern states.

Life History, Appearance, and Habits. Unlike the pea weevil and the broad bean weevil, these species breed continuously in the dry seeds, if they are stored in a warm place, and all stages may be found in winter. In the spring the adults escape from storerooms, and those that hibernated out-of-doors become active. They appear upon the plants as the latter come into bloom and feed on dew and honeydew and possibly on the foliage slightly. The adults are only half as large as the pea weevil, of a general light olive-brown color, mottled with darker brown and gray (Fig. 19.26,*a*). They are about ⅛ inch in length, the appendages are red-

---

[1] *Acanthoscelides* (= *Mylabris* or *Bruchus*) *obtectus* (Say).
[2] *Callosobruchus maculatus* (Fabricius) [= *Bruchus quadrimaculatus*].
[3] *Callosobruchus chinensis* (Linné).

dish, and the body narrows evenly toward the small head. In the cowpea weevil the elytra have a large rounded spot at mid-length, and the tips are black. There is a similar pair of black areas on the part of the abdomen exposed beyond the wing covers. The bean weevil has smaller black mottlings and can also be distinguished from the former species by having one large and two small teeth at the apex of the hind femur, while the cowpea weevil has only one such tooth. The whitish ellipsoidal eggs of the bean weevil are laid in loose groups of a dozen or more, in holes chewed by the female in the green pods along the seam where the two parts of the pod meet or in any natural crack that she finds in the pod. The cowpea weevil does not make a cavity for her eggs, but glues them securely to pods or exposed beans.

The minute, whitish, hairy grubs that hatch from the eggs, 3 to 30 days later, are equipped with short slender legs; and they scatter throughout the pod, seek out the developing seeds, and eat their way to the inside.

FIG. 19.27. Navy beans showing emergence holes of bean weevils. About natural size. (*From Fernald, "Applied Entomology."*)

On account of their very small size the entrance holes heal over and leave only a slight brown dot, so that growers and shippers rarely detect infestations of the weevils inside the beans. After feeding a few days the larvae molt and appear as white grubs with very small heads, no legs, and no long hairs. Feeding and growth continue until the larvae are about ⅛ inch long, nearly half as thick, wrinkled, and humpbacked (Fig. 19.26, *b*). The larval stage may occupy 2 weeks to 6 months or longer, depending upon the temperature and the moisture content of the beans. The pupal stage (*c*) is passed in the larval cell, which has been cemented over on the inside to exclude the larval excrement from contaminating the pupa, and which has a cylindrical extension to the outside but not penetrating the thin seed coat. Through these circular holes (Fig. 19.27) the adults commonly escape 2 to 8 weeks after the larvae entered the seeds. Adults emerging from seeds out-of-doors soon seek out growing beans and deposit their eggs upon them. If the weather remains warm, several generations may develop in the field. When beans are harvested, all stages may be taken into storage, and the adults as they transform continue to lay eggs on or among the beans. Breeding goes on steadily as long as there is any food left in the beans and the temperature is warm enough. Six or seven generations may be completed in a year, and as many as 28 weevils have been known to develop in one bean.

Control Measures. When beans are harvested, they should be at once sacked up tightly and, as soon as they are dry, either fumigated or heated as described for the pea weevil. Under no circumstances should weevily beans be planted. A simple method of killing the weevils and preventing their destructive increase during the winter is to treat the seeds with pyrethrins or premium-grade malathion as described on page 935. Bean vines and other refuse and leftover beans should be fumigated, fed, plowed under, or burned to prevent the weevils spreading into the fields in the spring.

References. U.S.D.A., *Farmers' Bul.* 1275, 1922; *Jour. Econ. Ento.*, **10**:74, 1917; *Ky. Agr. Exp. Sta. Bul.* 213, 1917; *Jour. Agr. Res.*, **24**:606, 1923; and **28**:347, 1924; *Tex. Agr. Exp. Sta. Bul.* 256, 1919; U.S.D.A., *Tech. Bul.* 593, 1938.

CHAPTER 20

INSECTS INJURIOUS TO DOMESTIC ANIMALS

For insects to attack living animals is a hazardous method of securing food. Unlike plants and stored foods, animals actively retaliate when insects attack them. Yet thousands of species of insects are specialized in structure or habit to secure all their food from the bodies of animals, and there is not a kind of wild or domesticated animal living that is not attacked by from one to many kinds of external or internal parasites.

So far as we know, no one has given us an accurate estimate of the loss that this attack upon our useful animals causes to man. We know that valuable animals, from the smallest chicks to the largest beef animals, are sometimes killed outright; that the flow of milk from dairy cattle decreases greatly during "fly time"; that beef and show cattle lose flesh and condition when flies or ticks are bad; that work horses are less efficient, often unmanageable, when annoyed by these pests; that lousy poultry, cattle, or sheep cannot be healthy; and that a number of deadly diseases are carried from animal to animal solely by the bites of insects and ticks. So it seems safe to say that this group of insects is deserving of much more careful study and active opposition than has been accorded to it.

General References. Can. Dept. Agr. Pub. 604, 1938; *U.S.D.A., Div. Ento. Bul.* 5(n.s.), 1896; *Wash. Agr. Ext. Ser. Bul.* 384, 1949; *U.S.D.A., Bur. Ento. Plant Quar., Cir.* E-762, 1949.

A. INSECTS ATTACKING HORSES, MULES, AND DONKEYS

FIELD KEY FOR THE IDENTIFICATION OF INSECTS INJURING HORSES, MULES, AND DONKEYS

A. *Free-flying insects that alight on the animals to suck blood, coming and going repeatedly:*
  1. Large, heavy-bodied, swift-flying, black, gray, or brownish, two-winged, often green-eyed flies, ½ to 1 inch long, with wings clear or banded, usually alight on head or shoulders and suck blood on warm sunny days. A drop or two of blood exudes from the puncture after they leave.............................
  ....*Horse flies* or *deer flies*, page 940.
  2. Flies about the size and general appearance of the house fly, suck blood, especially about the legs, causing animals to stamp their feet. Distinguished from the house fly by stiff, pointed beak or proboscis that projects forward from lower side of head, by broader black-spotted abdomen and grayer appearance, and by arista having hairs on upper side only. Palps less than ½ as long as proboscis. Sits generally with head up.................*Stable fly*, page 942.
  3. Small gnat-like humpbacked flies, not more than ¼ the size of the house fly, attack the animals, especially about the eyes, ears, and nostrils, and crawl into the hair to suck blood. Most abundant in spring.........................
  ....*Black flies* or *buffalo gnats*, page 944.
  4. Slender-bodied, scaly-winged, long-legged, two-winged insects, up to ½ inch long, especially abundant about lowlands and swamps at dusk or at night....
  .... *Mosquitoes*, page 998.

B. *Insects that stay on the animals all of the time:*
1. Wingless, flattened lice, with chestnut-colored head and thorax and yellowish black-banded abdomen, about $\frac{1}{12}$ inch long, that feed on the dry skin and hairs, but do not pierce the skin or draw blood. Head large, short, and broad, with chewing mouth parts; legs slender, abdomen not very broad............
....*Horse biting louse,* page 944.
2. Wingless, flattened, grayish bloodsucking lice, about $\frac{1}{8}$ inch long, with long, slender head, broad abdomen, and legs much thickened toward the end.......
....*Horse sucking louse,* page 945.
3. Leathery-skinned flattened long-legged two-winged flies, about $\frac{1}{3}$ inch long, cluster under hairs and suck blood......................................
Horse "tick" or *forest fly,* (see Hindle, "Flies and Disease: Blood-sucking Flies").
4. Animals rub their bodies vigorously, hairs stand erect, areas of skin with only scattered hairs remaining; skin with a fine eruption, scurfy, or covered with small dry yellow scabs. Minute, very long-legged, rounded mites, hardly visible without a lens, burrow into the skin, causing intolerable itching.......
....*Itch mite,* page 946.
5. Hair comes out in patches; skin with large moist scabs under which live myriads of minute, longer-legged oval-bodied mites, scarcely visible without a lens, that pierce the skin with their mouth parts but do not tunnel beneath it..........
....*Scab mite,* page 948.

C. *Short, chunky, spiny maggots that live in the alimentary canal and, during the spring, are passed in the droppings. Animals become run-down, unkempt, colicky:*
1. Elongate, whitish or yellowish eggs or "nits," glued by half their length to base of hairs, chiefly on the front legs. Light-brown bee-like flies, with faintly spotted wings, hover about forequarters of animals, laying these eggs in summer. Wing venation like that of the throat bot fly, and fly has a prominent spur on the trochanter of the hind legs. During winter months thick, spiny, yellowish or pinkish maggots, up to $\frac{2}{3}$ inch long, attach to walls of stomach or duodenum; circlets of spines double, the circlet on third from last segment (the tenth) lacking 1 or 2 pairs of spines at the middorsal line; next to last segment (the eleventh) with 1 to 5 spines visible from above at each side.
....*Common horse bot fly,* page 948.
2. Elongate whitish eggs, glued by nearly their full length to hairs, chiefly under the jaws, by a bee-like fly with rust-colored thorax and unspotted wings in which the discal cell ($1st\ M_2$) is only a little longer than the anterior basal cell (M) and the medial cross-vein is nearly in line with the radiomedial. During winter, short, thick, spiny, pale-yellowish maggots attach to walls of pharynx when small, sometimes cutting off breathing, and to stomach or duodenum; circlets of spines on larval segments in a single row..........*Throat bot fly,* page 948.
3. Elongate black eggs, glued to base of the minute dark hairs of horses' lips by a brightly marked bee-like fly with its abdomen white at base, black across the middle, and bright orange-red at the end; the thorax with a black, shining area between the wings; and unspotted wings in which the anterior basal cell (M) is not more than $\frac{1}{2}$ as long as the discal cell ($1st\ M_2$), and the medial cross-vein is distad of the radiomedial. During the winter, moderately spiny, pinkish maggots attach to stomach, duodenum, or rectum (often causing an obstruction) or cling exposed about margin of anus; circlets of spines on larval segments in double rows, but the circlet on third from last segment (the tenth) interrupted by at least half the width of the segment; next to last segment (the eleventh) without any spines visible from above...........................
....*Nose bot fly,* page 948.

General References. U.S.D.A., *Cir.* 148, 1930; *Ill. Agr. Exp. Sta. Cir.* 378, 1931.

313 (322, 333, 354). Horse Flies[1]

Importance and Type of Injury. Horses and mules strike with their heads, twitch the skin, shake their bodies, or otherwise evince sharp pain

[1] *Tabanus* and *Chrysops* spp., Order Diptera, Family Tabanidae.

as large heavy-bodied black or brown flies (Figs. 20.2 and 20.3) alight on the head, neck, shoulders, or back.   These insects fly alongside the animals, even when they are running swiftly, and bite repeatedly, but only in the daytime.   If they are not dislodged by the animal, they cut through the skin with their knife-like mouth parts and suck the blood for several minutes.   When they finally fly away, a drop or two of blood usually exudes from the hole they made.   In swampy, wooded sections animals in harness often become unmanageable and run away.   Webb and Wells estimated that, in areas where horse flies are prevalent, animals commonly lose an average of more than 3 ounces of blood a day.   They are forced to cease pasturing during most of the day, and as they congregate together, often wound each other by kicking or goring.   Horse flies are carriers of tularaemia, of a kind of filariasis (Calabar swellings) among human beings in West Africa, and of *el debab*, an Algerian disease of horses and camels.

F<small>IG</small>. 20.1. Larva of the black horse fly, *Tabanus atratus*. Line indicates natural size. (*From Ill. Natural History Surv.*)

They are also suspected carriers of anthrax, of equine infectious anemia or "swamp fever" of horses, and of surra.

Animals Attacked.   Horses, mules, cattle, hogs, man, dogs, deer, and other wild and domesticated animals.

Distribution.   Horse flies of various species occur throughout the world, being especially abundant in moist wooded areas and up to at least several thousand feet elevation in the mountains.

Life History, Appearance, and Habits.   The life histories of only a few of the many species have been studied.   Apparently the winter is generally passed as nearly full-grown larvae in the mud about lakes, streams, or wet areas of land.   These maggots (Fig. 20.1) are 2 inches or less in length.   They are pointed at each end, whitish or banded with black or brown, and with a fleshy elevated ring on each body segment.   They are very tough-skinned and in some sections are much prized as bait for fish. They become full-grown in late spring, when they pass through a pupal stage of several weeks in drier mud and then the flies begin to appear in early summer.   The eggs are laid on the leaves or stems of aquatic plants or trees; on stones and other objects that overhang water; or on grasses in moist swampy places, in dark-colored, wedge-shaped masses of several hundred.   Nearly a week later the very small maggots hatch, drop into the water, sink to the bottom, and bury themselves in the mud or sand. Their food is small animals such as other insects, earth worms, small crustacea, snails, and other horse fly larvae.   Some have a single generation and others apparently two each year.

Schwardt found that the big black horse fly, *Tabanus atratus* Fabricius, usually spends one winter and sometimes two in the larval stage, this larval period averaging 9 months but varying from 49 to 410 days, before producing the adults.   The striped horse fly, *Tabanus lineola* Fabricius, on the other hand, normally has at least two generations a year in Arkansas.   There is thus great variation in the life cycles, habits, and occurrence of different species in correlation with vegetation, drainage, and topography.

There are several hundred species of horse flies in the United States, ranging in size from about ⅓ inch long to nearly 1 inch long.   They are mostly black or brown, sometimes striped or spotted on the body, and many of the smaller species have the wings banded with brown.   The eyes are often brilliantly colored and banded, while alive.   Horse flies can usually be told from other flies by their antennae, which are divisible into three parts, the third being long and composed of five to eight rings and often with a short thumb-like projection at one side near its base.   Only the females bite, the males feeding on nectar, honeydew, and the like.

FIG. 20.2. Adult of the black horse fly, *Tabanus atratus*, male.   Natural size.   (*From Ill. Natural History Surv.*)

FIG. 20.3. Adult of the horse fly, *Tabanus sulcifrons*, female. Enlarged ½.   (*From Ill. Natural History Surv.*)

Control Measures.   No control has been discovered that is very effective.   Draining marshes and wet meadows where the flies develop is of the greatest value but should be done in such a way as to preserve the desirable wildlife of such areas, if possible.   Care should be taken to avoid overirrigation.   Oiling stagnant pools as for mosquitoes (page 1004) is of value in certain cases.   It may kill the adults which come to quiet pools and dip into them and also kills the young maggots as they hatch from the eggs and drop into the water.   Stabling animals during the day or covering them with light blankets, fly nets, or ear nets helps to ward off the attacks.   Spraying animals with synergized pyrethrins using 0.1 per cent pyrethrins and 1 per cent piperonyl butoxide, or equivalent synergist, as a water spray at 1 to 2 quarts per animal, or with 1 per cent pyrethrins and 10 per cent synergist as an oil spray applied as a mist at 1 to 2 ounces per animal, gives protection for 3 to 5 days.   Oil-based sprays of 3 to 5 per cent Lethane or Thanite and repellents such as butoxypolypropylene glycol, di-*n*-butyl succinate and di-*n*-propyl isocinchomeronate are also of value.   Granular applications of 5 per cent dieldrin at 0.6 pound per acre have given area-wide control when applied as larvicides to marshy breeding places.

References.   U.S.D.A., *Bur. Ento. Tech. Bul.* 12, Part II, 1906, and *Dept. Bul.* 1218, 1924; *Nev. Agr. Exp. Sta. Bul.* 102, 1921; *Ann. Ento. Soc. Amer.*, **24**:409–416, 1931, and **25**:631–637, 1932; *Ark. Agr. Exp. Sta. Bul.* 256, 1930; *La. Agr. Exp. Sta. Bul.* 93, 1907; *Ohio State Acad. Sci. Spcl. Paper* 5, 1903; *Jour. Econ. Ento.*, **30**:214, 1937; **44**:154, 1951; **49**:410, 1956; and **50**:379, 1957.

314 (322, 333, 355).   STABLE FLY[1]

Importance and Type of Injury.   The most injurious insect attacking horses and mules is a small fly, very similar in appearance to the common

[1] *Stomoxys calcitrans* (Linné), Order Diptera, Family Muscidae.

house fly, which bites the animals especially on the legs. It sometimes comes into houses, especially in stormy weather, and bites people about the ankles. Animals stamp their feet continually to dislodge these tormenting pests. The stable fly takes one or two drops of blood at a meal, several such meals in a day, so that each animal probably supplies hundreds, if not thousands, of fly meals a day when stable flies are abundant. As a result of the pain from the bites, the constant worry, and the loss of blood, animals lose weight, milk yield of dairy cattle is reduced, work animals become unmanageable, and sometimes animals are killed, either as the direct result of the flies or from disease induced by the flies. While the stable fly has been suspected of transmitting infantile paralysis, anthrax, leprosy, surra, and swamp fever (or infectious anemia of horses), it has not been proved to be the usual carrier of any animal disease in America.

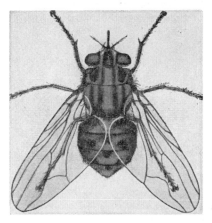

FIG. 20.4. The stable fly, *Stomoxys calcitrans*, adult. About 5 times natural size. (*From U.S.D.A., Farmers' Bul.* 1097.)

Animals Attacked. Horses, mules, cattle, hogs, dogs, cats, sheep, goats, guinea pigs, rabbits, rats, and man.

Distribution. The stable fly occurs in all parts of the United States and throughout most of the world. In the United States it appears to be most abundant "in the Central States from Texas to Canada, where grain is grown extensively" (Bishopp).

Life History, Appearance, and Habits. In the northern states the stable fly is believed to winter as larvae and pupae in wet straw piles or strawy manure. Farther south development continues throughout the year, and all stages may be found in winter. During the warm months of the year, breeding is continuous. The length of the several life stages has been given as follows: egg, commonly 2 or 3 days; larva, commonly 2 to 4 weeks; pupa, 1 to 3 weeks; and adult probably 3 weeks. Since the female is commonly 2 or 3 weeks old and must take several meals of blood before she begins laying eggs, the total average life cycle may be from 20 to 60 days, the longer period in cool weather. Five or six hundred elongate whitish eggs are deposited by the female in 4 or 5 batches. The stable fly (Fig. 20.4) is about ¼ inch long, of a general grayish color, like the house fly. It can be distinguished from the house fly by its habit of biting, and by its mouth parts, which stick forward from under the head as a stiff, somewhat pointed, slender beak, about twice as long as the head. The abdomen has seven rounded dark spots on the upper side, arranged in a figure 8 (Fig. 20.4). The cell nearest the tip of the wing (R_5) is open more than half its width. Both males and females suck blood as their chief food. They are active only during the day, either in the stable or in the field. The yellowish-white maggots of the stable fly develop in masses of straw, grain, hay, piles of grass, weeds, and other materials that have become water-soaked or contaminated with manure, and in the excrement of animals only if it contains much hay or straw. They have been found developing in great numbers on the trickling filters of sewage-disposal plants. In such cases flooding every 12 hours has been effective in cleaning out the larvae. Most serious outbreaks follow periods of excessive rainfall. The full-grown maggot is about ¾ inch long, tapering almost to a point at the head, and the posterior end is cut off squarely. It can be told from the house fly larva by looking at the spiracles on the last segment of the body. In the stable fly these are small, somewhat triangular, and the two separated by twice their own width; in the house fly they are almost touching, larger, and D-shaped (Fig. 21.15,*f*), with the three slits more sinuous. The insect passes a pupal stage of a week or two, in a brown puparium among the straw, and then emerges as the adult.

Control Measures. The destruction and avoidance of conditions in which the maggots thrive offer most promise, and, while this method cannot be expected to eliminate the fly, it should be possible to reduce its numbers. Straw from threshing should be scattered over the ground and plowed under, or burned, if it is not needed for bedding or feed. If it is to be preserved for use, it should be baled and stored in a dry place or else stacked carefully so that it does not become water-soaked and rotten. Masses of water-soaked feed should not be allowed to accumulate around stalls or feed troughs. The prompt disposal of manure and all other accumulations of fermenting organic matter, as explained for the house fly (page 1035), will also keep this species in check.

Animals may be covered with blankets, or old trousers may be pulled over the legs, or they may be allowed to run into darkened stables with nets, brush, or sacking so arranged over the doorway as to brush off the flies as the animals enter. About dairy barns and other stables, electrical or mechanical window traps may be used to catch myriads of the stable flies as well as house flies. Residual spraying of these areas as described on page 1036 will kill many of the resting flies. Animals can be protected from stable fly attack for several days by spraying with synergized pyrethrins, Lethane, or Thanite, or a fly repellent as described on page 389. The most satisfactory control with these materials will be gained by spraying the legs, belly, and sides of the animals once or twice per week.

References. U.S.D.A., Farmers' Bul. 1097, 1920; *Jour. Econ. Ento.,* **27**:1197, 1934; **48**:386, 1955; **50**:709, 1957; **51**:72, 269, 1958; and **52**:866, 1959.

315 (322, 352). BLACK FLIES OR BUFFALO GNATS[1]

These very small gnats (Fig. 21.4) hover about ears, eyes, and nostrils, alighting frequently and puncturing the skin with a very irritating bite. The young develop in the water, especially of rocky, swift-flowing streams. They are very difficult to destroy without destroying the fish in the stream and about the only means of control is to provide smudges, in the smoke of which animals can get relief from attack, to treat large areas with areosols, or to spray animals with repellents. These flies are more fully discussed under Insects That Attack Man (page 1005).

316. HORSE BITING LOUSE[2]

Importance and Type of Injury. At least two very different species of lice attack the horse, and it is important to recognize which is present in any case of infestation, since the control measures will be somewhat

---

[1] *Simulium* spp., Order Diptera, Family Simuliidae.

[2] *Bovicola* (= *Trichodectes*) *equi* (Linné); and *T. pilosus* Giebel, the European horse biting louse, Order Mallophaga, Family Trichodectidae.

different. This species does not pierce the skin or suck blood but runs freely about over the animal, nibbles at the dry skin and hairs, and causes great irritation. In the spring of the year, horses that have not wintered well rub against fences, stalls, and other objects. The coat, especially on the head, withers, and about the base of the tail becomes unkempt, hairs stand erect, and the skin is dry and full of scurf. When infested horses are brought from stables into the warm sunlight, the lice often are seen by hundreds clinging to the tips of the hairs.

Animals Attacked. Horses, mules, and donkeys.

Life History, Appearance, and Habits. These lice generally become noticeable in late winter or early spring, when all stages can usually be found on the animal. The full-grown ones (Fig. 20.5), are about $\frac{1}{10}$ inch long, of a chestnut-brown color except on the abdomen, which is yellowish with dark crossbands. The head is much broader and shorter than that of the following species and is rounded in front, forming a full semicircle in front of the antennae. The legs are slenderer than those of the blood-sucking horse louse. The eggs are glued to the hairs close to the skin, especially around the angle of the jaw and on the flanks. The eggs hatch in 5 to 10 days, into very small pale-colored lice, of the same general shape as the full-grown ones, and they become full-grown in 3 or 4 weeks. Breeding is continuous throughout the year, but the numbers become fewer in summer.

FIG. 20.5. The horse biting louse, *Bovicola equi*, female. About 25 times natural size. (*From U.S.D.A., Farmers' Bul.* 1493.)

Control Measures. Horses are not usually troubled with lice unless they have been neglected in feeding, stabling, and grooming. When present, the lice may be controlled by one of the measures suggested for cattle bloodsucking lice, page 959.

Reference. U.S.D.A., Farmers' Bul. 1493, 1926.

317. HORSE SUCKING LOUSE[1]

Importance and Type of Injury. This seems to be the most important louse infesting the horse, being commoner and also more irritating, because it feeds by piercing the skin and sucking the blood. The bites are painful and, when the lice become abundant, the loss of blood is a severe drain on the vitality of the host. The horse shows the same symptoms of scurfy skin, unkempt coat, and scratching or rubbing its body as in the case of the chewing louse. Often parts of the body will be rubbed raw, and the lice may feed about these wounds. Only an examination of the lice themselves will determine which species is present.

Animals Attacked. Horses, mules, and donkeys.

Life History, Appearance, and Habits. All sizes of lice and eggs will usually be found during the winter, when these insects are troublesome. The full-grown lice (Fig. 20.6) are of a dirty grayish or yellowish-brown color, about $\frac{1}{8}$ inch long, by half as broad at the middle of the abdomen. The thorax is only half as wide, and the head less than a third as wide, as the abdomen, distinctly narrowed toward the front. The legs are short

[1] *Haematopinus asini* (Linné) (= *H. macrocephalus* Burmeister), Order Anoplura, Family Haematopinidae.

and very clumsy, fitted for grasping about hairs.   The lice are commonest about head and neck and at the base of the tail.   The egg stage is normally from 11 to 20 days, but some may hatch as long as a month after they were laid.   The young lice are similar to the large ones except paler in color, and gradually grow to the size and color of the adults in 2 to 4 weeks.   There are several generations a year.

Control Measures.   See under Cattle Bloodsucking Lice, page 959.

Reference.   *U.S.D.A.*, *Farmers' Bul.* 1493, 1926.

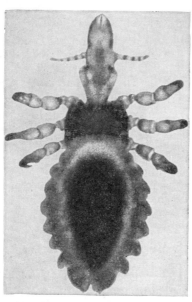

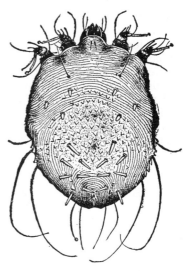

FIG. 20.6. The horse sucking louse, *Haematopinus asini*, female.   About 25 times natural size.   (*From U.S.D.A., Farmers' Bul.* 1493.)

FIG. 20.7. The mange or itch mite, *Sarcoptes scabiei*, female.   About 100 times natural size.   (*From U.S.D.A., Farmers' Bul.* 1493.)

318 (335, 366). ITCH OR MANGE MITE[1] OR SARCOPTIC MANGE

Importance and Type of Injury.   Animals rub and scratch their bodies vigorously.   Areas on the head, neck, back, or at the base of the tail become inflamed, pimply and scurfy, with the hairs bristling and only scattered hairs remaining.   Later the infestation may spread over the entire body, and large, dry, cracked scabs form on the thickened skin. To distinguish it from lousiness, some scrapings from the affected skin should be examined under a microscope for the mites which cause the trouble.   The mange of the horse may spread to man, causing "cavalryman's itch," but it does not persist on man, dying out in a few weeks (see also pages 973 and 1025.

Animals Attacked.   Horses, hogs, mules, man, dogs, cats, foxes, rabbits, squirrels, sheep, and cattle are attacked by different varieties of the same species.

Life History, Appearance, and Habits.   Mange is caused by a very small ovoid mite (Fig. 20.7), scarcely as big around as the cross-section

[1] *Sarcoptes scabiei* (De Geer), Order Acarina, Family Sarcoptidae.

of an ordinary pin ($\frac{1}{60}$ inch in diameter), and with very short legs that barely extend beyond the margins of the body. It is not a true insect, but an eight-legged form related to the ticks and spiders. The mites themselves will seldom be seen except by the specialist. They burrow beneath the skin on less-hairy parts of the body, making very slender winding tunnels, from $\frac{1}{10}$ inch to nearly 1 inch long. The serum discharged from the mouth of the burrows dries to form small dry papules. Within such a tunnel the female lays about 24 eggs and dies at the end of the tunnel. Within 3 to 10 days the eggs hatch to minute nymphs, which are at first six-legged. At the first molting they acquire an additional pair of legs, and after two more molts they are full-sized and ready to mate. After mating, the males die; the females again shed their skin and begin new tunnels 10 days to a month after hatching from the eggs. The mites secrete an extremely irritating poison, which, together with their tunneling, causes the excruciating itching. The trouble is most evident during the winter months, but some of the mites live on the animals the year round, unless treated. A generation may be completed in 2 weeks.

Control Measures. Mange is contagious, and the most important control is the isolation and quarantine of infected animals to prevent spread. The most effective treatments on horses or cattle consist in dipping or in high-pressure spraying, using about 2 gallons per animal, with one of the following formulations: (*a*) Lime-sulfur applied at 95 to 105°F., several times at weekly intervals. The dip can be made as follows:

Slake 10 pounds of fresh stone lime, thoroughly mix into the lime paste 24 pounds of sulfur, put through a screen to break up all the lumps. Add 30 gallons of boiling water and boil for 2 hours with frequent stirring. Allow to stand overnight, then siphon off the clear liquid without disturbing the sediment and add water to make 100 gallons ready to use. Care should be taken not to get any of the sludge or sediment into the dip.

Commercial lime-sulfur may be used at a dilution of 1 part to 15 parts of water. (*b*) Two sprayings or dippings, 10 to 14 days apart, with 0.075 per cent lindane or 0.5 per cent toxaphene. This treatment should not be used on dairy cows, and the animals should not be marketed before 1 month for toxaphene or 2 months for lindane. (*c*) Four sprayings at weekly intervals with 30 pounds of wettable sulfur in 100 gallons of water.

The walls and floors of stalls, curry combs, harness, and other objects about the horse should be treated with a good coal-tar-creosote disinfectant, since the mites may live on these objects for several weeks off the animal. Treated animals should not be returned to their former pasture for a month or two. Dogs and cats infested with mange mites may be treated as described above or may be clipped, subjected to a prolonged washing in warm water and green soap until all scabs are softened and loosened, and then an ointment consisting of lard, 8 ounces, sulfur, 1 ounce, and balsam of Peru, 1 dram, applied to the skin. It is sometimes recommended that one-fourth of the animal's body be treated on each of four succeeding days with this ointment. The bathing and ointment must be repeated at weekly intervals until all mites have been killed and all scabs have healed.

References. U.S.D.A., *Farmers' Bul.* 1493, 1926; *Jour. Econ. Ento.* **42**:444, 1949.

319 (326, 339). Scab Mite[1]

The scab mite is less serious on horses than on cattle and sheep. It is sometimes spoken of as the "wet mange," in contrast with the drier scabs that result from the attacks of the mange mite. It usually starts among the longer hair of the neck, withers, or base of tail. Treatment is the same as for mange mite (page 947).

320. Horse Bots

There are three kinds of bot flies that commonly molest horses in this country. They are known as the common horse bot fly,[2] the chin fly or throat bot fly,[3] and the lip or nose bot fly.[4] A fourth species, *Gasterophilus inermis* Brauer, was reported by Knipling in 1935.[5] The adult is very similar to the common horse bot, but lacks the spur on the hind trochanters. The larva is most like the lip bot, but, unlike that species, the circlet of spines on segment 3 is not interrupted on the ventral side and has as many spines as segment 4. The egg of this species is much shorter and thicker than any of the other bot flies and does not flare away from the hair to which it is attached any more at the head end than it does at the end toward the base of the hair.

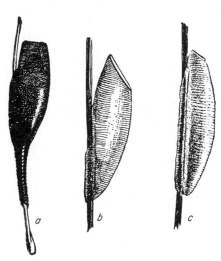

FIG. 20.8. Eggs of horse bots. *a*, The nose bot fly; *b*, common horse bot fly; and *c*, throat bot fly. About 35 times natural size. (*From U.S.D.A., Farmers' Bul.* 1503.)

Importance and Type of Injury. All three kinds of bot flies, but especially the nose bot fly, are dreaded by horses, which fight them viciously, although we know from the structure of the flies that they can neither bite nor sting the animals. Their sole object in flying about horses is to glue their eggs (Fig. 20.8) fast to the hairs. This act results in a slight pull on the hair, but it is believed that the alarm caused by the flies is mostly due to fear or nervous excitation caused by buzzing and striking. The most injurious stage of the flies is the maggot or larval stage, which lives in the digestive tract of the horse, causing mechanical injuries to the tongue, the lips, and the lining of the stomach and intestine, interfering with glandular activity, causing an inflamed ulcerous condition, absorbing food, which progressively starves the host, probably secreting toxic substances, and frequently causing complete obstruction to the passage of food substances from stomach to intestine (Fig. 20.9). A horse badly infested with bots presents a run-down condition, caused by digestive disturbances, and has a rough coat. The presence of bots may also be

[1] *Psoroptes equi* Gervais, Order Acarina, Family Sarcoptidae.
[2] *Gasterophilus intestinalis* (= *equi*) (De Geer), Order Diptera, Family Oestridae.
[3] *Gasterophilus nasalis* (= *veterinus*) (Linné).
[4] *Gasterophilus haemorrhoidalis* (Linné).
[5] *Ento. News,* **46**:105, 1935.

detected by finding the full-grown larvae in the feces in the spring months, although this occurs too late to apply effective control measures.

Animals Attacked.   Horses, mules, donkeys, and rarely dogs, rabbits, hogs, and man.

Distribution.   The common horse bot and the throat bot occur throughout the United States.   The throat bot is said to be especially abundant in the Rocky Mountain region.   The nose or lip bot fly is recorded from northern Illinois, westward to Idaho, and from northern Kansas and Colorado to Manitoba, Saskatchewan, and Alberta.   It has been spreading in all directions.

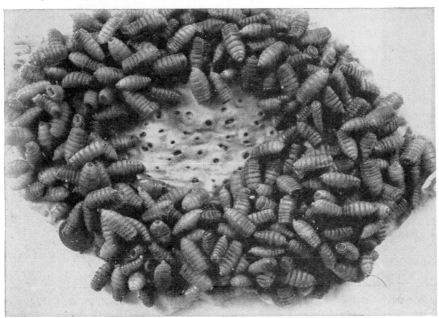

FIG. 20.9. Horse bots.   Larvae attached to lining of stomach of horse; at center, lesions where bots have been removed.   About ⅔ natural size.   (*From U.S.D.A., Dept. Bul.* 597.)

Life History, Appearance, and Habits.   The bots winter as larvae (Fig. 20.9) in the alimentary canal of the host, usually becoming full-grown by late winter or spring.   They may be found in the digestive tract every month of the year; but after the first of October only small larvae will be found, the mature larvae having all been passed by that date.   It is believed that the larvae feed upon the inflammatory products induced in the mucosa of the stomach by their presence.   The full-grown horse bots are thick tough-skinned maggots, blunt at the posterior end, tapering in front to the two strong mouth hooks, and with a circlet of prominent spines around each body segment, giving them a screw-like appearance. They are about ½ to ⅔ inch long when full-grown, yellowish white or pinkish in color.   The three kinds may be told by the characters given in the key.   When growth is complete, they release their hold on the walls of the stomach and pass to the ground, usually with the excrement, during the first half of May, in the case of the nose and throat bots, and somewhat later for the common horse bot.   This fact has been taken advantage of by quacks who recommend various treatments to be given in the late winter and cite the normal appearance of bots in the excrement as a result. The bot larvae burrow into the soil a short distance and inside their hardened skins form the pupae, from which adults appear 3 to 10 weeks later. The adults are abroad commonly, in the case of the nose bot, from early June until late summer in the northern states, being rare in the fall; while the throat bot appears early in June and the common horse bot about the last of June. These two species are abundant until heavy frosts, commonly about November 1. The adults look somewhat like bumblebees, being very hairy, two-winged, of a general brownish color, and about ⅔ inch long. They take no food and live only from 3 days to 3 weeks. The common horse bot may easily be told by the faint smoky spots on the wings. The throat bot has a rust-red thorax, clear wings, and a prominent band of black hairs at mid-length of the abdomen. The nose bot fly is somewhat smaller than the others, and the tip of the abdomen has reddish hairs.

The common horse bot[1] hovers about the animal without causing much excitement, bends her abdomen forward under the body, and, darting in, quickly glues an egg to a hair, usually on the front leg but also to a lesser extent on shoulders, belly, and hind legs. Egg after egg (Fig. 20.8, b) is attached in this way, so that the legs of horses on pasture or in corrals often assume a yellowish-gray cast from the large number of "nits" cemented to the hairs. The females average over 800 eggs. A very curious fact about the eggs of this fly is that they cannot hatch without some artificial aid. The necessary stimulus to hatching was long supposed to be moisture and friction. It has been discovered, however, that it is a sudden rise in temperature, the optimum temperature for hatching being between 114 and 118°F. The fully formed, tiny larva is ready to hatch 10 to 14 days after the egg was laid, but may remain alive in the eggshell for 100 to 140 days. At any time during this long period when the warm, moist lips or tongue of the horse pass over them in scratching or nibbling at the legs, they hatch very quickly and the larva clings to the lips or tongue. Upon entering the mouth, the first-instar larva burrows into the mucous membrane of the tongue and works its way backward toward the base of the tongue for 24 to 28 days, causing linear lesions. The molt to the second instar occurs when they leave the tongue. The small larvae then usually pass directly to the stomach, where they molt to the third instar after about 5 weeks. Unless the eggs on the hair of the horse are destroyed, larvae may continue to be taken into the mouth until about the middle of December. Having reached the stomach, 9 to 10 months are commonly spent there, during which growth is completed, there being but one generation a year.

The throat bot[2] hovers in front of the horse and, at intervals, darts upward and glues its eggs (Fig. 20.8,c), to the bases of hairs under the head or jaw of the horse or down on the throat. This causes the animals to nod their heads, drawing the nose in toward the breast and, when loose in pastures, to stand with their heads across the shoulders of another horse. These eggs are also whitish but can be distinguished from those of the common bot by the characters given in the key. The throat bot fly averages 400 to 500 eggs per female. These eggs do not require heat,

---

[1] *Gasterophilus intestinalis* (= *equi*) (De Geer), Order Diptera, Family Oestridae.
[2] *Gasterophilus nasalis* (= *veterinus*) (Linné).

moisture, or rubbing to cause hatching. After 10 to 12 days, the larvae emerge and are said to crawl down the hairs and wriggle through the hairs over the skin to enter between the lips. They are probably positively geotropic and hygrotropic and are said to reach the horse's mouth from eggs laid at least 8 inches away. The newly hatched larvae invade the spaces around and between the teeth, below the gum line, causing necrosis of the tissue and the formation of pus pockets. They remain in this location until several days after the first molt, a period of 17 to 30 days, when they pass into the duodenum. They tend to attach to the pharynx, paralyzing the muscles that control swallowing, or in the duodenum, and are less common in the stomach. The rest of the life cycle is similar to that of the common horse bot. The mature larvae pass from the host by the middle of May, the freshly dropped ones being whitish, whereas those of all the other species are reddish. Adults may be ovipositing by the first of June.

The lip or nose bot fly[1] is generally considered the worst species of the three. It lays its eggs, not about the nostrils, as the name nose bot fly wrongly implies, but attached very closely to the skin on the small hairs of the lips, mostly on the upper lip. Striking the horse on this sensitive spot causes it to jerk and toss its head, and repeated attacks often cause the animals to become frantic or to rub their lips over fences, or to stand with the lips appressed to another animal's back. The eggs (Fig. 20.8,a) are very different from the other species, being blackish and with a slender screw-like "tail" nearly as long as the egg. Only about 160 eggs per female are laid by this species. Moisture is apparently the stimulus to hatching of the eggs of the lip bot fly and temperature of little significance. Eggs attached to hairs of lips, where they are moistened by the saliva, hatch in 2 to 6 days; and it seems probable that those laid far from the lips never hatch. Immediately upon hatching the larvae are said to tunnel beneath the epidermis of the lips and migrate in this way into the mouth, thus causing the great irritation always noted when this fly is prevalent. Eventually they are swallowed with food or drink. From this point on, the life cycle is similar to that of the other species, except that these larvae, instead of passing with the excreta, have the habit of attaching to the walls of the rectum or to the skin about the anus for a time, after midwinter, before dropping to the soil, causing rubbing and switching of the tail. The adults begin to appear in early June, and some are present until frost, although any individual lives only a few days.

Control Measures. Bot flies are active only in the open and during the daytime, so that animals kept stabled during the day and pastured only at night, if at all, do not become infested. A piece of canvas stretched across the underside of the head and front of the neck, from the throatlatch or collar to the bit rings, effectively checks the egg-laying of throat bot flies. The lip fly may be similarly prevented from laying eggs by a belting lip protector, constructed as described by Bishopp and Dove in *U.S. Department of Agriculture Farmers' Bulletin* 1503. The application of equal parts of pine tar and lard will prevent egg-laying on the treated parts for about 4 days. By November first, in the latitude of central Illinois and Iowa, killing frosts will usually have destroyed all adult flies, so that no more eggs will be laid. The legs and even the underjaw and breast of all horses that bear eggs of bot flies should be

---

[1] *Gasterophilus haemorrhoidalis* (Linné).

promptly sponged, on a day when the temperature is below 60°F., with warm water (104 to 118°F.) containing 2 per cent phenol or coal-tar creosote. This causes the eggs to hatch and the larvae to die. No more larvae will then enter the body of the horse until the following summer. It is suggested that after delaying about 28 days more to allow any larvae burrowing in lips or tongue to reach the stomach, the digestive tract be fumigated in the following manner. After withholding all food for 24 hours and water for 5 hours, the horse is given a dose of carbon bisulfide, by means of a stomach tube or in a capsule that reaches the stomach unbroken. The proper dose is 2 to 2.5 cubic centimeters of carbon bisulfide for each 100 pounds of body weight of the horse or 6 drams per 1,000 pounds. It should be administered only by an experienced veterinarian and should not be attempted by inexperienced persons. Feeding and watering may be resumed 3 hours after the treatment. The carbon bisulfide forms a gas that kills the larvae, and they are passed in the feces beginning within 6 hours after treatment and continuing for 2 weeks or more. If the larvae are destroyed early in fall in this manner, the animals are spared most of the damage caused by the growth of the larvae in their stomachs, and, if communities cooperate in the treatment of all horses, the number of flies developing in successive years should be greatly reduced. In Illinois, 130,000 horses have thus been treated in a single year. This treatment also destroys the roundworm of the horse,[1] another serious intestinal parasite.

References. *U.S.D.A., Dept. Bul.* 597, 1918, and *Farmers' Bul.* 1503, 1926; *Bul. Ento. Res.* vol. 9, Part 2, pp. 91–106, September, 1918; *Iowa State Coll. Jour. Sci.*, **12**:181–203, 1938; *Jour. Econ. Ento.*, **33**:382, 1940.

B. INSECTS ATTACKING CATTLE

FIELD KEY FOR THE IDENTIFICATION OF INSECTS INJURING CATTLE

A. Free-flying insects that alight on the animals to suck blood, coming and going repeatedly:

1. Heavy-bodied, swift, brownish to black flies, ½ to 1 inch long, with 2 clear or banded wings, alight on animals and suck their blood, usually leaving a drop or two of blood exuding from the puncture. Especially active about marshes, swampy woods, or meadows and on warm sunny days.................................. ....*Horse flies*, pages 940 and 956.

2. Flies, of the size and general appearance of the house fly, suck blood especially from the legs, causing the animals to stamp their feet. Distinguished from the house fly by its stiff, slender beak, sticking straight forward from underside of head, by its broader black-spotted abdomen, grayer appearance, and by the bristle on the antenna having hairs on its upper side only. Palps less than half as long as proboscis. Usually rests with its head upward................... ....*Stable fly*, pages 942 and 956.

3. Flies similar to the house fly but slightly larger cluster on the faces of cattle, feeding on exudates of mucous membranes and causing great annoyance. Distinguished from house fly, in the male by the eyes, which almost meet instead of being separated by at least twice the width of the ocellar triangle, and in the female by the frontal stripe, which is less than twice as wide as that of one of the orbitals instead of 3 to 4 times as wide as in the house fly............. ....*Face fly*, page 956.

4. Flies similar to the stable fly, but only half as large, stand on their heads among the hairs especially of the back and sides of cattle and suck blood. Palps nearly as long as proboscis. They sometimes cluster in a mass about the base of the horns to rest.................................... ....*Horn fly*, page 954

[1] *Habronema* spp., Class Nematoda.

5. Very small, stout, black flies, less than ⅙ inch long, with a humped thorax, thick 11-segmented antennae as long as the head, 2 broad, delicately veined wings, and stout legs, pierce the skin and suck the blood, especially about the eyes, ears, and nostrils, causing extreme pain.............................. ....*Black flies, turkey gnats*, or *buffalo gnats*, page 1005.

6. Very slender-bodied scaly-winged bloodsucking flies swarming over the animals, especially at dusk and at night near swampy wooded sections............... ....*Mosquitoes*, pages 956 and 998.

B. *Insects that stay on the animals all the time:*

1. Very small lice, about ⅓ inch long, reddish in color, with distinct dark cross-bands on the abdomen; head broad, bluntly rounded in front. Chew hairs, epidermal scales, etc.; do not suck blood. Run about actively among the hairs. Eggs whitish, glued to hairs....................*Cattle biting louse*, page 957.

2. Slate-gray, wingless, broad, flat, bloodsucking lice, up to ⅛ inch long and half as broad; head pointed in front. The white eggs are glued to the hairs, especially about the fore part of the body.......*Short-nosed cattle louse*, page 959.

3. Similar to the short-nosed cattle louse but only about half as large when mature, often attaching in dense groups about head and neck. Eggs yellowish...... ....*A bloodsucking louse, Solenopotes capillatus* Enderlein, page 959.

4. Bluish, shiny, wingless lice, a little smaller, darker, much more slender, and with a longer head, than the preceding. Move about but little, usually stand on their heads among the hairs. Eggs nearly black, glued to the hairs...... ....*Long-nosed cattle louse*, page 958.

5. Hair comes out in patches; skin forms large scabs, under which may be found myriads of minute 8-legged oval-bodied pale mites, just visible to the naked eye, which pierce the skin with their mouth parts but do not tunnel beneath it...*Scab mite*, page 959.

6. Hard lumps, up to the size of a pea, form on skin of muzzle, head, and shoulders and burst open, disclosing a yellowish pus in which are many microscopic worm-like mites, with 8 short stubby legs toward one end of their short slender bodies...*Follicle mite*, page 960.

7. Eight-legged, leathery, glossy, ovate seed-like, brown to bluish-gray ticks, attached to the skin by their mouth parts and seldom moving about; up to ½ inch long. Drop off host to lay eggs in a cluster on the ground. Minute, first-instar nymphs or "seed ticks" have only 6 legs........................ *Cattle tick* (carries the disease known as Texas fever), *American dog tick*, and others, pages 960 and 962.

C. *Spiny worms or maggots that live in the flesh or in wounds:*

1. During the winter, tumorous swellings or "warbles" appear along the back, in which fat, wrinkled, white or blackish maggots, from ⅓ to 1 inch long, live; and from which they can be squeezed out. The full-grown larvae have minute spines on the posterior border of the ventral surface of segments 2 to 10, inclusive, *i.e.*, on all but the last larval segment. The surface of the stigmal plates is flat and the inner edges of the C-shaped stigmal plates are farther apart than in the bomb fly. These are the young of hairy, swift flies (about the size and appearance of a honeybee, ½ inch long, with hairy legs, uniformly white tegulae, and orange hairs at end of abdomen only) that glue small, slender, yellowish eggs in rows to hairs of legs, belly, etc., in early spring.... ....*Heel fly* or *common cattle grub*, page 963.

2. Like the above, but the last 2 segments of the larva without spines, *i.e.*, the ventral surface of segments 2 to 9 (but not 10) with small spines on the posterior border. The stigmal plates cup-shaped or funnel-shaped, and the inner edges of the C-shaped plates closer together than in the heel fly. Adults a little larger (⅔ inch long, with legs smoother, tegulae with a brown border, and yellowish hairs on front half of thorax as well as on base and tip of abdomen), appear later in spring and annoy cattle greatly, causing them to stampede or run madly, though the flies neither bite nor sting........................... ....*Bomb fly* or *northern cattle grub*, page 963.

3. Slender, whitish maggots, up to ¾ inch long, blunt behind, tapering in front.

and with an elevated ring at each segment, bearing short spines, burrow in wounds and *feed only in the living tissues of the animals.*   The eggs are laid about wounds and fly or tick punctures, by a metallic, bluish-green fly, larger than the house fly, with three black stripes on thorax and a reddish-yellow face. . . . . . .
. . . . *Screw-worm,* page 968.
4. Slender whitish maggots, blunt behind, tapering in front, and ¾ inch or less in length, work in the flesh about sores, especially following dehorning. . . . . . .
. . . . *Greenbottle fly* and *black blow fly,* page 984.

321. Horn Fly[1]

Importance and Type of Injury.   The horn fly is a close relative of the stable fly, and its harmful effects on cattle are very similar.   It pierces

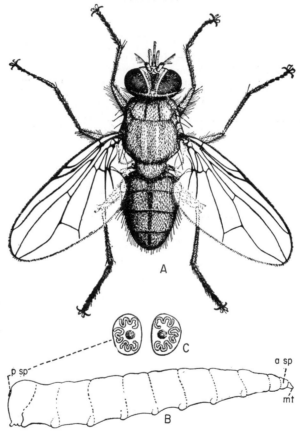

Fig. 20.10. Horn fly, *Haematobia irritans.* *A,* adult female, 10 times natural size; *B,* larva in side view; *a sp,* anterior spiracle; *mt,* mouth; *p sp,* posterior spiracles or stigmal plates; *C,* stigmal plates greatly magnified, showing median "button" and three sinuous spiracular slits.   (*Original.*)

the skin to suck blood, causing pain and annoyance and interfering with the feeding and resting of the animals, so that they lose weight, yield less milk, develop indigestion, and suffer other disorders.   These will be recognized as the small flies (Fig. 20.10), about half as big as the house fly

[1] *Haematobia* (= *Siphona*) *irritans* (Linné), Order Diptera, Family Muscidae.

or stable fly, that hover over the backs of cattle all summer long, especially out-of-doors, crawling down among the hairs on the withers, back, or belly and, with wings partly spread, suck blood until a swish of the head or tail scares them up, temporarily.   These flies have been suspected of transmitting anthrax.

Animals Attacked.   Cattle chiefly; also goats and sometimes horses, dogs, and sheep.   Annoying to people working about cattle.

Distribution.   This insect was first found in this country near Philadelphia in 1887 and is believed to have been brought to this country with shipments of cattle a year or two before.   Within 10 years it had spread over all of the United States east of the Rocky Mountains, and to California and Hawaii.   It is now generally distributed throughout America.

Life History, Appearance, and Habits.   Horn flies winter as larvae or pupae within puparia, in or beneath the droppings of cattle.   Toward the end of April in the latitude of central Illinois they begin to appear about cattle, and their numbers rapidly increase.   Apparently they develop only in fresh droppings of cattle, the flies darting from an animal to the fresh dung and depositing a few eggs on the surface of the mass, nearly always within a minute or two of the time it is dropped.   Within about 2 minutes the dung has lost its attraction for the egg-laying flies.   In masses of cattle dung dropped from 9 A.M. to 4 P.M., Mohr found an average of 150 larvae per mass, whereas dung dropped at night was uninfested.   The eggs are brown in color and not easy to see.   The maggots hatch from these eggs in a day or so, feed in the dung, and become full-grown (about $\frac{3}{8}$ inch long), in 3 to 5 days.   They then form pupae inside brown seed-like puparia, either in the soil or in the dung, and emerge as flies about a week later.   The entire life cycle may be completed in 10 days to 2 weeks.   The adults look like half-sized stable flies but do not have the spotted abdomen, and the palps at the sides of the beak are about two-thirds as long as the beak.   Although the flies are said to feed only once a day, they roost on the bodies of cattle both day and night.   When not feeding, they often rest about the head, especially on the base of the horn, if the animal has horns, sometimes so many of them that they make a black ring around the horn.   It should be clearly understood that no injury is done to the horn.   The flies remain abundant until frosts kill them, commonly toward the middle of October in central Illinois, when the immature stages in the dung go into hibernation.

Control Measures.   For beef cattle, thorough spraying of the back and flanks with the following materials, using 0.5 per cent in water at 2 quarts per animal, will give good control for about 3 weeks.   Sick animals and young calves should not be treated with the organophosphorus insecticides, and the indicated interval between treatment and marketing should be observed: methoxychlor, toxaphene (1 month), ronnel (2 months), and Co-Ral (2 months).   Back rubbers[1] treated with 5 per cent oil solutions of methoxychlor, DDT, or toxaphene are also effective.   For dairy cows, to avoid residues in milk, the neck and back should be rubbed at 2- to 3-week intervals with 1 teaspoonful of 50 per cent methoxychlor wettable or 3 teaspoonfuls of 5 per cent malathion dust.   Spraying daily with 1 per cent synergized pyrethrins or 3 to 5 per cent Lethane or Thanite in oil at 1 to 2 ounces per animal, or every 3 to 7 days with 0.05 per cent synergized pyrethrins in water at 1 to 2 quarts per animal, is also effective.   Fly

[1] *Jour. Econ. Ento.*, **45**:329, 1952; **52**:648, 1959.

repellents such as butoxypolypropylene glycol, di-*n*-butyl succinate, or di-*n*-propyl isocinchomeronate will protect animals for short periods. Darkened stables, with curtains or brush arranged over the entrance to brush the flies off as the cattle enter, give a measure of relief. If the cattle are being fed grain, hogs and chickens running with them help to control the flies by scattering the dung and destroying the maggots. Goats are annoyed much less if they are pastured apart from cattle.

References. *U.S.D.A., Bur. Ento. Cir.* 115, 1910; *Proc. La. Acad. Sci.,* **4**:129, 1938; *Jour. Econ. Ento.,* **21**:494, 1928; **31**:315, 1938; **47**:266, 1954; and **52**:1216, 1959.

321A. FACE FLY[1]

Importance and Type of Injury. The face fly, which closely resembles the house fly, congregates in clusters of 20 to 100 on the faces of cattle, where it feeds, without piercing the skin, upon the mucous membranes of the eyes, nose, and lips and upon fresh wounds or saliva deposits upon the shoulders, neck, briskets, and legs. The insect causes extreme annoyance, affecting milk and butterfat production.

Animals Attacked. Cattle chiefly, but other domestic animals and people working in the vicinity, are also annoyed.

Distribution. This insect, which is a native of Europe, was first found in Nova Scotia in 1952 and has now spread throughout the eastern and central portions of the United States.

Life History, Appearance, and Habits. The face fly very closely resembles the house fly in appearance (see distinguishing characters on page 952) and life history. The larvae breed in cow dung and other excrement and pupate in the adjacent soil. The adults are active from early spring to late autumn and congregate in large numbers in sunny spots, where they rest on buildings, fences, trees, etc., but avoid entering barns and shady areas.

Control Measures. The repellents di-*n*-butyl succinate and di-*n*-propyl isocinchomeronate will protect animals for about 1 day when thoroughly applied. Satisfactory control over a 1- to 5-day period may be obtained by painting the face of the animal with about 3 milliliters of a bait of 3 parts corn syrup and 1 part water containing 0.2 per cent DDVP and 1 per cent dimethoate.

References. *Can. Ento.,* **85**:422, 1953; *Jour. Econ. Ento.,* **53**:450, 1960.

322. OTHER BLOODSUCKING FLIES

The large bloodsucking horse flies or greenheads (Figs. 20.2 and 20.3) often punish cattle severely, especially if the pastures border woods or wet areas of land. Mosquitoes may dry up dairy cattle and cause loss of flesh in other animals, especially in wet seasons, when they come out of the water in swarms and bite the animals day and night. Draining wet land, screening stables, and providing smudges will give relief from horse flies and mosquitoes. Black flies (Fig. 21.4) attack about the eyes, nostrils, ears, and under the belly and may be sucked into the mouth or nostrils by snorting animals, which sometimes die as though suffocated by the innumerable gnats. Smudges or aerosol treatments appear to be the only practical help for this scourge. Further discussion of black flies is given on page 1005. The stable fly (Fig. 20.4), while generally less serious on cattle than the horn fly, in times of great epidemics may exceed the latter as a cattle pest. It bites chiefly about the legs. Control measures are the same as given under Horses.

[1] *Musca autumnalis* De Geer, Order Diptera, Family Muscidae.

323. Other Flies about Cattle

A number of flies, such as the house fly, that do not pierce the skin, visit the animals to suck up blood exuding from wounds made by horn flies, stable flies, horse flies, and others. They annoy the animals and may easily have a connection with the transmission of blood parasites from one animal to another. Other flesh flies besides the screw-worm sometimes deposit eggs or maggots in wounds where the maggots develop.

Cattle Lice

As in the case of the horse, two very different kinds of lice are common on cattle, some of them sucking the blood, and others, which cannot pierce the skin, living off the dry skin scales, hairs, and scabs. They are especially injurious to calves and to poorly fed, unhoused, old animals, during the winter months. The lack of oiliness of the skin of such animals makes conditions ideal for lice. Holsteins are said to be the worst infested and Jerseys the least so.

324. Cattle Biting Louse or Little Red Louse[1]

Importance and Type of Injury. When cattle rub and scratch against stanchions, fences, and other objects, and show areas of the skin which are full of scurf and partly denuded of hairs, or raw and bruised from rubbing, they are almost certainly infested either with lice or with scab mites. If lice are the cause of the irritation, parting the hairs and folds of the skin on head, neck, and shoulders will reveal the lice, while if the trouble is due to scab, no insects will be visible to the naked eye. The lice which suck blood (Fig. 20.12) are all a bluish slate color, while the biting louse (Fig. 20.11) is yellowish white with a reddish head and eight dark cross-bands on the abdomen, giving a somewhat ladder-like appearance to this part of the body. The biting lice crawl about freely over the skin between the hairs, irritating the skin both with their sharp claws and with their sharp chewing mandibles. When very abundant, they form colonies about the base of the tail or on the withers, which may become covered over with a light scurf, in patches as big as the hand.

Fig. 20.11. Cattle biting louse or little red louse, *Bovicola bovis,* female. About 20 times natural size. (*From U.S.D.A., Farmers' Bul.* 909.)

Under this scurf the lice are feeding on the raw skin. Such gross attacks weaken the animals, check growth, and predispose the animals to other diseases.

Animals Attacked. It is the general rule that each kind of animal has its own kinds of lice that do not feed on any other animal. Cattle lice have also been recorded from deer but do not attack other domestic animals.

Life History, Appearance, and Habits. Lice are most abundant during the winter, when the coat becomes thick and long and the skin is relatively dry of oil. At this season all sizes of lice and the eggs are to be found

---

[1] *Bovicola bovis* (Linné) (= *Trichodectes scalaris* Nitzsch), Order Mallophaga, Family Trichodectidae.

on an infested animal. The eggs are delicate, white, barrel-shaped objects, glued by one end to a hair while the other end has a slight rim within which fits a lid that is pushed off when the egg hatches. The young louse is paler in color but of the same form and structure as the old ones.   It appears from the egg within a week after laying, under favorable conditions, and within 3 weeks the young louse may have become full-grown (about $\frac{1}{12}$ inch long) and be laying eggs for another generation. Each female produces about 20 eggs.   Consequently they usually increase to great numbers during the winter and early spring, if not controlled. When the coat is shed in spring and during the heat of summer, the lice seem to disappear, but enough remain to carry the species over until favorable conditions permit them to increase again.   In one series of 692 lice collected from calves in Illinois, only 6 were males, and reproduction is usually parthenogenetic.

Control Measures.   These are given under Cattle Bloodsucking Lice, page 959.

References.   U.S.D.A., Farmers' Bul. 909, 1918; Conn. (Storrs) Agr. Exp. Sta. Bul. 97, 1918; Cornell Univ. Agr. Exp. Sta. Bul. 832, 1946.

325. Cattle Bloodsucking Lice

Importance and Type of Injury.   The symptoms of bloodsucking lice are similar to those shown when the little red louse is present, but the irritation is greater because of their habit of piercing the skin or "biting" to get the blood. The loss of blood keeps young animals runty and prevents normal production of milk or meat in older ones.   The animals scratch and rub persistently and patches of the skin become bare of hairs and are sore.

FIG. 20.12. Short-nosed cattle louse, *Haematopinus euryster-nus*, female. About 20 times natural size. *(From U.S.D.A., Farmers' Bul. 909.)*

Life History, Appearance, and Habits.   If an animal infested with bloodsucking lice is examined during the winter, dark-blue patches on the skin, often as big as a half dollar, which at first look like dirt, may be found on folds of the skin, on head, neck, withers, or along the inner surfaces of the legs.   Examining more closely will show that these spots are composed of clusters of little lice of all sizes (Fig. 20.12), standing on their heads, clinging by their claw-like legs to the hairs, and with noses appressed to the skin, from which they draw blood.   They move about very little except when laying their eggs.   The largest of them are only $\frac{1}{8}$ inch long, and successively smaller ones are generally present, down to the size of the egg. All stages are passed on the skin of the host, the eggs usually hatching in 10 days to 2 weeks.   The nymphs feed frequently, and the females begin to lay eggs when they are about 2 weeks old.   The eggs are glued fast to the base of the hairs by one end, and are somewhat keg-shaped.

The long-nosed cattle louse,[1] which is especially prevalent on young

---

[1] *Linognathus* (= *Haematopinus*) *vituli* (Linné), Order Anoplura, Family Haematopinidae.

cattle, is a slender species about $\frac{1}{10}$ inch long, the body one-third as wide as long, and the head nearly twice as long as broad and pointed in front. Its eggs are dark blue.   The short-nosed cattle louse[1] (Fig. 20.12) is the largest louse found on cattle, being $\frac{1}{8}$ inch long and much broader than the long-nosed louse, and the head is only a half longer than broad and bluntly pointed in front.   It is said to be more common on mature cattle. The eggs are white.   The small blue louse[2] is similar to the short-nosed cattle louse, with the head still shorter and rounded in front.   It is only about half as large as the short-nosed species, and its eggs, which are yellowish, are said not to hatch if removed from the host.   The cattle tail louse[3] is found in the long hair about the tail and also around the eyes and on the neck.   It resembles the short-nosed louse but is somewhat larger.

Control Measures.   Lice are most prevalent on poorly fed animals. Cattle should be well fed, kept in clean, light, well-ventilated stables, and not overcrowded.   As soon as they are brought off pasture in the fall, the neck and withers should be examined for the presence of lice. These will be especially easy to see on white or light-coated animals.   If even a few lice are found it is a good practice to treat all the animals in the herd by thoroughly spraying the entire body with one of the following materials, observing the indicated interval between treatment and marketing: 0.5 per cent methoxychlor, toxaphene (1 month), malathion, Co-Ral (2 months), ronnel (2 months), or 0.03 per cent lindane (1 month). A second application is usually needed after 2 to 3 weeks.   Sick animals or young calves should not be treated with the organophosphorus insecticides.   Dipping cattle in 0.5 per cent toxaphene, using a very stable emulsion especially prepared for the purpose, is very effective where large numbers of beef cattle are to be treated.   This should be done in the fall, before cold weather, and complete directions are given in *U.S. Department of Agriculture, Farmers' Bulletin* 909, 1953.   For dairy cows, to avoid residues in the milk, lice should be controlled by spraying with 0.025 per cent synergized pyrethrins or 0.0125 per cent rotenone or dusting with 1 per cent rotenone.

References.   *U.S.D.A., Farmers' Bul.* 909, 1953, *Bur. Ento. Cirs.* E-477, 1942, and E-762, 1949; *N.Y. (Cornell) Agr. Exp. Sta. Bul.* 832, 1946; *Jour. Econ. Ento.,* **40**:672, 1947; **48**:566, 1955; and **50**:618, 1957.

326 (319, 339).   SCAB MITE OR PSOROPTIC SCAB[4]

If animals that are rubbing and scratching, twitching the skin, and shaking their heads do not show lice upon examination, cattle scab should be suspected.   An examination of skin scrapings under a microscope will usually reveal the minute whitish eight-legged mites (Fig. 20.25) that cause this trouble by puncturing the skin with their sharp mouth stylets.   They are similar to the mange mite but can be distinguished by their longer legs and minor differences in the appendages.   They do not burrow into the skin but rest and feed upon the raw skin, completely covered over by scabs.   Infestations usually begin on thickly haired parts of the body.   The earliest symptoms are small reddened pimples that ooze pus.   As the mites increase, larger areas become covered with yellowish crusts filled with serum.   Large scabs form on the skin over the

---

[1] *Haematopinus eurysternus* (Nitzsch), Order Anoplura, Family Haematopinidae.

[2] *Solenopotes capillatus* Enderlein.

[3] *Haematopinus quadripertusus* Fahrenholz.

[4] *Psoroptes equi* Gervais, Order Acarina, Family Sarcoptidae.

mites and the hair comes out in great patches. This mite and its control are further discussed under Sheep, page 979. Valuable animals should be dipped a number of times.

327. Cattle Follicle Mite[1]

A minute worm-like mite sometimes attacks cattle, burrowing deep into the natural pores of the skin and causing lumps as large as peas to form about the head and over the shoulders. It injures the hides. The mites cannot easily be killed, and it is best to kill or market badly infected animals. Very valuable animals may be cured by persistent dipping, as for mange.

328. Cattle Tick[2]

Importance and Type of Injury. The most destructive parasite of domestic animals throughout the southern states is the cattle tick (Fig. 20.13). For more than a century this pest held back the cattle and

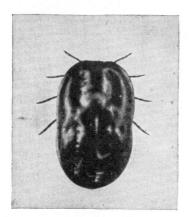

Fig. 20.13. Cattle tick or Texas-fever tick, *Boophilus annulatus.* At left, full-grown, engorged female tick, ready to drop from animal and deposit eggs; at right, female laying eggs on ground. Each female may lay as many as 5,000 eggs. Both figures 3 times natural size. (*From U.S.D.A., Farmers' Bul.* 498.)

dairy interests of the southern states and, indirectly, the entire agricultural development of these states. The cattle tick is a bloodsucking parasite and is injurious to cattle in the several ways described above for lice and flies. In addition to this it is the only means of spread of the disease known as cattle fever, tick fever, splenetic fever, or Texas fever, from sick animals to healthy ones. The principal symptoms of tick fever are a high fever, reddish discoloration of the urine, enlarged spleen, congested liver, loss of flesh and condition, dry muzzle, arched back, and drooping ears. Death results in from 10 per cent of the cases occurring during the summer months, in southern cattle, to 90 per cent of those during the autumn and early winter, especially among northern or imported stock. The cause of the disease is a minute parasite,[3] belonging to the phylum Protozoa, which lives inside the red corpuscles of the

[1] *Demodex bovis* Stiles, Order Acarina, Family Demodicidae.
[2] *Boophilus* (= *Margaropus*) *annulatus* (Say), Order Acarina, Family Ixodidae.
[3] *Babesia bigemina* (Smith and Killbourne).

blood and destroys them, thus causing the disease. It should be clearly understood that while the *direct* cause of the disease is a protozoon, the only method of spread of the disease is by the bite of the tick. If there are no ticks in a given section, there can be no Texas fever. Ticks do not pass from one animal to another, but spend their entire feeding period on a single animal. The germs of Texas fever pass through the eggs of diseased ticks to their young, and the young later get on new animals and so infect them. The presence of this disease has meant incalculable loss to the southland from the failure of diseased cattle to make beef production or dairying profitable, and especially from the "one-crop system" which has prevailed largely because of the difficulty and discouragements of cattle-raising in tick-infested territory. According to the Department of Agriculture, the average price of cattle in the tick territory is only about half that for similar grades in tick-free states. The hides are damaged from 50 cents to $1.25 each by the punctures of the ticks, and milk yield is reduced from 18 to 42 per cent, depending on the severity of the infestation. All in all, it is generally estimated that the tick costs southern agriculture about $2,500,000 a year. The punctures made by ticks attract the screw-worm fly (page 968) to lay her eggs upon the cattle.

Animals Attacked. The cattle tick is chiefly a pest of cattle but is found also on horses, mules, sheep, and deer. It does not transmit Texas fever to the other animals.

Distribution. The cattle tick occurs only in the southern half of North America, in the states south of West Virginia, Kentucky, Missouri, Kansas, Arizona, New Mexico, and Nevada. The disease that it transmits is prevalent also in Central and South America, Africa, Europe, and the Philippines, where it is carried by other kinds of ticks.

Life History, Appearance, and Habits. The cattle tick winters in two distinct ways in the South. Some individuals winter as eggs or minute six-legged "seed ticks," about half the diameter of a pinhead ($\frac{1}{32}$ inch). They may be found on the ground or on grass, weeds, and other objects in fields where infested cattle have pastured. Other individuals winter as nymphs or adults on the skin of the cattle. When the ticks reach maturity they mate on the animal, and then the female, having fed to repletion on blood, loosens her mouth parts from the skin and drops to the ground. At this time she is nearly $\frac{1}{2}$ inch long, the body olive-green to bluish-gray, bean-shaped, with hard, glossy, and somewhat wrinkled skin (Fig. 20.13, left). Within 2 to 6 weeks after dropping from the animal, each female ordinarily lays from a few hundred to 5,000 eggs, pushing them out in front of her head in a large brownish mass on the soil. The eggs

Fig. 20.14. Larvae or seed ticks of the cattle tick, just after hatching. 8 times natural size. (*From U.S.D.A., Farmers' Bul. 498.*)

usually hatch in from 13 days to 6 weeks, but eggs may remain viable for 7 or 8 months. The young that hatch from the eggs are known as larvae or "seed ticks," and have only 6 legs (Fig. 20.14). They crawl about actively on the ground and climb up on vegetation, often in large bunches, and remain there waiting for an animal to come along. If an animal brushes by them they cling fast and at once insert their mouth parts and feed in that one spot for a week or two, then shed their skins and appear as nymphs with an extra pair of legs, now having eight for the rest of their lives.   Another period of feeding for about a week enables them to molt and transform to adult males and females.   After mating, the female, still feeding, increases rapidly in size for a few days to a few weeks before she drops to the ground to lay eggs for another generation.   If the seed tick is not brushed off the vegetation, it must starve to death but may live 200 to 250 days before it perishes.   There are thus four stages in the life of a tick: the egg, the seed tick, the nymph, and the adult.   The life cycle is in two periods, a *parasitic* and a *nonparasitic* period.   The parasitic period begins when the tiny seed tick is picked up by an animal, and ends when the fully-engorged female drops from the host.   Part of the adult female lifetime, all the egg stage, and most of the seed tick's life are spent on or near the ground.   A life cycle requires about 60 days, at the least, and there are three generations throughout most of the South.

   Control Measures.   A quarantine maintained by state and federal governments prevents the movement of southern tick-infested cattle out of the quarantined area.   The cattle tick cannot survive the winter in the northern states.   Without the tick there can be no fever, so that cattle in the North, which are very susceptible, are protected from the disease by the quarantine.   In the South two measures of control have gone hand in hand: (1) Dipping or spraying animals with arsenic trioxide, toxaphene, or lindane.   It has been found that a farm can be completely freed of ticks by dipping all cattle, horses, and mules on the farm every 2 weeks, all spring and summer, and allowing them to range over the pastures and pick up the seed ticks between dippings.   (2) Rotating pasture lands, *i.e.*, keeping all animals off certain fields until the seed ticks will all certainly have starved, and then putting on such land only cattle freed of ticks by dipping.   Both these measures have been worked out with the greatest of care and are fully described in *U.S. Department of Agriculture, Farmers' Bulletin* 1057.   By a combination of quarantines, dipping, and pasture rotations, in operation since 1906, over 700,000 square miles of southern land (98 per cent of the area originally quarantined) have been freed of the curse of Texas fever.   With the exception of a buffer zone in Texas along the Mexican border, this disease has been eradicated from the United States.   The tick is controlled on beef cattle by dipping or thorough spraying with 0.025 per cent lindane or 0.5 per cent toxaphene, or by spraying with 0.5 per cent malathion or Co-Ral, or with 0.75 per cent ronnel (see cattle lice, page 959).   Dairy cattle may be dipped in 0.175 to 0.19 per cent arsenic trioxide, solubilized as sodium arsenite or sprayed with 0.1 per cent synergized pyrethrins or 0.15 per cent rotenone.

   References.   U.S.D.A., *Farmers' Buls.* 1057, 1932, and 1625, 1930, *Bur. Ento. Bul.* 72, 1907, and *Misc. Publ.* 2, 1927; *Tenn. Agr. Exp. Sta. Bul.* 113, 1915; *Jour. Econ. Ento.*, **41**:104, 1948; and **42**:276, 1949.

329 (340, 361, 362). OTHER TICKS

   Several other species of ticks attach to cattle and suck their blood, but none of them carry Texas fever or other cattle diseases so far as known.   The most serious species is the ear tick[1] which is prevalent in the semiarid sections of southwestern United States.   These ticks have the curious habit of entering the outer ears of cattle, horses, mules, dogs, sheep, and other animals, where they live and feed for

[1] *Otobius* (= *Ornithodoros*) *megnini* (Duges), Order Acarina, Family Argasidae.

from 1 to 7 months.   The ears may be literally packed full of these vermin, which range up to ⅓ inch or more in length.   After this period of feeding in the ear of the animal, the full-fed nymphs (Fig. 20.15), which have the skin covered with short heavy peg-like hairs, leave the host, crawl into cracks about stables, fences, or trees, and there shed their skins and become adults.   The adult stage is not found on animals but, after mating, the females lay eggs in dry places usually up off the ground.   The newly hatched "seed ticks" subsequently attach to the skin of the animals which they infest.   The Gulf Coast tick[1] has very similar habits.   These ticks are so irritating to

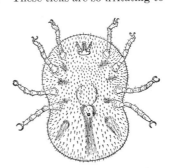

animals that they run about shaking their heads or roll on the ground until they become exhausted, ill, or deaf or die from the attack.   The wounds made by the tick in feeding often become infested by the screw-worm.   Nymphal Gulf Coast ticks are frequently found on quail, meadow larks, and other ground birds and may cause their death.   The ear tick[2] may be killed and reinfestation prevented for several weeks by injecting into each ear, by means of a syringe, ½ ounce of a mixture of 0.75 per cent lindane dissolved in 2 parts of xylene and 17 parts of pine oil by weight.   Spraying in the ears with 0.03 per cent lindane or 0.5 per cent toxaphene or dusting in the ears with 0.5 per cent Co-Ral has given effective control.   Good control of the Gulf Coast tick[1] has been obtained by treating the inside and outside of the ears and the base of the horns with a mixture of 5 parts of technical DDT dissolved in 15 parts by weight of dibutyl phthalate and mixed with 47 parts of rosin and 33 parts of hydrogenated methyl abietate.[3]   This should be repeated every 30 days to kill the ticks present and repel others.   The American dog tick,[4] which is often found on dogs, commonly attaches to cattle pastured in woodlands.

FIG. 20.15. Ear tick, *Otobius megnini*, nymph, ventral view, greatly enlarged. Note the mouth parts or capitulum, consisting of hypostome, chelicerae, and pedipalps on the underside of body between front legs and the covering of short peg-like setae. (*In part after Nuttall, Monograph of the Ixodidea.*)

The Rocky Mountain spotted-fever tick (page 1017) and the lone-star tick[5] also attack cattle, in regions where these ticks occur.   Area spraying of roadsides and woodlands with DDT, toxaphene, chlordane, or dieldrin at 1 to 2 pounds per acre has controlled these ticks.

References.   U.S.D.A., *Farmers' Buls.* 980, 1947, and 1150, 1920; *Jour. Econ. Ento.*, **29**:1068, 1936, and **40**:301, 303, 1947; *U.S.D.A., Bur. Ento. Plant Quar., Cirs.* E-719, 1947, and E-695, 1946.

330. Cattle Grubs or Ox Warbles

Two different species of bot flies attack cattle in America: the smaller one is known as the heel fly or common cattle grub;[6] the other is known as the bomb fly or northern cattle grub.[7]   There are very important differences in the habits of the two species, but their life histories and control are sufficiently similar to make their consideration together advantageous.

Importance and Type of Injury.   Tumors under the skin of the back, as big as the end of one's thumb, contain each a large fat maggot (Fig. 20.16),

[1] *Amblyomma maculatum* Koch.
[2] *Otobius* (= *Ornithodoros*) *megnini* (Duges), Order Acarina, Family Argasidae.
[3] Hercolyn.
[4] *Dermacentor variabilis* (Say), Order Acarina, Family Ixodidae.
[5] *Amblyomma americanum* (Linné), Order Acarina, Family Ixodidae.
[6] *Hypoderma lineatum* (De Villiers), Order Diptera, Family Oestridae.
[7] *Hypoderma bovis* (De Geer).

which may be squeezed out, but which emerges when "ripe" and falls to the ground. Hairy flies about as big as honeybees chase the animals in pastures, while laying their eggs on them. The cattle usually run wildly, with their tails in the air, and often injure themselves in their

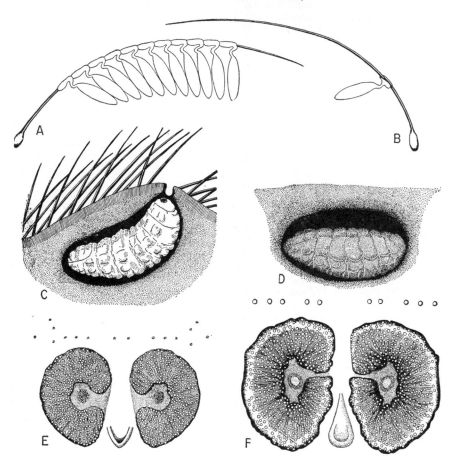

FIG. 20.16. Cattle grubs. *A*, eggs of the heel fly as often laid in a row on the hairs; note curved petiole arising near distal end of clamp. *B*, eggs of the bomb fly, as usually laid one to a hair; note straighter petiole arising near middle of clamp. *C*, larva in last instar in the tumor beneath skin of back; note hole through skin for respiration, with posterior end of larva bearing stigmal plates next to the hole. *D*, puparium in the soil in which the transformation from larva to fly takes place. *E*, posterior stigmal plates of the heel fly, *H. lineatum*, from fifth-instar larva; note that the plates are flat, with a rather wide gap on their inner margins. *F*, posterior stigmal plates of fifth-instar larva of the bomb fly, *H. bovis;* note that the plates are funnel-shaped and have a much narrower gap on their mesal margins. *A* and *B*, about 20 times natural size; *C* and *D*, enlarged about ⅓; *E* and *F*, greatly magnified. (*Drawn by Kathryn Sommerman.*)

attempts to get away. The small maggots from these eggs tunnel into the skin and migrate through the body for about 6 months before they find lodgment in the back, causing inflammation and suffering; milk production is reduced; growth and fattening checked; the quality of the meat is lowered, where the maggots tunnel through the flesh of the back; and the value of the hides for leather is greatly depreciated. The loss from this insect has been estimated to average $160,000,000 a year in the United States alone. Bishopp estimates that 75 per cent of the cattle in the United States are infested with from 1 to about 100 grubs each. In certain experiments a gain in milk flow of 25 per cent resulted when the grubs were removed from the back, and 5 per cent gain in weight was made by animals from which grubs were extracted, as compared with similar animals in which they were allowed to develop to maturity.

Animals Attacked. Cattle and bison and possibly some other wild ruminants and very rarely horses, goats, and man.

Distribution. The heel fly is the more prevalent and widely distributed species, occurring in every state in the union. The bomb fly does not occur in the southern states but is recorded as the more common species in the northeastern states. It ranges from coast to coast in southern Canada and the northern United States.

Life History, Appearance, and Habits. Many erroneous theories and statements about the life cycle of cattle grubs are current in print and among herdsmen. Consequently the true life cycle, which has now been established after more than 200 years of study, should be carefully noted. All cattle grubs winter as maggots (Fig. 20.16,C), in the backs of animals. In the latitude of central Illinois the first ones begin to appear under the skin of the back in late December. From that time on until April the swellings can easily be felt by passing the hand over the backs of infested cattle. The presence of the maggots causes the formation of a cyst about them; constant irritation by the spiny maggot keeps the walls of the cyst inflamed, and the grub lives on the secretions or suppuration inside the cyst wall. Soon after reaching the back, the larva cuts a small hole through the skin through which it takes air. Since maggots have spiracles at the posterior end of the body, this "tail" end lies against the hole in the hide. Any one larva spends about 2 months under the skin of the back, growing rapidly until it is about 1 inch long and very thick. It is significant that the larvae at this time live in that part of the body from which the most valuable steaks are secured and the holes they make are through that part of the skin which makes the best leather. When mature, the larva changes from white to dark brown or black and finally squeezes its body, tail foremost, through the hole in the skin and drops to the ground, usually in the morning. The larvae become mature and leave the back over a period of 5 or 6 months, from January to June in the North, and from November to March in the South. Having freed itself from the animal the maggot seeks protection on or in the soil and changes, through a hard-skinned pupal period, to the adult fly, which splits open the end of its case and emerges about 5 weeks later. The adults of the heel fly are found in Illinois from early April to the end of June and the bomb fly chiefly from early June to mid-August. Some of the flies (Fig. 20.17), therefore, may be found about cattle from April to September although any one fly probably lives only a few days and takes no food. In southwest Texas, Bishopp reports the heel fly as active throughout the warm periods of the winter.

The eggs of these flies are never laid on the back near the spot where the tumors appear, but mostly about the legs and lower part of the belly. The bomb fly is active chiefly on hot sunny days and is very tactless in laying its eggs. It darts at the animal with much buzzing and clings to the skin for a second while it glues an egg fast to the base of a hair. Then it retreats and in a few minutes strikes again. This quickly excites the animals, though it does not seem that it could hurt them, and they throw their tails in the air and start to run wildly. The fly follows until it loses its victim in the underbrush or in water. As many as 800 eggs may be laid by one fly, one to a hair here and there over the hind legs, especially near the hock and the knee, and also on the belly. This sudden stampeding of the herd is one of the serious phases of injury by this fly and the thing which suggested the name "bomb fly." The heel fly is sneaking in its egg-laying. It generally alights in the shadow near an animal, and if the animal is standing, quietly backs up and tucks its eggs among the

Fig. 20.17. Heel fly or common cattle grub fly, *Hypoderma lineatum;* male, about 4 times natural size. (*From U.S.D.A., Dept. Bul.* 1369.)

hairs of the "heels" or, if the animal is lying down, attaches them to such hairs as it can reach along the flanks, while standing on the ground. Egg-laying causes no pain and the fly is usually unnoticed, so that it may lay a dozen or more eggs on one hair in a row like the teeth of a comb (Fig. 20.16,*A*). In from 3 days to 1 week the eggs of either species hatch, and the minute larva crawls down the hair and bores into the skin near its base. Penetration of the skin takes several hours and causes much irritation to the animal. The tiny larvae then disappear from view, at a point usually on the legs, and the heel fly larvae are commonly next seen in butchered animals, lying in the walls of the pharynx, gullet and esophagus, especially from August to November. By this time they are second- and third-instar larvae, one-fourth to one-third grown. The maggots are constantly on the move, thus avoiding the attacks of white blood corpuscles, and there is every reason to believe their visit to the esophagus is only incidental to their general wandering about in the connective tissue of the body (Fig. 20.18). The larvae of the bomb fly are rarely found in the gullet, probably taking a more direct route to the back where they are frequently found inside the neural canal. At the approach of winter the heel fly maggots begin to disappear from the gullet and by the end of

December, the warbles begin to appear under the skin of the back as third-instar larvae about ½ inch long. In this stage the hole is cut through the skin, an operation requiring about 4 days. The larva then molts to the fourth instar and, 2 to 8 weeks later, to the fifth and final instar during which most of the growth takes place. According to Bishopp, the heel fly larva spends an average of 58 days in the larval stage after reaching the back, and the bomb fly larva an average of 73 days in this situation. There is but one generation a year.

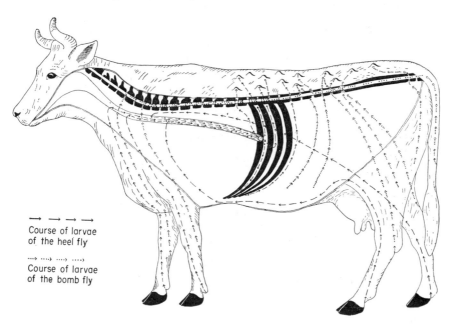

→ → → →
Course of larvae
of the heel fly

····⟩ ····⟩ ····⟩ ····⟩
Course of larvae
of the bomb fly

Fig. 20.18. Diagram showing the course of cattle grubs through the body of the animal, from the position where the eggs are laid on the legs or flanks of the animal to their final feeding place beneath the skin on the back. The solid arrows suggest the course of the larvae of the heel fly, *H. lineatum*, to the gullet, thence backward to the diaphragm and up to the back, or to the ventral ends of the ribs and thence beneath the pleura up to the back. The dotted arrows suggest the course of the larvae of the bomb fly, *H. bovis*, from the point where the eggs were laid, directly to the back, or through the neural canal inside the backbone and thence to the back. This species is not found in the gullet. (*Drawn by Kathryn Sommerman.*)

Control Measures. For many years the standard remedy involved attacking the grubs after they arrived in the back of the cattle, from December to June, using rotenone. This is still the only satisfactory treatment for dairy cattle. Rotenone should be applied as soon in midwinter as the first lumps appear under the skin, which will vary from early December in the South to the middle of February in the most northern areas, using (*a*) 1.5 per cent dust at 4 to 5 ounces per animal, thoroughly rubbed into the hair of the back, (*b*) a wash of 5 per cent rotenone powder, 12 ounces per gallon of water containing 0.5 ounce of wetting agent, brushed into the hair of the back at 1 to 2 pints per animal, or (*c*) a spray of 0.05 per cent rotenone applied to the back at 2 to 4 quarts per animal, using 350 to 400 pounds per square inch pressure. Several treatments will be required at 30- to 45-day intervals as long as the grubs appear in the back.   Ointment containing 0.5 to 2.5 per cent rotenone in lanolin, petroleum jelly, or olive oil has also been used, rubbing a small quantity into the top of each lump.   For beef cattle, treatment with the animal systemic insecticides Co-Ral and ronnel is more effective and prevents damage to the hide by killing the grubs before they reach the back. These materials should be applied from July to September, and the animals should not be marketed for 2 months after treatment.   Sick animals or calves should not be treated.   Co-Ral is applied as a 0.5 per cent spray, using from 2 to 4 quarts over the entire body.   Ronnel is applied orally with a balling gun or as a water drench, using one 37.5-gram bolus containing 40 per cent ronnel per 300 pounds of animal body weight.   Community cleanup programs where all cattle in a well-defined area are treated will greatly reduce the fly population the following summer.

Stabling of cattle during the daytime and running them on pasture only at night is a complete preventive measure, since the eggs are laid only in daytime and the flies do not enter stables.   Darkened sheds or brush shelters, into which the animals can retreat when flies are pestering them, are highly recommended.   It is a common practice to squeeze the larvae out of the backs of animals by pressure from the hands, or remove them with sterilized forceps, but this is likely to crush the maggots and it is considered dangerous to burst them under the skin because of the poisonous nature of the body fluids of the maggots, which may cause anaphylactic shock.   Suction pumps have been used to draw out both the maggots and the pus that surrounds them.   If the warbles are removed mechanically they must be killed before being thrown down.   No practical method of control is known for cattle on the range, but if herds can be kept on the move, as from lower to higher ranges as the season progresses, they will leave the larvae which drop from their backs far behind and escape attack by the egg-laying flies that develop from them.

References.   U.S.D.A., Dept. Bul. 1369, 1926, *Bur. Ento. Plant Quar. Cir.* E-623, 1944, and *Yearbook* **1952**:672, and *A.R.S.* 22–37, 1957; *Can. Dept. Agr. Health of Animals Branch Bul.* 16, 1912, and *Sci. Ser. Buls.* 22, 1916, and 27, 1919; *Jour. Agr. Res.,* **21**:439, 1921; *Jour. Econ. Ento.,* **40**:293, 928, 1947; **41**:779, 783, 1948; **51**:582, 876, 1958; and **52**:425, 1959.

331 (333, 369). SCREW-WORM[1]

Importance and Type of Injury.   The presence of the maggots of flies in the living bodies of man or other animals is called *myiasis.*   While a number of kinds of flies such as the horse bots and ox warbles attack perfectly healthy animals in this way, certain others parasitize animals only when there is a wound or diseased body opening to attract the egg-laying female flies.   Among the most serious of these pests is a dark, shiny blue-green blow fly, about twice as large as the house fly, with three black stripes on the back between the wings, and a reddish-yellow face (Fig. 20.20), which lays eggs only about the edges of wounds on animals, such as barbed-wire cuts, scratches from fighting, blood spots where ticks or other flies have bitten the animal, brand marks, sore eyes, and wounds from dehorning or castrating.   The maggots of these flies (Fig. 20.19)

[1] *Callitroga* (= *Cochliomyia*) *hominivorax* (Coquerel) (= *C. americana* C. & P.), Order Diptera, Family Calliphoridae.

start to feed in the wounds but soon invade the sound tissue, tearing it with their mouth hooks.   Wounds are prevented from healing, and the sickened animal hides away in the woods or brush, refusing to eat and usually dying if not found and treated.   The odor of an infested wound attracts additional flies to lay eggs about it and hundreds of maggots may

Fig. 20.19. Larval screw-worm, side view.   About 4 times natural size.   (*From U.S.D.A., Farmers' Bul.* 857.)

produce a terrible sore.   Other species, such as the secondary screw-worm[1] and blow flies (Fig. 1.11), which normally lay their eggs upon the carcasses of dead animals, may also attack wounds already infested by the screw-worm, but this species is said to cause 90 per cent of the primary invasions in the southern states.   How long the animal lives depends upon the number of larvae present and the location of the wound.   If the infestation is in the eyes or nasal passages or follows dehorning, meningitis frequently follows and kills the animal. Infestations about the navel of a newborn animal frequently result in peritonitis and death.   A very serious epidemic, beginning in 1932, resulted by 1934 in over 1,350,000 cases of infestation and the death of over 200,000 animals in the Gulf states alone.   Losses are estimated to reach $10,000,000 in certain years.

Animals Attacked.   Cattle, hogs, horses, mules, sheep, goats, man, dogs, and other domestic and wild animals.

Distribution.   Resident only in the most southern parts of the United States and southward to Argentina, this insect may range northward in summer to the Carolinas, Illinois, Kansas, and California and even

Fig. 20.20. Adult screw-worm fly, *Callitroga hominivorax.*   About 5 times natural size.   (*From U.S.D.A., Farmers' Bul.* 857.)

southern Canada.   It is probably always brought into these northern areas in infested animals, but may then breed there for several months, until killed by cold weather.

Life History, Appearance, and Habits.   The adults are active throughout the winter in extreme south Texas and Florida and begin to appear on the wing from early April to mid-June farther north.   The female screw-worm fly lays up to 3,000 eggs upon the dry skin near a wound or infected

[1] *Callitroga macellaria* (Fabricius).   This species normally breeds in unburied carcasses and transfers its attack to living animals only when it has become excessively abundant.   See *Ann. Trop. Med. Parasitology,* **27**:539, 1933.

body opening, in regular shingle-like masses of 200 to 400, firmly cemented to the skin and to each other. They hatch within 10 to 20 hours, and the small maggots tear out pockets in the healthy flesh adjacent to the wound with their sharp mouth hooks, severing minute blood vessels, and continually secreting a toxin that completely prevents healing and promotes contamination resulting in very foul-smelling pus-discharging wounds. The larvae are of typical maggot shape (Fig. 20.19) about ⅔ inch long when full-grown, with elevated spinose circlets at each segment, somewhat suggesting the ridges of a screw. Since they breathe through spiracles located at the large, blunt, posterior end of the body, they must frequently back out of the deeper recesses of a wound until the spiracles contact air. The larvae commonly become full-grown in from 4 to 10 days, when they drop to the ground and spend 3 to 14 days or longer in a brown seed-like puparium in the soil, transforming to adults. Under favorable conditions a generation averages about 3 weeks, and it is believed that there are commonly 8 to 10 generations during a summer.

Control Measures. The most important control measure is to prevent the breaking of the skin of animals or the flowing of blood during the warm periods of the year when the egg-laying flies are active. Dehorning, castrating, branding, earmarking, docking of lamb tails, and similar operations should not be performed in spring or summer. Dogs should not be allowed to bite livestock. Barbed wire, projecting nails, and similar snags that may tear the skin should be avoided. In areas where the flies hibernate in winter, animals should be bred so that the young will be born between mid-November and April first. In the most southern areas and for all summer births, the navels of the young and the vulvas of the dams should be treated lightly with one of the smears described below. Unavoidable wounds should also be treated, and every effort be made to avoid insect and tick bites, which are often followed by screw-worm attack. Screw-worm-infested wounds should be treated by applying one of the following smears, using a 1-inch paint brush to coat completely around the wound and work the material into any deep pockets made by the maggots. Treatment should be given twice the first week and weekly thereafter until the wound is healed.

Smear EQ 335		Smear 62	
Parts by Weight		*Parts by Weight*	
Lindane	3	Diphenylamine	3.5
Pine oil	35	Benzene	3.5
Mineral oil	40–44	Sulfonated castor oil	1
Emulsifier	8–12	Lampblack	2
Silica aerogel	8–12		

For beef cattle, spraying all wounds thoroughly and wetting the entire body with 0.5 per cent Co-Ral or ronnel as described under louse control (page 959) will protect the animals from infestation for 2 to 3 weeks by killing the newly hatched larvae. A smear of 5 per cent ronnel applied as suggested above is also very effective. The screw-worm was eradicated from Florida in 1960 by the release of sterile male flies as described on page 406.

References. U.S.D.A., Farmers' Bul. 857, 1926; Dept. Bul. 1472, 1927, Tech. Bul. 500, 1936, Circs. E-540, 1941, E-708, 1947, and E-813, 1951, Yearbook, **1952**:666; Fla. Agr. Ext. Bul. 123, Rev. 1944; Jour. Econ. Ento., **19**:536, 1926; **30**:735, 1937; **35**:70, 1942; **38**:66, 1945; **48**:462, 1955; **52**:106, 1217, 1959; and **53**:1110, 1960.

C. INSECTS ATTACKING HOGS

So far as its insect parasites are concerned, the hog sustains its reputation of being a hardy and healthy animal. It has only two serious insect parasites, the hog louse and the mange mite, besides which it is attacked to some extent by several kinds of flies.

FIELD KEY FOR THE IDENTIFICATION OF INSECTS INJURING HOGS

A. *Free-flying insects that alight on the animals to suck blood, coming and going repeatedly:*
  1. Flies of the size and appearance of the house fly often cluster about the ears and in other places on the body and suck blood from the animals, causing much pain. They can be recognized by the slender, stiff beak that projects forward from the lower side of the head; by the broad abdomen, that is gray with 4 to 6 black spots; and by the palps, that are less than half as long as the beak....
....*Stable fly*, page 942.
  2. Heavy-bodied, swift, brownish to black flies, $\frac{1}{2}$ to 1 inch long, with two clear or banded wings, alight on animals and suck their blood, usually leaving a drop or two of blood exuding from the puncture. Especially troublesome about marshes, swampy woods, or meadows, and on warm sunny days............
....*Horse flies*, page 940.
  3. Very small, stout, black flies, less than $\frac{1}{6}$ inch long, with a humped thorax, thick 11-segmented antennae as long as the head, 2 broad delicately veined wings, and stout legs, pierce the skin and suck the blood especially about the eyes, ears, and nostrils, causing extreme pain................................
........................*Black flies*, *turkey gnats* or *buffalo gnats*, page 1005.
B. *Wingless insects that crawl or jump upon the body to bite and suck blood, but do not spend their entire lives on the animal:*
  1. Small, brown, wingless insects, about $\frac{1}{16}$ inch long, very flat from side to side and with long hind legs, jump vigorously when disturbed...................
........................*Cat flea, dog flea, human flea*, and others, page 973.
C. *Wingless insects that stay on the animals all the time:*
  1. Hogs rub and scratch. Large, flattened, grayish, wingless lice, up to $\frac{1}{4}$ inch long, with head, thorax, abdomen, and legs bordered with black; head and legs long, the latter with a peculiar hook at the end to clasp about hairs; mouth parts withdrawn into the head when not sucking blood; especially abundant in folds of the skin. The large yellowish-white eggs are glued to the hairs on lower half of body. The only louse of the hog...........*Hog louse*, page 971.
  2. Hogs rub vigorously. The hair stands erect, the skin about the ears, on top of the neck, on the withers, and down the back to the base of the tail becomes cracked and scabby. Caused by minute short-legged rounded mites, just big enough to be seen, that burrow into the skin....*Itch or mange mite*, page 973.
  3. Tender skin about muzzle, eyes, or inner side of legs forms small hard pimples from the size of a pinhead to that of a small marble, which are filled with a yellowish pus......................................*Follicle mites*, page 974.
D. *Spiny worms or maggots that live in the flesh of wounds:*
  1. Slender, whitish maggots, up to $\frac{3}{4}$ inch long, blunt behind and tapering to a point in front, with elevated spiny rings about the segments giving them a screw-like appearance, and the dark, posterior tracheal trunks visible through the skin, burrow into the flesh of wounds and abrasions or into the body through its natural openings. The eggs are laid about scratches, wounds, or natural body openings by a dark bluish-green fly, about $\frac{3}{8}$ inch long, with three black stripes on thorax and with a reddish face....*Screw-worm*, pages 968 and 973.

332. Hog Louse[1]

Importance and Type of Injury. The only louse found on the body of hogs is a bloodsucking louse (Fig. 20.21), very similar in appearance to

---

[1] *Haematopinus suis* (= *adventicius*) (Linné), Order Anoplura, Family Haematopinidae.

the short-nosed cattle louse, but about twice as large when mature. It reaches a length of nearly ¼ inch and is the largest bloodsucking louse found on any farm animal. On account of its size, it is easily seen, although its color is a dirty gray-brown, almost matching the skin of the hog. The margins of the body and appendages are bordered with black. The lice torment the hogs by piercing the skin to suck the blood. This causes the animals to rub vigorously against feed troughs and fences and to scratch with their feet. The skin becomes thick, cracked, tender, and sore, the animals restless and unprofitable.

Animals Attacked. Hog lice do not infest other kinds of livestock.

Life History, Appearance, and Habits. The lice (Fig. 20.21) are most noticeable on hogs in cold weather. In winter they usually cluster in small clumps on the inside of the ears or in folds of skin about the neck or on the inside of the upper part of the legs. Big and little together, they cling to the hairs by their legs, which are adapted to clamp about the hair very securely. They feed frequently, puncturing the skin each time with very slender stylets, which are completely withdrawn into the head when not feeding. Egg-laying goes on all winter long, a female laying from three to six

Fig. 20.21. Female hog louse, *Haematopinus suis.* About 6 times natural size. (*From U.S.D.A., Farmers' Bul.* 1085.)

eggs a day, gluing each of them firmly to a hair close to the skin. The eggs are big enough to be seen, elongate, the smaller end glued by one side to the hair, the other end with a rounded cap. They are whitish in color when fresh, but after a few days become stained yellow or brownish. Most of the eggs are found on the lower half of the body. In 2 or 3 weeks the small louse pushes off the cap of the egg, seeks a tender place on the skin, and sucks blood until satisfied. It then withdraws its mouth parts and soon bites in another place. The young are of the same shape as the adults, but pale-colored. In 2 weeks, during which time they molt three times, they are full-grown, and mating and egg-laying begin again. The females live about 5 weeks, during the last 3 of which they lay eggs almost every day. There are probably six to a dozen or more generations a year. All stages are passed on the host. The lice never voluntarily leave a hog except when they can crawl directly upon the body of another hog. If dislodged from the animal they rarely live more than 3 days.

Control Measures. Complete control of hog lice may be obtained by spraying or dipping with one of the following materials, observing the indicated interval between treatment and marketing: 0.05 per cent lindane (1 month), 0.5 per cent toxaphene (1 month), 0.5 per cent malathion, 0.5 per cent methoxychlor, 0.5 per cent ronnel (6 weeks), and 0.5 per cent Co-Ral (7 weeks). The application should be repeated after 2 to 3 weeks if necessary. Sick animals should not be treated, and Co-Ral should not be used on young animals. Suggestions for dipping and detailed plans for a dipping vat are given in *U.S. Department of Agriculture, Farmers' Bulletin* 1085. The Department of Agriculture also suggests the use of medicated hog wallows. These should be made of concrete and contain

3 or 4 inches of water, to which is to be added, once a week, a quart of crude petroleum for each hog. A day or two after each application of oil, the wallow is drained, and the hogs are given water without oil the rest of the week. A 1 per cent solution of pine tar in water may be used instead of the petroleum on water. Such a wallow should be built in a shady place, all other wallows done away with, and the oil added in the evening, but not until the hogs are accustomed to using the wallow with water alone. The lice cannot live off the host more than a few days and do not breed in the bedding.

References. *U.S.D.A., Farmers' Bul.* 1085, 1952; *Tenn. Agr. Exp. Sta. Bul.* 120, 1918; *Jour. Econ. Ento.,* **40**:454, 1947.

333. FLIES

The stable fly (Fig. 20.4) bites hogs severely about the head and ears or at cracks or scratches in the skin. Besides the controls given on page 944, a deep, dusty furrow in which hogs may lie should be provided. Horse flies (Fig. 20.3), black flies (Fig. 21.4), and mosquitoes also bite hogs when abundant. These pests are discussed under Horses and Cattle. The screw-worm fly (Fig. 20.20) attacks all kinds of wounds, scratches, and sores on hogs, laying its eggs around them, and the maggots invade the sound flesh and prevent the wound from healing. Treatment is the same as given under Cattle.

334 (359). FLEAS

These well-known little pests (Fig. 21.9) often increase to great numbers about hog lots, doubtless making life very uncomfortable for the hogs, as they do also for any persons who come near. Mules and horses kept near infested hog lots are often annoyed. A thorough cleaning of the hog lots and spraying the soil with a dormant tree-spray oil, 1 part to 9 parts of water, or a good stock dip will be effective in destroying the fleas on the ground. If preferred, a coating of salt or crude flake naphthalene may be used. Fleas are further discussed as pests of man, page 1013.

335 (318, 366). ITCH OR MANGE MITE[1]

Importance and Type of Injury. When hogs are scratching and rubbing vigorously and their hair is standing erect, if an examination does not reveal the large gray hog lice, it is probable that the animals are infested with mange mites. If the skin about the eyes, ears, and along the top of the neck and back is inflamed, scurfy, scabby, with small pimples, or cracked and raw, scrapings from the skin should be made. It is necessary to scrape with a dull knife until the blood starts, to get the mites out of their burrows. Spread the scrapings out over some dark surface and examine under a hand lens or magnifying glass, in a warm place. If minute, pale-colored, nearly round, eight-legged mites (Fig. 20.7) are found crawling among the scrapings, the hog is infested with common mange, which is further discussed under the insects attacking horses. This mite may also attack man (page 1025).

Control Measures. The control of mange is the same as that suggested for horses (page 947): spraying or dipping with lime-sulfur, lindane, or toxaphene, but very thorough applications are necessary. Mange is very contagious, and all animals in the herd should be treated, whether they seem to be infested or not. Healthy animals must be kept so that they cannot touch infested ones until the latter are treated, and pens

[1] *Sarcoptes scabiei* (De Geer), Order Acarina, Family Sarcoptidae.

harboring infested animals must be disinfected, since the mites or their eggs may live off the animal for several weeks.

References. *U.S.D.A.*, *Farmers' Bul.* 1085, 1920; *Jour. Econ. Ento.*, **40**:451, 1947.

336. OTHER MITES AFFECTING HOGS, DOGS, CATS, RABBITS, AND FOXES

Sometimes hogs, dogs, cattle, and man become infested with the hog follicle mite[1] which lives in the hair follicles about the muzzle, eyes, base of tail, or on the tender skin on the inner sides of the legs. The skin becomes red and inflamed, and small hard pimples ranging in size from that of a pinhead to lumps as big as a small marble form and break, discharging a yellowish, cheesy pus. The trouble is caused by a very minute, slender, worm-like mite, with four pairs of very short legs. Secondary infection by *Staphylococcus* bacteria frequently adds to the seriousness of the infection by these mites. The dog follicle mite[2] causes the serious red mange of dogs, which is characterized by bare inflamed spots about the eyes, ears, and joints of the legs. The greatly elongate worm-like mites live deep in the hair follicles and cure is almost impossible. Ear mange mites,[3] affecting cats, dogs, rabbits, and foxes, attack the skin inside the ears, causing great irritation, deafness, and inability to coordinate movements. Sometimes the ears become packed with numerous laminae of dry scabs.

Control Measures. Because of the fact that the mites burrow so deeply in the skin, there is no practical way to cure an infested herd. Badly infested animals should be killed, the others fattened for butchering and disposed of as soon as practicable. Since the mites may live off the host for several days, the premises should be disinfected with a 1:15 lime-sulfur spray, kerosene emulsion, or one of the commercial dips before restocking with healthy hogs. For the ear mites, careful swabbing of the ears with glycerin containing 1 per cent phenol or sulfur in sweet oil or with 5 per cent DDT or 0.05 per cent lindane is suggested to destroy the mites.

D. INSECTS AFFECTING SHEEP AND GOATS
FIELD KEY FOR THE IDENTIFICATION OF INSECTS INJURING SHEEP AND GOATS

A. Free-flying insects that alight on the animals to suck blood, coming and going repeatedly:
    1. Flies of the size and general appearance of the house fly suck blood especially from the legs, causing the animals to stamp their feet. Distinguished from the house fly by its stiff, slender beak, sticking straight forward from the underside of the head, by its broader black-spotted abdomen and grayer appearance, and by the bristle on antenna having hairs on its upper side only. Palps less than half as long as proboscis. Usually rests with its head upward. . . . .
. . . .*Stable fly*, page 942.
    2. Very small, stout, black flies, less than 1/6 inch long, with a humped thorax, thick 11-segmented antennae, as long as the head, 2 broad, delicately veined wings, and stout legs, pierce the skin and suck the blood, especially about the eyes, ears, and nostrils, causing extreme pain. . . . .
. . . .*Black flies* or *buffalo gnats*, page 1005.
B. Insects that stay on the animal all the time:
    1. A very small brownish louse with a broad red head, only about 1/20 inch long, is found among the wool and on the skin of sheep, causing them to rub and scratch. . . . .*Sheep biting louse*, page 975.
    2. A louse, up to 1/12 inch long, the body tapering toward the front, the head as broad as long, and 2 transverse rows of hairs on each abdominal segment, is found among the short coarse hairs of the legs below the true wool. . . . .
. . . .*Bloodsucking foot louse*, page 975.

[1] *Demodex phylloides* Csokor, Order Acarina, Family Demodicidae.
[2] *Demodex canis* Leydig.
[3] *Otodectes cynotis* Hering, *Chorioptes cuniculi* Zurn, and others.

3. About the same size and appearance as *B,2*, but occurs over entire body and face under the wool, sometimes in clusters. Slightly more slender than the foot louse, and the head twice as long as broad. Wool about the clusters of lice generally discolored by the small pellets of excrement.....................
....*Bloodsucking body louse*, page 975.

4. Chestnut-brown wingless flattened tick-like insects, about ¼ inch long, with a leathery spiny skin, a sack-like unsegmented abdomen, short head, and 6 tapering, widespread legs, cling to the skin beneath the wool and suck blood. Often very severe on lambs. No exposed egg or larval stage; the rounded, brownish, ovate puparia, about ⅛ inch across, are often wrongly called "nits." The maggots are born full-grown and glued to the wool, especially about the neck...............................*Sheep-tick, louse fly* or *ked*, page 976.

5. The wool comes out in patches, "tagging" to weeds, fences, etc., on which the animals rub, leaving bare, scabby places on the skin. Under these scabs may be found thousands of minute 8-legged oval-bodied mites, just visible to the naked eye, which pierce the skin with their mouth parts, causing an exudation that hardens over them as a scab. They do not burrow into the skin.......
....*Scab mite*, page 979.

C. *Maggots that live in the bony cavities of the head or under the wool or in sores:*

1. Animals are nervous, push their noses between other sheep or into the dust, paw with their feet, shake their heads, run with heads held low, and sneeze. There is a copious, foul discharge from the nostrils. This trouble is caused by a grayish-brown fly, about ½ inch long, which deposits small active maggots in the nostrils. The maggots work upward through the nostrils and then tunnel through the nasal and frontal sinuses, frequently causing death. These larvae are fleshy, creamy-white, and brownish banded, finally reaching a length of about 1 inch..
........."Grub-in-the-head," "staggers," sheep bot, or sheep nose fly, page 982.

2. Slender, whitish maggots, blunt behind, tapering in front, up to ¾ inch long, burrow in matted, soiled wool, or into the flesh, especially at points injured by "ticks" or other piercing insects or by dogs. Eggs are laid by a brilliant-green or a blue, metallic fly, larger than the house fly, about ⅜ inch long..........
.....................*Greenbottle fly, screw-worm*, or *black blow fly*, page 983.

3. Slender, whitish maggots, up to ¾ inch long, pointed at the head end and truncate behind, with elevated, spiny rings about the segments giving them a screw-like appearance and the posterior tracheal trunks dark and showing through the skin, feed and eat into the living tissue next to wounds..........
....*Screw-worm*, page 968.

General References. U.S.D.A., Farmers' Bul. 1330, 1946; "Insect Parasites of Goats in the United States," *National Angora Record Jour.*, vol. 1, no. 1, September, 1922.

337. SHEEP AND GOAT LICE

Importance and Type of Injury. Three species of lice are known to live on sheep in this country: the bloodsucking body louse,[1] the bloodsucking foot louse,[2] and the red-headed or sheep biting louse.[3] They are by no means so prevalent as the sheep-tick. The usual symptoms of scratching and rubbing are caused by these lice biting and running over the skin. The biting lice eat off the wool fibers and tangle and soil the hair; the bloodsucking kinds rob the host of nutrition, and the bloodsucking body louse stains the wool with its small brown fecal spots. The sheep biting louse is said to be the most irritating, and the foot louse to be comparatively innocuous. Angora goats are injured and the clip of mohair and kid hair sometimes reduced as much as 10 to 25 per cent in

[1] *Haematopinus ovillus* (Neumann), Order Anoplura, Family Haematopinidae.
[2] *Linognathus pedalis* (Osborn), Order Anoplura, Family Haematopinidae.
[3] *Bovicola* (= *Trichodectes*) *ovis* (Linné), Order Mallophaga, Family Trichodectidae.

value by at least five different kinds of lice.   There are two kinds of blood-sucking lice,[1] which are blue-gray in color; and three species of chewing lice: a large yellowish one[2] and two reddish or orange species.[3]

Life History, Appearance, and Habits.   The sheep biting louse is the smallest and apparently the worst of the three species.   It is only $\frac{1}{20}$ inch long, pale-brownish in color, with a broad, reddish head, broadly rounded in front.   Each segment of the abdomen has a single transverse row of hairs.   These lice crawl about among the wool and on the skin, eating wool fibers and skin scales.   They do not suck blood but, when they cluster on the skin in great numbers, may cause raw sores.   The bloodsucking foot louse is similar in form to the short-nosed cattle louse but only about $\frac{1}{12}$ inch long and somewhat slenderer and paler colored. The head is about as wide as long, and each abdominal segment has two transverse rows of hairs.   The bloodsucking body louse is similar to the foot louse, but the head is twice as long as broad and bluntly pointed in front.   All these lice spend all stages, eggs, nymphs of various sizes and adults, on the body of the animal though they may live 3 days to a week if dislodged, and their eggs may hatch several weeks after being separated from the host.   The eggs are glued fast to the hairs, and hatch in 5 to 10 days in the case of the chewing louse, or 10 to 18 days in the other species.   The young are said to complete growth and begin laying eggs in about 2 weeks after hatching.

Control Measures.   These lice may be controlled by dipping with one of the following materials, observing the indicated interval between treatment and marketing: lindane at 0.025 per cent (1 month), toxaphene at 0.25 per cent (1 month), methoxychlor at 0.25 per cent, or rotenone at 1 pound of 5 per cent powder to 100 gallons (two applications at an interval of 2 to 3 weeks).   Spraying with lindane at 0.05 per cent, toxaphene at 0.5 per cent, methoxychlor at 0.5 per cent, malathion at 0.5 per cent, ronnel at 0.5 per cent (3 months), or Co-Ral at 0.25 per cent (2 months) is also effective.   Sick animals should not be treated, and only methoxychlor and rotenone should be used on milk goats.   Co-Ral should not be used on young animals.   All new animals should be examined before they are turned into a healthy flock, and especial attention should be given to males, which often infect a flock of ewes at breeding time.   There is also the danger of infestation from neighboring flocks along line fences.

References.   U.S.D.A., Leaflet 13, 1928, and *Bur. Ento. Cirs.* E-394, 1936, and E-762, 1949; and *Farmers' Bul.* 1330, 1946; *Jour. Econ. Ento.,* **39**:546, 1946.

338. Sheep-tick, Louse Fly, or Ked[4]

Importance and Type of Injury.   The sheep "tick" (Fig. 20.22) is one of the most remarkable insects known.   It is not a tick, like the Texas-fever tick or spotted-fever tick, but is a degenerate louse-like fly that has completely lost its wings.   It crawls about over the skin among the wool and feeds by thrusting its sharp mouth parts into the flesh and sucking blood (Fig. 20.23).   It causes sheep to rub, bite, and scratch at the wool,

---

[1] *Linognathus stenopsis* (Burmeister), the goat sucking louse, and *L. africanus* Kellogg and Paine, Order Anoplura, Family Haematopinidae.

[2] *Bovicola penicellata* Piaget, Order Mallophaga, Family Trichodectidae.

[3] *Bovicola caprae* (Gurlt), the goat biting louse, and *B. limbatus* (Gervais), the angora goat biting louse.

[4] *Melophagus ovinus* (Linné), Order Diptera, Family Hippoboscidae.

thus spoiling the fleece; when this tick is abundant, the animals are unthrifty and unprofitable. Estimates by a large number of sheep owners some years ago indicate that the presence of this insect causes a loss amounting to 20 to 25 cents a head a year, on the average, in weight and wool production. If this estimate is correct, the sheep-tick taxes the sheep industry of this country several million dollars a year on the approximately 50 million head of sheep kept by American farmers. They are

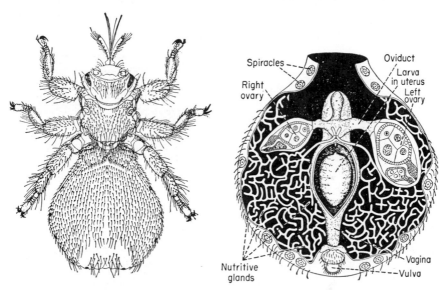

Fig. 20.22. The sheep-tick or ked. Left, dorsal view of adult, about 8 times natural size; note mouth parts, shielded by the maxillary palps, the compound eyes, and, in front of them, the concealed antennae; and along the sides of the thorax and abdomen, the spiracles. Right, dissection of the abdomen of the female, ventral view, to illustrate the viviparous reproduction. At center is a larva lying in the uterus, with its mouth toward the small nipple through which a nutritive fluid is passed from the elaborately branched, cylindrical glands that furnish the only food of the insect until the adult stage is reached. This larva originated from an egg produced in the right ovary (on the reader's left), while another egg is being produced by the left ovary, the two functioning in turn. When full-grown the larva will pass from the vulva in the condition shown in Fig. 20.24,A. (*Drawn by Kathryn Sommerman.*)

especially severe on lambs, to which they migrate readily from the ewes especially at shearing time.

Animals Attacked. Sheep and sometimes goats.

Distribution. Where sheep and goats are kept throughout the world.

Life History, Appearance, and Habits. The second remarkable thing about the sheep-tick is its method of reproduction. Unlike true ticks, which always drop to the ground to lay eggs, the sheep-tick spends its whole life on the animals. Two life stages are commonly found on sheep at any season of the year. The adults (Fig. 20.22), which are grayish-brown, wingless, and six-legged, have a broad, leathery, somewhat flattened, unsegmented, sac-like abdomen covered with short spiny hairs. The thorax and head are much narrower; the legs are widespread, the first pair appearing to come out at the sides of the head. The body is about $\frac{1}{4}$

inch long and covered with short, spiny hairs.   The other exposed stage of
the sheep tick is the so-called "nit."   It (Fig. 20.24,*E*) is a nearly round,
chestnut-brown, seed-like object that is glued fast to the hair, especially
about the neck, inside of thighs, and along the belly.   It is not an egg,
but a pupal case or puparium enclosing the pupal stage of the fly.   The
sheep-tick does not lay its eggs.   The maggots are nourished within the
body of the female until they are full-grown, never feeding externally.
When born, they are whitish in color, oval, about ⅛ inch long, and without appendages.   The female secretes a gelatinous glue which sticks
the larva to the hair near the skin, especially about the neck and belly.

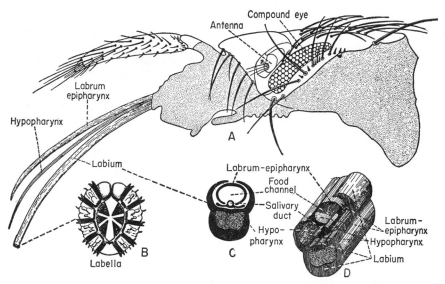

FIG. 20.23. Mouth parts of the sheep-tick—special biting-fly subtype, essentially like those
of the stable fly (*cf.* Fig. 4.8).   *A*, side view of the head with the mouth parts protruded
and separated to show the two stylets.   Note the curious concealed antenna in front of the
compound eye.   *B*, the labella greatly enlarged to show the prestomal, cutting teeth.
C, diagrammatic cross-section, and *D*, isometric projection of the proboscis with parts in
normal position, showing how the labrum-epipharynx and the hypopharynx form the
food channel, the salivary duct within the hypopharynx, and both stylets enclosed by the
labium.   (*Drawn by Kathryn Sommerman.*)

Within 12 hours the skin turns brown and forms a hard puparium about
the larva.   Within this case, pupation takes place, and in summer the
adult tick breaks out of its puparium from 19 to 23 days after it is born.
In cold weather 3 to 5 weeks, or even longer, may be spent in the puparium.   After emerging, the adults may mate in 3 or 4 days or not for
several months.   After mating, the females begin depositing full-grown
larvae in from 8 to 30 days.   Only one young is developed at a time, and
each female produces from 10 to 20 maggots, which are born at the rate
of about 1 a week.   The female may live as long as 5 or 6 months.   Sheep-
ticks never normally leave their hosts.   If separated from the animal,
they may live for a week or slightly longer, but most of them die in 3 or
4 days.   The pupae, however, may live 1 or 2 months apart from the host,
in warm weather.   Freezing kills this stage.   Breeding is continuous,

though slower in winter, and there are probably several generations a year.

Control Measures.   The most satisfactory control for the sheep-tick is to dip or spray the sheep with lindane, toxaphene, rotenone, malathion, Co-Ral, or ronnel, using the dosages and cautions suggested under sheep lice control (page 976).   Spraying with 0.1 per cent synergized pyrethrins or dusting with 1.5 per cent dieldrin, at least 3 months before marketing, has also been satisfactory.   Only rotenone and the pyrethrins should be used on milk goats.   Complete directions for dipping sheep are given in *N.Y. (Cornell) Agricultural Experiment Station Bulletin* 844, 1948.   The

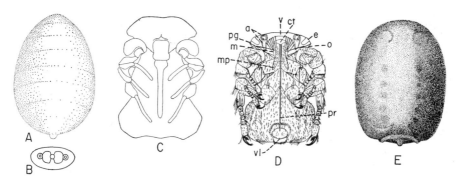

Fig. 20.24.   The sheep-tick.   *A*, new-born larva, as first glued to the wool of the sheep. *B*, the posterior stigmal plates of the larva.   *C*, an early-stage pupa, dissected from the puparium.   *D*, an adult dissected from its puparium shortly before emergence: *a*, antenna; *e*, compound eye; *mp*, maxillary palp; *o*, occiput; *pg*, postgena; *pr*, proboscis; *vl*, vulva. *E*, puparium, formed by hardening of the larval skin, within which the larva transforms from larval to adult stage.   All enlarged about 9 times.   (*A*, *B*, *and E drawn by Kathryn Sommerman; C and D by Carl Weinman.*)

cost of dipping should not exceed a few cents a head for each treatment. Dipping cannot be done in cold weather and should not be done following shearing until all cuts have healed.   After dipping, sheep should be kept from fields and pens previously used for from 1 month to 6 weeks. If sheep have not been dipped in the fall, and ticks become abundant in winter, they may be checked by several dustings of pyrethrum or derris powder sifted into the wool.   Since the ticks may crawl a considerable distance in search of animals, it is not wise to leave wool that has been sheared near the flock.   It should be stored at least 50 feet away.

References.   *U.S.D.A., Farmers' Bul.* 798, 1932; *Wyo. Agr. Exp. Sta. Buls.* 99, 1913, and 105, 1915; *Jour. Econ. Ento.*, **42**:410, 1949; *N.Y. (Cornell) Agr. Exp. Sta. Bul.* 844, 1948.

339 (319, 326).   SHEEP SCAB MITE[1]

Importance and Type of Injury.   One of the most injurious and contagious of sheep diseases is the trouble known as scab or scabies.   It is usually first indicated by "tagging," *i.e.*, the loss of bits of wool on weeds, fences, and other objects against which the animal has rubbed its body (Fig. 20.26).   The cause of this tagging or loss of hair is a very small eight-legged mite that punctures the skin with its sharp mouth parts until the

[1] *Psoroptes equi ovis* (Hering), Order Acarina, Family Sarcoptidae.

lymph exudes and flows over it. As this serum hardens to form a scab, the mites remain underneath on the raw skin where their continued feeding results in successive layers of scabs that eventually lift the hair out by the roots. The irritation and loss of blood due to the mites and an extremely irritating poison introduced as they feed cause the sheep to lose condition, become sickly, and, if not treated, to die. In less severe cases the loss of wool is a considerable item. The skin becomes first reddened, then white and glistening, then hardened and uniformly thickened over the infested part, and eventually bare of wool, cracked, and bleeding or oozing serum. This trouble is sometimes called "wet mange," in contrast with the dry mange caused by sarcoptic mites (page 946). Sheep which are rubbing or biting the skin or show a tangled condition of the fleece, should be examined for scab by a competent authority.

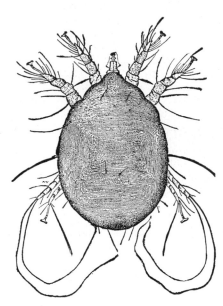

FIG. 20.25. The sheep scab mite, *Psoroptes equi ovis*, female. Greatly magnified. (*From U.S.D.A., Farmers' Bul.* 713.)

Animals Attacked. Sheep and cattle suffer most, while horses and goats are attacked rarely. Fine-wooled varieties of sheep are most seriously affected.

Distribution. Nearly cosmopolitan; but quarantines and dipping campaigns have eradicated it from many sections.

Life History, Appearance, and Habits. This mite feeds on the surface of the skin and does not burrow through the outer layers of the flesh, as does the mange mite. All stages of the scab mite will be found upon lifting a scab from an infested animal, the eggs, young, and adults swarming together under this protection. The largest mites (Fig. 20.25) are only as big as the cross-section of an ordinary pin or needle ($\frac{1}{50}$ inch in diameter), and may be seen as minute gray crawling specks against a black surface. A dozen or two of eggs are laid by the female in small clusters at the base of the hairs. They hatch in 2 to 10 days to six-legged nymphs. After the skin is shed the first time, the nymphs have an additional pair of legs. They grow rapidly and are mature in from 1 to 2 weeks. A female may live 1 month or more (may even survive away from a host for 2 or 3 weeks) and deposit up to 100 eggs. In wet cold weather generations succeed each other rapidly; the mites spread from sheep to sheep and whole flocks are soon affected.

Control Measures. For a number of years quarantine measures have been in effect against the shipment of scabby sheep, and the disease has been greatly reduced in prevalence. Dipping with lime-sulfur, nicotine, lindane, or toxaphene, as described on page 947, is the most effective way to stop injury by this mite. Many very dependable commercial dips are available. These should always be used strictly as directed by the manufacturer.   At least two dippings are necessary in cases of heavy infestation.   They should be made at 10- to 14-day intervals.   The sheep should be held in the dip for at least 2 minutes for lime-sulfur or nicotine and 1 minute with lindane and toxaphene, and the head ducked several times. Dipping is most effective shortly after shearing.   After dipping, the animals should be kept away from pens or pastures recently occupied, for at least 1 month.   The pens should be sprayed with a strong phenol

Fig. 20.26. A severe case of sheep scab due to the attacks of the scab mite.   (*From U.S.D.A., Bur. Animal Ind.*)

or creosote wash, such as 2 per cent cresol or 5 per cent carbolic acid, after having been thoroughly cleaned of all droppings, straw, and other refuse.

References.   U.S.D.A., Farmers' Bul. 713, 1916; *U.S.D.A., Farmers' Bul.* 1330, 1946, and *Bur. Ento. Plant Quar. Cir.* E-406, 1937.

340 (329, 361).   True Ticks

Although several species of ticks may attach to sheep to feed, they are not so injurious to sheep as to the less hairy animals.   In fact, pasturing infested ground with sheep has been suggested as a control for ticks, since the latter may become entangled in the wool and killed.   However, at least two species of ticks are injurious to sheep. The spotted-fever tick, which is discussed as a pest of man, page 1017, causes a kind of paralysis in sheep by biting and engorging along the backbone.   The removal of the large female ticks from the head, neck, and back, by hand, or the use of an arsenical dip results in rapid recovery from the disease.   The spinose ear tick (page 962) also attacks sheep, feeding especially inside the outer ear.   The lone-star tick[1] is very troublesome to goats, especially in the Southwest.

[1] *Amblyomma americanum* (Linné), Order Acarina, Family Ixodidae.

341. Sheep Bot or Nose Fly[1]

Importance and Type of Injury.   Sheep shake their heads, stamp their feet, and crowd together, holding their noses to the ground, especially in bare dusty places; or run away, with noses held low, in efforts to keep the fly from striking at their nostrils.   The presence of the maggots in the nostrils and head sinuses causes inflammation, and a copious "catarrhal" discharge.   The excess of mucus, together with irritating dust drawn into the air passages, causes sneezing and labored breathing.   The presence of the maggots may cause giddiness or "blind staggers."   Sometimes the flies may deposit their larvae in the eyes, nostrils, or mouth of man, sometimes causing blindness.

Animals Attacked.   Sheep, goats, wild deer, and, very rarely, man.

Distribution.   General throughout North America; especially troublesome from Idaho and Montana to New Mexico and Texas.

Life History, Appearance, and Habits.   The life history of this fly appears not to be well known.   In Canada both small and full-grown maggots have been found in the heads of sheep, both in winter and in summer, indicating that some larvae may remain in the head for more than a year.   In Russia the larvae are said to pass the winter in the second instar in a state of rest with little change, forming the third- and last-stage larvae in spring.   The larvae may complete their development in 2½ to 3½ months in lambs, but require much longer—at least 10 months—in older animals.   The full-grown larva is nearly 1 inch long by a third as thick, without definite head or legs, and the segments prominently marked with blackish crossbands.   Probably the typical cycle is for different larvae to become mature from early spring to late summer, when they retreat from the deeper tunnels they have made and drop or are sneezed out of the nostrils to the ground.   Here a transformation period of 20 to 60 days is passed in the hard, blackish puparia before the flies emerge.   In Texas adults are active every month of the year; farther north they are often present during 4 to 6 months in summer.   They are about ½ inch long, of a general grayish color, with minute dark spots peppered over the back and with eyes rather small and a broad space between them.   They take no food, but follow sheep on hot, still, sunny days and hide away in crevices of fences and walls, mornings, evenings, and in inclement weather.   The eggs hatch in the body of the female before they are laid, and a number of very active, small maggots, ½₀ inch long, enclosed in a little drop of sticky liquid, are deposited up the nostrils, as the female flies rapidly past the sheep's head without alighting.   The maggots work their way upward through the nostrils until they reach the sinus cavities between the bony plates of the skull in the region of the forehead (Fig. 20.27).   In other cases they find their way into the bronchi or into cavities in the horns or bones of the nose or jaw.   They lacerate the tissues, growing slowly and sometimes migrating far about in the head but probably do not penetrate through the skull into the brain, as often supposed.   As many as 80 larvae have been reported from one head, but the usual number is from 5 to 12.

Control Measures.   To prevent infestation by the sheep bot, the nostrils may be kept smeared with pine tar, applied weekly during the period when adults are active.   In small flocks the pine tar can be applied

---

[1] *Oestrus ovis* Linné, Order Diptera, Family Oestridae.

by hand with a brush. The first application should be made about mid-April in the latitude of Illinois. The sheep may be made to smear their own noses by an ingenious salt log, in which salt is placed in the bottom of 2- or 3-inch holes bored in the log and the edges of the holes smeared with pine tar. Hadwen recommends the provision of a dark shed with a curtain over the door, into which the sheep may retreat when the flies begin to attack them. Since the flies are short-lived and not

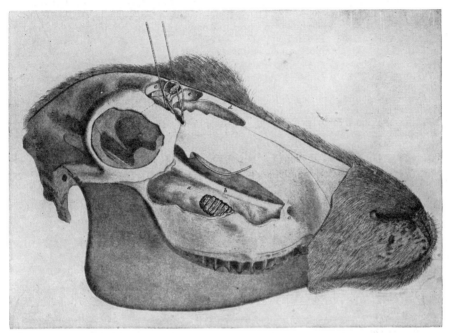

Fig. 20.27. Sheep bot fly, larvae (at *a* and *e e*) tunneling through sinuses of head of sheep. (*From U.S.D.A., Bur. Animal Ind.*)

capable of long flights, mobile herding or frequent change of pastures, which avoids concentrating the animals for grazing or watering in any place for more than a short time, is of some value in reducing infestations. Infested sheep may be treated by injecting 1 ounce of a 3 per cent solution of saponated cresol into each nostril. A skilled veterinarian may remove larvae from the heads of valuable animals by opening the sinuses above the nasal fossae with a trephine.

342. Fleeceworms or Wool Maggots

Importance and Type of Injury. When the wool of sheep becomes soggy from warm rains, or soiled with urine, feces, or blood from wounds or from lambing, certain blow flies are attracted to the animal and deposit their eggs in the dirty wool, most commonly about the rump, but also about the horns where wounds have resulted from fighting. The maggots feed in the wet wool and the adjacent skin, causing the latter to fester and the wool to loosen and become putrid; the inflamed, raw flesh, with the whitish maggots tunneling in it, is exposed. The dirty wounds readily become infected and the sheep may die of blood poisoning.

Animals Attacked. Sheep, goats, cattle, and other animals if they have putrid sores. Greasy fine-wooled sheep, such as Merinos, are more likely to be infested.

Life History, Appearance, and Habits. There are several kinds of flies that attack soiled wool, but among the most important are the greenbottle fly,[1] the black blow fly,[2] and the secondary screw-worm[3] (page 969). The first species is about twice the bulk of the common house fly and of a brilliant metallic, bluish-green color with bronze reflections, without stripes, and with a fine whitish bloom on the front of the thorax just behind the head. The black blow fly is a little larger, very dark greenish black in color all over, without stripes or grayish markings and not very bristly. Both of these flies probably winter as larvae or pupae in soil beneath carcasses, or in manure. At any rate the flies appear very early in spring, and from that time on, breeding is continuous except as checked by dry weather. The larvae of the black blow fly are said to live chiefly in carcasses or carrion; the greenbottle fly develops also in garbage.

Either fly may complete a generation from egg to egg in about 3 weeks. Their numbers increase as the season progresses and, during warm, rainy, or foggy weather, they are especially likely to lay eggs in the wool. The pupal stage is passed in the ground.

Control Measures. Since the flies attack animals chiefly after having become abundant by breeding in carcasses, all carrion should be destroyed promptly by burning or burying to a depth of at least 4 feet. If possible, the carcass should be sprinkled with lime before it is covered with earth. Trapping the adult flies with sweetened baits (page 1036) is of some value in reducing their numbers. Breeding hornless sheep, having the lambs come as early in the spring as practicable, shearing before lambing occurs, docking of lambs, "tagging" dirty sheep, and applying pine tar to wounds are sound preventive measures. When sheep are infested, the wool should be clipped close around the infested area, and this thoroughly wet with either Smear EQ 335 or ronnel smear (page 970) diluted 1 to 9 parts of water. Thorough spraying with 0.25 per cent Co-Ral is also effective, but should not be used within 2 months of marketing or on sick or young animals.

References. *U.S.D.A., Farmers' Bul.* 857, 1922; *U.S.D.A., Tech. Bul.* 270, 1931; *Jour. Econ. Ento.*, **41**:521, 1948.

E. INSECTS THAT ATTACK POULTRY

FIELD KEY FOR THE IDENTIFICATION OF INSECTS INJURING POULTRY

A. *Pests that visit the fowls only to secure food, coming and going repeatedly; or live on the fowls at night, hiding away in the daytime; or spend only part of the life cycle on the fowls, being free-living or intermittent parasites during other life stages:*

1. Mahogany-brown, broad, very flat or thin, oval, wingless bugs, of all sizes up to ⅕ inch long, live in nests, behind boards, and in cracks of houses during the daytime, and crawl out upon fowls at night and suck blood. Bugs have a bad odor. Small black spots of excreta from the bugs often seen on the eggs and about cracks......................................*Bed bug*, page 989.

2. Small gnat-like humpbacked flies, less than ⅙ inch long, with thick antennae, as long as the head; 2 broad delicate wings and stout legs, hover about the heads of fowls on the roost, suck blood, and appear to smother the birds, when very abundant in spring.......*Black flies, turkey gnats,* or *buffalo gnats,* page 1005.

3. A tiny, hard, long-legged, jumping insect, about 1/20 inch long and flattened from side to side; the females attach to, or burrow into, the skin about the eyes, comb, wattles, or vent in clusters, often forming dark areas visible from some distance; ulcers, blindness, and death, especially of young chicks, often result. Immature stages are passed in cracks of the henhouse or in the soil. A southern species......................*Sticktight* or *southern chicken flea,* page 989.

4. Small grayish to dark-red pear-shaped or ovate mites, from 1/40 to 1/20 inch long, with 8 slender legs, remain in cracks under the roosts or in nest boxes

---

[1] *Phaenicia* (= *Lucilia*) *sericata* (Meigen), Order Diptera, Family Calliphoridae.
[2] *Phormia regina* (Meigen), Order Diptera, Family Calliphoridae.
[3] *Callitroga macellaria* (Fabricius), Order Diptera, Family Calliphoridae.

during the day, except on sitting or laying hens in dark places. At night they swarm over the birds and suck the blood *.Poultry mite* or *roost mite*, page 991.

5. Similar to the poultry mite but females with tip of body slightly notched. The mites live on the fowls day and night, laying their eggs among the fluff feathers. Nymphs and adults suck blood.......*Northern fowl mite*, page 992.

6. In the South, poultry are attacked by a larger 8-legged oval-bodied brown tick, up to $\frac{1}{3}$ inch long, which in the adult stage attacks the host only at night, when it sucks blood in quantities, and hides in cracks during the day like the poultry mite. In its younger stages, however, it is a permanent parasite, remaining on the fowl day and night until ready to molt...............................
...............................*Fowl tick, adobe tick*, or *bluebug*, page 990.

B. *Small, wingless, flattened chewing lice that stay on the skin or feathers of the fowls all the time:*

1. Ovate yellow lice, less than $\frac{1}{16}$ inch long, with a single transverse row of hairs on each abdominal segment, above. Found along the shafts of the feathers of chickens rather than on the skin; when the feathers are parted, they run toward the body along the shaft. Chew at the feathers. Eggs glued to base of feathers. Not on young chicks...*Small body louse* or *shaft louse*, page 988.

2. Similar lice, from $\frac{1}{10}$ to $\frac{1}{8}$ inch long, darker yellow and more hairy; the hairs on upper side of abdomen in 2 transverse rows on each segment. Found running rapidly over the skin of chickens, turkeys, and pheasants in less feathered parts; not on the feathers. The most injurious species on grown chickens. Eggs attached especially to small feathers below the vent........
...............................*Large body louse*, page 988.

3. Lice about the size of the shaft louse, but dark grayish in color and with a longer head, found standing on their heads among the feathers of the head on chickens; move but little. Hairs on upper side of abdomen mostly confined to a wide median stripe. Antennae of male unusually large. The most injurious louse to young chicks. Eggs laid singly on the down of the head.......
...............................*Head louse*, page 987.

4. Similar to the head louse but more slender and darker in color. The only species found commonly on the large wing feathers of chickens, where it often lies between the barbules on the underside of the shaft, showing no signs of life. Eggs between barbules of large feathers..*Wing louse*, page 988.

5. A very large species, $\frac{1}{8}$ inch long and more than half as wide. Smoky gray to black in color; found on feathers of chickens and very active. Not common
...............................*Large chicken louse*, page 988.

6. A large chewing louse, $\frac{1}{8}$ inch long, with each hind angle of the head prolonged into a sharp process, at the end of which is a very long bristle; occurs on the feathers of the turkey, especially on the neck and breast...................
Large turkey louse, Chelopistes meleagridis (Linné) (= *Goniodes stylifer* Nitzsch).

7. A louse of equal length with the preceding, but only one-sixth as wide as long; pale yellowish, with a black margin around the body; especially common on the primary wing feathers of turkeys.................................
...............................*Slender turkey louse, Oxylipeurus polytrapezius* (Burm.).

8. A small species, only $\frac{1}{25}$ inch long, with head curiously expanded and rounded in front; dark red in color, with a white region in middle of abdomen; is common at the base of the large wing feathers of ducks and geese..............
...............................*Biting louse of ducks and geese, Docophorus icterodes* Nitzsch.

9. A larger species, $\frac{1}{6}$ inch long, slender, light yellow in color with a dark margin to the body and squarish dark spots on the abdomen. Infests ducks and geese, especially at the base of large wing feathers.......................
...............................*Squalid duck louse, Lipeurus squalidus* Nitzsch.

10. A short broad species, $\frac{1}{25}$ inch long, with abdomen squarish behind; whitish with a brown margin. Infests pigeons........*Small pigeon louse*, page 988.

11. An exceedingly slender louse about $\frac{1}{12}$ inch long, with dark abdomen and reddish-brown head and thorax; very abundant on old pigeons and partially feathered squabs......................*Slender pigeon louse*, page 988.

C. *Minute, almost invisible, 8-legged rounded mites that burrow into the skin beneath scales of legs or at base of feathers and feed and reproduce in the tunnels:*
1. Chickens, turkeys, pheasants, and other birds walk painfully or refuse to walk. Legs are encrusted with elevated scales from which a fine white powder and serum exude from the irritated and inflamed skin, and the legs become much swollen. Numerous minute, circular, very short-legged mites less than $\frac{1}{50}$ inch long burrow under the scales......................*Scaly leg mite,* page 993.
2. A similar but still smaller mite burrows into the skin at the base of the feathers of the rump, back, abdomen, head, and neck of chickens and pigeons, causing the feathers to fall or to be pulled out by the bird. If the stumps of such feathers are examined, an abundance of dry scales, crusts, and mites will be found......................................*Depluming mite,* page 994.

General References. *N.Y. (Cornell) Agr. Exp. Sta. Bul.* 359, 1915; *Ohio Agr. Exp. Sta. Bul.* 320, 1917; *Conn. (Storrs) Agr. Exp. Sta. Bul.* 86, 1916; *U.S.D.A., Farmers' Buls.* 801, 1931, and 1110, 1920.

343. Poultry Lice[1]

Importance and Type of Injury. Contrary to the belief of most poultrymen, the lice that live on fowls do not suck blood. They feed by nibbling or chewing the dry skin scales, feathers, or scabs on the skin. The irritation from the mouth parts, together with that of the sharp claws on their feet in running about over the skin, results in a nervous condition of the infested birds that prevents sleep, causes loss of appetite and diarrhea, and renders the weakened fowls easy prey for various poultry diseases. Young chickens and turkeys that are brooded by lousy hens are often killed in great numbers by the swarming of lice from the hen to them almost as soon as they hatch from the eggs. The most serious effect upon older fowls is a reduction in the number of eggs laid. Infested fowls are in a mopey, drowsy condition with droopy wings and ruffled feathers, refuse to eat, and gradually become emaciated. If the feathers of such a fowl are parted, the lice will often be found running about on the skin in great numbers, particularly below the vent, on the head, or under the wings.

Animals Attacked. Every kind of domestic fowl (and probably every kind of wild bird as well) has from one to several kinds of lice. In general, each species of bird has lice peculiar to it. The exceptions to this will be noted in discussing the different lice. At least a dozen kinds attack chickens and three to five different kinds are found on ducks, pigeons, and turkeys.

Distribution. Wherever fowls are kept.

Life History, Appearance, and Habits. Poultry lice generally breed faster and become more abundant in summer than in cold weather, but all stages can usually be found on the host in winter. All these chewing lice are permanent parasites, spending all life stages, generation after generation, on the same bird, and never normally leaving its body, except as they pass from one fowl to another, particularly from old to younger birds. The eggs are cemented fast to some part of the feathers. They are oval in shape, generally white in color, and often beautifully ornamented with spines and hairs (Fig. 5.1,*J*). While laid singly they may be abundant enough to form dense clusters on the fluffy feathers of badly infested chickens. In a few days or weeks the young nymph hatches from the egg in a form much like the parent lice only much smaller and paler in color. It at once begins running about and feeding, and in the course

[1] Order Mallophaga.

of the next few weeks passes through several molts, gradually assuming the size, form, and coloration of the adult.

Poultry lice are entirely wingless, six-legged insects with a much flattened body and broad head rounded in front. The mouth parts are near the middle of the underside of the head, the most prominent parts being two sharp-pointed teeth or mandibles. The legs are good-sized, and, in all of the species that live on birds, they have two claws at the end of the tarsus. Their relatives that live on hair-bearing animals, such as the

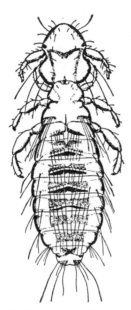

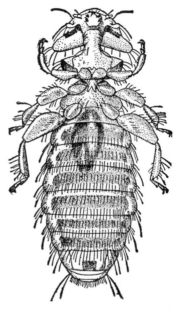

FIG. 20.28. Chicken head louse, *Cuclotogaster heterographus*, male. About 25 times natural size. (*From U.S.D.A., Farmers' Bul.* 801.)

FIG. 20.29. Chicken body louse, *Menacanthus stramineus*, female, underside. About 25 times natural size. (*From U.S.D.A., Farmers' Bul.* 801.)

little red louse of cattle or the chewing lice of horses and sheep, have only one tarsal claw, fitted for grasping about the hairs.

The chicken head louse[1] (Fig. 20.28) is especially noticeable and injurious on young chicks and turkeys. The dark-gray large-headed adults, about $\frac{1}{10}$ inch long, and the paler young ones are found standing head down along the base of the feathers on top of the head with their mouth parts against the skin. They constantly nibble at the skin scales but apparently never eat through the skin or into the flesh. Although they move about only a little, they pass very early from brooding hens to little chicks, which are often killed by them. The eggs are cemented to the barbs of the down or small feathers of the head or neck. They hatch in about 5 days and the young are full-grown in about 10 days more. In an incubator at a constant temperature of 33–34°C. the egg stage required 5 to 7 days, and there were 3 nymphal instars requiring a minimum of 6, 8,

---

[1] *Cuclotogaster* (= *Lipeurus*) *heterographus* (Nitzsch), Order Mallophaga, Family Liotheidae.

and 11 days, respectively, or a minimum of 30 days for a generation.   It is generally believed that the young become full-grown in about 10 days upon the host.[1]

The chicken body louse[2] (Fig. 20.29) lives most of the time on the skin of either chickens or turkeys, being especially abundant about the vent and under the wings, and is common on both young and old fowls.   When the feathers are parted, all sizes of the lice run rapidly to cover.   The smaller ones are pale yellowish white but the larger ones, which reach a length of nearly $\frac{1}{8}$ inch, appear brownish.   The body is covered with fine long hairs.   Bishopp and Wood consider this the most injurious louse of grown chickens, because it is constantly on the skin.   The eggs are fastened to the basal barbs from the shaft of the feathers, especially below the vent. They are said not to hatch on feathers dislodged from the host.   On the body they hatch in about a week, and 10 or 12 days of growth brings the nymphs to the adult stage.   They increase very rapidly: Lawson and Manter record having counted over 35,000 lice on one chicken, which they think was not half of those actually present.

The shaft louse, small body louse, or common body louse[3] is similar to the large body louse in appearance, but it is distinctly smaller ($\frac{1}{16}$ inch long), paler colored, and less hairy.   It has commonly been considered the most injurious louse of chickens, but Bishopp and Wood contend that it lives mostly on the feathers, lying along the shaft and running down the feathers to the skin when the feathers are parted.   It is very common about the vent, also on the back and breast.   It does not infest young chicks, presumably because of the lack of well-developed feathers.   The eggs are fastened to the base of feathers and hatch in 2 or 3 weeks.   These lice are very hardy, having been kept alive for 9 months.   This species occurs on ducks, turkeys, and guineas, at least when they are housed with chickens, and is sometimes troublesome to horses stabled near badly infested poultry.

The other lice on chickens are generally less abundant and live chiefly among the feathers, where their nibbling and crawling are not as annoying. These include the fluff louse,[4] which is found among the fluff under the vent and is about $\frac{1}{16}$ inch long; the chicken head louse[5] and the brown chicken louse,[6] both about $\frac{1}{10}$ inch long; the large chicken louse,[7] which is about $\frac{5}{32}$ inch long; and the wing louse,[8] a very slender species about $\frac{1}{10}$ inch long, which lives among the barbules of the wing feathers.   The lice that attack turkeys, geese, and ducks are said to be less abundant and generally not sufficiently injurious to require special treatment.   Pigeons, however, often become grossly overrun with the small pigeon louse[9] or the slender pigeon louse,[10] which live among the feathers of both old and young birds and doubtless interfere with the profitable raising of these fowls.

---

[1] *Jour. Parasitol.*, **6**:350, 1934.

[2] *Menacanthus* (= *Eomenacanthus*) *stramineus* (Nitzsch), Order Mallophaga, Family Menoponidae.

[3] *Menopon gallinae* (Linné), Family Menoponidae.

[4] *Goniocotes gallinae* (De Geer), Family Philopteridae.

[5] *Cuclotogaster heterographus* (Nitzsch), Family Philopteridae.

[6] *Goniodes dissimilis* Denny, Family Philopteridae.

[7] *Goniodes gigas* (Taschenberg), Family Philopteridae.

[8] *Lipeurus caponis* (Linné), Family Philopteridae.

[9] *Campanulotes* (= *Goniocotes*) *bidentatus compar* (Burm.), Family Philopteridae.

[10] *Columbicola columbae* (Linné), Family Philopteridae.

Control Measures. The most important factors in keeping these external parasites and the injurious effects from them in subjection are cleanliness, an abundance of fresh air and light, a good supply of drinking water, and provision of good clean dust baths.

Chicken lice may be controlled by (*a*) dusting with 4 to 5 per cent malathion or 1 per cent rotenone, using 1 pound to 100 birds, or with sulfur. Adding the malathion to dust-bath boxes provides a simple means of self-treatment. (*b*) Spraying the birds with 0.5 per cent malathion at 1 gallon to 100 birds or dipping them into 1 to 2 ounces of wettable sulfur in 1 gallon of water is also effective. (*c*) Treating nests and litter with 1 per cent malathion spray or 5 per cent malathion dust will give control. (*d*) Painting roosts with 40 per cent nicotine sulfate or 3 per cent malathion roost paint at about 1 pint to 150 linear feet is a simple method of control, which should be repeated 8 to 14 days later to destroy the lice hatching from eggs present at the first treatment.

References. Jour. Econ. Ento., **40**:918, 922, 1947; **51**:229, 1958; **52**:774, 1959; and **53**:328, 1960; *Jour. Kans. Ento. Soc.*, **29**:63, 1956.

344 (357) Bed Bug[1]

The common bed bug (Fig. 21.7) and several of its close relatives[2] are frequently pests in poultry houses. They hide, breed, and lay their eggs in nests, behind nest boxes, under loose boards, and in cracks about the walls, roosts, and roof of the building. At night the nymphs and adults find their way upon the sleeping hens and suck their blood. They are almost never found on the fowls in daytime. Sitting hens suffer especially from these pests and may be driven to leave the nests. The small black spots of excreta from the bed bugs may often be seen on the eggs and about cracks.

Control Measures. Because of the likelihood of carrying these bugs into the house, quite as much as on account of their injury to the fowls, vigorous control measures should be applied. Effective measures are given in the chapter on insects affecting man (page 1011). Chicken houses can be rid of these pests by spraying all cracks thoroughly with 5 per cent DDT or 2 per cent chlordane.

345 (359). Sticktight or Southern Chicken Flea[3]

Importance and Type of Injury. In the South and Southwest poultry sometimes show clusters of dark-brown objects about the face, eyes, ear lobes, comb, and wattles made by hundreds of small flattened fleas that have their heads embedded in the skin so that they cannot be brushed off. Young fowls are often killed, and egg-laying and growth are greatly checked by the loss of blood and the great irritation caused by the bites. The ears of dogs and cats often become lined with them.

Animals Attacked. Chickens, turkeys, and other poultry, and also cats, dogs, horses, and man.

Distribution. Southern, and southwestern United States from South Carolina to California.

Life History, Appearance, and Habits. Adult males and females are found, often *in copula*, on the heads of fowls. The females, at least, remain attached by their mouth parts in the same spot sometimes as long as 2 or 3 weeks. During this time the eggs are laid, being thrown with considerable force from the vagina of the female. The eggs hatch on the ground in from 2 days to 2 weeks, and the slender white larvae feed on the excreta of the adult fleas and possibly other filth in the cracks and litter about the floor of henhouses or on the ground in dry protected places. After a growing

---

[1] *Cimex lectularius* Linné, Order Hemiptera, Family Cimicidae.

[2] The poultry bug, *Haematosiphon inodorus* (Duges), the European pigeon bug, *Cimex columbarius* Jenyns, and the swallow bug, *Oeciacus vicarius* Horvath.

[3] *Echidnophaga gallinacea* (Westwood), Order Siphonaptera, Family Sarcopsyllidae.

period of 2 weeks to 1 month, they spin silken cocoons covered with dust and dirt, in which the pupal transformation occurs. The adults do not attach to the host for several days or 1 week and then a second period of about 1 week elapses before the females begin laying eggs. Only a few eggs, one to five, are laid at a time. The life cycle may be completed in from 1 to 2 months. The pest thrives best in dry cool weather, and the adults may live for several months under such conditions.

Control Measures. Thoroughly clean the house and yards, burning the dirt and litter on the spot, or soaking it with kerosene or creosote oil. Thorough dusting of poultry houses with 5 per cent malathion has given good control. Infested birds may be treated with malathion or rotenone. Keep dogs, cats, and rats away from poultry houses, as they may spread this flea. Exclude fowls from beneath buildings.

References. *Okla. Agr. Exp. Sta. Bul.* 123, 1919; *Jour. Agr. Res.,* **24**:1007, 1923; *Jour. Econ. Ento.,* **25**:164, 1932; and **39**:659, 1946; *U.S.D.A., Cir.* 388, 1934; *Jour. Agr. Res.,* **24**:1, 1923.

346. Fowl Tick or Bluebug[1]

Importance and Type of Injury. The injury and symptoms of attack by fowl ticks are much as in the case of poultry mites; weakness of the legs, droopy wings, pale comb and wattles, cessation of egg-laying and death from loss of blood. Small, rounded, reddish or dark-colored objects attach in clusters to the skin on neck, breast, thighs, or under wings, sucking blood and causing great irritation. Large, reddish or blue ticks up to $\frac{1}{2}$ inch long are found in daytime under bark of trees where the fowls roost or under loose boards about the henhouse or where roost poles meet the walls; black spots of excrement stain the woodwork near such cracks. These large ticks suck their fill of blood in about $\frac{1}{2}$ hour, and, when they are abundant, profitable poultry husbandry is out of the question unless control measures are applied. In addition to bleeding fowls, the fowl tick is the proved carrier of a highly fatal poultry disease known as fowl spirochaetosis in many parts of the world. There is danger that this disease may be introduced into North America.

Animals Attacked. Chickens are the preferred host, but all other domestic fowls and some wild fowls are attacked, and rarely domestic mammals and man.

Distribution. In the southwestern states from Texas westward to California, and in Florida. Apparently limited to warm, semiarid regions, around the world.

Life History, Appearance, and Habits. In cooler regions the ticks appear to winter chiefly as adults and half-grown nymphs. Throughout most of its range breeding may continue slowly even in winter. They thrive best, however, in hot dry weather. The adults are flattened, leathery, eight-legged, with thin edges to the body, egg-shaped in outline, and red to blue-black in color. They range in size from about $\frac{1}{4}$ to $\frac{1}{2}$ inch long. The brownish eggs are laid in cracks about the house. They hatch in from 10 days to 3 months, and the grayish six-legged nymphs seek a fowl, particularly at night, and attach in bunches to the skin, sucking blood from 3 to 10 days in one spot. They then release their hold and hide away in a protected place for about a week, molting and acquiring the fourth pair of legs. From this time on they suck blood only at night, hiding in cracks during daylight and alternately engorging and molting at intervals of a week or two until they become adult. The females deposit several lots of eggs, a total of 500 to 900. There is normally one generation a year, but if deprived of a host the adults may survive for 2 or 3 years without food or water.

Control Measures. The poultry house should be sprayed with 3 per cent malathion, applied with extreme thoroughness because these flattened ticks can hide deeply in cracks and are most hardy and resistant creatures. Roosts should be suspended by wires from the ceiling or supported entirely from the floor so that they do not touch the walls. Nests should be so constructed that they can be easily removed and treated. Poultry must not be allowed to roost in trees, barns, and other shelters which cannot be treated effectively. Any new fowls brought into the flock should first be isolated for 10 days to allow the seed ticks to drop off, and the temporary quarters then treated with 0.5 per cent chlordane or toxaphene or 5 per cent DDT spray. These sprays should also be applied to any outside resting places where the ticks are numerous.

References. *U.S.D.A., Farmers' Buls.* 1653, 1933, 1070, 1941.

[1] *Argas persicus* (Oken), Order Acarina, Family Argasidae.

347. Chicken Mite[1] or Poultry Mite

Importance and Type of Injury. These mites (Fig. 20.30) live in cracks about the roosts, floors, walls, or ceiling of the houses in the daytime and crawl upon the fowls at night or when they are on the nests. Only a very few mites are found on fowls during the daylight hours. Their only normal food is the blood of fowls, which they draw through their sharp piercing mouth parts at night. Since they rarely stay on poultry during the daytime, a flock may be badly run down by them without the owner's being aware of the cause of the trouble. Small areas about the roosts or elsewhere in the house show patches of gray, brown, or red mites or fine, black-and-white speckling as though dusted with pepper and salt—the

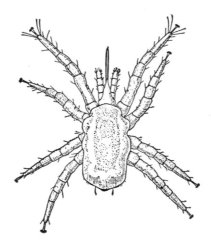

Fig. 20.30. The chicken mite, *Dermanyssus gallinae*, nymph before gorging on blood. Greatly magnified. (*From U.S.D.A., Farmers' Bul.* 801.)

excrement of the mites. When they become extremely abundant, the litter and manure may appear to be literally crawling with tiny gray-and-brown specks. The mites are especially abundant where the roost timbers rest on their supports. The fowls become droopy, pale about the head, and listless, and they stop laying. Sitting hens and chicks often die.

Animals Attacked. Chickens are preferred, and other poultry is not likely to be badly attacked if chickens can be reached. The mites often greatly irritate persons or animals about infested houses.

Distribution. Practically cosmopolitan; more troublesome in warmer and drier regions.

Life History, Appearance, and Habits. The mites become inactive and greatly reduced in numbers during cold weather, except in heated houses, although feeding and breeding to some extent may take place during warm spells, even in winter. With the advent of spring, activities are greatly accelerated. The females deposit their pearly-white elliptical eggs in dark protected places, such as cracks between timbers or in dry manure under the roosts. Two or three dozen eggs are laid over a period

---

[1] *Dermanyssus gallinae* (De Geer), Order Acarina, Family Dermanyssidae.

of several weeks. The eggs hatch in 2 to 4 days on the average. Several days more elapse, and the young six-legged nymph sheds its skin once and becomes eight-legged before it seeks a fowl and fills up on blood (Fig. 20.30). After engorging, it hides away in some dark place for a day or two or until it sheds its skin again. It then finds a host, engorges on blood a second time and again hides away for a day or two before it makes the third and final molt and becomes an adult. It requires only 1 week to 10 days to pass through the stages from egg to adult. The adults are $\frac{1}{40}$ to $\frac{1}{30}$ inch long, grayish in color but, when filled with blood, appearing bright red to nearly black. It is believed that the same adult does not feed every night, but after filling up on blood, leaves the fowl and during the next few days lays three to seven eggs, then feeds and repeats the process until all the eggs have been laid.

Control Measures. For the control of these mites attention must be given to the house rather than to the birds themselves. A very thorough cleanup of the poultry house is essential, since the mites abound in filth and infest any crack about the house big enough for them to crawl into. This should be followed by a thorough spraying with 1 per cent malathion, using 1 to 2 gallons to 1,000 square feet, or by dusting with 4 to 5 per cent malathion at 1 pound to 40 square feet of litter. The birds may be treated by spraying with 0.5 per cent malathion, using 1 gallon to 100 birds, or by dusting with 1 per cent malathion, using 1 pound to 100 birds. Painting the roosts with 40 per cent nicotine sulfate or 3 per cent malathion roost paint at about 1 pint to 150 linear feet is also effective. The mites may live 4 or 5 months in a house after all poultry has been removed, especially if surroundings are moderately moist. These pests may also stay on the bodies of fowls in small numbers for at most about 3 days. When moving fowls to a new building or introducing new birds to a flock, they should be isolated for 3 days and nights in pens so that the mites will all crawl away from them. Poultry houses should be so constructed as to receive plenty of light and air, to be dry, and to be cleaned easily.

References. *U.S.D.A., Dept. Buls.* 1652, 1933, 801, 1931, and 1228, 1924; *Jour. Econ. Ento.,* **40**:596, 918, 1947; **49**:447, 628, 1950; and **51**:158, 911, 1958.

348. Northern Fowl Mite or Feather Mite[1]

Importance and Type of Injury. Small reddish or brown eight-legged mites swarm over the skin, congregating about the vent, tail, and neck. The feathers become dirty from the eggs and excreta of the mites, and the skin becomes irritated and scabby. As a result, egg production is decreased and in severe infestations the fowls may be killed.

Animals Attacked. Chickens and other domestic fowls, English sparrow, starling, pigeon, and a variety of wild birds.

Distribution. Generally distributed throughout the United States.

Life History, Appearance, and Habits. The eggs are laid in the fluff feathers, where they adhere to the barbules, and are also found in the nest. Hatching occurs in 2 to 3 days, and the larval stage lasts about 9 hours. The mites become adult after about 4 days. The adult mite resembles the poultry mite in a superficial way but is slightly smaller, more hairy, and with smaller legs. It moves more rapidly, and the tip of the female abdomen has a slight notch. In habits, the northern fowl mite is very different, since it lives entirely on the body of the fowl instead of harboring in cracks about the poultry house.

[1] *Ornithonyssus* (= *Liponyssus*) *sylviarum* (Canestrini & Fanzago), Order Acarina, Family Dermanyssidae. The tropical fowl mite, *O. bursa* (Berlese), is very similar and generally more prevalent in warmer areas.

Control Measures. These are the same as for the poultry mite (page 992), but the northern fowl mite is harder to kill.

References. *U.S.D.A., Dept. Cir.* 79, 1920; *Can. Jour. Res.,* **D16**:230, 1938; *Ky. Agr. Exp. Sta. Bul.* 517, 1948; *Jour. Econ. Ento.,* **45**:926, 1952; **46**:827, 1953; **47**:943, 1954; **51**:188, 1958; and **52**:1195, 1959.

349. Scaly-leg Mite[1]

Importance and Type of Injury. This species and the following one are close relatives of the mites and ticks discussed earlier in this chapter, but their habits are quite different. They are more like mange or itch mites, since they remain on the body all the time and make tunnels in the skin, in which the eggs are laid. The scaly-leg mite attacks especially the feet and lower part of the legs but is also found about the comb and neck. The scales of the legs (Fig. 20.31) become elevated, and a fine white dust

Fig. 20.31. Rooster severely attacked by scaly-leg mite, *Knemidokoptes mutans.* (*From U.S.D.A., Farmers' Bul.* 1337.)

sifts from beneath them. Lymph and blood exude and red blotches form on the legs. The birds become crippled or even unable to walk at all. Great irritation must result from the burrowing of the mites.

Animals Attacked. Poultry and wild fowls; and rabbits, guinea pigs, and other animals housed near infested birds.

Life History, Appearance, and Habits. This trouble is caused by a tiny, eight-legged mite, measuring from $\frac{1}{100}$ to $\frac{1}{50}$ inch across, pale gray in color and nearly circular in outline. They are not seen by the naked eye. If the fowl's legs are soaked in warm soapy water, the scales may be lifted and the mites found by use of a microscope among the powder and lymph from beneath the scale. They have very short legs and the skin is traversed with fine lines as in the palm of one's hand. It is believed that the eggs are laid in the tunnels made by females beneath the skin scales. Like other mites, the young are at first six-legged and the development from that point on is a simple metamorphosis.

Control Measures. Poultrymen should be on the lookout for this trouble, especially on newly purchased birds, and should treat fowls as soon as the symptoms show up to prevent spread to others. The legs may be brushed with, or dipped in, crude petroleum or a mixture of raw linseed oil, 2 parts, and kerosene, 1 part. If the swollen scales are not largely shed within a month, the treatment should be repeated, but one treatment is usually sufficient.

[1] *Knemidokoptes mutans* (Robin & Lanquentin), Order Acarina, Family Sarcoptidae.

350. DEPLUMING MITE[1]

This mite is similar to the scaly-leg mite but is still smaller, and it burrows into the skin at the base of the feathers on the rump, back, head, abdomen, and legs. The oval mites may be found on the fallen feathers or among the dry powdery material in the skin. The irritation from the mites causes the fowls to pull out their feathers.

Control Measures. Sulfur ointment carefully applied to the affected parts, or a dip made by mixing 2 ounces of wettable sulfur and 1 ounce laundry soap to the gallon of water. Repeated applications will be necessary. The affected birds should be isolated.

[1] *Knemidokoptes gallinae* (Railliet), Order Acarina, Family Sarcoptidae.

CHAPTER 21

INSECTS THAT ATTACK AND ANNOY MAN AND AFFECT HIS HEALTH

The various ills man suffers from insects reach their climax in their attacks upon his person. "Bugs" are no respecters of persons. While for the most part clean, sanitary living and reasonable precautions about associating with less sanitary persons and surroundings will prevent these insects from attacking a given individual, nevertheless some of these unwelcome visitors are likely to come into any household at any time. In such a case, to be forewarned is to be forearmed. The housekeeper who knows when to ignore a newfound insect in her house as a creature of no real significance and when to fight one that may start an infestation which could cause trouble for months has the battle half won. Brues[1] says:

"The importance of insects as detrimental to public health is well known to professional zoologists, medical men, and laymen alike, but is usually emphasized only under the stress of particular circumstances, such as the safety of soldiers in war or of unusual outbreaks of diseases for which insects are directly responsible. Insect borne diseases present a *constant* menace to the world, and aside from the actual toll of lives which they exact, they impair its efficiency by enfeebling the health of its human population."

W. D. Hunter, in his address as President of the American Association of Economic Entomologists,[2] draws the conclusion that the losses caused by diseases transmitted by insects is approximately one-half as great as the losses caused to all farm products. He concludes: "Surely this is a sufficient argument for greater attention to medical entomology."

FIELD KEY FOR THE IDENTIFICATION OF INSECTS THAT ATTACK AND ANNOY MAN AND AFFECT HIS HEALTH

A. *Free-flying insects that alight on face, arms, and other exposed parts of the body to bite and to suck blood:*
  1. Slender-bodied long-legged insects, up to ½ inch long, with delicate wings fringed with scales, long slender mouth parts, and bushy antennae, make a high-pitched humming noise as they alight to suck blood. Especially abundant at dusk or at night and about swamps and woodlands...*Mosquitoes*, page 998.
  2. Small, chunky, humpbacked gnats, not over ⅕ inch long, with broad clear wings and short heavy mouth parts and antennae, alight on the body, crawl into eyes, ears, hair, or under the clothing and suck blood. They make comparatively little noise, and the bite at first is not very painful. Especially troublesome during the daytime.....................*Black flies*, page 1005.
  3. Very small midges, not over ¹⁄₁₀ inch long, with hairy but not scaly, sometimes mottled, wings, bite from early evening to early morning. They fly quietly and

[1] BRUES, C. T., "Insects and Human Welfare," Harvard Univ. Press, 1920.
[2] HUNTER, W. D., *Jour. Econ. Ento.*, **6**:27, 1913.

are seldom seen or heard until the hot, very painful bite is inflicted..........
.....................................*No-see-ums* or *punkies*, page 1007.

4. Large, heavy-bodied, brown, black, and orange, often green-eyed flies, about ½ inch long, with wings clear or banded, fly wildly about the head with much noise, and cause a very painful, bloody bite on arms, head, or neck. Encountered chiefly in woods or marshes on warm sunny days.....................
....*Deer flies* or *horse flies*, page 1008.

5. A fly about the size and general appearance of the house fly often comes indoors in lowering weather and bites especially about the ankles. It differs from the house fly in having a stiff, pointed beak that projects forward from lower side of the head, a broader black-spotted abdomen, grayer appearance, the apical cell (R_5) of wings more widely open, and hairs on upper side of arista only....
....*Stable fly*, pages 942 and 1008.

B. *Insects that crawl upon the body to bite and suck blood, not spending their entire life on the human host:*

1. Mahogany-brown, broad, very flat or thin, oval, wingless bugs, of all sizes up to ⅕ inch long, live in beds and cracks about the room and crawl over the body and bite at night. Bugs have a distinct disagreeable odor. *Bed bug*, page 1001.

2. Large, somewhat flattened, oval bugs, brownish or black, sometimes marked with pink or red; between ½ and 1 inch long, tapering to a slender head in front, with well-developed wings crossing over on the back, 2 or 3 large cells in the membrane of the wing, 4-segmented antennae, and 3-segmented labium. They bite when picked up and, in some localities, regularly visit the bodies of sleeping persons to suck their blood.................................
................*Poultry bug, kissing bug,* and other *assassin bugs*, page 1012.

3. Small, brown, wingless insects, about 1/16 inch long, very flat from side to side and with long hind legs, slip into the clothing or jump vigorously when disturbed. They bite especially about the legs and waist. Troublesome chiefly in basements or houses where dogs or cats are allowed....................
......................*Cat flea, dog flea, human flea,* and others, page 1013.

4. In southern states or tropical countries, painful, inflamed sores develop between the toes or under the toenails, each containing an ulcerated body that may become as big as a small pea. This is due to a flea that buries its body in the skin...*Chigoe*, page 1016.

5. Tough-skinned, 8-legged, reddish-brown to bluish-gray, seed-like bodies up to ½ inch long, often with silvery-white markings on the hard plate near the head, sink their mouth parts deeply and firmly in the skin and suck blood. No particular pain is felt at the time, but motor paralysis, infected sores, or spotted fever may follow the attack...
Rocky Mountain wood tick, castor-bean tick, dog tick, and others, pages 1017 and 1020.

6. An eruption on the skin like hives or chicken pox, accompanied by severe itching, and sometimes nausea, headaches, chills, or fever, results from handling straw or flour, dried fruits, meat, and other groceries, or follows a visit to grassy, brambly spots out-of-doors. Caused by nearly microscopic mites that crawl upon the skin and bite..
Chiggers, harvest mites, louse-like mites, or *flour* and *meal mites*, pages 1022 to 1025.

C. *Insects that live upon the body all their lives, generation after generation:*

1. Painful burning and itching bites, which become whitish scars ringed with brown pigmented skin, occur anywhere on parts of body covered by clothing. No "cause" found on the skin but examination of clothing about the neckband, armpits, waist, or crotch of trousers, reveals elongate, oval, flattened, wingless gray lice, up to ⅛ inch long, with 6 legs, each bearing a single curved claw at the end. Whitish keg-shaped eggs are laid in the seams of the clothing.....
.................................*Body louse, grayback,* or *cootie,* page 1026.

2. Hair of the head, especially back of the ears and at the nape of the neck, infested with crawling, grayish, 6-legged lice, similar to the above; the whitish eggs fastened to the hairs.............................*Head louse,* page 1026.

3. Painful burning and itching bites with small inflamed spots among the hairs

between the legs.   Small, broad, grayish, 6-legged lice up to $\frac{1}{16}$ inch long, that look something like miniature crabs and tend to remain fixed on one spot....
....*Crab louse*, page 1029.

4. Extreme itching of tender places on the skin, such as between the fingers, behind the knee, and inside the elbow, without visible cause.   Careful examination may reveal delicate, tortuous, gray thread lines just beneath the skin. Hard pimples as big as pinheads, containing a yellow matter, form on the skin. Scraping the affected spots to the "quick" and examining the scrapings under a lens reveals very small, whitish, 8-legged, nearly round mites..............
....*Itch mite*, page 1025.

D. *Short, whitish, legless, segmented maggots or "worms"*[1] *live under the skin in sores, or in the natural body cavities, or in the alimentary canal, and are not infrequently passed in the excreta:*

1. The maggots taper gradually from a blunt posterior end, which bears 2 rounded plates or short tubes for breathing, to a pointed head end which has short mouth hooks..
......Various *flesh flies, house flies, fruit flies,* and *root-maggot flies*, page 1031.

2. The maggots are thickest near mid-length and narrow strongly toward either end; the skin is very tough, and usually beset with many, minute, short, sharp spines..*Human bot flies*, page 1031.

3. The maggots are flattened, narrowing toward either end, each segment with prominent, fleshy, pointed processes, some of which have minute side spines...
...............................*Latrine fly* and *little house fly*, page 1031.

4. The maggots are nearly cylindrical, with head end rounded off and the opposite end prolonged into a long slender "tail" that is extensible like a telescope....
....*Rattailed maggots*, page 1031.

E. *Insects that frequent both filthy materials and human habitations:*

1. Flattened, oval, running, brown or black, nocturnal insects from $\frac{1}{2}$ to $1\frac{1}{2}$ inches long, with 6 long, very spiny legs, antennae longer than the body, and usually 4 finely veined wings that are seldom used; occur in kitchens, bakeshops, restaurants, public buildings, ships, and other moist warm places............
....*Cockroaches*, page 906.

2. Two-winged flies, about $\frac{1}{4}$ to $\frac{1}{3}$ inch long, of a general grayish color, with 4 equal black stripes on the back between the wings; the arista or antennal bristle feathered, *i.e.*, hairs coming off on two sides of it; the mouth parts soft, spongy, and retractile; the vein that ends nearest the wing bent forward so as nearly to meet the vein in front of it at the wing margin; and no large, bristly hairs on front segments of the abdomen......................*House fly*, page 1031.

F. *Insects and other arthropods that accidentally or occasionally hurt man, in defense of themselves or their nests, by stinging, biting, nettling, or blistering the skin:*

1. Bugs of various sizes and shapes up to 2 inches long, but all with a slender tubular beak projecting from the lower side of the head, 6 legs, and the front pair of wings thicker at base and thinner and overlapping at the tip; bite painfully when handled or when they fly against the face, especially at night.....
Assassin bugs, water-scorpions, electric-light bugs, and many other Hemiptera, pages 221 and 1012.

2. Elongate, rather soft-shelled, 6-legged beetles, commonly $\frac{1}{2}$ to $\frac{3}{4}$ inch long, blister or corrode the skin if crushed upon it............................
......................*Spanish fly,* and other *blister beetles,* pages 55 and 642.

3. Caterpillars ("worms") of a variety of sizes and colors, but always with 3 pairs of jointed legs and 5 pairs of fleshy prolegs, generally more or less spiny or woolly, nettle the skin when they brush against it, causing pain, itching, and inflamed spots.................*Various caterpillars* or *moth larvae*, page 23.

4. Four-winged, swift-flying, smooth or very hairy insects, usually conspicuously marked with yellow, insert the ovipositor or "stinger" from the posterior end

---

[1] Not to be confused with the parasitic *roundworms* and *flatworms*, which are usually much longer.   These, if cylindrical, are not segmented, and if segmented are much flattened.   Insect larvae never show more than 13 segments.

of the body, and inflict a painful sting..

..............................*Bees, wasps,* and *hornets,* pages 23 and 266.
5. Wingless, crawling, 6-legged, slender-waisted "ants," the hairy kinds often brilliantly colored, sting as in *F*,4.......................................

.........................*Stinging ants* and *velvet ants,* pages 23 and 272.
6. Wingless, many-colored, 8-legged creatures with the body in 2 regions, the part bearing the long legs separated from the abdomen by a slender pedicel or stalk, rarely puncture the skin and introduce a venom with the mouth parts........

........*Tarantulas* and other *spiders,* page 182; *black widow spider,* page 1008.
7. Elongate creatures, commonly 2 or 3 inches long, with a pair of "pincher legs" in front, eight long walking legs at the middle and a very long abdomen in two parts, the posterior slender part with a short swollen sting on the end; cause great pain by thrusting this sting into the flesh..........*Scorpions,* page 183.
8. Elongate, worm-like "hundred-legs," from 1 to 8 inches long, with distinct head and antennae, somewhat flattened body, and 15 or more pairs of legs, "bite" with a pair of poison claws just back of the head..*Centipedes,* page 181.

General References. HERMS, W. B., "Medical and Veterinary Entomology," 4th ed., Macmillan, 1947; RILEY, W., and O. JOHANNSEN, "Medical Entomology," McGraw-Hill, 1938; HERRICK, G., "Insects Injurious to the Household and Annoying to Man," Macmillan, 1926; MATHESON, R., "Medical Entomology," 2d ed., Comstock, 1950; EWING, "Manual of External Parasites," Thomas, 1929; PATTON, W. C., and A. M. EVANS, "Insects, Ticks, Mites, and Venomous Animals," Vols. I and II, Liverpool School of Tropical Medicine, 1929; HINDLE, E., "Flies in Relation to Disease: Blood-sucking Flies," Cambridge Univ. Press, 1914; GRAHAM-SMITH, G., "Flies in Relation to Disease: Non-blood-sucking Flies," Cambridge Univ. Press, 1914.

351 (322). Mosquitoes

Importance and Type of Injury. Besides the well-known painful bites inflicted by the females of mosquitoes, these insects are the proved carriers of five distinct human diseases. There is no other known method of acquiring malaria, yellow fever, dengue, and certain forms of filariasis except by the bites of mosquitoes which have previously bitten persons that had these diseases. Considering the entire world, malaria ("ague" or "chills and fever") has been said to be the most important disease. It causes a large percentage of the deaths among mankind. It has been estimated that in the United States in 1935 there were about 900,000 cases of malaria and perhaps 4,000 deaths, and for many years the prevalence of malaria retarded the proper development of the South. However, modern malaria-control methods and improved standards of living have made this disease a rarity in the United States today. In the tropics entire countries have been practically barred from civilization by this disease. As pointed out by Ross, these are unfortunately "more especially the fertile, well-watered and luxuriant tracts; precisely those which are of the greatest value to man." The several kinds of malaria are caused by microscopic organisms[1] that live in the blood, destroying the red corpuscles and causing anemia, accompanied by the characteristic alternating chills, fever, and sweating. Once introduced to a human body by a mosquito, the parasite may increase rapidly until as many as 3 billion are present in the blood of one patient; but it cannot pass from that person to another without the help of certain kinds of mosquitoes[2]

[1] *Plasmodium vivax, P. malariae, P. falciparum,* and *P. ovale,* Phylum Protozoa, Class Sporozoa.

[2] About 85 species of the genus *Anopheles,* Order Diptera, Family Culicidae. The important vectors of malaria in the United States are *A. quadrimaculatus* Say in eastern and southern United States and *A. freeborni* Aitken in the Pacific Coast region.

(Figs. 21.1 and 21.3), in the bodies of which a necessary part of its life cycle is completed. The sexual stages of the parasite, the *macro-* and *microgametocytes*, are taken into the stomach of the female Anopheles with a blood meal from a person suffering from malaria. Maturation into and union of the *gametes* occur in the mosquito stomach, forming a *zygote*, which penetrates into the stomach wall to form an *oöcyst*. Sporogony then occurs, and the oöcyst fills with minute *sporozoites*, which burst from the oöcyst into the body cavity and invade the salivary glands of the mosquito. When the mosquito bites, the sporozoites pass into their new host with the salivary fluid (Fig. 1.13). The entire process occurs in 1 to 2 weeks.

Until the end of the nineteenth century yellow fever was one of the most dreaded diseases in the world. Its cause was not known, and terrible epidemics swept tropical countries and our seaport towns such as New Orleans, Philadelphia, and Havana. This disease was the most potent factor in the failure of the French to build the Panama Canal, in the latter part of the last century. In 1900 American army surgeons working in Cuba discovered that the disease is spread by a particular kind of mosquito,[1] since known as the yellow-fever mosquito. Since this discovery the disease has been rapidly stamped out of country after country until it has now been nearly eradicated from many parts of the earth. Until 1929 it seemed certain that yellow fever could be annihilated from the entire world. Recently, however, a type of the disease known as "jungle fever" has been found in wild areas in South America and Africa, where it is believed that monkeys constitute a permanent reservoir or alternate hosts, and where it is spread by the bites of several other mosquitoes (Table 1.2). From this ineradicable source it may continually spread to towns and cities and start epidemics. There is continual grave danger of transporting infective mosquitoes or an incipient human case of yellow fever by airplane to many parts of the world where the mosquito carriers are present. The cause of the disease is a filterable virus. While its transformations in man and the mosquito are not known, there is evidence that it passes an essential part of its life cycle in each host. A method of vaccination against yellow fever has recently been perfected. The conquering of this disease constitutes one of the greatest triumphs of modern science.

Filariasis is caused by minute worms[2] that live in the blood and lymph. This disease is seldom fatal but is sometimes followed by enlargements or deformities of the legs, arms, genital organs, or other parts of the body, known as "elephantiasis." The embryo or larval worms are found in the lungs and deeper blood vessels during the daytime, but when the patient is resting, usually during the night, these worms swarm into the superficial blood vessels, and mosquitoes feeding at this time may draw in some of them with the diseased blood. The worms undergo a transformation in the muscles of the mosquito, and 2 or 3 weeks later, when about $\frac{1}{16}$ inch long, work out through the labium, as the mosquito is feeding on a new victim. They bore through the labella and into the skin of man, then migrate to the lymphatics where they grow to a length of about 3 inches and a diameter of $\frac{1}{100}$ inch. They mate, and

---

[1] *Aedes aegypti* (Linné), Order Diptera, Family Culicidae.
[2] *Wuchereria bancrofti* and *W. malayi*, Phylum Nemathelminthes, Class Nematoda, Family Filariidae.

the females produce the young ovoviviparously. Several kinds of mosquitoes[1] carry this disease. It occurs in many parts of the tropics.

The fourth disease of which mosquitoes are the known carriers is dengue or breakbone fever. It has occurred in our southern states, but is most prevalent in the tropics. It is seldom fatal but may attack practically the whole population of a village or community, so that temporarily great inconvenience and suffering are experienced. The symptoms are a very high, intermittent fever accompanied by terrible aches in the bones and joints and a skin rash. This virus disease is spread chiefly by the bites of the yellow-fever mosquito.[2] There were half a million cases of this disease in Texas in 1922.

The virus disease human encephalitis, which is widespread in North America, is transmitted by several species of mosquitoes.[3]

Animals Attacked. Mosquitoes bite all kinds of warm-blooded animals, especially domestic and wild mammals and even such creatures as snakes and turtles. While the diseases just discussed are known to be troublesome only in man, mosquitoes transmit diseases of other animals, such as equine encephalitis (page 32) and bird malaria.

Distribution. While yellow fever has been eradicated from many parts of the world, the mosquito which carried this disease is still common in the southern part of the United States and the tropics. Other mosquitoes range from the equator nearly to the poles and from sea level to at least 7,000 feet altitude.

Life History, Appearance, and Habits. Since there are about 140 species of mosquitoes in North America,[4] we shall attempt to give only certain general features of their life cycle and habits. Many species of mosquitoes go through the winter in the egg stage. Some winter as adult, fertilized females in washrooms, cellars, outbuildings, hollow trees, and other shelters, where they are often seen hiding in fall and spring. Others survive the winter in the larval stage, either freezing up with the water or remaining dormant at the bottom of ponds and puddles. Mosquitoes always develop in water, which contains microscopic plants and animals that serve as food for the larvae, and their eggs are laid on the water or in places where water is likely to accumulate, as on ice or snow or in dry depressions. The eggs of the common Culex mosquitoes are built in minute rafts that look like a bit of soot floating on the water but are really composed of several hundred eggs standing on end. The eggs of anopheline mosquitoes and yellow-fever mosquitoes are laid singly, the former having curious hollow expansions at the middle like a life belt, which keep them floating (Fig. 21.1).

The larvae of mosquitoes (Figs. 21.1 and 21.2) are the common "wrigglers" of rain barrels and quiet pools. The head is large, and has complex mouth brushes that, constantly in motion, waft food into the mouth. The mouth parts are of the chewing type, and they feed on algae and other small plant or animal life, either living or dead. The thorax is swollen

---

[1] Twenty or more species of *Aedes, Anopheles, Culex,* and *Mansonia* mosquitoes. *Culex pipiens quinquefasciatus* Say, which is an important vector of filariasis in some parts of the world, is the common southern house mosquito.

[2] *Aedes aegypti* (Linné). *A. albopictus* (Skuse) is also a vector.

[3] *Culex tarsalis* Coquillett, *C. pipiens pipiens* Linné, *C. pipiens quinquefasciatus* Say, *Aedes taeniorhynchus* (Wiedemann), and *A. sollicitans* (Walker).

[4] Matheson describes 139 species, including 13 of *Anopheles,* 58 of *Aedes,* 18 of *Culex,* and 18 of *Psorophora.*

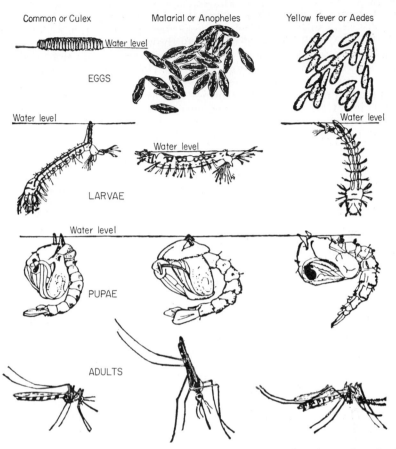

FIG. 21.1. Three important kinds of mosquitoes, showing egg, larval, pupal, and adult stages of each: in the left-hand column the life stages of a common house mosquito; in the center column, of a malaria mosquito; and in the right-hand column, of the yellow-fever mosquito.   5 or 6 times natural size.   (*From "Everyday Problems in Science," by Pieper and Beauchamp, copyright, 1925, Scott, Foresman & Co.*)

and appears as one segment, but has no trace of legs in this stage.   The abdomen is slenderer and bears on the eighth or next-to-last segment a short tube, known as a siphon, which the larva must thrust up into the air at intervals to breathe.   This supply of air is supplemented by four finger-like, tracheal gills attached to the last segment on the body, by which oxygen is taken from that dissolved in water.   The gills alone will not keep the larvae alive, and they must come often to the surface to breathe.   Indeed, they usually lie with the siphon

FIG. 21.2. Larva of a malaria mosquito, *Anopheles quadrimaculatus*, in resting position at surface of water. About 5 times natural size. (*From U.S.D.A., Farmers' Bul. 450.*)

projecting up through the surface film and the rest of the body hanging down at an angle in the water.   Most species descend to the bottom of shallow water to feed upon organic matter there. In the larvae of anopheline mosquitoes, the siphon is very short, and the larvae generally lie parallel to the surface just below the surface film feeding upon organisms, such as algae, floating on the surface of the water. When disturbed they swim down into the water by lashing the abdomen from side to side.

In as short a time as 2 days to 2 weeks the larvae may be full-grown, about ⅜ inch long in common species. The change to the pupal stage takes place quickly at the fourth molt. This is a very unusual kind of a pupa (Fig. 21.1). It swims about actively in the water, avoids enemies, and does nearly everything the larva does except to feed. It breathes through two trumpet-like tubes on the thorax. The eyes, legs, and wings can be seen developing through the body wall on the large combined head and thorax. The pupal stage is often called a "tumbler." After a few hours to a few weeks in this condition, the insect splits its skin down the back and the adult (Fig. 21.3) crawls out, balances for a few moments on the empty pupal shell until its wings spread and dry, and then flies away. Most species pass through a number of generations each year.

The larvae and pupae of many common species breed in stagnant water of large or small quantity—anything from a bit of rainwater, caught in a discarded tin can, to the acres and acres of marsh water along our coasts and streams. The larvae of anopheline mosquitoes breed in a great variety of situations, including especially slowly moving or standing water in which green algae abound. The yellow-fever mosquito breeds especially in and about dwellings and in cities wherever a bit of water, clean or foul, is left exposed long enough for it to complete its aquatic cycle. The adult of this species bites chiefly in the late afternoon and early morning. It is stealthy, does not make a very loud hum, is fond of crawling up under the clothing to bite, and prefers to stay in dwellings and other buildings. It has a peculiar, white, lyre-like pattern on the thorax. The anopheline mosquitoes bite chiefly in the evening and early morning. Many of them can be distinguished by the white-and-black or rusty-red spotting of the wings. The females also differ in having the palps about as long as the labium, whereas other mosquitoes have the palps, in the females, not over a third or fourth as long as the other mouth parts. When biting or resting, the long axis of the body in the anopheline mosquitoes is at an angle to the surface on which they are standing, while the culicine mosquitoes hold the body parallel to the surface. Other kinds of mosquitoes bite mostly at night, but there are enough that venture forth in daytime in most sections to make life uncomfortable at any hour of the day or night, especially in the woods or about lowlands.

Control Measures. Generally speaking, mosquito abatement should be carried out on a community-wide or large-area basis, under the direction of experts who will determine the exact species involved, their breeding periods, and the most effective and economical means of suppression. Satisfactory control depends upon an intimate knowledge of mosquito biology, especially such factors as life cycle, larval breeding habits, adult flight range, adult resting places, and adult food preferences. Because mosquitoes depend on an aquatic habitat for larval development, the most fundamental means of control lies in the elimination of breeding places by drainage, filling, pumping, or temporary dewatering. Domesticated species such as *Aedes aegypti* may be controlled by doing away with useless receptacles which may hold water, such as tin cans and urns, and by eliminating water which is standing in roof or street gutters, catch basins, drains, and the like.   Where drainage and dewatering methods are impossible, as in lakes, rivers, streams, and water-storage reservoirs, natural control is often possible.   The larval stages of most species of mosquitoes depend upon aquatic vegetation for protection and development of food organisms.   Especially effective anopheline control can thus be secured by keeping the water surface free of such vegetation by either mechanical or natural means.   Management of water levels achieves a similar purpose in keeping water out of marginal vegetation and is also

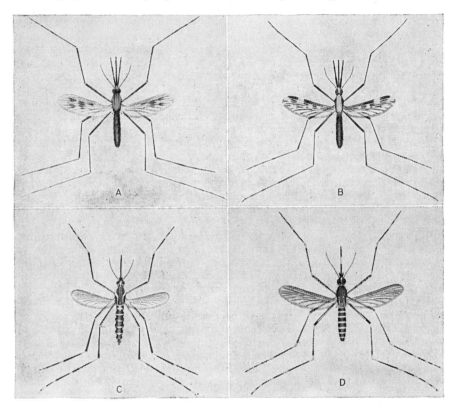

FIG. 21.3. Four important North American mosquitoes.   *A, Anopheles quadrimaculatus; B, Anopheles punctipennis; C, Aedes aegypti; D, Culex tarsalis.   (From Communicable Disease Center, U.S. Public Health Service.)*

effective in stranding larvae on dry land.   Water-level control is especially effective in rice fields, which must be alternately flooded and dewatered, and in storage reservoirs, where cyclical fluctuation and seasonal recession can be used to retard the invasion of marginal plants into the flooded areas.   Larvivorous fish such as Gambusia can occasionally be employed to keep ornamental pools and small ponds mosquito-free.   In spite of their usefulness, however, such natural approaches rarely achieve complete mosquito control and must be reinforced by other measures such as house screening and the application of insecticides.

The choice of larvicides for the control of mosquitoes is dependent upon the species and habits of the mosquito; hazards to human beings, animals, fish, aquatic organisms, and other wildlife; the presence of insecticide-resistant strains of mosquitoes; and cost factors. Paris green dust at 1 to 2 pounds per acre has given good control of anopheline larvae, which are surface feeders, but has not been very effective against culicine larvae. Kerosene or fuel oil applied at 20 to 40 gallons per acre is a simple and effective larvicide for small bodies of water. Pyrethrins emulsion applied at 0.1 per cent is effective but costly. Most larviciding operations are conducted, using emulsions or granular formulations of DDT, dieldrin, chlordane, heptachlor, or lindane at 0.05 to 0.1 pound per acre. Where resistant strains are encountered, EPN, parathion, or Baytex at 0.05 to 0.1 pound per acre or malathion at 0.5 pound per acre has been widely used. These insecticides are applied to small breeding areas by knapsack sprayer, at 1 quart to 1 gallon per acre, or by hand duster, at 1 to 5 pounds per acre. Larger areas are treated by power equipment or by aircraft.

Adult mosquitoes are most readily controlled in human habitations. Space-spraying with 0.2 per cent synergized pyrethrins at 1 ounce to 1,000 cubic feet will give rapid knockdown and relief from annoyance. Weekly treatments of this type have been demonstrated to give effective malaria control by preventing the survival of infected mosquitoes, and thus interrupting the chain of malaria transmission between gametocyte and sporozoite formation (Fig. 1.13). Space sprays of 3 to 5 per cent Lethane or Thanite have also been used for mosquito control. Residual spraying of habitations with persistent insecticides (page 442) is widely used in eradication programs against malaria and filariasis, but is not so effective in preventing the annoyance caused by the mosquitoes. Residual treatments of DDT at 200 milligrams per square foot of wall surface remain effective for 6 to 12 months, dieldrin at 50 milligrams per square foot for 6 to 12 months, and lindane at 50 milligrams per square foot for 3 months. Malathion at 200 milligrams per square foot is somewhat less persistent but is effective against strains of mosquitoes resistant to the other materials. Area treatments using ground aerosols or aircraft sprays of 5 per cent DDT, 2.5 per cent chlordane, 0.25 per cent lindane, or 5 per cent malathion in fuel oil are effective against adult mosquitoes when applied at 0.2 to 1 pound per acre and have been used for malaria and encephalitis control operations. The liberal use of space sprays or aerosols may be employed to free outdoor gatherings from the annoyance of mosquito biting for periods of several hours.

The careful and complete mosquitoproofing of dwellings or beds is essential to comfort and health in many parts of the world. Sixteen-mesh screen will prevent the entrance of all North American malaria vectors but will not always protect against the entrance of *Aedes aegypti* and is ineffective against sand flies or punkies, for which smaller-mesh screen should be used. In mosquitoproofing, especial attention should be paid to fireplace covers, to covering cracks with heavy kraft paper, and to filling holes with wood or metal patches.

Where individuals must be exposed to mosquito attack, very effective protection from biting can be obtained by the use of repellents such as dimethyl phthalate, 2-ethyl-1,3-hexanediol, diethyltoluamide, and dimethyl carbate, applied as described on page 389. The use of fine-mesh veils, gloves, and tightly woven clothing is also of benefit. For relief from the bites of mosquitoes and the bites and stings of other insects, the first rule is not to break the skin by scratching. In many cases, prompt relief from pain or itching has been secured by the local application of an antihistamine ointment such as 5 per cent 2-methyl-9-phenyl-2,3,4,9-tetrahydro-1-pyridindene hydrogen tartrate.[1]

References. HOWARD, L., H. DYAR, and F. KNAB, "The Mosquitoes of North and Central America and the West Indies," 4 vols., Carnegie Institution, Washington, 1912; *U.S.D.A., Bur. Ento. Bul.* 88, 1910, and *Farmers' Bul.* 1354, 1923; *N.J. Agr. Exp. Sta. Buls.* 276, 1915, and 348, 1921; *Jour. Econ. Ento.*, **30**:10, 1937; MATHESON, R., "Mosquitoes of North America," 2d ed., Comstock, 1944; HERMS, W., and H. GRAY, "Mosquito Control," 2d ed., Commonwealth Fund, 1944; "Malaria Control on Impounded Waters," U.S. Government Printing Office, 1947; HORSFALL, W., "Mosquitoes," Ronald, 1955; *U.S.D.A., Agr. Handbook* 152, 1959; *W.H.O. Tech. Rept.* 191, 1960.

352 (315, 322). BLACK FLIES, BUFFALO GNATS, OR TURKEY GNATS[2]

Importance and Type of Injury. During the daytime, out-of-doors, small, clear-winged, humpbacked, chunky, blackish gnats (Fig. 21.4) hover about the eyes, ears, nostrils, and other parts of the body. They make little noise but promptly alight, run greedily over the skin and suck blood through their short, sharp mouth parts. They often appear in great numbers and may be drawn into the air passages of animals, as poultry and large animals are apparently sometimes smothered by the swarms of flies. The bites are very irritating, and in many sections of our northern states and Canada these little bloodsucking gnats make life unendurable for a definite season, mostly in spring. They bite especially about the face and neck but also on arms and any other part of the body that is exposed; they do not hesitate to squeeze under the clothing or into the hair to bite. The bites are not especially painful when made but become increasingly itching, swollen, and irritating for some days. The venom has a specific effect upon the glands about the ears and neck causing symptoms similar to mastoiditis. They are carriers of a filarial parasite[3] causing "Mexican blindness."

Animals Attacked. All warm-blooded animals.

Distribution. Some species occur in nearly all parts of the United States and Canada. Especially troublesome in the northern woods and mountains. They have a definite season after which little trouble is experienced for the rest of the summer.

Life History, Appearance, and Habits. There are many species of black flies which differ considerably in life cycle. They generally winter as maggots below the water surface in swift, rocky streams, and the adult flies in many sections are most abundant in spring. The adults are from $\frac{1}{25}$ to $\frac{1}{5}$ inch long, with broad clear wings, short, stiff horn-like antennae, and sooty black, chunky body, from the front end of which the head hangs downward, giving them a curious humpbacked appearance. The yellowish to black eggs are deposited on the surface of rocks, sticks, or vegetation at, or actually below, water in swiftly flowing streams. The eggs are often subtriangular in outline and laid, a single layer deep, in large patches covering many square inches. The eggs may hatch in 4 to 12 days and the grayish-black, legless maggots attach themselves to rocks, sticks, or other obstructions in streams where the water churns or boils over them, or to the slightly submerged leaves of sedges or trees. Here they may be

[1] Thephorin. See *Jour. Amer. Med. Assoc.*, **140**:603, 1949.

[2] Various species of *Simulium* and related genera, Order Diptera, Family Simuliidae.

[3] *Onchocerca volvulus*, Phylum Nemathelminthes, Class Nematoda, Order Spirurida.

seen attached by their tail ends, standing erect or stretched downstream like a banner in the wind, or writhing over their submerged support by bending the body from side to side.   From a single stone, 10 inches in diameter, 2,880 larvae have been taken.   The individual larva is less than ½ inch long, shaped like an Indian club or bowling pin, with a brownish head at the smaller end that bears a pair of marvelous mouth rakes each consisting of 30 to 60 long curved bristles, palmately arranged and constantly in motion, straining microscopic plants and animals from the current and scooping or fanning them toward the chewing mouth parts. The larvae hang on to the rocks or loop about by means of a complex disk, bearing hundreds of minute hooked hairs, near each end of the body, and also by clinging to silken threads that they spin over the surface of the rocks or sticks.   They respire through three retractile lobes or clusters of soft, white, blood gills near the tail.   The food is minute plant and

FIG. 21.4. Adult buffalo gnat, *Prosimulium pecuarum*. About 5 times natural size. (*From H. Garman, Ky. Agr. Exp. Sta.*)

animal life such as algae, protozoa, and diatoms. The larvae are therefore entirely harmless creatures that develop in streams often far from human habitations.   After 2 to 6 weeks in this stage, or in some species not until the following spring, growth is completed, and the larva makes a slipper- or vase-shaped silken cocoon, wide open at the downstream end, fastened to the rock where the larva lived.   In this pocket the pupa is hooked, with its 4 to 60 long thread-like tracheal gills which attach to the prothorax.   From about 2 days to 3 weeks later the adults emerge, float to the surface of the water, and quickly take wing before being drowned in the current.   Simultaneously with their emergence from the pupae they change from harmless aquatic curiosities to bloodthirsty plagues (Fig. 21.4) that seek warm-blooded animals of all kinds, pierce the skin and draw the blood, leaving at the same time an irritating venom that causes extreme pain.   It is only the females that suck blood.   Sometimes black flies appear in sections remote from swift rocky streams.   This may be due to migration (aided by winds) from a considerable distance, but it must be remembered that there are many species of these flies, and the life habits of most of them are imperfectly known. Their favorite breeding grounds are the ripples of cold swift streams. Some kinds develop in large streams, ditches, and other slowly moving water, but they cannot live in standing water.

Control Measures.   When one must venture into territory preoccupied by these gnats, the clothing should be securely closed at boot tops, neck, and wrists, and gloves and head veils worn if possible.   A double layer of clothing, however light, is desirable; and long-sleeved full-length underwear is recommended.   Repellents such as dimethyl phthalate, dimethyl carbate, diethyltoluamide, and 2-ethyl-1,3-hexanediol will protect against biting for as long as 5 hours (page 389).   Clothing impregnated with formulation M-1960 (page 393) gives very effective protection against black flies.   The locating of tents or camps on high dry spots, away from trees and underbrush, and where the wind strikes, will avoid the worst of the attacks.   One of the most important measures for the protection of livestock is to provide smudges in the fields or before the doors of barns or poultry houses by burning bark, moist punky wood, old leather, or green grass.   The flies will not attack in the smoke.   Applications of DDT aerosols have been found to be very effective in controlling adult black flies when applied over large areas.   Most treatments have been made with 5 per cent DDT in fuel oil applied at about 0.2 pound of DDT per acre from the ground or by helicopter (Fig. 8.15). The larvae of black flies have been controlled by dripping DDT emulsion into streams at the rate of 0.1 to 1 part of DDT per million parts of water. This treatment, however, may also kill fish and other aquatic organisms. Cleaning and deepening channels to remove logs, roots, stones, and other obstructions that cause ripples and waterfalls helps to reduce the numbers of these flies.   The outlets of lakes, dams, and spillways should have a clear unobstructed drop into a deep pool.

References.   U.S.D.A., Bur. Ento. Bul. 5 (n.s.), 1896, and *Dept. Bul.* 329, 1916, and *Bur. Ento., Tech. Bul.* 26, 1914; *N.Y. State Museum Bul.* 289, 1932; *Jour. Econ. Ento.,* **38**:671, 694, 1945; and **39**:227, 1946.

353. Punkies, No-see-ums, and Sand Flies[1]

Importance and Type of Injury.   These bloodsucking midges (Fig. 21.5) are not uniformly distributed but occur locally in numbers sufficient to make them almost intolerable.   They bite chiefly in the evening and very early in the morning.   Most

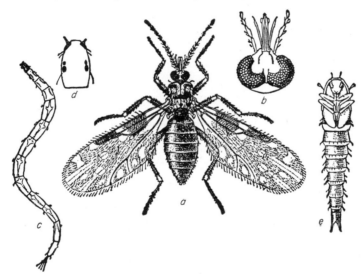

Fig. 21.5. A punkie or no-see-um, *Culicoides guttipennis.   a,* Adult, 15 times natural size; *b,* head of adult, more enlarged; *c,* larva; *d,* head of larva; *e,* pupa.   (*From Riley and Johannsen, "Handbook of Medical Entomology," Comstock Pub. Co.*)

species are attracted to artificial lights.   They are most prevalent from mid- to late summer.   In some species the bite is very burning and painful, and the victim is likely to be astonished when he notes the extremely small size of the creature that inflicted it.

Life History, Appearance, and Habits.   Punkies are very small two-winged midges, commonly about $\frac{1}{15}$ to $\frac{1}{25}$ inch long, the wings hairy and sometimes pictured, but never with flattened scales.   Two long veins near the front of the wings are distinct, the others very faint.   The mouth parts are short, the body moderately heavy, and the legs rather stout.   The larvae develop in decaying leaves and silt, where salty

[1] *Culicoides* spp. Order Diptera, Family Chironomidae.

tidal water backs up into fresh-water streams, along the margins of ponds, streams, and pools of various kinds, and in rot holes of trees. They are extremely slender and have a small brown head and a tuft of hairs at the opposite end. They breathe by means of blood gills after the manner of fish. The pupae of some species are said to float in the water, breathing from above the surface film.

Control Measures. Tree holes near residences should be drained, repaired by tree surgery, or saturated with creosote. Near cities where the problem is acute, the building of dikes or bulkheads, with pumps and tide gates to prevent flooding by tides or surface drainage, may be the only method of permanent relief. Satisfactory control has been obtained by treating breeding areas with 10 per cent DDT dust at 12 to 15 pounds per acre or with 5 per cent DDT emulsion at 3 to 5 gallons per acre. Such applications should be made from 3 to 5 days before periods of maximum populations. Some control of the adult midges over large areas can be obtained by airplane spraying with 5 per cent DDT solution at $\frac{1}{2}$ to 1 gallon per acre or by aerosol treatments with DDT at 0.2 pound per acre. Painting of window and door screens with 5 per cent DDT in kerosene will largely prevent the entrance of the midges for periods of several weeks, while similar applications of pyrethrum extract stabilized with 20 parts of lubricating oil (S.A.E. 5) or with an activator such as piperonyl butoxide give good protection for several days. Sixty-mesh silk bolting cloth over doors and windows is required to exclude the smallest of these pests. Fair personal protection can be obtained by skin applications of dimethyl phthalate or 2-ethyl-1,3-hexanediol or their mixtures (page 389).

References. U.S.D.A., Cir. E-441, 1938; Jour. Parasitol., **20**:162–172, 1934; N.Y. State Museum Bul. 289, 1932; Jour. Econ. Ento., **40**:472–475, 805–814, 1947; and **39**:227–235, 1946.

354 (313). HORSE FLIES OR DEER FLIES[1]

Many of the horse flies do not attack man, but a number of species of banded-winged smaller deer flies descend upon him when he ventures into their haunts in woods or marshes. They fly threateningly about the head in wild circles and, if not constantly warded off, alight and sink their mouth blades into the exposed skin. The bite is instantly very painful and considerable blood is drawn. Horse flies are the known carriers of *Loa loa*, tularaemia, and anthrax, which they may spread by biting a person or animal soon after feeding on a diseased animal or carcass.

Control Measures. General measures are discussed under Horses (page 942). The same repellents suggested for mosquitoes are useful in preventing the bites of horse flies.

355 (314, 322, 333). STABLE FLY[2]

This insect, which has been fully discussed under Horses (page 942), commonly comes indoors or on porches, especially before a storm, and bites people but is also troublesome out-of-doors. It selects especially the legs and ankles on which to feed and rarely bites elsewhere. It is noiseless and stealthy. Its bite is instantly very painful but has no prolonged after-effects. It is often called "the biting house fly" but should not be confused with the true house fly. The two species can be distinguished by the characters given in the key.

356. BLACK WIDOW SPIDER[3]

Importance and Type of Injury. This is the most seriously venomous spider native to North America. Its bites commonly give rise to very severe symptoms, but death results in only about 5 per cent of the known cases, and recovery is usually complete in from 2 to 5 days. The bite itself is usually a mere pinprick, but excruciating pains usually begin in a few minutes and spread from the point of the bite to arms, legs, chest,

---

[1] *Chrysops* spp., Order Diptera, Family Tabanidae.

[2] *Stomoxys calcitrans* (Linné), Order Diptera, Family Muscidae.

[3] *Latrodectus mactans* (Fabricius), Order Araneida, Family Theridiidae. The closely related *L. geometricus* Koch occurs in the subtropical parts of the United States.

back and abdomen; and within a few hours symptoms, such as chills, vomiting, difficult respiration, profuse perspiration, delirium, partial paralysis, violent abdominal cramps, pains, and spasms, frequently result. The pain is so severe as to lead frequently to a diagnosis as appendicitis, colic, or food poisoning. The venom is described as about fifteen times as potent as rattlesnake venom.

Food. A great variety of insects and other very small animals.

Distribution. General throughout most of the Western Hemisphere; extremely common in many of the southern states.

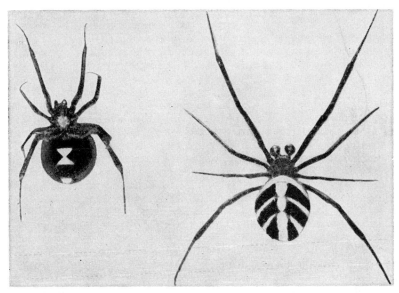

Fig. 21.6. The black widow spider, *Latrodectus mactans.* At left, the female from the underside, showing the characteristic hourglass-shaped spot, which is red and on the *underside* of the abdomen. At right, the male, showing the markings that often occur upon the upper side of the male but not the female abdomen; note his enlarged pedipalps. Female, enlarged about ½; male, several times natural size. (*From Baerg, Ark. Agr. Exp. Sta.*)

Life History, Appearance, and Habits. Young adults, which have not yet laid eggs, winter in buildings, rodent burrows, and other sheltered places, becoming mature in late spring. The female is a shiny, coal-black, eight-legged, slender-waisted creature, nearly ½ inch long (Fig. 21.6), the slim glossy black legs having a reach of about 1½ inches. The best recognition mark is an hourglass-shaped bright-reddish spot on the *under*side of the globular abdomen. There may be red or yellow marks on the upper side of the abdomen also, especially in the male. There is a "comb" or row of toothed setae on the tarsus of the fourth pair of legs. The male is very much smaller than his mate and has greatly swollen pedipalps. During the warm months of the year the spiders are found in sheltered, dimly lighted places such as barns, garages, basements, outdoor toilets, hollow stumps, rodent holes, and among trash, brush and dense vegetation. Cold and drought drive them into buildings. The female constructs an irregular, tangled web of tough silk with a funnel-shaped retreat extending to or toward the ground, from which she forages out to entangle, paralyze, and devour such prey as blunder into it. After a prolonged courtship the male transfers sperms from his gonopore to that of the female by means of his pedipalps. It has been said that the female often kills and eats her mate after mating, hence the name black "widow." During spring and summer, the female lays many eggs, enclosing them in grayish silken egg balls attached to her web. Each ball contains from 200 to 900 eggs, and a female may construct from 5 to 15 such egg cases. The rate of increase is greatly curtailed by the cannibalistic habits of the young, so that from 1 to 12 young may be all that survive from each egg case. From 10 to 30 days after the egg ball is made, the young spiders emerge from it. Growth requires 2 or 3 months, during which the male molts three to six times and the female six to eight times. The old ones apparently die in summer or autumn after having laid their eggs. In late summer young males and females live together in the same web.

Control Measures. The spiders can best be killed by residual applications of 5 to 10 per cent DDT, 2 per cent chlordane, or 1 to 2 per cent lindane, sprayed directly on the spider webs and liberally about basements, outdoor toilets, outbuildings, and other areas harboring spiders. Such treatments will remain effective for periods as long as a month or more. In case one has the misfortune to be bitten by one of the spiders, a physician should be called at once and advised to administer intravenous injections of a 10 per cent solution of calcium gluconate in 10-cubic-centimeter doses. Sodium amytal has also been recommended to be used in the same way and, if neither of the above is available, magnesium sulfate solution has been used with varying success. Perhaps the best remedy to use while awaiting the arrival of a physician is frequent baths as hot as can be endured. The poisons spread so rapidly that the application of a tourniquet and making incisions through the site of the bite and sucking out the venom are of doubtful value. The sterilizing of the skin about the bite with tincture of iodine should, however, be performed. The physician might be advised that strychnine may be useful as a heart stimulant and sedatives may be employed, although hypodermics of morphine have frequently been found to be practically useless in relieving the pain. It should be clearly understood that this spider is not an aggressive one but is really very shy and retiring. There appear to be two circumstances under which they bite people: (1) when they are squeezed, as by picking them up or while donning shoes or clothing in which they are hiding, or when they are encountered among sheets or blankets in the bed; (2) especially when the female is guarding her egg case, if she is disturbed in her retreat, she may rush out and bite the intruder. This seems to be a reaction largely to a vibration of her web and is the response which she would normally give when some insect blunders against her web. Such prey is usually entangled with the silk, paralyzed by the bite, and subsequently devoured. In a large number of cases which have been studied, it has been found that by far the majority of the bites have been inflicted while the person was using an outside toilet. In such cases apparently the web spun across the toilet seat had been vibrated and the spider had rushed out and inflicted the bite in the typical manner. In some areas and in some seasons the black widow spider has become so abundant among crops such as tomatoes and grapes that there is some hazard in picking the fruit. Under such circumstances workmen should wear gloves while working among the plants where the insects are hiding.

References.   Calif. Agr. Exp. Sta. Bul. 591, 1935; *Ark. Agr. Exp. Sta. Bul.* 325, 1936; *Ore. Agr. Exp. Sta. Cir.* 112, 1935; *Quart. Rev. Biol.*, **11**:123–160, 1936.

357 (344). Bed Bug[1]

Importance and Type of Injury.   The bite of a bed bug is not generally felt immediately, but the venom introduced soon causes itching, burning, and swelling to a variable degree, depending on individual susceptibility.   On many people the bites become increasingly painful for a week or more.   Bed bugs thrive under crowded and squalid living conditions and are often associated with army barracks, labor and prison camps, and similar situations where they may readily contact a variety of hosts. Because of the ease with which the insect is carried in baggage and on the clothing, hotelkeepers and theater managers find it a serious problem, and anyone who travels extensively will eventually encounter this bug, not only in the cheap but in the best of hotels and in sleeping cars.   Under such conditions the bugs probably feed on different persons nearly every time they draw blood, and this circumstance would seem favorable for the spread of blood-infesting disease organisms.   However, no direct proof has been advanced that bed bugs are common vectors of any known diseases.

Animals Attacked.   Besides feeding on man, the bed bug will attack mice, rabbits, guinea pigs, horses, cattle, and poultry.   They are often abundant in neglected poultry houses.

Distribution.   The true bed bug is found all over North America and is nearly cosmopolitan.

Life History, Appearance, and Habits.   In warmer regions and in rooms that are kept uniformly heated, the bed bug probably breeds throughout the year; in rooms where the temperature lowers perceptibly in winter it apparently winters chiefly as adults and nymphs, and egg-laying is suspended until spring.   The eggs may be found at any time during the warm months of the year.   They are laid in cracks of furniture, behind baseboards, under loose edges of wall paper and all such crevices, where the adults hide during the day.   They are elongate, whitish, big enough to be easily seen, and have a distinct cap at one end.   Each female may lay from 75 to over 500 eggs, at the rate of 3 or 4 a day, gluing them to wood or fabrics in their hiding places.   They hatch in 6 to 17 days.   The young bugs are similar to the adult when they hatch, but paler, yellow in color.   They molt five times, and at the last molt the abbreviated, useless, rudimentary wings appear and the insect is adult.   Any one bug probably does not feed every night but, at intervals of several days to a week, usually once before each molt.   They may become adult within a month or two after hatching, but ordinarily growth is much slower, and there are from one to four generations a year.   They may live long periods (4 to 12 months) without food.   They may also feed on mice and other animals than man, so that empty houses may remain infested for long periods.   Both males and females live on blood alone.

Any very flat brown bug is likely to be mistaken for the bed bug. There are many species living under bark of trees[2] that look something like bed bugs, but do not bite, and there are several other species that live in the nests of birds and bats, sucking their blood.   Many false notions

[1] *Cimex lectularius* Linné, Order Hemiptera, Family Cimicidae.
[2] Flat bugs, Order Hemiptera, Family Aradidae.

about bed bugs are current because people do not distinguish these similar-looking insects.   The true bed bug (Fig. 21.7) may be recognized by the very slender third and fourth segments of its antennae (the second segment being shorter than the third), by its almost lunate-shaped prothorax into which the head is sunken and by being covered with very short, curved, serrate hairs.   The adults are about ⅕ inch long.   They emit an oily substance which has a very offensive odor.

Control Measures.   Bed bugs may be readily controlled by residual spraying of houses, barns, poultry barns, barracks, etc., with 5 per cent DDT, 2 per cent chlordane, or 0.5 per cent lindane.   These materials may be applied as wettables, emulsions, or solutions and will give good control for several months or more.   Particular attention should be given to applying the spray to cracks and crevices in the walls and to bedsteads, springs, and mattresses.   If the bugs are believed to be localized in some one room or piece of furniture, they may be killed by spraying all hiding places with a deodorized kerosene spray containing either DDT, chlordane, lindane, an organic thiocyanate, or 0.2 per cent synergized pyrethrins, applied in sufficient amounts to wet the articles thoroughly.   Where strains of the bug resistant to these insecticides are found, treatment with 0.5 per cent Diazinon or 0.5 to 1.0 per cent malathion is effective.   Power sprayers will force these sprays into the remote hiding places of the bugs more effectively.   Mattresses, rugs, and upholstery can be treated by steam cleaning or by having them vacuum fumigated.   New, and especially secondhand furniture, laundry, traveling bags, and similar articles delivered to the home should be watched so that the bed bug is not introduced with them.   Where spray treatments are not desirable, infestations may be eliminated by fumigation with hydrogen cyanide, using 10 ounces of sodium cyanide, or the equivalent of calcium cyanide or liquid hydrogen cyanide, to each 1,000 cubic feet of space (page 375).   In congested places individual rooms or apartments can more safely be fumigated with methyl bromide, or methyl formate or ethylene oxide mixed with carbon dioxide (page 384), but community cooperation should be enlisted because the bugs will migrate from house to house.   If one is traveling extensively, it is advisable to carry a small bottle of pyrethrum powder to be sprinkled over the bed under the sheets, if these pests are found or suspected in the room.   Such treatment will usually keep them from attacking for the single night.

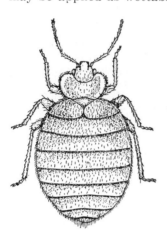

FIG. 21.7. The bed bug, *Cimex lectularius.* 8 times natural size. (*Original.*)

References.   U.S.D.A., *Farmers' Bul.* 754, 1916, and *Leaflet* 146, 1938; *Jour. Econ. Ento.,* **38**:265, 606, 1945; *U.S.D.A., Misc. Publ.* 606, 1946.

358. ASSASSIN BUGS OR KISSING BUGS

Importance and Type of Injury.   The bites inflicted by the Mexican bed bugs or bloodsucking conenoses,[1] China bed bugs,[2] and the so-called "kissing bugs" or masked

[1] *Triatoma sanguisuga* (Leconte), Order Hemiptera, Family Reduviidae.
[2] *Triatoma protracta* (Uhler), Order Hemiptera, Family Reduviidae.

hunter[1] are scarcely exceeded in severity by any other insect. They have been likened in effects to snake bites. The pain is intense and usually affects a considerable part of the body; swelling generally follows; and in the worst cases faintness, vomiting, and other ill effects are experienced that may last weeks or even months. These bites may be experienced when one picks up the bugs or when they fly against the face. In the South and Southwest, the *Triatoma* spp. mentioned above are aggressive and come into houses and bite at night to secure a meal of blood. It is probable that many of the painful bites of which spiders are accused are caused by these bugs. In South and Central America *Triatoma megista* (Burm.) and several other species are carriers of a highly fatal human disease known as Chagas' disease (Table 1.2). In some of these species which are habitual bloodsuckers, the bites are entirely painless.

Animals Attacked. Man, domestic mammals, and poultry.

Life History, Appearance, and Habits. The eggs of these bugs are mostly laid out-of-doors under stones, logs, or other shelter or, in some species, on plants. The young bugs probably feed chiefly on other insects but may attack warm-blooded animals. One species is known as the masked bed bug hunter because the nymph has a sticky secretion all over the body to which dust and lint adhere so that, as it crawls along the floor, it looks like a bit of lint being blown along. It is believed to catch bed bugs and suck the blood from them. So far as these species have been studied, it appears that most species have but one generation a year.

These bugs (Fig. 21.8), which are from $\frac{2}{3}$ to 1 inch long, have the characteristics of the order Hemiptera. The head is long, somewhat conical, the prothorax narrows in front, the wings cross over flat on the back and have 2 or 3 large cells in the membrane; and the edges of the abdomen are produced as thin flat plates at the sides of the wings. These plates are in some species marked with pink or red bars, though the general color is dark-brown or black.

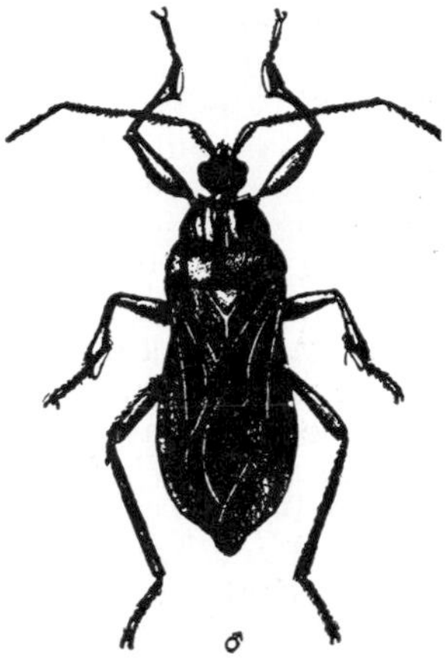

Fig. 21.8. An assassin bug or "kissing bug," *Melanolestes picipes*, male. About 3 times natural size. (*From Ill. Natural History Surv.*)

Control Measures. Satisfactory control of severe infestations about dwellings has been obtained by residual spraying with lindane at 50 milligrams per square foot or dieldrin at 125 milligrams per square foot. Breeding places should be eliminated, and the bugs prevented from entering houses, by screening. If they alight on the face flip them off quickly and do not take hold of them.

350 (334, 360). FLEAS

Importance and Type of Injury. Everyone knows that fleas commonly infest dogs and cats, and many have experienced the painful irritating bites that result when they suck the blood of man. The bites are likely not to be felt immediately but become increasingly irritating and sore for several days to a week or more afterward. It is rather characteristic of flea bites that there are frequently 2 or 3 in a row. They bite mostly about the legs. Unlike lice and ticks which, having found a host, cling to it for dear life, fleas shift from host to host and feed indifferently on several kinds of animals. The cat flea[2] is nearly as likely to be found on a dog or a man as on a cat, and the so-called human flea[3] has been taken from dogs, skunks, rats, mice, hogs, and deer. The rat flea and those

---

[1] *Reduvius personatus* (Linné) and *Melanolestes picipes* Herrick-Schaeffer.

[2] *Ctenocephalides felis* (Bouché), Order Siphonaptera, Family Pulicidae.

[3] *Pulex irritans* Linné, Order Siphonaptera, Family Pulicidae.

of ground squirrels also bite man. This promiscuous-feeding habit makes possible the most serious injury that fleas inflict, *viz.*, the transmission of bubonic plague from man to man. This is not the only way that plague may be contracted, but, in the great epidemics, fleas have been the most important factor in the spread of the disease. Plague is a bacterial disease caused by *Pasteurella* (= *Bacillus*) *pestis*. This organism causes a fatal disease in rats as well as in man. In the fourteenth century a great epidemic swept the Old World during which it was estimated that about 25 million people died of the disease. Bubonic plague has been present in the western part of the United States since 1900 and has become established in at least 17 genera of wild rodents as sylvatic plague. It has spread as far east as Kansas, Nebraska, and Texas. The oriental rat flea[1] is by far the most important vector in the spread of plague from rats to man, but at least 24 other species of fleas found on rats and rodents have been proved to be capable vectors. These include the northern rat flea,[2] the mouse flea,[3] the cat flea,[4] the dog flea,[5] and the human flea.[6] The flea transmits the infection by regurgitation of the bacilli in the stomach contents through the proboscis during feeding. This is caused by a temporary blocking of the proventriculus of the flea by a gelatinous mass of the multiplying bacilli in the ingested blood.[7] The oriental rat flea is also the principal vector in the spread of murine or endemic typhus from rats to man. This disease, which should not be confused with epidemic or Old World typhus, is caused by *Rickettsia typhi*. It is widespread in rats and in other rodents, and over 90 per cent of the human cases in the United States occur in the southern states and in southern California. The rickettsia multiply in the epithelial cells of the stomach and in the Malpighian tubes and are spread to humans principally by rubbing infected flea feces into wounds made by the biting flea or by scratching the flea bite. These fleas also serve as intermediate hosts for the dog tapeworm, *Dipylidium caninum*, and the rodent tapeworm, *Hymenolepis diminuta*, which occasionally parasitize man.

Animals Attacked. Man, hogs, dogs, cats, rabbits, foxes, and other fur-bearing animals, and rats, mice, and all other rodents as well as many other animals are bitten by fleas. Plague and murine typhus occur in man, rats, and many other rodents.

Distribution. Fleas as a group are cosmopolitan. In the eastern United States, the cat and dog fleas are most often found in dwellings. In the western states the human flea is most often encountered in houses. The oriental rat flea is widely distributed in all sections of the country and is most abundant in the southern states and southern California, where it comprises from 40 to 60 per cent of the fleas collected from rats. The northern rat flea is most widely distributed in the northern states, where it may comprise up to 68 per cent of the fleas collected from rats.

Life History, Appearance, and Habits. Fleas, like most of the parasites of the large animals, continue their activity and reproduction throughout the winter, but breeding and all life processes are somewhat

---

[1] *Xenopsylla cheopis* (Rothschild), Order Siphonaptera, Family Pulicidae.

[2] *Nosopsyllus* (= *Ceratophyllus*) *fasciatus* (Bosc), Family Dolichopsyllidae.

[3] *Leptopsylla* (= *Ctenopsyllus*) *segnis* (Schönherr), Family Hystrichopsyllidae.

[4] *Ctenocephalides* (= *Ctenocephalus*) *felis* (Bouché), Family Pulicidae.

[5] *Ctenocephalides canis* (Curtis), Family Pulicidae.

[6] *Pulex irritans* Linné, Family Pulicidae.

[7] JELLISON, W., *Ann. Rev. Ento.*, **4**:389, 1959.

slowed down by cold weather. The adults are hard-skinned, very spiny insects, extraordinarily thin from side to side, with piercing-sucking mouth parts retaining both pairs of palps, concealed antennae, no trace of wings, and very large legs fitted for jumping. Their only food is blood. Although the adults are the only stage most of us see, fleas pass through all the stages of a complete metamorphosis. The eggs (Fig. 21.9,b) are either deposited in the dust, dirt, or bedding of the host or laid while the female is on an animal. They are never glued fast to the hairs, and those laid on the host usually sift readily through the hairs to the ground. The eggs are white, relatively large, short ovoid, $\frac{1}{50}$ inch long. Only a few are laid at a time, though the total number may be several hundred. The length of the egg stage is given as 2 to 14 days. The young flea (Fig. 21.9,c) is a very slender whitish larva, about $\frac{1}{6}$ to $\frac{1}{4}$ inch long, with a small pale-brown head but without legs or eyes. The body is plainly segmented and is covered with scattered, very long hairs. The larva has chewing mouth parts and feeds on a variety of dry organic

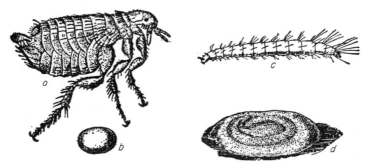

Fig. 21.9. A flea. *a*, Adult; *b*, egg; *c*, larva; *d*, cocoon. All much enlarged. (*From U.S.D.A.*)

matter such as the excreta of the adult fleas or that of mice, rats, and other rodents. Larvae can be reared successfully on the dirt scraped from the cracks of a wooden floor. The larval stage occupies from 1 to 5 weeks or more, and two molts have been recorded. However, since the larvae generally eat their molted skins, there may be additional instars that have not been observed. When full-grown, the larva forms a small oval cocoon of white silk to which adhere particles of dust and trash so as to give it a dirty, obscure appearance. From this cocoon (*d*) the flea emerges in 5 days to 5 weeks although it may pass the winter in this way. Thus 2 or 3 weeks to 2 or 3 months and rarely as long as 2 years are required for a generation of fleas of different species and under varied conditions. It thus happens that a house or basement in which cats or dogs have been permitted may be closed during an absence of the occupants and found upon their return to be overrun with these pests, although all animals have been excluded for a considerable time. Even a stray cat or dog, sleeping on one's doorstep or under a porch, may peddle enough fleas or flea eggs to start an epidemic. Adult fleas, after having fed, have been kept alive for 50 to 100 days without further food, while unfed adults may remain alive from 1 to 2 years.

Control Measures. The habits of fleas make it plain that either to rid an animal of the pests or to clean up infested premises, the control measures must be of two kinds: (1) the destruction of rats and other rodents and the treatment of any cats, dogs, hogs, or other infested animals, to kill the adults on them; (2) thorough and vigorous treatment of the kennels, pens, or sleeping quarters and even of the dry soil, dry manure, and other litter in hog lots and barnyards, under porches, and in or under buildings to which the infested animals have had access.   The best treatment of pets is frequent dusting with 0.5 to 1 per cent rotenone, 0.2 per cent synergized pyrethrins, or 4 to 5 per cent malathion.   Other materials which may be used with dogs and other animals that do not lick themselves are 5 per cent DDT, 2 per cent chlordane, or 1 per cent lindane. Malathion as a 0.25 per cent dip or 0.5 per cent spray is generally effective. Infested yards or premises may be treated by spraying or dusting with 5 per cent DDT, 2 per cent chlordane, or 1 per cent lindane.   Where fleas are resistant to these materials, spraying with a 1 per cent emulsion of Diazinon, malathion, or Dipterex at 1 gallon to 1,000 square feet is highly effective.   In living quarters light sprays of these materials may be applied to beds and bedding.   Dusting wooden or earthen floors, pet houses, and mats or rugs on which animals sleep with dusts of the above materials will also eliminate the fleas.   In addition to the above measures, houses or basements in which fleas are established may be treated by fumigation with hydrogen cyanide (see page 375) or by sprinkling flake naphthalene over the floors of tightly closed rooms at 1 pound to each 100 square feet for 24 to 48 hours.   After such treatment the floors should be thoroughly scrubbed with hot soapsuds to kill the eggs, or with an oil mop wet in kerosene.   It should be realized that cats and dogs cannot be allowed regular access to the house without starting an infestation of fleas sooner or later, unless they are treated at regular intervals with a good flea powder or bath of flea soap.   Where fleas have become established about barns, hog lots, and outbuildings, the infestations may be cleaned up by removing the manure and litter and spreading it in the fields at some distance from the buildings and thoroughly treating the ground in and around the infested buildings with one of the insecticides mentioned above.   Personal protection from fleas can be obtained by treating trouser legs and socks with diethyltoluamide or benzyl benzoate or applying these materials to the skin (page 389).

Considerable success in the control of murine typhus and plague has been obtained by controlling rat fleas with applications of 5 to 10 per cent DDT dust, blown into all areas suspected of harboring rats, such as holes and runways and between double walls and floors, and applied in patches laid about holes and along runways.   Such treatments, when carefully organized on a city- or area-wide basis, have resulted in initial control of rat fleas as high as 99 per cent and in more than 80 per cent control 4 months after treatment.[1]

References.   U.S.D.A., Dept. Bul. 248, 1915, and Farmers' Bul. 897, 1917; Jour. Econ. Ento., **39**:417, 767, 1946; U.S.D.A., Cir. 977, 1955; "Rat-borne Disease Prevention and Control," U.S. Public Health Service, 1949; W.H.O. Tech. Rept. 191, 1960.

360 (359). CHIGOE FLEA[2]

Importance and Type of Injury.   In the southern part of the United States, Mexico, the West Indies, and other tropical regions, a small reddish-brown flea, about $\frac{1}{25}$ inch

---

[1] U.S. Pub. Health Repts., **63**:129–141, 1948.

[2] *Tunga* (= *Dermatophilus*) *penetrans* (Linné), Order Siphonaptera, Family Sarcopsyllidae.

long, has the despicable habit of burrowing into the skin, especially between the toes and under the toenails.   It causes much pain and itching and, as the female enlarges beneath the skin, a pus-filled ulcer is formed.   The sore is very likely to become infected with bacteria, and the entire toe, foot, or limb may be lost by blood poisoning. This insect must not be confused with the chigger mites, since both are called chiggers.

Animals Attacked.   Man; hogs, and other domestic animals.

Life Cycle, Appearance, and Habits.   Chigoes, like other fleas, develop in the soil or in filth, through a slender larval stage and a pupal period spent in a cocoon.   When they become adult, they attach to warm-blooded animals and suck blood.   After mating, the female, aided by her long mouth parts, works her way into the skin, and her body becomes enormously enlarged, as her eggs develop, until she may be as large as a small pea.   The eggs are generally extruded through the entrance hole and drop to the ground, where they develop to adults in about 1 month.

Control Measures.   Infested sores should be opened with sterile instruments, the fleas removed, and the wound given an antiseptic dressing.   Shoes or boots should always be worn in the infested territory.   Where the pest abounds, pigs and all other animals should be kept away from dwellings.

361 (340). Rocky Mountain Wood Tick or Spotted-fever Tick[1]

Importance and Type of Injury.   The Rocky Mountain wood tick is the most important tick in the United States.   It is very annoying to domestic and wild animals and is commonly found attached securely to the skin of children, hunters, and other persons who are much out-of-doors, especially in brush- and scrub-covered areas.   The bites are not felt at the time the tick attaches but cause more or less inflammation later.   If the tick is forcibly removed, its mouth parts usually remain in the skin and cause an ulcer that is in danger of bacterial infection and serious complications.   Another somewhat mysterious injury caused by ticks is a paralysis of the motor nerves affecting first the legs, and a few days later the arms, and gradually spreading, if not checked, until death may result. This disease, which is known as tick paralysis, results apparently only when the tick attaches at the back of the neck or base of the skull and feeds very greedily, a week or so after it first attaches.   The careful removal of the tick usually results in a speedy recovery.   It is believed to be due to a poisonous substance introduced by the mouth parts of the tick.

The most serious injury to man is the transmission of Rocky Mountain spotted fever.   This is a highly fatal, continuous fever that begins a few days to a week after the attachment of an infected tick to the skin and often results in death within 2 weeks.   A peculiar skin rash of grayish or brownish spots usually appears on the arms, legs, and other parts of the body a few days after the fever begins.   There have been about 1,000 cases a year in the United States.   The fatality in the Bitter Root valley of Montana runs from 70 to 90 per cent.   In the other western states and in the states east of the Rockies (where the American dog tick is the principal carrier) the mortality is less than 25 per cent.   The cause of the disease is a minute, intracellular rickettsia, discovered by Wolbach in 1916 and called *Rickettsia* (= *Dermacentroxenus*) *rickettsii*, which attacks especially the peripheral blood vessels.   Another offence by the Rocky Mountain wood tick is the dissemination of tularaemia, which, like spotted fever, is a rodent disease that occasionally attacks man and domestic animals.   This disease is said to have killed 5,000 sheep in a single year

---

[1] *Dermacentor andersoni* Stiles (= *D. venustus* Banks), Order Acarina, Family Ixodidae.

in Idaho.   It also affects cattle, goats, rabbits, many rodents and certain game birds.   Many kinds of insects and other ticks transmit the organism, *Pasteurella* (= *Bacterium*) *tularensis*, among wild animals and from the wild animal hosts to man and domestic animals (Table 1.2).

Animals Attacked.   Spotted fever affects man and many kinds of rodents.   Tularaemia is also prevalent among men, domestic and wild animals, and certain game birds.   Tick paralysis has been noted only in man, sheep, and cattle, although it has been produced experimentally also in horses, dogs, rabbits, and guinea pigs.   The Rocky Mountain wood tick feeds in its nymphal stages upon small wild animals, mostly rodents, such as ground squirrels, woodchucks, rabbits, rats, and mice and, when adult, upon large animals such as men, dogs, horses, cows, mules, sheep, deer, mountain goats, and many others.

Distribution.   The Rocky Mountain wood tick occurs in the eight or nine states centering around southern Idaho, and rarely in the states bordering these.   Spotted fever and tularaemia, however, are now known to occur throughout the United States (page 1020).

Life History, Appearance, and Habits.   The Rocky Mountain wood tick has four distinct life stages: the egg, the "seed tick" or "larva," the nymph, and the adult.   The adults of the two sexes are also very different in appearance.   The females (Fig. 21.10, upper) are dark reddish brown, the anterior third of the body covered with a hard shield which is white, splotched with several small reddish dashes and dots each side of the center.   In the male (Fig. 21.10, lower) the shield or *scutum* covers the entire back, and the males are grayish white, splotched with bluish-gray markings.   Before feeding, the males and females are about $\frac{1}{16}$ inch long, but, when fully engorged, the female is much larger, $\frac{1}{2}$ inch long by $\frac{1}{3}$ inch wide.   The larvae or seed ticks, upon hatching, are only about $\frac{1}{40}$ inch long, have only three pairs of legs, and are pale yellow in color, but they increase greatly in size and become slate gray after feeding.   The nymphs are easily distinguished from the larvae by their larger size ($\frac{1}{16}$ to $\frac{1}{6}$ inch long), darker color, and four pairs of legs.   Adults can always be told from nymphs by the presence of the genital opening at the anterior third of the body on the underside.

Each tick during its lifetime requires three hosts, which are probably always three different individuals and commonly three different species of animals.   Only the adults feed on man and the large animals.   The larvae and nymphs feed on small wild animals, especially rodents.   Each stage remains anchored by its mouth parts to one spot on its host for a week or two, then drops to the ground and spends a week or two, or sometimes the entire winter, resting, digesting the blood it has swallowed, and molting.   Two, three, or four years may be occupied completing a generation, 2 years being the most common.   The winter is passed as unfed males and females or as nymphs, among grass and leaves on the ground.   The adults and nymphs feed only during the spring months, from about the middle of March to the middle of July; and this is the only period when man contracts spotted fever.   During the first warm days of spring they climb up on the brush and may attach to any passing object.   If this is a man, horse, cow, or other large animal, they insert their mouth parts and draw blood for the next week or two.   During this time the males visit the females, mating occurs on the host animal, and, when fully engorged, the female drops off the host.   During the next month she normally deposits 4,000 to 7,000 small brown eggs, all in one mass, under stones and other trash on the ground, laying several hundred each day. From 2 to 7 weeks later the eggs hatch, especially during June and July.   These "larvae" have only six legs and are much smaller than pinheads.   They climb up on grass, brush, and other objects, and the fortunate few that are brushed off by some small wild rodent may suck its blood for a week, then drop off the host and spend 1 to 4 weeks on the ground

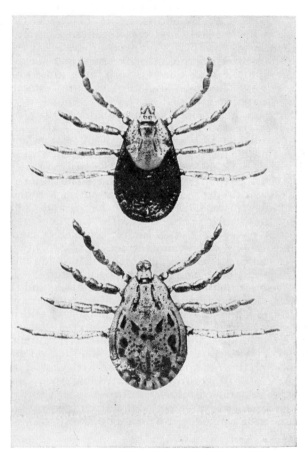

Fig. 21.10. The Rocky Mountain wood tick, *Dermacentor andersoni*.   Adult, unengorged female, above; adult male, below.   About 5 times natural size.   (*From U.S.D.A., Bur. Ento. Bul.* 105.)

before they molt to the eight-legged nymphal stage.   By this time hot weather usually causes the nymphs to go into a period of rest under the shelter of grass or leaves where they remain until the following spring. The first warm spring days bring them up on the vegetation again, where, if fortunate, they attach to a rabbit, mouse, ground squirrel, or woodchuck, suck its blood for a week, drop to the ground, and devote the next 6 or 8 weeks to the change to the adult stage.   The hot weather of summer now prevents these adults from seeking a host and they remain in estivation and hibernation until the following spring.   Such is the normal life cycle of the lucky ones—nymphs and adults being present the year round, but seeking hosts and feeding only in the first half of the growing season, and the larvae feeding only during midsummer. The larvae from late-laid eggs or ones that do not find a host perish during the winter. Nymphs and adults, however, which fail to find a host before midsummer may survive an additional year before feeding, and the adults even a third and fourth year before they find a host or finally perish.

Control Measures. Every effort should be made to discover the presence of ticks on the body and the clothing by careful examination, especially of the head, promptly after exposure to tick-infested brush. A change of clothing after walking in ticky areas, and keeping the outdoor clothing away from sleeping quarters, are important. Repellents such as indalone, diethyltoluamide, dimethyl carbate, dimethyl phthalate, and benzyl benzoate are fairly effective when applied to socks and outer clothing. The U.S. Army clothing impregnant M-1960 (page 393) is the most satisfactory treatment. If ticks are found attached to the skin, they should be touched with a few drops of chloroform, gasoline, or turpentine or a hot needle to cause them to relax their mouth parts, or grasped firmly with tweezers or fingers and removed by a steady pull so as not to break off the mouth parts in the skin. The point of attachment should then be sterilized by dipping the point of a round toothpick in carbolic acid, silver nitrate, or tincture of iodine and drilling it into the hole made by the mouth parts. Since the tick does not transmit the disease unless it feeds for 4 to 8 hours, thorough examination of the body, morning, noon, and evening when ticks are prevalent, is very important. A vaccine is available which will provide immunity to the disease with the aid of seasonal "booster" injections. The very complicated life cycle of the tick, its many diverse hosts, and the wild nature of much of the infested territory make destruction extremely difficult. Control measures suggested include the dipping or spraying of horses, cattle, and sheep, three times at 10-day intervals, during the spring months, as directed for the cattle tick; restricted grazing of domestic animals in tick-free pastures, the use of repellent oils on grazing animals; the destruction of the rodents which serve as hosts for the larvae and nymphs, with poisoned grain scattered around their burrows or with calcium cyanide introduced into the burrows. Area spraying or dusting with DDT, chlordane, toxaphene, or dieldrin at 1 to 2 pounds per acre or with lindane at 0.5 pound per acre is effective.

References. *U.S.D.A., Bur. Ento. Buls.* 105, 1911, and 106, 1912; *Mont. State Bd. Ento. Seventh Bien. Rept.,* 1929; *Arch. Path.,* **15**:389–429, 1932; *Jour. Med. Res.,* **11**:1–197, 1919; *Jour. Econ. Ento.,* **30**:51, 1937; *U.S. Pub. Health Repts.,* **62**:1162, 1947; and **63**:339, 1948; *W.H.O. Tech. Rept.* 191, 1960; *U.S.D.A., Cir.* 977, 1955.

362. AMERICAN DOG TICK OR WOOD TICK[1]

Importance and Type of Injury. Previous to 1931 this tick was considered merely an annoying parasite of dogs and human beings who roamed woody places; but with the discovery that this common wood tick is the carrier in the central and eastern states of Rocky Mountain spotted fever, caused by *Rickettsia rickettsii*, it has taken on real significance. Although there have been fewer than 150 cases a year of this disease east of the Rocky Mountains, and the death rate is only about 25 per cent, it is a

---

[1] *Dermacentor variabilis* (Say), Order Acarina, Family Ixodidae.

dreaded disease and the tick itself a repulsive bugbear of the out-of-doors. This tick is also a carrier of tularaemia and of bovine anaplasmosis, as well as a troublesome parasite of dogs and horses.

Animals Attacked.   Dogs, man, cattle, horses, hogs, sheep, cats, rabbits, mice, and many other domestic and wild mammals.

Distribution.   Throughout most of the United States except the Rocky Mountain area; especially abundant in areas of considerable humidity which are covered with grass or underbrush, where its mice hosts are abundant.

Life History, Appearance, and Habits.   In the most southern states all stages of the tick may be found the year round, although reproduction is slowed down by both the cool weather of winter and the hot dry weather of midsummer.   In the North all stages, except the eggs, apparently winter successfully; the larvae and nymphs on mice and the unengorged adults in clumps of grass and similar shelter.   The adult, which is the stage that attacks man, is most abundant in spring and early summer and is rare after the first of August.   This is the only stage that attacks man, the dog, and the domestic animals.   They are flattened, chestnut brown to blue gray, very tough-skinned creatures with eight large legs. When unengorged, they measure about $\frac{3}{16}$ inch in length.   The male has a hard white-marked shield all over his back, whereas in the female this scutum reaches only half the length of the body and scarcely $\frac{1}{10}$ its length after engorgement.   The fully fed female reaches a length of $\frac{1}{2}$ inch and becomes bluish gray in color.   If they do not feed, adults may live in moist situations nearly 3 years but normally die in from 3 weeks to 3 months.   Females may remain attached in one spot to the host for 5 to 13 days.   On the fourth to sixth day the female may be visited by a male, mating occurs, and the female then rapidly fills with blood to repletion, releases her mouth parts, and drops to the ground.   She may spend 2 weeks to a month laying from 4,000 to 6,500 round brownish eggs, in some protected place on the ground, in an elongate mass in front of her, and then dies in a few days to several weeks.   The eggs hatch in 1 to 2 months.   The tiny six-legged, yellow to grayish larvae, $\frac{1}{40}$ inch long, climb up to the ends of grass blades and other leaves, ready to attach to some passing animal.   They may survive in moist places for a year or more if they do not find a host sooner; but, if fortunate, they may attach to a host a few days after hatching and be full fed in 3 to 12 days.   They then drop from the host, and after 1 to 30 weeks on the ground (the longer interval in cold weather) they shed their skins and become eight-legged nymphs.   They are $\frac{1}{16}$ inch long before feeding, may attach, if they find a host, in from 4 days to a year or more, feed for 3 to 10 days, and again drop to the ground where they may undergo the final molt to the adult stage in from 20 to over 200 days.   The nymphs may live more than 1 year, if they do not find a host sooner.   Many individuals probably spend 2 or 3 years completing their life cycle.

Control Measures.   In most cases avoidance of moist areas where grass and underbrush abound during spring and early summer or the careful examination of the body and especially the hair of the head at least twice a day and removal of any ticks as promptly as possible after exposure will prevent serious consequences.   Tincture of iodine should be forced into the minute hole made by the tick's mouth parts.   Care should be taken not to get the blood of crushed ticks into the eyes or into scratches on the skin.   Dogs should be kept from roaming in ticky places or treated with 4 to 5 per cent rotenone dust to kill the ticks they collect. A wash or dip of 2 to 4 ounces of rotenone powder and 1 ounce of soap to 1 gallon of water may be applied every 5 days to dogs that run out-of-doors. Spraying or dusting the vegetation along roads, paths, and trails where the ticks congregate with DDT, toxaphene, chlordane, or dieldrin at 1 to 3 pounds or lindane at 0.5 pound per acre has resulted in almost complete eradication of the ticks. If they are not susceptible to these materials, spraying with 0.5 per cent Diazinon is effective. Where the ticks invade houses and outbuildings, residual spraying with 5 per cent DDT, 3 per cent chlordane, 0.5 per cent dieldrin or lindane, or 0.5 per cent Diazinon is effective, treating floors, cracks, and other areas where the ticks congregate. Repellents such as indalone, diethyltoluamide, dimethyl carbate, dimethyl phthalate, and benzyl benzoate or the U.S. Army clothing impregnant M-1960 (page 393), applied to socks and outer clothing, will prevent tick attachment. Clearing away underbrush, keeping grass closely mowed, poisoning or trapping meadow mice, and wearing boots laced up over trouser legs and clothing securely fastened about neck and wrists are aids in control.

References. *U.S.D.A.*, *Circs.* 478, 1938, and 977, 1955, and E-454, 1948, *Tech. Bul.* 905, 1946, and *Bur. Ento. Bul.* 106, 1912; *W.H.O. Tech. Rept.* 191, 1960; *Jour. Econ. Ento.*, **30**:51, 1937; **38**:553, 1945; **39**:235, 1946; **40**:303, 1947; and **41**:427, 1948; *Arch. Path.*, **15**:389, 1933.

MITES

A number of species of mites are troublesome to man, among them being the chigger or jigger (not to be confused with the chigoe flea, see page 1016), harvest mites, louse-like mites, and flour and meal mites. Mites are close relatives of ticks and spiders and are not true insects. They represent distinct families of the class Arachnida and order Acarina, the larger ones of which are called ticks and the smaller ones mites.

363. Chiggers, Jiggers, or Red Bugs[1]

Importance and Type of Injury. From 12 to 24 hours after one has been on an outing in summer or autumn, particularly to spots where tall grass, weeds, and brambles abound, the skin may become inflamed in spots, especially where the clothing closely pressed the skin. Scattered red blotches of varied size appear, accompanied by a most intense itching that may not subside for a week or more. Some persons become feverish, extremely nervous, and seriously disturbed by such infestations. Closely related species[2] are vectors of scrub typhus or tsutsugamushi fever, caused by the organism *Rickettsia tsutsugamushi* and occurring in Japan, Malay States, and the South Pacific islands.

Animals Attacked. Man, domestic animals, poultry, certain ground-nesting birds, various snakes, land turtles, frogs, toads, squirrels, rabbits, and possibly other rodents.

Distribution. The distribution of chiggers is not uniform and is somewhat peculiar. Some local areas are badly infested, while others apparently similar and not far distant are practically free.

Life History, Appearance, and Habits. The causes of this trouble are small "larval" or six-legged mites, less than $\frac{1}{150}$ inch in diameter and

[1] *Eutrombicula alfreddugesi* (Oudemans) [= *Trombicula irritans* (Riley)], *E. splendens* (Ewing), *E. batatas* (Linné) and others, Order Acarina, Family Trombidiidae.
[2] *Trombicula akamushi* (Brumpt) and *T. deliensis* Walch.

hence almost invisible to the naked eye.   According to Miller[1] at least
one species of chigger[2] spends the winter in earthen cells ½ to 1 inch
below ground in the adult stage.   The following spring they come out
of the ground and lay their eggs.   The young six-legged mites that hatch
from the eggs normally live on land turtles or on snakes, under the overlapping scales of the back, according to Miller,[1] or on rabbits, according
to Ewing,[3] and it is only these first-stage "larvae" that
attack man.   These are rounded, oval, bright orange-
yellow in color, blind, and run very rapidly.   When
one walks through grass and underbrush, these young
chiggers may swarm over the body for several hours but
are not felt until they settle down and begin to feed (Fig.
21.11).   They undoubtedly introduce a definite poison
that causes the irritation.   It has often been stated that
chiggers burrow into the skin and suffer a speedy death
as a penalty for trespassing upon man; but we are
apparently to be deprived of this consolation, for Ewing[3]
contends that they do not burrow beneath the skin but
only insert the mouth parts, sometimes in a skin pore
or hair follicle.   When full-fed they drop off man.
According to Miller, the full-fed, first-stage larvae fall
from their hosts in late September, in southern Ohio,
and go into the loose soil to pass the winter.   After
molting and spending 2 or 3 weeks as quiescent nymphs,
they molt to the adult stage but remain in their earthen
cells until spring.   There is only one generation a year.
Ewing states that about 10 months are spent in the
adult stage.   The adults do not attack man but are
scavengers on the excrement of other arthropods or on
decaying wood.

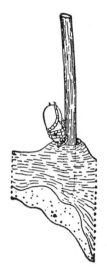

Fig. 21.11. Chigger
engorging, or feed-
ing, at the base
of a hair.   (*From
U.S.D.A., Dept.
Bul.* 896.)

   Control Measures.   If one anticipates a visit to the
domains of these little tormentors, almost complete
freedom from attack can be obtained by treating the
clothing with dimethyl phthalate, dibutyl phthalate, or
benzyl benzoate.   These materials can be applied (*a*) as a ½-inch barrier
around sleeves, cuffs, neck, waist, and fly; (*b*) by rubbing 4 to 5 ounces
or spraying 1 to 2 ounces over the clothing; or (*c*) by impregnating the
entire garments in a 5 per cent solution of the material in acetone or
ethyl alcohol, or in a 5 per cent emulsion of the material with about 0.5
per cent emulsifier.[4]   The U.S. Army formulation M-1960 is very suitable for this purpose.   Clothing treated in this manner will continue to
protect even after several washings.   (*d*) Dusting the arms and legs and
openings of clothing with 5 per cent benzil is effective in preventing
chigger attack for several days, and a similar application of finely powdered sulfer has limited effectiveness.   A soapy bath within a few hours
after exposure, allowing the soap to dry on the skin, will usually prevent
infection.   After itching begins, little can be done except by applying

   [1] MILLER, A. E., *Science*, **61** (no. 1578): 345–346, 1925.
   [2] *Eutrombicula alfreddugesi* (Oudemans).
   [3] EWING, H. E., *Science, Supplement* 2, **59**: xiv, 1924.
   [4] Suggested emulsifiers are Tweens 60 and 80, Triton X-100, Stearate 61-C-2280,
or if these are unavailable, soft soap.

cooling ointments, such as ammonia or salicylic acid in alcohol with a little olive oil.   Premises infested with these mites may be freed by spraying or dusting with lindane at 0.25 to 0.5 pound, dieldrin at 0.5 to 1 pound, or chlordane or toxaphene at 2 to 4 pounds per acre.   Such treatments may remain effective for several months.   Other suggested treatments include dusting with sulfur at 50 to 100 pounds per acre, cutting out brush and keeping grass closely trimmed, and pasturing with sheep.

References. U.S.D.A., Dept. Bul. 986, 1921, and *Cir.* 977, 1955; *W.H.O. Tech. Rept.* 191, 1960; *Jour. Econ. Ento.*, **37**:283, 1944; **40**:790, 1947; and **41**:731, 936, 1948.

364. Straw Itch or Harvest Mites[1]

Importance and Type of Injury.   Workmen while threshing wheat and other small grains, or handling beans or cotton, and occasionally people that sleep on straw-filled mattresses, may be overrun with microscopic mites that produce symptoms much like chiggers and are often called by that name. Within a day after exposure a hive-like eruption appears over much of the body (Fig. 21.12). These spots itch intolerably for several days to a week, and vomiting, headache, and fever may occur.

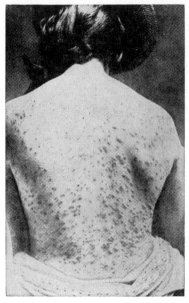

FIG. 21.12. Eruptions caused by bites of the straw itch mite, *Pyemotes ventricosus*. (*From U.S.D.A., Bur. Ento. Cir.* 118.)

Animals Attacked.   Man and insects only; so far as known.

Distribution.   This mite has been most troublesome in the central grain-growing states.

Life History, Appearance, and Habits.   This mite is usually predaceous on other insects, among which are the wheat jointworm, granary and rice weevils, and the Angoumois grain moth.   Like the chiggers, they are practically microscopic, except that the female after mating becomes greatly enlarged with developing young so that her body, originally about $\frac{1}{125}$ inch long, then measures nearly $\frac{1}{16}$ inch.   The young mites hatch from the eggs, and the nymphs develop to the adult condition within the abdomen of the female. The adults are then born viviparously; they mate soon after birth, and within a week may have matured a second generation in the same manner.   Each female may produce as many as 200 or 300 adult mites.   The predaceous activities of the mites and their attacks on man, unlike that of chiggers, are therefore confined to the adult stage.

Control Measures.   The repellent measures suggested for chiggers (page 1023) may be of value.   The free use of sulfur dust in the clothing or the application of a greasy ointment to the body before working among infested straw, stubble, or seeds, followed by a soapy bath and change of clothes promptly afterward, should prevent any unpleasant results.   Infested mattresses and other material may be cleaned by fumigation, by heating to at least 130°F. for about 6 hours, or by steaming.   After the irritation has begun, slight relief may be secured from the application of cooling ointments.

Reference.   U.S.D.A., Bur. Ento., Cir. 118, 1910.

[1] *Pyemotes* (= *Pediculoides*) *ventricosus* (Newport), Order Acarina, Family Pediculoididae.

365 (299). FLOUR, MEAL, AND CHEESE MITES[1]

Trouble similar to that from chiggers and harvest mites is sometimes experienced by persons who work with flour, meal, sugar, dried fruits, copra, cheese, hams, and the like, and is called "grocer's itch." This is caused by small mites (Fig. 19.15), which normally feed on these stored products (page 919), crawling upon the body and inserting their mouth parts. The cheese mites are often eaten and sometimes cause digestive troubles. The dried powdered bodies or excrement of these mites may cause skin irritations or allergic disturbances.

They develop very rapidly under favorable conditions and have a remarkable non-feeding stage known as a *hypopus* that sometimes intervenes between nymphs and adults. In this stage the mouth parts are wanting, and there is a minute sucker on the underside of the body with which they attach to the bodies of such insects as occur in their feeding places, and probably to mice, and may thus be transported considerable distances to start a new infestation.

Control Measures. When workmen are suffering from grocer's itch, the source of the mites should be found and the infested material or storeroom freed of them by fumigation (pages 370 and 920) or by superheating (page 405) and thorough cleaning.

366 (318, 335). ITCH MITE[2]

Importance and Type of Injury. The human itch mite (Fig. 20.7) is the same creature which, in horses and dogs, causes mange. There are several strains or varieties, each somewhat adapted to its particular host. The itch mite differs from the mites just discussed in spending its entire life cycle on the host. The eggs are laid in tunnels made by the burrowing females beneath the skin. Because of their internal position and very small size, the cause of itch was for centuries unknown, the trouble was attributed to improper living, immorality, or "bad blood." Since no proper treatment was known, the mites multiplied for generations and years, on the body of their victim, thus giving rise to the expression "seven-year itch." The female mites (Fig. 20.7) are only about $\frac{1}{60}$ inch in diameter, the males half as large, with four pairs of short stubby legs, the third pair in the male and the third and fourth pairs in the female ending in a bristle nearly as long as the body, the other pairs terminating in small stalked suckers. The skin is pale colored and striated like the palm of one's hand and with scattered tack-like spines on the upper side. The mites have neither eyes nor tracheae. The newly mated females dig into the skin, making tunnels about $\frac{1}{50}$ inch in diameter and up to 1 inch in length, in which about 24 eggs are laid. The egg stage, the six-legged "larval" stage, and two eight-legged nymphal stages each last only 2 or 3 days. The young females start to dig new burrows in which mating takes place, and egg-laying may begin in 10 days to 2 weeks after the female hatched from the egg. The adults are said to live a month or more. This burrowing and more especially the feeding of the mites cause the extreme itching that is the chief symptom of the disease. The tunnels are parallel with the surface of the skin and not very deep. They can often be seen as delicate gray thread lines beneath the skin, between the fingers and toes, behind the knee, on the external genitalia, and, in prolonged cases, over most of the body except the head. Hard pimples about the size of pinheads containing a yellow fluid form over the affected skin, and, as these are scratched, they usually become infected and cause large ugly sores and scabs. The nervous strain and the venom introduced

[1] Order Acarina, Family Tyroglyphidae.

[2] *Sarcoptes scabiei* (De Geer), Order Acarina, Family Sarcoptidae.

by the feeding of the mites may greatly depress the individual and disturb the health.

Animals Attacked. Man, dogs, rabbits, and ferrets. Varieties of the same species cause mange of many other animals (pages 946, 973).

Control Measures. Since the mites crawl about over the body chiefly at night, infection usually occurs by occupying the same bed with one who has the itch. The use of the same towels or clothing with a victim may also spread the disease, since the mites can live for 10 days or more off the body in moist conditions. Infected individuals should be isolated from the rest of the household as much as possible. Treatment consists in applying the following formulation[1] diluted 1 part to 5 parts of water.

	Parts
Benzyl benzoate	68 (by weight)
DDT	6
Benzocaine (ethyl *p*-aminobenzoate)	12
Sorbitan monooleate polyoxyalkalene derivative[1]	14

[1] Tween 80.

About 75 cubic centimeters of diluted emulsion should be applied to the entire body, except the head, by using a sponge or by spraying. The individual should not bathe for 24 hours following treatment. A single treatment will usually eliminate the infestation, but a second may occasionally be necessary after 1 week. Underwear, outer clothing, bed clothing, and towels used by the patient should be sterilized by superheating or otherwise. Applications of 1 per cent lindane or 10 per cent sulfur ointments have also been suggested.

Reference. U.S.D.A., Cir. 977, 1955.

367. HUMAN BODY LOUSE[2] AND HEAD LOUSE[3]

Importance and Type of Injury. Human lice are often the inseparable companions of those who dwell in unsanitary situations and have been responsible for a vast amount of human misery and disease (page 28). In time of war, troops have suffered unspeakably from these pests. During the American Civil War they were known as "graybacks," but during the First World War they became notorious under the name of "cooties." In times of peace, lice are largely confined to jails, prisons, labor camps, slums of cities, or other places where unsanitary living prevails. All the lice found on the body of man are of the bloodsucking kind or Anoplura. The bites of these lice, although scarcely felt at the time, are as irritating as those of fleas and bed bugs. Their crawling on the skin is also very annoying. Scratching is almost inevitable, and there is danger of streptococcic infection. The skin becomes scarred, thickened, and bronze-colored, with brownish spots. A generally tired, "grippy" feeling, fever, and an irritable and pessimistic state of mind are attributed to the feeding of lice. The most serious injury, however, is the role of the body louse as the vector of several very serious human diseases.[4] The most serious of these is epidemic typhus caused by *Rickettsia prowazekii.* The rickettsia are ingested by the louse and invade the cells of the mucous membrane of its stomach, where they propagate so intensively that after a few days the cells burst and liberate the rick-

[1] U.S. Army NBIN Formulation.
[2] *Pediculus humanus humanus* Linné, Order Anoplura, Family Pediculidae.
[3] *Pediculus humanus capitis* De Geer.
[4] WEYER, F., *Ann. Rev. Ento.*, **5**:405, 1960.

ettsia into the lumen of the alimentary canal. They are then carried away by the feces and invade human beings through skin injuries or through the mucous membranes. Typhus transmission commonly occurs through the crushing of an infected louse near a feeding puncture or by scratching with fingernails infected with the body fluids or feces of the louse. Infected lice usually die in 8 to 12 days. Epidemic typhus is a continuous fever accompanied by a spotted skin eruption. The mortality is ordinarily 5 to 25 per cent, but under war conditions, more often 50 to 70 per cent. Typhus is in general a disease of cool climates and of winter weather and occurs every year in the higher and cooler parts of Mexico. There have been outbreaks in New York and Philadelphia, but the United States has never experienced the devastating epidemics which occurred in Europe during the First World War. Endemic or murine typhus caused by *Rickettsia typhi* is spread among rodents and from rodents to man by fleas (page 1013) and possibly by the tropical rat mite.[1] In addition to typhus, the body louse is also the vector of trench fever, caused by *Rickettsia quintana*, which reproduces in the lumen of the stomach and infects man through the feces. The body louse and the head louse are also vectors of relapsing fevers, which are caused by the spirochetes *Borrelia recurrentis, B. duttoni*, and others. These microorganisms invade the hemolymph and are transmitted only when this is lost through injury to the louse. With both trench and relapsing fevers the louse is unaffected and remains infected over its lifetime.

Animals Attacked. These lice live only on the various races of man, except that apes or monkeys in confinement are occasionally infested.

Distribution. Lice occur in all countries and upon all races of people. The body louse is less abundant in the tropics, probably because, as Nuttall suggests, of the lesser amount of clothing worn as well as of temperatures too high for them.

Life History, Appearance, and Habits. Head lice and body lice are almost indistinguishable in appearance, but their habits are very distinct. They have usually been considered two different species; but, since Bacot[2] has shown that they produce fertile offspring when crossed, and Nuttall[3] has shown that there are no constant morphological differences, and Keilin and Nuttall[4] have shown that the head lice lose all their usual differences when reared for a few generations under conditions that the body louse normally experiences, we must consider that they are only races or varieties of one species.

Human lice (Fig. 21.13) are grayish, flattened, wingless six-legged insects, $\frac{1}{16}$ to $\frac{1}{6}$ inch long by nearly one-half as wide, with short five-segmented antennae, simple eyes, rather heavy legs that terminate in a single, sharp, curved claw for grasping about hairs, and piercing mouth parts that, when not in use, disappear completely into the front of the head (Fig. 4.10). The body louse is typically larger than the head louse, lighter in color, with slightly longer antennae and longer, more slender legs, and the constrictions between the abdominal segments are less conspicuous than in the head louse. In spite of these differences it is usually difficult to tell from an isolated specimen which form it is. Normally, however, the habits are very distinct. The head louse lives *on the skin*

---

[1] *Ornithonyssus* (= *Liponyssus*) *bacoti* (Hirst), Order Acarina, Family Dermanyssidae.

[2] BACOT, A., *Parasitology*, **9**:228, 1917.

[3] NUTTALL, G. H. F., *Parasitology*, **11** (nos. 3, 4): 329, 1919.

[4] KEILIN, D., and G. H. F. NUTTALL, *Parasitology*, **11** (nos. 3, 4): 279, 1919.

and among the hairs of the head and glues its eggs to hairs in the typical louse manner.   The body louse lives *in the clothing*, only going upon the skin to feed, and even then retaining its hold on the adjoining part of the clothing; *it lays its eggs in seams of the clothing.*   Both these races occur associated with man the year round but become most prevalent in winter or at any time when their victims crowd together in warm, poorly ventilated houses, which furnish conditions best suited for the increase of lice. All stages may be found at any season of the year.   The head louse lays its eggs chiefly glued to the hairs of the head, back of the ears, and at the back of the neck.   The body louse is

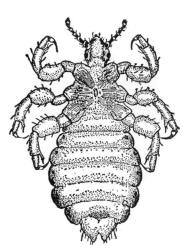

remarkable in being the only louse that lays its eggs among the seams of clothing, to the fibers of which they adhere.   The eggs are elongate oval in outline, about $\frac{1}{25}$ inch long, with a distinct pebbled lid at one end and whitish in color.   They hatch in about a week under favorable conditions, and the very tiny, slender lice at once begin feeding.   They continue to feed from two to six or more times a day for the next 1 to 4 weeks, growing meantime and molting three times before they become adult.   The adults live for about 1 month more, the females laying an average of 8 to 10 eggs a day until a total of 50 to 100 are deposited by the head louse and 200 to 300 by the body louse. Under conditions where the clothing is not, or cannot be, changed for long periods, the number of lice on one person may sometimes range from 1,000 to 10,000.   The body louse may survive as adults, eggs, and nymphs a total of a month or more in moist clothing off the host.

Fig. 21.13. Human body louse, *Pediculus humanus humanus*, underside. 20 times natural size. (*From Herrick, "Insects Injurious to the Household," copyright, 1914, Macmillan. Reprinted by permission.*)

Control Measures.   On account of the abundance of lice in Europe during the First World War and their announced connection with the spread of typhus and trench fever, the most extensive investigations were undertaken and hundreds of remedies and very elaborate plans for delousing troops were devised.   For an excellent discussion of this subject, the reader is referred to the article by Nuttall.[1]   It was not until the outbreak of the Second World War, however, that effective louse powders were developed by entomologists of the U.S. Department of Agriculture. The materials were so effective that an epidemic of 1,403 cases of typhus in Naples, Italy, in 1944, was completely arrested by mass delousing with over 3,000,000 applications of the powders.[2]   In peacetime most persons will be interested chiefly in methods of avoiding infestation by lice and possibly in helping some unfortunate to get rid of an infestation. The indiscriminate use of public rooms or sleeping cars and all contact with unclean persons should be avoided.

For the control of head lice, the infested head should be treated by two applications of 10 per cent DDT powder in pyrophyllite or of the following

[1] NUTTALL, G. H. F., *Parasitology*, **10** (no. 4): 411, 1918.
[2] *Amer. Jour. Hygiene*, **45**:305, 1947.

formulation,[1] about 7 to 10 days apart.   Treatment with 0.2 per cent lindane in hair oil has given good results.

	Per Cent
Pyrethrins	0.2
Sulfoxide	2.0
2,4-dinitroanisole	2.0
Mixed isopropyl cresols	0.1
Pyrophyllite	95.7

For the control of body lice, the DDT powder is preferred as it is longer lasting.   The clothing need not be removed for treatment, but dust should be applied to hat, sleeves, the back through the neck opening, under the shirt, and to the crotch from front and rear, using a hand duster.   From 1.5 to 2 ounces of powder is required per adult, fully clothed.   Where DDT-resistant strains are encountered, 1 per cent lindane or 1 per cent malathion dusts have given satisfactory results.   The DDT–benzyl benzoate formulation described under Itch Mite Control (page 1026) is very effective for the control of body lice, head lice, pubic lice, and scabies.   This concentrate is to be emulsified to 5 per cent in water before use.   It is then applied by hand or by spraying to the head, pubic and anal regions, armpits, and other hairy areas of the body.   About 20 cubic centimeters are needed per person, and the individual should not bathe for at least 24 hours.   The impregnation of clothing with DDT is an effective control measure for body lice.   This can readily be accomplished by washing the garments in 1 to 2 per cent DDT emulsion or by applying an equivalent solution of DDT in a volatile solvent such as petroleum naphtha or gasoline.   The dosage of DDT applied should be about 2 per cent of the weight of the garments treated.   Such treatment will eliminate lice for 6 to 8 weeks in garments worn continuously but washed every week.

When the infestation is of body lice, especial attention must be given to killing the lice and their eggs in clothing and baggage.   The surest and safest methods are vacuum fumigation with hydrogen cyanide or methyl bromide fumigation.[2]   This fumigation treatment should be applied to the clothing of infested persons in conjunction with the delousing measures described above.   Methyl bromide should be used at 9 pounds per 1,000 cubic feet above 60°F. and 12 pounds below 60°F., in a vault or a tight pit dug in the ground.   Small gas-tight neoprene coated bags with an ampoule of methyl bromide have been used by the U.S. Army Quartermaster Corps.   Destruction may also be accomplished by steam sterilization, or by a dry heat of 140 to 160°F. for 6 to 12 hours, or by fumigation with carbon bisulfide or carbon tetrachloride in a tight container.   Sleeping quarters of infested persons may be disinfested by live steam or by fumigation as for bed bugs.

References.   Parasitology, 9:228, 1917; 10:411, 1918; and 11:279, 1919; Res. Publ., Univ. Minn., vol. 8, no. 4, 1919; *U.S.D.A., Cir.* 977, 1955.

368. Crab Louse or Pubic Louse[3]

Importance and Type of Injury.   Although this species is very closely related to the head louse and body louse, it presents a very different

---

[1] U.S. Army MYL Louse Powder.

[2] *U.S.D.A., Cir.* 745, 1946.

[3] *Phthirus pubis* (Linné), Order Anoplura, Family Pediculidae.

appearance (Fig. 21.14) and its habits are quite distinct. It is almost entirely limited to the part of the body at the crotch of the legs and to the armpits, but it rarely occurs among the eyelashes, eyebrows, and beard. Nuttall[1] shows that the reach of the two extended legs of the louse corresponds closely to the distance apart of the coarse hairs in these places and concludes that this louse so seldom infests the head because the hairs there are finer and closely crowded together. Severe itching, especially in the hairy parts of the pubic region, accompanied by inflamed spots, is caused by the feeding of these small lice. When these spots are scratched, a more or less severe eczema may develop. According to Nuttall,[2] faint bluish-gray spots of varying size characteristically appear beneath the skin where these lice have fed. The same author finds that fever, headaches, and other disturbances may occasionally result from the poison introduced by the feeding of the lice. This louse is not known to transmit any disease.

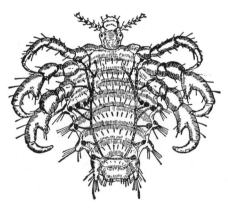

FIG. 21.14. Crab louse, *Phthirus pubis*, underside. About 20 times natural size. (*From Herrick, "Insects Injurious to the Household," copyright, 1914, Macmillan. Reprinted by permission.*)

Animals Attacked and Distribution. Known to attack only the white and Negro races of man and probably occurs wherever these races live. It has been taken from dogs.

Life History, Appearance, and Habits. The eggs are attached to the coarse hairs of the body among which the lice feed. Nuttall[1] thinks that as many as 50 eggs are laid by one female, although other authors record only about a dozen. The same author finds that the egg stage lasts about a week and the nymphs molt three times and are mature in 2 or 3 weeks after hatching. The full-grown female louse is about $\frac{1}{12}$ to $\frac{1}{16}$ inch long by about two-thirds as broad, the male about half as large, but the legs sticking out at the sides give them a still broader appearance. They are grayish white in color with darker legs and shoulders. The general appearance of the louse is that of a minute gray speck which, when magnified, suggests a miniature crab. The legs are similar to those of the body louse except that those of the first pair lack the "thumb" against which the curved claw closes in the other legs. The thorax is very broad and the margins of the abdomen have short, finger-like projections, the ends of which are provided with bristles.

Control Measures. One should use care to avoid contamination with this pest in public toilet rooms, baths, and unclean rooming houses, and especially by close association with infested persons. The most satisfactory and simple treatment is to dust the affected areas with 10 per cent DDT powder. The pubic and anal regions, the armpits, and other areas of dense body hair should be thoroughly treated twice within 7 to 10 days and the powder well distributed by rubbing. About 10 grams of

[1] NUTTALL, G. H. F., *Parasitology*, **10** (3):383, 1918.
[2] *Ibid.*, 375–382, 1918.

powder should be applied at each treatment and left undisturbed for at least 24 hours. The MYL louse powder (page 1029) is also effective and should be applied in the same manner. Two similar applications of the DDT–benzyl benzoate formulation (page 1026) are also an effective means of control.

References. *Parasitology,* **10**:375, 1918; *U.S.D.A., Cir.* 977, 1955.

369. INTERNAL INSECT PARASITES

Importance and Type of Injury. While there are in the tropics several species of maggots that live in the flesh of the human body, such as the Congo floor maggot,[1] the tumbu fly,[2] and the human bot fly,[3] myiasis (as the presence of fly maggots in the body is called) is really exceptional and accidental under normal conditions in temperate America.

About 106 different species of flies have been recorded as occasionally infesting the human body in their larval stages, 19 of them in North America. They penetrate the unbroken skin, occur in wounds, in the cavities leading from the mouth and nose, in the eyes, in the alimentary tract, and very rarely in the urinogenital passages. They reach these positions by being swallowed with water or food containing the eggs or young maggots, or the latter are deposited in wounds or body openings by the adult flies. In either case the larvae of the species concerned find conditions such within the body that they can live for a long time and often until full-grown. The results of the infection are apt to be very pronounced, and the condition is so offensive that it is likely to assume undue importance from the vivid accounts of the sporadic cases in medical literature.

The screw-worm fly, which has been discussed as a pest of cattle (page 968), sometimes attacks men, especially through the nostrils or in wounds, with dreadful results. A flesh fly known as *Wohlfahrtia vigil* Walker has been taken working under the skin. Horse bots (page 948) and ox warbles (page 963) have been found living beneath the skin of man, and the sheep bot (page 982) may cause blindness by penetrating the eye. The little house fly[4] has been recorded a number of times from the urinary tract, especially of women. The largest number of cases is recorded from the intestines. These are caused by the little house fly, the latrine fly,[5] the pomace fly,[6] the house fly, several kinds of flesh flies, and species of rattailed maggots.[7] The usual symptoms are nausea, vomiting, severe abdominal pains, a bloody diarrhea, and sooner or later the passage of maggots with the excrement.

Control Measures. Reasonable care about eating uncooked meats, vegetables, and fruits or using poorly cleaned milk bottles, and the protection of the mouth, nostrils, and other body openings, especially in the case of children and others when they are afflicted with catarrh, bad breath, or otherwise unclean, is important in avoiding the attacks of these flies. Control of the screw worm in animals is the best safeguard from attack upon man. The presence of fly larvae in any part of the body should receive the immediate attention of a skilled physician to avoid serious injury and possibly a horrible death.

References. *Jour. Econ. Ento.* **30**:29, 1937; *U.S.D.A., Misc. Pub.* 631, 1947.

370. HOUSE FLY[8]

Importance and Type of Injury. The common house fly (Fig. 21.15) is a thoroughly cosmopolitan and domesticated insect and is present at one

---

[1] *Auchmeromyia luteola* (Fabricius), Order Diptera, Family Muscidae.
[2] *Cordylobia anthropophaga* (Blanch.), Order Diptera, Family Muscidae.
[3] *Dermatobia hominis* (Linné, Jr.), Order Diptera, Family Oestridae.
[4] *Fannia canicularis* (Linné), Order Diptera, Family Anthomyiidae.
[5] *Fannia scalaris* (Fabricius), Order Diptera, Family Anthomyiidae.
[6] *Drosophila* spp., Order Diptera, Family Drosophilidae.
[7] Several genera of Order Diptera, Family Syrphidae.
[8] *Musca domestica* Linné, Order Diptera, Family Muscidae.

time or another in nearly every habitation in the world. Its numbers generally represent 98 per cent or more of the flies commonly collected in dwellings. In addition to its disagreeable presence and habits, the house fly has long been suspected as a vector of human and animal diseases. Its disgusting habits of walking and feeding on garbage and excrement and also on the human person and food make it an ideal agent for the transfer of disease organisms. Indeed, it has been shown that house flies are naturally infected with the pathogens of more than 20 human diseases, and many authorities believe that the fly is an important vector of typhoid fever, epidemic or summer diarrhea, amoebic and bacillary dysentery, cholera, poliomyelitis, and various parasitic worms. However, adequate epidemiological evidence is available only for bacillary

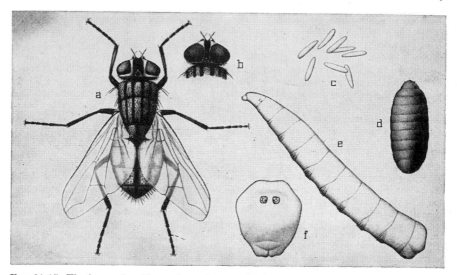

FIG. 21.15. The house fly, *Musca domestica*. *a*, Adult female, about 5 times natural size; *b*, head of male; *c*, eggs, about 6 times natural size; *d*, puparium, about 4 times natural size; *e*, larva or maggot, about 7 times natural size; *f*, last segment of larva to show spiracles. (*From Ill. Natural History Surv.*)

dysentery, where it has been shown that the incidence of the disease can be reduced by fly control.[1] House flies also serve as intermediate hosts of roundworms[2] and transfer them to sores or to lips and eyes of horses; and of tapeworms which infest poultry when these eat the flies or their maggots. It must be understood that the house fly is not the sole vector of any of these diseases in the way that mosquitoes are related to malaria or lice to typhus. In all these cases the germs of the diseases are picked up by the fly from human excrement, sputum, the carcasses of diseased animals, manure, and other filth, which the fly visits; or are taken up by the larvae and remain in a living condition in its body during the pupal stage and may then be scattered about by the deposits from the body of the adult fly. That the hairy body of the house fly is an admirable carrier of bacteria is attested by Esten and Mason who found an average of 1,250,000 bacteria on each of 414 flies examined, the maximum on a single fly being 6,600,000. The house fly carries disease germs

[1] *U.S. Pub. Health Repts.*, **63**:1319–1334, 1948.
[2] *Habronema* spp., Phylum Nemathelminthes, Class Nematoda.

either on the outside of the body clinging to its mouth parts, on its sticky foot pads (pulvilli), wings, and other body surfaces, or in its alimentary canal where they resist digestion and may remain in a living condition for many days. If carried externally, they may be dropped or rubbed off or washed off by the fly, on foods, drinks, wounds, or the eyes or lips of man. If carried internally, they are deposited either by regurgitation or in the feces ("fly specks") in such situations that they may be taken into the mouth or upon some delicate body surface, like the eye or a sore, and start disease. A minor but not unimportant injury from the house fly is as a nuisance about men and other animals where rest and sleep are disturbed by its ubiquitous buzzing and alighting on the skin.

Animals Attacked. Man is not the only animal injured by the house fly. Although it does not bite, it annoys domestic animals by alighting on their bodies and serves to transmit certain diseases by sipping blood from wounds made by other insects.

Distribution. The house fly is practically cosmopolitan and is the commonest fly found in houses throughout most of the world. Explorers in parts of the earth rarely frequented by man have been greatly annoyed by the house fly.

Life History, Appearance, and Habits. In the more northern parts of the United States, houses flies winter chiefly as larvae or pupae in their puparia (Fig. 21.15,*d*), beneath manure piles, and in other breeding grounds. In heated buildings some adults probably survive, and in warm dairy barns and in warmer parts of the earth it continues to breed slowly throughout the winter. The house fly combines a large family with one of the shortest life cycles known among insects. From 100 to 150 eggs (Fig. 21.15,*c*) are laid at a time, and from two to seven such batches may be deposited by one female, so that a total of 500 eggs from one fly is probably a normal production. The record so far as known is 21 batches, or a total of 2,387 eggs, from one fly. The whole life cycle from egg to adult may be completed in from 6 to 20 days and a new generation started in from 2 to 20 days more. The length of the various stages under favorable conditions is somewhat as follows:

| | | | *Adult to Eggs* |
Egg	*Larva*	*Pupa*	*(Preoviposition Period)*
8 to 30 hours	5 to 14 days	3 to 10 days	$2\frac{1}{4}$ to 23 days

In one series of experiments, the average adult life of 3,000 flies kept in cages was 19 days. Another investigator found 30 days to be a normal lifetime. One fly was kept alive 70 days.

Most house flies come from animal manure, as it is in this substance that they prefer to lay their whitish elongate eggs, 25 of which end to end reach 1 inch. A young house fly (Fig. 21.15,*e*) is not a miniature of its parents but is the legless, footless, and almost headless, white maggot familiar to every boy who has cleaned the manure from stables. The larvae of house flies are believed to feed upon the microorganisms which cause fermentation and decay and hence live only in moist masses of organic matter which are warm enough to promote the growth of these organisms. Feeding on this material, the maggot reaches a length of $\frac{1}{3}$ to $\frac{1}{2}$ inch in a few days, and then forms a seed-like, chestnut-colored puparium (*d*) of its third larval skin, within which the change to adult takes place. Before forming the puparium, the larva migrates to some drier part of the substance on which it fed. Every house fly thus passes a wingless egg, larval, and pupal existence in some fermenting material before it can appear in the familiar winged condition.   While horse manure is preferred, they can develop in any moist, warm, fermenting, organic matter, ranging from piles of grass, decaying fruits or vegetables, garbage, and waste about feed troughs, to human and animal excrement of all kinds.   While it was estimated a few years ago that 90 per cent of our flies develop in horse manure under average conditions, and the replacement of horses by motor cars has greatly reduced their numbers about cities and villages, we may be sure that the complete extermination of the horse would not remove the house fly pest from our midst.

The thing above all else that makes the house fly such a dangerous messmate is its inordinate curiosity, which leads it to go everywhere and to feed on almost anything.   From manure piles, sewage, garbage, sputum, and carcasses of animals to food of all kinds in dining room, kitchen, restaurant, and grocery, and to the lips, eyes, and nursing bottles of sleeping children is "all in the day's work" for a busy house fly.   If they stayed in filthy places or remained all the time in our houses, the objection to them would not be so great.   But their promiscuous habits result in polluting their bodies with filth and shedding it in their paths, wherever they go.   House flies are characteristically day-active, showing preference for direct sunshine.   The adults are almost omnivorous but, because of their sponging mouth parts (Fig. 4.15), can take food only in the liquid state.   Solids must be dissolved in their saliva or the regurgitated contents of their stomachs.   They have a keen sense of smell and thirst for liquids of nearly every kind.

The point as to how far a house fly can fly becomes an important one when the question of control in restricted areas arises.   In a number of experiments marked flies have been recovered at distances ranging from 50 yards to 13 miles from the point where they were released.   In general it seems that few flies will range farther than is necessary to secure food and a place to lay their eggs.   When abundant in a city, it is fairly certain that they are breeding in the same or an adjoining block.   Their roving habits, however, point to the necessity in general of community action to control them.   It must not be assumed that any fly caught in a house is the house fly.   Only a checking by *all* the characteristics given in the key (page 997) will determine it certainly.

There are a number of other species of domestic flies which frequent filth and refuse and often become very annoying.   These include:

1. The black blow fly[1] (page 984) and the secondary screw-worm fly[2] (page 969), which breed primarily in animal carcasses, and the greenbottle flies[3] (page 984) and the bluebottle flies,[4] which breed in garbage, excrement, and decomposing flesh and animal matter.

2. The little house fly[5] and the latrine fly[6] are often mistaken for small house flies. The adults frequent houses and barns, especially during the early summer, and exhibit a characteristic hovering and jerky flight.   The larvae, which have flattened bodies and lateral processes, breed in excrement and decaying matter.

[1] *Phormia regina* (Meigen), Family Calliphoridae.

[2] *Callitroga macellaria* (Fabricius), Family Calliphoridae.

[3] *Phaenicia* (= *Lucilia*) *sericata* (Meigen), *P. pallescens* (Shannon), and *P. caeruleiviridis* (Macquart), Family Calliphoridae.

[4] *Calliphora erythrocephala* (Meigen), *C. vomitoria* (Linné), and *Cynomyopsis cadaverina* (Desv.), Family Calliphoridae.

[5] *Fannia canicularis* (Linné), Family Anthomyiidae.

[6] *Fannia scalaris* (Fabricius), Family Anthomyiidae.

3. The stable fly[1] (page 942) is very similar in appearance to the house fly and is often referred to as the biting house fly.

4. The flesh flies[2] are large grayish flies with checkerboard patterns on the abdomens. The eggs hatch within the body of the female and living larvae are deposited upon the breeding medium. Many species are parasitic on invertebrate animals, while others infest wounds or breed in carrion or excrement.

5. The eye gnat[3] is a very annoying pest of man and animals, especially in areas of sandy or mucky soil undergoing cultivation. The larvae breed in freshly cultivated soil, excrement, and decaying matter, while the adults, which are about $\frac{1}{12}$ inch long, congregate and feed at the eyes, nostrils, and mouth. This insect has been suspected of spreading acute contagious conjunctivitis, the so-called "pink eye" disease.

6. The vinegar fly[4] is a small yellowish fly about $\frac{1}{8}$ inch in length which breeds in and is commonly found about decaying fruits and garbage cans.

Control Measures. The presence of house flies is an indication of our failure to dispose properly of manure, garbage, sewage, food wastes, human excrement, dead animals, or other organic waste. Therefore proper environmental sanitation is fundamental to successful fly control, and fly breeding can be prevented by the simple practices of burying such organic matter, as in the cut-and-fill method of garbage disposal, or by drying it so that its moisture content is below that required for larval development, as is the ultimate result in modern sewage-disposal facilities. However, under situations such as cattle feed lots and poultry farms where birds are caged above the ground, the accumulation of manure is so great as to make satisfactory fly control a major problem. A $\frac{1}{2}$-ton manure pile, after being exposed for only 4 days, was found to contain an estimated 400,000 fly larvae, or 400 flies to the pound. Manure removed from such localities should be hauled away every 2 or 3 days and scattered thinly over fields. Manure is most valuable as a fertilizer when fresh, and the common practice of storing it in the barnyard results in the loss of 25 to 65 per cent of its fertilizing value in 6 months' time, because of leaching and hot fermentation. More important, flies cannot breed in the thinly scattered material because it dries out and there is no fermentation. After manure is well rotted, flies are not attracted to it to lay their eggs. In instances where it is not practical to remove manure, methods which promote its rapid drainage and drying are of value. The breeding of flies in stored manure can be largely prevented by making piles with straight sides and clean margins and covering the entire pile with heavy building or roofing paper so that the heat developed underneath kills the larvae.

Open outdoor toilets are the greatest menace from the standpoint of fly-borne human diseases. Where privies cannot be avoided, they should be built as flyproof as possible (see *U.S. Department of Agriculture, Farmers' Bulletin* 463) and the refuse kept covered by daily applications of waste crankcase oil (if not used as fertilizer) or liberal amounts of lime. The application of 0.5 per cent borax solution so as to thoroughly wet manure is an effective larvicidal measure. Diazinon, chlordane, lindane, and dieldrin are the most efficient larvicides against susceptible flies when applied at 50 to 100 milligrams per square foot. However, larviciding has not in general been a practical means of fly control because of the difficulty in obtaining adequate penetration of the breeding media. *Para-*

---

[1] *Stomoxys calcitrans* (Linné), Family Muscidae.

[2] *Sarcophaga haemorrhoidalis* (Fallen) and others, Family Sarcophagidae.

[3] *Hippèlates pusio* (Loew) and *H. collusor* Loew, Family Chloropidae.

[4] *Drosophila melanogaster* Meigen, Family Drosophilidae.

dichlorobenzene has been suggested for application to garbage cans at 2 ounces per container.

In addition to these control measures, which go to the heart of the problem and prevent the fly from reaching its vexatious winged condition, successful fly control depends upon the use of supplementary practices to destroy the adult. The electrified screens for doors and windows and electrified fly traps are valuable about dwellings and barns. Various homemade traps are described in *U.S. Department of Agriculture, Farmers' Bulletin* 734. Bananas and milk or brown sugar and cheese, equal parts, kept wet and fermenting, make attractive bait for traps. Sanitary practices seldom entirely prevent the breeding of flies, and supplementary control with insecticides is usually necessary. Space-spraying (page 429) with 0.1 per cent synergized pyrethrins or allethrin gives rapid knockdown and temporary relief from adult flies in dwellings, barns, etc. Other insecticides which may be used for this purpose include 5 per cent DDT or methoxychlor, 2.5 per cent chlordane or 1 per cent lindane, as emulsions or oil solutions. Where the flies are resistant to the chlorinated hydrocarbons, space-spraying with 0.1 per cent Diazinon or 2 per cent ronnel or malathion is effective.

Residual spraying is an effective way to control flies in enclosures, and where they are susceptible to the chlorinated hydrocarbons, emulsions or suspensions of 5 per cent DDT or methoxychlor, 2.5 to 5 per cent chlordane, or 0.5 to 1 per cent lindane or dieldrin are very effective and provide lengthy residual action. Only methoxychlor and lindane are suitable for use in dairies or other places where food may be contaminated. Where the flies are resistant to the chlorinated hydrocarbons, residual sprays of 1 per cent Diazinon or ronnel or 5 per cent malathion are usually effective and may be used in dairy barns. Such treatments remain effective for 1 to 2 months, and their efficiency is often extended by the addition of 2.5 parts of sugar to 1 part of toxicant. The use of 5 to 7.5 parts of DDT with 1 part of synergist such as DMC or *p*-chlorophenyl *N*,*N*-dibutylsulfonamide has given satisfactory control of DDT-resistant flies.

Another effective and inexpensive way to employ the organophosphorus insecticides is as baits. One to two per cent Diazinon, Dipterex, DDVP, or Dibrom in granulated sugar sprinkled over floors at about 2 to 4 ounces per 1,000 square feet and applied daily has given excellent control. Liquid baits of 0.1 to 0.2 per cent of these toxicants and 10 per cent sugar may be applied by sprinkling.

The use of cotton cords about $\frac{3}{32}$ inch in diameter impregnated with parathion, Diazinon, or ronnel by immersing in 10 to 25 per cent xylene solution, and then suspended near the ceilings of dairies, barns, pigsties, chicken sheds, porches, etc., has given effective control for 2 to 4 months. The cords should be installed at the rate of about 30 linear feet to 100 square feet of floor area. Because of the presence of house fly strains resistant to the various chlorinated hydrocarbons and certain organophosphorus insecticides, the chemical control of flies is an exacting operation requiring the imaginative usage of the insecticides and methods mentioned above.

References. HEWITT, C. G., "The House Fly," Manchester Univ. Press, 1910; GRAHAM-SMITH, G. ,"Flies and Disease," Cambridge Univ. Press, 1914; WEST, L., "The Housefly," 2d ed., Comstock, 1950; *U.S.D.A., Farmers' Buls.* 734, 1936, and 1408, 1926, and *Dept. Buls.*, 118, 1914, 245, 1915, and 408, 1916; *Iowa Agr. Exp. Sta. Bul.* 345, 1936; *Ill. Agr. Exp. Sta. Cir.* 626, 1948; *Ann. Rev. Ento.*, **1**:323, 1956; *U.S.D.A., Cir.* 977, 1955; *W.H.O. Tech. Rept.* 191, 1960.

INDEX

Page references in **boldface** type indicate principal discussions

Moths, codling, 131, 689, **727**
  diamond-back, 662, **666**
  fig, 929
  flannel, 254
  giant silkworm, 263
  grape berry, 772, 775
  gypsy, 6, 317, 420, 454, 685, 818, **830**
  hag, 25
  hawk, 258, 657
  hummingbird, 656
  Indian-meal, 892, **931**
  io, 25
  leaf-roller, 257
  leopard, 828, **851**
  mandibulate, 253
  meal, 892
  Mediterranean flour, 892, **929**
  oriental fruit, 749, **763**
  pea, 622, **627**
  potato tuber, 640, **648**
  royal, 262
  satin, 818
  silkworm, 47, 263
  Sphinx, 258
  tapestry (*see* Carpet moth)
  tiger, 261
  tobacco, 890, **911**
  tussock, 260
  webbing clothes, 915
  white-marked tussock, 685, 818, **828**
  yucca, 254
Mountain pine beetle, 847
Mouse flea, 1014
Mouse nests, 289
Mouth, 127–148
Mouth hooks, 148, 294
Mouth parts, 127–148
  anchoring subtype, 144
  anoplurous subtype, 143, 227
  brushing subtype, 141
  chelate subtype, 144
  chewing type, 130, 140
  chewing-lapping type, 146
  common biting-fly subtype, 143
  degenerate types, 148
  dipterous subtype, 143, 148
  flea subtype, 144
  grasping subtype, 141
  grasping-sucking subtype, 141
  grinding subtype, 141
  hemipterous subtype, 143
  louse subtype, 143
  maggot, 148
  mandibulo-suctorial subtype, 141, 244
  masticatory subtype, 141
  mite, 144
  muscid subtype, 138, 144
  piercing-sucking type, 133, 142, 294
  predaceous subtype, 141
  rasping-sucking type, 141, 212
  scraping subtype, 141
  siphonapterous type, 144
  siphoning type, 146, 249
  spatulate subtype, 141
  special biting-fly subtype, 138, 144
  sponging type, 145, 293

Mouth parts, tick, subtype, 144
Mud-dauber, 277, 278
Muirixenos dicranotropidis, 241
Mules, 939–952
Mundulea spp. 334
Murgantia histrionica, 668
Murine typhus, 1014
Musca autumnalis, 956
  domestica, 77, 93, **1031**
Muscid subtype of mouth parts, 138, 144
Muscidae, 32, 309
Muscles, 91, 112–114, 136
  metabolism in, 112
  of mouth parts, 136, 139, 140
  number of, in insect body, 91
  wing-shaped, 101, 104
Museum beetles, 890, 912
  (*See also* Carpet beetles)
Museum pests, 890, 912
Mushroom fly, 880
Mushroom mite, 919
Mushrooms, pests of, 199, 300, 880, 919
Mutillidae, 280
Muzzles for flies, 407
Mycetophilidae, 300
Myiasis, 1031
Myiophagus sp., 803
MYL louse powder, 334, 1029
Mylabridae, 240
Myoneural junction, 118
Myrmeleonidae, 243
Myzus circumflexus, 19
  persicae, 18, 19, 22, 676, **754**, 873
  rosarum, 873

Nabam, 389
Nabidae, 224
Nadata gibbosa, 259
Nagana, 32
Names, insect, approved list of, 178
  common, 178
  scientific, 179
   how to write, 179
Nanus mirabilis, 19
Naphthalene, 386, 865, 914
1-Naphthyl N-methyl carbamate, 356
Narcissus bulb fly, 860, 880
Nasutitermes triodiae, 2–3
Natural classification, 175
Natural control, 422
Natural enemies, control by, 424
Natural selection, 76
NBIN formulation, 1026
Necrobia rufipes, 918
Necrophorus marginatus, 234
Negro bug, 673, 677
Nemathelminthes, 176
Nematoda, 416
Nematodes, 860, 886
Nematus ribesii, 782
Nemertinea, 176
Nemocera, 295
Neodiprion lecontei, larva of, 170
Neophylax concinnus, pupal case of, 172
Neoplectana glasseri, 416
Neoquassin, 338

Plum leafhopper, 18
Plum web-spinning sawfly, 765
Plum wilt, 16
Plumose antenna, 86
Plutella maculipennis, 666
Pnyxia scabiei, 640
Pod weevil, 548, **578**
Podosesia syringae fraxinae, 852
 syringae syringae, 852
Poecilocapsus lineatus, 784
Poison baits, 323–324, 469
 bran, 469
Poisoned bordeaux, 387
Poisoned grain, 1020
Poisonous insects, 23–25
Polar bodies, 152
Poliomyelitis, 1032
Polistes, 282, 283
Pollen, 56–60
Pollen basket, 56, 88, 273
Pollen brush, 56, 88, 273
Pollination, 56–60
Pollinizers, insects as, 56–60, 249
Polycentropidae, 247
Polyembryony, 269
Polyhedrosis disease, 417
Polypeptides, 99
Polyphaga, 231
Polyphenols in cuticle, 81
Pomace fly, 45, 75
Pompilidae, 278
Popillia japonica, 77, 749
Poplar borer, 819, **842**
 and willow borer, 820
Poplar leaf beetle, 817
Porcellio laevis, 883
Pore canals in cuticle, 80
Porthetria dispar, 115, 121, **830**
Postalar callosity, 293
Postclypeus, 128, 186
Postembryonic development, 153, 154
Posterior cells, 268, 293
Posterior orbit, 293
Postgena, 128, 147
Postscutellum, 84
Potassium cyanide, 376
Potato, 17, 18, 41, 639–649
 (*See also* Sweetpotato)
Potato aphid, 17, 134, 639, 646, **655**
Potato beetle, Colorado, 5, 131, 423,
 639, **640**
Potato blight, 641
Potato bug (*see* Potato beetle, Colorado)
 old-fashioned (*see* Blister beetles)
Potato flea beetle, 18, 131, **603**, 604
Potato leaf roll, 18
Potato leafhopper, 547, 621, **643**
Potato psyllid, 639, 655
Potato scab, 17
Potato scab gnat, 639
Potato stalk borer, 639, **647**
Potato tuber moth, 639, **647**
Potato tuberworm, 591, **597**, 639, 647,
 655
Potherb butterfly, 667
Potter wasp, 281

Poultry bug, 996, **1012**
Poultry lice, 150, **986**
Poultry mite, 144, 985, **991**
Powder post beetle, 36, 890, **903**
Prairie tent caterpillar, 828
Praying mantis, 88, 203
Predaceous beetles, 64–66, 141, 230
 diving, 88, 141, 232
Predaceous bugs, 224
Predaceous insects, 45, 64–67
Predaceous subtype of mouth parts, 141
Predaceous vertebrates, 416
Predators, 45, 64–67
 importation of, 414
Preoral cavity, 141
Prepupal period, 174
Prescutum, 84
Prestomal teeth, 138
Pretarsus, 88
Prickly pear, 417
Prionoxystus macmurtrei, 850
 robiniae, 850
Prionus, corn, 464
Prionus laticollis, larva, 171
Priophorus rubivorus, 786
Proboscis of Lepidoptera, 145–148,
 250–251
Proctodeum, 97
Proctotrupidae, 276
Prodenia eridania, 106, **479**
 ornithogalli, **479**, 559
 praefica, 559
Prognathous head, 129
Prolan, 344
Prolegs, 87, 100, 251
Prominents, 252
Pronotum, 216, 219, 268
Pronuba yuccasella, 254
Propane, 448
Propleura, 219
Propyl-*N,N*-diethyl succinamate, 391
Propyl isome, 333
Propylene oxide, 387
Prosimulium pecuarum, 1006
Prosopidae, 284
Prospaltella aurantii, 805
Protease, 99
Protective behavior, 93–95
Protective coloration, 94
Protective construction, 93–95
Protective mimicry, 94
Protective positions and reactions, 93–95
Protective resemblance, 94, 202
Protective size, 93
Prothoracic glands, 116, 117, 169
Prothorax, 84, 85
Protoaphin, 82
Protoparce quinquemaculata, 594, 656
 sexta, 594, 656
Protoveratridine, 337
Protoveratrine, 337
Protozoal diseases, 14–22, 28–35, 417
 of insects, 417
Protractor muscles, 139
Proventriculus, 97
Proxate, 386

Two-spotted lady beetle, 66
Two-spotted spider mite, 616, 688, 717, 786, 859, **875**
Two-striped grasshopper, 467
Tylodermia fragariae, 796
Tylus, 219
Tympanal organs, 120, 121
Tympanum, 84, 88
Types, of antennae, 85–86
  of mouth parts, 130–148
  of reproduction, 157, 158
Typhlocyba pomaria, 713
Typhlodromus bellinus, 877
  reticulatus, 877
Typhoid, 43, 1032
Typhus, endemic, 34, 1014
  epidemic, 34, 1027
  murine, 34, 1014
Typophorus nigritus viridicyaneus, 650
Tyrophagus castellanii, 919
  lintneri, 919
Tyrosinase, 82
Tyrosine, 82

Udea rubigalis, 675
Ugly-nest caterpillars, 257
Ultraviolet, 406
Ultrawet, 401
Unarmored scales, 805–807
Ungues, 84, 89
Unslaked lime, 365, 387
Unsulfonatable oils, 368
Upholstered furniture, 912–917
Upper lip, 127–148
Urea, 108
Uric acid, 108
Uricase, 108
Urinary products, 108
Urino-genital diseases, 56
Uterine glands, 297

Vacuum fumigation, 374, 934
Vagina, 97, 125
Value, of insect-pollinated crops, 56–58
  of insect products, 44–73
  of insects, 44–75
    aesthetic, 74
    commercial, 58
Valve, cardiac, 98
  pyloric, 98
Varied carpet beetle, 914
Variegated cutworm, **478**, 559, 594
Variety, 180
Vas deferens, 98, 125, 126
Vasates lycopersici, 657
Vat for dipping, 972
Vatsol, 401
Vegetable flies, baits for, 324
Vegetable gardens, insects injurious to, 598–683, 749
Vegetable-oil soap, 370
Vegetable oils, 367
Vegetable weevil, 599, **608**
Veins, 90, 91, 267, 268, 293
Velsicol AR-60, 401
Velvet ants, 189, 272, 998

Venomous insects, 23
Venoms, 23–25, 270–271
  methods of applying, 23–25, 270–**271**
  mosquito, 114
Venter, 219
Ventral nerve cord, 97, 98, 118
Ventriculus, 97
Veratridine, 337
Veratrimine, 337
Veratrine, 337
Veratrum album, 336
  viride, 336
Verruga peruana, 32
Vertex, 84, 128, 136, 250
Vertical triangle, 293
Vesicles, seminal, 125, 126
Vespa maculata, 281, 282
Vespidae, 272, 281
Vespoid digger wasp, 280
Vetchworm, 499
Viceroy butterfly, 266
Vine borer, squash (*see* Squash vine borer)
Vinegar fly, 1035
Violet aphid, 873
Viruses, 14, 17, 20, 21, 33
Vision, insect, 121
  mosaic, 123
Visual end organs, 123
Vitamins, 102
Vitelline membrane, 152, 154
Viviparous reproduction, 157
Volunteer plants, destruction of, 410, **533**
Volunteer wheat, 533, 541

Walking-leaves, 202
Walking-sticks, 202
Walking upside down, 88, 89
Wall, body, 79–81
Wallows, hog, 972
Walnut caterpillar, 817, **826**
Walnut husk fly, 812
Warble flies, 26, 27, **963**
Warbles, ox (cattle grubs), 26, 27, **963**
  injury by, 23, 26, 27, 963
Warehouses, insects in, 910–938
Warning, of gas, 378
  of insect outbreaks, 425
Warning coloration, 81
Washington's lady beetle, 66
Wasps, 148, 241, **266–292**, 998
  braconid, 67, 70, 72, **275**
  chalcid, 67, 70, **276**
  cuckoo, 279
  digger, 170, **277**
  dryinid, 278
  egg-parasite, 69, 70, **276**
  ensign, 276
  gall, 277
  ichneumon, 67, 70, **274**
  mud-dauber, 277, **278**
  paper-nest, 281
  parasitic, 61, 67, 242, 267
  spider, 278
  vespoid, 280
Water as control measure, 423, 875
Water boatmen, 221